Encyclopedia of Metalloproteins

Robert H. Kretsinger • Vladimir N. Uversky
Eugene A. Permyakov
Editors

Encyclopedia of Metalloproteins

Volume 1

A–C

With 1109 Figures and 256 Tables

Editors
Robert H. Kretsinger
Department of Biology
University of Virginia
Charlottesville, VA, USA

Vladimir N. Uversky
Department of Molecular Medicine
College of Medicine
University of South Florida
Tampa, FL, USA

Eugene A. Permyakov
Institute for Biological Instrumentation
Russian Academy of Sciences
Pushchino, Moscow Region, Russia

ISBN 978-1-4614-1532-9   ISBN 978-1-4614-1533-6 (eBook)
ISBN 978-1-4614-1534-3 (print and electronic bundle)
DOI 10.1007/ 978-1-4614-1533-6
Springer New York Heidelberg Dordrecht London

Library of Congress Control Number: 2013931183

© Springer Science+Business Media New York 2013
This work is subject to copyright. All rights are reserved by the Publisher, whether the whole or part of the material is concerned, specifically the rights of translation, reprinting, reuse of illustrations, recitation, broadcasting, reproduction on microfilms or in any other physical way, and transmission or information storage and retrieval, electronic adaptation, computer software, or by similar or dissimilar methodology now known or hereafter developed. Exempted from this legal reservation are brief excerpts in connection with reviews or scholarly analysis or material supplied specifically for the purpose of being entered and executed on a computer system, for exclusive use by the purchaser of the work. Duplication of this publication or parts thereof is permitted only under the provisions of the Copyright Law of the Publisher's location, in its current version, and permission for use must always be obtained from Springer. Permissions for use may be obtained through RightsLink at the Copyright Clearance Center. Violations are liable to prosecution under the respective Copyright Law.
The use of general descriptive names, registered names, trademarks, service marks, etc. in this publication does not imply, even in the absence of a specific statement, that such names are exempt from the relevant protective laws and regulations and therefore free for general use.
While the advice and information in this book are believed to be true and accurate at the date of publication, neither the authors nor the editors nor the publisher can accept any legal responsibility for any errors or omissions that may be made. The publisher makes no warranty, express or implied, with respect to the material contained herein.

Printed on acid-free paper

Springer is part of Springer Science+Business Media (www.springer.com)

Preface

Metal ions play an essential role in the functioning of all biological systems. All biological processes occur in a milieu of high concentrations of metal ions, and many of these processes depend on direct participation of metal ions. Metal ions interact with charged and polar groups of all biopolymers; those interactions with proteins play an especially important role.

The study of structural and functional properties of metal binding proteins is an important and ongoing activity area of modern physical and chemical biology. Thirteen metal ions – sodium, potassium, magnesium, calcium, manganese, iron, cobalt, zinc, copper, nickel, vanadium, tungsten, and molybdenum – are known to be essential for at least some organisms. Metallo-proteomics deals with all aspects of the intracellular and extracellular interactions of metals and proteins. Metal cations and metal binding proteins are involved in all crucial cellular activities. Many pathological conditions are correlated with abnormal metal metabolism. Research in metallo-proteomics is rapidly growing and is progressively entering curricula at universities, research institutions, and technical high schools.

Encyclopedia of Metalloproteins is a key resource that provides basic, accessible, and comprehensible information about this expanding field. It covers exhaustively all thirteen essential metal ions, discusses other metals that might compete or interfere with them, and also presents information on proteins interacting with other metal ions. *Encyclopedia of Metalloproteins* is an ideal reference for students, teachers, and researchers, as well as the informed public.

Acknowledgements

We extend our sincerest thanks to all of the contributors who have shared their insights into metalloproteins with the broader community of researchers, students, and the informed public.

About the Editors

Robert H. Kretsinger Department of Biology, University of Virginia 395, Charlottesville, VA 22904, USA

Robert H. Kretsinger is Commonwealth Professor of Biology at the University of Virginia in Charlottesville, Virginia, USA. His research has addressed structure, function, and evolution of several different protein families. His group determined the crystal structure of parvalbumin in 1970. The analysis of this calcium binding protein provided the initial characterization of the helix, loop, helix conformation of the EF-hand domain and of the pair of EF-hands that form an EF-lobe. Over seventy distinct subfamilies of EF-hand proteins have been identified, making this domain one of the most widely distributed in eukaryotes.

Dr. Kretsinger has taught courses in protein crystallography, biochemistry, macromolecular structure, and history and philosophy of biology, and has served as chair of his department and of the University Faculty Senate.

His "other" career as a sculptor began, before the advent of computer graphics, with space filling models of the EF-hand (www.virginiastonecarvers.com). He has also been an avid cyclist for many decades.

Vladimir N. Uversky Department of Molecular Medicine, College of Medicine, University of South Florida, Tampa, FL 33612 USA

Vladimir N. Uversky is an Associate Professor at the Department of Molecular Medicine at the University of South Florida (USF). He obtained his academic degrees from Moscow Institute of Physics and Technology (PhD in 1991) and from the Institute of Experimental and Theoretical Biophysics, Russian Academy of Sciences (DSc in 1998). He spent his early career working mostly on protein folding at the Institute of Protein Research and Institute for Biological Instrumentation, Russia. In 1998, he moved to the University of California, Santa Cruz, where for six years he studied protein folding, misfolding, protein conformation diseases, and protein intrinsic disorder phenomenon. In 2004, he was invited to join the Indiana University School of Medicine as a Senior Research Professor to work on intrinsically disordered proteins. Since 2010, Professor Uversky has been with USF, where he continues to study intrinsically disordered proteins and protein folding and misfolding processes. He has authored over 450 scientific publications and edited several books and book series on protein structure, function, folding and misfolding.

Eugene A. Permyakov, Institute for Biological Instrumentation, Russian Academy of Sciences, Pushchino, Moscow Region, Russia

Eugene A. Permyakov received his PhD in physics and mathematics at the Moscow Institute of Physics and Technology in 1976, and defended his Doctor of Sciences dissertation in biology at Moscow State University in 1989. From 1970 to 1994 he worked at the Institute of Theoretical and Experimental Biophysics of the Russian Academy of Sciences. From 1990 to 1991 and in 1993, Dr. Permyakov worked at the Ohio State University, Columbus, Ohio, USA. Since 1994 he has been the Director of the Institute for Biological Instrumentation of the Russian Academy of Sciences. He is a Professor of Biophysics and is known for his work on metal binding proteins and intrinsic luminescence method. He is a member of the Russian Biochemical Society. Dr. Permyakov's primary research focus is the study of physico-chemical and functional properties of metal binding proteins. He is the author of more than 150 articles and 10 books, including *Luminescent Spectroscopy of Proteins* (CRC Press, 1993), *Metalloproteomics* (John Wiley & Sons, 2009), and *Calcium Binding Proteins* (John Wiley & Sons, 2011). He is an Academic Editor of the journals *PLoS ONE* and *PeerJ*, and Editor of the book *Methods in Protein Structure and Stability Analysis* (Nova, 2007).

In his spare time, Dr. Permyakov is an avid jogger, cyclist, and cross country skier.

Section Editors

Sections: Physiological Metals: Ca; Non-Physiological Metals: Pd, Ag

Robert H. Kretsinger Department of Biology, University of Virginia, Charlottesville, VA, USA

Section: Physiological Metals: Ca

Eugene A. Permyakov Institute for Biological Instrumentation, Russian Academy of Sciences, Pushchino, Moscow Region, Russia

Sections: Metalloids; Non-Physiological Metals: Ag, Au, Pt, Be, Sr, Ba, Ra

Vladimir N. Uversky Department of Molecular Medicine, College of Medicine, University of South Florida, Tampa, FL, USA

Sections: Physiological Metals: Co, Ni, Cu

Stefano Ciurli Laboratory of Bioinorganic Chemistry, Department of Pharmacy and Biotechnology, University of Bologna, Italy

Sections: Physiological Metals: Cd, Cr

John B. Vincent Department of Chemistry, The University of Alabama, Tuscaloosa, AL, USA

Section: Physiological Metals: Fe

Elizabeth C. Theil Children's Hospital Oakland Research Institute, Oakland, CA, USA
Department of Molecular and Structural Biochemistry, North Carolina State University, Raleigh, NC, USA

Sections: Physiological Metals: Mo, W, V

Biswajit Mukherjee Department of Pharmaceutical Technology, Jadavpur University, Kolkata, India

Sections: Physiological Metals: Mg, Mn

Andrea Romani Department of Physiology and Biophysics, Case Western Reserve University, Cleveland, OH, USA

Sections: Physiological Metals: Na, K

Sergei Yu. Noskov Institute for BioComplexity and Informatics and Department for Biological Sciences, University of Calgary, Calgary, AB, Canada

Section: Physiological Metals: Zn

David S. Auld Harvard Medical School, Boston, Massachusetts, USA

Sections: Non-Physiological Metals: Li, Rb, Cs, Fr

Sergei E. Permyakov Protein Research Group, Institute for Biological Instrumentation, Russian Academy of Sciences, Pushchino, Moscow Region, Russia

Sections: Non-Physiological Metals: Lanthanides, Actinides

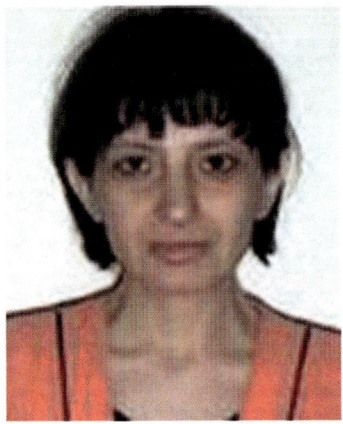

Irena Kostova Department of Chemistry, Faculty of Pharmacy, Medical University, Sofia, Bulgaria

Sections: Non-Physiological Metals: Sc, Y, Ti, Zr, Hf, Rf, Nb, Ta, Tc, Re, Ru, Os, Rh, Ir

Chunying Chen Key Laboratory for Biological Effects of Nanomaterials and Nanosafety of CAS, National Center for Nanoscience and Technology, Beijing, China

Sections: Non-Physiological Metals: Hg, Pb

K. Michael Pollard Department of Molecular and Experimental Medicine, The Scripps Research Institute, La Jolla, CA, USA

Sections: Non-Physiological Metals: Al, Ga, In, Tl, Ge, Sn, Sb, Bi, Po

Sandra V. Verstraeten Department of Biological Chemistry, University of Buenos Aires, Buenos Aires, Argentina

List of Contributors

Satoshi Abe Department of Biomolecular Engineering, Graduate School of Bioscience and Biotechnology, Tokyo Institute of Technology, Yokohama, Japan

Vojtech Adam Department of Chemistry and Biochemistry, Faculty of Agronomy, Mendel University in Brno, Brno, Czech Republic

Central European Institute of Technology, Brno University of Technology, Brno, Czech Republic

Olayiwola A. Adekoya Pharmacology Research group, Department of Pharmacy, Institute of Pharmacy, University of Tromsø, Tromsø, Norway

Paul A. Adlard The Mental Health Research Institute, The University of Melbourne, Parkville, VIC, Australia

Magnus S. Ågren Department of Surgery and Copenhagen Wound Healing Center, Bispebjerg University Hospital, Copenhagen, Denmark

Karin Åkerfeldt Department of Chemistry, Haverford College, Haverford, PA, USA

Takashiro Akitsu Department of Chemistry, Tokyo University of Science, Shinjuku-ku, Tokyo, Japan

Lorenzo Alessio Department of Experimental and Applied Medicine, Section of Occupational Health and Industrial Hygiene, University of Brescia, Brescia, Italy

Mamdouh M. Ali Biochemistry Department, Genetic Engineering and Biotechnology Division, National Research Centre, El Dokki, Cairo, Egypt

James B. Ames Department of Chemistry, University of California, Davis, CA, USA

Olaf S. Andersen Department of Physiology and Biophysics, Weill Cornell Medical College, New York, NY, USA

Gregory J. Anderson Iron Metabolism Laboratory, Queensland Institute of Medical Research, PO Royal Brisbane Hospital, Brisbane, QLD, Australia

Janet S. Anderson Department of Chemistry, Union College, Schenectady, NY, USA

João Paulo André Centro de Química, Universidade do Minho, Braga, Portugal

Claudia Andreini Magnetic Resonance Center (CERM) – University of Florence, Sesto Fiorentino, Italy

Department of Chemistry, University of Florence, Sesto Fiorentino, Italy

Alexey N. Antipov A.N. Bach Institute of Biochemistry Russian Academy of Sciences, Moscow, Russia

Tayze T. Antunes Kidney Research Center, Ottawa Hospital Research Institute, University of Ottawa, Ottawa, ON, Canada

Varun Appanna Department of Chemistry and Biochemistry, Laurentian University, Sudbury, ON, Canada

Vasu D. Appanna Department of Chemistry and Biochemistry, Laurentian University, Sudbury, ON, Canada

Cristina Ariño Department of Analytical Chemistry, University of Barcelona, Barcelona, Spain

Vladimir B. Arion Institute of Inorganic Chemistry, University of Vienna, Vienna, Austria

Farukh Arjmand Department of Chemistry, Aligarh Muslim University, Aligarh, UP, India

Fabio Arnesano Department of Chemistry, University of Bari "Aldo Moro", Bari, Italy

Joan L. Arolas Proteolysis Lab, Department of Structural Biology, Molecular Biology Institute of Barcelona, CSIC Barcelona Science Park, Barcelona, Spain

Afolake T. Arowolo Department of Biochemistry, Microbiology & Biotechnology, Rhodes University, Grahamstown, Eastern Cape, South Africa

Nebojša Arsenijević Faculty of Medical Sciences, University of Kragujevac, Centre for Molecular Medicine, Kragujevac, Serbia

Samuel Ogheneovo Asagba Department of Biochemistry, Delta State University, Abraka, Delta State, Nigeria

Michael Aschner Department of Pediatrics, Division of Pediatric Clinical Pharmacology and Toxicology, Vanderbilt University Medical Center, Nashville, TN, USA

Center in Molecular Toxicology, Vanderbilt University Medical Center, Nashville, TN, USA

Center for Molecular Neuroscience, Vanderbilt University Medical Center, Nashville, TN, USA

The Kennedy Center for Research on Human Development, Vanderbilt University Medical Center, Nashville, TN, USA

Michael Assfalg Department of Biotechnology, University of Verona, Verona, Italy

William D. Atchison Department of Pharmacology and Toxicology, Michigan State University, East Lansing, MI, USA

List of Contributors

Bishara S. Atiyeh Division of Plastic Surgery, Department of Surgery, American University of Beirut Medical Center, Beirut, Lebanon

Sílvia Atrian Departament de Genètica, Facultat de Biologia, Universitat de Barcelona, Barcelona, Spain

Christopher Auger Department of Chemistry and Biochemistry, Laurentian University, Sudbury, ON, Canada

David S. Auld Harvard Medical School, Boston, MA, USA

Scott Ayton The Mental Health Research Institute, The University of Melbourne, Parkville, VIC, Australia

Eduard B. Babiychuk Department of Cell Biology, Institute of Anatomy, University of Bern, Bern, Switzerland

Petr Babula Department of Natural Drugs, Faculty of Pharmacy, University of Veterinary and Pharmaceutical Sciences Brno, Brno, Czech Republic

Damjan Balabanič Ecology Department, Pulp and Paper Institute, Ljubljana, Slovenia

Wojciech Bal Institute of Biochemistry and Biophysics, Polish Academy of Sciences, Warsaw, Poland

Graham S. Baldwin Department of Surgery, Austin Health, The University of Melbourne, Heidelberg, VIC, Australia

Cynthia Bamdad Minerva Biotechnologies Corporation, Waltham, MA, USA

Mario Barbagallo Geriatric Unit, Department of Internal Medicine and Medical Specialties (DIMIS), University of Palermo, Palermo, Italy

Juan Barceló Lab. Fisiología Vegetal, Facultad de Biociencias, Universidad Autónoma de Barcelona, Bellaterra, Spain

Khurram Bashir Graduate School of Agricultural and Life Sciences, The University of Tokyo, Tokyo, Japan

Partha Basu Department of Chemistry and Biochemistry, Duquesne University, Pittsburgh, PA, USA

Andrea Battistoni Dipartimento di Biologia, Università di Roma Tor Vergata, Rome, Italy

Mikael Bauer Department of Biochemistry and Structural Biology, Lund University, Chemical Centre, Lund, Sweden

Lukmaan Bawazer School of Chemistry, University of Leeds, Leeds, UK

Carine Bebrone Centre for Protein Engineering, University of Liège, Sart–Tilman, Liège, Belgium

Institute of Molecular Biotechnology, RWTH–Aachen University, c/o Fraunhofer IME, Aachen, Germany

Konstantinos Beis Division of Molecular Biosciences, Imperial College London, London, South Kensington, UK

Membrane Protein Lab, Diamond Light Source, Harwell Science and Innovation Campus, Chilton, Oxfordshire, UK

Research Complex at Harwell, Harwell Oxford, Didcot, Oxforsdhire, UK

Catherine Belle Département de Chimie Moléculaire, UMR-CNRS 5250, Université Joseph Fourier, ICMG FR-2607, Grenoble, France

Andrea Bellelli Department of Biochemical Sciences, Sapienza University of Rome, Rome, Italy

Gunes Bender Department of Biological Chemistry, University of Michigan Medical School, Ann Arbor, MI, USA

Stefano Benini Faculty of Science and Technology, Free University of Bolzano, Bolzano, Italy

Stéphane L. Benoit Department of Microbiology, The University of Georgia, Athens, GA, USA

Tomas Bergman Department of Medical Biochemistry and Biophysics, Karolinska Institutet, Stockholm, Sweden

Lawrence R. Bernstein Terrametrix, Menlo Park, CA, USA

Marla J. Berry Department of Cell & Molecular Biology, John A. Burns School of Medicine, University of Hawaii at Manoa, Honolulu, HI, USA

Ivano Bertini Magnetic Resonance Center (CERM) – University of Florence, Sesto Fiorentino, Italy

Department of Chemistry, University of Florence, Sesto Fiorentino, Italy

Gerd Patrick Bienert Institut des Sciences de la Vie, Universite catholique de Louvain, Louvain-la-Neuve, Belgium

Andrew N. Bigley Department of Chemistry, Texas A&M University, College Station, TX, USA

Luis M. Bimbo Division of Pharmaceutical Technology, University of Helsinki, Helsinki, Finland

Ohad S. Birk Head, Genetics Institute, Soroka Medical Center Head, Morris Kahn Center for Human Genetics, NIBN and Faculty of Health Sciences, Ben Gurion University, Beer Sheva, Israel

Ruth Birner-Gruenberger Institute of Pathology and Center of Medical Research, Medical University of Graz, Graz, Austria

Cristina Bischin Department of Chemistry and Chemical Engineering, Babes-Bolyai University, Cluj-Napoca, Romania

Florian Bittner Department of Plant Biology, Braunschweig University of Technology, Braunschweig, Germany

Jodi L. Boer Department of Biochemistry and Molecular Biology, Michigan State University, East Lansing, MI, USA

Judith S. Bond Department of Biochemistry and Molecular Biology, Pennsylvania State University College of Medicine, Hershey, PA, USA

Martin D. Bootman Life, Health and Chemical Sciences, The Open University Walton Hall, Milton Keynes, UK

Bhargavi M. Boruah CAS Key Laboratory of Pathogenic Microbiology and Immunology, Institute of Microbiology, Chinese Academy of Sciences, Beijing, China

Graduate University of Chinese Academy of Science, Beijing, China

Sheryl R. Bowley Division of Hemostasis and Thrombosis, Beth Israel Deaconess Medical Center, Harvard Medical School, Boston, MA, USA

Doreen Braun Institute for Experimental Endocrinology, Charité-Universitätsmedizin Berlin, Berlin, Germany

Davorka Breljak Unit of Molecular Toxicology, Institute for Medical Research and Occupational Health, Zagreb, Croatia

Leonid Breydo Department of Molecular Medicine, Morsani College of Medicine, University of South Florida, Tampa, FL, USA

Mickael Briens UPR ARN du CNRS, Université de Strasbourg, Institut de Biologie Moléculaire et Cellulaire, Strasbourg, France

Joan B. Broderick Department of Chemistry and Biochemistry, Montana State University, Bozeman, MT, USA

James E. Bruce Department of Genome Sciences, University of Washington, Seattle, WA, USA

Ernesto Brunet Dept. Química Orgánica, Facultad de Ciencias, Universidad Autónoma de Madrid, Madrid, Spain

Maurizio Brunori Department of Biochemical Sciences, Sapienza University of Rome, Rome, Italy

Susan K. Buchanan Laboratory of Molecular Biology, National Institute of Diabetes and Digestive and Kidney Diseases, US National Institutes of Health, Bethesda, MD, USA

Gabriel E. Büchel Institute of Inorganic Chemistry, University of Vienna, Vienna, Austria

Živadin D. Bugarčić Faculty of Science, Department of Chemistry, University of Kragujevac, Kragujevac, Serbia

Melisa Bunderson-Schelvan Department of Biomedical and Pharmaceutical Sciences, Center for Environmental Health, The University of Montana, Missoula, MT, USA

Jean-Claude G. Bünzli Center for Next Generation Photovoltaic Systems, Korea University, Sejong Campus, Jochiwon–eup, Yeongi–gun, ChungNam–do, Republic of Korea

École Polytechnique Fédérale de Lausanne, Institute of Chemical Sciences and Engineering, Lausanne, Switzerland

John E. Burke Medical Research Council, Laboratory of Molecular Biology, Cambridge, UK

Torsten Burkholz Division of Bioorganic Chemistry, School of Pharmacy, Saarland State University, Saarbruecken, Germany

Bruce S. Burnham Department of Chemistry, Biochemistry, and Physics, Rider University, Lawrenceville, NJ, USA

Ashley I. Bush The Mental Health Research Institute, The University of Melbourne, Parkville, VIC, Australia

Kunzheng Cai Institute of Tropical and Subtropical Ecology, South China Agricultural University, Guangzhou, China

Iván L. Calderón Laboratorio de Microbiología Molecular, Universidad Andrés Bello, Santiago, Chile

Glaucia Callera Kidney Research Center, Ottawa Hospital Research Institute, University of Ottawa, Ottawa, ON, Canada

Marcello Campagna Department of Public Health, Clinical and Molecular Medicine, University of Cagliari, Cagliari, Italy

Mercè Capdevila Departament de Química, Facultat de Ciències, Universitat Autònoma de Barcelona, Cerdanyola del Vallés (Barcelona), Spain

Fernando Cardozo-Pelaez Department of Pharmaceutical Sciences, Center for Environmental Health Sciences, University of Montana, Missoula, MT, USA

Bradley A. Carlson Molecular Biology of Selenium Section, Laboratory of Cancer Prevention, National Cancer Institute, National Institutes of Health, Bethesda, MD, USA

Silvia Castelli Department of Biology, University of Rome Tor Vergata, Rome, Italy

Tommy Cedervall Department of Biochemistry and Structural Biology, Lund University, Chemical Centre, Lund, Sweden

Sudipta Chakraborty Department of Pediatrics and Department of Pharmacology, and the Kennedy Center for Research on Human Development, Vanderbilt University Medical Center, Nashville, TN, USA

Henry Chan Department of Molecular Biology, Division of Biological Sciences, University of California at San Diego, La Jolla, CA, USA

N. Chandrasekaran Centre for Nanobiotechnology, VIT University, Vellore, Tamil Nadu, India

Loïc J. Charbonnière Laboratoire d'Ingénierie Moléculaire Appliquée à l'Analyse, IPHC, UMR 7178 CNRS/UdS ECPM, Strasbourg, France

Malay Chatterjee Division of Biochemistry, Department of Pharmaceutical Technology, Jadavpur University, Kolkata, West Bengal, India

Mary Chatterjee Division of Biochemistry, Department of Pharmaceutical Technology, Jadavpur University, Kolkata, West Bengal, India

François Chaumont Institut des Sciences de la Vie, Universite catholique de Louvain, Louvain-la-Neuve, Belgium

Juan D. Chavez Department of Genome Sciences, University of Washington, Seattle, WA, USA

Chi-Ming Che Department of Chemistry, State Key Laboratory of Synthetic Chemistry and Open Laboratory of Chemical Biology of the Institute of Molecular Technology for Drug Discovery and Synthesis, The University of Hong Kong, Hong Kong, China

Elena Chekmeneva Department of Chemistry, University of Sheffield, Sheffield, UK

Di Chen The Developmental Therapeutics Program, Barbara Ann Karmanos Cancer Institute, and Departments of Oncology, Pharmacology and Pathology, School of Medicine, Wayne State University, Detroit, MI, USA

Hong-Yuan Chen National Key Laboratory of Analytical Chemistry for Life Science, School of Chemistry and Chemical Engineering, Nanjing University, Nanjing, China

Jiugeng Chen Laboratory of Plant Physiology and Molecular Genetics, Université Libre de Bruxelles, Brussels, Belgium

Sai-Juan Chen State Key Laboratory of Medical Genomics, Shanghai Institute of Hematology, Rui Jin Hospital Affiliated to Shanghai Jiao Tong University School of Medicine, Shanghai, China

Zhu Chen State Key Laboratory of Medical Genomics, Shanghai Institute of Hematology, Rui Jin Hospital Affiliated to Shanghai Jiao Tong University School of Medicine, Shanghai, China

Robert A. Cherny The Mental Health Research Institute, The University of Melbourne, Parkville, VIC, Australia

Yana Chervona Department of Environmental Medicine, New York University Medical School, New York, NY, USA

Christopher R. Chitambar Division of Hematology and Oncology, Medical College of Wisconsin, Froedtert and Medical College of Wisconsin Clinical Cancer Center, Milwaukee, WI, USA

Hassanul Ghani Choudhury Division of Molecular Biosciences, Imperial College London, London, South Kensington, UK

Membrane Protein Lab, Diamond Light Source, Harwell Science and Innovation Campus, Chilton, Oxfordshire, UK

Research Complex at Harwell, Harwell Oxford, Didcot, Oxfordshire, UK

Samrat Roy Chowdhury Department of Pharmaceutical Technology, Jadavpur University, Kolkata, West Bengal, India

Stefano Ciurli Department of Agro-Environmental Science and Technology, University of Bologna, Bologna, Italy

Stephan Clemens Department of Plant Physiology, University of Bayreuth, Bayreuth, Germany

Nansi Jo Colley Department of Ophthalmology and Visual Sciences, UW Eye Research Institute, University of Wisconsin, Madison, WI, USA

Gianni Colotti Institute of Molecular Biology and Pathology, Consiglio Nazionale delle Ricerche, Rome, Italy

Giovanni Corsetti Division of Human Anatomy, Department of Biomedical Sciences and Biotechnologies, Brescia University, Brescia, Italy

Max Costa Department of Environmental Medicine, New York University Medical School, New York, NY, USA

Jos A. Cox Department of Biochemistry, University of Geneva, Geneva, Switzerland

Adam V. Crain Department of Chemistry and Biochemistry, Montana State University, Bozeman, MT, USA

Ann Cuypers Centre for Environmental Sciences, Hasselt University, Diepenbeek, Belgium

Martha S. Cyert Department of Biology, Stanford University, Stanford, USA

Sabato D'Auria National Research Council (CNR), Laboratory for Molecular Sensing, Institute of Protein Biochemistry, Naples, Italy

Verónica Daier Departamento de Química Física/IQUIR-CONICET, Facultad de Ciencias Bioquímicas y Farmacéuticas, Universidad Nacional de Rosario, Rosario, Argentina

Charles T. Dameron Chemistry Department, Saint Francis University, Loretto, PA, USA

Subhadeep Das Division of Biochemistry, Department of Pharmaceutical Technology, Jadavpur University, Kolkata, West Bengal, India

Nilay Kanti Das Department of Dermatology, Medical College, Kolkata, West Bengal, India

Rupali Datta Department of Biological Sciences, Michigan Technological University, Houghton, MI, USA

Benjamin G. Davis Chemistry Research Laboratory, Department of Chemistry, University of Oxford, Oxford, UK

Dennis R. Dean Department of Biochemistry, Virginia Tech University, Blacksburg, VA, USA

Kannan Deepa Biochemical Engineering Laboratory, Department of Chemical Engineering, Indian Institute of Technology Madras, Chennai, Tamil Nadu, India

Claudia Della Corte Unit of Liver Research of Bambino Gesù Children's Hospital, IRCCS, Rome, Italy

Simone Dell'Acqua REQUIMTE/CQFB, Departamento de Química, Faculdade de Ciências e Tecnologia, Universidade Nova de Lisboa, Caparica, Portugal

Dipartimento di Chimica, Università di Pavia, Pavia, Italy

Hakan Demir Department of Nuclear Medicine, School of Medicine, Kocaeli University, Umuttepe, Kocaeli, Turkey

Sumukh Deshpande Institute for Biocomplexity and Informatics, Department of Biological Sciences, University of Calgary, Calgary, AB, Canada

Alessandro Desideri Department of Biology, University of Rome Tor Vergata, Rome, Italy

Interuniversity Consortium, National Institute Biostructure and Biosystem (INBB), Rome, Italy

Patrick C. D'Haese Laboratory of Pathophysiology, University of Antwerp, Wilrijk, Belgium

José Manuel Díaz-Cruz Department of Analytical Chemistry, University of Barcelona, Barcelona, Spain

Saad A. Dibo Division Plastic and Reconstructive Surgery, American University of Beirut Medical Center, Beirut, Lebanon

Pavel Dibrov Department of Microbiology, University of Manitoba, Winnipeg, MB, Canada

Adeleh Divsalar Department of Biological Sciences, Tarbiat Moallem University, Tehran, Iran

Ligia J. Dominguez Geriatric Unit, Department of Internal Medicine and Medical Specialties (DIMIS), University of Palermo, Palermo, Italy

Delfina C. Domínguez College of Health Sciences, The University of Texas at El Paso, El Paso, TX, USA

Rosario Donato Department of Experimental Medicine and Biochemical Sciences, University of Perugia, Perugia, Italy

Elke Dopp Institute of Hygiene and Occupational Medicine, University of Duisburg-Essen, Essen, Germany

Melania D'Orazio Dipartimento di Biologia, Università di Roma Tor Vergata, Rome, Italy

Q. Ping Dou The Developmental Therapeutics Program, Barbara Ann Karmanos Cancer Institute, and Departments of Oncology, Pharmacology and Pathology, School of Medicine, Wayne State University, Detroit, MI, USA

Ross G. Douglas Zinc Metalloprotease Research Group, Division of Medical Biochemistry, Institute of Infectious Disease and Molecular Medicine, University of Cape Town, Cape Town, South Africa

Annette Draeger Department of Cell Biology, Institute of Anatomy, University of Bern, Bern, Switzerland

Gabi Drochioiu Alexandru Ioan Cuza University of Iasi, Iasi, Romania

Elzbieta Dudek Department of Biochemistry, University of Alberta, Edmonton, AB, Canada

Todor Dudev Institute of Biomedical Sciences, Academia Sinica, Taipei, Taiwan

Henry J. Duff Libin Cardiovascular Institute of Alberta, Calgary, AB, Canada

Evert C. Duin Department of Chemistry and Biochemistry, Auburn University, Auburn, AL, USA

R. Scott Duncan Vision Research Center and Departments of Basic Medical Science and Ophthalmology, School of Medicine, University of Missouri, Kansas City, MO, USA

Michael F. Dunn Department of Biochemistry, University of California at Riverside, Riverside, CA, USA

Serdar Durdagi Institute for Biocomplexity and Informatics, Department of Biological Sciences, University of Calgary, Calgary, AB, Canada

Kaitlin S. Duschene Department of Chemistry and Biochemistry, Montana State University, Bozeman, MT, USA

Ankit K. Dutta School of Molecular and Biomedical Science, University of Adelaide, Adelaide, South Australia, Australia

Naba K. Dutta Ian Wark Research Institute, University of South Australia, Mawson Lakes, South Australia, Australia

Paul J. Dyson Institut des Sciences et Ingénierie Chimiques, Ecole Polytechnique Fédérale de Lausanne (EPFL) SB ISIC-Direction, Lausanne, Switzerland

Brian E. Eckenroth Department of Microbiology and Molecular Genetics, University of Vermont, Burlington, VT, USA

Niels Eckstein Federal Institute for Drugs and Medical Devices (BfArM), Bonn, Germany

David J. Eide Department of Nutritional Sciences, University of Wisconsin-Madison, Madison, WI, USA

Thomas Eitinger Institut für Biologie/Mikrobiologie, Humboldt-Universität zu Berlin, Berlin, Germany

Annette Ekblond Cardiology Stem Cell Laboratory, Rigshospitalet University Hospital, Copenhagen, Denmark

Jean-Michel El Hage Chahine ITODYS, Université Paris-Diderot Sorbonne Paris Cité, CNRS UMR 7086, Paris, France

Alex Elías Laboratorio de Microbiología Molecular, Departamento de Biología, Universidad de Santiago de Chile, Santiago, Chile

Jeffrey S. Elmendorf Department of Cellular and Integrative Physiology and Department of Biochemistry and Molecular Biology, and Centers for Diabetes Research, Membrane Biosciences, and Vascular Biology and Medicine, Indiana University School of Medicine, Indianapolis, IN, USA

Sanaz Emami Department of Biophysics, Institute of Biochemistry and Biophysics (IBB), University of Tehran, Tehran, Iran

Vinita Ernest Centre for Nanobiotechnology, VIT University, Vellore, Tamil Nadu, India

Miquel Esteban Departmte of Analytical Chemistry, University of Barcelona, Barcelona, Spain

Christopher Exley The Birchall Centre, Lennard-Jones Laboratories, Keele University, Staffordshire, UK

Chunhai Fan Laboratory of Physical Biology, Shanghai Institute of Applied Physics, Shanghai, China

Marcelo Farina Departamento de Bioquímica, Centro de Ciências Biológicas, Universidade Federal de Santa Catarina, Florianópolis, SC, Brazil

Nicholas P. Farrell Department of Chemistry, Virginia Commonwealth University, Richmond, VA, USA

Caroline Fauquant iRTSV/LCBM UMR 5249 CEA-CNRS-UJF, CEA/Grenoble, Bât K, Université Grenoble, Grenoble, France

James G. Ferry Department of Biochemistry and Molecular Biology, Eberly College of Science, The Pennsylvania State University, University Park, PA, USA

Ana Maria Figueiredo Instituto de Pesquisas Energeticas e Nucleares, IPEN-CNEN/SP, Sao Paulo, Brazil

David I. Finkelstein The Mental Health Research Institute, The University of Melbourne, Parkville, VIC, Australia

Larry Fliegel Department of Biochemistry, University of Alberta, Edmonton, AB, Canada

Swaran J. S. Flora Division of Pharmacology and Toxicology, Defence Research and Development Establishment, Gwalior, India

Juan C. Fontecilla-Camps Metalloproteins; Institut de Biologie Structurale J.P. Ebel; CEA; CNRS; Université J. Fourier, Grenoble, France

Sara M. Fox Department of Pharmacology and Toxicology, Michigan State University, East Lansing, MI, USA

Ricardo Franco REQUIMTE FCT/UNL, Departamento de Química, Faculdade de Ciências e Tecnologia, Universidade Nova de Lisboa, Caparica, Portugal

Stefan Fränzle Department of Biological and Environmental Sciences, Research Group of Environmental Chemistry, International Graduate School Zittau, Zittau, Germany

Christopher J. Frederickson NeuroBioTex, Inc, Galveston Island, TX, USA

Michael Frezza The Developmental Therapeutics Program, Barbara Ann Karmanos Cancer Institute, and Departments of Oncology, Pharmacology and Pathology, School of Medicine, Wayne State University, Detroit, MI, USA

Barbara C. Furie Division of Hemostasis and Thrombosis, Beth Israel Deaconess Medical Center, Harvard Medical School, Boston, MA, USA

Bruce Furie Division of Hemostasis and Thrombosis, Beth Israel Deaconess Medical Center, Harvard Medical School, Boston, MA, USA

Roland Gaertner Department of Endocrinology, University Hospital, Ludwig-Maximilians University Munich, Munich, Germany

Sonia Galván-Arzate Departamento de Neuroquímica, Instituto Nacional de Neurología y Neurocirugía Manuel Velasco Suárez, Mexico City, DF, Mexico

Livia Garavelli Struttura Semplice Dipartimentale di Genetica Clinica, Dipartimento di Ostetrico-Ginecologico e Pediatrico, Istituto di Ricovero e Cura a Carattere Scientifico, Arcispedale S. Maria Nuova, Reggio Emilia, Italy

Carolyn L. Geczy Inflammation and Infection Research Centre, School of Medical Sciences, University of New South Wales, Sydney, NSW, Australia

Emily Geiger Department of Biological Sciences, Michigan Technological University, Houghton, MI, USA

Alayna M. George Thompson Department of Chemistry and Biochemistry, University of Arizona, Tucson, AZ, USA

Charles P. Gerba Department of Soil, Water and Environmental Science, University of Arizona, Tucson, AZ, USA

Miltu Kumar Ghosh Department of Pharmaceutical Technology, Jadavpur University, Kolkata, West Bengal, India

Pramit Ghosh Department of Community Medicine, Medical College, Kolkata, West Bengal, India

Saikat Ghosh Department of Pharmaceutical Technology, Jadavpur University, Kolkata, West Bengal, India

Hedayatollah Ghourchian Department of Biophysics, Institute of Biochemistry and Biophysics (IBB), University of Tehran, Tehran, Iran

Jessica L. Gifford Department of Biological Sciences, Biochemistry Research Group, University of Calgary, Calgary, AB, Canada

Danuta M. Gillner Department of Chemistry, Silesian University of Technology, Gliwice, Poland

Mario Di Gioacchino Occupational Medicine and Allergy, Head of Allergy and Immunotoxicology Unit (Ce.S.I.), G. d'Annunzio University, Via dei Vestini, Chieti, Italy

Denis Girard Laboratoire de recherche en inflammation et physiologie des granulocytes, Université du Québec, INRS-Institut Armand-Frappier, Laval, QC, Canada

F. Xavier Gomis-Rüth Proteolysis Lab, Department of Structural Biology, Molecular Biology Institute of Barcelona, CSIC Barcelona Science Park, Barcelona, Spain

Harry B. Gray Beckman Institute, California Institute of Technology, Pasadena, CA, USA

Claudia Großkopf Department Chemicals Safety, Federal Institute for Risk Assessment, Berlin, Germany

Thomas E. Gunter Department of Biochemistry and Biophysics, University of Rochester School of Medicine and Dentistry, Rochester, NY, USA

Dharmendra K. Gupta Departamento de Bioquímica, Biología Celular y Molecular de Plantas, Estación Experimental del Zaidin, CSIC, Granada, Spain

Nikolai B. Gusev Department of Biochemistry, School of Biology, Moscow State University, Moscow, Russian Federation

Mandana Haack-Sørensen Cardiology Stem Cell Laboratory, Rigshospitalet University Hospital, Copenhagen, Denmark

Bodo Haas Federal Institute for Drugs and Medical Devices (BfArM), Bonn, Germany

Hajo Haase Institute of Immunology, Medical Faculty, RWTH Aachen University, Aachen, Germany

Fathi Habashi Department of Mining, Metallurgical, and Materials Engineering, Laval University, Quebec City, Canada

Alice Haddy Department of Chemistry and Biochemistry, University of North Carolina, Greensboro, NC, USA

Nguyêt-Thanh Ha-Duong ITODYS, Université Paris-Diderot Sorbonne Paris Cité, CNRS UMR 7086, Paris, France

Jesper Z. Haeggström Department of Medical Biochemistry and Biophysics (MBB), Karolinska Institute, Stockholm, Sweden

James F. Hainfeld Nanoprobes, Incorporated, Yaphank, NY, USA

Sefali Halder Department of Pharmaceutical Technology, Jadavpur University, Kolkata, West Bengal, India

Boyd E. Haley Department of Chemistry, University of Kentucky, Lexington, KY, USA

Raymond F. Hamilton Jr. Department of Biomedical and Pharmaceutical Sciences, Center for Environmental Health, The University of Montana, Missoula, MT, USA

Heidi E. Hannon Department of Pharmacology and Toxicology, Michigan State University, East Lansing, MI, USA

Timothy P. Hanusa Department of Chemistry, Vanderbilt University, Nashville, TN, USA

Edward D. Harris Department of Nutrition and Food Science, Texas A&M University, College Station, TX, USA

Todd C. Harrop Department of Chemistry, University of Georgia, Athens, GA, USA

Andrea Hartwig Department Food Chemistry and Toxicology, Karlsruhe Institute of Technology, Karlsruhe, Germany

Robert P. Hausinger Department of Biochemistry and Molecular Biology, Michigan State University, East Lansing, MI, USA

Department of Microbiology and Molecular Genetics, 6193 Biomedical and Physical Sciences, Michigan State University, East Lansing, MI, USA

Hiroaki Hayashi Department of Dermatology, Kawasaki Medical School, Kurashiki, Japan

Xiao He CAS Key Laboratory for Biomedical Effects of Nanomaterials and Nanosafety & CAS Key Laboratory of Nuclear Analytical Techniques, Institute of High Energy Physics, Chinese Academy of Sciences, Beijing, China

Yao He Institute of Functional Nano & Soft Materials, Soochow University, Jiangsu, China

Kim L. Hein Centre for Molecular Medicine Norway (NCMM), University of Oslo Nordic EMBL Partnership, Oslo, Norway

Claus W. Heizmann Department of Pediatrics, Division of Clinical Chemistry, University of Zurich, Zurich, Switzerland

Michael T. Henzl Department of Biochemistry, University of Missouri, Columbia, MO, USA

Carol M. Herak-Kramberger Unit of Molecular Toxicology, Institute for Medical Research and Occupational Health, Zagreb, Croatia

Christian Hermans Laboratory of Plant Physiology and Molecular Genetics, Université Libre de Bruxelles, Brussels, Belgium

Griselda Hernández New York State Department of Health, Wadsworth Center, Albany, NY, USA

Akon Higuchi Department of Chemical and Materials Engineering, National Central University, Jhongli, Taoyuan, Taiwan

Department of Reproduction, National Research Institute for Child Health and Development, Setagaya–ku, Tokyo, Japan

Cathay Medical Research Institute, Cathay General Hospital, Hsi–Chi City, Taipei, Taiwan

Russ Hille Department of Biochemistry, University of California, Riverside, CA, USA

Alia V. H. Hinz Department of Chemistry, Western Michigan University, Kalamazoo, MI, USA

John Andrew Hitron Graduate Center for Toxicology, University of Kentucky, Lexington, KY, USA

Miryana Hémadi ITODYS, Université Paris-Diderot Sorbonne Paris Cité, CNRS UMR 7086, Paris, France

Christer Hogstrand Metal Metabolism Group, Diabetes and Nutritional Sciences Division, School of Medicine, King's College London, London, UK

Erhard Hohenester Department of Life Sciences, Imperial College London, London, UK

Andrij Holian Department of Biomedical and Pharmaceutical Sciences, Center for Environmental Health, The University of Montana, Missoula, MT, USA

Richard C. Holz Department of Chemistry and Biochemistry, Loyola University Chicago, Chicago, IL, USA

Charles G. Hoogstraten Department of Biochemistry and Molecular Biology, Michigan State University, East Lansing, MI, USA

Ying Hou Key Laboratory for Biomechanics and Mechanobiology of the Ministry of Education, School of Biological Science and Medical Engineering, Beihang University, Beijing, China

Mingdong Huang Division of Hemostasis and Thrombosis, Beth Israel Deaconess Medical Center, Harvard Medical School, Boston, MA, USA

David L. Huffman Department of Chemistry, Western Michigan University, Kalamazoo, MI, USA

Paco Hulpiau Department for Molecular Biomedical Research, VIB, Ghent, Belgium

Amir Ibrahim Plastic and Reconstructive SurgeryBurn Fellow, Massachusetts General Hospital / Harvard Medical School & Shriners Burn Hospital, Boston, USA

Mitsu Ikura Ontario Cancer Institute and Department of Medical Biophysics, University of Toronto, Toronto, Ontario, Canada

Andrea Ilari Institute of Molecular Biology and Pathology, Consiglio Nazionale delle Ricerche, Rome, Italy

Giuseppe Inesi California Pacific Medical Center Research Institute, San Francisco, CA, USA

Hiroaki Ishida Department of Biological Sciences, Biochemistry Research Group, University of Calgary, Calgary, AB, Canada

Vangronsveld Jaco Centre for Environmental Sciences, Hasselt University, Diepenbeek, Belgium

Claus Jacob Division of Bioorganic Chemistry, School of Pharmacy, Saarland State University, Saarbruecken, Germany

Sushil K. Jain Department of Pediatrics, Louisiana State University Health Sciences Center, Shreveport, LA, USA

Peter Jensen Department of Dermato-Allergology, Copenhagen University Hospital Gentofte, Hellerup, Denmark

Klaudia Jomova Department of Chemistry, Faculty of Natural Sciences, Constantine The Philosopher University, Nitra, Slovakia

Raghava Rao Jonnalagadda Chemical Laboratory, Central Leather Research Institute (Council of Scientific and Industrial Research), Chennai, Tamil Nadu, India

Hans Jörnvall Department of Medical Biochemistry and Biophysics, Karolinska Institutet, Stockholm, Sweden

Olga Juanes Dept. Química Orgánica, Facultad de Ciencias, Universidad Autónoma de Madrid, Madrid, Spain

Sreeram Kalarical Janardhanan Chemical Laboratory, Central Leather Research Institute (Council of Scientific and Industrial Research), Chennai, Tamil Nadu, India

Paul C. J. Kamer School of Chemistry, University of St Andrews, St Andrews, UK

Tina Kamčeva Laboratory of Physical Chemistry, Vinča Institute of Nuclear Sciences, University of Belgrade, Belgrade, Serbia

Laboratory of Clinical Biochemistry, Section of Clinical Pharmacology, Haukeland University Hospital, Bergen, Norway

ChulHee Kang Washington State University, Pullman, WA, USA

Kazimierz S. Kasprzak Chemical Biology Laboratory, Frederick National Laboratory for Cancer Research, Frederick, MD, USA

Jane Kasten-Jolly New York State Department of Health, Wadsworth Center, Albany, NY, USA

Jens Kastrup Cardiology Stem Cell Laboratory, Rigshospitalet University Hospital, Copenhagen, Denmark

The Heart Centre, Cardiac Catheterization Laboratory, Rigshospitalet University Hospital, Copenhagen, Denmark

Prafulla Katkar Department of Biology, University of Rome Tor Vergata, Rome, Italy

Fusako Kawai Center for Nanomaterials and Devices, Kyoto Institute of Technology, Kyoto, Japan

Jason D. Kenealey Department of Biomolecular Chemistry, University of Wisconsin, Madison, WI, USA

Bernhard K. Keppler Institute of Inorganic Chemistry, University of Vienna, Vienna, Austria

E. Van Kerkhove Department of Physiology, Centre for Environmental Sciences, Hasselt University, Diepenbeek, Belgium

Kazuya Kikuchi Division of Advanced Science and Biotechnology, Graduate School of Engineering, Osaka University, Suita, Osaka, Japan

Immunology Frontier Research Center, Osaka University, Suita, Osaka, Japan

Michael Kirberger Department of Chemistry, Georgia State University, Atlanta, GA, USA

Masanori Kitamura Department of Molecular Signaling, Interdisciplinary Graduate School of Medicine and Engineering, University of Yamanashi, Chuo, Yamanashi, Japan

Rene Kizek Department of Chemistry and Biochemistry, Faculty of Agronomy, Mendel University in Brno, Brno, Czech Republic

Central European Institute of Technology, Brno University of Technology, Brno, Czech Republic

Nanne Kleefstra Diabetes Centre, Isala clinics, Zwolle, The Netherlands

Department of Internal Medicine, University Medical Center Groningen, Groningen, The Netherlands

Langerhans Medical Research Group, Zwolle, The Netherlands

Judith Klinman Departments of Chemistry and Molecular and Cell Biology, California Institute for Quantitative Biosciences, University of California, Berkeley, Berkeley, CA, USA

Michihiko Kobayashi Graduate School of Life and Environmental Sciences, Institute of Applied Biochemistry, The University of Tsukuba, Tsukuba, Ibaraki, Japan

Ahmet Koc Department of Molecular Biology and Genetics, Izmir Institute of Technology, Urla, İzmir, Turkey

Sergey M. Korotkov Sechenov Institute of Evolutionary Physiology and Biochemistry, The Russian Academy of Sciences, St. Petersburg, Russia

Peter Koulen Vision Research Center and Departments of Basic Medical Science and Ophthalmology, School of Medicine, University of Missouri, Kansas City, MO, USA

Nancy F. Krebs Department of Pediatrics, Section of Nutrition, University of Colorado, School of Medicine, Aurora, CO, USA

Zbigniew Krejpcio Division of Food Toxicology and Hygiene, Department of Human Nutrition and Hygiene, The Poznan University of Life Sciences, Poznan, Poland

The College of Health, Beauty and Education in Poznan, Poznan, Poland

Robert H. Kretsinger Department of Biology, University of Virginia, Charlottesville, VA, USA

Artur Krężel Department of Protein Engineering, Faculty of Biotechnology, University of Wrocław, Wrocław, Poland

Aleksandra Krivograd Klemenčič Faculty of Health Sciences, University of Ljubljana, Ljubljana, Slovenia

Peter M. H. Kroneck Department of Biology, University of Konstanz, Konstanz, Germany

Eugene Kryachko Bogolyubov Institute for Theoretical Physics, Kiev, Ukraine

Naoko Kumagai-Takei Department of Hygiene, Kawasaki Medical School, Okayama, Japan

Anil Kumar CAS Key Laboratory for Biomedical Effects of Nanoparticles and Nanosafety, National Center for Nanoscience and Nanotechnology, Chinese Academy of Sciences, Beijing, China

Graduate University of Chinese Academy of Science, Beijing, China

Thirumananseri Kumarevel RIKEN SPring-8 Center, Harima Institute, Hyogo, Japan

Valery V. Kupriyanov Institute for Biodiagnostics, National Research Council, Winnipeg, MB, Canada

Wouter Laan School of Chemistry, University of St Andrews, St Andrews, UK

James C. K. Lai Department of Biomedical & Pharmaceutical Sciences, College of Pharmacy and Biomedical Research Institute, Idaho State University, Pocatello, ID, USA

Maria José Laires CIPER – Interdisciplinary Centre for the Study of Human Performance, Faculty of Human Kinetics, Technical University of Lisbon, Cruz Quebrada, Portugal

Kyle M. Lancaster Department of Chemistry and Chemical Biology, Cornell University, Ithaca, NY, USA

Daniel Landau Department of Pediatrics, Soroka University Medical Centre, Ben-Gurion University of the Negev, Beer Sheva, Israel

Albert Lang Department of Molecular and Cell Biology, California Institute for Quantitative Biosciences, University of California, Berkeley, Berkeley, CA, USA

Alan B. G. Lansdown Faculty of Medicine, Imperial College, London, UK

Jean-Yves Lapointe Groupe d'étude des protéines membranaires (GÉPROM) and Département de Physique, Université de Montréal, Montréal, QC, Canada

Agnete Larsen Department of Biomedicine/Pharmacology Health, Aarhus University, Aarhus, Denmark

Lawrence H. Lash Department of Pharmacology, Wayne State University School of Medicine, Detroit, MI, USA

David A. Lawrence Department of Biomedical Sciences, School of Public Health, State University of New York, Albany, NY, USA

Laboratory of Clinical and Experimental Endocrinology and Immunology, Wadsworth Center, Albany, NY, USA

Peter A. Lay School of Chemistry, University of Sydney, Sydney, NSW, Australia

Gabriela Ledesma Departamento de Química Física/IQUIR-CONICET, Facultad de Ciencias Bioquímicas y Farmacéuticas, Universidad Nacional de Rosario, Rosario, Argentina

John Lee Department of Biochemistry and Molecular Biology, University of Georgia, Athens, GA, USA

Suni Lee Department of Hygiene, Kawasaki Medical School, Okayama, Japan

Silke Leimkühler From the Institute of Biochemistry and Biology, Department of Molecular Enzymology, University of Potsdam, Potsdam, Germany

Herman Louis Lelie Department of Chemistry and Biochemistry, University of California, Los Angeles, CA, USA

David M. LeMaster New York State Department of Health, Wadsworth Center, Albany, NY, USA

Joseph Lemire Department of Chemistry and Biochemistry, Laurentian University, Sudbury, ON, Canada

Thomas A. Leonard Max F. Perutz Laboratories, Vienna, Austria

Alain Lescure UPR ARN du CNRS, Université de Strasbourg, Institut de Biologie Moléculaire et Cellulaire, Strasbourg, France

Solomon W. Leung Department of Civil & Environmental Engineering, School of Engineering, College of Science and Engineering and Biomedical Research Institute, Idaho State University, Pocatello, ID, USA

Bogdan Lev Institute for Biocomplexity and Informatics, Department of Biological Sciences, University of Calgary, Calgary, AB, Canada

Aviva Levina School of Chemistry, University of Sydney, Sydney, NSW, Australia

Huihui Li School of Chemistry and Material Science, Nanjing Normal University, Nanjing, China

Yang V. Li Department of Biomedical Sciences, Heritage College of Osteopathic Medicine, Ohio University, Athens, OH, USA

Xing-Jie Liang CAS Key Laboratory for Biomedical Effects of Nanoparticles and Nanosafety, National Center for Nanoscience and Nanotechnology, Chinese Academy of Sciences, Beijing, China

Patrycja Libako Faculty of Veterinary Medicine, Wroclaw University of Environmental and Life Sciences, Wrocław, Poland

Carmay Lim Institute of Biomedical Sciences, Academia Sinica, Taipei, Taiwan

Department of Chemistry, National Tsing Hua University, Hsinchu, Taiwan

Sara Linse Department of Biochemistry and Structural Biology, Lund University, Chemical Centre, Lund, Sweden

John D. Lipscomb Department of Biochemistry, Molecular Biology, and Biophysics, University of Minnesota, Minneapolis, MN, USA

Junqiu Liu State Key Laboratory of Supramolecular Structure and Materials, College of Chemistry, Jilin University, Changchun, China

Qiong Liu College of Life Sciences, Shenzhen University, Shenzhen, P. R. China

Zijuan Liu Department of Biological Sciences, Oakland University, Rochester, MI, USA

Marija Ljubojević Unit of Molecular Toxicology, Institute for Medical Research and Occupational Health, Zagreb, Croatia

Mario Lo Bello Department of Biology, University of Rome "Tor Vergata", Rome, Italy

Yan-Chung Lo The Genomics Research Center, Academia Sinica, Taipei, Taiwan

Institute of Biological Chemistry, Academia Sinica, Taipei, Taiwan

Lingli Lu MOE Key Laboratory of Environment Remediation and Ecological Health, College of Environmental & Resource Science, Zhejiang University, Hangzhou, China

Roberto G. Lucchini Department of Experimental and Applied Medicine, Section of Occupational Health and Industrial Hygiene, University of Brescia, Brescia, Italy

Department of Preventive Medicine, Mount Sinai School of Medicine, New York, USA

Bernd Ludwig Institute of Biochemistry, Goethe University, Frankfurt, Germany

Quan Luo State Key Laboratory of Supramolecular Structure and Materials, College of Chemistry, Jilin University, Changchun, China

Jennene A. Lyda Department of Pharmaceutical Sciences, Center for Environmental Health Sciences, University of Montana, Missoula, MT, USA

Charilaos Lygidakis Regional Health Service of Emilia Romagna, AUSL of Bologna, Bologna, Italy

Jiawei Ma Key Laboratory for Biomechanics and Mechanobiology of the Ministry of Education, School of Biological Science and Medical Engineering, Beihang University, Beijing, China

Jian Feng Ma Plant Stress Physiology Group, Institute of Plant Science and Resources, Okayama University, Kurashiki, Japan

Megumi Maeda Department of Biofunctional Chemistry, Division of Bioscience, Okayama University Graduate School of Natural Science and Technology, Okayama, Japan

Axel Magalon Laboratoire de Chimie Bactérienne (UPR9043), Institut de Microbiologie de la Méditerranée, CNRS & Aix-Marseille Université, Marseille, France

Jeanette A. Maier Department of Biomedical and Clinical Sciences L. Sacco, Università di Milano, Medical School, Milano, Italy

Robert J. Maier Department of Microbiology, The University of Georgia, Athens, GA, USA

Masatoshi Maki Department of Applied Molecular Biosciences, Graduate School of Bioagricultural Sciences, Nagoya University, Nagoya, Japan

R. Manasadeepa Department of Pharmaceutical Technology, Jadavpur University, Kolkata, West Bengal, India

David J. Mann Division of Molecular Biosciences, Department of Life Sciences, Imperial College London, South Kensington, London, UK

G. Marangi Istituto di Genetica Medica, Università Cattolica Sacro Cuore, Policlinico A. Gemelli, Rome, Italy

Wolfgang Maret Metal Metabolism Group, Diabetes and Nutritional Sciences Division, School of Medicine, King's College London, London, UK

Bernd Markert Environmental Institute of Scientific Networks, in Constitution, Haren/Erika, Germany

Michael J. Maroney Department of Chemistry, Lederle Graduate Research Center, University of Massachusetts at Amherst, Amherst, MA, USA

Brenda Marrero-Rosado Department of Pharmacology and Toxicology, Michigan State University, East Lansing, MI, USA

Christopher B. Marshall Ontario Cancer Institute and Department of Medical Biophysics, University of Toronto, Toronto, Ontario, Canada

Dwight W. Martin Department of Medicine and the Proteomics Center, Stony Brook University, Stony Brook, NY, USA

Ebany J. Martinez-Finley Department of Pediatrics, Division of Pediatric Clinical Pharmacology and Toxicology, Vanderbilt University Medical Center, Nashville, TN, USA

Center in Molecular Toxicology, Vanderbilt University Medical Center, Nashville, TN, USA

Jacqueline van Marwijk Department of Biochemistry, Microbiology & Biotechnology, Rhodes University, Grahamstown, Eastern Cape, South Africa

Pradip K. Mascharak Department of Chemistry and Biochemistry, University of California, Santa Cruz, CA, USA

Anne B. Mason Department of Biochemistry, University of Vermont, Burlington, VT, USA

Hidenori Matsuzaki Department of Hygiene, Kawasaki Medical School, Okayama, Japan

Jacqueline M. Matthews School of Molecular Bioscience, The University of Sydney, Sydney, Australia

Andrzej Mazur INRA, UMR 1019, UNH, CRNH Auvergne, Clermont Université, Université d'Auvergne, Unité de Nutrition Humaine, Clermont-Ferrand, France

Paulo Mazzafera Departamento de Biologia Vegetal, Universidade Estadual de Campinas/Instituto de Biologia, Cidade Universitária, Campinas, SP, Brazil

Michael M. Mbughuni Department of Biochemistry, Molecular Biology, and Biophysics, University of Minnesota, Minneapolis, MN, USA

Joseph R. McDermott Department of Biological Sciences, Oakland University, Rochester, MI, USA

Megan M. McEvoy Department of Chemistry and Biochemistry, University of Arizona, Tucson, AZ, USA

Astrid van der Meer Interfaculty Reactor Institute, Delft University of Technology, Delft, The Netherlands

Petr Melnikov Department of Clinical Surgery, School of Medicine, Federal University of Mato Grosso do Sul, Campo Grande, MS, Brazil

Gabriele Meloni Division of Chemistry and Chemical Engineering and Howard Hughes Medical Institute, California Institute of Technology, Pasadena, CA, USA

Ralf R. Mendel Department of Plant Biology, Braunschweig University of Technology, Braunschweig, Germany

Mohamed Larbi Merroun Departamento de Microbiología, Facultad de Ciencias, Universidad de Granada, Granada, Spain

Albrecht Messerschmidt Department of Proteomics and Signal Transduction, Max-Planck-Institute of Biochemistry, Martinsried, Germany

Marek Michalak Department of Biochemistry, University of Alberta, Edmonton, AB, Canada

Faculty of Medicine and Dentistry, University of Alberta, Edmonton, AB, Canada

Isabelle Michaud-Soret iRTSV/LCBM UMR 5249 CEA-CNRS-UJF, CEA/Grenoble, Bât K, Université Grenoble, Grenoble, France

Radmila Milačič Department of Environmental Sciences, Jožef Stefan Institute, Ljubljana, Slovenia

Glenn L. Millhauser Department of Chemistry and Biochemistry, University of California, Santa Cruz, Santa Cruz, CA, USA

Marija Milovanovic Faculty of Medical Sciences, University of Kragujevac, Centre for Molecular Medicine, Kragujevac, Serbia

Shin Mizukami Division of Advanced Science and Biotechnology, Graduate School of Engineering, Osaka University, Suita, Osaka, Japan

Immunology Frontier Research Center, Osaka University, Suita, Osaka, Japan

Cristina Paula Monteiro Physiology and Biochemistry Laboratory, Faculty of Human Kinetics, Technical University of Lisbon, Cruz Quebrada, Portugal

Augusto C. Montezano Kidney Research Center, Ottawa Hospital Research Institute, University of Ottawa, Ottawa, ON, Canada

Pablo Morales-Rico Department of Toxicology, Cinvestav-IPN, Mexico city, Mexico

J. Preben Morth Centre for Molecular Medicine Norway (NCMM), Nordic EMBL Partnership, University of Oslo, Oslo, Norway

Jean-Marc Moulis Institut de Recherches en Sciences et Technologies du Vivant, Laboratoire Chimie et Biologie des Métaux (IRTSV/LCBM), CEA–Grenoble, Grenoble, France

CNRS, UMR5249, Grenoble, France

Université Joseph Fourier–Grenoble I, UMR5249, Grenoble, France

Isabel Moura REQUIMTE/CQFB, Departamento de Química, Faculdade de Ciências e Tecnologia, Universidade Nova de Lisboa, Caparica, Portugal

José J. G. Moura REQUIMTE/CQFB, Departamento de Química, Faculdade de Ciências e Tecnologia, Universidade Nova de Lisboa, Caparica, Portugal

Mohamed E. Moustafa Department of Biochemistry, Faculty of Science, Alexandria University, Alexandria, Egypt

Amitava Mukherjee Centre for Nanobiotechnology, VIT University, Vellore, Tamil Nadu, India

Biswajit Mukherjee Department of Pharmaceutical Technology, Jadavpur University, Kolkata, West Bengal, India

Balam Muñoz Department of Toxicology, Cinvestav-IPN, Mexico city, Mexico

Francesco Musiani Department of Agro-Environmental Science and Technology, University of Bologna, Bologna, Italy

Joachim Mutter Naturheilkunde, Umweltmedizin Integrative and Environmental Medicine, Belegarzt Tagesklinik, Constance, Germany

Bonex W. Mwakikunga Council for Scientific and Industrial Research, National Centre for Nano–Structured, Pretoria, South Africa

Department of Physics and Biochemical Sciences, University of Malawi, The Malawi Polytechnic, Chichiri, Blantyre, Malawi

Chandra Shekar Nagar Venkataraman Condensed Matter Physics Division, Materials Science Group, Indira Gandhi Centre for Atomic Research, Kalpakkam, Tamil Nadu, India

Hideaki Nagase Kennedy Institute of Rheumatology, Nuffield Department of Orthopaedics, Rheumatology and Musculoskeletal Sciences, University of Oxford, London, United Kingdom

Sreejayan Nair University of Wyoming, School of Pharmacy, College of Health Sciences and the Center for Cardiovascular Research and Alternative Medicine, Laramie, WY, USA

Manuel F. Navedo Department of Physiology and Biophysics, University of Washington, Seattle, WA, USA

Tim S. Nawrot Centre for Environmental Sciences, Hasselt University, Diepenbeek, Belgium

Karel Nesmerak Department of Analytical Chemistry, Faculty of Science, Charles University in Prague, Prague, Czech Republic

Gerd Ulrich Nienhaus Institute of Applied Physics and Center for Functional Nanostructures (CFN), Karlsruhe Institute of Technology (KIT), Karlsruhe, Germany

Department of Physics, University of Illinois at Urbana–Champaign, Urbana, IL, USA

Crina M. Nimigean Department of Anesthesiology, Weill Cornell Medical College, New York, NY, USA

Department of Physiology and Biophysics, Weill Cornell Medical College, New York, NY, USA

Department of Biochemistry, Weill Cornell Medical College, New York, NY, USA

Yasumitsu Nishimura Department of Hygiene, Kawasaki Medical School, Okayama, Japan

Naoko K. Nishizawa Graduate School of Agricultural and Life Sciences, The University of Tokyo, Tokyo, Japan

Research Institute for Bioresources and Biotechnology, Ishikawa Prefectural University, Ishikawa, Japan

Valerio Nobili Unit of Liver Research of Bambino Gesù Children's Hospital, IRCCS, Rome, Italy

Nicholas Noinaj Laboratory of Molecular Biology, National Institute of Diabetes and Digestive and Kidney Diseases, US National Institutes of Health, Bethesda, MD, USA

Aline M. Nonat Laboratoire d'Ingénierie Moléculaire Appliquée à l'Analyse, IPHC, UMR 7178 CNRS/UdS ECPM, Strasbourg, France

Sergei Yu. Noskov Institute for Biocomplexity and Informatics, Department of Biological Sciences, University of Calgary, Calgary, AB, Canada

Wojciech Nowacki Faculty of Veterinary Medicine, Wroclaw University of Environmental and Life Sciences, Wrocław, Poland

David O'Connell University College Dublin, Conway Institute, Dublin, Ireland

Masafumi Odaka Department of Biotechnology and Life Science, Graduate School of Technology, Tokyo University of Agriculture and Technology, Koganei, Tokyo, Japan

Akira Ono Department of Material & Life Chemistry, Faculty of Engineering, Kanagawa University, Kanagawa-ku, Yokohama, Japan

Laura Osorio-Rico Departamento de Neuroquímica, Instituto Nacional de Neurología y Neurocirugía Manuel Velasco Suárez, Mexico City, DF, Mexico

Patricia Isabel Oteiza Departments of Nutrition and Environmental Toxicology, University of California, Davis, Davis, CA, USA

Takemi Otsuki Department of Hygiene, Kawasaki Medical School, Okayama, Japan

Rabbab Oun Strathclyde Institute of Pharmacy and Biomedical Sciences, University of Strathclyde, Glasgow, UK

Vidhu Pachauri Division of Pharmacology and Toxicology, Defence Research and Development Establishment, Gwalior, India

Òscar Palacios Departament de Química, Facultat de Ciències, Universitat Autònoma de Barcelona, Cerdanyola del Vallès (Barcelona), Spain

Maria E. Palm-Espling Department of Chemistry, Chemical Biological Center, Umeå University, Umeå, Sweden

Claudia Palopoli Departamento de Química Física/IQUIR-CONICET, Facultad de Ciencias Bioquímicas y Farmacéuticas, Universidad Nacional de Rosario, Rosario, Argentina

Tapobrata Panda Biochemical Engineering Laboratory, Department of Chemical Engineering, Indian Institute of Technology Madras, Chennai, Tamil Nadu, India

Lorien J. Parker Biota Structural Biology Laboratory, St. Vincent's Institute of Medical Research, Fitzroy, VIC, Australia

Department of Biochemistry and Molecular Biology, Bio21 Molecular Science and Biotechnology Institute, The University of Melbourne, Parkville, VIC, Australia

Michael W. Parker Biota Structural Biology Laboratory, St. Vincent's Institute of Medical Research, Fitzroy, VIC, Australia

Department of Biochemistry and Molecular Biology, Bio21 Molecular Science and Biotechnology Institute, The University of Melbourne, Parkville, VIC, Australia

Marianna Patrauchan Department of Microbiology and Molecular Genetics, College of Arts and Sciences, Oklahoma State University, Stillwater, OK, USA

Sofia R. Pauleta REQUIMTE/CQFB, Departamento de Química, Faculdade de Ciências e Tecnologia, Universidade Nova de Lisboa, Caparica, Portugal

Evgeny Pavlov Department of Physiology & Biophysics, Faculty of Medicine, Dalhousie University, Halifax, NS, Canada

V. Pennemans Biomedical Institute, Hasselt University, Diepenbeek, Belgium

Harmonie Perdreau Centre for Molecular Medicine Norway (NCMM), Nordic EMBL Partnership, University of Oslo, Oslo, Norway

Alice S. Pereira Departamento de Química, Faculdade de Ciências e Tecnologia, Requimte, Centro de Química Fina e Biotecnologia, Universidade Nova de Lisboa, Caparica, Portugal

Eulália Pereira REQUIMTE, Departamento de Química e Bioquímica, Faculdade de Ciências da Universidade do Porto, Porto, Portugal

Eugene A. Permyakov Institute for Biological Instrumentation, Russian Academy of Sciences, Pushchino, Moscow Region, Russia

Sergei E. Permyakov Protein Research Group, Institute for Biological Instrumentation of the Russian Academy of Sciences, Pushchino, Moscow Region, Russia

Bertil R. R. Persson Department of Medical Radiation Physics, Lund University, Lund, Sweden

John W. Peters Department of Chemistry and Biochemistry, Montana State University, Bozeman, MT, USA

Marijana Petković Laboratory of Physical Chemistry, Vinča Institute of Nuclear Sciences, University of Belgrade, Belgrade, Serbia

Le T. Phung Department of Microbiology and Immunology, University of Illinois, Chicago, IL, USA

Roberta Pierattelli CERM and Department of Chemistry "Ugo Schiff", University of Florence, Sesto Fiorentino, Italy

Elizabeth Pierce Department of Biological Chemistry, University of Michigan Medical School, Ann Arbor, MI, USA

Andrea Pietrobattista Unit of Liver Research of Bambino Gesù Children's Hospital, IRCCS, Rome, Italy

Thomas C. Pochapsky Department of Chemistry, Rosenstiel Basic Medical Sciences Research Center, Brandeis University, Waltham, MA, USA

Ehmke Pohl Biophysical Sciences Institute, Department of Chemistry, School of Biological and Biomedical Sciences, Durham University, Durham, UK

Joe C. Polacco Department of Biochemistry/Interdisciplinary Plant Group, University of Missouri, Columbia, MO, USA

Arthur S. Polans Department of Ophthalmology and Visual Sciences, UW Eye Research Institute, University of Wisconsin, Madison, WI, USA

K. Michael Pollard Department of Molecular and Experimental Medicine, The Scripps Research Institute, La Jolla, CA, USA

Charlotte Poschenrieder Lab. Fisiología Vegetal, Facultad de Biociencias, Universidad Autónoma de Barcelona, Bellaterra, Spain

Thomas L. Poulos Department of Biochemistry & Molecular Biology, Pharmaceutical Science, and Chemistry, University of California, Irivine, Irvine, CA, USA

Richard D. Powell Nanoprobes, Incorporated, Yaphank, NY, USA

Ananda S. Prasad Department of Oncology, Karmanos Cancer Center, Wayne State University, School of Medicine, Detroit, MI, USA

Walter C. Prozialeck Department of Pharmacology, Midwestern University, Downers Grove, IL, USA

Qin Qin Wise Laboratory of Environmental and Genetic Toxicology, Maine Center for Toxicology and Environmental Health, Department of Applied Medical Sciences, University of Southern Maine, Portland, ME, USA

Thierry Rabilloud CNRS, UMR 5249 Laboratory of Chemistry and Biology of Metals, Grenoble, France

CEA, DSV, iRTSV/LCBM, Chemistry and Biology of Metals, Grenoble Cedex 9, France

Université Joseph Fourier, Grenoble, France

Stephen W. Ragsdale Department of Biological Chemistry, University of Michigan Medical School, Ann Arbor, MI, USA

Frank M. Raushel Department of Chemistry, Texas A&M University, College Station, TX, USA

Frank Reith School of Earth and Environmental Sciences, The University of Adelaide, Centre of Tectonics, Resources and Exploration (TRaX) Adelaide, Urrbrae, South Australia, Australia

CSIRO Land and Water, Environmental Biogeochemistry, PMB2, Glen Osmond, Urrbrae, South Australia, Australia

Tony Remans Centre for Environmental Sciences, Hasselt University, Diepenbeek, Belgium

Albert W. Rettenmeier Institute of Hygiene and Occupational Medicine, University of Duisburg-Essen, Essen, Germany

Rita Rezzani Division of Human Anatomy, Department of Biomedical Sciences and Biotechnologies, Brescia University, Brescia, Italy

Marius Réglier Faculté des Sciences et Techniques, ISM2/BiosCiences UMR CNRS 7313, Aix-Marseille Université Campus Scientifique de Saint Jérôme, Marseille, France

Oliver-M. H. Richter Institute of Biochemistry, Goethe University, Frankfurt, Germany

Agnes Rinaldo-Matthis Department of Medical Biochemistry and Biophysics (MBB), Karolinska Institute, Stockholm, Sweden

Lothar Rink Institute of Immunology, Medical Faculty, RWTH Aachen University, Aachen, Germany

Alfonso Rios-Perez Department of Toxicology, Cinvestav-IPN, Mexico city, Mexico

Rasmus Sejersten Ripa Cardiology Stem Cell Laboratory, Rigshospitalet University Hospital, Copenhagen, Denmark

Cluster for Molecular Imaging and Department of Clinical Physiology, Nuclear Medicine and PET, Rigshospitalet University Hospital, Copenhagen, Denmark

Marwan S. Rizk Deptartment of Anesthesiology, American University of Beirut Medical Center, Beirut, Lebanon

Nigel J. Robinson Biophysical Sciences Institute, Department of Chemistry, School of Biological and Biomedical Sciences, Durham University, Durham, UK

João B. T. Rocha Departamento de Química, Centro de Ciências Naturais e Exatas, Universidade Federal de Santa Maria, Santa Maria, RS, Brazil

Juan C. Rodriguez-Ubis Dept. Química Orgánica, Facultad de Ciencias, Universidad Autónoma de Madrid, Madrid, Spain

Harry A. Roels Louvain Centre for Toxicology and Applied Pharmacology, Université catholique de Louvain, Brussels, Belgium

Andrea M. P. Romani Department of Physiology and Biophysics, School of Medicine, Case Western Reserve University, Cleveland, OH, USA

S. Rosato Struttura Semplice Dipartimentale di Genetica Clinica, Dipartimento di Ostetrico-Ginecologico e Pediatrico, Istituto di Ricovero e Cura a Carattere Scientifico, Arcispedale S. Maria Nuova, Reggio Emilia, Italy

Barry P. Rosen Department of Cellular Biology and Pharmacology, Florida International University, Herbert Wertheim College of Medicine, Miami, FL, USA

Erwin Rosenberg Institute of Chemical Technologies and Analytics, Vienna University of Technology, Vienna, Austria

Amy C. Rosenzweig Departments of Molecular Biosciences and of Chemistry, Northwestern University, Evanston, IL, USA

Michael Rother Institut für Mikrobiologie, Technische Universität Dresden, Dresden, Germany

Benoît Roux Department of Pediatrics, Biochemistry and Molecular Biology, The University of Chicago, Chicago, IL, USA

Namita Roy Choudhury Ian Wark Research Institute, University of South Australia, Mawson Lakes, South Australia, Australia

Jagoree Roy Department of Biology, Stanford University, Stanford, USA

Kaushik Roy Division of Biochemistry, Department of Pharmaceutical Technology, Jadavpur University, Kolkata, West Bengal, India

Marian Rucki Centre of Occupational Health, Laboratory of Predictive Toxicology, National Institute of Public Health, Praha 10, Czech Republic

Anandamoy Rudra Department of Pharmaceutical Technology, Jadavpur University, Kolkata, West Bengal, India

Giuseppe Ruggiero National Research Council (CNR), Laboratory for Molecular Sensing, Institute of Protein Biochemistry, Naples, Italy

Kelly C. Ryan Department of Chemistry, Lederle Graduate Research Center, University of Massachusetts at Amherst, Amherst, MA, USA

Lisa K. Ryan New Jersey Medical School, The Public Health Research Institute, University of Medicine and Dentistry of New Jersey, Newark, NJ, USA

Janusz K. Rybakowski Department of Adult Psychiatry, Poznan University of Medical Sciences, Poznan, Poland

Ivan Sabolić Unit of Molecular Toxicology, Institute for Medical Research and Occupational Health, Zagreb, Croatia

Kalyan K. Sadhu Division of Advanced Science and Biotechnology, Graduate School of Engineering, Osaka University, Suita, Osaka, Japan

Anita Sahu Institute of Pathology and Center of Medical Research, Medical University of Graz, Graz, Austria

P. Ch. Sahu Condensed Matter Physics Division, Materials Science Group, Indira Gandhi Centre for Atomic Research, Kalpakkam, Tamil Nadu, India

Milton H. Saier Jr. Department of Molecular Biology, Division of Biological Sciences, University of California at San Diego, La Jolla, CA, USA

Jarno Salonen Laboratory of Industrial Physics, Department of Physics, University of Turku, Turku, Finland

Abel Santamaría Laboratorio de Aminoácidos Excitadores, Instituto Nacional de Neurología y Neurocirugía Manuel Velasco Suárez, Mexico City, DF, Mexico

Luis F. Santana Department of Physiology and Biophysics, University of Washington, Seattle, WA, USA

Hélder A. Santos Division of Pharmaceutical Technology, University of Helsinki, Helsinki, Finland

Dibyendu Sarkar Earth and Environmental Studies Department, Montclair State University, Montclair, NJ, USA

Louis J. Sasseville Groupe d'étude des protéines membranaires (GÉPROM) and Département de Physique, Université de Montréal, Montréal, QC, Canada

R. Gary Sawers Institute for Microbiology, Martin-Luther University Halle-Wittenberg, Halle (Saale), Germany

Janez Ščančar Department of Environmental Sciences, Jožef Stefan Institute, Ljubljana, Slovenia

Marcus C. Schaub Institute of Pharmacology and Toxicology, University of Zurich, Zurich, Switzerland

Sara Schmitt The Developmental Therapeutics Program, Barbara Ann Karmanos Cancer Institute, and Departments of Oncology, Pharmacology and Pathology, School of Medicine, Wayne State University, Detroit, MI, USA

Paul P. M. Schnetkamp Department of Physiology & Pharmacology, Hotchkiss Brain Institute, University of Calgary, Calgary, AB, Canada

Lutz Schomburg Institute for Experimental Endocrinology, Charité – University Medicine Berlin, Berlin, Germany

Gerhard N. Schrauzer Department of Chemistry and Biochemistry, University of California, San Diego, La Jolla, CA, USA

Ruth Schreiber Department of Pediatrics, Soroka University Medical Centre, Ben-Gurion University of the Negev, Beer Sheva, Israel

Ulrich Schweizer Institute for Experimental Endocrinology, Charité-Universitätsmedizin Berlin, Berlin, Germany

Ion Romulus Scorei Department of Biochemistry, University of Craiova, Craiova, DJ, Romania

Lucia A. Seale Department of Cell & Molecular Biology, John A. Burns School of Medicine, University of Hawaii at Manoa, Honolulu, HI, USA

Lance C. Seefeldt Department of Chemistry and Biochemistry, Utah State University, Logan, UT, USA

William Self Molecular Biology & Microbiology, Burnett School of Biomedical Sciences, University of Central Florida, Orlando, FL, USA

Takashi Sera Department of Applied Chemistry and Biotechnology, Graduate School of Natural Science and Technology, Okayama University, Okayama, Japan

Aruna Sharma Laboratory of Cerebrovascular Research, Department of Surgical Sciences, Anesthesiology & Intensive Care medicine, University Hospital, Uppsala University, Uppsala, Sweden

Hari Shanker Sharma Laboratory of Cerebrovascular Research, Department of Surgical Sciences, Anesthesiology & Intensive Care medicine, University Hospital, Uppsala University, Uppsala, Sweden

Honglian Shi Department of Pharmacology and Toxicology, University of Kansas, Lawrence, KS, USA

Xianglin Shi Graduate Center for Toxicology, University of Kentucky, Lexington, KY, USA

Satoshi Shinoda JST, CREST, and Department of Chemistry, Graduate School of Science, Osaka City University, Sumiyoshi-ku, Osaka, Japan

Maksim A. Shlykov Department of Molecular Biology, Division of Biological Sciences, University of California at San Diego, La Jolla, CA, USA

Siddhartha Shrivastava Rensselaer Nanotechnology Center, Rensselaer Polytechnic Institute, Troy, NY, USA

Center for Biotechnology and Interdisciplinary Studies, Rensselaer Polytechnic Institute, Troy, NY, USA

Sandra Signorella Departamento de Química Física/IQUIR-CONICET, Facultad de Ciencias Bioquímicas y Farmacéuticas, Universidad Nacional de Rosario, Rosario, Argentina

Amrita Sil Department of Pharmacology, Burdwan Medical College, Burdwan, West Bengal, India

Radu Silaghi-Dumitrescu Department of Chemistry and Chemical Engineering, Babes-Bolyai University, Cluj-Napoca, Romania

Simon Silver Department of Microbiology and Immunology, University of Illinois, Chicago, IL, USA

Britt-Marie Sjöberg Department of Biochemistry and Biophysics, Stockholm University, Stockholm, SE, Sweden

Karen Smeets Centre for Environmental Sciences, Hasselt University, Diepenbeek, Belgium

Stephen M. Smith Departments of Molecular Biosciences and of Chemistry, Northwestern University, Evanston, IL, USA

Małgorzata Sobieszczańska Department of Pathophysiology, Wroclaw Medical University, Wroclaw, Poland

Young-Ok Son Graduate Center for Toxicology, University of Kentucky, Lexington, KY, USA

Martha E. Sosa Torres Facultad de Quimica, Universidad Nacional Autonoma de Mexico, Ciudad Universitaria, Coyoacan, Mexico DF, Mexico

Jerry W. Spears Department of Animal Science, North Carolina State University, Raleigh, NC, USA

Sarah R. Spell Department of Chemistry, Virginia Commonwealth University, Richmond, VA, USA

Christopher D. Spicer Chemistry Research Laboratory, Department of Chemistry, University of Oxford, Oxford, UK

St. Hilda's College, University of Oxford, Oxford, UK

Alessandra Stacchiotti Division of Human Anatomy, Department of Biomedical Sciences and Biotechnologies, Brescia University, Brescia, Italy

Jan A. Staessen Study Coordinating Centre, Department of Cardiovascular Diseases, KU Leuven, Leuven, Belgium

Unit of Epidemiology, Maastricht University, Maastricht, The Netherlands

Maria Staiano National Research Council (CNR), Laboratory for Molecular Sensing, Institute of Protein Biochemistry, Naples, Italy

Anna Starus Department of Chemistry and Biochemistry, Loyola University Chicago, Chicago, IL, USA

Alexander Stein Hubertus Wald Tumor Center, University Cancer Center Hamburg (UCCH), University Hospital Hamburg-Eppendorf (UKE), Hamburg, Germany

Iryna N. Stepanenko Institute of Inorganic Chemistry, University of Vienna, Vienna, Austria

Martin J. Stillman Department of Biology, The University of Western Ontario, London, ON, Canada

Department of Chemistry, The University of Western Ontario, London, ON, Canada

Walter Stöcker Johannes Gutenberg University Mainz, Institute of Zoology, Cell and Matrix Biology, Mainz, Germany

Barbara J. Stoecker Department of Nutritional Sciences, Oklahoma State University, Stillwater, OK, USA

Edward D. Sturrock Zinc Metalloprotease Research Group, Division of Medical Biochemistry, Institute of Infectious Disease and Molecular Medicine, University of Cape Town, Cape Town, South Africa

Minako Sumita Department of Biochemistry and Molecular Biology, Michigan State University, East Lansing, MI, USA

Kelly L. Summers Department of Biology, The University of Western Ontario, London, ON, Canada

Raymond Wai-Yin Sun Department of Chemistry, State Key Laboratory of Synthetic Chemistry and Open Laboratory of Chemical Biology of the Institute of Molecular Technology for Drug Discovery and Synthesis, The University of Hong Kong, Hong Kong, China

Claudiu T. Supuran Department of Chemistry, University of Florence, Sesto Fiorentino (Florence), Italy

Hiroshi Suzuki Department of Biochemistry, Asahikawa Medical University, Asahikawa, Hokkaido, Japan

Q. Swennen Biomedical Institute, Hasselt University, Diepenbeek, Belgium

Ingebrigt Sylte Medical Pharmacology and Toxicology, Department of Medical Biology, University of Tromsø, Tromsø, Norway

Yoshiyuki Tanaka Graduate School of Pharmaceutical Sciences Tohoku University, Sendai, Miyagi, Japan

Shen Tang Department of Chemistry, Georgia State University, Atlanta, GA, USA

Akio Tani Research Institute of Plant Science and Resources, Okayama University, Kurashiki, Okayama, Japan

Pedro Tavares Departamento de Química, Faculdade de Ciências e Tecnologia, Requimte, Centro de Química Fina e Biotecnologia, Universidade Nova de Lisboa, Caparica, Portugal

Jan Willem Cohen Tervaert Clinical and Experimental Immunology, Maastricht University, Maastricht, The Netherlands

Tiago Tezotto Departamento de Produção Vegetal, Universidade de São Paulo/Escola Superior de Agricultura Luiz de Queiroz, Piracicaba, SP, Brazil

Elizabeth C. Theil Children's Hospital Oakland Research Institute, Oakland, CA, USA

Department of Molecular and Structural Biochemistry, North Carolina State University, Raleigh, NC, USA

Frank Thévenod Faculty of Health, School of Medicine, Centre for Biomedical Training and Research (ZBAF), Institute of Physiology & Pathophysiology, University of Witten/Herdecke, Witten, Germany

David J. Thomas Pharmacokinetics Branch – Integrated Systems Toxicology Division, National Health and Environmental Research Laboratory, U.S. Environmental Protection Agency, Research Triangle Park, NC, USA

Ameer N. Thompson Department of Anesthesiology, Weill Cornell Medical College, New York, NY, USA

Department of Physiology and Biophysics, Weill Cornell Medical College, New York, NY, USA

Rüdiger Thul School of Mathematical Sciences, University of Nottingham, Nottingham, UK

Jacob P. Thyssen National Allergy Research Centre, Department of Dermato-Allergology, Copenhagen University Hospital Gentofte, Hellerup, Denmark

Milon Tichy Centre of Occupational Health, Laboratory of Predictive Toxicology, National Institute of Public Health, Praha 10, Czech Republic

Dajena Tomco Department of Chemistry, Wayne State University, Detroit, MI, USA

Hidetaka Torigoe Department of Applied Chemistry, Faculty of Science, Tokyo University of Science, Tokyo, Japan

Rhian M. Touyz Kidney Research Center, Ottawa Hospital Research Institute, University of Ottawa, Ottawa, ON, Canada

Institute of Cardiovascular & Medical Sciences, BHF Glasgow Cardiovascular Research Centre, University of Glasgow, Glasgow, UK

Chikashi Toyoshima Institute of Molecular and Cellular Biosciences, The University of Tokyo, Tokyo, Japan

Lennart Treuel Institute of Applied Physics and Center for Functional Nanostructures (CFN), Karlsruhe Institute of Technology (KIT), Karlsruhe, Germany

Institute of Physical Chemistry, University of Duisburg–Essen, Essen, Germany

Shweta Trivedi Department of Animal Science, North Carolina State University, Raleigh, NC, USA

Thierry Tron iSm2/BiosCiences UMR CNRS 7313, Case 342, Aix-Marseille Université, Marseille, France

Chin-Hsiao Tseng Department of Internal Medicine, National Taiwan University College of Medicine, Taipei, Taiwan

Division of Endocrinology and Metabolism, Department of Internal Medicine, National Taiwan University Hospital, Taipei, Taiwan

Tsai-Tien Tseng Center for Cancer Research and Therapeutic Development, Clark Atlanta University, Atlanta, GA, USA

Samantha D. Tsotsoros Department of Chemistry, Virginia Commonwealth University, Richmond, VA, USA

Petra A. Tsuji Department of Biological Sciences, Towson University, Towson, MD, USA

Hiroshi Tsukube JST, CREST, and Department of Chemistry, Graduate School of Science, Osaka City University, Sumiyoshi-ku, Osaka, Japan

Sławomir Tubek Institute of Technology, Opole, Poland

Raymond J. Turner Department of Biological Sciences, University of Calgary, Calgary, AB, Canada

Toshiki Uchihara Laboratory of Structural Neuropathology, Tokyo Metropolitan Institute of Medical Science, Tokyo, Japan

Takafumi Ueno Department of Biomolecular Engineering, Graduate School of Bioscience and Biotechnology, Tokyo Institute of Technology, Yokohama, Japan

Christoph Ufer Institute of Biochemistry, Charité – Universitätsmedizin Berlin, Berlin, Germany

İrem Uluisik Department of Molecular Biology and Genetics, Izmir Institute of Technology, Urla, İzmir, Turkey

Balachandran Unni Nair Chemical Laboratory, Central Leather Research Institute (Council of Scientific and Industrial Research), Chennai, Tamil Nadu, India

Vladimir N. Uversky Department of Molecular Medicine, University of South Florida, College of Medicine, Tampa, FL, USA

Joan Selverstone Valentine Department of Chemistry and Biochemistry, University of California, Los Angeles, CA, USA

Marian Valko Department of Chemistry, Faculty of Natural Sciences, Constantine The Philosopher University, Nitra, Slovakia

Faculty of Chemical and Food Technology, Slovak Technical University, Bratislava, Slovakia

J. David van Horn Department of Chemistry, University of Missouri-Kansas City, Kansas City, MO, USA

Frans van Roy Department for Molecular Biomedical Research, VIB, Ghent, Belgium

Department of Biomedical Molecular Biology, Ghent University, Ghent, Belgium

Marie Vancová Institute of Parasitology, Biology Centre of the Academy of Sciences of the Czech Republic and University of South Bohemia, České Budějovice, Czech Republic

Jaco Vangronsveld Centre for Environmental Sciences, Hasselt University, Diepenbeek, Belgium

Antonio Varriale National Research Council (CNR), Laboratory for Molecular Sensing, Institute of Protein Biochemistry, Naples, Italy

Milan Vašák Department of Inorganic Chemistry, University of Zürich, Zürich, Switzerland

Claudio C. Vásquez Laboratorio de Microbiología Molecular, Departamento de Biología, Universidad de Santiago de Chile, Santiago, Chile

Oscar Vassallo Department of Biology, University of Rome Tor Vergata, Rome, Italy

Claudio N. Verani Department of Chemistry, Wayne State University, Detroit, MI, USA

Nathalie Verbruggen Laboratory of Plant Physiology and Molecular Genetics, Université Libre de Bruxelles, Brussels, Belgium

Sandra Viviana Verstraeten Department of Biological Chemistry, IQUIFIB (UBA-CONICET), School of Pharmacy and Biochemistry, University of Buenos Aires, Argentina, Buenos Aires, Argentina

Ramon Vilar Department of Chemistry, Imperial College London, South Kensington, London, UK

John B. Vincent Department of Chemistry, The University of Alabama, Tuscaloosa, AL, USA

Hans J. Vogel Department of Biological Sciences, Biochemistry Research Group, University of Calgary, Calgary, AB, Canada

Vladislav Volarevic Faculty of Medical Sciences, University of Kragujevac, Centre for Molecular Medicine, Kragujevac, Serbia

Anne Volbeda Metalloproteins; Institut de Biologie Structurale J.P. Ebel; CEA; CNRS; Université J. Fourier, Grenoble, France

Eugene S. Vysotski Photobiology Laboratory, Institute of Biophysics Russian Academy of Sciences, Siberian Branch, Krasnoyarsk, Russia

Anne Walburger Laboratoire de Chimie Bactérienne (UPR9043), Institut de Microbiologie de la Méditerranée, CNRS & Aix-Marseille Université, Marseille, France

Andrew H.-J. Wang Institute of Biological Chemistry, Academia Sinica, Taipei, Taiwan

Jiangxue Wang Key Laboratory for Biomechanics and Mechanobiology of the Ministry of Education, School of Biological Science and Medical Engineering, Beihang University, Beijing, China

Xudong Wang Department of Pathology, St Vincent Hospital, Worcester, MA, USA

John Wataha Department of Restorative Dentistry, University of Washington HSC D779A, School of Dentistry, Seattle, WA, USA

David J. Weber Department of Biochemistry and Molecular Biology, University of Maryland School of Medicine, Baltimore, MD, USA

Nial J. Wheate Faculty of Pharmacy, The University of Sydney, Sydney, NSW, Australia

Chris G. Whiteley Graduate Institute of Applied Science and Technology, National Taiwan University of Science and Technology, Taipei, Taiwan

Roger L. Williams Medical Research Council, Laboratory of Molecular Biology, Cambridge, UK

Judith Winogrodzki Department of Microbiology, University of Manitoba, Winnipeg, MB, Canada

John Pierce Wise Sr. Wise Laboratory of Environmental and Genetic Toxicology, Maine Center for Toxicology and Environmental Health, Department of Applied Medical Sciences, University of Southern Maine, Portland, ME, USA

Pernilla Wittung-Stafshede Department of Chemistry, Chemical Biological Center, Umeå University, Umeå, Sweden

Bert Wolterbeek Interfaculty Reactor Institute, Delft University of Technology, Delft, The Netherlands

Simone Wünschmann Environmental Institute of Scientific Networks in Constitution, Haren/Erika, Germany

Robert Wysocki Institute of Experimental Biology, University of Wroclaw, Wroclaw, Poland

Shenghui Xue Department of Biology, Georgia State University, Atlanta, GA, USA

Xiao-Jing Yan Department of Hematology, The First Hospital of China Medical University, Shenyang, China

Xiaodi Yang School of Chemistry and Material Science, Nanjing Normal University, Nanjing, China

Jenny J. Yang Department of Chemistry, Georgia State University, Atlanta, GA, USA

Natural Science Center, Atlanta, GA, USA

Vladimir Yarov-Yarovoy Department of Physiology and Membrane Biology, Department of Biochemistry and Molecular Medicine, School of Medicine, University of California, Davis, CA, USA

Katsuhiko Yokoi Department of Human Nutrition, Seitoku University Graduate School, Matsudo, Chiba, Japan

Vincenzo Zagà Department of Territorial Pneumotisiology, Italian Society of Tobaccology (SITAB), AUSL of Bologna, Bologna, Italy

Carla M. Zammit School of Earth and Environmental Sciences, The University of Adelaide, Centre of Tectonics, Resources and Exploration (TRaX) Adelaide, Urrbrae, South Australia, Australia

CSIRO Land and Water, Environmental Biogeochemistry, PMB2, Glen Osmond, Urrbrae, South Australia, Australia

Lourdes Zélia Zanoni Department of Pediatrics, School of Medicine, Federal University of Mato Grosso do Sul, Campo Grande, MS, Brazil

Huawei Zeng United States Department of Agriculture, Agricultural Research Service, Grand Forks Human Nutrition Research Center, Grand Forks, ND, USA

Cunxian Zhang Department of Pathology, Women & Infants Hospital of Rhode Island, Kent Memorial Hospital, Warren Alpert Medical School of Brown University, Providence, RI, USA

Chunfeng Zhao Institute for Biocomplexity and Informatics and Department of Biological Sciences, University of Calgary, Calgary, AB, Canada

Anatoly Zhitkovich Department of Pathology and Laboratory Medicine, Brown University, Providence, RI, USA

Boris S. Zhorov Department of Biochemistry and Biomedical Sciences, McMaster University, Hamilton, ON, Canada

Sechenov Institute of Evolutionary Physiology and Biochemistry, Russian Academy of Sciences, St. Petersburg, Russia

Yubin Zhou Department of Chemistry, Georgia State University, Atlanta, GA, USA

Division of Signaling and Gene Expression, La Jolla Institute for Allergy and Immunology, La Jolla, CA, USA

Michael X. Zhu Department of Integrative Biology and Pharmacology, The University of Texas Health Science Center at Houston, Houston, TX, USA

Marcella Zollino Istituto di Genetica Medica, Università Cattolica Sacro Cuore, Policlinico A. Gemelli, Rome, Italy

A

A Kinase Anchoring Protein 150 (AKAP150)

► Calcium Sparklets and Waves

AAP, Aminopeptidase from *Vibrio Proteolyticus (Aeromonas Proteolytica)*

► Zinc Aminopeptidases, Aminopeptidase from Vibrio Proteolyticus (Aeromonas proteolytica) as Prototypical Enzyme

Acclimatization

► Cadmium Exposure, Cellular and Molecular Adaptations

Accumulation

► Thallium, Distribution in Animals

ACD, N-Acetylcitrulline Deacetylase

► Zinc Aminopeptidases, Aminopeptidase from Vibrio Proteolyticus (Aeromonas proteolytica) as Prototypical Enzyme

Acireductone Dioxygenase

Thomas C. Pochapsky
Department of Chemistry, Rosenstiel Basic Medical Sciences Research Center, Brandeis University, Waltham, MA, USA

Synonyms

1,2-Dihydroxy-3-keto-5-thiomethylpent-1-ene dioxygenase; ARD; E2; E2'; EC 1.13.11.54 & 1.13.11.53

Definition

Acireductone dioxygenase (ARD) catalyzes the penultimate reaction in the methionine salvage pathway (MSP), the oxidative addition of O_2 to 2-hydroxy-3-keto-thiomethylpent-1-ene. The ARD-catalyzed reaction represents a branch point in the MSP. The on-pathway reaction (Scheme 1) results in the formation of formate and the keto-acid precursor of methionine, 4-thiomethyl-2-ketobutyrate. An off-pathway reaction (Scheme 2) results in the production of formate, carbon monoxide and 3-thiomethylpropionate. It has been shown that the identity of the metal bound in the ARD active site controls which of these two reactions occur. In the case of the enzyme isolated from *Klebsiella oxytoca*, KoARD, the first ARD to be characterized, the on-pathway enzyme (KoARD') binds Fe^{+2} and that catalyzing the off-pathway reaction (KoARD) binds Ni^{+2}. Co^{+2} and Mn^{+2} can replace

V.N. Uversky et al. (eds.), *Encyclopedia of Metalloproteins*, DOI 10.1007/978-1-4614-1533-6,

Acireductone Dioxygenase, Scheme 1 On-pathway reaction catalyzed by Fe-bound ARD' in the methionine salvage pathway

Acireductone Dioxygenase, Scheme 2 Off-pathway oxidation of acireductone catalyzed by Ni-bound ARD with carbon monoxide (CO) as product

Ni^{+2} in generating off-pathway activity, while Mg^{+2} confers partial on-pathway activity (Dai et al. 2001). The enzyme has since been identified in both prokaryotic and eukaryotic organisms, although catalytic activity and metal-binding preferences have only been thoroughly characterized for the *K. oxytoca* enzyme.

Structure

ARD is a member of the structural class of proteins known as cupins, or "jellyrolls." The core of the cupin structure is a β-helix formed from two antiparallel β-strands. If one were to twist a hairpin into a helix, one would obtain the cupin-folding topology. Several α-helices are arranged in pseudosymmetric fashion around the central β-helix in ARD (Fig. 1). The active site of ARD is located at the wide end of the β-helix, with access controlled by nearby helix and loop structures. The active site incorporates a three-histidine one-carboxylate pseudo-octahedral metal ligation scheme, with two His residues in adjacent equatorial positions around the metal, and the axial positions occupied by the third His and a glutamyl carboxylate group. The two equatorial positions closest to the opening of the active site are presumed to be the site of substrate binding to the metal. The ligation is reminiscent of and clearly related to the 2-His

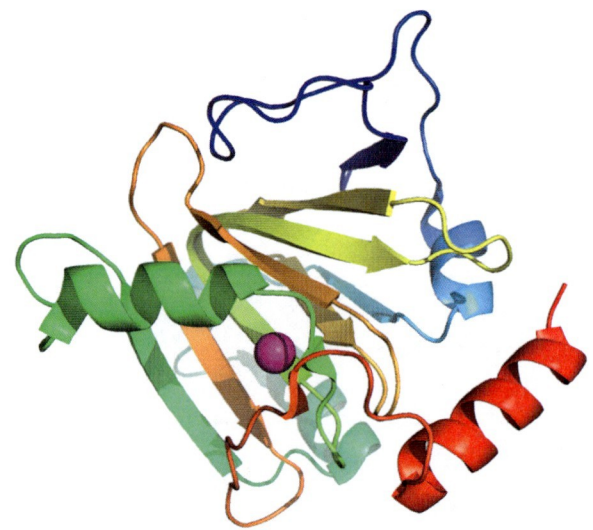

Acireductone Dioxygenase, Fig. 1 Crystallographic structure of MmARD, 1VR3, with active site in foreground. Presumed nickel ion is shown as *magenta sphere*

1-carboxylate facial Fe(II) ligation found in many nonheme dioxygenases of the cupin superfamily, including those dependent on α-ketoglutarate.

Three structures of ARD enzymes are currently available in the RCSB Protein Data Base. One is a crystallographic structure (Fig. 1), that of mouse ARD (MmARD, PDB ID 1VR3) (Xu et al. 2006), and two are NMR-derived, that of the iron-containing

KoARD' (2HJI) (Ju et al. 2006) and nickel-containing KoARD (1ZRR) from *Klebsiella oxytoca* (Pochapsky et al. 2002; Pochapsky et al. 2006). In the case of the MmARD structure, the ligand(s) occupying the presumed substrate binding site on the metal (which is presumed but not known to be Ni(II)) are unidentified, while the KoARD structures required XAS and homology modeling to characterize metal binding due to the paramagnetism of the metal in either case.

Comparison of the KoARD and KoARD' structures provide some insight into the origin of both the differential chemoselectivity of the two enzymes as well as their chromatographic separability (Ju et al. 2006). The ligation of Fe(II) and Ni(II) in both forms of the enzyme is similar, as determined by mutagenesis and XAS investigations (Chai et al. 2008). However, nickel-ligand bond lengths in KoARD are slightly shorter than the corresponding iron-ligand bonds in KoARD'. As the positions of three of the protein-based ligands (two histidine residues and one glutamate) are fixed within the β-helix, the differential bond lengths between the two forms are amplified in the position of the fourth ligand, a histidine that resides on a flexible loop. In KoARD, this loop is involved in several hydrogen bond–salt bridge interactions that stabilize a partially occluded active site entrance and permit formation of a 3,10 helix near the C-terminus of the polypeptide that sits atop the β-helix (Fig. 2). In KoARD', the active site is more open, the top helix is absent (Fig. 3), and the C-terminal peptide is disordered. This has been termed the result of a structural entropy switch, in that some parts of the iron-binding KoARD' are in fact more ordered than in the nickel-containing KoARD (Ju et al. 2006).

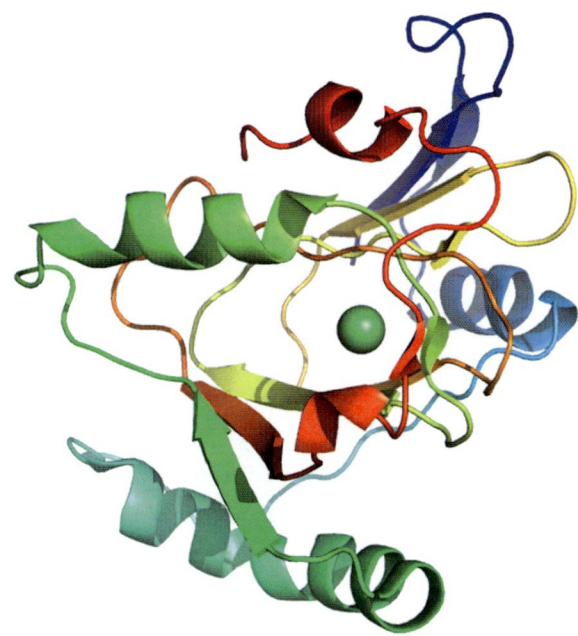

Acireductone Dioxygenase, Fig. 2 KoARD structure 1ZRR. *Green sphere* indicates position of Ni^{+2} ion. Note 3,10 helix on *top (red)*

Mechanism of Catalysis

The substrate of ARD, 1,2-dihydroxy-3-keto-5-thiomethylpent-1-ene, has not been synthesized de novo. However, the desthio analog, 1,2-dihydroxy-3-keto-5-hex-1-ene (acireductone) can be generated in situ by treatment of synthetically prepared 1-phosphonooxy-2,2-dihydroxy-3-oxohexane with E1 enolase phosphatase. This analog is turned over by both ARD and ARD' and is appropriate for activity assays and mechanistic studies. The acireductone substrate is quite reactive, and oxidative decomposition occurs in air with exposure to trace amounts of metal

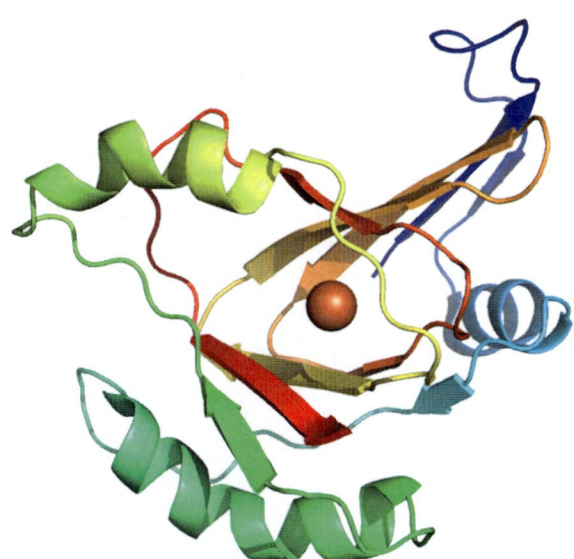

Acireductone Dioxygenase, Fig. 3 KoARD' structure 1HJI. *Orange sphere* indicates the position of Fe^{+2}

ions and/or peroxides, leading to the on-pathway products of keto acid and formate. Because of its reactivity, it was initially assumed that the on-pathway oxidation of acireductone was uncatalyzed in vivo, and it was

Acireductone Dioxygenase, Scheme 3 Proposed mechanisms for regioselectivity in the oxidation of acireductone by Fe-ARD' (left) and Ni-ARD (right)

only upon discovery of ARD that it was shown that the reaction is in fact catalyzed. However, it is clear that acireductone reacts readily with oxygen, and no evidence has yet been found for direct interaction between O_2 and the metal center in either form of ARD, as would be expected for oxygen activation (Dai et al. 2001). Nor is there evidence for even transient changes in metal oxidation state. As such, it is proposed that the role of the metal is to direct the regiochemistry of substrate oxidation by selective Lewis acid activation of either C-2 or C-3 on the substrate (Scheme 3). In the proposed mechanism, the first step for either on- or off-pathway reaction would be chelation of the metal by substrate at the two equatorial positions that do not bear proteinaceous ligands. Ligation would be to the oxygens at C-1 and C-3 in the case of the off-pathway reaction and C-1 and C-2 for on-pathway chemistry. In either case, substrate binding would be followed by initial electron transfer to molecular oxygen, generating a nucleophilic superoxide radical anion. The Lewis acidity of the metal ion would direct the attack of superoxide to form either the five-membered (in the case of ARD) or four-membered (ARD') cyclic peroxide (Ju et al. 2006).

The cyclic peroxides thus formed would undergo electrocyclic rearrangements to yield either the on-pathway (ARD') or off-pathway products (ARD).

Note that in Scheme 3, there is a requirement for isomerization of 1-ene double bond in order to obtain the geometry required for ARD-type (C-1,C-3 oxygens) ligation. The mechanism of this isomerization is unclear. However, model studies of Ni-acireductone complexes support both the geometry of the ligation and demonstrate the production of carbon monoxide from such complexes (Rudzka et al. 2010).

Finally, the role of the structural rearrangement between ARD and ARD' in determining product distribution is unclear. The model complexes that have been synthesized to test the ARD reaction model are sterically quite hindered, suggesting that sterics may be important in controlling the type of chelation observed (Rudzka et al. 2010). Given that the ARD' active site is more accessible than that of ARD, it is possible that substrate is able to bind to ARD' a more extended conformation, yielding the C-1,C-2-attached oxygen ligation preferentially. However, this hypothesis is not supported experimentally as yet (Ju et al. 2006).

Alternative Functions of ARD

The fact that the on-pathway chemistry catalyzed by ARD' occurs spontaneously in vitro suggests that the enzyme may have more complex roles to play than simply directing the last step of the MSP. Given that carbon monoxide has been implicated in a variety of biological signaling pathways, including neurotransmission, apoptosis suppression/induction, and bacterial quorum sensing, it is conceivable that the off-pathway functions of ARD are more significant than the on-pathway function in vivo. Some of these functions may not even require enzymatic activity: The human ARD ortholog (HsARD) was first identified as a permissive trigger for hepatitis infection in nonpermissive cell lines, not as an enzyme. ARD has since been implicated in carcinogenesis in testicular cancer (Oram et al. 2007) and regulation of activity of matrix metalloproteinase I (Hirano et al. 2005). While the role of the metal center in such activity is as yet unclear, we know that Ni(II) binding in KoARD represents a kinetic trap, since the protein must be denatured in order to remove nickel ion, while the iron form loses metal rather easily, and without denaturation of the protein. This would provide a potentially useful switching mechanism for controlling any regulatory binding functions that ARD might provide in nonenzymatic roles.

Cross-References

▶ Iron Proteins, Mononuclear (non-heme) Iron Oxygenases
▶ Nickel-Binding Sites in Proteins

References

Chai SC, Ju TT, Dang M, Goldsmith RB, Maroney MJ, Pochapsky TC (2008) Characterization of metal binding in the active sites of acireductone dioxygenase isoforms from Klebsiella ATCC 8724. Biochemistry 47:2428

Dai Y, Pochapsky TC, Abeles RH (2001) Mechanistic studies of two dioxygenases in the methionine salvage pathway of Klebsiella pneumoniae. Biochemistry 40:6379

Hirano W, Gotoh I, Uekita T, Seiki M (2005) Membrane-type 1 matrix metalloproteinase cytoplasmic tail binding protein-1 (MTCBP-1) acts as an eukaryotic aci-reductone dioxygenase (ARD) in the methionine salvage pathway. Genes Cells 10:565

Ju TT, Goldsmith RB, Chai SC, Maroney MJ, Pochapsky SS, Pochapsky TC (2006) One protein, two enzymes revisited: A structural entropy switch interconverts the two isoforms of acireductone dioxygenase. J Mol Biol 363:523

Oram SW, Ai J, Pagani GM, Hitchens MR, Stern JA, Eggener S, Pins M, Xiao W, Cai X, Haleem R, Jiang F, Pochapsky TC, Hedstrom L, Wang Z (2007) Expression and function of the human androgen-responsive gene ADI1 in prostate cancer. Neoplasia 9:643

Pochapsky TC, Pochapsky SS, Ju TT, Mo HP, Al-Mjeni F, Maroney MJ (2002) Modeling and experiment yields the structure of acireductone dioxygenase from Klebsiella pneumoniae. Nat Struct Biol 9:966

Pochapsky TC, Pochapsky SS, Ju TT, Hoefler C, Liang J (2006) A refined model for the structure of acireductone dioxygenase from Klebsiella ATCC 8724 incorporating residual dipolar couplings. J Biomol NMR 34:117

Rudzka K, Grubel K, Arif AM, Berreau LM (2010) Hexanickel Enediolate Cluster Generated in an Acireductone Dioxygenase Model Reaction. Inorg Chem 49:7623

Xu QP, Schwarzenbacher R, Krishna SS, McMullan D, Agarwalla S, Quijano K, Abdubek P, Ambing E, Axelrod H, Biorac T, Canaves JM, Chiu HJ, Elsliger MA, Grittini C, Grzechnik SK, DiDonato M, Hale J, Hampton E, Han GW, Haugen J, Hornsby M, Jaroszewski L, Klock HE, Knuth MW, Koesema E, Kreusch A, Kuhn P, Miller MD, Moy K, Nigoghossian E, Paulsen J, Reyes R, Rife C, Spraggon G, Stevens RC, Van den Bedem H, Velasquez J, White A, Wolf G, Hodgson KO, Wooley J, Deacon AM, Godzik A, Lesley SA, Wilson IA (2006) Crystal structure of acireductone dioxygenase (ARD) from Mus musculus at 2.06 A resolution. Proteins Struct Funct Bioinform 64:808

Actinide and Lanthanide Systems, High Pressure Behavior

P. Ch. Sahu and Chandra Shekar Nagar Venkataraman
Condensed Matter Physics Division, Materials Science Group, Indira Gandhi Centre for Atomic Research, Kalpakkam, Tamil Nadu, India

Synonyms

$4f$ systems – lanthanide elements; $5f$ systems – actinide elements; f-electron systems – lanthanide and actinide elements; Lanthanides – rare-earths

Definition

In the periodic table, the 15 elements starting from lanthanum (La) to lutetium (Lu) are called lanthanides and the other 15 elements starting from actinium (Ac) to lawrencium (Lr) are called actinides. Pressures above one atmosphere are called high pressure.

The effects of high pressure on these elements leading to changes in their physical behavior, especially structural behavior, are discussed in this entry.

Introduction

The actinide and lanthanide elements occupy a special position in the periodic table. They exhibit very interesting physical and chemical properties and are of great technological importance.

The Lanthanides

The lanthanides comprise the 15 elements (La–Lu) in the periodic table. This series begins with the filling of the 4f electron shell having the electronic configuration: $[Xe]4f^05d^{0-1}6s^2$. The 4f orbitals in the lanthanide series are inner and do not participate in chemical bonding. They are core-like (localized) in nature and the chemistry of the lanthanide ions are in general very similar (Temmerman et al. 2009; Johansson and Brooks 1993). The s and d valence electrons thus dominate in bonding and leave their respective atomic cores in a trivalent state, except for Eu and Yb, which exhibit divalency. This is because of the tendency of the 4f shell to be half-filled or filled, the lone d electron finds energetically more favorable to occupy the 4f shell. Thus the $[Xe]4f^{7/14}5d^06s^2$ configuration in these two elements lead to divalency. The trivalent lanthanide elements are called regular lanthanides.

The atomic volumes of the lanthanides are shown in Fig. 1. In lanthanides, a gradual decrease of atomic volume across the series is found except for Eu and Yb, which are divalent. This gradual decrease is due to the increase in the nuclear charge and partial screening of the electronic charge by the inner 4f electrons seen by the nucleus due to the other outer electrons, leading to the well-known "lanthanide contraction." Their bulk properties also are expected to vary in a gradual manner across the series. For example, as we proceed from La to Lu with increasing atomic number (or decreasing atomic volume), the crystal structure sequence fcc → dhcp → Sm-type (Hexagonal) → hcp is observed at ambient conditions (Johansson and Rosengren 1975; Benedict and Holzapfel 1993; Holzapfel 1995).

The Actinides

The actinide series comprises the 15 elements (Ac–Lr) in the periodic table and begins with the filling of the 5f

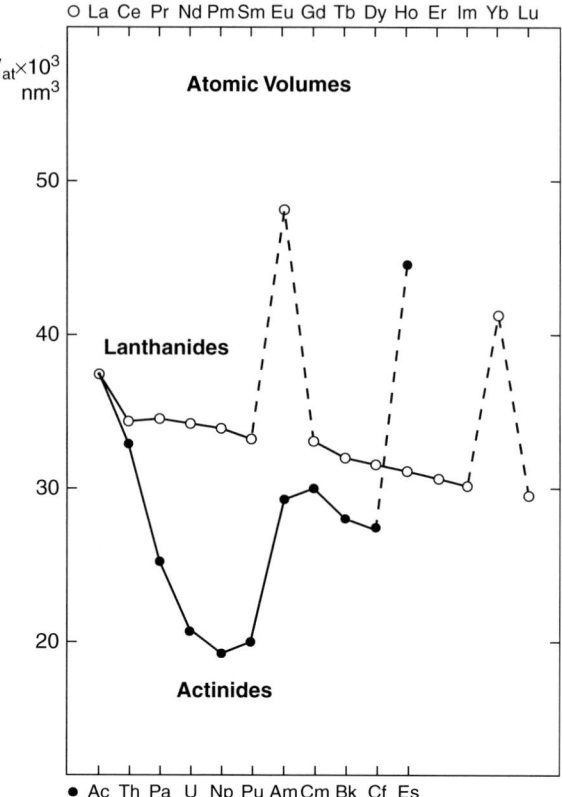

Actinide and Lanthanide Systems, High Pressure Behavior, Fig. 1 The equilibrium atomic volumes of the lanthanide and actinide elements (Benedict and Holzapfel 1993). The deviations for Eu, Yb, and Es are due to their divalency

electron shell having the electronic configuration: $[Rn]5f^06d^{0-1}7s^2$. In actinides, since there is an inner 4f orbital, the electrons in the 5f orbital are pushed out due to core orthogonalization (5f orbitals orthogonal to 4f). The 5f orbitals are thus more extended compared to the 4f and are very close in energy to the 6d. However, as the atomic number increases, the spatial extension of the 5f orbitals shrink and become more localized. In fact, the 5f orbitals in early actinides (up to Pu) are itinerant (delocalized) and participate in bonding like the d orbitals in transition metals, and those in the late actinides (Am onward) are localized like the 4f orbitals in lanthanides (Johansson and Brooks 1993; Johansson and Li 2007; Moore and van der Laan 2009). Thus, the 5f orbitals in actinides exhibit dual nature, and the itinerant to localization transition takes place as a function of atomic number at Am. This can be clearly seen in their atomic volumes in Fig. 1. In early actinides, atomic volumes decrease in a parabolic manner like the 3d/4d transition metal series because the itinerant 5f electrons are less effective in screening the nuclear charge. The late actinides (Am onward) exhibit a gradual decrease in their atomic volumes similar to the lanthanide series.

In actinide systems, most of the physical properties are governed by the nature of their 5f-electron states. For instance, the systems with itinerant 5f-electron states exhibit complex anisotropic physical properties and higher bulk modulus and stabilize in low symmetry crystal structures (tetragonal, orthorhombic, monoclinic, etc.). On the other hand, localized 5f-electron-based systems exhibit local magnetic moments and low bulk modulus and adopt high symmetry crystal structures (dhcp, fcc, etc.) (Johansson and Brooks 1993; Moore and van der Laan 2009).

High Pressure and Its Effect on Materials

Nowadays, high-pressure investigations up to pressures of few megabars (1 megabar = 100 GPa) can be carried out with the help of diamond anvil cells (DACs) using a variety of experimental techniques such as x-ray diffraction, neutron diffraction, Raman spectroscopy, optical absorption, and reflectivity, Mossbauer spectroscopy, NMR, and electrical resistivity (Jayaraman 1983, 1986). The basic principle of a DAC is shown in Fig. 2. It consists of a pair of gem quality diamond anvils with a metal gasket sandwiched between the two diamond flat culets. The metal gasket compressed to a thickness of about 50 μm has a central hole of diameter 100–200 μm, which contains a tiny sample immersed in a pressure-transmitting fluid. An external force is applied on the anvils to compress the fluid and thereby transmitting hydrostatic pressure to the sample. The diamond culets are small enough so that a few kN force is sufficient to generate pressures in the range of 100 GPa. Diamond being transparent to almost the entire electromagnetic spectrum, either an x-ray or laser beam can be focused on the sample through the diamond window to carry out a variety of in situ high-pressure investigations on the pressurized sample. Also, diamond, being the hardest material, enables achieving the highest static pressure (>500 GPa or 5 megabar) in a laboratory.

Application of external pressure on materials brings the constituent atoms closer and closer. This leads to

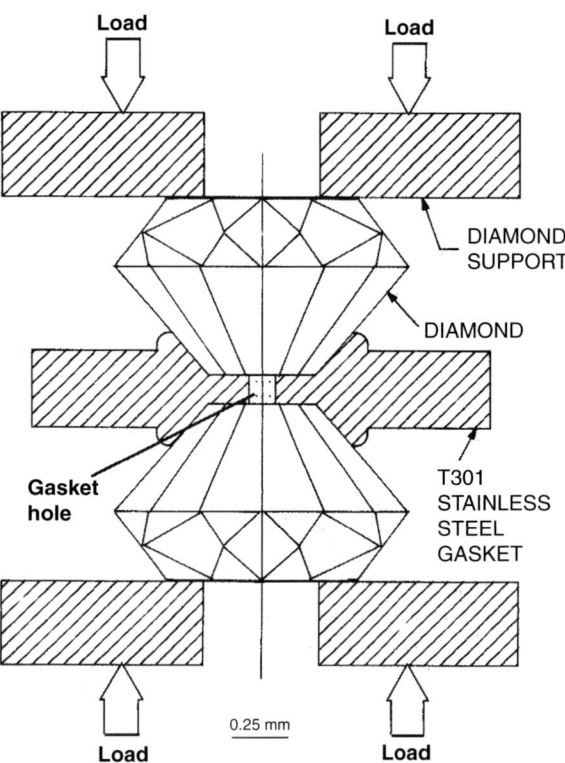

Actinide and Lanthanide Systems, High Pressure Behavior, Fig. 2 Basic principle of a diamond anvil cell for generating very high pressures

phenomenal changes in the materials which can be broadly classified into two categories: the lattice compression and the electronic structure change (Sahu and Chandra Shekar 2007). However, these changes are not totally independent, and often, one is associated with the other. The decrease in interatomic distances or increase in the density leading to changes in the phonon spectra, increase in the free energy (G) and the associated phase transitions stabilizing compact structures constitute the "lattice effects." As the interatomic distances decrease, the overlap of outer electronic orbitals in a solid increases. This leads to the following three principal effects in its electronic structure: (1) broadening of the electronic energy bands (or increase in the energy band widths), (2) shifting of the energy bands with respect to the Fermi energy E_F, and (3) shifting of the Fermi level itself to higher values. All these electronic effects lead to dramatic changes in their physical and chemical properties. For instance, closing of energy gaps lead to metal–insulator transitions, shift in energy bands lead to interband electron and valence transitions, change in the topology of the Fermi surface lead to Lifshitz type transitions and so on.

Effect of Pressure on Lanthanides and Actinides

Lanthanides

We have seen that the 4f orbitals in lanthanides are localized, do not participate in bonding, and contribute to the formation of local magnetic moments. The 5f orbitals in actinides on the other hand exhibit dual nature: itinerant in the first half of the series (up to Pu) and contribute to the bonding, localized in the next half (Am onward) resembling the lanthanides. The nature of these 4f/5f orbitals (i.e., whether itinerant or localized) depends on their f-orbital overlaps, which in turn depend on the interatomic distance in the solid. These interatomic distances or the orbital overlaps can be tuned in a controlled manner by applying external pressure. Thus in systems with localized f orbitals to start with, can be transformed into delocalized states by the application of pressure. Once the f orbitals become delocalized and participate in bonding, the systems exhibit very interesting changes in their crystal structures and physical and chemical properties. Generally, they transform to low-symmetry crystal structures associated with large volume collapse across the transition (Johansson and Rosengren 1975; Benedict and Holzapfel 1993; Holzapfel 1995). The stabilization of low-symmetry crystal structures is the direct manifestation of the highly complicated shaped f orbitals, and the volume collapse is due to their participation in bonding. Their physical properties also become highly anisotropic and the chemical nature changes due to increase in their valency. Also, their local magnetic moments get destroyed due to delocalization of the f orbitals.

In regular lanthanides, we have seen the structure sequence: hcp $\rightarrow$ Sm-type $\rightarrow$ dhcp $\rightarrow$ fcc as a function of decreasing atomic number at ambient conditions. The same structural sequence has been observed in lanthanide elements as function of increasing pressure (Johansson and Rosengren 1975; Benedict and Holzapfel 1993; Holzapfel 1995). This sequence is seen most completely for the heavier lanthanides, while the lighter elements start already at ambient pressure with the high-pressure structures of the heavier elements. However, this pressure-induced structural sequence should not be attributed to the effect of 4f electron states, because similar pressure-induced sequence has been seen in La which does not have any 4f occupancy. Later, energy band calculations show that the d-band contribution to the total energy is what actually drives the lattice through the observed crystal structure sequence (Duthie and Pettifer 1977). Lu, which crystallizes into an hcp structure at STP, has a d-band occupation number of about 1.5. La on the other hand, which crystallizes into a dhcp structure at STP, has a d-band occupation number of the order of 2.7. Pressure has the effect of increasing the energy of electrons in the s-band relative to the d-band, which initiates the s $\rightarrow$ d transfer and the number of d electrons per atom in the conduction band tends toward 3. A similar sequence has been observed in yet another non-4f element yttrium. Subsequent high-pressure investigations have revealed yet another high-pressure phase, known as the distorted fcc phase in La, Y, Pr, Nd, Pm, Sm, Gd, and Yb. This phase transition also is not because of the 4f effect, as both La and Y do not have any 4f occupancy. Hence, the correct high-pressure sequence in lanthanides is: hcp $\rightarrow$ Sm-Type $\rightarrow$ dhcp $\rightarrow$ fcc $\rightarrow$ dist. fcc. At still higher pressures, a number of anomalous low-symmetry phases (monoclinic, tetragonal, α-U type orthorhombic) have been found outside this sequence and their origins have been largely attributed to be the result of 4f-electron delocalization and/or hybridization (Benedict and Holzapfel 1993; Johansson and Brooks 1993). The 4f delocalization transitions are generally accompanied with large volume collapse. Ce with fcc structure at STP and having only one 4f electron on the verge of delocalization, in fact, undergoes an isostructural transition with a volume collapse of about 16% at 0.7 GPa due to delocalization of its 4f electron. However, it does not stabilize in any low-symmetry structure characteristic of delocalized 4f due to the dominant 5d states, which favors fcc structure. It transforms to low-symmetry bcm and then to bct phase at 5.3 and 12.5 GPa, respectively, due to the increased delocalization of the 4f states. The structural transitions observed in lanthanides under pressure are summarized in Fig. 3 (Schiwek et al. 2002). Some recent results on Pr, Eu, Gd, Tb, Dy, and Tm reporting more structural transitions at multimegabar pressures are not shown in Fig. 3 and it is beyond the scope of this entry to include these results and cite the relevant references.

metal. Although, Pm has been found to exhibit the structural sequence: dhcp → fcc → trigonal under pressure and falls in line with other trivalent lanthanides (Benedict and Holzapfel 1993; Holzapfel 1995), it is not shown in this diagram.

The divalent lanthanide elements Eu and Yb, which stabilize in the bcc and fcc structure, respectively, at STP, are expected to exhibit normal behavior and enter the rare-earth crystal structure sequence at higher pressures. In fact, Eu transforms to the hcp structure at 12.5 GPa and Yb shows the structural sequence: fcc → bcc → hcp → fcc → hp3 at 4, 26, 53, and 98 GPa, respectively. The hp3 phase has been found to be stable up to 202 GPa. The divalent to trivalent transition in Yb takes place gradually and behaves like the regular trivalent lanthanides Sm and Nd (Chesnut and Vohra 1999).

Actinides

The complex structures at atmospheric pressure of the lighter actinides are a consequence of itinerant 5f electrons. These metals are expected to be less compressible (i.e., exhibit larger bulk moduli) and exhibit few phase transitions under pressure. In contrast, the transplutonium metals (Am onward) have localized 5f electrons not contributing to bonding at atmospheric pressure, and are therefore expected to be "soft" (lower bulk moduli). Under pressure, the localized 5f electrons are expected to become itinerant due to more orbital overlap and acquire structures displayed by the light actinides (Benedict and Holzapfel 1993; Holzapfel 1995; Lindebaum et al. 2003; Moore and van der Laan 2009).

The structural transitions observed in actinides under pressure are summarized in Fig. 5. The light actinides with low-symmetry structures almost remain stable and show high bulk modulus (120–150 GPa). Pu, however, has a low bulk modulus of 40 GPa, as its 5f states are on the verge of localization and exhibit complex behavior. It transforms from its monoclinic to hexagonal phase at about 40 GPa. Pa also transforms from bct to α-U structure at 77 GPa. Th, which adopts the fcc structure at STP, has no 5f occupancy. However, its 5f level is located very close (about 1 eV above) to the Fermi level (E_F). Under high pressure, the 5f level gets lowered below E_F, giving rise to finite occupancy of the 5f states, and contributes to bonding. It transforms from its high-symmetry fcc to a low-symmetry bct structure at ~60 GPa and behaves like other light actinide metals.

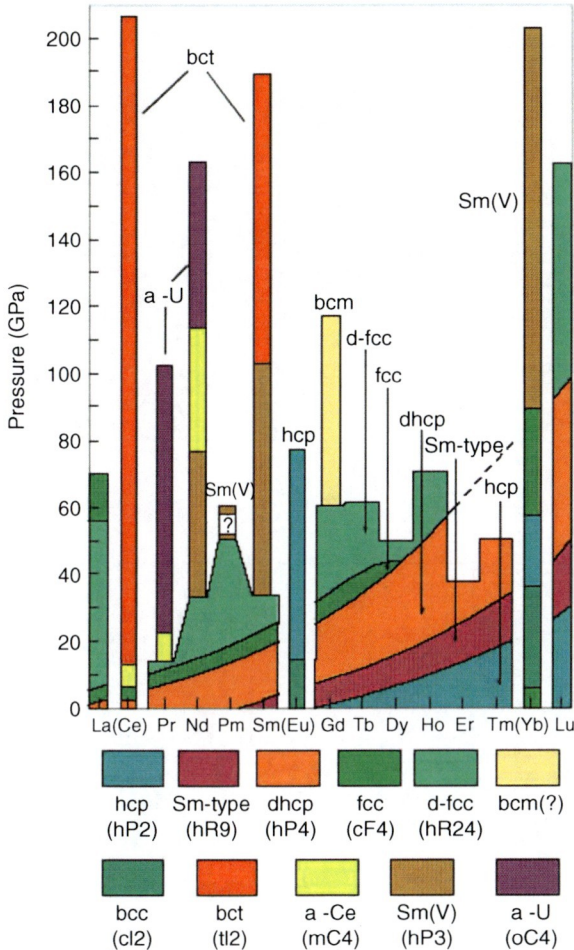

Actinide and Lanthanide Systems, High Pressure Behavior, Fig. 3 Generalized high-pressure phase diagram of the lanthanides (Schiwek et al. 2002)

Johansson and Rosengren (1975) had constructed an interesting generalized P–T phase diagram for the lanthanides as shown in Fig. 4. The very fact that such a diagram could be constructed, based on the individual P–T phase diagrams of the lanthanides, gives very strong evidence for the close similarity between the lanthanide elements. In fact, this generalized phase diagram clearly demonstrates the inertness of the 4f electrons as regards their participation in the bonding. One can also clearly notice the absence of Eu and Yb in this diagram, the reason being their divalent metallic behavior in contrast to the common trivalent lanthanides. Also, the absence of Ce is due to the reason that high pressure destroys the inertness of its 4f electron, and therefore, it is no longer an adequate trivalent

Actinide and Lanthanide Systems, High Pressure Behavior, Fig. 4 A generalized P–T phase diagram of the trivalent lanthanides (Johansson and Li 2007). The empty spaces are for the missing elements Ce, Eu, Yb, and Pm (reasons given in section "Lanthanides")

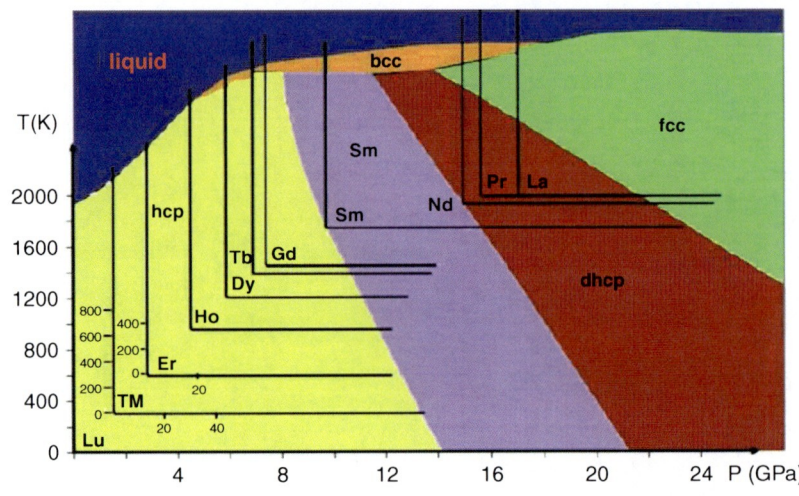

Actinide and Lanthanide Systems, High Pressure Behavior, Fig. 5 A "pseudobinary" phase diagram of the 5f actinide metals as a function of pressure (Moore and van der Laan 2009). The pressure behaviors of Np and Pu are not shown, but the ground-state crystal structure of each metal is indicated. Under high pressure, Pu transforms to a hexagonal structure at 37 GPa, and Np remains stable in its orthorhombic phase up to 52 GPa. The area indicated by "structure not determined" represents α-U type structures

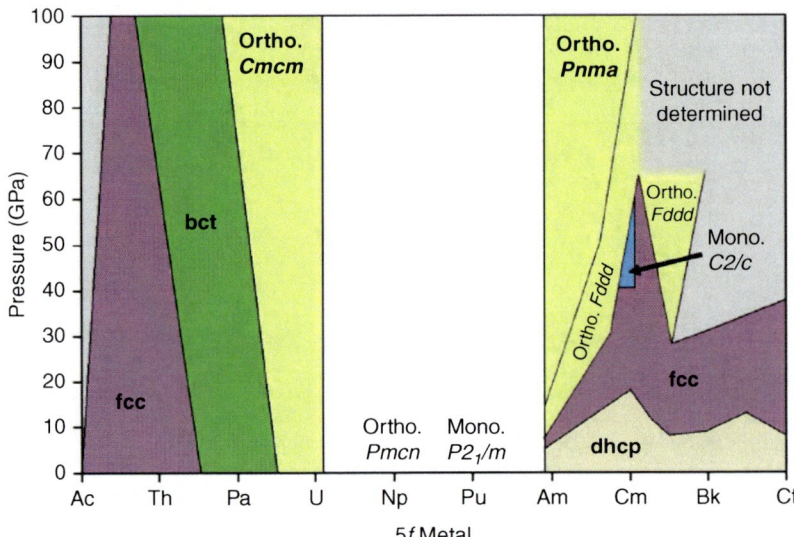

The late actinides, as expected, behave like the lanthanides under pressure. They transform from their dhcp to fcc/dfcc and eventually to low-symmetry (trigonal, α-U type orthorhombic) structures due to delocalization of their 5f states. The 5f delocalization with volume collapse takes place in one or two steps. For example, in Am, the fcc → fco (face centered oprthorhombic) transition at 10 GPa with a volume collapse of 2% takes place due to partial delocalization of 5f states, then it transforms to the α-U type structure at 16 GPa with a volume collapse of 7%. Similarly, in Cm and Cf, the 5f delocalization takes place in two steps, where as in Bk, it takes place in one step. Their bulk moduli, which are of the ~40 GPa and are similar to that of the lanthanides, dramatically increase to 120–150 GPa on delocalization of their 5f states and behave like the early actinides.

Their local magnetic moments also get destroyed due to delocalization of the 5f electrons. Some of the actinide elements like Th, Pa, and U also exhibit superconductivity at very low temperatures (Moore and van der Laan 2009). High pressure has very interesting effect on the superconducting transition temperature (T_C) of these metals. Whereas the T_C decreases with pressure in Th, it increases in case of Pa and U, and this difference in behavior has been attributed to the presence of 5f states.

Conclusions

The lanthanides having localized 4f states exhibit the structural sequence hcp → Sm-type → dhcp → fcc → dist. fcc as a function of increasing pressure or decreasing atomic number. At higher pressures, they exhibit volume collapse transitions due to delocalization of their 4f states and stabilize in low-symmetry structures. Their bulk moduli increase to very high values across the 4f delocalization transitions. In case of actinides, the light actinides with low-symmetry structures show high bulk modulus (120–150 GPa) and few structural transitions. The late actinides (Am onward) with localized 5f states and dhcp structure have very low bulk moduli (~40 GPa) like the lanthanides. They transform from their dhcp structure to fcc/dfcc and eventually to low symmetry α-U type structures due to delocalization of their 5f states and exhibit very high bulk moduli, similar to the early actinides.

Cross-References

▶ Actinides, Physical and Chemical Properties
▶ Lanthanides, Physical and Chemical Characteristics

References

Benedict U, Holzapfel WB (1993) High pressure studies and structural aspects. In: Gschneidner KA Jr, Eyring L, Lander GH, Choppin GR (eds) Handbook on the physics and chemistry of rare-earths, vol 17, Lanthanide and actinide physics – I. Elsevier, Amsterdam
Chesnut GN, Vohra YK (1999) Structural and electronic transitions in ytterbium metal to 202 GPa. Phys Rev B 82:1712–1715
Duthie JC, Pettifer DG (1977) Correlation between d-band occupancy and crystal structure in the rare earths. Phys Rev Lett 38:564–567
Holzapfel WB (1995) Structural systematics of 4f and 5f elements under pressure. J Alloys Compd 223:170–173
Jayaraman A (1983) Diamond anvil cell and high pressure physical investigations. Rev Mod Phys 55:65–108
Jayaraman A (1986) Ultrahigh pressures. Rev Sci Instrum 57:1013–1031
Johansson B, Brooks MSS (1993) Theory of cohesion in rare earths and actinides. In: Gschneidner KA Jr, Eyring L, Lander GH, Choppin GR (eds) Handbook on the physics and chemistry of rare-earths, vol 17, Lanthanide and actinide physics – I. Elsevier, Amsterdam
Johansson B, Li S (2007) The actinides – a beautiful ending of the periodic table. J Alloys Compd 444–445:202–206
Johansson B, Rosengren A (1975) Generalised phase diagram for the rare-earth elements: calculations and correlations of the bulk properties. Phys Rev B 11:2836–2857
Lindebaum A, Heathman S, Le Bihan T, Haire RG, Idiri M, Lander GH (2003) High pressure crystal structures of actinide elements to 100 GPa. J Phys Condens Matter 15: S2297–S2303
Moore KT, van der Laan G (2009) Nature of the 5f states in actinide metals. Rev Mod Phys 81:235–298
Sahu P Ch, Chandra Shekar NV (2007) High pressure research on materials – 2. Experimental techniques to study the behaviour of materials under high pressure. Resonance 12:49–64
Schiwek A, Porsch F, Holzapfel WB (2002) Phase diagrams for cerium – lanthanide alloys under pressure. In: Krell U, Schneider JR, von Zimmermann M (eds) Scientific Contributions. Part 1, HASYLAB Annual Report 2002, Hamburg
Temmerman WM, Petit L, Svane A et al (2009) The dual, localized or band-like character of the 4f-states. In: Gschneidner KA Jr, Bunzli JCG, Pencharsky VK (eds) Handbook on the physics and chemistry of rare-earths, vol 39. Elsevier, Amsterdam

Actinides, Interactions with Proteins

Mohamed Larbi Merroun
Departamento de Microbiología, Facultad de Ciencias, Universidad de Granada, Granada, Spain

Synonyms

Complexation of actinides by proteins; Coordination of actinides to proteins; Uptake of actinides by proteins

Definition

The coordination of actinides to proteins largely through functional groups including carboxyl, hydroxyl, phosphate, etc.

Introduction

Actinides are chemical elements with atomic numbers ranging from 89 (actinium) to 103 (lawrencium) (Connelly et al. 2005). Five actinides have been found in nature: thorium, protoactinium, uranium, neptunium, and plutonium. Uranium is a widely distributed and occurs in almost all soils. Thorium is present at low levels in rocks and soils.

Small quantities of persisting natural plutonium have also been identified in the environment. Transients amounts of protactinium and atoms of neptunium are produced from the radioactive decay of uranium and from transmutation reactions in uranium ores, respectively (Asprey 2012). The other actinides (curium, americium, etc.) are synthetic elements (Darling 2011). Naturally occurring uranium and thorium and synthetically produced plutonium are the most abundant actinides on earth.

The distribution of actinides in the environment is continuously incremented due to human activities including nuclear weapon production and nuclear energy production, the storage of radioactive wastes, etc. The risk assessment of the distribution of these radioactive elements in the environment on human health is needed and implies the understanding of the mechanisms of interaction of these toxic elements with living organisms including humans, animals, microbes, and their cellular components (e.g., proteins, ADN, and ARN).

Proteins are one of the most important biomolecules involved in the complexation of heavy metals, including actinides. Ionizable functional groups such as carboxyl, phosphoric, amine, and hydroxyl represent potential binding sites for the sequestration of metal ions. Of these groups, carboxyl and phosphate are the main binding sites of actinides (Merroun et al. 2005).

A good characterization of the actinide/protein system implies the determination of the structural parameters of the formed complexes (nature of neighboring atoms to actinide, bond distances, etc.). Among the analytical techniques used to determine these parameters, synchrotron-based techniques such as X-ray absorption spectroscopy (XAS) are of great importance. This technique has been used to determine the oxidation state (X-ray absorption near edge structure, XANES) and to identify the number of atoms and their distances in the local structural environment (extended X-ray absorption fine structure, EXAFS) of actinides within a variety of proteins (Merroun et al. 2005; Merroun and Selenska-Pobell 2008). In addition, XAS is among the few analytical methods that can provide information on the chemical environment of actinides in biological samples at dilute metal concentrations. XAS is a nondestructive method, and no sample reduction or digestion is required which would alter the chemistry of the element of interest.

The molecular scale characterization of actinide/protein interactions is of great interest to understand the potential transport of radionuclide inside living organism (Jeanson et al. 2009). In addition, the structural information obtained will help to design and synthesize potential specific chelating agents for actinides and their use in the elimination of incorporated radionuclides (Jeanson et al. 2009). It also provides insights on the role played by these biomolecules, particularly of microbial origin, on the mobility of these inorganic contaminants in the environment.

Interactions of Actinides with Microbial Proteins

Microbial cells may synthesize a variety of metal-binding peptides and proteins, e.g., metallothioneins and phytochelatins, which regulate metal ion homeostasis and affect toxic responses (Avery 2001). In addition, most archaea and a large number of bacteria possess proteinaceous layers external to their cell walls (Sleytr et al. 1996) which play a major role in the coordination of metal ions.

From the point of view of environmental impact, the interactions of actinides with microbial proteins are poorly studied, and most of these studies were focused on U. The best example of the proteins/uranium coordination studies is represented by bacterial surface layers proteins (S-layer).

Microbial S-Layers and Actinides

The crystalline bacterial S-layer, as probably the most abundant bacterial cellular proteins, represents the outermost cell envelope component of diverse types of bacteria and archaea (Sára and Sleytr 2000). S-layers are generally composed of identical protein or glycoprotein subunits, and they completely cover the cell surface during all stages of bacterial growth and division. Most S-layers are 5–15-nm thick and possess pores of identical size and morphology in the range of 2–6 nm (Beveridge 1994). As porous lattices completely covering the cell surface, the S-layers can provide prokaryotic cells with selective advantages by functioning as protective coats, as structures involved in cell adhesion and surface recognition, and as molecule or ion traps (Sleytr 1997). Surface layer sheets are, in some cases, glycosylated (Sleytr et al. 1996),

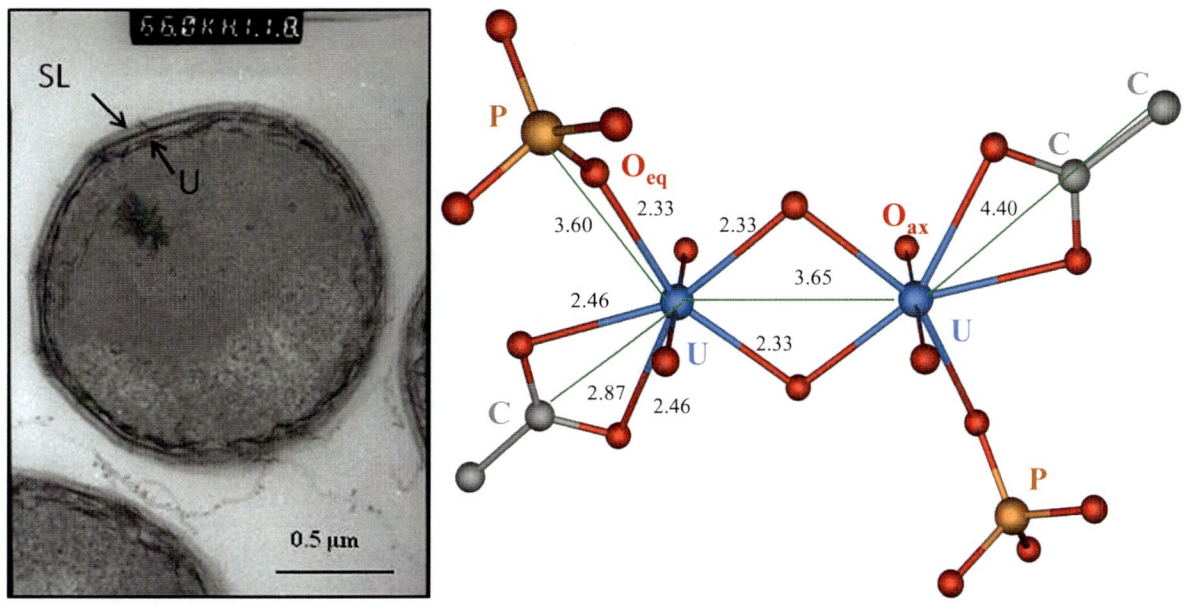

Actinides, Interactions with Proteins, Fig. 1 Transmission electron micrograph (*left*) of thin section of *L. sphaericus* JG-7B cells treated with uranium (U), where the metal is accumulated mainly within the inner side of S-layer protein (SL) of this bacterium. Model (*right*) used for the fit of the EXAFS spectrum of U complexes formed the S-layer of the strain JG-7B (Reprinted with permission from Merroun et al. (2005) American Society of Microbiology)

possess a large number of carboxylated amino acids (Sleytr et al. 1996), and can be also phosphorylated (Merroun et al. 2005). Johnson and Bardford (1993) reported that one-third of proteins in cells are phosphorylated. The phosphorylation of proteins is needed for the many cellular functions including regulation of gene transcription, signal transduction throughout the cell, regulation of enzyme activities, etc. (Johnson and Bardford 1993). These functions would be damaged by the coordination of proteins with heavy metals. Therefore, the interactions of heavy metals with proteins are of great importance (Fahmy et al. 2006).

The coordination of uranium to S-layer proteins of two uranium mining waste bacterial isolates, *Lysinibacillus sphaericus* (formerly *Bacillus sphaericus*) JG-A12 and JG-7B, has been extensively studied using molecular scale techniques (Merroun et al. 2005; Merroun and Selenska-Pobell 2008). X-ray absorption (XAS) studies showed that, in both bacterial S-layer proteins, U(VI) is coordinated to carboxyl groups in a bidentate fashion with an average distance between the U atom and the C atom of 2.88 ± 0.02 Å and to phosphate groups in a monodentate fashion with an average distance between the U atom and the P atom of 3.62 ± 0.02 Å (Merroun et al. 2005). The local coordination of U within the S-layer protein of these two bacteria is illustrated in the Fig. 1.

Analyses of the amino acid composition of the S-layer proteins of the strain JG-A12 demonstrated a high content of glutamic acid and aspartic acid and of other amino acids such as serine and threonine. The C-terminal part especially consists of stretches of glutamic acid and aspartic acid (both residues with carboxyl groups) and of serine and threonine (both residues with hydroxyl groups), the latter being potential phosphorylation sites. The carboxyl and phosphate groups of these amino acids are probably implicated in the complexation of uranium. Acidic amino acids (Asp and Glu) or the free carboxylate terminus has been demonstrated to be the binding sites for uranium in most of the structure of uranyl protein complexes found in the Protein Data Bank (PDB) (Van Horn and Huang 2006).

Compared to the S-layer protein of the closely related strain *L. sphaericus* NCTC 9602, the S-layer protein of the uranium mining waste pile isolate *L. sphaericus* JG-A12 contains about six times more phosphorus (Merroun et al. 2005). This difference may reflect an adaptation of *L. sphaericus* JG-A12 to its highly with uranium-contaminated environment.

Actinides, Interactions with Proteins, Fig. 2 (a) Transmission electron micrograph of thin section of the cells of *Lysinibacillus sphaericus* JG-A12 treated with europium (III). The europium is coordinated mainly to S-layer proteins of this bacterium. (b) Spectrum of energy dispersive X-ray analysis of the Eu/S-layer protein accumulates

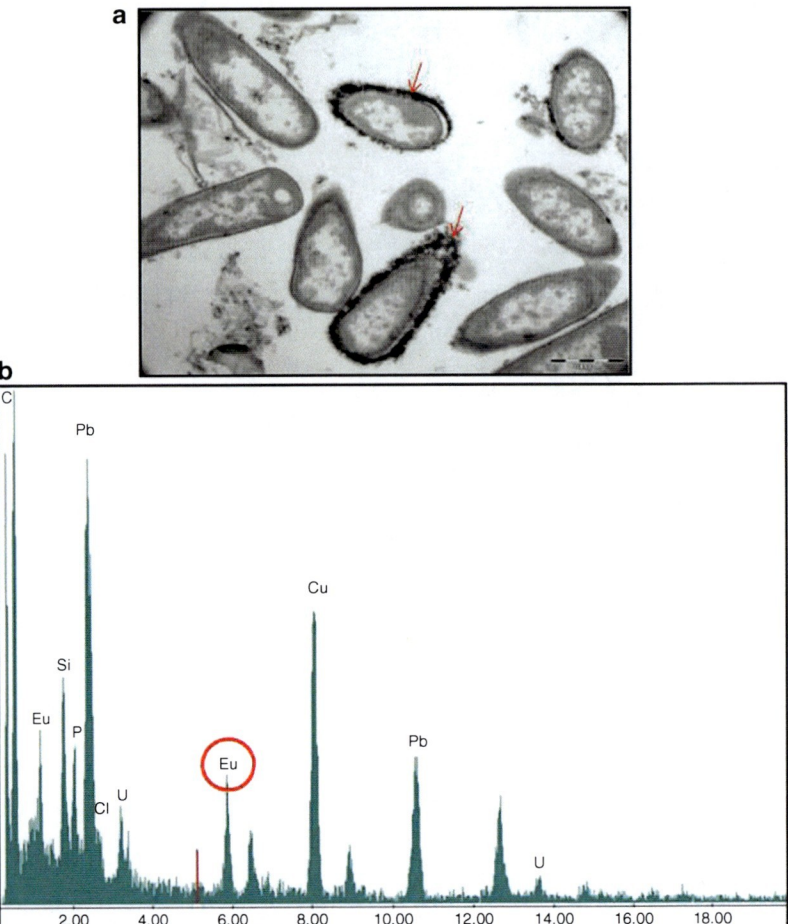

The high uranium affinity of phosphate groups on the surface of the cell may allow *L. sphaericus* JG-A12 to bind selectively large amounts of uranium before getting damaged by this toxic radionuclide. Figure 1 shows the cellular localization of the deposited U on the internal side of the S-layer protein, where the negatively charged groups are localized, and also into the cross-linked peptidoglycan layers of the cell wall.

The S-layer of *L. sphaericus* JG-A12 is involved in interactions not only with uranium but also with other radionuclides. Figure 2 shows, for instance, an example of accumulation of the inactive analog of the trivalent actinide curium(III), namely, the trivalent lanthanide europium(III), onto the surface of the strain. In this case, as well as in the case of U, the S-layer proteins may act as a molecular sieve, ion trap, or protective shell (Sleytr 1997). Because of the ability of S-layer to replace the "older" S-layer sheets on the cell surface, one can speculate about the mechanism of its protective function against uranium and other toxic metals. The saturation with metals (e.g., uranium) may lead to denaturation of the S-layer lattice, which is then replaced by freshly synthesized protein monomers (Merroun et al. 2005; Merroun and Selenska-Pobell 2008).

Coordination of Human/Animal Proteins to Actinides

In humans, the accidental release of actinides results in an internal contamination with these elements including uranium, thorium, neptunium, plutonium, and americium. Independently of the route of contamination, the radionuclide is absorbed into and then transported by blood (Jeanson et al. 2009). Several authors investigated the actinide/blood protein interactions in order to understand the behavior of actinides in biological tissues. Most of these studies were focused on uranium, and few investigations were performed to elucidate the mechanisms of coordination of proteins to other actinides like Th, Np, Pu, Am, etc.

The coordination of tetravalent actinides (actinides (IV)) (thorium, neptunium, plutonium) by transferrin (the iron transport glucoprotein) was studied using a combination of spectroscopic techniques in including UV/Vis spectroscopy, X-ray absorption spectroscopy, etc. (Jeanson et al. 2010). The main objective of the later work is to compare the behavior of these elements to that of iron(III), the endogenous transferrin cation, from a structural point of view. There are some similarities between the behavior of Fe(III) and Np(IV)/Pu(IV). At least one tyrosine residue probably participate in the actinide coordination sphere, forming a mixed hydroxy-transferrin complex in which actinides are bound with transferrin both through actinide–tyrosine and through actinide–OH bonds (Jeanson et al. 2010). Grossmann et al. (1992) suggested that transferrin binds actinides(IV) in the iron complexation site which include two tyrosine, one histidine, and one aspartate. Th(IV) was not coordinated to ferritin under the experimental conditions of this work. In the case of calmodulin, Seeger et al. (1997) showed that plutonium (III) is coordinated to calcium site of the this protein.

In the case of uranium in the serum, the uranyl cation, a linear dioxo cation with an overall +2 charge (UO_2^{2+}), occurs as uranyl bis- and tris-carbonate complexes and UO_2/protein/carbonate complexes with human serum albumin (HSA, the most abundant serum protein implicated in the transport of essential divalent cations including Ni(II), Cu(II), Zn(II), etc.) and transferrin (TF). Of the two fractions, the uranyl–carbonate complexes are more diffusible into tissue (liver, kidneys, bones, etc.) while the portion bound to protein is the portion cleared from the serum via transport to and elimination from the kidneys (Van Horn and Huang 2006). The U(VI)–TF complexation constant values were slightly higher than those observed for U(VI)–HSA complex (Michon et al. 2010) which explain the high U binding capacities of the HSA protein(10 mol of U(VI) in comparison to that of the TF protein (5 mol of U(VI)).

Phosphovitin, from egg yolk, is another protein chosen as ideal model compound to study the complexation of organic phosphate residues of proteins with actinides. This protein is slightly glycosylated and highly phosphorylated being one of the most phosphorylated proteins in nature where 8–10% of the molecular weight composes of phosphorus (Byrne et al. 1984). The complexation of phosphovitin with U(VI) studied at different phosphate/uranium ratios using attenuated total reflection Fourier transform infrared (ATR FT-IR), indicated the implication of two phosphate groups in the coordination of this radionuclide (Li et al. 2010).

In addition to proteins, small peptides were used as models for studying the nature of proteins functional groups implicated in the coordination of actinides. Thus, a linear pentapeptide, acetyl-diaspartyl-prolyl-diaspartyl-amide, was studied as potential chelating ligand for thorium(IV), neptunium(IV), and plutonium(IV) cations forming polynuclear species with oxy- or hydroxo-bridged cations (Jeanson et al. 2009).

Conclusion

This entry described the molecular scale characterization of interactions between actinides and proteins of different origin (human, animal, and microbial). Carboxyl and phosphates groups are the main functional groups involved in the binding of actinides. Protein/actinides interaction research is an inherently multidisciplinary and requires a combination of wet chemistry, spectroscopy, microbiology, biochemistry, and radiochemistry. This emerging area of sciences poses many opportunities and challenges for microbiologists, biochemists, and toxicologists. Proteins could be used as biosorbent for bioremediation of actinide contaminated waters. A more immediate use of research results is in predicting the environmental and public health consequences of nuclear waste isolation.

Cross-References

▶ Toxicity

References

Asprey LB (2012) "actinoid element." Encyclopædia Britannica. Encyclopædia Britannica Online. Encyclopædia Britannica Inc., 2012. Web. 11 Jan. 2012. http://www.britannica.com/EBchecked/topic/4354/actinoid-element

Avery SV (2001) Metal toxicity in yeast and the role of oxidative stress. Adv Appl Microbiol 49:111–142

Beveridge T (1994) Bacterial S-layers. Curr Opin Struct Biol 4:204–212

Byrne BM, Schip ADV, Vandeklundert JAM, Arnberg AC, Gruber M, Ab G (1984) Amino acid sequence of phosphovitin derived from nucleotide sequence of part of the chicken vitellogenin gene. Biochemistry 23: 4275–4279

Connelly NG et al (2005) "Elements" nomenclature of inorganic chemistry. Royal Society of Chemistry, London

Darling D (ed) (2011) Encyclopedia of science online

Fahmy K, Merroun ML, Pollmann K, Raff J, Savchuk O, Hennig C, Selenska-Pobell S (2006) Secondary structure and Pd(II) coordination in S-layer proteins from *Bacillus sphaericus* studied by infrared and X-ray absorption spectroscopy. Biophys J 91:996–1007

Grossmann G, Neu M, Pantos E et al (1992) X-ray solution scattering reveals conformational changes upon iron uptake in lactoferrins, serum and ovaltransferrins. J Mol Biol 225:811–819

Jeanson A, Berthon C et al (2009) The role of aspartyl-rich pentapeptides in comparative complexation of actinide(IV) and iron(III). New J Chem 33:976–985

Jeanson A, Ferrand M, Funke H, Hennig C, Moisy P, Solari PL, Vidaud C, Den Auwer C (2010) The role of transferrin in actinide(IV) uptake: comparison with iron(III). Chem A Eur J 16:1378–1387

Johnson LN, Bardford D (1993) The effect of phosphorylation on the structure and function of proteins. Ann Rev Biophys Biomol Struct 22:199–232

Li B, Raff J, Barkleit A, Bernhard G, Foerstendorf H (2010) Complexation of U(VI) with highly phosphorylated protein, phosvitin a vibrational spectroscopic approach. J Inorg Biochem 104:718–725

Merroun ML, Selenska-Pobell S (2008) Bacterial interactions with uranium: and environmental perspective. J Cont Hydrol 102:285–295

Merroun ML, Raff J, Rossberg A, Hennig C, Reich T, Selenska-Pobell S (2005) Complexation of uranium by cells and S-layer sheets of *Bacillus sphaericus* JG-A12. Appl Environ Microbiol 71:5532–5543

Michon J, Frelon S, Garnier C, Coppin F (2010) Determinations of uranium(VI) binding properties with some metalloproteins (transferring, albumin, metallothionein and ferritin) by fluorescence quenching. J Fluoresc 20: 581–590

Pollmann K, Raff J, Merroun ML, Fahmy K, Selenska-Pobell S (2006) Metal binding by bacteria from uranium mining waste piles and its technological applications. Biotechnol Adv 24:58–68

Sára M, Sleytr UB (2000) S-layer proteins. J Bacteriol 182:859–868

Seeger PA et al (1997) Neutron resonance scattering shows specific binding of plutonium to the calcium binding sites of the protein calmouldin and yields precise distance information. J Am Chem Soc 119:5118–5125

Sleytr UB (1997) I. Basic and applied S-layer research: an overview. FEMS Microbiol Rev 20:5–12

Sleyter UB, Messner P, Pum D, Sára MRG (1996) Crystalline bacterial cell surface proteins. Academic/RG Landes, London

Van Horn JD, Huang H (2006) Uranium(VI) bio-coordination chemistry from biochemical, solution and protein structure data. Coord Chem Rev 250:765–775

Actinides, Physical and Chemical Properties

Fathi Habashi
Department of Mining, Metallurgical, and Materials Engineering, Laval University, Quebec City, Canada

Actinides are a group of metals following actinium in the periodic table. The first member is thorium (discovered in 1829) followed by protactinium (discovered in 1917), then uranium (discovered in 1789). The trans-uranium elements do not occur in nature and were prepared artificially. They became an important group of metals after the discovery of the phenomenon of fission in 1939.

Element	Symbol	Atomic number	Atomic weight	Occurrence or preparation
Actinium	Ac	89	227	Relative abundance 3×10^{-14}
Thorium	Th	90	232.038	Relative abundance 1.1×10^{-3}
Protactinium	Pa	91	231	Relative abundance 8×10^{-11}
Uranium	U	92	238.03	Relative abundance 4×10^{-4}
Neptunium	Np	93	237	Bombarding ^{238}U by neutrons
Plutonium	Pu	94	242	Bombarding ^{238}U by deuterons
Americium	Am	95	243	Bombarding ^{239}Pu by neutrons
Curium	Cm	96	247	Bombarding ^{239}Pu by α-particles
Berkelium	Bk	97	247	Bombarding ^{241}Am by α-particles
Californium	Cf	98	249	Bombarding ^{242}Cm by α-particles
Einsteinium	Es	99	254	Product of nuclear explosion
Fermium	Fm	100	253	Product of nuclear explosion
Mendelevium	Md	101	256	Bombarding ^{253}Es by α-particles

(*continued*)

Actinides, Physical and Chemical Properties

Element	Symbol	Atomic number	Atomic weight	Occurrence or preparation
Nobelium	No	102	254	Bombarding ^{243}Am by ^{15}N or ^{238}U with α-particles
Lawrencium	Lw	103	257	Bombarding ^{252}Cf by ^{10}B or ^{11}B and of ^{243}Am with ^{18}O

Only thorium, uranium, and plutonium are of technical importance for nuclear energy.

Position in the Periodic Table

																	1 H	2 He
3 Li	4 Be											5 B	6 C	7 N	8 O	9 F	10 Ne	
11 Na	12 Mg	13 Al										14 Si	15 P	16 S	17 Cl	18 Ar		
19 K	20 Ca	21 Sc	22 Ti	23 V	24 Cr	25 Mn	26 Fe	27 Co	28 Ni	29 Cu	30 Zn	31 Ga	32 Ge	33 As	34 Se	35 Br	36 Kr	
37 Rb	38 Sr	39 Y	40 Zr	41 Nb	42 Mo	43 Tc	44 Ru	45 Rh	46 Pd	47 Ag	48 Cd	49 In	50 Sn	51 Sb	52 Te	53 I	54 Xe	
55 Cs	56 Ba	57 La	72 Hf	73 Ta	74 W	75 Re	76 Os	77 Ir	78 Pt	79 Au	80 Hg	81 Tl	82 Pb	83 Bi	84 Po	85 At	86 Rn	
87 Fr	88 Ra	89 Ac	90 Th	91 Pa	92 U	93 Np	94 Pu	95 Am	96 Cm	97 Bk	98 Cf	99 Es	100 Fm	101 Md	102 No	103 Lw		

Physical and Nuclear Properties

	Thorium 232	Uranium 238	Plutonium 239
Meting point, °C	1,842	1,150	639.5
Density	11.7	19.1	19.8
Crystal structure	Face-centered cubic	Body-centered cubic	Body-centered cubic
Isotope percentage, %	100	99.28	
Half life, years	2.2×10^{10}	4.5×10^9	2.4×10^4
Type of decay	α-emission	α-emission	α-emission

Uranium 235 undergoes fission when bombarded by thermal neutrons; it breaks apart into two smaller elements and at the same time emitting several neutrons and a large amount of energy:

$$^{235}_{92}U + ^1_0n \rightarrow ^{89}_{36}Kr + ^{144}_{56}Ba + 3^1_0n + 200 \text{ MeV}$$
$$\downarrow \qquad \qquad \downarrow$$
$$^{89}_{39}Y + 3e^- \qquad ^{144}_{60}Nd + 4e^-$$

Uranium 238 absorbs neutrons forming uranium 239 which is a beta emitter with a short half-life; its daughter neptunium 239 also emits an electron to form plutonium 239.

$$^{238}_{92}U + ^1_0n \rightarrow ^{239}_{92}U \rightarrow ^{239}_{93}Np + e^-$$
$$\downarrow$$
$$^{239}_{94}Pu + e^-$$

Plutonium 239 undergoes fission with the emission of several neutrons and can maintain a chain reaction. Thus, a nuclear reactor using uranium as a fuel, not only produces energy, but also produces another nuclear fuel. Under certain conditions it is possible to generate fissionable material at a rate equal to or greater than the rate of consumption of the uranium. Such a reactor is known as a *breeder reactor*.

Thorium absorbs neutrons and is transformed to uranium 233:

$$^{232}_{90}Th + ^1_0n \rightarrow ^{233}_{90}Th \rightarrow ^{233}_{91}Pa + e^-$$
$$\downarrow$$
$$^{233}_{92}U + e^-$$

Uranium 233 also undergoes fission and can be used to operate a breeder reactor in the presence of a thorium blanket according to the scheme:

$$^{233}_{92}U + ^1_0n \rightarrow \text{Fission products} + \text{Energy}$$

One neutron that is captured by $^{232}_{90}Th$ / One neutron to continue chain reaction

$$^{232}_{90}Th$$
$$\downarrow$$
$$^{233}_{90}Th$$
$$\downarrow$$
$$^{233}_{91}Pa + e^-$$
$$\downarrow$$
$$^{233}_{92}U + e^-$$

The electronic configuration of actinides is similar to the lanthanides in that the two outermost shells are the same (with minor exceptions due to quantum chemistry considerations).

Ac	Th	Pa	U	Np	Pu	Am	Cm	Bk	Cf	Es	Fm	Md	No	Lw
2	2	2	2	2	2	2	2	2	2	2	2	2	2	2
8	8	8	8	8	8	8	8	8	8	8	8	8	8	8
18	18	18	18	18	18	18	18	18	18	18	18	18	18	18

(continued)

Ac	Th	Pa	U	Np	Pu	Am	Cm	Bk	Cf	Es	Fm	Md	No	Lw
32	32	32	32	32	32	32	32	32	32	32	32	32	32	32
19	20	21	22	23	24	25	26	27	29	30	31	32	32	32
9	9	9	9	9	9	9	9	9	9	8	8	8	8	9
2	2	2	2	2	2	2	2	2	2	2	2	2	2	2

Main Compounds

- Thorium oxide, ThO_2, from which the metal is produced by reduction with calcium.
- Uranium tetrafluoride, UF_4, from which the metal is produced by reduction with magnesium
- Uranium hexafluoride, UF_6, a gas prepared from UF_4 by reaction with fluorine. Used for U 235 isotope separation
- Plutonium oxide, PuO_2, used for the production of the metal by reduction with calcium

References

Habashi F (ed) (1997) Handbook of extractive metallurgy, vol 3, Part 9-radioactive metals. Wiley-VCH, Weinheim

Habashi F (2003) Metals from ores: an introduction to extractive metallurgy. Métallurgie Extractive Québec, Québec City, Distributed by Laval University Bookstore, www.zone.ul.ca

Activate: Stimulate

▶ Monovalent Cations in Tryptophan Synthase Catalysis and Substrate Channeling Regulation

Activation of Thyroxine (T4) by Deiodination to Triiodothyronine (T3)

▶ Selenoproteins and the Biosynthesis and Activity of Thyroid Hormones

Active Site Zinc

▶ Zinc Alcohol Dehydrogenases

Acute Phase

▶ Magnesium and Inflammation

Acute Promyelocytic Leukemia

▶ Arsenic in Therapy
▶ Promyelocytic Leukemia–Retinoic Acid Receptor α (PML–RARα) and Arsenic

ADAM

▶ Zinc Adamalysins

ADAMTS

▶ Zinc Adamalysins

AdcR

▶ Zinc Sensors in Bacteria

ADH1/ADC1

▶ Zinc Storage and Distribution in S. cerevisiae

AdoMet:Arsenic(III) Methyltransferase

▶ Arsenic Methyltransferases

Adsorption Studies

▶ Colloidal Silver Nanoparticles and Bovine Serum Albumin

Adverse Health Effects Related to Mn

▶ Manganese Toxicity

Aequorin

▶ Sarcoplasmic Calcium-Binding Protein Family: SCP, Calerythrin, Aequorin, and Calexcitin

Ag, MT

▶ Metallothioneins and Silver

AI-2 – Autoinducer 2

▶ Boron, Biologically Active Compounds

Al and Polyphosphoinositides

▶ Aluminum and Phosphatidylinositol-Specific-Phospholipase C

Alkali Cation Binding Sites and Structure–Functions Relations

▶ Potassium in Biological Systems

Alkali Cation Binding Sites and Structure-Functions Relationships

▶ Potassium-Binding Site Types in Proteins

Alkali Cations

▶ Potassium, Physical and Chemical Properties
▶ Sodium, Physical and Chemical Properties

Alkali Metals

▶ Potassium, Physical and Chemical Properties
▶ Sodium, Physical and Chemical Properties

Alkaline Earth Metal

▶ Strontium Binding to Proteins

Alkaline Metals

▶ Cesium, Therapeutic Effects and Toxicity

Alkaline-Earth Metal

▶ Barium, Physical and Chemical Properties
▶ Strontium, Physical and Chemical Properties

Allergy

▶ Chromium and Allergic Reponses

Allocation on Earth's Surface

▶ Lithium in Biosphere, Distribution

Allosteric: Other Site

▶ Monovalent Cations in Tryptophan Synthase Catalysis and Substrate Channeling Regulation

Alpha Rays

▶ Polonium and Cancer

Alpha-Radioactivity

▶ Polonium and Cancer

ALS

▶ Copper-Zinc Superoxide Dismutase and Lou Gehrig's Disease

ALS, Amyotrophic Lateral Sclerosis

▶ Zinc Aminopeptidases, Aminopeptidase from Vibrio Proteolyticus (Aeromonas proteolytica) as Prototypical Enzyme

Aluminum

▶ Aluminum in Plants

Aluminum and Bioactive Molecules, Interaction

Xiaodi Yang and Huihui Li
School of Chemistry and Material Science, Nanjing Normal University, Nanjing, China

Synonyms

Aluminum complexes with low molecular mass substances; Aluminum speciation

Definition

Interaction of aluminum and low molecular mass substances of biological interest is a phenomenon that may explain the effects of aluminum in the environment and organisms.

Aluminum [Al(III)] is a potential neurotoxic agent implicated in the pathogenesis of many neurological disorders (Yokel et al. 2001). The possible role of Al(III) in the development of pathologies, such as Alzheimer's disease, has often been investigated. The interaction of this element with biologically related sites, represented by proteins, enzymes, and coenzymes as well as enzyme-active sites and low molecular mass organic substrates, is prevalent in biological fluids.

Al-α-KG: Constitutional Aspect

Alpha-Ketoglutate (α-KG) is common in many microorganisms, and can also be isolated from higher plants.

The α-KG was found to bind Al(III) in a bidentate manner at the carboxylate and carbonyl moieties. The mononuclear 1:1 ($AlLH_{-1}$, AlL^+, $AlHL^{2+}$) and 2:1 (AlL_2^-, $AlL_2H_{-2}^{3-}$) species, and dinuclear 2:1 (Al_2L^{4+}) species were found in acidic aqueous solution. Meanwhile, Al(III) can promote α-KG tautomerize to its enolic structure in solutions. Figure 1 shows α-KG and its Al(III) complexes in equilibrium between the *keto* and *enol* forms. There exists an equilibrium between the two forms: *keto*-form ↔ *enol*-form. The equilibrium is, however, shifted to the left or right, depending on the solvent type, solution acidity, and the concentration of Al–α-KG. The affinity of Al(III) for the enol-form is stronger than for the keto-form, especially at high Al(III) concentrations and high-pH solutions. Therefore, Al(III) can promote the keto–enol tautomerization of α-KG in aqueous solutions (Yang et al. 2003a, 2007).

Al-L-Glu: Configurational Aspect

Amino acids, like glutamic acid (Glu) and aspartic acid (Asp), are important acidic neurotransmitters in the central nervous system (CNS) (Delonce et al. 2002). Glu rather acts as a bidentate ligand, displaying similarity with simple carboxylate group.

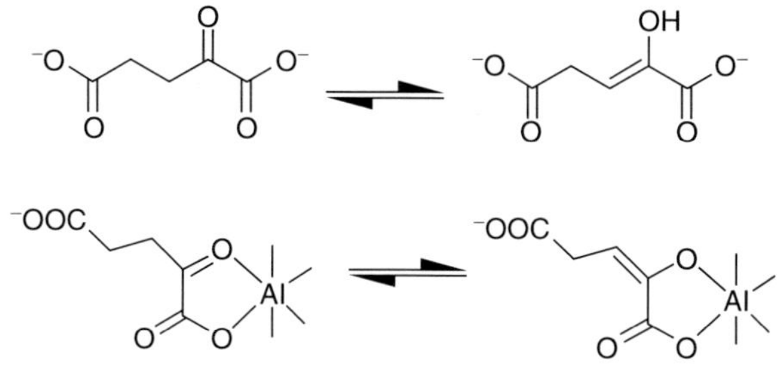

Aluminum and Bioactive Molecules, Interaction, Fig. 1 Tautomerization of α-KG and its aluminum complexes (Yang et al. 2003a)

There is an electrostatic repulsive effect of the $-NH_3^+$ group and the Al^{3+} ion. The two carboxylate groups and one deprotonated amino group of L-glutamate are proposed binding sites for Al, and are thought to form the mononuclear 1:1 ($AlLH^{2+}$, AlL^+, and $AlLH_{-1}$) species and dinuclear 2:1 (Al_2L^{4+}) species in acidic aqueous solutions (Fig. 2). The tridentate binding mode of the likely five- + seven-membered joint chelate also existed in acidic aqueous solutions. The 1:1 species can be ascribed to only one reasonable tridentate binding mode ($-COO^-$, $-NH_2$, $-COO^-$) with two OH^- groups around Al(III) under the physiological condition. The formation of the tridentate Al–Glu complexes explains the above possible mechanism that Al(III) promotes the racemization from L-amino acid to D-amino acid, which is also very important in the food chemistry (Yang et al. 2003d).

Al-NAD, NADH, NADP: Conformational Aspect

NAD^+, and its reduced form, NADH, in the glutamate dehydrogenase reaction system, act as a coenzyme ubiquitously involved in biological redox processes and play an important role in the conversion of chemical energy to metabolic energy. Both of the coenzymes play crucial roles in oxidations, which are accomplished by the removal of hydrogen atoms in cells. The diphosphate bridge oxygen atoms of NAD^+ are the potential binding sites (Fig. 3). It was obvious that the conformation of NAD^+ not only depends on the Al(III) binding but also depends on the solvent effect.

Al(III) can coordinate with NAD^+ through the following binding sites: N_7' of adenine- and pyrophosphate-free oxygen (O_A^1, O_N^1, O_A^2) to form various mononuclear 1:1 ($AlLH_2^{3+}$, $AlLH^{2+}$) and 2:1 (AlL_2^-) species, and dinuclear 2:2 ($Al_2L_2^{2+}$) species. The conformations of NAD^+ and Al-NAD^+ depend on the solvents and different species in the complexes. The occurrence of an Al-linked complexation causes structural changes at the primary recognition sites and secondary conformational alterations for coenzymes (Yang et al. 2003c).

Al(III) interacts with NADH by occupying the binding sites of pyrophosphate oxygen atoms and locks the adenine moiety of coenzyme in an *anti*-folded conformation. Meanwhile, the weak attractive interactions ("association") may occur between Al(III) and the hydroxyl groups of ribose rings through the intramolecular hydrogen bonding. The occurrence of Al(III)-linked conformational changes in these flexible molecules brings structural changes at the primary (by occupying the binding sites of phosphate oxygens O_{N1} and O_{A1}, hydroxyls of ribose rings) and secondary (restrict folded form of NADH in plant cell physiological pH and Al(III) concentration) recognition sites for substrates and enzyme. Furthermore, at biologically relevant pH and concentrations of Al(III) and NADH, Al(III) could increase the amount of folded forms of NADH, which will result in reducing the coenzyme NADH activity in hollow-dehydrogenases reaction systems. The presence of possible competing ligands such as citrate, oxalate, and tartate could detoxify the Al(III) toxic effect. Organic acids protecting the coenzyme NADH from Al(III) lesions are of a biological relevant value in this respect (Yang et al. 2003b).

β-Nicotinamide adenine dinucleotide phosphate ($NADP^+$) and its reduced form NADPH is an important coenzyme in both plants and animals. Like other phosphate molecules of biological importance, such as adenosine monophosphate (AMP), diphosphate (ADP), and triphosphate (ATP), $NADP^+$ shows strong binding capability with Al(III) via the complexation of the nonbridging phosphate group (Murakami and Yoshino 2004). At μM concentration level and in neutral pH aqueous solutions, the 1:1 species, Al($NADP^+$), predominate.

Al, Al_{13}-GSH, GSSG: Species Aspect

Reduced glutathione (γ-L-glutamyl-L-cysteinylglycine, GSH) is a fundamental low molecular mass antioxidant that serves several biological functions. It is an essential

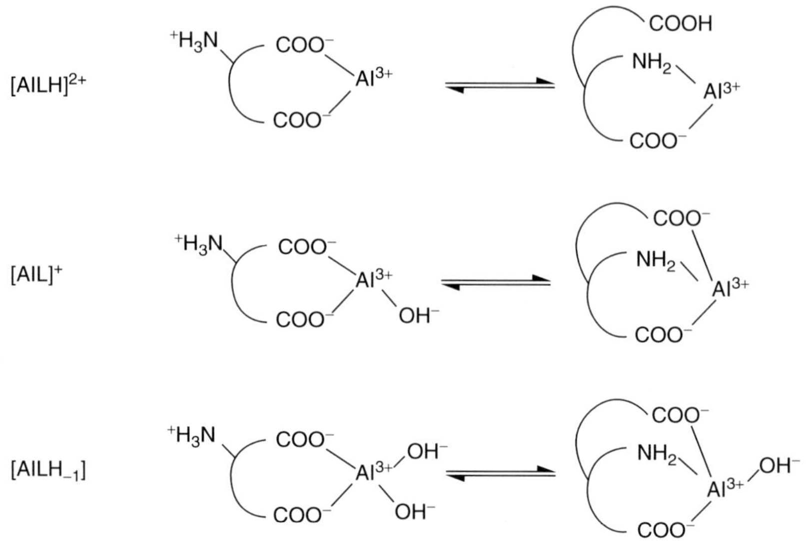

Aluminum and Bioactive Molecules, Interaction, Fig. 2 Bidentate and tridentate binding modes of the 1:1 Al-Glu species in acidic aqueous solutions (Al:Glu = 0.01M:0.01M) (Yang et al. 2003d)

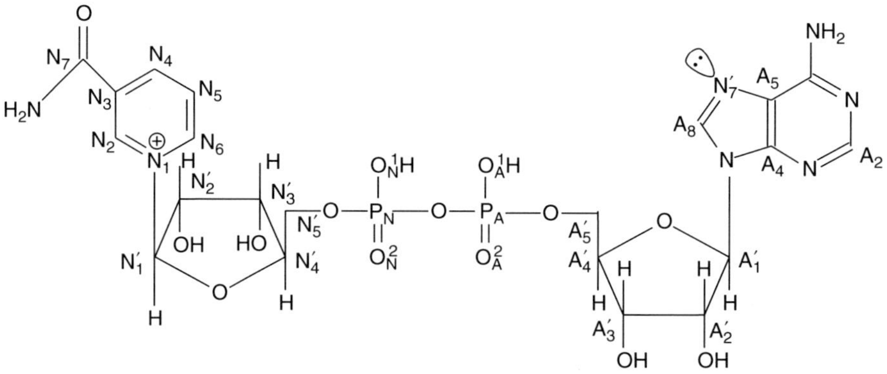

Aluminum and Bioactive Molecules, Interaction, Fig. 3 The structure and atomic numbering of NAD$^+$

constituent of all living cells and usually the most abundant intracellular nonprotein thiol source. Upon enzymatic and non-enzymatic oxidation, GSH forms glutathione disulfide (GSSG).

GSH is expected to display a greater affinity for Al^{3+} ions and thus interfere with aluminum's biological role more significantly than simple natural amino acids. Al^{3+} coordinates with the important biomolecule GSH through carboxylate groups to form various mononuclear 1:1 (AlHL, AlH$_2$L and AlH$_{-1}$L), 1:2 (AlL$_2$) complexes, and dinuclear (Al$_2$H$_5$L$_2$) species, where H^{4}L$^+$ denotes totally protonated GSH (Table 1). Besides the monodentate complexes through carboxylate groups, the amino groups and the peptide bond imino and carbonyl groups may also be involved in binding with Al^{3+} in the bidentate and tridentate complexes (Wang et al. 2009).

Oxidized glutathione (GSSG) exists in animal cells and in the form of phytochelatins in plant cells. It is, therefore, widely available in all living things. The coordination chemistry of Al(III) with glutathione is of vital importance, since it serves as a model system for the binding of this metal ion by larger peptide and protein molecules. Al(III) can coordinate with the important biomolecule GSSG through the following binding sites: glycyl and glutamyl carboxyl groups to form various mononuclear 1:1 (AlLH$_4$, AlLH$_3$, AlLH$_2$, AlLH, AlL, AlLH$_{-1}$, AlLH$_{-2}$) and several binuclear 2:1 (Al$_2$LH$_4$, Al$_2$LH$_2$, Al$_2$L) species (where H$_6$L^{2+} denotes the totally protonated oxidized glutathione) in acidic aqueous solutions. The carboxylate groups are effective binding sites for Al(III). The possible binding sites are the negative charged C-terminal Gly-COO$^-$ and Glu-COO$^-$ groups. The coordination

Aluminum and Bioactive Molecules, Interaction, Table 1 Proton and aluminum(III) complexes formation constants (logβ) of the hydroxo and low molecular mass substances at $T = 25 \pm 0.1\,°C$ and $I = 0.10$ M KCl

Ligand species	logβ [a]	Hydroxo species	logβ	Complexation species	logβ [a]
α-KG		AlH$_{-1}$	−5.33	AlLH^{2+}	6.55
H$_2$L	2.41	AlH$_{-2}$	−10.91	AlL$^+$	3.83
HL$^-$	7.04	AlH$_{-3}$	−16.64	AlLH$_{-1}$	−0.87
		AlH$_{-4}$	−23.46	AlL$_2^-$	5.75
		Al$_2$H$_{-2}$	−7.15	AlL$_2$H$_{-2}^{3-}$	−4.58
		Al$_3$H$_{-4}$	−13.13	Al$_2$L^{4+}	6.32
		Al$_{13}$H$_{-32}$	−107.41		
L-Glu				AlLH^{2+}	10.68
H$_2$L	2.17			AlL$^+$	7.42
HL$^-$	6.15			AlLH$_{-1}$	2.46
L^{2-}	15.11			Al$_2$L^{4+}	9.56
NAD$^+$				AlLH$_2^{3+}$	15.60
H$_2$L	3.46			AlLH^{2+}	12.31
HL$^-$	10.95			Al$_2$L$_2^{2+}$	19.82
GSH		AlH$_{-1}$	−5.33	AlH$_2$L	21.85 ± 0.05
H$_4$L	24.12 ± 0.04	AlH$_{-2}$	−10.91	AlHL	17.70 ± 0.03
H$_3$L	21.10 ± 0.03	AlH$_{-3}$	−16.64	AlH$_{-1}$L	7.76 ± 0.08
H$_2$L	18.00 ± 0.02	AlH$_{-4}$	−23.46	AlL$_2$	24.72 ± 0.05
HL	9.53 ± 0.02	Al$_2$H$_{-2}$	−7.15	Al$_2$H$_5$L$_2$	49.6 ± 0.2
		Al$_3$H$_{-4}$	−13.13		
		Al$_{13}$H$_{-32}$	−107.41		
GSSG		AlH$_{-1}$	−5.33	AlLH$_4$	26.82 ± 0.03
H$_6$L	28.95 ± 0.02	AlH$_{-2}$	−10.91	AlLH$_3$	23.56 ± 0.03
H$_5$L	27.35 ± 0.01	AlH$_{-3}$	−16.64	AlLH$_2$	20.32 ± 0.03
H$_4$L	25.32 ± 0.02	AlH$_{-4}$	−23.46	AlLH	15.83 ± 0.03
H$_3$L	21.99 ± 0.02	Al$_2$H$_{-2}$	−7.15	AlL	11.92 ± 0.04
H$_2$L	18.23 ± 0.02	Al$_3$H$_{-4}$	−13.13	AlLH$_{-1}$	4.96 ± 0.01
HL	9.54 ± 0.02	Al$_{13}$H$_{-32}$	−107.41	AlLH$_{-2}$	−5.06 ± 0.01
				Al$_2$L	14.07 ± 0.04
				Al$_2$LH$_2$	21.93 ± 0.03
				Al$_2$LH$_4$	28.86 ± 0.05

[a] Averages (± standard deviations)

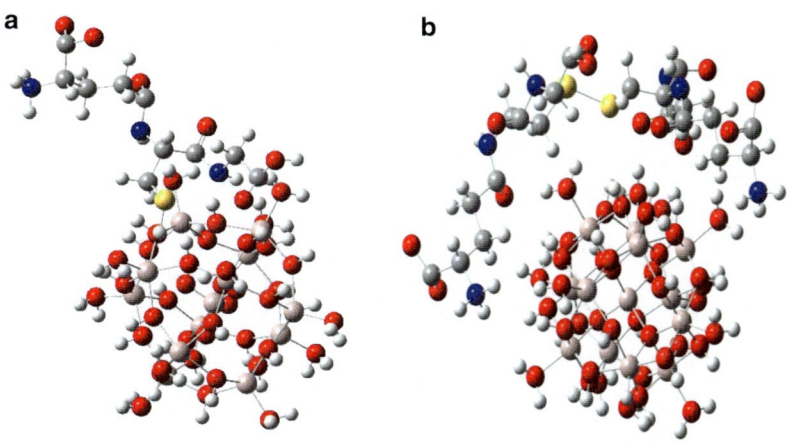

Aluminum and Bioactive Molecules, Interaction, Fig. 4 Binding mode of substances on the Al$_{13}$ clusters. (**a**) GSH and (**b**) GSSG. Color codes: C, *gray*; H, *light gray*; O, *red*; N, *blue*; S, *yellow*; Al, *pink*

of Al(III) induces small conformational changes of peptide GSSG (Yang et al. 2008).

Nanometer-sized tridecameric aluminum polycation (nano-Al_{13}) is known as a new type clarifying agent, showing more effective coagulation effects and rapid aggregation velocity in a relatively wider pH range, and widely applied in the water treatment. Al_{13} clusters could exert strong effect to the conformation of GSH and GSSG, due to the significant interactions between the ligands and cluster surface (Fig. 4).

Cross-References

▶ Aluminum in Biological Systems
▶ Aluminum in Plants
▶ Aluminum Speciation in Human Serum
▶ Aluminum, Biological Effects
▶ Lead and Alzheimer's Disease
▶ Mercury and Alzheimer's Disease
▶ Zinc in Alzheimer's and Parkinson's Diseases

References

Delonce R, Fauconnrau B, Piriou A, Huguet F, Guillard O (2002) Aluminum L-glutamate complex in rat brain cortex: in vivo prevention of aluminum deposit by magnesium D-aspartate. Brain Res 946:247–252

Murakami K, Yoshino M (2004) Aluminum decreases the glutathinoe regeneration by the inhibition of NADP-isocitrate dehydrogenase in mitochondria. J Cell Biochem 93:1267–1271

Wang XL, Li K, Yang XD, Wang LL, Shen RF (2009) Complexation of Al(III) with reduced glutathione in acidic aqueous solutions. J Inorg Biochem 103:657–665

Yang XD, Bi SP, Wang XL, Liu J, Bai ZP (2003a) Multimethod characterization of the interaction of aluminum ion with α-ketoglutaric acid in acidic aqueous solutions. Anal Sci 19:273–279

Yang XD, Bi SP, Yang L, Zhu YH, Wang XL (2003b) Multi-NMR and fluorescence spectra study the effects of aluminum (III) on coenzyme NADH in aqueous solutions. Spectrochim Acta A 59:2561–2569

Yang XD, Bi SP, Yang XL, Yang L, Hu J, Liu J (2003c) NMR spectra and potentiometry studies of aluminum(III) binding with coenzyme NAD^+ in acidic aqueous solutions. Anal Sci 19:815–821

Yang XD, Tang YZ, Bi SP, Yang GS, Hu J (2003d) Potentiometric and multi-NMR studies of aluminum(III) complex with L-glutamate in acidic aqueous solutions. Anal Sci 19:133–138

Yang XD, Zhang QQ, Li LF, Shen RF (2007) Structural features of aluminium(III) complexes with bioligands in glutamate dehydrogenase reaction system – a review. J Inorg Biochem 101:1242–1250

Yang XD, Zhang QQ, Chen RF, Shen RF (2008) Speciation of aluminum(III) complexes with oxidized glutathione in acidic aqueous solutions. Anal Sci 24:1005–1012

Yokel RA, Rhineheimer SS, Sharma P, Elmorce D, McNamara PJ (2001) Entry, half-life, and desferrioxamine-accelerated clearance of brain aluminum after a single (26)Al exposure. Toxicol Sci 64:77–82

Aluminum and Health

▶ Aluminum, Biological Effects

Aluminum and Phosphatidylinositol-Specific-Phospholipase C

Sandra Viviana Verstraeten[1] and Patricia Isabel Oteiza[2]
[1]Department of Biological Chemistry, IQUIFIB (UBA-CONICET), School of Pharmacy and Biochemistry, University of Buenos Aires, Argentina, Buenos Aires, Argentina
[2]Departments of Nutrition and Environmental Toxicology, University of California, Davis, Davis, CA, USA

Synonyms

Al and polyphosphoinositides

Definitions

Lipid bilayer: Spontaneous arrangement of polar lipids, such as phospholipids, when in contact with water-containing solutions. Given that phospholipids have in their structure a hydrophilic headgroup and two hydrophobic tails, they adopt a disposition of a flat two-layered sheet with the headgroups facing the aqueous milieu and the hydrophobic tails grouped toward the inside of the sheet, isolated from water (Scheme 1).

Liposomes: Artificial membranes composed of lipids arranged as a bilayer (Scheme 1). Liposomes are custom made; thus by selecting their lipid and/or protein composition, liposomes become a useful tool

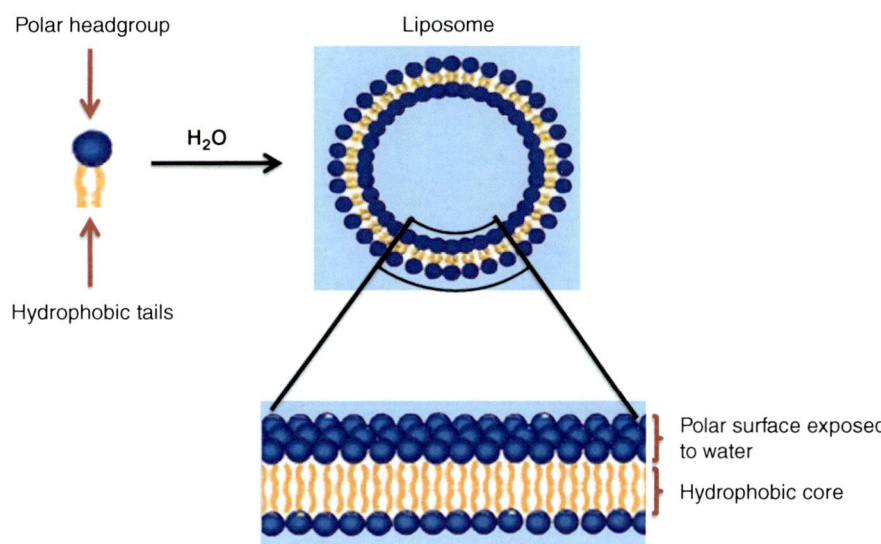

Aluminum and Phosphatidylinositol-Specific-Phospholipase C, Scheme 1 Amphiphilic lipids (such as phospholipids) spontaneously arrange into bilayers when dissolved in water-containing solutions

to assess lipid behavior under controlled conditions (such as changes in aqueous media composition, temperature, pressure, interaction with proteins or nucleic acids, etc.). Given that liposomes contain an internal water space, it is also possible to entrap water-soluble molecules. This property makes them useful as carriers (e.g., drug delivery).

Biological membrane: Naturally occurring membrane composed of lipids and proteins, which determines the boundaries of a cell. Biological membranes serve as selective barriers and prevent the leakage of molecules located within the cell. The outermost membrane of a cell, called the plasma membrane, participates in cell recognition and response, in cell-cell adhesion and interaction with other cellular and extracellular structures. It is also involved in cell signaling and contains identity markers. In addition to the plasma membrane, eukaryotic cells also have internal membranes that delimit the different intracellular compartments or organelles such as the mitochondria, nuclei, lysosomes, peroxysomes, and chloroplasts. The membrane of each organelle has characteristic lipid and protein compositions that create the adequate environment for them to exert their biological functions. Supporting a relevant role of biological membranes in cell function, alterations in membranes lipid or protein composition lead to a wide range of pathologies.

Membrane biophysical properties: Group of measurable properties that define the dynamics of a lipid bilayer. They include: (1) Membrane fluidity that is related to the viscosity of the bilayer. This property is determined by both the length of the acyl chains and the number of unsaturations of membrane fatty acids, and by the temperature. (2) Lateral phase separation (or lateral diffusion), a property closely related to membrane fluidity that reflects the velocity of lipids displacement along the membrane. (3) Membrane hydration reflects membrane water content. It is closely associated with membrane fluidity given that a more fluid membrane will allow water molecules to penetrate into deeper regions at the hydrophobic core of the bilayer. On the contrary, rigid membranes will extrude water from the hydrophobic core. (4) Membrane permeability, which determines how easily a given molecule will cross the bilayer. This property relies not only on the membrane composition but also on the chemical nature of the molecule permeating the bilayer. (5) Surface potential, determined by the different charged polar headgroup of phospholipids. (6) Topology, a property determined by the characteristic that not all phospholipids have the same length. Therefore, mixtures of relatively short phospholipids and long ones will determine membrane regions of different widths.

Cell signal transduction: Cascade of events that transmit the information from the environment to the inside of the cell. These events are frequently triggered by the interaction of a stimulus (e.g., proteins,

Aluminum and Phosphatidylinositol-Specific-Phospholipase C, Fig. 1 Chemical structure of phosphoinositides and phosphatidylcholine. The hydroxyl groups of phosphatidylinositol (*PI*) that can be phosphorylated by specific kinases to generate the different PPIs are indicated in *red*

Introduction

Communication among cells, regardless if they are adjacent or distant, requires an intricate network of molecules that carry the chemical or electrical information to the target cell, where they trigger a biological action. The chemical nature of these molecules (stimuli) is diverse, from physical (e.g., light) or chemical (e.g., oxygen) molecules to peptides and proteins. At the target cell, they interact with specific receptors and drive the generation of one or more intracellular second messengers, an effect that depends on both the characteristics of the receptor and the cell type involved. In general, receptors can be classified in two groups, ionotropic and metabotropic. The first group is associated with the modulation of the opening/closing of ionic channels, leading to the controlled influx/efflux of inorganic ions. On the other hand, metabotropic receptors are associated with enzymes chemical and physical stimuli) with cell surface or intracellular receptors. This interaction will determine the recruitment and activation of different intracellular proteins, which in turn, will lead to a biological action (e.g., changes in cell metabolism, changes in gene transcription).

Polyphosphoinositides (PPIs): Family of phospholipids that contain an inositol moiety in their headgroup. The simplest one is phosphatidylinositol (Fig. 1). Although this molecule has five hydroxyl groups (the sixth being involved in the O-P bond that connects phosphoinositol with diacylglycerol), only those at positions 3, 4, and 5 of the inositol ring can be phosphorylated by the action of specific kinases (Fig. 1). Each phosphoinositide has been associated with a variety of biological actions, being phosphatidylinositol (4, 5) bisphosphate (PIP_2, Fig. 1) the most studied. Research on the biological actions of this family of lipids is a growing field of interest in the areas of cell biology and metabolic regulation.

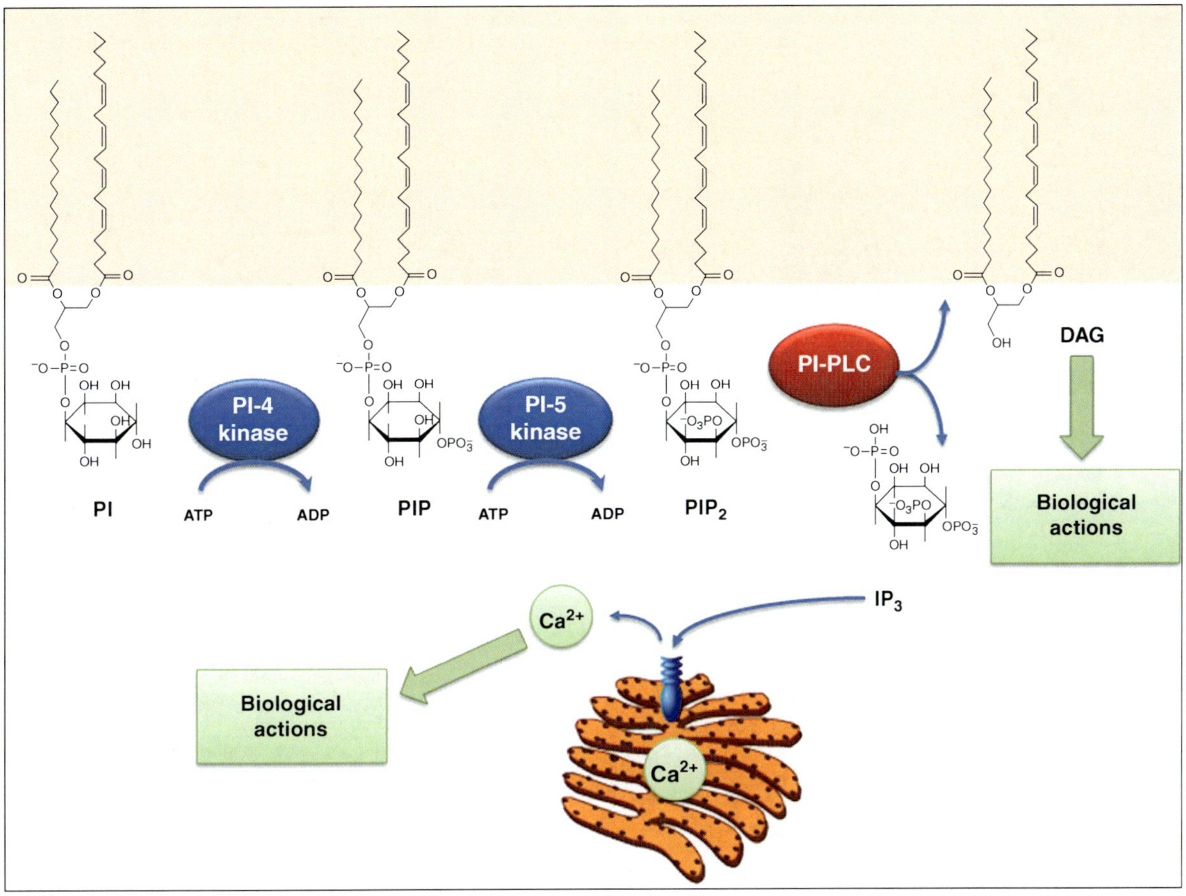

Aluminum and Phosphatidylinositol-Specific-Phospholipase C, Scheme 2 Summary of PPIs metabolism

that catalyze the generation of intracellular second messengers, each one having defined mechanisms of action.

A major signaling pathway in biological systems involves the binding of certain hormones, growth factors, and neurotransmitters to their corresponding membrane receptor which activates the turnover of polyphosphoinositides (PPIs). PPI comprises a group of phosphorylated species derived from phosphatidylinositol (PI). PI is a substrate for different kinases that act on the hydroxyl groups at the positions 3, 4, or 5 of the inositol ring (Fig. 1). Upon the proper stimuli, phosphatidylinositol (4, 5) bisphosphate (PIP$_2$, Fig. 1) is hydrolyzed by a specific phospholipase C (PI-PLC) rendering inositol trisphosphate (IP$_3$) and diacylglycerol, two molecules with relevant second-messenger function. Due to its lipidic nature, diacylglycerol remains within the bilayer and activates membrane-associated enzymes, such as protein kinase C (PKC) while the water-soluble IP$_3$ is released to the cytosol. IP$_3$ activates a family of IP$_3$ receptors (IP$_3$Rs) which are present in the plasma membrane, endoplasmic reticulum, Golgi apparatus, and nuclear membranes (Corry and Hool 2007). At the endoplasmic reticulum, binding of IP$_3$ to the IP$_3$R leads to the opening of the channel and to the consequent Ca^{2+} mobilization from this intracellular reservoir to the cytosol where it activates a number of signaling cascades (Scheme 2).

The capacity of aluminum (Al^{3+}) to inhibit the PPI-dependent signaling pathway has been reported several years ago (Shafer and Mundy 1995;

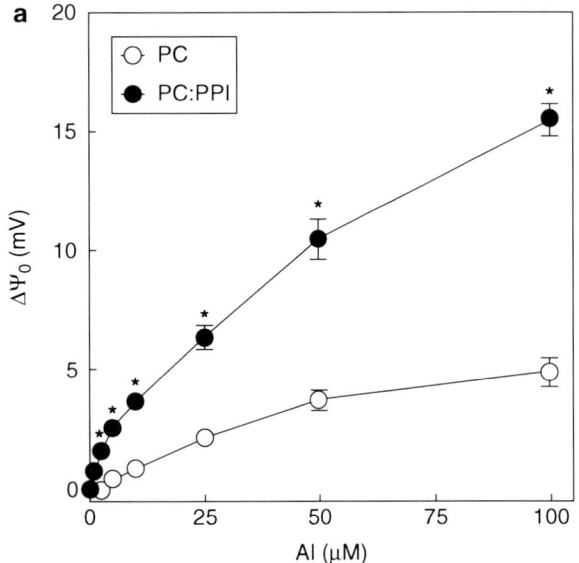 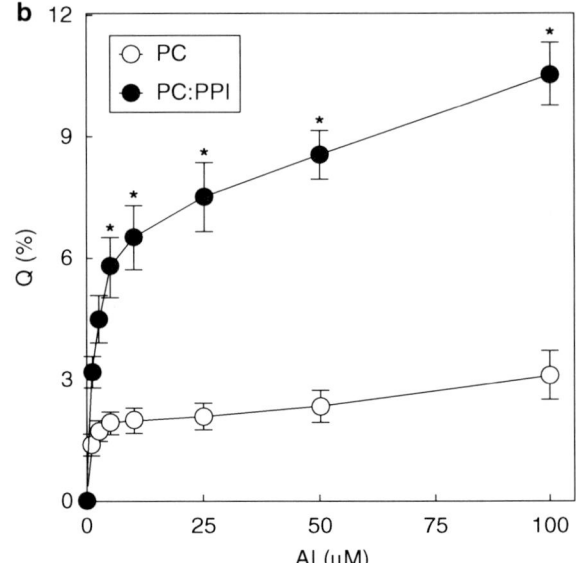

Aluminum and Phosphatidylinositol-Specific-Phospholipase C, Fig. 2 Al^{3+} modifies liposome surface potential and generates PPI-enriched domains. PC (○) or PC:PPI (60:40 molar ratio) (●) liposomes were incubated in the presence of variable amounts of Al^{3+}. (**a**) Liposome surface potential was evaluated using the fluorescent probe ANS. (**b**) Formation of lipid domains was assessed using the fluorescent probe C6-NBD-PC. Results are shown as the mean ± SEM of at least three independent experiments. * Significantly different from the value measured in PC liposomes at the same Al^{3+} concentration ($P < 0.001$, two-way ANOVA) (Adapted from Arch Biochem Biophys, 408, Verstraeten SV and Oteiza PI, "Al^{3+}-mediated changes in membrane physical properties participate in the inhibition of polyphosphoinositide hydrolysis", 263–271, 2002, with permission from Elsevier)

Nostrandt et al. 1996). This inhibition could be ascribed to the interaction of the metal with at least one of the four major components of this pathway: the membrane receptor, the G protein coupled to the receptor that activates the enzyme PI-PLC, PI-PLC itself, and its substrate PIP_2. So far, neither a direct interaction of Al^{3+} with the receptor nor the inhibition of the G protein have been proven to cause Al^{3+}-mediated inhibition of PIP_2 hydrolysis (Shafer and Mundy 1995). Therefore, the mechanism underlying this inhibition should reside in the interaction of the metal with PI-PLC and/or PIP_2. In the following sections, we will review current evidence showing that Al^{3+} can inhibit PI-PCL mainly due to the capacity of Al^{3+} to interact with PIP_2-containing membranes in a manner that impedes the availability of the substrate to enzyme, rather than to a metal-protein interaction which would directly alter PI-PLC activity.

Aluminum Interaction with PPI Causes Major Alterations in Membrane Physical Properties

In vitro, the binding of Al^{3+} to acidic phospholipids alters the dynamics of lipids at the bilayer (Verstraeten et al. 1997a). Given that phosphate is the main biological ligand for Al^{3+} (Martin 1986), metal interactions with phosphate groups present in the inositol moiety of PPI may enhance this ability of Al^{3+} and modify the biophysical properties of PPI-containing membranes. To investigate this hypothesis, a series of in vitro experiments using liposomes composed of a mixture of phosphatidylcholine (PC) (Fig. 1) and PPI in a 60:40 molar ratio were performed.

The binding of Al^{3+} to lipid bilayers occurs through the ionic interaction between the positively charged metal and the negative charges in the polar headgroup of phospholipids. Therefore, an overall increase in the superficial charge of liposomes is evidence of such metal-lipid interactions. In the absence of Al^{3+}, PC:PPI

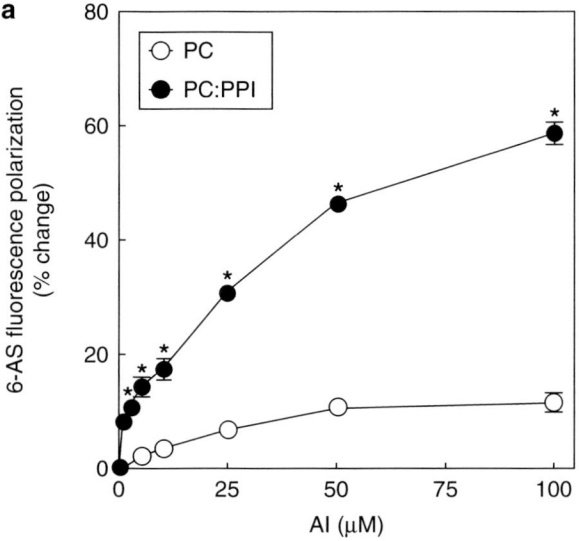 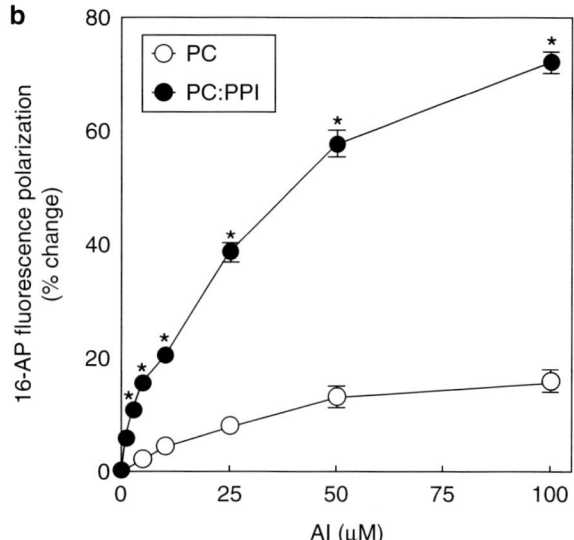

Aluminum and Phosphatidylinositol-Specific-Phospholipase C, Fig. 3 Al^{3+} decreases membrane fluidity in PPI-containing liposomes. PC (○) or PC:PPI (60:40 molar ratio) (●) liposomes were incubated in the presence of variable amounts of Al^{3+} and the fluidity at the surface level (**a**) and at the hydrophobic core of the bilayer (**b**) was evaluated from the changes in the fluorescence polarization of the probes 6-AS and 16-AP, respectively. Results are shown as the mean ± SEM of at least three independent experiments. * Significantly different from the value measured in PC liposomes at the same Al^{3+} concentration ($P < 0.001$, two-way ANOVA) (Adapted from Arch Biochem Biophys, 408, Verstraeten SV and Oteiza PI, "Al^{3+}-mediated changes in membrane physical properties participate in the inhibition of polyphosphoinositide hydrolysis", 263–271, 2002, with permission from Elsevier)

liposomes are approximately 15 mV more negative than PC liposomes due to the presence of PPI phosphate groups (Verstraeten et al. 2003). In the presence of Al^{3+}, PC liposomes experience a slight increase in the surface membrane potential ($\Delta\Psi_0$) (Fig. 2a), indicating a weak ionic interaction between the metal and the membrane. This result is expected, based on the chemical structure of PC (see Fig. 1). PC is a zwitterionic phospholipid, with a negative charge conferred by the phosphate group esterifying the diacylglycerol molecule, and a positive charge from the quaternary ammonium of the choline residue. On the other hand, the presence of one or more phosphate groups in PPI positively affects the interaction of Al^{3+} with the membrane, as evidenced from the marked increase in the surface potential of PPI-containing liposomes (Fig. 2a).

Multivalent cations such as Al^{3+} have the tendency to interact with more than one binding site, with 1:2 or 1:3 stoichiometries (Martin 1986). In the case of PPI-containing membranes, Al^{3+} may bind to either phosphate groups located within a single phospholipid or to those located in neighbor molecules. In the case of PIP, the existence of only one phosphate group in the inositol ring favors the interaction of the metal with more than one phospholipid. On the other hand, PIP_2 contains two phosphate groups which, in principle, will favor the binding of one molecule of Al^{3+} with the two phosphates groups of a single phospholipid molecule. However, the different spatial disposition of those residues, facing the opposite sides of the inositol ring, would lead to the interaction of Al^{3+} with more than one PIP_2 molecule. If the latter occurs, PPI will segregate in the lateral phase of the bilayer, generating PPI-enriched domains. To assess this possibility, the generation of membrane domains was evaluated with the fluorescent probe C6-NBD-PC. When C6-NBD-PC molecules are in close proximity, fluorescence quenching occurs. In PC liposomes, the separation of lipids into domains (fluorescence quenching, Q%) was minimal upon their interaction with Al^{3+} (Fig. 2b). However, when liposomes contained PPI, a marked and Al^{3+} concentration-dependent lateral phase separation was observed (Fig. 2b) (Verstraeten et al. 2003).

Membrane fluidity is another important biophysical property that has a profound impact on the functionality of membrane-associated proteins. Al^{3+} has been proven to promote a decrease in membrane fluidity both in diverse in vitro and in vivo models (Verstraeten et al. 1997a, b, 1998, 2002; Verstraeten and Oteiza 2000, 2002). It is important to point out that, even when the interaction of Al^{3+} with phospholipids occurs at the superficial level of the bilayer, membrane rigidification propagates from the surface to the deepest region of the bilayer (Fig. 3). However, the magnitude of the effect will depend on the nature of the phospholipid involved. In this way, in PC liposomes the effect of Al^{3+} promoting the rigidification of the membrane was almost negligible when evaluated either at the superficial level (probe 6-AS, Fig. 3a) or at the deepest hydrophobic region (probe 16-AP, Fig. 3b). The presence of PPI in the membrane resulted in an enhanced rigidifying effect of Al^{3+}. The magnitude of the effect was significantly higher at the hydrophobic portion than at the surface of the bilayer (Verstraeten et al. 2003). In accordance to the higher rigidity found in PPI-containing liposomes treated with Al^{3+}, membranes became partially dehydrated (Verstraeten and Oteiza 2002).

Aluminum-Mediated Changes in Membrane Physical Properties Determine the Ability of PI-PLC to Hydrolyze PIP_2

The findings discussed above raised the question if the binding of Al^{3+} to phosphate groups in PPI was the sole factor responsible for the previously described PI-PLC inhibition, or if Al^{3+}-mediated changes in membrane physical properties could contribute to such inhibition. To answer that question, PC:PPI liposomes were pre-incubated with Al^{3+} and subsequently submitted to the action of two PI-PLCs from different origin. One of the enzymes assessed was isolated from bovine brain and corresponds to the catalytic active subunit of the delta isoform (Verstraeten et al. 2003). The other was obtained commercially and was purified from *Bacillus cereus* (EC 3.1.4.10).

After the incubation of the samples in the presence of brain PI-PLC, the only specie hydrolyzed by the enzyme was PIP_2 (Fig. 4a) a result that is in accordance with the substrate specificity of this enzyme (Rebecchi and Pentyala 2000). A similar specificity was observed when the bacterial PI-PLC was assessed (Fig. 4b). It is important to stress that the extent of PIP_2 hydrolysis by both enzymes was similar in intact liposomes as well as in the micelles that were obtained by disrupting liposomes with the detergent Triton X-100. This finding indicates that both enzymes had a complete accessibility to their substrates even when they were incorporated in a bilayer.

When liposomes were pre-incubated with 50 µM Al^{3+}, PI-PLC activity was abolished (brain PI-PLC) or significantly impaired (bacterial PI-PLC). Surprisingly, when the bilayer structure of liposomes was disrupted by the addition of Triton X-100, the activity of the enzymes was fully recovered (Fig. 4). However, it is possible to speculate that this treatment may have displaced Al^{3+} from its binding sites, increasing the accessibility of PI-PLC to its substrates. To investigate this hypothesis, the amount of Al^{3+} bound to lipids was measured. As expected from the weakness of Al^{3+} interaction with the zwitterionic phospholipid PC, the amount of Al^{3+} bound to PC liposomes was very low and close to the detection limit of the method. In PC:PPI liposomes the amount of Al^{3+} bound was 40-fold higher than in PC liposomes (Fig. 5). The disruption of liposomes with Triton X-100 did not affect Al^{3+}-phospholipid binding. These findings stress the strength of the ionic bonds established between Al^{3+} and the phosphate moieties in the inositol headgroups of PIP and PIP_2.

The catalytic domain of eukaryotic isoforms of PI-PLC is organized in a α/β TIM barrel that lodges the conserved amino acids His_{311}, His_{356}, Glu_{341}, Asp_{343}, and Glu_{390} involved in the binding and hydrolysis of the O-P bond that connects phosphoinositol with diacylglycerol (Rebecchi and Pentyala 2000). Surrounding the active site there is a ridge of hydrophobic residues that allows the enzyme to insert into the lipid bilayer, locating the active site close to the substrates. This is a common mechanism (called the interfacial activation mechanism) for several hydrophilic enzymes that act on hydrophobic substrates such as lipases and phospholipases (Aloulou et al. 2006). For the specific case of PI-PLC, it has been demonstrated that the insertion of the hydrophobic ridge into the bilayer strictly depends on the lateral pressure of the bilayer (reviewed in Rebecchi and

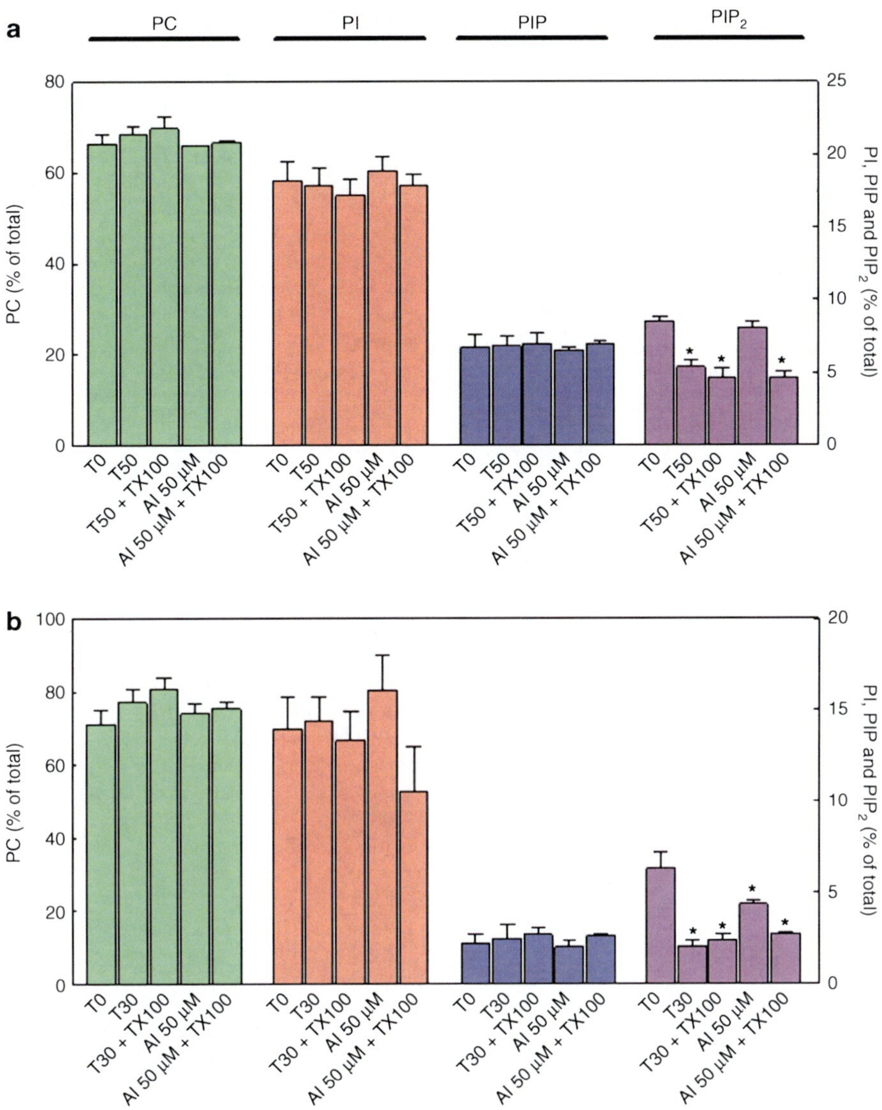

Aluminum and Phosphatidylinositol-Specific-Phospholipase C, Fig. 4 Al^{3+} inhibition of PI-PLC-mediated PIP_2 hydrolysis depends on membrane integrity. (**a**) PC:PPI (60:40) or (**b**) PC:PPI (90:10) liposomes were incubated for 10 min in the presence of 50 μM Al^{3+}. After incubation, portions of the samples were treated with or without 0.5% (v/v) Triton X-100 in order to disrupt membranes, and further incubated for (**a**) 60 min (T60) at 37°C in the presence of brain PI-PLC, or (**b**) 30 min (T30) at 37°C in the presence of PI-PLC from *Bacillus cereus*. Lipids were resolved by high-performance thin-layer chromatography, and PC, PI, PIP and PIP_2 were quantified. Results are shown as the mean ± SEM of at least three independent experiments. * Significantly different from the value measured in control liposomes (T0) ($P < 0.01$, one-way ANOVA) (Adapted from Arch Biochem Biophys, 408, Verstraeten SV and Oteiza PI, "Al^{3+}-mediated changes in membrane physical properties participate in the inhibition of polyphosphoinositide hydrolysis", 263–271, 2002, and Chem Phys Lipids, 122, Verstraeten SV and Oteiza PI, "Al^{3+}-mediated changes on membrane fluidity affects the activity of PI-PLC but not of PLC", 159–163, 2003, with permission from Elsevier)

Pentyala 2000). Therefore, in rigid membranes, where the lipids are tightly packed, PI-PLC is not able to penetrate the bilayer and unable to reach its substrates. This might be the mechanism operative in the inhibition of PI-PLC activity by Al^{3+}, which would act creating a lipid environment unfavorable for the docking of the enzyme rather than causing a direct inhibition.

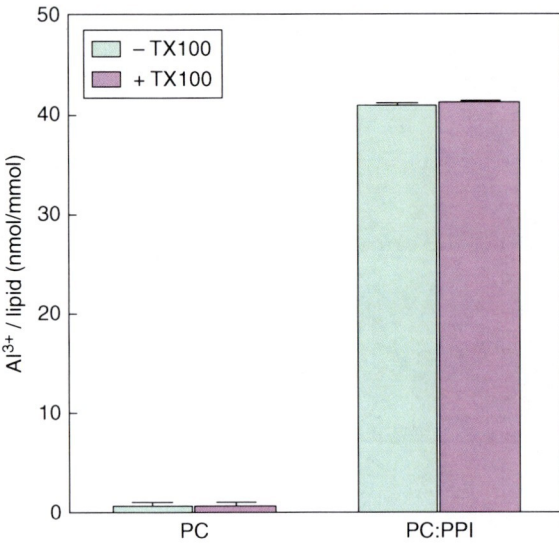

Aluminum and Phosphatidylinositol-Specific-Phospholipase C, Fig. 5 The presence of PPI in the membrane enhances Al^{3+} binding. PC or PC:PPI (60:40 molar ratio) liposomes were incubated in the presence of 50 μM Al^{3+}. After incubation, portions of the samples were treated without (▫) or with (▪) 0.5% (v/v) Triton X-100 in order to disrupt membranes. The binding of Al^{3+} to the membrane was measured with the reagent lumogallion. Results are shown as the mean ± SEM of four independent experiments (Adapted from Arch Biochem Biophys, 408, Verstraeten SV and Oteiza PI, "Al^{3+}-mediated changes in membrane physical properties participate in the inhibition of polyphosphoinositide hydrolysis", 263–271, 2002, with permission from Elsevier)

Conclusions

Multivalent cations are documented to modify the activity of a myriad of enzymes. When evaluating the underlying mechanisms, it is important to consider that metal-protein interactions may not be the only cause for an altered functioning of membrane enzymes. Metals can interact with enzyme lipid substrates and alter their molecular organization and dynamics, making them less available to the enzyme.

Cross-References

- ▶ Aluminum and Bioactive Molecules, Interaction
- ▶ Aluminum, Biological Effects
- ▶ Calcium in Biological Systems
- ▶ Lipases

References

Aloulou A, Rodriguez JA, Fernandez S, van Oosterhout D, Puccinelli D, Carrière F (2006) Exploring the specific features of interfacial enzymology based on lipase studies. Biochim Biophys Acta 1761:995–1013

Corry B, Hool L (2007) Calcium Channels. In: Chung SH, Andersen O, Krishnamurthy V (eds) Biological membrane ion channels: dynamics, structure, and applications. Springer, New York

Martin RB (1986) The chemistry of aluminum as related to biology and medicine. Clin Chem 32:1797–1806

Nostrandt AC, Shafer TJ, Mundy WR, Padilla S (1996) Inhibition of rat brain phosphatidylinositol-specific phospholipase C by aluminum: regional differences, interactions with aluminum salts, and mechanisms. Toxicol Appl Pharmacol 136:118–125

Rebecchi MJ, Pentyala SN (2000) Structure, function, and control of phosphoinositide-specific phospholipase C. Physiol Rev 80:1291–1335

Shafer TJ, Mundy WR (1995) Effects of aluminum on neuronal signal transduction: mechanisms underlying disruption of phosphoinositide hydrolysis. Gen Pharmacol 26:889–895

Verstraeten SV, Oteiza PI (2000) Effects of Al^{3+} and related metals on membrane phase state and hydration: correlation with lipid oxidation. Arch Biochem Biophys 375:340–346

Verstraeten SV, Oteiza PI (2002) Al^{3+}-mediated changes in membrane physical properties participate in the inhibition of polyphosphoinositide hydrolysis. Arch Biochem Biophys 408:263–271

Verstraeten SV, Nogueira LV, Schreier S, Oteiza PI (1997a) Effect of trivalent metal ions on phase separation and membrane lipid packing: role in lipid peroxidation. Arch Biochem Biophys 338:121–127

Verstraeten SV, Golub MS, Keen CL, Oteiza PI (1997b) Myelin is a preferential target of aluminum-mediated oxidative damage. Arch Biochem Biophys 344:289–294

Verstraeten SV, Keen CL, Golub MS, Oteiza PI (1998) Membrane composition can influence the rate of Al^{3+}-mediated lipid oxidation: effect of galactolipids. Biochem J 333:833–838

Verstraeten SV, Erlejman AG, Zago MP, Oteiza PI (2002) Aluminum affects membrane physical properties in human neuroblastoma (IMR-32) cells both before and after differentiation. Arch Biochem Biophys 399:167–173

Verstraeten SV, Villaverde MS, Oteiza PI (2003) Al^{3+}-mediated changes on membrane fluidity affects the activity of PI-PLC but not of PLC. Chem Phys Lipids 122:159–163

Aluminum Complexes with Low Molecular Mass Substances

▶ Aluminum and Bioactive Molecules, Interaction

Aluminum in Biological Systems

Christopher Exley
The Birchall Centre, Lennard-Jones Laboratories,
Keele University, Staffordshire, UK

Definition

Aluminum is the third most abundant element in the Earth's crust and the most abundant metal (Exley 2003). Aluminum is neither required by biological systems nor is it known to participate in any essential biological processes. While today all living organisms contain some aluminum, there is no scientific evidence that any organism uses aluminum for any biological purpose. There is similarly no evidence from the proteome or genome that any organism has utilized aluminum at any time in the evolutionary record. Aluminum's abundance and paradoxical lack of biological function remains a biochemical enigma.

It is argued that aluminum's absence from biochemical processes can be best explained in terms of its "historical" absence from biochemical evolution (Exley 2009a). In spite of its abundance in the Earth's crust, aluminum was not biologically available for the greater part of biochemical evolution. It was not available to participate in the natural selection of the elements of life. It is shown that silicon, as silicic acid, is a geochemical control of the biological availability of aluminum and has probably played a significant role in keeping aluminum out of biota (Exley 1998).

If aluminum had been biologically available during early biochemical evolution then this encyclopedia entry would have been listing and describing myriad proteins and biochemical systems which are either built around or require aluminum. This can be anticipated as it is now known that when aluminum is biologically available it is a serious ecotoxicant and can act so at to displace or nullify biologically essential metals and, in particular, magnesium, calcium, and iron. It is known that aluminum is the principal antagonist in fish death in acid waters, in forest decline in acidified catchments, in poor crop productivity on acid soils, and it is also known that aluminum is both acutely and chronically toxic in humans (Exley 2009b). The scientific evidence indicates that biologically available aluminum is biologically reactive and not always to the detriment of the biochemical processes concerned. For example, while some phosphoryl-transferring enzymes, such as hexokinase, are potently inhibited by aluminum, others are not and appear at least to use Al-ATP as effectively as Mg-ATP as a biochemical substrate (Furumo and Viola 1989; Korchazhkina et al. 1999). In addition, while aluminum is not known to play any role in any enzyme it has been shown that when aluminum is engineered into the active site of a purple acid phosphatase the activity of this enzyme was not hindered by the presence of aluminum. This was described by the authors as the first example of an active aluminum enzyme (Merkx and Averill 1999). The lack of biochemical essentiality of aluminum is best explained as being due to it being excluded from biochemical evolution (Exley 2009a). If biologically available aluminum had been present, then aluminum would either be an essential element in modern biochemistry (the phenotype of which would not necessarily be similar to what is known today) or there would still be clues in the evolutionary record pertaining to the selection of aluminum out of biochemical processes. It is argued that today aluminum is a silent visitor to contemporary biota including human beings and is only now participating in the evolutionary process. These "visits" are not inconsequential and are already being manifested as both acute and chronic toxicity (Exley 2009b). The latter may prove to be an unrecognized burden upon biological systems including human health unless steps are taken to protect the environment and the body from its burgeoning presence in modern life and its increasing participation in modern biochemistry (Exley 1998).

References

Exley C (1998) Silicon in life: a bioinorganic solution to bioorganic essentiality. J Inorg Biochem 69:139–144

Exley C (2003) A biogeochemical cycle for aluminium? J Inorg Biochem 97:1–7

Exley C (2009a) Darwin, natural selection and the biological essentiality of aluminium and silicon. Trends Biochem Sci 34:589–593

Exley C (2009b) Aluminium and medicine. In: Merce ALR, Felcman J, Recio MAL (eds) Molecular and supramolecular bioinorganic chemistry: applications in medical sciences. Nova Science, New York, pp 45–68

Furumo NC, Viola RE (1989) Inhibition of phosphoryl-transferring enzymes by aluminium-ATP. Inorg Chem 28:820–823

Korchazhkina OV, Wright G, Exley C (1999) No effect of aluminium upon the hydrolysis of ATP in the coronary circulation of the isolated working rat heart. J Inorg Biochem 76:121–126

Merkx M, Averill BA (1999) Probing the role of the trivalent metal in phosphate ester hydrolysis: preparation and characterisation of purple acid phosphatases containing (AlZnII)-Zn-III and (InZnII)-Zn-III active sites, including the first example of an active aluminium enzyme. J Am Chem Soc 121:6683–6689

Aluminum in Plants

Charlotte Poschenrieder and Juan Barceló
Lab. Fisiología Vegetal, Facultad de Biociencias, Universidad Autónoma de Barcelona, Bellaterra, Spain

Synonyms

Aluminum

Definition

Aluminum is a chemical element with atomic number 13 and atomic weight 26.98. In acid environments Al solubility increases and Al ions can be taken up by plants.

Presence of Aluminum in Plant's Life

Aluminum is an amphoteric element without any established biological function. Aluminum ranges third in abundance of the chemical elements in the lithosphere. In nature Al does not occur as free metal, but mainly in the form of only lightly soluble oxides and silicates. The availability of Al and, in consequence, the possibility of Al to interact with plants is mostly restricted to acid environments. In aqueous media with pH below 5, Al $(H_2O)_6^{3+}$ is the predominant monomeric Al species. This Al form, usually written simplified as Al^{3+}, is thought to be the main toxic Al species (Kochian et al. 2004). In the chemically complex soil solutions different inorganic (e.g., fluoride, sulfate, silicon) and organic ligands (e.g., organic acids, phenolics, hydroxamates, humic and fulvic acids) compete with OH to complex Al. Binding of Al to these ligands reduces the toxicity. Especially organic Al complexes are considered practically nontoxic to plants. Polymeric Al, especially in the tridecameric form ($[AlO_4Al_{12}(OH)_{24}(H_2O)_{12}]^{7+}$; simplified known as Al_{13}, has been claimed to be extremely toxic to plants and microorganisms. The environmental relevance of this polycation, however, is still under debate. In conclusion, acid environments are the main scenarios for potential Al toxicity. Among those acidic freshwaters, forests affected by acid deposition due to air pollution, and acid tropical soils have focused most research (Poschenrieder et al. 2008).

Phytotoxicity of Aluminum

Aluminum toxicity is among the most important abiotic stress factors limiting crop production on acid soils, especially in tropical regions. Plant roots explore the soils for acquiring water and essential mineral nutrients. In acid mineral soils roots are the first plant organ to contact with toxic Al^{3+} and inhibition of root elongation is a fast consequence in Al-sensitive plant species. The most Al-sensitive root part is the transition zone, the region where cells leaving the meristem (cell division zone) are about to enter the elongation zone (cell expansion zone). However, Al not only reduces root length but changes the entire root architecture. Aluminum-stressed roots show enhanced lateral root initiation. The elongation of these newly initiated laterals is quickly inhibited. The result is a stunted root system of bushy appearance. Affected plants are more sensitive to seasonal draught and low phosphorous availability.

As for other ions, the mechanisms of Al^{3+} toxicity are related to its binding to the organic molecules required for the proper functioning of the plant cells. The trivalent Al^{3+} is among the hardest Lewis acid and has high affinity for oxygen donor ligands. Aluminum strongly binds to phosphatidyl, carboxyl, and phenyl groups. This implies multiple toxicity targets both in cell walls and plasmalemma as well as inside the cells.

Cell Walls: Cell walls are main sites for Al binding in roots. Cross-linking by Al of the carboxyl groups of cell wall pectins has been made responsible for Al-induced cell wall stiffening leading to inhibition of root cell elongation. Cross-linking of other polar wall constituents such as the hydroxyproline-rich glycoprotein (HRGP) extensin can also contribute to inhibition of root cell elongation. It remains to be established if such cross-linking of HRGPs is caused by direct binding of Al to the protein or is mediated by Al-induced production of reactive oxygen species (ROS) (Poschenrieder et al. 2008).

▸ *Plasma Membrane*: Carboxyl and phosphatidyl groups are potential binding sites for Al in the plasma membrane. Al binding to the plasmalemma can change key properties of this membrane affecting fluidity, lateral lipid phase separation, membrane potential, and ion channel activity (Ma 2007). Aluminum-induced alteration of Ca^{2+} homeostasis (Rengel and Zhang 2003), inhibition of proton adenosine triphosphatase (H^+-ATPase), and lipid peroxidation are characteristic features of the Al toxicity syndrome (Ma 2007). Aluminum by interacting with plasmamembrane components can alter different signaling pathways. ▸ Calcium signaling and inositol 1,4,5 triphosphate–mediated signal transduction are targeted by Al (Rengel and Zhang 2003; Ma 2007). Moreover, aluminum fluoride (AlF_4^-) is a well-known activator of G-proteins. Trimeric G proteins, which couple extracellular signals to internal effectors, are essential in plant signal transduction.

Symplastic Al Targets: In contrast to the classical view that the plasma membrane is almost impermeable to trivalent cations, it is now well established that Al penetrates into root tip cells within minutes upon exposure. Inside the cells Al^{3+} is expected to be immediately bound to the multiple ligands with high affinity for this metal ion. Binding to organic acids or phenolic substances renders nontoxic Al complexes. However, toxicity targets inside the cells are, among others, ATP, GTP, nucleic acids, and glutamate. Aluminum even at nanomolar concentrations can efficiently compete with Mg for binding to ATP (Ma 2007). Staining of root tips with the Al-specific fluorescent dye lumogallion revealed the presence of Al in the nucleoli of root tip cell nuclei after a few hours of Al exposure.

Root Growth Inhibition: In the view of both the multiple target molecules of Al and the fast Al effects on different cell signaling pathways, the primary mechanism or mechanisms of Al-induced inhibition of root growth is still under debate. Two basic mechanisms should be distinguished: the initial fast inhibition of root elongation that is reversible if Al is removed after short exposure times and the irreversible inhibition. The reversible inhibition is most probably caused by Al cross-linking of cell wall pectins causing stiffening of the walls and reduced elongation of root tip cells.

Contrastingly, the irreversible inhibition implies arrest of both root cell division and root cell elongation. Aluminum alters the patterning of root tip cells. Cell division in the apical meristem is inhibited, while stimulation of cell division in more mature, subapical root zones promotes the emergence of new lateral roots. This change in root architecture seems related to Al-induced inhibition of polar auxin transport in the transition zone of the root. Auxin is a growth-stimulating phytohormone that promotes lateral root formation. The inhibition of auxin transport may cause a local increase of auxin concentrations close to the transition zone inducing the formation of new lateral roots. The polar, cell-to-cell transport of auxin is mediated by intracellular vesicle trafficking. Vesicle trafficking, in turn is closely controlled by the actin cytoskeleton (Poschenrieder et al. 2009).

The cytoskeleton has for long been recognized as a target of Al toxicity both in plant root tip cells and astrocytes of the human brain. In vitro, Al induced rigor in plant actin filament network and aggregation of bovine brain cytoskeleton proteins has been observed. It remains to be established whether Al-induced cytoskeleton alterations in vivo are directly caused by Al binding, or are indirect consequences of other primary toxicity mechanisms of Al (e.g., alterations of the energy metabolism, prooxidant activity, and interference of protein phosphorylation).

Aluminum Resistance in Plants

There are considerable differences in Al resistance among crop species and varieties. Exclusion of Al from the root tips, the most Al sensitive plant part, is a fundamental strategy to avoid Al toxicity. In many species this is brought about by exudation of organic acids from the root tips into the rhizosphere (Kochian et al. 2004). The organic acids like malate, citrate, or oxalate bind Al in a nontoxic form that is not taken up by the roots. There are two different patterns of organic

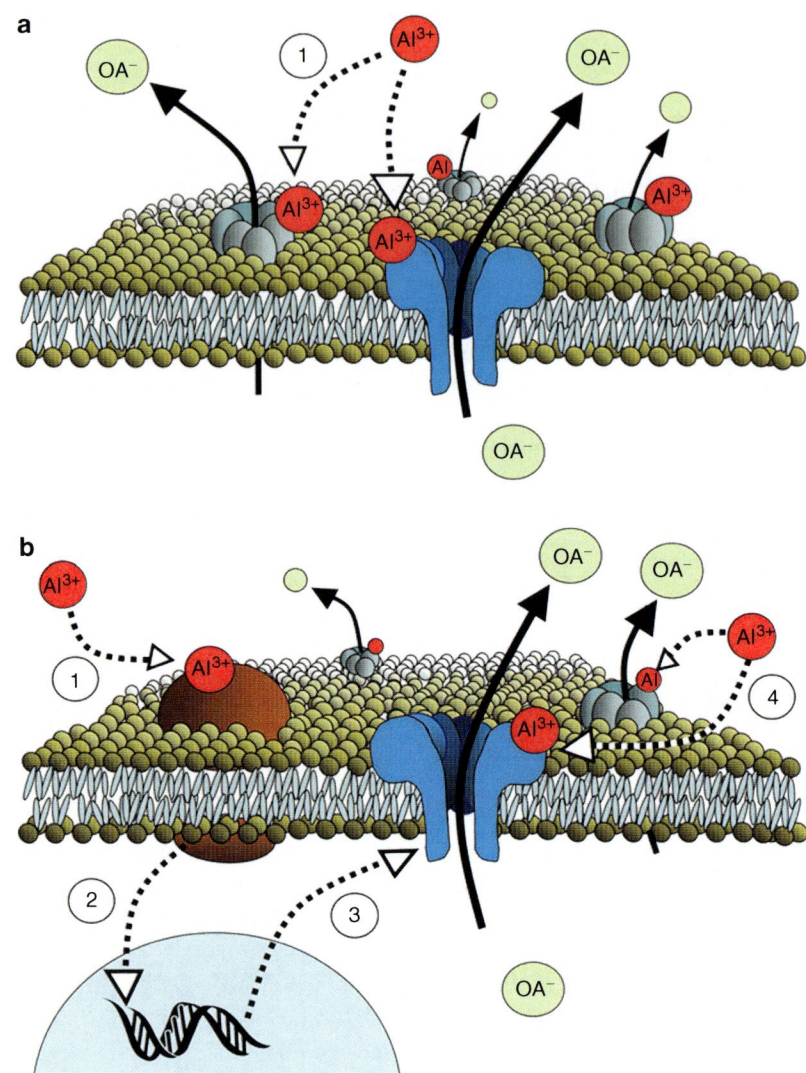

Aluminum in Plants, Fig. 1 Hypothetical models for the Al^{3+}-activated efflux of organic anions by members of the ALMT and MATE families of proteins.
(**a**) Pattern I: The protein is expressed constitutively in root apices with Al^{3+}-resistant genotypes showing greater expression than Al^{3+}-sensitive genotypes. Al^{3+} (*red cycles*) activates organic anion (OA^-) efflux by interaction directly with the preexisting protein in the plasma membrane.
(**b**) Pattern II: Al^{3+} first induces the expression of proteins through a signal transduction pathway possibly involving a specific receptor (*arrows 1, 2* and *3*) or nonspecific stress responses. Al^{3+} then activates organic anion efflux by interacting with the newly synthesized proteins (*arrow 4*) (Reproduced with permission from Delhaize et al. 2007)

acid exudation. Pattern I where the exudation of organic acids is triggered immediately upon Al exposure and pattern II where a lag time of several hours between start of exposure and organic acid release is observed (Fig. 1) (Delhaize et al. 2007). Wheat is the best studied pattern I plant; examples of pattern II behavior are rye, triticale, *Cassia tora,* and bean. Most genes responsible for Al resistance that have been isolated so far are genes encoding for organic acid efflux proteins. In cereals, proteins responsible for Al-activated efflux of organic acids are encoded by genes from two different families: *ALMT* encoding an Al-activated malate permease (*Al*uminum-induced *m*alate *t*ransporter) and *MATE* (*M*ultidrug *a*nd *T*oxic compound *E*xtrusion).

TaALMT1 is a major gene for Al resistance in wheat. Its constitutive expression in root tips is consistent with the pattern I type of Al-induced efflux of malate in Al-resistant wheat varieties (Fig. 1a). ALMT proteins, however, form a large, diverse family of plant proteins with transport functions not only restricted to Al resistance.

MATE proteins have been identified in plants, humans, animals, and microorganisms. In plants, different members of this large family of efflux proteins have been related to Al resistance, Fe efficiency, vacuolar anthocyanin transport, and transport of secondary metabolites including alkaloids. Alt_{Sb}, the Al resistance gene of sorghum (*Sorghum bicolor*), encodes a citrate efflux protein of the MATE family. Aluminum exposure activates the expression of this Al resistance gene in root tips of sorghum. This is consistent with the pattern II type of Al-induced citrate efflux from root tip cells in this species (Fig. 1b) (Delhaize et al. 2007).

Not in all plant species Al-induced exudation of organic acids from root tips fully accounts for varietal differences in Al resistance and other mechanisms can play a role. Mutational analyses in the model plant *Arabidopsis thaliana* have identified two Al^{3+}-sensitive mutants, *als1* and *als3*. ALS1 and ALS3 are ABC-type transporters. ALS1 is located in the tonoplast of the vacuole of root tips and in the vascular system. ALS3 is a plasma membrane transporter found in root cortex cells, in the phloem, and in hydatodes. Hydatodes are special cells located in the leaf borders where plants can extrude water droplets under conditions of high root pressure, a process called gutation (Ryan et al. 2011). It has been suggested that these ALS proteins may transport Al in chelated form contributing to Al tolerance by detoxification of Al inside the plants.

The high constitutive Al tolerance in rice (*Oryza sativa*) is related to two genes, *OsSTAR1* and *OsSTAR2*, encoding ABC transporters. Both proteins STAR1 and STAR2 associate to form a protein complex bound to vesicle membranes. Not organic acids, but UDP-glucose is the transported molecule. Vesicle-mediated transport of UDP-glucose to the apoplast is a fundamental process required for the formation of the cell wall. It is still not clear whether the STAR proteins are involved either in Al exclusion by conferring special cell wall properties or in tolerance of high internal Al concentrations by promoting Al compartmentation inside the plant (Ryan et al. 2011). Still unknown are the mechanisms of the extraordinarily high Al resistance in *Brachiaria decumbens*, a forage grass species of African origin. Neither organic acid exudation nor root-induced changes in the rhizosphere pH seem responsible for hyperresistance to Al in *Brachiaria*.

Aluminum in Plants, Table 1 Aluminum concentrations in shoots of selected accumulator species from different botanical families. Potential Al ligands, present in high concentrations in the different families, are given

Family/species	Leaf Al concentrations (mg/kg)
Polygonaceae	Oxalate, malate, citrate, quercetin anthraquinones
Fagopyrum esculentum	480–15,000
Theaceae	Catechin epigallocatechin gallate
Camellia sinensis	1,000–30,000
Eurya acuminata	19,600
Hydrangeaceae	Delphidin (flowers) citrate, hydrangenol (leaves)
Hydrangea sp.	400–3,000
Melastomataceae	Citrate, oxalate, tannate
Melastoma malabathricum	590–10,000
Miconia lutescens	6,800
Miconia ferruginata	4,300
Acisanthera uniflora	20,200
Pternande caerulescens	16,600
Memecylaceae	Flavonoids
Memecylon laurinum	12,700–30,500
Rubiaceae	Quercetin, kaempferol
Urophyllum macrophyllum	23,100

Modified from Barceló and Poschenrieder (2002) and Jansen et al. (2002)

Hyperaccumulation of Aluminum

While Al exclusion is the main strategy for Al resistance in most crop plants, there is a considerable number of plant species, native to acid soils with high Al^{3+} availability, which accumulate extremely high concentrations of Al in their shoots (Table 1). Most of these plant species from different botanical families are not used as commercial crops. Notable exceptions are the tea plant (*Camellia sinensis*), buckwheat (*Fagopyrum esculentum*), and the ornamental *Hydrangea* species. Common characteristics of Al hyperaccumulator plants are high degree of mycorrhizal colonization and high internal concentrations of either or both organic acids and phenolic substances. The mycorrhizal colonization seems crucial for phosphorus acquisition by these plants growing in soils with low P availability. To what extent mycorrhiza may contribute to the high accumulation of Al in nontoxic form is still unclear. Citrate seems to be the main ligand for Al in the xylem of *Melastoma*, tea, and buckwheat. After ligand exchange, Al can be stored in the leaves in the form of Al-oxalate. High concentrations of phenolic compounds can contribute to Al tolerance in these hyperaccumulating species both by binding Al in nontoxic form and by their antioxidant properties.

Localization studies of Al in leaves of hydroponically grown tea plants using low-energy X-ray fluorescence microscopy (LEXRF) have shown that Al mainly accumulates in the cell walls of leaf epidermal cells. Moreover, Al was identified in xylem and phloem tissue. Taken together, these results suggest that in tea plants Al preferentially moves apoplastically with the transpiration stream to the epidermal cells. Here water is evaporated to the air and the Al is stored in the cell walls. However, the Al signal detected in the phloem region indicates that some Al can move symplastically through the leaves.

Despite the high Al concentrations in tea leaves, no health risk for consumption of tea by humans has been shown. The high amount of phenolic substances present in tea-based beverages binds Al in nontoxic form and more than 99% of the Al passes the digestive tract unabsorbed. The small absorbed fraction is excreted by the kidney.

Beneficial Effects of Aluminum

Although most investigations on Al in plants address toxicity and resistance mechanisms, Al-induced stimulation of plant performance has also been reported. Aluminum is not an essential nutrient for plants and no specific function for Al has been established. In acid mineral soils, however, plants may benefit from the presence of Al^{3+} in the soil solution because the trivalent Al can ameliorate the toxicity of high proton concentrations (Kinraide 1993; Kidd and Proctor 2001).

Aluminum-induced stimulation of growth in Al hyperaccumulating plants seems related to enhanced phosphorous availability and uptake. Moreover, metal hyperaccumulation can be beneficial by inhibition of plant pathogens or as a feeding deterrent.

Cross-References

▶ Aluminum Speciation in Human Serum
▶ Aluminum and Bioactive Molecules, Interaction
▶ Aluminum in Biological Systems
▶ Aluminum, Genes Involved in Novel Adaptive Resistance in Rhodotorula glutinis

References

Barceló J, Poschenrieder C (2002) Fast root growth responses, root exudates, and internal detoxification a clues to the mechanisms of aluminium toxicity and resistance: a review. Environ Exp Bot 48:75–92

Delhaize E, Gruber BD, Ryan PR (2007) The roles of organic anion permeases in aluminium resistance and mineral nutrition. FEBS Lett 581:2255–2262

Jansen S, Watanabe T, Smets E (2002) Aluminium accumulation in leaves of 127 species in Melastomataceae, with comments on the order Myrtales. Ann Bot 90:53–64

Kidd PS, Proctor J (2001) Why plants grow poorly on very acid soils: are ecologists missing the obvious. J Exp Bot 52:791–799

Kinraide TB (1993) Aluminium enhancement of plant growth in acid rooting media. A case of reciprocal alleviation of toxicity by two toxic cations. Physiol Plant 88:619–625

Kochian LV, Hoekenga OA, Piñeros MA (2004) How do crop plants tolerate acid soils? Mechanisms of aluminium tolerance and phosphorus efficiency. Annu Rev Plant Biol 55:459–493

Ma FJ (2007) Syndrome of aluminum toxicity and diversity of aluminum resistance in higher plants. Int Rev Cytol 264:225–252

Poschenrieder C, Gunsé B, Corrales I, Barceló J (2008) A glance into aluminum toxicity and resistance in plants. Sci Total Environ 400:356–368

Poschenrieder C, Amenós M, Corrales I, Doncheva S, Barceló J (2009) Root behavior in response to aluminum toxicity. In: Baluška F (ed) Plant-environment interactions. From sensory plant biology to active plant behavior. Springer, Berlin, pp 21–43

Rengel Z, Zhang W-H (2003) Role of dynamics of intracellular calcium in aluminium-toxicity syndrome. New Phytol 159:295–314

Ryan PR, Tyerman SD, Sasaki T, Furuichi T, Yamamoto Y, Zhang WH, Delhaize E (2011) The identification of aluminium-resistance genes provides opportunities for enhancing crop production on acid soils. J Exp Bot 62:9–20

Aluminum Speciation

▶ Aluminum and Bioactive Molecules, Interaction

Aluminum Speciation in Human Serum

Janez Ščančar and Radmila Milačič
Department of Environmental Sciences, Jožef Stefan Institute, Ljubljana, Slovenia

Synonyms

High molecular mass aluminum species in serum; Hyphenated techniques; Low molecular mass aluminum species in serum; Speciation analysis

Definition

The role of trace elements and their impact on the environment and living organisms depend not only on their total concentration but also on chemical forms in which they are actually present. Individual chemical species of trace elements in different samples are quantitatively determined by speciation analysis.

A fundamental tool for speciation analysis of aluminum (Al) is the combination of separation technique coupled to element-specific detector. For the time being the most powerful hyphenation is coupling of liquid chromatography to inductively coupled plasma mass spectrometry (ICP-MS). For identification of Al-binding ligands, mass spectrometry (MS)-based techniques exhibit the greatest potential.

Introduction

Al is the most abundant metal in the lithosphere, comprising 8% of the Earth's crust. The sparingly soluble nature of most Al compounds and its low bioavailability considerably decreases the probability of an Al body load in humans from environmental sources. The main route of entry of Al into the human body is through the consumption of food, beverages, and drinking water with higher Al concentrations. Due to effective urinary clearance, the whole body Al burden in adult person is low, about 35 mg. Also the physiological serum Al concentrations are low, around 2 ng Al mL^{-1} or less. Al is not considered to be an essential element in humans, but its toxic effects are known. In past decades Al represented potential overload in dialysis patients. Consequently, due to elevated Al burden, many clinical disorders were observed such as renal osteodystrophy, microcytic anemia, and dialysis encephalopathy. Concern over maintaining the high quality of water and fluids used for dialysis considerably prevented intoxication of dialysis patients with Al. Nevertheless, the absorption of Al via consumption of Al-based drugs and its accumulation in target organs is still a possible source of Al overload in renal patients. Al is also accumulated in brain of Alzheimer's patients. Risk for Al intoxication represent parenterally administered nutrients, contaminated with Al, particularly in infants with insufficient kidney function. Since Al salts are common adjuvants in vaccines, neurotoxic effects due to frequent vaccination may also develop. The outcomes of several epidemiological studies indicated on possible connection between elevated Al concentrations in drinking water and increased risk for different neurotoxic disorders (Milačič 2005; Yokel 2004).

To better understand the toxicity of Al in humans, it is essential to identify and quantify the chemical species in which Al is transported and stored in the body. For these reasons, among a variety of body fluids and tissues, Al speciation was the most frequently investigated in human serum (Milačič 2005; Harrington et al. 2010). Due to the complex chemistry of Al in serum, its low total concentration and the high risk of contamination by extraneous Al, speciation of Al in serum is a difficult task for analytical chemists. In the last two decades, analytical techniques for the determination of the amount and composition of high molecular mass Al (HMM-Al) as well as low molecular mass Al (LMM-Al) species in human serum has been progressively developed. For separation of Al species microultrafiltration, size exclusion chromatography (SEC) and anion-exchange chromatography were applied (Sanz-Medel et al. 2002; Milačič et al. 2009; Murko et al. 2011). Separated Al species were determined either by electrothermal atomic absorption spectrometry (ETAAS) or inductively coupled plasma mass spectrometry (ICP-MS). Identification of Al-binding ligand in separated serum fractions was performed by sodium dodecyl sulfate polyacrylamide gel electrophoresis (SDS-PAGE) and mass spectrometry (MS)-based techniques (Kralj et al. 2004; Bantan et al. 1999; Murko et al. 2009).

Al-Binding Ligands in Human Serum and the Role of Al Speciation

Blood serum is a very complex matrix with pH 7.4, containing HMM and LMM compounds and high concentration of salts. LMM metal complexes are the most active in terms of bioavailability. One of the most important LMM serum constituents is citrate. It is considered to be one of the major LMM binding ligands in mammalian serum and as an important LMM ligand for metals transport in human body. In human serum citrate occurs in concentrations about 0.1 mmol L^{-1}. Al-citrate complex is formed through coordination binding of Al with hydroxyl group and the two terminal carboxylates of citrate, thus leaving a free carboxylate that is dissociated at physiological pH. The complex is very stable. In addition to citrate, the important LMM compound in serum is phosphate, present in concentration about 2 mmol L^{-1}. Al^{3+} readily forms sparingly soluble species with phosphate anion, which may precipitate from body fluids. Due to the sparingly soluble nature of phosphate species, soluble Al-phosphate complexes in serum occur in much lesser extent than Al-citrate, while other LMM constituents of blood serum like lactate, oxalate, and amino acids have significantly lower affinities for Al^{3+} than citrate and phosphate and were not considered to contribute in complex formation. The presence of Al$(OH)_4^-$ was predicted in human serum, but in negligible concentrations (Milačič 2005). The majority of Al^{3+} in human serum is bound to HMM proteins. The potential HMM Al-binding ligands of serum proteins are albumin (molecular mass 66,000 Da, average serum concentration about 0.6 mmol L^{-1}) and transferrin (Tf) (molecular mass 795,000 Da, average serum concentration about 37 µmol L^{-1}). Albumin is too weak as a metal ion binder at physiological pH values to be able to effectively compete for Al^{3+} with other much stronger carriers such as Tf and citrate. Al readily binds to the two high-affinity iron-binding sites, the N-terminal and C-terminal domains. Since each Tf molecule has two metal-binding sites, the total binding capacity of Tf is about 74 µmol L^{-1}. Under normal conditions Tf is only about 30% saturated with Fe, leaving about 50 µmol L^{-1} of unoccupied metal-binding sites. This excess is more than enough to complex Al in human serum at normal concentration levels (Al concentrations about 0.1 µmol L^{-1}) and even in serum of dialysis patients (Al concentrations up to 5 µmol L^{-1}). The Tf is characterized by the unique requirement that Al is only bound as a ternary complex between the metal, the protein, and a carbonate anion, which is referred to as the synergistic anion (Milačič 2005).

The chemical speciation of Al in human serum is of crucial importance in assessing which individual compounds contribute to Al neurotoxicity. Tf and citrate may facilitate Al transport across the blood–brain barrier, while Tf seems to mediate Al transport into neurons and glial cells (Yokel 2004). The speciation analysis is also very important in the evaluation of the efficiencies of chelation therapies in patients overloaded with Al (Murko et al. 2011). For these reasons efforts of many investigators were oriented to the development of reliable analytical procedures for the determination of Al species in human serum.

Al Speciation in Human Serum

Physiological concentrations of Al in human serum are very low, around 2 ng mL^{-1}. A decade ago elevated concentrations of Al up to 150 ng mL^{-1} were found in dialysis patients. Presently, these concentrations are appreciably lower due to intensive care to prevent Al overload in renal patients. Complexity of the serum constituents, low total concentration of Al, the high risk of contamination, and the tendency of Al to be adsorbed by different chromatographic supports make speciation of Al in human serum a difficult task for analytical chemists. In the following sections, problems related to Al speciation in human serum are discussed and the progress that was made through the development of the analytical procedures is presented.

Contamination: Sources and Elimination

To obtain reliable analytical data in Al speciation, it is of extreme importance to prevent all possible sources of contamination. The only appropriate material for laboratory ware is high-density polyethylene or Teflon that should be cleaned before use with 10% HNO_3. Traces of Al should be removed also from eluents, reagents, columns, and filtering devices. For effective cleaning of membranes of the filtering devices, NaOH was found to be the most convenient. To remove Al from eluents used in chromatographic separations,

CLEANING PROCEDURES

Laboratory ware and tubes
- Polyethylene or Teflon
- Cleaning with 10% HNO_3 (soaking for 24 h, rinsing with Mili Q water)

Eluents
- Chelating ion-exchange chromatography (Chelex 100, Na^+ form, batch procedure) followed by passing of eluent through silica based reversed-phase C_{18} HPLC column
or
- Kelex 100-impregnated silica C_{18} scavenger column used "on-line" to clean the eluents

Microultrafiltration membranes
- 0.1 mol L^{-1} NaOH or 2 mol L^{-1} citric acid

Analytical procedures performed under clean room conditions class 10000

CLEANING OF SEC AND FPLC COLUMNS

Cleaning is carried out at a flow rate of 1 mL min^{-1}
   15 min rinsing with water
- 1 mL (SEC) or 0.5 mL (FPLC) of 1 mol L^{-1} NaOH injected onto the column resin
- 15 min rinsing with water
- 10 min gradient elution from 100% water to 100% 2 mol L^{-1} citric acid
- 10 min rinsing with 2 mol L^{-1} citric acid
- 10 min gradient elution from 100 % 2 mol L^{-1} citric acid to 100% water
- 15 min rinsing with water
- 15min rinsing with 0.05 mol L^{-1} Tris-HCl buffer (pH 7.4)

↓

LOW BLANKS < 0.5 ng Al mL^{-1} (ETAAS detection)

CLEANING OF CIM MONOLITHIC DEAE-8 COLUMN

Cleaning is carried out at a flow rate of 5 mL min^{-1}
   5 min rinsing with 1 mol L^{-1} NaOH
- 20 min rinsing with 0.2 mol L^{-1} Tris-HCl buffer (pH 7.4)
- 20 min rinsing with 0.05 mol L^{-1} Tris-HCl buffer (pH 7.4)
- 8 min rinsing with 2 mol L^{-1} citric acid
- 30 min rinsing with 0.05 mol L^{-1} Tris-HCl buffer (pH 7.4)

↓

VERY LOW BLANKS < 0.15 ng Al mL^{-1} (ICP-MS detection)

Aluminum Speciation in Human Serum, Fig. 1 The most frequently applied cleaning procedures in Al speciation

different chelating resins in combination with silica-based columns, which have strong affinity to adsorb Al, were applied (Sanz-Medel et al. 2002). Chromatographic supports used in the speciation procedure were cleaned by application of citric acid, acetic acid or sodium citrate, and NaOH (Milačič et al. 2009). In order to obtain reliable analytical data in speciation of Al in serum samples, the overall cleaning procedure and careful handling of samples to prevent contamination is of crucial importance. The most frequently applied cleaning procedures in Al speciation are schematically presented in Fig. 1.

Fractionation Procedures

Because of low Al concentration the speciation of Al in serum of healthy subjects was in general performed in spiked samples or in serum of renal patients with elevated Al concentrations. To separate Al bound to proteins (HMM-Al) from Al bound to LMM compounds, fractionation procedures based on microultrafiltration or size exclusion chromatography (SEC) were commonly applied. The microultrafiltration procedures were generally more liable to contamination problems than SEC procedures (Milačič et al. 2009). Microultrafiltration and SEC procedures in general gave the same results on the amount of HMM-Al and LMM-Al species. Studies of many researchers confirmed that about 90% of Al is bound to proteins, while the remaining Al corresponded to the LMM serum fraction (Sanz-Medel et al. 2002; Kralj et al. 2004; Milačič et al. 2009). Recently, a fractionation procedure was developed for rapid separation of proteins from the LMM-Al species. The fractionation was performed on HiTrap desalting SEC column coupled on-line to ICP-MS in 10 min (Murko et al. 2011). By this simple SEC procedure the partitioning of Al between serum proteins and the LMM-Al compounds may be quantified in both the HMM serum fraction (first 5 min of the separation procedure) and the LMM serum fraction (the following 5 min of the separation procedure). Typical elution profile of Al for separation of serum proteins and LMM species by HiTrap desalting SEC-ICP-MS procedure is presented in Fig. 2. In comparison to ultrafiltration and SEC procedures that take 0.5–1 h (Sanz-Medel et al. 2002, Milačič et al. 2009), the separation time of the HiTrap desalting SEC procedure is much shorter. In addition, ultrafiltration procedures enable quantification of Al

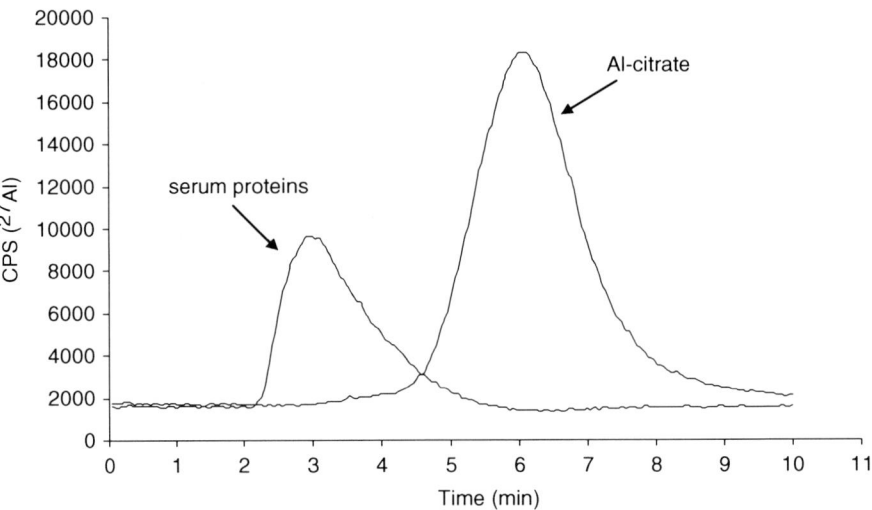

Aluminum Speciation in Human Serum, Fig. 2 Rapid separation of serum proteins and LMM species by HiTrap desalting SEC-ICP-MS procedure (Reproduced with permissions from The Royal Society of Chemistry (Murko et al. 2011))

only in the LMM serum fraction (Sanz-Medel et al. 2002; Milačič et al. 2009), while by the use of the conventional SEC procedures the quantification of Al is in general possible only in the protein fractions (Sanz-Medel et al. 2002; Kralj et al. 2004; Milačič et al. 2009). It is of great importance that the HiTrap desalting SEC fractionation procedure does not influence the chemical speciation of the serum constituents and therefore enables further speciation procedures to be applied in the separated HMM and LMM serum fractions. The fast fractionation procedure is also reliable and sensitive enough to perform fractionation of serum samples containing Al at physiological concentrations levels. It represents also a promising tool for studies of the efficiencies of the chelation therapies in patients overloaded with Al. Such therapies have been already performed in dialysis patients (Sanz-Medel et al. 2002) and are presently again actual in modalities for treatment of other clinical disorders. The latest discoveries on relationship between brain metal dishomeostasis and progression of Alzheimer's disease rendered chelation therapy as a promising option for curing this neurological disease (Hedge et al. 2009).

Chromatographic Procedures in Speciation of LMM-Al Species in Human Serum

Data obtained from computer modeling on the distribution of LMM-Al species in human serum predicted a variety of LMM-Al compounds, of which citrate was considered as one of the most important Al-binding ligand. Therefore, there was a need to experimentally determine the composition and content of LMM-Al species in human serum. This was rather difficult task for analytical chemists, since Al tends to adsorb on a variety of chromatographic supports. A big problem represented also contamination with Al arising from fillings of chromatographic columns (Sanz-Medel et al. 2002). The most powerful analytical procedure was developed by the use of strong anion-exchange Mono Q fast protein liquid chromatography (FPLC) column. It enabled quantitative separation of Al-citrate and soluble Al-phosphate species, applying gradient elution from water to 4 mol L^{-1} NH_4NO_3 in 10 min (Bantan et al. 1999). The use of appropriate cleaning procedures (NaOH and citric acid) lowered the column blanks below 1 ng mL^{-1} Al. In the analysis of serum samples, the microultrabiltrable fraction of spiked human serum and serum of dialysis patients was injected onto the column resin. 0.2 mL fractions were collected throughout the chromatographic run and Al determined by ETAAS. In fractions that contained Al, the binding ligand was presumed on the basis of the retention time and identified also by electrospray ionization tandem mass spectrometry (ESI-MS-MS). Based on the mass spectra and the corresponding daughter ion spectra, the presence of citrate (peak m/z 191 and the corresponding daughter ion spectra with m/z 111, 87 and 85), phosphate (peak m/z 97 and the corresponding daughter ion spectra with m/z

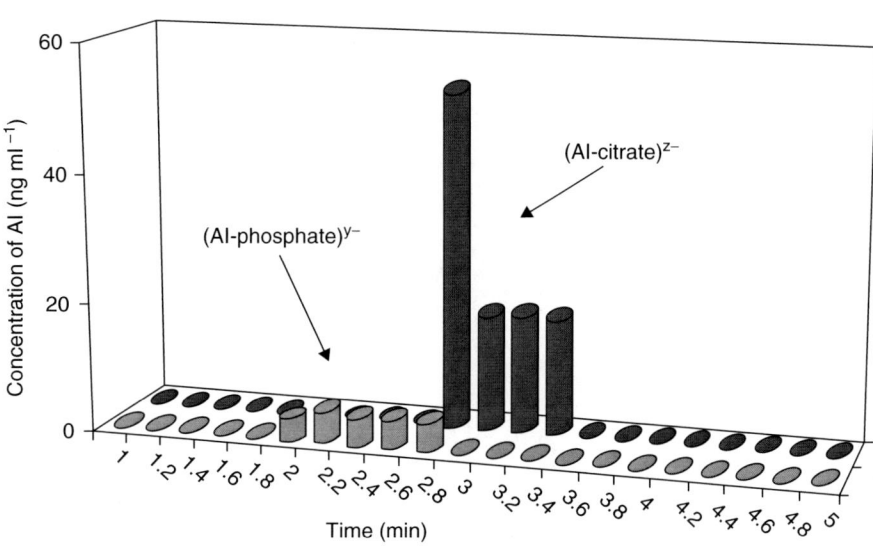

Aluminum Speciation in Human Serum, **Fig. 3** Typical chromatograms of Al-citrate and Al-phosphate (100 ng mL^{-1} Al) at pH 7.4. Separation was performed on a Mono Q HR 5/5 anion-exchange FPLC column and separated species were detected by ETAAS (Reproduced with permissions from Elsevier (Bantan Polak et al. 2001))

97 and 79), and both phosphate- and citrate-binding ligands was confirmed in separated serum fractions. Typical chromatograms for separation of Al-citrate and Al-phosphate at pH 7.4 are presented in Fig. 3. An example of ESI-MS-MS spectra in separated serum fractions containing Al is presented in Fig. 4. By combining data from speciation of Al by FPLC-ETAAS procedure and further identification of the Al-binding ligands, it was experimentally proven that Al-citrate, Al-phosphate, and ternary Al-citrate-phosphate species were present in serum samples of healthy persons (serum spiked with Al) and dialysis patients. These data were in agreement with the modeling calculations in human serum. In serum of healthy adults, the percentage of LMM-Al species ranged from 14% to 20%, while in dialysis patients the LMM-Al fraction was higher, up to 50%.

Chromatographic Procedures in Speciation of HMM-Al Species in Human Serum

SEC procedures were not selective enough to separate albumin and transferrin (Sanz-Medel et al. 2002; Milačič et al. 2009). Therefore, more powerful chromatographic procedures in combination with various specific detection techniques were needed to characterize and quantify the Al-binding proteins in human serum. To obtain higher resolution in separation of serum proteins, anion-exchange chromatographic columns were applied. One of the first promising chromatographic supports represented a polymeric anion-exchange Protein Pak DEAE-5-PW column (Sanz-Medel et al. 2002). Separated proteins were followed by UV detection, while Al in separated fractions was determined by ETAAS. The selected column fractions that contained Al were also characterized by SDS-PAGE (Fig. 5). Although immunoglobulin G (IgG) and Tf were not quantitatively separated, the results indicated that Al was eluted under the Tf peak. To obtain better separation of proteins, an anion-exchange FPLC Mono Q HR 5/5 column coupled to UV and element-specific ETAAS detectors was later used by several researchers. Results revealed that Al was bound exclusively to Tf. In order to lower the limits of detection, the same chromatographic column was further applied, while for detection of separated Al species quadruple ICP-MS and high-resolution ICP-MS instruments were used. Special attention was focused on efficient cleaning of eluents, while the cleaning of column supports remained the limiting factor for speciation of Al in unspiked human serum at physiological concentration levels (Milačič et al. 2009). The results of different researchers again confirmed that about 90% of Al in human serum was bound to Tf. Since there are no reference materials for speciation analysis of Al, the use of complimentary analytical procedures is desirable. As an alternative to

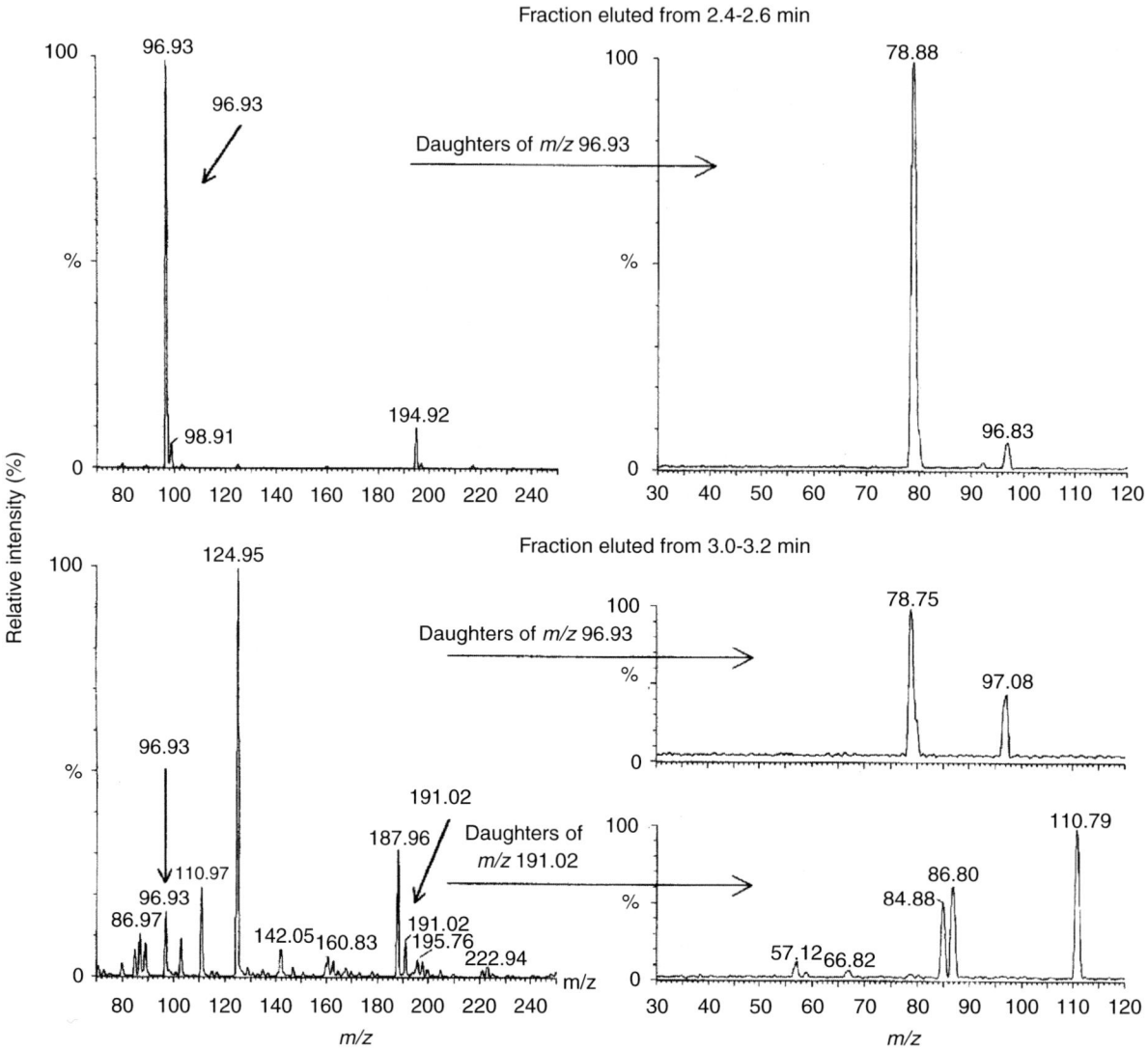

Aluminum Speciation in Human Serum, Fig. 4 ES-mass spectra and corresponding daughter ion mass spectra of m/z 97 for eluted fractions from 2.4 to 2.6 and m/z 97 and 191 for eluted fractions from 3.0 to 3.2 min, respectively, on an anion-exchange FPLC column for serum sample (Reproduced with permissions from The Royal Society of Chemistry (Bantan et al. 1999))

ion-exchange FPLC columns, ion-exchange separation supports based on convective-interaction media (CIM) were further applied. Al speciation was performed on CIM methacrylate-based monolithic anion-exchange diethyl amino ethane (DEAE) column. A great advantage of CIM DEAE column is its ability to sustain rigorous cleaning of the chromatographic support with NaOH (as presented in Fig. 1). Efficient cleaning of the column support and eluents used in the chromatographic procedure resulted in extremely low blanks (below 0.15 ng Al mL^{-1}). By coupling the CIM DEAE column to highly sensitive element

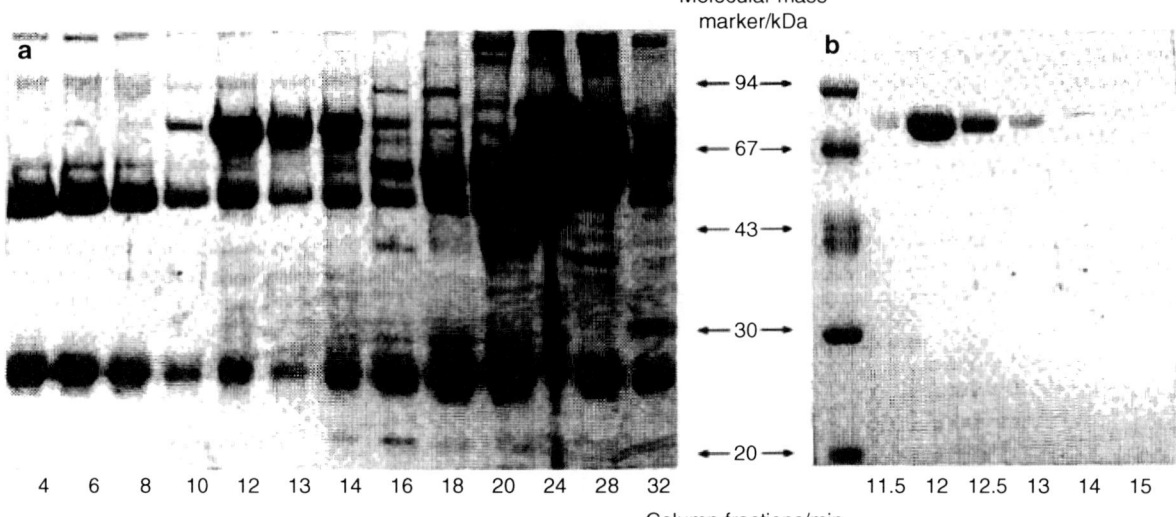

Aluminum Speciation in Human Serum, Fig. 5 SDS-PAGE of selected column fraction collected for elution of 500 μL of: (**a**) diluted spiked serum sample, and (**b**) Tf standard solution. Molecular mass markers are given in the center (Reproduced with permissions from The Royal Society of Chemistry (Wróbel et. al. 1995))

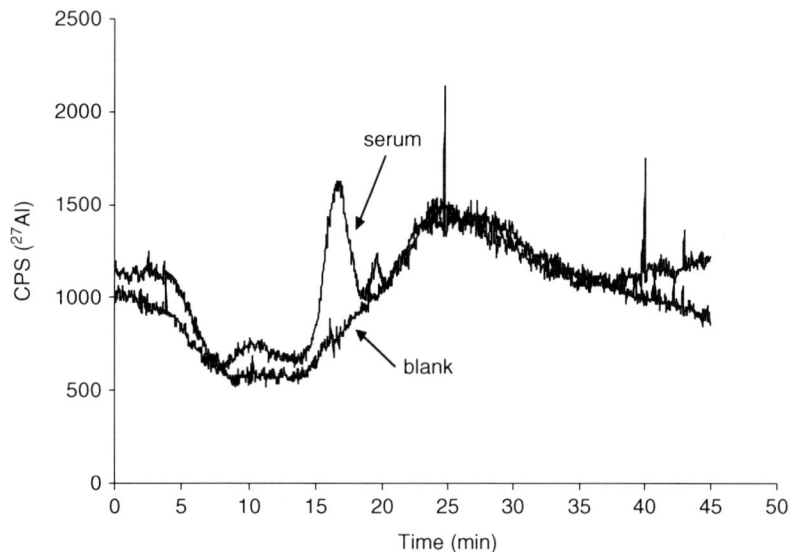

Aluminum Speciation in Human Serum, Fig. 6 The Al elution profiles for the separation of unspiked serum (1 + 4) and blank sample after overall cleaning procedure. The peak of Al in unspiked serum sample (1 + 4) corresponds to 1.04 ng mL^{-1} of Al (Reproduced with permissions from American Chemical Society (Murko et al. 2009))

specific detector such as ICP-MS, speciation of Al was possible at physiological concentration levels from 0.5 to 2 ng Al mL^{-1} (Fig. 6) (Murko et al. 2009). Data demonstrated that about 90% of Al in unspiked human serum was separated at the retention volume characteristic for Tf. To identify the Al-binding ligand, the column fraction that contained Al was further characterized by acquity ultra performance liquid chromatography–electrospray ionization mass spectrometry (UPLC-ESI-MS) (Fig. 7). The results again confirmed that Al-binding ligand was Tf. The data on the speciation of Al at physiological concentration levels represent an important basis for studies of Al distribution and its fate in human body.

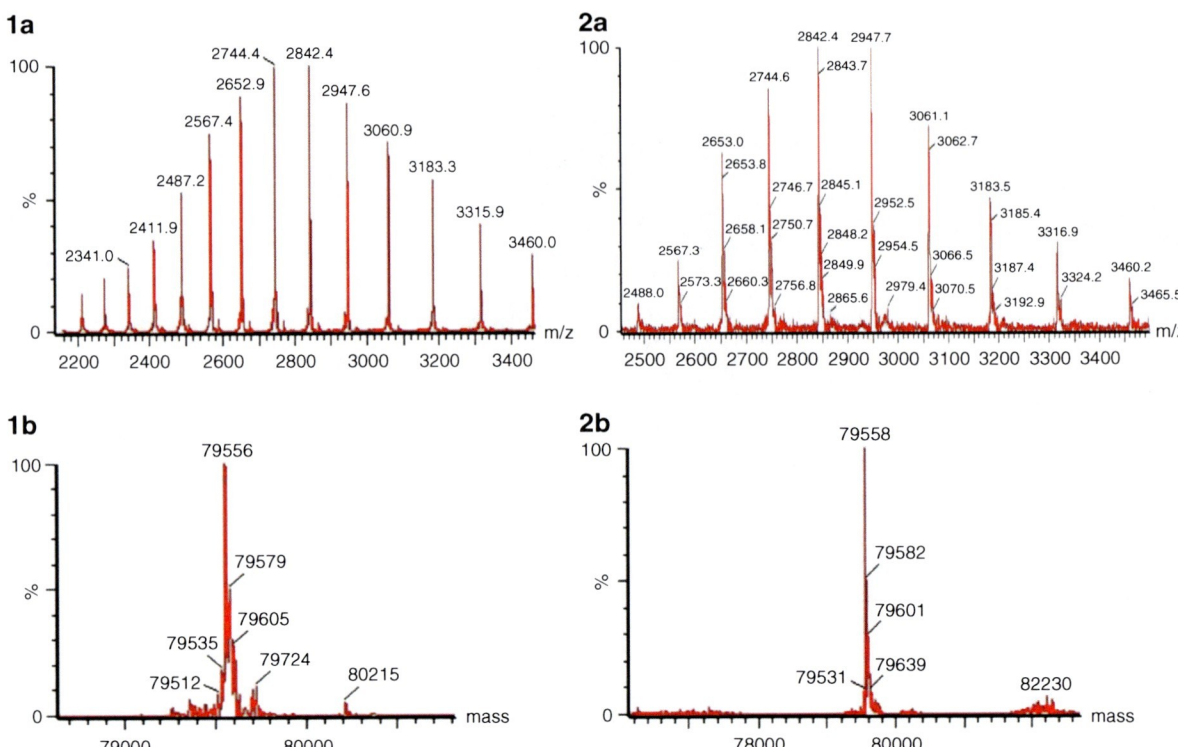

Aluminum Speciation in Human Serum, Fig. 7 UPLC-ESI-MS analysis of transferrin standard (1a and 1b) and human serum transferrin after separation on CIM DEAE column (2a and 2b). (a) ESI-mass spectrum, (b) deconvoluted mass spectrum (Reproduced with permissions from American Chemical Society (Murko et al. 2009))

Conclusions

Al is involved in many clinical disorders. To understand Al transport through the human body and its accumulation in target organs, it is necessary to obtain information on the chemical species of Al in human serum. In last decades analytical procedures for speciation of Al have been progressively developed. Due to the very low physiological concentrations of Al in serum of healthy subjects and the environmental abundance of Al, there is a high risk of contamination during speciation analysis. To avoid contamination by extraneous Al, appropriate handling of samples and efficient cleaning procedures should be applied. Ultrafiltration and SEC procedures enable fractionation of Al in serum. With these procedures it is possible to distinguish only between the amount of LMM-Al and HMM-Al species. However, they can be successfully applied as an appropriate analytical tool, to follow the effectiveness of chelation therapies in patients overloaded with Al. For identification and quantification of Al chemical species present in human serum, more powerful analytical procedures are required. Anion-exchange FPLC columns in combination with element-specific detectors, mass spectrometry–based techniques, UV, and SDS-PAGE were found to be powerful analytical tools for speciation both LMM-Al as well as HMM-Al species in spiked human serum, or in serum of dialysis patients with elevated Al concentrations. The combination of microultrafiltration and speciation analysis of ultrafiltrable Al by anion-exchange FPLC and ESI-MS-MS techniques demonstrated that the LMM-Al fraction is composed of Al-citrate, Al-phosphate, and ternary Al-citrate-phosphate complexes. Anion-exchange FPLC columns were also applied in separation of serum proteins. Investigations were performed on human serum spiked with Al. As element-specific detectors ETAAS and ICP-MS techniques were used. In combination with UV detection and identification of Al-binding ligand by SDS-PAGE, it has been demonstrated that about 90% of Al in spiked human serum is bound to Tf. Later, CIM DEAE monolithic column was applied in separation of serum proteins. By the use of rigorous cleaning of the column support and adequate cleaning of eluents used in the chromatographic procedure, the significant lowering of blanks and quantitative speciation of HMM-Al compounds at physiological concentration levels was possible. The identification of Al-binding ligand was performed by UPLC-ESI-MS. Data revealed that at physiological concentration levels the same amount of Al (about 90%) as in spiked human serum is bound exclusively to Tf. The possibility to perform speciation of Al at physiological concentration levels is of great importance in investigations of Al transportation and deposition in the target organs of the human body.

Cross-References

▶ Aluminum Complexes with Low Molecular Mass Substances
▶ Aluminum in Biological Systems
▶ Aluminum, Biological Effects

References

Bantan Polak T, Milačič R, Mitrović B, Benedik M (2001) Speciation of low molecular weight Al complexes in serum of CAPD patients. J pharm Biomed Anal 26:189–201

Bantan T, Milačič R, Mitrović B, Pihlar B (1999) Investigation of low molecular weight Al complexes in human serum by fast protein liquid chromatography (FPLC)-ETAAS and electrospray (ES)-MS-MS techniques. J Anal At Spectrom 14:1743–1748

Harrington CF, Clough R, Hansen HR, Hill SJ, Tyson JF (2010) Atomic spectrometry update. Elemental speciation. J Anal At Spectrom 25:1185–1216

Hedge ML, Bharathi P, Suram A, Venugopal C, Jagannathan R, Poddar P, Srinivas P, Sambamurti K, Rao KJ, Ščančar J, Messon L, Zecca L, Zatta P (2009) Challenges associated with metal chelation therapy in Alzheimer's disease. J Alzheimers Dis 17:457–468

Kralj B, Ščančar J, Križaj I, Benedik M, Bukovec P, Milačič R (2004) Determination of high molecular mass Al species in serum and spent CAPD fluids of dialysis patients combining SEC and anion-exchange FPLC with ETAAS detection. J Anal At Spectrom 19:101–106

Milačič R (2005) Speciation of aluminum in clinical aspects: (health & disease). In: Cornelis R (ed) Handbook of elemental speciation II: species in the environment, food, medicine & occupational health. Wiley, New York/Chichester, pp 27–39

Milačič R, Murko S, Ščančar J (2009) Problems and progresses in speciation of Al in human serum: an overview. J Inorg Biochem 103:1504–1513

Murko S, Milačič R, Kralj B, Ščančar J (2009) Convective interaction media monolithic chromatography with ICPMS and ultraperformance liquid chromatography-electrospray ionization MS detection: a powerful tool for speciation of aluminum in human serum at normal concentration levels. Anal Chem 81:4929–4936

Murko S, Ščančar J, Milačič R (2011) Rapid fractionation of Al in human serum by the use of HiTrap desalting size exclusion column with ICP-MS detection. J Anal At Spectrom 26:86–93

Sanz-Medel A, Soldado Cabezuelo A, Milačič R, Bantan Polak T (2002) The chemical speciation of aluminium in human serum. Coord Chem Rev 228:373–383

Wróbel K, Blanco González E, Sanz-Medel A (1995) Aluminium and silicon speciation in human serum by ion-exchange high-performance liquid chromatography–electrothermal atomic absorption spectrometry and gel electrophoresis. Analyst 120:809–815

Yokel RA (2004) Aluminum. In: Merian E, Anke M, Inhant M, Stoeppler M (eds) Elements and their compounds in the environment, 2nd edn. Wiley-VCH, Veinheim, pp 635–658

Aluminum, Biological Effects

Patrick C. D'Haese
Laboratory of Pathophysiology,
University of Antwerp, Wilrijk, Belgium

Synonyms

Aluminum and health; Behavior of aluminum in biological systems

Definition

This entry refers to the ways of exposure of various biological systems to aluminum and to the mechanisms underlying the interactions of the element with these systems. Concerns on aluminum toxicity on ecosystems gain interest in view of the continuously increasing use of aluminum compounds in industry and acid rain.

Introduction

Despite the ubiquity of aluminum (Al) in the environment and its presence in living organisms, be it in small amounts (ppb-ppm range), no biological function has so far been attributed to the element. For this reason aluminum is considered a nonessential metal. It has long been thought that aluminum was inert for living organisms and as such was not regarded as a toxic element until the 1960s. At that time, only a few reports dealt with the toxic effects of aluminum in humans and animals, which, however, did not receive much attention. In the 1970s, aluminum exposure was linked to particular disease states occurring in patients with end-stage renal failure particularly those treated by dialysis. Although aluminum toxicity is mainly concerned with dialysis patients and the element is adequately removed by the kidneys, occupationally exposed workers (e.g., welders) are also at risk for its deleterious effects. In the latter population some evidence has been provided for the element to cause pulmonary and neurological lesions. Aluminum exposure has also been linked to the development of Alzheimer's disease, an issue which is still controversial.

Occurrence

Aluminum is the most abundant metallic element in nature (8% of the earth crust) and the third most abundant element preceding iron (4.7%) but less abundant than oxygen and silicon. Aluminum exists primarily associated with silicates and oxides in minerals, which are rather insoluble, explaining the relatively low aluminum concentrations in rivers, lakes, and sea water (<1 μg/L). Acidification of lakes as a result of acid rain enhanced aluminum solubility and hence toxicity for fish and even birds living in the immediate surroundings.

Human activities have also in other ways enhanced exposure of living organisms to aluminum since it is being used in industrial applications such as, for example, food additive, flocculant in water treatment, pharmaceuticals, etc. Especially, exposure to aluminum dust by welders and workers in aluminum-producing industries has resulted in neurological problems and lung diseases (Buchta et al. 2005; Donoghue et al. 2011).

Aqueous Chemistry

Aluminum, electronic structure: [Ne] $3s^2 3p^1$, MM 27, exists only in the oxidation state III. It is amphoteric, combining with both acids and bases to form, respectively, aluminum salts and aluminates. It has a small ionic radius (54 pm) closer to iron (65 pm) than most other elements. From the combination of a small ionic radius with a high charge, it follows that in aqueous solutions the free Al^{3+} concentration is very small due to the formation of aluminum hydroxide complexes. In a solution with a pH below 5 aluminum exists as $Al(H_2O)_6^{3+}$. With rising pH, this complex deprotonates under the formation of the insoluble $Al(OH)_3$ at neutral pH, which redissolves at higher pH as the $Al(OH)_4^-$ complex.

Exposure to Aluminum

Because of its abundance in the environment, aluminum is frequently consumed as an incidental component of water or food, including infant formula. Aluminum is also intentionally added to food as a caking or emulsifying agent. As a result, bread made with aluminum-based baking powder can contain up to 15 mg aluminum per slice, and processed American cheese can contain as much as 50 mg aluminum per slice. It has been shown recently that exposure to aluminum in humans may also occur through vaccination as certain vaccines may contain specific aluminum salts (primarily aluminum hydroxide and aluminum phosphate) as an adjuvant. Aluminum adjuvants are important components of vaccines, since they stimulate the immune system to respond more effectively to protein or polysaccharide antigens that have been adsorbed to the surface of insoluble aluminum particles (Mitkus et al. 2011).

Till the early 1980s patients with chronic renal failure were exposed to high amounts of aluminum as at that time aluminum hydroxide was frequently used for phosphate control and was given in doses of grams/day during years. Moreover, when treated by hemodialysis the use of aluminum-contaminated dialysis fluids further increased the risk for aluminum overload and serious toxicological effects. Due to replacement of aluminum-based phosphate binders by nonaluminum-containing compounds together with the introduction of adequate water treatment installations caricatural aluminum overload nowadays is rarely seen although moderate exposure still frequently occurs in less developed countries and with the "accidental" use of aluminum-leaking instrumental devices related to the dialysis treatment (D'Haese and De Broe 1999).

Aluminum Metabolism in Humans

Inhalation
A fraction of the aluminum present in dust remains indefinitely in the lungs after inhalation, thus without entering the systemic blood compartment. Especially in welders and workers in the aluminum industry, uptake of the element via this route has caused health problems related to lungs and brain.

Gastrointestinal Absorption of Aluminum
The gastrointestinal absorption of aluminum is importantly affected by the speciation of the element and values between 0.001% and 27% have been reported. This wide variation is mainly due to analytical difficulties, contamination, and differences in experimental protocols. With the application of the ^{26}Al radioisotope and its detection by Accelerator Mass Spectrometry (AMS) more reliable results on fractional gastrointestinal absorption varying between 0.04% and 1% could be obtained.

At least as important as the knowledge on the amounts of aluminum absorbed after oral intake is the exact mechanism that governs this process. No unified explanation has emerged till today. It is believed that intestinal absorption of aluminum includes both (1) paracellular pathways along enterocytes and through junctions by passive processes and (2) transcellular pathways through the enterocytes involving both passive and active processes.

Local factors altering aluminum absorption include citrate which has been reported to significantly enhance the element's absorption by opening the epithelial tight junctions although effects to a much lesser extent have been reported more recently by others using more sophisticated techniques. Other factors reported to alter aluminum absorption are fasting state, gastric pH, and silicon. With regard to the latter element, recent studies reported silicon to have therapeutic potential to prevent oral absorption and retention in mammals, particularly in the setting of Alzheimer's disease.

General factors which have been reported to enhance aluminum absorption are the uremic state, diabetes mellitus, hyperparathyroidism, and Down syndrome. During the last decades, the potentially increased bioavailability of aluminum resulting from an increased anthropogenic acidification of soils has become a matter of concern.

Tissue Distribution and Excretion of Aluminum
Following enteral absorption, aluminum is transported mainly in the plasma in association with the iron-binding protein transferrin. In adults with normal renal function, the total aluminum burden is estimated to be 30 mg with the highest levels found in the lungs and skeletal muscles. In patients with renal insufficiency, chronic accumulation will occur as the major elimination route, that is, the kidney is partially or even totally absent, and in the past, aluminum body burdens up to 1,660 mg have been reported in dialysis patients. In these patients also a different tissue distribution pattern is observed as compared to individuals with normal renal function with bone and liver as the major storage sites.

As only a very small fraction of ingested aluminum is absorbed, the feces represent the major excretion pathway. In the presence of a normal renal function, minute fractions that are absorbed are subsequently removed via the kidney. As for the absorption of aluminum, the exact mechanism of aluminum excretion by the kidney has not been unraveled up to now although it has been inferred from micropuncture experiments and other techniques that a significant fraction of aluminum filtered by the glomerulus is reabsorbed. Somewhat in contrast herewith is the statement made by some that the aluminum clearance is about 5–10% of the glomerular filtration rate, which in combination with the finding that in the blood compartment aluminum is about 90–95% protein bound would suggest that the unbound serum aluminum is filtered and excreted without tubular reabsorption.

Aluminum Toxicity

General
The chemistry and biochemistry of the aluminum ion (Al^{3+}) dominate the pathways that lead to toxic outcomes. The molecular targets of aluminum toxicity are multifaceted and appear to involve the disruption of essential metal homeostasis such as calcium,

magnesium, and iron. Aluminum has been demonstrated to replace calcium within the bone and interferes with calcium-based signaling events. Magnesium has been observed to be replaced by aluminum for binding to phosphate groups on the cell membrane, on ATP, and on DNA. Perhaps, the main targets of aluminum toxicity are iron-dependent biological processes. Interference of aluminum with iron homeostasis also leads to the production of reactive oxygen species (ROS), which in turn may also lead to toxicological effects.

Overall, experimental animal studies have failed to demonstrate carcinogenicity attributable solely to aluminum compounds and aluminum has thus not been classified with respect to carcinogenicity. *Aluminum production*, however, has been classified as carcinogenic by the International Agency for research on Cancer (IARC). Occupational limits exist in several countries for exposures to aluminum dust and aluminum oxide. For nonoccupational environments, limits have been set for intake in foods and drinking water.

Humans

Subjects with Normal Renal Function

Neurological Effects In subjects with normal renal function, toxicity of aluminum has been mainly associated with neurological and lung diseases. Although in this population a definite role for aluminum in the development of neurological diseases remains elusive, the link between aluminum and Alzheimer's disease has been the subject of scientific debate for several decades. Following the findings from an epidemiological study of Martyn et al. (1989) showing a high incidence of Alzheimer's disease in areas with a high level of aluminum in the drinking water in England and Wales similar observations were made by others, while other studies revealed increased brain aluminum levels in patients with Alzheimer's disease. The discovery of aluminum in the brain of Alzheimer's patients and the association of aluminum with amyloid plaques provided some further evidence to the connection between the metal toxin and Alzheimer's disease, although it should be mentioned that such associations could not be confirmed by some other groups. Mounting evidence has suggested the significance of oligomerization of β-amyloid protein and neurotoxicity in the molecular mechanism of Alzheimer's disease pathogenesis. In their excellent overview, Kawahara and Kato-Negishi (Kawahara and Kato-Negishi 2011) revisit the link between aluminum and Alzheimer's disease and integrate aluminum and amyloid cascade hypotheses in the context of β-amyloid oligomerization.

Amyotrophic lateral sclerosis (ALS) along with Parkinsonism dementia has been seen in certain populations with great frequency, such as in the Chamorros of Guam, in the Kii Peninsula of Japan and in southern West New Guinea, three regions of the Western pacific region. Studies later on revealed an abnormal mineral distribution in the soil and drinking water of these areas consisting of a virtual lack of calcium and magnesium coupled with high levels of aluminum and manganese. Based on experimental studies in rats it was hypothesized that long-term intake of low calcium and high aluminum water and/or diet reduces magnesium in the body and may cause degenerative neurological disorders by altering the normal biological effects of magnesium.

Increased exposure to aluminum by the use of aluminum hydroxide as an adjuvant in the anthrax vaccine has also been linked to the Gulf War Syndrome, a multisystem disorder afflicting many veterans of Western armies in the 1990–1991 Gulf War. A number of those showed neurological deficits including various cognitive dysfunctions and motor neuron disease, the latter expression virtually indistinguishable from classical amyotrophic lateral sclerosis (ALS) except for the age of onset as the great majority of diseased individuals were less than 45 years of age.

Pulmonary Effects Following inhalation exposure, the effects of aluminum are mainly exerted on the respiratory system. Reports have been published dealing with the development of asthma, cough, or decreased pulmonary function (Nayak 2002). To which extent these effects are solely due to aluminum or rather are the consequence of multielement exposure or have a multifactorial etiology remains questionable. A small minority exposed to powdered metallic aluminum appear to develop an alveolitis that progresses to fibrosis, a lung disease known as aluminosis, which was already described in the 1930s and 1940s. Although it was assumed that under today's working conditions lung fibrosis induced by aluminum dust could not occur anymore, several severe cases of aluminum-induced lung fibrosis have been reported in

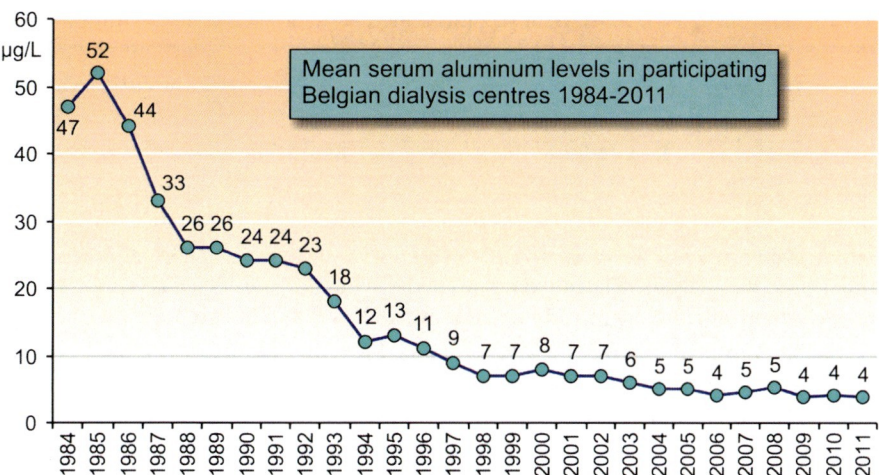

Aluminum, Biological Effects, Fig. 1 Evolution of the mean serum aluminum levels of dialysis patients from different centers in Belgium

more recent years. Experimental studies provided some evidence for a causal role of the element be it that in most of these studies exposure was much higher than that encountered in environmental or even occupational settings and, thus, have to be interpreted with caution.

Subjects with Impaired Renal Function, in Particular Dialysis Patients
As the urinary excretion route is the major pathway of elimination of aluminum, impairment of the kidney will be an important determination of the element's accumulation in the human body. In the past, aluminum-based phosphate binders were frequently used in these subjects, which led to the development of severe aluminum-related diseases such as osteomalacia and dialysis encephalopathy (De Broe and Coburn 1990), particularly when patients were treated by hemodialysis. Indeed, according to the concentration gradient between the ultrafiltrable amount of aluminum in the blood (i.e., serum-) compartment (<5% of the total concentration), a transfer of the element from the dialysis fluid toward the patient thus may already occur at dialysate aluminum levels as low as 2 µg/L (D'Haese and De Broe 1999).

It is now well recognized that the critical localization of aluminum at the mineralization front underlies the mineralization defect in aluminum-related osteomalacia. While this type of bone disease is accompanied by distinctly elevated levels of the element in bone (>15 µg/g wet weight vs <2 µg/g in normals), much lower levels were reported in the brain of dialysis patients with encephalopathy (2.3 ± 0.83 µg/g vs <0.60 ± 0.37 µg/g in dialysis patients without dementia) indicating the brain to be much more vulnerable to the toxic effects of aluminum as compared to other organs.

Due to the introduction of adequate water treatment systems and aluminum-free phosphate-binding agents together with the establishment of regular monitoring programs chronic, caricatural aluminum overload nowadays is rarely seen and in most developed countries serum values now vary around 3 µg/L (Fig. 1). Nevertheless, the risk for accumulation/intoxication should not be neglected (D'Haese and De Broe 1999). Indeed, as demonstrated in a recent survey, may moderate aluminum accumulation still occur in chronic renal failure patients, either or not treated by dialysis, from particular regions in the world. With such levels more subtle disorders at the level of the parathyroid gland function and bone turnover, resistance to erythropoietin therapy, and anemia may be observed. The anemia is typically microcytic and the presence of microcytosis with normal serum ferritin levels suggests that aluminum intoxication may be the causative factor. It is thought that aluminum might induce a defect in iron utilization or interfere with the bioavailability of stored iron for erythropoiesis.

Research on the speciation and protein-binding characteristics of aluminum has led to a better insight in the mechanisms underlying the element's tissue distribution and it has been hypothesized that iron depletion due to the widespread use of erythropoietin may result in an increased binding of aluminum to transferrin (Van Landeghem 1997). Consequently, the uptake of the aluminum-transferrin complex by cells via endocytosis of transferrin receptors increases. This will cause a preferential accumulation of aluminum in transferrin receptor expressing tissues such as, for example, the parathyroid gland and the osteoblast which may result in a reduced PTH secretion and cellular proliferation. Insight in the speciation of aluminum in cerebrospinal fluid of severely intoxicated dialysis patients also put some light on the mechanisms underlying the dialysis-related encephalopathy which, as mentioned above, may already occur at much lower levels than those seen in bone or other organs. Indeed, it has been shown that whereas in serum of dialysis patients up to 90% of the total concentration appears bound to transferrin, the remaining fraction most probably occurring as (a) low molecular mass compounds such as citrate or silicate, in the brain evidence has been presented for the element not to circulate as a protein-bound complex. This could be explained by the fact that the molar concentration of citrate in cerebrospinal fluid is 900-fold higher than that of transferrin by which aluminum in the latter biological fluid will preferentially bind to citrate in contrast to the situation in serum, where the molar citrate/transferrin ratio is only 4 (Van Landeghem 1997). To a certain extent, this may also provide an answer to the question as to why the incidence of neurological diseases such as Alzheimer's disease and cognitive impairment is not increased in dialysis patients despite the relatively higher exposure of this population to aluminum, an issue which often has been used as a major argument for disregarding such an association. Indeed, literature data have shown silicon levels in both CSF and serum of dialysis patients to be increased up to 100-fold as compared to subjects with normal renal function. Case studies furthermore demonstrated that also in the CSF, dialysis patients have much higher silicon levels as compared to subjects with normal renal function; CSF and serum silicon levels are quite identical. This together with data attributing a protective effect of silicon against aluminum toxicity by either affecting the bioavailability of aluminum or reversing the aluminum-induced conformational changes of neurofibrillary tangles characteristic of Alzheimer's disease may put this controversial issue in another perspective.

Due to its high protein binding, aluminum can hardly be removed during dialysis even when serum aluminum levels are high (>100 µg/L). Therefore, in cases of aluminum intoxication/overload, chelation therapy using desferrioxamine is applied. With this treatment aluminum stored in bone and tissues is picked up by the chelator under the formation of the aluminum-desferrioxamine complex, that is, aluminoxamine (MW 583 Da), which by its relatively low molecular mass can be removed efficiently from the body during dialysis (or the urine in subjects with intact renal function). Care should be taken during desferrioxamine therapy, however, as in severely intoxicated patients (serum aluminum levels >300 µg/L) a neurologic syndrome similar to aluminum-related encephalopathy has been reported. The exact mechanism is unknown but is theorized to be due to redistribution of aluminum mobilized by desferrioxamine into the brain, hereby assuming that the formed aluminoxamine complex is able to pass the blood-brain barrier (D'Haese and De Broe 2007).

Aquatic Systems and Plants

Aluminum toxicity has also been reported in aquatic systems. So has it been demonstrated that aluminum is a strong inhibitor of the mullet cytochrome P450 reductase activity while exposure to environmentally relevant concentrations of aqueous aluminum at neutral pH was reported to have a negative effect on the immunocompetence of the crayfish *Pacifastacus leniusculus*, specifically in the ability of the hemocytes to remove bacteria from the circulation. It is likely that the impairment of immunocompetence is due to hypoxia rather than direct toxicity of aluminum. Prolonged exposure to aluminum abolished this effect, indicating that the crayfish is able to adapt to exogenous aluminum. More or less in line with observations in humans has a protective role against aluminum toxicity been attributed in these organisms also. While in most cases this has been described to be due to ex vivo silicon–aluminum interactions by the formation of inert hydroxyaluminosilicates, evidence has been presented more recently for the existence of a silicon-specific intracellular mechanism for aluminum detoxification in aquatic snails involving regulation of orthosilicic acid (White et al. 2008).

Concerns on aluminum toxicity on ecosystems gain in interest in view of the continuously increasing use of aluminum compounds in industry and acid rain which will result in an increased solubility of aluminum in aquatic systems and in soil. Aluminum toxicity is considered a major constraint for crop production in acidic soil worldwide. When the soil pH is lower than 5, aluminum enters the root tip and causes inhibition of cell elongation and cell division leading to root stunting as well as a deep change in the entire root architecture.

Cross-References

▶ Aluminum and Bioactive Molecules, Interaction
▶ Aluminum in Biological Systems
▶ Aluminum in Plants
▶ Aluminum Speciation in Human Serum
▶ Aluminum, Physical and Chemical Properties
▶ Iron Homeostasis in Health and Disease
▶ Iron Proteins, Transferrins and Iron Transport
▶ Lead and Alzheimer's Disease
▶ Magnesium

References

Buchta AM, Kiesswetter BE, Schäper BM, Zschiesche CW, Schaller DKH, Kuhlmann AA, Letzel AS (2005) Neurotoxicity of exposures to aluminium welding fumes in the truck trailer construction industry. Environ Toxicol Pharmacol 19(3):677–685

D'Haese PC, De Broe ME (1999) Trace metals in chronic renal failure patients treated by dialysis. Trace Elem Electrolytes 16(4):163–174 (Review)

D'Haese PC, De Broe ME (2007) Aluminum, lanthanum and strontium. In: Daugirdas JT, Blake PG, Ing TS (eds) Handbook of dialysis, 4th edn. Wolters Kluwer, Philadelphia

De Broe ME, Coburn JW (eds) (1990) Aluminum and renal failure. Kluwer, Dordrecht

Donoghue AM, Frisch N, Ison M, Walpole G, Capil R, Curl C, Di Corleto R, Hanna B, Robson R, Viljoen D (2011) Occupational asthma in the aluminum smelters of Australia and New Zealand: 1991–2006. Am J Ind Med 54(3):224–231

Kawahara M, Kato-Negishi M (2011) Pathogenesis of Alzheimer's disease: the integration of the aluminum and amyloid cascade hypotheses. Int J Alzheimers Dis, Article ID 276393, 17 pages, doi:10.4061/2011/276393

Martyn CN, Osmond C, Edwardson JA, Barker DJP, Harris EC, Lacey RF (1989) Geographical relation between Alzheimer's disease and aluminium in drinking water. Lancet 1(8629):59–62

Mitkus RJ, King DB, Hess MA, Forshee RA, Walderhaug MO (2011) Updated aluminum pharmacokinetics following infant exposures through diet and vaccination. Vaccine 29(51):9538–9543

Nayak P (2002) Aluminum: impacts and disease. Environ Res 89(2):101–115 (Review)

Van Landeghem GF (1997) Aluminum speciation in biological fluids. Implications for patients with end stage renal failure. Ph.D. thesis, University of Leiden, Leiden (ISBN 90-9010801-7)

White KN, Ejim AI, Walton RC, Brown AP, Jugdaohsingh R, Powell JJ, McCrohan CR (2008) Avoidance of aluminum toxicity in freshwater snails involves intracellular silicon-aluminum biointeraction. Environ Sci Technol 42(6):2189–2194

Aluminum, Genes Involved in Novel Adaptive Resistance in *Rhodotorula glutinis*

Fusako Kawai[1] and Akio Tani[2]
[1]Center for Nanomaterials and Devices, Kyoto Institute of Technology, Kyoto, Japan
[2]Research Institute of Plant Science and Resources, Okayama University, Kurashiki, Okayama, Japan

Synonyms

Aluminum-resistant genes related to cation homeostasis

Definition

Three genes probably related to cation homeostasis are involved in adaptive acquirement of novel heritable aluminum (Al) resistance in *Rhodotorula glutinis* IFO1125.

So far, no nutrient role is known for aluminum (Al), which has toxic effects not only on plants and microorganisms, but also on human beings. The strong toxicity of Al is caused by inorganic monomeric ion (Al^{3+}, $Al(OH)^{2+}$, and $Al(OH)_2^{+}$), especially Al^{3+}, which is maintained at low pH below 4.5, but at higher pH, it is gradually converted to harmless forms. In what follows, Al toxicity is associated with Al^{3+} below pH 4.0. The acidification of soil causes the dissolution of Al salts from soils and even a micromolar range of cationic Al ions shows severe inhibition of plant growth. Resistant plant species exist and, in some, the lowering of toxicity is caused by complexation of Al with organic acids originating from the metabolic pathways. Since the acidification of soil and concomitant dissolution of Al causes serious agricultural and environmental problems, extensive studies have been carried out on the mechanism of plant resistance to Al. On the other hand, screening of highly Al-resistant microorganisms resulted in isolation of super-resistant microorganisms, belonging to fungi, *Aspergillus flavus*, *Penicillium* sp., *Penicillium janthinellum*, and *Trichoderma asperellum* (Kawai et al. 2000). Usually such filamentous fungi are known to resistant to acidic pH and Al. The other isolates belong to yeasts, *Rhodotorula glutinis* and *Cryptococcus humicola*. Since yeasts of high resistance to Al are not known, we focused on one of the isolates, *Rhodotorula glutinis* Y-2a. Strain Y-2A was resistant up to 200 mM Al, the concentration of which does not occur in natural environments. A normal counterpart *R. glutinis* in culture collections (IFO1125) was, however, sensitive to Al toxicity, and it showed retarded growth in the presence of 50–100 μM Al in liquid culture. Curiously, supplementation of Al resulted in longer lag phase, but not in decreased growth rate. Namely, when the growth started in the presence of Al after long lag phase, the growth ratio was comparable to that in the absence of Al. It seemed that the strain adapted to Al during the lag phase. When the strain was pre-grown in liquid culture in the absence of Al and spread onto Al-containing agar plates, colony-forming unit (CFU) varied depending on Al concentration, but the colonies once appeared on Al-containing plate showed variable CFU to Al concentrations added to agar plates. These results suggested that the resistance obtained by one round of Al treatment was somewhat unstable, and that this resistance cannot be explained by simple mutation. When the strain was cultivated in the presence of Al at more than 50 μM repeatedly by ten times in liquid culture, the acquired resistance was heritable and stable, which was the first report on the novel adaptive Al resistance (Tani et al. 2004). In addition, when the stable resistant cells were repeatedly exposed to stepwise increments in Al concentration, the resistant level increased up to more than 5 mM. The growth yield of resistant strains decreased as their resistance increased. It is known that Al ions inhibit Mg-uptake systems, but resistant cells obtained the improved Mg-uptake system under Al stress, since they could grow in the presence of Al and low Mg. When the wild type was repeatedly treated by Cu, heritable and stable Cu resistance (50 μM to 1.6 mM) similar to Al resistance was acquired, but the Cu-resistant cells did not show Al resistance and vice versa. These results suggested that the underlying resistance mechanisms of Al and Cu resistance are different, but the strain is capable of adapting to such increased metal stress.

The crucial role of mitochondrial regulation in adaptive Al resistance was found (Tani et al. 2008). Resistant cells contained Al when grown in the presence of Al, but the content of which was significantly low compared to Al added to the medium, suggesting that the cells have a barrier to Al. The concentration and ionic form of Al (Al^{3+}) in the media did not change during the cultivation, suggesting that the cells neither chelate Al^{3+} nor make a complex with it to detoxify. Transmission electron microscopic analyses revealed a greater number of mitochondria in resistant cells. The formation of small mitochondria with simplified cristae structures was observed in the wild type strain grown in the presence of Al and in resistant cells grown in the absence of Al. Addition of Al to cells resulted in high mitochondrial membrane potential and concomitant generation of reactive oxygen species (ROS). Exposure to Al also resulted in elevated levels of oxidized cellular proteins and lipids. Addition of the antioxidants such as α-tocopherol and ascorbic acid alleviated the Al toxicity. These results suggested that ROS generation is the main cause of Al toxicity. Differential display analysis indicated upregulation of mitochondrial genes in the resistant cells. Resistant cells were found to have 2.5- to 3-fold more mitochondrial DNA (mtDNA) than the wild type strain. Analysis of tricarboxylic acid cycle and respiratory-chain enzyme activities in wild type and resistant cells revealed significantly reduced cytochrome *c* oxidase activity and resultant high ROS production in the resistant cells. The adaptive increased Al resistance resulted from an increased number of mitochondria and increased mtDNA content, which is a compensatory response to reduced respiratory activity caused by a deficiency in complex IV function.

Three nuclear genes (*RgFET3*, *RgGET3* and *RgCMK*) were also found to be upregulated in the resistant cells. They code for proteins homologous to *Saccharomyces cerevisiae* FET3p (ferrooxidase), GET3p (guanine nucleotide exchange factor for Gpa1P), and CMK1p and CMK2p (calmodulin-dependent protein kinases), respectively. Their expression was promoted in the presence of Al. These three genes were cloned from the wild type *R. glutinis* and introduced into *S. cerevisiae* BY4741 and its derivatives (*fet3Δ*, *get3Δ*, *cmkΔ* and *cmk2Δ*) (Tani et al. 2010).

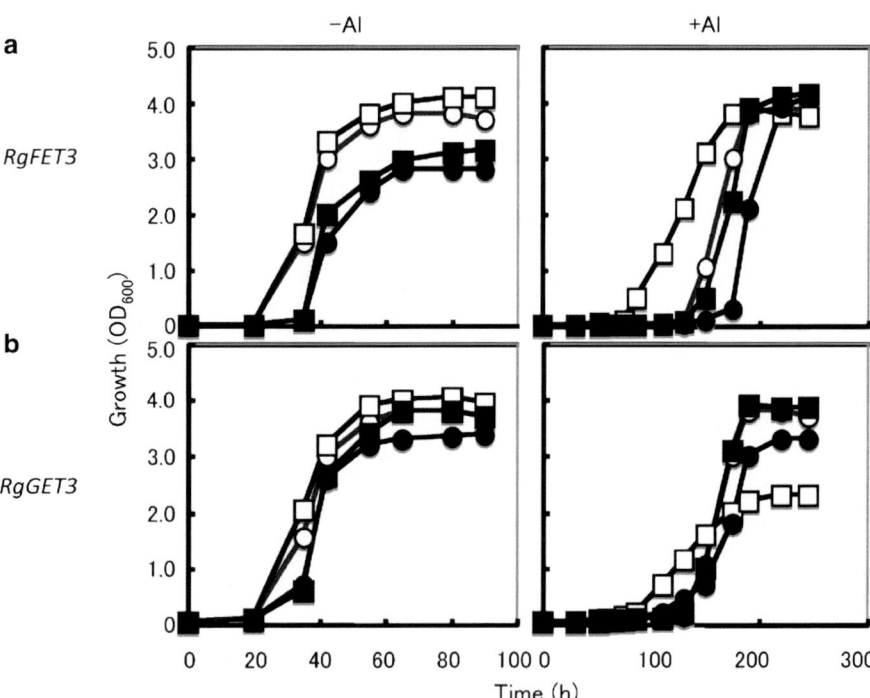

Aluminum, Genes Involved in Novel Adaptive Resistance in *Rhodotorula glutinis*, Fig. 1 Growth of *S. cerevisiae* and its transformants with *RgFET3* (*top panel*) and *RgGET3* (*lower panel*). The wild type *S. cerevisiae* and its derivatives *fet3Δ* and *get3Δ* were grown in the absence (*left panels*) and presence (*right panels*) of Al (50 μM). Open symbols, wild type *S. cerevisiae*; closed symbols, mutant derivatives; *circles*, vector control (*p*YES2); *squares*, transformants (Cited from Tani et al. 2010)

The introduction of *RgFET3* improved the growth of the wild type and *fet3Δ* of *S. cerevisiae* in the presence of Al (Fig. 1a). This result indicated that RgFet3p functioned as high-affinity Fe uptake system as Fet3p (Stearman et al. 1996). Fet3p receives copper in a late or post-Golgi compartment into which Ccc2p, a P-type copper-translocating ATPase, and Gef1p, CLC chloride-transport protein, supply copper ions. CCC2 expression is regulated by iron and Gef1p is involved in cellular cation homeostasis. The RgFet3p-GFP protein is localized at the cell periphery in *S. cerevisiae*. The supplementation of Fe (III) promoted *R. glutinis* growth under Al stress. From these results, the role of RgFet3p in *R. glutinis* would be iron uptake, and the high expression level in resistant cells under Al stress would reflect cellular demand for iron. This is in accordance with the idea that iron ion should be of great importance for its mitochondrial activity under Al stress and that iron competes with Al. Thus, RgFet3p from *R. glutinis* is considered to have relevance to Al resistance through the cellular Fe requirement (Fig. 1a).

RgGet3p is similar to bacterial ArsA ATPase, and *GET3* mutant of *S. cerevisiae* was found to be sensitive to various metal ions and temperature. Later, it was shown that Get3p is involved in Golgi-to-ER trafficking (Schuldiner et al. 2005). Get3p binds to Gef1p, which is required for efficient copper supply to the Golgi, and cytosolic copper supports this interaction. Furthermore, in *gef1Δ*, Fet3p does not mature normally because of copper shortage. In addition, genes involved in the retrograde transport (*COG6*, *COG8*, *RGP1*, *RIC1*, *TLG2* and *YPT6*) were identified as Al-tolerance genes (Kakimoto et al. 2005), with which Get3p interacts in *S. cerevisiae*. Thus, Get3p is important for copper loading to Golgi, and copper is necessary for Fet3p maturation. *RgGET3* complemented the slow growth rate of *get3Δ* under Al (Fig. 1b). The RgGet3p had a tendency to form punctate bodies in the cytosol under Al stress, which was very similar to the localization of Get3p in *S. cerevisiae*. These results suggested that RgGet3p functions as Get3p. Thus, RgGet3p was considered to localize at Golgi-ER trafficking, which is important for Al tolerance. RgGet3p may enable copper loading to RgFet3p, through as yet unidentified Gef1 homologue.

The roles of calmodulin-dependent protein kinases (Cmk1p and Cmk2p) in *S. cerevisiae* have not been fully understood yet. They are considered to be involved in stress responses for weak organic acids such as sorbate or benzoate, and in the acquisition of thermotolerance. The wild type yeast shows growth retardation followed by adaptation to weak organic acids, but *cmk1* cells are resistant. The wild type yeast can tolerate exposure to high temperature, but *cmk1*Δ mutants show decreased levels of induced thermotolerance. Cmk2p becomes independent of Ca^{2+} and calmodulin via autophosphorylation, but Cmk1p does not. *CMK2* is induced in a calcineurin-dependent manner. Calcineurin controls cell wall biosynthesis and cation homeostasis. One of the phosphorylation target of calcineurin is Crz1p, whose binding site is 5′-GNGGC(G/T)CA-3′. Interestingly, in *C. neoformans*, calcineurin activity is dependent on cyclophilin A, and cyclophilin A is required for laccase (Fet3p homologue) and is also the target of cyclosporin A. Furthermore, calmodulin from chicken was shown to bind aluminum. *RgCMK* is induced in both the wild type and resistant cells of *R. glutinis* in the presence of Al. *S. cerevisiae cmk1*Δ and *cmk2*Δ mutants exhibited no clear deficiency in growth under Al stress, and introduction of *RgCMK* into them did not show any distinct effect. But the response to calcium ion and cyclosporin A was clearly different between the wild type and the resistant cells of *R. glutinis*. These results suggested altered Ca ion homeostasis in resistant cells and the involvement of calmodulin-dependent signaling pathway in Al resistance. The adaptation of *Rhodotorula* cells to Al necessitates repeated cultivation and its resistance was not stable in earlier treatment. Such behavior of tolerance is quite similar to the induced thermotolerance and adaptation to weak organic acids seen in *S. cerevisiae*, which is mediated by calmodulin-dependent kinases. The complete growth inhibition caused by cyclosporin A in the wild type and no effect in the resistant cells in the presence of Al will be the key to understand the Ca-mediated mechanism of adaptive Al resistance, and this difference may further be linked to RgFet3p again through cyclophilin A. The consensus sequence for Crz1p binding was found in the promoter region for *RgFET3* (5′-GAGGCGCA-3′, 169 bp upstream from transcription initiation site) and *RgCMK* (5′-CTGGCGCA-3′ [possible] and 5′-GAGGCGGA-3′, 548 and 393 bp upstream, respectively). This also implies the link between RgCmkp and RgFet3p mediated by calcineurin, although FET3 is not the target of Crz1p in *S. cerevisiae*.

Thus, at least two genes (*RgFET3* and *RgGET3*) were considered to have relevance to Al resistance in *R. glutinis*. The relevance of these genes to Al resistance has not been reported previously. The relevance of *RgCmk* has not been clarified yet, but interestingly, RgGet3p and RgCmkp may have links to RgFet3p, through as yet unidentified mechanisms. Although these three genes were not recognized as Al-tolerance gene in the work of mutant screening of *S. cerevisiae*, RgGet3p and RgCmkp are supposed to interact with these tolerance genes directly or indirectly, which suggests the common tolerance mechanisms between *S. cerevisiae* and *R. glutinis*. The difference is that the tolerance level of *R. glutinis* became higher and higher by repetitive cultivation under increased concentration of Al. In the process of this adaptation, cellular cation homeostasis and the regulation of mitochondrial activity and amount have been considerably changed.

In conclusion, the novel adaptive Al resistance found in *R. glutinis* IFO1125 may be attributable to altered cation homeostasis and the regulation of mitochondrial activity and amount in the resistant cells. Besides, Mg-uptake system and calcium signaling probably play significant roles in this adaptation. The inference may be reasonable because Al competes with Mg, Fe, and Ca in various cellular systems. In other words, the microorganisms able to alter their cation homeostasis and mitochondrial regulation in response to Al may become resistant to increased Al stress. The novel adaptive Al resistance found in *R. glutinis* may be widespread in various yeasts, as yet undiscovered.

References

Kakimoto M, Kobayashi A, Fukuda R, Ono Y, Ohta A, Yoshimura E (2005) Genome-wide screening of aluminum tolerance in *Saccharomyces cerevisiae*. Biometals 18: 467–474

Kawai F, Zhang D, Sugimoto M (2000) Isolation and characterization of acid- and Al-tolerant microorganisms. FEMS Microbiol Lett 189:143–147

Schuldiner M, Collins SR, Thompson NJ, Denic V, Bhamidipati A, Punna T, Ihmels J, Andrews B, Boone C, Greenblatt JF, Weissman JS, Krogan NJ (2005) Exploration of the function and organization of the yeast early secretory pathway through an epistatic miniarray profile. Cell 123:507–519

Stearman R, Yuan DS, Yamaguchi-Iwai Y, Klausner RD, Dancis A (1996) A permease-oxidase complex involved in high-affinity iron uptake in yeast. Science 271:1552–1557

Tani A, Zhang D, Duine JA, Kawai F (2004) Treatment of the yeast *Rhodotorula glutinis* with $AlCl_3$ leads to adaptive acquirement of heritable aluminum resistance. Appl Microbiol Biotechnol 65:344–348

Tani A, Inoue C, Tanaka Y, Yamamoto Y, Kondo H, Hiradate S, Kimbara K, Kawai F (2008) The crucial role of mitochondrial regulation in adaptive aluminium resistance in *Rhodotorula glutinis*. Microbiology 154:3437–3446

Tani A, Kawahara T, Yamamoto Y, Kimbara K, Kawai F (2010) Genes involved in novel adaptive aluinum resistance in *Rhodotorula glutinis*. J Biosci Bioeng 109:453–458

Aluminum, Physical and Chemical Properties

Fathi Habashi
Department of Mining, Metallurgical, and Materials Engineering, Laval University, Quebec City, Canada

Aluminum is the most abundant metallic element in the earth's crust. It occurs in a variety of minerals combined with oxygen, silicon, the alkali and alkaline-earth metals, and fluorine, and as hydroxides, sulfates, and phosphates. Many applications of aluminum are based upon its low density, high electrical and thermal conductivities, and resistance to corrosion. Pure aluminum is soft but it can be alloyed with other elements to increase strength and impart a number of useful properties.

Physical Properties

Atomic number	13
Atomic weight	26.98
Relative abundance, %	8.13
Melting point, °C	660.5
Boiling point, °C	2,494
Heat of fusion, J/g	397
Heat of vaporization, kJ/g	10.8
Heat capacity, $J\ g^{-1}\ K^{-1}$	0.90
Density, g/cm^3	2.699
Density of liquid, g/cm^3	
at 700°C	2.357
at 900°C	2.304
Crystal structure	Face-centered cubic

(*continued*)

Atomic diameter, m	2.86×10^{-10}
Lattice constant (length of unit cube) at 25°C, m	4.0496×10^{-10}
Coefficient of expansion at 20°C, K^{-1}	23×10^{-6}
Thermal conductivity at 25°C, $W\ cm^{-1}\ K^{-1}$	2.37
Electrical resistivity of pure aluminum at 25°C, $\mu\Omega\cdot cm$	2.5
Surface tension at melting point, N/cm	8.68×10^{-3}
Viscosity at melting point, Pa·s	0.0012
Magnetic susceptibility at 25°C, $mm^{-3}\ mol^{-1}$	16×10^{-3}
Thermal neutron cross section, cm^2 Barn	$(2.32 \pm 0.03) \times 10^{-25}$ 0.232 ± 0.003
Nuclear magnetic moment, $A\cdot m^2$	1.84×10^{-26}

Chemical Properties

Aluminum is a typical metal in the sense that when it loses its three outermost electrons, it will have the electronic structure of an inert gas. Although it is one of the most reactive metals, it is widely used as a material of construction because of the continuous adherent thin oxide film that rapidly forms on its surface. Aluminum is used to store nitric acid, concentrated sulfuric acid, organic acids, and many other reagents. However, it dissolves in alkaline solutions:

$$2Al + 2OH^- + 6H_2O \rightarrow 2[Al(OH)_4]^- + 3H_2$$

Aluminum is amphoteric; it reacts with mineral acids to form soluble salts with hydrogen evolution:

$$2Al + 6H^+ \rightarrow 2Al^{3+} + 3H_2$$

Aluminum powder reduces many oxides when heated to produce metals and alloys:

$$3MO + 2Al \rightarrow 3M + Al_2O_3$$

where M is a divalent metal.

All aluminum compounds are colorless. Alumina is a major aluminum compound produced mainly from bauxite. It is the starting material for producing the metal by fused salt electrolysis. It is also used as a catalyst, as an adsorbent, as abrasive, in ceramics, etc. Aluminum sulfate is an important flocculating agent for purifying water. Anhydrous aluminum chloride is an important *Friedel–Crafts catalyst* in the chemical and petrochemical industries. Aluminum compounds are considered nontoxic.

References

Frank WB et al (1997) Aluminum. In: Habashi F (ed) Handbook of extractive metallurgy. Wiley, Weinheim, pp 1093–1127

Habashi F (1994) Aluminum and its position in the periodic table. Educ Chem (Bombay) 11(2):18–24. http://www.meta-ynthesis.com/webbook/35_pt/pt.html#hab

Aluminum-Resistant Genes Related to Cation Homeostasis

▶ Aluminum, Genes Involved in Novel Adaptive Resistance in *Rhodotorula glutinis*

Amine N-Ethylcarbamoylborane

▶ Amine-Boranes

Amine Oxidase (Copper-Containing)

▶ Copper Amine Oxidase

Amine-Boranes

Bruce S. Burnham
Department of Chemistry, Biochemistry, and Physics, Rider University, Lawrenceville, NJ, USA

Synonyms

Amine *N*-ethylcarbamoylborane; Amine-carbamoylborane; Amine-carbomethoxyborane; Amine-carboxyborane; Amine-cyanoborane; Borane; Borano-amine; Boronated amine

Definition

Amine-boranes are complexes of a Lewis acid borane (e.g., BH_3, BH_2CN, BH_2COOH, BH_2COOCH_3, $BH_2CONHCH_2CH_3$) and an amine (e.g., R,R′,R″N) where the lone electron pair on the nitrogen atom forms a coordinate covalent bond by donating both of its electrons into the vacant orbital of borane.

Introduction

Chemistry of Boron

Boron is in column three of the periodic table and has three valence electrons. This leaves the boron atom with a vacant orbital capable of accepting a pair of electrons acting as a Lewis acid. Typically, molecules with atoms containing lone pairs of electrons will form complexes with boron such as nitrogen, oxygen, and phosphorus (Spielvogel 1988).

When coordinated to amino groups, the borane becomes tetrahedral making it isosteric to a tetrahedral carbon atom. Since the borane is more lipophilic, this type of isosteric replacement is attractive in the design of therapeutic agents since this property enhances their ability to cross cell membranes (Spielvogel 1988).

Synthesis

The most common borane functionalities present in the pharmacologically active compounds are borane (BH_3), cyanoborane (BH_2CN), carboxyborane (BH_2COOH), carbomethoxyborane (BH_2COOCH_3), and *N*-ethyl carbamoylborane ($BH_2CONHCH_2CH_3$). The cyanoborane can be conveniently prepared by two methods, see Figs. 1 and 2: (1) by refluxing the amine hydrochloride and sodium cyanoborohydride in an inert solvent such as tetrahydrofuran (THF) or (2) via a Lewis acid exchange reaction, either by using equimolar or excess amounts of a cyanoborane and the amine heated to at least 60–70°C in an inert solvent. In the Lewis acid exchange reaction, a weakly basic or bulky amine or phosphine, as its substituted borane adduct, is exchanged for a more basic or less bulky amine. The Lewis acid exchange reactions must be carried out under anhydrous conditions since water would complex with the borane. The Lewis acid exchange method is also a general route, which has also been used in the preparation of other borane

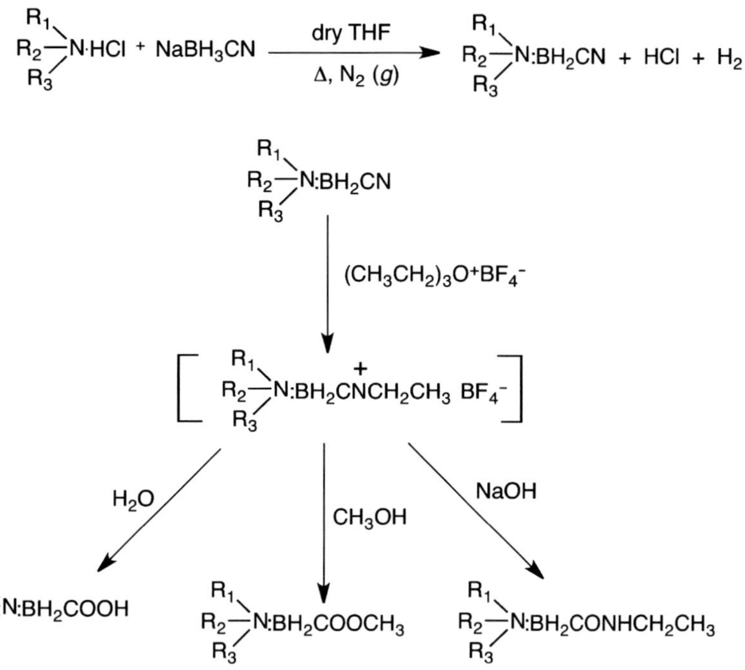

Amine-Boranes, Fig. 1 Synthesis of the cyano-, carboxy-, and N-ethylcarbamoylborane adducts of amines

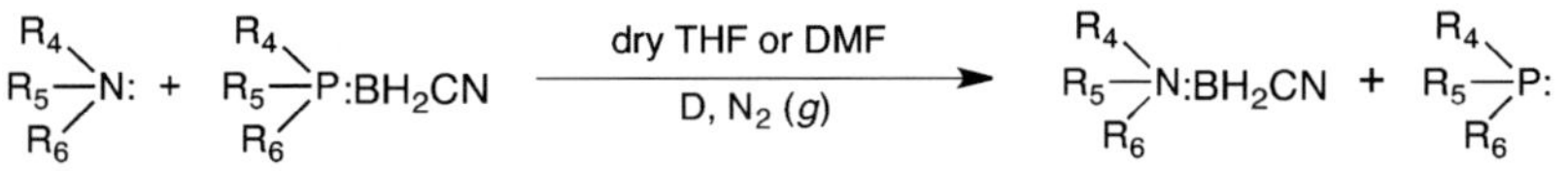

Amine-Boranes, Fig. 2 Synthesis of the cyanoborane adducts of amines via Lewis acid exchange

adducts (e.g., BH_3 and BH_2COOH). Subsequently, the cyanoborane can be converted to the carboxyborane through the formation of the nitrilium salt by refluxing the cyanoborane with triethyloxonium tetrafluoroborate in dichloromethane. Then, addition of water hydrolyzes the nitrilium salt to the carboxyborane. The nitrilium salt can alternatively be treated with methanol to yield the carbomethoxyborane or with a hydroxide base to yield the N-ethylcarbamoylborane. Standard coupling conditions (e.g., a carbodiimide with triethylamine or CCl_4 with Ph_3P) have also been used to prepare the ester and amide derivatives from the carboxyborane (Burnham 2005).

The aforementioned isosteric replacement makes amine-boranes attractive functional groups in designing drugs, and such compounds have demonstrated anticancer, antiviral, hypolipidemic, and anti-inflammatory activities (Burnham 2005). One of the reasons amine-boranes were investigated as therapeutic agents was their potential use in boron neutron capture therapy (BNCT), which is a treatment for inoperable tumors. BNCT is based on the principle that ^{10}B, when exposed to thermal neutrons, will split and therefore emit locally ionizing radiation. The natural abundance of the stable ^{10}B isotope is 19.8%. When a ^{10}B atom is bombarded with low kinetic energy thermal neutrons of approximately 0.025 eV, it splits into ^7Li and ^4He (α-particles). In 94% of the neutron captures, a 0.48-MeV gamma ray is emitted, and the remaining 2.31 MeV is the average kinetic energy between the ^7Li (0.84 MeV) and the ^4He particles (1.46 MeV). The path length of the ^7Li and ^4He particles produced in situ in tissue are 5 and 9 μm, respectively, which is less than the diameter of a mammalian cell (~10 μm). Therefore, this process can selectively deliver high-energy ionizing radiation to a single cell while minimizing collateral damage to the surrounding cells that do not contain ^{10}B atoms. For BNCT to achieve therapeutic success, the boronated compounds must be selectively distributed into the neoplastic tissue. It has been determined that

5–30 μg ^{10}B per gram of tissue is needed for effective neutron capture and that the response is enhanced twofold if the ^{10}B is localized in the cell nucleus (Perks et al. 1988).

Amino Acid Derivatives

Anticancer Activity

An initial strategy toward employing amine-borane isosteres was the synthesis of analogues of the amino acids glycine (H_2NCH_2COOH) or betaine (($CH_3)_3N^+CH_2COO^-$) where the amino methylene (H_2N-CH_2- or ($CH_3)_3N^+CH_2$-) would be replaced by an amine-borane (H_2N:BH_2- or ($CH_3)_3N$:BH_2-) (Hall et al. 1979). These derivatives of amino acids were designed to inhibit protein biosynthesis or other cell processes utilizing glycine or betaine, hence stopping cell replication. The most potent of the early series was trimethylamine-cyanoborane (($CH_3)_3N$:BH_2CN). However, it was quite toxic to normal mice with an LD_{50} of 70 mg/kg, I.P. Of the series of aliphatic amine-borane, amine-cyanoborane, amine-carboxyborane, amine-carbomethoxyborane, and amine N-ethylcarbamoylborane, trimethylamine-carboxyborane proved to be the most potent and least toxic of the compounds synthesized. The substitution of the methylene for the borane resulted in an increase in the pK_a of the carboxylic acid group from 1.83 to 8.14–8.38 making it the weakest known simple carboxylic acid of zero net charge further illustrating the enhanced lipophilicity of the amine-borane derivatives (Spielvogel 1988). Trimethylamine-carboxyborane demonstrated in vitro cytotoxicity against suspended murine and human-cultured cells including L_{1210}, Tmolt^3, and HeLa-S^3 cell lines. An in vivo Ehrlich ascites antitumor screen in CF_1 mice demonstrated that Me_3N:BH_2CO_2H inhibited tumor growth of 82% at 20 mg/kg/day, I.P. Mixed Walker-246 carcinosarcoma ascitic tumor showed a treated-to-control (T/C) value of 174% at 2.5 mg/kg/day, I.P. Some activity was observed in the B-16 melanoma with a T/C value of 134%, and in the Lewis lung, with a T/C value of 144% at 20 mg/kg/day, I.P., in $C_{57}BL/6$ mice. No acute toxicity was observed in mice at 100 mg/kg/day, I.P., and the LD_{50} value in CF_1 mice was determined to be 1,800 mg/kg, I.P. In L_{1210} cell culture, the inhibition of DNA, RNA, and protein syntheses by Me_3N:BH_2CO_2H occurred at 1, 2, and 3 times the ED_{50} value after incubation for 60 min. Mode of action (MOA) studies in L_{1210} cells showed that the decrease in DNA synthesis occurred due to inhibition of the following enzyme activities: thymidine kinase, TDP kinase, PRPP amidotransferase, IMP dehydrogenase, dihydrofolate reductase, and ribonucleotide reductase. The combined effects of inhibiting regulatory pathways in de novo purine biosynthesis (PRPP amidotransferase, IMP dehydrogenase, and dihydrofolate reductase), together with inhibition of the activities of thymidine kinases and ribonucleotide reductase, will lead to the overall inhibition of DNA synthesis (Hall et al. 1990). The amine-carboxyborane did demonstrate an inhibition of the phosphorylation of topoisomerase II by protein kinase C (PKC) resulting in decreased activity of topoisomerase II, which results in DNA strand scission after 8 h. The inhibition of PKC phosphorylation was shown to be mediated in a similar manner to TNF-α since $Me_3NBH_2CO_2H$ was shown to competitively bind with [^{14}C]-TNF-α to high-affinity TNF-a receptors on L_{1210} and L_{929} cells (Miller et al. 1998). In vivo boron distribution studies with Me_3N:BH_2CO_2H, in which the agent was administered at an I.P. dose of 40 mg ^{n}B/kg to BALB mice with Harding-Passey melanoma, demonstrated tumor boron concentrations of 45.7 and 17.5 μg ^{n}B/g of tissue. These concentrations are adequate for BNCT, but the tumor:blood and tumor:brain ratios were only ~1:1, which is not sufficient selectivity for BNCT use (Spielvogel 1988).

Hypolipidemic Activity

A number of the amine-borane amino acid analogues were shown to have inhibitory activity against HMG-CoA reductase, the rate-limiting enzyme in de novo cholesterol biosynthesis. The inhibition of HMG-CoA reductase is a target for the statins (e.g., atorvastatin, simvastatin, etc.), which are clinically used to lower cholesterol concentrations in the blood. It was found that amine-boranes were found to lower serum lipid in rodents. Of the amino acid derivatives, the most effective agents were trimethylamine-carboxyborane, trimethylamine-carbomethoxyborane, trimethylamine-N-n-octyl-carbamoylborane, and N,N-dimethyl-N-octadecylamine-borane. The most potent agent was the methyl ester trimethylamine-carbomethoxyborane, which lowered serum cholesterol and triglyceride levels to 44% and 77% of control values, respectively, in mice at a dose of 10 mg/kg/day, I.P. In vitro MOA studies on CF_1 mouse liver homogenates revealed that this compound inhibited the activities of *sn*-glycerol-3-phosphate acyl transferase, ATP-dependent citrate lyase, and acyl-CoA cholesterol acyl transferase. In studies on lipid levels of serum lipoproteins from rats, cholesterol and triglyceride levels were reduced in the chylomicron, VLDL, and LDL fractions. Importantly, the largest reduction occurred in the LDL fraction. In this fraction, cholesterol levels were decreased to 57% of control values; triglyceride levels were decreased to 44% of control values. Also, this carbomethoxyborane adduct increased serum HDL cholesterol levels by 95% (Hall et al. 1987).

Anti-inflammatory Activity

Potent anti-inflammatory activity was observed in rodents for the glycine and betaine analogues of amine-boranes including trimethylamine-borane, trimethylamine-cyanoborane, and ammonia-cyanoborane were potent anti-inflammatory agents in rodents. These agents demonstrated activity in the carrageenan-induced edema of mouse foot pads (Winter's test), writhing reflex, and rat chronic adjuvant arthritis screen. Key enzymes involved in inflammation whose activities were inhibited included acid phosphatase, cathepsin, and prostaglandin synthetase. The derivatives with the best activities were ammonia-cyanoborane and trimethylamine-cyanoborane; however, those compounds were quite toxic with LD_{50} values of 30 and 70 mg/kg, I.P., respectively, in mice (Hall et al. 1980).

Amine-Borane Derivatives of Peptides

The logical extension of amino acid analogues was the use of the amine-borane adducts to prepare peptides. Di- and tripeptides with $Me_3N:BH_2CO_2H$ or $H_3N:BH_2CO_2H$ at the *N*-terminus have been prepared using standard carbodiimide coupling procedures (Sood et al. 1990). These peptides showed weaker antineoplastic activity than the previously discussed α-amino acid analogues. The dipeptides' cytotoxic effects were due to the inhibition of DNA synthesis, primarily affecting de novo purine biosynthesis inhibiting the activities of PRPP amidotransferase and IMP dehydrogenase.

Boron-containing dipeptides also demonstrated hypolipidemic activity in mice (Sood et al. 1990). The active peptides were L-serine or L-leucine coupled to trimethylamine-carboxyborane as the *N*-terminus (Miller et al. 1999).

Dipeptides in the class mentioned above demonstrated anti-inflammatory activity but were less potent than the simple amino acid analogues. The more potent derivatives were those containing a tyrosine or tryptophan at the C-terminus (Miller et al. 1999).

Amine-Borane Derivatives of Proteins

Early attempts to attach *p*-boronophenylalanine to monoclonal antibodies (mAb) specific for tumor cells were successful but did not yield a high enough concentration of ^{10}B. Borane cage compounds ($B_{10}H_{10}$) have been attached to mAb's yet rendered the mAb ineffective. A strategy to use the Merrifield solid-phase peptide synthesis was successful in polymerizing *closo* and *nido* borane cages with subsequent coupling to a mAb. Again, while the chemical synthesis was successful, the immunogenicity on the mAb was lost. Another method employed was boronating a DL-polylysine polymer with trimethylamino-octahydrodecaborane via an isocyanato linkage. These conjugated mAb did retain 40% to 90% of their immunogenicity; however, the boronated mAb had increased liver uptake over the nonboronated mAb (Soloway et al. 1993).

Other Biomolecule Amine-Borane Derivatives

Since amino acids are used as building blocks for other biomolecules, amine-cyano-, carboxy-, carbomethoxy-, and carbamoylborane adducts of simple heterocyclic amines (e.g., morpholine, piperidine, piperazine, and imidazole) and nucleosides have been synthesized and evaluated for pharmacological activity. The most potent compounds in vitro and in vivo were the heterocycles piperidine-carboxyborane and *N*-methylmorpholine-carboxyborane, the nucleosides 2′-deoxycytidine-N^3-cyanoborane and 3′,5′-*O-bis* (triisopropylsilyl)-2′-deoxyguanosine-N^7-cyanoborane, and the deprotected 2′-deoxyguanosine-N^7-cyanoborane. The heterocyclic compounds as well as the nucleosides targeted DNA synthesis and de novo purine biosynthesis at PRPP amidotransferase and IMP dehydrogenase. 2'-Deoxycytidine-N^3-cyanoborane showed increased uptake in vitro in Tmolt3 cells over Bg-9 fibroblast cells. However, in vitro BNCT studies with V-79 Chinese hamster cells showed no uptake of the unlabelled compound into the cells at concentrations up to 100 mM (Spielvogel et al. 1992).

Summary

Borane (BH_3) and its Lewis acid derivatives (BH_2CN, BH_2COOH, BH_2COOCH_3, and $BH_2CONHC_2H_5$) can form adducts with amines forming an amine-borane where the B–N bond is a coordinate covalent bond. Such compounds can be easily synthesized from $NaBH_3CN$ and Lewis acid exchange reactions. Amine-borane adducts resembling the amino acids glycine and betaine demonstrated anticancer activity as well as hypolipidemic activity. Thus far, their use in BNCT has not been established due to either low or nonselective tumor cell uptake. Some anti-inflammatory effects were observed but for the more toxic cyanoborane adducts.

References

Burnham B (2005) Synthesis and pharmacological activity of amine-boranes. Curr Med Chem 12:1995–2010

Hall I, Starnes C, Spielvogel B et al (1979) Boron betaine analogues: antitumor activity and effects on Ehrlich ascites tumor cell metabolism. J Pharm Sci 68:685–688

Hall I, Starnes C, McPhail A et al (1980) Anti-inflammatory activity of amine cyanoboranes, amine carboxyboranes, and related compounds. J Pharm Sci 69:1025–1029

Hall I, Spielvogel B, Sood A et al (1987) Hypolipidemic activity of trimethylamine-carbomethoxyborane and related derivatives in rodents. J Pharm Sci 76:359–365

Hall I, Spielvogel B, Sood A (1990) The antineoplastic activity of trimethylamine carboxyboranes and related esters and amides in murine and human tumor cell lines. Anticancer Drugs 1:133–141

Miller M, Woods C, Murthy M et al (1998) Relationship between amine-carboxyboranes and TNF-alpha for the regulation of cell growth in different tumor cell lines. Biomed Pharmacother 52:169–179

Miller M, Sood A, Spielvogel B et al (1999) The hypolipidemic and anti-inflammatory activity of boronated aromatic amino acids in CF1 male mice. Met Based Drugs 6:337–344

Perks C, Mill A, Constantine G et al (1988) A review of boron neutron capture therapy (BNCT) and the design and dosimetry of a high-intensity, 24 keV, neutron beam for BNCT research. Br J Radiol 61:1115–1126

Soloway A, Barth R, Carpenter D (eds) (1993) Advances in neutron capture therapy. Springer, New York

Sood A, Sood C, Spielvogel B et al (1990) Boron analogues of amino acids IV. Synthesis and characterization of di- and tripeptide analogues as antineoplastic, anti-inflammatory and hypolipidemic agents. Eur J Med Chem 25:301–308

Spielvogel B (1988) Pharmacologically active boron analogues of amino acids. In: Liebman J, Greenberg A (eds) Advances in boron and the boranes. VCH Publishers, New York

Spielvogel B, Sood A, Shaw B et al (1992) In: Allen B et al (eds) Progress in neutron capture therapy for cancer. Springer, New York

Amine-Carbamoylborane

▶ Amine-Boranes

Amine-Carbomethoxyborane

▶ Amine-Boranes

Amine-Carboxyborane

▶ Amine-Boranes

Amine-Cyanoborane

▶ Amine-Boranes

Amyotrophic Lateral Sclerosis

▶ Copper-Zinc Superoxide Dismutase and Lou Gehrig's Disease

Anaerobic Prokaryotes

▶ Zinc and Iron, Gamma and Beta Class, Carbonic Anhydrases of Domain Archaea

Analytical and High Resolution Electron Microscopy

▶ Palladium, Colloidal Nanoparticles in Electron Microscopy

ANCA

▶ Silicon Exposure and Vasculitis

Anemia

▶ Iron Homeostasis in Health and Disease

Angiotensin I-Converting Enzyme

Ross G. Douglas and Edward D. Sturrock
Zinc Metalloprotease Research Group, Division of Medical Biochemistry, Institute of Infectious Disease and Molecular Medicine, University of Cape Town, Cape Town, South Africa

Synonyms

Carboxycathepsin; Dipeptide hydrolase and peptidase P; Dipeptidylcarboxypeptidase I; Kininase II

Definition

Angiotensin-converting enzyme (ACE) is a metallopeptidase that utilizes a zinc ion as a crucial cofactor in its catalytic mechanism. ACE has wide human tissue distribution and plays a key role in the regulation of blood pressure, cardiovascular function, electrolyte balance, hematology, and tissue fibrosis. Potent inhibitors of ACE have been developed and are clinically useful in the treatment of hypertension, post-myocardial infarction and diabetic nephropathy.

Background

ACE is a zinc metallopeptidase that is responsible for the hydrolysis of the penultimate peptide bond of several substrates of physiological consequence. This dipeptidyl peptidase is best known for its conversion of inactive decapeptide angiotensin I to active octapeptide hormone angiotensin II, leading to sodium retention, aldosterone release, and ultimately vasoconstriction. The central role that ACE plays in blood pressure regulation emphasizes both the enzyme's physiological importance and its therapeutic importance as a target for the treatment of cardiovascular disease.

ACE is widely expressed in many human tissues, especially on vascular endothelial cells. It is anchored to the cell surface by means of a single hydrophobic transmembrane region. While ACE is targeted for extracellular surface localization, a region proximal to the transmembrane region can be proteolytically cleaved and therefore result in a soluble form of the protein. This posttranslational processing is generally referred to as "ectodomain shedding." Two isoforms of ACE exist: a widely expressed somatic form (somatic ACE) and a truncated form, testis ACE, that is expressed in male germinal cells and plays an important role in fertility.

The *Ace* gene is located on the 17q23 locus and is 21 kilobases in length with 26 variably sized exons. Both ACE isoforms are the products of the same gene, with the truncated testis ACE arising from a testis-specific intronic promoter. Thus, testis ACE is approximately half the size of somatic ACE. Somatic ACE is expressed as a 1, 277 amino acid mature protein while testis ACE is 701 amino acids in length. Molecular cloning of the somatic isoform revealed an unusual feature for enzymes: The enzyme consisted of two homologous domains (designated N- and C-domains depending on their location on the polypeptide chain), each containing a fully functional active site. In contrast, testis ACE contained only one domain and was identical to the C-domain of somatic ACE with the exception of a 36-amino acid region at the N-terminus (Fig. 1). Comparison of the two somatic ACE domains reveals high sequence identity between the two domains (approximately 60% identity). Analysis of residues present in the active sites shows even higher position similarity (approximately 90% identity). Elucidation of the three-dimensional

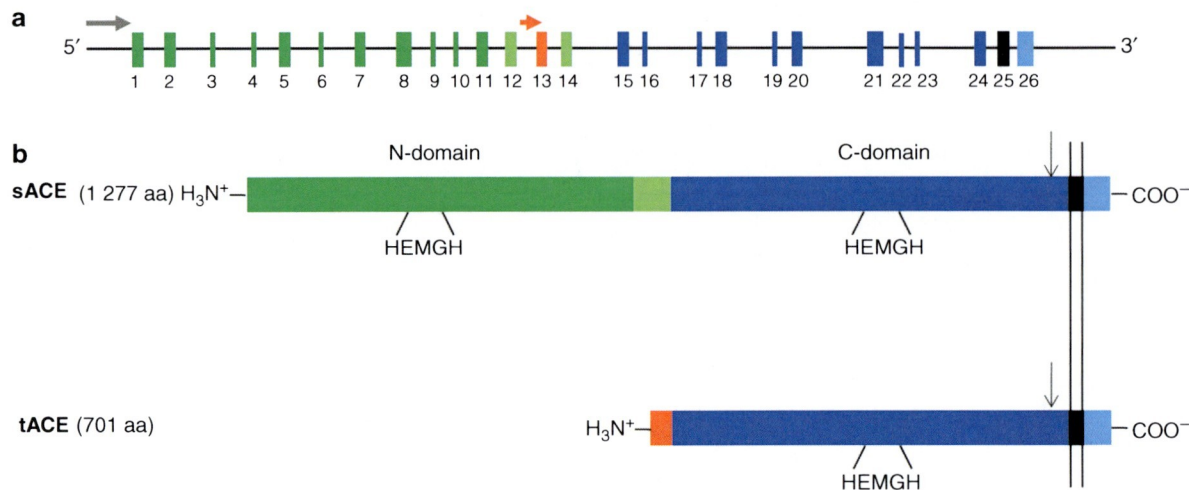

Angiotensin I-Converting Enzyme, Fig. 1 Schematic representation of the Ace gene and the two isoform products. (**a**) The Ace gene consists of 26 exons all of which are transcribed for somatic ACE with the exception of exon 13. The promoter for somatic ACE expression is indicated as a *gray arrow*. Testis ACE arises from the presence of a germinal cell–specific promoter in intron 12 (*orange arrow*) and translation of exon 13 results in a unique 36 amino acid N-terminus of testis ACE whereas the remainder of the protein is identical to the C-domain of sACE. (**b**) Somatic ACE N- and C-domains with catalytic HEMGH motifs are shown in *green* and *blue,* respectively, the inter-domain linker region in *light green*, transmembrane region in *black*, and cytoplasmic C-terminus in *light blue*. The *black arrow* indicates the approximate cleavage site of ACE to be solubilized into the surrounding medium

structures of each domain has further revealed highly conserved structural topology between the two domains.

Another noteworthy feature of ACE biochemistry is the presence of glycosylation (carbohydrate or glycan chains) on the protein surface. The presence of these glycans is important for proper folding of the protein and thus key in the production of active enzyme (Acharya et al. 2003).

Biological Function

ACE is one of the central proteases in the renin-angiotensin-aldosterone system (RAAS), an important system for blood pressure regulation and electrolyte homeostasis. In the classical linear system, 55-kDa plasma protein angiotensinogen is N-terminally cleaved by the aspartyl protease renin to yield angiotensin I. ACE is then responsible for the conversion of this inactive decapeptide into the active octapeptide hormone by cleavage of the penultimate C-terminal peptide bond. The product of ACE hydrolysis, angiotensin II, mediates effects through the angiotensin receptor to induce a signaling cascade to ultimately increase sodium retention, aldosterone release, and vasoconstriction. While the classical system appears relatively simple, research performed in recent decades underlines the complexities of the RAAS (see (Fyhrquist and Saijonmaa 2008) for additional information). The physiological significance of ACE action is emphasized in studies involving transgenic mice. ACE null mice display low blood pressure, renal defects, anemia, reduced ability to concentrate urine, and decreased fertility. Overexpression of ACE in heart tissue specifically resulted in mice with atrial enlargement, cardiac arrhythmia, and a tendency to die of sudden cardiac death due to ventricular fibrillation (Bernstein et al. 2005).

In addition to angiotensin I hydrolysis, ACE is able to cleave a variety of physiologically relevant peptides. These include vasodilator peptides bradykinin, angiotensin1–7, and substance P as well as other non-vasoactive peptides such as amyloid β-peptide and anti-fibrotic peptide *N*-acetyl-Ser-Asp-Lys-Pro (AcSDKP) (Table 1). ACE can also operate outside of its classical dipeptidyl peptidase function in vitro. However, the in vivo actualities

Angiotensin I-Converting Enzyme, Table 1 An overview of selected ACE substrates. Cleavage sites by ACE are indicated by an arrow. The broken arrow indicates the sequential cleavage site after the removal of the first dipeptide

Substrate name	Biological action of substrate	Substrate peptide sequence and ACE cleavage sites	Biological action of product
Angiotensin I	Inactive	↓ D – R – V – Y – I – H – P – F – H – L	Vasoconstriction Hypertrophy Fibrosis
Bradykinin	Vasodilation	↓ ↓ R – P – P – G – F – S – P – F – R	Inactive
Angiotensin (1–7)	Vasodilation	↓ D – R – V – Y – I – H – P	Inactive
Substance P	Vasodilation Pain response	↓ ↓ R – P – K – P – Q – Q – F – F – G – L – M -NH$_2$	Inactive
Gonadotropin-releasing hormone	Sexual development	↓ ↓ ↓ pyro-E – H – W – S – Y – G – L – R – P – G -NH$_2$	No in vivo data published
N-acetyl-SDKP	Anti-fibrosis	↓ Ac-S – D – K – P	Inactive

of this property remain to be properly elucidated (Bernstein et al. 2011).

In addition to its role in hormone cleavage, ACE has also been shown to play a role in immunological function. Studies suggest that ACE has a physiological function in editing the carboxyl termini of proteasome produced major histocompatibility complex class 1 antigenic peptides (Shen et al. 2011). Further, modification of sACE domain activities results in marked changes of inflammatory cytokine levels, implying a function in cytokine regulation.

While the exact details of the mechanism of ACE substrate hydrolysis are not currently fully characterized, it is presumed to be a general base-type mechanism similar to thermolysin. The substrate is positioned and stabilized through interactions with residues in the enzyme active site (particularly with the C-terminal carboxylate of the substrate and Gln, Lys, and Tyr residues) and the peptide bond carbonyl oxygen coordinating the zinc ion. Proper positioning of the substrate displaces the zinc ion–associated water molecule, resulting in a nucleophilic attack of the carbonyl carbon of the peptide bond by water. After proton exchange, the resultant tetrahedral intermediate promptly collapses to form the product pair (Sturrock et al. 2004).

While ACE is classically a metalloenzyme, other works have shown it to play an interesting "noncatalytic" role. Binding of substrate and inhibitors tended to result in phosphorylation of the cytoplasmic tail of ACE which led to an induction of the JNK/c-Jun pathway. Other works have shown that angiotensin II binding specifically resulted in Ca^{2+} release while other inhibitors and substrates did not elicit this effect. Therefore, while ACE is popularly referred to as such due to its documented conversion of angiotensin I, ACE has the ability not only to process other physiologically relevant substrates but to function in a receptor-like role as well (Lambert et al. 2010).

Domain Substrate Specificities

Somatic ACE contains two homologous domains that have high sequence identity and conserved structural topology between their active sites. Despite this marked similarity, the two domains show different in vivo substrate cleavage abilities. As examples, the C-domain has been shown to be the prominent site for the conversion of angiotensin I. Meanwhile, the N-domain is the major domain for the cleavage of anti-fibrotic peptide AcSDKP. Both domains cleave bradykinin with similar efficiencies (Bernstein et al. 2011).

The two domains, while having discreet substrate cleavage abilities, do not operate in a completely independent manner. Biochemical analysis indicates that in an in vitro system, the domains tend to exhibit a negative cooperative effect on the other domain. That is, the domains in isolation have improved

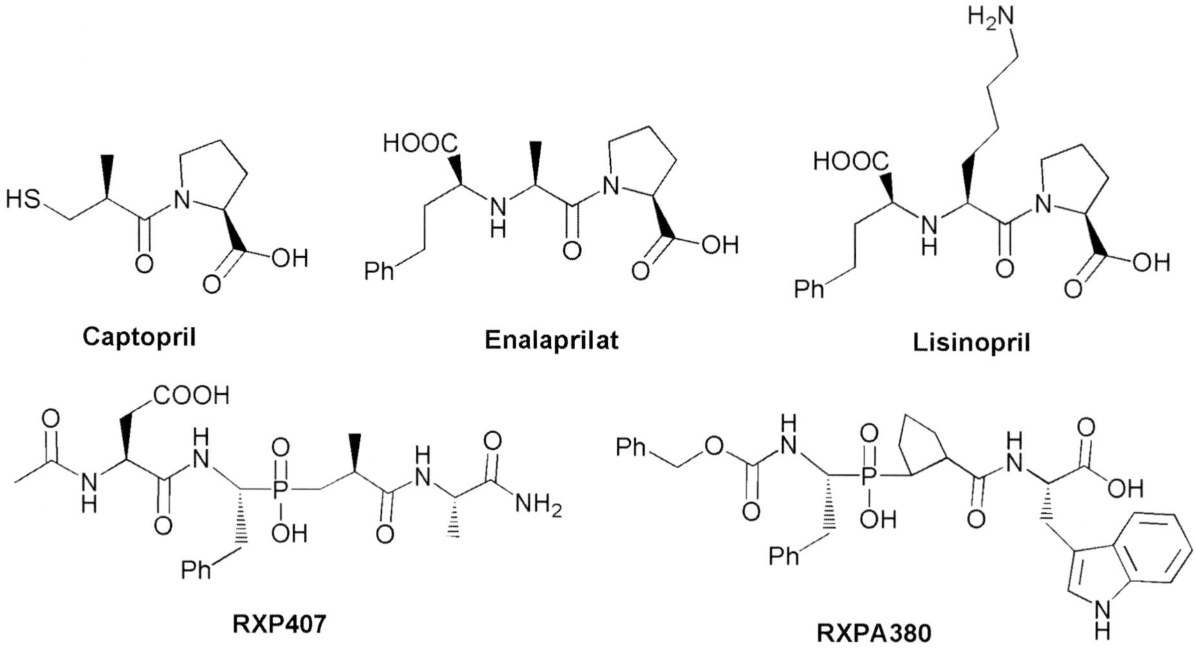

Angiotensin I-Converting Enzyme, Fig. 2 Chemical structures of a selection of ACE inhibitors. Captopril, enalaprilat, and lisinopril are clinically approved ACE inhibitors for the treatment of hypertension, heart failure, and diabetic nephropathy. RXP407 and RXPA380 are phosphinic peptidomimetic inhibitors that are selective for the N- and C-domains, respectively

activities compared to the full-length somatic ACE. The total activity of somatic ACE is therefore approximately the mean of the two isolated domain activities. The exact physiological significance of this observation is not yet known.

ACE Inhibitors

With ACE playing a central role in cardiovascular functioning, it is perhaps not surprising that ACE inhibitors are successfully used in the treatment of hypertension, congestive heart failure, post-myocardial infarction, left ventricular dysfunction, and diabetic nephropathy. First-generation ACE inhibitors were based on bradykinin-potentiating peptides from the snake venom of *Bothrops jararaca* and, with modifications to the zinc-binding group (to contain a sulfhydryl as a binding group), led to approval of the first inhibitor captopril for clinical use in 1981. Since the finding that ACE inhibition significantly contributes to reduction of blood pressure in the human system, a total of 17 ACE inhibitors containing different zinc-binding groups and functionalities have been designed, synthesized, and approved for the clinic (Fig. 2) (Redelinghuys et al. 2005).

It is important to note that first-generation current clinical ACE inhibitors were designed with neither the knowledge of the two-domain ACE composition nor the crystal structures. Since these findings, research has been carried out, and continues to be, on attempting to develop inhibitors that are selective for one domain. Such efforts have resulted in the production of phosphinic peptidomimetic inhibitors RXP407 and RXPA380 that are selective for the N- and C-domains, respectively (Fig. 2). These developments are important since adverse drug events (such as persistent dry cough and angioedema) associated with clinical ACE inhibitor treatment are possibly due to elevated plasma bradykinin levels. Currently, all small molecule ACE inhibitors block both domains with approximately equal binding affinities. Thus, inhibitors that are selective for the C-domain could allow for treatment of blood pressure with lower elevations of bradykinin levels (and therefore reduced incidence of side effects), while N-domain-selective inhibitors could result in the treatment of conditions related to tissue fibrosis due to AcSDKP buildup without affecting blood pressure regulation (Bernstein et al. 2011).

Three-Dimensional Structures of ACE

The overall structure of both the N- and C-domains of ACE represents an ellipsoid divided into two subdomains by a deep central cleft. The catalytic zinc-containing active site is located deep within this cleft, and investigations have shown that substrates could perhaps gain access through hinge movement by twisting the N- and C-termini of the protein. The structures are predominantly α-helical with only six β-sheets seen in the resolved structures. The determination of such structures has allowed for an understanding of ACE inhibitor binding at the atomic level and allowed researchers to identify obligatory binding sites within the active site (Fig. 3). Side chains of substrates or inhibitors have traditionally carried the nomenclature of $P_N...P_2, P_1, P_1', P_2'...P_N'$ based on the insertion and interaction with the corresponding $S_N...S_2, S_1, S_1', S_2'...S_N'$ subsites. The resolution of both domain structures with ACE inhibitor lisinopril revealed a very similar overall binding mode within the active sites. The inhibitors' carboxyl-alkyl group in both cases is coordinating the catalytic zinc ion. The P_1 phenylalanine extends into the corresponding S_1 subsite to interact with residues Thr496/Val518 of the N- and C-domains respectively, while the P_1' lysyl moiety interacts with unique C-domain residues Glu162 and Asp377 (replaced by Asp140 and Gln355 in the N-domain). In the S_2' subsite, the P_2' prolyl carboxylate has significant interactions with conserved residues Gln259/281, Lys489/511, and Tyr498/520 (N/C numbering) and these residues have been suggested as important for substrate and inhibitor positioning.

In addition to an appreciation of overall topology and inhibitor binding mode, these structures allowed for the identification of amino acids that were present in the active sites but differed in chemical nature between the N- and C-domains. This finding has been useful in assessment of the contribution of these unique amino acids to the selective binding of inhibitors and processing of specific substrates. In particular, residues Tyr369 and Arg381 in the N-domain (replaced by a Phe and Glu in the C-domain) appear to be important

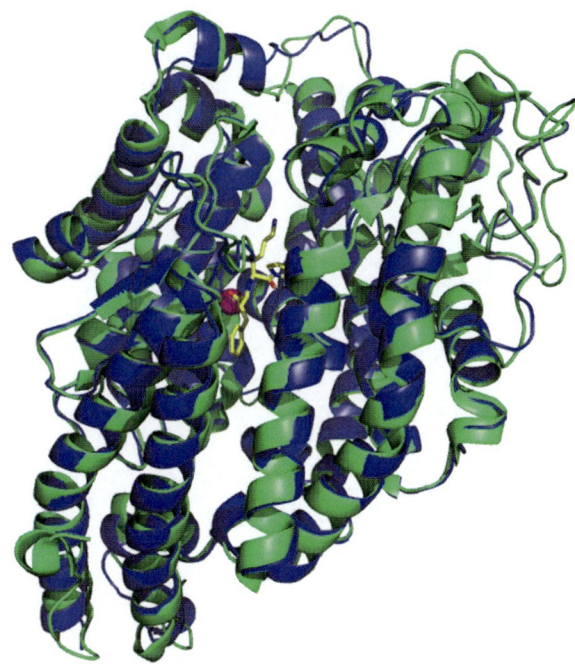

Angiotensin I-Converting Enzyme, Fig. 3 Three-dimensional structures of the somatic ACE N- and C-domains. The N- (*green ribbon*) and C-domains (*blue ribbon*) display marked overall structural topology. Potent ACE inhibitor lisinopril and the catalytic zinc ion are shown in *yellow sticks* and *magenta sphere*, respectively. A deep central cleft can be noted down the center of the molecules resulting in two subdomains

in the selective binding of RXP407 while the S_2' subsite and Phe391 (replaced by more bulky hydrophilic residues and Tyr respectively in the N-domain) seem to be the major contributors toward the selective binding of RXPA380. The crystal structures have therefore provided important insights as to the structure-function relationship of this significant enzyme (Anthony et al. 2012, Watermeyer et al. 2010).

Concluding Remarks

ACE is an important metalloenzyme in its physiological contribution to cardiovascular function and is therefore a good drug target for the improvement of human health. Furthermore, the presence of two homologous catalytic domains that have different properties and play diverse functional roles in immunology, hematology, and fibrosis adds another dimension to the function of this ubiquitous enzyme. The role of the zinc ion as crucial to the catalytic mechanism emphasizes the importance of metal ions in biological function and stresses the need for thorough understanding of the role of bioinorganics in health and disease.

Cross-References

▶ Angiotensin I-Converting Enzyme
▶ Thermolysin
▶ Zinc Carboxypeptidases

References

Acharya KR, Sturrock ED, Riordan JF et al (2003) Ace revisited: a new target for structure-based drug design. Nat Rev Drug Discov 2:891–902
Anthony CS, Masuyer G, Sturrock ED et al (2012) Structure based drug design of angiotensin-I converting enzyme inhibitors. Curr Med Chem 19:845–855
Bernstein KE, Xiao HD, Frenzel K et al (2005) Six truisms concerning ACE and the renin-angiotensin system educed from the genetic analysis of mice. Circ Res 96:1135–1144
Bernstein KE, Shen XZ, Gonzalez-Villalobos RA et al (2011) Different in vivo functions of the two catalytic domains of angiotensin-converting enzyme (ACE). Curr Opin Pharmacol 11:105–111
Fyhrquist F, Saijonmaa O (2008) Renin-angiotensin system revisited. J Intern Med 264:224–236
Lambert DW, Clarke NE, Turner AJ (2010) Not just angiotensinases: new roles for the angiotensin-converting enzymes. Cell Mol Life Sci 67:89–98
Redelinghuys P, Nchinda AT, Sturrock ED (2005) Development of domain-selective angiotensin I-converting enzyme inhibitors. Ann NY Acad Sci 1056:160–175
Shen XZ, Billet S, Lin C et al (2011) The carboxypeptidase ACE shapes the MHC class I peptide repertoire. Nat Immunol 12:1078–1085
Sturrock ED, Natesh R, van Rooyen JM et al (2004) Structure of angiotensin I-converting enzyme. Cell Mol Life Sci 61:2677–2686
Watermeyer J, Kroger WL, Sturrock ED et al (2010) Angiotensin-converting enzyme – new insights into structure, biological significance and prospects for domain-selective inhibitors. Curr Enzym Inhib 5:134–147

Annexin

▶ Calcium-Binding Proteins, Overview

Annexins

Annette Draeger and Eduard B. Babiychuk
Department of Cell Biology, Institute of Anatomy, University of Bern, Bern, Switzerland

Synonyms

Calpactins; Chromobindins; Endonexins; Lipocortins

Definition

Calcium binding sites: Structural motifs, which enable the binding of a calcium molecule. Type I is defined as EF-hand motif, which consists of a 12-residue loop between two helices. Type II and type III binding sites describe structural propensities for calcium binding which do not correspond to the EF-hand motif and consist of shorter loops. The type II calcium binding site differs in its conformational contribution of individual or paired amino acids from the type III site, which is considered to be a "minor" calcium binding site.

Introduction

The annexins are a multigene family of Ca^{2+}-dependent membrane-binding proteins. They are structurally related and expressed in most phyla and species of eukaryotes. In vertebrates, 12 annexins are expressed (A1–11 and A13), displaying different splice versions (Gerke and Moss 2002). The annexins have been implicated in physiological processes, which are generally concerned with the regulation of plasma membrane organization and membrane-related signaling events, such as the control of vesicle trafficking during endo- and exocytosis (Raynal and Pollard 1994; Gerke and Moss 2002). In addition, they have been credited with extracellular functions, among them prominently featuring anticoagulative and anti-inflammatory properties (Dassah et al. 2008; Babbin et al. 2008).

In general, the annexins act as intracellular Ca^{2+} sensor/effector proteins; however, they have also been assigned Ca^{2+}-independent functions

(Raynal and Pollard 1994; Gerke and Moss 2002). In this entry, only the Ca^{2+}-dependent functions of this protein family will be considered: Intracellular interactions of the annexins which occur in a Ca^{2+}-independent way, as well as extracellular functions of annexins, which take place at saturating Ca^{2+} concentrations, are beyond the scope of this entry.

Annexins and Molecular Structure

The annexins share a core structure made up of four homologous domains, each of approximately 70 amino acids in length (except for annexin A6, which, as a result of gene duplication, possesses eight). Each of the four highly conserved domains consists of five α-helices, wound in a tight right-handed superhelix and connected by short loops (Raynal and Pollard 1994). In contrast, the NH_2-terminal domains of the annexins are highly variable and harbor phosphorylation sites, which are important for the interaction with other proteins (Gerke and Moss 2002). It is this variability which is thought to influence the specificity of individual annexins, whereas the conserved core domain unites the annexins as a family of structurally related proteins.

The annexin core shows both intramolecular and intermolecular homology (Gerke and Moss 2002) and harbors type II and type III Ca^{2+} binding sites. Originally identified by its resistance to proteolysis, the core forms a curved disk with a central hydrophobic pore. Ca^{2+} also coordinates carbonyl and carboxyl groups of the protein and phosphoryl groups at the glycerol backbone of membrane phospholipids. Annexin-membrane interactions occur via the formation of a ternary complex between the convex surface of the annexin core, Ca^{2+}, and the negatively charged phospholipids (rev by Gerke and Moss 2002). Ca^{2+} binding leads to the conformational changes in the core, particularly in repeat III (rev by Gerke and Moss 2002).

The Ca^{2+} Sensitivity of Annexins

Biochemical analysis in vitro revealed that the annexins have individual Ca^{2+} requirements for their interaction with negatively charged phospholipids (rev by Raynal and Pollard 1994; Monastyrskaya et al. 2009).

In vivo, at low $[Ca^{2+}]_i$ in resting cells, annexins are diffusely distributed throughout the cytoplasm. After stimulation, they translocate to the plasmalemma. Direct comparison of the Ca^{2+} sensitivity of several annexins by live cell imaging techniques established that annexin A2 shows the highest Ca^{2+} sensitivity of plasmalemmal translocation, followed by annexin A6, A4, and A1 (rev Monastyrskaya et al. 2009). Recent data have shown that the half maximal $[Ca^{2+}]_i$ required for annexin A2 translocation in vivo was 0.5 µM (Potez et al. 2011), whereas that of annexin A6 and annexin A1 was 5 and 10 µM, respectively (rev by Draeger et al. 2011).

Annexin A2 translocates to the plasma membrane in response to physiological stimuli, which lead to a transient Ca^{2+} release from intracellular stores (rev by Monastyrskaya, et al. 2009). Simultaneous Ca^{2+} and annexin imaging demonstrated that the duration and timing of this association precisely corresponded to the elevation in $[Ca^{2+}]_i$. Thus, it is highly likely that this molecule takes part in membrane-associated signaling processes. Annexin A2 is particularly abundant in endothelial and in smooth muscle cells, cell types which have a highly varied and intricate relationship with each other and with their surroundings (Babiychuk and Draeger 2000). Annexin A6 translocates to the plasmalemma only after a sustained influx of Ca^{2+}, such as is induced by the activation of store-operated channels during cellular stress responses (rev by Monastyrskaya et al. 2009). The expression of plasma membrane–anchored annexin A6 led to a downregulation of store-operated Ca^{2+} entry, implicating that this protein may play a role in the maintenance of intracellular Ca^{2+} homeostasis in the environmentally stressed cells (Monastyrskaya et al. 2009). Annexin A1 translocates to the plasmalemma only after a massive elevation in $[Ca^{2+}]_i$ such as occurs during plasmalemmal injury. Its role in the repair of plasmalemmal lesions caused either by mechanical damage or by pore-forming toxins has recently been established (rev by Draeger et al. 2011). Hence, the intracellular translocation of annexins might represent an important mechanism for the control of numerous physiological and pathological cellular functions.

For the investigator, annexins are useful tools which allow the monitoring of changes in $[Ca^{2+}]_i$. Especially in damaged cells, their differential Ca^{2+} sensitivity of membrane binding makes them more reliable Ca^{2+}

sensors than fluorescent Ca^{2+} dyes which diffuse out of the cytoplasm once the plasmalemma is perforated (rev by Draeger et al. 2011).

The annexins' Ca^{2+} sensitivity of membrane binding can be modulated via the NH_2-terminal interaction with other proteins (rev by Gerke and Moss 2002; Monastyrskaya et al. 2009). Especially well investigated in this respect are annexins A1 and A2. These annexins exist either as monomers or heterotetramers in which two molecules of annexin A1 are associated with two molecules of S100A11 (p10) or, in the case of annexin A2, with dimers of S100A10 (p11). NH_2-terminal cleavage of annexins A1 and A2 abrogates the formation of heterotetramers leading to opposite effects on Ca^{2+} sensitivity in annexin A1 and A2: It increases the Ca^{2+} sensitivity of annexin A1 and decreases that of annexin A2 (rev by Monastyrskaya et al. 2009). Binding to S100A10 has also been demonstrated to enhance annexin A2's properties to aggregate membranes at micromolar concentrations (rev by Gerke and Moss 2002; Monastyrskaya et al. 2009). Seven members of the S100 protein family (S100A1, S100A4, S100A6, S100A10, S100A11, S100A12, and S100B) are known to interact with at least one of the 12 annexin proteins (rev by Rintala-Dempsey et al. 2008). In addition, some S100 proteins, i.e., S100A6, have been shown to form complexes with several annexins (A2, A5, A6, and A11) (rev by Rintala-Dempsey et al. 2008). Thus, it is possible that the Ca^{2+} sensitivity not only of annexins A1 and A2 but also of other annexins is modulated accordingly.

An additional layer of complexity is added by the observation that annexins are able to adapt their individual Ca^{2+} sensitivity to match changes in the lipid composition within the plasmalemma. All annexins associate with negatively charged phospholipids, but several members of this protein family prefer sites of distinct lipid composition (rev by Monastyrskaya et al. 2009; Draeger et al. 2011). Apart from their association with distinct lipids, they can specifically bind to – or specifically avoid – distinct lipid microdomains or certain assemblies of lipids. One example is the association of annexin A1 with ceramide during apoptosis. As a consequence of cellular stress, an elevation in $[Ca^{2+}]_i$ leads to an increased hydrolysis of plasmalemmal sphingomyelin to ceramide, which self-associates thereby coalescing into large membrane platforms. At these conditions, annexin A1 preferentially associates with the newly formed ceramide platforms (rev by Draeger et al. 2011). Another example is annexin A2, which was identified as a phosphatidylinositol (4,5)-bisphosphate-binding protein being recruited to sites of actin assembly at the plasma membrane and to endocytic vesicles (rev by Gerke et al. 2005).

Thus, several annexins present within any one cell, endowed with a distinct Ca^{2+} threshold for membrane translocation and specific lipid targeting properties, can be considered as a broad-range Ca^{2+} and lipid sensing/effecting system (rev by Monastyrskaya et al. 2009; Draeger et al. 2011). The advantages of this mechanism are obvious: By expressing a unique set of annexins, each cell type can develop its characteristic annexin "profile" and adjust its Ca^{2+} homeostasis to its specific functional requirements. Nonetheless, it is conceptually difficult to understand why – apart from annexins A2 and A6 – all other annexins investigated so far (annexins A1, A4, A5, A7, and A11) have Ca^{2+} sensitivities, which are too low to be compatible with cell survival.

In order to resolve this, one needs to consider yet another of the annexins' functional properties: their ability to fuse and aggregate membranes.

Aggregation and Fusion of Membranes

Membrane fusion is of vital importance for the life of cells; it regulates the contacts with the extracellular environment via the exo- and endocytosis of substances, and it is instrumental in intracellular protein transfer. Annexin A7 (formerly "synexin") was initially identified in a search for proteins that promote the aggregation of chromaffin granules of the adrenal medulla (rev by Raynal and Pollard 1994).

Vesicle association and involvement in vesicular transport have since been shown for multiple annexins (rev by Raynal and Pollard 1994; Futter and White 2007; Monastyrskaya et al. 2009). Annexin A1 is targeted to endosomes and contributes to EGF-mediated membrane inward vesiculation. Annexin A2 contributes to intracellular vesicle movement and regulates endosomal functions. Annexin A3 has been credited with a role in phagosomal aggregation, annexin A5 has been colocalized with late endosomes, and annexin A6, which is associated with endocytic transport, is capable of binding to two adjacent phospholipid membranes. Annexins A1, A2, and A6 play

Annexins, Fig. 1 Annexin-mediated plasma membrane repair and damage control mechanisms. Annexins are diffusely distributed within the cytoplasm and translocate to the plasmalemma in response to Ca^{2+} influx via a toxin- or complement-induced pore. The injured region is isolated by annexin molecules, which fuse the adjoining membranes which are then released in the form of a microvesicle or microparticle. Alternatively, cellular subcompartments (blebs) can be sealed off from the cell body. This mechanism allows the regional compartmentalization of $[Ca^{2+}]$ and facilitates membrane repair

a role in the regulation of endocytic membrane traffic and in the biogenesis of multivesicular bodies.

In fact, it is notable that those annexins, which have been probed for membrane aggregation properties, all appear to support vesicle aggregation. However, the mechanism by which annexins link two membranes is not completely understood. Recent reports, which have closely monitored membrane fusion, have largely focused on annexins A1 and A2 heterotetramers. A structural model of membrane binding and membrane fusion has implicated the amphipathic helices of the annexin A1 NH_2-terminus (Gerke et al. 2005). They interact directly with one membrane while the core domain binds to the second membrane (Gerke et al. 2005). However, at least in vitro, isolated annexin cores have the ability to aggregate membranes without the N-terminal domain (Raynal and Pollard 1994). Cryo-electron microscopy of the annexin A2 heterotetramer suggested that in the presence of Ca^{2+} and at physiological pH, the complex forms membrane bridges in chromaffin granules and large unilamellar vesicles.

Annexins in Membrane Injury

The combination of Ca^{2+} sensing with fusogenic properties shared by the proteins of the annexin family are of considerable practical importance for a cell as soon as it suffers membrane injury. The integrity of the plasma membrane can be compromised by mechanical or chemical injury, by pore-forming toxins or even by the organism's internal defenses such as blood complement complexes or perforins (rev by McNeil and Steinhardt 2003).

Depending on the size of the lesion and on the nature of the injury, the cell has to adopt different strategies for resealing the lesion. Cells suffering mechanical damage, which causes large physical defects (>0.2 μm), cannot reseal their plasma membrane spontaneously but have to cover the defect with vesicles which originate from the cell's cytoplasmic compartment. During this so-called exocytotic repair mechanism, membrane patches are created by the fusion of lysosomes into large lipid segments, which are transported to the plasma membrane and inserted into the lesion (Idone et al. 2008). It is likely that – apart from lysosomes – other intracellular membrane reservoirs can also be recruited for this purpose. Well investigated in plasma membrane repair are the protein families of the SNAREs, synaptotagmins, and ferlins (rev by Draeger et al. 2011). A potential cooperation of the annexins with these proteins has frequently been invoked.

In contrast to the repair of a defect characterized by "free lipid edges," the protein-lined pores caused by toxins or complement need first to be quarantined and subsequently excised and discarded. Depending on the nature of the toxin or on the cell type, the pore-containing membrane segments are either taken up by endocytosis and neutralized intracellularly or shed extracellularly in the form of microvesicles or microparticles (rev by Draeger et al. 2011) (Fig. 1).

In damaged cells, the sudden influx of extracellular Ca^{2+} constitutes the initial trigger for the activation of cellular membrane repair responses (rev by Draeger et al. 2011). A limited increase in $[Ca^{2+}]_i$ is thought to induce transcriptional activation and numerous physiological activities (rev by McNeil and Steinhardt 2003). However, a surfeit of Ca^{2+} is detrimental to the cell and ultimately leads to cell death.

Plasmalemmal repair appears to be most effective at Ca^{2+} concentrations between 5 and 10 μM (rev by Draeger et al. 2011). If $[Ca^{2+}]$ rises above 10 μM, the cells are unable to complete membrane repair, and persistent changes in the architecture of the plasmalemma are brought about by the externalization of phosphatidylserine, which leads to the loss of membrane lipid asymmetry; simultaneously, sphingomyelin is hydrolyzed to ceramide, which self-associates into large membrane platforms (rev by Monastyrskaya et al. 2009; Draeger et al. 2011). Coinciding with the shift of phosphatidylserine from the inner to the outer leaflet of the plasma membrane, newly assembled ceramide platforms from the outer leaflet move into the opposite direction – being internalized by so-called massive endocytosis as the steric properties of the plasmalemma are significantly altered (Lariccia et al. 2011).

These profound changes in plasma membrane architecture are initiated by an unrestrained influx of Ca^{2+} and ultimately cause cell death. At the "hotspots" of Ca^{2+} entry, the local $[Ca^{2+}]$ rises to above 10 μM, leading to a surge of annexins, which translocate from the cytoplasm to the injured plasma membrane. An essential role of annexin A1 in membrane repair was demonstrated by laser damage to selective regions of the plasmalemma of HeLa cells, which triggered a translocation of annexin A1 followed by membrane repair. Membrane repair could be specifically blocked by antibodies against annexin A1 or by a dominant-negative annexin A1 protein mutant, which was incapable of Ca^{2+} binding (McNeil et al. 2006). This was the first experimentally confirmed demonstration of a member of the annexin protein family being involved in membrane repair (McNeil et al. 2006). More recently, annexin A1 has been shown to be instrumental in the repair of membrane lesions after an attack by streptolysin O (SLO), a bacterial pore-forming toxin. The downregulation of annexin A1 by siRNA significantly decreased the ability of human embryonic kidney cells (HEK 293) to withstand a SLO attack (rev by Draeger et al. 2011). By fusing membrane sites on either side of the SLO pore, annexin A1 imprisoned the lesion, which was subsequently pinched off and discarded into the extracellular space as a microvesicle or microparticle (rev by Draeger et al. 2011) (Fig. 1). These remnants of the plasma membrane consist of uni- or multilamellar vesicles and are largely devoid of cytoplasm. They have been detected in all bodily fluids investigated to date and are considered to be a sequel of cell stress or cell injury, often in the wake of systemic disease. They contain annexins A1 (rev by Draeger et al. 2011) and A6 (Potez et al. 2011) in addition to distinct cell surface proteins. Microvesicles are known to be released in response to Ca^{2+} influx into blood cells and platelets and have been shown to contain annexin A7 (rev by Draeger et al. 2011).

In addition to the diverse membrane repair mechanisms described above, early warning systems are operative in cells that are about to suffer a toxin or a complement attack on their plasma membrane. Sublytic concentrations of toxins elicit the release of Ca^{2+} from intracellular stores and trigger the detachment of the subcortical cytoskeleton from the plasma membrane in the form of blebs (rev by Draeger et al. 2011). Blebbing is described as the appearance of multiple membrane protrusions in a cell, which increase the cell surface and create additional intracellular compartments. If the membrane of such a compartment is perforated and flooded with extracellular Ca^{2+}, annexin A1 will move into the bleb and effect a fusion of the membranes of the neck of the bleb. The cell is thus able to save its life by sacrificing the bleb, which can be locally detached from the membrane or – if repair is successful, meaning that physiological $[Ca^{2+}]_i$ is restored – annexin A1 will back-translocate to the cytoplasm and the connection between cell body and bleb will be reestablished (rev by Draeger et al. 2011) (Fig. 1).

Cells can exhibit widely differing phenotypes. Displaying multiple protrusions, cell types such as glia and neurons are prone to rapid changes in $[Ca^{2+}]_i$, in particular within their thin-necked appendages. They are thus at high risk of experiencing an uncontrollable surge of Ca^{2+} after a membrane perforation. Since $[Ca^{2+}]_i$ can change very rapidly, membrane protection must be realized under different conditions. The annexin protein family is ideally suited to provide such a shield: Several annexins work in concert in order to protect the plasma membrane. Depending on the site of injury and the extent of the rise in local $[Ca^{2+}]_i$, the translocation of different annexins is being triggered: Initially, highly Ca^{2+}-sensitive ones, and – if the attack persists or if initial damage limitation mechanisms fail – annexins with lower Ca^{2+} sensitivity will translocate to the injured site and undertake membrane repair.

Conclusion

An incidental appearance of annexins in biochemical assays has been recorded by many scientists and frequently led to a lasting interest in these proteins. The annexins are a family of Ca^{2+}-regulated phospholipid-binding proteins. Presumably attributable to their involvement in numerous membrane-associated processes, they have been credited with a bewildering range of functions. It appears that their structural similarity and ubiquitous expression have obscured rather than furthered the understanding of their functional role.

Annexins display distinct Ca^{2+} sensitivities of plasmalemmal translocation and differential lipid affinities. Their capability to react to a surge in $[Ca^{2+}]_i$ by translocating to the plasma membrane in addition to their fusogenic properties enable them to perform a variety of membrane surgical operations: excising and shedding a protein-lined perforation or patching a membrane hole by fusing the "sticky lipid edges."

In order to prevent a lethal outcome of plasmalemmal perforation, several members of the annexin family working in concert can provide rapid and efficient protection. Functioning as a membrane repair emergency team also explains why most annexins investigated so far operate in a range of $[Ca^{2+}]_i$ which can be considered nonphysiological and indeed lethal for normal cellular function. Annexins A2 and A6, which display a higher Ca^{2+} sensitivity, are presumably fulfilling additional functions in cell signaling, membrane transport, and regulation of intracellular Ca^{2+} homeostasis (Fig. 1).

References

Babbin BA, Laukoetter MG, Nava P et al (2008) Annexin A1 regulates intestinal mucosal injury, inflammation, and repair. J Immunol 181:5035–5044

Babiychuk EB, Draeger A (2000) Annexins in cell membrane dynamics: Ca^{2+}-regulated association of lipid microdomains. J Cell Biol 150:1113–1123

Dassah M, Deora AB, He K, Hajjar KA (2008) The endothelial cell annexin A2 system and vascular fibrinolysis. Gen Physiol Biophys 28:F20–F28

Draeger A, Monastyrskaya K, Babiychuk EB (2011) Plasma membrane repair and cellular damage control: the annexin survival kit. Biochem Pharmacol 81:703–712

Futter CE, White IJ (2007) Annexins and endocytosis. Traffic 8:951–958

Gerke V, Moss SE (2002) Annexins: from structure to function. Physiol Rev 82:331–371

Gerke V, Creutz CE, Moss SE (2005) Annexins: linking Ca^{2+} signaling to membrane dynamics. Nat Rev Mol Cell Biol 6:449–461

Idone V, Tam C, Andrews NW (2008) Two-way traffic on the road to plasma membrane repair. Trends Cell Biol 18:552–559

Lariccia V, Fine M, Magi S et al (2011) Massive calcium-activated endocytosis without involvement of classical endocytic proteins. J Gen Physiol 137:111–132

McNeil PL, Steinhardt RA (2003) Plasma membrane disruption: repair, prevention, adaptation. Annu Rev Cell Dev Biol 19:697–731

McNeil AK, Rescher U, Gerke V, McNeil PL (2006) Requirement for annexin A1 in plasma membrane repair. J Biol Chem 281:35202–35207

Monastyrskaya K, Babiychuk EB, Draeger A (2009) The annexins: spatial and temporal coordination of signaling events during cellular stress. Cell Mol Life Sci 66:2623–2642

Potez S, Luginbühl M, Monastyrskaya K, Hostettler A, Draeger A, Babiychuk EB (2011) Tailored protection against plasmalemmal injury by annexins with different Ca^{2+}-sensitivities. J Biol Chem 286:17982–17991

Raynal P, Pollard HB (1994) Annexins: the problem of assessing the biological role for a gene family of multifunctional calcium- and phospholipid-binding proteins. Biochim Biophys Acta 1197:63–93

Rintala-Dempsey AC, Rezvanpour A, Shaw GS (2008) S100-annexin complexes–structural insights. FEBS J 275:4956–4966

Anti-apoptotic

▶ Lithium, Neuroprotective Effect

Antiarrhythmics

▶ Sodium Channel Blockers and Activators

Anticancer Characteristics of Gold(III) Complexes

▶ Gold(III) Complexes, Cytotoxic effects

Anticancer Drug

▶ Gold(III), Cyclometalated Compound, Inhibition of Human DNA Topoisomerase IB

Anticancer Metallodrugs

▶ Zinc, Metallated DNA-Protein Crosslinks as Finger Conformation and Reactivity Probes

Antimicrobial Action of Silver

▶ Silver as Disinfectant

Antimicrobial Action of Titanium Dioxide

▶ Titanium Dioxide as Disinfectant

Antimony, Impaired Nucleotide Excision Repair

Claudia Großkopf[1] and Andrea Hartwig[2]
[1]Department Chemicals Safety, Federal Institute for Risk Assessment, Berlin, Germany
[2]Department Food Chemistry and Toxicology, Karlsruhe Institute of Technology, Karlsruhe, Germany

Synonyms

Inhibition of DNA repair

Definition

Antimony interferes with the repair of a specific DNA lesion which is induced by UVC radiation. The nucleotide excision repair, a major DNA repair pathway, is responsible for the removal of this lesion and is accomplished by the concerted interaction of more than 30 proteins. Two of these proteins, XPE and XPA, both playing a crucial role in the recognition of the DNA lesions, appear to be affected by antimony via different mechanisms.

Antimony

Essential biological functions for the metalloid antimony have not been identified so far. Nevertheless, the toxicity is in the center of interest, especially since antimony shares toxicological features with arsenic, a human carcinogen. Epidemiological data indicate that antimony might be carcinogenic in humans, too, sufficient evidence, however, exists only in experimental animals. Thus, in two inhalation studies, an increased number of lung tumors were observed in rats after long-term exposure toward antimony trioxide dust (for review see De Boeck et al. 2003).

The toxicity of arsenic and antimony highly depends on their oxidation state: The trivalent species are much more cytotoxic than the pentavalent ones and cause DNA damage while the pentavalent do not. Since both elements are not mutagenic, it is supposed that the trivalent species do not directly interact with DNA but mediate their genotoxic properties via an indirect mechanism such as inhibition of DNA repair. Since DNA lesions are continuously generated, the inhibition of their repair gives rise to an accumulation of DNA damage, too. Several toxic metals such as cadmium, cobalt, or arsenic are known to inhibit the repair of DNA lesions and antimony seems to act in this way as well. In the case of antimony, an inhibition of the repair of DNA double strand breaks was shown, even though at comparatively high concentrations (reviewed in Beyersmann and Hartwig 2008).

A striking feature of the trivalent species is their high affinity toward thiol groups of peptides and proteins. It has been therefore speculated that antimony might interact with proteins involved in the DNA repair process (Schaumlöffel and Gebel 1998; De Boeck et al. 2003).

The Nucleotide Excision Repair Pathway

In response to various types of DNA lesions, highly specialized repair systems have evolved to maintain genomic stability. One of them, the nucleotide excision repair (NER) pathway, is responsible for the removal of bulky, helix-distorting DNA lesions typically induced by environmental mutagens such as UVC radiation or benzo[a]pyren. The removal of these lesions comprises several steps including lesion

Inhibition of Nucleotide Excision Repair

Diverse methods exist to measure the removal of lesions by nucleotide excision repair. UVC radiation is frequently used as a model. It specifically generates two kinds of lesions: the 6-4 photoproducts and the cyclobutane pyrimidine dimers both formed by covalent binding between two adjacent pyrimidine bases. An elegant way to detect these lesions is immunofluorescence labeling: Cultured human cells previously radiated with UVC are fixed in formaldehyde solution and exposed to antibodies directed against the respective lesion. The fluorescence of the antibodies is finally visualized under the microscope and quantified with special software. By choosing different repair times (the time between irradiation of the cells with UVC and fixation), the decline in signal intensity, respectively lesion number, can be traced. With respect to the well-documented inhibitory effect of arsenic on nucleotide excision repair, a possible impact of antimony was of high interest. Therefore, the test system was used to quantify the removal of lesions in the presence of the trivalent antimony compound $SbCl_3$. Interestingly, the removal of the 6-4 photoproducts was not affected while the removal of the cyclobutane pyrimidine dimers was. In comparison to cells not treated with antimony, significantly, more pyrimidine dimers remained in the presence of $SbCl_3$ 24 h after lesion induction with UVC, consistent with a repair inhibition of up to 50 %. Most notably, the effect on the repair of cyclobutane pyrimidine dimers was already observed at noncytotoxic concentrations of antimony.

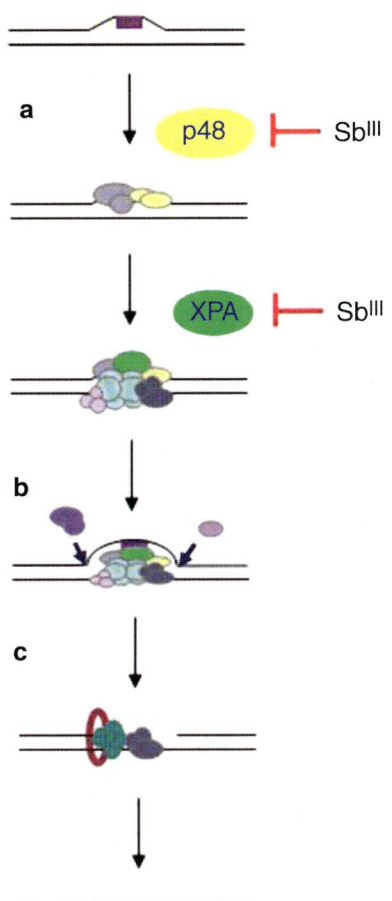

Antimony, Impaired Nucleotide Excision Repair, Fig. 1 Molecular targets for trivalent antimony within nucleotide excision repair, which comprises of (**a**) lesion recognition and formation of a pre-incision complex, (**b**) removal of the damaged DNA strand by dual incision up- and downstream of the DNA lesion and (**c**) gap-filling by polymerization and ligation

recognition, excision of the damaged oligonucleotide, polymerization, and ligation of the new fragment (Fig. 1).

Altogether more than 30 proteins are involved in this process, sequentially assembling and disassembling at the lesion site. Of special importance are the so-called XP proteins A-G. Their name is derived from the severe human genetic disorder xeroderma pigmentosum, which is caused by defects in these proteins and is characterized by extreme UV sensitivity and enhanced risk for skin cancer (Friedberg et al. 2006).

Lesion Recognition

The lesion-specific impact of antimony might be explained by an interference with lesion recognition, a very crucial step in nucleotide excision repair. Most NER lesions present in nontranscribed DNA regions are directly recognized by the repair protein XPC, which binds to the DNA damage and recruits further repair proteins. In case of UVC-induced DNA lesions, however, a further protein complex, named UV-DDB complex, appears to be involved in this process and acts as an initial sensor. Especially the cyclobutane pyrimidine dimers, which cause only a very subtle distortion of the DNA helix, are recognized and repaired more efficiently in the presence of this complex. It has been shown that XPC does not bind to the pyrimidine dimers in the absence of XPE, which is part of the UV-DDB complex, whereas the recognition of the 6-4 photoproducts is not impaired. Due to its tight transcriptional and proteasomal regulation, XPE is assumed to be a limiting factor for the recognition of these lesions (Sugasawa 2010). Actually, quantification of the cellular XPE level revealed that treatment with trivalent antimony caused a decrease in the amount of XPE protein which was due to a reduction of the XPE gene expression. Also, antimony diminished the UVC-induced expression of XPE. In contrast to that, no impact was observed on the XPC protein level.

Experimental results indicate that besides XPE, also XPA might become limiting for the repair of cyclobutane pyrimidine dimers (Cleaver et al. 1995; Shell et al. 2009). Exposure toward trivalent antimony did not alter the expression and protein level of XPA; nevertheless, antimony impaired the association of XPA at the DNA lesion. The association of XPA was examined by generating local spots of DNA damage via UVC irradiation through a porous filter. Labeling of the XPA protein with fluorescent antibodies revealed a diminished assembly of the protein at the lesion site 10 min after irradiation in the presence of antimony trichloride. After 60 min, a time point at which half of the protein normally had already dissociated from the lesion, more XPA remained at the DNA lesion when co-treated with antimony, indicating a delayed association of XPA. Again, no effect was observed on XPC, which binds to the damaged DNA prior to XPA. Also an impairment of XPE as underlying mechanism can be ruled out since effects on XPA were observed after a short treatment with antimony for only 2 h while effects on the XPE protein level were restricted to later time points.

Zinc-Binding Motif of XPA

Like many other proteins involved in maintaining genomic integrity, e.g., transcription factors and repair proteins, XPA contains a zinc-binding motif in its structure. Zinc-binding motifs are formed by cysteine or histidine side chains coordinating one or more zinc ions and mediate protein-protein interactions or DNA binding. Metal ions known for their affinity toward thiols and imidazoles are able to interact with zinc-binding motifs and thus pose a risk for the correct conformation and function of the protein (Hartwig 2001). With respect to these data, the zinc-binding domain of XPA appears to be a promising target for trivalent antimony species, too.

A direct interaction of antimony with the repair protein was investigated by measuring zinc release from a 37 amino acids containing peptide saturated with zinc, which resembles the zinc-binding domain of human XPA. Liberated zinc was quantified via formation of a colored complex with 4-(2-pyridylazo)-resorcinol (PAR) and spectrometric detection. Zinc release was provoked in a dose-dependent manner already at a less than equimolar ratio of antimony trichloride and the peptide. In great excess, the antimony compound was even as effective as hydrogen peroxide which was used as positive control. Although no direct conclusions can be drawn from the isolated zinc-binding domain to what might happen to the whole protein in the cell, these results support the assumption that antimony interacts with XPA and impairs its function within nucleotide excision repair.

Conclusions

DNA repair inhibition is a common feature of toxic metals and offers at least one explanation for their mutagenicity and carcinogenicity. Regarding first results on nucleotide excision repair, this also seems to be true for antimony. Remarkable is, however, the lesion specificity of the repair inhibition observed in case of UVC-induced DNA damage. This might be explained by two different mechanisms affecting lesion recognition (Fig. 1). On the one hand, the association of the repair protein XPA is altered in the presence of trivalent antimony. With respect to its high affinity toward thiol and imidazole groups, antimony may directly interact with the zinc-binding domain of XPA and impair its function within nucleotide excision repair; investigations with the isolated zinc-binding domain of XPA strongly support this assumption. On the other hand, the expression of XPE, a limiting factor for the recognition of UVC-induced DNA lesions, is diminished after treatment with antimony. Interestingly, XPE is transcriptionally regulated by the tumor suppressor protein p53, which also contains a zinc-binding motif. It is known from literature that the conformation of the protein as well as its transcriptional activity is sensitive toward zinc-chelating compounds, oxidation, and also metals as demonstrated for cadmium (Meplan et al. 1999). Therefore, it can be assumed that although different target proteins within lesion recognition might be responsible for the inhibitory effect of antimony on NER, both interactions may be due to the interaction with thiol and imidazole groups of zinc-binding motifs. This issue needs to be further investigated.

Cross-References

▶ Arsenic
▶ Zinc-Binding Proteins, Abundance
▶ Zinc-Binding Sites in Proteins
▶ Zinc Cellular Homeostasis

References

Beyersmann D, Hartwig A (2008) Carcinogenic metal compounds: recent insight into molecular and cellular mechanisms. Arch Toxicol 82:493–512

Cleaver JE, Charles WC, McDowell ML, Sadinski WJ, Mitchell DL (1995) Overexpression of the XPA repair gene increases resistance to ultraviolet radiation in human cells by selective repair of DNA damage. Cancer Res 55:6152–6160

De Boeck M, Kirsch-Volders M, Lison D (2003) Cobalt and antimony: genotoxicity and carcinogenicity. Mutat Res 533:135–152

Friedberg EC, Walker GC, Siede W, Wood RD, Schultz RA, Ellenberger T (2006) DNA repair and mutagenesis. ASM Press, Washington, DC

Hartwig A (2001) Zinc finger proteins as potential targets for toxic metal ions: differential effects on structure and function. Antioxid Redox Signal 3:625–634

Meplan C, Verhaegh G, Richard MJ, Hainaut P (1999) Metal ions as regulators of the conformation and function of the tumor suppressor protein p53: implication for carcinogenesis. Proc Nutr Soc 58:565–571

Schaumlöffel N, Gebel T (1998) Heterogeneity of the DNA damage provoked by antimony and arsenic. Mutagenesis 13:281–286

Shell SM, Li Z, Shkriabai N, Kvaratskhelia M, Brosey C, Serrano MA, Chazin WJ, Musich PR, Zou Y (2009) Checkpoint kinase ATR promotes nucleotide excision repair of UV-induced DNA damage via physical interaction with xeroderma pigmentosum group A. J Biol Chem 284:24213–24222

Sugasawa K (2010) Regulation of damage recognition in mammalian global genomic nucleotide excision repair. Mutat Res 685:29–37

Antimony, Physical and Chemical Properties

Fathi Habashi
Department of Mining, Metallurgical, and Materials Engineering, Laval University, Quebec City, Canada

Antimony is a metalloid; its outermost electrons are not free to move in the crystal structure because they are fixed in position in a covalent bond. It has metallic luster but is brittle with no useful mechanical properties. Sulfide ores with antimony contents between 5% and 25% are roasted to give volatile Sb_2O_3, which can be reduced directly to the metal. The oxide forms between 290°C and 340°C in an oxidizing atmosphere:

$$2Sb_2S_3 + 9O_2 \rightarrow 2Sb_2O_3 + 6SO_2.$$

If too much oxygen is available, the nonvolatile antimony (V) oxide is formed:

$$Sb_2S_3 + 5O_2 \rightarrow Sb_2O_5 + 3SO_2.$$

Therefore, oxygen supply must be kept low to inhibit its formation. Metallic antimony may form when Sb_2O_3 reacts with the sulfide:

$$2Sb_2O_3 + Sb_2S_3 \rightarrow 6Sb + 3SO_2.$$

Antimony metal is used mainly in alloys with lead or other metals. Nearly 50% of the total demand is accounted for by storage batteries, power transmission devices, communications equipment, type metal, solder, and ammunition. The compounds of antimony have a wide range of industrial uses, including uses in flame retardants, industrial chemicals, rubber, plastics, ceramics, and glass.

Physical Properties

Atomic number	51
Atomic weight	121.76
Relative abundance in the Earth's crust, %	1×10^{-4}
Isotopes	121, 123
Density	
Solid at 20°C, g/cm^3	6.688

(continued)

Liquid at 630.5°C, g/cm³	6.55
Melting point, °C	630.5
Boiling point at 101.3 kPa, °C	1,325
Heat of fusion, kJ/mol	10.49
Tensile strength, N/mm²	10.8
Modulus of elasticity, N/mm²	566
Hardness, on the Mohs scale	3
Crystal form	Rhombohedral
Surface tension, mN/m	
Solid at 432°C	317.2
Liquid at 630°C	349
Liquid at 1,200°C	255
Molar heat capacity at 630.5°C, $J\ mol^{-1}\ K^{-1}$	
Solid	30.446
Liquid	31.401
Coefficient of linear expansion [0–100°C], $°C^{-1}$	10.8×10^{-6}
Thermal conductivity, $W\ m^{-1}\ K^{-1}$	
At 0°C	18.51
At 100°C	16.58

There are unstable forms of metallic antimony: yellow antimony, black amorphous antimony, and what is known as explosive antimony. Yellow antimony is formed when air or oxygen is passed through liquid stibine. Black amorphous antimony is obtained by rapidly cooling antimony vapors and is also formed from yellow antimony at −90°C. At room temperature, black antimony slowly reverts to metallic antimony, and at 400°C, reversion is spontaneous. Black amorphous antimony ignites spontaneously in air. Explosive antimony is obtained by electrolysis of antimony (III) chloride solution in hydrochloric acid at a high current density with an antimony anode and a platinum cathode. It consists of black amorphous antimony contaminated with antimony trichloride. In the vapor state, antimony exists as Sb_4; at higher temperature, it splits into Sb_2.

Chemical Properties

When antimony loses its outermost electrons, it forms trivalent compounds, as in $SbCl_3$, or pentavalent, as in $SbCl_5$. It can also add electrons in the outermost shell forming trivalent compounds as in SbH_3. Antimony tetroxide, Sb_2O_4, is according to x-ray diffraction a double oxide, Sb_2O_3 and Sb_2O_5, or antimony antimonate, $Sb^{III}Sb^{V}O_4$. Pure antimony does not change in air at room temperature, and it is not tarnished in humid air or pure water. If heated in air, the molten metal ignites. Above 750°C, steam oxidizes liquid antimony to antimony trioxide, and hydrogen is evolved. Fluorine, chlorine, bromine, and iodine react violently with antimony at room temperature to form trihalides. Antimony (III) sulfide is the product of the reaction with sulfur, hydrogen sulfide, or dry sulfur dioxide. It is soluble in alkali sulfide solution.

Antimony is resistant to concentrated hydrofluoric, dilute hydrochloric, and dilute nitric acids. It is readily soluble in a mixture of nitric and tartaric acids and in aqua regia. At room temperature, it is not attacked by dilute or concentrated sulfuric acid. It is attacked at 90–95°C by concentrated sulfuric acid, and sulfur dioxide is evolved. Pure antimony is resistant to solutions of ammonium and alkali-metal hydroxides and to molten sodium carbonate. If heated to redness, it reacts with molten sodium or potassium hydroxide to form hydrogen gas and antimonites. Poisoning with antimony and its compounds can result from exposure to airborne particles in the workplace.

References

Habashi F (2001) Arsenic, antimony, and bismuth production. In: Encyclopedia of materials: science & technology, pp 332–336

Hanusch K, Herbst K-A, Rose G, Wolf HU (1997) In: Habashi F (ed) Handbook of extractive metallurgy. Wiley, Weinheim, pp 823–844

Antimony-Based Therapy of Leishmaniases, Molecular and Cellular Rationale

Gianni Colotti and Andrea Ilari
Institute of Molecular Biology and Pathology,
Consiglio Nazionale delle Ricerche, Rome, Italy

Synonyms

Pentavalent antimonial therapy against leishmaniases, molecular bases

Definition

Pentavalent antimonials have been used for decades as first-line compounds against leishmaniasis. The search for the molecular basis of antimony-based treatment against the parasite has long time eluded all investigators' efforts. In the last two decades, many causes of both antimonials action and resistance toward these drugs on a molecular level have been identified. Both mechanisms of action and of resistance are multifactorial, the first depending on the effect on different targets, and the second on different strategies used by the parasite to escape the killing actions of antimonials.

Leishmaniasis

The Leishmaniases are a group of vector-borne diseases caused by the infection with protozoans belonging to the at least 21 species of parasite of the genus Leishmania. These include the *Leishmania (L.) donovani* complex with two main species (*L. donovani* and *L. infantum*), the *L. mexicana* complex with three main species (*L. mexicana, L. amazonensis*, and *L. venezuelensis*), *L. tropica, L. major, L. aethiopica*, and the subgenus *Viannia (V.)* with four main species (*Leishmania (V.) braziliensis, L. (V.) guyanensis, L. (V.) panamensis*, and *L. (V.) peruviana*). Animals and humans themselves are the reservoirs of the disease. Leishmaniasis is transmitted through the bite of phlebotomine sand flies infected with the protozoan. The parasite can then be internalized via macrophages in the liver, spleen, and bone marrow of the mammalian host. Leishmania parasites are dimorphic organisms, that is, with two morphological forms in their life cycle: promastigotes in the digestive organs of the insect vector, and amastigotes in the mononuclear phagocytic system of the mammalian host.

Leishmaniasis affects about 13 million people in at least 88 countries, and is a poverty-related disease. The clinical manifestations range from cutaneous leishmaniasis (CL), characterized by ulcers which can heal spontaneously, often leaving scars, to the destruction of cutaneous and subcutaneous tissues in mucocutaneous forms, to the involvement of liver and many organs in the most severe form, visceral leishmaniasis (VL) or kala-azar, caused by *L. donovani* and *L. infantum*. In VL, patients develop fever, splenomegaly, hypergammaglobulinemia, and pancytopenia; VL is almost always fatal if untreated, and causes about 500,000 cases and 80,000 deaths per year.

Anti-Leishmanial Drugs

No vaccine is available yet against leishmaniasis, and chemotherapy is the main option against the disease. The existing weapons against leishmaniasis are limited, because companies invest only a limited budget on the research against neglected diseases, which affect mostly people in the developing world. Pentavalent antimonials are in use since the first half of the twentieth century, and constitute the first-line drugs in many countries. However, the increase of clinical resistance, especially in the region of Bihar (India) prevents the use of antimonials in these regions. Pentavalent antimonial drugs are administered parenterally, at doses of 20–30 mg of Sb/(kg day) for at least 20 days, and present several side effects, which include nausea, vomiting, weakness and myalgia, abdominal colic, diarrhea, skin rashes, and hepatotoxicity. Pancreatitis and cardiotoxicity can be important problems of antimonial treatment of leishmaniasis.

Second-line drugs include pentamidine and lipid formulations of amphotericin B, a less toxic antifungine. Though amphotericin B preparations have high cure rate against leishmaniasis, their high cost impairs their extensive use in developing countries. Miltefosine, originally developed as anticancer drug, is the first oral drug against leishmaniasis; it is effective also against antimony-resistant leishmaniases, but high costs, severe side effect, long treatment schemes, long half-life, and potential teratogenicity limit their use. Paromomycin, an aminoglycoside antibiotic, is a promising drug that is currently in clinical trial; it is cheap and effective, but needs long-term parenteral treatments for VL, and it may generate resistance (Croft and Coombs 2003).

Antimony-Based Drugs Against Leishmaniasis

Potassium antimony tartrate (tartar emetic), a trivalent antimonial, was used against leishmaniasis for the first time by Vianna in 1913, but Sb(III) compounds have low efficacy and stability, while being highly toxic, with serious side effects.

Antimony-Based Therapy of Leishmaniases, Molecular and Cellular Rationale, Fig. 1 Structural formula in aqueous solution of 1:1 Sb-NMG and Sb-SG complexes in Glucantime (**a**) and Pentostam (**b**), according to Frézard et al. (2008)

Pentavalent antimonials have been used against leishmaniasis since the 1920s, when Brahmachari successfully used urea stibamine against VL. Subsequently, other Sb(V) compounds have been synthesized and used, such as stibosan, neostibosan, antimony gluconate (Solustibosan) in 1937; and sodium stibogluconate (SSG, or Pentostam) in 1945.

Nowadays the two main pentavalent antimonials in clinical use are the highly water-soluble complexes of Sb(V) with N-methyl-D-glucamine (NMG, in Glucantime) and sodium gluconate (SG, in Pentostam). The structure of these complexes has been unknown for half a century, because of their amorphous state. Both compounds are a mixture of oligomeric structures with the general formula (sugar-Sb)n-sugar, with the prevalence of 1:1 Sb-NMG and Sb-SG complexes in diluted samples. They are mostly zwitterionic in solution (Fig. 1).

Antimony Metabolism in Leishmania

Both Sb(V) and Sb(III) accumulate, in antimony-sensitive and antimony-resistant strains of different *Leishmania* species. Entry of antimonials in the parasite appears to take place via different routes for Sb(V) and Sb(III). Sb(III) uptake is competitively inhibited in parasite cells by As(III), belonging to the same group, and depends on a parasitic transporter, aquaglyceroporine (AQP1), involved in the entry of several metabolites.

The route of entry of pentavalent antimonials is currently unknown. Since gluconate competitively inhibits the uptake of Sb(V), while As(V) and phosphate, which are known to enter via a phosphate transporter, do not compete with Sb(V) transport, pentavalent antimonials are believed to enter *Leishmania* cells by means of a transporter which recognizes the sugar moiety of the drug.

Sb(V)-containing compounds are more effective and about tenfold less toxic than trivalent antimony derivatives. According to the recent literature, pentavalent antimonials are considered to be mainly prodrugs. Sb(V) can be reduced by the host macrophages or by the parasite to the more toxic Sb(III) form, both enzymatically and non-enzymatically (Ashutosh et al. 2007; Haldar et al. 2011).

Leishmania parasites reduce Sb(V) to Sb(III) in a stage-specific fashion. Amastigotes reduce Sb(III) more efficiently than promastigotes, and are therefore more susceptible to pentavalent antimonials.

Nonenzymatic reduction is carried out by thiols. Reduced glutathione, the main thiol present in the host cytosol, can reduce Sb(V) to Sb(III) within the phagolysosome containing *Leishmania* amastigotes. In vitro, such conversion is quite slow, although pentavalent antimony reduction is favored by the acidic pH and temperature of the macrophage organelle. Cysteine and cysteinyl-glycine, which are the predominant thiols within lysosomes, also reduce Sb(V) to Sb(III) at 37 °C, with a higher rate, compared with the glutathione-mediated reaction. However, reduction of Sb(V) to Sb(III) in the macrophage cannot be very high in vivo, because Sb(III) is very toxic, and concentrations of 25 μg/mL kill 50% of the macrophages. Probably the majority of the highly toxic Sb(III) is generated inside the parasite.

Antimony-Based Therapy of Leishmaniases, Molecular and Cellular Rationale, Fig. 2 Structure of reduced trypanothione (T(SH)$_2$), formed by conjugation of two glutathione molecules (in *black*, with the two -SH groups colored *green*) and a spermidine (*red*). In the oxidized form of trypanothione, the two -SH groups are oxidized to form a -S-S- disulfide bridge

The most important thiol involved in Sb(V) reduction in *Leishmania* is trypanothione (N1,N8-bis (glutathionyl)spermidine, T(SH)$_2$), which is the predominant thiol within the parasite (Fig. 2).

T(SH)$_2$ has been found to rapidly reduce Sb(V) to Sb(III), especially under acidic conditions and at slightly elevated temperature. Such reaction occurs preferentially in the intracellular amastigotes, which have a lower intracellular pH and live at a higher temperature than promastigotes, although promastigotes contain higher intracellular amounts of T(SH)$_2$ and glutathione than amastigotes.

Two enzymes have been demonstrated to catalyze the reduction of Sb(V) in *Leishmania*. The thiol-dependent reductase TDR1 is a parasite-specific tetrameric protein found highly abundant in the amastigotes, which uses glutathione as a source of reducing power to reduce Sb(V) to Sb(III). LmACR2, homologue to yeast arsenate reductase, identified in *L. major*, is also able to reduce Sb(V), by using reduced glutathione and glutaredoxin as cofactors. LmACR2 is inhibited by trivalent antimony and As(III), and its overexpression increases sensitivity to Pentostam.

Reduction of Sb(V) to Sb(III) is a critical event for the parasite, and a decrease of reductive activity in the parasite can be associated with antimony resistance.

Antimony Targets

Sb(III), as other thiophilic metals such as As(III) and Bi(III), strongly bind thiols. When sulfhydryls bind these metals, they form completely deprotonated and highly stable thiolate anions, whose nucleophilicity is attenuated upon formation of metal complexes with high thermodynamic stability, which dissociate with very slow rates.

The mechanism of action of Sb(III) against leishmaniasis is mainly due to its interaction with low-molecular thiols, cysteine-containing peptides and, in particular, thiol-dependent enzymes.

Trypanothione Metabolism: T(SH)$_2$ and Trypanothione Reductase

The trypanosomatids (*Leishmania* spp., *Trypanosoma* spp., and other parasites) possess a unique redox metabolism, based on the trypanothione and on thiol-dependent enzymes, which is the main target of pentavalent antimonials (Fig. 3) (Colotti and Ilari 2011).

The trypanothione-dependent redox metabolism is considered as a highly potential drug target against leishmaniases, because of the absence of the T(SH)$_2$ system in mammals, the lack of a functional redundancy within the redox metabolism, and the sensitivity of Trypanosomatids against oxidative stress (i.e., most enzymes of the pathway are essential for the survival of all *Leishmania* species). In addition, *Leishmania* parasites lack mammalian redox enzymes such as glutathione reductase, selenocysteine-containing glutathione peroxidases and thioredoxin reductase, and catalase.

T(SH)$_2$ is synthesized by trypanothione synthetase, by conjugation of two glutathione molecules with a spermidine, and its concentration is very high (1–2 mM) in all parasite forms and growth phases. T(SH)$_2$ is the electron donor in several metabolic

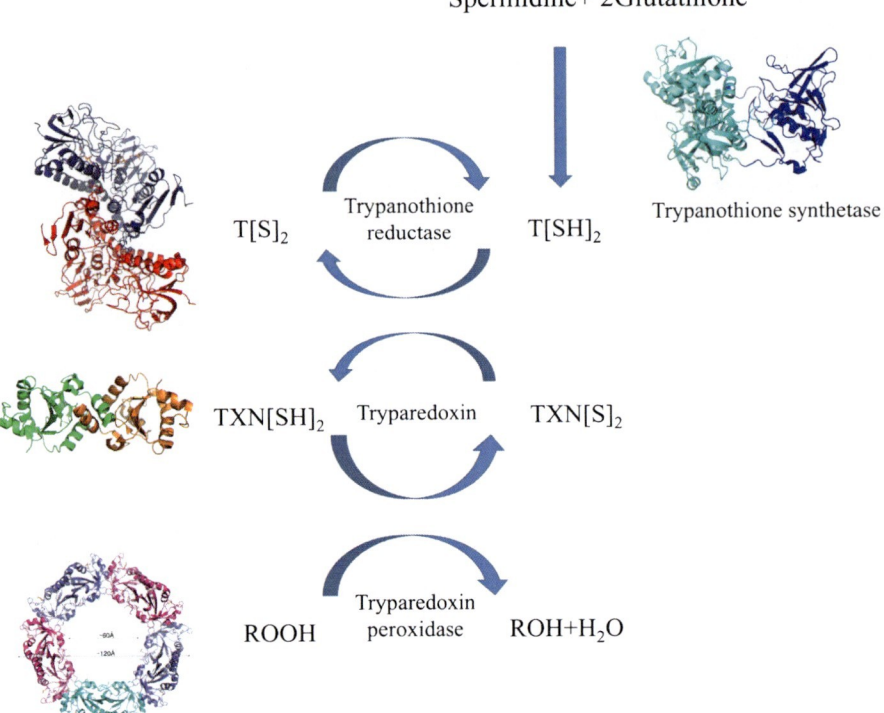

Antimony-Based Therapy of Leishmaniases, Molecular and Cellular Rationale, Fig. 3 The trypanothione metabolism in *Leishmania*. Trypanothione is synthesized by Trypanothione synthetase, is kept reduced by Trypanothione reductase (TR), and provides reducing equivalents via Tryparedoxin to Tryparedoxin-dependent peroxidase for the reduction of the peroxides

pathways, where it delivers reducing equivalents to oxidized glutathione, dehydroascorbate, and in particular to the dithiol tryparedoxins (TXN1, cytosolic; and TXN2, mitochondrial) which, in turn, reduce different enzymes and participates in hydroperoxides elimination, together with the tryparedoxin-dependent peroxidases (TDPXs).

Tryparedoxins and TDPXs are homologues of thioredoxin and thioredoxin peroxidase, respectively. Tryparedoxins, reduced by T(SH)$_2$, operate as sources of reducing electrons during removal of peroxides by the TDPXs, which catalyze the reduction of H$_2$O$_2$ and organic hydroperoxides to water or alcohols, thereby detoxifying the *Leishmania* parasites.

T(SH)$_2$ is kept reduced by trypanothione reductase (TR), thereby maintaining the oxidoreductive balance in *Leishmania* parasite, protecting the parasites from oxidative damage and toxic heavy metals, and allowing delivery of reducing equivalents for DNA synthesis. The T(SH)$_2$/TR system replaces many of the antioxidant and metabolic functions of the glutathione/glutathione reductase and thioredoxin/thioredoxin reductase systems present in mammals and is necessary for the survival of all *Leishmania* species.

TR, as well as glutathione reductase, is a member of the family of FAD-dependent NAD(P)H oxidoreductases, that is, dimeric flavoenzymes that catalyze the transfer of electrons between pyridine nucleotides and dithiol compounds. Each TR monomer is formed by three different domains, such as the FAD-containing domain (residues 1–160 and 289–360, *L. infantum* TR numbering), the NADPH-binding domain (residues 161–289), and the interface domain (residues 361–488). TR and human glutathione reductase share structural and mechanistic similarities, although they are mutually exclusive for their dithiol substrates. Substrate specificity is largely determined by only five amino acids at the substrate binding site: the active site pocket of TR is wider, more hydrophobic, and negatively charged than that of glutathione reductase, and can, therefore, select oxidized trypanothione, bulkier and more positively charged, with respect to oxidized glutathione.

The first step of the reaction catalyzed by TR is the binding of NADPH, which transiently reduces the flavin. Reduction of the Cys52-Cys57 disulfide in the active site by the reduced flavin follows, with the formation of a short-lived covalent intermediate with Cys57, and subsequently of a stable charge-transfer

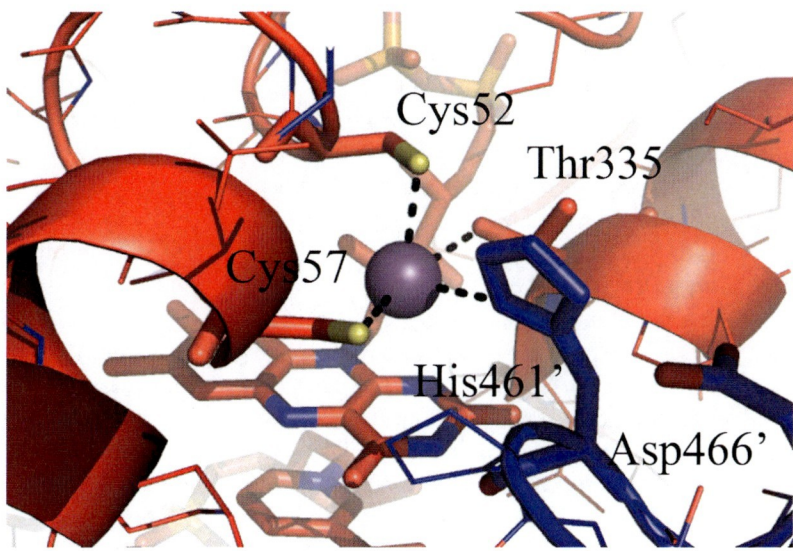

Antimony-Based Therapy of Leishmaniases, Molecular and Cellular Rationale, Fig. 4 Antimony (Sb(III), in *purple*) binds to the active site of Trypanothione reductase. Sb(III) inhibits TR activity by binding to the two active cysteine residues (Cys52 and Cys57), to Thr335 and to His461′, belonging to the second monomer of the enzyme dimer. The residues belonging to the two TR monomers are drawn in *red* (first monomer) and in *blue* (second monomer) (Authors: Colotti Gianni, Ilari Andrea)

complex between the flavin and the Cys57 thiolate. Following the generation of the charge-transfer complex, NADP$^+$ dissociates and is replaced by another NADPH. Cys52, which is activated similarly to serine and cysteine proteases by the His461′-Glu466′ pair (belonging to the second subunit of the TR dimer) can then react with the oxidized trypanothione to produce a mixed disulfide, followed by attack of Cys57 on Cys52 and full Trypanothione dithiol reduction.

Fairlamb and collaborators demonstrated that the antimonials interfere with the trypanothione metabolism in two ways: Sb(III) can form stable complexes with T(SH)$_2$ and glutathione, in 1:1 and 1:3 antimony-thiol ratios, respectively, or ternary Sb(III)- T[SH]$_2$-glutathione complexes. These complexes can be sequestered inside a vacuole or are rapidly extruded by ATP-binding cassette (ABC) transporters, thereby decreasing the thiol buffering capacity of the parasite. Antimonials also strongly inhibit TR in vivo and in vitro, resulting in the lack of reduction and regeneration of the thiols and the consequent accumulation of the disulfide forms of trypanothione and glutathione. Together, these two mechanisms impair the thiol redox potential and all thiol-dependent processes in both amastigote and promastigote stages of the life cycle of *Leishmania* (Wyllie et al. 2004).

Colotti, Ilari et al. solved X-ray crystal structure of the reduced *Leishmania* TR in complex with Sb(III) and NADPH, thus disclosing the molecular basis of TR inhibition (Baiocco et al. 2009).

Sb(III) inhibits TR by binding strongly to the catalytic pocket of the enzyme, engaging the residues involved in the catalytic mechanism in the formation of a complex. Sb(III) binds to Cys52, to Cys57, to Thr335, and to His461′ of the twofold symmetry–related subunit, thus disallowing the hydride transfer from the protein to the trypanothione and therefore, T(SH)$_2$ reduction (Fig. 4).

Trivalent antimony may also bind to and inhibit other enzymes of the trypanothione metabolism, but only preliminary results are available on the interaction with tryparedoxin and TDPXs and their possible inhibition.

The T(SH)$_2$ metabolism, compromised by antimonials, is essential for many different cellular functions, such as the detoxification of peroxides via the tryparedoxin-dependent peroxidase system, the biosynthesis of deoxyribonucleotides via ribonucleotide reductase, the detoxification of methylglyoxal to D-lactate via T(SH)$_2$-dependent glyoxalase I and glyoxalase II, and the detoxification of xenobiotics via trypanothione *S*-transferase. All these functions are strongly impaired upon pentavalent antimonials administration. In addition, oxidized trypanothione inhibits ribonucleotide reductase that is essential for deoxyribonucleotides synthesis.

Further, survival of amastigotes within host macrophages requires them to withstand oxidant stress from potentially lethal macrophage responses, such as increased production of superoxide, hydrogen peroxide, hydroxyl radicals, and nitric oxide.

Impaired antioxidant defenses caused by inhibition of trypanothione reductase in amastigotes and by thiol efflux are probably the main determinants of parasite's death upon antimony treatment.

Zinc Finger Proteins

Other targets for antimonials have been identified by Frézard, Demicheli, and coworkers. They have shown that Sb(III) binds to CCHC and CCCH zinc finger peptide models, determining the ejection of Zn(II) and stabilizing the structure of the motif. These structural elements are contained in many proteins, interacting with nucleic acids and proteins and involved in diverse functions, including DNA recognition, RNA packaging, protein folding and assembly, lipid binding, transcriptional activation, cell differentiation and growth, and regulation of apoptosis. Several zinc finger proteins containing CCCH or CCHC motifs are involved in different cellular functions in trypanosomatids. In *L. major*, the protein HEXBP, containing nine CCHC motifs, binds to the hexanucleotide repeat sequence found in the intervening region of the GP63 gene cluster, the most abundant surface glycoprotein of the parasite, and it is likely to be involved in DNA replication, structure, and repair. The interaction between Sb(III) and these motifs may, therefore, modulate or mediate the pharmacological action of antimonial drugs (Frézard et al. 2012).

Other Targets of Sb(III)

Treatment of *L. infantum* amastigotes with Sb(III) induces an increase in intracellular Ca^{2+} via nonselective cation channels in the host and in the parasite, and apoptosis (programmed cell death), with the presence of DNA fragmentation, a marker of late events of apoptosis.

Studies on the effect of arsenite, chemically similar to Sb(III), show important alterations in the parasites microtubules, and possibly in several signal-transduction pathways, upon As(III) treatment. In particular, expression of α- and β-tubulin in promastigotes is altered upon arsenite administration, while it remains unaltered in arsenite-resistant parasites. Microtubule polymerization can be regulated by α- and β-tubulins phosphorylation, which is highly increased in the arsenite-resistant mutant.

Sb(V) Targets

Sb(V) has also been shown to possess intrinsic anti-leishmanial activity, that is, to have drug-like effects without the need for reduction to Sb(III). Pentavalent sodium stibogluconate inhibits glycolysis and fatty acid beta oxidation, but the specific targets in these pathways have not been identified.

Sodium stibogluconate, but not Sb(III), specifically inhibits type I DNA topoisomerase from *L. donovani*, thus inhibiting DNA unwinding and cleavage. The formation of this metal-protein complex has been shown to correlate with the levels of in vivo sensitivity toward antimonial drugs.

Formation of Sb(V)-ribonucleoside complexes, both in the ratio of 1:1 and 1:2 has also been reported. NMR studies suggest that antimony binds to ribose hydroxyl groups, probably via ring chelation at C2′ and C3′. Formation of such complex is kinetically favored in acidic conditions, such as those present in the macrophage phagolysosome, while its dissociation is very slow at neutral pH. Sb(V)-ribonucleoside complexes might inhibit the *Leishmania* purine transporters, or might enter the parasite, where it may interfere with the purine salvage pathway.

Effects on the Host Macrophages: Activation of the Host Immune Responses

Antimonials also contribute to parasite elimination by activating both innate as well as adaptive host immune responses (Haldar et al. 2011).

Microbicidal molecules, like nitric oxide (NO) or hydrogen peroxide (H_2O_2) generated by the host macrophage upon antimony treatment, play a role in the leishmanicidal activity of antimonials.

Treatment with Pentostam induces reactive oxygen species (ROS) generation and NO production in murine macrophages infected by *L. donovani* amastigotes, by promoting two waves of killing. The first wave is due to the induction of ROS, while the second wave is mediated by NO generation.

ROS generation is mediated by phosphorylation of ERK via PI3K-PKC-Ras/Raf pathway, while NO production depends on activation of the PI3K/Akt pathway and p38MAPK. In particular, p38MAPK induces TNFα production, which in turn induces iNOS2 (nitric oxide synthase) expression and subsequent NO generation. Parasite killing is inhibited by treatment with antiTNFα antibodies.

Leishmania infection can increase PTPase activity, with the consequent dysregulation of PTK-dependent signaling in the macrophages. Pentostam inhibits SHP1 and SHP2 classes of PTPases, such that dephosphorylation of ERKs, a mechanism by which *Leishmania* parasites can escape activation of PI3K-PKC-Ras/Raf-ERK1/2 pathway for ROS generation and the PI3K-Akt-p38 MAPK pathway for NO generation, is not available.

A Pentostam-dependent increase of IL-12 production has been observed in macrophages: IL-12 induces Th cells to produce IFN-γ, which activates TNF-α synthesis and, consequently, NO production. Efficacy of Pentostam treatment of VL depends on the endogenous levels of IL-2, IL-4, and IL-12.

Pentavalent antimonials also contribute to activation of cell-mediated response. This effect, together with the generation of ROS and NO, potentiates the ability of macrophages to eliminate *Leishmania* parasites and also protect from relapse.

Pentostam induces T-cell response, while B-cell response is not activated. Cytotoxic T lymphocyte response is increased by pentavalent antimonials, which increase the expression and the presentation of MHC I class-mediated antigens. Th1 cells are essential for a complete antileishmanial effect of pentavalent antimonials. Patients with HIV-VL co-infections have poor response upon antimony treatment.

Resistance Toward Antimonials

Pentavalent antimonials have been used for decades for the treatment of leishmaniases, and remain the first choice drugs in Africa, Asia (except the Bihar district, in India), and South America.

In the last three decades, however, widespread resistance toward these compounds has been reported for *L. donovani*–dependent VL in the Bihar district, where in the late 1970s the original low-dose treatment based on the administration of 10 mg/kg/day for 6–10 days was found to be ineffective for 30% of the patients.

Although several revisions of antimony-based treatment recommendations have been emanated, with increases in daily regimens and in treatment length, and temporary increases in cure rates, resistance toward Sb(V)-based drugs has continued to spread, such that in some villages in the Bihar district 100% resistance toward antimonials has been observed.

Pentavalent antimonials, however, continue to be effective in the rest of the world, although small areas of resistance are spreading in small regions of Nepal and of India, surrounding Bihar. The reasons for resistance are not fully understood. Misuse and overuse of antimonial drugs and the presence of high amount of arsenic in the water have determined long-term exposure of the parasite (and of the hosts) to the metals. Another reason for the spread of resistance to antimonials is the high number of persons infected in Bihar, that makes the transmission of the parasite in that region mostly anthroponotic (humans are the reservoirs of the parasite), rather than zoonotic (in most part of the world, animals such as dogs and horses are the main reservoirs). In zoonotic transmission, drug-resistant parasites kill the animals and are eliminated preferentially, while in anthroponotic transmission, drug-sensitive parasites are eliminated while drug-resistant parasites continue to circulate, and resistance continue to develop in patients with relapses (Perry et al. 2011; Haldar et al. 2011).

Resistance can be generated in vitro by exposing the parasite cultures to increasing concentrations of drugs.

Mechanisms of Resistance Toward Pentavalent Antimonials

Many Gene Expression Profiling studies and some genomic studies have been carried out on antimony-resistant and antimony-sensitive strains of several *Leishmania* species, reporting differential expression for many different proteins, in accordance with the multiplicity of mechanisms leading to antimony resistance. At least four mechanisms of resistance to antimonial drugs have been identified (Carter et al. 2006; Ashutosh et al. 2007; Haldar et al. 2011).

1. Restricted entry of antimonials into the parasite cell AQP1, responsible for entry of Sb(III) into *Leishmania* cells, has been associated with resistance toward antimonials. Overexpression of AQP1 increases sensitivity toward both Sb(III) and pentavalent antimonials in *Leishmania* cells. Further, AQP1 has been found to be downregulated in both promastigote and amastigote stages in antimony-resistant clinical isolates and mutants.
2. Low reduction of Sb(V) to Sb(III).
   Both alterations in enzymatic and in nonenzymatic Sb(V) reduction have been associated to antimony

resistance. Transfection of LmACR2 in *L. infantum* promastigotes has been found to increase sensitivity to Pentostam in amastigotes. In addition, the levels of thiols in the infected host macrophage and in the parasite are related to antimony resistance. The expression of γ-glutamylcysteine synthetase (γ-GCS), the enzyme which catalyzes the rate-limiting step in glutathione synthesis, is downregulated in the macrophages of antimony-resistant isolates, determining the reduction of glutathione concentrations in the host macrophage and promoting a less reducing environment, with a consequent low ability to reduce Sb(V) to the more toxic Sb(III).

3. Increased expression of proteins of redox metabolism in *Leishmania*.

   Overexpression of TR has been observed in antimony-resistant strains, and high levels of tryparedoxin and TDPX, enzymes in hydroperoxide detoxification, have been reported in parasites resistant to antimonials. Such increased levels determine an increased metabolism of peroxides, and link clinical resistance with enhanced antioxidant protection.

4. Sequestration and/or efflux of the drug.

   The ABC transporter MRPA has been found to be amplified in Leishmania mutants resistant to Sb(III), Sb(V), and As(III). Its overexpression seems to confer resistance by sequestration of metal-thiol conjugates, rather than by increasing metal efflux. At least eight proteins belong to the MRP1 family and may be associated in antimony resistance dependent on thiol-based efflux, but no precise identification is available.

   Increase of metal efflux has been observed in *Leishmania* strains expressing high amounts of glutathione and trypanothione, for example, overexpressing ornithine decarboxylase, responsible for spermidine synthesis, and γ-GCS, responsible for glutathione synthesis. Increases in the intracellular level of $T(SH)_2$ induce resistance toward Sb(III), since an increased level of the Sb-trypanothione complexes can be sequestered or can be extruded.

   Other genes and proteins able to confer antimony resistance have also been identified. These enclose genes coding for proteins that belong to the superfamily of the leucine-rich repeat proteins, histone H2A, and kinetoplastid membrane protein-11 (KMP-11).

Cross-References

► Antimony, Impaired Nucleotide Excision Repair
► Antimony, Physical and Chemical Properties
► Arsenic and Aquaporins
► Arsenic
► Arsenic in Nature
► Arsenic, Mechanisms of Cellular Detoxification
► As
► Ribonucleotide Reductase

References

Ashutosh SS, Goyal N (2007) Molecular mechanisms of antimony resistance in *Leishmania*. J Mol Microbiol 56:143–153

Baiocco P, Colotti G, Franceschini S, Ilari A (2009) Molecular basis of antimony treatment in Leishmaniasis. J Med Chem 52:2603–2612

Carter KC, Hutchison S, Henriquez FL, Légaré D, Ouellette M, Roberts CW, Mullen AB (2006) Resistance of *Leishmania donovani* to sodium stibogluconate is related to the expression of host and parasite γ-glutamylcysteine synthetase. Antimicrob Agents Chemother 50(1):88–95

Colotti G, Ilari A (2011) Polyamine metabolism in *Leishmania*: from arginine to trypanothione. Amino Acids 40(2):269–285

Croft SL, Coombs GH (2003) Leishmaniasis- current chemotherapy and recent advances in the search for novel drugs. Trends Parasitol 19(11):502–508

Frezard F, Martins PS, Barbosa MCM, Pimenta AMC, Ferreira WA, de Melo JE, Mangrum JB, Demicheli C (2008) New insights into the chemical structure and composition of the pentavalent antimonial drugs meglumine antimonate and sodium stibogluconate. J Inorg Biochem 102:656–665

Frézard F, Silva H, Pimenta AMC, Farrell N, Demicheli C (2012) Greater binding affinity of trivalent antimony to a CCCH zinc finger domain compared to a CCHC domain of kinetoplastid proteins. Metallomics 4(5):433–440

Haldar AK, Sen P, Roy S (2011) Use of antimony in the treatment of Leishmaniasis: current status and future directions. Mol Biol Int 2011:1–23

Perry MR, Wyllie S, Prajapati VK, Feldmann J, Sundar S, Boelaert M, Fairlamb AH (2011) Visceral leishmaniasis and arsenic: an ancient poison contributing to antimonial treatment failure in the Indian subcontinent? PLoS Negl Trop Dis 5(9):1–5, e-1227

Wyllie S, Cunningham ML, Fairlamb AH (2004) Dual action of antimonial drugs on thiol redox metabolism in the human pathogen *Leishmania donovani*. J Biol Chem 279(38): 39925–39932

Antineoplastic

► Tin Complexes, Antitumor Activity

Antioxidant Enzymes

▶ Manganese and Catalases

Antiproliferative Characteristics of Gold(III) Compounds

▶ Gold(III) Complexes, Cytotoxic Effects

Antiproliferative Compounds

▶ Tin Complexes, Antitumor Activity

Antithrombotic

▶ Nanosilver, Next-Generation Antithrombotic Agent

Antitumor Drug – Anticancer Drug

▶ Osmium Complexes with Azole Heterocycles as Potential Antitumor Drugs

Antitumor Effects of Metal-Based Drugs

▶ Gold(III) Complexes, Cytotoxic Effects

Antiviral Metallodrugs

▶ Zinc, Metallated DNA-Protein Crosslinks as Finger Conformation and Reactivity Probes

Apoferritin, Activation by Gold, Silver, and Platinum Nanoparticles

Afolake T. Arowolo[1], Jacqueline van Marwijk[1] and Chris G. Whiteley[2]
[1]Department of Biochemistry, Microbiology & Biotechnology, Rhodes University, Grahamstown, Eastern Cape, South Africa
[2]Graduate Institute of Applied Science and Technology, National Taiwan University of Science and Technology, Taipei, Taiwan

Definition

The effect of silver, gold, and platinum nanoparticles on the ferroxidase activity of apoferritin showed a respective increase in specific activity of 110-fold, ninefold, and ninefold over the control at the molar ratio of 1,000:1 (500:1 for Au:apoferritin). Typical color change, from light yellow to orange (Ag), light yellow to a dark brown/black (Au), or light orange to dark brown (Pt) when apoferritin was mixed with $AgNO_3$, $AuCl_3$, or K_2PtCl_4 followed by sodium borohydride to afford respective M:apoferritin nanoparticle complexes in a ratio of between 250:1 and 4,000:1. These complexes were characterized by UV/Vis, inductively coupled plasma optical emission spectroscopy (ICP-OES), Fourier transform infra red (FTIR), transmission electron microscopy (TEM), and energy dispersive X-ray spectroscopy. TEM revealed that the size of nanoparticles increased as the molar ratio of metal to apoferritin increased with an average size of 3–6 nm generated with M:apoferritin molar ratios of 250:1 to 4,000:1. FTIR showed no structural changes of the apoferritin when the nanoparticles were attached to the protein.

Introduction

The multidisciplinary field of "nanotechnology" embraces applications of nanoscale(10^{-9} m) functional materials including particles in cross disciplinary studies such as health care, environmental biology, chemical and material sciences, and communications (Rao et al. 2004; Rai et al. 2009). One of the fundamental principles that sets this type of technology apart from traditional standard protocols is the fact that as the particle itself becomes smaller and smaller its association with biomacromolecules changes which, inevitably, leads to these molecules behaving differently to the native ones. The biological syntheses (Cao 2004) of noble metal nanoparticles through the bioreduction of metal salts by both prokaryotic and eukaryotic organisms (Whiteley et al. 2011; Fendler 1998) is the most cost-effective, simple, and eco-friendly process even though a limitation still remains with control of the mechanism that determines the particle size and shape (Riddin et al. 2006, 2009, 2010; Rashamuse and Whiteley 2007; Whiteley et al. 2011). A successful synthesis of platinum nanoparticles from both a eukaryote (*Fusarium oxysporum*) and a prokaryote (*Desulphovibrio*) (Riddin et al. 2006; Rashamuse and Whiteley 2007; Rashamuse et al. 2008; Govender et al. 2009, 2010), showed no specific control on their shape and size and a mixed "bag" of triangles, squares, spheres, trapezoids, and pyramids of varying size was produced. The shape-and size-controlled synthesis of functional metal nanoparticles (Konetzka 1977) has attracted great attention since these particles have unique electronic, magnetic, catalytic, and optical properties. One promising approach of size-controlled nanomaterials synthesis is a biomimetic one using organic assemblies as templates for directed fabrication. The use of a protein cavity or cage as a limited growth field for nanoparticles is not novel but serves as an ideal template to confine particle growth in a homogeneous distribution as well as a stabilizer against particle aggregation. One such cage is apoferritin, a globular protein of approximately 440 kDa in 24 identical subunits that has ferroxidase activity and catalyzes the oxidation of Fe^{2+} to Fe^{3+} in the presence of molecular oxygen as an electron acceptor. The subunits form a spherical protein shell of 12 nm with an internal aqueous cavity of 8 nm. Apart from Fe^{3+} several other zero-valent apoferritin encapsulated metal nanoparticles have been prepared within this protein cage and these include palladium, platinum, copper, cobalt, nickel, cadmium, chromium, and zinc. Apoferritin from horse spleen (HSAF) was the first protein cage to be used as a template for synthesis of inorganic nanoparticles, and it was probably the most intensively studied and best understood example. One of the most important features of this protein cage is the perforation of the protein shell by small channels (0.5 nm diameter), found at the junctions of the subunits, which allow the introduction of several metal cations and metal complexes into the cage.

Although several studies of protein-nanoparticle interactions have appeared in the past few years none have actually addressed what actually occurs at the interface between the protein and the nanoparticle. Perhaps the most promising suggestion that has materialized is an understanding of the effective "unit" as that of being a dynamic layer or "corona." The binding between nanoparticles and protein changes the structure of the latter and as the particle gets smaller not only does its interaction with the protein change but the composition of the protein itself changes. Consequently, the "nanoparticle-enzyme corona" may have totally different biological properties in comparison to a native protein or enzyme.

In view of the foregoing discussion regarding the anticipated different properties exhibited by enzymes/proteins in the presence of nanoparticles and since no reports had appeared in the literature there was special interest in the activity of apoferritin in the presence of nanoparticles. It was expected that apoferritin associated with platinum nanoparticle may have a wide range of affinities and activities dependent on the size of the nanoparticle.

This preparation of nanoparticles within a biomacromolecular apoferritin cage has lead to some unprecedented enhancement of its ferroxidase activity.

M:HSAF Nanoparticle Complex

Synthesis

M:HSAF nanoparticle complexes of silver, gold, or platinum were formed after HSAF was incubated with varying concentrations of respective metal salt [$AgNO_3$, $AuCl_3$, or K_2PtCl_4] followed by treatment with $NaBH_4$ (Deng et al. 2009). The preparation of the complex is carried out at pH 8 in order to increase the electronegativity of the interior of HSAF thereby improving electrostatic attraction between positive metal cations and the core of HSAF. The change in nanoparticle size was determined by TEM analysis and a variation in molar ratios of M:HSAF was used to determine the saturation point of HSAF in terms of the amount of metal that the protein was capable of encapsulating.

Nanoparticle formation was monitored visibly by color change from light yellow to orange (Ag), light yellow to a dark brown/black (Au) or light orange to dark brown (Pt). Color change is the initial evidence of nanoparticle formation. The intensity of color increased as the amount of metal salt added to the aqueous solution of HSAF increased. Since more particles were stabilized as the precursor platinum to HSAF increased it suggested an increase in the encapsulation of metal atoms with increasing metal concentration that was available in solution. Studies on metals nanoparticles stabilized by HASF have suggested divalent precursor metal salts were preferably accommodated within the core of HSAF. This may be due to the fact that apoferritin in its normal biological function takes up iron in its Fe^{2+} state which is later converted to Fe^{3+} within its core.

The M:HSAF complex was purified by size exclusion chromatography and the elution profile indicated a single peak coincidental with a peak of pure apoferritin which indicated that HSAF remained intact after incubation with metal salt and that the metal was bound to HSAF. Analysis with native PAGE (5 %) showed a co-migration of HSAF only and M:HSAF nanoparticles complexes implying that the metal nanoparticles were attached to HSAF.

Characterization

Synthesized nanoparticles were characterized by UV-visible spectroscopy, ICP-OES, FTIR, TEM, and EDX analysis.

The synthesized nanoparticles showed similar absorption spectra for HSAF and for all M:HSAF ratios typical of surface plasmon resonance bands consistent with metal nanoparticles encapsulated by protein shells. There was an increase in absorption maxima with increasing platinum nanoparticles around 230–280 nm due to the attachment of the platinum nanoparticles to HSAF.

With Au:HSAF nanoparticle complexes there were two absorbance maxima, one at 280 nm, corresponding to pure HSAF, and the other at 520 nm that represented the characteristic surface plasmon resonance band of a spherical gold nanoparticle <20 nm in diameter. The intensity of these two absorbance peaks increased with increasing gold nanoparticle concentration.

With Ag:HSAF nanoparticle complexes there was, apart from the obvious protein absorbance peak present at 280 nm, a rather prominent surface plasmon resonance band at 414 nm the intensity of which increased with increasing silver nanoparticle concentration.

ICP-OES showed an increase in platinum nanoparticle concentration but a decrease in gold and silver nanoparticle concentration with increase in molar ratio of metal salt, respectively, to a fixed HSAF concentration. The M:HSAF ratio based on protein content of HSAF before and after synthesis, respectively, was also estimated. The greatest number of platinum atoms stabilized was 404; that of gold and silver were 357 and 125, respectively, based on the starting HSAF concentration and 920 based on the final concentration of HSAF after synthesis (gold and silver were 661 and 202, respectively).

The biologically synthesized M:HSAF nanoparticle complexes were analyzed by FTIR spectroscopy in order to investigate if there were any functional groups on HSAF that were responsible for the stabilization and/or coordination of the nanoparticles as well as to study the structural integrity of HSAF after the synthesis of the nanoparticles. Noticeable findings were seen between 1,200–1,800 cm^{-1} indicating that the integrity of HSAF remained intact. The presence of amide I (1,590–1,650 cm^{-1}) and amide II (1,500–1,560 cm^{-1}) bands which are characteristic of proteins and peptides, in both spectra of HSAF and M:HSAF nanoparticle complexes indicated protein stability. Furthermore the presence of an amide I peak (1,630–1,650 cm^{-1}), often ascribed to predominantly α-helical structured protein (like HSAF) in aqueous solutions, all pointed to the fact that the synthesis of metal nanoparticles in the presence of HSAF may not have compromised the overall structure of HSAF. With a more detailed analysis of the FTIR spectra of the M:HSAF nanoparticle complexes it was noticed that additional peaks appeared in the amide I region at 1,737–1,740 cm^{-1}, which were not seen in the spectrum of HSAF, suggesting the binding of the metal (before reduction to nanoparticles) to the carboxylate side chains of acidic amino acids. These residues predominantly line the interior of HSAF and give it its net negative charge at physiological pH, implying a possibility of metal nanoparticle synthesis within the cavity of HSAF.

M:HSAF nanoparticle complexes were characterized by TEM in order to determine particle size and distribution, position of the nanoparticle in the HSAF and to confirm the synthesis of metal nanoparticles in the presence of HSAF. The nanoparticles were predominantly spherical that agreed, obviously, with the spherical shape of the HSAF interior. Analysis revealed a general increase in the size of the nanoparticles with an increase in M:HSAF ratio. It was noticed, however, that there was no significant increase in the size of the nanoparticles as the M:HSAF increased implying that the cavity of HSAF restricted metal nanoparticle nucleation within its core. The average size of the metal nanoparticles obtained with different molar concentration of metal salt was between 2 and 6 nm and since the size obtained was below that of the interior diameter of HSAF (8 nm), it gave further evidence that the metal nanoparticles formed within the core of HSAF.

Confirmation of the presence and stabilization of metal in M:HSAF nanoparticle complexes was made by EDX analysis.

Activation

There is a ninefold increase in ferroxidase activity with an increase in the molar ratios of Pt:HSAF nanoparticles seen at a molar ratio of 1,000:1 [Pt:HSAF]. At a Pt:HSAF ratio of 500:1 there is a fivefold increase while at a ratio of 250:1 there is only a twofold increase. Platinum is a widely used catalyst that exhibits increased catalysis with increased surface area and so it is not surprising that this metal, at the nanoscale level, showed enhanced activity of ferroxidase.

Furthermore, the release of platinum nanoparticles after the uptake of iron further confirmed that platinum nanoparticles were initially present within the core of the HSAF. It may also suggest the ability of the interior of HSAF to accommodate more than one type of metal atom (Fe and Pt) within its cavity. The functional groups responsible for the coordination of both platinum and iron within the core of HSAF may not be involved in the storage and oxidation of iron in the HSAF core. A different mechanisms of Fe^{2+} uptake and oxidation by apoferritin proposed that the basic amino acids residues within the core of apoferritin were responsible for its ferroxidase activity. This site was different from that containing acidic amino residues reported to be implicated in the uptake, nucleation, and stabilization of nanoparticles.

There is also a ninefold increase in ferroxidase activity with an increase in the molar ratio of Au:HSAF nanoparticles seen at a molar ratio of 500:1 [Au:HSAF]. At Au:HSAF ratios of 250:1 and 1,000:1, respectively, there is a six- and eightfold increase, while at a ratio of 2,000:1 there is a 2.7-fold increment.

As far as silver nanoparticles were concerned there is a 110-fold increase in ferroxidase activity with an increase in the molar ratio of Ag:HSAF nanoparticles of 1,000:1. At Ag:HSAF ratio of 500:1 there is a 90-fold increase, while at a ratio of 2,000:1 there is a 67-fold increase; at a ratio of 250:1 there is only a 29-fold increase.

Future Directions

Furthermore novelty manifests itself with possible clinical applications in the treatment of diseases associated with poor iron absorption and possible decrease in oxidative stress associated with the toxic levels of iron in biological systems.

Gold and silver nanoparticles have shown extensive popularity in nanomedicine for their anticancer/antitumor and antimicrobial properties. From a structural and mechanistic point of view it is not clear why there is this considerable increase in activity of ferroxidase in the present study. Though it may be considered speculation to propose tentative suggestions it is probable that one, or more, of the following may occur:

1. Noble metallic nanoparticles, especially Au and Ag have catalytic properties.
2. The distances between the glutamic acid, aspartic acid, glutamine, and water molecule within the active site of ferroxidase are about 8–10 Å – well within the range for the Au/Ag nanoparticles to bind. This would, in turn, enhance the negativity of the amino acids in the core increasing the binding of Fe^{2+}.
3. The Au/Ag nanoparticles interact with the di-ferrous [Fe–Fe] complex in the enzyme reactive core increasing the rate of removal of electrons and oxidation to $[Fe^{3+}-Fe^{3+}]$.
4. The nanoparticles facilitate the addition of a water molecule to enhance the formation of the hydrolysis product [FeOOH] and in so doing interact with the released protons to drive the reaction.
5. The nanoparticles interact with polar amino acids at the reactive site changing the conformation of the reactive core increasing the rate of addition of oxidant.

6. The nanoparticles chelate to the cysteine sulfur atoms and/or the zinc at the subunit interface disrupting any structural conformations (cysteine bridges) forming and potentially blocking the access of oxidant through channels to the active region.

Cross-References

▶ Gold and Nucleic Acids

References

Cao G (2004) Nanostructures and nanomaterials: synthesis, properties and applications. Imperial College Press, London

Deng QY, Yang B, Wang JF, Whiteley CG, Wang XN (2009) Biological synthesis of platinum nanoparticles with apoferritin. Biotechnol Lett 31:1505–1509

Fendler JH (1998) Nanoparticles and nanostructured films: preparation, characterisation and applications. Wiley, New York

Govender Y, Riddin TL, Gericke M, Whiteley CG (2009) Bioreduction of platinum salt into nanoparticles: a mechanistic perspective. Biotechnol Lett 31:95–100

Govender Y, Riddin TL, Gericke M, Whiteley CG (2010) On the enzymatic formation of platinum nanoparticles. J Nanoparticle Res 12:261–271

Konetzka WA (1977) Microbiology of metal transformations. In: Weinberg ED (ed) Microorganisms and minerals. Marcel Dekker, New York, pp 317–342

Rai M, Yadav P, Bridge P, Gade A (2009) Myconanotechnology: a new and emerging science. In: Rai M, Bridge PD (eds) Applied mycology. CABI, UK, pp 258–267

Rao CNR, Muller A, Cheetham AK (2004) The chemistry of nanomaterials: synthesis, properties and applications, vol 1. Wiley-VCH, Weinheim

Rashamuse KJ, Mutambanengwe CCZ, Whiteley CG (2008) Enzymatic recovery of Platinum (IV) from industrial wastewater using a Biosulphidogenic hydrogenase. Afr J Biotech 7(8):1087–1095

Rashamuse K, Whiteley CG (2007) Bioreduction of platinum (IV) from aqueous solution using sulphate reducing bacteria. Appl Microbiol Biotechnol 75:1429–1435

Riddin TL, Gericke M, Whiteley CG (2006) Analysis of the inter- and extracellular formation of platinum nanoparticles by *Fusariumoxysporum f. splycopersici* using response surface methodology. Nanotechnol 17:3482–3489

Riddin TL, Govender Y, Gericke M, Whiteley CG (2009) Two different hydrogenase enzymes from sulphate reducing bacteria are responsible for the bioreductive mechanism of platinum into nanoparticles. Enzyme Microbial Technol 45:267–273

Riddin TL, Gericke M, Whiteley CG (2010) Biological synthesis of platinum nanoparticles: effect of initial metal concentration. Enzyme Microbial Technol 46:501–505

Whiteley CG, Govender Y, Riddin T, Rai M (2011) Enzymatic synthesis of platinum nanoparticles: prokaryote and eukaryote systems. In: Rai M, Duran N, Southam G (eds) Metal nanoparticles in microbiology. Springer, New York, pp 103–134. ISBN 978-3-642-18311-9

Apoptosis

▶ Arsenic and Primary Human Cells
▶ Tin Complexes, Antitumor Activity

Apoptosis, Programmed Cell Death

▶ Calcium and Viruses
▶ Chromium(VI), Oxidative Cell Damage

Apoptotic Cell Death

▶ Selenium and Glutathione Peroxidases

Aquaporin-Mediated Arsenic Transport

▶ Arsenic and Aquaporins

Aquaporin-Mediated Boron Transport

▶ Boron and Aquaporins

Aquaporin-Mediated Selenite Transport

▶ Selenium and Aquaporins

Aquaporin-Mediated Silicon Transport

▶ Silicon and Aquaporins

Aquaporins and Transport of Metalloids

François Chaumont and Gerd Patrick Bienert
Institut des Sciences de la Vie, Universite catholique de Louvain, Louvain-la-Neuve, Belgium

Synonyms

Major Intrinsic Proteins; Metalloid transport; Solute channels; Water channels

Definition

Aquaporins or major intrinsic proteins (MIPs) are channels facilitating the passive diffusion of small uncharged solutes across biological membranes. In addition to water, several aquaporin isoforms were shown to transport gases (NH_3 and CO_2) or other uncharged solutes including metalloid compounds (arsenite, boric acid, silicic acid, selenite, and antimonite).

Aquaporin Discovery and Phylogeny

Aquaporins are membrane proteins belonging to the ancient major intrinsic protein (MIP) family whose members are found in all kingdoms of life. Although the term aquaporin was initially restricted to MIPs transporting water molecules, it is now used to describe all MIPs.

The first demonstration of the water channel activity of a MIP protein was published in 1992 when Peter Agre and coworkers heterologously expressed in Xenopus oocytes a 28-kDa protein (CHIP28) from human erythrocyte membrane (Preston et al. 1992). This protein was renamed AQP1 and became the archetypal membrane water channel. Peter Agre shared the 2003 Nobel Prize in chemistry for the discovery of water channels and their molecular and physiological characterization of AQP1. While 13 aquaporin genes have since been identified in mammals, higher plants contain a larger number of isoforms with 33, 35, 36, 37, and 55 homologs in rice, *Arabidopsis*, maize, tomato, and poplar, respectively (reviewed in Gomes et al. 2009; Hachez and Chaumont 2010; Maurel et al. 2008). This large multiplicity of isoforms probably offers better adaptive advantages for plants to grow in different environmental conditions as a result of divergent substrate specificity, localization, and regulation.

Based on their amino acid sequences, aquaporins can be divided into three classes, as proposed for mammalian aquaporins (Ishibashi et al. 2009). Class I originally regroups water-specific channels but also includes aquaporins shown to be permeable to other solutes such as carbon dioxide (CO_2) and nitric oxide (NO). Class II aquaporins, also named aquaglyceroporins, can facilitate the transport of solutes such as glycerol and urea, as well as water and, for some of them, metalloids. Class III regroups the most divergent aquaporins (less than 20% identity with class I and class II aquaporins) and includes the more recently discovered mammalian AQP11 and AQP12 and the plant SIPs (small basic intrinsic proteins). However, as the number of available aquaporin sequences is increasing, their phylogenetic analyses become more complex. Nowadays, a total of 13 subfamilies can be distinguished whereof five are plant-specific (Danielson and Johanson 2010).

Aquaporin Topology and Pore Selectivity

Aquaporins are channels with a molecular mass around 30 kDa, typically containing six transmembrane domains (TM1–TM6) linked by five loops (A–E), with the N- and C-termini facing the cytosol (reviewed in Chaumont et al. 2005; Gomes et al. 2009). The two halves of the protein show an obvious structural symmetry. The cytosolic loop B and extracytosolic loop E both form short hydrophobic α-helices that dip halfway into the membrane from opposite sides forming two hemipores. These two loops contain the generally conserved asparagine-proline-alanine (NPA) motifs, which contribute to the substrate selectivity and proton exclusion mechanism. Aquaporin molecular structures have been determined with a high resolution (5–2.2 Å) by electron microscopy and X-ray crystallography using 2D and 3D crystals. Aquaporins assemble as tetramers in the membrane in which each monomer forms an independent channel (Murata et al. 2000). The common channel structure consists in three topological parts, an extracellular and a cytoplasmic vestibule connected by a narrow pore acting as a selectivity filter. In spite of this highly conserved structure, differences in channel specificity derive from size exclusion mechanisms at two constriction regions (the NPA motif and the aromatic/arginine (ar/R) constriction) as well as from stereoscopic interaction between the substrate and the amino acid residues facing the pore.

The narrow NPA region is positively charged because of the dipole moments of the two short α-helices formed by the loops B and E, and it induces a reorientation of the water molecules within the constriction by promoting an interaction of these latter with the asparagine residues (Murata et al. 2000). The ar/R constriction, located ~7–8 Å from the NPA motifs toward the extracellular vestibule, also constitutes a steric barrier which measures ~2 Å in diameter in most of the water-selective aquaporins and is ~1 Å larger in aquaglyceroporins. This constriction is formed by interaction of four amino acid residues, two from the TM2 and TM5 and two from the loop E, that play a key role in controlling the passage of uncharged molecules. Substitutions at these positions and in their neighborhood modulate the polarity and the size of the pore and therefore the specificity of the channel.

Aquaporins and Metalloids

As mentioned above, aquaporins are not solely water channels but are involved in the transport of a wide range of small uncharged solutes including metalloids. Aquaporins channel activity and specificity are generally tested by heterologous expression in Xenopus oocytes or in yeast strains (reviewed in Hachez and Chaumont 2010). Transport assays in oocytes include cell swelling measurements to determine the water or solute channel activity and measurements of the accumulation of radioactively labeled substrates into the cell according to time. In yeast, growth tests in the presence of a toxic solute, complementation assays of strains deleted in specific transporters, or membrane permeability measurement using stopped-flow spectrometry allow the determination of the intrinsic permeability of a single aquaporin isoform. These systems have been used to demonstrate that several aquaporins facilitate the diffusion of metalloids.

Aquaporins facilitating the transport of metalloids in animals, plants, prokaryotes, and protozoan generally belong to class II (aquaglyceroporins). In plants, most of them are members of the nodulin-26-like intrinsic protein (NIP) family and of the recently discovered X intrinsic protein (XIP) subfamilies, although some plasma membrane intrinsic proteins (PIPs) belonging to the class I (aquaporin) were also found to be involved in this process (reviewed in Ishibashi et al. 2009; Bienert and Chaumont 2011; Bienert et al. 2011; Danielson and Johanson 2010). In general, NIP aquaporins are physiologically essential metalloid transporters in plants, responsible for the uptake, translocation, or extrusion of uncharged metalloid species including boron, silicon, arsenic, and antimony (Bienert et al. 2008) (see ▶ Arsenic and Aquaporins, ▶ Boron and Aquaporins, ▶ Selenium and Aquaporins, ▶ Silicon and Aquaporins). Selenium is occasionally included in the list of metalloid as selenite shares chemical features with other hydroxylated metalloid species. This is in accordance with the observation that these compounds share a common NIP-facilitated transmembrane pathway (see ▶ Selenium and Aquaporins).

Interestingly, the properties of the four amino acid residues of the ar/R filter essential for the pore specificity appear to be an important feature for metalloid transport. Accordingly, plant NIPs are further subdivided into three subgroups, the NIPIs, NIPIIs, and NIPIIIs (Danielson and Johanson 2010; Ma 2010). NIPI proteins have been reported to transport water, hydrogen peroxide, glycerol, lactic acid, or the metalloids antimonite and arsenite (Bienert and Chaumont 2011). NIPII proteins are permeable to solutes such as water, ammonia, urea, or the metalloids boric acid, antimonite, and arsenite (Bienert and Chaumont 2011). Finally, NIPIII subgroup isoforms have a distinct selectivity filter consisting of relatively small amino acid residues and forming a potentially larger constriction region compared with other NIP subgroups or microbial and mammalian aquaglyceroporins (Danielson and Johanson 2010; Ma 2010). The pore diameter and channel path of NIPIIIs were modeled to be among the widest of all aquaporins. This is in accordance with the fact that NIPIII proteins allow the passage of silicic acid which is larger than typical MIP substrates such as water, glycerol, or other transported metalloid species (Ma 2010).

Aquaporin Regulation

Water and/or solutes movement through aquaporins is mainly dependent on the concentration gradients of these compounds across the membranes, as aquaporins only facilitate their diffusion and do not act as active transporters. Therefore, the amount of active channels in the membranes has to be tightly regulated according to the metabolic demand (reviewed in Chaumont et al. 2005; Gomes et al. 2009; Maurel et al. 2008). The first regulation step is through regulation of mRNA and protein expression level according to the cell type, developmental stage, and environmental cues. For instance, expression of different boron transporters belonging to the NIP subfamily is upregulated in plants grown under boron-limited conditions, and this seems to be cell-type specific (see ▶ Boron and Aquaporins). Once translated, aquaporins have to be sent to their target membrane, a mechanism that depends on their phosphorylation status and on physical interactions with different aquaporin isoforms as well as with regulatory proteins of the secretory pathway. These trafficking regulation mechanisms probably explain the polarization of metalloid NIP transporters in specific plasma membrane regions in specialized root cells (see ▶ Boron and Aquaporins, ▶ Silicon and Aquaporins). Finally, gating of aquaporins through reversible phosphorylation, pH, osmotic gradient, and hydrostatic gradients has been reported for different isoforms and represents a rapid way to control the membrane permeability to different solutes in response to developmental and environmental constraints. To our knowledge, the gating of metalloid-transporting aquaporins has not yet been studied in detail.

Concluding Remarks

The discovery that aquaporins are channels facilitating the diffusion of a wide range of small uncharged solutes including metalloids has definitely opened new perspectives concerning their roles in cell physiology. Further characterization of the molecular basis defining the pore selectivity of metalloid-transporting aquaporins through the generation of 2D or 3D crystals would definitely reveal essential amino acid residues. Several aquaporins from very different organisms have been shown to transport metalloids, but their involvement in cell and/or organism physiology has still to be determined for most of them. A better understanding of their function and regulation is therefore required before envisioning their use to improve metalloid homeostasis in living cells.

Funding

This work was supported by grants from the Belgian National Fund for Scientific Research (FNRS), the Interuniversity Attraction Poles Programme–Belgian Science Policy, and the "Communauté française de Belgique–Actions de Recherches Concertées". GPB was supported by a grant from the FNRS.

Cross-References

▶ Arsenic and Aquaporins
▶ Boron and Aquaporins
▶ Selenium and Aquaporins
▶ Silicon and Aquaporins

References

Bienert GP, Chaumont F (2011) Plant aquaporins: roles in water homeostasis, nutrition, and signaling processes. Springer, Berlin/Heidelberg

Bienert GP, Thorsen M, Schüssler MD et al (2008) A subgroup of plant aquaporins facilitate the bidirectional diffusion of As (OH)$_3$ and Sb(OH)$_3$ across membranes. BMC Biol 6:26

Bienert GP, Bienert MD, Jahn TP et al (2011) Solanaceae XIPs are plasma membrane aquaporins that facilitate the transport of many uncharged substrates. Plant J 66:306–317

Chaumont F, Moshelion M, Daniels MJ (2005) Regulation of plant aquaporin activity. Biol Cell 97:749–764

Danielson JAH, Johanson U (2010) Phylogeny of major intrinsic proteins. Landes Bioscience-Springer, New York

Gomes D, Agasse A, Thiébaud P et al (2009) Aquaporins are multifunctional water and solute transporters highly divergent in living organisms. Biochim Biophys Acta 1788:1213–1228

Hachez C, Chaumont F (2010) Aquaporins: a family of highly regulated multifunctional channels. Landes Bioscience-Springer, New York

Ishibashi K, Koike S, Kondo S et al (2009) The role of a group III AQP, AQP11 in intracellular organelle homeostasis. J Med Invest 56:312–317

Ma JF (2010) Silicon transporters in higher plants. Landes Bioscience-Springer, New York

Maurel C, Verdoucq L, Luu DT et al (2008) Plant aquaporins: membrane channels with multiple integrated functions. Annu Rev Plant Biol 59:595–624

Murata K, Mitsuoka K, Hirai T et al (2000) Structural determinants of water permeation through aquaporin-1. Nature 407:599–605

Preston GM, Carroll TP, Guggino WB et al (1992) Appearance of water channels in Xenopus oocytes expressing red cell CHIP28 protein. Science 256:385–387

ARD

▶ Acireductone Dioxygenase

ArgE, ArgE-Encoded N-Acetyl-L-ornithine Deacetylase

▶ Zinc Aminopeptidases, Aminopeptidase from Vibrio Proteolyticus (Aeromonas proteolytica) as Prototypical Enzyme

Argyrophilia

▶ Silver Impregnation Methods in Diagnostics

Arsen

▶ Arsenic in Pathological Conditions
▶ Arsenic, Free Radical and Oxidative Stress

Arseniasis

▶ Arsenicosis

Arsenic

Damjan Balabanič[1] and Aleksandra Krivograd Klemenčič[2]
[1]Ecology Department, Pulp and Paper Institute, Ljubljana, Slovenia
[2]Faculty of Health Sciences, University of Ljubljana, Ljubljana, Slovenia

Synonyms

Arsenic metabolic pathway; Arsenic toxicity; Human health effects of arsenic exposure

Definition

Arsenic (As) is a naturally occurring element widely distributed in the Earth's crust. Exposure to excess As, principally from contaminated drinking water, is considered one of the top environmental health threats worldwide. Most of this exposure is from natural geological sources of As that contaminate groundwater (Mukherjee et al. 2006).

As has the common oxidation states -3, $+3$, and $+5$. Trivalent As is more soluble and mobile than the pentavalent form. The log K_{ow} value for As is 1.3, indicating lipophilicity. For the oxidation and reduction of As can be responsible also microorganisms such as *Bacillus* and *Pseudomonas*. As fungi can produce toxic and highly volatile As. Methylation of As to organoarsenicals occurs in vertebrates (including humans) and invertebrates.

Exposure to As

In the environment, As is combined with oxygen, chlorine, and sulfur to form inorganic As compounds. For the general human population, the main exposure to organic and inorganic As is through ingestion. Both organic and inorganic As are present in varying amounts in water and food (meat, poultry, fish, shellfish, dairy products, and cereals) (Chakraborti 2011). Inorganic forms of As can exist as either arsenate or arsenite. Although arsenate is less toxic, it can be converted to arsenite in humans through metabolism. The major metabolic pathway for inorganic As in humans is methylation. After absorption, inorganic As is transported by blood to other organs. A single oral dose of inorganic As results in increased As concentrations in liver, kidneys, lungs, and intestinal mucosa. Some As, predominantly in the pentavalent form is excreted directly in feces. Once absorbed, pentvalent As is excreted also by the kidneys. Trivalent As is readily excreted in bile to the intestines, where it is available for reabsorption or feces elimination.

Effects of As

As affects breathing, hearth rhythm, and increases the risk of bladder cancer. Its exposure can also cause gastrointestinal disturbances, liver and renal diseases, reproductive health effects, and dermal changes (Balabanič et al. 2011; Chakraborti 2011). As causes oxidative damage by the production of reactive oxygen species which produces different bioactive molecules (Flora et al. 2007; Zhang et al. 2011). Chronic exposure to As has been associated with increased numbers of spontaneous abortions in contaminated water users and women working at or living in close proximity to smelters, as well as women whose partners worked at the smelters (Balabanič et al. 2011). The rate of spontaneous abortion was even higher when both partners worked at the smelters. Chronic exposure to high level of As elevated serum lipid peroxide levels and lowered nonprotein sulfhydryl levels suggesting oxidative stress (Flora et al. 2007).

Tests for genotoxicity have indicated that As compounds inhibit DNA repair and induce chromosomal aberrations, sister-chromatid exchanges, and micronuclei formation in both human and rodent cells in culture and in cells of exposed humans (Chakraborti 2011; Kesari et al. 2012). The mechanism of genotoxicity is not known, but may be due to the ability of arsenate to inhibit DNA replicating or repair enzymes, or the ability of arsenate to act as a phosphate analog.

Cells accumulate As by using an active transport system normally used in phosphate transport. Once absorbed, As toxicity is generally attributed to the trivalent form. Toxic effects are exerted by reacting with sulfhydryl enzyme systems. The tissue rich in oxidative metabolism, such as the alimentary tract, liver, kidney, lung, and epidermis, are therefore most affected (Chakraborti 2011).

One of the mechanisms by which As exerts its toxic effect is through impairment of cellular respiration by the inhibition of various mitochondrial enzymes and the uncoupling of oxidative phosphorylation. Most toxicity of As results from its ability to interact with sulfhydryl groups of proteins and enzymes and to substitute phosphorous in a variety of biochemical reactions (Zhang et al. 2007).

Most of the inorganic As is metabolized to dimethylarsinic acid [DMA (V)] and monomethylarsonic acid [MMA (V)] before excretion in the urine. Methylation of As involves a two electron reduction of pentavalent [As (V) and MMA (V)] to trivalent [As (III) and MMA (III)] As species followed by the transfer of a methyl group from a methyl donor, such as S-adenosylmethionine (Thompson 1993; Tchounwou et al. 2003). This methylation mechanism has been widely accepted and the metabolites MMA (V) and DMA (V) have been consistently observed in human urine. A key intermediate for the methylation of MMA (V) to DMA (V) is the MMA (III) species (Suzuki et al. 2004; Reichard et al. 2007).

The generally held view of As carcinogenesis in the past was that arsenite was the most likely cause of carcinogensis and that methylation of As species was a detoxification pathway. The present view of As carcinogenesis is that there are many possible chemical forms of As that may be causal in carcinogenesis and that methylation of As may be a toxification pathway. MMA (III) has been found in urine of humans exposed to As without and with concomitant treatment with chelators (Aposhian et al. 2000). Some of the biological activities that MMA (III) is known to possess in various experimental systems include enzyme inhibition, cell toxicity, and genotoxicity (Kesari et al. 2012; Lin et al. 2012). MMA (III) is an excellent choice as a cause of As carcinogenesis. DMA (III) has been detected in human urine of As-exposed humans administered a chelator.

Various hypotheses have been proposed to explain the carcinogenicity of inorganic As. Nevertheless, molecular mechanisms by which this arsenical induces cancer are still poorly understood. Results of previous studies indicated that inorganic As does not act through classic genotoxic and mutagenic mechanisms, but rather may be a tumor promoter that modifies signal transduction pathways involved in cell growth and proliferation (Simeonova and Luster 2000). Inorganic As (III) has been shown to modulate expression and/or DNA-binding activities of several key transcription factors, including nuclear factor kappa B, tumor suppressor 53 (*p53*), and activating protein-1 (AP-1) (Hu et al. 2002; Chang et al. 2009). Mechanisms of AP-1 activation by trivalent inorganic As include stimulation of the mitogen-activated protein kinase (MAPK) cascade with a consequent increase in the expression and/or phosphorylation of the two major AP-1 constituents, *c-Jun* and *c-Fos* (Hu et al. 2002; Sanchez et al. 2009). Several studies have examined *p53* gene expression and mutation in tumors obtained from subjects with a history of As ingestion. The *p53* participates in many cellular functions, cell-cycle control, DNA repair, differentiation, and apoptosis (Yu et al. 2008; Kim et al. 2011). The tumor suppressor protein *p53* is one component of the DNA damage response pathway in mammalian cells. Some of these normal cellular functions of *p53* can be modulated and sometimes inhibited by interactions with either cellular proteins or oncoviral proteins of certain DNA viruses (Yu et al. 2008; Qu et al. 2009).

Conclusions

Integration of the available scientific information on As indicates that geogenic and anthropogenic contamination of natural resources represents a major public health problem in many countries of the world. Exposure to As occurs via ingestion, inhalation, and dermal contact. Such exposure has been associated with a significant number of systemic health effects in various organs and tissue systems including skin, lung, liver, kidney, bladder, gastrointestinal tract, reproduction system, respiratory system, and hematopoietic system. Evidence from recent studies has linked As consumption in drinking water with two noncancer health conditions (hypertension and diabetes mellitus) that are a major cause of morbidity and mortality.

Experiments in animals and in vitro have demonstrated that As has many biochemical and cytotoxic effects at low doses of human exposure. Those effects include induction of oxidative damage, altered DNA methylation and gene expression, changes in intracellular levels of *p53* protein, inhibition of thioredoxin reductase, inhibition of pyruvate dehydrogenase, induction of protein-DNA cross-links, altered regulation of DNA-repair genes, glutathione reductase and other stress-response pathways, and induction of apoptosis. Additionally, the evidence of carcinogenicity in humans is very strong, especially for cancer of the skin, lung, liver, kidney, and bladder. Further epidemiological studies are highly recommended to investigate the dose-response relationship between As ingestion and noncancer endpoints. Because of the very large populations exposed, these endpoints are common causes of morbidity and mortality, even small increases in relative risk at low doses of As exposure could be of considerable public health significance. Such information is also important in developing a comprehensive risk assessment and management program for As.

Cross-References

▶ Arsenic in Nature
▶ Arsenic in Pathological Conditions
▶ Arsenic Methyltransferases
▶ Arsenic, Free Radical and Oxidative Stress
▶ Arsenic-Induced Stress Proteins

References

Aposhian HV, Zheng B, Aposhian MM, Le XC, Cebrian ME, Cullen W, Zakharyan RA, Ma M, Dart RC, Cheng Z, Andrewes P, Yip L, O'Malley GF, Maiorino RM, Van Voorhies W, Healy SM, Titcomb A (2000) DMPS-arsenic challenge test II. Modulation of arsenic species, including monomethylarsonous acid (MMA (III)), excreted in human urine. Toxicol Appl Pharmacol 165:74–83

Balabanič D, Rupnik M, Krivograd Klemenčič A (2011) Negative impact of endocrine-disrupting compounds on human reproductive health. Reprod Fertil Dev 23:403–416

Chakraborti D (2011) Arsenic: Occurrence in Groundwater. In: Nriagu J (ed) Encyclopedia of Environmental Health. Elsevier Science, Oxford, pp 165–180

Chang JS, Gu MK, Kim KW (2009) Effect of arsenic on p53 mutation and occurrence of teratogenic salamanders: their potential as ecological indicators for arsenic contamination. Chemosphere 75:948–954

Flora SJS, Bhadauria S, Kannan GM, Singh N (2007) Arsenic-induced oxidative stress and the role of antioxidant supplementation during chelation: a review. J Environ Biol 28:333–347

Hu Y, Jin X, Snow ET (2002) Effect of arsenic on transcription factor AP-1 and NF-κB DNA binding activity and related gene expression. Toxicol Lett 133:33–45

Kesari VP, Kumar A, Khan PK (2012) Genotoxic potential of arsenic at its reference dose. Ecotoxicol Environ Safety 80:126–131

Kim YJ, Chung JY, Lee SG, Kim JY, Park JE, Kim WR, Joo BS, Han SH, Yoo KS, Yoo YH, Kim JM (2011) Arsenic trioxide-induced apoptosis in TM4 Sertoli cells: the potential involvement of p21 expression and p53 phosphorylation. Toxicology 285:142–151

Lin PY, Lin YL, Huang CC, Chen SS, Liu YW (2012) Inorganic arsenic in drinking water accelerates N-butyl-N-(4-hydroxylbutyl)nitrosamine induced bladder tissue damage in mice. Toxicol Appl Pharmacol 259:27–37

Mukherjee A, Sengupta MK, Hossain MA, Ahamed S, Das B, Nayak B, Lodh D, Rahman MM, Chakraborti D (2006) Arsenic contamination in groundwater: a global perspective with emphasis on the Asian scenario. J Health Popul Nutr 24:142–163

Qu J, Lin J, Zhang S, Zhu Z, Ni C, Zhang S, Gao H, Zhu M (2009) HBV DNA can bind to p53 protein and influence p53 transactivation in hepatoma cells. Biochem Biophys Res Commun 386:504–509

Reichard JF, Schnekenburger M, Puga A (2007) Long term low-dose arsenic exposure induces loss of DNA methylation. Biochem Biophys Res Commun 352:188–192

Sanchez Y, Calle C, de Blas E, Aller P (2009) Modulation of arsenic trioxide-induced apoptosis by genistein and functionally related agents in U937 human leukaemia cells. Regulation by ROS and mitogen-activated protein kinases. Chem Biol Interact 182:37–44

Simeonova PP, Luster MI (2000) Mechanisms of arsenic carcinogenicity: genetic or epigenetic mechanisms? J Environ Pathol Toxicol Oncol 19:281–286

Suzuki KT, Katagiri A, Sakuma Y, Ogra Y, Ohmichi M (2004) Distributions and chemical forms of arsenic after intravenous administration of dimethylarsinic and monomethylarsonic acids to rat. Toxicol Appl Pharmacol 198:336–344

Tchounwou BP, Patlolla AK, Centeno JA (2003) Carcinogenic and systemic health effects associated with arsenic exposure-a critical review. Toxicol Pathol 31:575–588

Thompson DJ (1993) A chemical hypothesis for arsenic methylation in mammals. Chem Biol Interact 88:89–114

Yu X, Robinson JF, Gribble E, Hong SW, Sidhu JS, Faustman EM (2008) Gene expression profiling analysis reveals arsenic-induced cell cycle arrest and apoptosis in p53-proficient and p53-deficient cells through differential gene pathways. Toxicol Appl Pharmacol 233:389–403

Zhang X, Yang F, Shim JY, Kirk KL, Anderson DE, Chen X (2007) Identification of arsenic-binding proteins in human breast cancer cells. Cancer Lett 225:95–106

Zhang Z, Wang X, Cheng S, Sun L, Yo S, Yao H, Li W, Budhraja A, Li L, Shelton BJ, Tucker T, Arnold SM, Shi X (2011) Reactive oxygen species mediate arsenic induced cell transformation and tumorigenesis through Wnt/β-catenin pathway in human colorectal adenocarcinoma DLD1 cells. Toxicol Appl Pharmacol 256:114–121

Arsenic Accumulation in Yeast

▶ Arsenic and Yeast Aquaglyceroporin

Arsenic and Alcohol, Combined Toxicity

Honglian Shi
Department of Pharmacology and Toxicology,
University of Kansas, Lawrence, KS, USA

Synonyms

Combined toxicity of arsenic and alcohol drinking; Synergistic toxicity of arsenic and ethanol

Definition

Combined toxicity of arsenic and alcohol refers to harmful effects resulting from exposures to arsenic and alcohol (or ethanol) at the same time. In some cases, the combined toxicity is synergistic. The most common form of the exposures is intake of water and alcohol.

Introduction

Historically, co-exposures to arsenic and alcohols caused severe diseases. In the early 1900s, there were over 6,000 cases and 70 deaths from heart diseases attributed to drinking arsenic-contaminated beer in England (Cullen 2008). Wine fermented from grapes treated with arsenical fungicides resulted in peripheral vascular disease and cardiomyopathies in a group of German vintners in the 1920s (Engel et al. 1994). A recent study reported the presence of As in sweet little Gabonese palm wine in Gabon (Mavioga et al. 2009). In the USA, arsenic-contaminated moonshine was implicated as the cause of a dozen cases of cardiovascular diseases in the state of Georgia (Gerhardt et al. 1980). In these incidents, had the patients taken the amount of arsenic in their drinks alone or the amount of alcohol alone, such severe damages would not have occurred. This indicates that the co-exposure made alcohol and arsenic toxic at their nontoxic concentrations.

To date, there are no epidemiological studies on the population who are co-exposed to arsenic and alcohol and on how severe a public health problem it is. Nonetheless, high prevalence of arsenic exposure and the alcohol epidemic make it highly possible that certain populations are co-exposed to arsenic and alcohol (Bao and Shi 2010).

Alcohol and Arsenic Pharmacokinetics

It is well established that alcohol consumption on a regular basis increases deposition of arsenic and copper in liver and kidney. Higher total urinary arsenic has been reported in alcohol consumers than nonalcoholics (Hsueh et al. 2003). It has also experimentally proven that ethanol increases the uptake and retention in the liver and kidneys (Flora et al. 1997).

Alcohol may facilitate the uptake of arsenic through cell membranes that are damaged and changed in their molecular composition by the direct effect of alcohol. Alcohol consumption may also alter the methylation of arsenic, affecting its distribution and retention in tissues (Tseng 2009).

Action of Alcohol and Arsenic on Proteins and Their Activities

Kinases. Experimentally, co-exposure, but not exposure to arsenic or alcohol alone, increased active Fyn tyrosine kinase, phosphorylation of important proteins such as PKCδ, membrane localization of phospholipase Cγ1, vascular endothelial cell growth factor, and insulin-like growth factor-1 (Klei and Barchowsky 2008).

Redox proteins. Arsenite activates NADPH oxidase (Lynn et al. 2000), which is a major source of intracellular superoxide anion radical ($O_2^{\cdot-}$). Methylated arsenic species can release redox-active iron from ferritin and have synergic effect with ascorbic acid to do so while alcohol elevate serum ferritin levels (Ioannou et al. 2004). Ethanol induces cytochrome P450 2E1 (CYP2E1), which has a high NADPH oxidase activity and produces $O_2^{\cdot-}$, H_2O_2, and hydroxyethyl radicals. Chronic ethanol consumption can result in a 10–20-fold increase in hepatic CYP2E1 in animals and humans. In addition, NO production is increased by the effect of ethanol as well as arsenic on inducible nitric oxide synthase (Kao et al. 2003).

Arsenic reduces plasma and cellular GSH levels. For instance, arsenic drinking (12 mg/kg) for 12 weeks depletes GSH levels in the liver and the brain of rats (Flora 1999). Similarly, possible contributions of impaired antioxidant defenses to ethanol-induced oxidative stress have been extensively investigated. Besides impairing catalase and superoxide dismutase (SOD) activities, alcohol depletes cellular GSH. Fernandez-Checa et al. have found that alcohol intake can lower the hepatic mitochondrial GSH content by 50–85% in rats (Fernandez-Checa et al. 1991). A combined exposure decreased hepatic and plasma GSH more markedly than ethanol or arsenic alone (Flora et al. 1997).

Mitochondrial proteins. Mitochondria are important targets of arsenic and ethanol. Arsenic is accumulated in mitochondria via phosphate transport proteins and the dicarboxylate carrier. Once accumulated in mitochondria, arsenic uncouples the oxidative phosphorylation because ATP synthase undergoes oxidative arsenylations (Crane and Lipmann 1953). Ethanol metabolism promotes a substantial reduction of both cytosolic and mitochondrial NAD, substantially increasing the NADH level. This increase poses an acute metabolic challenge for energy metabolism. Meanwhile, ethanol increases utilization of oxygen mainly through ethanol oxidation, resulting in localized and transient hypoxia. In recent studies, the mitochondria permeability transition has been identified as a key step for the induction of mitochondrial cytochrome c release and caspase activation by ethanol (Kurose et al. 1997; Higuchi et al. 2001). The harmful effects of arsenic on the enzymes of antioxidant defense systems such as thioredoxin reductase will potentiate ethanol's effect on membrane permeability, mtDNA damage, and mitochondrial dysfunction, which will cause more mitochondrial dysfunction and damage.

Proteins in DNA methylation, damage, and repair. DNA methylation is an important determinant in controlling gene expression whereby hypermethylation has a silencing effect on genes and hypomethylation may lead to increased gene expression. Ethanol can interact with one carbon metabolism and DNA methylation. It reduces the activity of methionine synthase which remethylates homocysteine to methionine with methyltetrahydrofolate as the methyl donor. It inhibits the activity of DNA methylase which transfers methyl groups to DNA in rats.

Arsenic is also able to induce DNA hyper- and hypomethylation (Lee et al. 1985). Arsenic interferes with DNA methyltransferases, resulting in inactivation of tumor suppressor genes through DNA hypermethylation. When co-exposed to alcohol and arsenic, cells are exposed to combined effects on modification of DNA methylation.

Acetaldehyde is the immediate product of the ethanol metabolism by alcohol dehydrogenase. Acetaldehyde causes point mutations in the hypoxanthine phosphoribosyltransferase 1 locus in human lymphocytes, and induces sister chromatid exchanges and gross chromosomal aberrations. It can bind to proteins such as *O*-6-methylguanine-DNA methyltransferase.

It is well known that arsenic interferes with DNA repair system. A link between the enhancing effects and inhibition of DNA repair processes has been documented by Okui and Fujiwara (1986). The interaction of arsenite with the removal of DNA damage induced by *N*-methyl-*N*-nitrosourea (MNU) has been characterized by Li and Rossman (1989). They observed an accumulation of DNA strand breaks after MNU treatment in the presence of arsenite in permeabilized V79 cells, indicating the inhibition of a later step of base excision repair.

Summary

Alcohol consumption and arsenic are two high risk factors for human health. Co-exposure to alcohol and Arsenic has potential synergistic effects on many protein activities, which play important role in maintaining cellular functions such as phosphorylation, redox homeostasis, mitochondrial functions, and DNA repair. Besides necessary epidemiological studies to assess the prevalence of the co-exposure, future studies are needed to clarify the exact mechanisms and identify potential pathways for prevention and therapeutic treatments for co-exposure of arsenic and alcohol.

Cross-References

▶ Arsenic
▶ Arsenic in Nature
▶ Arsenic Methyltransferases
▶ Arsenic and Primary Human Cells
▶ Arsenic, Biologically Active Compounds
▶ Arsenic, Free Radical and Oxidative Stress

References

Bao L, Shi H (2010) Potential molecular mechanisms for combined toxicity of arsenic and alcohol. J Inorg Biochem 104:1229–1233

Crane RK, Lipmann F (1953) The effect of arsenate on aerobic phosphorylation. J Biol Chem 201:235–243

Cullen WR (2008) Is arsenic an aphrodisiac. In: Is arsenic an aphrodisiac. Royal Society of Chemistry, Cambridge, pp 120–124

Engel RR, Hopenhayn-Rich C, Receveur O, Smith AH (1994) Vascular effects of chronic arsenic exposure: a review. Epidemiol Rev 16:184–209

Fernandez-Checa JC, Garcia-Ruiz C, Ookhtens M, Kaplowitz N (1991) Impaired uptake of glutathione by hepatic mitochondria from chronic ethanol-fed rats. Tracer kinetic studies in vitro and in vivo and susceptibility to oxidant stress. J Clin Invest 87:397–405

Flora SJ (1999) Arsenic-induced oxidative stress and its reversibility following combined administration of N-acetylcysteine and meso 2,3-dimercaptosuccinic acid in rats. Clin Exp Pharmacol Physiol 26:865–869

Flora SJ, Pant SC, Malhotra PR, Kannan GM (1997) Biochemical and histopathological changes in arsenic-intoxicated rats coexposed to ethanol. Alcohol 14:563–568

Gerhardt RE, Crecelius EA, Hudson JB (1980) Moonshine-related arsenic poisoning. Arch Intern Med 140:211–213

Higuchi H, Adachi M, Miura S, Gores GJ, Ishii H (2001) The mitochondrial permeability transition contributes to acute ethanol-induced apoptosis in rat hepatocytes. Hepatology 34:320–328

Hsueh YM, Ko YF, Huang YK, Chen HW, Chiou HY, Huang YL, Yang MH, Chen CJ (2003) Determinants of inorganic arsenic methylation capability among residents of the Lanyang Basin, Taiwan: arsenic and selenium exposure and alcohol consumption. Toxicol Lett 137:49–63

Ioannou GN, Dominitz JA, Weiss NS, Heagerty PJ, Kowdley KV (2004) The effect of alcohol consumption on the prevalence of iron overload, iron deficiency, and iron deficiency anemia. Gastroenterology 126:1293–1301

Kao YH, Yu CL, Chang LW, Yu HS (2003) Low concentrations of arsenic induce vascular endothelial growth factor and nitric oxide release and stimulate angiogenesis in vitro. Chem Res Toxicol 16:460–468

Klei LR, Barchowsky A (2008) Positive signaling interactions between arsenic and ethanol for angiogenic gene induction in human microvascular endothelial cells. Toxicol Sci 102:319–327

Kurose I, Higuchi H, Kato S et al (1997) Oxidative stress on mitochondria and cell membrane of cultured rat hepatocytes and perfused liver exposed to ethanol. Gastroenterology 112:1331–1343

Lee TC, Huang RY, Jan KY (1985) Sodium arsenite enhances the cytotoxicity, clastogenicity, and 6-thioguanine-resistant mutagenicity of ultraviolet light in Chinese hamster ovary cells. Mutat Res 148:83–89

Li JH, Rossman TG (1989) Inhibition of DNA ligase activity by arsenite: a possible mechanism of its comutagenesis. Mol Toxicol 2:1–9

Lynn S, Gurr JR, Lai HT, Jan KY (2000) NADH oxidase activation is involved in arsenite-induced oxidative DNA damage in human vascular smooth muscle cells. Circ Res 86:514–519

Mavioga EM, Mullot JU, Frederic C, Huart B, Burnat P (2009) Sweet little Gabonese palm wine: a neglected alcohol. West Afr J Med 28:291–294

Okui T, Fujiwara Y (1986) Inhibition of human excision DNA repair by inorganic arsenic and the co-mutagenic effect in V79 Chinese hamster cells. Mutat Res 172:69–76

Tseng CH (2009) A review on environmental factors regulating arsenic methylation in humans. Toxicol Appl Pharmacol 235:338–350

Arsenic and Aquaporins

Gerd Patrick Bienert and François Chaumont
Institut des Sciences de la Vie, Universite catholique de Louvain, Louvain-la-Neuve, Belgium

Synonyms

Aquaporin-mediated arsenic transport; Arsenic channels; Major intrinsic proteins and arsenic transport

Definition

Aquaporins or major intrinsic proteins (MIPs) are transmembrane channel proteins, which facilitate the passive and bidirectional diffusion of water and a variety of small and noncharged compounds across biological membranes. Aquaporins are found in organisms of all kingdoms of life and are present in all main subcellular membrane systems. The substrate specificity/spectra of aquaporins are highly isoform-dependent. Several isoforms from bacteria, protozoa, yeasts, mammals, fishes, and plants were shown to facilitate the transmembrane diffusion of diverse arsenic species (i.e., arsenite, monomethylarsonous acid, monomethylarsonic acid, and dimethylarsinic acid). All organisms face the challenge to handle considerable variations in the concentration of metalloids which they are exposed to, in terms of either the demand to acquire sufficient amounts for their metabolism or, conversely, the necessity to extrude them to prevent toxicity. This is achieved through homeostatic processes that require, among others, aquaporin-mediated transport across membranes at the cellular level.

Chemical Prerequisites for Metalloid Substrates to Pass Through Aquaporins

Aquaporins are membrane proteins known as diffusion facilitators for water and small uncharged solutes (see ▶ Aquaporins and Transport of Metalloids). Among the latter, the inorganic undissociated arsenite (As(OH)$_3$) molecule fulfills all requirements to pass through certain aquaporins. The molecular volume, electrostatic charge distribution, polarity, and capacity to form hydrogen bonds strikingly resemble the glycerol molecule, the typical substrate of aquaglyceroporins (Porquet and Filella 2007). Furthermore, with pKa$_1$, pKa$_2$, and pKa$_3$ values of 9.2, 12.1, and 13.4, respectively, arsenite is mostly undissociated at normal physiological pH conditions (>94% undissociated molecules at pH < 8) and, therefore, represents a typical substrate for aquaporins (Zhao et al. 2010b; Bienert et al. 2008b). Antimonite (Sb(OH)$_3$) is highly similar to arsenite both in terms of (bio)chemical reactivity and in physicochemical parameters (Porquet and Filella 2007).

Aquaporin-Mediated Arsenic and Antimony Transport in Bacteria

Important contributions to metalloid and in particular arsenic transport pathways and detoxification systems in bacteria were made by Rosen and collaborators. In a screen aiming at identifying metalloid transporters in *Escherichia coli*, they isolated a mutant that was resistant to extracellularly applied antimonite. The gene carrying the mutation turned out to be the GlpF aquaglyceroporin. EcGlpF is the major player for arsenite and antimonite uptake into *E. coli* (Rosen and Tamas 2010). As aquaglyceroporins are widespread in prokaryotes, it is likely that bioavailable arsenite is adventitiously taken up by these channels. Consequently, active bacterial arsenic efflux pumps have to transport it out of the cells again. It is also obvious that aquaglyceroporins can be used to detoxify bacteria from arsenite, especially when grown in oxidizing conditions. In these conditions, arsenate is the predominant arsenic species and is taken up by phosphate transporters, which cannot distinguish between arsenate and phosphate molecules due to chemical similarity. Arsenate is reduced to arsenite by arsenate reductases, which then has to be pumped out actively in a process costing metabolic energy (Rosen and Tamas 2010). However, export could occur without the cost of metabolic energy via passive diffusion through aquaglyceroporins driven by the naturally occurring chemical gradient towards the outside of the cells. That bacterial aquaglyceroporins might be used for arsenic detoxification stayed a hypothesis until Rosen and coworkers discovered that an aquaglyceroporin (AqpS) is encoded by a gene found in the arsenic resistance (*ars*) operon in specific bacteria such as *Sinorhizobium meliloti*. In this bacterial strain, it replaces the typically occurring *arsB* gene, which encodes an active arsenite extrusion transporter (Rosen and Tamas 2010). The combination of AqpS and the arsenate reductase ArsC forms an effective pathway for cell arsenate detoxification. This provides a clear evidence that aquaporin-mediated arsenic transport is not only adventitious and responsible for arsenic influx but plays a physiological role in arsenite efflux and detoxification. The exchange of *arsB*, encoding an ATP-driven arsenite extrusion pump found in many bacterial *ars* operons, by *aqp*S, encoding an aquaglyceroporin, is an interesting example of how bacteria develop genetic and metabolic strategies to be optimally adapted to all kinds of ecological niches and environments. Another exciting discovery of such an evolutional optimization event in bacteria with respect to arsenic detoxification was made by Beitz and coworkers. They identified that actinobacteria from soil (*Frankia alni*) and marine environments (*Salinispora tropica*) genetically combined the two-step process of arsenate reduction by ArsC and the extrusion by the ATP-driven arsenite extrusion pump (ArsB) to a one-step resistance process mediated by a single gene (Wu et al. 2010). This gene is translated into a protein which consists of the fusion of an aquaglyceroporin to a C-terminal arsenate reductase domain of a phosphotyrosine-phosphatase origin. These bacteria engineered a functional system, which is constituted by one protein enzymatically generating the substrate of its transport capacity. This allows increasing the concentration of the toxic substrate only locally at the place of transport, preventing toxic cellular accumulation (Wu et al. 2010). This genetic fusion of a metabolic arsenate-reducing enzyme to an arsenic-permeable aquaglyceroporin is another example for the physiological involvement of aquaporins in arsenic detoxification and transport.

Aquaporin-Mediated Arsenic Transport in Yeast

The best-characterized aquaglyceroporin from yeast is the glycerol facilitator Fps1 from *Saccharomyces cerevisiae*, which is the functional homolog of the GlpF aquaglyceroporin from *E. coli*. The major physiological role of Fps1p is the regulation of intracellular levels of glycerol in response to changing osmotic conditions and stresses (Rosen and Tamas 2010). In addition, based on both genetic and direct transport measurements, it was shown that Fps1p mediates the bidirectional transmembrane transport of arsenite and antimonite (Rosen and Tamas 2010; Wysocki and Tamás 2010; Maciaszczyk-Dziubinska et al. 2012). Interestingly, regulation of Fps1 at the transcriptional and posttranscriptional levels physiologically controls its channel function with respect to different substrates. When *S. cerevisiae* is externally exposed to either antimonite or arsenite, the transcription of the *Fps1* gene is rapidly downregulated to decrease the total amount of proteins and reduce the entry of metalloids. Furthermore, the activity of plasma membrane-localized Fps1 proteins is inhibited through phosphorylation of an amino acid residue located in the N-terminus, resulting in channel closure (Rosen and Tamas 2010; Wysocki and Tamás 2010). It is assumed that metalloids activate the mitogen-activated protein kinase Hog1p which subsequently phosphorylates and closes Fps1p. Altogether, these data indicate that Fps1p is involved in the physiological regulation of metalloid transport processes in *S. cerevisiae* (Rosen and Tamas 2010; Wysocki and Tamás 2010).

Aquaporin-Mediated Arsenic and Antimony Transport in Protozoan Species

The adventitious metalloid transport capacity of microbial aquaglyceroporins represents one of the molecular bases for an effective use of arsenic and antimony-containing drugs. Such drugs have been used against protozoan parasitic infections since the pre-antibiotic era. The arsenic-containing drug Melarsoprol and the antimony-containing drugs Pentostam and Glucantime are still used as the first-line treatments against leishmaniasis and sleeping sickness caused by trypanosomes and have not been replaced by other therapeutic agents since more than 40 years (Mukhopadhyay and Beitz 2010). All three aquaglyceroporins identified in *Trypanosoma brucei* are proposed to be involved in osmoregulation and glycerol transport and have been demonstrated to channel antimonite and arsenite (Mukhopadhyay and Beitz 2010). Whether the metalloid transport capacity is responsible for the uptake of metalloid-containing drugs and whether drug resistance is due to a reduced drug uptake into the parasites are not known yet. However, this link has been revealed for aquaglyceroporins from *Leishmania* species. For example, antimony-resistant isolates of *Leishmania donovani* showed a downregulation of one of their aquaglyceroporins, resulting in a reduced uptake of antimony (Mukhopadhyay and Beitz 2010). *Leishmania major* possesses five aquaporins including LmAQP1, which is permeable to arsenite and antimonite. Disruption of *LmAQP1* gene conferred increased antimony resistance while *Leishmania* overexpressing this isoform became hypersensitive to antimony-containing drugs (Mukhopadhyay and Beitz 2010). Altogether, these data suggest that adventitious metalloid transport by parasitic aquaglyceroporins crucially contributes to their high sensitivity to metalloid-containing drugs. On the other hand, parasites may develop drug resistance by suppressing or minimizing channel abundance under stress conditions or by changing channel properties. This latter adaptation might be achieved by maintaining selectivity of the channel for physiologically essential substrates while making it impermeable to detrimental metalloid species.

Aquaporin-Mediated Arsenic Transport in Mammals

Human and other mammalian genomes encode 13 aquaporins (AQP0–AQP12). AQP3, AQP7, AQP9, and AQP10 are aquaglyceroporins (Verkman 2011). The ability of AQP7 and AQP9 from rat, mouse, and human to transport arsenite and antimonite species was first demonstrated in transport assays performed in heterologous expression systems (*S. cerevisiae* and *Xenopus laevis* oocytes) (Liu 2010). In contrast, human AQP3 and AQP10 are not able to significantly transport metalloids in these systems. AQP9 displays the highest metalloid transport capacity and became a convenient model to further study metalloid transport of mammalian aquaglyceroporins (Liu 2010). Biochemical transport assays and mutagenesis approaches revealed that metalloids and glycerol use the same translocation pathway through AQP9. Moreover, AQP9 transports the organic methylated arsenic species monomethylarsonous acid (MMAs(III)), monomethylarsonic acid (MMAs(V)), and dimethylarsinic acid (DMAs(V)) (Liu 2010). In humans, arsenite is methylated in the liver to MMAs(V), MMAs(III), DMAs(V), and dimethylarsinous acid DMAs(III); and these compounds are then excreted into the bile and urine and cleared from the body (Liu 2010). This led to the hypothesis that, in addition to its role in arsenite uptake from the blood into the liver, AQP9 is also involved in the export of methylated derivatives of arsenic out of the liver back to the blood (Liu 2010). The characterization of AQP9-null mice, which do not express AQP9 anymore, revealed that these mice are more sensitive to arsenite exposure, accumulate more arsenic, and excrete less arsenic in the feces and urine compared to the wild-type mice (Carbrey et al. 2009), supporting the hypothesis that aquaglyceroporins are involved in arsenic detoxification in mammals.

Arsenic- and antimony-containing drugs have been used as pharmaceuticals for more than 2,000 years and were developed and applied systematically since the nineteenth century. Despite being proven human carcinogens, arsenicals are still used nowadays in therapies against a wide range of tumors which are resistant to other treatments. Arsenic trioxide (As_2O_3) is used as first-line therapeutics against acute promyelocytic leukemia in which an abnormal accumulation of immature leucocytes takes place in the bone marrow (Mukhopadhyay and Beitz 2010). As in the case of microbial aquaglyceroporins, the adventitious transport capacity of mammalian aquaglyceroporins for metalloids might be one of the reasons why arsenite trioxide represents such a powerful antitumor agent both in vitro and in vivo (Mukhopadhyay and Beitz 2010; Bienert et al. 2008b). Human leukemia cells overexpressing AQP9 accumulate more metalloids and concomitantly become hypersensitive to arsenic and antimony agents (Mukhopadhyay and Beitz 2010). AQP9 expression in leukemia cells of different cell lines correlated directly to arsenic trioxide sensitivity (Mukhopadhyay and Beitz 2010). Downregulation of aquaglyceroporins in cancer cells might turn them insensitive to arsenic-containing drugs. As an example, a human lung adenocarcinoma cell line, which became significantly more drug-resistant compared to the parental cell line, accumulated less arsenic which was correlated to a concomitant downregulation of AQP3 (Mukhopadhyay and Beitz 2010). The development of pharmacological compounds able to increase the expression or activity of aquaglyceroporins could increase the efficiency of arsenic-containing drugs. Two such substances have already been described: vitamin D and all-trans-retinoic acid, which both increase the expression of AQP9 in cancer cells and, concomitantly, their hypersensitivity to arsenic trioxide due to an increased arsenic uptake rate (reviewed in Bienert et al. 2008b; Mukhopadhyay and Beitz 2010).

Aquaporin-Mediated Arsenic and Antimony Transport in Plants

Physiological studies in plants suggested that channel-mediated passive diffusion is a major uptake mechanism for the noncharged arsenite molecule, and it was suggested that aquaporins are responsible for this transport capacity (Meharg and Jardine 2003). Among the five subfamilies of aquaporins found in higher plants (the plasma membrane intrinsic proteins (PIPs), the tonoplast intrinsic proteins (TIPs), the small-basic intrinsic proteins (SIPs), the nodulin26-like intrinsic proteins (NIPs), and the X-intrinsic proteins (XIPs) (▶ Aquaporins and Transport of Metalloids), NIPs represent the structural and functional analogs of microbial and mammalian aquaglyceroporins in terms of steric channel properties and substrate spectra (reviewed in Bienert et al. 2008b). NIPs mediate the transport of a wide range of small uncharged molecules including ammonia, hydrogen peroxide, urea, lactic acid, glycerol, boric acid, and silicic acid (Bienert and Chaumont 2011) and, therefore, were obvious candidates for testing their ability to channel arsenite and antimonite across membranes. When expressed in *S. cerevisiae* mutants which are highly resistant to elevated (externally applied) arsenite levels, NIPs from *Arabidopsis thaliana* (AtNIP5;1, AtNIP6;1, and AtNIP7;1), *Oryza sativa* (OsNIP2;1, OsNIP2;2, and OsNIP3;2), and *Lotus japonicus* (LjNIP5;1 and LjNIP6;1) significantly increased the cell sensitivity to arsenite and antimonite (Bienert et al. 2008a). It was further shown that this increased sensitivity is directly linked to an increased uptake of the toxic metalloids.

Other transport studies using Xenopus oocytes demonstrated a permeability to arsenite of NIPs from *A. thaliana* (AtNIP1;1, AtNIP1;2, AtNIP5;1, and AtNIP7;1) and *O. sativa* (OsNIP1;1, OsNIP2;1, OsNIP2;2, and OsNIP3;1) (Bienert and Jahn 2010; Zhao et al. 2010b). The impact of NIPs on arsenite uptake, accumulation, and tolerance was demonstrated to be of major biological relevance using *nip* knockout mutants in both Arabidopsis and rice. In Arabidopsis, a forward genetic screen testing the growth behavior of mutagenized lines on medium containing root growth-inhibiting concentrations of arsenite led to the isolation of three independent arsenite-tolerant lines, all of them having a mutation in the *AtNIP1;1* gene (Kamiya et al. 2009). Further experiments clearly confirmed the dominant role of AtNIP1;1 in arsenite uptake.

In rice, mutated lines for OsNIP2;1 (an aquaporin which functions as an essential silicon influx channel in roots (▶ Silicon and Aquaporins)) were isolated and exhibited a significantly decreased arsenite uptake capacity. This indicated that OsNIP2;1 represents a major uptake pathway for arsenite. At the same time, in arsenate stress condition, OsNIP2;1 promoted arsenic detoxification by channeling arsenite out of the roots (Ma et al. 2008; Zhao et al. 2010a). Information on arsenite entry routes into rice plants is highly valuable for strategies aiming at breeding or engineering low-arsenic-accumulating cultivars. Arsenic contamination of rice-based food products is a public health issue as rice actually accounts for the largest contribution to dietary intake of inorganic arsenic for the population (Zhao et al. 2010b; Bienert and Jahn 2010). High arsenic accumulation in rice is probably promoted by the following facts: (1) rice is cultivated in regions where bioavailable arsenic is present at high levels in soils. (2) Rice is grown in paddy soils, in which arsenite is the predominant arsenic species due to the reductive conditions. Arsenite is more mobile and more bioaccessible than arsenate, which is predominant in oxidative field conditions. (3) Last but not least, rice uses NIP-mediated uptake pathways to ensure assimilation of large amounts of silicon, a physiologically highly beneficial metalloid (▶ Silicon and Aquaporins). However, adventitious uptake of toxic arsenite molecules through these channels cannot be prevented. Moreover, other known metalloid transporters are present and active such as Lsi2 protein which in addition to silicon also transports arsenic species (▶ Silicon and Aquaporins).

OsNIP2;1 also mediates the uptake of undissociated pentavalent MMA and DMA (Zhao et al. 2010b). The *lsi1* rice mutant (standing for "*low silicon 1*") carrying a loss-of-function mutation in OsNIP2;1 took up less than 50% of these organic arsenic species compared to wild-type plants. In nature, these methylated arsenic species derive most probably from soil microorganisms.

In summary, permeability to arsenite (and maybe to some organic arsenic species) seems to be a common feature to all three different subclasses of plant NIPs (NIPI, NIPII, and NIPIII) even though they differ in their amino acid residue composition of their selectivity filter (Danielson and Johanson 2010). Whether some NIPs are involved in arsenic detoxification mechanisms in certain plant species has to be resolved. Differential regulations of NIPs at the transcriptional and post-translational level in response to arsenic and/or other physiologically relevant substrates would suggest such a role.

When heterologously expressed in Xenopus oocytes, three members of the PIP subfamily of rice aquaporins (OsPIP2;4, OsPIP2;6, and OsPIP2;7) significantly increase arsenic uptake compared to control oocytes (Mosa et al. 2012). These PIP isoforms are highly expressed both in the roots and the shoots in standard conditions. Overexpression of these PIPs in Arabidopsis resulted in an increased arsenite tolerance and higher biomass accumulation. Moreover, the lines overexpressing OsPIP2;6 exhibited both influx and efflux capacity of arsenite depending on the metalloid concentration gradient between the external medium and the root (Mosa et al. 2012). Whether other PIPs or plant aquaporin isoforms have the ability to transport arsenic species needs to be investigated as well as their role in plant arsenic homeostasis.

Concluding Remarks

Aquaporins constitute indispensable and high-capacity transport systems for reduced and uncharged forms of the toxic metalloids arsenic and antimony. These channels fulfill physiologically relevant roles in arsenic homeostasis, which is supported by several findings: (1) aquaporin-mediated arsenite transport is conserved in all kingdoms of life, (2) mutants lacking aquaporins are impaired in uptake or extrusion of various arsenic species, (3) aquaporins are regulated at the transcriptional and post-translational level in response to arsenic stress, and (4) aquaporins are part of arsenic resistance operons and form fusion proteins with arsenate reductases in some bacteria, observations providing genetic evidence for their role in arsenic resistance mechanisms.

Funding

This work was supported by grants from the Belgian National Fund for Scientific Research (FNRS), the Interuniversity Attraction Poles Programme–Belgian Science Policy, and the "Communauté française de Belgique–Actions de Recherches Concertées." GPB was supported by a grant from the FNRS.

Cross-References

▶ Aquaporins and Transport of Metalloids
▶ Boron and Aquaporins
▶ Selenium and Aquaporins
▶ Silicon and Aquaporins

References

Bienert GP, Chaumont F (2011) Plant aquaporins: roles in water homeostasis, nutrition, and signalling processes. Springer, Berlin

Bienert GP, Jahn TP (2010) Major intrinsic proteins and arsenic transport in plants: new players and their potential role. Landes Bioscience-Springer, New York

Bienert GP, Thorsen M, Schüssler MD et al (2008a) A subgroup of plant aquaporins facilitate the bi-directional diffusion of As(OH)$_3$ and Sb(OH)$_3$ across membranes. BMC Biol 6:26

Bienert GP, Schüssler MD, Jahn TP (2008b) Metalloids: essential, beneficial or toxic? Major intrinsic proteins sort it out. Trends Biochem Sci 33:20–26

Carbrey JM, Song L, Zhou Y et al (2009) Reduced arsenic clearance and increased toxicity in aquaglyceroporin-9-null mice. Proc Natl Acad Sci USA 106:15956–15960

Danielson JAH, Johanson U (2010) Phylogeny of major intrinsic proteins. Landes Bioscience-Springer, New York

Kamiya T, Tanaka M, Mitani N et al (2009) NIP1;1, an aquaporin homolog, determines the arsenite sensitivity of Arabidopsis thaliana. J Biol Chem 284:2114–2120

Liu Z (2010) Roles of vertebrate aquaglyceroporins in arsenic transport and detoxification. Landes Bioscience-Springer, New York

Ma JF, Yamaji N, Mitani N et al (2008) Transporters of arsenite in rice and their role in arsenic accumulation in rice grain. Proc Natl Acad Sci USA 105:9931–9935

Maciaszczyk-Dziubinska E, Wawrzycka D, Wysocki R et al (2012) Arsenic and antimony transporters in eukaryotes. Int J Mol Sci 13:3527–3548

Mehard AA, Jardine L (2003) Arsenite transport into paddy rice (Oryza sativa) roots. New Phytol 157:39–44

Mosa KA, Kumar K, Chhikara S et al (2012) Members of rice plasma membrane intrinsic proteins subfamily are involved in arsenite permeability and tolerance in plants. Transgenic Res (in press)

Mukhopadhyay R, Beitz E (2010) Metalloid transport by aquaglyceroporins: consequences in the treatment of human diseases. Landes Bioscience-Springer, New York

Porquet A, Filella M (2007) Structural evidence of the similarity of Sb(OH)$_3$ and As(OH)$_3$ with glycerol: implications for their uptake. Chem Res Toxicol 20:1269–1276

Rosen BP, Tamas MJ (2010) Arsenic transport in prokaryotes and eukaryotic microbes. Landes Bioscience-Springer, New York

Verkman AS (2011) Aquaporins at a glance. J Cell Sci 124:2107–2112

Wu B, Song J, Beitz E (2010) Novel channel enzyme fusion proteins confer arsenate resistance. J Biol Chem 285: 40081–40087

Wysocki R, Tamás MJ (2010) How *Saccharomyces cerevisiae* copes with toxic metals and metalloids. FEMS Microbiol Rev 34:925–951

Zhao FJ, Ago Y, Mitani N et al (2010a) The role of the rice aquaporin Lsi1 in arsenite efflux from roots. New Phytol 186:392–399

Zhao FJ, McGrath SP, Meharg AA (2010b) Arsenic as a food chain contaminant: mechanisms of plant uptake and metabolism and mitigation strategies. Annu Rev Plant Biol 61:535–559

Arsenic and Primary Human Cells

Denis Girard
Laboratoire de recherche en inflammation et physiologie des granulocytes, Université du Québec, INRS-Institut Armand-Frappier, Laval, QC, Canada

Synonyms

Apoptosis; Arsenical compounds; Mode of action

Definition of the Subject

The subject of arsenic and primary human cells refers to the role of arsenical compounds in the physiology of cells of primary origin. The importance of this subject resides in the fact that most of the studies reported in the literature focus on the role of arsenic in cancer cells or in immortalized cancer cell lines, but very few deal with the arsenic-primary cell interaction. Primary cells are those that are directly isolated from blood (immune origin) or isolated from a tissue or organ (nonimmune origin) like epithelial or hepatocyte cells. Because arsenical compounds may be toxic or, paradoxically, could be used in the treatment of diverse diseases, it is important to better understand how these compounds alter primary cells of different origins.

Generalities

Metalloids are elements that are neither metals nor nonmetals and this group includes boron, silicon, germanium, tellurium, antimony, and arsenic. Arsenic (As) is among the most intensively studied elements in the field of metal toxicology. The term "arsenic" is, however, frequently erroneously used in the literature since it includes several distinct arsenical compounds, including the potent anticancer drug arsenic trioxide (ATO). Humans can be exposed to arsenic via air and food, but the major exposure route is through the contaminated drinking water where millions of humans are exposed worldwide to this environmental toxicant (Yoshida et al. 2004). It is well established that arsenic poisoning via groundwater is a worldwide problem; arsenic-contaminated groundwater has been found in aquifers in Bangladesh, China, India, Nepal, Argentina, Mexico, and Taiwan (Nicolli et al. 2012; Yoshida et al.2004). In addition to causing significant problems in the provision of safe drinking water, elevated concentrations of arsenic in water raised also concern regarding food safety. Long-term exposure to arsenic is associated with cancers of the lungs, liver, skin, urinary tract, kidneys, etc. Epidemiologic and experimental evidences support the conjecture that arsenic could play a role in hypertension and cardiovascular diseases (Chen et al. 1995). For example, a positive relationship between inorganic arsenic exposure from drinking water and hypertension was reported in epidemiologic studies conducted in arsenic-endemic areas in Taiwan and Bangladesh (Chen et al. 1995, 2007; Rahman et al. 1999).

Paradoxically, arsenic compounds have been used in traditional oriental medicine to treat a variety of diseases other than cancers, such as inflammatory diseases (Sears 1988). Experimental studies have indicated that arsenic exposure may be involved in the development of hypertension through the promotion of inflammation, oxidative stress, and endothelial dysfunction (Aposhian et al. 2003; Balakumar et al. 2008). However, it was only in the 1970s that arsenic agents were found to be highly effective for treating several types of leukemia (Sears 1988). To date, ATO is considered to be one of the most potent drugs for chemotherapy of cancers. In addition, arsenic was recently reported to be a potential candidate for the treatment of a variety of diseases, including arthritis (Mei et al. 2011) and asthma (Chu et al. 2010).

Route of Transport of Arsenic Through the Cell Membrane

Like other metalloids, arsenic passes through cell membrane via channels involved in passive transport, such as aquaporins and aquaglyceroporins (Zangi and Filella 2012). While aquaporins selectively conduct water molecules, aquaglyceroporins will transport also other small uncharged molecules. They are bidirectional channels operating via passive diffusion. The direction of the transport depends on the concentration gradient across the membrane; if the intracellular concentration of the solute is too high and causes damage, the channel will facilitate elimination of the compound out of the cell if the extracellular concentration is lower. Therefore, there are no specific receptors identified for arsenical compounds. Nevertheless, the role of aquaporin 9 in accumulation of arsenic and its cytotoxicity in primary mouse hepatocytes was reported (Shinkai et al. 2009). Aquaporin 9 is a member of the aquaglyceroporin subfamily of aquaporins involved in the transfer of water and small solutes, including arsenite. Since arsenic toxicity is largely associated with intracellular accumulation of this metalloid inside the cells, pretreatment with sorbitol, a competitive inhibitor of aquaporin 9, and siRNA-mediated knockdown of aquaporin 9 were found to significantly decrease the arsenite uptake in the cell and its cytotoxicity, suggesting that aquaporin 9 is a channel for arsenite in primary mouse hepatocytes.

Toxicity and Mode of Action

The toxicity of arsenic compounds is partly explained by their ability to bind and inactivate several sulfhydryl-containing proteins and enzyme systems. Oxidative stress is certainly among the more documented mechanisms of arsenic toxicity. Arsenic acts by causing a marked imbalance between production of reactive oxygen species (ROS) and antioxidant defense resulting in alteration of the cellular redox status. This is explained by the fact that it is the balance between ROS production and antioxidant defenses that determines the degree of oxidative stress. However, the precise mechanisms by which arsenic induces oxidative stress are not yet fully understood but it is well accepted that ROS exert their effects by modulating cell apoptosis. Thus, the principal mode of action of arsenic is via its ability to induce cell apoptosis via ROS but also via caspases activation (Miller et al. 2002). The cytoskeleton, especially the microtubules, is an important cellular target for arsenic, probably due to the high sulfhydryl protein content of cytoskeletal components. However, in addition to microtubules, intermediate filaments and microfilaments are also important targets of arsenic. For example, ATO induces the cleavage of several microfilament-associated proteins, including myosin, paxillin, gelsolin, and actin as well as of the two intermediate filament proteins lamin B1 and vimentin. The cleavage of these cytoskeletal proteins occurs via caspase activation since treatment of cells with caspase inhibitors reversed the ability of arsenic to induce the cleavage of these proteins (Binet et al. 2006). Of note, ATO induces cell differentiation when used at a low concentration (ranging between 0.5 and 5 µM), while it induces cell apoptosis at concentrations ranging from 5 to 20 µM in a variety of cell types. Treatment of acute promyelocytic leukemia (APL) with ATO is associated with the disappearance of the PML-RARα fusion transcript, the characteristic APL gene product of the chromosomal translocation t(15;17). ATO can affect the PML portion of the aberrant protein since the PML is characterized by the presence of a cysteine-rich region.

Cell Signaling

Arsenic is known to activate the mitogen-activated protein kinase (MAPK) pathway in a variety of cancer cells. For example, ATO-induced U937 cell apoptosis was found to be dependent on activation of p38 and inactivation of extracellular signal-regulated kinases-1/2 (ERK-1/2). The p38 mitogen-activated protein kinase pathway was also involved in ATO-induced human acute promyelocytic leukemia (APL) NB-4 cells, K562 CML-blast crisis cell line as well as in the MCF-7 human breast carcinoma and the LNKAP prostate carcinoma cell lines (Binet et al. 2009). Activation of c-jun N-terminal kinase (JNK) is also an event occurring in ATO-induced apoptosis in APL. Recently, it was demonstrated that ATO could downregulate survivin, a member of the inhibitor of apoptosis family that is highly expressed in various cancer cells, via activation of both p38 and JNK in apoptotic human lung adenocarcinoma cell line

H1355. In neutrophils, ATO was found to recruit p38 mitogen-activated protein kinase and/or c-jun NH$_2$-terminal MAPK but not ERK-1/2 (Binet and Girard 2008). Therefore, ATO can act by modulating all the three major MAPK pathways, namely, p38, Erk-1/2, and JNK, but a certain selectivity exists depending on the cell types since the activation of these three MAPKs is not necessarily systematically observed.

Arsenic and Primary Cells

Although the vast majority of studies investigating the role of arsenic compounds in cell physiology have been conducted with cancer cell lines or in immature cancer cells isolated from patients, their roles on primary cells have retained less attention (Binet et al. 2009). Due to its high efficiency for the treatment of APL, ATO is certainly one of the most studied arsenical compounds. Cells of immune origin, including B or T lymphocytes, monocytes, macrophages, and, more recently, neutrophils, are important targets of arsenical compounds.

As for a variety of cancer cell lines, the main mechanism of action of arsenic in primary immune cells is also via induction of cell apoptosis. Inorganic arsenic was found to significantly repress major functions of human T-lymphocytes and macrophages (Lemarie et al. 2006, 2008). Arsenic blocks the differentiation of blood monocytes into functional macrophages by inhibiting survival signaling pathways and, at noncytotoxic concentrations, arsenic reverses, at least partially, the phenotypic and genotypic features of mature macrophages (Bourdonnay et al. 2009). While both ATO and ascorbic acid mediate cytotoxicity in chronic lymphocytic leukemia B lymphocytes when used alone, the efficacy of ATO is enhanced by ascorbic acid (Biswas et al. 2010). In vitro experiments conducted with peripheral blood mononuclear cells from healthy individuals indicated that low concentrations of arsenic tended to increase the number of natural T regulatory lymphocytes. In contrast, higher concentrations (>5.0 µM) decreased the cell number (Hernandez-Castro et al. 2009). In neutrophils, ATO induced apoptosis by a mechanism involving caspase activation, cytoskeletal breakdown, ROS production, and de novo protein synthesis. ATO also induced an endoplasmic reticulum stress in these cells (Binet et al. 2006, 2009, 2010). Using pharmacological inhibitors, the proapoptotic activity of ATO was found to occur by a MAPK-independent mechanism. In contrast, the ability of ATO to enhance adhesion, migration, phagocytosis, release, and activity of gelatinase, but not azurophilic granules, is dependent upon activation of p38 and/or JNK.

Recently, the role of arsenic in a variety of cells of primary origin other than immune cells has been documented. Arsenite (sodium meta-arsenite, NaAsO$_2$) was found to decrease the expression of tumor necrosis factor-alpha and vascular endothelial growth factor, two important inflammatory mediators, in primary human hepatocytes (Noreault-Conti et al. 2012). In another study, this arsenical compound inhibited glutamate metabolism in human primary cultured astrocytes probably responsible or partially involved in arsenic-induced neurotoxicity (Zhao et al. 2012). Investigating the differential sensitivity of primary human cultured chorion and amnion cells prepared from fetal membranes to As^{3+}, aquaporin 9 and multidrug resistance-associated protein were found to be involved in the control of cellular As^{3+} accumulation (Yoshino et al. 2011). While ATO is known to induce apoptosis in various cancer cells including lung cancer cells, little is known about the toxicological effects of ATO on normal primary lung cells. In one study, the cellular effects of ATO on human pulmonary fibroblast cells in relation to cell growth inhibition and death were reported (Park and Kim 2012). ATO was found to inhibit the cell growth with an IC(50) of $\sim$30–40 µM at 24 h and induced cell death. This was accompanied by the loss of mitochondrial membrane potential. Human pulmonary fibroblasts were considered to be highly resistant to ATO. The mode of action involved an increased expression of p53 and decrease of the antiapoptotic Bcl-2 protein. Interestingly, ATO activated caspase-8 but not caspase-3 in these cells and administration of caspase-8 siRNA attenuated pulmonary fibroblast cell death whereas caspase-3 siRNAs did not, indicating that ATO induced the growth inhibition and death in these cells via caspase-8 and not caspase-3. A possible involvement of arsenic in Alzheimer's disease has been recently proposed (Gong and O'Bryant 2010). In this respect, a study reported that sodium arsenite and its main metabolite, dimethylarsinic acid, affected the expression and processing of the amyloid precursor protein, using primary neuronal cells. It was demonstrated that sodium arsenite and dimethylarsinic acid up-regulated the expression and processing of amyloid precursor protein in vitro. After proteolysis, this precursor protein generates beta amyloid, the primary component of amyloid plaques found in the brains of Alzheimer's disease patients (Zarazua et al. 2011). In an immunotoxicological study, ATO was administered in mice (inhalation exposure) and spleen was used for several assays. No spleen cell cytotoxicity was observed and there were no changes in spleen cell surface marker expression for B and T lymphocytes, macrophages, and natural killer cells (Burchiel et al. 2009). Also, the cell proliferation of both B and T lymphocytes in response to mitogens was not affected by arsenic treatment, and no changes were found in the natural killer-mediated lysis of Yac-1 target cells. However, the primary T-dependent antibody response was highly susceptible to arsenic suppression. Because arsenic is associated with cardiac toxicity, a study was conducted in order to evaluate the cytotoxic effect of ATO on cardiac myocytes (Raghu and Cherian 2009). To do so, primary culture of rat myocytes was treated with different concentrations of ATO for various periods of time and the cardiac toxicity was assessed by monitoring cell viability, mitochondrial and deoxyribonucleic acid integrity, ROS generation, calcium overload, and cell apoptosis. ATO was found to alter mitochondrial integrity, generation of ROS, calcium overload, and apoptosis in a concentration- and time-dependent manner. However, no DNA fragmentation was observed. Therefore, it was concluded that ATO induces apoptosis in cardiomyocytes by generation of ROS and the induction of calcium overload. Exposure of mouse fetal liver cells to sodium arsenite ($NaAsO_2$) was found to induce adaptive responses and aberrant gene expression (Liu et al. 2008). For example, expression of genes related to steroid metabolism, such as 17beta-hydroxysteroid dehydrogenase-7 and Cyp2a4, was increased approximately twofold. This study indicates that the aberrant gene expression observed in response to arsenic insults could alter genetic programming very early in life, potentially contributing to tumor formation later in life. Primary rat vascular smooth muscle cells were isolated from aortic explants from adult Sprague Dawley rats and were exposed to arsenic ($NaAsO_2$) in order to elucidate cell signaling events occurring in response to arsenic insults (Pysher et al. 2008). It was observed that arsenic can alter focal adhesion protein co-association leading to activation of downstream pathways. More specifically, arsenic caused a sustained increase in focal adhesion kinase-Src association and activation, and stimulation of downstream PAK, ERK and JNK pathways known to be involved in cellular survival, growth, proliferation, and migration in vascular smooth muscle cells.

All of the above studies clearly indicated that arsenical compounds can alter the physiology of primary cells of a variety of origins. This includes cells of immune origin like B and T lymphocytes, natural killer cells, monocytes, macrophages and neutrophils and in cells of nonimmune origin like hepatocytes, astrocytes, chorion and amnion cells prepared from fetal membranes, pulmonary fibroblasts, neuronal cells, myocytes, fetal liver cells, and vascular smooth muscle cells. Also, these studies demonstrate the importance to pursue investigations of primary cells in response to arsenic insults in order to better control potential arsenic toxicity in cells of primary origin and, in parallel, to better elucidate the complex mode of action of arsenical compounds, including the cell signaling pathways involved in a biological response.

Utilization of Arsenical Compounds in Combination with Other Agents

In addition to its direct effect on cells of primary origin, the utilization of arsenic in combination with other agents is becoming a very interesting avenue of research for the development of therapeutic strategies. The main objective is to decrease the concentration as much as possible of one arsenical compound and to mix it with another agent, limiting potential toxic effects. For example, ATO has been recently used in combination with silibinin, a natural polyphenolic flavonoid, in glioblastoma multiform cell line, U87MG and the results showed that ATO, in some cases, improved and/or complemented the anticancer effects, suggesting a new combination therapy for the highly invasive human glioma treatment (Dizaji et al. 2012). Combined administration of suberoylanilide hydroxamic acid and ATO was recently found to be an effective approach to the treatment of lung cancer (Chien et al. 2011). In combination with blocking monoclonal antibodies against CD154 and LFA-1, ATO was found to prolong heart allograft survival in allo-primed T cells–transferred mice (Lin et al. 2011). Genistein, an isoflavone known to inhibit tyrosine kinases, was reported to potentiate the effect of ATO in human hepatocellular carcinoma (Ma et al. 2011).

Interestingly, one study reported that Imatinib Mesylate induced mainly the intrinsic pathway of cell apoptosis, whereas ATO induced the endoplasmic reticulum stress-mediated pathway of cell apoptosis and that the combination of these two anticancer drugs was more effective for inducing the intrinsic, extrinsic, and ER stress-mediated pathways of cell apoptosis, resulting in a more effective and efficient induction of apoptosis in K562 cancer cells (Du et al. 2006). Therefore, in cancer therapy, it could be possible to further accelerate cell apoptosis using drugs that when combined together will activate different cell apoptotic pathways. In human peripheral blood lymphocytes, a combination of curcumin with arsenic was found to ameliorate the toxic effect of arsenic when used alone by reducing the frequency of structural aberrations, hypoploidy and primary DNA damage (Tiwari and Rao 2010).

This strategy is not restricted to ATO, since other arsenical compounds are known to act in synergy with different drugs. Arsenic disulfide (As_2S_2) was reported to synergize with a phosphoinositide 3-kinase inhibitor (PI-103) to eradicate acute myeloid leukemia stem cells by inducing their differentiation (Hong et al. 2011). Tetra-arsenic oxide (As_4O_6), in combination with paclitaxel, was found to increase apoptosis in vitro, in gastric, cervix and head and neck cancer cell lines (Chung et al. 2009). A combination of arsenic sulfide (As_4S_4) and Imatinib was found to possess more profound therapeutic effects than As_4S_4 or Imatinib used alone in a mouse model of chronic myeloid leukemia (Zhang et al. 2009).

Cross-References

▶ Arsenic and Aquaporins
▶ Arsenic in Nature
▶ Arsenic in Therapy
▶ Arsenic, Free Radical and Oxidative Stress

References

Aposhian HV, Zakharyan RA, Avram MD, Kopplin MJ, Wollenberg ML (2003) Oxidation and detoxification of trivalent arsenic species. Toxicol Appl Pharmacol 193:1–8

Balakumar P, Kaur T, Singh M (2008) Potential target sites to modulate vascular endothelial dysfunction: current perspectives and future directions. Toxicology 245:49–64

Binet F, Antoine F, Girard D (2009) Interaction between arsenic trioxide and human primary cells: emphasis on human cells of myeloid origin. Inflamm Allergy Drug Targets 8:21–27

Binet F, Cavalli H, Moisan E, Girard D (2006) Arsenic trioxide (AT) is a novel human neutrophil pro-apoptotic agent: effects of catalase on AT-induced apoptosis, degradation of cytoskeletal proteins and de novo protein synthesis. Br J Haematol 132:349–358

Binet F, Chiasson S, Girard D (2010) Arsenic trioxide induces endoplasmic reticulum stress-related events in neutrophils. Int Immunopharmacol 10:508–512

Binet F, Girard D (2008) Novel human neutrophil agonistic properties of arsenic trioxide: involvement of p38 mitogen-activated protein kinase and/or c-jun NH_2-terminal MAPK but not extracellular signal-regulated kinases-1/2. J Leukoc Biol 84(6):1613–1622

Biswas S, Zhao X, Mone AP, Mo X, Vargo M, Jarjoura D, Byrd JC, Muthusamy N (2010) Arsenic trioxide and ascorbic acid demonstrate promising activity against primary human CLL cells in vitro. Leuk Res 34:925–931

Bourdonnay E, Morzadec C, Sparfel L, Galibert MD, Jouneau S, Martin-Chouly C, Fardel O, Vernhet L (2009) Global effects of inorganic arsenic on gene expression profile in human macrophages. Mol Immunol 46:649–656

Burchiel SW, Mitchell LA, Lauer FT, Sun X, McDonald JD, Hudson LG, Liu KJ (2009) Immunotoxicity and biodistribution analysis of arsenic trioxide in C57Bl/6 mice following a 2-week inhalation exposure. Toxicol Appl Pharmacol 241:253–259

Chen CJ, Hsueh YM, Lai MS, Shyu MP, Chen SY, Wu MM, Kuo TL, Tai TY (1995) Increased prevalence of hypertension and long-term arsenic exposure. Hypertension 25:53–60

Chen Y, Factor-Litvak P, Howe GR, Graziano JH, Brandt-Rauf P, Parvez F, van Geen A, Ahsan H (2007) Arsenic exposure from drinking water, dietary intakes of B vitamins and folate, and risk of high blood pressure in Bangladesh: a population-based, cross-sectional study. Am J Epidemiol 165:541–552

Chien CW, Yao JH, Chang SY, Lee PC, Lee TC (2011) Enhanced suppression of tumor growth by concomitant treatment of human lung cancer cells with suberoylanilide hydroxamic acid and arsenic trioxide. Toxicol Appl Pharmacol 257:59–66

Chu KH, Lee CC, Hsin SC, Cai BC, Wang JH, Chiang BL (2010) Arsenic trioxide alleviates airway hyperresponsiveness and eosinophilia in a murine model of asthma. Cell Mol Immunol 7:375–380

Chung WH, Sung BH, Kim SS, Rhim H, Kuh HJ (2009) Synergistic interaction between tetra-arsenic oxide and paclitaxel in human cancer cells in vitro. Int J Oncol 34:1669–1679

Dizaji MZ, Malehmir M, Ghavamzadeh A, Alimoghaddam K, Ghaffari SH (2012) Synergistic effects of arsenic trioxide and silibinin on apoptosis and invasion in human glioblastoma U87MG cell line. Neurochem Res 37:370–380

Du Y et al (2006) Coordination of intrinsic, extrinsic, and endoplasmic reticulum-mediated apoptosis by imatinib mesylate combined with arsenic trioxide in chronic myeloid leukemia. Blood 107:1582–1590

Gong G, O'Bryant SE (2010) The arsenic exposure hypothesis for Alzheimer disease. Alzheimer Dis Assoc Disord 13:13

Hernandez-Castro B, Doniz-Padilla LM, Salgado-Bustamante M, Rocha D, Ortiz-Perez MD, Jimenez-Capdeville ME, Portales-Perez DP, Quintanar-Stephano A,

Gonzalez-Amaro R (2009) Effect of arsenic on regulatory T cells. J Clin Immunol 29:461–469

Hong Z et al (2011) Arsenic disulfide synergizes with the phosphoinositide 3-kinase inhibitor PI-103 to eradicate acute myeloid leukemia stem cells by inducing differentiation. Carcinogenesis 32:1550–1558

Lemarie A, Bourdonnay E, Morzadec C, Fardel O, Vernhet L (2008) Inorganic arsenic activates reduced NADPH oxidase in human primary macrophages through a Rho kinase/p38 kinase pathway. J Immunol 180:6010–6017

Lemarie A, Morzadec C, Merino D, Micheau O, Fardel O, Vernhet L (2006) Arsenic trioxide induces apoptosis of human monocytes during macrophagic differentiation through nuclear factor-kappaB-related survival pathway down-regulation. J Pharmacol Exp Ther 316:304–314

Lin Y, Dai H, Su J, Yan G, Xi Y, Ekberg H, Chen J, Qi Z (2011) Arsenic trioxide is a novel agent for combination therapy to prolong heart allograft survival in allo-primed T cells transferred mice. Transpl Immunol 25:194–201

Liu J, Yu L, Tokar EJ, Bortner C, Sifre MI, Sun Y, Waalkes MP (2008) Arsenic-induced aberrant gene expression in fetal mouse primary liver-cell cultures. Ann N Y Acad Sci 1140:368–375

Ma Y, Wang J, Liu L, Zhu H, Chen X, Pan S, Sun X, Jiang H (2011) Genistein potentiates the effect of arsenic trioxide against human hepatocellular carcinoma: role of Akt and nuclear factor-kappaB. Cancer Lett 301:75–84

Mei Y, Zheng Y, Wang H, Gao J, Liu D, Zhao Y, Zhang Z (2011) Arsenic trioxide induces apoptosis of fibroblast-like synoviocytes and represents antiarthritis effect in experimental model of rheumatoid arthritis. J Rheumatol 38:36–43

Miller WHJ, Schipper HM, Lee JS, Singer J, Waxman S (2002) Mechanisms of action of arsenic trioxide. Cancer Res 62:3893–3903

Nicolli HB, Garcia JW, Falcon CM, Smedley PL (2012) Mobilization of arsenic and other trace elements of health concern in groundwater from the Sali River Basin, Tucuman Province, Argentina. Environ Geochem Health 34:251–262

Noreault-Conti TL, Fellows A, Jacobs JM, Trask HW, Strom SC, Evans RM, Wrighton SA, Sinclair PR, Sinclair JF, Nichols RC (2012) Arsenic decreases RXRalpha-dependent transcription of CYP3A and suppresses immune regulators in hepatocytes. Int Immunopharmacol 4:4

Park WH, Kim SH (2012) Arsenic trioxide induces human pulmonary fibroblast cell death via the regulation of Bcl-2 family and caspase-8. Mol Biol Rep 39:4311–4318

Pysher MD, Chen QM, Vaillancourt RR (2008) Arsenic alters vascular smooth muscle cell focal adhesion complexes leading to activation of FAK-src mediated pathways. Toxicol Appl Pharmacol 231:135–141

Raghu KG, Cherian OL (2009) Characterization of cytotoxicity induced by arsenic trioxide (a potent anti-APL drug) in rat cardiac myocytes. J Trace Elem Med Biol 23:61–68

Rahman M, Tondel M, Ahmad SA, Chowdhury IA, Faruquee MH, Axelson O (1999) Hypertension and arsenic exposure in Bangladesh. Hypertension 33:74–78

Sears DA (1988) History of the treatment of chronic myelocytic leukemia. Am J Med Sci 296:85–86

Shinkai Y, Sumi D, Toyama T, Kaji T, Kumagai Y (2009) Role of aquaporin 9 in cellular accumulation of arsenic and its cytotoxicity in primary mouse hepatocytes. Toxicol Appl Pharmacol 237:232–236

Tiwari H, Rao MV (2010) Curcumin supplementation protects from genotoxic effects of arsenic and fluoride. Food Chem Toxicol 48:1234–1238

Yoshida T, Yamauchi H, Fan SG (2004) Chronic health effects in people exposed to arsenic via the drinking water: dose-response relationships in review. Toxicol Appl Pharmacol 198:243–252

Yoshino Y, Yuan B, Kaise T, Takeichi M, Tanaka S, Hirano T, Kroetz DL, Toyoda H (2011) Contribution of aquaporin 9 and multidrug resistance-associated protein 2 to differential sensitivity to arsenite between primary cultured chorion and amnion cells prepared from human fetal membranes. Toxicol Appl Pharmacol 257:198–208

Zangi R, Filella M (2012) Transport routes of metalloids into and out of the cell: a review of the current knowledge. Chem Biol Interact 25:25

Zarazua S, Burger S, Delgado JM, Jimenez-Capdeville ME, Schliebs R (2011) Arsenic affects expression and processing of amyloid precursor protein (APP) in primary neuronal cells overexpressing the Swedish mutation of human APP. Int J Dev Neurosci 29:389–396

Zhang QY et al (2009) A systems biology understanding of the synergistic effects of arsenic sulfide and Imatinib in BCR/ABL-associated leukemia. Proc Natl Acad Sci USA 106:3378–3383

Zhao F, Liao Y, Jin Y, Li G, Lv X, Sun G (2012) Effects of arsenite on glutamate metabolism in primary cultured astrocytes. Toxicol In Vitro 26:24–31

Arsenic and Vertebrate Aquaglyceroporins

Joseph R. McDermott and Zijuan Liu
Department of Biological Sciences,
Oakland University, Rochester, MI, USA

Synonyms

Arsenic channels; Arsenic transporters

Definitions

Arsenic (As): A column VI element sharing similar properties with phosphate (P) and antimony (Sb). Dissolved arsenic usually exists in trivalent (As^{III}) and pentavalent forms (As^V) which are common contaminants in aquatic environments and human agriculture. Arsenic is widely and unevenly distributed globally and poses a major health concern, with drinking water and food sources in many areas far surpassing the 10 ppb (parts per billion) arsenic limit set by the World Health Organization.

Aquaglyceroporin: A highly conserved, evolutionarily ancient member of the major facilitator superfamily (MFS) and subgroup of the aquaporin family of membrane channels that permeate water, glycerol, and other noncharged molecules (Liu 2010). Channels in this family present in all domains and play important roles in osmolarity regulation and nutrient uptake. Recently, aquaglyceroporins were identified as major means of access for metalloids into cells and play critical roles in metalloid uptake and detoxification.

Arsenic chemical forms and toxicology: Most immobile arsenic coexists with other elements in ores. Soluble arsenic exists naturally or is produced by microbial activity. Arsenic can have -3, $+3$, and $+5$ valence and exists in many forms of compounds. The inorganic soluble forms include trivalent arsenite (As^{III}, $As(OH)_3$, or As_3O_3), and pentavalent arsenate (As^V, AsO_4^{3-}). Metabolism of these inorganic species by prokaryotes and eukaryotes produces a profile of organic species including MMA^{III} (monomethylarsonous acid, $CH_4As(OH)_2$), MMA^V (monomethylarsonic acid, $CH_4AsO(OH)_2$), DMA^{III} (dimethylarsonous acid, $(CH_3)_2AsO(OH)$), TMA (trimethylarsine, $(CH_3)_3As$ (gas)) and TMAO (trimethylarsine oxide, $(CH_3)_3AsO$). Marine organisms can metabolize inorganic arsenic into the same species in addition to nontoxic arsenobetaine (ArsB) (Thomas 2007). Each arsenic species has different structures, cellular targets, carcinogenicity, and overall toxicity. In general, the cellular toxicity is ranked from most to least toxic as DMA^{III}, $MMA^{III} > As^{III} > As^V > DMA^V$, MMA^V (Liu 2010). TMAO and ArsB are not considered to cause notable toxicity (Mandal 2002).

Inorganic arsenic causes toxicity through causing dysfunction in many cellular targets. As^V can inactivate ATPases by substituting with the chemically similar phosphate (ref). As^{III} can strongly bind vicinal cysteines, which are present in the active site of many enzymes (Mandal 2002). Inorganic arsenic causes oxidative stress by creating reactive oxygen species (ROS). This can occur directly during its metabolism, and perhaps most significantly, by inhibiting mitochondrial enzymes and leading to increases in the ROS leakage which occurs with oxidative phosphorylation. As the mechanisms of arsenic toxicity are so numerous, a key factor in determining overall arsenic toxicity becomes the capacity to metabolize inorganic arsenic to other forms and the ability of the organism to extrude arsenic from the body.

Depending on the type of arsenic exposure and dosage, arsenic can either manifest acute toxicity and lethality or increase risk of cancer and other pathologies. Acute exposure leads to vomiting and cardiac abnormalities. At the cellular level, arsenic induces apoptosis. Chronic arsenic exposure frequently occurs in areas with contaminated drinking water and agriculture, particularly in Bangladesh and southern India. It is also elevated in developing countries with increased industrial mining activity. Chronic arsenic exposure is linked to hyperpigmentation of skin and extensive cancer risk, such as skin and bladder cancers.

Inorganic arsenite (believed to be As^{III}) was applied in traditional Chinese medicine and other cultures have also used arsenic to treat infections. Currently, this old drug has new applications with promising results in multiple cancer treatments. The trivalent form is currently an FDA-approved drug clinically applied to treat acute promyelocytic leukemia. Anthropogenic sources of arsenic contamination in the environment have been common but are decreasing out of concern for toxicity to animals. Pentavalent arsenate (as chromated copper arsenate) has been used in wood preservation. Additionally, a variety of arsenic compounds have been widely applied as herbicides and pesticides in past decades.

Arsenic detoxification in living organisms: Due to the widespread presence of arsenic, all living organisms – from prokaryotes to eukaryotes – developed systems for arsenic detoxification throughout evolution. The general steps involved in cellular arsenic metabolism are (a) membrane transporter–mediated uptake, (b) cellular metabolism, and (c) efflux. Among these steps, membrane transport is the first and likely rate-limiting step for downstream responses.

Previous studies of metalloid transport and aquaglyceroporins: Transporter-mediated uptake for As^V (AsO_4^{3-}) has been established since the early 1950s where it was found As^V can be assimilated by microbes and mammalian cells via phosphate transporters. AsO_4^{3-} and PO_4^{3-} share very close structural similarity and compete for PO_4^{3-} in enzymes and biological molecules (Ballatori 2002). The transport pathway for the more toxic As^{III} has been found only relatively recently. In 1997, using a transposon mutation system in *E. coli*, mutation of aquaglyceroporin GlpF was identified by leading to an antimonite resistance phenotype. Antimonite shares very close physical and chemical properties with arsenite, and it soon followed that GlpF was identified as the first membrane transporter to facilitate As^{III} uptake. This has been proved by later experiments that showed GlpF deletion leads to decreased As^{III} accumulation in *E. coli*. Following this study, a yeast homolog, FPS1, was shown to facilitate As^{III} transport. The genetic deletion of *FPS1* leads to a dramatic increase in As^{III} tolerance as well as a decrease in As^{III} accumulation (Liu 2010). In 2002, the mammalian homologues, AQP7 and AQP9, were overexpressed in yeast and oocytes from *X. laevis*, and both were found to transport As^{III} and Sb^{III}, with AQP9 having the greatest As^{III} transport capacity. Studies using mammalian cell lines showed that AQP3 is also an efficient transporter for As^{III} (Liu 2010). Using AQP9 knockout mice, AQP9 was shown to be a critical transporter and to be required for overall As^{III} clearance. The AQP9 knockout mice exhibited higher arsenic retention in tissues and less tolerance for As^{III}. Thus, AQP9 was found to serve as the major As^{III} transporter in mammals and many homologs in other organisms were found to share this function (Carbrey et al. 2009). Studies showed that in addition to As^{III}, the internally generated arsenic metabolite, MMA^{III}, is a popular substrate for aquaglyceroporins. Mammalian AQP9 and many homologs exhibit robust transport for MMA^{III}, and AQP9 also transports other As^{III} metabolites. As aquaglyceroporins are bidirectional channels, they likely provide efflux of As^{III} metabolites rather than only having function in uptake.

Experimental methods used to study metalloid transport by aquaglyceroporins: The most popular and direct method to study aquaglyceroporin function is by transport assays to measure activity of functionally expressed aquaglyceroporins. Aquaglyceroporin expression can be modified for overexpression and reduced expression in in vitro systems. Cell lines can be transfected to overexpress target proteins or silence endogenous expression with siRNA. This system allows assessment of the relative contribution of aquaglyceroporins to cellular transport and their ability to function in the presence of a relatively normal cell environment. Another effective expression system is the oocytes of the African clawed frog, *Xenopus laevis*, which express high amounts of proteins following injection of the proteins RNA. Compared to cell lines, the *X. laevis* expression system often allows clearer interpretation of results as the oocytes have almost no background transporters or other activities which cause indirect effects. To perform transport assays in these systems, the cells are exposed to the substrate of interest in a buffered system to simulate various conditions, followed by quantification of internalized substrates. Precise quantification of metalloids may be achieved by using inductively coupled plasma mass spectrometry (ICP-MS) which is capable of measuring metalloids in parts-per-trillion range to parts-per-billion range. ICP-MS can also be coupled with HPLC in the front end to determine the chemical form of metalloids. To assay non-metalloid substrates (such as glycerol), radiolabeled isotopes are incorporated into the substrate as a tracer and the accumulation of substrate can be quantified by scintillation counting (Liu 2010).

As most metalloids are toxic, cell lines are also useful to observe phenotypes and detailed mechanisms of toxicity. In the case of arsenic, it is common that the overexpression of aquaglyceroporins will increase arsenic sensitivity and deletion or silencing of endogenous aquaglyceroporins will increase arsenic tolerance. Studies not directly related to function, such as mechanisms of aquaglyceroporin regulation, localization, activation, and specialized roles are also studied in these models.

Cellular roles of aquaglyceroporins: Aquaglyceroporins belong to the aquaporin family. Orthodox aquaporins are known as water channels which allow very rapid diffusion of H_2O in response to osmotic gradients, while aquaglyceroporins have larger channel entrances and pore diameters which allow transport of H_2O and a variety of small, uncharged molecules which are often nutrients.

Aquaporins are ubiquitously present to maintain cellular osmolarity and for various other roles that are tissue specific. Thirteen aquaporins have been identified in mammals, designated AQP0-12. Four out of 13 human aquaporins (AQP3, AQP7, AQP9, and AQP10) are aquaglyceroporins and share 37–45% sequence identity with each other (Liu 2010). These AQPs are highly conserved among vertebrates and respective AQPs of one species have closer identity with orthologous AQPs than they do with other family members in their own species. Aquaglyceroporins exhibit wide expression patterns. AQP3 is detected in skin, kidney, testis, and erythrocytes. AQP7 expresses in testis, adipose tissue, heart, and kidney. In rat and human, AQP9 is abundantly expressed in liver. In contrast with rat, human AQP9 is also expressed in peripheral leukocytes and in tissues that accumulate leukocytes, such as lung, spleen, and bone marrow. Studies also found that human AQP9 is present in duodenum, jejunum, and ileum as well as in brain astrocytes. Human AQP10 is expressed in the duodenum and jejunum. Much of the in-depth expression patterns are based on rodent studies and these have mostly been consistent with human expression patterns, but with some variability.

All aquaglyceroporins appear capable of glycerol and H_2O transport but vary in the efficiency of water transport and range of other substrates. Cell swelling assays show that aquaglyceroporins transport H_2O much less rapidly than orthodox aquaporins; thus, a common opinion is that while H_2O is transported, this is not their normal physiological role. However, despite the lower efficiency in H_2O transport, AQP9 activation and water transport has been shown to be required in the volume-dependent motility of neutrophils (Karlsson et al. 2011). Other substrates are considerably broad and aquaglyceroporins vary in selectivity and efficiency of transport. AQP9 has the most identified substrates which include polyols, purines, pyrimidines, carbamides, monocarboxylic acids, and gaseous CO_2.

The known physiological roles of aquaglyceroporins are understood from their tissue expression and substrate specificity, and confirmation by genetic ablation and rescue experiments. AQP3 has been shown to transport H_2O_2, which enables AQP3-dependent local responses to H_2O_2 signaling (Miller et al. 2010). AQP7, located in adipocytes, mediates efflux of glycerol produced under fasting conditions, which can coordinate with uptake into hepatocytes mediated by AQP9 (which is upregulated by fasting) for gluconeogenesis. The AQP9 knockout mice showed a significant malfunction in glycerol metabolism (Rojek et al. 2007; Jelen et al. 2011). Given the wide substrate specificity of these channels and their complex expression patterns, their physiological roles are multifunctional and quite diverse. Though numerous roles have been found, complete roles remain under investigation. Many aquaglyceroporin roles are been difficult to pinpoint as there is some redundancy in their functions and many roles may only be noticed in specialized models. Many new roles of aquaglyceroporins are still under ongoing investigations.

Aquaglyceroporin structure and translocation mechanisms: The three-dimensional structures of several aquaporins including human aquaporin-1 (AQP1) and *E. coli* aquaglyceroporin GlpF have been solved with water in the channels and with or without glycerol within GlpF. All aquaporins are homotetramers and each monomer consists of six transmembrane alpha-helical segments. Aquaglyceroporin entrances are lined with exposed residues capable of hydrogen bonding to reduce the energy costs of desolvating molecules which have water hydration shells. These residues also play a stereoselective role by aligning substrates as they proceed further down the channel to several restriction and filtering regions. The narrowest region of the pore contains a highly conserved aromatic histidine residue closely oriented to the positively charged hydrogen of a conserved charged arginine. This tight region only allows one molecule to pass at a time while also creating an energy barrier for protons. Another conserved feature is two asparagine-proline-alanine (NPA) signature motifs which also act as filters, first by imposing size restrictions and second, creating an electrostatic field forming an energy barrier to protons and ions. Comparing the structures of aquaporins and aquaglyceroporins, it is clear that they have different pore diameters. The narrowest region of AQP1 is 2.8 Å, which is just large enough for water molecules, while a 4.0 Å diameter is observed in GlpF, accommodating transport of larger molecules (Liu 2010).

Aquaglyceroporins have sites for serine phosphorylation to regulate their activity. In some cases, aquaglyceroporins are normally trafficked to the plasma membrane and are always active, but in others, they may remain docked in scaffolds until intracellular signals cause phosphorylation and membrane trafficking. Phosphorylation sites may also affect localization and colocalization with other membrane proteins.

Metalloid transport by aquaglyceroporins: Inorganic metalloids are commonly found in tri- and tetra-hydroxylated forms, such as with arsenite (As(OH)$_3$), antimonite (Sb(OH)$_3$), and silicate (Si(OH)$_4$). These small, polar, and neutral molecules are expected to favorably interact with hydrogen bonding residues at the pore entrance and pass through channel restriction regions. Because of this chemical similarity in terms of size, solvation, and neutral charge, such metalloids are classified to be transported as molecular mimics of normal substrates in vertebrates. There is an exception to this case known in plants, as aquaglyceroporin homologs known as NIPs (nodulin-like intrinsic proteins) fulfill a need in boron and silicon transport. In vertebrates, transport of boron and silicon by AQPs does not appear to be essential. In support of molecular mimicry and a shared transport mechanism between metalloid and nutrient substrates, arsenite transport by AQP9 can be reduced by mutating residues which show correspondingly negative effects on glycerol transport. However, molecular mimicry does not completely explain As^{III} transport, as AQP3, AQP7, AQP9, and AQP10 all transport glycerol, but AQP10 does not transport As^{III} and AQP3 and AQP7 transport As^{III} at a lesser rate than AQP9.

Impact of AQP9 arsenic transport on human health and cancer treatment: Aquaglyceroporins appear to be the major cellular transporters of inorganic metalloids as well as several organic forms. It is widely believed this transport of metalloid substrates is purely adventitious, as there appears to be no cellular benefit in allowing uptake of such toxins. Many cell culture experiments show increased AQP9 expression sensitizes cells to As^{III}. However, while clearly increasing toxicity on the cellular level, arsenic transport by AQP9 may have some benefit to arsenic exposure on a physiological scale. AQP9 expression in the liver may help arsenic processing, as the liver is more equipped to methylate As^{III} and perform other metabolic modifications which may detoxify As^{III} and improve its clearance. AQP9 also transports the products of As^{III} methylation, MMA^V and MMA^{III}, and may efflux them from hepatocytes after their synthesis. In support of a beneficial role in arsenic exposure, AQP9 knockout mice exhibited less As^{III} tolerance as well as greater arsenic accumulation than wild-type mice.

Arsenic cellular toxicity has made As^{III} a suitable chemotherapeutic agent in certain types of cancers. As^{III} (under the trade name Trisenox (As_2O_3)) is used to treat acute promyelocytic leukemia (APL) with great success, being capable of causing complete remission with tolerable levels of side effects. A critical property of As^{III} effectiveness comes from its ability to be concentrated in AQP9-expressing myelocytes. As AQP9 efficiently transports arsenite, it follows that activity and expression of AQP9 in this cancer type impacts the degree of selective toxicity of arsenite to these cells. Western analysis of clinical APL samples shows that AQP9 expression is a critical predictor of Trisenox therapeutic effectiveness in this cancer type. In the future, As^{III} may become a viable treatment for other tumor types and AQP9 expression should be considered prior to drug application (Agre and Kozono 2003).

Perspectives: The adventitious uptake of toxic metalloids via nutrient transporters represents a popular phenomenon in many organisms. The major role of aquaglyceroporins in metalloid uptake and subsequent cellular toxicity has been firmly established. It is less clear how much aquaglyceroporins contribute to metalloid detoxification or the exact reason of AQP9 knockout mice experiencing higher lethality in As^{III} exposure. The simplest hypothesis is that AQP9 allows arsenic extrusion by efflux of arsenic metabolites, but future studies are needed to verify the exact mechanisms responsible. A better understanding of the role of aquaglyceroporins in this process may allow design of intervention strategies to modulate aquaglyceroporin expression to alleviate metalloid toxicity.

Cross-References

- Arsenic
- Arsenic and Aquaporins
- Arsenic in Nature
- Arsenic in Pathological Conditions
- Arsenic in Therapy
- Arsenic, Free Radical and Oxidative Stress
- Arsenic, Mechanisms of Cellular Detoxification
- Boron and Aquaporins
- Silicon and Aquaporins

References

Agre P, Kozono D (2003) Aquaporin water channels: molecular mechanisms for human diseases. FEBS Lett 555(1):72–78

Ballatori N (2002) Transport of toxic metals by molecular mimicry. Environ Health Perspect 110(Suppl 5):689–694

Carbrey JM, Song L et al (2009) Reduced arsenic clearance and increased toxicity in aquaglyceroporin-9-null mice. Proc Natl Acad Sci USA 106(37):15956–15960

Jelen S, Wacker S et al (2011) Aquaporin-9 protein is the primary route of hepatocyte glycerol uptake for glycerol gluconeogenesis in mice. J Biol Chem 286(52):44319–44325

Karlsson T et al (2011) Aquaporin 9 phosphorylation mediates membrane localization and neutrophil polarization. J Leukoc Biol 90(5):963–973

Liu Z (2010) Roles of vertebrate aquaglyceroporins in arsenic transport and detoxification. In: Jahn T, Bienert G (eds) MIPs and their role in the exchange of metalloids. Landes Bioscience, Austin

Mandal BK (2002) Arsenic round the world: a review. Talanta 58(1):201–235

Miller EW et al (2010) Aquaporin-3 mediates hydrogen peroxide uptake to regulate downstream intracellular signaling. Proc Natl Acad Sci USA 107(36):15681–15686

Rojek AM, Skowronski MT et al (2007) Defective glycerol metabolism in aquaporin 9 (AQP9) knockout mice. Proc Natl Acad Sci USA 104(9):3609–3614

Thomas DJ (2007) Molecular processes in cellular arsenic metabolism. Toxicol Appl Pharmacol 222(3):365–373, Epub 2007 Feb 23. Review

Arsenic and Yeast Aquaglyceroporin

Robert Wysocki
Institute of Experimental Biology, University of Wroclaw, Wroclaw, Poland

Synonyms

Arsenic accumulation in yeast; Glycerol channel Fps1; Glycerol facilitator Fps1

Definition

Aquaglyceroporins are ubiquitous integral membrane proteins which belong to the MIP (major intrinsic protein) superfamily and mediate passive transport of water and/or small solutes, like glycerol and urea. In the budding yeast *Saccharomyces cerevisiae*, there are two aquaglyceroporins: Fps1 (fdp1 suppressor 1) involved in osmoregulation by controlling the intracellular level of glycerol and Yfl054 of unknown physiological role. In addition, Fps1 constitutes a major uptake route for trivalent inorganic arsenic and antimony metalloids into the yeast cells and plays a role in arsenic export out of the cells down the concentration gradient of metalloid.

Arsenic and the Role of Yeast as a Model Organism to Study Metalloid Transport

Arsenic is a highly toxic metalloid which is ubiquitously present in the environment, sometimes in quite high concentrations. What is more, arsenic is easily accumulated by living organisms and imposes a major health and agriculture problem in many areas worldwide. On the other hand, arsenic is also used in the modern therapy as an anticancer drug. Thus, knowledge of cellular uptake pathways of arsenic is crucial for understanding how to minimize arsenic accumulation in normal cells as well as how to increase arsenic intake into cancer cells or organisms used for remediation of regions polluted by this metalloid.

Baker's yeasts are unicellular fungi which serve as an excellent eukaryotic model organism to study various biological processes, including mechanisms of arsenic transport. Studies in yeast allowed to identify and characterize major routes of arsenic uptake and efflux (Wysocki and Tamás 2010, 2011). Thanks to the fact that arsenic transport pathways are conservative, this helped to understand how other organisms, including humans, acquire tolerance to metalloids. Importantly, the yeast mutant cells devoid of arsenic uptake and efflux systems are used as a biological tool to identify plant and human genes involved in arsenic transport (Maciaszczyk-Dziubinska et al. 2012).

Routes for Arsenic Uptake into Yeast Cells

Arsenic is able to enter the cells in a pentavalent form of arsenate anion AsO_4^{3-} and in a trivalent state as arsenite, which exists as an uncharged form of $As(OH)_3$ at the neutral pH in solution. Arsenate structurally resembles phosphate anion and permeates the cells via transporters dedicated for phosphate uptake. $As(OH)_3$ is similar to glycerol that enables arsenite uptake through the glycerol transporters. In addition, $As(OH)_3$ was also demonstrated to enter yeast and mammalian cells using glucose transporters (Maciaszczyk-Dziubinska et al. 2012). In yeast, at least two phosphate permeases, Pho84 and Pho87, are involved in accumulation of arsenate. In the absence of glucose, arsenite uptake is catalyzed by any of 18 glucose transporters present in the yeast cell. However, in the presence of glucose, the main route for arsenite uptake in yeast is constituted by the glycerol

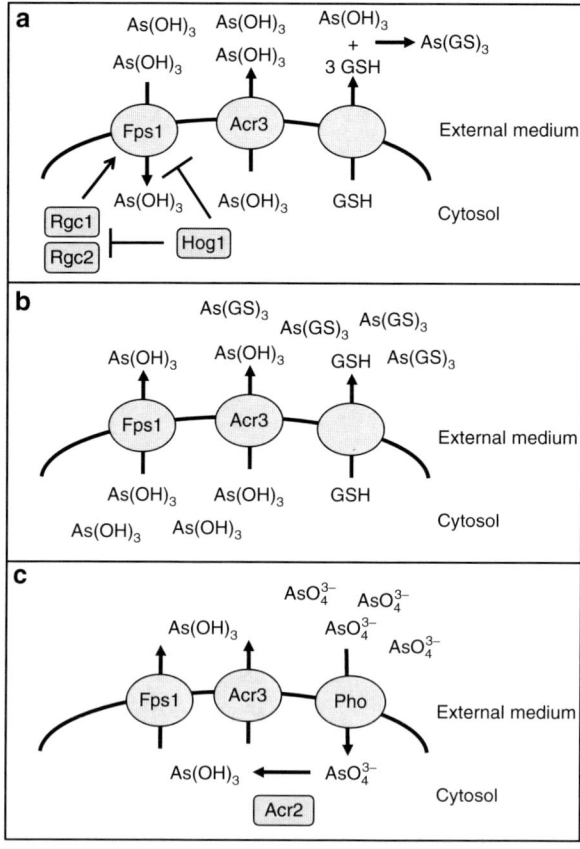

Arsenic and Yeast Aquaglyceroporin, Fig. 1 A dual role of the aquaglyceroporin Fps1 in accumulation and extrusion of arsenite in the yeast cells. (**a**) When there is a high concentration of arsenite [As(OH)$_3$] in the environment, the aquaglyceroporin Fps1 is a major route for As(OH)$_3$ accumulation into the yeast cell. To prevent As(OH)$_3$ uptake, the Fps1 channel closes as a result of Hog1-dependent phosphorylation of its N-terminal tail and inhibition of the positive regulators of Fps1, Rgc1, and Rgc2. (**b**) When the external concentration of free As(OH)$_3$ drops due to extrusion of glutathione (GSH) which forms As(GS)$_3$ complexes, Fps1 facilitates As(OH)$_3$ export from the cytosol to the environment. (**c**) In the case of pentavalent arsenic (AsO$_4^{3-}$) treatment, arsenate enters the cell through the phosphate transporters (Pho). In the cytosol, arsenate is rapidly reduced to arsenite by the arsenate reductase Acr2. Then, arsenite is transported out of the cell via the Fps1 channel down the concentration gradient. (**a–c**) In all cases, arsenite efflux is also mediated by a specific arsenite transporter Acr3

Fps1 is a Glycerol Channel Involved in Osmoregulation

The yeast glycerol channel Fps1 is a member of the MIP superfamily which comprises integral membrane proteins including a large family of water-transporting aquaporins and glycerol facilitators called also aquaglyceroporins (Benga 2009). The MIP channels are usually less than 300 amino acid long and contain six transmembrane domains connected by five loops named A, B, C, D, and E. The loops B and E form two half-transmembrane domains and carry a highly conserved MIP channel motif asparagine-proline-alanine (NPA). In the membrane, MIP channels exist as homotetramers but each subunit constitutes an individual pore. Fps1 is an unusual example of MIP channel as the NPA motif is changed into asparagine-proline-serine (NPS) in the loop B and asparagine-leucine-alanine (NLA) in the loop E. In addition, the length of Fps1 is extended to 669 amino acids due to the presence of large N- and C-terminal tails, about 250 and 150 amino acid long, respectively (Hohmann 2002). Fps1 is located in the plasma membrane and catalyzes both uptake and efflux of glycerol. The physiological role of Fps1 is regulation of intracellular glycerol content in response to changes in the external osmolarity (Tamás et al. 1999). In microorganisms, accumulation of compatible solutes, like glycerol, increases the internal osmolarity. Thus, upon a hyperosmotic shock, yeast cells inhibit glycerol transport out of the cell as well as increase biosynthesis of glycerol. On the contrary, in response to a hypo-osmotic stress, yeast cells rapidly release glycerol to the external medium to prevent bursting. Fps1-mediated transport of glycerol plays a crucial role during osmoadaptation (Hohmann 2002). In response to a hyperosmotic stress, Fps1 channel closes to maintain a high concentration of glycerol in the cytosol and opens to allow glycerol release when the external osmolarity drops. Thus, the yeast deletion mutant lacking *FPS1* gene (*fps1Δ*) loses viability upon transfer from high to low osmolarity conditions. It has been demonstrated that the Fps1 mutants with the truncated N-terminal cytosolic domain (fps1Δ-1) exhibit unregulated transport of glycerol, cannot close in response to hyperosmotic shock, and thus are unable to grow in the presence of high concentrations of salts (Tamás et al. 1999). Further studies showed that two short stretches of amino acids, 222–238 in the N-terminal tail and 535–546 in the C-terminal

channel Fps1. More importantly, arsenite transport via Fps1 is bidirectional and regulated at both transcriptional and posttranscriptional levels either to reduce arsenite intake or increase arsenite export out of the cell (Fig. 1) (Wysocki and Tamás 2010, 2011).

extension, are required for the regulation of Fps1 transport activity and deletion of any of these sequences results in a constitutive open channel.

Identification of Fps1 as a Major Uptake System for Antimonite and Arsenite

First, it has been observed that mutation in the *Escherichia coli glpF* gene encoding for the glycerol channel, which is similar to Fps1, renders bacteria highly resistant to antimony, suggesting that the glycerol transporters might be involved in antimony uptake. Next, a phenotypic analysis of the yeast mutant lacking the aquaglyceroporin Fps1 revealed that both antimonite and arsenite permeate the yeast cells via the glycerol channel (Fig. 1a) (Wysocki et al. 2001). The *fps1Δ* mutant is highly resistant to antimonite and arsenite and accumulates less arsenite as demonstrated by direct transport studies using a radioactive arsenite. Moreover, applying a hyperosmotic stress to wild type cells in order to close the Fps1 channel also results in resistance to these metalloids. On the other hand, expression of a constitutive open mutant form of Fps1 (fps1Δ-1) increases sensitivity to antimonite and arsenite. It is important to mention that the yeast aquaglyceroporin Fps1 is the first reported eukaryotic transporter involved in antimony and arsenite accumulation into the cells (Wysocki et al. 2001).

Aquaglyceroporin Fps1 is a Bidirectional Metalloid Channel

The MIP channels mediate a passive diffusion of solutes across the membranes down the concentration gradient. Thus, the aquaglyceroporin Fps1 should be able to facilitate arsenite transport out of the cell when the cytosolic concentration of arsenite is higher than outside the cell. Interestingly, arsenate, which enters the yeast cells via the phosphate transporters, is rapidly reduced to the trivalent form by the action of arsenate reductase Acr2 present in the cytosol (Fig. 1c). Reduction of arsenate is an important step of arsenic detoxification because the trivalent arsenic is the only substrate for the plasma membrane transporter Acr3 which efficiently extrudes arsenite from the yeast cells against its concentration gradient (Fig. 1a) (Wysocki and Tamás 2010, 2011). However, during arsenate poisoning as a result of arsenate reduction, the concentration of arsenite in the cytosol becomes higher than in the surroundings and arsenite could diffuse via the aquaglyceroporin Fps1 out of the cell. Recent studies proved that Fps1 is indeed the bidirectional channel involved in both uptake and efflux of arsenite and antimonite (Maciaszczyk-Dziubinska et al. 2010). The yeast cells containing multiple copies of *FPS1* gene and thus overexpressing the Fps1 protein exhibit an elevated extrusion of arsenic compared to normal control cells. In contrast, the yeast mutant lacking the aquaglyceroporin Fps1 shows a significant decrease of arsenic transport out of the cell and is more sensitive to arsenate than wild type cells. This demonstrates that the aquaglyceroporin Fps1 contributes to metalloid tolerance of yeast cells by extrusion of arsenite formed after arsenate reduction (Fig. 1c). Surprisingly, it was observed that overexpression of Fps1 causes increased resistance to arsenite due to a lesser accumulation of arsenite within the cells. This strongly indicates that the aquaglyceroporin Fps1 facilitates arsenite efflux in the presence of high concentration of this metalloid in the medium. The question arises how the Fps1 channel can mediate arsenite transport against its concentration gradient. It was shown that in response to arsenite treatment, the yeast cells secrete a tripeptide called glutathione to bind arsenite outside the cells and limit arsenite influx as glutathionylated arsenite is not able to enter the cells (Wysocki and Tamás 2010, 2011). Over time the concentration of free arsenite in the vicinity of the cells becomes lower than in the cytosol which allows a passive diffusion of arsenite via the aquaglyceroporin Fps1 out of the cell (Fig. 1b).

Regulation of Arsenic Transport via the Aquaglyceroporin Fps1

Although toxic metals usually permeate the cells via the membrane proteins developed for accumulation of essential metals and nutrients using molecular mimicry, cells are not defenseless and employ several strategies to downregulate such entry pathways and prevent accumulation of unwanted substances. First, cells can reduce the number of transporter molecules in the plasma membrane by limiting the production of transporter proteins at the level of transcription or by inducing endocytic removal of transporters from the plasma membrane followed by their degradation in the vacuole. The alternative and much faster response involves inhibition of transport activities of these proteins which are already present in the plasma membrane. At least two such responses have been described for the aquaglyceroporin Fps1 during arsenite stress. A short-term exposure to arsenite leads to decrease of *FPS1* mRNA production (Wysocki et al. 2001). The mechanism of negative regulation of *FPS1* gene transcription is not known. However, in the presence of arsenite, the level of Fps1 protein remains constant and Fps1 is not removed from the plasma membrane (Maciaszczyk-Dziubinska et al. 2010). This suggests that accumulation of arsenite into the yeast cells is blocked by inhibition of Fps1 transport activity. Indeed, in response to arsenite stress, Fps1 undergoes phosphorylation on threonine 231 located in the N-terminal tail which leads to closing of Fps1 channel (Thorsen et al. 2006). Arsenite-induced phosphorylation of Fps1 depends on a mitogen-activated protein kinase (MAPK) Hog1 (High osmolarity glycerol 1), which is involved in response to several stresses by regulating various aspects of metabolism to adapt the cells to unfavorable conditions. By a yet unknown mechanism, arsenite activates the HOG signaling pathway to transduce a stress signal by phosphorylating and activating the Hog1 kinase within 15–30 min from arsenite addition to the yeast cells. Then activated Hog1 remains localized in the cytoplasm to phosphorylate and close the aquaglyceroporin Fps1 in order to restrain arsenite accumulation. That is why the cells lacking the Hog1 kinase or expressing Fps1 with threonine 231 mutated into alanine (Fps1-T231A) are highly sensitive to arsenite and accumulate more arsenite than normal cells. Interestingly, deletion of two genes *RGC1* and *RGC2* (Regulator of the Glycerol Channel 1 and 2), which encode for two pleckstrin homology (PH) domain proteins of unknown function, suppresses the arsenite sensitivity of *HOG1* deletion mutant (Beese et al. 2009). In addition, cells lacking Rgc1 and Rgc2 are defective for glycerol efflux via the aquaglyceroporin Fps1, suggesting that they are involved in a positive regulation of Fps1 by promoting the open state of Fps1 channel. It was shown that in response to arsenite and hypo-osmotic stress, Rgc2 undergoes hyperphosphorylation, which is markedly reduced in the *HOG1* deletion mutant (Beese et al. 2009). Thus, Hog1 may downregulate the activity of Fps1 channel not only by a direct phosphorylation of N-terminal tail of Fps1 but also indirectly by affecting Rgc2, a positive regulator of Fps1 (Fig. 1a). However, arsenite-induced phosphorylation of Hog1 is only maintained up to 180 min after arsenite addition to the cells, suggesting that negative regulation of Fps1 is released after this time despite the presence of metalloid in the medium. This coincides with the upregulation of *FPS1* gene expression and accumulation of Fps1 mRNA (Maciaszczyk-Dziubinska et al. 2010). Such switch from the negative to the positive regulation of Fps1 channel activity is in a good agreement with a dual role of the aquaglyceroporin Fps1 in both uptake and efflux of metalloids in the yeast cells.

Yeast as a Tool for Identification of Arsenic Transporters

Yeast are often used as a host for heterologous expression and functional analysis of proteins from other organisms. Detailed understanding of arsenic transport pathways, including the aquaglyceroporin Fps1, the plasma membrane arsenite transporter Acr3, and the ABC (ATP-Binding Cassette) pump Ycf1 mediating a vacuolar accumulation of arsenite complexed with glutathione, made the yeast *Saccharomyces cerevisiae* a perfect model for studies of foreign proteins involved in arsenic transport and tolerance. The yeast triple mutant lacking *ACR3*, *FPS1*, and *YCF1* genes, and thus exhibiting no arsenic transport activity and high resistance to arsenite due to the lack of arsenite uptake, was successfully used for identification of plant and mammalian aquaglyceroporins involved in metalloid accumulation (Wysocki and Tamás 2010, 2011). For example, based on the reversion of arsenite resistance and the increased accumulation of radioactive arsenite in the *FPS1* deletion strain upon expression of the rat AQP9 on a plasmid, it could be concluded that aquaglyceroporins constitute a major uptake pathway for arsenic also in mammals. A similar approach was used to identify plant aquaporins involved in arsenic accumulation. Importantly, it was noticed that expression of plant aquaporins not only sensitizes yeast cells to arsenite but also improves the growth of yeast in the presence of arsenate. This observation allowed to propose that aquaglyceroporins are bidirectional channels which are capable of both uptake and efflux of arsenic depending on its concentration gradient.

Cross-References

▶ Arsenic
▶ Arsenic and Aquaporins

- Arsenic and Vertebrate Aquaglyceroporins
- Arsenic in Nature
- Arsenic in Therapy
- Arsenic, Biologically Active Compounds
- Arsenic, Mechanisms of Cellular Detoxification

References

Beese SE, Negishi T, Levin DE (2009) Identification of positive regulators of the yeast Fps1 glycerol channel. PLoS Genet 5: e1000738

Benga G (2009) Water channel proteins (later called aquaporins) and relatives: past, present, and future. IUBMB Life 61: 112–133

Hohmann S (2002) Osmotic stress signaling and osmoadaptation in yeasts. Microbiol Mol Biol Rev 66:300–372

Maciaszczyk-Dziubinska E, Migdal I, Migocka M et al (2010) The yeast aquaglyceroporin Fps1p is a bidirectional arsenite channel. FEBS Lett 584:726–732

Maciaszczyk-Dziubinska E, Wawrzycka D, Wysocki R (2012) Arsenic and antimony transporters in eukaryotes. Int J Mol Sci 13:3527–3548

Tamás MJ, Luyten K, Sutherland FC et al (1999) Fps1p controls the accumulation and release of the compatible solute glycerol in yeast osmoregulation. Mol Microbiol 31:1087–1104

Thorsen M, Di Y, Tängemo C et al (2006) The MAPK Hog1p modulates Fps1p-dependent arsenite uptake and tolerance in yeast. Mol Biol Cell 17:4400–4410

Wysocki R, Tamás MJ (2010) How *Saccharomyces cerevisiae* copes with toxic metals and metalloids. FEMS Microbiol Rev 34:925–951

Wysocki R, Tamás MJ (2011) *Saccharomyces cerevisiae* as a model organism for elucidating arsenic tolerance mechanisms. In: Bánfalvi G (ed) Cellular effects of heavy metals. Springer, Berlin, pp 87–112

Wysocki R, Chéry CC, Wawrzycka D et al (2001) The glycerol channel Fps1p mediates the uptake of arsenite and antimonite in *Saccharomyces cerevisiae*. Mol Microbiol 40:1391–1401

Arsenic Black

- Arsenic in Pathological Conditions
- Arsenic, Free Radical and Oxidative Stress

Arsenic Channels

- Arsenic and Aquaporins
- Arsenic and Vertebrate Aquaglyceroporins

Arsenic in Nature

Sai-Juan Chen[1], Xiao-Jing Yan[2] and Zhu Chen[1]
[1]State Key Laboratory of Medical Genomics, Shanghai Institute of Hematology, Rui Jin Hospital Affiliated to Shanghai Jiao Tong University School of Medicine, Shanghai, China
[2]Department of Hematology, The First Hospital of China Medical University, Shenyang, China

Synonyms

Arsenicals; Organic and inorganic arsenic

Definition

Arsenic is a chemical element with the symbol As on the periodic table along with nitrogen, phosphorus, antimony, and bismuth. The atomic number of arsenic is 33 and the relative atomic mass is 74.92. Naturally occurring arsenic is composed of one stable isotope, ^{75}As. Many radioisotopes of arsenic have also been synthesized. Arsenic was first documented by Albertus Magnus in 1,250. The term "arsenic" probably originates from the Persian word az-zarnikh or other modifications of its root word, "zar," which refers to yellow or gold orpiment. Arsenic is ubiquitous in nature and ranks twentieth among the elements in abundance in the Earth's crust, fourteenth in seawater, and twelfth in the human body (Jomova et al. 2011).

Arsenic Compounds

Arsenic is a metalloid element, meaning that it displays some properties of both a metal and a nonmetal. It is widely distributed in the biosphere with inorganic or organic forms. The most common valence states of arsenic are -3 (arsine), 0 (elemental arsenic), $+3$ (arsenite), and $+5$ (arsenate). Under reducing conditions, arsenite (As^{III}) is the dominant form; arsenate (As^V) is generally the stable form in oxygenated environments. Although very rare in nature, elemental arsenic may form in hydrothermal deposits at low temperatures (50–200°C) under very anoxic and low-sulfur conditions. Solid samples of elemental arsenic tend to be brittle, nonductile, and insoluble in water.

Arsenic in Nature, Fig. 1 Arsenic minerals

Three most common solid forms of elemental arsenic are metallic gray, yellow, and black arsenic (Gorby 1988). Gray arsenic is the most common and important for use in industry. It has a metallic sheen and conducts electricity. Yellow arsenic is metastable, is a poor electrical conductor, and does not have a metallic sheen. Black arsenic is glassy, brittle, and a poor electrical conductor. Arsenic can exist in many different chemical forms in combination with other elements. Most pure arsenic compounds have no smell or special taste and are white or colorless powders that do not evaporate. Arsenic salts exhibit a wide range of solubilities depending on pH and the ionic environment. Inorganic arsenic is found usually to combine with sulfur, oxygen, halogen elements, and metals (e.g., copper, nickel, cobalt, iron, lead). It occurs naturally in the minerals and ores (Fig. 1), such as realgar (As_4S_4, "red arsenic"), orpiment (As_2S_3, "yellow arsenic"), arsenolite (As_2O_3, "white arsenic"), arsenopyrite (FeAsS), cobaltite (CoAsS), and niccolite (NiAs). Organic arsenic is an arsenic compound containing one or more arsenic-carbon bonds that can be found in nature, in water, natural gas, and shale oil, with the most important general types being those which contain methyl groups. Examples of organic arsenic are methylarsine (CH_3AsH_2), dimethylarsine [$(CH_3)_2AsH$], trimethylarsine [$(CH3)3As$], monomethylarsonic acid (MMA^V) [$CH_3AsO(OH)_2$], monomethylarsenous acid (MMA^{III}) [$CH_3As(OH)_2$], dimethylarsinic acid (DMA^V) [$(CH_3)_2AsO(OH)$], dimethylarsenous acid (DMA^{III}) [$(CH_3)_2AsOH$], trimethylarsinic oxide (TMAO) [$(CH_3)_3AsO$], tetramethylarsonium ion (TMA^+)[$(CH_3)_4As^+$], and others (United States Environmental Protection Agency 2000; Ng 2005).

Occurrence and Source

The Earth's crust is an abundant natural source of arsenic. It is present in more than 320 different minerals, the most common of which is arsenopyrite (Foster 2003). Arsenic may enter the air, water, and soil from wind-blown dust and may get into water from runoff and leaching. Volcanic eruptions are the most important source of arsenic in atmosphere. Arsenic may enter the environment during the mining and smelting of arsenic-containing ores. Small amounts of arsenic also may be released into the atmosphere from coal-fired power plants and incinerators because coal and waste products often contain some arsenic. Many common arsenic compounds can dissolve in water. Thus, arsenic can get into lakes, rivers, or underground water by dissolving in rain or snow or through the discharge of industrial wastes. Some of the arsenic will stick to particles in the water or sediment on the bottom of lakes or rivers, and some will be carried along by the water. Ultimately, most arsenic ends up in the soil or sediment. Arsenic is also ubiquitous in the plant kingdom and all living organisms. The main dietary forms for human are seafood, rice, mushrooms, and poultry. Although some fish and shellfish take in more arsenic, which may build up in tissues, most of this arsenic is in an organic form called arsenobetaine (commonly called "fish arsenic") that is much less harmful.

Environmental levels of arsenic vary. Mean total arsenic concentrations in air from remote and rural areas range from 0.02 to 4 ng/m^3. Mean total arsenic concentrations in urban areas range from 3 to about 200 ng/m^3; much higher concentrations (>1,000 ng/m^3) have been measured in the vicinity of industrial sources, although in some areas, this is decreasing because of pollution abatement measures. In water, levels of arsenic are lowest in seawater, typically 1–2 μg/L. It is higher in rivers and lakes where the concentrations are generally below 10 μg/L. Arsenic levels in groundwater can range up to 3 mg/L in areas with volcanic rock and sulfide mineral deposits. Background concentrations in soil range from 1 to 40 mg/kg, with mean values often around 5 mg/kg, which increase if there are natural and/or man-made sources of arsenic contamination present (Scientific Facts on Arsenic 2004).

Applications

Industrial processes can contribute to the presence of arsenic in air, water, and soil. Environmental contamination of arsenic also occurs because it is used in agricultural pesticides and in chemicals. In the past, inorganic arsenic compounds were predominantly used as wood preservatives, pesticides, herbicides, and paints, and now can no longer be used in agriculture for its environmental issues. However, organic arsenic compounds, namely, cacodylic acid, disodium methylarsenate (DSMA), and monosodium methylarsenate (MSMA), are still used as pesticides.

Some organic arsenic compounds are used as additives in animal feed. Small quantities of elemental arsenic are added to other metals to form metal mixtures or alloys with improved properties. In addition, arsenic is used in alloys (primarily in lead-acid batteries for automobiles) and in semiconductors and light-emitting diodes.

Toxicity

Arsenic is an essential trace element for some animals; however, arsenic and many of its compounds are especially potent poisons. The toxicity of an arsenic-containing compound depends on its valence state, its form (inorganic or organic), its solubility, and the physical aspects governing its absorption and elimination. Generally, inorganic arsenic species are more toxic than organic forms to living organisms, and trivalent arsenite is more toxic than pentavalent arsenic (arsenate). The reported lethal dose of arsenic ranges from 120 to 200 mg in adults and is 2 mg/kg in children (Ellenhorn et al. 1997). Since arsenic is found naturally in the environment, people may be exposed to some arsenic by eating food, drinking water, or breathing air. For drinking water, the World Health Organization (WHO) has set the International Drinking Water Standard for arsenic concentration at 10 μg/L in 1993. For air, the Occupational Safety and Health Administration has established limits of 0.01 mg/m^3 for inorganic and organic arsenic compounds.

Arsenic can cause skin lesions, hepatic injury, hemorrhagic gastroenteritis, cardiac arrhythmia, cancers and psychiatric disease (Hughes 2002; Jomova et al. 2011). Intensive studies have been carried out to elucidate the underlying mechanisms (Chen et al. 2011). The results show that trivalent arsenic (As^{III}) can firmly bind the sulfhydryl groups of biomolecules such as glutathione and lipoic acid and the cysteinyl residues of many proteins and enzymes and therefore inhibits the activities of enzymes such as glutathione reductase, glutathione peroxidases, thioredoxin reductase, and thioredoxin peroxidase. Arsenic also affects flavin enzymes such as NAD(P)H oxidase and NO synthase isozymes. Consequently, arsenic exposure leads to production of reactive oxygen species (ROS). Arsenic interferes with many signal transduction cascades and activates (or inactivates) transcription factors by alteration of global histone H3 methylation. Another mechanism of arsenic toxicity involves substitution of pentavalent arsenic (As^V) for phosphorus in many biochemical reactions, leading to rapid hydrolysis of high-energy bonds in compounds such as ATP.

Cross-References

▶ Arsenic
▶ Arsenic in Pathological Conditions
▶ Arsenicosis
▶ As

References

Chen SJ, Zhou GB, Zhang XW et al (2011) From an old remedy to a magic bullet: molecular mechanisms underlying the therapeutic effects of arsenic in fighting leukemia. Blood 117:6425–6437
Ellenhorn MJ, Schonwald S, Ordog G et al (1997) Ellenhorn's medical toxicology, 2nd edn. Williams & Wilkins, Baltimore
Foster AL (2003) Spectroscopic investigations of arsenic species in solid phases. In: Welch AH, Stollenwerk KG (eds) Arsenic in ground water. Kluwer, Boston
Gorby MS (1988) Arsenic poisoning. West J Med 149:308–315
Hughes MF (2002) Arsenic toxicity and potential mechanisms of action. Toxicol Lett 133:1–16
Jomova K, Jenisova Z, Feszterova M et al (2011) Arsenic: toxicity, oxidative stress and human disease. J Appl Toxicol 31:95–107
Ng JC (2005) Environmental contamination of arsenic and its toxicological impact on humans. Environ Chem 2:146–160
Scientific Facts on Arsenic (2004). http://www.greenfacts.org/en/arsenic/
United States Environmental Protection Agency (2000) Arsenic occurrence in public drinking water supplies. U.S. Environmental Protection Agency, Office of Water, Washington, DC

Arsenic in Pathological Conditions

Swaran J. S. Flora
Division of Pharmacology and Toxicology, Defence Research and Development Establishment, Gwalior, India

Synonyms

Arsen; Arsenic black; Arsenicals; Colloidal arsenic; Gray or grey arsenic; Metallic arsenic; Sodium arsenate; Sodium arsenite

Definition

Arsenic (As), the 33rd element of the periodic table, is classified as a metalloid and is also ubiquitous and highly abundant in nature. Being odorless and colorless, its presence is not immediately obvious and thus, serious human health hazard exists. The use of arsenic as a poison has been known and reported for many years. Arsenic has long been used worldwide as poison, in medicines and pesticides, which still prevails in many countries. The very name of arsenic is thus synonymous with poison. Arsenic exists in the environment as pentavalent (As^{5+}, arsenate) and trivalent (As^{3+}, arsenite) forms, and arsenite is considered as more toxic than arsenate. Human exposure to arsenic may occur from inhalation, skin absorption, and primarily by water ingestion and food like rice, mushrooms, seafood, and poultry, which are reported to have the highest concentrations of arsenic. Environmental contamination of arsenic, particularly in drinking water sources mainly because of anthropogenic activities, is a major cause of concern for human arsenic exposure (Jomova et al. 2011). Small amounts of arsenic absorbed over a period of time may result in chronic poisoning which may produce nausea, headache, coloration and scaling of the skin, hyperkeratosis, anorexia, and white lines across the fingernails. These common symptoms are followed by significant pathological abruptions with or without involvement of internal organs. Ingestion of large amounts (acute poisoning) of arsenic results in severe gastrointestinal pain, diarrhea, vomiting, and swelling of the extremities. Acute poisoning causes renal failure and shock which may ultimately be fatal. Arsenic-induced generation of free radicals and oxidative stress can cause DNA damage, and inhibition of various proteins mainly including transcription factors, regulatory proteins, and induction and/or inhibition of apoptosis. These oxidative stress biomarkers correlated with determination of arsenic concentration in urine, hair, or fingernails of the exposed population that are common diagnostic tools for epidemiological studies. Chronic arsenic exposure has been reported to .induce several pathological conditions such as abnormal skin pigmentation, vasculopathy resulting in dry gangrene of extremities, keratosis, ischemic heart disease, respiratory disease, diabetes, gastrointestinal disturbances, splenomegaly, and neurological defects.

Arsenic (III) binds to sulfhydryl groups leading to the inhibition of a number of enzymes in cellular energy pathway (including pyruvate dehydrogenase), and DNA synthesis and repair, etc. Arsenic is metabolized by reduction and methylation, which was earlier considered as a detoxification mechanism. These reactions are catalyzed by glutathione-S-transferase omega-1 and arsenic (III) methyltransferase. Highly reactive and carcinogenic trivalent arsenic intermediates (monomethylarsonous and dimethylarsonous), are present in urine of arsenic-exposed human subjects more than the pentavalent forms confirming toxicity-mediated through biotransformation. Arsenic is also known to induce cancer of skin, lung, kidney and bladder. Cancer initiation involves the promotion of oxidative stress, in which the antioxidant capacity of the living organism is overwhelmed by arsenic-induced reactive oxygen species (ROS). Latter results in molecular damage to proteins, lipids, and most significantly DNA. Although the toxic effects of arsenic on humans and environment have been well documented, the mechanism by which arsenic induces adverse health effects is not well characterized. The antidotes for arsenic poisoning are dimercaprol, succimer, and recently proposed monoester of succimer.

Arsenic-Induced Clinicopathological Effects
Dermal Effects
Clinical cases of arsenic exposure are most commonly known and identified with dermal toxicity. Appearance of skin pigmentation following exposure is an early indication of initial stage of arsenic poisoning. Chronic exposure of arsenic results in the development of skin lesions, including hyperkeratosis and hyperpigmentation. These symptoms are often used as a diagnostic feature for arsenicosis. Skin cancer induced by arsenic may take a long time to appear, sometimes taking several decades to develop symptoms. Arsenic penetrates through epithelium and causes allergic contact dermatitis in subjects exposed to contaminated water.

Mees' lines: Formation of single, solid white transverse bands of about 1 or 2 mm in width completely crossing the nail of all fingers at the same relative distance from the base.

Melano-Keratosis
Melanosis: It is dark pigmentation of skin surface which initially is visible in palms and spreads to the whole body gradually.

Keratosis: In this condition, rough, dry, and spotted nodules are observed in palms and/or soles.

Both melanosis and keratosis are the chief symptoms of arsenical dermatitis (ASD) and combination of these two features – melanosis and keratosis – in the same patient points to the diagnosis of arsenical dermatitis.

Spotted melanosis (*Spotted or raindrop pigmentation*) is commonly defined as "raindrops in the dust," and is usually seen on chest, back, or limbs.

Spotted and Diffuse Keratosis of palms indicates moderate to severe toxicity. This condition is characterized by rough, dry, spotted nodules which appear after 5–10 years of arsenic exposure. Later (after 10 years), the skin appears dry and gets thickened. This stage is called diffuse keratosis. Gradual thickening of soles may lead to cracks and fissures, a condition termed as "hyperkeratosis."

Leucomelanosis: It is the advanced stage of the disease in which development of pigmentation and depigmented spots (white and black in color) occurs on legs or trunk. Leucomelanosis is common in advanced stage of arsenicosis or condition where arsenic consumption is stopped but had spotted melanosis earlier.

Dorsal keratosis: In severe case of long-term arsenic exposure, skin becomes rough and dry often with the development of palpable nodules (spotted keratosis) on the dorsal skin of hands, feet, legs, or even other parts of the skin, resulting in whole body keratosis.

Gastrointestinal Disturbances
Gastrointestinal symptoms are observed more during acute exposure to inorganic arsenic, which occurs within 30 min of exposure, than chronic exposure. Clinically, acute arsenic poisoning occurs in two distinct forms: acute paralytic syndrome and acute gastrointestinal syndrome. While acute gastrointestinal syndrome leads to dry mouth and throat, heartburn, nausea, abdominal pain and cramps, and moderate diarrhea; chronic arsenic ingestion results in symptomatic gastrointestinal irritation or may produce gastritis, esophagitis, or colitis. Symptoms of acute arsine gas exposure are nonspecific and include headache, weakness, nausea, vomiting, accompanied with abdominal pain. Within a few hours of exposure, dark red urine is seen and within 1–2 days, jaundice is evident. The trio abdominal pain, hematuria, and jaundice are characteristic of arsine gas poisoning.

The gastrointestinal tract appears to be the critical target of toxicity following oral exposure to monomethylarsenous acid (MMA). A dose level of 72.4 mg MMA kg^{-1} per day led to a thickened wall; edema; and hemorrhagic, necrotic, ulcerated, or perforated mucosa in the large intestine and a significant increase in the incidence of squamous metaplasia of the epithelial columnar absorptive cells in the colon and rectum.

Hematological Effects

Acute and chronic arsenic exposure leads to alterations in hematopoietic system. Anemia (normochromic, normocytic, aplastic, and magaloblastic) and leucopoenia (granulocytopenia, thrombocytopenia, myeloid, myelodysplasia) are common effects of arsenic poisoning which may be due to a direct hemolytic or cytotoxic effect on the blood cells. High levels of arsenic have been linked with bone marrow depression in humans and it also leads to red blood cell hemolysis which can cause death within hours. Blood cells undergo hemolysis in the presence of arsenic. Arsenic lowers the GSH levels, which leads to oxidation of sulfhydryl groups in the hemoglobin. The formed hemocyanin then combines with arsenic, which reduces oxygen uptake by cells.

Hepatotoxic Effects

Long-term exposure to arsenic leads to liver damage and the exposed subject may report bleeding esophageal varices, ascites, jaundice, or simply an enlarged liver. Clinical examination reveals swollen, tender liver and elevated levels of hepatic enzymes. These effects are often observed after chronic arsenic exposures to as little as 0.02–0.1 mg As/kg/day.

A correlation between chronic arsenic exposure and abnormal liver function, namely, hepatomegaly, hepatoportal sclerosis, liver fibrosis, and cirrhosis, is well known.

- *Abnormal liver function* is manifested by gastrointestinal disturbances like abdominal pain, indigestion, loss of appetite, and by clinical elevation of serum enzymes following chronic exposure to arsenic.
- *Hepatomegaly* is defined as the abnormal enlargement of liver. More than 75% of the arsenic-exposed subjects in West Bengal, India, report hepatomegaly which is also positively correlated with hepatic arsenic content and arsenic concentration in drinking water.
- *Hepatoportal sclerosis* (*Noncirrhotic portal hypertension*), also known as noncirrhotic portal fibrosis, idiopathic portal hypertension, and Banti syndrome, is a rare arsenic-related condition characterized by portal hypertension but without liver cirrhosis. Initial clinical symptoms of hepatoportal sclerosis are manifested by splenomegaly, anemia, and episodes of gastrointestinal hemorrhage.
- *Liver fibrosis and cirrhosis* is generally present in cirrhotic patients who consume "homemade brew" made with water highly contaminated with arsenic. Liver cirrhosis is one of the major causes of arsenic-related mortality in Guizhou, China, and is potentially associated with hepatocellular carcinoma (HCC).

The susceptibility of liver to arsenic is a consequence of its primary role in arsenic metabolism. In liver, conversion of As^{3+} to As^{5+} results in ROS generation, which leads to mitochondria-mediated liver cell death. Arsenic causes mitochondrial damage and impairs mitochondrial functions. Additionally, toxic methylated arsenic species such as MMA^{III} show greatest binding with hepatocytes. Arsenic involves several toxic mechanisms such as ROS-mediated oxidative stress, inflammatory response, or metabolic hindrances to induce its hepatotoxic effect.

Cardiovascular Effect

Arsenic is also associated with various cardiovascular diseases such as Raynaud's disease, myocardial infarction, myocardial depolarization, cardiac arrhythmias, hypertension, carotid atherosclerosis, ischemic heart disease, and vascular disease (States et al. 2009). Black Foot Disease (BFD) causing ischemic heart disease (ISHD) is one of the major complications following arsenic exposure. This peripheral neuropathy and vascular disease is characterized by severe systemic atherosclerosis and dry gangrene in the lower extremities featuring blackening of feet and hands at end stages (Mazumder 2008). Although cardiovascular effects of arsenic are well defined, the mechanism requires exploration. Arsenic via excessive ROS generation alters the regulation of gene expression, inflammatory responses, and endothelial nitric oxide homeostasis which is important in maintaining vascular tone, leading to cardiovascular endpoints. Arsenic may also induce cardiovascular effects in infants

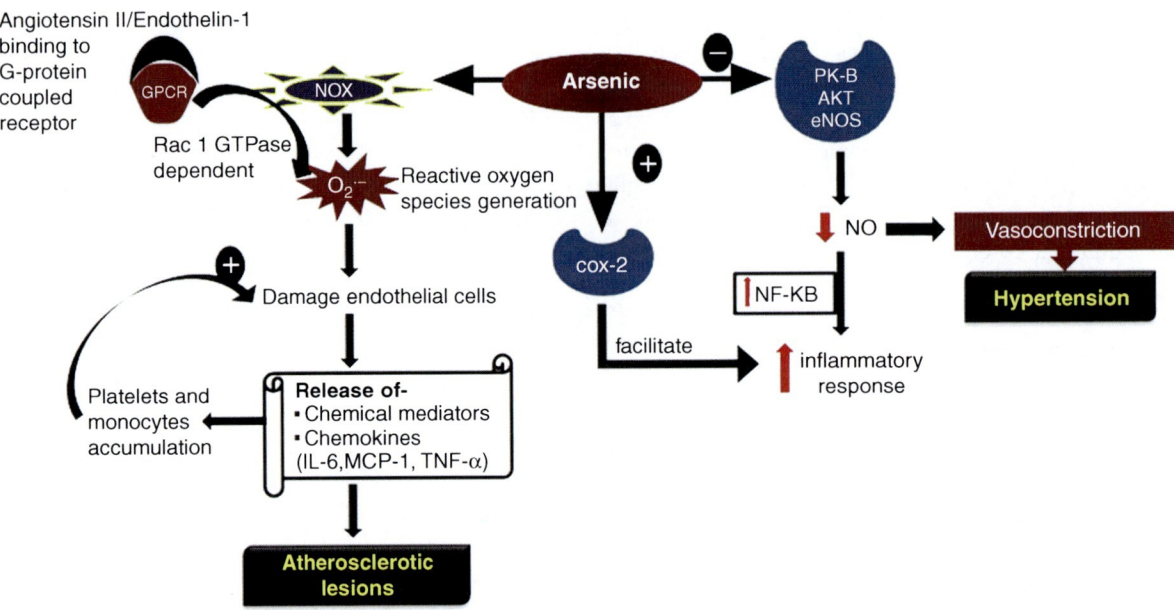

Arsenic in Pathological Conditions, Fig. 1 Arsenic-induced cardiovascular pathologies

following in utero exposure. Although with limited evidence, human data and animals studies suggest that arsenic triggers the onset of atherosclerosis in infants. Arsenic-induced alteration in plasma triglycerides and impaired vasorelaxation response are major observations reported.

Mechanism Involved for Cardiovascular Lesions

Arsenic induces cardio-toxicity that is a result of both cardiac and more importantly vascular effects is mediated through multiple mechanisms. Chronic arsenic exposure causes hypertension due to elevated peripheral resistance resulting from stiffness and lower compliance of the vessel wall. These altered vascular functions are caused by perturbed regulation of vasomotor function and/or structural remodeling of blood vessels. Arsenic-induced ROS or altered redox signaling in both the vascular endothelial and the smooth muscle cell forms the foundation of underlying toxic mechanisms. NADPH oxidase (NOX) enzyme complex involved in hypertension and other vascular disease by superoxide generation is stimulated by arsenic. In the process, various endogenous hypertensive peptides, such as angiotensin II or endothelin-1, interact with membrane linked G-protein-coupled receptors (GPCR) to initiate Rac1-GTPase–dependent superoxide generation by NOX enzymes. Arsenic-induced ROS generation damages the endothelial cells, thereby affecting the endothelial nitric oxide homoeostasis, which play an important role in maintaining vascular tone. Nitric oxide (NO) is a vascular endothelial factor involved in vasodilatation, anti-inflammation, inhibition of platelet adhesion and aggregation, smooth muscle cell proliferation and migration. Exposure to sodium arsenite inactivates protein kinase B/Akt and eNOS, resulting in endothelial cytotoxicity and reduced generation of NO subsequently decreasing the endothelium-mediated vasorelaxation (Balakumar and Kaur 2009). Arsenic also increases the phosphorylation of myosin light chain and calcium flux in the blood vessel. Further, arsenic disrupts the blood pressure regulatory mechanisms by virtue of its effect on hepatic, renal, and neurological system (Fig. 1).

Atherosclerosis is another cardio-pathology closely associated with arsenic exposure. Arsenic-mediated atherosclerosis is initiated by ROS-induced oxidative insult to endothelial vascular cells, instigating activation of inflammatory chemical mediators, adhesion molecules, and chemokines. These chemokines and proinflammatory cytokines including monocyte

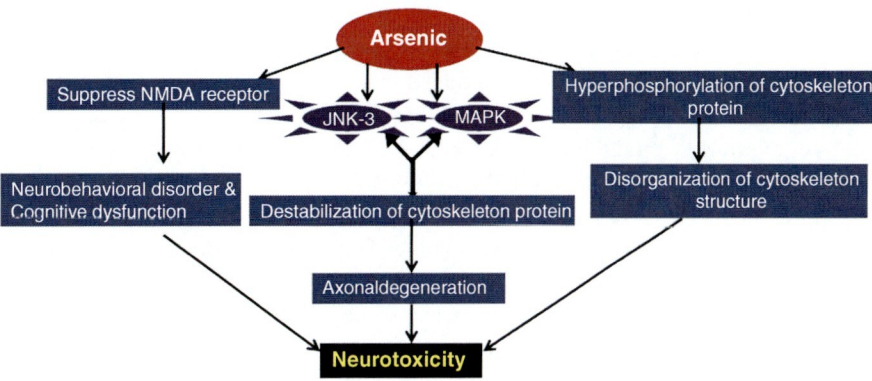

Arsenic in Pathological Conditions, Fig. 2 Role of arsenic in neuropathology

chemoattractant protein-1 (MCP-1), interleukin-6 (IL-6), and tumor necrosis factor alpha, presence of which was detected at high concentrations in atherosclerotic lesions. Arsenic is known to induce the overexpression of MCP-1 and IL-6 in vascular and smooth muscle cells, thereby leading to generation of atherosclerotic lesions. Release of these mediators attract platelets and monocytes which facilitates their activation, forming a continuous process. Foam cell formation by scavenging of oxidized low-density lipoprotein (LDL) by macrophages initiates vascular plaque formation. Physiologically these oxidized LDLs modulate intracellular signal transduction but also generate ROS and peroxides. Arsenic increases accumulation of protein adducts of malondialdehyde (MDA) and 4-hydroxy-*trans*-2-nonenal (HNE) in vascular lesions. Arsenic-induced low NO bioavailability supports the inflammatory processes by increased transcription factors like NF-κB and plaque formation by vasoconstriction. Arsenic is also known to upregulate cyclooxygenase-2 (COX-2) expression in endothelial cells facilitating inflammatory processes. Finally, arsenic promotes the coagulation processes by inhibiting tissue type plasminogen activator, thus reducing fibrinolysis. Progressive development of atherosclerosis which is either initiated or facilitated by arsenic ultimately continues till congestion of cardiac blood vessel by plaque formation (Flora 2011).

Neurological Effects

Ingestion of inorganic arsenic results in neural injury and damage to both peripheral and central nervous system. During acute exposure, encephalopathy with symptoms such as persistent headache, mental confusion, hallucination, seizures, coma, diminished recent memory, distractibility, abnormal irritability, lethargy restlessness, loss of libido, and increased urinary urgency is observed. This is often associated with anxiety, panic attacks, somatizations, and secondary depression. Repeated arsenic exposure in individuals causes polyneuropathy of sensorimotor. Arsenic exposure results in the polyneuropathy, particularly in the lower limbs which is characterized by tingling, numbness, burning soles, and weakness.

Arsenic-induced neurotoxicity causes changes in cytoskeletal protein composition and hyperphosphorylation which may lead to disorganization of the cytoskeletal structure, a potential cause of neurotoxicity (Fig. 2). Peripheral neuropathy and the cytoskeletal defects are typical neurological features (Mathew et al. 2010). Arsenic-induced neurotoxicity involves induction of apoptosis in the cerebral neuron by activating p38 mitogen-activated protein kinase and JNK3 pathway. Destabilization and disruption of nerve cytoskeletal proteins leading to axonal degeneration is another mechanism involved in arsenic-induced neurotoxicity. These nerve cytoskeletal proteins serves as flexible scaffold for cells that is responsible for the communication between cell parts and other functions. The major components of myelinated neurons are neurofilaments (NF-H, NF-M, NF-L). Among them NF-L is crucial and is required by both NF-H and NF-M to form a heteropolymer in the cytoskeleton of the neuronal cell. Within the peripheral nerves, arsenic by virtue of its affinity for -SH moiety, binds and subsequently degrade neurofilaments (NF-L). This ultimately leads to arsenic recirculation, and rendering it available for further targeting NF-L. Arsenic and its metabolite (monomethylarsonic acid and monomethylarsonous) suppress the NMDA receptors in hippocampus, leading to neurobehavioral disorders and cognitive dysfunction.

Renal Effects

Arsenic causes acute renal failure and chronic renal dysfunction. Pathologically, arsenic-induced manifestations are due to hypotensive shock, hemoglobinuric or myoglobinuric tubular injury. Arsenic during its excretion via kidneys gets converted into less soluble but more toxic trivalent arsenic. The capillaries, tubules, and glomeruli of kidneys are involved in this process and, thus, are also the sites of arsenic induced damage. Arsenic induces lesions such as loss of capillary integrity and increase in glomerular capillary permeability. Mitochondrial damage and arsenic-induced hemolysis leads to tubular necrosis with a high risk of renal failure. The damaged tubular cells lead to proteinuria, oligouria, hematuria, acute tubular necrosis (ATN), renal insufficiency, or frank renal failure. Dimethylarsenic acid (DMA) also causes renal damage which is characterized by increased volume and pH of urine, urinary calcium level, increase in water consumption, and decreased levels of electrolytes. Increased kidney weights and minimal tubular epithelial cell degeneration, tubular casts, and focal mineralization have also been associated with arsenic exposure. The urinary system seems to be a more recessive target for DMA than for MMA (Jomova et al. 2011).

Respiratory Effects

Acute arsenic exposure leads to the development of pulmonary edema, adult respiratory distress syndrome (ARDS), along with respiratory failure from muscle weakness, and phrenic nerve damage leading to apnea. Laryngitis, bronchitis, pharyngitis, rhinitis, tracheobronchitis, shortness of breath, nasal congestion, conjunctiva congestion, redness of the eyes, chronic cough, chronic asthmatic bronchitis and asthma, and perforation of the nasal septum have been associated with exposure to arsenic.

Arsenic accumulates in lungs along with other target sites following ingestion and biotransformation by liver. Although mechanisms for arsenic-induced lung toxicity and cancer are not well defined, one of the hypothesis is that high oxygen partial pressure in lungs facilitates arsenic-induced oxidative stress. In the presence of high oxygen, a DMA metabolite, dimethylarsine which is a gas and thus excreted by lungs reacts with oxygen to generate free radical. Arsenic exposure through inhalation forms an important route to understand arsenic-induced respiratory defects. Respiratory tract absorbs more arsenic than gastrointestinal tract. The fate of inhaled arsenic particles is different from ingested arsenic. Arsenic particles have low solubility and thus show slow body clearance retaining for longer time compared to ingested arsenic. Inhaled arsenic causes inflammation of the lung tissue due to mechanical interaction along with other mechanisms such as oxidative stress, insult to biomolecules especially DNA, suppression of p53 leading to manifestations like cancer (Celik et al. 2008).

Reproductive and Developmental Effects

Arsenic-induced reproductive toxicity has not been widely studied, thus limited information is available. Arsenic-induced male reproductive toxicity includes low sperm count, postmeiotic spermatogenesis, abnormal hormonal secretion, and altered enzyme activity. Female reproductive system is also adversely affected by arsenic exposure. Major toxic effects include ovarian steroidogenesis, prolonged diestrus, degradation in ovarian, follicular, and uterine cells; and increased meiotic aberrations in oocyte. Sporadic human, while contradictory animal data suggest arsenic and its methylated metabolites can cross the placenta. Arsenic exposure increases the incidence of preeclampsia in pregnant women; decreases birth weight of newborn infants; and increases the risk of malfunctions and stillbirths, spontaneous abortions, preterm births, and infant mortality. On the other hand, laboratory studies suggest an increase in malformations and stillbirths in animals. All these data are indicative but not conclusive. Thus, a properly designed epidemiologic study in a sufficiently large population is required in order to assess the potential adverse effects of arsenic in human reproduction (Jomova et al. 2011; Flora et al. 2011).

Arsenic causing oxidative insult to the placental cells which interferes with the communication between the mother and the developing embryo and adversely affects the nutritional translocation through placenta. Arsenic generates ROS and reduces important placental antioxidants such as thioredoxin reductase. Arsenic induces anemia in pregnant female due to it hematological adverse effects (Flora et al. 2011).

Diabetes Mellitus

Diabetes mellitus observed in patients exposed to arsenic is similar to Type II diabetes mellitus characterized

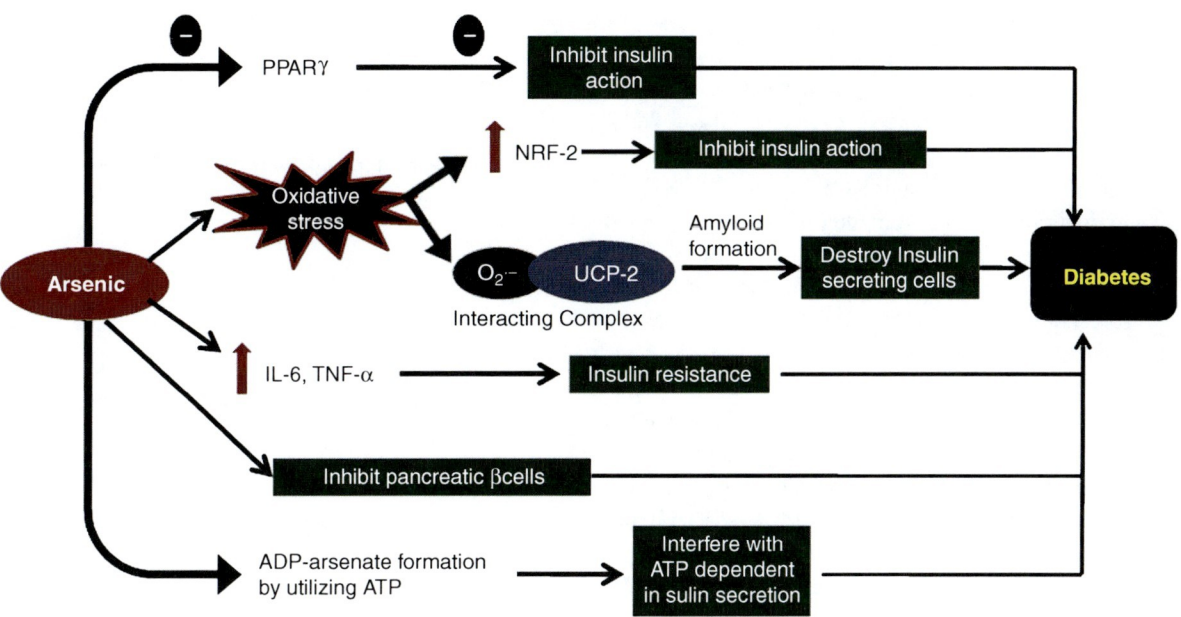

Arsenic in Pathological Conditions, Fig. 3 Role of arsenic in inducing diabetes mellitus

by both insulin resistance and a relative deficiency in insulin secretion (Flora 2011). Diabetogenic effects of arsenic are known; however, the exact mechanism remains unclear. Arsenic substitutes phosphate in the formation of adenosine triphosphate (ATP) and other phosphate intermediates involved in glucose metabolism. This slows down the normal metabolism of glucose, thus interrupting the production of energy and interfering with the ATP-dependent insulin secretion (Fig. 3).

Chronic arsenic exposure leads to increased oxidative stress and upregulation of tumor necrosis factor-a (TNF-α) and interleukin-6 (IL-6). These cytokines have been known for their effect on the induction of insulin resistance. Arsenic also shows inhibitory effect on the expression of peroxisome proliferator-activated receptor-g (PPAR-g), a nuclear hormone receptor important for activating insulin action. A major pathogenic link between insulin resistance and cell dysfunction is oxidative stress which exerts its action through activation of nuclear factor-kB (NF-kB). Superoxide production induced by arsenic can impair insulin secretion by interaction with uncoupling protein-2 (UCP-2). This may lead to amyloidal formation in the pancreas, which could progressively destroy the insulin-secreting cells (Tseng 2004). Arsenic inhibits glucose-stimulated insulin secretion (GSIS) by pancreatic β cells, besides increasing Nrf_2-mediated oxidative response, and decreasing glucose-stimulated peroxide accumulation. These observations suggest that low levels of arsenic triggers a cellular adaptive response, impairing ROS signaling involved in GSIS, and thus disturbs β-cell function.

Carcinogenic Effects

Inorganic arsenic exposure induces cancer in human lungs, urinary bladder, skin, kidney, and liver, with the majority of deaths from lung and bladder cancer. Trivalent form of arsenic is more genotoxic than pentavalent form. Arsenic induces cancer by various mechanisms: (1) arsenic causes impairment of cellular respiration by the inhibition of various mitochondrial enzymes, and the uncoupling of oxidative phosphorylation, (2) arsenic potentially interacts with sulfhydryl groups of proteins and enzymes, and it substitutes phosphorous in a variety of biochemical reactions, and (3) arsenic can potentially induce the release of iron from Ferritin, an iron storage protein, which regulates intracellular concentration of iron. Once released free iron may catalyze the decomposition of hydrogen peroxide via the Fenton reaction, thereby generating the reactive hydroxyl radical which causes DNA damage. 8-Hydroxy-2′-deoxyguanosine (8-OhdG) is one of the major ROS-induced DNA damage products and used as biomarker of oxidative stress to DNA. Hepatic 8-OhdG levels increase on exposure to organic and inorganic arsenic, suggesting that arsenic elevates rate of free radical attack on DNA. Some clastogenic effects of arsenic are also mediated via free radicals (e.g., peroxynitrite, superoxide, hydrogen peroxide, and possibly free iron). Arsenic methylation leads to formation of more toxic, reactive, and carcinogenic trivalent methylated arsenicals (MMAIII and DMAIII). No carcinogenic effect of MMA (V) has been observed so far in in vivo studies. DMA(V) and MMA(V) are stored and reduced to form DMA (III) and MMA(III) in the lumen of the bladder; this organ is much more susceptible to cancer than liver and kidney (Cohen et al. 2006; Jomova et al. 2011). DMA (III) does not directly interact with DNA to exert genotoxicity but follows indirect pathway, i.e., formation of ROS. ROS formation activates a cascade of transcription factors (e.g., AP-1, c-fos, and NF-kB), and oversecretion of pro-inflammatory and growth-promoting cytokines, resulting in increased cell proliferation and ultimately carcinogenesis. In summary, it can be suggested that the exact mechanism for arsenic-induced carcinogenicity is still unclear; however, genetic and epigenetic changes, the role of oxidative stress, enhanced cell proliferation, and modulation of gene expression are some current recommendations.

Therapeutic Strategies

British Anti-Lewisite (BAL) or 2,3-dimercaptopropanol (Dimercaprol) is one of the oldest chelating agents for treating acute arsenic poisoning following ingestion, inhalation, or absorption. BAL however, is rather considered most toxic chelator available that restricts its application to few acute poisoning cases. Later, few derivatives of BAL were introduced which were safe and effective. Meso 2,3-dimercaptosuccinic acid (DMSA) and sodium 2,3-dimercaptopropane 1-sulfonate (DMPS) are water-soluble dithiols with safer drug profiles (Fig. 4). Both these drugs show predominantly extracellular distribution with DMPS

Arsenic in Pathological Conditions, Fig. 4 Chemical structures of DMSA and DMPS

Arsenic in Pathological Conditions, Fig. 5 Chemical structure of MiADMSA

showing some intracellular distribution also (Flora 2011). A major drawback associated with DMSA is its extracellular distribution, since it is unable to cross the cell membrane. Thus, esters of DMSA were synthesized to enhance chelation from intracellular compartments. Monoisoamyl DMSA (MiADMSA), a C5 branched chain alkyl monoester of DMSA, is one such investigational chelator which exhibits lipophilicity as compared to the parent DMSA. MiADMSA thus has been identified as a promising drug against arsenic-induced pathological lesions in experimental animals. Safety of MiADMSA has been well established in preclinical in vitro and in vivo models with copper depletion as the only prominent reversible side effect (Fig. 5). Combination therapy, administering a chelating agent either with an antioxidant or another structurally different chelator formulates effective therapeutic strategies against chronic arsenic toxicity. The advantages include providing an immediate and additional antioxidant effects and more pronounced arsenic excretion from different body compartments, respectively. Such combination therapy with a lipophilic and lipophobic (MiADMSA and DMSA) chelating agents would limit drawbacks like metal redistribution and form a safer therapy regime. Experimental studies revealed that such strategies result not only in better reduction of body arsenic burden but also more effective recoveries in biomarkers, and neurological defects (Flora and Pachauri 2010). Thus, advanced therapeutic strategies may exhibit superior results as compared to traditional chelation monotherapy.

Cross-References

▶ Arsenic
▶ Arsenic and Aquaporins
▶ Arsenic in Nature
▶ Arsenic in Therapy
▶ Arsenic in Tissues, Organs, and Cells
▶ Arsenic Methyltransferases
▶ Arsenic, Free Radical and Oxidative Stress
▶ Arsenic, Mechanisms of Cellular Detoxification
▶ Arsenic-Induced Diabetes Mellitus
▶ Arsenicosis
▶ As

References

Balakumar P, Kaur J (2009) Arsenic exposure and cardiovascular disorders: an overview. Cardiovasc Toxicol 9:169–176
Celik I, Gallicchio L, Boyd K et al (2008) Arsenic in drinking water and lung cancer: a systematic review. Environ Res 108:48–55
Cohen SM, Arnold LA, Eldan M et al (2006) Methylated arsenicals: the implications of metabolism and carcinogenicity studies in rodents to human risk assessment. Crit Rev Toxicol 36:99–133
Flora SJS (2011) Arsenic induced oxidative stress and its reversibility. Free Radic Biol Med 51:257–281
Flora SJS, Pachauri V (2010) Chelation in metal intoxication. Int J Environ Res Pub Health 7:2745–2788
Flora SJS, Pachauri V, Saxena G (2011) Arsenic, cadmium and lead. In: Gupta RC (ed) Reproductive and developmental toxicity. Academic Press, Boston, pp 415–438
Jomova K, Jenisova J, Feszterova M et al (2011) Arsenic: toxicity, oxidative stress and human disease. J Appl Toxicol 31:95–107
Mathew L, Vale A, Adcock JE (2010) Arsenical peripheral neuropathy. Pract Neurol 10:34–38
Mazumder GDN (2008) Chronic arsenic toxicity and human health. Ind J Med Res 128:436–447
States JC, Shrivastava S, Chen Y, Barchowsky A (2009) Arsenic and cardiovascular disease. Toxicol Sci 107:312–323
Tseng CH (2004) The potential biological mechanisms of arsenic-induced diabetes mellitus. Toxicol Appl Pharmacol 197:67–83

Arsenic in Therapy

Sai-Juan Chen[1], Xiao-Jing Yan[2] and Zhu Chen[1]
[1]State Key Laboratory of Medical Genomics, Shanghai Institute of Hematology, Rui Jin Hospital Affiliated to Shanghai Jiao Tong University School of Medicine, Shanghai, China
[2]Department of Hematology, The First Hospital of China Medical University, Shenyang, China

Synonyms

Acute promyelocytic leukemia; Arsenic sulfide; Arsenic trioxide; Darinaparsin; Melarsoprol

Definition

Arsenic is a natural substance and a traditional poison. The toxicity of arsenic is a double-edged sword. In fact, it has also been used as a drug with appropriate application for over 2,400 years in both traditional Chinese medicine and the Western world to treat many diseases from syphilis to cancer. However, the medical use of arsenic faced an embarrassment partly due to its low effect compared to modern therapy, but mostly because of concerns about the toxicity and potential carcinogenicity. In the modern era, interest in arsenic as a chemotherapy is rekindled after reports from China describing a high proportion of hematologic responses in patients with acute promyelocytic leukemia (APL) treated with arsenic trioxide. Numerous studies in the last two decades have confirmed arsenic trioxide as a successful treatment for APL. These results provided new insights into the pathogenesis of this malignancy and raised hopes that arsenicals might be useful in treating other cancers.

An Old Remedy

Although elemental arsenic was isolated about 700 years ago, arsenic compounds have been used as medicines by Greek and Chinese healers since more than 2,400 years ago (Fig. 1). Despite its carcinogenicity and the toxic effects associated with long-term exposure, scientists and physicians have used the

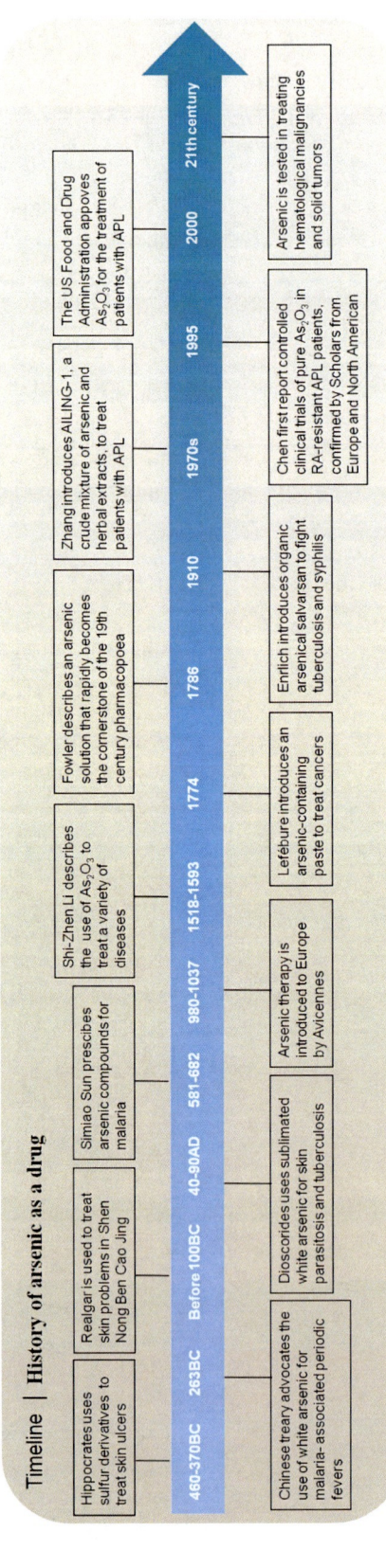

Arsenic in Therapy, Fig. 1 History of arsenic as a drug

poison successfully in practice to treat numerous ailments and diseases (Chen et al. 2011; Zhu et al. 2002). The first mention of arsenicals was made by Hippocrates (460–370 BC), who used realgar and orpiment pastes to treat ulcers. In China, arsenic pills for the treatment of malaria-associated periodic fever were recorded in the Chinese Nei Jing Treaty (263 BC). Realgar-containing pastes were used in the treatment of certain skin problems like carbuncle and were historically recorded in Shen Nong Ben Cao Jing (before 100 BC). Dioscorides (40–90 AD) later noted that arsenicals could cause hair loss but clear scabies, lice, and many skin growths that might have been cancers. Si-Miao Sun (581–682 AD) purified a medicine composed of realgar, orpiment, and arsenic trioxide in treating malaria. Later, the pharmacopoeia of Shi-Zhen Li (1518–1593 AD) in the Ming Dynasty described the use of As_2O_3 to treat a variety of diseases. Arsenic therapy was introduced to Europe by Avicenna (980–1037 AD) and Paracelsus (1493–1541 AD).

During the eighteenth, nineteenth, and twentieth centuries, a number of arsenic compounds have been used as medicines, including arsphenamine, neosalvarsan, and arsenic trioxide (Antman 2001; Zhu et al. 2002). In the 1700s, Lefébure introduced an arsenic-containing paste proposed to be an "established remedy to radically cure all cancers." Later, English inventor Thomas Fowler developed a solution of arsenic trioxide in potassium bicarbonate (KH_2AsO_3, 1% w/v), aptly named "Fowler's solution," that was used to treat asthma, chorea, eczema, pemphigus, and psoriasis. In the nineteenth century, arsenides and arsenic salts were used in the form of external pastes to treat ulcers and cancer. They were also prescribed as antiperiodics, antipyretics, antiseptics, antispasmodics, caustics, cholagogues, depilatories, hemantinics, sedatives, and tonics. The drugs were in liquid or solid form and could be inhaled as vapor, injected, administered intravenously, or given as enemas to treat systemic illnesses. In 1878, Fowler's solution was used to lower white blood cell counts in patients with chronic myelogenous leukemia (CML) and became a mainstay in the treatment of leukemia until it was succeeded by radiation in the twentieth century. Nevertheless, Fowler's solution resumed treating CML in 1931 after a report showing the remarkable response in nine patients to the treatment. However, this period was short because of the reports of arsenic's chronic toxicity and the efficacy of busulfan in 1950s. In 1910, a Nobel laureate, physician, and founder of chemotherapy, Paul Ehrlich introduced salvarsan, an organic arsenic-based product which was screened from 500 organic arsenic compounds. This compound was shown to be effective in treating tuberculosis and syphilis. Other organic arsenicals, such as melarsoprol, are still used to treat trypanosomiasis. Since this time, the use of the arsenicals in medicine declined because newer drugs had a greater therapeutic index and the toxicity of arsenic was concerned.

A New Trick

In the early 1970s, a group from Harbin Medical University in China studied arsenic oxide in a number of cancer types (Chen et al. 2011; Wang and Chen 2008). They found that intravenous infusions of Ailing-1 containing 1% As_2O_3 and trace amounts of mercury chloride could be used to treat patients with APL and achieved complete clinical remission (CR) in two-thirds of patients. From 1994, clinical trials with pure As_2O_3 were conducted by our group in Shanghai Second Medical University, in collaboration with the group from Harbin (Chen et al. 2011; Wang and Chen 2008). The efficacy of pure As_2O_3 has been demonstrated in patients with not only newly diagnosed APL but also APL cases relapsed after retinoic acid (RA) plus chemotherapy. A molecular remission is obtainable in a relatively high proportion of the patients, from 70% to 90% in different studies. Further studies conducted in Japan, Europe, and North America reported similar results. In 2000, the US Food and Drug Administration (FDA) approved As_2O_3 for the treatment of patients with APL that are resistant to ATRA. The efficacy of As_4S_4 in APL was also investigated (Chen et al. 2011; Wu et al. 2011). In traditional Chinese medicine, combination therapy containing multiple drugs with distinct but related mechanisms, called formulas, have been used for more than 2,000 years to amplify therapeutic efficacies of each agent and to minimize adverse effects. On the basis of traditional Chinese medicine (TCM) theories, a patented Realgar-Indigo naturalis formula (RIF) was designed in the 1980s, in which a mined ore realgar was the principal element, whereas Indigo naturalis, Salvia miltiorrhiza, and radix pseudostellariae were adjuvant components to assist the effects of realgar. Multicenter clinical trials showed that a CR rate of more than 90% and a 5-year overall survival (OS) rate of about 85% were achieved in patients with APL receiving the RIF.

The remarkable therapeutic efficacy of As_2O_3 in APL rekindled the enthusiasm of the experts in hematology and oncology. The therapeutic efficacy of arsenicals in other hematologic cancers and solid tumors is under intensive investigation (Cui et al. 2008; Emadi and Gore 2010; Hu et al. 2005). Arsenic has been tested in treating hematological malignancies such as CML, multiple myeloma (MM), myelodysplastic syndrome, non-APL acute leukemia, myelofibrosis/myeloproliferative disorder, and lymphoid malignancies (including non-Hodgkin lymphoma) and has displayed beneficial effects in some cases, but the therapeutic efficacy was limited as compared to the APL miracle. Solutions containing arsenicals have been used as therapeutic agents since ancient times for CML, a malignant myeloproliferative disease of pluripotential hematopoietic stem cells that is characterized by BCR–ABL fusion protein with constitutively activated tyrosine kinase activity, and the results suggested that arsenic exhibits some inhibitory effects on proliferation of BCR–ABL–expressing cells. Clinical trials are underway to test the efficacy of arsenicals (As_2O_3 and As_4S_4) in combination with tyrosine kinase inhibitor imatinib for CML, because these regimens have showed synergetic effects in vitro and in vivo. Arsenic holds therapeutic promise in the treatment of MM, with data showing growth inhibitory and apoptotic effects of arsenic on myeloma cells and mice model. Clinical trials have demonstrated arsenic trioxide's activity in advanced refractory or high-risk MM. Several trials have demonstrated promising results in human T cell lymphotropic virus type I (HTLV-I)-associated adult T cell leukemia-lymphoma (ATL). A high synergistic effect between IFN-α and As_2O_3 in ATL-derived cell lines has been demonstrated. Similar results have been obtained with fresh leukemia cells derived from an ATL patient. Accordingly, an association between IFN-α and As_2O_3 has shown therapeutic effects in patients with ATL that have become refractory to other treatments.

There are more than 100 recently completed or ongoing clinical trials listed on www.clinicaltrials.gov evaluating As_2O_3 alone or in combination with other agents for treatment of cancers, excluding APL.

As_2O_3 is under investigation as treatment for a variety of solid tumors including bladder cancer, glioma, breast cancer, hepatocellular carcinoma, pancreatic adenocarcinoma, cervical cancer, colorectal cancer, esophageal cancer, germ cell tumors, liver cancer, lung cancer, and melanoma. Limited clinical activity as a single agent has been reported in a small number of patients with hepatocellular carcinoma, melanoma, and renal cell carcinoma; As_2O_3 in combination with chemotherapy has shown promising activity in osteosarcoma and Ewing sarcoma.

Besides As_2O_3 and As_4S_4, some newly developed organic arsenic compounds have demonstrated anticancer efficacy in vitro, such as melarsoprol, dimethylarsinic acid, GSAO, and darinaparsin (Elliott et al. 2012; Mann et al. 2009). GSAO, 4-(N-(S-glutathionylacetyl)amino) phenylarsonous acid, has potential antiangiogenic capability with application in cancer where tumor metastasis relies on neovascularization. The phase I clinical study of GSAO is ongoing in patients with solid tumors refractory to standard therapy. Darinaparsin (ZIO-101) is an organic arsenical composed of dimethylated arsenic linked to glutathione and has significant activity in a broad spectrum of hematologic and solid tumors in preclinical models. The phase I/II clinical studies of darinaparsin are ongoing in patients with refractory solid tumors and demonstrate antitumor activity.

Cross-References

▶ Arsenic in Nature
▶ Arsenic, Biologically Active Compounds
▶ Promyelocytic Leukemia–Retinoic Acid Receptor α (PML–RARα) and Arsenic

References

Antman KH (2001) Introduction: the history of arsenic trioxide in cancer therapy. Oncologist 6(Suppl 2):1–2

Chen SJ, Zhou GB, Zhang XW et al (2011) From an old remedy to a magic bullet: molecular mechanisms underlying the therapeutic effects of arsenic in fighting leukemia. Blood 117:6425–6437

Cui X, Kobayashi Y, Akashi M et al (2008) Metabolism and the paradoxical effects of arsenic: carcinogenesis and anticancer. Curr Med Chem 15:2293–2304

Elliott MA, Ford SJ, Prasad E et al (2012) Pharmaceutical development of the novel arsenical based cancer therapeutic GSAO for phase I clinical trial. Int J Pharm 426:67–75

Emadi A, Gore SD (2010) Arsenic trioxide – an old drug rediscovered. Blood Rev 24:191–199

Hu J, Fang J, Dong Y, Chen SJ et al (2005) Arsenic in cancer therapy. Anticancer Drugs 16:119–127

Mann KK, Wallner B, Lossos IS et al (2009) Darinaparsin: a novel organic arsenical with promising anticancer activity. Expert Opin Investig Drugs 18:1727–1734

Wang ZY, Chen Z (2008) Acute promyelocytic leukemia: from highly fatal to highly curable. Blood 111:2505–2515

Wu J, Shao Y, Liu J et al (2011) The medicinal use of realgar (AsS) and its recent development as an anticancer agent. J Ethnopharmacol 135:595–602

Zhu J, Chen Z, Lallemand-Breitenbach V (2002) How acute promyelocytic leukaemia revived arsenic. Nat Rev Cancer 2:705–713

Arsenic in Tissues, Organs, and Cells

Sai-Juan Chen[1], Xiao-Jing Yan[2] and Zhu Chen[1]
[1]State Key Laboratory of Medical Genomics, Shanghai Institute of Hematology, Rui Jin Hospital Affiliated to Shanghai Jiao Tong University School of Medicine, Shanghai, China
[2]Department of Hematology, The First Hospital of China Medical University, Shenyang, China

Synonyms

Arsenic metabolism; Arsenic methyltransferases; Arsenic pharmacokinetics; Biomonitoring

Definition

As contained in water, soil, food, or atmosphere, the primary routes of arsenic entry into the body are ingestion and inhalation. Very little internal exposure to arsenic occurs via the material passing through the skin into the body. After absorption, arsenic is widely distributed by the blood stream throughout the body. Most tissues rapidly clear arsenic, except for skin, hair, and nails. The absorbed As undergoes biomethylation primarily by the liver. Inorganic As in the body is metabolized by the reduction of arsenate (As^{III}) to arsenite (As^V), followed by sequential methylation to monomethylarsonic acid (MMA) and dimethylarsenic acid (DMA). These methylation reactions have traditionally been regarded as a detoxification mechanism since the methylated metabolites exert less acute toxicity and reactivity with tissue constituents than inorganic As. Approximately 70% of arsenic is excreted, mainly in urine. Smaller amounts may be eliminated in the feces, sweat, hair, or nails. The metabolic process and distribution of arsenic in human body are closely associated with the paradoxical effects of arsenic: poison and medicine.

Sources and Route of Exposure

Humans are exposed to arsenic in the environment through ingestion of food and water, inhalation of polluted air, and, to a much lesser extent, dermal absorption. Food is usually the largest source except in areas where drinking water is naturally contaminated with arsenic. The daily intake of arsenic from food and beverages is generally between 20 and 300 μg/day (http://www.greenfacts.org/en/arsenic/ 2004). Arsenic in food is mainly in the form of organic arsenic. About one-quarter of the arsenic present in the diet is inorganic arsenic, mainly from foods such as meat, poultry, dairy products, and cereals. In drinking water, arsenic is present in the more toxic, inorganic form. The predominant form in water from deep wells is arsenite. Surface water contains mostly the arsenate, and the airborne pollutant is predominantly the unhydrated trivalent compound As_2O_3 (Rossman and Klein 2011).

Absorption

Correspondingly, the primary routes of arsenic entry into the body are ingestion and inhalation. Dermal absorption also occurs, but to a lesser extent. In the gastrointestinal (GI) tract, soluble arsenic compounds from food and beverages are rapidly and extensively absorbed into the blood stream. The amount of arsenic absorbed into the body from inhaled airborne particles is highly dependent on two factors, the size of particles and their solubility. The amount of arsenic absorbed by inhalation has not been determined precisely, but it is thought to be within 60–90% (ATSDR 2009).

The majority of arsenic enters the body in the trivalent inorganic form via a simple diffusion

Arsenic in Tissues, Organs, and Cells, Fig. 1 The methylation process of arsenic metabolism in humans

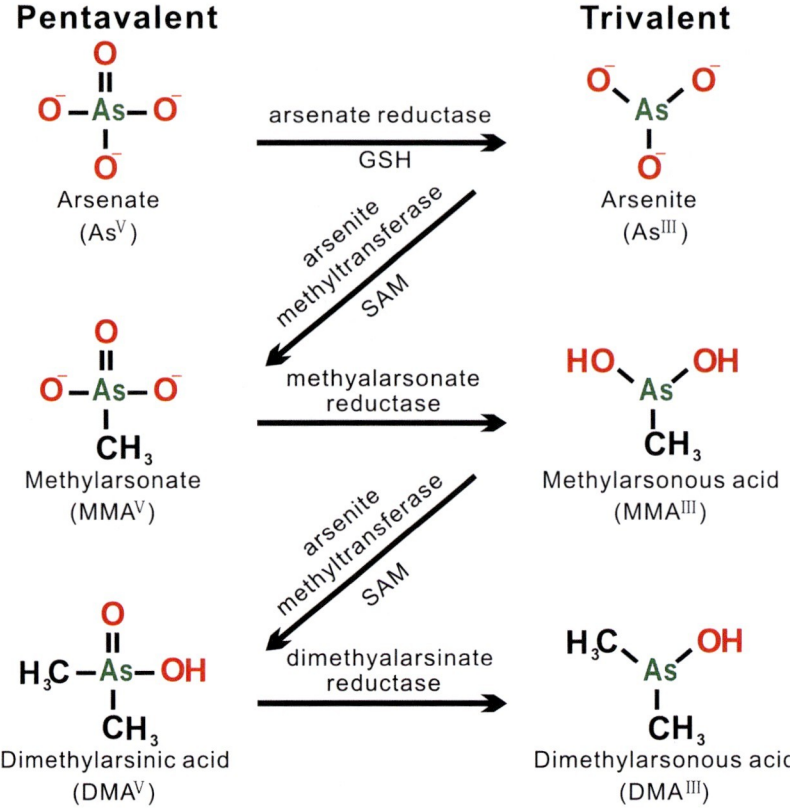

mechanism. Only a small amount of pentavalent inorganic arsenic can cross cell membranes via an energy-dependent transport system, after which it is immediately reduced to trivalent arsenic. Arsenic is absorbed into the blood stream at the cellular level and is taken up by red blood cells, white blood cells, and other cells that reduce arsenate to arsenite (ATSDR 2009). Since cells would have no reason to evolve uptake systems for toxic elements, both trivalent arsenite and pentavalent arsenate are taken up adventitiously by existing transport systems. Arsenate is taken up by phosphate transporters, while arsenite is taken up by aquaglyceroporins (AQP7 and AQP9 in mammals) and hexose permeases (GLUT1 and GLUT4 in mammals) (Rosen and Liu 2009).

Distribution and Metabolism

After absorption through the lungs or GI tract, arsenic is widely distributed by the blood throughout the body. Most tissues rapidly clear arsenic, except for keratin-rich tissues such as skin, hair, nails, and, to a lesser extent, in bones and teeth. The absorbed As is metabolized primarily by the liver and excreted by the kidneys into the urine. Smaller amounts may be eliminated in the feces, hair, nails, sweat, and skin desquamation. The metabolism of arsenic is a particularly complex pathway. In human, arsenic metabolism is characterized by two main types of reactions (Jomova et al. 2011; Nicolis et al. 2009; Rossman and Klein 2011): (1) reduction reactions of pentavalent to trivalent arsenic and (2) oxidative methylation reactions in which trivalent forms of arsenic are sequentially methylated to form methylated products using S-adenosyl methionine (SAM) as the methyl donor and glutathione (GSH) as an essential cofactor. The postulated scheme is as follows: $As^V \rightarrow As^{III} \rightarrow MMA^V \rightarrow MMA^{III} \rightarrow DMA^V \rightarrow DMA^{III}$ (Fig. 1). The enzyme catalyzing the biomethylation has been identified as arsenite methyltransferase (also named Cyt19, encoded by the gene *AS3MT*). Several factors have been suggested to influence the metabolism of arsenic, such as age, gender, genetic polymorphisms in genes coding for enzymes involved in arsenic metabolism, exposure level, nutrition, smoking, alcohol use, and diseases (Cui et al. 2008). The metabolism and disposition of inorganic arsenic may be influenced by its valence state. In vitro studies have shown that the cellular uptake of As^{III} is greater and faster than that of As^V (Cui et al. 2008). Ingested organoarsenicals such as MMA, DMA, and arsenobetaine are much less extensively metabolized and more rapidly eliminated in urine than inorganic arsenic in both laboratory animals and humans. Studies show that, regardless of the As concentration, methylated metabolites can be detected in the liver cancer cell line HepG2, whereas no such metabolites are found in the leukemia cell lines NB4 and U937 (Cui et al. 2008). This observation suggests that leukemia cells have special metabolic system of arsenic which may exert antileukemia effects. Methylation has long been considered the main route of arsenic detoxification, but more recently, there has been a growing body of literature supporting other detoxification mechanisms. For example, a number of animal species lack arsenic methylation and excrete inorganic arsenic. The implication is that there may be other important arsenic detoxification mechanisms in mammals.

Excretion

Absorbed As is rapidly cleared from the blood, most of which is cleared with a half-life of about 1 h. In samples collected from patients for three weeks after the last intravenous administration of 0.15 mg As_2O_3/kg body wt, the arsenic content in blood cells is measured 6–10 times higher than plasma levels. Pentavalent arsenic is found in blood only transiently at the end of therapy and rapidly disappears. Trivalent arsenic concentration of blood rapidly decreases after the end of administration falling below detection limits 10 h after end of infusion. Methylated metabolites appear in blood 6 h after arsenic infusion and decreased more slowly over a few days. An elimination half-life of 70 h is reported for As_4S_4 oral administration. DMA is the primary As species in blood plasma and red blood cells with lower concentrations of inorganic As and MMA. Arsenic concentrations in blood from people with no unusual exposures range from 0.3 to 2 µg/L, much lower than those in urine, hair, or nails (Nicolis et al. 2009; Orloff et al. 2009).

The main route of arsenic excretion is in the urine through the kidneys. During the first 2–4 h after intravenous administration of As_2O_3, trivalent inorganic arsenite is the main compound found in urine, while pentavalent metabolites monomethylarsonic and dimethylarsinic acids become the major urine arsenic species after the first 24 h, dimethylarsinic acid being generally the most important one in percentage. Only small amounts of pentavalent inorganic arsenate are detected in urine. Approximately 50% of excreted arsenic in human urine is dimethylated and 25% is monomethylated, with the remainder being inorganic. However, there may be individual variations in percentage. After a single intravenous injection of radiolabeled trivalent inorganic arsenic in human volunteers, most of the arsenic are cleared through urinary excretion within 2 days, although a small amount of arsenic is found in the urine up to 2 weeks later. Fish arsenic is largely not biotransformed in vivo, but it is rapidly excreted unchanged in the urine. The biologic half-life of ingested fish arsenic in humans is estimated to be less than 20 h, with total urinary clearance in approximately 48 h. The concentration of metabolites of inorganic arsenic in urine reflects the absorbed dose of inorganic arsenic on an individual level. Generally, it ranges from 5 to 20 µg/L, but may even exceed 1,000 µg/L (Nicolis et al. 2009; Orloff et al. 2009).

Arsenic in the blood is in equilibrium with As in the hair root, so that As is incorporated into the hair shaft as the hair grows out. Arsenic (As^{III}) has a high affinity for sulfhydryl groups that are prevalent in keratin and other proteins in hair. The binding of As to keratin may be regarded as an excretory pathway since once As is incorporated into the hair matrix, it is no longer biologically available. Arsenic in hair is predominantly inorganic and less than 10% is organic. Arsenic in water or dust that comes into contact with the hair can also bind to the sulfhydryl groups in keratin and other proteins in hair. In individuals with no known exposure to As, the concentration of As in hair from multiple studies is generally in the range of 0.02–0.2 µg/g (Orloff et al. 2009; Rossman and Klein 2011).

The nail bed has a rich blood supply, and As is incorporated into the nail as it grows out. As with hair, inorganic As^{III} binds to sulfhydryl groups in keratin proteins in the nail matrix. External contamination of nails with As is a potential problem for biomonitoring, as it is with hair. Inorganic As is the primary As species in fingernails, although lower concentrations of DMA and MMA are also detected.

Normal As concentrations in nails are reported to range from 0.02 to 0.5 μg/g (Orloff et al. 2009; Rossman and Klein 2011).

Cross-References

- Arsenic and Aquaporins
- Arsenic and Primary Human Cells
- Arsenic and Vertebrate Aquaglyceroporins
- Arsenic in Nature
- Arsenic in Pathological Conditions
- Arsenic Methyltransferases
- Arsenic, Biologically Active Compounds
- Arsenic, Mechanisms of Cellular Detoxification

References

ATSDR (2009) Case studies in environmental medicine. Arsenic toxicity. http://www.atsdr.cdc.gov/csem/arsenic/docs/arsenic.pdf

Cui X, Kobayashi Y, Akashi M et al (2008) Metabolism and the paradoxical effects of arsenic: carcinogenesis and anticancer. Curr Med Chem 15:2293–2304

Jomova K, Jenisova Z, Feszterova M et al (2011) Arsenic: toxicity, oxidative stress and human disease. J Appl Toxicol 31:95–107

Nicolis I, Curis E, Deschamps P et al (2009) Arsenite medicinal use, metabolism, pharmacokinetics and monitoring in human hair. Biochimie 91:1260–1267

Orloff K, Mistry K, Metcalf S (2009) Biomonitoring for environmental exposures to arsenic. J Toxicol Environ Health B Crit Rev 12:509–524

Rosen BP, Liu Z (2009) Transport pathways for arsenic and selenium: a mini review. Environ Int 35:512–515

Rossman TG, Klein CB (2011) Genetic and epigenetic effects of environmental arsenicals. Metallomics 3:1135–1141

Scientific Facts on Arsenic (2004). http://www.greenfacts.org/en/arsenic/

Arsenic Metabolic Pathway

- Arsenic

Arsenic Metabolism

- Arsenic in Tissues, Organs, and Cells

Arsenic Methyltransferase

- Arsenic Methyltransferases

Arsenic Methyltransferases

David J. Thomas[1] and Barry P. Rosen[2]
[1]Pharmacokinetics Branch – Integrated Systems Toxicology Division, National Health and Environmental Research Laboratory, U.S. Environmental Protection Agency, Research Triangle Park, NC, USA
[2]Department of Cellular Biology and Pharmacology, Florida International University, Herbert Wertheim College of Medicine, Miami, FL, USA

Synonyms

AdoMet:arsenic(III) methyltransferase; Arsenic methyltransferase; Arsenite methyltransferase; Arsenite S-adenosylmethionine methyltransferase; Cyt19; Monomethylarsonous acid methyltransferase; S-adenosyl-L-methionine:arsenic(III) methyltransferase

Definition

Arsenic methyltransferase (EC # 2.1.1.137) is an enzyme that catalyzes reactions that convert inorganic arsenic into mono-, di-, and tri-methylated products. Examples of arsenic methyltransferases have been identified in archaea, prokaryotes, and eukaryotes. Conversion of inorganic arsenic into methylated metabolites affects the environmental transport and fate of arsenic and its metabolism and disposition at cellular and systemic levels.

The metalloid arsenic enters the environment by natural processes (volcanic activity, weathering of rocks) and by human activity (mining, smelting, herbicides, and pesticides). Although arsenic has been exploited for homicidal and suicidal purposes since antiquity, its significance as a public health issue arises from its potency as a human carcinogen. In addition, considerable epidemiological evidence shows that chronic exposure to inorganic arsenic contributes to increased risk of other diseases (Hughes et al. 2011). Interest in the biomethylation of arsenic as a factor in its environmental fate and its actions as a toxicant and a carcinogen originated in the nineteenth century with observations that microorganisms converted inorganic arsenicals used as wallpaper pigments into Gosio gas, a volatile species that is released into the atmosphere and that Gosio gas was trimethylarsine. Subsequent detection of methylated arsenicals in natural waters and in human urine suggested that biomethylation of arsenic was a widespread phenomenon (Cullen 2008).

These studies piqued interest in understanding the molecular basis of biomethylation of arsenic. Studies in mammals demonstrated methylation of arsenic to be enzymatically catalyzed and culminated in isolation and purification of an arsenic methyltransferase from rat liver cytosol and cloning and expression of cognate genes from rat, mouse, and human (Thomas et al. 2007). This gene was initially identified as *cyt19* but is now designated as arsenic (+3 oxidation state) methyltransferase (*As3mt*). The human *AS3MT* gene (accession number NP_065733.2) encodes a 375 amino-acid protein (41,745Da, EC # 2.1.1.137) that contains sequence motifs commonly found in non-DNA methyltransferases. Putative *As3mt* genes have been identified in genomes of deuterostomes ranging in complexity from purple sea urchin (*Strongylocentrotus purpuratus*) to *Homo sapiens*. The central role of As3mt in methylation of arsenic has been demonstrated by *As3mt* gene silencing by RNA interference and by heterologous expression of the *As3mt* gene in a human cell line that does not methylate arsenic (Thomas et al. 2007). Studies in *As3mt* knockout mice show capacity for arsenic methylation to be greatly diminished, albeit not eliminated, in these animals. Residual capacity for arsenic methylation in *As3mt* knockout mice may reflect methylation of arsenic by gut microbiota or arsenic methyltransferase activity of other methyltransferases. Arsenic methylation catalyzed by As3mt is central to the processes that control the distribution, retention, and clearance of arsenicals in tissues after exposure to inorganic arsenic. For example, reduced methylation of arsenic in *As3mt* knockout mice is associated with increased accumulation of inorganic arsenic in urinary bladder and exacerbation of injury to this organ. Conversely, there is evidence that methylated arsenicals containing trivalent arsenic formed during enzymatically catalyzed methylation of arsenic are more potent than inorganic trivalent arsenic (arsenite) as enzyme inhibitors, cytotoxins, and genotoxins (Thomas et al. 2007). Hence, methylation of arsenic in higher organisms has aspects that can be considered simultaneously as pathways for both activation and detoxification of this metalloid.

Parallel studies have demonstrated arsenic biomethylation in bacteria, archaea, fungi, and eukaryotic algae. In these organisms, *ar*senite S-adenosyl*m*ethyltransferase (ArsM) orthologues of As3mt catalyze arsenic methylation. Typically, prokaryotic and archaeal *arsM* genes are downstream of an *arsR* gene that encodes the archetypal arsenic-responsive transcriptional repressor that controls expression of *ars* operons suggesting that ArsM genes evolved to confer arsenic resistance. The first identified *arsM* gene was cloned from the soil bacterium *Rhodopseudomonas palustris* (Qin et al. 2006) and encodes for a 283-residue ArsM protein (29,656 Da) (accession number NP_948900.1). Its expression in an arsenic hypersensitive strain of *Escherichia coli* conferred resistance to arsenite. *E. coli* cells expressing recombinant *arsM* biotransformed medium arsenic into the methylated species, dimethylarsinic acid, trimethylarsine oxide, and trimethylarsine. These results clearly demonstrate that *arsM* gene expression is necessary and sufficient for arsenic detoxification. Purified recombinant ArsM catalyzed transfer of methyl groups from S-adenosylmethionine to arsenite, forming di- and trimethylated species, with the final product being trimethylarsine gas.

Commonalities and differences have been identified in the structure and catalytic function of arsenic methyltransferases from highly divergent species. Figure 1 shows that amino acid sequences are highly conserved in arsenic methyltransferases from species ranging from human (hAS3MT) to the thermophile *Cyanidioschyzon* strain 5,508, an environmental isolate of the eukaryotic red alga *Cyanidioschyzon merolae* from Yellowstone National Park (CmArsM). In all known arsenic methyltransferase orthologues, three cysteine residues are strictly conserved. These conserved cysteine residues (61, 156, and 206 in hAS3MT and 72, 174, and 224 in CmArsM7 from *C. merolae*) are required for catalytic function (Song et al. 2009; Marapakala et al. 2012). Loss of catalytic

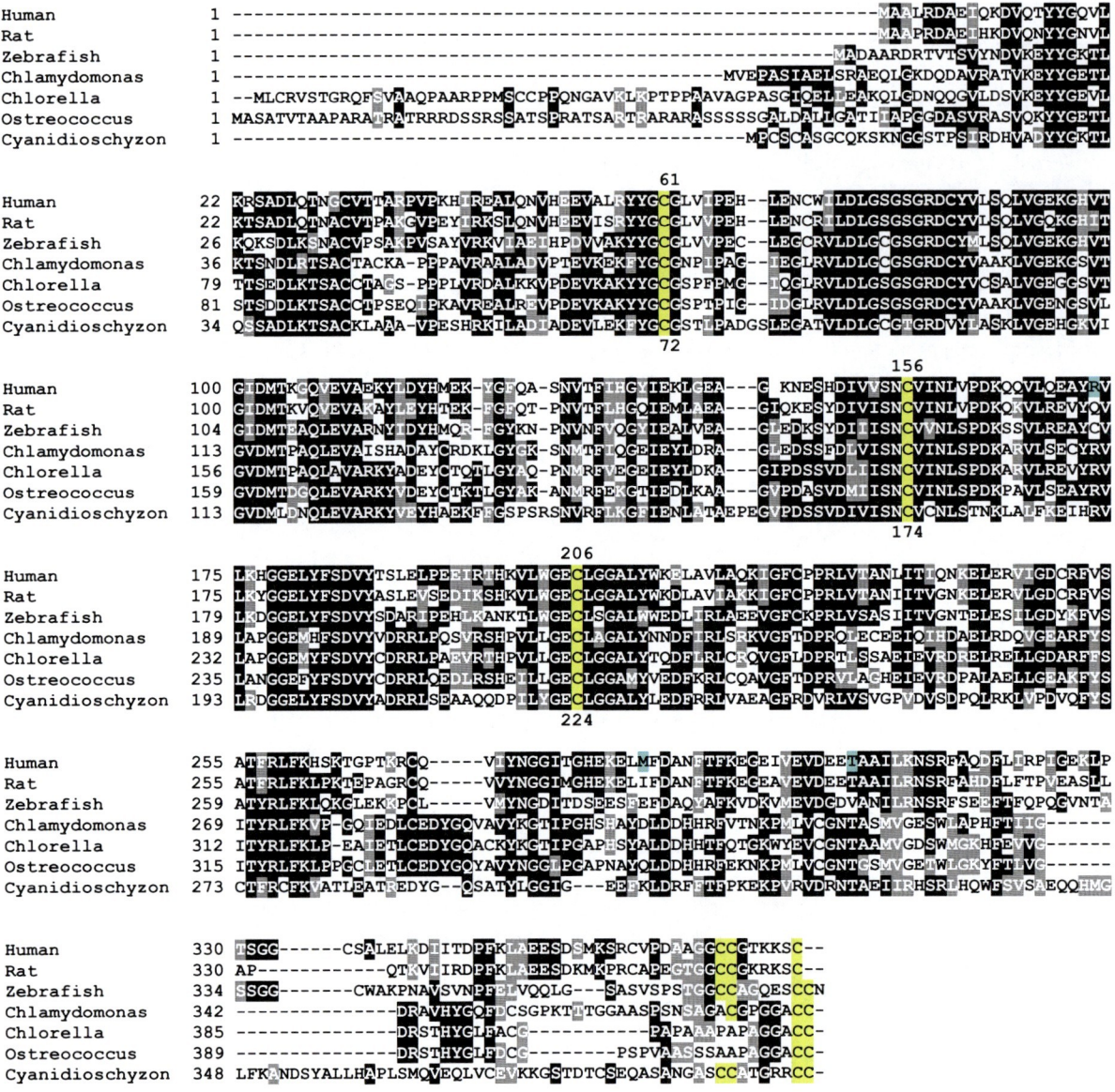

Arsenic Methyltransferases, Fig. 1 Conserved cysteine residues are highlighted in *yellow*

activity in As3mt in which these critical cysteine residues have been replaced has been attributed to disruption of critical disulfide bonds between these cysteine residues (Fomenko et al. 2007). In contrast, formation of disulfide bonds is not observed in CmArsM during catalysis (K. Marapakala and B.P. Rosen, unpublished results). A final judgment on the role of critical cysteine residues will depend on structural characterization of the enzymes and elucidation of the catalytic process. In addition to these three strictly conserved cysteine residues, all arsenic methyltransferase orthologues have multiple C-terminal cysteine residues and pairs that are not absolutely conserved. These cysteine residues could form a site for initial binding of arsenite before its transfer to the catalytic cysteine residues or function to regulate enzyme activity.

There are notable differences between As3mt and ArsM in reductants that support catalytic function. Arsenic methylation catalyzed by recombinant rat

As3mt (rAs3mt) shows a strong dependence on the presence of dithiol reductants, with activity maximized by addition of a coupled system containing the endogenous reductant thioredoxin (Tx), thioredoxin reductase (TxR), and NADPH. The monothiol glutathione (GSH) does not support methylation catalyzed by rAs3mt but does increase the overall rate of methylation when added to reaction mixtures containing the Tx/TxR/NADPH coupled systems and reduces the capacity of rAs3mt to catalyze formation of trimethylated arsenicals (Thomas et al. 2007). In contrast, studies with hAS3MT have shown GSH supports conversion of inorganic arsenic into mono- and dimethylated products (Hayakawa et al. 2005). Methylated products were detected in the presence of at least 2 mM GSH, and there was a positive relation between the percentage of arsenite present in an arsenite-$(GSH)_3$ complex and the extent of hAS3MT-catalyzed methylation. Although direct comparisons of the results with rAs3mt and hAS3MT are problematic due to differences in assay conditions and analytical procedures, it is notable that yields of methylated metabolites were much higher in rAs3mt-catalyzed reactions with the Tx/TxR/NADPH coupled systems than in hAS3MT-catalyzed reactions with GSH. In contrast, GSH alone is sufficient to support methylation of arsenic catalyzed by CmArsM7 (Marapakala et al. 2012). Arsenite-$(GSH)_3$ or methylarsonous acid-$(GSH)_2$ are bound to the enzyme faster than are arsenite or methylarsonous acid, suggesting that complexation of substrate by GSH accounts for its role in CmArsM7-catalyzed reactions. Because GSH is the only thiol present in cells at millimolar concentrations, it is possible that spontaneously formed $As(GS)_3$ is a direct donor of metalloid to CmArsM. In CmArsM-catalyzed reactions, the first round of methylation converts arsenite to methylarsonous acid and a second round of methylation converts methylarsonous acid to dimethylarsinous acid faster than methylarsonous acid dissociates from the enzyme. The latter product with a relatively lower affinity for the enzyme dissociates faster than it undergoes the third round of methylation. This model posits binding of a glutathionylated trivalent arsenical to the enzyme during repeated cycles of methylation and predicts dimethylarsinous acid to be the principal metabolic product.

Differences in requirements for specific reductants in reactions catalyzed by arsenic methyltransferases suggest two distinct models for the pathway of arsenic methylation. The Challenger scheme for arsenic methylation was originally developed as a chemically plausible pathway for the conversion of inorganic arsenic to methylated products based on studies in microorganisms (Challenger 1951). As shown in Fig. 2a, this pathway involves alternating rounds of oxidative methylation in which a methyl group is added to a trivalent arsenical. The resulting pentavalent arsenical is reduced to trivalency for additional cycles of oxidative methylation. An alternative pathway proposed by Hirano and coworkers involves reductive methylation of arsenic (Hayakawa et al. 2005). Here, trivalent arsenicals present as GSH conjugates or bound to protein thiols undergo repeated rounds of methyl group addition, and the appearance of methylated products containing pentavalent arsenic in urine is due to nonenzymatic oxidation of the glutathionylated intermediates or products (Fig. 2b). This pathway is supported by recent studies with CmArsM7 (Marapakala et al. 2012). Further studies of molecular interactions between reductants and arsenic methyltransferases or of the role of arsenothiol complexes as preferred substrates are essential to delineating the relevant pathway for arsenic methylation.

Variation in capacity for enzymatically catalyzed arsenic methylation can have practical consequences. A common genetic variant in human *AS3MT* (NCBI rs11191439) that occurs at a frequency of about 10% of most populations results in the replacement of a methionyl residue in position 287 with a threonyl residue (AS3MT/M287T). In population-based studies of the effect of AS3MT genotype on urinary arsenical profiles, AS3MT/M287T genotype has been associated with a higher concentration of methylated arsenic than was found with the wild-type AS3MT genotype (Engstrom et al. 2011) and with increased disease susceptibility (Hsieh et al. 2011) Thus, a specific AS3MT genotype may be linked to an arsenic methylation phenotype and to a disease susceptibility phenotype through increased production of methylated arsenic, a reactive and toxic metabolite of inorganic arsenic.

The bacterial *arsM* gene has the potential use for improving food safety. Rice, the primary source of nutrition for more than half of the world's population, is a natural arsenic accumulator. Because consumption of arsenic-containing rice could increase cancer risk

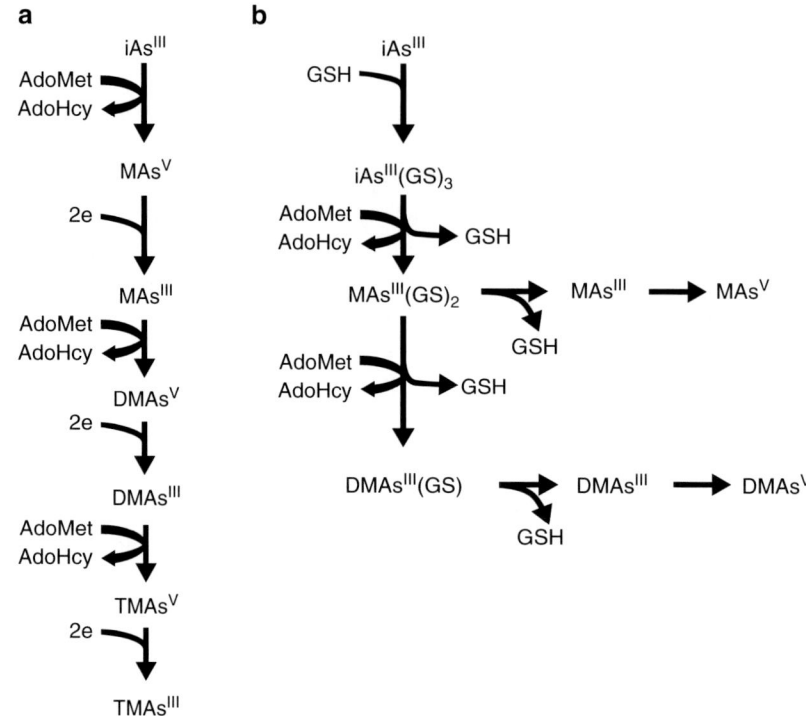

Arsenic Methyltransferases, Fig. 2 Alternative pathways for biomethylation of arsenic. (**a**) Oxidative methylation of arsenic converts trivalent arsenicals to methylated species containing pentavalent arsenic. Donation of a methyl group donor from S-adenosylmethionine (AdoMet) yields S-adenosylhomocysteine (AdoHcy). Pentavalent arsenic in methylated arsenicals is reduced to trivalency before each round of methylation (**b**). Reductive methylation converts a trivalent oxyarsenical into a glutathione (GSH) complex. This complex is the substrate for repeated rounds of methylation. Dissociation of arsenical-GSH complexes yields trivalent arsenicals that can be oxidized to pentavalency. iAs^{III} arsenite, $iAs^{III}(GS)_3$ arsenite-triglutathione, MAs^V methylarsonic acid, MAs^{III} methylarsonous acid, $MAs^{III}(GS)_2$ methylarsonous acid-diglutathione, $DMAs^V$ dimethylarsinic acid, $DMAs^{III}$ dimethylarsinous acid, $DMAs^{III}(GS)$ dimethylarsinous acid-glutathione, $TMAs^V$ trimethylarsine oxide, $TMAs^{III}$ trimethylarsine

(Williams et al. 2007), it would be useful to reduce arsenic accumulation in this food crop. One approach to reducing arsenic levels has been to clone the *R. palustris arsM* gene into rice to create transgenic plants that methylate arsenic (Meng et al. 2011). Although the process requires optimization, it confirms the validity of the approach and demonstrates that rice plants can be engineered to biotransform arsenic. This process might be optimized by using a plant gene rather than a bacterial one to transform rice. Although higher plants do not have *arsM* genes and do not methylate arsenic, *arsM* genes from lower plants, especially from eukaryotic algae, are candidates for use in rice modification (Yin et al. 2011). With the goal of increasing biomethylation and volatilization of arsenic, future experiments will be conducted with rice transformed with the *CmarsM* gene from the red alga *C. merolae*.

Disclaimer

This manuscript has been reviewed in accordance with the policy of the National Health and Environmental Effects Research Laboratory, U.S. Environmental Protection Agency, and approved for publication. Approval does not signify that the contents necessarily reflect the views and policies of the Agency, nor does mention of trade names or commercial products constitute endorsement or recommendation for use. BPR acknowledges support from United States Public Health Service NIH Grant R37 GM55425.

Cross-References

► Arsenic
► Arsenic and Aquaporins
► Arsenic and Vertebrate Aquaglyceroporins
► Arsenic in Nature
► Arsenic in Tissues, Organs, and Cells
► Arsenic Methyltransferases
► Arsenic, Biologically Active Compounds
► Arsenic, Mechanisms of Cellular Detoxification

References

Challenger F (1951) Biological methylation. Adv Enzymol Relat Subj 12:429–491
Cullen WR (2008) Is arsenic an aphrodisiac? The sociochemistry of an element. RSC Publishing, Cambridge
Engstrom K, Vahter M, Mlakar SJ et al (2011) Polymorphisms in arsenic(+III oxidation state) methyltransferase (AS3MT) predict gene expression of AS3MT as well as arsenic metabolism. Environ Health Perspect 119:182–188
Fomenko DE, Xing W, Adair BM et al (2007) Large-scale identification of catalytic redox-active cysteines by detecting sporadic cysteine/selenocysteine pairs in homologous sequences. Science 315:387–389
Hayakawa T, Kobayashi Y, Cui X et al (2005) A new metabolic pathway of arsenite: arsenic-glutathione complexes are substrates for human arsenic methyltransferase Cyt19. Arch Toxicol 79:183–191
Hsieh Y-C, Lien L-M, Chung W-T, Hsieh FI et al (2011) Significantly increased risk of carotid atherosclerosis with arsenic exposure and polymorphisms in arsenic metabolism genes. Environ Res 111:804–810
Hughes MF, Beck B, Chen Y et al (2011) Arsenic exposure and toxicology: a historical perspective. Toxicol Sci 123:305–332
Marapakala K, Qin J, Rosen BP (2012) Identification of catalytic residues in the As(III) S-adenosylmethionine methyltransferase. Biochemistry 51:944–951
Meng XY, Qin J, Wang LH et al (2011) Arsenic biotransformation and volatilization in transgenic rice. New Phytol 191:49–56
Qin J, Rosen BP, Zhang Y et al (2006) Arsenic detoxification and evolution of trimethylarsine gas by a microbial arsenite S-adenosylmethionine methyltransferase. Proc Natl Acad Sci USA 103:2075–2080
Song X, Geng Z, Zhu J et al (2009) Structure-function roles of four cysteine residues in the human arsenic (+3 oxidation state) methyltransferase (hAS3MT) by site-directed mutagenesis. Chem Biol Interact 179:321–328
Thomas DJ, Li J, Waters SB et al (2007) Arsenic (+3 oxidation state) methyltransferase and the methylation of arsenicals. Exp Biol Med 232:3–13
Williams PN, Raab A, Feldmann J et al (2007) Market basket survey shows elevated levels of as in South Central U.S. processed rice compared to California: consequences for human dietary exposure. Environ Sci Technol 41: 2178–2183
Yin XX, Chen J, Qin J et al (2011) Biotransformation and volatilization of arsenic by three photosynthetic cyanobacteria. Plant Physiol 156:1631–1638

Arsenic Pharmacokinetics

► Arsenic in Tissues, Organs, and Cells

Arsenic Sulfide

► Arsenic in Therapy
► Promyelocytic Leukemia–Retinoic Acid Receptor α (PML–RARα) and Arsenic

Arsenic Toxicity

► Arsenic

Arsenic Transporters

► Arsenic and Vertebrate Aquaglyceroporins

Arsenic Trioxide

► Arsenic in Therapy
► Promyelocytic Leukemia–Retinoic Acid Receptor α (PML–RARα) and Arsenic

Arsenic, Biologically Active Compounds

Leonid Breydo
Department of Molecular Medicine, Morsani College of Medicine, University of South Florida, Tampa, FL, USA

Synonyms

DMA^{III} – dimethylarsenite; DMA^{V} – dimethylarsenate; MMA^{III} – monomethylarsenite; MMA^{V} – monomethylarsenate

Definition

Arsenic is widespread in the environment and many of its derivatives are highly toxic and carcinogenic. Toxicity of arsenic-containing compounds is primarily derived from their ability to form protein adducts via a reaction with cysteine residues and to deplete the cellular glutathione levels. Arsenic in living organisms is reduced and methylated to form less toxic organoarsenic compounds. Arsenic is also incorporated into arsenolipids, arsenosugars, and other types of organic compounds.

Inorganic and Organic Arsenic Compounds in the Environment

Arsenic is the 20th most abundant element on Earth and is often present together with sulfide minerals (Rezanka and Sigler 2008). In inorganic compounds, arsenic is present in a variety of oxidation states with As^{3+} and As^{5+} being the most common. Arsenic is released into the environment during burning of coal and other fossil fuels and during smelting of arsenic-containing ores. Arsenic derivatives have been routinely used as pesticides, wood preservers, and fungicides further contributing to release of this element although they are being phased out due to their relatively high toxicity. In the beginning of the twentieth century, organoarsenic compounds have been used as antibiotics although their use in this capacity has been discontinued due to availability of more effective and less toxic alternatives.

Arsenic is considered an essential element for humans and other mammals (Fig. 1). It is present in serum at the concentrations around 10 ng/ml. Lower levels of arsenic in blood (5–7 ng/ml) were observed in patients with CNS and cardiovascular diseases (Mayer et al. 1993). Low arsenic diet (0.1 mg/kg or less) leads to slower growth and higher risk of birth defects in several types of experimental animals. Arsenic is believed to be important in promotion of glutathione biosynthesis and metabolism of several amino acids including arginine and methionine (Uthus et al. 1983). However, the exact mechanism of its action is unknown.

At higher concentrations, many arsenic derivatives are both highly acutely toxic and carcinogenic. Arsenates and elemental arsenic are not particularly toxic but toxicity of arsenites and many organoarsenic compounds (especially As^{3+} organoarsenics) is substantial. High abundance of arsenic in soil leads to its persistent presence in water at the level of several ppb (part per billion). WHO has determined that levels of arsenic in water above 10 ppb are unsafe for human health (Fawell and Nieuwenhuijsen 2003). However, in some water sources, arsenic levels are significantly higher, exposing over 100 million people in the world to the risk of increased incidence of cancer and other diseases. Several organoarsenic compounds (diphenylchloroarsine, diphenylcyanoarsine, phenyldichloroarsine, and lewisite) have been used as chemical weapons, and dimethylarsinic acid has been used as a defoliant (Fig. 2). Arsenic derivatives (especially arsenites) have also been used as poisons since at least the Middle Ages.

The most important mechanism of toxicity of arsenites and trivalent organoarsenic compounds involves their reactions with thiols. Arsenic derivatives rapidly react with cysteine residues of proteins such as enzymes and metal carrier proteins to form the As adducts (Fig. 3). These adducts are highly stable, and their formation inhibits normal function of enzymes and other proteins. Especially stable adducts are formed when As complexes with two neighboring sulfhydryl groups in a single protein. While excess glutathione can restore the protein function, prolonged exposure to arsenites depletes glutathione levels and leads to increased ROS production. Confirming its effect on ROS production, As^{3+} (1–25 µM) increased mRNA levels of antioxidant enzymes such as heme oxygenase-1, thioredoxin peroxidase 2, NADPH dehydrogenase, and glutathione S-transferase P subunit (Hirano et al. 2003). These enzymes are overexpressed as a defense mechanism against oxidative stress.

Toxicity of arsenate occurs via a different mechanism and stems from its ability to replace inorganic phosphate and form unstable arsenyl esters that hydrolyze spontaneously (Fekry et al. 2011; Nielsen and Uthus 1984). During glycolysis, this spontaneous hydrolysis of arsenate esters prevents formation of highly energetic phosphate esters that are later converted to ATP. Thus, energy produced during glycolysis is lost if arsenic substitutes for phosphorus.

Arsenic, Biologically Active Compounds

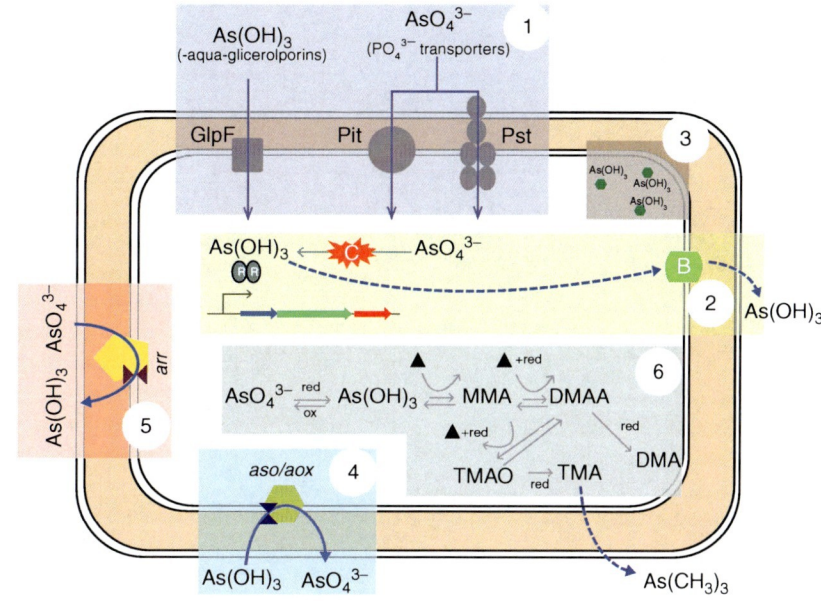

Arsenic, Biologically Active Compounds, Fig. 1 Diagram of the different processes involved in microbial arsenic metabolism. (*1*) Arsenic enters the cells. (*2*, *5*) Reduction of arsenate to arsenite and extrusion of arsenite. (*3*) Reaction of arsenite with thiols. (*4*) Oxidation of arsenite to arsenate. (*6*) Methylation of arsenate and arsenite (Reproduced with permission from Paez-Espino et al. (2009))

Arsenic, Biologically Active Compounds, Fig. 2 Common arsenic-containing compounds in vivo: arsenate, arsenite, trimethylarsine, diphenylchloroarsine, phenyldichloroarsine, diphenylcyanoarsine, dimethylarsinic acid, arsenobetaine, arsenocholine

Metabolism of Arsenate and Arsenite

Metabolism of inorganic arsenate and arsenite in mammals involves four steps. A two-electron reduction of arsenate to arsenite is followed by methylation of arsenite to form monomethylarsonic acid (MMA^V). Another two-electron reduction of MMA^V to form monomethylarsenic acid (MMA^{III}) occurs, which is methylated to form dimethylarsonic acid (DMA^V) (Fig. 4).

The first step is the reduction of arsenate to arsenite. It is catalyzed by arsenate reductases. There are three types of this enzyme with two found exclusively in bacteria. Prokaryotic arsenate reductases (ArsC) utilize reduced glutathione and a thiol transfer protein glutaredoxin as reducing

Arsenic, Biologically Active Compounds, Fig. 3 Reaction of arsenite with thiol

Arsenic, Biologically Active Compounds, Fig. 4 Reductive methylation of arsenite and decomposition of DMA^V to dimethylarsine

agents. ArsC has a single catalytic Cys residue that forms a covalent intermediate with arsenate. Glutathione reacts with this intermediate, forming a mixed disulfide and displacing arsenite. Glutaredoxin reduces the mixed disulfide, regenerating the active site cysteine and reduced glutathione. Eukaryotic arsenate reductase has been identified in yeast. It is not closely related to bacterial arsenate reductases but instead is similar to tyrosine phosphatases. It has the same cofactor requirements as ArsC and appears to operate via the same mechanism. Glutathione and other thiols can also reduce arsenate to arsenite nonenzymatically but this reaction is quite slow in physiological conditions.

In addition, some phosphorylases are capable of reducing arsenate to arsenite. Purine nucleoside phosphorylase, glyceraldehyde-3-phosphate dehydrogenase (GAPDH), glycogen phosphorylase, and ornithine carbamoyl transferase (Nemeti and Gregus 2009) have been identified as such enzymes although it is likely that other phosphorylases can catalyze this reaction as well. Phosphorylases add phosphate to their substrates, resulting in the cleavage of substrate into two products, one of which is phosphorylated. Arsenate can replace phosphate in these reactions, leading to an arsenylated metabolite. For example, purine nucleoside phosphorylase cleaves inosine with arsenate to yield hypoxanthine and ribose-1-arsenate. The resulting arsenate ester is rapidly reduced to arsenite by thiols in nonenzymatic fashion (Gregus et al. 2009).

Reverse reaction (oxidation of arsenite to arsenate) is catalyzed by the arsenite oxidase. It is a bacterial enzyme that consists of two subunits (aoxA and aoxB). This enzyme contains both molybdenum bound to pterin cofactor and iron-sulfur clusters (Santini and vanden Hoven 2004) and is similar to other molybdenum-containing redox enzymes. Azurin is used as an electron donor. Since arsenate is significantly less toxic compared to arsenite, this enzyme is common in the bacteria living in arsenic-rich environments (Santini and vanden Hoven 2004). For some of those bacteria, arsenite oxidation serves as a source of energy.

Conversion of arsenite to organoarsenic derivatives starts from its methylation to yield monomethylarsonic acid (MMA^V). This reaction is catalyzed by arsenite methyltransferase with S-adenosyl methionine (SAM) serving as a methyl donor (Fig. 4). Arsenate can also be methylated by this enzyme but the reaction is significantly slower. This enzyme is common in bacteria and animals but it is not present in plants. In some mammalian species, the enzyme is present in inactive truncated form. Methylation likely evolved as detoxification pathway as As(V) methylated species are relatively nontoxic and are rapidly cleared (Hall and Gamble 2012; Thomas et al. 2004).

MMA^V can be reduced to form monomethylarsonic acid (MMA^{III}). This reaction is catalyzed by MMA^V reductase, a member of glutathione S-transferase superfamily (Zakharyan et al. 2001). MMA^{III} can be methylated further to dimethylarsonic acid (DMA^V) by arsenite methyltransferase. DMA^V is reduced to DMA^{III} by MMA^V reductase although a specific DMA reductase may be present as well. DMA^{III} may be methylated further to trimethylarsine oxide although this reaction is inhibited by the presence of thiols and is thus uncommon in vivo. DMA^V is a major arsenic metabolite in humans (Mandal et al. 2001). While it does not react with thiols as rapidly as As^{3+} derivatives, both DMA^V and DMA^{III} can decompose to produce dimethylarsine (Fig. 4). Dimethylarsine rapidly reacts with oxygen to produce the dimethylarsenic peroxyl radical and other ROS (Yamanaka et al. 2009). These ROS are capable of damage to DNA and proteins and may be carcinogenic.

In general, organoarsenic compounds are significantly less toxic compared to arsenite. For example, LD_{50} for arsenite is 0.014 g/kg while that for MMA^V is 0.7 g/kg and for DMA^{III} is 3.3 g/kg (Paez-Espino et al. 2009). Formation of organoarsenic compounds is clearly a part of detoxification process for arsenic.

Bacterial Arsenite Extrusion System

Many bacteria are resistant to arsenic toxicity. In addition to the methylation/reduction process described above, arsenite extrusion system contributes to this resistance. This system is composed of several proteins encoded by the *ars* operon: transcriptional repressor ArsR, arsenite efflux pump ArsB, and arsenate reductase ArsC (Paez-Espino et al. 2009). Additional components of the arsenite extrusion system (ArsA, ArsD, ArsH) are present in some bacteria as well. ArsA is an ATPase that serves as a source of energy for the arsenic pump ArsB and ArsD is an arsenic chaperon for this pump (Lin et al. 2007). ArsH is oxidoreductase that contributes to As reduction in some bacteria.

In its resting state, ArsR is bound to the promoter of *ars* operon inhibiting its transcription. Upon binding to As(III) derivatives, ArsR is activated and dissociates from DNA, allowing the transcription of arsenic extrusion proteins to occur. ArsR repressor is activated by covalent interaction of As(III) with three cysteines. ArsA ATPase is activated in the same manner. The role of ArsD is to transfer As(III) species to ArsA using the same three cysteine arrangement (Paez-Espino et al. 2009). The role of ArsH in the system is unclear.

Arsenic in Plants, Fungi, and Marine Organisms

Many species of mushrooms and other fungi have elevated concentrations of arsenic. In many of these

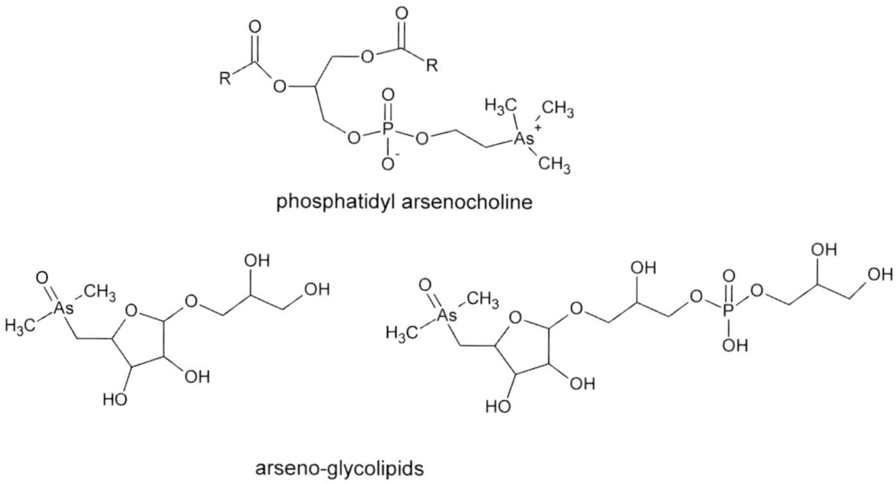

Arsenic, Biologically Active Compounds,
Fig. 5 Arsenolipids: phosphatidyl arsenocholine, dimethylarsenylribose derivatives

phosphatidyl arsenocholine

arseno-glycolipids

species, arsenic is primarily present in the form of arsenobetaine (up to 60–90% of total arsenic). In marine animals, arsenic is primarily present as arsenobetaine as well (Fig. 2). In addition, other arsenic-containing compounds have been found in marine organisms including arsenocholine, MMA, DMA, and a variety of arsenolipids (Fig. 5). Arsenolipids can be divided into two structural classes: hydrocarbon-based lipids and glycerolipids. Hydrocarbon-based arsenolipids usually contain dimethylarsinoxide moiety attached to a hydrocarbon chain. The mechanism for the biosynthesis of these compounds is unclear, and it has been suggested that dimethylarsinoyl-propionic acid might be a substrate for a fatty acid synthase. Arsenic-containing glycolipids include phosphatidylarsenocholine (presumably derived from arsenobetaine) and various derivatives of ribose with dimethylarsinoxide moiety in a 5′ position (Rezanka and Sigler 2008). These lipids, widespread in marine organisms, are believed to be synthesized by algae. Complex organoarsenic compounds do not react with thiols and are thus essentially nontoxic. Their formation is presumably a detoxification pathway for these organisms that live in the environment with relatively high arsenic concentrations.

macromolecules including DNA and RNA and small molecule metabolites (Wolfe-Simon et al. 2011). The authors showed that these bacteria can grow at very low phosphate concentrations (0.02% of dry weight). In addition, EXAFS data indicated that arsenic is present in these bacteria as arsenate with geometry resembling that of phosphate in nucleic acids. These findings were hotly disputed with other researchers pointing out the half-life of the model arsenate triester is 0.06 s under physiological conditions. In addition, it was pointed out that arsenate esters are easily reduced to arsenite (Fekry et al. 2011). Other researchers have shown that very low phosphate concentrations observed in GFAJ-1 may be sufficient for synthesis of nucleic acids, given very slow propagation of these bacteria. In addition, mass spectroscopy showed that DNA isolated from GFAJ-1 bacteria contained no covalently bound arsenic. Growth of these bacteria at low phosphate levels was independent of arsenic levels and stability of DNA was inconsistent with the presence of highly labile arsenate ester bonds (Reaves et al. 2012).

Overall, it can be concluded that there is no evidence that arsenic is incorporated into the bacterial DNA.

Can As Replace Phosphorus in DNA?

A recent publication claimed that a strain of *Halomonadacea* bacteria (GFAJ-1) isolated from the arsenic-rich waters of Mono Lake, California, is able to substitute arsenic for phosphorus in its

Conclusions

Arsenic is a ubiquitous element, and wide variety of arsenic compounds is present in the environment. The most important feature of many arsenic compounds is their toxicity due to formation of stable thiol adducts.

Metabolism of arsenic derivatives is dominated by detoxification pathways attempting to convert toxic arsenite to less toxic organoarsenic derivatives. Despite their toxicity, arsenic compounds are important components of the food chains of many organisms and may play important biological roles.

Cross-References

▶ Arsenic
▶ Arsenic in Nature
▶ Arsenic Methyltransferases
▶ Arsenic, Mechanisms of Cellular Detoxification
▶ As

References

Fawell J, Nieuwenhuijsen MJ (2003) Contaminants in drinking water. Br Med Bull 68:199–208

Fekry MI, Tipton PA, Gates KS (2011) Kinetic consequences of replacing the internucleotide phosphorus atoms in DNA with arsenic. ACS Chem Biol 6:127–130

Gregus Z, Roos G, Geerlings P, Nemeti B (2009) Mechanism of thiol-supported arsenate reduction mediated by phosphorolytic-arsenolytic enzymes: II. Enzymatic formation of arsenylated products susceptible for reduction to arsenite by thiols. Toxicol Sci 110:282–292

Hall MN, Gamble MV (2012) Nutritional manipulation of one-carbon metabolism: effects on arsenic methylation and toxicity. J Toxicol, 595307

Hirano S, Cui X, Li S, Kanno S, Kobayashi Y, Hayakawa T, Shraim A (2003) Difference in uptake and toxicity of trivalent and pentavalent inorganic arsenic in rat heart microvessel endothelial cells. Arch Toxicol 77:305–312

Lin YF, Yang J, Rosen BP (2007) ArsD: an As(III) metallochaperone for the ArsAB As(III)-translocating ATPase. J Bioenerg Biomembr 39:453–458

Mandal BK, Ogra Y, Suzuki KT (2001) Identification of dimethylarsinous and monomethylarsonous acids in human urine of the arsenic-affected areas in West Bengal, India. Chem Res Toxicol 14:371–378

Mayer DR, Kosmus W, Pogglitsch H, Mayer D, Beyer W (1993) Essential trace elements in humans. Serum arsenic concentrations in hemodialysis patients in comparison to healthy controls. Biol Trace Elem Res 37:27–38

Nemeti B, Gregus Z (2009) Glutathione-supported arsenate reduction coupled to arsenolysis catalyzed by ornithine carbamoyl transferase. Toxicol Appl Pharmacol 239:154–161

Nielsen FH, Uthus E (1984) Arsenic. In: Frieden E (ed) Biochemistry of the essential ultratrace elements. Plenum Press, New York, pp 319–340

Paez-Espino D, Tamames J, de Lorenzo V, Canovas D (2009) Microbial responses to environmental arsenic. Biometals 22:117–130

Reaves ML, Sinha S, Rabinowitz JD, Kruglyak L, Redfield RJ (2012) Absence of detectable arsenate in DNA from arsenate-grown GFAJ-1 cells. Science 337:470–473

Rezanka T, Sigler K (2008) Biologically active compounds of semi-metals. Phytochemistry 69:585–606

Santini JM, Vanden Hoven RN (2004) Molybdenum-containing arsenite oxidase of the chemolithoautotrophic arsenite oxidizer NT-26. J Bacteriol 186:1614–1619

Thomas DJ, Waters SB, Styblo M (2004) Elucidating the pathway for arsenic methylation. Toxicol Appl Pharmacol 198:319–326

Uthus EO, Cornatzer WE, Nielsen FH (1983) Consequences of arsenic deprivation in laboratory animals. In: Lederer WH (ed) Arsenic symposium, production and use, biomedical and environmental perspectives. Van Nostrand Reynhold, New York, pp 173–189

Wolfe-Simon F, Switzer Blum J, Kulp TR, Gordon GW, Hoeft SE, Pett-Ridge J, Stolz JF, Webb SM, Weber PK, Davies PC et al (2011) A bacterium that can grow by using arsenic instead of phosphorus. Science 332:1163–1166

Yamanaka K, Kato K, Mizoi M, An Y, Nakanao M, Hoshino M, Okada S (2009) Dimethylarsine likely acts as a mouse-pulmonary tumor initiator via the production of dimethylarsine radical and/or its peroxy radical. Life Sci 84:627–633

Zakharyan RA, Sampayo-Reyes A, Healy SM, Tsaprailis G, Board PG, Liebler DC, Aposhian HV (2001) Human monomethylarsonic acid (MMA(V)) reductase is a member of the glutathione-S-transferase superfamily. Chem Res Toxicol 14:1051–1057

Arsenic, Free Radical and Oxidative Stress

Swaran J. S. Flora and Vidhu Pachauri
Division of Pharmacology and Toxicology,
Defence Research and Development Establishment,
Gwalior, India

Synonyms

Arsen; Arsenic black; Arsenicals; Free radicals; Hydroxyl radicals; Metallic arsenic; Oxidative stress; Reactive oxygen species; Sodium arsenate; Sodium arsenite; Superoxides anions

Definition

Arsenic, a naturally occurring metalloid, is known to generate free radicals that play a fundamental role in inducing oxidative stress–mediated toxicity.

Oxidative stress refers to serious imbalance between level of free radicals generated and cellular antioxidant defense that ultimately alters cellular redox status. Thus, oxidative stress may be induced either by accelerating the pro-oxidant species (direct) or by depleting the antioxidant defense (indirect). Free radicals are chemical entities with unpaired electrons or an open shell configuration. Most of these are highly reactive chemically, and play important role in normal physiology and in pathologic manifestations. Arsenic-induced oxidative stress is a multifactorial phenomenon produced by both direct and indirect mechanisms. Directly arsenic causes free radical generation (ROS) during its biotransformation processes or by facilitating through cellular pathway disruption. Arsenic also inhibits the antioxidant mechanisms of the body, thereby indirectly causing elevated ROS and oxidative stress. Arsenic-induced ROS results in the damage to biomolecules that are important cell components and modulates various cell signaling pathways which ultimately results in apoptosis or proliferative defects like cancer.

Arsenic: The Toxicant

Arsenic is a toxic metalloid acquiring most common oxidation states of +3, +5, and −3 existing in both organic and inorganic forms. Inorganic arsenic (iAs) is most abundant in nature existing as trivalent (As^{III}) or pentavalent forms (As^V). Arsenic in Earth's crust forms complex with pyrite which under favorable conditions, subjected to oxygen, pH, redox conditions, etc., gets oxidized to mobilize arsenic and release to groundwater. Elevated concentrations of arsenic above the WHO guideline level of 10 mg/l are present in most countries, although the prevalence and concentrations vary considerably. Southeast Asia is among the most severely affected regions, with highest ground water arsenic concentration reported from Bangladesh. Inorganic arsenic is a well-documented potent human carcinogen, causing cancer in skin, lungs, urinary bladder, kidney, and, possibly liver. Chronic exposure to arsenic through drinking water is associated with detectably increased risk of several non-cancer diseases (e.g., hyperkeratosis, pigmentation, cardiovascular diseases, hypertension, respiratory, neurological, liver and kidney disorders, and diabetes mellitus).

Arsenic exposure induces diverse adverse effects at cellular and subcellular/molecular levels. This entry focuses on: (1) arsenic-induced free radical generation and oxidative stress, (2) role of methylation in arsenic-induced oxidative stress and toxicity, (3) cellular pathways, proteins, and biomolecules affected, and (5) the role of antioxidant in therapy against arsenic toxicity.

Direct Oxidative Stress

Arsenic-Induced Free Radicals Generation

Arsenic-induced free radical generation forms the foundation for producing oxidative stress. The most important players in the process of oxidative signaling include superoxide anion ($O_2^{\bullet-}$), hydroxyl radical ($^{\bullet}OH$), hydrogen peroxide (H_2O_2), singlet oxygen ($^1O_2$), and peroxyl radical ($LOO^{\bullet}$). Thus oxygen based radicals form the most important class and superoxide anions are considered the "primary" reactive oxygen species (ROS). These radicals may further interact through enzyme- or metal-catalyzed reactions to form "secondary" ROS such as H_2O_2 and $^{\bullet}OH$ by $O_2^{\bullet-}$ dismutation and Fenton reaction. Mitochondria being directly involved in arsenic-induced ROS generation forms a crucial site for induction as well as important target for ROS-induced oxidative insult. Respiratory or the Electron Transport Chain (ETC) present at mitochondria physiologically functions to generate mitochondrial membrane potential (MMP). The MMP is a proton gradient generated by pumping out protons by utilizing energy produced during transfer of electron through ETC. At ETC, electron is transferred from NADH and succinate through enzymatic series of donors and acceptors to ultimately reach oxygen that is reduced to water. MMP is then utilized to generate ATP by oxidative phosphorylation. Free radical superoxide is generated when some electrons leak to directly reach oxygen, bypassing the complete ETC transactions. These radicals then attack the phospholipids of mitochondrial membrane damaging it to further disrupt the respiratory chain, thus forming a vicious circle. Arsenic promotes deflection of electrons from ETC, thus inducing ROS generation and a decline of MMP that ultimately leads to apoptosis (see later sections). Arsenite increases the free radical generation especially at the ubiquinone site (Complex I) of the respiratory chain. Increase in cellular superoxide anions following arsenic exposure also involves cytosolic, peroxidase enzymes like cytochrome P-450, that sequentially transfer two electrons from NADPH to molecular oxygen. Flavin enzymes such as NAD(P)H oxidase and NO synthase isoforms also mediate arsenic-induced ROS generation. NAD(P)H oxidase physiologically activates during respiratory outburst for generation of superoxide anions for phagocytosis in neutrophils. Arsenic increases the NAD(P)H oxidase activity by either upregulating the enzyme at different levels like gene expression and translocation of its subunit p22$_{phox}$ and Rac1 respectively, or by enzyme activation. NO synthase isoforms physiologically catalyze generation of NO from substrate L-arginine involving essential cofactors. Under stress conditions like limited availability of substrate or cofactor, NO synthase results in reduced NO production along with superoxide anion generation. Arsenic depletes heme tetrahydrobiopterin (BH$_4$), one of its five cofactors possibly due to destruction by superoxide anions. In view of increased superoxide generation, depleted levels of NO during arsenic exposure may also be attributed to peroxynitrite (ONOO−) formation following reaction between NO and $O_2^{\bullet-}$. Peroxynitrite is not a free radical but rather a highly reactive anion, a powerful oxidant and nitrating agent. In addition to mitochondria, other important sources of arsenic-induced ROS generation include (1) superoxide anion production in the process of arsenic biotransformation

$$H_3AsO_3 + H_2O + O_2 \rightarrow H_3AsO_4$$
$$+ H_2O_2 \left(\Delta rG\theta = -40.82 \text{ kcalmol}^{-1} \right)$$

$$Fe(II) + H_2O_2 \rightarrow Fe(III) + \bullet OH$$
$$+ OH^- \text{ (Fenton reaction)};$$

(2) during oxidation of arsenic from its As^{III} to As^{V} form, the intermediary free radical arsenic species are generated such as dimethylarsinic peroxyl [(CH$_3$)$_2$AsOO$^\bullet$] and dimethylarsinic radicals [(CH$_3$)$_2$As$^\bullet$]; (3) methylated arsenic species especially dimethylarsinous acid (DMA^{III}) release redox-active iron from ferritin. Arsenic induces heme oxygenase isoform 1 (HO-1) leading to production of additional free iron, CO, and biliverdin. Free iron plays a central role in ROS generation via Fenton type and/or Haber–Weiss reaction that combines a Fenton reaction, such as facilitating conversion of $O_2^{\bullet-}$ and H_2O_2 into highly reactive $^\bullet$OH.

$$O_2^{\bullet-} + H_2O_2 \rightarrow O_2 + {^\bullet}OH + OH^-$$

Arsenic increases oxygen consumption by cell, thereby contributing to an increased metabolism and ultimately ROS generation. These effects are due to uncoupled oxidative phosphorylation. ROS is critically associated with various tightly regulated physiological pathways and processes essential for integrity of cell and synchronization of biological system. ROS acts as a secondary messenger and, by modulating intracellular redox status, induces signaling pathways, downstream gene expression, and cell proliferation or death. Thus, arsenic by induction of ROS disrupts various physiological processes (Flora 2011; Jomova et al. 2011).

Arsenic Methylation and Toxicity

Virtually every organism possesses an arsenic detoxification/biotransformation mechanism involving transport systems and enzymatic pathways possibly due to its ubiquitous occurrence in nature. Since arsenic species have no physiological role, they utilize the existing cellular transports by mimicking electrochemical characteristics of essential ions rather than any dedicated system. In pH range of 4–10, pentavalent arsenate species are negatively charged while trivalent arsenic compounds remain neutral, and thus they readily cross the cell membrane. The pentavalent arsenate is reduced to trivalent arsenite in blood stream before entering cell for further metabolism. The reduction of arsenate to arsenite was initially believed to occur nonenzymatically utilizing GSH which is now known as enzymatic reaction through arsenate reductase. Arsenate reductase however requires inosine and thiol compound for the reaction. Reduction of arsenate can also be catalyzed by human liver MMA^V reductase/ hGSTO-1. Arsenite was initially believed to enter cell by simple diffusion, being a neutral species at physiological pH. However, it is now known to be transported inside cell via aquaporin isozyme 7 or 9 (AQP7/9), a member of aquaglyceroporins. Arsenate (iAs^V) enters cell by phosphate carrier system and competes with phosphate inside the cell, for example, binding with polyphosphates like ADP.

Arsenic metabolism follows two possible pathways. The classically known pathway follows reduction and oxidative methylation and the second novel pathway involves GSH conjugation reactions and methylation. In the classical pathway, pentavalent arsenate or its methylated species such as monomethylarsenate (MMA^V) or dimethylarsenate (DMA^V) is reduced to its corresponding arsenite through arsenate reductases. These arsenate reductases are omega isoform of GSH S-transferase (GSTO-1) and purine nucleoside phosphorylase (PNP). Subsequently, the trivalent arsenite (iAs^{III}) or its methylated species, monomethylarsenite (MMA^{III}), then undergoes oxidative methylation through arsenite methyltransferase and Cyt19 to form pentavalent monomethylarsenate (MMA^V) or dimethylarsenite (DMA^V), respectively. The MMA^V reductase is the rate-limiting enzyme for the inorganic arsenic methylation/classical pathway. MMA^V reductase has an absolute requirement for GSH not replaceable by any other thiol. In the novel pathway, inorganic trivalent arsenite (iAs^{III}) forms GSH conjugates to produce arsenite triglutathione [As(SG)$_3$] and subsequent methylation by Cyt 19 to yield a monomethylarsonic diglutathione [MMS(SG)$_2$] followed by dimethylarsinic glutathione [DMS(SG)]. The GSH-conjugated species however can undergo spontaneous degradation to form intermediates of the classical pathways. For example, the MMA(SG)$_2$ may either (1) be degraded to MMA^{III}, followed by oxidation to MMA^V, or (2) methylate to DMA(SG) through Cyt 19, followed by DMA^{III} to DMA^V. DMA^V is the major metabolite excreted in urine. In the process enzymatic methylation of arsenic requires S-adenosylmethionine as the methyl donor and reducing agents such as GSH and cysteine. Unlike most xenobiotics arsenic biotransformation/methylation does not result in its detoxification. The trivalent arsenic species especially the MMA^{III} are most toxic among its methylated forms and represent obligatory intermediates to ultimate end products like MMA^V and DMA^V. Therefore, Cyt19 polymorphism forms an important factor capable of influencing arsenic toxicity in humans.

The iAs^{III} possesses unshared 4s electron pair that renders it capable to interact with biological molecules, which is lacking in the iAs^V. The methylated arsenite possesses one, two, or three methyl substituents that significantly alter the chemical behavior of the moiety, in addition to the unshared electrons. Quantum mechanical calculations predict that addition of methyl group to MMA^V increases the dipole moment and the ionization energy and decreases the molecular volume available for delocalization of electron density. Since dipole moment is directly correlated to log P, it is speculated that log P will increase following methylation of MMA^V. Although iAs^{III} species are not free radicals themselves, they may give rise to free radical species. Dimethylarsine, a trivalent arsenic species, can react with molecular oxygen to produce dimethylarsenic [$(CH_3)_2As^{\bullet}$] radical and superoxide anion. Addition of another oxygen molecule to the dimethylarsenic radical forms dimethylarsenic peroxyl [$(CH_3)_2AsOO^{\bullet}$] radical (Fig. 1). Superoxide and hydrogen peroxide, generated through superoxide dismutase from superoxide anion, produces hydroxyl radicals (Cohen and Arnold 2006; Kumagai and Sumi 2007; Aposhian et al. 2004).

Indirect Oxidative Stress

Other than accelerating the pro-oxidant mechanism, arsenic induces oxidative stress also by inhibiting or depleting the antioxidant defense. The major components of antioxidant defense system are nonenzymatic and enzymatic free radical scavengers, inhibitors to free radical generating enzyme and metal chelators. Reduced glutathione (GSH) is an important nonenzymatic antioxidant that plays an important role in maintaining cellular redox status. Arsenic alters cellular GSH concentration to induce oxidative stress. Arsenic-induced oxidative stress is mediated through decrease in GSH concentration by three important mechanisms: (1) GSH is utilized during arsenic metabolism, (2) arsenic by virtue of thiol-binding property shows high affinity for GSH, and (3) arsenic-induced free radicals oxidize GSH which gets converted to glutathione disulfide (GSSG). However, arsenic-induced GSH modulation depends on arsenic concentration and cellular response to arsenic attack. During acute arsenic exposures, increase in GSH concentration occurs possibly as cellular response to combat arsenic-induced elevated

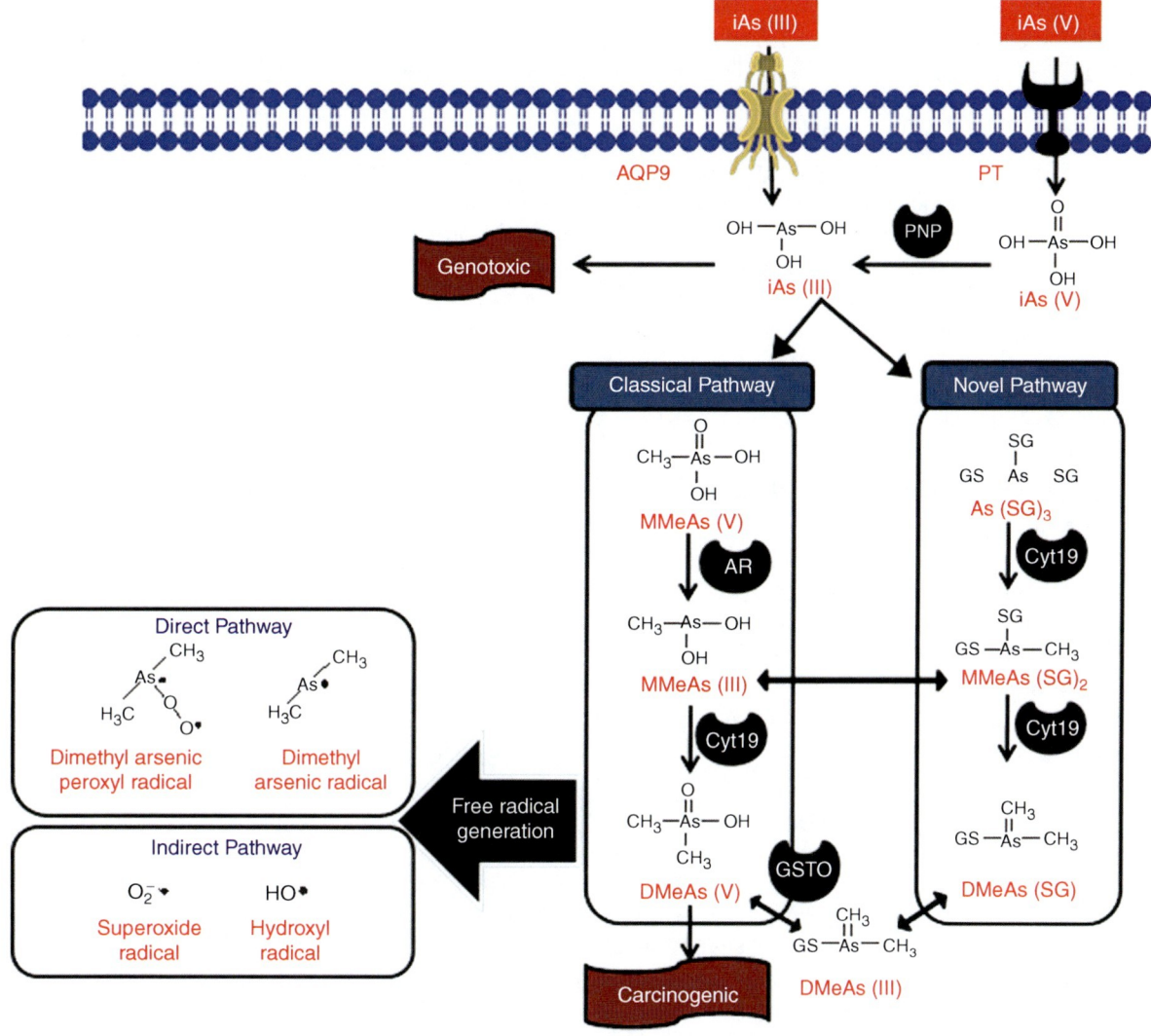

Arsenic, Free Radical and Oxidative Stress, Fig. 1 Role of biotransformation in arsenic toxicity

ROS levels. However, GSH depletion occurs following chronic arsenic exposures (Shi et al. 2004). Other nonenzymatic cellular antioxidants affected by arsenic include vitamin C and E, flavonoids, carotenoids, essential elements (zinc, selenium), and amino acids such as taurine, methionine, cysteine, and thiol compounds. Thiols constitute the most important class of compounds involved in arsenic-induced toxic mechanism. These include N-acetylcysteine, thioredoxin, α-lipoic acid other than GSH. Enzymatic antioxidants such as superoxide dismutase (SOD), catalase, and glutathione peroxidase also show immediate increase followed by decrease in their activity following arsenic exposure subjected to dose and duration of exposure. Arsenic alters the antioxidant status of the cell by its action either at the molecular level, i.e., inhibition of antioxidant activity by structural alterations, such as binding to thiol groups and displacement of essential metal cofactors, and/or at the genetic level by affecting the antioxidant expression. For example, arsenic induces downregulation of SOD2 expression in patients with arsenic-induced skin lesions. These subjects were also indicated of deficient wound healing. Arsenic possibly causes suppression of chemokine response pathway by downregulation of TNF and CCL20 expression.

Arsenic also affects other important cellular enzymes such as thioredoxin reductase which is depleted in a concentration-dependent manner especially by methylarsine(III), ultimately resulting in induction of apoptosis. Thioredoxin reductase physiologically functions to catalyze reduction of many disulfide-containing substrates and is NADPH dependent. Heme oxygenase (HO-1) is a cytoprotective enzyme, an oxidative stress protein, expression of which is regulated by Nrf-2. Arsenic induces upregulation of HO-1 expression which is suggested as cellular adaptive mechanism to combat its toxicity (Vizcaya-Ruiza et al. 2009).

Arsenic Disrupts Signal Transduction Pathways

Cell signal transduction pathways are tightly regulated orchestra that operate crucial physiological functions and dictate paraphrase like cell differentiation, apoptosis, and proliferations. Intrinsic interaction of arsenic with the components of cell signaling pathways either directly or through ROS determines its toxicity and fate of the cell. Major pathways affected by arsenic include tyrosine phosphorylation system, mitogen-activated protein kinases (MAPKs), and transcription factor families such as NF-κB and AP-1 (Flora 2011). Tyrosine phosphorylation system modulates and participates in various cell physiological processes and signal transduction pathways. Activation of enzyme protein tyrosine kinases and/or inhibition of tyrosine phosphatases induces tyrosine phosphorylation reaction. These kinases are either receptor mediated, i.e., receptor tyrosine kinases (RTKs) or nonreceptor tyrosine kinases (NTKs). RTKs include several growth factors such as epidermal growth factor receptor (EGFR), platelet-derived growth factor receptor (PDGFR), vascular endothelial growth factor receptor (VEGFR). Activation of RTKs either by ligand binding or receptor cross-linking leads to autophosphorylation on tyrosine residue, inducing conformational change that initiates cascade of reactions. Arsenic causes increased total cellular tyrosine phosphorylation, mainly affecting EGFR. Arsenic either binds directly with EGFR or causes dimerization for its activation. Tyrosine phosphorylation–induced conformation change in the EGFR relays the signal to downstream cascade pathway. EGFR tyrosine phosphorylation induced by arsenic is also responsible for activation of MAPK via Ras protein–mediated signaling. Arsenic induces EGFR also via NTKs such as Src in human epithelial cell line either by direct interaction with its vicinal thiol groups or integrin rearrangements, or indirectly via ROS. Arsenic-induced Src causes tyrosine phosphorylation of EGFR at tyrosine residues that are unique from its autophosphorylation sites. Src also activates MAPKs which is mediated by two parallel pathways, i.e., EGFR-dependent or EGFR-independent pathways.

MAPKs induction by arsenic is largely dependent on arsenic species, dose, duration, and cell type exposed. MAPK is a family of serine/threonine phosphorylating proteins that in response to various extracellular stress–mediated response regulates gene expression. Three major classes of MAPK, namely, extracellular signal regulated kinases (ERKs), c-Jun N-terminal kinases (JNKs), and p38 kinases, are differentially affected by arsenic exposure. ERKs are responsible for cell differentiation, transformation, and proliferation, while JNKs and p38 kinase are involved in cell growth arrest and apoptosis stimulated in response to stress. Thus, arsenic-induced carcinogenic effect involves the activation of ERKs, while apoptotic and anticarcinogenic effects are mediated through JNKs and p38 kinases. Ras/Raf/Mek pathway mediates ERK and p38 kinases activation and Rac, Rho, and MEKK3-4 activate the JNKs. Arsenic-induced ERK activation is mediated by various signal pathways most defined through RTKs such as EGFR. ERK in turn activates AP-1 (c-Jun + c-Fos) and ATF-2 transcription factors, resulting in its proliferating response. Mechanism for arsenic-induced JNK-mediated apoptosis is unclear. JNK upon activation is translocated to mitochondria and interacts with Bcl-$_{XL}$ and Bcl-2 where it phosphorylates and inactivates Bcl-2, ultimately resulting in release of cytochrome c and apoptosis proteases (Fig. 2).

MAPKs activation by arsenic influences the activation of transcription factors such as NK-κB and AP-1. The transcriptional pathways closely are involved in cell transformation and apoptosis. NK-κB is a dimeric transcription factor(s) belonging to Rel family. This is activated rapidly in response to stress and functions as a transcription factor for gene controlling cytokines, growth factors, and acute response proteins. Arsenic may pose differential and dose-dependent effect on NK-κB, i.e., at low,

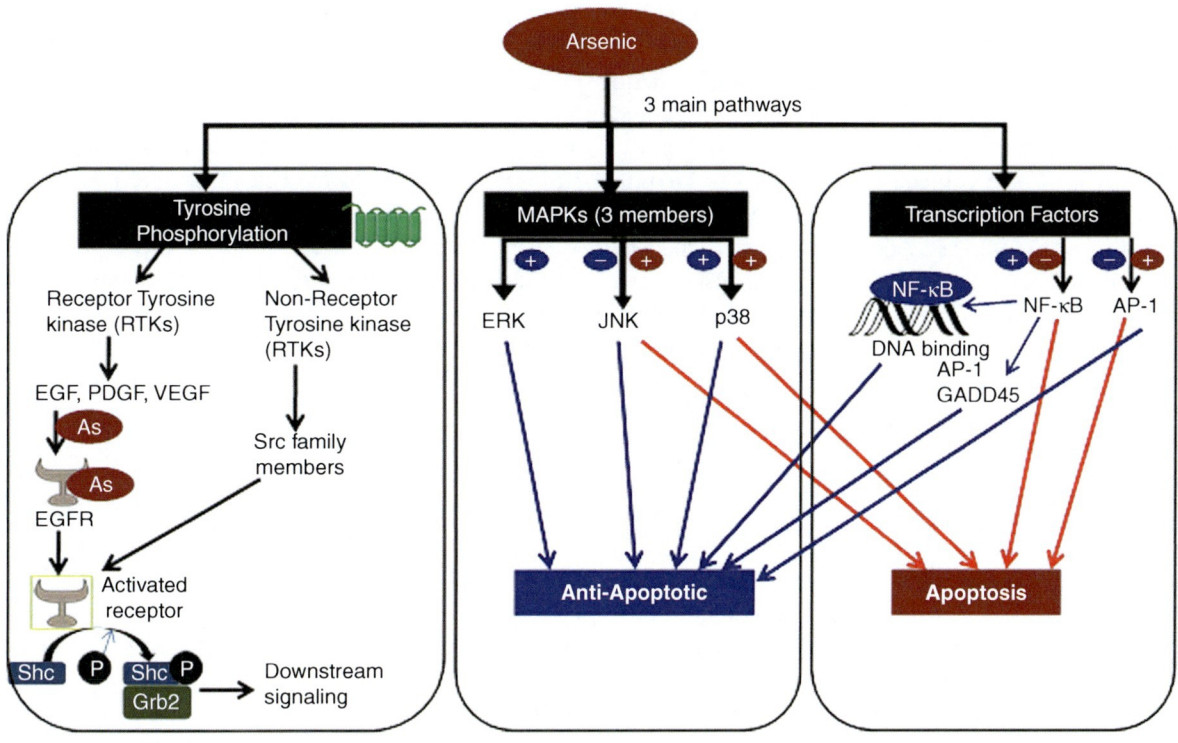

Arsenic, Free Radical and Oxidative Stress, Fig. 2 Effect of arsenic on cellular signaling pathways

non-cytotoxic dose it activates NK-κB while at higher dose inhibits it. Activation of NK-κB generally involves arsenic-induced ROS or via degradation of its inhibitor IκB. Additionally, arsenic-induced NK-κB activation involves cross talk with ERKs and JNKs. Inhibition of NK-κB involves repression of IκB and inhibition of phosphorylation and degradation of IκB. Arsenic induces AP-1 transactivation and expression of c-jun and c-fos gene (Qian et al. 2003; Kitchin and Ahmad 2003).

Arsenic on Cellular Pathways: Protein Interface

As discussed in the previous section iAs^{III} has high affinity for sulfhydryl groups, and iAs^{V} mimics phosphate and is an uncoupler of oxidative phosphorylation. Thus, both iAs^{III} and iAs^{V} have potential to produce oxidative stress through uncoupling of mitochondrial oxidative phosphorylation, and/or increased cellular production of H_2O_2. Arsenic-induced oxidative stress leads to cell damage and death through activation of oxidative sensitive signaling pathways that affects a large number of genes and proteins involved in different physiologic and pathologic pathways.

Proteins involved with oxidative stress, such as heat shock proteins, metallothionein, heme oxygenase, ubiquitin, aldose reductase, and ferritin light chain, are induced by arsenic. Arsenic induces DNA damage, lipid peroxidation, and protein modification and apoptosis, through ROS-mediated reactions. In addition, the expression of genes encoding for several drug-metabolizing enzymes, such as CYP1B1, CYP2B9, CYP7A1, CYP7B1, CYP3A11, is also affected/altered (suppressed) during acute arsenic toxicity.

Heat Shock or Stress Proteins

Heat shock or stress proteins (HSPs/SPs) are constitutively expressed redox-sensitive proteins, that are synthesized in almost all organisms following induction due to exposed to a range of stressors, including oxidative stress, heat shock, free radicals, UV radiation, and heavy metals. Physiologically present in low concentrations, these function as "molecular chaperones" contributing to maintenance of the cellular homeostasis and defense. These support the proper folding, assembly, and distribution during protein synthesis. These are also involved in modulation of immune system. HSPs through epigenetic repair also called heat shock response protect and repair vulnerable protein targets and DNA repair. HSPs are classified as metalloproteins (MT), ubiquitin (Ub), proteins of 15–40kDa such as HO, proteins of 60, 70, 90, and 110kDa. Additionally, stress-induced synthesis of HSP compared to heat shock-induced proteins is classified as analogous, subset, and specific induction. The specific induction causes synthesis of proteins exclusively by metals but not by heat such as metallothionein (MT). These metalloproteins are involved directly in the cellular detoxification of metal or protection and repair of metal-induced metabolic dysfunction. HSP induction by arsenic constitutes a defensive mechanism of cells against oxidative stress–induced damage. iAs^{III} is an effective inducer of a number of stress-related proteins including HSP27, HSP32, HSP60, HSP70, and HSP90 (Razo et al. 2001).

HSP27 belongs to small heat shock or stress proteins (sHSPs) family. Under stress condition such as arsenic exposure, these large oligomers dissociate into small oligomers as a result of phosphorylation by kinases. iAs^{III} increases the HSP27 phosphorylation by inhibiting protein phosphatase activity. HSP27 prevents apoptosis by regulating the activation of the phosphoinositide 3-kinase/protein kinase B pathway and inhibiting cytochrome c-dependent activation of procaspase 9. It can also decrease the intracellular level of ROS by modulating the metabolism of glutathione to maintain it in a reduced state. Thus, these have the potential to protect cell against iAs^{III} toxicity.

The Hsp60 synthesis, during stress, in cells is rapidly upregulated. Hsp65 is involved in autoimmunity by stimulating antibodies directed against it. Arsenic modulates cytokine secretion and increases the production of many cytokines including the tumor necrosis factor-a (TNFα), the proinflammatory cytokine interleukin 1a (IL-1a), the neurotrophic cytokine IL-6, the neutrophil chemotactic cytokine IL-8 as well as cellular growth factors such as the transforming growth factor-a (TGFα) and the granulocyte macrophage-colony stimulating factor (GM-CSF) as immunopathological response. HSP 60 and HSP 70 mediate inflammatory responses to arsenic as the expression of IL-1b, IL-6, TNF-a, and IFN-d mRNAs has been associated with the expression of Hsp60, Hsp47, Hsp70, and Hsp70/Hsp86 mRNAs. These proteins are involved in the induction/progression of hypertension, atherosclerosis and are markers for early cardiovascular disease associated with chronic arsenic exposure. Arsenic (iAs^{III}) also induces proteins that are functional homologue of HSP60 such as TCP1 involved with microfilaments.

HSP70 physiologically functions as chaperone by assisting in the membrane translocation of new synthesized structural or secretory proteins. These also prevent the misfolding and translocation during new protein synthesis by binding to them until process completion. The Hsp70 family of proteins plays a key role in reproductive physiology, particularly in gametogenesis and embryogenesis, and encodes the major inducible Hsp expressed in a wide range of tissues. Trivalent arsenic induces HSP70 differentially in various tissues. DMA exposure in vivo, for example, causes higher accumulation of HSP72 in lung compared to kidney. Presence of the stress-inducible *Hsp70–1* and *Hsp70–3* helps protect embryos from the effects of As^{III}-induced neural tube defects and related toxicity.

Hsp 90 is induced in stress responses that result in glucose depletion, intracisternal calcium, or other disruption of glycoprotein trafficking. iAs^{III}-induced stress response causes induction of HSP90 and HSP70 which is mediated through increased progesterone receptor (PR) activity. Homologous to HSP70 and HSP90 are "glucose-regulated proteins" (Grps) (74–78 and 94 kDa), another class of stress proteins which are involved in glucose-related stress. iAs^{III} induces Grp74 and Grp94; however, elevation of Grp94 in arsenic-induced diabetes needs to be investigated.

Metallothionein (MT)

MT is a sulfhydryl-rich, low-molecular-weight (6–7-kDa) metal-binding protein that provides protection against metal-induced oxidative cellular injury. Induction of metallothioneins (MT I and II) represents a major means of metal detoxification in the body to maintain trace elements within a physiological range or to protect the body from the damage by metal overload. Arsenic-induced MT greatly depends upon the arsenic species. iAs^{III} is the strongest inducer which is followed by its pentavalent forms, iAs^{V}, $MMAs^{V}$, and $DMAs^{V}$. Arsenic-induced MT induction is mediated by ROS, signaling pathways such as inflammatory response and induction of transcription factors such as MTF-1, rather than iAs chelation.

MTs regulate metal homeostasis by sequestering metals in protein-bound forms, providing a zinc reserve, and serving as a scavenger to quench ROS and other free radicals. The increased cellular tolerance to arsenic may result from an increase in the efficiency of transcription through an activation of the involved regulatory factors.

Heme Oxygenase (HO)

HO, an enzyme of 32 kDa, is responsible for the physiological breakdown of heme into equimolar amounts of biliverdin, carbon monoxide, and Fe using O_2 and electrons donated by NADPH-cytochrome P450 reductase. Induction of HO-1 both in vivo and in vitro is a hallmark of arsenic-induced oxidative stress. HO-1 induction is dependent on the level of arsenic, its redox status, and target organ. Trivalent arsenicals (iAs^{III}) form the most effective inducers, while pentavalent methylated forms ($MMAs^V$ or $DMAs^V$) are ineffective. Since distribution and biotransformation of arsenic is organ specific, the kinetics and dynamics of arsenicals dictate the induction. For example, liver shows highest HO-1 induction due to predominance of iAs form. After arsenic exposure, induction of HO-1 may confer cytoprotection from oxidative damage by promoting the conversion of pro-oxidant metalloporphyrins, such as heme, to bile pigments (biliverdin and bilirubin) with free radical scavenging capabilities. The levels of the *HO-1* gene are transcriptionally controlled. The expression of HO-1 is under the regulation of AP-1 and NF-kB transcription factors and is mediated through activation of MAP kinase pathway.

Ubiquitin (Ub)

Ub is a low-molecular-weight (7–8-kDa) protein involved in the nonlysosomal degradation of intracellular proteins. Under stressful conditions, Ub acts as the cofactor for cytosolic degradation of abnormal proteins to maintain chromosome structure by a set of proteins induced under different environmental stress factors. iAs^{III} at high concentrations causes its inhibition while following low level exposures results in an increase in the Ub-conjugation enzymes, inducing the accumulation of Ub-conjugated proteins. Therefore, when As^{III} inhibits the deubiquitination, it contributes to increase in its toxicity by proteolysis of aged, damaged, or dysfunctional proteins. In contrast, when As^{III} increases the level of Ub, the protein allows the cell to enhance the repair process of tagged proteins; it also targets selective molecules for degradation, repair of accumulated proteins and damaged DNA, and provides transient elevation of cellular defenses.

Aldose Reductase (AR)

Aldose reductase is a member of the aldoketoreductase family that acts as an antioxidative stress protein. Following arsenite-induced oxidative insult, these metabolize several aldehyde product compounds including 4-hydroxy *trans*-2-nonenal, a major toxic product of lipid peroxidation. Arsenite upregulates AR in vitro.

Arsenic-Mediated ROS-Induced DNA Damage

Arsenic-induced oxidative stress leads to DNA damage, DNA hypo- or hyper-methylation forming an important mechanism underlying its apoptotic, necrotic, or carcinogenic effects. Acute inorganic arsenic exposure in vivo results in induction of DNA damage-inducible proteins GADD45 and GADD153 and the DNA excision repair protein ERCC1, as well as other DNA damaging-repair genes. Enzymes involved in nucleotide excision repair (NER) and base excision repair (BER) are also affected by arsenic. Trivalent arsenicals inhibit BER and NER activity by interacting with zinc finger motifs of proteins in these two DNA repair systems. Arsenic unlike other carcinogens does not induce mutations. It does not directly covalently bind to DNA but causes damage indirectly through ROS, leading to DNA adducts, strand break, cross-link, and chromosomal aberrations. Arsenic-induced hydroxyl radicals are mainly responsible for DNA damage through oxidation of any of the four bases especially thymidine. DNA methylation state is also altered by arsenic exposure by modifying the activity of DNA methylation enzymes. Mechanisms involved include either direct interaction of arsenic-thiol enzyme or indirectly through ROS-enzyme interaction. Depletion of S-adenosyl-methionine(s) required for DNA methylation via arsenic may also contribute to DNA effects. Arsenic also inhibits DNA ligase, thus inhibiting DNA repair. Arsenic also causes strand break via superoxide radicals and related secondary radicals by activation of NADH oxidase.

Arsenic-Mediated ROS-Induced Apoptosis

Apoptosis is a critical cellular response to maintain normal cell development and proper function of multicellular organisms. There are two major signaling pathways of apoptosis: the death receptor pathway (extrinsic) and the mitochondrial pathway (intrinsic). Arsenic-induced ROS-mediated apoptosis is well documented in the literature, in which the mitochondria-driven apoptotic pathway is probably a more favorable mechanism for arsenic-induced cell death. Excess ROS increases the mitochondrial membrane permeability and damages the respiratory chain, resulting in increased ROS production. The disruption in the mitochondrial membrane causes the release of cytochrome c from the mitochondria. The release of cytochrome c then activates caspase-9, which initiates the activation of caspases-3, -6, and -7, leading to apoptosis. Caspases, which are aspartate-specific cysteine proteases, are cytoplasmic proenzymes that play an important role in initiation and effector phases of apoptosis. A large percentage of the altered genes and proteins are known to be regulated by redox-sensitive transcription factors, (SP1, NF-κB, AP-1), suggesting that, at environmentally relevant levels of chronic exposure, arsenic may be acting through alteration of cellular redox status.

Role of Antioxidants in Arsenic-Induced Oxidative Stress

Various synthetic and natural antioxidants show beneficiary effects in arsenic-induced toxicity. Although chelation therapy forms the mainstay in therapy against arsenic toxicity, antioxidants are recommended as adjuvant for the additional benefits. There is no therapy defined yet for arsenic toxicity post chronic exposure. Since oxidative stress forms the basis for arsenic-induced toxicity, antioxidant therapy is recommended as the primary and immediate strategy to combat toxicity in clinical cases. Antioxidants have shown promising results both as monotherapy and as adjuvant to the primary chelation under investigation. Conventionally known antioxidants, such as vitamin C, E, N-acetyl cysteine (NAC), etc., provide significant benefit in arsenic toxicity. NAC is a precursor of L-cysteine and reduced glutathione. It is a thiol-containing mucolytic agent. Thus, mechanisms involved in NAC-mediated protection against arsenic include ROS scavenging, elevate cellular antioxidant, GSH, and mild chelation by virtue of sulfhydryl-containing group. Additionally antioxidants of natural origin and crude forms like the plant extracts have also gained success in lowering arsenic-induced oxidative stress. Some of the most promising candidates include *Curcumin, Hippophae rhamnoides, Aloe Vera barbadensis, Centella asiatica* and *Allium sativum*. Certain antioxidants like NAC or α-lipoic acid are associated with chelation ability and show specific protection against arsenic toxicity. *Moringa oleifera* is one such natural antioxidant that protects against arsenic-induced oxidative stress additionally by lowering the body arsenic burden. The chelation property of *Moringa oleifera* is attributed to the interaction between cysteine- and methionine-rich proteins that are present in high amount in its seeds. Similarly, organosulfur compounds present in garlic form the active agents effective in reducing the clastogenic effects of arsenic with possible role of p53 and heat shock proteins. Arsenic chelation ability of garlic is due to its thiosulfur components which act as Lewis acids and interact with Lewis base (arsenic) to form stable components. Natural polyphenols such as flavonoids are another class of compounds that have shown beneficiary effects. α-Lipoic acid (1,2-dithiolane-3-pentanoic acid, 1,2-dithiolane-3-valeric acid or thioctic acid, LA,) and its reduced form, the dihydrolipoic acid (6,8-dimercaptooctanoic acid or 6,8-thioctic acid, DHLA), are established synthetic antioxidants which also possess chelation property due to its thiol moiety. Lipoic acid protects against arsenic toxicity by its dual ability of lowering free radicals and arsenic from the system (Flora 2011).

Cross-References

- ▶ Arsenic
- ▶ Arsenic and Aquaporins
- ▶ Arsenic in Nature
- ▶ Arsenic in Pathological Conditions
- ▶ Arsenic in Therapy
- ▶ Arsenic in Tissues, Organs, and Cells
- ▶ Arsenic Methyltransferases
- ▶ Arsenic, Mechanisms of Cellular Detoxification
- ▶ Arsenic-Induced Diabetes Mellitus
- ▶ Arsenicosis
- ▶ As

References

Aposhian HV, Zakharyan RA, Avram MD et al (2004) A review of the enzymology of arsenic metabolism and a new potential role of hydrogen peroxide in the detoxication of the trivalent arsenic species. Toxicol Appl Pharmacol 198:327–335

Cohen SM, Arnold LL (2006) Methylated arsenicals: the implications of metabolism and carcinogenicity studies in rodents to human risk assessment. Crit Rev Toxicol 36:99–133

Flora SJS (2011) Arsenic-induced oxidative stress and its reversibility. Free Radic Biol Med 51:257–281

Jomova K, Jenisova J, Feszterova M et al (2011) Arsenic: toxicity, oxidative stress and human disease. J Appl Toxicol 31:95–107

Kitchin KT, Ahmad S (2003) Oxidative stress as a possible mode of action for arsenic carcinogenesis. Toxicol Lett 137:3–13

Kumagai Y, Sumi D (2007) Arsenic: signal transduction, transcription factor, and biotransformation involved in cellular response and toxicity. Ann Rev Pharmacol Toxicol 47:243–262

Qian Y, Castranova V, Shi X (2003) New perspectives in arsenic-induced cell signal transduction. J Inorg Biochem 96:271–278

Razo LMD, Quintanilla-Vega B, Brambila-Colombres E et al (2001) Stress proteins induced by arsenic. Toxicol Appl Pharmacol 177:132–148

Shi H, Shi X, Liu KJ (2004) Oxidative mechanism of arsenic toxicity and carcinogenesis. Mol Cell Biochem 255:67–78

Vizcaya-Ruiza AD, Barbiera O, Ruiz-Ramosb R et al (2009) Biomarkers of oxidative stress and damage in human populations exposed to arsenic. Mutat Res 674:85–92

Arsenic, Mechanisms of Cellular Detoxification

Ebany J. Martinez-Finley[1,2] and Michael Aschner[1,2,3,4]
[1]Department of Pediatrics, Division of Pediatric Clinical Pharmacology and Toxicology, Vanderbilt University Medical Center, Nashville, TN, USA
[2]Center in Molecular Toxicology, Vanderbilt University Medical Center, Nashville, TN, USA
[3]Center for Molecular Neuroscience, Vanderbilt University Medical Center, Nashville, TN, USA
[4]The Kennedy Center for Research on Human Development, Vanderbilt University Medical Center, Nashville, TN, USA

Synonyms

As

Definition

Arsenic (As) is a naturally occurring metalloid, found in water, air, and food; a frequent source of human exposure to inorganic As is from groundwater contamination. As has several oxidation states but those of greatest toxicologic importance are the inorganic arsenite (III) and arsenate (V) states.

As in the Environment

Arsenic (As) is a naturally occurring metalloid found in the Earth's crust and present in the environment. It can leach into the groundwater and its presence is the result of industrial emissions and natural sources such as surrounding sediment especially in places with volcanic tuff. The highest natural levels of As in groundwater are found in the West, Midwest, and Northeast regions of the United States and in India, China, Chile, Taiwan, and Bangladesh. Arsenic and arsenic compounds have been used in many insecticide sprays, weed killers, fungicides, wood preservatives, waste incineration, disinfectant compounds, paints, drugs, dyes, soaps, semiconductors, agricultural products, prints, and enamels. Its use has also been documented in homeopathic remedies and most notably historically as a homicide and suicidal agent. It is found in rice, poultry, and seafood and arsenic trioxide (ATO) has been used in the treatment of acute promyelocytic leukemia (APL) relapse.

Human Health Impact

A common source of human exposure to inorganic As is through groundwater and drinking water contamination. The maximum contaminant level (MCL) of As in drinking water is regulated by the US Environmental Protection Agency (EPA) and the World Health Organization (WHO) at 10 parts per billion (ppb 0.01mg/L) (Lievremont et al. 2009). Due to its widespread nature, As poses a significant risk to human health. Inorganic As has been associated with skin, lung and bladder cancers, vascular diseases, hypertension, genotoxicity, cellular disruption, developmental anomalies, decreased intellectual function, and diabetes (NRC, 2001). Arsenic is considered a Class I human carcinogen by the International Agency for Research on Cancer due to its permissiveness to skin, lung and bladder cancers.

Elemental and Inorganic As

Organic arsenic is formed by the combination of arsenic with carbon and hydrogen. As can form inorganic complexes with oxygen, chlorine, and sulfur. It has several oxidation states but those of greatest toxicologic importance are inorganic trivalent arsenite (AsIII) and pentavalent arsenate (AsV), the two forms that are also most commonly found in drinking water (Smedley and Kinniburgh 2002).

Fate of As in the Human Body

Inorganic As species are well absorbed by both the oral and inhalation routes but can also enter the body through dermal exposure and injection. Once absorbed by the body As can be distributed to all systems. Arsenic is stored in the liver, kidney, intestines, spleen, lymph nodes, and bones and can pass both the blood-brain and placental barriers. The rate of absorption following oral and inhalational routes ranges from ~75% to 95%, with the rate of As sulfide, lead arsenate and other insoluble forms is much lower than the soluble forms. Several studies indicate that the absorption of As in ingested dust or soil is likely to be less than absorption of As from ingested salts. If there are differences in absorption rates in children versus adults these have not been documented (U.S. Agency for Toxic Substances and Diseases Registry 2005).

Transport of As is efficient due to its increased mobility resulting from its enzymatic methylation. The primary detoxification pathway of As is through its biomethylation, although it is well established that many of the metabolites exert their own toxicities. Inorganic arsenite (III) is more toxic than arsenate (V). Inorganic arsenic is primarily biotransformed in the liver by oxidation/reduction, followed by a series of methylation steps with S-adenosylmethionine (SAM) as the donor of methyl groups and reduced glutathione as a cofactor and excretion in the urine and feces as either monomethylarsonate (MMA) or dimethylarsinate (DMA). Metabolism of As does not vary by route of exposure, although there is tissue variation in the rate of methylation and the availability of methyl groups such as methionine, and cysteine is not a rate limiting step. There is some evidence for differences in methylation capacity among individuals. The main route of excretion is urinary, accounting for ~55–87% of daily oral or inhalation intakes of arsenate or arsenite. There also appears to be an upper-dose limit to excretion in the urine, which can account for the toxicity of As in tissues. Fecal excretion accounted for less than 5% of eliminated As. As is cleared from the blood within a few hours and most As absorbed from the lungs or the gastrointestinal tract is excreted within 1–2 days. An alternative metabolism pathway consists of nonenzymatic formation of glutathione complexes with arsenite resulting in the formation of As triglutathione followed by a series of methylation steps (U.S. Agency for Toxic Substances and Diseases Registry 2005).

Diagnosis of As exposure is difficult as renal excretion and blood levels may not give an accurate picture of total exposure. Renal excretion levels may be indicative of As exposure within 24 h with a half-life of 4 days of exposure. In cases where intoxication may have occurred weeks prior to medical evaluation, hair and fingernail samples can be used to determine levels for up to 3–6 months. The use of chelating agents for treatment of As poisoning has been advantageous if administered within hours of As absorption, their efficacy is probably due to the sulfhydryl groups they contain. Emetics, cathartics, lavages, and activated charcoal have also been used to remove As from the gastrointestinal tract (Flora and Pachauri 2010).

Cellular Handling of As

The gamut of health effects produced by As exposure can be attributed to whole body organ distribution of arsenic. Several mechanisms of As-induced damage have been proposed including oxidative stress, induction of chromosomal abnormalities through indirect effects on DNA, genotoxic damage, epigenetic mechanisms resulting in altered DNA methylation status, and disruption of protein activity by binding to and inactivation of sulfhydryl groups, especially disruption of metalloproteins, altered gene expression, and/or cell signaling. As generally does not induce point mutations.

Because arsenic is a nonessential metal it does not have a designated transport system, therefore cellular uptake of inorganic arsenite is mediated by the aquaglyceroporins (AQPs) which are normally responsible for transporting small uncharged molecules across cell membranes. AQP9, the major subtype found in liver and astrocytes, has been shown to handle both inorganic arsenite and its intermediate MMA. The major kidney, testis, and adipose tissue aquaglyceroporin, AQP7, has been implicated in uptake of arsenite. Inorganic arsenate is a phosphate analogue and therefore can be taken into the cell via phosphate transporters. Glucose transporter isoform 1 (GLUT1) is a glucose permease found in erythrocytes and epithelial cells forming the blood-brain barrier that has also been implicated in arsenite and MMA uptake (Martinez-Finley et al. 2012).

Arsenite has a high affinity for thiol (–SH) groups and in cells can form a complex with the amino acid cysteine, which contains many thiol groups. Toxic effects of arsenic are a result of its binding to vicinal sulfhydryl groups in key enzymes and interfereing with biological processes such as gluconeogenesis and DNA repair. The molecular mechanism(s) through which As causes oxidative stress in cells is related to binding of As to the thiol proteins, with some of these targets being metalloproteins. It is hypothesized that As binds thiol groups of enzymes that are stabilized by Zn^{2+} and displaces Zn from the protein binding site. Arsenate must be reduced in the body to arsenite so it is thought that it interferes with proteins by binding sulfhydryl groups, however, it may also act as a phosphate analogue, affecting other processes such as ATP production, bone formation, and DNA synthesis.

Arsenic Health Effects

Skin cancer and skin lesions are a major concern following exposure to As. These can include hyperkeratosis, hyperpigmentation, and depigmentation. Most often the presence of As in the lung leads to lung cancer as an endpoint. Additionally, effects of As on the cardiovascular system include cardiac arrhythmias, hypertrophy of the ventricular wall, Blackfoot disease, and altered myocardial depolarization. Other associations include cardiovascular effects due to chronic consumption of As-contaminated drinking water in pregnant mothers, leading to congenital heart disease in children, as well occurrence of hypertension (U.S. Agency for Toxic Substances and Diseases Registry 2005).

Spontaneous abortion, stillbirth, reduced birth weight, and infant mortality are reportedly associated with As exposure. Following fetal exposure, other studies have reported a slight increase in cancers overall, increased chronic obstructive pulmonary disease, increased cognitive deficits, increased acute myocardial infarction deaths, increased liver cancer mortality, and an increase in adult diseases following early life exposure. Experimental animal models have shown teratogenic effects including cleft palate, delayed bone hardening, and neural tube defects following in utero exposure to As. Arsenic levels in breast milk are negligible even when maternal blood As levels are high.

Inorganic As might act as an estrogen-like chemical in vivo and induce tumors in mice according to experimental animal and cell culture data which also shows that As is a potent endocrine disruptor, altering gene regulation by the closely related glucocorticoid, mineralocorticoid, progesterone, and androgen steroid receptors. Arsenic has shown positive correlation with diabetes. Deficits in cognitive function following As exposure have been reported and neurological symptoms following exposure include neuropathy, hallucinations, agitation, emotional changes, and memory loss (U.S. Agency for Toxic Substances and Diseases Registry 2005).

Conclusions

As is an environmental toxicant that affects many of the body's systems and has a high affinity for thiol groups, making it particularly detrimental to many cellular proteins.

Acknowledgments The authors wish to acknowledge funding by grants from the National Institutes of Environmental Health Sciences (R01 ES07331, P30 ES000267 and T32 ES007028).

Cross-References

▶ As

References

Flora, Pachauri (2010) Chelation in Metal Intoxication. Int J Eniron Res Public Health 7(7):2745–2788

Lievremont D et al (2009) As in contaminated waters: biogeochemical cycle, microbial metabolism and biotreatment processes. Biochimie 91(10):1229–1237

Martinez-Finley et al (2012) Cellular transport and homeostasis of essential and nonessential metals. Metallomics

National Research Council (2001) Arsenic in Drinking Water 2001 Update. National Academy Press

Smedley, Kinniburgh (2002) A review of the source, behaviour and distribution of arsenic in natural waters. Applied Geochemistry 17:517–569

U.S. Agency for Toxic Substances and Diseases Registry (ATSDR) (2005) Toxicological profile for Arsenic. http://www.atsdr.cdc.gov/toxprofiles/tp2.pdf. Accessed 14 Nov 2011

Arsenic, Physical and Chemical Properties

Fathi Habashi
Department of Mining, Metallurgical, and Materials Engineering, Laval University, Quebec City, Canada

Arsenic is a bright silver-gray metalloid; its outermost electrons are not free to move in the crystal structure because they are fixed in position in a covalent bond. It has metallic luster but is brittle with no useful mechanical properties. In addition to the metallic-like form, there are other modifications which change into the metallic-like form above 270°C:

- Black arsenic, formed as a mirror when arsenic hydride is passed through an incandescent glass tube and also when the vapor is rapidly cooled.
- Yellow arsenic, formed by the sudden cooling of arsenic vapor and consists of transparent, waxy crystals. It is unstable and changes into metallic-like arsenic on exposure to light or on gentle heating.
- Brown arsenic, is either a special modification or simply a more finely divided form is obtained in the reduction of solutions of arsenic trioxide in hydrochloric acid with tin(II) chloride or hypophosphorous acid.

Physical Properties

Atomic number	33
Atomic weight	24.92
Relative abundance in the Earth's crust, %	5×10^{-4}
Density, g/cm^3 at 20°C	
Metallic arsenic	5.72
Yellow arsenic	2.03
Brown arsenic	3.7–4.1
Melting point at 3.7 MPa, °C	817
Sublimation point at 0.1 MPa, °C	613
Linear coefficient of thermal expansion: K^{-1}	ca. 5×10^6
Specific heat capacity, J g^{-1} K^{-1}	
At 18°C	0.329
0–100°C	0.344
Electrical resistivity at 0°C, $\Omega \cdot$ cm	24×10^{-6}
Hardness, Mohs scale	3–4
Crystalline form	Triagonal
Potential with respect to the normal hydrogen electrode, V	ca. 0.24

Chemical Properties

The surface of elemental arsenic tarnishes in humid air. When heated in air, it burns with a bluish-white flame, forming dense vapors of arsenic trioxide. In compounds, it has oxidation states of +3, +5, and −3. Concentrated nitric acid and aqua regia oxidize arsenic to arsenic acid; arsenic is oxidized to the +3 state by dilute nitric acid or concentrated sulfuric acid and by boiling alkali hydroxides in air. Hydrochloric acid has little effect on arsenic. Chlorine combines fierily with arsenic to form arsenic trichloride. When the metal is heated with sulfur, AsS, As$_2$S$_3$, or As$_2$S$_5$ is obtained, depending on the ratios used.

Arsenic combines with metals to form arsenides. When subjected to oxidizing roasting, arsenides give partly metal oxide and arsenous acid and partly basic arsenates. With oxygen, arsenic forms three oxides: arsenic trioxide, As$_2$O$_3$, arsenic pentoxide, As$_2$O$_5$, and As$_2$O$_4$, which apparently contains trivalent and pentavalent arsenic alongside. Arsenous acid is derived from arsenic trioxide and can only exist in aqueous solution. Its salts are the arsenates(III) (formerly arsenites). Arsenic acid is derived from arsenic pentoxide. The highly poisonous arsenic hydride, AsH$_3$, is formed from arsenic compounds in acidic solution in the presence of strong reducing agents, e.g., zinc or from suitable arsenides, e.g., As_2Zn_3 and acids:

$$As_2Zn_3 + 6HCl \rightarrow 2AsH_3 + 3ZnCl_2$$

References

Grossman H, Hanusch K, Herbst K-A, Rose G, Wolf HU (1997) Arsenic. In: Habashi F (ed) Handbook of extractive metallurgy. Wiley-VCH, Weinheim, pp 795–822

Habashi F (2001) Arsenic, antimony, and bismuth production. In: Buschow KHJ et al (eds) Encyclopedia of materials: science and technology. Elsevier, Amsterdam/New York, pp 332–336

Arsenical Compounds

▶ Arsenic and Primary Human Cells

Arsenicals

▶ Arsenic in Nature
▶ Arsenic in Pathological Conditions
▶ Arsenic, Free Radical and Oxidative Stress

Arsenic-Induced Diabetes Mellitus

Chin-Hsiao Tseng
Department of Internal Medicine, National Taiwan University College of Medicine, Taipei, Taiwan
Division of Endocrinology and Metabolism, Department of Internal Medicine, National Taiwan University Hospital, Taipei, Taiwan

Synonyms

Chronic arsenic intoxication impairs glucose homeostasis; Environmental arsenic exposure and diabetes

Definition

Human beings can be exposed to arsenic from a variety of sources. Chronic arsenic exposure can cause skin lesions, neuropathy, cardiovascular disease, and cancer. Since the mid-1990s, many epidemiologic studies have shown a potential link between arsenic exposure and diabetes mellitus. Studies from animals and cell cultures have provided a possibility of diabetogenic effects of arsenic by mechanisms resulting in impaired insulin secretion, induction of insulin resistance and reduced cellular glucose transport. Most of these effects may be explained by the basic biochemical properties of arsenic including (1) functional similarity between arsenate and phosphate, (2) high affinity of arsenite to sulfhydryl groups, (3) increased oxidative stress, and (4) interference with gene expression. Factors related to individual susceptibility to arsenic-induced diabetes mellitus may include arsenic exposure dosage, nutritional status, genetic predisposition, antioxidant supplementation, and interaction with other trace elements.

Introduction

Arsenic can occur in the +5, +3, 0, and −3 valence states and can form alloys with metals. It is ubiquitous and can be found in soil, air, water, food, or some medications. Inorganic forms of arsenic usually present in well water are much more toxic than organic forms found in crustacean seafood. Inorganic arsenic is readily absorbed through the gastrointestinal tract, and mainly metabolized in the liver to organic forms of methylated arsenic, which can be excreted in the urine. It is also now known that both arsenite (inorganic trivalent form) and arsenate (inorganic pentavalent form) can be actively transported into cells via aquaglyceroporins and phosphate transporters, respectively (Tseng 2004, 2009).

Arsenic is well known for its acute toxicity and has been used for hundreds of years as a poisonous substance. With chronic exposure to arsenic, skin lesions, neuropathy, and cancer originating from the lung, liver, kidney, urinary bladder, and skin can develop. Arsenic is also atherogenic and may increase the risk of hypertension, ischemic heart disease, stroke, and peripheral arterial disease (Tseng 2008). In recent years, epidemiologic studies have demonstrated that exposure to arsenic may induce diabetes mellitus in humans; and many studies have been done on animals and tissue cultures to investigate the potential mechanisms.

This entry first briefly reviews the pathophysiology of diabetes and the epidemiologic evidence linking arsenic exposure and diabetes; and finally provides potential biological explanations for the pathogenic links based on the biochemical properties of arsenic and evidence from current literature.

Key Proteins and Pathways Related to Glucose Homeostasis

Insulin, a hormone secreted by the islet β cells of the pancreas, is the principal hormone to lower blood glucose by suppressing gluconeogenesis and glycogenolysis in the liver and by stimulating the uptake of glucose into skeletal muscle and fat. Insulin molecule contains two peptide chains referred to as the A chain and B chain. Two disulfide bonds link these two chains together and insulin exerts its physiological actions by binding to its receptor on cell membrane.

There are at least 14 glucose transporters (GLUT) in the body. Among them, GLUT2 is expressed highly in pancreatic β cells, which is not insulin-dependent and functions as a glucose sensor for the glucose-stimulated insulin secretion. The molecular mechanisms of glucose-stimulated insulin secretion include the following steps which occur in the pancreatic β cells: (1) glucose entry into β cells via GLUT2, (2) phosphorylation of glucose by glucokinase and activation of glycolysis, (3) production of adenosine triphosphate (ATP) with reduced adenosine diphosphate (ADP) concentration leads to a rise in the ATP:ADP ratio, (4) closure of ATP-sensitive K-channel (Kir6.2) leads to cell membrane depolarization, (5) opening of the voltage-sensitive L-type Ca^{2+} channel, and (6) influx of calcium ions leading to insulin secretion (Polakof et al. 2011).

GLUT4 is insulin-dependent and is the major transporter for glucose to enter the cells of the adipose, muscle, and cardiac tissues (Augustin 2010). The insulin receptor complex contains two α- and two β-subunits, linked together by interchain disulfide bridges. Insulin binding to the α-subunits causes a cascade of signaling events including autophosphorylation of tyrosine residues on the β-subunits, tyrosine phosphorylation of the insulin receptor substrates, and activation of the phosphatidylinositol 3-kinase (PI 3-kinase) pathway, which triggers a series of downstream events involving protein kinase C (PKC), leading to insulin-stimulated translocation of GLUT4 to the plasma membrane and subsequent glucose transport via the transporters (Choi and Kim 2010).

Some other pathways not related to insulin signaling may also lead to the expression of GLUT4 and its translocation to the cell membrane. For examples, exercise may be associated with an increase in GLUT4 expression (Hussey et al. 2012) and an amplified signal transduction which is independent of proximal insulin signaling and leads to GLUT4 translocation to the cell membrane, probably mediated by atypical PKC (Maarbjerg et al. 2011). The mitogen-activated protein kinases (serine/threonine protein kinases), which are activated in response to a variety of external stimuli, may also be involved in exercise-induced glucose transport (Widegren et al. 2001). Metabolic stress can induce glucose transport via the 5′ adenosine monophosphate-activated protein kinase pathway, which is also independent of insulin-activated PI-3 kinase.

The disulfide bonds of insulin and insulin receptors are essential for maintaining the normal function and structure of the proteins. The exofacial sulfhydryl groups of the GLUT4 also play an important role in the maximal activity of the transporters, and are crucial for the regulation of transport rates by insulin.

Pathophysiology of Type 2 Diabetes Mellitus

The pathophysiology leading to hyperglycemia in T2DM is very complicated. Any gene mutation or metabolic disturbance leading to a defect in insulin secretion, insulin transport, insulin action, glucose transport, or enzymes associated with glucose metabolism can theoretically result in hyperglycemia or clinical diabetes.

Overexpression of tumor necrosis factor α (TNFα) and interleukin-6 (IL-6) has been found to play an important role on the induction of insulin resistance leading to the development of diabetes. On the other hand, activation of peroxisome proliferator-activated receptor γ (PPARγ) plays an important role in glucose homeostasis by increasing insulin sensitivity. PPARγ is downregulated in adipocytes with the treatment of IL-6, and activation of PPARγ reduces TNFα expression. The expression of TNFα and IL-6 are regulated by nuclear factor-κB (NF-κB), and reactive oxygen species (ROS) resulting from hyperglycemia or other stress stimuli can lead to insulin resistance through its interactions with cytokines and other mediators involving the activation of NF-κB pathway (Henriksen et al. 2011).

Another mechanism leading to hyperglycemia in patients with T2DM involves the inability of insulin to inhibit hepatic glucose production. Enhanced phosphoenolpyruvate carboxykinase (an enzyme catalyzing the rate-limiting step in gluconeogenesis) activity leading to increased gluconeogenesis has been shown to be a major source of increased hepatic glucose production in patients with diabetes.

Pancreatic β-cell dysfunction has also been demonstrated in patients with diabetes. Progressive formation of amyloidosis with loss of β cells is always a major pathological change found in patients with diabetes (Westermark et al. 2011). ROS can induce rapid polymerization of monomeric pancreatic islet amyloid polypeptide into amylin-derived islet amyloid, which is extremely resistant to proteolysis.

Epidemiologic Evidence for Arsenic-Induced Diabetes

There are approximately 25 published papers in epidemiologic studies related to arsenic and diabetes mellitus (Chen et al. 2007; Del Razo et al. 2011; Huang et al. 2011; Navas-Acien et al. 2006, 2008; Tseng et al. 2000, 2002). The first studies were published in the mid-1990s and were mainly conducted in populations exposed to relatively high levels of inorganic arsenic from drinking water in Taiwan and Bangladesh. In these regions, a higher risk of diabetes mellitus associated with arsenic exposure can be demonstrated in a dose-responsive pattern. The evidence for such a link among residents with lower dose of arsenic exposure from drinking water in the USA and Mexico is not conclusive. Some studies conducted among copper smelters and art glass workers with arsenic exposure from the air in Sweden also suggest a potential role of arsenic on diabetes.

Animal Studies

There are approximately 25 animal studies conducted mainly in rats or mice aiming at investigating the effects of arsenic on diabetes mellitus or outcomes relevant to diabetes (Tseng 2004). Most studies used sodium arsenite or arsenic trioxide, administered via routes such as drinking water, intraperitoneal or subcutaneous injection, gavage, or oral intake from diet or as capsule. Hyperglycemia, impaired glucose tolerance, impaired insulin sensitivity, and pancreatic toxicity have been reported. Some studies also observe an attenuation of the toxic effects of arsenic on pancreas and on hyperglycemia when methyl donors or antioxidants are treated simultaneously. However, it should be noted that mice and rats may be less susceptible to arsenic toxicity than human beings.

Biochemical Properties of Arsenic That May Interfere with Glucose Metabolism

The following biochemical properties of arsenic may interfere with glucose metabolism and may be responsible for the mechanisms of arsenic-induced diabetes mellitus (Table 1) (Tseng 2004):

Substitution of Arsenate for Phosphate
Arsenate can replace phosphate in energy transfer phosphorylation reactions, resulting in the formation of ADP-arsenate instead of ATP. ADP-arsenate can serve as a substrate for hexokinase resulting in the formation of glucose-6-arsenate instead of glucose-6-phosphate. At high concentrations of arsenate, the activity of hexokinase is also inhibited.

Arsenite Reaction with Sulfhydryl Groups
Arsenite has high affinity for sulfhydryl groups of proteins and can form stable cyclic thioarsenite complexes with vicinal or paired sulfhydryl groups of cellular proteins. One example of such reaction is the complex formation of arsenite with dihydrolipoamide, a cofactor of the enzymes pyruvate dehydrogenase and α-ketoglutarate dehydrogenase. This chemical change can cause aberration in structures of proteins and inactivate many enzymes and receptors. Inhibition of pyruvate dehydrogenase can impair the production of ATP by blocking the processing of citric acid cycle,

Arsenic-Induced Diabetes Mellitus, Table 1 Biochemical properties of arsenic related to its induction of diabetes

Biochemical property	Explanations
Functional similarity between arsenate and phosphate	Arsenate may substitute phosphate and form ADP-arsenate and glucose-6-arsenate, leading to impaired glucose metabolism and inefficient energy production.
High affinity of arsenite to sulfhydryl groups	Arsenite forms cyclic thioarsenite complex with paired sulfhydryl groups in proteins (insulin, insulin receptor, glucose transporters), and enzymes (pyruvate dehydrogenase and α-ketoglutarate dehydrogenase) could lead to impaired glucose transport and metabolism.
Increased oxidative stress	Oxidative stress can lead to formation of amyloid in pancreatic islet cells, leading to progressive β-cell loss. Superoxide may impair insulin secretion by interaction with uncoupling protein 2. Insulin resistance can also be induced by oxidative stress.
Interference with gene expression	Induction of insulin resistance by enhancing the expression of NF-κB, TNFα, and IL-6 and by inhibiting the expression of PPARγ.

ADP adenosine diphosphate, *NF-κB* nuclear factor-κB, *TNFα* tumor necrosis factor α, *IL-6* interleukin-6, *PPARγ* peroxisome proliferator-activated receptor γ

which is critical for providing reducing equivalents to the mitochondria needed for electron transport.

Although a high concentration of arsenite ($IC_{50} > 100$ μM) is required for the inhibition of pyruvate dehydrogenase, recent studies suggest that trivalent methylated metabolites can exhibit inhibitory effect on pyruvate dehydrogenase with a two- to sixfold higher potency than arsenite.

Oxidative Stress

During the metabolism of arsenic, oxidative stress can be generated by the production of ROS and free radicals like hydrogen peroxide, hydroxyl radical species, nitric oxide, superoxide anion, dimethylarsinic peroxyl radical, and dimethylarsinic radical. A variety of normal physiological functions can be disrupted through the production of ROS and induction of oxidative stress.

Oxidative stress induced by arsenic was not only demonstrated in animal or cell biology studies, studies in human beings also disclosed an increased oxidative stress in subjects with chronic arsenic exposure. Therefore, oxidative stress can play an important role in the pathogenesis of atherosclerosis, cancer, and diabetes related to chronic arsenic exposure.

Gene Expression

Arsenic can influence the expression of a variety of proteins involving signal transduction and gene transcription. These include a variety of cytokines and growth factors associated with inflammation including IL-6. Arsenic has also been shown to upregulate TNFα and to inhibit the expression of PPARγ.

Arsenic Effects on Glucose Transport and Potential Link with Insulin Resistance

Studies suggest that arsenite or its methylated trivalent metabolites can cause insulin resistance by inhibiting insulin signaling, impairing the translocation of glucose transporters, and impeding glucose uptake into the cells.

Sulfhydryl groups on proteins are crucial for their maximal activity. Trivalent arsenicals may form stable cyclic thioarsenite complexes with sulfhydryl groups of cellular proteins including insulin, insulin receptors, and glucose transporters, thereby may impede the normal function of these proteins, leading to insulin resistance.

Arsenic may also induce a status of insulin resistance through its molecular effects on inducing gene expression of some cytokines including TNFα and IL-6. On the other hand, sodium arsenite at physiologically relevant concentration of 6 μM prevents adipocyte differentiation through a mechanism of inhibiting the expression of both PPARγ and CCAAT/enhancer-binding protein α. An interaction between PPARγ and its coactivator retinoid X receptor alpha is associated with an improvement in insulin sensitivity. Therefore, disruption of the expression of PPARγ will surely lead to a status of insulin resistance.

A growing number of reports in recent years have suggested a link between increased ROS production or oxidative stress and the development of insulin resistance and β cell dysfunction in humans. This stress-sensitive pathway may involve the transcription factor NF-κB induced by arsenite, which may in turn induce the expression of TNFα and IL-6, followed by insulin resistance.

Evidence also suggests that arsenite may reduce the myogenic differentiation from myoblasts to myotubes by repressing the transcription factor myogenin. This may also lead to impairment in glucose uptake into the cells.

Therefore, arsenic effects on the functional groups of proteins and on the expression of genes involving ROS, cytokines, PPARγ, and NF-κB can work together in the induction of insulin resistance and diabetes mellitus associated with chronic exposure to arsenic (Tseng 2004).

Arsenic Effects on Glucose Metabolism and Energy Production

Because of its biochemical properties, arsenic may theoretically impair glucose metabolism by acting as an uncoupler of oxidative phosphorylation, as an inhibitor of sulfhydryl containing enzymes such as α-ketoglutarate dehydrogenase and pyruvate dehydrogenase, and as a competitor for phosphate-binding sites on glycolytic enzymes. The formation of ADP-arsenate instead of ATP can cause an inefficient production of energy and results in generalized inhibition of the metabolic pathways that require ATP. Glucose-6-phosphate is not only important as a mediator for glycolysis, gluconeogenesis, glycogenesis, and glycogenolysis, it is also important as an initiator for the pentose phosphate pathway, which generates nicotinamide adenine dinucleotide phosphate (NADPH, an important cofactor in the reduction of glutathione [GSH]) and provides the cell with ribose-5-phosphate for the synthesis of nucleotides and nucleic acids. Substitution of phosphate in the formation of glucose-6-phosphate by yielding glucose-6-arsenate may lead to an inefficient metabolism of glucose. Insufficient production of NADPH from the pentose phosphate pathway further disrupts the ability of the cells to deal with oxidative stress.

Therefore, a slowdown of the metabolic pathways induced by arsenic or its metabolites may contribute partly to an impairment in glucose metabolism (Tseng 2004).

Arsenic Effects on Insulin Biosynthesis and Secretion

Arsenic may theoretically cause impairment in insulin secretion by mechanisms involving functional or structural changes. Because arsenic may cause formation of ADP-arsenate instead of ATP, it is possible that the ATP-dependent insulin secretion could be impaired in the absence of sufficient energy supply.

Interaction between arsenite or its methylated metabolites and the sulfhydryl groups of insulin or proinsulin and competition with zinc in the formation of hexamers during the synthetic stages of insulin are also theoretically possible.

Arsenic may also indirectly cause functional impairment in insulin secretion through the generation of free radicals. Uncoupling protein 2 (UCP2) is a negative regulator of insulin secretion. It mediates proton leak across the inner mitochondrial membrane. A superoxide-UCP2 pathway has been suggested to cause impairment in insulin secretion in pancreatic β cells. Arsenic is well known for its ability to induce the production of superoxide. If excess superoxide is produced in the pancreatic β-cells, an impairment of insulin secretion is expected. Recent studies suggest that Nrf2-mediated antioxidant response may be responsible for the impairment in glucose-stimulated insulin secretion in pancreatic β cells induced by arsenic.

The increased oxidative stress induced by arsenic can theoretically cause structural damages to the pancreatic islets with the formation of amyloidosis, which not only prevents the release of insulin into the circulation, but also destroys the insulin-secreting β cells insidiously after prolonged exposure to arsenic.

Therefore, with prolonged arsenic exposure, insulin secretion may be impaired by either a disruption in insulin secretion function or a progressive loss of β cell mass (Tseng 2004).

Individual Susceptibility to Arsenic-Induced Diabetes Mellitus

Individual susceptibility is likely to vary based on duration and cumulative dosage of exposure, genetic variation, metabolic condition, nutritional status, health status, and other factors (Tseng 2004, 2007, 2009). The variation in individual susceptibility may influence the toxic effect of arsenic on target organs and determine the clinical development of diabetes mellitus.

Nutritional status and supplement with antioxidants are important factors determining the chronic toxicity associated with arsenic exposure. Thus, good nutritional status with sufficient intake of antioxidants can reduce oxidative stress induced by arsenic and can possibly prevent the onset of arsenic-induced diabetes mellitus. GSH is essential for arsenic detoxification and is important for insulin action. Therefore, protein-restricted diets or deficiency in selenium, which is required in the biosynthesis of GSH, may aggravate the hyperglycemia induced by arsenic.

Most of the trace elements do not work in isolation and the interactions between arsenic and other trace elements from environmental co-contamination and oral intake could modulate the chronic toxicity and clinical manifestations associated with arsenic exposure. These interactions may include zinc, selenium, magnesium, and chromium. Deficiency of these trace elements per se may induce hyperglycemia. In low phosphate conditions, arsenate uptake will be accelerated and its toxicity will probably be promoted. Although there is not much published research, the interactions between arsenic and other trace elements that have an effect on glucose homeostasis, such as lithium and vanadium, can also contribute to the hyperglycemic effect of arsenic.

Summary

Epidemiologic studies suggest a potential link between arsenic exposure and diabetes mellitus in humans. Although the etiological links have not yet been fully elucidated, studies from experimental animals and cell biology can provide pathophysiological mechanisms for arsenic-induced diabetes mellitus. The mechanisms may involve one of the following biochemical properties of arsenic: (1) functional similarity between arsenate and phosphate, (2) high affinity of arsenite with sulfhydryl groups, (3) increased oxidative stress, and (4) effects on gene expression. As a result, insulin resistance, β cell dysfunction, and diabetes mellitus can be induced by chronic arsenic exposure. Individual variability in detoxification capability, nutritional status, and interactions with other trace elements may also influence the susceptibility of arsenic-exposed subjects to develop diabetes mellitus.

Cross-References

▶ Arsenic
▶ Arsenic and Alcohol, Combined Toxicity
▶ Arsenic and Primary Human Cells
▶ Arsenic and Vertebrate Aquaglyceroporins
▶ Arsenic in Pathological Conditions
▶ Arsenic Methyltransferases
▶ Arsenic, Free Radical and Oxidative Stress
▶ Arsenic, Mechanisms of Cellular Detoxification
▶ As

References

Augustin R (2010) The protein family of glucose transport facilitators: it's not only about glucose after all. IUBMB Life 62:315–333

Chen CJ, Wang SL, Chiou JM, Tseng CH, Chiou HY, Hsueh YM, Chen SY, Wu MM, Lai MS (2007) Arsenic and diabetes and hypertension in human populations: a review. Toxicol Appl Pharmacol 222:298–304

Choi K, Kim YB (2010) Molecular mechanism of insulin resistance in obesity and type 2 diabetes. Korean J Intern Med 25:119–129

Del Razo LM, García-Vargas GG, Valenzuela OL, Castellanos EH, Sánchez-Peña LC, Currier JM, Drobná Z, Loomis D, Stýblo M (2011) Exposure to arsenic in drinking water is associated with increased prevalence of diabetes: a cross-sectional study in the Zimapán and Lagunera regions in Mexico. Environ Health 10:73

Henriksen EJ, Diamond-Stanic MK, Marchionne EM (2011) Oxidative stress and the etiology of insulin resistance and type 2 diabetes. Free Radic Biol Med 51:993–999

Huang CF, Chen YW, Yang CY, Tsai KS, Yang RS, Liu SH (2011) Arsenic and diabetes: current perspectives. Kaohsiung J Med Sci 27:402–410

Hussey S, McGee S, Garnham A, McConell G, Hargreaves M (2012) Exercise increases skeletal muscle GLUT4 gene expression in patients with type 2 diabetes. Diabetes Obes Metab. doi:10.1111/j.1463-1326.2012.01585.x, Epub ahead of print

Maarbjerg SJ, Sylow L, Richter EA (2011) Current understanding of increased insulin sensitivity after exercise – emerging candidates. Acta Physiol (Oxf) 202:323–335

Navas-Acien A, Silbergeld EK, Streeter RA, Clark JM, Burke TA, Guallar E (2006) Arsenic exposure and type 2 diabetes: a systematic review of the experimental and epidemiological evidence. Environ Health Perspect 114:641–648

Navas-Acien A, Silbergeld EK, Pastor-Barriuso R, Guallar E (2008) Arsenic exposure and prevalence of type 2 diabetes in US adults. JAMA 300:814–822

Polakof S, Mommsen TP, Soengas JL (2011) Glucosensing and glucose homeostasis: from fish to mammals. Comp Biochem Physiol B Biochem Mol Biol 160:123–149

Tseng CH (2004) The potential biological mechanisms of arsenic-induced diabetes mellitus. Toxicol Appl Pharmacol 197:67–83

Tseng CH (2007) Arsenic methylation, urinary arsenic metabolites and human diseases: current perspective. J Environ Sci Health C Environ Carcinog Ecotoxicol Rev 25:1–22

Tseng CH (2008) Cardiovascular disease in arsenic-exposed subjects living in the arseniasis-hyperendemic areas in Taiwan. Atherosclerosis 199:12–18

Tseng CH (2009) A review on environmental factors regulating arsenic methylation in humans. Toxicol Appl Pharmacol 235:338–350

Tseng CH, Tai TY, Chong CK, Tseng CP, Lai MS, Lin BJ, Chiou HY, Hsueh YM, Hsu KH, Chen CJ (2000) Long-term arsenic exposure and incidence of non-insulin-dependent diabetes mellitus: a cohort study in arseniasis-hyperendemic villages in Taiwan. Environ Health Perspect 108:847–851

Tseng CH, Tseng CP, Chiou HY, Hsueh YM, Chong CK, Chen CJ (2002) Epidemiologic evidence of diabetogenic effect of arsenic. Toxicol Lett 133:69–76

Westermark P, Andersson A, Westermark GT (2011) Islet amyloid polypeptide, islet amyloid, and diabetes mellitus. Physiol Rev 91:795–826

Widegren U, Ryder JW, Zierath JR (2001) Mitogen-activated protein kinase signal transduction in skeletal muscle: effects of exercise and muscle contraction. Acta Physiol Scand 172:227–238

Arsenic-Induced Stress Proteins

Alfonso Rios-Perez, Pablo Morales-Rico and Balam Muñoz
Department of Toxicology, Cinvestav-IPN, Mexico city, Mexico

Synonyms

Arsenic; Heat shock proteins; HSPs; Stress

Definition

Arsenic is a ubiquitous element in the Earth's crust. Exposure to this metalloid may induce a variety of responses including oxidative stress, metabolic alterations, growth inhibition, and eventually cell death. A complex enzyme network metabolizes arsenic, inducing the expression of stress proteins. Stress proteins are molecules produced after exposure to a physical or chemical agent. The most common stress proteins induced by arsenic exposure are metallotionein, ubiquitin, and heat shock proteins (HSPs).

Introduction

In nature, arsenic is found in several forms including oxides, chlorine, or sulfur compounds. Water is the most common medium for environmental distribution of arsenic (Del Razo et al. 2001). Exposure to arsenic produces a variety of stress responses in cells including morphological, physiological, biochemical, and molecular alterations. Arsenic (As) may enter an organism through epithelial exposure or ingestion, where consumption of contaminated drinking water is the most common mean among humans (ATSDR 2007). Although the World Health Organization (WHO) has determined As concentrations of 10 μg/L as acceptable, higher levels have been measured in parts of Argentina, Mexico, Bangladesh, India, China, Canada, Chile, and the USA, among others. Furthermore, since 2007 arsenic has been classified as a Class I human carcinogen by the International Agency for Research on Cancer (IARC), verifying the existence of substantial evidence for its carcinogenicity to humans. For this reason further investigation into arsenic toxicity and its carcinogenic effects should be a priority among health organizations.

Arsenic Exposure

Chronic exposure to arsenic is mainly connected with intake from natural sources, however, higher risk occupations such as mining and smelting may also be involved in disease development (ATSDR 2007). At present, millions of individuals worldwide are considered chronically exposed to arsenic via drinking water, with consequences ranging from acute toxicities, such as stomachache, nausea, vomiting, and diarrhea to the development of malignancies, such as skin, bladder, and lung cancer. Due to its indiscernible physical characteristics (no odor, color, or flavor), arsenic exposure often goes unnoticed, especially when ingested through contaminated food or drinking water. In this context, long-term effects are a major health concern in affected areas.

Metabolism of Arsenic

Arsenic enters the cells via transport through the membrane by aquaporin and hexosepermease transporters. In humans, as in many mammals, arsenic is readily absorbed and distributed to many organs, although the liver is the main target and the site of most of its biotransformation. The reduction of arsenate (iAsV)

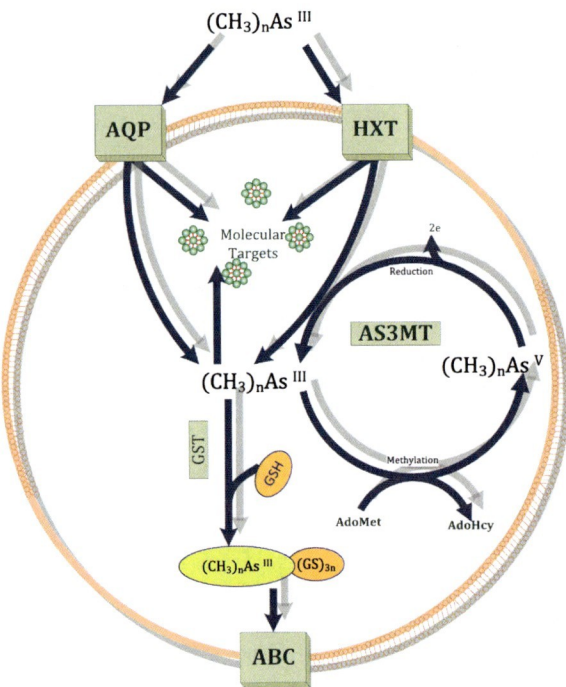

Arsenic-Induced Stress Proteins, Fig. 1 *Model of arsenic metabolism.* Arsenic enters through membrane transporters such as aquaporin (AQP) or the hexose permease transporter (HXT) and may interact with specific molecules. Trivalent arsenic may be conjugated via glutathione-S-transferase (GST) and excreted through the ATP dependent transporter (ABC) or methylated via arsenic-3-methyl transferase (AS3MT) (Adapted from Thomas 2007)

processes. Stress responses that result from As exposure include the induction of certain proteins that maintain homeostasis in the cell.

Induction of Stress Proteins at mRNA and Protein Levels by Arsenic Exposure

Stress response is an adaptation event by prokaryotes and eukaryotes, regulated at the transcriptional, translational, and posttranslational levels. Altered patterns of protein synthesis induced by heat, UV radiation, or chemical exposure activate the expression of stress proteins (SPs) (Nriagu 2000). However, in normal conditions, several of the major SPs are constitutively expressed at low levels and work as "chaperones," facilitating the correct folding, assembly, and distribution of recently synthesized proteins (Kato et al. 1997). The presence of higher concentrations of As in the environment induces an expression of SPs that resembles that of the heat shock response (Ahsan et al. 2010). To date, the specific SPs related with exposure to As include: metallotioneins (MTs), ubiquitin (Ub), alpha B-crystallin, and heat shock proteins (HSPs) of various sizes including HSP25, HSP27, HSP30, HSP60, HSP70, HSP90, and HSP105 (Del Razo et al. 2001). MTs are small thiol-rich proteins typically related with Zn homeostasis; however, MTs are also induced in the presence of a wide range of metals, including As^{3+}. In mice, a lack of MTs increases toxicity of arsenic to the kidneys. However, the induction of MT by arsenic is probably a response to the generation of reactive species. Nevertheless, As modulation of MT expression may result from diverse interactions. For instance, As could act directly by an interaction with metal response elements (MRE) in the DNA such as MTF-1 or indirectly by glucocorticoids, the antioxidant response element, or through its effect on Zn regulation (Del Razo et al. 2001). The stress protein Ub, also expressed in the presence of As, binds to proteins and acts as a signal for degradation via proteosomes. Exposure to arsenite at low concentrations causes an increase in Ub-conjugation enzymes, resulting in a large number of Ub-conjugated proteins. Alpha B-crystallin, a major protein in the vertebrate lens, also has sequence similarity with small HSPs; arsenite exposure in lung cells induces this protein in response. Although alpha B-crystallin is constitutively expressed in cells, its expression increases in response to other

to arsenite (iAsIII) is required for the enzymatic methylation of arsenic to yield mono-, di-, and possibly trimethylated metabolites. This process is catalyzed by an As^{3+} methyltransferase (As3MT), in which S-adenosylmethionine acts as the donor of methyl groups and produces trivalent and pentavalent methylated arsenicals (Thomas 2007). Arsenic is typically conjugated (Fig. 1) with glutathione via glutathione-S-transferase or with betaine, producing arsenobetaine and arsenocholine. In addition, arsenic generates ROS and free radicals such as hydrogen peroxide, nitric oxide, or hydroxyl radical (OH) species. This metabolic activity may lead to stress responses, growth inhibition, and cell death. Furthermore, As^{3+} may interact with proteins and inhibit DNA repairing

types of stress such as higher temperatures, drugs, and toxic metals (Wijeweera et al. 1995).

Under normal conditions HSPs are expressed at basal levels; however, As exposure is capable of triggering the oxidation of glutathione, leading to the activation of heat shock transcription factor (HSF). This transcription factor increases expression of HSPs such as Hsp70 and Hsp27. High concentrations of arsenite increase the phosphorylation level of Hsp27, inhibiting protein phosphatase activity and acting as a protective mechanism against oxidative damage. Hsp70 is a large, multigene family that includes both constitutively expressed Hsc70 and stress-inducible Hsp70 forms (Das et al. 2010). Cells that accumulate inducible Hsp70 after As exposure demonstrate protection, enhanced recovery, and resistance to subsequent exposures (Han et al. 2005). Hsp70 may guard cells from As-induced chromosome alterations through several mechanisms that facilitate repair of DNA damage (Barnes et al. 2002). Hsp70 and Hsp27 are associated with base excision repair enzymes that reduce alterations in DNA induced by As. Considering all stress proteins, Hsp70 is shown to be one of the most conserved stress proteins, whereby many studies have shown accumulation of Hsp70 in cells or tissues after As treatment. For this reason Hsp70 response may act as a biomarker to inorganic arsenic (iAs) exposure and damage. Hsp90 synthesis is also induced by arsenite in various cell types and although it is abundant in the cytosol of normal cells its expression increases with stress conditions (Lau et al. 2004). The Hsp90 chaperone is involved in several signal transduction pathways by stabilizing protein complexes. Specific inhibitors of Hsp90 increase the toxic effect of arsenic trioxide by inducing apoptosis, arresting the cell cycle, and blocking cell signaling. High molecular weight HSPs such as Hsp105 or Hsp110, are also expressed after arsenite treatment in different cells such as keratinocytes, embryonic, epithelial, and tumor-derived cells. These HSPs are suggested to play a protective role following stress induced by arsenite exposure (Liu et al. 2001).

Hemeoxygenase 1 (HO-1), an enzyme involved in heme degradation, is another protein that increases its expression after exposure to As. HO-1 is considered to be an inducible stress protein that confers protection against oxidative stress. The most useful HO-1 inducers are trivalent arsenicals, mainly iAsIII, in which neither MMAsV nor DMAsV are effective HO-1 inducers. Acute and subchronic exposure to iAs increases bilirubin excretion; HO-1 catabolizes heme into three products: carbon monoxide (CO), biliverdin (which is rapidly converted to bilirubin), and free iron radicals. For this reason, the inducibility of HO-1 by iAs in cells could be considered a potential biomarker (Lau et al. 2004).

Regulation of Stress Proteins Induced by Arsenic

It is well established that HSP may control a wide variety of cell processes including environmental response, adaptation, growth stimulation and inhibition, differentiation, and cell death. For this reason, As exposure may activate transcriptional responses that control the processes mentioned above (Pirkkala et al. 2001).

Arsenite stimulates the binding of transcriptional regulators to DNA such as the heat shock factor HSF1 through phosphorylation and subsequent activation of signaling pathways. Phosphorylation is an important determinant of the transactivating potency of HSF1. This protein is a member of the heat shock transcription factor family (the HSF family has been found in vertebrates (HSF1–4) and plants) and contains two distinct carboxyl-terminal activation domains, AD1 and AD2, which are controlled by a centrally located, heat-responsive regulatory domain (RD). Constitutive phosphorylation of two specific serine/proline motifs, S303 and S307, is important for the function of the RD and may be critical for the negative regulation of HSF1 (Fig. 2). These phosphorylation sites are targeted by glycogen synthase kinase 3ß (GSK-3ß) and extracellular signal-regulated kinase (ERK-1), whereas S363 may be a phosphorylation site by c-Jun N-terminal kinase (JNK) (Liu et al. 2010). The phosphorylation site S230 appears to be a suitable substrate for calcium/calmodulin-dependent protein kinase II (CaMK II). Protein-damaging stress induced by arsenic leads to the activation of HSF1, which then binds to upstream regulatory sequences in the promoters of heat shock genes that enhance heat shock gene expression. The activation of HSF1 proceeds through a multistep pathway, involving multiple protein modifications. HSF1 activity is modulated at different levels by the HSPs resulting in a regulatory feedback

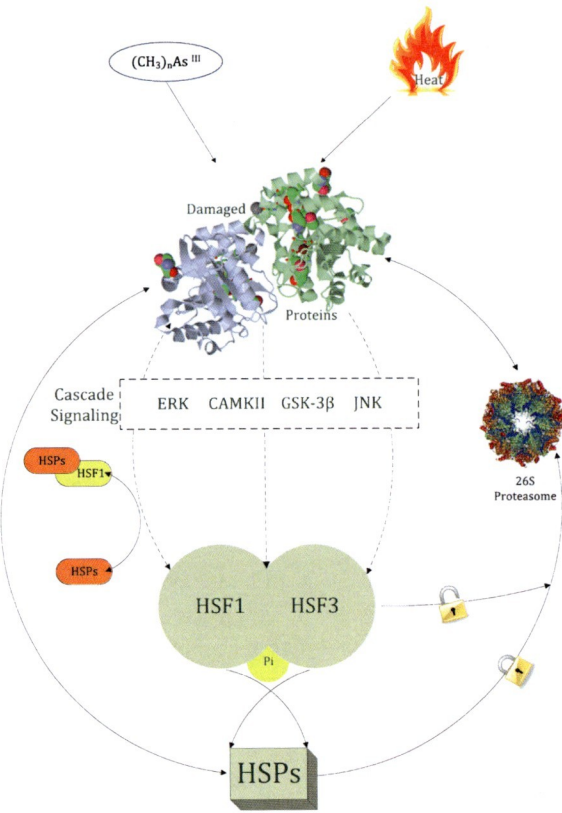

Arsenic-Induced Stress Proteins, Fig. 2 *Regulation of stress protein by arsenic or heat.* Arsenic or heat induce changes that activate several signal pathways and result in the transcriptional activity of HSFs. HSFs activate the expression of HSPs thereby mediating regulation via a feedback loop

conclusively demonstrated to be a carcinogen in humans although some reports suggest that As may act as a carcinogen in other animals in early life and with transplacental exposure (Tokar et al. 2010). Currently, sufficient evidence supports the observation that inhalation exposure to iAs leads to respiratory irritation, nausea, skin, and neurological effects, and an increased risk of developing lung cancer. Dermal exposure to iAs on the other hand, shows little risk of adverse effects with the exception of local irritation. In addition, hyperkeratinization, gangrene, cardiovascular effects, skin, lung, prostate, and bladder cancer have been associated with the ingestion of drinking water contaminated with As. From the various sources of As in the environment, long-term exposure of arsenic in drinking water likely poses the greatest threat to human health. The capacity of As^{3+} species to bind biologically important proteins and to alter DNA methylation profiles and oxidative stress account for the most plausible proposed mechanisms of arsenic carcinogenesis.

Summary

Study of heat shock proteins induced by arsenic may lead to the understanding of how cells respond to environmental stress and how chronic effects result in the development of diseases. It is obvious that the effects of arsenic exposure represent classical stress responses of cells with the participation of the "standard" heat shock mechanism. Arsenicals may activate the expression of the major HSP families and other stress response proteins such as metallotionein and hemoxygenase (OH-1). Among the different arsenic species, iAsIII has been shown to be the most powerful stress inducer. Additional work is necessary to completely uncover the general sequence of molecular events in the signal pathway induced by arsenic.

mechanism. However to date, regulation remains relatively unknown as do the phosphorylation sites and kinases/phosphatases responsible for HSF1 phosphorylation.

Another possible mechanism of regulation induced by As is through the glucocorticoid receptor (GR) where As simulates the effect of a hormone binding to the receptor. Glucocorticoids and other steroid hormones regulate many of the genes affected by low-dose As because As acts as a endocrine disruptor, altering gene regulation (Nriagu 2000).

Chronic Diseases Resulting from Arsenic Exposure

Ingestion of As may occur via food (mainly in fish, seafood, algae, and cereals), air (coal-fired power generation and smelting), or water. Long-term exposure of iAs has been associated with several human diseases (Goering et al. 1999). To date, iAs has only been

Cross-References

▶ Arsenic
▶ Arsenic and Aquaporins
▶ Arsenic in Nature
▶ Arsenic in Pathological Conditions
▶ Arsenic in Tissues, Organs, and Cells
▶ Arsenic Methyltransferases

- Arsenic, Biologically Active Compounds
- Arsenic, Free Radical and Oxidative Stress
- Arsenic, Mechanisms of Cellular Detoxification
- As
- Phosphatidylinositol 3-Kinases

References

Ahsan N, Lee DG, Kim KH et al (2010) Analysis of arsenic stress-induced differentially expressed proteins in rice leaves by two-dimensional gel electrophoresis coupled with mass spectrometry. Chemosphere 78:224–231

ATSDR (2007) Draft toxicological profile for arsenic. Agency for Toxic Substances and Disease Registry, Atlanta

Barnes JA, Collins BW, Dix DJ et al (2002) Effects of heat shock protein 70 (Hsp70) on arsenite-induced genotoxicity. Environ Mol Mutagen 40:236–242

Das S, Pan D, Bera AK et al (2010) Stress inducible heat shock protein 70: a potent molecular and toxicological signature in arsenic exposed broiler chickens. Mol Biol Rep 37:3151–3155

Del Razo LM, Quintanilla-Vega B, Brambila-Colombres E et al (2001) Stress proteins induced by arsenic. Toxicol Appl Pharmacol 177:132–148

Goering PL, Aposhian HV, Mass MJ et al (1999) The enigma of arsenic carcinogenesis: role of metabolism. Toxicol Sci 49:5–14

Han SG, Castranova V, Vallyathan V (2005) Heat shock protein 70 as an indicator of early lung injury caused by exposure to arsenic. Mol Cell Biochem 277:153–164

Kato K, Ito H, Okamoto K (1997) Modulation of the arsenite-induced expression of stress proteins by reducing agents. Cell Stress Chaperones 2:199–209

Lau ATY, He Q-Y, Chiu J-F (2004) A proteome analysis of the arsenite response in cultured lung cells: evidence for in vitro oxidative stress-induced apoptosis. Biochem J 382:641–650

Liu J, Kadiiska MB, Liu Y et al (2001) Stress-related gene expression in mice treated with inorganic arsenicals. Toxicol Sci 61:314–320

Liu J, Zhang D, Mi X et al (2010) p27 suppresses arsenite-induced Hsp27/Hsp70 expression through inhibiting JNK2/c-Jun- and HSF-1-dependent pathways. J Biol Chem 285:26058–26065

Nriagu LBJ (2000) Molecular aspects of arsenic stress. J Tox Environ H Part B 3:293–322

Pirkkala L, Nykanen P, Sistonen L (2001) Roles of the heat shock transcription factors in regulation of the heat shock response and beyond. FASEB J 15:1118–11131

Thomas DJ (2007) Molecular processes in cellular arsenic metabolism. Toxicol Appl Pharmacol 222:365–373

Tokar EJ, Benbrahim-Tallaa L, Ward JM et al (2010) Cancer in experimental animals exposed to arsenic and arsenic compounds. Crit Rev Toxicol 40:912–927

Wijeweera JB, Thomas CM, Gandolfi AJ et al (1995) Sodium arsenite and heat shock induce stress proteins in precision-cut rat liver slices. Toxicology 104:35–45

Arsenicism

- Arsenicosis

Arsenicosis

Nilay Kanti Das[1], Pramit Ghosh[2] and Amrita Sil[3]
[1]Department of Dermatology, Medical College, Kolkata, West Bengal, India
[2]Department of Community Medicine, Medical College, Kolkata, West Bengal, India
[3]Department of Pharmacology, Burdwan Medical College, Burdwan, West Bengal, India

Synonyms

Arseniasis; Arsenicism; Arsenism; Chronic arsenicosis ("arsenicosis" needs to be differentiated from "acute arsenic poisoning")

Definition

Arsenicosis is defined by the World Health Organization (WHO) working group as a "chronic health condition arising from prolonged ingestion (not less than 6 months) of arsenic above a safe dose, usually manifested by characteristic skin lesions, with or without involvement of internal organs" (WHO Regional Office for South-East Asia 2003). The *maximum permissible limit* recommended by WHO in groundwater is 10 μg/L; however, in India, Bangladesh, the accepted level is <50 μg/L in the absence of an alternative source of potable water in the affected area (Smedley and Kinniburgh 2002).

Magnitude of the Problem

Arsenicosis is a global problem with some local predilection. Asian countries are the worst affected with major disease burden reported from countries like *Bangladesh, India*, and *China*; however, people are also affected in parts of Mongolia, Hungary,

Arsenicosis, Table 1 Country profiles (in decreasing order of population at risk) in relation to arsenic contamination of groundwater

Countries	Area affected (sq. km)	Concentration range (µg/L)	Population at risk (in millions)	Sources of contamination	First reported case
Bangladesh	**1,18,849**	**<0.5–2,500**	**30**[a]	**Geogenic**	**1995**
India/West Bengal	**38,865**	**<10–3,200**	**6**[a]	**Geogenic + Anthropogenic**	**1983**
Argentina	10,00,000	1–7,500	2	Geogenic	1917
Chile	35,000	100–1,000	0.5	Geogenic + Anthropogenic	1962
Mexico	32,000	8–620	0.4	Geogenic + Anthropogenic	
China/Taiwan	4,000	10–1,820	0.1	Geogenic + Anthropogenic	1968 in Taiwan
Mongolia	4,300	1–2,400	0.1	Geogenic	
Thailand	100	1 to <5,000	0.015	Anthropogenic	1996

[a]Percentage of the total population at risk is about 25% in Bangladesh and 6% in West Bengal (India)
Source: Ghosh et al. (2008) (Permission obtained from editor-in-chief IJDVL)

Mexico, Argentina, Chile, Canada, and USA (Table 1) (Ghosh et al. 2008; Smith et al. 2000).

Approximately 40,000-sq. km area in West Bengal (12 out of 19 districts), 118,000-sq. km area in Bangladesh (50 out of 64 districts), and 4,000 sq. km in China (8 provinces and 37 counties) are reported to be affected (Ghosh et al. 2008). Apart from the Asian countries, reports from Hungary (Great Hungarian Plain) reveal that 4,263-sq. km area is having arsenic in groundwater in the range of 25–50 µg/L (Kapaj et al. 2006). With increasing scope of field testing and screening, these figures are on a rising trend, and increasing number of areas affected with arsenic contamination and at-risk population are identified with every new survey conducted. Reports from community-based surveys revealed that one in every five to ten persons in the endemic areas of south eastern Asian countries is actually manifesting the disease (Smith et al. 2000; Ghosh et al. 2008).

Epidemiological Determinants

Arsenicosis is a disease with multifactorial determinants, but the key factor is exposure occurring through naturally contaminated groundwater used for drinking or in food preparation. Sources such as mining and some pesticides and wood preservatives may contribute to human exposure and should be controlled in order to prevent environmental contamination.

Agent Factor

The metalloid arsenic is the obvious agent factor, and the trivalent (*arsenite*, As^{+3}) and pentavalent (*arsenate*, As^{+5}) states are principally implicated in arsenicosis. Arsenic is perhaps unique among the heavy metalloids and oxyanion-forming elements (e.g., arsenic, selenium, antimony, molybdenum, vanadium, chromium, uranium, rhenium) in its sensitivity to mobilization at the pH values typically found in groundwaters (pH 6.5–8.5) and under both oxidizing and reducing conditions (Smedley and Kinniburgh 2002).

Most oxyanions including arsenate tend to become less strongly sorbed as the pH increases. Under some conditions at least, these anions can persist in solution at relatively high concentrations (tens of µg/L) even at near-neutral pH values. Therefore, the oxyanion-forming elements are some of the most common trace contaminants in groundwaters. However, relative to the other oxyanion-forming elements, arsenic is among the most problematic in the environment because of its relative mobility over a wide range of redox conditions. It can be found at concentrations in the mg/L range when all other oxyanion-forming metals are present in the µg/L range (Smedley and Kinniburgh 2002).

Highest concentration of arsenic in water is found in groundwater, and range of concentration is also pretty wide, hence making it difficult to derive a typical or "usual" value. Moreover, majority of the researches focused on areas endemic for arsenicosis, resulting in extreme values, and these are possibly unrepresentative of global pattern.

Arsenicosis, Table 2 Postulated mechanisms of *geochemical processes* for leaching of arsenic in groundwater

Postulations	Principles and proposed mechanism
"Pyrite oxidation" hypothesis	*Mechanism*: Excessive withdrawal of groundwater for the purpose of irrigation → The heavy groundwater usage leads to lowering of the water table → gap created is being filled up by air (containing oxygen) → subsequent oxidation of As-containing pyrites on the wall of the aquifer → after rainfall, water table recharges → as in its oxidized form, i.e., arsenate (As^{5+}) leaches out of the sediment into the aquifer
	Limitations: Fails to explain the presence of reduced form, arsenite (As^{3+}), in the groundwater
"Oxyhydroxide reduction" hypothesis	*Principle*: As in pore, water is controlled by the solubility of iron and manganese oxyhydroxides in the oxidized zone and metal sulfides in the reduced zone Digenetic sulfides act as an important sink for As in reducing sulfidic sediment
	Mechanism: In ground condition (with finely grained surface layers impeding the penetration of air into the aquifer and microbial oxidation of organic carbon depleting the dissolved oxygen), a greatly reduced environment is created → under reducing condition, oxyhydroxides of iron and manganese dissolve → arsenic sulfides precipitate and arsenite (As^{3+})
"Oxidation-reduction theory"	*Mechanism*: Arsenic originates from the arsenopyrite oxidation (as proposed in "pyrite oxidation" hypothesis) → arsenic thus mobilized forms the minerals → gets reduced to arsenite (As^{3+}) underground in favorable Eh conditions

Source: Ghosh et al. (2008) (Permission obtained from editor-in-chief IJDVL)

Baseline concentration of arsenic in atmospheric precipitation, river water, rainwater, or seawater is found to be quite low. The concentration of arsenic may increase to manifolds in river water or surface water where concentration is usually very low, if it gets contaminated by industrial pollution, by increased geothermal activities, or if water is flowing through bedrocks (with increased alkalinity and pH). Seasonal variation may influence geothermal activity or leaching of arsenic in surface water/river water. In low-flow period during summer months, it gets increased, and this may also be linked to temperature-controlled microbial reduction of As^{+5} to As^{+3} with consequent increased mobility of As^{+3} (Smedley and Kinniburgh 2002; Appelo 2006).

Host Factor

Propensity to develop arsenicosis in a community is variable. In large scale studies, it was noted that manifestation of arsenicosis was present among 11–20% of people living in arsenic contaminated areas of West Bengal, India, and Bangladesh (Ghosh et al. 2008). Understandably the susceptibility of individuals to develop full-blown disease varies in the same community using same water source.

The disease manifestation is strongly associated with *malnutrition*, especially with deficiency of proximate principle like protein or deficiency of trace element like *selenium*. People belonging to *lower socioeconomic strata* are more vulnerable to develop this disease. *Age* of the individual is also an important parameter which determines the expression of the disease and it usually takes *5–20 year of exposure* to develop the disease. Nevertheless, children with <10 year of exposure have developed this disease (Smith et al. 2000).

Genetic factors, viz., *decreased capacity of methylation of arsenic and genetic polymorphism of enzymes XPD and XRCC1*, having *null or variant genotype* of at least one of the three (M1/T1/P1) *glutathione S-transferases* are risk factors for carcinogenesis in arsenicosis patients (Ghosh et al. 2008).

Environmental Factors (Sources of Arsenic)

Arsenicosis is a disease, principally linked to environmental situation. People can get exposed to arsenic due to its presence in water (principal route of entry), soil, air, or less commonly through other miscellaneous agents as detailed below. Arsenic can enter these routes either though natural (geogenic) processes or by man-made (anthropogenic) activities, or both may contribute in certain situations (Ghosh et al. 2008).

Arsenic can gain entry in underground water as result of *geochemical processes* (vide Table 2), or it may be of anthropogenic origin, e.g., use of *pesticide, insecticide,* or *industrial pollution*. Presence of arsenic in soil is thought to be the result of use of *phosphate fertilizer* which could result increased phosphate concentration in soil and consequently increase in sediment biota leading to increased resorption of arsenic from the sediment (Ghosh et al. 2008).

Volcanic eruption contributes to the majority of atmospheric flux of arsenic (as As_2O_3). Natural low-temperature bio-methylation and microbial reduction are the other natural sources of arsenic in air. Of the anthropogenic causes, *smelting of nonferrous metal* and *combustion of fossil fuels* are important. The arsenic returns to the earth by wet or dry precipitation and contributes the arsenic load of water and soil (Ghosh et al. 2008).

Organic form of arsenic is primarily found in marine organisms in the form of *arsenobetaine* in marine animals and *dimethyl arsenoyl ribosides* in marine algae and can serve as important arsenic source for individuals consuming seafood (Ghosh et al. 2008).

Exposure to arsenic through medications (e.g., *Fowler's solutions* or 1% potassium arsenite) containing arsenic is of historical importance, but even in present days, *homeopathic medicine* containing arsenic is still being used in some countries. Recently, *arsenic trioxide* (Trisenox) has obtained FDA approval for use in acute promyelocytic leukemia, but its impact in causing arsenicosis is yet to be evaluated.

Gallium arsenide, used as a silicone substitute in computer microchips, poses an emerging threat arising due to the expanding opto-electric and micro-electronic industries (Ghosh et al. 2008).

Bioaccumulation of arsenic is still not an established threat, but case reports and studies of presence of arsenic in crops (particularly rice) irrigated with arsenic contaminated water or milk and meat sample obtained from animal husbandry of arsenic contaminated areas point toward this phenomenon (Ghosh et al. 2008).

Pathogenesis

Arsenic gaining entry in the body undergoes sequential reduction and oxidative methylation and is converted to less toxic methylated metabolite, monomethyl arsenic acid (MMA), and dimethyl arsenic acid (DMA). This *bio-methylation* is considered as *bio-inactivation* process, and persons differing in this capability suffer due to accumulation of As^{+3} and As^{+5} (Sengupta et al. 2008; Gomez-Caminero et al. 2001a).

As^{+3} has the ability to bind with the sulfhydryl groups present in various essential compounds (e.g., glutathione, cysteine) and can lead to inactivation of many thiol group-containing enzymes (e.g., pyruvate dehydrogenase) and other functional alteration (e.g., prevention of binding of steroid to glucocorticoid receptors). As^{+5}, on the other hand, mostly mediates its toxicity after conversion to the trivalent state but in certain situation, can render direct toxicity by replacing phosphate during glycolysis (phenomenon known as "*arsenolysis*") leading to ineffective adenosine triphosphate (ATP) production (Sengupta et al. 2008; Gomez-Caminero et al. 2001a).

Inorganic arsenic is capable of causing *teratogenic* effects, and animal studies have demonstrated that it can lead to neural tube defect, inhibition of development of limb-buds, pharyngeal arch defect, anophthalmia, etc. Arsenic can also induce *genotoxicity* by inhibiting DNA excision repair of thymine dimer, DNA ligase, and tubulin polymerization and altering the activity of tumor suppressor gene p53 by DNA methylation. Arsenic is also thought to build up oxidative stress by the production of oxygen free radical, which in turn can cause chromatid exchange and chromosomal aberration. The genotoxicity in turn is responsible for carcinogenicity. The free radicals generation also leads to increased *apoptosis* (programmed cell death) seen in arsenicosis (Sengupta et al. 2008; Gomez-Caminero et al. 2001b).

The cutaneous toxicity seen with arsenicosis is influenced by various growth factors and transcription factors. Expression of keratin 16 (marker for hyperproliferation) and keratin 8 and 18 (marker for less-differentiated epithelial cells) is found to be increased by arsenic exposure. Arsenic has shown to reduce transcription factor AP1 and AP2 which in turn reduces keratinocyte differentiation marker, involucrin. Arsenic is shown to trigger production of interleukin (IL)-8 and also stimulate the secretion of IL-1, tumor necrosis factor-alpha (TNFα), transforming growth factor (TGF) beta, and granulocyte monocyte colony-stimulating factor (GM-CSF) by the keratinocytes. The changed cytokine milieu is believed to exert the cutaneous changes (Sengupta et al. 2008; Gomez-Caminero et al. 2001a). The pathogenesis of other systemic manifestations is highlighted in Table 3.

Arsenicosis, Table 3 Proposed mechanism of pathogenesis for systemic manifestations in arsenicosis

Systemic involvements	Proposed mechanism of pathogenesis
1. Atherosclerosis	• Monoclonal expansion of smooth muscle cell
	• Production of reactive oxygen species (H_2O_2 & –OH radical) → endothelial cell proliferation and apoptosis
	• Upregulation of inflammatory signal → release of TNFα from mononuclear cells or stimulates cyclooxygenase II pathway
	• Enhances arterial thrombosis and platelets aggregation
2. Hepatotoxicity (fibrosis)	• Predominant lesion of hepatic fibrosis appears to be induced by oxystress and elevation of cytokines (TNFα and IL-6) associated with increasing level of collagen in the liver
	• Reduction/weakening of hepatic glutathione and enzymes of antioxidant defense system of liver → free radical accumulation → peroxidative damage of lipid membrane
3. Respiratory system	• Pathophysiology is not well understood, but oxidative damage is thought to play a role
	• Arsenic is potent respiratory toxicant, and ingested arsenic can reach respiratory tract to damage lung tissue
4. Neurological	• Predominantly sensory with distal axonopathy due to axonal degeneration
5. Genitourinary	• Renal tubular necrosis, nephritis, and nephrosis
6. Diabetes mellitus	• As^{3+} suppress insulin stimulated glucose uptake by interfering the mobilization of glucose transporters in adipose cell
	• Interfering the transcription factor involved in insulin-related gene expression

Source: Sengupta et al. (2008) (Permission obtained from editor-in-chief IJDVL)

Clinical Features

Arsenicosis is a multisystem disorder, but the hallmark clinical manifestations are its cutaneous changes. The cutaneous changes are so pathognomonic that the condition can be clinically confirmed by looking at the skin changes. The skin lesions can take the form of *melanosis* (pigmentary changes/dyspigmentation), *keratosis* (thickening of skin), or *malignant/ premalignant lesions*. Respiratory, hepatobiliary, vascular, and nervous systems are among the major systems affected by arsenicosis, apart from constitutional symptoms, including weakness, headache, pedal edema, etc. (Sengupta et al. 2008). Though the systemic features of arsenicosis are nonspecific, but at times they may serve as taletell signs especially in early suspected cases. The cutaneous changes and neurological defects produced by arsenicosis are irreversible, but respiratory, hepatobiliary, gastrointestinal, and vascular effects can be reversed (World Bank technical report vol II). The manifestations of arsenicosis vary according to the cumulative exposure of arsenic, and the description of the exposure dose for different manifestations is detailed in Fig. 1.

Cutaneous Manifestations

The cutaneous lesions are typically bilateral and are the earliest and commonest changes seen in arsenicosis patients. It is generally agreed that there is latency of approximately 10 years for development of skin lesion (i.e., exposure of arsenic and development of manifestations), though there are reports of arsenicosis in children less than 10 years (Smith et al. 2000).

Melanosis is found to be the earliest and commonest of all the cutaneous manifestation of arsenicosis and is most pronounced on trunk. The most characteristic arsenical dyspigmentation takes the form of guttate hypopigmented macule on hyperpigmented background, referred to as *leucomelanosis* (Fig. 2). In some cases, it can also take the form of freckle-like spotty pigmentation, quoted as *raindrop pigmentation* by some authors (Fig. 3). There can also be appearance of *diffuse pigmentation* of skin and/or *pigmentation of mucosa* (e.g., on tongue, buccal mucosa) (Sengupta et al. 2008; GuhaMajumder 2008).

Arsenical keratosis mostly affects the palms and soles, though in long standing cases can affect other areas, including trunk and dorsa of feet or hands (Fig. 3). The keratosis is symmetric and graded

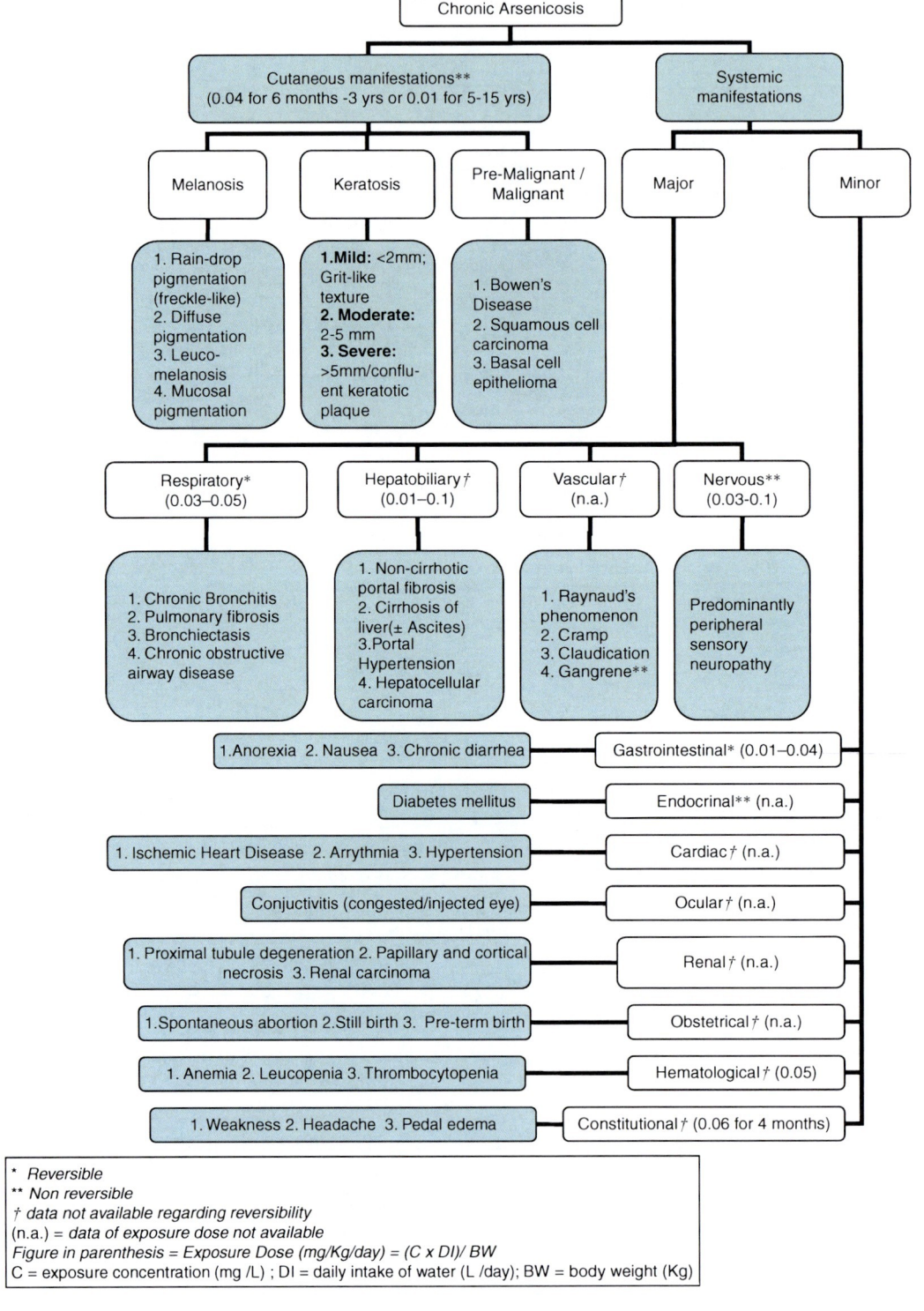

Arsenicosis, Fig. 1 Clinical manifestations in arsenicosis (Source: Sengupta et al. (2008) (Permission obtained from editor-in-chief IJDVL))

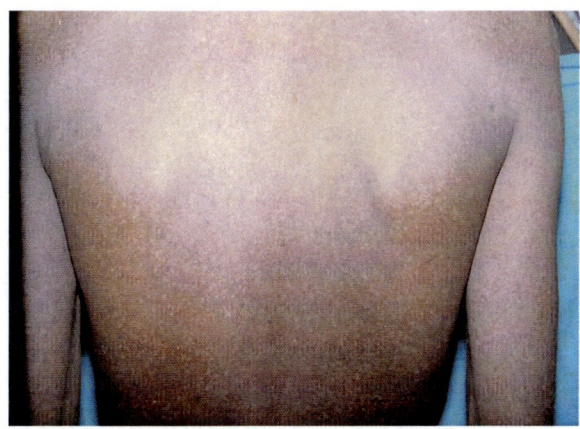

Arsenicosis, Fig. 2 Leucomelanosis affecting the trunk (Source: Sengupta et al. (2008) (Permission obtained from editor-in-chief IJDVL))

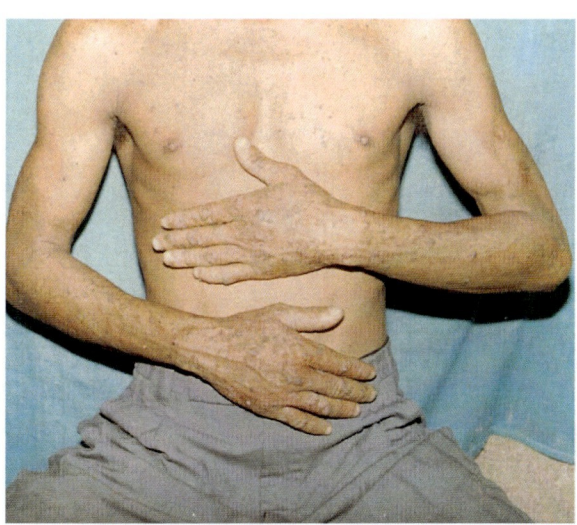

Arsenicosis, Fig. 3 Raindrop pigmentation on the trunk, along with arsenical hyperkeratosis affecting the dorsa of hands, fingers, and forearm (Source: Das and Sengupta (2008) (Permission obtained from editor-in-chief IJDVL))

according to its severity and extent into mild (grit-like texture or papule less than 2 mm) (Fig. 4), moderate (wart-like excrescence of size 2–5 mm) (Fig. 5), or severe (keratotic elevation of more than 5 mm or diffuse and confluent keratosis) (Fig. 6). Arsenical keratosis is often the forerunner of cutaneous malignancy and sometimes regarded as early clinical marker of carcinogenicity of internal organs (especially carcinoma of lungs and bladder) (Sengupta et al. 2008; GuhaMajumder 2008).

Carcinogenicity

Arsenic-induced carcinogenicity mostly affects the skin arising usually on keratotic lesions, though it may arise on apparently normal skin too. There is a dose-response relationship between skin cancers and arsenic consumption with a lifetime risk of developing skin cancer ranging from 1 to 2 per 1,000 with the intake of 1-μg/Kg body weight/day (approximately 1 l of water with arsenic concentration of 50 μg/L) (Smith et al. 2000). Bowen's disease is the most common form of precancerous lesions, and any hyperkeratotic plaque in arsenicosis suspect showing fissures and crusts at places should raise the suspicion for this premalignant form. *Multicentric Bowen's disease* affecting both covered and sun-exposed skin is a characteristic feature of arsenicosis (Fig. 7). Untreated Bowen's disease can transform into *squamous cell carcinoma* (Fig. 7) and can contribute significantly to the morbidity and mortality associated with the condition. *Basal-cell carcinoma* (Fig. 8) may also appear as a result of arsenicosis as well as other less-reported malignancies including *Merkel cell carcinoma* and *melanoma* (Sengupta et al. 2008; GuhaMajumder 2008).

Cancer risk for arsenicosis patients is not limited to skin malignancies, and there are reports of *visceral malignancies* including carcinoma of *liver, lungs, kidney, and urinary bladder*. There are also reports of malignancies arising from prostrate, uterus, and lymphatic tissues in cases of arsenicosis (Sengupta et al. 2008). It has been estimated that lifetime risk of dying from internal malignancy is high (around 13 per 1,000 on consumption of 1 l of water with arsenic concentration of 50 μg/L) than that of skin cancers because the later is detected early and not fatal if treated adequately (Smith, Lingus and Rahman, Smith et al. 2000). Based on the dose-response relationship, it is suggestive that if the patients with skin manifestations continue to drink arsenic contaminated water, the risk of mortality due to internal malignancies would increase when sufficient latency has been reached. Arsenic is regarded as class I human carcinogen by the International Agency for Research on Cancer (International Agency for Research on Cancer 2004).

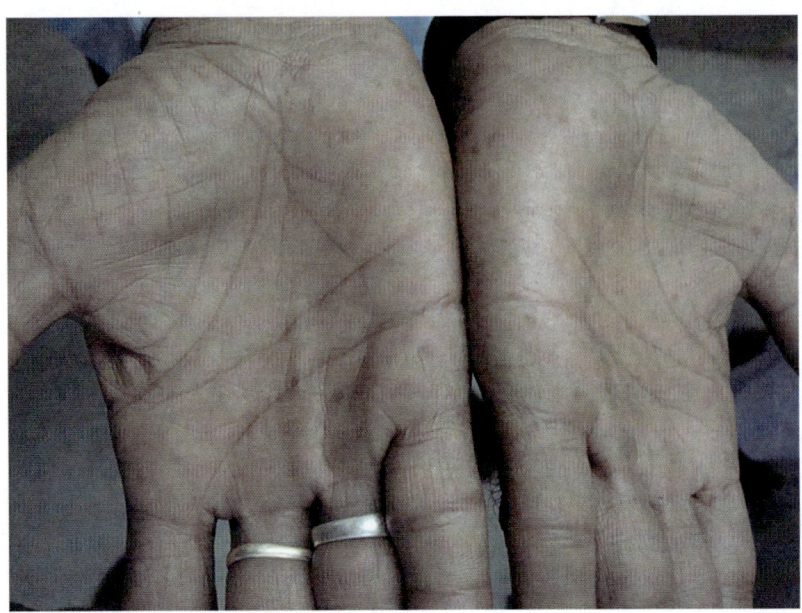

Arsenicosis, Fig. 4 Mild variety of arsenical keratosis producing grit-like texture

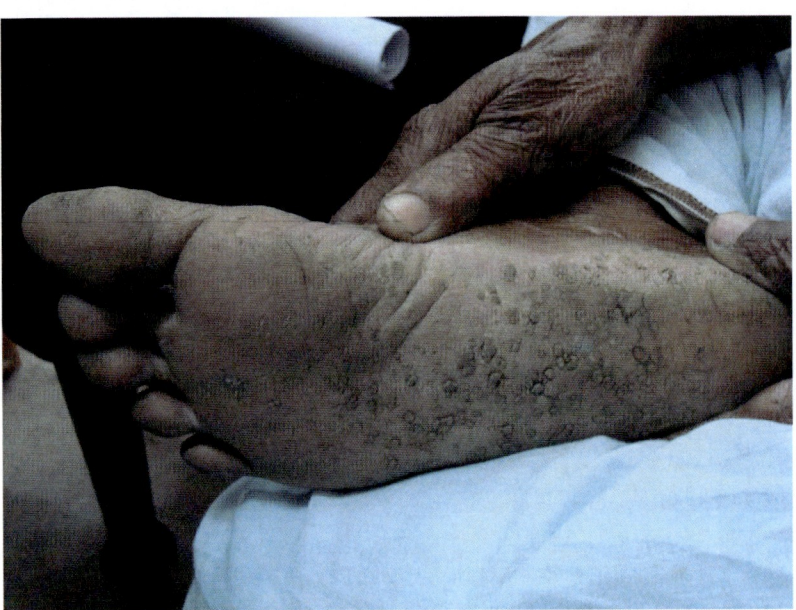

Arsenicosis, Fig. 5 Moderate variety of arsenical keratosis

Respiratory System

Respiratory symptoms include *chronic cough* and *respiratory distress*, which may result from *restrictive lung disease, obstructive lung disease, combination of both, interstitial lung disease, bronchitis,* or *bronchiectasis*. Crepitation and rhonchi are found on auscultation, and the pulmonary function test shows reduced forced expiratory volume in 1 s (FEV_1) and forced vital capacity (FVC) (Sengupta et al. 2008; GuhaMajumder 2008).

Hepatobiliary System

Arsenicosis produces hepatocellular toxicity mostly the form of *non-cirrhotic portal fibrosis* associated with *portal hypertension*, though *cirrhosis of liver* is rarely reported too. The portal fibrosis is characterized

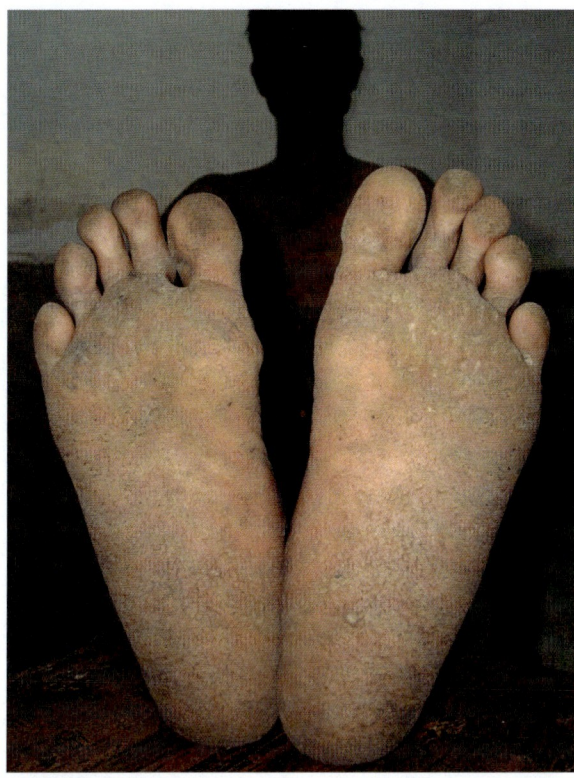

Arsenicosis, Fig. 6 Severe variety of arsenical keratosis (Source: Sengupta et al. (2008) (Permission obtained from editor-in-chief IJDVL))

by expansion of portal zones with streak fibrosis containing leash of blood vessels. Prevalence of *hepatomegaly* is found to be significantly higher in patients with arsenicosis, and degree of liver enlargement shows a dose-response relationship (Sengupta et al. 2008; GuhaMajumder 2008).

Cardiovascular System

Vascular effects are mostly seen in southwest coast of Taiwan, where *Blackfoot disease* is endemic. This is characterized by dry gangrene and spontaneous amputation resulting from severe systemic arteriosclerosis. Signs of ischemia, including intermittent claudication, rest pain, and ischemic neuropathy, along with diminished or absent arterial pulsation, pallor on elevation, and rubor on dependency, are diagnostic of blackfoot disease. Histology shows either arteriosclerosis obliterans or thromboangiitis obliterans in this unique peripheral vascular disease.

Apart from peripheral vasculature, chronic arsenic poisoning can also affect cardio- and cerebrovascular systems, leading to *hypertension, ischemic heart disease, cerebrovascular disease*, and *carotid atherosclerosis* (Sengupta et al. 2008; GuhaMajumder 2008).

Neurological System

Peripheral sensory neuropathy is characteristic of arsenicosis and is manifested by paraesthesia (e.g., burning and tingling sensation), numbness, pain, etc. Motor involvement is rare and if occurs, can result in distal limb weakness and atrophy and diminished or absent tendon reflexes. The neuropathy is due to distal axonopathy with axonal degeneration, especially the large myelinated fibers. Decreased nerve conduction amplitude is found in arsenic-induced sensory neuropathy with little/no change in nerve conduction velocity and visual/auditory evoked potential (Sengupta et al. 2008; GuhaMajumder 2008).

Other Systems

Arsenicosis can give rise to gastrointestinal symptoms in the form of *dyspepsia, anorexia, nausea, vomiting, diarrhea*, and *abdominal pain*. Anuria and dysuria are also reported in arsenicosis, arising due to *renal tubular necrosis*. Potential of arsenic to cross the placental barrier is thought to be the reason behind bad obstetrical outcome with increased incidence of *spontaneous abortion, still birth, preterm birth*, and *infant mortality* that are reported with arsenicosis. Anemia, *leucopenia*, and *thrombocytopenia* are usual accompaniment of arsenicosis and are the result of decreased erythropoiesis. There are also reports of *diabetes mellitus, conjunctival congestion, and non-pitting edema of feet* in arsenicosis, apart from constitutional symptoms including *headache and weakness* (Sengupta et al. 2008; GuhaMajumder 2008).

Diagnosis

Diagnosis of arsenicosis relies both on clinical manifestations (as detailed above) and laboratory estimation of arsenic content in drinking water and biological samples (Table 4). In countries where the laboratory backup cannot be made available in field situation, the cases can be clinically confirmed by virtue of its characteristic clinical features. WHO has proposed an algorithmic approach for diagnosing arsenicosis (Fig. 9) with the locally available resources. This approach has shown to have both the sensitivity and specificity of

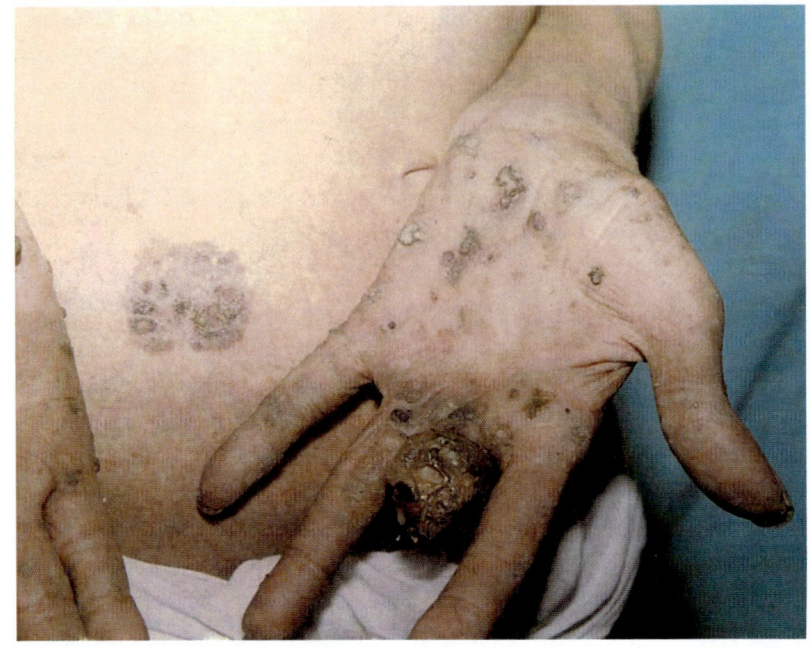

Arsenicosis, Fig. 7 Squamous cell carcinoma affecting the hands, with Bowen's disease on the abdomen same patient with arsenical hyperkeratosis affecting the palm and forearm (Source: Sengupta et al. (2008) (Permission obtained from editor-in-chief IJDVL))

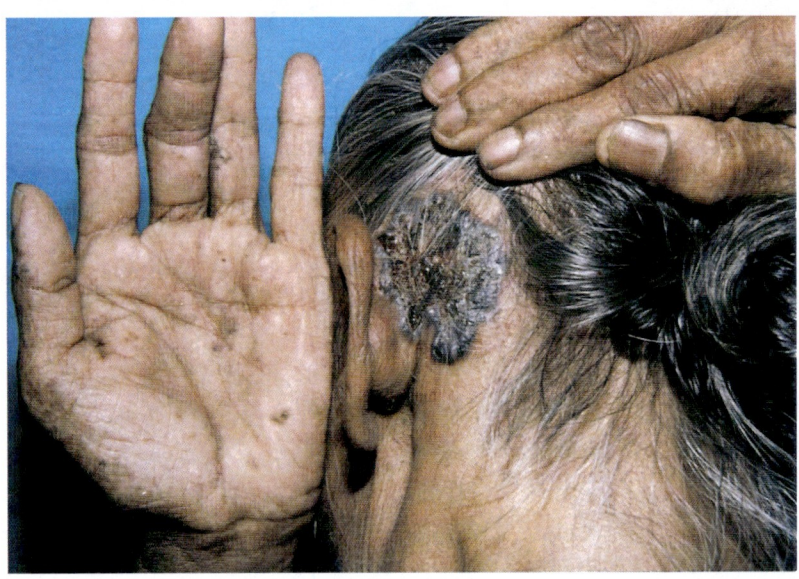

Arsenicosis, Fig. 8 Basal-cell carcinoma behind the ear, with arsenical keratosis on the palms (Source: Sengupta et al. (2008) (Permission obtained from editor-in-chief IJDVL))

more than 80% (WHO Regional Office for South-East Asia 2003; Das and Sengupta 2008).

High index of suspicion for arsenicosis is important while assessing persons hailing from areas documented to have high-arsenic content in drinking water, especially if history of similar skin lesions in persons (family members or neighbors) drinking water from the same source. At the same time, it is also important to thoroughly examine the suspected/probable cases to rule out clinical mimickers of arsenicosis (Table 5) (Das and Sengupta 2008).

Determination of Arsenic Content in Water and Biological Samples (Hair, Nail, and Urine)

Atomic absorption spectrometry is considered as the standard reference methods ("gold standard") for determination of arsenic content due to its high

Arsenicosis, Table 4 Characteristic clinical and laboratory criteria for diagnosis of arsenicosis

Clinical (cutaneous) manifestations		
1. Melanosis	(a) Fine-freckled or spotted pigmentation (raindrop pigmentation)	
	(b) Diffuse or generalized hyperpigmentation	
	(c) Guttate hypopigmentation on normal or hyperpigmented background (leucomelanosis)	
	(d) Mucosal pigmentation (esp. oral mucosa)	
2. Keratosis	(a) Mild: Minute papules (<2 mm) with slight thickening of palms and soles associated with grit-like texture detected primarily by palpation	
	(b) Moderate: Multiple keratotic papules (2–5 mm) present symmetrically on palms and soles	
	(c) Severe: Large discrete papule (>5 mm) or confluent keratotic elevation with nodular, wart-like, or horny appearance	
3. Malignant/ premalignant lesions	(a) Bowen's disease (squamous cell carcinoma in situ): multicentric Bowen's disease in non sun-exposed areas	
	(b) Squamous cell carcinoma and basal-cell epithelioma	
Laboratory criteria and collection method for establishment of exposure to arsenic		
1. Water (for at least 6 months)	(a) WHO guideline: >10 µg/L (also expressed as 10 parts per billion)	
	(b) Indian standard: >50 µg/L (also expressed as 50 parts per billion)	
	Collection method:	
	50 ml of sample in acid washed plastic container, which is completely filled to avoid oxidation from air in the bottle	
	Samples thus collected can be stored in room temperature for a week, at -20°C for 6 months and at -80°C for longer periods	
2. Hair & Nail (evidence of past exposure within last 9 months)	(a) Hair: >1 mg/kg of dry hair	
	Collection method:	
	Collected after washing with arsenic-free shampoo and ensuring that it is free of coloring agent containing arsenic	
	For females, 30 hairs of at least 6-cm length and for males, 60 short hairs from the base	
	(b) Nail: >1.5 mg/kg of nail	
	Collection method:	
	Clipping every finger and toenail after they have grown for 1 month	
	Both hair and nail shipped at room temperature to the laboratory, where it can be stored at 4°C till tested	
3. Urine (evidence of recent exposure)	>50 µg/L in subjects who has not consumed seafood in the last 4 days	
	Collection method:	
	Same as for water. Furthermore, concentrated hydrochloric acid (1 in 100 ml of urine) is added to prevent the bacterial growth in urine, to it	

Source: Das and Sengupta (2008) (Permission obtained from editor-in-chief IJDVL)

sensitivity and specificity. Other methods of arsenic estimation including colorimetric, inductive-coupled plasma, radiochemical methods, voltammetry, x-ray spectroscopy, hyphenated techniques, etc. are also described, but they suffer the inadequacy of being semiquantitative methods with low sensitivity. The development for *test kits for arsenic detection* is a much felt need for use in field situations, but unfortunately, none of the presently available test kit has proven reliable (Das and Sengupta 2008).

Maximum permissible limit for arsenic content in drinking water was reduced to 10 from 50 µg/L in 1993 by WHO. It has to be kept in mind that this limit is set depending on the analytical capability of available infrastructure to detect arsenic, and this limit would have been much lower if standard basis of risk assessment for industrial chemical is followed for arsenic. Though many countries, like United States and Japan, have adopted the new standard, developing nations which are worst affected by arsenicosis, like India and Bangladesh, are still operating under the national water quality standard of 50 µg/L due to inadequate testing facility (Smedley and Kinniburgh, 2002). This is making the residents more vulnerable to

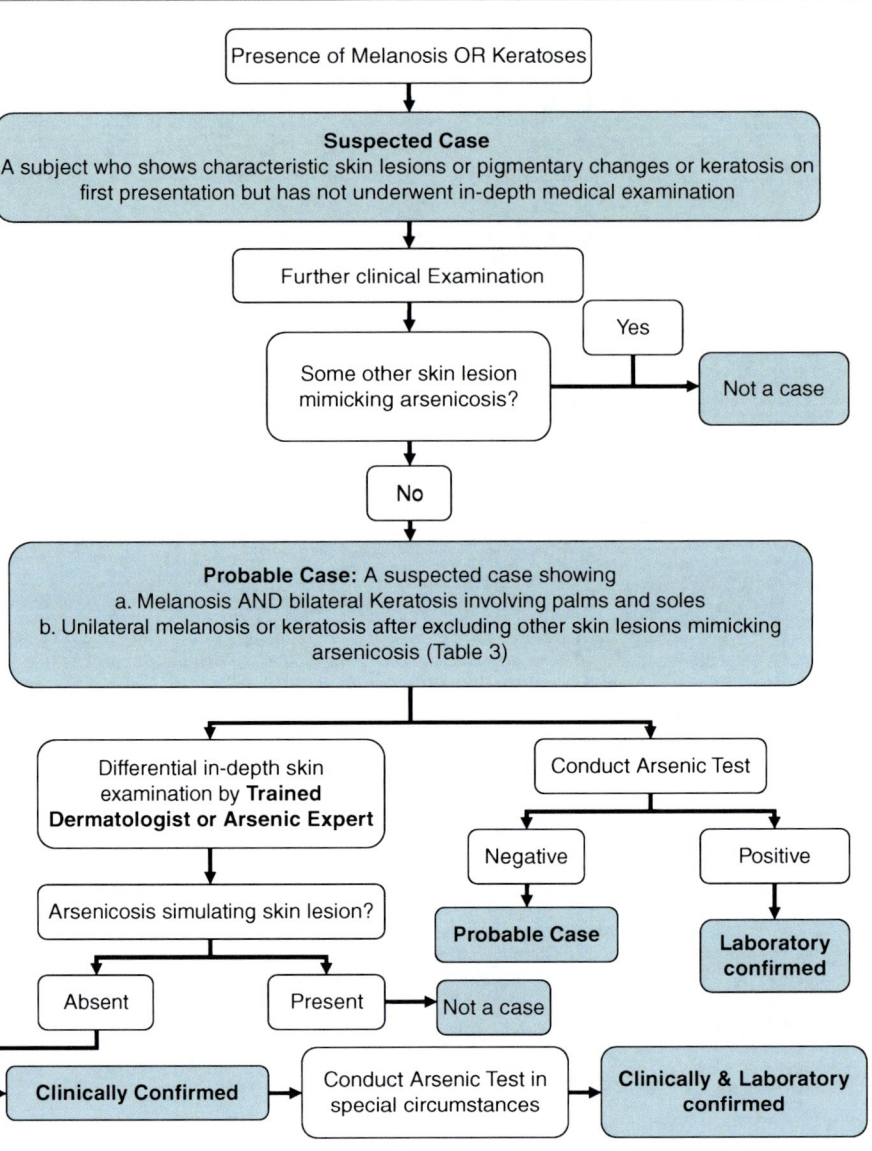

Arsenicosis, Fig. 9 Algorithmic approach toward diagnosis of arsenicosis (Adopted with modification from World Health Organization, regional office of Southeast Asia, A Field Guide for Detection, Management and Surveillance of Arsenicosis Cases; Editor, Deoraj Caussy) (Source: Das and Sengupta (2008) (Permission obtained from editor-in-chief IJDVL))

develop the disease by making them drink unsafe water under the false impression of safe water.

Newer Biomarkers of Arsenic Exposure

Biomarkers that reflect the exposure to arsenic in preclinical stage are being explored to take remedial measures before full-blown expression of disease has occurred. In this respect, raised *urinary uroporphyrin-III* and *coproporphyrin-III* have shown promise to detect arsenic exposure. Lowered *metallothionein* (metal-binding protein that protects against metal intoxication) in blood and tissue has also shown association in arsenic-exposed group and can be utilized in future (Das and Sengupta 2008).

Histology as Diagnostic Clues

Histology is invaluable in diagnosing the malignant changes and for early detection of premalignant lesions. *Dysplastic keratinocytes*, with presence of *anisocytosis* and *mitotic figures*, and *dyskeratosis* are important histological features of malignant changes (Fig. 10). Loss of cellular architecture and polarity with wind-blown appearance and intact basement membrane suggests Bowen's disease. *Hyperkeratosis, parakeratosis, acanthosis, and papillomatosis* are features of keratotic lesions but are not specific for arsenical keratosis. It is important to evaluate the histology of keratotic lesions because of their malignant potential. Histology of pigmented lesions has fallen out of

Arsenicosis, Table 5 Clinical mimickers of arsenicosis

Clinical types	Other dermatological conditions to be differentiated
Melanosis	
Spotted(raindrop)	Pityriasis versicolor, freckels, lentigines, telangiectasia macularis eruptiva perstans
Diffuse/generalized	Photopigmentation, ashy-grey dermatosis, lichen planus pigmentosus, Addison's disease, hemochromatosis
Leucomelanosis	Pityriasis versicolor, epidermodysplasia verruciformis, pityriasis lichenoides chronica, xeroderma pigmentosa, pigmented xerodermoid, idiopathic guttate hypomelanosis, leprosy, PKDL, salt and pepper pigmentation in scleroderma
Mucosal	Drug induced, Peutz-Jeghers syndrome, racial pigmentation, pigmented nevus, Addison's disease
Keratoses	
Punctate keratoses (on palms, soles, as well as on body)	Verruca vulgaris, corn/calluses, pitted keratolysis, occupational keratoses, seborrheic keratoses, lichen amyloidoses, lichen planus hypertrophicus
Diffuse keratoses (on palms and soles)	Psoriasis, pityriasis rubra pilaris, occupational keratoses, hyperkeratotic eczema, hereditary keratoderma, tinea pedis, mechanic's hand sign in dermatomyositis

Source: Das and Sengupta (2008) (Permission obtained from editor-in-chief IJDVL)

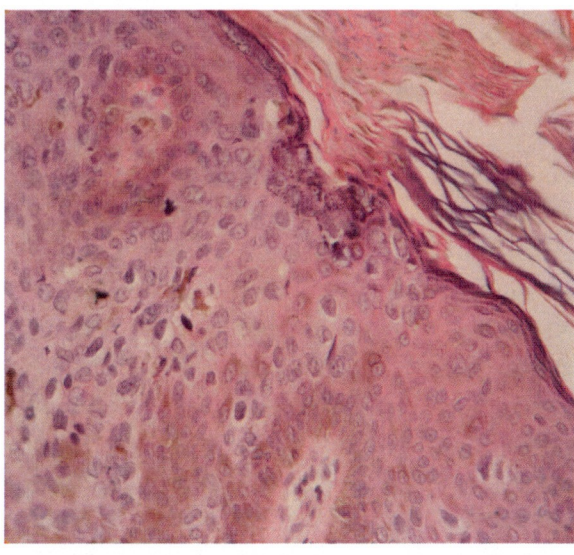

Arsenicosis, Fig. 10 Parakeratotic hyperkeratosis with focal hypergranulosis, increased basal layer pigmentation, and early dysplastic changes (H&E stain, × 400) (Source: Sengupta et al. (2008) (Permission obtained from editor-in-chief IJDVL))

grace because of their nonmalignant potential (Sengupta et al. 2008).

Strategies for Control of Arsenicosis

Basic principle for arsenic mitigation program should include the following (WHO Regional consultation meeting 1997):

- Identification of the affected people
- Provision of symptomatic management including nutritional supplement
- Medical management of seriously ill patients in health centers
- Improved and rapid diagnostic facility in the regional centers
- Provision of safe drinking water
- Capacity building of health care providers
- Awareness generation in susceptible community
- Implement comprehensive and integrated studies for better alternatives

In addition to these, comprehensive program implementation plan, evaluation of existing technologies, research and development for newer approaches, networking with stakeholders, and better convergence should be part of long-term policy.

Provision of Safe Drinking Water

Of all sources of arsenic exposure to human beings, drinking water remains the most crucial one. Safe drinking water sources either can be identified (*rainwater harvesting/surface water/safe aquifers*, etc.) or engineered (*arsenic treatment units/arsenic removal plants*) (Table 6) (Ghosh et al. 2008; Yamamura et al. 2000).

It is technically feasible to achieve arsenic concentrations of 5 μg/L or lower using any of several possible treatment methods. However, this requires

Arsenicosis, Table 6 Details of the methods for obtaining arsenic-free safe water

Source of water	Methods	Technical details	Countries adopting the method
Surface water	Pond sand filter	Uses bed of fine sand (porous medium) through which water slowly percolates downward	Myanmar
		The pond should not be used for other purposes, e.g., bathing, washing, and fishing	
	Rainwater harvesting	Rainwater collected using sheet material rooftop or plastic sheet → then diverted to a storage container	Bangladesh, Cambodia, and Taiwan
		Cannot be used in areas with a prolonged dry season	
	Piped water supply of surface water after simple treatment	Treatment needed to reduce turbidity, chlorination to protect against bacteriological contamination of surface water	Bangladesh and India
Groundwater	Dug well	Wells are excavated below the water table till the incoming water exceeds the digger's bailing rate → lined with stones, bricks, or tiles to prevent collapse or contamination	Bangladesh, Nepal, and Myanmar
	Deep tube well (from safe aquifer)	Groundwater is mapped to identify at which depth arsenic-safe groundwater is located, i.e., shallow groundwater is separated from the deep groundwater by clay layer	Bangladesh, India, and Nepal
		Monitoring is important because arsenic may percolate into deep aquifer in future	
	Arsenic removal technologies	Basic principles of arsenic removal from water are based on conventional techniques of oxidation, coprecipitation and adsorption on coagulated flocs, adsorption onto sorptive media, ion exchange, and membrane filtration	1. Community arsenic removal plants: Bangladesh, India, Vietnam, and Taiwan
		Routine monitoring of arsenic safety of the water is needed	2. Household arsenic removal plants: Bangladesh, Nepal, Vietnam, and India
		If the source is surface water, bacteriological checkup also needs to be performed	

Source: Ghosh et al. (2008) (Permission obtained from editor-in-chief IJDVL)

careful process optimization and control, and a more reasonable expectation is that 10 μg/L should be achievable by conventional treatment (e.g., coagulation) (Ghosh et al. 2008; Yamamura et al. 2000).

The ideal solution is to use alternative sources of water that are low in arsenic. However, it is important that this does not result in *risk substitution*. Especially tropical countries most affected by arsenicosis area also bear the major burden of gastrointestinal disease. A large group of population is exposed to water source contaminated with enteric pathogen. Water safety frameworks should be used during planning, installation, and management of all new water points, especially ones based on surface water and very shallow groundwater, to minimize risks from fecal and other non-arsenic contamination. Screening for arsenic and other possible chemical contaminants of concern that can cause problems with health or acceptability, including fluoride, nitrate, iron, and manganese, is also important to ensure that new sources are acceptable. Periodic screening may also be required after a source is established to ensure that it remains safe (Ghosh et al. 2008; Yamamura et al. 2000; Johnston et al. 2001).

Many of the major problems lie in rural areas, where there are many small supplies, sometimes to the household level. At this level, water availability and financial and technical resources are all limited. There are several available approaches, but a basic requirement for education can never be overemphasized. In particular, there is a need to understand the risks of high-arsenic exposure and the sources of arsenic exposure.

To achieve safe arsenic concentration in water for human consumption, a number of approaches have been successfully used in rural areas, including *source substitution* and the use of both high- and low-arsenic sources blended together. *Pond water supply, piped water supply of surface water, rainwater harvesting, deep tube well from safe aquifers, and community-/household-based arsenic removal plants* are there as alternatives. Best suited one or combination of these methods has been tried in different parts of the world with variable success. These sources may be used to provide drinking water and cooking water or to provide water for irrigation. High-arsenic water can still be used for bathing and clothes washing or other requirements that do not result in contamination of food, in situation where no better alternative is available (Ghosh et al. 2008; Ahmed et al. n.d.).

Arsenic removal process can be of four basic types: *Oxidation sedimentation, coagulation filtration, sorption techniques*, and *membrane techniques*. Decision for selecting the most suitable alternative depends upon effectiveness of the process in removing arsenic, installation and running cost, water output, etc. A report revealed that most cost effective method for long-term solution is piped water supply, whereas for medium-term solution at community level, oxidation and filtration or coagulation filtration are better alternatives. For household purpose, iron-filings/iron-coated sand filters are preferable choices for getting arsenic-free water (Ghosh et al. 2008; Ahmed et al. n.d.).

Health Communication

General community should be made aware of relevant health effects of arsenic poisoning, available alternatives for health protection, safe water options, testing facilities for detection of arsenic contamination, dispelling myths, and misconception about the disease. Communication should be a continuous process, and ultimate aim is to involve the community in planning, implementing, and evaluating arsenicosis control program at the local level to make it a sustainable one. Translational research should be carried out to bring the fruits of scientific research and development to the doorsteps of those who need it most (Galway n.d.).

Treatment of Arsenicosis

The treatment of arsenicosis is essentially symptomatic since no effective remedial measures are known till date. The symptomatic relief that can be offered is also very limited because most of the changes are irreversible and for some clinical manifestations even supportive therapy are not available.

Hyperkeratosis can be treated to some extent by using different keratolytic agents. Five to ten percent salicylic acid or 10–20% urea is recommended by WHO for treatment of arsenical keratosis. Retinoids are also tried in arsenicosis, but their use is limited by their cost and side effect profile. The *malignant and premalignant lesions* are to be treated by surgical excision at the earliest (Das and Sengupta 2008).

Dyspeptic symptoms are treated by H_2 blockers or proton pump inhibitors along with prokinetic agents. For manifestations of portal hypertension, like *esophageal varices*, sclerotherapy or banding may be done. *Peripheral vascular diseases* can be managed by vasodilator drugs (e.g., pentoxifylline or calcium channel blocker) with limited success. Tricyclic antidepressant (e.g., amitriptyline) is sometimes used for relieving the painful dysesthesia of arsenical *peripheral neuropathy*. Bronchodilators are used for controlling symptoms of *obstructive lung diseases*; also, it is important to avoid smoking and dusty environment (Das and Sengupta 2008).

Role of *chelation therapy* using dimercaptosuccinic acid (DMSA), dimercaptopropane succinate (DMPS), or D-penicillamine has not shown any favorable reports in recent studies. Presently, they are not considered for the treatment in spite of early success stories.

In arsenicosis, it is important to be vigilant to detect malignancies (either cutaneous or internal) at the earliest so as to offer the benefit of early intervention. Hence, constant follow-up and surveillance is needed for all who reside in arsenic contaminated zone.

Cross-References

▶ Arsenic in Nature
▶ Arsenic in Therapy
▶ Arsenic Methyltransferases
▶ Arsenic, Free Radical and Oxidative Stress
▶ Arsenic-Induced Diabetes Mellitus

References

Ahmed F, Minnatullah K, Talbi A (n.d.) Arsenic contamination of ground water in South & East Asian countries [pdf]. Technical report, Paper 3, Arsenic mitigation technologies in South & East Asia. World Bank document, vol II. Available at http://siteresources.worldbank.org/INTSAREGTOPWATRES/Resources/ArsenicVolII_PaperIII.pdf. Accessed 15 Apr 2012

Appelo T (2006) Arsenic in ground water – a world problem [pdf]. Netherlands national committee of the IAH. Available at http://www.iah.org/downloads/occpub/arsenic_gw.pdf. Accessed 15 Apr 2012

Das NK, Sengupta SR (2008) Arsenicosis: diagnosis and treatment. Ind J Dermatol Venereol Leprol 74(6):571–581

Galway M (n.d.) Communication for development [pdf]. World Health Organization. Available at http://www.who.int/water_sanitation_health/dwq/arsenicun7.pdf. Accessed 15 Apr 2012

Ghosh P, Roy C, Das NK, Sengupta SR (2008) Epidemiology and prevention of chronic arsenicosis: an Indian perspective. Ind J Dermatol Venereol Leprol 74(6):582–593

Gomez-Caminero A et al (2001a) Kinetics and metabolism in laboratory animals and human. Arsenic and arsenic compound. Environmental health criteria 224. International programme on chemical safety. World Health Organization, Geneva, pp 112–167

Gomez-Caminero A et al (2001b) Effects on laboratory mammals and in vitro test systems. Environmental health criteria 224. International programme on chemical safety. World Health Organization, Geneva, pp 168–233

GuhaMajumder DN (2008) Chronic arsenic toxicity & human health. Indian J Med Res 128:436–447

International Agency for Research on Cancer (2004) Some drinking-water disinfectants and contaminants, including arsenic. IARC monographs on the evaluation of carcinogenic risks to humans, vol 84. Available at http://monographs.iarc.fr/ENG/Monographs/vol84/mono84-1.pdf. Accessed 15 Apr 2012

Johnston R, Heijnen H, Wurzel P (2001) Safe water technology [pdf]. World Health Organization. Available at http://www.who.int/water_sanitation_health/dwq/arsenicun6.pdf. Accessed 15 Apr 2012

Kapaj S, Ptereson H, Liber K, Bhattacharya P (2006) Human health effects from chronic arsenic poisoning: a review. J Environ Sci Health 41:2399–2428

Sengupta SR, Das NK, Datta PK (2008) Pathogenesis, clinical features and pathology of chronic arsenicosis. Ind J Dermatol Venereol Leprol 74(6):559–570

Smedley PL, Kinniburgh DG (2002) Source and behaviour of arsenic in natural waters [pdf]. World Health Organization. Available at http://www.who.int/water_sanitation_health/dwq/arsenicun1.pdf. Accessed 15 Apr 2012

Smith AH, Lingus EO, Rahman M (2000) Contamination of drinking water by Arsenic in Bangladesh: a public health emergency. Bull WHO 78(9):1093–1103

WHO Regional Consultation Meeting (1997) Arsenic in drinking water and resulting arsenic toxicity in India and Bangladesh: recommendation for action. WHO, New Delhi. Available at http://www.searo.who.int/LinkFiles/Health_Topics_recommend.pdf. Accessed 15 Apr 2012

WHO Regional Office for South-East Asia (2003) Arsenicosis case-detection, management and surveillance. Report of a regional consultation. WHO, New Delhi

World Bank technical report vol II. An overview of current operational responses to the arsenic issue in the South and East Asia. Arsenic contamination of ground water in South and East Asian countries. Available at http://siteresources.worldbank.org/INTSAREGTOPWATRES/Resources/ArsenicVolII_PaperII.pdf. Accessed 15 Apr 2012

Yamamura S et al (2000) Drinking water guidelines and standards [pdf]. World Health Organization. Available at http://www.who.int/water_sanitation_health/dwq/arsenicun5.pdf. Accessed 15 Apr 2012

Arsenism

▶ Arsenicosis

Arsenite Methyltransferase

▶ Arsenic Methyltransferases

Arsenite S-Adenosylmethionine Methyltransferase

▶ Arsenic Methyltransferases

Artificial Metalloenzyme

▶ Palladium, Coordination of Organometallic Complexes in Apoferritin

Artificial Metalloprotein

▶ Palladium, Coordination of Organometallic Complexes in Apoferritin

Artificial Selenoproteins

Quan Luo and Junqiu Liu
State Key Laboratory of Supramolecular Structure and Materials, College of Chemistry, Jilin University, Changchun, China

Synonyms

Incorporation of selenium into protein; Selenium-containing protein mimic

Definition

Through incorporation of selenium into existing protein scaffolds, artificial selenoproteins can be generated by chemical or biological strategies to mimic the behaviors of selenoproteins in natural processes. The implementation of biomimetic functions in these artificial systems will help to clarify the biological and biomedical effects of selenium and increase the understanding for the structures and functions of selenoproteins, which may lead to their valuable applications in human health.

Introduction

Selenium was discovered by the Swedish chemist Jöns Jacob Berzelius in 1817 and has been recognized as an essential micronutrient since 1957. However, selenium biochemistry really emerged after the bacterial enzymes formate dehydrogenase and glycine reductase were reported to contain selenium in 1973. At the same time, the biochemical role of selenium began to be established when it was verified as a vital part of the active site of the antioxidant enzyme glutathione peroxidase (Rotruck et al. 1973).

Selenium exerts its physiological functions in vivo mainly due to its presence in proteins as the 21st amino acid, selenocysteine (Sec, U). Unlike the other amino acids, Sec is encoded by the opal stop codon UGA and undergoes a very complicated biosynthetic pathway to be incorporated into protein by using its own tRNA^{Sec} to recognize UGA. Both cis- and trans-acting factors are involved in decoding UGA for Sec insertion rather than translation termination (Gladyshev and Hatfield 2010). Sec is structurally similar to cysteine (Cys), except that it contains selenium instead of sulfur. However, the lower pKa value (5.2) and stronger nucleophilicity of Sec make it much more reactive, resulting in the unique redox behavior.

Selenoproteins, a class of proteins that contain one or more Sec residues, exist in archaea, bacteria, and eukaryotes. So far, more than 40 selenoprotein families have been identified using genomic and bioinformatic approaches, however, with about half of them still having unknown functions. The majority of the characterized selenoproteins are involved in various redox reactions, and their Sec residue is a key functional group for biological functions by reversibly changing its redox state. Selenoproteins can be classified into three major groups (Gladyshev 2006). The largest selenoprotein group, including glutathione peroxidases (GPxs); thyroid hormone deiodinases (DIs); selenoproteins H, M, N, T, V, and W; Sep15; and selenophosphate synthetase 2 (SPS2), contains Sec in the N-terminal regions and in most cases employs

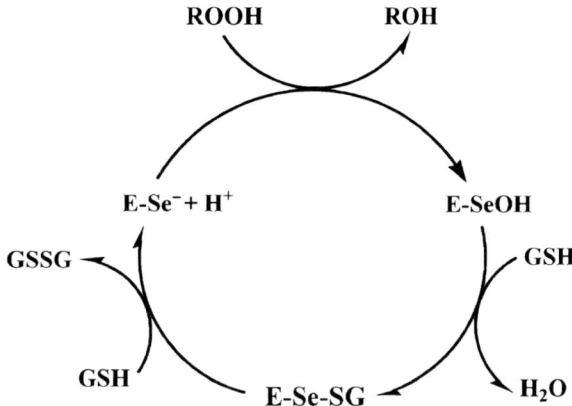

Artificial Selenoproteins, Fig. 1 The catalytic mechanism of GPx

a similar thioredoxin folds. The second selenoprotein group has Sec located in C-terminal sequences, such as thioredoxin reductases (TRs) and selenoproteins S, R, O, I, and K. Selenoproteins in the third group utilize Sec to coordinate a redox-active metal in the active sites. Hydrogenase, formate dehydrogenase (FDH), and formylmethanofuran dehydrogenase are examples of such proteins, which are found only in prokaryotes. Among these selenoproteins, the former two groups of mammalian selenoproteins are of particular interest and importance. In vitro and in vivo studies indicate that selenoproteins in mammals are important members of enzymes, and their expression or activity can influence the risk of a range of diseases, such as cancer, Keshan disease, virus infection, male infertility, impaired immunity, and so on (Lu and Holmgren 2009).

GPx is the first known mammalian selenoprotein and also perhaps the best studied. It is a highly efficient antioxidant enzyme that catalyzes the reduction of hydroperoxides, including hydrogen peroxides, by reducing glutathione (GSH) and functions to protect the cell from oxidative damage (Fig. 1). Over the past few decades, considerable effort has been devoted to mimic the behaviors of selenium in GPx, and a great number of artificial GPx models (e.g., ebselen, ebselen analogs, selenenamides, diselenides, α-phenylselenoketones, selenium-containing cyclodextrins, polymers, proteins) were reported to exhibit GPx-like functions (Huang et al. 2011; Liu et al. 2012). This has promoted researches in the well-known field, selenoprotein biomimetic chemistry, through the reasonable design and synthesis of a series of selenoprotein mimics, ranging from small molecular compounds to macromolecular ones. As a new development in this regard, the protein-based GPx mimics have attracted much attention. Various approaches have been employed to create such molecules by incorporation of Sec into proteins. These artificial selenoproteins show structural and functional properties similar to GPxs and exhibit excellent catalytic efficiencies and specificities rivaling natural ones. All studies aim at exploring the structure-function relationships of selenoproteins and finding their valuable applications in the treatment of human diseases. In this chapter, we summarize recent advances in artificial selenoproteins prepared by chemical or biological strategies, including semisynthetic selenoproteins, selenium-containing catalytic antibodies, bioimprinted selenoproteins, and genetically engineered selenoproteins. How to improve the properties of these artificial systems has also been described.

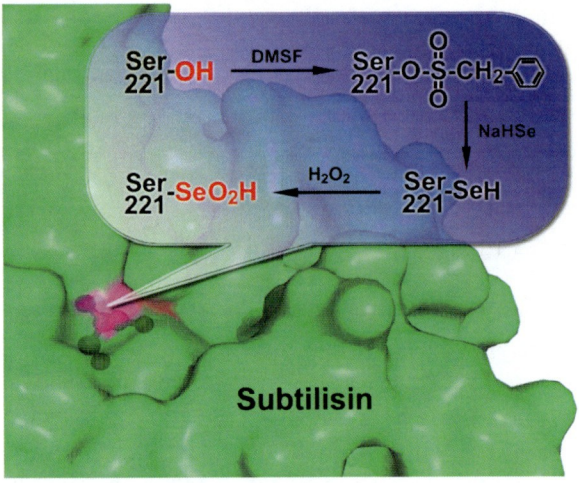

Artificial Selenoproteins, Fig. 2 The preparation of selenosubtilisin by site-specific chemical modification of the active site Ser221 of subtilisin to Sec

Semisynthetic Selenoproteins

Naturally existing proteins are ideal precursor molecules for the preparation of artificial selenoproteins, whereas the incorporation of Sec into proteins remains a challenging problem for protein engineering. In nature, this process is highly regulated, with the UGA stop codon being used to specify the insertion of Sec. Therefore, it is difficult to obtain selenoproteins by using traditional recombinant DNA technology. As an alternative strategy, semisynthesis is an attractive technology that utilizes well-defined proteins as preformed intermediates for the resynthesis of the protein to avoid much synthetic effort, and Sec incorporation can be achieved by the chemical modification of protein scaffolds at the desired position.

Site-specific modification of proteins was developed by the groups of Polgar and Bender, who created a new, active enzyme by chemical transformation of the active site serine residue of subtilisin to cysteine. In subsequent work, Yokosawa et al. successfully prepared a thiol-trypsin using a similar chemical method, and Clark et al. converted the active site thiol of papain into a hydroxyl group to obtain the hydroxyl-papain, as a complementary for the work mentioned above. Despite these semisynthetic enzymes proved to retain only limited catalytic activity with respect to the typical substrates of the wild-type enzymes, these studies are significant because the method that can be used to produce new semisynthetic enzymes with novel catalytic activities was established (Qi et al. 2001).

Inspired by earlier work on thiol-subtilisin, the first artificial selenoprotein was prepared by selective activation of Ser[221] with phenylmethanesulfonyl fluoride (PMSF) to form a sulfonylated subtilisin, which was reacted with NaSeH to produce selenosubtilisin by nucleophilic substitution (Fig. 2) (Wu and Hilvert 1990). The alteration of the active site of subtilisin yielded meaningful results that the semisynthetic selenoenzyme exhibited significant GPx-like redox activity and can catalyze the reduction of a variety of hydroperoxides by an aryl thiol. The rate for the reduction of tert-butyl hydroperoxide (t-BuOOH) in the presence of selenosubtilisin is over 70,000 times higher than that for diphenyl diselenide (PhSeSePh, a well-studied antioxidant) when using 3-carboxy-4-nitrobenzenthiol (TNB) as substrate. Kinetic investigation revealed that the selenosubtilisin-catalyzed reaction is a typical ping-pong mechanism, analogous to that described for natural GPx. These studies indicated that the redox activity and kinetics of selenosubtilisin can be attributed to the effect of an active site S/Se substitution. In addition, crystallographic analysis and substrate screening with selenosubtilisin suggested that this chemical modification did not disturb the overall structural integrity of the protein and the substrate selectivity of selenosubtilisin was similar to that of wild-type subtilisin, leading to numerous possibilities in the design and even prediction of its substrate selectivity. As a result, an enrichment of novel racemic alkyl aryl hydroperoxides were chemically synthesized and reported to be catalyzed by selenosubtilisin with high efficiency and selectivity, which may complement the set of naturally available biocatalysts for enantioselective synthesis.

Along a similar line, trypsin was chemically converted into selenotrypsin by replacement of the active serine with Sec. The modified enzyme exhibited considerable GPx activity and was found to have a second-order rate constant about 10^3-fold lower than GPx. This indicated that the chemical approach can be also applied in the site-specific modification of other serine proteases, and GSH is not a suitable substrate for selenotrypsin.

Although chemical modification of active-site residue dramatically changes the function of subtilisin and trypsin, the novel artificial selenoenzymes generated in this way are usually not optimally efficient when using GSH as the thiol substrate. To improve this problem, the active site of protein scaffolds should be modified and regulated to accommodate GSH. Recent years have seen some progress in the design of the artificial selenoenzymes following this routine. Luo and coworkers enhanced the GSH specificity of the semisynthetic selenoezyme by chemically modifying naturally occurring glutathione transferases (GSTs) (Huang et al. 2011). Taking advantage of the structure similarities between seleno-GST and GPx in the specific GSH-binding sites and the perfect location of Sec in the correct vicinity to reactive group of GSH, the resulting selenoenzymes displayed a significantly high efficiency for catalyzing the reduction of H_2O_2 by GSH. As examples, the GPx activity of rat theta-class-derived seleno-GST (Se-rGST T2-2) was found to be 102 $\mu mol \cdot min^{-1} \cdot mg^{-1}$ that surpasses the activities of some natural GPxs (e.g., rabbit liver GPx, bovine liver GPx, human hepatoma HepG 2 cell giGPx, and human plasma pGPx), and the GPx activity of a new modified human glutathione transferase zeta1-1 (Se-hGSTZ1-1) was 8,602 ± 32 U/μM, about 1.5 times higher than that of rabbit liver GPx (5,780 U/μM). These results demonstrated that the general principle of combining a functional group involved in catalysis with a specific binding site for the substrate could be applied to the generation of other efficient semisynthetic biocatalysts.

Selenium-Containing Catalytic Antibodies

Antibodies are immune system-related proteins that selectively recognize a large number of structurally diverse ligands, called antigens, with the goal of eliciting an immune response. Each antibody recognizes a specific antigen, similar to that of enzymes in specifically binding substrate molecules. The possibility of the induction of catalytic antibodies (abzyme) was initially suggested from Pauling's enzyme hypothesis (Nevinsky et al. 2000). He hypothesized that enzymes achieve catalysis by stabilizing the transition state through binding interactions. According to this hypothesis, if the antigen is a transition state analog (hapten) of a corresponding reaction, the binding energy provided by the antibody could stabilize the transition state and facilitate the transformation. This prediction was subsequently confirmed by Lerner and Schultz's groups when the first anti-hapten monoclonal catalytic antibodies were obtained. At present, the technology that induces antibody production by haptens was widely applied in the design of abzymes.

The first known attempt to produce a selenium-containing catalytic antibody (seleno-abzymes) was made by Ding et al. The commercially available anti-human IgM (Fcμ) (rabbit immunoglobulin fraction) was adopted for incorporation of the catalytic group via selective chemical modification of Ser into Sec (Ding et al. 1994). Due to the lack of substrate binding site, the modified antibody was confirmed to have relatively low GPx activity as compared with natural enzymes.

Subsequently, Luo et al. succeeded in the preparation of monoclonal antibodies (McAbs) with GSH-specific binding sites using a series of hydrophobically modified GSH and GSSG as haptens, and the catalytic group Sec was then incorporated into the McAb by the semisynthetic method mentioned above to generate GPx mimics (Fig. 3) (Huang et al. 2011). In this process, substrate analogs rather than transition state analogs were used as haptens, because the hapten design based on transition state structure does not usually work well, especially for complex chemical reactions with unknown transition states (such as GPx). Intriguingly, these catalytic antibodies exhibited remarkably high catalytic efficiency. The GPx activity of seleno-abzyme Se-4A4 was 1,097 U/μM, approaching the magnitude level of native enzyme activity. Moreover, three new selenium-containing murine catalytic antibodies, including

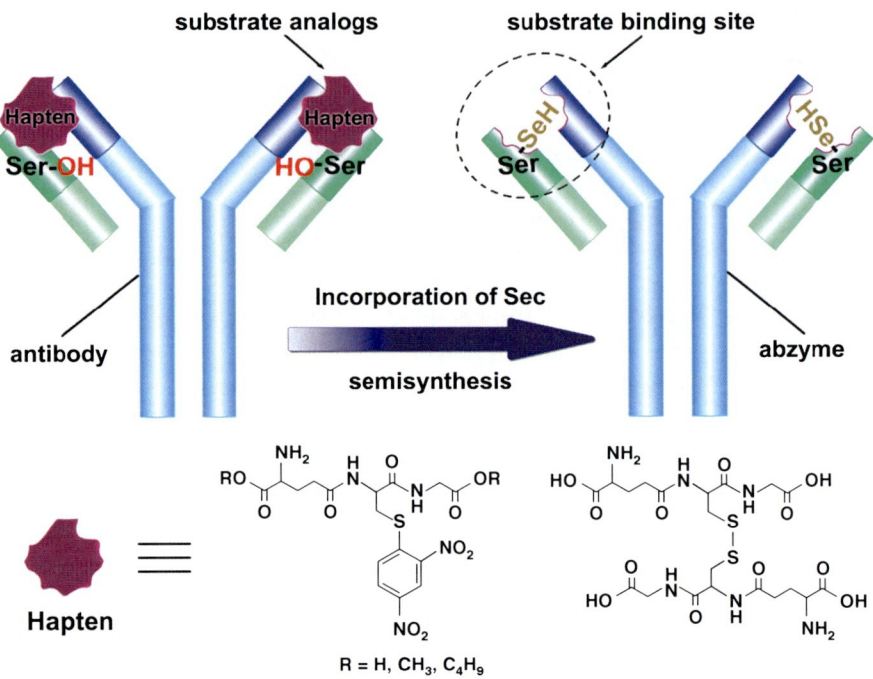

Artificial Selenoproteins, Fig. 3 The combination of monoclonal antibody preparation technique and semisynthetic method to generate selenium-containing catalytic antibodies with GPx activities

Se-3G5, Se-2F3, and Se-5C9, were reported to display remarkable activities (12,900, 24,300, and 21,900 U/μM, respectively) rivaling natural GPx from rabbit liver (Huang et al. 2011). These results demonstrated that an extremely high degree of precision in the match between the complementary structures of antibody binding sites and substrates is crucial for the efficient catalytic activities of artificial seleno-abzymes.

Recently, instead of an immunization protocol, the traditional genetic engineering technology has been explored as an efficient and more manipulative strategy to acquire single-chain variable region fragments of antibodies (scFv) for pharmacological use. Compared to McAb, scFv has a smaller size and less immunogenicity, without sacrificing its ability to bind antigen. Therefore, Ren et al. prepared the first selenium-containing single-chain abzyme (Se-2F3-scFv) from the 2F3 monoclonal antibody by successful construction of a high-level expression vector, and the produced seleno-abzyme shows GPx activity of 3,394 U/μM, which approaches that of rabbit liver GPx but was lower than that of Se-2F3 (Huang et al. 2011). The investigation of mitochondrial damage model induced by reactive oxygen species (ROS) indicated that Se-2F3-scFv can effectively inhibit the swelling of mitochondria, leading to new treatments for oxidative stress-related diseases.

To further improve the immunogenicity of seleno-abzyme, in vitro evolutionary method using a phage display system has been applied for protein engineering of human antibody fragments. As a result, several human antibodies B8, H6, C1, and B9 with the GSH-binding site were screened out for direct incorporation of Sec to form the corresponding seleno-abzymes (Liu et al. 2012), and most of them were detected to reach the level of native enzyme activity. Moreover, some human antibodies were also explored to enhance the GPx activity by the combination of computer-assisted modeling and site-directed mutagenesis. On the basis of modeling analysis, an Ala/Ser180 substitution at the binding site of Se-B3-scFv could greatly increase the enzyme-substrate interaction, resulting in 2.16 times enhancement in the GPx activity (Liu et al. 2012). It is suggested that attractive noncovalent interactions between enzyme and substrate can also play a key role in catalysis through the stabilization of substrate conformation within the active site.

Bioimprinted Selenoproteins

Molecular recognition is an important feature for enzyme-mediated catalysis, and the binding specificity

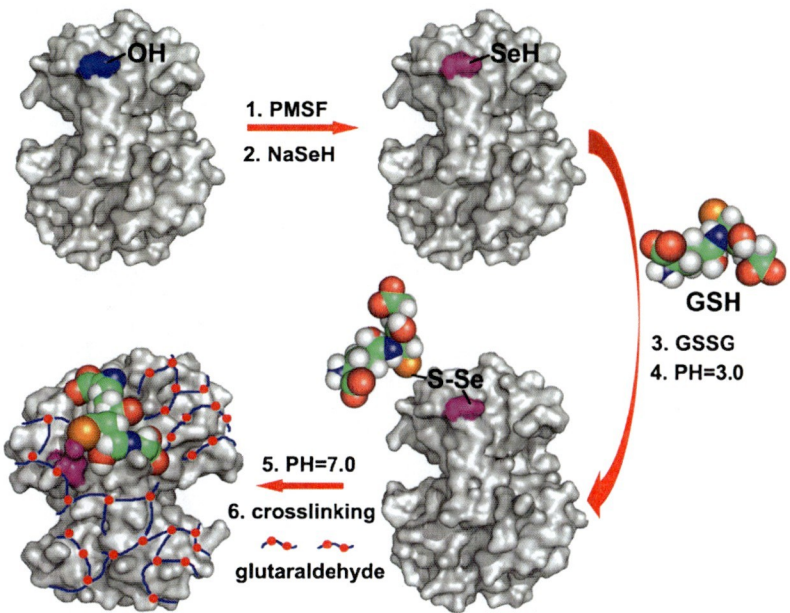

Artificial Selenoproteins, Fig. 4 Schematic representation of the production process of the GSH-imprinted selenoprotein

depends on attractive forces and complementary structures between enzyme and substrate. In the preparation of enzyme mimics, a number of recognition elements, both natural (e.g., antibodies) and synthetic (e.g., modified cyclodextrins), have been described to mimic the catalytic behaviors of enzymes, whereas these receptor molecules can only selectively bind their respective type of substrates. This means that each recognition problem requires a novel solution.

As an alternative strategy for the development of new and efficient catalysts, molecular imprinting is a more generic method that allows for the preparation of receptor scaffolds with specific binding sites for a target molecule through the copolymerization of functional and cross-linking monomers in the presence of a template molecule and followed by the removal of the template to yield a binding site with size and shape complementary to the target molecule in the cross-linked network polymer. Based on the same principle, the technique of bioimprinting has been developed using biopolymers for this imprinting procedure.

The design of bioimprinted proteins with GPx activity was proposed by Liu et al. N,S-bis-2,4-dinitrophenyl-glutathione (GSH-2DNP), a GSH derivative, was synthesized to imprint egg albumin (Liu et al. 2004). In this process, the structural change of egg albumin was induced by the noncovalent interactions (such as hydrogen bonds, ion pairing, and hydrophobic interactions) between protein and template molecule in mild denaturing condition. The new conformation was then fixed using a biofunctional cross-linker glutaraldehyde. After removal of the template by dialysis, an imprinted GPx enzyme model with template-shaped cavity capable of binding GSH was produced by incorporation of Sec using semisynthetic method, and its GPx activity was found to be 101–817 U/μM, approaching the level of selenium-containing catalytic antibody Se-4A4. These results proved the feasibility of bioimprinting to prepare an artificial recognition system with high selectivity toward target molecule for the design of selenoenzyme.

Later, the imprinting procedure was optimized by the same group using a covalent Se-S bond to link the template molecule GSH and semisynthesized selenosubtilisin to form GSH-selenosubtilisin (Fig. 4). The advantage of this improvement is that the catalytic group Sec is located in close proximity to the reactive thiol of the bound substrate GSH, a perfect position inside the active site to catalyze the reaction. Thus, the second-order rate constant of bioimprinted selenosubtilisin for H_2O_2 is larger than that of bioimprinted selenoalbumin (Liu et al. 2008). Moreover, the imprinted selenosubtilisin was found to have no activity against aromatic substrate TNB, suggesting the bioimprinting technique completely

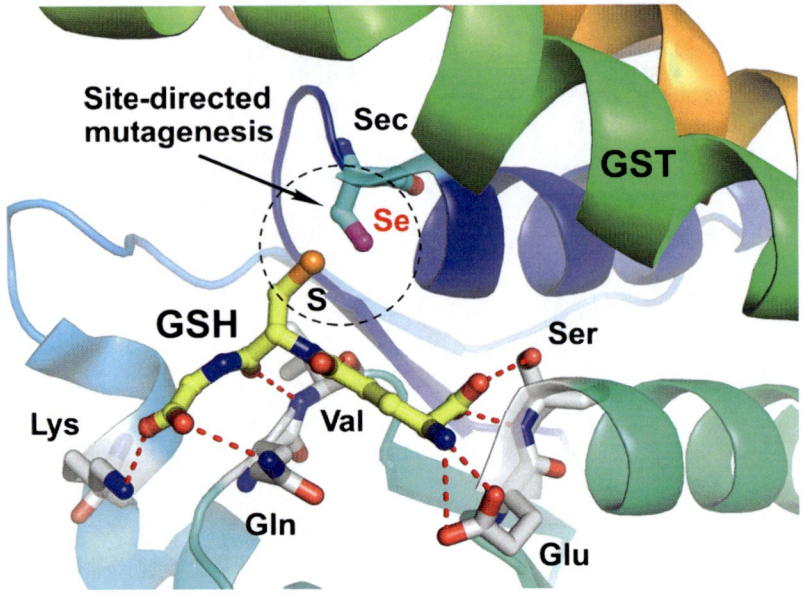

Artificial Selenoproteins, **Fig. 5** Schematic representation of genetically engineered seleno-GST with an active site Sec and a specific GSH-binding site

rearranged the active site of subtilisin to accommodate the native GPx substrate GSH.

Genetically Engineered Selenoproteins

Genetic engineering of proteins that contain noncanonical amino acid residues is a powerful tool for the redesign of natural proteins with novel properties. It utilizes the endogenous cellular machinery to translate the target protein, in which the unnatural amino acid can be selectively incorporated with high translation fidelity and efficiency. Cowie and Cohen pioneered the use of an auxotrophic bacterial host that starved for the natural amino acid and supplemented with a close structural analog to incorporate selenomethionine (SeMet) instead of Met. Budisa et al. extended the idea to create a series of atomic mutations ($-CH_2- \rightarrow -S- \rightarrow -Se- \rightarrow -Te-$) in proteins by substitution of Met by its noncanonical isosteres norleucine, selenomethionine, and telluromethionine (Beatty and Tirrell 2009). Several decades later, this strategy was successfully used to solve the problem of Sec biosynthesis.

The possibility of using a cysteine auxotrophic strain for the efficient substitution of cysteine residues in a protein by selenocysteine was first reported by Böck et al. (Muller et al. 1994). The similar method for incorporation of Sec was then exploited to substitute Cys149 residue in gene-phosphorylating glyceraldehyde 3-phosphate dehydrogenase (GAPDH). The obtained seleno-GAPDH exhibits high GPx activity against the aromatic substrate, which is similar to that of the first semisynthetic selenosubtilisins but remains low compared to selenoglutathione peroxidase.

Following this work, GST was chosen as protein scaffold to address the low GPx activity occurring on seleno-GAPDH toward GSH (Fig. 5). It is well known that GST belongs to the thioredoxin superfamily and shares a similar structural fold to GPx, known as thioredoxin fold, which is an ideal scaffold to create GPx mimics with high affinity for GSH. After the catalytically essential residue selenocysteine was bioincorporated into GSH-specific binding scaffold by replacing the active site serine 9 with a cysteine and then substituting it with selenocysteine in the cysteine auxotrophic system, the genetically engineered seleno-GST exhibited a remarkable activity of 2,957 U/μM rivaling native GPx. Furthermore, another member of the thioredoxin superfamily, glutaredoxin (Grx), was reported to offer a smaller protein scaffold to endow GPx properties by engineering a catalytic Sec residue (Huang et al. 2011). The novel seleno-Grx showed a GPx activity of 2,723 U/μM for the reduction of H_2O_2 with GSH, displaying obvious advantage for potential medical applications. These studies indicated that the redesign of evolutionarily related enzymes may be a common idea to design efficient biocatalysts.

Summary

The field of artificial selenoprotein is a rapidly evolving subject, especially for the antioxidant selenoprotein GPx. So far, several strategies that combine expertise from chemistry and biology have been developed to design the protein-based GPx mimics by site-specific incorporation of Sec. These studies do not simply reproduce naturally occurring selenoproteins, but offer an alternative approach to explore the biological role of selenium and the structure-function relationships of selenoprotein. Characterization of these artificial selenoprotein systems has provided valuable insight into the nature of molecular recognition and catalysis, which may promote the ability to create more efficient artificial selenoproteins for application in chemistry, biology, and medicine.

Cross-References

▸ Selenium, Biologically Active Compounds

References

Beatty KE, Tirrell DA (2009) Noncanonical amino acids in protein science and engineering. In: Köhrer C, RajBhandary UL (eds) Protein engineering. Springer, Berlin/Heidelberg, pp 127–153

Ding L, Zhu Z, Luo GM et al (1994) Artificial imitation of glutathione peroxidase by using chemical mutation method. Chin J Biochem Mol Biol 10:296–299

Gladyshev VN (2006) Selenoproteins and selenoproteomes. In: Hatfield DL, Berry MJ, Gladyshev VN (eds) Selenium: its molecular biology and role in human health, vol 2. Springer, New York

Gladyshev VN, Hatfield DL (2010) Selenocysteine biosynthesis, selenoproteins and selenoproteomes. In: Atkins JF, Gesteland RF (eds) Recoding: expansion of decoding rules enriches gene expression, vol 1. Springer, New York

Huang X, Liu XM, Luo Q, Liu JQ, Shen JC (2011) Artificial selenoenzymes: designed and redesigned. Chem Soc Rev 40:1171–1184

Liu JQ, Zhang K, Ren XJ et al (2004) Bioimprinted protein exhibits glutathione peroxidase activity. Anal Chim Acta 504:185–189

Liu L, Mao SZ, Liu XM et al (2008) Functional mimicry of the active site of glutathione peroxidase by glutathione imprinted selenium-containing protein. Biomacromolecules 9:363–368

Liu JQ, Luo GM, Mu Y (eds) (2012) Selenoproteins and mimics. Springer, Heidelberg.

Lu J, Holmgren A (2009) Selenoproteins. J Biol Chem 284:723–727

Muller S, Senn H, Gsell B, Vetter W, Baron C, Böck A (1994) The formation of diselenide bridges in proteins by incorporation of selenocysteine residues: biosynthesis and characterization of (Se)$_2$-thioredoxin. Biochemistry 33:3404–3412

Nevinsky GA, Kanyshkova TG, Buneva VN (2000) Natural catalytic antibodies (abzymes) in normalcy and pathology. Biochemistry (Moscow) 65:1245–1255

Qi D, Tann CM, Haring D, Distefano MD (2001) Generation of new enzymes via covalent modification of existing proteins. Chem Rev 101:3081–3111

Rotruck JT, Pope AL, Ganther HE et al (1973) Selenium: biochemical role as a component of glutathione peroxidase. Science 179:588–590

Wu ZP, Hilvert D (1990) Selenosubtilisin as a glutathione peroxidase mimic. J Am Chem Soc 112:5647–5648

As

▸ Arsenic, Mechanisms of Cellular Detoxification

Ascorbate Oxidase

Albrecht Messerschmidt
Department of Proteomics and Signal Transduction, Max-Planck-Institute of Biochemistry, Martinsried, Germany

Synonyms

EC 1.10.3; L-ascorbate; Oxygen oxidoreductase

Definition

Ascorbate oxidase is a multicopper oxidase that catalyses the four-electron reduction of dioxygen with concomitant one-electron oxidation of the substrate.

Ascorbate oxidase, laccase, and ceruloplasmin known as blue oxidases are typical members of the multicopper oxidase (MCO) family (Messerschmidt 1997). They contain at least one type-1 copper center and a trinuclear copper site. Ascorbate oxidase has been reviewed several times (Messerschmidt 1994, 1997, 2001, 2010). It occurs in higher plants. The main source for investigations has been zucchini squash. The immunohistochemical localization of

Ascorbate Oxidase,
Fig. 1 Ribbon diagram of the monomer structure of ascorbate oxidase (PDB-code: 1AOZ), domain 1 in *turquoise*, domain 2 in *green*, and domain 3 in *pink* (Reproduced from Messerschmidt 2010, with permission from Elsevier Inc.)

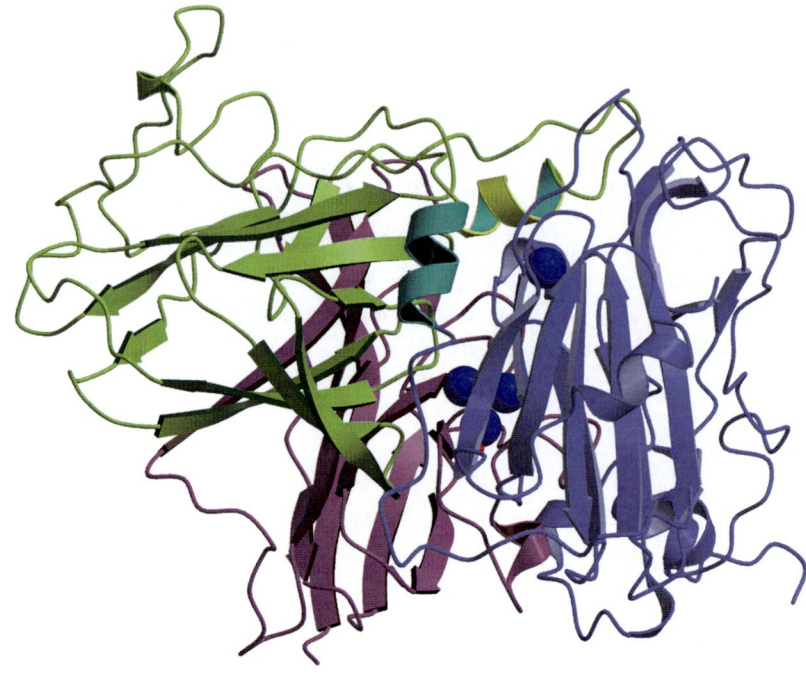

ascorbate oxidase in green zucchini shows that it is distributed ubiquitously over vegetative and reproductive organs. Its in vivo role has not been fully elucidated. It might be involved in processes such as fruit ripening, growth promotion, or in susceptibility to disease. The best and probably the physiological substrate is ascorbate, but catechols and polyphenols are also substrates in vivo.

The spectral properties of the blue oxidases like ascorbate oxidase, laccase, and ceruloplasmin are very similar. The copper ions of copper proteins have been classified according to their spectroscopic properties. Type-1 Cu^{2+} shows high absorption in the visible region (generally $>3,000$ M^{-1} cm^{-1} at 600 nm) and an EPR spectrum with $A_{\parallel} < 95 \times 10^{-4}$ cm^{-1}. Type-2, or normal, Cu^{2+} has undetectable absorption and the EPR line shape of the low-molecular-mass copper complexes ($A_{\parallel} > 140 \times 10^{-4}$ cm^{-1}). Type-3 (Cu^{2+}) is characterized by strong absorption in the near-ultraviolet region ($\lambda = 330$ nm) and by the absence of an EPR signal. The type-3 center consists of a pair of copper ions that are antiferromagnetically coupled. The above-mentioned signals disappear upon reduction.

Ascorbate oxidases have eight coppers with two type-1, two type-2, and two type-3 copper centers per homotetramer. The X-ray structure (Messerschmidt et al. 1992) indicated that the non-blue EPR-active type-2 copper together with the type-3 copper pair form an integral part of the trinuclear active copper center. There has been experimental evidence in earlier studies that the type-2 copper is close to the type-3 copper. Solomon and coworkers (Spira-Solomon et al. 1986), based on absorption, EPR, and low-temperature magnetic CD spectroscopy of azide binding to laccase, classified this metal-binding site as a trinuclear copper site.

Ascorbate oxidase was the first blue oxidase whose crystal structure was determined (Messerschmidt et al. 1992). It is a homodimeric enzyme with a molecular mass of 70 kDa and 552 amino acid residues per subunit (zucchini). The three-domain structure and the location of the type-1 and trinuclear copper centers in the ascorbate oxidase monomer as derived from the crystal structure are shown in Fig. 1. The folding of all three domains is of a similar β-barrel type. The structure is distantly related to the small blue copper proteins like plastocyanin or azurin. The mononuclear type-1 copper site is located in domain 3 and has the four canonical type-1 copper ligands (His, Cys, His, Met) also found in the small blue copper proteins plastocyanin and azurin. The trinuclear copper species is bound between domains 1 and 3 and has eight histidine ligands symmetrically supplied by both

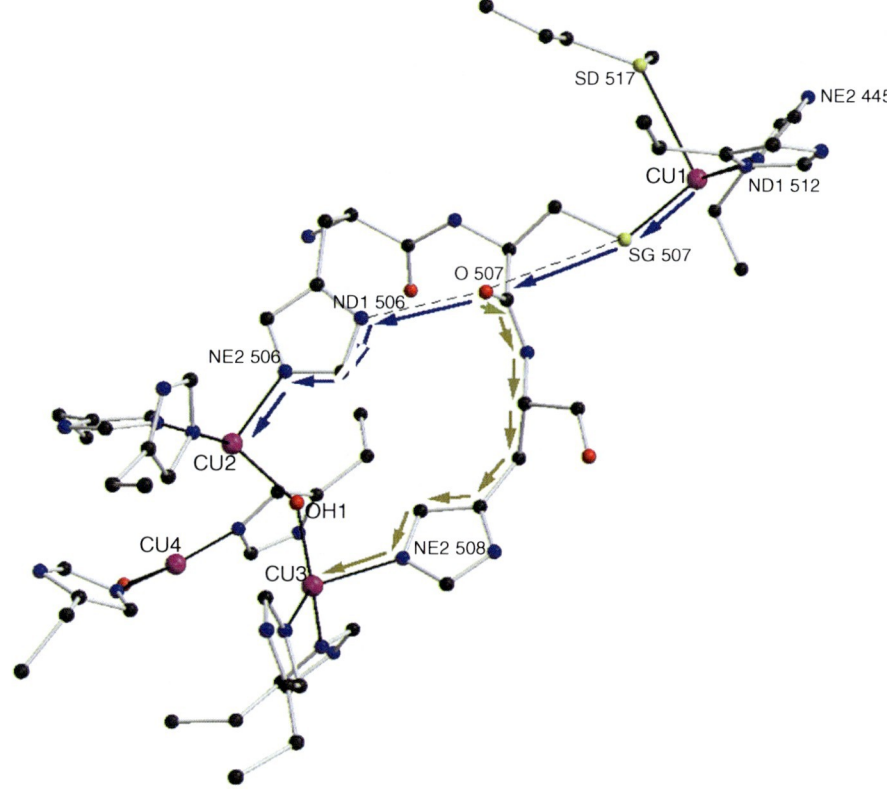

Ascorbate Oxidase, Fig. 2 Drawing of the region in the ascorbate oxidase monomer between the type-1 copper center and the trinuclear copper site (PDB-code: 1AOZ) (Reproduced from Messerschmidt (2010), with permission from Elsevier Inc.)

domains and two oxygen atoms. Seven histidines are coordinated by their NE2 atoms to the copper ions, whereas His62 is ligated to CU3 by its ND1 atom. The region of the molecule between the two centers is depicted in Fig. 2 showing the respective copper coordinations. The distances between the type-1 copper and the three coppers of the trinuclear center are 12.20, 12.69, and 14.87 Å, respectively. A binding pocket for the reducing substrate that is complementary to an ascorbate molecule is located near the type-1 copper site and accessible from solvent. A broad channel providing access from solvent to the trinuclear copper species, which is the binding and reaction site for the dioxygen, is present in ascorbate oxidase. During catalysis, an intramolecular electron transfer between the type-1 copper and the trinuclear copper cluster must occur. Intramolecular electron transfer from the type-1 copper to the type-3 copper pair of the trinuclear copper site may be through-bond, through-space, or a combination of both. A through-bond pathway is available for both branches each with 11 bonds (see Fig. 2). The alternative combined through-bond and through-space pathway from the type-1 copper Cu1 to Cu2 of the trinuclear center that involves a transfer from the SG atom of Cys507 to the main-chain carbonyl of Cys507 and through the hydrogen bond of this carbonyl to the ND1 atom of His506. Electron transfer processes in blue oxidases have been discussed in detail (Farver and Pecht 1997).

The crystal structure of the reduced form of ascorbate oxidase shows the type-1 copper site geometry virtually unchanged whereas the trinuclear site displays considerable structural changes (Messerschmidt et al. 1993). The bridging oxygen ligand OH1 is released, and the two coppers, Cu2 and Cu3, move toward their respective histidines and become three-coordinated in a trigonal-planar arrangement. The copper–copper distances increase from an average of 3.7–5.1 Å for Cu2–Cu3, 4.4 Å for Cu2–Cu4, and 4.1 Å for Cu3–Cu4. In the crystal structure of the peroxide form (Messerschmidt et al. 1993), the bridging oxygen ligand OH1 is released as well, and the peroxide is bound end on to the copper Cu2 of the trinuclear copper cluster. Solomon and coworkers (see e.g., Solomon et al. 1997) concluded from their spectroscopic data obtained from ascorbate oxidase

**Ascorbate Oxidase,
Fig. 3** Reaction scheme for laccase and ascorbate oxidase as proposed by Solomon and coworkers (Lee et al. 2002) (Reproduced from Lee et al. (2002), with permission from American Chemical Society)

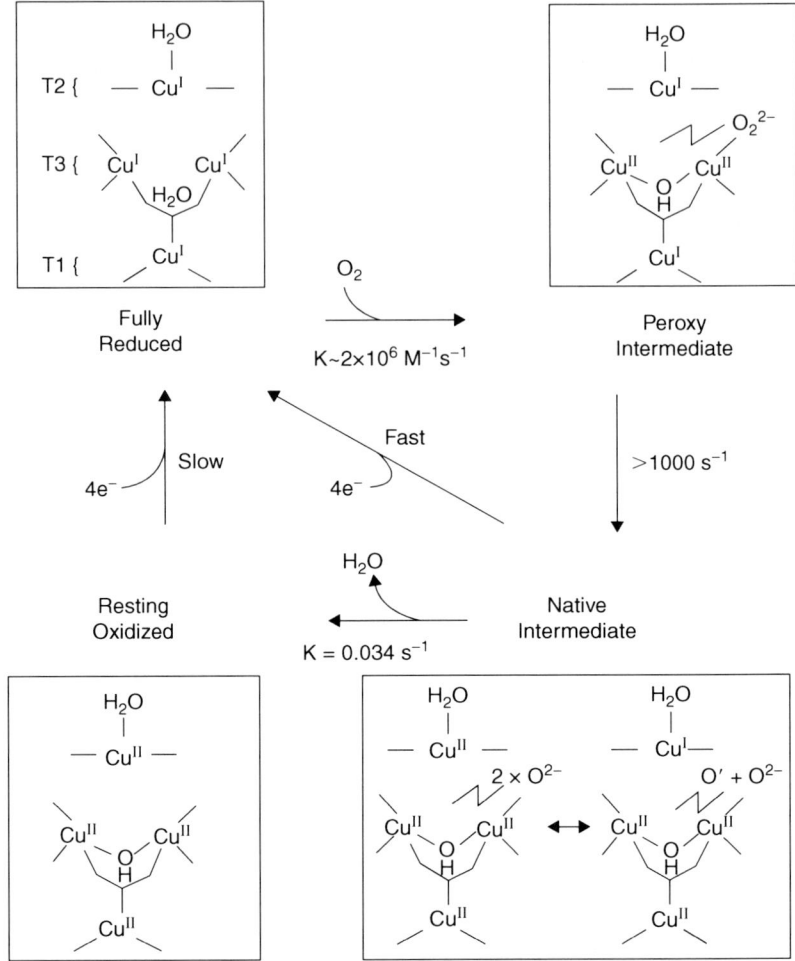

and laccase that their reoxidation intermediate binds as 1,1-μ hydroperoxide between either Cu2 and Cu4 or Cu3 and Cu4.

A "two-site ping-pong bi bi" mechanism has been deduced for tree laccase from steady-state kinetics. This will be valid for ascorbate oxidase as well because both enzymes are structurally and mechanistically closely related. A reaction scheme for ascorbate oxidase has been proposed based on the available spectroscopic, kinetic, and structural information (Messerschmidt 2001) that should also be valid for laccase or ceruloplasmin in the main features. A somewhat different view of the catalytic mechanism has been presented by Lee et al. (2002) based on spectroscopic and computational chemical techniques. This reaction scheme is shown in Fig. 3. The reaction starts from the resting oxidized state (lower left panel in Fig. 3). This state has been observed in the X-ray structure of fully oxidized ascorbate oxidase and is also the starting point in the scheme proposed by Messerschmidt (2001). The addition of four reduction equivalents leads to the fully reduced state (upper left panel in Fig. 3). The main distinction between both schemes is the presence of a bridging oxygen ligand between the type-3 copper pair, which is a water in the fully reduced state and a hydroxyl in the other three states. This central oxygen species could not be observed in the various ascorbate oxidase X-ray structures and is therefore not part of the mechanism reported in Messerschmidt (2001). The native intermediate state is split up into two substates. The right one is similar to the state of release of the first water molecule and formation of an $O^{\bullet-}$ radical in the scheme reported in Messerschmidt (2001) with the difference that the

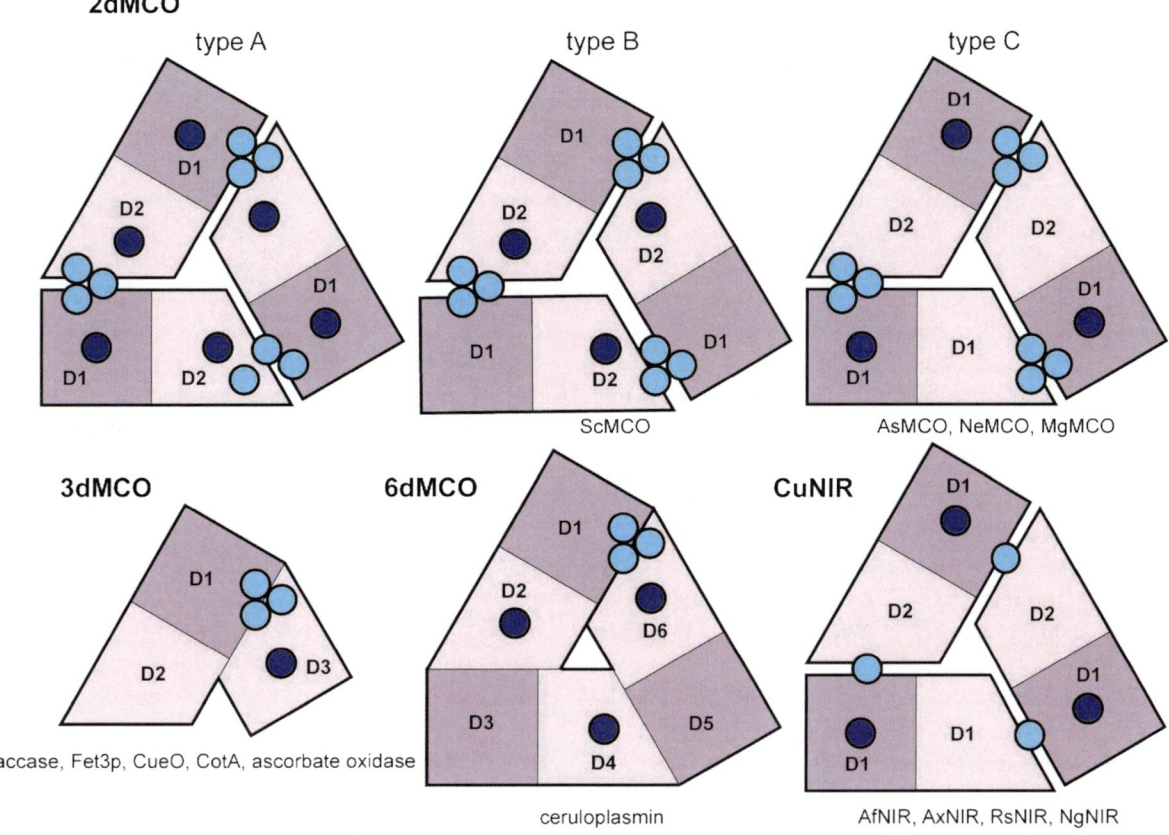

Ascorbate Oxidase, Fig. 4 MCO classification scheme. The *cyan* circles indicate the trinuclear copper site. Single *dark blue* circles represent the type-1 copper site. The CiNIR domain structure is shown for comparison (Reproduced from Lawton and Rosenzweig 2011, with permission from John Wiley and Sons)

type-2 copper is reduced instead of one copper from the type-3 copper pair. After transfer of the fourth electron to the oxygen radical, the second water molecule is released (left side in the lower right panel of Fig. 3) and returns to the oxidized resting form. In the scheme reported in Messerschmidt (2001), the enzyme moves back to the fully reduced state without traversing the resting state. However, in the main features, both reaction schemes are very similar and reflect the spectroscopic, kinetic, and structural known data. Bento et al. (2006) have presented a reaction scheme that does not involve the fully reduced state in the catalytic cycle and does not contain the spectroscopically proven oxygen radical intermediate. As it is very unlikely that dioxygen binds to the fully oxidized enzyme and the oxygen radical intermediate had not been included, this scheme seems to be very implausible.

According to the current state of knowledge, the MCO family comprises now members that consist of two domains (2dMCO), three domains (3dMCO), and six domains (6dMCO).

Although not being an oxidase, copper-dependent nitrite reductase (CuNIR) is closely related to the MCO family. 2dMCOs have recently been reviewed, and their relationship among each other and to CuNIR is discussed (Lawton and Rosenzweig 2011). The corresponding MCO classification scheme is shown in Fig. 4. The most surprising finding is that 2dMCOs occur as active enzymes in a trimeric form resembling CuNIRs. The trinuclear copper site is located between domains 1 and 2 of symmetry-related molecules. This is similar to CuNIRs where the type-2 copper site is also situated between domains 1 and 2 of symmetry-related molecules. 2dMCOs can be subdivided into three different types as deduced from the amino acid sequences. Type A has type-1 copper centers in both domains, type B in domain 2, and type C in domain 1. X-ray structures are known from type B (*Streptomyces coelicolor* MCO, ScMCO) and type C (*Arthrobacter* sp. MCO, AsMCO; *Nitrosomonas europaea* MCO, NeMCO; *Uncultured bacteria* MCO, MgMCO). Despite the similarities between 2dMCOs and CuNIRs, there are significant differences at the termini. In all currently characterized CuNIRs, the N-terminal residues extend from the first cupredoxin domain into the crevice between domains 1 and 2 whereas the C-terminal b-strand hydrogen bonds with domain 1 of the adjacent monomer. In contrast, the C-termini of 2dMCOs either occupy the space between domains 1 and 2 (NeMCO and MgMCO) or do not extend into the neighboring monomer (AsMCO). In AsMCO, the domain–domain interface is filled by the N-terminus, similar to CuNIRs, but in a slightly different fashion. It is important to note that a significant portion of the N-terminus of AsMCO is not modeled.

The trimeric structure of 2dMCOs results in a domain arrangement that is surprisingly similar to the domain arrangement of 3dMCOs. Structural conservation is especially evident at the 2dMCO monomer–monomer interface and the analogous interface between domains 1 and 3 in 3dMCOs. This similarity results in the trinuclear copper site being coordinated in the same way in 2dMCOs and 3dMCOs and suggests that formation of the trinuclear copper site is dependent on precise positioning of the domains. Thermal denaturation studies on ferrous transport 3 protein (Fet3p), a 3dMCO involved in iron metabolism, show that binding of the trinuclear copper ions decreases the stability of domains 1 and 3. It was suggested that domain 2 compensates for this effect by reducing the degrees of freedom for domains 1 and 2. Thus, domain 2 likely plays a large role in positioning and stabilizing the interface between domains 1 and 3 in 3dMCOs. In 2dMCOs, this interface forms as a result of trimerization. Therefore, oligomerization in 2dMCOs and the presence of an additional domain in 3dMCOs likely serve a similar purpose.

Ceruloplasmin has all six domains on a single amino acid sequence chain, but its domains are similarly arranged to the 2dMCOs and CuNIRs. The trinuclear copper site is located between domains 1 and 6 and domains 2, 4, and 6 each hold a type-1 copper.

Models for MCO evolution are all consistent that the evolution began with a single domain duplication, followed by a trimerization event and metal binding at the monomer–monomer interfaces. The existence of functional 2dMCOs confirms the plausibility of these models. In light of the conserved domain placement between 2dMCOs and 3d/6dMCOs, it is extremely likely that modern day 2dMCOs and 3d/6dMCOs share a common 2dMCO ancestor.

In addition, the 2dMCO structures highlight their similarities to CuNIRs. Strikingly, the only major difference between CuNIR and the type-C 2dMCOs, NeMCO and ScMCO, is the presence of the type-3 copper ions. Two pathways for CuNIR evolution have been proposed: The first involves evolution of CuNIR following the initial domain duplication event, and the second involves evolution of CuNIR from a type-C 2dMCO. Current structural and sequence analysis cannot exclude either of these possibilities, and it is possible that both pathways may be relevant to multicopper blue protein evolution.

Cross-References

▶ Biological Copper Transport
▶ Catechol Oxidase and Tyrosinase
▶ Copper-Binding Proteins
▶ Copper, Biological Functions
▶ Laccases
▶ Monocopper Blue Proteins

References

Bento I, Carrondo MA, Lindley PF (2006) Reduction of dioxygen by enzymes containing copper. J Biol Inorg Chem 11:539–547

Farver O, Pecht I (1997) Electron transfer reactions in multi-copper oxidases. In: Messerschmidt A (ed) Multi-copper oxidases. World Scientific, Singapore

Lawton TJ, Rosenzweig AM (2011) Two-domain multicopper oxidase. In: Messerschmidt A (ed) Handbook of metalloproteins, vol 5. Wiley, Chichester, pp 591–599

Lee SK, George SD, Antholine WE, Hedman B, Hodgson KO, Solomon EI (2002) Nature of the intermediate formed in the reduction of O_2 to H_2O at the trinuclear copper cluster active site in native laccase. J Am Chem Soc 124:6180–6193

Messerschmidt A (1994) Blue copper oxidases. Adv Inorg Chem 40:121–185

Messerschmidt A (ed) (1997) Multi-copper oxidases. World Scientific, Singapore

Messerschmidt A (2001) Ascorbate oxidase. In: Messerschmidt A, Huber R, Poulos T, Wieghardt K (eds) Handbook of metalloproteins, 1st edn. Wiley, Chichester

Messerschmidt A (2010) Copper metalloenzymes. In: Mander, L., Lui, H.-W. (eds) Comprehensive natural products II, chemistry and biology, vol 8. Elsevier, Oxford, pp 489–545

Messerschmidt A, Ladenstein R, Huber R, Bolognesi M, Avigliano L, Petruzzelli R, Rossi A, Finazzi-Agro A (1992) Refined crystal structure of ascorbate oxidase at 1.9 Å resolution. J Mol Biol 224:179–205

Messerschmidt A, Luecke H, Huber R (1993) X-ray structures and mechanistic implications of three functional derivatives of ascorbate oxidase from zucchini: reduced, peroxide and azide forms. J Mol Biol 230:997–1014

Solomon EI, Machokin TE, Sundaram UM (1997) Spectroscopy of multi-copper oxidases. In: Messerschmidt A (ed) Multi-copper oxidases. World Scientific, Singapore

Spira-Solomon DJ, Allendorf MD, Solomon EI (1986) Low-temperature magnetic circular dichroism studies of native laccase: confirmation of a trinuclear copper active site. J Am Chem Soc 108:5318–5328

Ascorbic Acid, Vitamin C

▶ Chromium(VI), Oxidative Cell Damage

Aspartic Acid: Aspartate, Asp, D

▶ Magnesium Binding Sites in Proteins

Aspirin, Acetylsalicylic Acid

▶ Chromium(VI), Oxidative Cell Damage

Assimilation of Cadmium

▶ Cadmium Absorption

Association/Binding Constant

▶ Calcium Ion Selectivity in Biological Systems

Astacin

▶ Zinc-Astacins

Astacus Protease

▶ Zinc-Astacins

Asymmetric Allylic Alkylation/Amination

▶ Palladium-catalysed Allylic Nucleophilic Substitution Reactions, Artificial Metalloenzymes

Atomic Number 38

▶ Strontium, Physical and Chemical Properties

Atomic Number 56

▶ Barium, Physical and Chemical Properties

ATP Phosphohydrolase (Na$^+$,K$^+$, Exchanging)

▶ Sodium/Potassium-ATPase Structure and Function, Overview

ATP2A1-3 (Human Genes for SERCA1-3)

▶ Calcium ATPases

ATP2B1-4 (PMCA1-4)

▶ Calcium ATPases

ATR1 Paralogs in Yeast

▶ Boron Stress Tolerance, YMR279c and YOR378w

Atr1, Boron Exporter in Yeast

Ahmet Koc and İrem Uluisik
Department of Molecular Biology and Genetics, Izmir Institute of Technology, Urla, İzmir, Turkey

Synonyms

Boric acid transport; Boron efflux pump; Boron toxicity; Boron transport in yeast

Definition

Boron is a micronutrient which is essential for plants and beneficial for animals. However, excess amount of it is toxic for organisms. Yeast Atr1 is a multidrug efflux pump and its major function is to export boron to outside of the cell. This entry will basically focus on boron transport and the role of Atr1 protein in boron stress response paths.

Boron in Living Organisms

In biological systems, boron is found mostly as boric acid [$B(OH)_3$] especially at neutral pH. Boric acid is a small molecule and is known as a weak Lewis acid which has a pKa of 9.24. Boric acid and borate can both bind and make strong interactions with biological compounds containing hydroxyl groups in *cis* position. These compounds include nucleic acids, nicotinamide, phosphoinositides, and sugars such as ribose. The finding of ribose stabilization achieved by borate puts forward a new vision to the "RNA world" hypothesis.

Boron is an essential metalloid for many organisms. Boron essentiality for plants was first established in 1923 by the finding that the growth of the plant *Vicia faba* was inhibited when boron was not present in the medium, however, when boron was supplied the growth was rescued (Warington 1923). Since that time boron has been known to be a necessary micronutrient for all vascular plants. Boron is also important for the life cycle of diatoms and algae, for the stabilization of heterocyst and nitrogen fixation in cyanobacteria and in some actinomycetes (*Frankia*, *Anabaena*). These organisms were dominant in ancient times, thus, boron may have special roles in the evolution of early life (Tanaka and Fujiwara 2008). Newly, a bacterial species *Bacillus boroniphilus* was isolated from soils of Kütahya, Turkey. The bacterium has a need for boron for its growth and can tolerate more than 450 mM boron (Ahmed et al. 2007). Also animals such as zebra fish (*Danio rerio*), rainbow trout (*Oncorhynchus mykiss*), and frogs (*Xenopus laevis*) required boron for their embryonic development. In addition, an interesting role of boron was found in the year 2002. It is involved in bacterial quorum sensing mechanism via appearing in the conserved structure of autoinducer AI-2 which is produced by various bacteria (Chen et al. 2002). Another important study related to one of the primary functions of boron comes from the year 1996 and revealed that boron acts as a crosslinker in plant cell walls. This crosslinking of rhamnogalacturonan II by borate cis-diol ester bonds is crucial for plants. Rhamnogalacturonan II is pectin and the formation of its network was found to be important especially for normal leaf expansion (Kobayashi et al. 1996).

When boron is present at high concentrations it becomes toxic. The optimum amount of boron differs from one organism to another. While 10 mM boric acid can cause lethality in *Arabidopsis*, the yeast *Saccharomyces cerevisiae* can grow in media containing 80 mM boric acid (Kaya et al. 2009).

Both deficiency and toxicity of boron can have some unwanted side effects on organisms. In plants, for example, usually expanding organs are firstly affected by boron deficiency. Abnormal cell wall formations, altered cytoskeleton polymerization, inhibition of leaf expansion and root elongation, as well as flower and fruit development are other examples of the results of boron deprivation (Tanaka and Fujiwara 2008). Boron toxicity symptoms occur mostly in the margins of leaves and other observed consequences of boron toxicity in plants are reduced growth of shoots and roots, lower photosynthetic rates, chlorosis, and necrosis of leaves. There are also several studies about the consequences of boron deprivation in human and animals including impairment of growth, abnormal bone development, the decrease in blood steroid hormones, and the increase in calcium excretion. In animals, excessive boron causes neurological effects, weight loss, diarrhea, anorexia, testicular atrophy, cardiovascular defects, and skeletal malformation. However, the molecular mechanisms behind these observed defects caused by boron deficiency or toxicity remain unresolved.

Boron Transporters

Basically three mechanisms have been proposed for the detoxification of the metals and metalloids. First mechanism includes export of the metal or metalloid from the cytosol; second mechanism includes the compartmentalization of metal or metalloid from the cytosol to the vacuole via the transporters located on the vacuolar membrane; and third mechanism involves the chelation (Wysocki and Tamas 2010).

Boron toxicity is a worldwide problem and causes a decrease in agricultural yield and concomitant increase in economical loss. Since plants are directly exposed to boron in the soil, many of the studies come from the plant systems. However, the yeast *Saccharomyces cerevisiae* is also one of the widely used model organisms for the characterization of boron tolerance genes in plants.

There are three ways of boron uptake and transport; passive diffusion, active transport by boron transporters, and facilitated transport by NIP channels (Tanaka and Fujiwara 2008). Boron has a very high permeability coefficient and it is an uncharged molecule so that it can easily pass the lipid bilayer. The first boron efflux transporter *BOR1* was found in *Arabidopsis* in 2002. This transporter is localized to plasma membrane and responsible for the xylem loading against concentration gradient under low boron (Takano et al. 2002). However, when plants are exposed to high amounts of boron, *BOR1* is degraded in the vacuole. *BOR1* paralogs and homologs were found in many organisms, except prokaryotes. For instance, *Arabidopsis thaliana* genome carries seven homologs of *BOR1* which are probably expressed in different tissues or have different structural properties. *YNL275W* is the only homolog of *A.thaliana BOR1* in the yeast *Saccharomyces cerevisiae*. It was found that yeast Bor1 located on the plasma membrane and responsible for the boron export (Takano et al. 2002). In addition to yeast *BOR1* gene, there are two other genes *FPS1* (glycerol channel) and *DUR3* (urea transporter) suggested to influence the boron tolerance of yeast cells (Nozawa et al. 2006). Human *BOR1* homolog is the *HsNaBC1* which stands for human Na^+-coupled borate transporter providing the borate uptake into the cell coupled with the Na^+ (Park et al. 2004). In mammals, boron is rapidly excreted in the urine therefore it will not surprising if mammals would have sophisticated systems for boron efflux. The regulation of these transporters and channels is important to avoid the deficiency or toxicity of boron.

Yeast Atr1 Function as a Boron Exporter

In a 2009 study, a yeast genomic DNA library was screened for the genes that provide resistance to yeast cells and *ATR1* isolated as a boron resistance gene (Kaya et al. 2009). The deletion of *ATR1* was shown to cause sensitivity to boron. When boron tolerance of *atr1Δ* deletion mutant was compared to the other gene deletion mutants that have been known as boron tolerance genes in yeast, such as *BOR1*, *DUR3*, and *FPS1* (Nozawa et al. 2006), *atr1Δ* mutant was found to be more sensitive to boron than other mutants. It cannot tolerate even 50 mM boric acid, whereas other mutants showed a wildtype-like growth phenotype in the presence of boric acid. However, it was also shown that overexpression of *ATR1* gene both in wild-type and *atr1Δ* mutant provided strong boron resistance to these cells. *ATR1*-overexpressing cells could grow in media containing 225 mM boric acid and this is the highest boron concentration that can be tolerated by a eukaryote.

To find out where Atr1 is localized in the cell, the researchers created an *ATR1-GFP* fusion protein. By using a confocal microscope it was shown that the fusion protein is localized specifically to the plasma membrane and vacuole.

In order to determine how Atr1 provides boron resistance to the cells, intracellular boron measurements were performed both in wild-type and *atr1Δ* mutants. The *atr1Δ* cells were found to accumulate more boron than wild-type cells; however, when *ATR1* gene was overexpressed in wild-type and *atr1Δ* cells, intracellular boron levels decreased by 25% and 47%, respectively.

It was shown that Gcn4 transcription factor is required for *ATR1* expression in response to boron. It has been also found that boron shows its toxicity especially by inhibiting protein synthesis (Uluisik et al. 2011). Gcn2 is known as a protein kinase which is activated in response to starvation and stress conditions and phosphorylates eIF2α, leading the induction of GCN4 and general amino acid control (GAAC) pathway. Boron stress induces Gcn2-dependent phosphorylation of eIF2α and inhibits general protein synthesis (Uluisik et al. 2011).

Cross-References

▶ Boron Stress Tolerance, YMR279c and YOR378w

References

Ahmed I, Yokota A et al (2007) A novel highly boron tolerant bacterium, Bacillus boroniphilus sp nov., isolated from soil, that requires boron for its growth. Extremophiles 11(2):217–224

Chen X, Schauder S et al (2002) Structural identification of a bacterial quorum-sensing signal containing boron. Nature 415(6871):545–549

Kaya A, Karakaya HC et al (2009) Identification of a novel system for boron transport: Atr1 is a main boron exporter in yeast. Mol Cell Biol 29(13):3665–3674

Kobayashi M, Matoh T et al (1996) Two chains of rhamnogalacturonan II are cross-linked by borate-diol ester bonds in higher plant cell walls. Plant Physiol 110(3):1017–1020

Nozawa A, Takano J et al (2006) Roles of BOR1, DUR3, and FPS1 in boron transport and tolerance in *Saccharomyces cerevisiae*. FEMS Microbiol Lett 262(2):216–222

Park M, Li Q et al (2004) NaBC1 is a ubiquitous electrogenic Na$^+$-coupled borate transporter essential for cellular boron homeostasis and cell growth and proliferation. Mol Cell 16(3):331–341

Takano J, Noguchi K et al (2002) Arabidopsis boron transporter for xylem loading. Nature 420(6913):337–340

Tanaka M, Fujiwara T (2008) Physiological roles and transport mechanisms of boron: perspectives from plants. Pflugers Archiv-Eur J Physiol 456(4):671–677

Uluisik I, Kaya A et al (2011) Boron stress activates the general amino acid control mechanism and inhibits protein synthesis. PLoS One 6(11):e27772

Warington K (1923) The effect of boric acid and borax on the broad bean and certain other plants. Oxford University Press, Oxford

Wysocki R, Tamas MJ (2010) How *Saccharomyces cerevisiae* copes with toxic metals and metalloids. FEMS Microbiol Rev 34(6):925–951

[Au(C^N^C)(IMe)]CF$_3$SO$_3$

▶ Gold(III), Cyclometalated Compound, Inhibition of Human DNA Topoisomerase IB

AuNPs

▶ Gold Nanoparticle Platform for Protein-Protein Interactions and Drug Discovery

Autoimmune Diseases as the Complication of Silicosis

▶ Silica, Immunological Effects

Avian Thymic Hormone

▶ Parvalbumin

B

Ba

▶ Barium(II) Transport in Potassium(I) and Calcium(II) Membrane Channels

Bacterial Calcium Binding Proteins

Delfina C. Domínguez[1] and Marianna Patrauchan[2]
[1]College of Health Sciences, The University of Texas at El Paso, El Paso, TX, USA
[2]Department of Microbiology and Molecular Genetics, College of Arts and Sciences, Oklahoma State University, Stillwater, OK, USA

Synonyms

Calerythrin; Calmodulin; Calsymin; EF-hand proteins

Definition

Bacterial proteins that bind calcium and are implicated in the regulation of various cellular events. These proteins appear to be involved in a wide variety of functions including chemotaxis, heat shock, pathogenesis, transport (influx and efflux), cell differentiation, and cell signaling.

Background

The calcium ion (Ca^{2+}) is perhaps the most important intracellular messenger in eukaryotes and regulates many cellular processes including cell differentiation, movement, cell cycle, transport mechanisms, and gene expression (Celio et al. 1996). Many effects of Ca^{2+} are mediated by ▶ calcium-binding proteins (CaBPs). Some of these proteins act as Ca^{2+} reservoirs or buffers. More specialized CaBPs such as calmodulin, act as signal transducers activating phosphorylation cascades leading to regulation in gene expression, and control of Ca^{2+} channel activity.

In prokaryotes, a similar role for Ca^{2+} has been proposed but is still not well defined. Ca^{2+} has been implicated in various bacterial physiological processes such as spore formation, chemotaxis, heterocyst differentiation, transport, and virulence. Several reports have shown that bacteria are capable of maintaining intracellular ▶ Ca^{2+} homeostasis, and Ca^{2+} transients are produced in response to adaptation to nitrogen starvation, and environmental stress (reviewed in Dominguez 2004). These findings as well as the demonstrated effect of Ca^{2+} on gene expression suggest a regulatory role of Ca^{2+} in bacterial physiology.

Sequence analyses of prokaryotic genomes indicate the presence of CaBPs proteins with different Ca^{2}-binding motifs (Zhou, et al. 2006). EF-hand containing proteins have been documented in several genera of bacteria (Dominguez 2004; Norris et al. 1996). While the genes of some CaBPs have been cloned and partially characterized, their functional activity remains to be investigated. Although bacterial cells appear to be equipped with the appropriate components (efflux and influx mechanisms, proteins kinases, Ca^{2+} homeostasis) of Ca^{2+} signaling networks, a physiologically integrated signal transduction system(s) remains to be characterized.

V.N. Uversky et al. (eds.), *Encyclopedia of Metalloproteins*, DOI 10.1007/978-1-4614-1533-6,
© Springer Science+Business Media New York 2013

Bacterial Calcium Binding Proteins, Table 1 Bacterial proteins containing EF-hand- and EF-hand-like motifs

Protein	Organism	Motif	Deviation from EF-hand	References
Calerythrin	*Saccharopolyspora erythrea*	Helix-loop-helix	None	Michiels et al. (2002)
Calsymin	*Rhizobium etli*	Helix-loop-helix	None	Michiels et al. (2002)
Glucanotransferase	*Thermotoga maritime*	Helix-loop-helix	Shorter loop	Zhou et al. (2006)
Slt35	*Escherichia coli*	Helix-loop-helix	Longer loop	Zhou et al. (2006)
Protective antigen	*Bacillus anthracis*	Loop-loop-helix	Entering helix missing	Zhou et al. (2006)
Dockerin	*Clostridium thermocellum*	Loop-loop-helix	Entering helix missing	Zhou et al. (2006)
Galactose-binding protein	*Salmonella typhimurium*	Helix-loop-strand	Exiting helix missing	Zhou et al. (2006)
Alginate-binding protein	*Sphingomonas* sp	Helix-loop-strand	Exiting helix missing	Zhou et al. (2006)
Alkaline protease	*Pseudomonas aeruginosa*	Strand-loop-strand	Entering and exiting helix missing	Zhou et al. (2006)

Structure

EF-Hand Motif and EF-Hand Like Motif Proteins

Sequence analyses of prokaryotic genomes have revealed the presence of 397 putative EF-hand proteins. Most of these proteins with a few exceptions (Calerythrin from *Saccharopolyspora erythrea*, Calsymin from *Rhizobium etli*, the *Brucella abortus* Asp24 and *Streptomyces coelicolor* CabD) are hypothetical proteins (Michiels et al. 2002) and have not been functionally characterized. According to Zhou et al. (2006), most of the predicted EF-hand bacterial proteins contain a single EF-hand motif, while only 39 bacterial proteins have 2–6 EF-hand motifs, 16 of which were previously described in (Michiels et al. 2002).

The superfamily of EF-hand proteins is the largest and best characterized group of CaBPs. Since the description of the EF-hand motif in 1973, more than 66 subfamilies have been reported (Kretsinger 1976: Zhou et al. 2006). All these proteins share a common structural motif that consists of a Ca^{2+}-binding loop flanked by two α-helices. Acidic amino acids in the loop play a very important role in Ca^{2+} binding (Kretsinger 1976).

The canonical EF-hand loop binds Ca^{2+} via side chain oxygen atoms. Residues 1, 3, 5, 7, 9, and 12 of the loop provide the ligands for Ca^{2+} binding in a pentagonal bipyramidal fashion. Residues 1, 3, and 5 act as monodentate ligands, and residue 12, usually a Glu or Asp, is a bidentate Ca^{2+} ligand (Kretsinger 1976). The prokaryotic EF-hand domain shows great sequence and structural diversity. Some of these proteins differ in the length of the Ca^{2+}-binding loop, which may be shorter or longer than 12 residues, while other proteins deviate in the secondary structure flanking the Ca^{2+}-binding loop. However, the structural geometry of these proteins resembles the classical EF-hand motif (Zhou et al. 2006).

Five classes of EF-hand and EF-hand-like motifs have been reported in bacteria (Table 1). They include (1) the typical helix-loop helix EF-hand structure seen in Calerythrin and Calsymin, (2) the extracellular Ca^{2+}-binding region (Excalibur) with a shorter loop containing 10 residue motif DxDxDGxxCE found in various bacteria, (3) the longer 15-residue Ca^{2+}-binding loop seen in the *E. coli* lytic transglycosylase B, and (4) and (5) classes lacking the first or second helix as described in the *C. thermocellum* dockerin and the *Sphingomonas* sp. alginate-binding protein, respectively (Zhou et al. 2006).

Ca^{2+}-Binding ß-Roll Motif

Many Gram-negative bacteria secrete proteins with a ß-roll or parallel ß-helix structure, which contains multiple Ca^{2+}-binding and glycine-rich sequence motifs. These proteins are secreted by the Type 1 secretion system (T1SS) and often have been linked to pathogenesis. Many of these proteins contain a so called repeats-in-toxin (RTX) domain located upstream of the C-terminal uncleaved secretion signal. The RTX is found in several families of proteins including extracellular lipases, proteases, epimerases, and hemolytic toxins.

The RTX domain consists of tandem repeating nonamers of the sequence GGXGXDXUX, in which U is an aliphatic amino acid and X is any amino acid. The first six amino acids bind Ca^{2+} and the last three residues form a ß-strand. The number of tandem repeats varies from 5 to 45. The highly conserved aspartate is required for Ca^{2+} binding and without Ca^{2+} the beta roll structure is not formed. It is important to point out that Ca^{2+} is essential for RTX proteins folding, which takes place outside the cell (Michiels et al. 2002).

Completion of the genome sequence in *Mycobacterium tuberculosis* revealed the presence of a family of proteins with the sequence signature PE_PGRS. The PE domain is linked to a glycine-rich sequence known as the polymorphic GC-rich repetitive sequence (PGRS) region, which has been used in fingerprinting of *M. tuberculosis* clinical isolates. These proteins appear to have multiple Ca^{2+}-binding and glycine-rich motifs (GGXGXD/NXUX) as found in the RTX toxins. It is suggested that these proteins may be involved in host-pathogen interactions (Michiels et al. 2002).

Ca^{2+}-Binding Greek Key Motif

The ß-γ-crystallin superfamily includes Ca^{2+}-binding proteins found in various taxa including eukaryotes, eubacteria, and archea. In eukaryotes, ß-γ-crystallins are the major structural proteins of the lens. The common topological feature of these proteins is a double Greek key motif arranged as four adjacent antiparallel ß-strands sharing the third ß-strand with the opposite motif and forming a domain (Sharma and Balasubramanian 1996). ß-γ-crystallins from all of the three kingdoms showed that Ca^{2+} coordination is conserved in the form of N/D-N/D-#-I-S/T-S, in which the residue occupying the position # provides a main chain carbonyl for direct Ca^{2+} coordination, and the nonpolar residue I contributes to the hydrophobic core of the domain. In contrast to the EF-hand superfamily, ß-γ-crystallins showed Ca^{2+}-binding affinities within the μM range (as often seen in extracellular CaBPs), whereas Ca^{2+}-sensors of the EF-hand superfamily exhibit higher Ca^{2+}-binding affinities in the lower μM to nM range (Kretsinger 1976). The binding of Ca^{2+} produces a variety of physicochemical responses ranging from sequestration to stabilization.

Bacterial proteins with sequence conservation and three-dimensional structure closely related to ß-γ-crystallins include protein S from *Myxococcus xanthus*, *Yersinia pestis*, and the M-crystallin from the archaeon *Methanosarcina acetivorans*. Protein S (PS) is a monomeric CaBP that has two Ca^{2+}-binding domains, each domain consisting of two dissimilar Greek key motifs. PS is a spore-coat-forming protein, whose expression increases during starvation. This protein protects the organism over a long period of desiccation in a Ca^{2+}-dependent manner.

The *Y. pestis* CO92 crystallin is a putative secreted protein. The protein has a unique AA and BB type combination instead of the typical AB or BA found in other members of the ß-γ-crystallin superfamily. In the absence of Ca^{2+} the domains are intrinsically unstructured and only in the presence of Ca^{2+}, acquire their ß-γ-crystallin fold.

The M-crystallin from the archaeon *M. acetivorans* has been recently added to the ß-γ-crystallin superfamily. The protein appears to have a remarkable structural similarity with vertebrate ß-γ-crystallins. M-crystallin has a typical AB-type Greek key motif arrangement with Ca^{2+}-binding properties (packing and conformational changes) more similar to vertebrate lens proteins than to their microbial homologues. Similar to other lens ß- and γ- crystallins, binding of Ca^{2+} produce no noticeable conformational changes (Barnwal et al. 2009). The presence of the Ca^{2+}-binding motifs needs to be tested for functional necessity if not for viability of the organisms.

Evolution

Based on the discovery of bacterial proteins with similarity to calmodulin, the idea that CaBPs originated from prokaryotes was first proposed in 1987 by Swan and coworkers. Later, the prediction and identification of various EF-hands in the genus *Streptomyces* lead to the assumption that CaBPs evolved from Gram positive bacteria. The continued increase of genomic information and computational analyses during the past years have shown that prokaryotic EF-hands exhibit great diversity and high structural variability within the Ca^{2+} loop (Zhou et al. 2006). Based on these observations, it was proposed that the evolution of EF-hand and EF-hand-like proteins in bacteria is far more complex than previously anticipated, and they postulated that the canonical EF-hand motif could be the most visible, but perhaps not the most ancient.

The recent findings of single-handed EF motifs in prokaryotes raised the possibility that the EF-hand motif could be a structural "mobile unit" for Ca^{2+} binding that may have undergone modifications during evolution and had been incorporated into host proteins. Since pseudo-EF-hand proteins are absent from bacterial genomes, they are likely to be phylogenetically younger than canonical EF-hand motifs (Zhou et al. 2006).

Functions

Calcium Transport and Transporters

Three major types of ▶ Ca^{2+} transport systems have been described in prokaryotes: Ca^{2+} exchangers, ▶ Ca^{2+} ATPases, and polyhydroxybutyrate – polyphosphate (PHB-PP) complexes. However, only a few of these have been characterized biochemically, and their physiological significance has yet to be demonstrated.

Calcium Exchangers

In most bacteria Ca^{2+} is exported by Ca^{2+} exchangers, Ca^{2+}/H^+ or Ca^{2+}/Na^+ antiporters. These are low-affinity Ca^{2+} transport systems that use the energy stored in the electrochemical gradient of ions. Depending on the gradient, exchangers can also operate in the reverse (Ca^{2+} entry) direction. Ca^{2+} exchangers differ in ion specificity and have been identified in a number of bacterial genera (Norris et al. 1996). Major examples include *E. coli* proteins ChaA, YrbG, and PitB that have been reported as Ca^{2+}/H^+, Ca^{2+}/Na^+ antiporters (Saaf et al. 2001) and Ca^{2+}/PO_4^{3-} symporter (van Veen et al. 1994), respectively. ChaA may also exhibit Na^+/H^+, and K^+/H^+ antiport activity and play a role in tolerance to high concentrations of Ca^{2+} and Na^+. An acidic motif, EHEDDSDDDD-209 in ChaA is strikingly similar to a motif in ▶ calsequestrin and was proposed to be used as a signature motif for identifying genes encoding Ca^{2+}/H^+ antiporter activity or other Ca^{2+}-dependent activities. YrbG represents a large family of Ca^{2+}/Na^+ exchangers that includes both prokaryotic and eukaryotic proteins. Finally, PitB transports Ca^{2+} as a metal phosphate ($MeHPO_4$) complex in symport with H^+ (Norris et al. 1996).

Ca^{2+} ATPases

Ca^{2+} ATPases are mostly high-affinity Ca^{2+} pumps that export Ca^{2+} from the cytosol to the extracellular environment by using the energy stored in ATP. Several P-type Ca^{2+} translocating ATPases have been described in bacteria and also in the archaeon *Methanobacterium thermoautotrophicum*. These proteins are homologous to the SERCA Ca^{2+} transporter from sacroplasmic reticulum of eukaryotes, and transiently form a phosphorylated intermediate by the transfer of the γ-phosphate of ATP to an aspartic acid residue of the protein (reviewed in Dominguez 2004).

Polyhydroxybutyrate – Polyphosphate (PHB-PP) Complexes

PHB-PP complexes form non-proteinaceous voltage-gated ion channels in planar lipid bilayer. The channels are highly selective for Ca^{2+} over Na^+ at physiological pH (reviewed in Shemarova and Nesterov 2005) and can coexport Ca^{2+} and PP. The complexes are highly abundant in stationary growth phase cells and display many characteristics of protein Ca^{2+} channels: voltage-activated, selective for divalent cations, permeant to Ca^{2+}, Sr^{2+} and Ba^{2+}, and blocked in a concentration-dependent manner by La^{3+}, Co^{2+}, Cd^{2+} and Mg^{2+}.

In addition to translocating Ca^{2+} and maintaining Ca^{2+} homeostasis, Ca^{2+} transporters may play other physiological roles including chemotaxis, sporulation, pathogenesis and survival in a host. These findings provide further support for the idea that Ca^{2+} transport is regulated in bacteria even if the specific transporters and their function(s) remains to be demonstrated unequivocally. Moreover, the data suggest that the Ca^{2+} flux systems may present new targets for future antimicrobial agents.

Calcium Signaling

Ca^{2+} signaling in eukaryotes has long been understood; however, in prokaryotes, it is less well defined. Earlier it has been shown that Ca^{2+} is implicated in the regulation of a number of physiological processes including the expression of virulence factors, chemotaxis, cell division, competence, and autolysis, spore germination, and acid tolerance (reviewed in Dominguez 2004; Shemarova and Nesterov 2005). More recent studies provide further evidence that Ca^{2+} is involved in the regulation of gene expression. The responding processes include iron acquisition, biosynthesis of pyocyanin and proteases, quinolone signaling, nitrogen metabolism, oxidative, and general stress responses. However, the mechanisms of such regulation are not clear.

Sensing extracellular signals may occur via two different mechanisms: two-component regulators or by signal uptake followed by signal transduction or relay. A two-component *ca*lcium-*r*egulated system *carSR* has been shown to negatively regulate polysaccharide production and biofilm formation in *Vibrio cholera* (Bilecen and Yildiz 2009). The AtoS-AtoC two-component system in *E. coli* was shown to regulate the effect of extracellular Ca^{2+} on biosynthesis of poly-(R)-3-hydroxybutyrate. On the other hand, increased cytosolic Ca^{2+} was shown to alter gene transcription *E. coli* (Naseem et al. 2009).

CaBPs are most likely candidates for binding Ca^{2+} and transducing or relaying Ca^{2+} signal. Thus, for example, a four-EF-hand CaBP CabC has been shown to bind Ca^{2+} and regulate aerial hyphal formation in *Streptomyces coelicolor* (Wang et al. 2008).

Finally, a variety of external stimuli affect cytosolic maintenance of Ca^{2+} in bacteria (reviewed in Shemarova and Nesterov 2005). These include photosensitization in *Propionibacterium acnes* and oxidative stress in *B. subtilis*. Also, butane 2,3-diol, a glycerol fermentation product commonly produced in a human gut, activates Ca^{2+} transients in *E. coli*, thus suggesting the role of Ca^{2+} in bacteria-host cell signaling.

Concluding Remarks

It is clear that evidence in support of the importance of calcium in bacteria is accumulating. However, much work needs to be done to identify and characterize the physiologically important CaBPs. A full assessment of the precise regulatory mechanisms that convert changes in cytosolic free Ca^{2+} into physiological functions awaits further analysis.

Cross-References

▶ Biological Copper Transport
▶ Calcium ATPase
▶ Calcium-Binding Proteins
▶ Calmodulin
▶ Calsequestrin
▶ Penta-EF-Hand Calcium-Binding Proteins

References

Barnwal RP et al (2009) Solution structure and calcium-binding properties of M-crystallin, a primordial betagamma-crystallin from archaea. J Mol Biol 386:675–689

Bilecen K, Yildiz FH (2009) Identification of a calcium-controlled negative regulatory system affecting *Vibrio cholerae* biofilm formation. Environ Microbiol 11:2015–2029

Celio MR, Tomas P, Shawler B (1996) Guidebook to the calcium-binding proteins. Sambrook & Tooze, Oxford University Press, Oxford

Dominguez DC (2004) Calcium signalling in bacteria. Mol Microbiol 54:291–297

Kretsinger RH (1976) Calcium-binding proteins. Annu Rev Biochem 45:239–266

Michiels J et al (2002) The functions of Ca(2+) in bacteria: a role for EF-hand proteins? Trends Microbiol 10:87–93

Naseem R et al (2009) ATP regulates calcium efflux and growth in *E. coli*. J Mol Biol 391:42–56

Norris et al (1996) "Bacterial calcium-binding proteins". In: Celio MR, Pauls TL, Schwaller B (eds) Sambrook & Tooze. Oxford University Press, pp 209–212

Saaf A et al (2001) The internal repeats in the Na+/Ca2+ exchanger-related Escherichia coli protein YrbG have opposite membrane topologies. J biol chem 276:18905–18907

Sharma Y, Balasubramanian D (1996) "Crystallins". In: Celio MR, Pauls TL, Schwaller B (eds) Sambrook & Tooze. Oxford University Press, pp 225–228

Shemarova IV, Nesterov VP (2005) Evolution of mechanisms of calcium signaling: the role of calcium ions in signal transduction in prokaryotes. Zh Evol Biokhim Fiziol 41:12–17

van Veen HW et al (1994) Translocation of metal phosphate via the phosphate inorganic transport system of Escherichia coli. Biochemistry 33:1766–1770

Wang SL et al (2008) CabC, an EF-hand calcium-binding protein, is involved in Ca2+-mediated regulation of spore germination and aerial hypha formation in *Streptomyces coelicolor*. J Bacteriol 190:4061–4068

Zhou Y et al (2006) Prediction of EF-hand calcium-binding proteins and analysis of bacterial EF-hand proteins. Proteins 65:643–655

Bacterial Mercury Resistance Proteins

Simon Silver and Le T. Phung
Department of Microbiology and Immunology,
University of Illinois, Chicago, IL, USA

Synonyms

Microbial toxic mercury resistance; Proteins with mercury as "natural" substrate

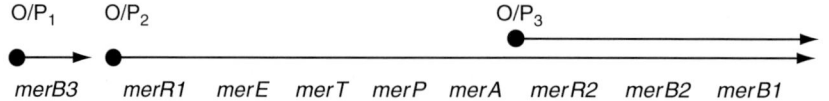

Bacterial Mercury Resistance Proteins, Fig. 1 The genes of the *Bacillus mer* gene complex, i.e., the *mer* operon. OP, operator/promoter mRNA initiation sites; *merR* regulatory gene; *merE* and *merT* genes for membrane transport proteins; *merP* gene for outer surface mercury binding protein; *merA* gene for mercuric reductase enzyme; and *merB* genes for organomercurial lyase enzymes; → direction of mRNA synthesis. GenBank accession AB066362, GI 15076639 (Chen et al. 2008)

Definition

Bacteria carry out chemical transformations of mercury compounds. Often these transformations result from bacterial resistance systems to inorganic mercury (Hg^{2+}) and to organomercurials (such as methylmercury and phenylmercury). There are four types of mercury-specific proteins: detoxifying enzymes producing less toxic products, membrane transport proteins that bring Hg^{2+} into the cells to be detoxified, mercury-binding proteins at the outer cell surface, and DNA-binding transcriptional regulatory proteins governing mRNA synthesis (Barkay et al. 2003; Silver and Phung 2005; Silver and Hobman 2007).

Mercury Resistance (the mer Operon)

The genes for bacterial mercury resistance are clustered in a "*mer* operon" of contiguous genes that are regulated together and are found widely on the plasmids and the chromosomes of both gram-positive and gram-negative bacteria. The number of genes (and therefore corresponding protein products) in different bacteria range from fewer to as many as 9, but in all cases include related protein products (Silver and Phung 2005). Figure 1 shows the genes involved in mercury resistance in *Bacillus*, the most complex of such determinants to have been studied in detail. It includes 3 *merB* genes for organomercurial lyase enzymes, 2 *merR* genes for regulatory proteins, and three operator/promoter mRNA transcriptional control sites (Chen et al. 2008).

Mercuric Reductase Enzyme

Mercuric reductase is a homodimeric protein, whose two active sites lie at the interface between the subunits (Fig. 2a), and include a redox-active cysteine pair, closely homologous to that in the paralogous proteins glutathione reductase and lipoamide dehydrogenase (Ledwidge et al. 2005). The cysteine pair carries electrons transferred from a loosely bound NAD(P)H first to a tightly bound flavin adenine dinucleotide (FAD) next and finally to the active site cysteines, where the substrate Hg^{2+} is reduced to Hg^0 (which is then spontaneously released, Fig. 3) (Silver and Phung 2005). This structural and functional homology raises the question of which enzyme came first in early cell evolution; and we hypothesize that toxic mercury was present early after the origin of life, over three billion years ago, perhaps before the appearance of glutathione and lipoamide. Mercuric reductase is the oldest.

With changing scientific communication, the Wikipedia report (http://en.wikipedia.org/wiki/Mercury%28II%29_reductase) is a good source of current information. Mercuric reductase contains a number of functionally different regions (or domains), starting with a Hg^{2+}-binding domain of approximately 70 amino acids in length (with a cysteine pair dithiol; Fig. 2a) at the N-terminus and ending with a second Hg^{2+}-binding cysteine pair dithiol at the C-terminus of the polypeptide (Ledwidge et al. 2005). In between are the binding determinants of NADPH and FAD, as well as the active site sequence, including a third conserved cysteine pair (Figs. 2a and 3). The substrate Hg^{2+} is initially bound to the N-terminal cysteine pair of one subunit, then transferred to the C-terminal cysteine pair of the opposite subunit, and then to the active site cysteine pair at the interface of the first subunit, for reduction to Hg^0. The currently understood intermediate steps in the binding of Hg^0, electron transfer, and reduction are shown in Fig. 3.

Organomercurial Lyase Enzyme

Organomercurial lyases are small monomeric proteins (just over 200 amino acids in length; determined by the *merB* gene) that function via a biomolecular S_E2 electrophilic substitution reaction mechanism. H^+ originating from aspartic acid residue Asp99 adds to the carbon of the organomercurial (either alkyl such as in methyl- or ethyl-mercuric compounds or aromatic such as in phenylmercury) from the same side as the mercury (Barkay et al. 2003; Miller 2007). Hg-C bond cleavage

Bacterial Mercury Resistance Proteins, Fig. 2 (a) Mercuric reductase showing dimeric structure with bound NADP and FAD and cysteine pair thiols (Silver and Phung 2005) (cartoon adapted from RCSB PDB structure 1ZK7 and 1ZX9; see also Fig. 3) with Hg^{2+} bound at both active sites, although probably only one functions at a time; and (b) organomercurial lyase ribbon structure showing active site cysteine pair Cys96 Cys159 thiols and proton donor Asp99, with substrate-binding surface (Lafrance-Vanasse et al. 2009; Miller 2007; Parks et al. 2009) (Adapted from RCSB PDB database ID: 3F0O, viewed in Jmol; see also Fig. 4)

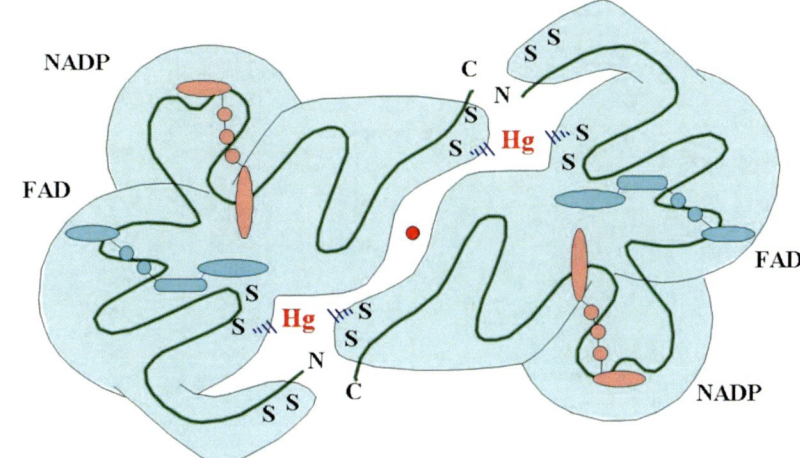

a Mercuric reductase

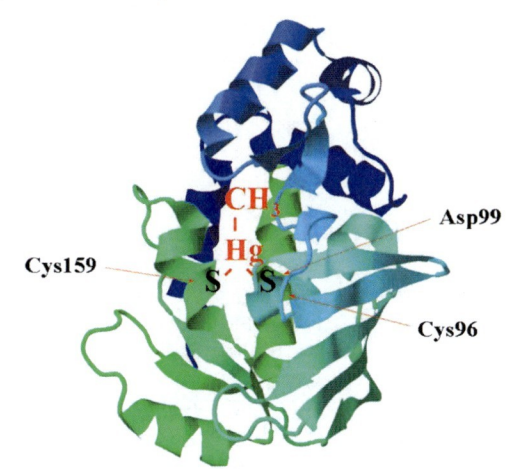

b Organomercurial Lyase

and H^+ addition occur. The currently understood steps of the reaction cycle are shown in Fig. 4 and the protein structure for MerB (which has been independently solved by NMR spectroscopy and by X-ray diffraction from crystalline MerB) is shown in Fig. 2b. An active site cysteine pair (Cys96 and Cys159) covalently binds Hg during the reaction. As a first step, the organomercurial compound enters an internalized active site and forms a Hg-S bond with the thiol from Cys96; the proton released from the Cys96 thiol protonates Asp99 (Fig. 4). Cys159 then forms a second Hg-S bond resulting in a trigonal $R-HgS_2$ intermediate. Finally, the H donated by Asp99 attacks and releases the RH product, with the addition of a water molecule at the position on the Hg^{2+} previously occupied by the organic R group.

Structural studies provide support for the mechanism proposed in Fig. 4 (Parks et al. 2009; Lafrance Vanasse et al. 2009), in addition to demonstrating the close proximity of Asp99, Cys96, and Cys159 in the protein molecule (Fig. 2b). The binding surface for the organomercurial substrate lies on a hydrophobic surface within the protein molecule. The N-terminal region of the MerB protein appears to function as a "lid" opening to accommodate the organomercurial substrate and then closing over the complex. How the organomercurial lyase substrate-binding surface accommodates different organomercurial groups

Bacterial Mercury Resistance Proteins, Fig. 3 The steps of mercuric reductase (MerA) activity (Silver and Phung 2005; Barkay et al. 2003). (1) Hg-dithiol substrate binding, *in vivo* directly from MerT protein or in vitro from glutathione or mercaptoethanol. (2) Release of second thiol linkage, and formation of di-cysteine bound Hg^{2+} at C-terminal of second subunit. (3) Transfer of Hg^{2+} to active site Cys135 of one subunit. (4) Formation of Cys135 Cys140 dithiol-bound Hg^{2+} at active site. (5) Reduction of Hg^{2+} to Hg^0 (with electrons from NADPH to FAD) and release of Hg^0. (6) Binding of NADPH to regenerate an active enzyme

[for example, small (methyl) and larger (phenyl)] is not understood. The physical basis for different versions of this MerB having different ranges of organomercurial substrates is also not known.

Whereas mercuric reductase constitutes a "sub-branch" paralogous group within the larger family of enzymes including glutathione reductase and lipoamide dehydrogenase, organomercurial lyases are essentially novel, lacking close homologs (Silver and Phung 2005). Nevertheless, both mercuric reductase and organomercurial lyase enzymes are found broadly among bacterial types, both Gram positive and Gram negative, indicating an ancient origin. There are also less-studied versions of mercuric reductase and organomercurial lyase genes found in a few Archaea; however, these genes and enzymes are not found in eukaryote organisms.

Mercury Transport and Binding Proteins

It seems counter-intuitive for bacteria to have a membrane transport system whose function is to bring the toxic cation Hg^{2+} from a relatively harmless position outside of the bacterial cell to inside the cytoplasm. However, uptake of Hg^{2+} via small membrane proteins is a well-studied process (Silver and Hobman 2007), and is part of all bacterial mercury resistance systems. A functional Hg^{2+} uptake system in the absence (due to mutation) of mercuric reductase causes the cells to be more sensitive ("hypersensitive") to Hg^{2+} than are bacteria lacking all mercury-related genes. The overall logic appears to be that the required electron donor NADPH is a high-energy molecule (like ATP) that cannot be released from the cell. Therefore, evolution resulted in an orderly tight pathway, functioning as a molecular "bucket brigade," starting

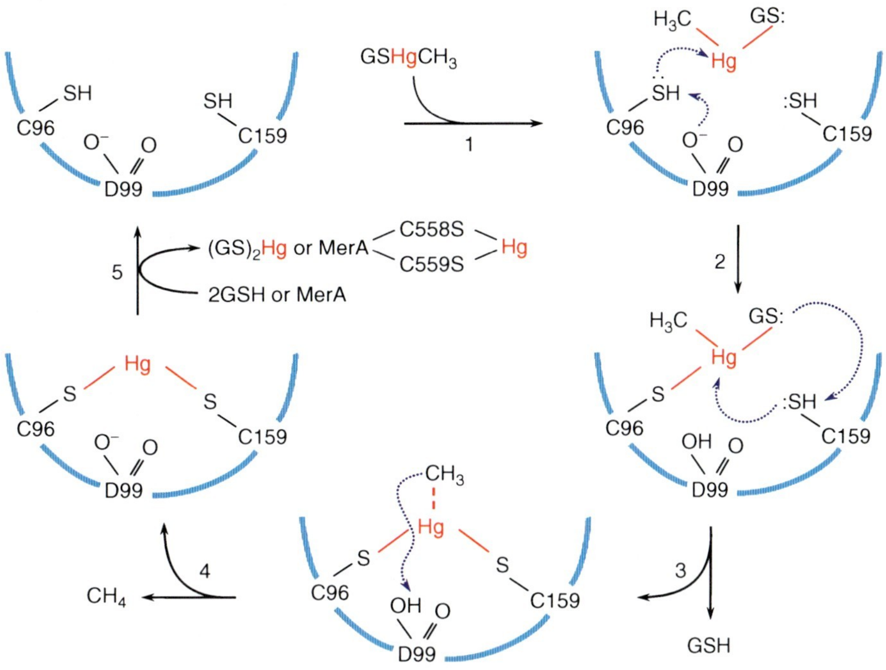

Bacterial Mercury Resistance Proteins, Fig. 4 The steps of organomercurial lyase (MerB) activity. (1) Binding substrate methylmercury coupled to glutathione (GS) to Cys96, and transfer of H$^+$ to Asp99. (2) Binding to second active site Cys159 thiol. (3) Release of reduced GSH. (4) Transfer of H$^+$ from Asp99 and release of CH$_4$. (5) Transfer of Hg^{2+} to GSH or directly to the MerA protein (Miller 2007; Silver and Phung 2005; Parks et al. 2009)

with the binding of Hg^{2+} at the cell surface, to movement across the cell membrane, and ending with binding of Hg^{2+} on mercuric reductase, without freely diffusible Hg^{2+} at any stage. Transport proteins constitute the initial stages of a cysteine pair to cysteine pair cascade, so that Hg^{2+} initially bound outside the cell surface is passed on from one cysteine pair to the next (the exchange frequency can be quite rapid) without ever being released free at any intermediate stage (as the release frequency would be low).

There are three modes of membrane-embedded Hg^{2+}-transport proteins found in the inner membranes of gram-negative bacteria and homologous proteins in gram-positive bacteria. The first one (identified from the genes of transposons Tn*21* and Tn*501*) is MerT, an unusual protein that spans the membrane three times (Fig. 5a) (Silver and Hobman 2007; Brown et al. 2002). There is a closely adjacent di-cysteine dithiol pair in the N-terminal alpha-helical region of MerT and a second di-cysteine dithiol pair in the cytoplasmic loop between the second and third alpha helices (Fig. 5a). All four cysteines are required for function and it is thought that Hg^{2+} is transferred initially from the small periplasmic MerP protein (see below) to the intramembrane cysteine pair, and then to the cytoplasmic cysteine pair in MerT, and finally directly to the N-terminal cysteine pair of a MerP-like domain at the N-terminus of mercuric reductase. The alternative membrane transport proteins, MerF (Howell et al. 2005) and MerC, have respectively two or four transmembrane alpha helices (Fig. 5a). However, both have an intramembrane N-terminal cysteine pair and a cytoplasmic loop cysteine pair, similar to MerT. The functional differences between these three membrane proteins, for example, binding of Hg^{2+} and movement from the membrane protein to mercuric reductase, are not known.

With mercury resistance systems of gram-negative bacteria, Hg^{2+} binding begins with highly specific binding in the periplasm to the small MerP protein (Steele and Opella 1997; Serre et al. 2004), a 72-amino-acid monomer required for transport of MerP across the membrane. The sequence and structure of MerP are unrelated to those of other periplasmic binding proteins. MerP has a tight β1α1β2β3α2β4 structure (Fig. 5b) with a conserved GMTC14X$_2$C17 Hg^{2+}-binding determinant in the exposed loop between the α1 and β1 segments. Surprisingly a closely homologous approximately 70-amino-acid Hg^{2+}-binding domain is found also at the N-terminus of the larger mercuric reductase enzymes (Silver and Phung 2005), whose subunit contains 560 amino acids with one copy (as with transposons in gram-negative bacteria; Fig. 2a) or 630 amino acids with two tandem

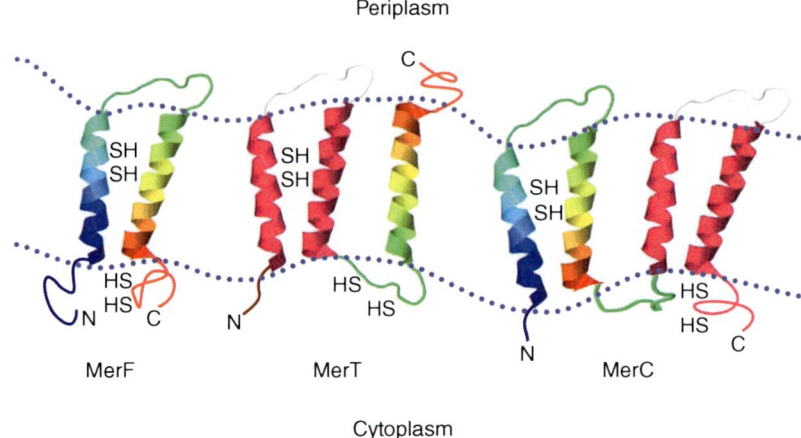

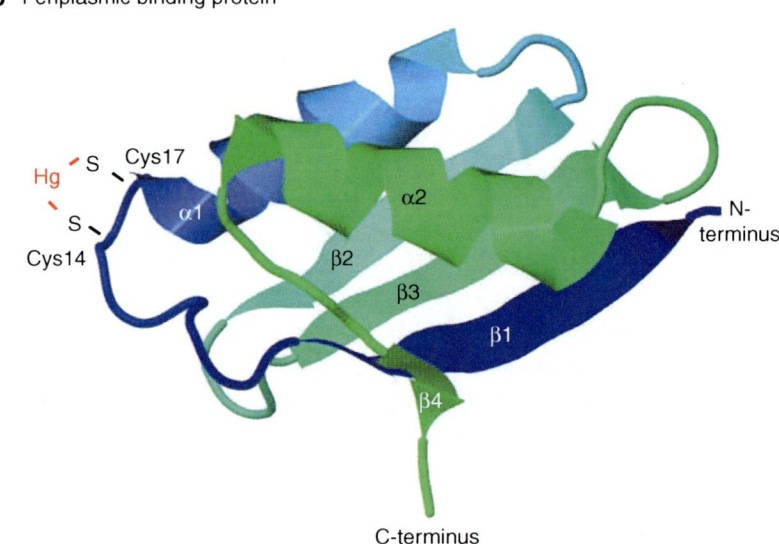

Bacterial Mercury Resistance Proteins, Fig. 5 (a) Mercury transport membrane proteins MerT, MerC, and MerF with cysteine pairs (SH) (Brown et al. 2002; Silver and Hobman 2007). The MerF structure is adapted from RCSB PDB database ID: 1WAZ, viewed in Jmol, and used to generate models for MerT and MerC. (b) The periplasmic MerP mercury-binding protein with the Cys14 Cys17 Hg^{2+} binding pair (Steele and Opella 1997), adapted from RCSB PDB database ID: 1OSD, viewed in Jmol

Hg^{2+}-binding domains (in gram-positive *Bacillus*). An exception is MerA of *Streptomyces*, which lacks this domain (and has about 480 amino acids). Closely related MerP-like cation-binding domains are found at the N-terminus of Cd^{2+}-resistance P-type ATPases in bacteria and Cu^+-transporting P-type ATPases in bacteria and humans (Silver and Phung 2005). The binding of Hg^{2+} by MerP has been well documented in protein structures for MerP with or without Hg^{2+} (Steele and Opella 1997). The basis of cation specificity with Hg^{2+} for MerP or Cd^{2+} or Cu^{2+} for homologs is, however, not known.

Transcriptional Regulatory Protein MerR

MerR, the transcriptional regulatory protein that binds to the DNA (operator/promotor region, OP; Fig. 1), was the first studied of what has subsequently been recognized as a larger class of such regulatory proteins (Barkay et al. 2003; Brown et al. 2003). MerR binds to the OP region of the DNA in both the absence and in the presence of Hg^{2+}. In the absence of Hg^{2+} (Fig. 6a), the −35 and −10 RNA polymerase–binding motifs in the DNA sequence are on opposite faces of the DNA (and in an unusually long nonfunctional distance) making it impossible for RNA polymerase to bind in

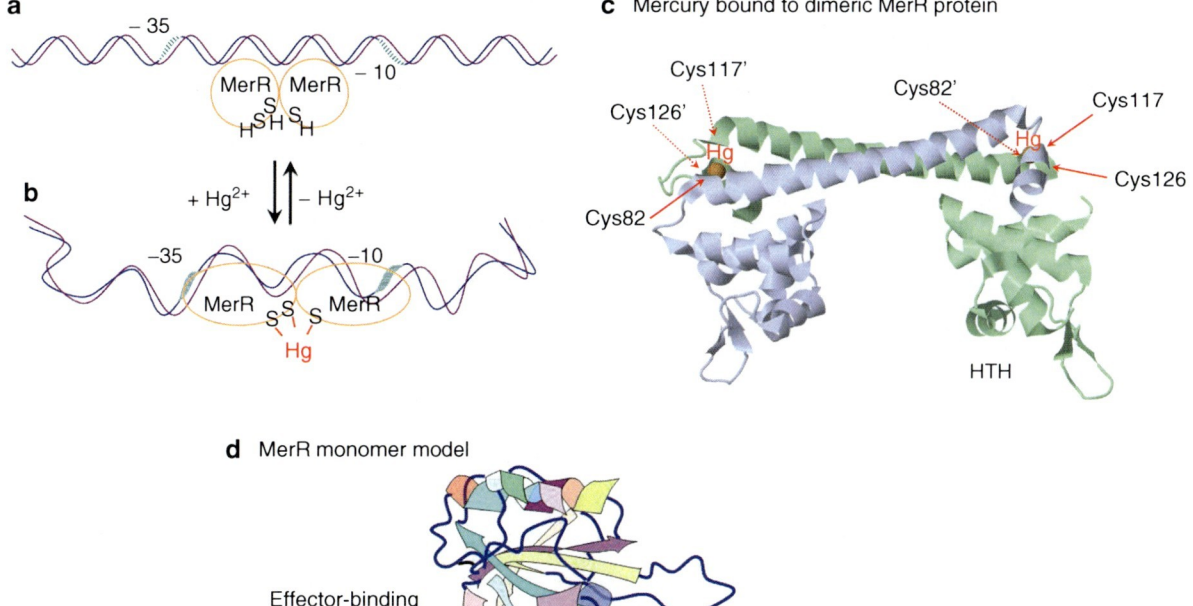

Bacterial Mercury Resistance Proteins, Fig. 6 (**a**) The transcriptional activator protein MerR bound to OP operator/promoter region DNA region without Hg^{2+} and (**b**) with Hg^{2+} with -35 and -10 RNA polymerase–binding sites on the same surface of the DNA with bending and twisting the DNA to "open" for RNA synthesis. (**c**) The structure of the MerR dimer with the three Hg^{2+}-binding cysteines (cysteines 117 and 126 of one monomer and cysteine 82' of the second monomer). One DNA-binding helix-turn-helix site is apparent in the structure (Modified from RCSB PDB database ID: 1Q05 biological assembly, viewed in Jmol for CueR, a close homolog of MerR) (Changela et al. 2003). (**d**) Structure of MerR monomer showing separate domains for Hg^{2+} binding and for binding to DNA (HTH, helix-turn-helix) connected by the long α5 alpha helix

a transcriptionally functional manner. DNA-bound MerR in the absence of Hg^{2+} therefore represses mRNA synthesis about 100-fold. Adding Hg^{2+} to MerR rotates the structure of the protein and results in a twisting and shortening of the DNA so that the -35 and -10 polymerase-binding motifs are properly positioned on the same surface, allowing RNA polymerase binding in a functional manner (Fig. 6b) (Utschig et al. 1995). The presence of Hg^{2+} stimulates mRNA synthesis about 100-fold – together making for an unusual 10,000-fold range of mRNA synthesis, from off to on. In the absence of Hg^{2+}, the MerR protein functions as a repressor, reducing transcription effectively to 0%; in the presence of Hg^{2+}, the unusual "twist and turn" of both the MerR protein and the DNA results in 100% maximum transcription.

Homodimeric MerR is a helix-turn-helix transcriptional regulator with amino acids 9 through 28 from the N-terminus forming an $\alpha 1$ helix-turn-$\alpha 2$ helix (HTH) DNA-binding motif (Fig. 6c and d) (Guo et al. 2010). A pair of long α helices ($\alpha 5$ and $\alpha 5$', one from each subunit) form the dimer subunit interface; and the MerR Hg^{2+}-binding sites are at the ends of helices $\alpha 5$ and $\alpha 5$', away from the DNA-binding HTH (Fig. 6c and d). Three cysteine residues, Cys82 of one subunit together with Cys117' and Cys126' of the other subunit, form one Hg^{2+}-binding site (Fig. 6c), but only one of the two sites binds Hg^{2+} at a time (Kaisa et al. 2011).

Mercury Oxidation by Catalase

The familiar enzyme hydroperoxidase-catalase, whose usual physiological function is to accelerate the rate of conversion of H_2O_2 to H_2O, can also catalyze the oxidation of relatively unreactive mono-atomic Hg^0 (which is present in the cell as a gas) to more reactive Hg^{2+} (Smith et al. 1998). This enzyme is found broadly from E. coli (encoded by the katG gene) to humans and in the environment converts Hg^0 gas to water-soluble Hg^{2+}, completing the mercury cycle. Abiotic nonenzymatic oxidation of Hg^0 is probably responsible for atmospheric Hg^0 being deposited in the sea and lakes (Barkay et al. 2003). However, but catalase in animals is probably responsible for converting the nontoxic Hg^0 (breathed into the lungs and transferred into the red blood cells in the lungs) to Hg^{2+}, which remains in the body and can damage metabolism as a protein thiol poison.

Mercury Methylation

The methylating of inorganic Hg^{2+} to CH_3Hg^+ (methyl mercury) and then to some extent to dimethyl mercury, $(CH_3)_2Hg$, appears to be nonenzymatic and therefore not carried out directly by proteins. Methylation appears to result from cell-released methyl-cobalamin, as an inadvertent side reaction and not a favored process for that abundant cofactor (Silver and Phung 2005; Barkay et al. 2003). Mercury is not methylated with S-adenosylmethionine (SAM), as is inorganic arsenic.

Cross-References

▶ Arsenic-Induced Stress Proteins

References

Barkay T, Miller SM, Summers AO (2003) Bacterial mercury resistance from atoms to ecosystems. FEMS Microbiol Rev 27:355–384

Brown NL, Shih Y-C, Leang C, Glendinning KJ, Hobman JL, Wilson JR (2002) Mercury transport and resistance. Biochem Soc Trans 30:715–718

Brown NL, Stoyanov JV, Kidd SP, Hobman JL (2003) The MerR family of transcriptional regulators. FEMS Microbiol Rev 27:145–163

Changela A, Chen K, Xue Y, Holschen J, Outten CE, O'Halloran TV, Mondragon A (2003) Molecular basis of metal-ion selectivity and zeptomolar sensitivity by CueR. Science 301:1383–1387

Chen CY, Hsieh JL, Silver S, Endo G, Huang CC (2008) Interactions between two MerR regulators and three operator/promoter regions in the mercury resistance module of Bacillus megaterium. Biosci Biotechnol Biochem 72:2403–2410

Guo H-B, Johs A, Parks JM, Olliff L, Miller SM, Summers AO, Liang L, Smith JC (2010) Structure and conformational dynamics of the metalloregulator MerR upon binding of Hg (II). J Mol Biol 398:555–568

Howell SC, Mesleh MF, Opella SJ (2005) NMR structure determination of a membrane protein with two transmembrane helices in micelles: MerF of the bacterial mercury detoxification system. Biochemistry 44:5196–5206

Kaisa M, Hakkila KM, Nikander PA, Junttila SM, Lamminmäki UJ, Virta MP (2011) Cd-specific mutants of mercury-sensing regulatory protein MerR, generated by directed evolution. Appl Environ Microbiol 77:6215–6224

Lafrance-Vanasse J, Lefebvre M, Di Lello P, Sygusch J, Omichinski JG (2009) Crystal structures of the organomercurial lyase MerB in its free and mercury-bound forms. J Biol Chem 284:938–944

Ledwidge R, Patel B, Dong A, Fiedler D, Falkowski M, Zelikova J, Summers AO, Pai EF, Miller SM (2005) NmerA, the metal binding domain of mercuric ion reductase, removes Hg^{2+} from proteins, delivers it to the catalytic core, and protects cells under glutathione-depleted conditions. Biochemistry 44:11402–11416

Miller SM (2007) Cleaving C-Hg bonds: two thiolates are better than one. Nat Chem Biol 3:537–538

Parks JM, Guo H, Momany C, Liang L, Miller SM, Summers AO, Smith JC (2009) Mechanism of Hg–C protonolysis in the organomercurial lyase MerB. J Am Chem Soc 131: 13278–13285

Serre L, Rossy E, Pebay-Peyroula E, Cohen-Addad C, Coves J (2004) Crystal structure of the oxidized form of the periplasmic mercury-binding protein MerP from Ralstonia metallidurans CH34. J Mol Biol 339:161–171

Silver S, Hobman J (2007) Mercury microbiology: resistance systems, environmental aspects, methylation and human health. In: Nies DH, Silver S (eds) Molecular microbiology of heavy metals. Springer, Heidelberg, pp 357–370

Silver S, Phung LT (2005) A bacterial view of the Periodic Table: genes and proteins for toxic inorganic ions. J Ind Microbiol Biotechnol 32:587–605

Smith T, Pitts K, McGarvey JA, Summers AO (1998) Bacterial oxidation of mercury metal vapor, Hg(0). Appl Environ Microbiol 64:1328–1332

Steele RA, Opella SJ (1997) Structures of the reduced and mercury-bound forms of MerP, the periplasmic protein from the bacterial mercury detoxification system. Biochemistry 36:6885–6895

Utschig LM, Bryson JW, O'Halloran TV (1995) Mercury-199 NMR of the metal receptor site in MerR and its protein-DNA complex. Science 268:380–385

Bacterial Response to Gallium

▶ Gallium in Bacteria, Metabolic and Medical Implications

Bacterial Tellurite Processing Proteins

Raymond J. Turner
Department of Biological Sciences,
University of Calgary, Calgary, AB, Canada

Synonyms

Chalcogen resistance; Metal resistance; Metalloid resistance; Tellurite tolerance

Definition

Tellurite processing proteins from bacteria are those that have been shown to have some biochemical activity and or transformational effect on the tellurite oxyanion (TeO_3^{2-}).

Introduction

Tellurite resistance determinants (groups of genes or an operon specifically responsible) have been isolated and characterized by a number of groups. The Te^r genes first appeared associated with conjugated plasmids and were described in the 1970s (reviewed by Walter and Taylor 1992). However, there have now been many physiological studies that have identified genes, and/or their gene product, that display a basal activity responsible for tellurite oxyanion processing and in many cases detoxification (Zannoni et al. 2008; Chasteen et al. 2009; Turner et al. 2011). It is also not unusual to find multiple genes within a given bacterium that upon deletion, or overexpression, display a tellurite-dependent phenotype. The observations suggest that these genes provide some selective advantage in natural environments (Chasteen et al. 2009), and that these enzymes, although not specific for tellurite, are able to process it and in some cases other oxyanions (Zannoni et al. 2008).

Genes Involved in Basal Tellurite Resistance Levels

Upon studying various genes in bacteria for their physiological contributions, several have been identified to be necessary for the basal resistance to tellurite. Their deletion results in higher sensitivity to tellurite, and in many cases general oxidative stress (reviewed by Zannoni et al. 2008; recent work cited below). Most of this work has been explored in *Escherichia coli*, yet many other organisms are focus of study for tellurite resistance. Such chromosomal determinants in this category include:

NarGHI, NapA; These proteins are defined as nitrate reductase. In *E. coli*, NarGHI and its homologue NarXYZ are cytoplasmically localized membrane associated bioenergetic enzymes. These enzymes have demonstrated selenate reductase activity but also tellurite reduction activity (Avazeri et al. 1997). The loss of these enzymes shows a 200-fold decrease in resistance to tellurite, presenting an MIC of 0.03 µg/mL from the wild-type level of 2. Additionally, it has been shown that the periplasmically localized nitrate reductase, NapA, also has this activity.

Trx, Grx, Gor, Gsh; The genes are involved in thiol redox buffering and contribute to the redox poise with the cytoplasm of bacteria. They include thioredoxin (*trx*), glutaredoxin (*grx*), and glutathione reductase (*gor*) and glutathione biosynthetic enzymes (*gsh*). Deletion of any of these genes can lead to a loss of resistance to an MIC of 0.25–0.5 from 2 µg/ml (Turner et al. 1995). It is assumed that a Painter-like reaction, similar to thiol reaction with selenite, is occurring where $2RSH + TeO_3^{2-} \rightarrow RSTeRS$. This reaction is thought to continue to RSSR and Te(0) (Turner et al. 1998; Zannoni et al. 2008). Consistent with this observation is *iscS* (cysteine desulfurase) which was found to confer some of the resistance (Tantalean et al. 2003).

DsbAB; The loss of either *dsbA* or *dsbB* genes coding for disulfide-bond catalyzing enzymes in the periplasm showed a remarkable loss of tellurite tolerance down to 0.008–0.0015 μg/ml from the base level of 2 in wild type. Unexpectedly, the same observation was not observed for the other *dsb* genes (C and D) (Turner et al. 1999). The biochemical mechanism for this observation has been suggested through studies in *Rhodobacter capsulatus* as a reverse electron sink (Borsetti et al. 2007).

Sod; The loss the superoxide dismutase genes of *sodA* and/or *sodB* in *E. coli* displayed a reduction of basil resistance to a level of 0.125–0.25 μg/ml (Turner et al. 1995). This is generally thought to be due to the need of Sod activity to recover from the general oxidative response from the reduction reactions of tellurite to elemental tellurium.

BtuE; a peroxidase (Arenas et al. 2011) displayed a role in tellurite resistance when in a catalase/peroxidase deletion strain. *KatG*; *E. coli* mutants in this catalase/peroxidase gives increased sensitivity and an extract enzyme showed a NAD(P)H-dependent reduction of tellurite (Calderon et al. 2006).

ActP; The acetate permease of *Rhodobacter capsulatus* was defined as the tellurite uptake transporter in this organism and Δ*actP* strains showed a 25-fold increase in tellurite resistance (Borghese and Zannoni 2010). It was concluded that tellurite entrance into the cells is via this monocarboxylate transporter.

LepA; The translational GTPase *lepA* is well conserved in bacteria. It was observed that a Δ*lepA* *E. coli* strain became sensitive to 0.05 μg/ml tellurite (Shoji et al. 2010).

AceEF/LpdA; The pyruvate dehydrogenase (*aceE*), dihydrolipoamide transacetylase (*aceF*), and dihydrolipoamide dehydrogenase (*lpdA*) from *Aeromonas caviae ST* have been shown to give rise to tellurite resistance when cloned and overexpressed. It is considered that all three of these enzymes have the ability to reduce TeO_3^{2-} to Te(0) (overviewed in Chasteen et al. 2009).

CobA; *Geobacillus stearothermophilus V cobA* gene encoding uroporphyrinogen-III C-methyltransferase utilizes S-adenosyl-L-methionine and mediates moderate levels of tellurite resistance when expressed (Araya et al. 2009). However, similar to specific tellurite resistance determinant TehB, no methylated volatile derivatives are found as products suggesting additional reactions must occur.

YqhD; This gene provides tolerance to compounds that generate membrane lipid peroxidation and has a NADPH-dependent aldehyde reduction activity (Perez et al. 2008). As expression of this *yqhD* in *E. coli* provides for tellurite resistance, it implies that peroxidation is a consequence of tellurite exposure.

Summary

Overall the different tellurite processing proteins identified to date are very different at the genetic and protein functional level. The common phenotype of bacteria exposed to tellurite is the reduction of TeO_3^{2-} to Te(0) crystals which is displayed by a blackening of broth cultures and colonies. This activity has directed researchers toward redox biochemistries. Overall, there appears to be many different enzymatic activities involving tellurite, yet it has been difficult to separate direct versus indirect affects. There is a clear dependence on the physiological state of the cells and that the thiol–redox balance is important as tellurite can react with reduced thiol compounds readily at physiological conditions (Turner et al. 1995, 1999). This parallels the observation of general reduction activities that are observed by various metabolic systems as well as the oxidative stress enzymes. However, this is not a consistent observation and likely is species dependent as little reactive oxygen species were observed when *E. coli* is exposed to tellurite compared to other metals (Harrison et al. 2009). A metabolomic study evaluating the hyper-resistance to tellurite in *Pseudomonas pseudoalcaligenes* KF707 is correlated with the induction of oxidative stress response, resistance to membrane perturbation, and general reconfiguration of cellular metabolism (Tremaroli et al. 2009). These observations suggest a pleiotropic biochemistry for the tellurium oxyanion.

Cross-References

▶ Bacterial Tellurite Resistance
▶ Tellurium and Oxidative Stress
▶ Tellurium in Nature
▶ Tellurite-Detoxifying Protein TehB from *Escherichia coli*

References

Araya MA, Tantaleán JC, Pérez JM, Fuentes DE, Calderón IL, Saavedra CP, Burra R, Chasteen TG, Vásquez CC (2009) Cloning, purification and characterization of *Geobacillus stearothermophilus* V uroporphyrinogen-III C-methyltransferase: evaluation of its role in resistance to potassium telluirte in *Escherichia coli*. Res Microbiol 160:125–133

Arenas FA, Covarrubias PC, Sandovai JM, Perez-Donoso JM, Imlay JA, Vasquez CC (2011) The *Escherichia coli* BtuE protein functions as a resistance determinant against reactive oxygen species. PLoS One 6:e15979

Avazeri C, Turner RJ, Pommier J, Weiner JH, Giordano G, Vermeglio A (1997) Tellurite and selenate reductase activity of nitrate reductases from *Escherichia coli:* correlation with tellurite resistance. Microbiology 143:1181–1189

Borghese R, Zannoni D (2010) Acetate permease (ActP) is responsible for tellurite (TeO_3^{2-}) uptake and resistance in cells of the facultative phototroph *Rhodobacter capsulatus*. Appl Environ Microbiol 76:942–944

Borsetti F, Francia F, Turner RJ, Zannoni D (2007) The thiol: disulfide oxidoreductase DsbB mediates the oxidizing effects of the toxic metalloid tellurite (TeO_3^{2-}) on the plasma membrane redox system of the facultative phototroph *Rhodobacter capsulatus*. J Bacteriol 189:851–859

Calderon IL, Arenas FA, Perez JM, Fuentes DE, Araya MA, Saavedra DP, Tantalean JC, Pichuantes SE, Youderian PA, Vasquez CC (2006) Catalases are NAD(P)H-dependent tellurite reductases. PLoS One 20:e70

Chasteen TG, Fuentes DE, Tantalean JC, Vasquez CC (2009) Tellurite: history, oxidative stress, and molecular mechanisms of resistance. FEMS Microbiol Rev 33:820–832

Harrison JJ, Tremaroli V, Stan MA, Chan CS, Vacchi-Suzzi C, Heyne BJ, Parsek MR, Ceri H, Turner RJ (2009) Chromosomal antioxidant genes have metal ion-specific roles as determinants of bacterial metal tolerance. Environ Microbiol 11:2491–2509

Perez JM, Arenas FA, Pradenas GA, Sandoval JM, Vasquez CC (2008) Escherichia coli YqhD exibits aldehyde reductase activity and protects from the harmful effect of lipid peroxidation-derived aldehydes. J Biol Chem 283:7346–7353

Shoji S, Janssen BD, Hayes CS, Fredrick K (2010) Translation factor LepA contributes to tellurite resistance in *Escherichia coli* but plays no apparent role in the fidelity of protein synthesis. Biochimie 92:157–163

Tantalean JC, Araya MA, Saavedra CP, Fuentes DE, Perez JM, Calderon IL, Youderian P, Vasquez CC (2003) The *Geobacillus stearothermophilus* V iscS gene, encoding cysteine desulfurase, confers resistance to potassium tellurite in *Escherichia coli* K-12. J Bacteriol 185:5831–5837

Tremaroli V, Workentine ML, Weljie AM, Vogel HJ, Ceri H, Viti C, Tatti E, Zhang P, Hynes AP, Turner RJ, Zannoni D (2009) Metabolomic investigation of the bacterial response to a metal challenge. Appl Environ Microbiol 75:719–728

Turner RJ, Weiner JH, Taylor DE (1995) The tellurite resistance determinants *tehAtehB* and *klaAklaBtelB* have different biochemical requirements. Microbiology 141:3133–3140

Turner RJ, Weiner JH, Taylor DE (1998) Selenium metabolism in *Escherichia coli*. Biometals 11:223–227

Turner RJ, Weiner JH, Taylor DE (1999) Tellurite-mediated thiol oxidation in *Escherichia coli*. Microbiology 145:2549–2557

Turner RJ, Borghese R, Zannoni D (2011) Microbial reduction of tellurium metalloids as a tool in biotechnology. Biotech Adv. doi 10.1016/j.biotechadv.2011.08.018 (in press)

Walter EG, Taylor DE (1992) Plasmid-mediated resistance to tellurite: expressed and cryptic. Plasmid 27:52–64

Zannoni D, Borsetti F, Harrison JJ, Turner RJ (2008) The bacterial response to the chalcogen metalloids Se and Te. Adv Microbial Physiol 53:1–71

Bacterial Tellurite Resistance

Raymond J. Turner
Department of Biological Sciences, University of Calgary, Calgary, AB, Canada

Synonyms

Chalcogen resistance; Metal resistance; Metalloid resistance; Tellurite tolerance

Definition

Bacteria resistance to tellurite is defined here as specific genes, cluster of genes, or specific genetic operons referred to as resistance determinants. Tellurite (TeO_3^{2-}), redox state Te(IV), is the oxyanion of the chalcogen tellurium (Te; atomic number 52) in group 16 with elements O, S, Se, and Po. In addition to specific determinants a number of bacteria species have been identified to have high levels of resistance to tellurite yet the specific genetics has not been defined; these are referred to as tellurite tolerant.

Introduction

Bacteria treat tellurite quite differently from other antibacterial metal ions. Evolution has seen that a biochemical mechanism of resistance is uniform across species for most all other metal ions other than tellurite. At least five genetically different resistance determinants have been identified with little to no homology between them. Genes responsible for tellurite resistance (Te^r) in various organisms have been isolated and characterized by a number of groups. The Te^r genes first appeared associated with conjugated plasmids and were described by Anne Summer and Diane Taylor toward the end of 1970s (reviewed by Walter and Taylor 1992); however, several determinants and plasmid homologues have now been found associated with the chromosome as well (Taylor 1999; Zannoni et al. 2008; Chasteen et al. 2009; Turner et al. 2011). Plasmid-encoded resistance determinants are generally associated with plasmids of the H and P incompatibility groups. There are also several Te^r determinants emerging from various bacterial families, suggesting that these determinants provide some selective advantage in natural environments (Chasteen et al. 2009). Such advantages may be unrelated to the Te^r phenotype, as the levels of resistance demonstrated in the laboratory do not always correlate with the levels of tellurium ion species present in the ecological or pathogenic environment. An interesting characteristic of the genes encoding Te^r is they often confer other phenotypes as well, which suggests tellurite-processing proteins could be a form of moonlighting enzymatic activity.

Tellurite is typically sold and used as potassium tellurite; K_2TeO_3. It is difficult to clearly define the ionic species that the bacteria would experience in various growth medium or in the environment. Te can exist in a number of redox states: Telluride (Te^{2-}) → elemental (Te^0) → tellurite (TeO$_3^{2-}$) → tellurate (TeO$_4^{2-}$). In aqueous conditions in water Te(IV) at pH 7.0 exists at a ratio of HTeO$_3^-$/TeO$_3^{2-}$ of ~10^4/1. Te (VI) would likely be tellurate, TeO$_4^{2-}$. Thus the standard reduction potential of the Te/TeO$_3^{2-}$ couple (−0.42 V) at basic pHs would be raised to -0.12 V for the couple HTeO$_3^-$/TeO$_3^{2-}$ at pH 7.0, with no Te^{4+} present due to its instability in water. Although one cannot rule out that the oxyanion of Te could be complexed with organo or metallo cations, the primary species bacteria likely experience for the toxic form that they have developed resistance to is HTeO$_3^-$ (Zannoni et al. 2008).

Tellurite Resistance Determinants

There are now six well-defined Te^r determinants including the Ter, Teh, Tel, Tpm, CysK, and Ars. Their resistance levels toward K_2TeO_3 are summarized in Table 1.

Bacterial Tellurite Resistance, Table 1 Resistance levels mediated by specific tellurite resistance determinants

Te^r	MIC (μg K$_2$TeO$_3$/ml)	Mechanism[a, b]
NONE	1–4[c]	
Ter	512–1,024	Reductase[b]
TehAB	128	Methylation[a] and efflux[b]
KlaABTelB	256	Unknown
TpmT	256	Methyltransferase[b]
CysK	1,000	Reductase[b]
ArsABC	64	ATP-dependent efflux[a]

[a]Known mechanism
[b]Proposed or hypothesized
[c]The typical resistance level of most bacteria is in this range. However, highly resistant species do exist and their resistance does not utilize these specific determinants

Ter

Plasmids within the incompatibility group HI-2 and HII confer protection against colicins and resistance to potassium tellurite (Taylor 1999). These resistances are associated with a large cluster of genes (*terZABCDEF*) referred to as the *ter* Te^r determinant (Walter and Taylor 1992). This determinant was also found to be associated with the pathogenicity island of *Escherichia coli* H157:O7, and other pathogenic organisms such as *Yersinia pestis* and *Klebsiella pneumoniae*. A *terZABCDE* operon was also identified in *Proteus mirabilis*, and is common in the *Proteus* genus. Further, *ter* gene homologues are found on the chromosomes of a wide range of bacteria yet little clues are provided to their function. Little is known about the function of each of the genes in the operon. Overall, there is considerable homology between the *ter* genes of the different plasmids and chromosomes, yet the *ter* operon appears to be differently regulated in different organisms.

The biochemical mechanism of the Ter determinant remains unknown. It has been suggested to be involved in tellurite transport, tellurite reduction to Te(0), reduced protein, and glutathione thiol oxidation. The TerD structure has been determined by NMR (PBP: 2KXV), which revealed that this protein binds Ca^{2+} in a common fold that are also thought to be found in TerE and TerZ. TerB NMR structure is also available (PDB 2JXU) yet no clues on the mechanism are provided. There are conflicting reports whether there is increased tellurite uptake or not, yet it is clear that all studies show the accumulation of Te(0) nanocrystals (see Zannoni et al. 2008 and Turner et al. 2011).

TehAB

The *tehAB* genes were first described as a Te^r that was believed to have originated from the IncHII plasmid pHH1508a (Walter and Taylor 1992). However, follow-up studies demonstrated that it was actually on the *E. coli* genome (Taylor et al. 1994), and now homologues are found on many bacteria genomes. The genes mediate resistance only upon overexpression. TehA appears to have sequence characteristics of a transport protein and falls into the C4-dicarboxylate transporter/malic acid family. TehB associates weakly with the membrane. It contains three conserved motifs found in S-adenosylmethionine (SAM)-dependent nonnucleic acid methyltransferase (Liu et al. 2000) and has demonstrated SAM binding and methylase activity. Pairs of cysteines in TehA and TehB have been shown to be required for full resistance. Additionally, the resistance is very dependent on many other host metabolic systems (Turner et al. 1995). TehB has the ability to mediate resistance on its own and is partially responsible for the natural resistance of a number of organisms including *Streptococcus*. TehB has been recently crystallized (PDB: 2XVM) and suggests an SN2 nucleophilic attack between the SAM to the telluro oxyanion (Choudhury et al. 2011). Although TehB appears to be a SAM-dependent telluro methylase, methyl telluride does not appear to be the final product in vivo. It is hypothesized that further reactions may occur in the cell before a final organotelluro compound is effluxed out (Zannoni et al. 2008).

kilAtelAB/klaABtelB

The RK2 plasmids have a complex network of co-regulated genes known as the *kil-kor* regulon. A normally cryptic Te^r was identified on some isolates as RK2Te^R and was mapped to the *kilA* locus. The operon is comprised of three genes, *klaA*, $-B$, $-C$, which in the Te^r versions are referred to as *kilABTelB*. A single Ser125 to Cys mutation in the integral membrane protein TelB mediates the resistance, and this mutation is solely responsible for the resistance; yet all three of the genes are required for the resistance. The mutation generates a cysteine pair (Cys 125: 132) in an extramembrane loop. Both cysteines are required for resistance and the presence of the determinant protects against cell glutathione oxidation (Turner et al. 2001). The data to date suggests some form of thiol chemistry may be involved in the resistance mechanism. As opposed to the *tehAB* determinant, *kilAtelAB* is much less dependent on the physiological state of the cell to mediate full resistance (Turner et al. 1995). Although there is apparently less "blackening" of cultures' reduced uptake or efflux of tellurite has been ruled out for this determinant as the biochemical mechanism (Turner et al. 1995).

There is some confusion in the literature in that KlaA homologues are found on the chromosome of many organisms, however, it is referred to as TelA and is incorrectly annotated as putative toxic anion resistance protein. It was found that a chromosomal homologue of TelA in *Listeria monocytogenes* demonstrated resistance to antibiotics like nisin, yet no mechanism was proposed. Due to the lethality phenotype associated with the kilA(klaA) gene, very little microbial and biochemical studies are available and thus the biochemical mechanism of this determinant remains elusive.

TpmT

The *tpm* gene was cloned from the tellurite resistant *Pseudomonas syringae* pathovar *pisi* (Cournoyer et al. 1998). Since this observation the gene has been found on a number of other *Pseudomonas spp. tpmT* encodes a SAM-dependent thiopurine methyltransferase enzyme, which led the authors to propose that the resistance likely occurs through a volatilization of tellurite into dimethyl telluride.

CysM/CysK

Chromosomally encoded genes, homologous to those involved in cysteine biosynthesis, have been found to mediate tellurite resistance. The *cysM* gene from *Staphylococcus aureus* SH1000 was found to be functionally homologous to the *O*-acetyl serine (thiol)-lyase B family of cysteine synthase proteins. A deletion in this gene gives increased sensitivity to tellurite and could mediate Te^R when transformed into *E. coli*. A homologue of this gene was also identified on an IncHI3 plasmid and was designated *cysK*. This enzyme is a pyridoxal 5′-phosphate-dependent enzyme and catalyzes the transformation of *O*-acetyl-L serine and S^{2-} to L-cysteine and acetate (Ramirez et al. 2006). Somehow this reductase-like enzyme mediates resistance to both pore forming colicins and tellurite, which suggests some parallels toward the Ter determinant yet no homology exists. Homologues of *cysK* have been identified in

Azospririllum brasilense and *Geobacter stearothermophilus*. In the second case, a NADH-dependent reduction of tellurite and an additional gene of *iscS* (cysteine disulfurase) were found to confer some of the resistance (Tantalean et al. 2003). CysK was considered to be a key determinant responsible for the tellurite resistance in *Rhodobacter sphaeroides*, however, other genes such as *trgAB* and a *telA* homologue were also involved (O'Gara et al. 1997). This above work suggests that CysK likely does not mediate the resistance on its own but in concert with other universal systems found in other bacteria as they demonstrate resistance to *E. coli* when cloned.

ArsABC

The arsenate/arsenite resistance determinant *arsABC* which is an arsenate reductase (ArsC) with an ATP-dependent arsenite efflux pump (ArsAB) was also found to mediate moderate levels of tellurite resistance (Turner et al. 1992). It is likely that the pump also recognizes the $HTeO_3^-$ ion. Although the *arsC* gene was required to mediate full resistance, its role is unknown.

Tellurite Tolerant Bacteria

Intrinsic low-level tellurite resistance has been reported for a few Gram positive organisms such as *Corynebacterium diphtheriae, Streptococcus faecalis*, and some *Staphylooccus aureus* strains. Yet increasing number of organisms have been described to have high levels of tellurite resistance and/or processing (reduction), however, the specific genetics and/or the biochemical processes are not completely resolved. Some examples include: *Stenotrophomonas maltophilia Sm777, Pseudomonas pseudoalcaligenes* KF707, *Rhodotorula mucilaginosa*. Additionally, there are some "super" tellurite resistant bacteria such as *Pseudoalteromonas telluritireducens, Pseudoalteromonas spiralis, Bacillus* sp STG-83 and *Paenibacillus sp.* TeW which is tolerant to ~300–500 μg/ml. A number of obligately aerobic photosynthetic bacteria have been defined with very high levels of tellurite resistance (MIC range from 750 to 2,300 μg/mL) and include: *Erythrobacter litoralis, Erythromicrobium hydrolyticum, E. ursincola, E. ramosum, E. sibiricum, E. ezovicum, Roseococcus thiosulfatophilus* (Yurkov et al. 1996).

Summary

Overall the different determinants are very different at the genetic and protein functional level. However, many of the resistance determinants have at least one gene that is an integral membrane protein (TerC, TehA, TelA). Another common theme is that resistant organisms display increased reduction of tellurite to Te (0) crystals which is displayed by a blackening of broth cultures and colonies. For the most part, the Te^r determinants *ter, klaABtelB, teh*, and *ars* do not mediate resistance to tellurate (Te(VI), TeO_4^{2-}). However, some of the specific determinants and other general metabolite genes involved with tellurite tolerance also mediate some resistance to selenite (SeO_3^{2-}). An example is TpmT. Overall the different Te^r appear to have different dependence on the physiological state of the cells and that the thiol–redox balance is important as tellurite can react with reduced thiol compounds readily at physiological conditions (Turner et al. 1995, 1999). This parallels the observation of general reduction activities that are observed by various metabolic systems as well as the oxidative stress enzymes. Unfortunately, the nature of the chemistry of tellurite reduction makes for challenging biochemistry experiments with the determinants. This and the apparent pleiotropic nature of the physiological targets has not allowed for a clear view of microbiological and biochemical specific mechanisms.

Cross-References

▶ Bacterial Tellurite Resistance
▶ Catalases as NAD(P)H-Dependent Tellurite Reductases
▶ Tellurite-Resistance Protein TehA from *Escherichia coli*
▶ Tellurium and Oxidative Stress
▶ Tellurite-Detoxifying Protein TehB from *Escherichia coli*

References

Chasteen TG, Fuentes DE, Tantalean JC, Vasquez CC (2009) Tellurite: history, oxidative stress, and molecular mechanisms of resistance. FEMS Microbiol Rev 33: 820–832

Choudhury HG, Cameron AD, Iwata S, Beis K (2011) Structure and mechanism of the chalcogen-detoxifying protein TehB from *Escherichia coli*. Biochem J 435:85–91

Cournoyer B, Watanabe S, Vivian A (1998) A tellurite-resistance genetic determinant from phytopathogeneic pseudomonads encodes a thiopurine methyltransferase: evidence of a widely conserved family of methyltransferases. Biochim Biophys Acta 1297:161–168

Liu M, Turner RJ, Winstone TL, Saetre A, Dyllick-Brenzinger M, Jickling G, Tara LW, Weiner JH, Taylor DE (2000) *Escherichia coli* TehB requires S-adenosylmethionine as a cofactor to mediate tellurite resistance. J Bacteriol 182:6509–6513

O'Gara JP, Gomelsy M, Kaplan S (1997) Identification and molecular genetic analysis of multiple loci contributing to high-level tellurte resistance in *Rhodobacter sphaeroides* 2.4.1. Appl Environ Microbiol 63:4713–4720

Ramirez A, Castaneda M, Xiqui ML, Sosa A, Baca BE (2006) Identification, cloning and characterization of *cysK*, the gene encoding O-acetylserine (thiol)-lyase from *Azospirillum brasilense*, which is involved in tellurite resistance. FEMS Microbiol Lett 261:272–279

Tantalean JC, Araya MA, Saavedra CP, Fuentes DE, Perez JM, Calderon IL, Youderian P, Vasquez CC (2003) The *Geobacillus stearothermophilus* V *iscS* gene, encoding cysteine desulfurase, confers resistance to potassium tellurite in *Escherichia coli* K-12. J Bacteriol 185:5831–5837

Taylor DE (1999) Bacterial tellurite resistance. Trends Microbiol 7:111–115

Taylor DE, Hou Y, Turner RJ, Weiner JH (1994) Location of a potassium tellurite resistance operon (*tehAtehB*) within the terminus of *Escherichia coli* K-12. J Bacteriol 176:2740–2742

Turner RJ, Hou Y, Weiner JH, Taylor DE (1992) The arsenical ATPase efflux pump mediates tellurite resistance. J Bacteriol 174:3092–3094

Turner RJ, Weiner JH, Taylor DE (1995) The tellurite resistance determinants *tehAtehB* and *klaAklaBtelB* have different biochemical requirements. Microbiology 141:3133–3140

Turner RJ, Weiner JH, Taylor DE (1999) Tellurite-mediated thiol oxidation in *Escherichia coli*. Microbiology 145:2549–2557

Turner RJ, Aharonowitz Y, Weiner JH, Taylor DE (2001) Glutathione is a target in bacterial tellurite toxicity and is protected by tellurite resistance determinants in *Escherichia coli*. Can J Microbiol 47:33–40

Turner RJ, Borghese R, Zannoni D (2011) Microbial reduction of tellurium metalloids as a tool in biotechnology. Biotechnol Adv. doi:10.1016/j.biotechadv.2011.08.018

Walter EG, Taylor DE (1992) Plasmid-mediated resistance to tellurite: expressed and cryptic. Plasmid 27:52–64

Yurkov V, Jappe J, Vermeglio A (1996) Tellurite resistance and reduction by obligately aerobic photosynthetic bacteria. Appl Environ Microbiol 62:4195–4198

Zannoni D, Borsetti F, Harrison JJ, Turner RJ (2008) The bacterial response to the chalcogen metalloids Se and Te. Adv Microbial Physiol 53:1–71

Barium

▶ Barium(II) Transport in Potassium(I) and Calcium(II) Membrane Channels

Barium and Protein–RNA Interactions

Thirumananseri Kumarevel
RIKEN SPring-8 Center, Harima Institute,
Hyogo, Japan

Synonyms

Metal-ion-mediated protein–nucleic acid interactions; Role of metal ions in protein–nucleic acid complexes

Definition

Genomic studies are providing researchers with a potentially complete list of the molecular components present in living systems. It is now obvious that several metal ions are essential to life. More specifically, biological macromolecules (proteins and nucleic acids) that require metal ions to perform their physiological functions are widespread in all organisms. Here, we explored the importance and involvement of one of the alkali earth metals, barium, in the biological system. Based on structural and functional analyses, we clearly demonstrated how the divalent metal ions produce the structural rearrangements that are required for *hut* mRNA recognition. The applications and health risks of barium metal ions are also discussed.

General Background on Barium Metal

The chemical element barium is the 56th element in the chemical periodic table with the symbol of *Ba*, atomic number of 56, and weight of the 137.327. It is the fifth element in group 2, a soft silvery white metallic, and one of the alkaline earth metals. Barium is never found in nature in its pure form due to its reactivity with air. The metal oxidizes very easily and reacts with water or alcohol. The most commonly occurring minerals containing this element are the very insoluble barium sulfate (BaSO$_4$) and barium carbonate (BaCO$_3$).

The name barium originates from the Greek word "barys," meaning "heavy," which refers to the high density of some common barium ores. It was first discovered by electrolysis of molten barium salts by Sir Humphrey Davy in 1808 in England. Prior to this discovery, Carl Scheele had *distinguished* baryta (barium oxide, BaO) from lime (calcium oxide, CaO) in 1774 but could not isolate barium itself (Mark Winter, "Webelements," www.webelements.com).

Physical and Chemical Properties of Barium Metal

The general, physical, and other important properties of barium are provided in Table 1. Barium is a soft and ductile metal. The barium compounds are notable for their relatively high specific gravity and are also called heavy spar due to their high density. Barium is a highly reducing metal. Thus, the surface of barium metal is covered with a thin layer of oxide that helps protect the metal from attack by air. Once ignited, barium metal burns in air and yields a mixture of white barium oxide (BaO) and barium nitride (Ba$_3$N$_2$). In the periodic table, barium is placed three positions below magnesium, indicating that barium is more reactive with air than magnesium.

For example, $2Ba + O_2 \rightarrow 2BaO$

$$B2 + O_2 \rightarrow BaO_2$$

$$3Ba + N_2 \rightarrow Ba_3N_2$$

Barium reacts readily with water to form barium hydroxide and hydrogen gas. The reaction is violent if barium is powdered. Barium also reacts violently with dilute acids, alcohol, and water. It also combines with several metals, including aluminum, zinc, lead, and tin, to form intermetallic compounds and alloys:

$$Ba + 2H_2O \rightarrow Ba(OH)_{2+}H_2 \uparrow$$

The abundance of barium in different environments is shown in Table 2. Naturally occurring barium is a mixture of seven stable isotopes, and the most abundant among these is ^{138}Ba (~72%). There are 22 barium isotopes known, but most of these are highly radioactive and have half-life periods in the range of several milliseconds to several days. The only notable exceptions are ^{133}Ba, which has a half-life of 10.51 years, and ^{137m}Ba with a half-life of 2.55 min. Barium isotopes are used in a variety of applications. Notably, ^{130}Ba is used in the production of ^{131}Ba/^{131}Cs, which is used in internal radiotherapy to treat tumors/cancers in the cervix, prostate, breast, skin, or other body sites. ^{133}Ba is a standard gamma-ray reference source, which has been used in nuclear physics experimental studies ("Barium in Webelements & Wikipedia" www.webelements.com; http://en.wikipedia.org/wiki/Barium).

Other applications of barium are given below:
1. The most important application of elemental barium is as a scavenger to remove the last traces of oxygen and other gases in vacuum tubes such as TV cathode ray tubes (CRT).
2. An alloy of barium with nickel is used in spark plug wires.
3. Barium sulfate, as a permanent white or blanc fixe, is used in X-ray diagnostics (barium meals or barium enemas).
4. Barium sulfate is important to the petroleum industry, e.g., as drilling mud, a weighting agent in drilling new oil wells.
5. It is also a filler in a variety of products such as rubber.
6. Lithopone, a pigment that contains barium sulfate and zinc sulfide, is a permanent white that has good covering power and does not darken when exposed to sulfides. It is used in interior paints and in some enamels.
7. Barium carbonate is used in glass making and cement. Being a heavy element, barium increases the refractive index and luster of the glass.
8. Barium carbonate is also used as a rat poison.
9. Barium nitrate and chlorates are used to give green color in pyrotechnics.
10. Barium peroxide can be used as a catalyst to start an aluminothermic reaction when welding rail tracks together. It is an oxidizing agent, which is used for bleaching.
11. Barium titanate is a potential dielectric ceramic used for capacitors.
12. Barium fluoride is used as a source material in the manufacture of optical components such as lenses. Barium fluoride is also a common and one of the best scintillators for the detection of X-rays, gamma rays, or other high-energy particles.

Barium and Protein–RNA Interactions, Table 1 Physical, chemical, and other important factors of barium metal

Barium appearance

Silver Gray

Natural Barium Sulfate

General properties	
Name, symbol, number	Barium, Ba, 56
Pronunciation	BAIR-ee-əm
Element category	Alkaline earth metals
Group period block	2, 6, s
Standard atomic weight	137.33 g·mol^{-1}
Electron configuration	[Xe] 6 s2
Electrons per shell	2, 8, 18, 18, 8, 2
Physical properties	
Phase	Solid
Density	(Near r.t.) 3.51 g·cm^{-3}
Liquid density at mp	3.338 g·cm^{-3}
Melting point	1,000 K 727°C 1,341°F
Boiling point	2,170 K 1,897°C 3,447°F
Heat of fusion	7.12 kJ·mol^{-1}
Heat of vaporization	140.3 kJ·mol^{-1}
Specific heat capacity	(25°C) 28.07 J·mol^{-1}·K^{-1}

Vapor pressure

P (Pa)	1	10	100	1 k	10 k	100 k
at T (K)	911	1,038	1,185	1,388	1,686	2,170

Atomic properties	
Oxidation states	2 (strongly basic oxide)
Electronegativity	0.89 (Pauling scale)
Ionization energies	1st: 502.9 kJ·mol^{-1}
	2nd: 965.2 kJ·mol^{-1}
	3rd: 3,600 kJ·mol^{-1}
Atomic radius	222 pm
Covalent radius	215±11 pm
Van der Waals radius	268 pm
Other properties	
Crystal structure	body-centered cubic
Magnetic ordering	paramagnetic
Electrical resistivity	(20°C) 332 nΩ·m
Thermal conductivity	(300 K) 18.4 W·m^{-1}·K^{-1}

(*continued*)

Barium and Protein–RNA Interactions, Table 1 (continued)

Other properties	
Thermal expansion	(25°C) 20.6 μm·m^{-1}·K^{-1}
Speed of sound (thin rod)	(20°C) 1,620 m/s
Young's modulus	13 GPa
Shear modulus	4.9 GPa
Bulk modulus	9.6 GPa
Mohs hardness	1.25
CAS registry number	7440-39-3
Most stable isotopes	

Main article: isotopes of barium

Isotope	Z(p)	N(n)	Mass	Half-life
^{130}Ba	56	74	129.9063208	PRIMORDIAL radioactive [7E+13 a]
^{132}Ba	56	76	131.9050613	STABLE [>300E+18 a]
^{133}Ba	56	77	132.9060075	10.51(5) a
^{134}Ba	56	78	133.9045084	Stable
^{135}Ba	56	79	134.9056886	Stable
^{136}Ba	56	80	135.9045759	Stable
^{137}Ba	56	81	136.9058274	Stable
^{138}Ba	56	82	137.9052472	Stable

Barium and Protein–RNA Interactions, Table 2 Abundance of barium in different environments

Location	ppb^a by weight	ppb by atoms
Universe	10	0.09
Sun	10	0.1
Meteorite	2,800	410
Crustal rocks	340,000	51,000
Seawater	30	1.4
Stream	25	0.2
Human	300	14

^{a}ppb – parts per billion (=10^9), both in terms of weight and in terms of number of atoms are given in the table above

Barium Metal in Biological Systems

Although barium has no biological role, all of its compounds that are soluble in water or acid are toxic. Inhaled dust containing barium compounds can also accumulate in the lungs, causing a benign condition called baritosis. Since barium sulfate is highly insoluble in water as well as stomach acids, it can be ingested as a suspension (barium meal) for body imaging. It is eliminated completely from the digestive tract and does not bioaccumulate, unlike other heavy metals.

The British Pharmaceutical Codex from 1907 indicates that barium chloride [*barii chloridum*, BaCl$_2$.2H$_2$O] has a stimulating action on the heart and other muscles. It was said that it "raises blood pressure by constricting the vessels and tends to empty the intestines, bladder and gall bladder." Its poisonous nature was also pointed out. Barium sulfide (BaS) was used as a depilatory agent (removes hair). At low doses, barium acts as a muscle stimulant, whereas higher doses affect the nervous system, causing cardiac irregularities, tremors, weakness, anxiety, dyspnea, and paralysis. This may be due to its ability to block potassium ion channels, which are critical to the proper function of the nervous system. However, individual responses to barium salts vary widely, with some being able to handle barium nitrate casually without problems and others becoming ill from working with it in small quantities.

Metal Barium-Ion-Binding Proteins

Barium Interactions with Proteins or Bound Ligands

Barium is less notable in biological systems compared to metals such as iron (Fe) or zinc (Zn). However, there are useful interactions involving barium in biological systems, such as the binding of barium ions to phosphoinositide-specific phospholipase. Barium and calcium are used as mediators by bonding to the active

Barium and Protein–RNA Interactions, Table 3 Metal barium in proteins

Sl. no	PDB code	Name	Ref.	No. Ba^{2+}	Interactions with
1	1DJH	Phospholipase C-delta1	Essen et al. (1997) Biochemistry 36: 2753	3	N312, D343, E390, I651, D653, N677, D706, D708
2	1SOF	Bacterioferritin	Liu et al. (2004) Febs Lett 573:93	2	N148, Q151
3	1WRO	HutP	Kumarevel et al. (2005) Nucleic Acids Res. 33:5494	6	His ligand, H77, H73, E90, H138
4	2ADI	Monoclonal anti-cd4 antibody q425	Zhou et al. (2005) PNAS 102:14575	1	D32, E50, N100
5	2B5E	Disulfide isomerase	Tian et al. (2006) Cell 124:61	1	E194
6	2BOU	EGF domains	Abbott et al. (to be published)	2	D43, N63, D95, E98, N114
7	2CIS	D-hexose-6-phosphate mutarotase	Graille et al. (2006) J Biol Chem 281:30175	1	Glucose-6-phosphate
8	2DNS	D-amino acid amidase	Okazaki et al. (2007) J Mil Biol 368:79	4	D122, nonspecific
9	2ITD	Potassium channel KcsA–Fab complex	Lockless et al. (2007) Plos Biol 5: e121	2	T175
10	2QDE	Mandelate racemase	Agarwal et al. (to be published)	1	D195, N197, E221, D246, E247
11	2 V02	Calmodulin	Kursula and Majava (2007) Acta Cryst. F 63:653	1	N60, D56, D58, E67
12	2V2F	Penicillin-binding protein (PBP1A)	Job et al (2008) J. Biol. Chem. 283:4886	1	T364, S472
13	2W2F	p-Coumaric acid decarboxylase	Rodriguez et al. (2010) Proteins 78:1662s	1	–
14	2 W54	Xanthine dehydrogenase	Dietzel et al. (2009) 284:8764	1	E172, H173, T266
15	2WTR	Beta arrestin-1	Zhou et al. (to be published)	2	–
16	2X6B	Potassium channel	Clarke et al. (2010) Cell 141:1018	1	T96
17	3AA6	Actin capping protein	Takeda et al. (2010) Plos Biol 8: e1000416	1	D38
18	3E8F	NaK channel	Alam & Jiang (2009) Nat. Struct. Mol. Biol. 16,35	2	–
19	3LR4	Sensor protein	Edwards et al. (to be published)	5	D136
20	3NKV	Ras-related protein Rab-1b	Muller et al. (2010) Science 329,946	2	D44, D132, T134
21	3NS4	Vacuolar protein sorting-associated protein 53	Vasan et al. (2010) PNAS 107:14176	3	N620, T720, N727
22	1VBZ	Hepatitis delta virus ribozyme	Ket et al. (2004) Nature 429:201	2	Ura120, Ura163
23	1ZQB	DNA polymerase	Pelletier & Sawaya (1996) Biochemistry 35:12778	3	D190, D192, T101, V102, I106, Gua7, Thy6

sites of the enzymes in the reaction as phospholipase connects phospholipids together to form a membrane. In another example, barium salts were used to determine the packing density of phospholipids in the cellular membrane of certain species of bacteria. Because of its interactions with phospholipids, a third function of barium in biological systems is the modeling of phospholipid head groups. Barium diethyl phosphate is a model compound used to analyze the head groups of phospholipids. From these examples, it is clear that barium is often used in the arrangement and connection of phospholipids in cellular membranes (Essen et al. 1997; Snyder et al. 1999; Herzfeld et al. 1978). Recently, the number of barium metal ion containing crystal structures available is rapidly increasing. We searched the protein databank (Berman et al. 2000, 2002) (PDB www.pdb.org) and found 23 nonhomologous protein structures containing barium metals (Table 3). Based on the analysis of protein–metal ion interactions, barium metal ions appear to prefer to interact with aspartic acid followed by asparagine, glutamic acid, and threonine.

Metal (Barium)-Ion-Mediated Protein–RNA Interactions

General Introduction to HutP

Regulating gene expression directly at the mRNA level represents a novel approach in the control of cellular processes in all organisms. In this respect, RNA-binding proteins, while in the presence of their cognate ligands, play a key role by targeting the mRNA to regulate its expression through attenuation or antitermination mechanisms. The distinction between the attenuation and antitermination pathways is the end result of interactions between the terminator and the activated protein that decides whether transcription is terminated or allowed to continue. In the attenuation process, the regulatory protein, activated by the regulatory molecule, pauses the transcription at the terminator structure, which otherwise permits the readthrough of the transcription apparatus. The best example for this kind of regulation is the tryptophan biosynthetic operon (Yanofsky 2000; Gollnick and Babitzke 2002; Antson et al. 1995, 1999). In contrast to this mechanism, the antitermination process requires the activated protein to bind to the preexisting terminator structure to allow the RNA polymerase to transcribe the downstream genes. An example of this is the BglG/SacY family of antitermination proteins. The target sequences for these antiterminator proteins comprise either single- or double-stranded regions of their respective mRNAs.

HutP (16.2 kDa, 148 aa) is an RNA-binding antiterminator protein that regulates the expression of the *h*istidine *ut*ilization (*hut*) operon in *Bacillus subtilis* by binding to *cis*-acting regulatory sequences on *hut* mRNA. In the *hut* operon, HutP is located just downstream from the promoter, while the five other subsequent structural genes, *hutH*, *hutU*, *hutI*, *hutG*, and *hutM*, are positioned far downstream from the promoter. In the presence of L-histidine and divalent metal ions, HutP binds to the nascent *hut* mRNA leader transcript. This allows the antiterminator to form, thereby preventing the formation of the terminator and permitting transcriptional readthrough into the *hut* structural genes. In the absence of L-histidine and divalent metal ions or both, HutP does not bind to the *hut* mRNA, thus allowing the formation of a stem-loop terminator structure within the nucleotide sequence located between *hutP* and the structural genes. Similar to HutP, many regulatory proteins that involve allosteric regulation by small molecules to modulate their binding to the cognate mRNA have been described for various operons. These proteins must initially be activated by their specific ligands before they can function as antiterminators/attenuators (Oda et al. 1992, 2000; Kumarevel et al. 2002, 2003, 2004a, b).

Requirement of L-Histidine

HutP requires L-histidine ($\sim$10 mM) for binding to the *hut* mRNA. To obtain additional insights into the requirement of L-histidine and its important functional groups responsible for activation, we analyzed 15 different L-histidine analogs. Among the analogs tested, L-histidine β-naphthylamide (HBN) and L-histidine benzyl ester showed higher affinity (10-fold) over L-histidine. L-histidine methyl ester and L-β-imidazole lactic acid showed similar affinity as L-histidine (Kd $\sim$300 nM), and urocanic acid, histamine, and L-histidinamide showed only weak activation. D-Histidine, imidazole-4-acetic acid, L-histidinol, α-methyl-DL-histidine, 1-methyl-L-histidine, 3-methyl-L-histidine, and 3-(2-thienyl)-L-alanine failed to show any activation. Based on the analysis of the active analogs, we found that the imidazole group as well as the backbone moiety of L-histidine is essential for activation. Moreover, HutP activation by L-histidine is highly stereospecific since D-histidine prevented *hut* mRNA binding to HutP. This suggests that the correct positioning of the α-amino and carboxy moieties of L-histidine is essential for HutP activation (Kumarevel et al. 2003, 2004a).

Requirement of Divalent Metal Ions for the Activation of HutP

Our previous analyses suggested that HutP binds to its cognate RNA only in the presence of L-histidine. To analyze the ability of HutP to bind to mRNA, the reactions are carried out in the presence of L-histidine (10 mM) and Mg^{2+} ions (5 mM). However, we do not know what function the metal ions play in antitermination complex formation. In order to understand the role of metal ions in the formation of this complex, we used a gel mobility shift assay, and these studies suggested that Mg^{2+} ions are important for the HutP–RNA interactions. When $MgCl_2$ was omitted from the binding reactions, HutP failed to bind to *hut* mRNA. When 0.5 mM of $MgCl_2$ was incorporated in the binding buffer, we clearly observed formation of an antitermination complex. The level of complex formation was concentration-dependent, increasing further

Barium and Protein–RNA Interactions, Table 4 Selected divalent and monovalent metal ions and its properties

	Metal ion (compound used)	Ionic radii	Preferred coordination	Concentration in bacterial cells (mg/kg)
Divalent	Mg (MgCl$_2$)	0.72	6	7 × 10^3
	Ca (CaCl$_2$)	0.99, 1.12	6, 8	5.1 × 10^3
	Mn (MnCl$_2$)	0.83	6	260
	Cu (CuCl$_2$)	0.57, 0.73	4, 6	150
	Zn (ZnCl$_2$)	1.02, 1.08	4, 6	83
	Co (CoCl$_2$)	0.74	6	7.9
	Cd (CdCl$_2$)	0.95	4–7	0.31
	Ba (BaCl$_2$)	1.35, 1.38	6, 7	–
	Sr (SrCl$_2$)	1.13	6	–
	Yb (YbCl$_2$)	1.02–1.14	6–8	–
	Ni (NiCl$_2$)	0.69	6	–
	Pb (PbCl$_2$)	1.19–1.49	4–12	–
	Ag (AgNO$_3$)	0.94	6	–
	Hg (Hg (CN)$_2$)	0.69–1.14	2,4,6,8	–
	Pt (K$_2$PtCl$_4$)	0.80	6	–
Monovalent	Na (NaCl)	0.99–1.39	4–12	4.6 × 10^3
	K (KCl)	1.37–1.64	4–12	115 × 10^3

with higher metal ion and L-histidine concentrations. The metal ion Kd (489 μM) value for the HutP–RNA interactions appeared to be more efficient (>10-fold) compared to the metal ions for other protein–RNA interactions, suggesting the existence of an efficient metal-ion-binding pocket (Kumarevel et al. 2004c; 2005a).

Biochemical Analysis of HutP–RNA Interactions in the Presence of Various Metal Ions

From the aforementioned studies, it is clear that divalent metal ions, Mg^{2+} ions, are essential for mediating the HutP–RNA interactions. To substantiate the requirement for divalent metal ions and also to identify the best divalent metal ions that support the interactions, we performed binding reactions in the presence of various divalent metal ions. The properties of the divalent metal ions used in the present study including their ionic radii, preferred coordination, and concentrations present within the bacterial cells are summarized in Table 4. Of the 15 different divalent metal ions tested, 12 (Mg^{2+}, Ca^{2+}, Mn^{2+}, Zn^{2+}, Co^{2+}, Cd^{2+}, Ba^{2+}, Sr^{2+}, Ni^{2+}, Pb^{2+}, Ag^{2+}, Pt^{2+}) were able to mediate the HutP–RNA interactions. The only metal ions that failed to support the interactions were Cu^{2+}, Yb^{2+}, and Hg^{2+} (Fig. 1). Among the 12 divalent ions that participated in the interactions, Mn^{2+}, Zn^{2+}, and Cd^{2+} were more efficient, followed by Mg^{2+}, Co^{2+}, and Ni^{2+} ions (Fig. 1). Interestingly, the divalent metal ions that are less abundant in the bacterial cell, such as Mn^{2+}, Zn^{2+}, and Cd^{2+}, were the active divalent metal ions, whereas the more commonly found divalent metal ions, Mg^{2+}, Ag^{2+}, and Ca^{2+}, were weakly efficient (Kumarevel et al. 2005b).

Interestingly, Ni^{2+} ions, which are reportedly not present within bacterial cells, also mediate the complex formation more efficiently than Mg^{2+}, and other metals that are not found in bacteria, such as Ba^{2+}, Pb^{2+}, Pt^{2+}, and Sr^{2+}, also participate in the complex. We compared the atomic radii of the metal ions that support the HutP–RNA interactions in order to study the pocket that accommodates the metal ions. When we compared the ionic radii of the metal ions, based on their preferred hexameric coordination, we found that the radii for Cu^{2+}, Mg^{2+}, and Co^{2+} were nearly the same size (Table 4). However, Cu^{2+} failed to interact with the complex. Similarly, although the radii between the Yb^{2+} and Zn^{2+} ions were essentially the same size, Yb^{2+} was noninteractive, whereas Zn^{2+} showed the highest interactions with the protein–RNA complex. The Ba^{2+} ion, which has the longest ionic radius, showed the lowest support for the HutP–RNA interactions. Although these studies suggested that the atomic radii of the divalent metal ions are important for their support in mediating the HutP–RNA interactions, it is possible that the underlying mechanism may be much more complicated.

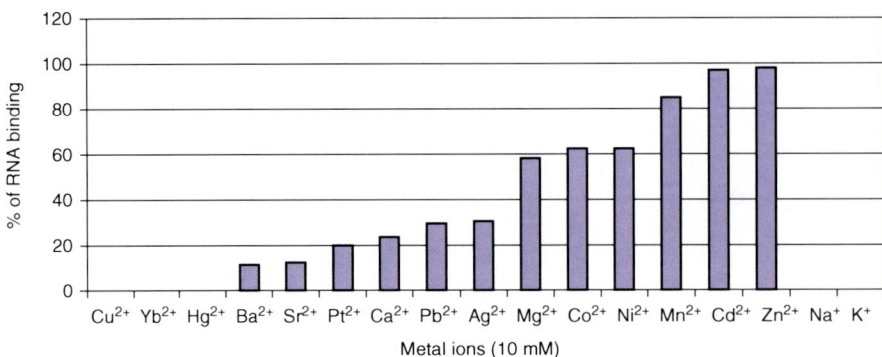

Barium and Protein–RNA Interactions, Fig. 1 Analysis of the abilities of various metal ions to mediate the HutP–RNA interactions. Fifteen divalent and two monovalent cations were analyzed by a gel shift assay. The amounts of complexed and free RNA were used to calculate the percentage of metal ion interactions involved in making the ternary complex

Both tested monovalent cations (Na$^+$, K$^+$) failed to mediate the HutP–RNA interactions even at high (10,100 mM) concentrations in the absence of divalent cations. From these analyses, it can be concluded that divalent cations are mandatory for the interactions between HutP and its RNA and cannot be replaced with monovalent cations.

Structural Analysis of HutP–RNA Interactions in the Presence of Ba^{2+} Metal Ions

To clarify the Ba^{2+} metal ion coordinations within the HutP and L-histidine ligand complex, crystallization trials were carried out with HutP, L-histidine, and BaCl$_2$, as well as with other ions, such as MnCl$_2$ or MgCl$_2$. All of the complexes were crystallized successfully, and the structures HutP–L-histidine–Ba^{2+}, HutP–L-histidine–Mn^{2+}, and HutP–L-histidine–Mg^{2+} were solved by molecular replacement with HutP–HBN (PDB id, 1VEA) as a search model (Kumarevel et al. 2004a, 2005a, b; Gopinath et al. 2008). The overall structures of the HutP–L-histidine–Ba^{2+} are shown in Fig. 2a. Each asymmetric unit contains three molecules of HutP and three L-histidines related by non-crystallographic threefold axis, forming a tight trimer, with each monomeric HutP molecule consisting of four α-helices and four β-strands, arranged in the order α- α -β- α - α - β- β- β in the primary structure, and the four antiparallel β-strands form a β-sheet in the order β1- β2- β3- β4, with two α-helices each on the front and the back (Fig. 2a–c). Consistent with previous X-ray and biochemical analyses, these complexes form a hexameric structure along the twofold axis (Fig. 2b).

The exceptional quality of both the experimentally phased electron density map and that obtained with the calculated phases and 2|F$_{obs}$|-|F$_{calc}$| amplitudes allowed us to identify three specific divalent metal binding sites unambiguously in the complexes. These three specific binding sites were consistent in the present HutP–L-histidine–Ba^{2+} and HutP–L-histidine–Mn^{2+} and our other reported HutP–L-histidine–Mg^{2+} and HutP–L-histidine–Mg^{2+}–21mer RNA complexes. The bound metal ions (Ba^{2+}, Mn^{2+}, Mg^{2+}) were located at the dimer or dimer–dimer interface with a similar recognition motif, critically forming the hexacoordination with the L-histidine ligand and the histidine cluster of the HutP protein (Figs. 2c, 3a). Out of the six coordinations, two were with the amino and carboxyl groups of the L-histidine ligand, and three were with the imidazole nitrogens of His138, His73, and His77. The sixth coordination of the Mg^{2+} ion was that with a water molecule (Fig. 3a). This typical hexacoordination with the L-histidine and histidine cluster may not be possible with the monovalent metal ions (K$^+$, Na$^+$), and hence, these metal ions could not mediate the protein–RNA interactions (Figs. 1 and 3a). In the case of the HutP–L-histidine–Ba^{2+} complex, we found three additional nonspecific Ba^{2+} ions located at the center of the HutP trimer (Figs. 2a–b and 3b). However, these three Ba^{2+} are related by a non-crystallographic threefold axis and may occur as alternative positions of a disordered Ba^{2+} ion. A close view of these Ba^{2+} positions and their interactions (Fig. 3b) shows that each Ba^{2+} is bonded with two coordinations derived from the two molecules, i.e., Gly89 of one molecule and Glu90 from the other adjacent molecule.

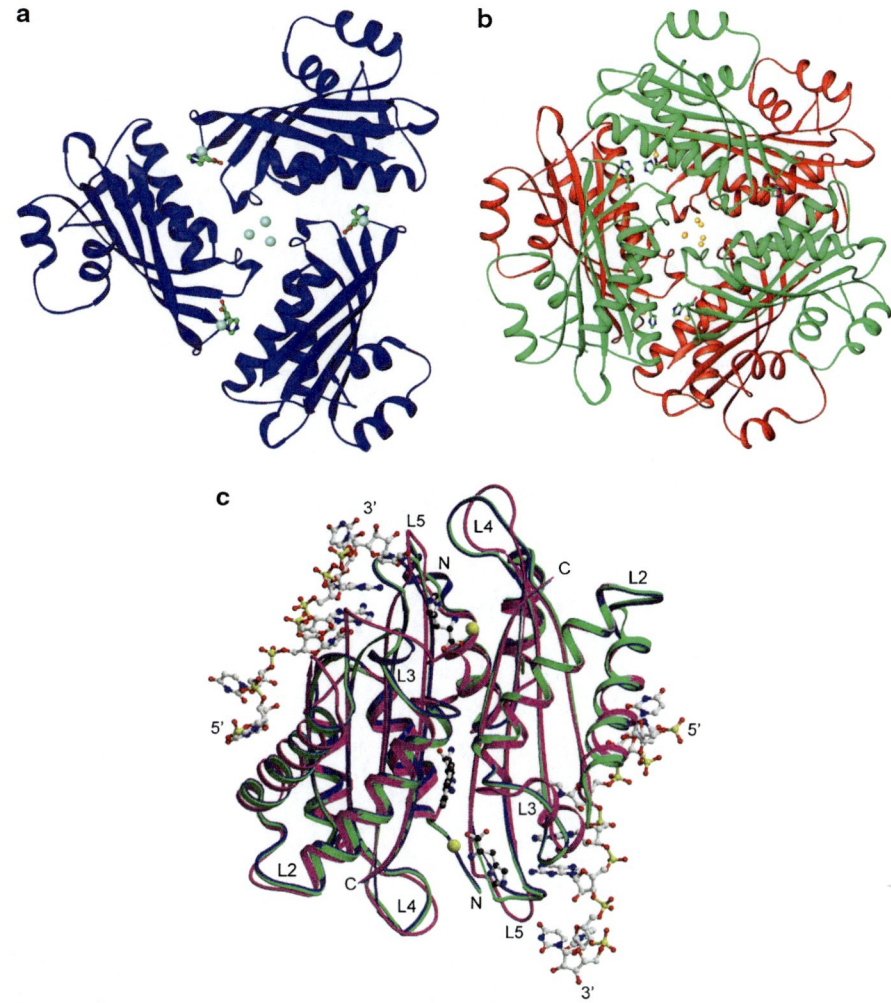

Barium and Protein–RNA Interactions, Fig. 2 Crystal structure of the HutP with divalent metal ion and RNA complexes. (**a**) Structure of the HutP–L-histidine–Ba^{2+} complex shown in a *ribbon* diagram. The bound L-histidines are shown as a *ball-and-stick*, and the divalent metal ion is represented by CPK model. (**b**) Hexameric formation of the divalent metal complex (HutP–L-histidine–Ba^{2+}). One trimer is shown in *red*, and the other is *green*. The divalent metal ions represented by the CPK model. (**c**) The crystal structure of the HutP complexes. HutP-histidine β-naphthylamide (HBN, an L-histidine analog) complex (magenta) superimposed along with the divalent metal ion complex (*blue*) and the HutP quaternary (HutP–L-histidine, Mg^{2+}, and RNA) complex (*green*) shown in a *ribbon* diagram with labels for N-, C-terminals, loop regions, and RNA directions. The bound HBN is shown as a *ball-and-stick* model colored by atom type (nitrogen, *blue*; carbon, *green*; oxygen, *red*). The L-histidines and HBN are represented by *ball-and-stick* models in green, and the Mg^{2+} ions are represented by CPK models, colored in *yellow*

However, this nonspecific Ba^{2+} binding site is not required for protein–RNA interactions, because Mn^{2+} and Mg^{2+} can also mediate the complex formation even without this metal ion interaction.

Site-directed mutational analyses showed that all of the histidines (His73, His77, and His138) involved in the metal ion interactions were required for the coordinations as observed recently with Mg^{2+} ions. When we analyze all three of the divalent metal ion (Ba^{2+}, Mn^{2+}, and Mg^{2+}) complexes, we found that although the metal ion coordinations were essentially the same in all of the cases, however, the coordination distance changed slightly (2.28 Å for Mg^{2+}, 2.26 Å for Mn^{2+}, and 2.60 Å for Ba^{2+}), depending on the metal ion radius (Fig. 4). The maximum difference in coordination distances observed for Ba^{2+} was 0.33 Å, in comparison with Mg^{2+}, and this

Barium and Protein–RNA Interactions, Fig. 3 Stereo view of divalent metal ion coordinations in the complex structures. (**a**) A close-up view of the Ba^{2+} ion binding site in the HutP–L-histidine–Ba^{2+} complex. Hydrogen bonds are indicated by *broken lines*. The L-histidine ligand and the protein residues are represented by *ball-and-stick* models colored by atom type, as shown in Fig. 2c. The Ba^{2+} and water molecules are represented by CPK models in *blue* and *red*, respectively. (**b**) A close-up view of the nonspecific Ba^{2+} ion binding site and its interactions. Hydrogen bonds and the color scheme are described in Fig. 3a

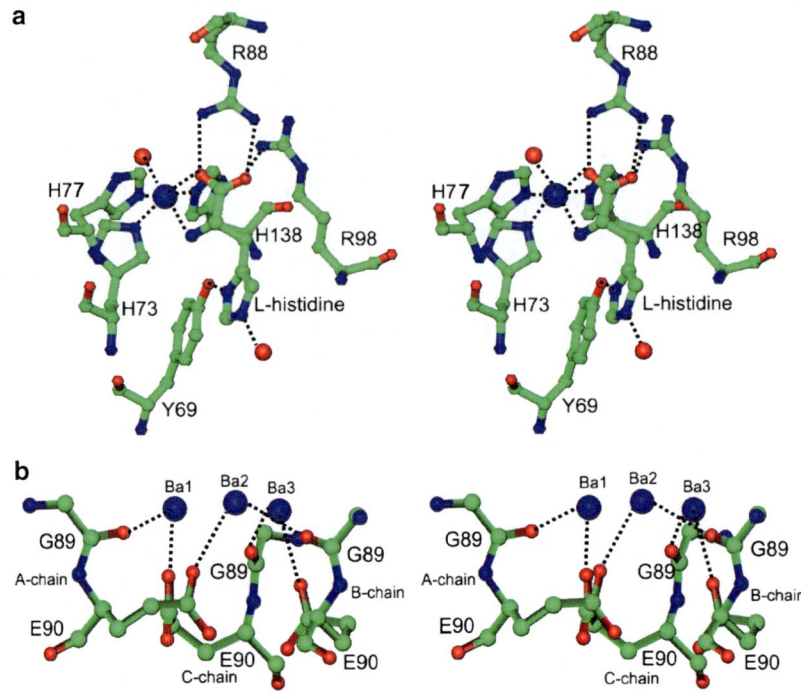

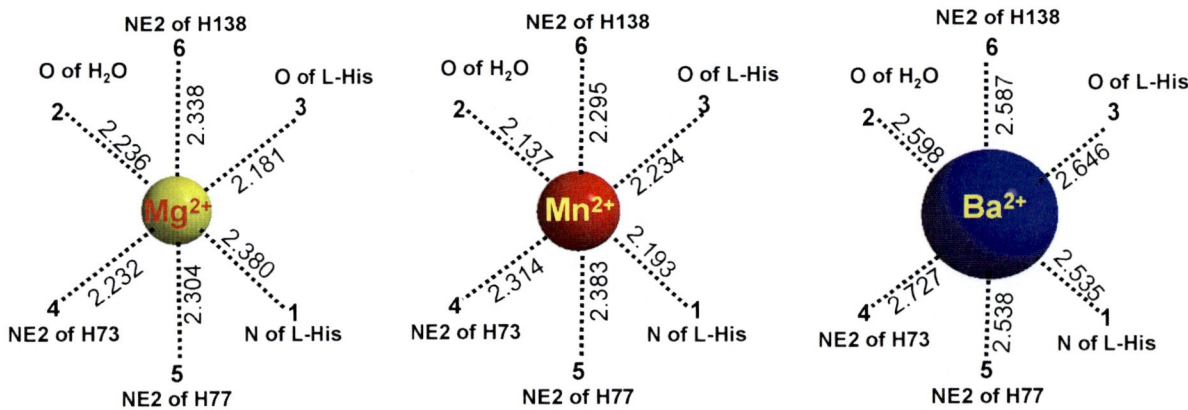

Barium and Protein–RNA Interactions, Fig. 4 Divalent metal ion coordination distance comparison for different metal ions observed in the complex structures. A schematic hexacoordination of the metal ions, drawn and numbered as in Fig. 3a

may be due to the difference in ionic radii (Table 4). Due to the larger ionic radius of the Ba^{2+} ions, all six of the metal-coordinating functional groups displaced around (0.3–0.4 Å) from the center of the ion, as compared with other metal ions (Mn^{2+} and Mg^{2+}). From these analyses, it seems that the metal ion resides in between the HutP monomers and that interface may accept a wide range of divalent metal ions with different ionic radii.

Role of Divalent Metal Ions in the HutP–RNA Interactions

When we analyzed our reported structures of HutP (apo HutP, HutP–Mg^{2+}, HutP–HBN complex, HutP–L-histidine–Mg^{2+}, HutP–L-histidine–Ba^{2+}, HutP–L-histidine–Mn^{2+}, and HutP–L-histidine–Mg^{2+}–21mer RNA complex, HutP–L-histidine–Mg^{2+}–55mer RNA complex), it was quite clear that the specific metal ion coordinations with the histidine cluster of HutP (His73, His77, His138) and the L-histidine ligand are indispensable for the structural rearrangements, which is required to recognize the RNA. The apo HutP, HutP–Mg^{2+}, and HutP–HBN complexes adopted similar conformations, indicating that the Mg^{2+} ion or L-histidine ligand alone may not be sufficient for the activation of HutP. Two metal ion interactions are lost in the HutP–Mg^{2+} complex in the absence of L-histidine, providing an explanation for why the metal ion itself may not reside in the complex to facilitate the required structural rearrangements. When we introduced the divalent metal ion into the HutP–L-histidine complex, the metal ion bound to the L-histidine and moved approximately 12 Å on either side of the dimer interface (Fig. 2c). This movement of L-histidine along with the metal ions is apparently linked to the many local conformational changes, especially in the loop regions, L3, L4, and L5, in addition to the L-histidine binding site (Figs. 2c and 3a). These conformational changes become evident upon a comparison of the inactive/preactive conformations and the divalent metal ion complexes. The divalent metal ion complex represents the activated conformation of HutP which is required for the recognition of RNA, since there are no further conformational changes observed of the activated HutP protein upon RNA binding. Therefore, we suggest that the divalent metal ions play a major role in activating the HutP protein through the structural rearrangements specifically required for RNA recognition. Thus, HutP represents the first example of a single-stranded RNA-binding protein that requires metal ions for mediating RNA–protein interactions (Kumar et al. 2006; Kumarevel 2007).

Metal Barium Ion and Its Interactions with Nucleic Acids

Metal ions are unavoidably involved in almost every aspect of nucleic acid chemistry. Due to their polyanionic nature, the natural nucleic acids DNA and RNA always occur in combination with cations. In principle, each nucleotide carries one negative charge associated with its phosphate groups, and it needs to be shielded to enable the formation of a stable secondary structure. This shielding can be provided by a variety of cations (polyamines, mono-, or divalent cations). The maintenance of nucleic acid structural integrity is the main role of these metal ions.

However, metal ions are also important for the biological action of nucleic acids such as RNA folding, catalytic cofactors in ribozymes, and mediation of protein–RNA interactions.

The most common intracellular metal ion is the potassium ion. Magnesium and calcium are also two other important metal ions for the function of nucleic acids. Although the barium ion is not biologically important for the functions of nucleic acids, it is used as a divalent metal ion during the crystal screening process. When we searched the PDB (Berman et al. 2000, 2002; (www.pdb.org)) for barium metals containing nucleic acid complexes, we found 23 nonhomologous structures containing one or more barium ions (Table 5). Most of the bound barium ions interact with the nucleic acids (DNA/RNA) either through phosphate backbones or predominantly with guanines (Table 5).

Environmental and Health Effects of Barium Metal

The environmental background level of barium is low, ~0.0015 parts per billion (ppb) in air. Most surface water contains about 0.38 ppb or less of barium, and some areas of underground water wells may have up to 1 ppb of barium. High amounts of barium may be found in soils (100 ~ 3,000 ppm) and in some foods, such as nuts, seaweed, fish, and certain plants. The amount of barium found in water and food usually is not high enough to cause any health concern, or at least no related ailments have been registered thus far. However, ongoing investigations are being conducted to determine if long-term exposure to low levels of barium may cause any health concerns.

Extensive use of barium in industries, such as mining processes, refining processes, and the production of barium compounds, can pollute the environment by releasing great amounts of barium waste into the air. As a result, barium concentrations in air, water, and soil may be higher than those naturally occurring at many locations. Additionally, people with the greatest known risk of exposure to high levels of barium are those working in the industries that make or use barium compounds. Most of the health risks that they undergo are caused by breathing in air that contains barium sulfate or barium carbonate. Many hazardous waste sites contain certain amounts of barium compounds,

Barium and Protein–RNA Interactions, Table 5 Metal barium in nucleic acids

Sl. no	PDB code	Name	Ref.	No. Ba^{2+}	Interactions with
DNA					
1	1IHH	Dodecamer duplex	Spingler et al. (2001) Inorg Chem 40:5596	1	ADE10, GUA11
2	1QYK	DNA quadruplex	Cardin et al. (to be published)	2	Only backbones
3	1XCS	Oligonucleotide/drug complex	Valls et al. (2005) J Biol Inorg Chem 10:476	1	Only backbones
4	220D	DNA decamer	Gao et al. (1995) Biophys J 69:559	1	–
5	284D	Four-stranded DNA	Salisbury et al. (1997) PNAS 94:5515	11	–
6	3FQB	Z-DNA	Mandal et al (to be published)	2	Gua4
7	3GOJ	Holliday junction	Naseer & Cardin (to be published)	4	Gua4 (1)
8	3GOM	Holliday junction	Naseer & Cardin (to be published)	5	Gua4 (1)
9	2DQO	DNA:RNA duplex	Juan et al (2007) Nucleic Acids Res. 35,1969	2	rGua13, rAde17
10	1Y7F	Oligonucleotide racemase	Egli et al. (2005) Biochemistry 44:9045	1	Gua3
11	1IOF	Decamer complex	Tereshko et al. (2001) Nucleic Acids Res 29:1208	1	Gua3
12	2GPX	A-DNA octamer	Jiang et al (2007) Nucleic Acids Res 35:477	1	Gua3
RNA					
13	1J6S	RNA tetraplex	Pan et al. (2003) Structure 11:815	8	Gua8, Gua10, Gua11, Gua16, Gua17, Gua20, Gua22, Gua23
14	1KD4	Polyribonucleotide	Kacer et al. (2003) Acta Cryst Sec D 59:423	2	Gua2, Ura3, Gua8
15	1NTA	Streptomycin RNA-aptamer	Tareshko et al. (2003) Chem Biol 10:175	3	Gua6, Ura10, Ura11, Gua114, Gua115
16	1U9S	Ribonuclease P	Krasilnikov et al. (2004) Science 306:104	21	Gua79, Gua98, Gua99, Gua100, Gua110, Gua123, Gua130, Gua141, Gua157, Ura158, Gua159, Gua161, Gua162, Gua167, Gua171, Cyt175, Gua182, Ura183, Ade192, Ade197, Gua218, Cyt219
17	1Y6S	HIV-1 DIS RNA	Ennifar et al. (2003) Nucleic Acids Res. 31:2671	6	Ura3, Ura6, Gua4, Gua7, Gua9, Gua10
18	2HOL	Thi-box riboswitch	Edwards & Ferre-D'Amare (2006) Structure 14:1459	9	Gua36, Gua40, Ade41, Gua60, Gua66, Gua78
19	3F2W	FMN riboswitch	Serganov et al. (2009) Nature 458:233	16	Gua9, Gua16, Gua19, Gua28, Gua33, Gua41, Gua62, Gua97, Gua101, Ade102 Gua36, Ura52, Gua70, Ura89, Gua93
20	3IQN	SAM-I riboswitch	Stoddard et al. (2010) Structure 18:787	5	–
21	3OXJ	Glycine riboswitch	Huang et al. (2010) Mol Cell 40:774	24	Gua1, Gua8, Ura19, Gua60, Ura61, Gua79, Gua80
22	3P4B	Polyribonucleotide	Pallan et al.(2010) Nucleic Acids Res. (In press)	2	–
23	439D	Polyribonucleotide	Perbandt et al. (2001) Acta Cryst Sec D 57:219	1	Gua81, Gua82

and these sites may be a source of exposure for people living and working near them. Exposure near hazardous waste sites may occur by breathing dust, eating soil or plants, or drinking water that is polluted with barium. Skin contact may also cause health problems. Barium that enters the body by breathing, eating, or drinking is removed mainly in feces and urine. Most of the barium that enters our body is removed within a few days, and almost all of it is excreted within 1–2 weeks. Most of the barium that remains in the body goes to the bones and teeth. However, the long-term health effects of the barium that stays in the body are not known.

The health effects of different barium compounds depend on how well the specific barium compound dissolves in water. For example, barium sulfate does not dissolve well in water and has few adverse health effects. Doctors sometimes give barium sulfate orally or by placing it directly in the rectum of patients for purposes of obtaining X-rays of the stomach or intestines. The use of this particular barium compound in this type of medical test is not harmful to people. However, barium compounds such as barium acetate, barium carbonate, barium chloride, barium hydroxide, barium nitrate, and barium sulfide that dissolve in water can cause adverse health effects. Most of what we know about health risks from barium comes from studies in which a small number of individuals were exposed to fairly large amounts of barium for short periods. Eating or drinking very large amounts of barium compounds dissolved in water may cause paralysis or death in a few individuals. Some people who eat or drink somewhat smaller amounts of barium for a short period may potentially have difficulties in breathing, increased blood pressure, changes in heart rhythm, stomach irritation, minor changes in blood pressure, muscle weakness, changes in nerve reflexes, swelling of the brain, and damage to the liver, kidney, heart, and spleen. One study showed that people who drank water containing as much as 10 ppm of barium for 4 weeks did not have increased blood pressure or abnormal heart rhythms. We have no reliable information about the possible health effects in humans who are exposed to barium by breathing or by direct skin contact. However, many of the health effects may be similar to those seen after eating or drinking barium. There is no information available on the ability of barium to cause birth defects or affect reproduction in humans. Barium has not been shown to cause cancer in humans.

However, the health effects of barium are more well studied in experimental animals than in humans. Rats that ate or drank barium over short periods had buildup of fluid in the trachea (windpipe), swelling and irritation of the intestines, changes in organ weights, decreased body weight, and increased rates of death. Rats that ate or drank barium over long periods had increased blood pressure and changes in the function and chemistry of the heart. Mice that ate or drank barium over a long period had shortened life spans. There is no reliable information about the health effects in experimental animals that are exposed to barium by breathing or by direct skin contact.

Cross-References

▶ Barium Binding to EF-Hand Proteins and Potassium Channels
▶ Barium, Physical and Chemical Properties
▶ Cobalt Proteins, Overview
▶ Copper-Binding Proteins
▶ Magnesium
▶ Magnesium Binding Sites in Proteins
▶ Magnesium in Biological Systems
▶ Nickel Ions in Biological Systems
▶ Nickel-Binding Proteins, Overview
▶ Nickel-Binding Sites in Proteins
▶ Zinc-Binding Proteins, Abundance
▶ Zinc-Binding Sites in Proteins

References

Antson AA, Dodson EJ, Dodson G, Greaves RB, Chen XP, Gollnick P (1999) Structure of the *trp* RNA-binding attenuation protein, TRAP, bound to RNA. Nature 401:235–242

Antson AA, Otridge J, Brzozowski AM, Dodson EJ, Dodson GG, Wilson KS, Smith TM, Yang M, Kurecki T, Gollnick P (1995) The structure of trp RNA-binding attenuation protein. Nature 374:693–700

Barium (2001) In wikipedia, the free encyclopedia. Retrieved 20 Apr 2011, from http://en.wikipedia.org/wiki/Barium

Berman HM, Battistuz T, Bhat TN, Bluhm WF, Bourne PE, Burkhardt K, Feng Z, Gilliland GL, Iype L, Jain S, Fagan P, Marvin J, Padilla D, Ravichandran V, Schneider B, Thanki N, Weissig H, Westbrook JD, Zardecki C (2002) The protein data bank. Acta Crystallogr D Biol Crystallogr 58:899–907

Berman HM, Westbrook J, Feng Z, Gilliland G, Bhat TN, Weissig H, Shindyalov IN, Bourne PE (2000) The protein data bank. Nucleic Acids Res. 28:235–242

Essen L-O, Perisic O, Lynch DE, Katan M, Williams RL (1997) A ternary metal binding site in the C2 domain of phosphoinositide-specific phospholipase C-δ1. Biochemistry 36:2753–2762

Gollnick P, Babitzke P (2002) Transcription attenuation. Biochim Biophys Acta 1577:240–250

Gopinath SCB, Balasubramanian D, Kumarevel TS, Misono TS, Mizuno H, Kumar PKR (2008) Insights into anti-termination regulation of the *hut* operon in *Bacillus subtilis*: importance of the dual RNA-binding surfaces of HutP. Nucleic Acids Res 36:3463–3473

Herzfeld J, Griffin RG, Haberkorn RA (1978) ^{31}P chemical-shift tensors in barium diethyl phosphate and urea-phosphoric acid: model compounds for phospholipid head-group studies. Biochemistry 17:2711–2718

Kumar PKR, Kumarevel TS, Mizuno H (2006) Structural basis of HutP-mediated transcription anti-termination. Curr Opin Struct Biol 16:18–26

Kumarevel TS (2007) Structural insights of HutP-mediated regulation of transcription of the *hut* operon in *Bacillus subtilis*. Biophys Chem 128:1–12

Kumarevel TS, Fujimoto Z, Padmanabhan B, Oda M, Nishikawa S, Mizuno H, Kumar PKR (2002) Crystallization and preliminary X-ray diffraction studies of HutP protein: an RNA-binding protein that regulates the transcription of *hut* operon in *Bacillus subtilis*. J Struct Biol 138:237–240

Kumarevel TS, Mizuno H, Kumar PKR (2003) Allosteric activation of HutP protein, that regulates transcription of hut operon in *Bacillus subtilis*, mediated by various analogs of histidine. Nucleic Acids Res Suppl 3:199–200

Kumarevel TS, Fujimoto Z, Karthe P, Oda M, Mizuno H, Kumar PKR (2004a) Crystal structure of activated HutP; an RNA binding protein that regulates transcription of the hut operon in *Bacillus subtilis*. Structure 12:1269–1280

Kumarevel TS, Fujimoto Z, Mizuno H, Kumar PKR (2004b) Crystallization and preliminary X-ray diffraction studies of the metal-ion-mediated ternary complex of the HutP protein with L-histidine and its cognate RNA. BBA- Prot Proteomics 1702:125–128

Kumarevel TS, Gopinath SCB, Nishikawa S, Mizuno H, Kumar PKR (2004c) Identification of important chemical groups of the *hut* mRNA for HutP interactions that regulate the *hut* operon in *Bacillus subtilis*. Nucleic Acids Res 32:3904–3912

Kumarevel TS, Mizuno H, Kumar PKR (2005a) Structural basis of HutP-mediated anti-termination and roles of the Mg^{2+} ion and L-histidine ligand. Nature 434:183–191

Kumarevel TS, Mizuno H, Kumar PKR (2005b) Characterization of the metal ion binding site in the anti-terminator protein, HutP, of *Bacillus subtilis*. Nucleic Acids Res 33:5494–5502

Oda M, Katagai T, Tomura D, Shoun H, Hoshino T, Furukawa K (1992) Analysis of the transcriptional activity of the hut promoter in Bacillus subtilis and identification of a *cis*-acting regulatory region associated with catabolite repression downstream from the site of transcription. Mol Microbiol 6:2573–2582

Oda M, Kobayashi N, Ito A, Kurusu Y, Taira K (2000) *cis*-acting regulatory sequences for antitermination in the transcript of the *Bacillus subtilis hut* operon and histidine-dependent binding of HutP to the transcript containing the regulatory sequences. Mol Microbiol 35:1244–1254

Snyder S, Kim D, McIntosh TJ (1999) Lipopolysaccharide Bilayer structure: effect of chemotype, core mutations, divalent cations, and temperature. Biochemistry 38:10758–10767

Winter M, The periodic table on the web "WebElements™." Source: WebElements http://www.webelements.com/

Yanofsky C (2000) Transcription attenuation: once viewed as a novel regulatory strategy. J Bacteriol 182:1–8

Barium Binding to EF-Hand Proteins and Potassium Channels

Vladimir N. Uversky
Department of Molecular Medicine, University of South Florida, College of Medicine, Tampa, FL, USA

Synonyms

Ca^{2+}-activated K^+ channels; Calcium-modulated potassium channels; Calmodulin; EF-hand calcium-binding proteins; Potassium channels

Definition

Barium can effectively interact with several EF-hand calcium-binding proteins. This metal ion is also considered as a potent inhibitor of the regulated potassium channels. It also can inactivate some calcium channels.

Some Physicochemical Properties of Barium

The chemical element barium is the 56th element in the chemical periodic table with the symbol of Ba, atomic number of 56, and atomic mass of the 137.33. It belongs to the group 2 (of which it is a fifth element) in the periodic table. Therefore, barium, which is a soft, silvery-white metal, with a slight golden shade when ultrapure, and highly reactive metal, is the heaviest nonradioactive member of the alkaline-earth elements. Barium is a relatively common element in Earth (it has a concentration of 0.0425% in the Earth's crust and 13 µg/L in sea water, being 14th in natural abundance in the Earth's crust), being naturally found only in combination with other elements in minerals. Since ultrapure barium is hard to prepare, its major physical properties are not well characterized as of yet.

Chemically, barium is rather similar to magnesium, calcium, and strontium, being even more reactive. It always exhibits the oxidation state of +2 and easily and exothermally reacts with chalcogens. Its reactions with water and alcohols are also very exothermic. Due to its ability to react with oxygen and air, barium is stored under oil or inert gas atmosphere.

The most common industrial use of barium metal or barium-aluminum alloys is gettering, that is, the removal of unwanted gases from the vacuum tubes. Barium sulfate is used in the petroleum industry (as the drilling fluid), paint, or as a filler for plastics and rubbers. Due to its low toxicity and relatively high density, barium sulfate is widely used as the radiocontrast agent in the X-ray imaging of the digestive tract. Barium nitrate can be added to fireworks to provide them with a green color.

There is no information on the toxicity of the metallic barium due to its high reactivity. Barium is not carcinogenic and does not bioaccumulate. Although insoluble barium sulfate is not toxic, water-soluble barium compounds are poisonous. At low doses, barium ions act as a muscle stimulant, whereas higher doses affect the nervous system, causing cardiac irregularities, tremors, weakness, anxiety, dyspnea, and paralysis. These effects were attributed to the capability of Ba^{2+} to block potassium ion channels, which are critical to the proper function of the nervous system (Werman and Grundfest 1961; Gilly and Armstrong 1982; Armstrong and Taylor 1980).

Barium and EF-Hand Proteins

Calmodulin and Barium

Since calcium ions are important second messengers in various signaling processes of almost all organisms, cells contain various Ca^{2+}-binding proteins that can sense calcium concentration and often undergo the calcium ion binding-induced conformational changes which alter their ability to interact with other proteins. An illustrative example of such calcium sensor is calmodulin (CaM, an abbreviation for CALcium-MODULated proteIN), which is a prototypic EF-hand calcium-binding protein that acts by sensing calcium levels and binding to target proteins in a regulatory manner. The importance of this protein for all the vertebrates is reflected in its 100% sequence conservation. Calmodulin is a small protein comprising of 148 amino acids long with the molecular mass of 16.7 kDa. It contains four EF-hand motifs, each of which binds a single Ca^{2+} ion. Structurally, CaM has two approximately symmetrical globular domains (the N- and C-domains, each containing two EF-hands), separated by a flexible linker region. EF-hands are characterized by the presence of specific electronegative environment necessary for calcium coordination. Calcium binding induces noticeable conformational changes, where hydrophobic methyl groups of the key methionine residues become solvent exposed, generating specific hydrophobic surfaces, which can in turn bind to basic amphiphilic helices (BAA helices) on the target protein.

In addition to be able to sense calcium, CaM is able to be activated by other metal ions, although the CaM affinities of many metal ions are noticeably lower than that of calcium. For example, the CaM-regulated activation of cerebellar nitric oxide synthase requires an over 200-fold higher concentration of Ba^{2+} than of Ca^{2+}, and the interaction between CaM and caldesmon is weakened by the exchange of Ca^{2+} for Ba^{2+} (Kursula and Majava 2007). Based on the X-ray crystallographic analysis of the calmodulin crystals soaked in barium chloride, it was concluded that Ba^{2+} is able to substitute Ca^{2+} in EF-hand 2 but not in any of the other EF-hands. The coordination of the Ba^{2+} ion was similar to that of Ca^{2+}, and the coordination distances between the ion and the O atoms of CaM EF-hand 2 ranged between 2.5 and 2.8 Å, whereas the coordination distances for Ca^{2+} in CaM are usually 2.3 Å (Kursula and Majava 2007). In addition to the canonical binding within the EF-hand 2, CaM possesses a second Ba^{2+}-binding site in the close vicinity of the EF-hand 2 (Kursula and Majava 2007). The differences in the Ca^{2+} and Ba^{2+} binding to calmodulin were explained by the much larger ionic radius of Ba^{2+} (1.35 Å) when compared with Ca^{2+} (0.99 Å) (Kursula and Majava 2007).

Barium and FhCaBP4

In addition to calmodulin, different organisms are known to contain some additional calcium sensors. In trematodes (which are flatworm parasites), these sensors are presented by a family of proteins which combine EF-hand-containing domains with dynein light chain (DLC)-like domains. Functional characterization of the member of this protein family, FhCaBP4 from liver fluke (*Fasciola hepatica*), that has an N-terminal domain containing two imperfect EF-hand sequences and a C-terminal dynein light chain-like domain revealed that in addition to calcium this protein can bind barium (Orr et al. 2012).

Barium and KChIPs

KChIPs are Kv channel-interacting proteins that bind to the cytoplasmic N-terminus of Kv4 α-subunits and regulate the ion current of Kv channels. There are four KChIP families, KChIP1, KChIP2, KChIP3, and KChIP4, each is encoded by different genes and contains several splice variants. Although all KChIPs share a high degree of sequence homology at their C-terminal domains that contain four EF-hands, their N-terminal domains are notably diversified (Chang et al. 2003). A representative member of this family of proteins, KChIP1, was shown to possess two types of Ca^{2+}-binding sites, high-affinity and low-affinity Ca^{2+}-binding sites. However, only low-affinity-binding site for Mg^{2+}, Sr^{2+}, and Ba^{2+} was observed in this protein (Chang et al. 2003). Deletion-mutation analysis revealed that EF-hand 4 of KChIP1 is likely to represent the high-affinity-binding site, whereas the intact low-affinity-binding site for metal ions is likely to be located at EF-hand 2 or EF-hand 3. Although in its apo-form KChIP1 has a propensity to form dimers, binding of metal ions promoted structural changes in this protein leading to its tetramerization (Chang et al. 2003).

Ba^{2+}-Dependent Inactivation of L-Type α_{1C} Calcium Channel

The voltage-dependent calcium channels control the calcium ion influx that plays a major role in the excitation-contraction coupling of cardiac myocytes. In the cardiac L-type calcium channel, there are seven to nine distinct α_1-subunits (α_{1A}, α_{1B}, α_{1C}, α_{1D}, α_{1E}, α_{1F}, α_{1G}, α_{1H}, and α_{1S}). The C-terminal region of the α_{1C}, with its EF-hand binding motif, is especially important for a negative feedback mechanism related to the regulating voltage-dependent calcium influx in cardiac cells. This EF-hand binding motif is mostly conserved between the C-termini of six of the seven α_1-subunit Ca^{2+} channel genes. Mutational analysis of the role of the glutamate residue E1537 located in the EF-hand binding motif of the cardiac α_{1C} channel revealed that the whole-cell Ba^{2+} and Ca^{2+} currents proceeded more slowly as the glutamate residue was replaced by more hydrophobic residues in the following order: E > Q > G ≈ S > A (Bernatchez et al. 1998).

Barium as a Modulator of c-Fos Expression and Posttranslational Modification

The analysis of the effect of the exogenous barium on PC12 rat pheochromocytoma cells revealed the existence of the transient induction of the *c-fos* gene and showed that the barium-induced c-fos protein underwent less extensive posttranslational modifications (Curran and Morgan 1986). These effects were specific to barium and were not shared by a range of di- and trivalent cations examined (such as cadmium, cesium, cobalt, magnesium, manganese, lanthanum, strontium, tin, or zinc ions in concentrations up to 10 mM). The unique ability of barium to boost the *c-fos* gene expression was due to the capability of this cation to enter the cell through a voltage-dependent calcium channel and interact with calmodulin to stimulate c-fos expression (Curran and Morgan 1986).

Barium and Potassium Channels

Potassium Channels and Structural Determinants of Their Specificity

Many fundamental biological processes including electrical signaling in the nervous system depend on the rapid diffusion of potassium ions across cell membranes. This diffusion across the cell membranes is determined by the existence of specific proteins, K^+ channels, which use diverse mechanisms of gating (the processes by which the pore opens and closes), but they all exhibit very similar ion permeability characteristics (Doyle et al. 1998). In fact, all K^+ channels show a selectivity sequence of $K^+ \approx Rb^+ > Cs^+$, whereas their permeability for the smallest alkali metal ions Na^+ and Li^+ is very low (Doyle et al. 1998). In fact, all potassium channels are at least 10,000 times more permeable to K^+ than to Na^+. On the other hand, K^+ channels exhibit a throughput rate approaching the diffusion limit (Doyle et al. 1998). This unusual combination of a high K^+ selectivity with a high throughput rate suggested the existence of some specific properties of the ion conductance pore.

The X-ray analysis of the potassium channel from *Streptomyces lividans* (KcsA K^+ channel, which is an integral membrane protein with high sequence similarity to all known K$^+$ channels, particularly in the pore region) revealed that the four identical subunits of this protein create an inverted cone that contained the selectivity filter of the pore in its outer end (Doyle et al. 1998). The narrow selectivity filter is only 12 Å long, whereas the remainder of the pore is wider and lined with hydrophobic amino acids.

Structurally, this selectivity filter was shown to be characterized by two essential features. First, the main chain atoms of regions involved in the formation of the selectivity filter create a stack of sequential oxygen rings. This stack of oxygen rings represents numerous closely spaced sites of suitable dimensions for coordinating a dehydrated K$^+$ ion that therefore has only a very small distance to diffuse from one site to the next within the selectivity filter. The second important structural feature of the selectivity filter is the protein packing around it. Here is how this structural feature is described in the original article (Doyle et al. 1998): "The Val and Tyr side chains from the V-G-Y-G sequence point away from the pore and make specific interactions with amino acids from the tilted pore helix. Together with the pore helix Trp residues, the four Tyr side chains form a massive sheet of aromatic amino acids, twelve in total, that is positioned like a cuff around the selectivity filter. The hydrogen bonding, for example, between the Tyr hydroxyls and Trp nitrogens, and the extensive van der Waals contacts within the sheet offer the immediate impression that this structure behaves like a layer of springs stretched radially outward to hold the pore open at its proper diameter." This intricate spatial organization of the selectivity filter and a region around it in the potassium channel produce structural constraints to coordinate K$^+$ ions but not smaller Na$^+$ ions. Since in the crystal structure, the selectivity filter contained two K$^+$ ions about 7.5 Å apart, it was proposed that such configuration promoted ion conduction by exploiting electrostatic repulsive forces to overcome attractive forces between K$^+$ ions and the selectivity filter (Doyle et al. 1998).

Structural Basis for the Barium-Induced Inhibition of the Ca^{2+}-Activated K$^+$ Channels

The effect of barium on the potassium channels is determined by its physicochemical properties. In fact, the atomic, covalent, and van der Waals radii of barium are of 222, 215 ± 11, and 268 pm, respectively. These values compare favorably with those of potassium (227, 203 ± 12, and 275 pm, respectively). Therefore, stereochemically and geometrically, a potassium channel "sees" the Ba^{2+} cation as a divalent K$^+$ ion (Jiang and MacKinnon 2000). However, although the size allows barium to fit into the selectivity filter of the channel, its doubled charge defines tighter binding. Therefore, binding of Ba^{2+} to the channel's filter prevents the rapid flow of K$^+$ (Jiang and MacKinnon 2000). In fact, the Ca^{2+}-activated K$^+$ channel was shown to be forced to close with a single Ba^{2+} ion inside the pore. Furthermore a Ba^{2+} ion inside the closed channel was trapped and cannot escape until the channel was opened (Miller 1987).

Since the trapping of barium inside the channel and its ability to exit to the external side were strongly dependent on the potassium concentration on both sides of the membrane, the existence of three K$^+$ ion sites (the external lock-in site, the enhancement site that is located closer to the barium-binding site, and the internal lock-in site located between the position of Ba^{2+} and the internal entryway) was proposed (Neyton and Miller 1988). The presence of K$^+$ at the external lock-in site obstructs the outward movement of barium, whereas occupancy of both, the external lock-in and the enhancement, sites by K$^+$ would destabilize Ba^{2+} and speed its exit to the inside (Neyton and Miller 1988). The internal lock-in site is occupied by K$^+$ with the affinity of ∼10 mM, being also not very selective for K$^+$ when compared with Na$^+$ (Neyton and Miller 1988). In agreement with this model, structural analysis of the Rb$^+$-saturated KcsA K$^+$ channel equilibrated with solutions containing barium chloride revealed that there were four distinct Rb$^+$-binding sites inside the channel. These Rb$^+$-binding sites contained one outer, two inner, and one cavity ions. The outer ion was near the extracellular entryway and the inner ion was closer to the cavity and was shown to be present at either of two positions, closer to the inner ion or closer to the outer ion. Ba^{2+} resides at a single location within the selectivity filter, being located at the location of the inner ion closest to the central cavity (Jiang and MacKinnon 2000).

Inhibition of the Na+/K(+)-Pump in Mast Cells

When mast cells are stimulated (both in a IgE-directed and non-IgE-directed way), they release potent mediators of inflammation (e.g., histamines), thereby playing an important role in the pathogenesis of disease states in which the inflammatory response contributes to the development of the clinical symptoms (Knudsen 1995). Mast cells contain Na+/K(+)-pump that is able to respond to changes in the intracellular sodium concentration and that is inhibited by lanthanides and several divalent cations, including barium (Knudsen 1995).

Barium Inhibition of Kir2 Channels

Inward rectifier K^+ (Kir2) channel is present in a number of excitable and non-excitable cell types, such as cardiac muscle, skeletal muscle, some neurons, epithelial cells, and vascular endothelial cells. In cardiac, neurons, and other excitable tissues, these channels stabilize the membrane potential until a threshold potential can be reached. The Kir2 conductance, which is qualitatively dependent on the membrane potential and extracellular K^+ concentration, is inhibited by barium (Park et al. 2008).

Barium Activation of the Ca^{2+}-Activated and Voltage-Dependent BK-Type K^+ Channel

The Ca^{2+}-activated and voltage-dependent big K^+ channel (BK-type channel also known as Maxi-K or slo1 channel) is characterized by the large conductance of K^+ ions through the cell membrane and encodes negative feedback regulation of membrane voltage and Ca^{2+} signaling. Here, depolarization of the membrane voltage and increased intracellular Ca^{2+} levels both cause BK channels to open, which hyperpolarizes the membrane and closes voltage-dependent channels, including Ca^{2+} channels, reducing Ca^{2+} influx into the cell. BK channels are found virtually in all excitable and non-excitable tissues, with the exception of heart, and play a number of crucial roles in regulation of smooth muscle tone and neuronal excitability, for example, being involved in controlling the contraction of smooth muscles and electrical tuning of hair cells in the cochlea. BK channel represents a homo-tetramer of the channel forming α-subunits (each being the product of the *KCNMA1* gene) decorated by modulatory β-subunits (encoded by *KCNMB1*, *KCNMB2*, *KCNMB3*, or *KCNMB4* genes). BK α-subunit is a large modular protein containing seven transmembrane segments (residues 1–343) and a large intracellular C-terminal domain (residues 344–1113). In the transmembrane N-terminal part, there are several functional segments, such as a unique transmembrane region (S0) that precedes the 6 transmembrane segments (S1–S6) conserved in all voltage-dependent K^+ channels and comprising a voltage-sensing domain (S1–S4) and a K^+ channel pore domain (S5, selectivity filter, and S6). A cytoplasmic C-terminal domain (CTD) consisting of a pair of RCK regions (RCK1, residues 344–613, and RCK2, residues 718–1056) that are connected by a long linker and assemble into an octameric gating ring on the intracellular side of the tetrameric channel, in addition to regulating the pore directly, may also modulate the voltage sensor (Yuan et al. 2010). The CTD contains a high-affinity binding site for Ca^{2+}, called "calcium bowls," encoded within the RCK2 domain of each monomer (Zhou et al. 2012). Furthermore, RCK1 contain a second high-affinity Ca^{2+}-binding site. Therefore, one tetrameric BK channel contains eight Ca^{2+}-binding sites that contribute to regulation by Ca^{2+}.

Early analysis of the divalent cation selectivity of the BK-channel Ca^{2+}-binding sites revealed that Ca^{2+} and Sr^{2+} were effective at the Ca^{2+} bowl, and Ca^{2+}, Cd^{2+}, and Sr^{2+} acted through the RCK1 domain, whereas divalent cations of smaller ionic radius, such as Mn^{2+}, Co^{2+}, Mg^{2+}, and Ni^{2+}, were ineffective or only weakly effective at either of the higher-affinity sites (Zeng et al. 2005). It was shown recently that Ba^{2+}, which is typically considered as a specific blocker of potassium channels, is also able to activate BK channel via specific binding to the Ca^{2+}-bowl site (Zhou et al. 2012). Since ionic radius of Ba^{2+} is larger than that of Sr^{2+} and Ca^{2+}, this finding provided a valuable support to the hypothesis that ionic radius is an important determinant of selectivity differences among different divalent cations observed for each Ca^{2+}-binding site of the BK channels (Zhou et al. 2012).

Cross-References

▶ Barium, Physical and Chemical Properties
▶ Calmodulin
▶ EF-Hand Proteins
▶ Potassium Channel Diversity, Regulation of Potassium Flux across Pores
▶ Potassium Channels, Structure and Function

References

Armstrong CM, Taylor SR (1980) Interaction of barium ions with potassium channels in squid giant axons. Biophys J 30:473–488

Bernatchez G, Talwar D, Parent L (1998) Mutations in the EF-hand motif impair the inactivation of barium currents of the cardiac alpha1C channel. Biophys J 75:1727–1739

Chang LS, Chen CY, Wu TT (2003) Functional implication with the metal-binding properties of KChIP1. Biochem Biophys Res Commun 311:258–263

Curran T, Morgan JI (1986) Barium modulates c-fos expression and post-translational modification. Proc Natl Acad Sci USA 83:8521–8524

Doyle DA, Morais Cabral J, Pfuetzner RA, Kuo A, Gulbis JM, Cohen SL, Chait BT, MacKinnon R (1998) The structure of the potassium channel: molecular basis of K^+ conduction and selectivity. Science 280:69–77

Gilly WF, Armstrong CM (1982) Divalent cations and the activation kinetics of potassium channels in squid giant axons. J Gen Physiol 79:965–996

Jiang Y, MacKinnon R (2000) The barium site in a potassium channel by x-ray crystallography. J Gen Physiol 115:269–272

Knudsen T (1995) The Na+/K(+)-pump in rat peritoneal mast cells: some aspects of regulation of activity and cellular function. Dan Med Bull 42:441–454

Kursula P, Majava V (2007) A structural insight into lead neurotoxicity and calmodulin activation by heavy metals. Acta Crystallogr Sect F Struct Biol Cryst Commun 63:653–656

Miller C (1987) Trapping single ions inside single ion channels. Biophys J 52:123–126

Neyton J, Miller C (1988) Potassium blocks barium permeation through a calcium-activated potassium channel. J Gen Physiol 92:549–567

Orr R, Kinkead R, Newman R, Anderson L, Hoey EM, Trudgett A, Timson DJ (2012) FhCaBP4: a *Fasciola hepatica* calcium-binding protein with EF-hand and dynein light chain domains. Parasitol Res 111:1707–1713

Park WS, Han J, Earm YE (2008) Physiological role of inward rectifier K(+) channels in vascular smooth muscle cells. Pflugers Arch 457:137–147

Werman R, Grundfest H (1961) Graded and all-or-none electrogenesis in anthropod muscle. J Gen Physiol 44:997–1027

Yuan P, Leonetti MD, Pico AR, Hsiung Y, MacKinnon R (2010) Structure of the human BK channel Ca^{2+}-activation apparatus at 3.0 A resolution. Science 329:182–186

Zeng XH, Xia XM, Lingle CJ (2005) Divalent cation sensitivity of BK channel activation supports the existence of three distinct binding sites. J Gen Physiol 125:273–286

Zhou Y, Zeng XH, Lingle CJ (2012) Barium ions selectively activate BK channels via the Ca^{2+}-bowl site. Proc Natl Acad Sci USA 109:11413–11418

Barium(II) Transport in Potassium(I) and Calcium(II) Membrane Channels

J. Preben Morth and Harmonie Perdreau
Centre for Molecular Medicine Norway (NCMM),
Nordic EMBL Partnership, University of Oslo,
Oslo, Norway

Synonyms

Ba; Barium

Definition

Barium is an alkaline earth metal that can be used as a substrate analogue of other alkaline earth metals (e.g., calcium) and alkali metals (e.g., potassium) in electrophysiology and membrane transport researches.

Background

Barium is a chemical element from the group 2 of alkaline earth metals, with symbol Ba and atomic number 56. The name barium originates from Greek *barys*, meaning "heavy," and though the chemical has been known for centuries, it was initially isolated in 1808 by the English chemist Sir Humphrey Davy.

Barium is the fourteenth most abundant element in the Earth's crust and is estimated to constitute about 0.05% of the elemental mass. Nevertheless, Ba is not found in its pure form (a soft silvery-white metal) in nature due to its high reactivity with oxygen and water. The most common naturally occurring barium minerals are barite (barium sulfate, $BaSO_4$) and witherite (barium carbonate, $BaCO_3$).

Physical and Chemical Properties

Barium belongs to the same family as the well-known magnesium (Mg) and calcium (Ca) metals with which it shares certain ▸ physical and chemical properties:

– Relatively soft metal with a shiny silvery-white color
– Harder, denser, and higher melting point than sodium and potassium
– High reactivity with oxygen and water (but less than alkali metals from group 1)
– Oxidation number of +2

The biological applications of barium are limited since $BaSO_4$ have low solubility in water (0.0024 g/100 mL at 20°C). The free Ba^{2+} ions are highly toxic and have been used as rodenticides and pesticides. Incidences of human poisonings by soluble barium salts are rare; however, isolated cases have been reported, in particular for people living or working near high-risk areas like heavy industrial sites. Barium poisoning can notably induce a rapid hypokalemia, resulting in nausea, diarrhea, cardiac problems, and muscular spasms/paralysis. Patients are then treated by potassium administration (Smith and Gosselin 1976).

Despite its toxicity, barium chloride ($BaCl_2$) is the most commonly used barium salt as its solubility in water reaches 35.8 g/100 mL at 20°C. $BaCl_2$ is particularly used in brine purification, pigment synthesis, and salt manufacturing. In laboratories, this barium salt is used as a test for sulfate ions, as it reacts with these ions to produce a white precipitate of barium sulfate.

There are seven naturally occurring barium isotopes, the most abundant being ^{138}Ba (71.7%). About a dozen radioactive isotopes of barium are also known, but they present very short half-lives (from several milliseconds to several seconds), and none of them has any practical commercial application. The only notable exception is ^{133}Ba which has a half-life of 10.51 years and is used as a standard source for gamma-ray detectors in nuclear physics studies.

Clinical Applications

Barium compounds are usually toxic and not used in any clinical studies. The only exception is barium sulfate, used for its density, insolubility, and X-ray opacity as a radiocontrast agent for imaging the human gastrointestinal tract. This form of barium, insoluble in water or human fluids, does not bioaccumulate. "Barium meals" and "barium enemas" are then used for radiologic examination of patients with swallowing disorders, polyps, inflammatory bowel diseases, colorectal cancers, etc. (Ott 2000).

Scientific Applications

Because of its toxicity, the use of barium in scientific approaches is somewhat limited. Barium chloride is yet utilized in electrophysiological studies as a substrate analogue of other alkaline earth metals (e.g., calcium) and alkali metals (e.g., potassium). The common stock and working solutions are 1 M $BaCl_2$ and 10–100 nM $BaCl_2$, respectively.

Study of Potassium (K) Channels

Ba^{2+} ions are well-known efficient K^+ channel (▶ Potassium Channels, Structure and Function) blockers, and they are widely used in electrophysiological and structural studies as a probe to investigate the permeation mechanisms of these channels.

Neyton and Miller showed in 1988 (Neyton and Miller 1988a, b), through single-channel studies of Ca^{2+}-activated K^+ (BK) channels, that barium ions have a size that allow them to enter inside the pore. Because of their charges, Ba^{2+} ions bind tightly and prevent the flow of K^+ in the narrow pore of BK channels. They also showed that barium is a reversible blocker effective from either side of the membrane and sensitive to the presence of K^+ ions in the pore. Indeed, it is observed that in the presence of low extracellular concentrations of K^+ ions (0.01 mM), Ba^{2+} ions mainly dissociate to the external solution, while in the presence of higher K^+ ion concentrations (500 mM), dissociation to the internal solution is favored. These properties were attributed to two distinct K^+ ion binding sites and classified the K^+ channels as "long-pore channels," constituted of multiple ions in single file inside a long and narrow pore.

For scientific experimental usage today, the blockade properties of Ba^{2+} ions are used to understand the ion selectivity of K^+ channels. Indeed, K^+ channels are able to conduct K^+ ions at nearly diffusion-limited rates and, at the same time, to prevent Na^+ ions from conducting. A number of studies are about the necessary properties of ions (i.e., size, charge) and channels to recognize each other.

A recent structural and thermodynamic study of *Streptomyces lividans* K^+ channel (KcsA) revealed that two Ba^{2+} ions can bind the channel by an exothermic reaction (Lockless et al. 2007). They proposed that the ion/channel selectivity is achieved by the size of the ion and the structure of the ion-binding site. K^+ and Ba^{2+} ions have indeed different electric field strengths at their surface but nearly the same size: The ionic radius of Ba^{2+} (1.35 Å) is very close to that of K^+ (1.33 Å), whereas the radii of Ca^{2+} (0.99 Å) and Na^+ (0.95 Å) are smaller. Furthermore, the ion-binding sites of KcsA are created by eight oxygen ligands and appropriately sized for K^+/Ba^{2+} ions and not for Na^+ ions. Figure 1 presents the crystal structure of KcsA in complex with Ba^{2+} ions.

Study of Calcium (Ca) Channels

The basic mechanism of Ca^{2+} channels (▶ Bacterial Calcium Binding Proteins) is complex because it

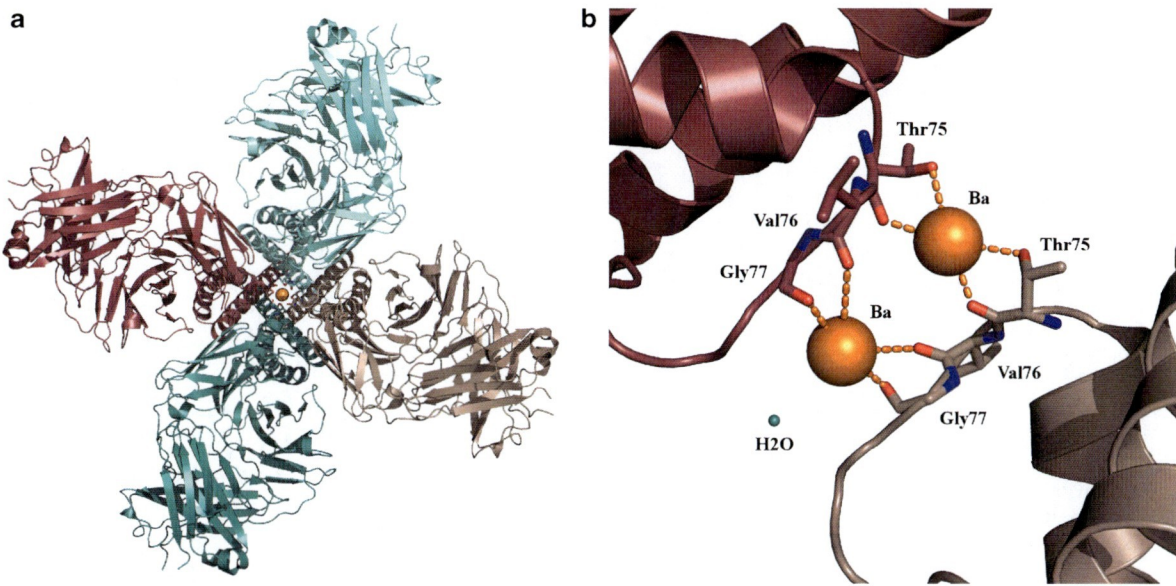

Barium(II) Transport in Potassium(I) and Calcium(II) Membrane Channels, Fig. 1 Biological assembly of Streptomyces lividans K$^+$ channel (KcsA) in complex with barium chloride. (**a**) Assumed biological molecule composed of four partners. (**b**) Interacting residues in the metal binding site, only two partners are represented. The figure was generated with Protein Data Bank identification code: 2ITD. Using the PyMOL Molecular Graphics System, Version 1.3, Schrödinger, LLC

depends on the presence or absence of extracellular Ca^{2+} ions:

- When Ca^{2+} ions are present (physiological conditions), the channels are selective (▶ Calcium Ion Selectivity in Biological Systems) and exclude other cations.
- When Ca^{2+} ions are absent, the channels become permeable to Ba^{2+} and other divalent and monovalent cations.

In contrast to the K$^+$ channel, the Ca^{2+} channel pore has a relatively wide diameter of 6 Å which allows, in absence of the high-affinity Ca^{2+} ions, the conduction of other ions, mainly depending on their size (McCleskey and Almers 1985).

Thereby, Ba^{2+} ions can be used, according to the experimental conditions:

- To reduce the contamination by K$^+$ currents and to obtain a more specific result in the study of Ca^{2+} channels. For example, Almog et al. replaced Ca^{2+} ions (2 mM) in the application solution with Ba^{2+} ions (5 mM) for the characterization of voltage-gated Ca^{2+} channels, because of the similarity of the activation curves with Ca^{2+} or Ba^{2+} (Almog and Korngreen 2009).
- To differentiate Ca^{2+} channel types. Indeed, inward currents are 2-fold larger with Ba^{2+} than Ca^{2+} for L-channels, whereas they are quite comparable for T-channels (Hess and Tsien 1984; Huguenard 1996).
- To unravel the ion selectivity mechanism of the different Ca^{2+} channels. For instance, subtle differences can be shown between Ca^{2+} and Ba^{2+} permeations in T-type channels, more particularly in the presence of Mg^{2+} ions (Serrano et al. 2000).

Cross-References

- ▶ Barium, Physical and Chemical Properties
- ▶ Calcium Ion Selectivity in Biological Systems
- ▶ Calcium-Binding Proteins, Overview
- ▶ Potassium Channels, Structure and Function

References

Almog M, Korngreen A (2009) Characterization of voltage-gated Ca(2+) conductances in layer 5 neocortical pyramidal neurons from rats. PLoS One 4(4):e4841

Hess P, Tsien RW (1984) Mechanism of ion permeation through calcium channels. Nature 309(5967):453–456

Huguenard JR (1996) Low-threshold calcium currents in central nervous system neurons. Annu Rev Physiol 58:329–348

Lockless SW, Zhou M, MacKinnon R (2007) Structural and thermodynamic properties of selective ion binding in a K$^+$ channel. PLoS Biol 5(5):e121

McCleskey EW, Almers W (1985) The Ca channel in skeletal muscle is a large pore. Proc Natl Acad Sci USA 82(20):7149–7153

Neyton J, Miller C (1988a) Discrete Ba^{2+} block as a probe of ion occupancy and pore structure in the high-conductance Ca^{2+} – activated K$^+$ channel. J Gen Physiol 92(5):569–586

Neyton J, Miller C (1988b) Potassium blocks barium permeation through a calcium-activated potassium channel. J Gen Physiol 92(5):549–567

Ott DJ (2000) Accuracy of double-contrast barium enema in diagnosing colorectal polyps and cancer. Semin Roentgenol 35(4):333–341

Serrano JR, Dashti SR, Perez-Reyes E, Jones SW (2000) Mg(2+) block unmasks Ca(2+)/Ba(2+) selectivity of alpha1G T-type calcium channels. Biophys J 79(6):3052–3062

Smith RP, Gosselin RE (1976) Current concepts about the treatment of selected poisonings: nitrite, cyanide, sulfide, barium, and quinidine. Annu Rev Pharmacol Toxicol 16:189–199

Barium, Physical and Chemical Properties

Timothy P. Hanusa
Department of Chemistry, Vanderbilt University, Nashville, TN, USA

Synonyms

Alkaline-earth metal; Atomic number 56

Definition

The heaviest nonradioactive member of the alkaline-earth elements (atomic number 56), barium is a soft, silvery, highly reactive metal. Its compounds have various uses, particularly in drilling fluids, paints, glasses, and pyrotechnics, but all soluble compounds of barium are toxic to mammals. The insoluble barium sulfate is used as a radiocontrast agent in medical imaging.

Background

Barium is a member of the alkaline-earth family of metals (Group 2 in the periodic table). It is a relatively common element, ranking approximately 14th in natural abundance in the Earth's crust; it is more plentiful than, for example, sulfur or zinc. The free metal does not occur in nature, but is found combined in minerals such as barite (barium sulfate, $BaSO_4$) (Fig. 1), or less commonly, in witherite (barium carbonate, $BaCO_3$). The density of such compounds (i.e., $BaSO_4$, 4.5 g cm^{-3}; $BaCO_3$, 4.3 g cm^{-3}) has long been reflected in their names; barite, for example, was at one time known as *terra ponderosa*, heavy spar, or barytes (from the Greek βαρύς *barys*, meaning "heavy"). Barium oxide (BaO, 5.7 g cm^{-3}) was similarly called *baryta*, and barium hydroxide (Ba(OH)$_2$), baryta water. When Sir Humphry Davy first isolated the metal by electrolysis of a mixture of barium oxide and mercuric oxide in 1808, he used the stem of the word baryta to form the name of the element. (Ironically, the name is not as appropriate for barium itself; the density of the metal (3.51 g cm^{-3}) is less than half that of iron, and it is only 30% more dense than aluminum.) The most important commercial source of barium is barite; its production in recent decades has varied between 6 and 8 million tons/year. Over half the world's supply comes from China, with India, the USA, and Morocco together providing about one quarter of the total.

Isotopes

Naturally occurring barium comprises six stable isotopes; ^{138}Ba is the most common (72%), followed by ^{137}Ba (11%) and ^{136}Ba (7.9%) (Table 1). One other naturally occurring isotope (^{130}Ba, 0.11%) undergoes decay by double electron capture; its half-life (7×10^{13} year) is many times the age of the universe, however, and can be considered stable for practical purposes. The synthetic isotopes are radioactive with far shorter half-lives, with mass numbers ranging from 114 to 153. The isotope with the longest half-life (^{133}Ba, 10.5 year) is used as a gamma reference source. The next longest-lived isotope (^{140}Ba) has a half-life of only 13 days, and it, along with ^{141}Ba (18 min) and ^{139}Ba (83 min), played a key role in the discovery of nuclear fission by Hahn, Strassman, and Meitner in 1938. They found that after the ^{235}U nucleus split under neutron bombardment, the three barium isotopes were detectable among the fission products. Some use has been made of ^{131}Ba/^{131}Cs (half-lives of 11.5 and 9.7 days, respectively) in radiation treatments (prostate brachytherapy).

Barium, Physical and Chemical Properties, Fig. 1 Sand-encrusted blades of barite ($BaSO_4$) form *rose rock*, the official state rock of Oklahoma

Barium, Physical and Chemical Properties, Table 2 Atomic and physical properties of barium

Atomic number	56	$E°$ for M^{2+}(aq) + $2e^- \rightarrow M(s)$	−2.91 V
Number of naturally occurring isotopes	7	Melting point	727°C
Atomic mass	137.33	Boiling point	1,870°C
Electronic configuration	[Xe]$6s^2$	Density (20°C)	3.51
Ionization energy (kJ mol^{-1})	502.7 (1st); 965 (2nd)	ΔH_{fus} (kJ mol^{-1})	7.8
Metal radius	2.22 Å	ΔH_{vap} (kJ mol^{-1})	136
Ionic radius (6-coordinate)	1.35 Å	Electrical resistivity (20°C)/ μohm cm	34

Barium, Physical and Chemical Properties, Table 1 Selected isotopes of barium

Nuclide	Nat. abund (%)	Nuclear spin	Half-life
^{130}Ba	0.106%	0	7×10^{13} year
^{131}Ba	–	+1/2	11.5 days
^{132}Ba	0.101%	0	stable
^{133}Ba	–	+1/2	10.51 year
^{134}Ba	2.42%	0	stable
^{135}Ba	6.59%	+3/2	stable
^{136}Ba	7.85%	0	stable
^{137}Ba	11.2%	+3/2	stable
^{138}Ba	71.7%	0+	stable
^{139}Ba	–	−7/2	83.1 min
^{140}Ba	–	0	12.75 days
^{141}Ba	–	−3/2	18.3 min
^{142}Ba	–	0	10.6 min

Properties of the Metal

Various physical and chemical properties of barium are listed in Table 2. Barium is a lustrous, silvery metal that is relatively soft (Mohs hardness of 1.5). It crystallizes in a body-centered cubic lattice with $a = 5.028$ Å. It readily oxidizes when exposed to air, and must be protected from oxygen during storage. One of the few applications for elemental barium takes advantage of this property; thin films of the metal serve as getters in vacuum and cathode ray tubes, and chemically remove traces of oxygen that otherwise could cause tube failure. Barium also forms an alloy with nickel that finds use in automobile spark plug electrodes, and it is a component of Frary metal, an alloy of lead, calcium, and barium that has been used as a bearing metal.

General Properties of Compounds

Barium exclusively displays the +2 oxidation state in its compounds. The metal is highly electropositive ($\chi = 0.89$ on the Pauling scale; 0.97 on the Allred-Rochow scale; cf. 0.93 and 1.01, respectively for sodium), and with a noble gas electron configuration for the Ba^{2+} ion ([Xe]$6s^0$), the metal-ligand interactions are usually viewed as electrostatic. To a first approximation, the bonding can be considered as nondirectional, and strongly influenced by ligand packing. The structures of barium compounds with multidentate and sterically bulky ligands can be highly irregular.

It has been clear since the 1960s that a more sophisticated analysis of bonding than that provided by simple electrostatics must be used with some compounds of the heavy alkaline-earth metals. The gaseous Group 2 dihalides (MF_2 (M = Ca, Sr, Ba), MCl_2 (M = Sr, Ba), $BaBr_2$, BaI_2) (Hargittai 2000), for example, are nonlinear, contrary to the predictions of electrostatic bonding. An argument based on the "reverse polarization" of the metal core electrons by the ligands has been used to explain their geometry, an analysis that makes correct predictions about the ordering of the bending for the dihalides (i.e., Ca < Sr < Ba; F > Cl > Br > I). The "reverse polarization" analysis can be recast in molecular orbital terms; that is, bending leads to a reduction in the antibonding character in the HOMO.

An alternative explanation for the bending in ML_2 species has focused on the possibility that metal d orbitals might be involved. Support for this is provided by calculations that indicate a wide range of small molecules, including MH_2, MLi_2, $M(BeH)_2$, $M(BH_2)_2$, $M(CH_3)_2$, $M(NH_2)_2$, $M(OH)_2$, and MX_2 (M = Ca, Sr, Ba) should be bent, owing partially to the effect of metal d-orbital occupancy. The energies involved in bending are sometimes substantial (e.g., the linearization energy of $Ba(NH_2)_2$ is placed at ca. 28 kJ mol^{-1}) (Kaupp and Schleyer 1992). Spectroscopic confirmation of the bending angles in most of these small molecules is not yet available, however.

The standard classification of alkali and alkaline-earth ions as hard (type a) Lewis acids leads to the prediction that ligands with hard donor atoms (e.g., O, N, halogens) will routinely be preferred over softer (type b; P, S, Se) donors. This is generally true, but studies have demonstrated that the binding of s-block ions to "soft" aromatic donors can be quite robust; for example, the gas-phase interaction energy of two barium atoms with benzene to form the $[Ba(C_6H_6)Ba]$ complex is calculated to be exothermic with a bond energy $D_e[Ba–(C_6H_6)Ba]$ of 162.0 kJ mol^{-1} (Diefenbach and Schwarz 2005). Furthermore, it has been suggested that the toxicity of certain barium compounds may be related to the ability of Ba^{2+} to coordinate to "soft" disulfide linkages, even in the presence of harder oxygen-based residues (Murugavel et al. 2001).

The Ba^{2+} ion is large (the 6-coordinate radius is 1.35 Å) and approximately the same size as potassium (1.38 Å) and polyatomic cations such as NH_4^+ and PH_4^+. Coordination numbers in barium compounds are typically high; a 12-coordinate barium center exists in the $[Ba(NO_3)_6]^{4-}$ ion, for example. In the presence of sterically compact ligands (e.g., -NH_2, -OMe, halides), extensive oligomerization or polymerization will occur, leading to the formation of nonmolecular compounds of limited solubility or volatility. However, sterically bulky ligands can be used to produce compounds with low coordination numbers and improved solubility; barium is only 3-coordinate in $\{Ba[N(SiMe_3)_2]_2\}_2$, for example (Fig. 2) (Westerhausen 1998).

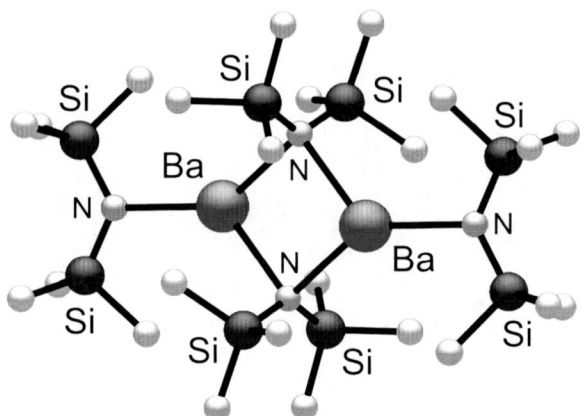

Barium, Physical and Chemical Properties, Fig. 2 Solid state structure of $\{Ba[N(SiMe_3)_2]_2\}_2$

Barium, Physical and Chemical Properties, Table 3 Bond energies in diatomic barium molecules

Ba–E	Enthalpy (kJ mol^{-1})
Ba–H	176 ± 15
Ba–O	562 ± 13
Ba–S	400 ± 19
Ba–F	587 ± 7
Ba–Cl	436 ± 8
Ba–Br	363 ± 8
Ba–I	321 ± 6

The strengths of common barium-element bonds are listed in Table 3. These are for gas-phase species, and so must be used with caution in comparison to the solid state. Bonds to oxygen and fluorine are the strongest that barium forms, and are comparable to those of some early transition metals (e.g., Mo–O = 560 kJ mol^{-1}; V–F = 590 kJ mol^{-1}).

Common Reactions and Compounds of Barium

With Barium-Oxygen Bonds

In addition to its reaction with oxygen at room temperature, barium burns in air to yield a mixture of white barium oxide, BaO, barium nitride, Ba_3N_2, and the peroxide, BaO_2 (1–3). Barium oxide is cleanly made by heating barium carbonate, $BaCO_3$, and is a high melting solid (1,923°C) with a rock salt lattice that reacts with CO_2 in the reverse of the reaction used to prepare it (4). For a time in the nineteenth century, the reaction of barium oxide with air at 500°C was used to produce the peroxide on an industrial scale (the Brin process); BaO_2 could be subsequently heated at 700°C to release oxygen (5). Prior to the introduction of low-temperature air fractionation, this was a major method for producing oxygen. The oxide finds use today in fluorescent lamps as an electrode coating to enhance electron release.

$$2Ba(s) + O_2(g) \rightarrow 2BaO(s) \quad (1)$$

$$Ba(s) + O_2(g) \rightarrow BaO_2(s) \quad (2)$$

$$3Ba(s) + N_2(g) \rightarrow Ba_3N_2(s) \quad (3)$$

$$BaO(s) + CO_2(g) \rightarrow BaCO_3(s) \quad (4)$$

$$2BaO(s) + O_2(g) \rightleftarrows 2BaO_2(s) \quad (5)$$

About 75% of all barium carbonate produced goes into the manufacturing of specialty glass. The addition of barium increases the refractive index of glass, but unlike lead oxide, it does not increase the dispersion. Before the introduction of lanthanide-containing glasses in the 1930s, this property was critically important in the design of lenses that would minimize astigmatic aberrations. Barium-containing glass is also used to provide radiation shielding in cathode ray and television tubes, although the voltage in such devices must be controlled to avoid secondary emission of X-rays.

Barium and barium oxide react with water to form the hydroxide, $Ba(OH)_2$, and in the case of barium, also hydrogen gas (H_2) (6, 7). Barium hydroxide, which crystallizes as the octahydrate, $Ba(OH)_2 \cdot 8H_2O$, is readily soluble in water (38 g L^{-1}), and is similar to the alkali metal hydroxides in its base strength.

$$Ba(s) + 2H_2O(g) \rightarrow Ba(OH)_2(aq) + H_2(g) \quad (6)$$

$$BaO(s) + H_2O(l) \rightarrow Ba(OH)_2(aq) \quad (7)$$

Barium sulfate (barita) is the main commercial source of barium, although the metal will react with sulfuric acid to form the sulfate (8), and it can be formed from an exchange reaction (e.g., 9). The latter works because of the sulfate's low solubility in water ($K_{sp} = 1.1 \times 10^{-10}$), which causes it to precipitate from solution. Barium sulfate itself is not the direct source of most barium compounds, but it is converted to the soluble sulfide BaS by heating with carbon (10) or hydrogen. The sulfide is then used to prepare other barium derivatives.

$$Ba(s) + H_2SO_4(l) \rightarrow BaSO_4(s) + H_2(g) \quad (8)$$

$$BaCl_2(aq) + Na_2SO_4(aq) \rightarrow BaSO_4(s) + 2NaCl(aq) \quad (9)$$

$$BaSO_4 + 2C(s) \rightarrow BaS(s) + 2CO_2(g) \quad (10)$$

The density of $BaSO_4$ is exploited in its largest industrial use (approx. 80% of world consumption), that of a component in water-based drilling fluid. The sulfate is also used as a pigment in paints (where it is known as *blanc fixe*, i.e., "permanent white"), either alone or in combination with titanium dioxide or zinc sulfide. It finds some use as a filler for rubber or paper, and its insolubility makes it valuable as a radiocontrast agent in medical imaging (see Section "General Properties of Compounds").

Barium nitrate, and to a lesser extent barium chlorate, is used to create a green color in fireworks and other pyrotechnics. Mixed with thermite (aluminum/metal oxide mixture), a small percentage of sulfur and a binder, barium nitrate forms Thermate-TH3, which is used in incendiary grenades.

The technique of chemical vapor deposition (CVD; sometimes abbreviated as MOCVD [metalorganic chemical vapor deposition]) has been under intensive development for the s-block elements, and particularly the alkaline-earth metals, since the late 1980s (Pierson 1999). The production of complex oxides of barium, such as the perovskite-based titanate $BaTiO_3$ and superconducting cuprates (e.g., $YBa_2Cu_3O_{7-x}$) has been one focus of this research. Barium titanate is a ferroelectric ceramic that is used in capacitors, as a piezoelectric material, and in nonlinear optical applications. The high-temperature superconductors have a variety of applications, including use in Josephson junctions and in current-carrying cable.

Simple metal alkoxides and acac derivatives are usually unsuitable as precursors to electronic materials, and even specially modified compounds have difficulties with deposition. For example, fluorinated

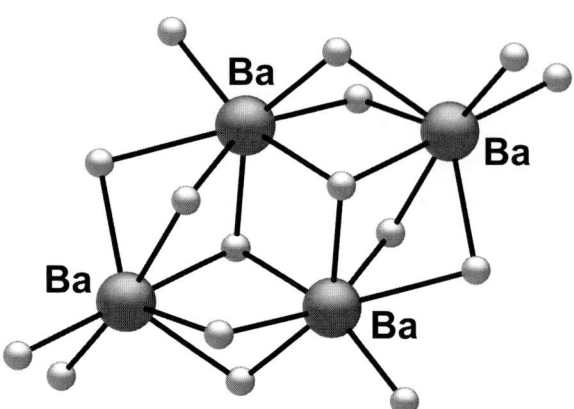

Barium, Physical and Chemical Properties, Fig. 3 Barium-oxygen core of Ba(tmhd)$_2$; in the solid state, the material is a tetrametallic compound. Carbon and hydrogen atoms have been omitted for clarity

compounds are favored for their increased volatility, but their use can lead to the deposition of metal fluorides that contaminate the deposited oxides and require extra processing to remove. Early reports on the thermal behavior of the widely used barium oxide precursor Ba(tmhd)$_2$ (tmhd = 2,2,6,6-tetramethylheptane-3,5-dionate anion, Fig. 3) indicated that ca. 25–40% of the material remained unsublimed under oxide-forming conditions. Such problems were later found to be the result of using partially decomposed or impure material; with the pure compound, only a 5–6% residue remains after heating to 410°C (Drake et al. 1993).

With Barium-Sulfur Bonds

Barium sulfide can be formed from the reaction of barium sulfate with hydrogen (11), but the original and most important method uses carbon as the reducing agent (10). Like the oxide, it possesses a rock salt crystal structure, and is soluble in water. A particular crystalline form of barite found near Bologna, Italy in the early seventeenth century was dubbed "Bologna stone." When strongly heated with charcoal (which formed the sulfide), the material was phosphorescent, and glowed for a time after exposure to bright light. The phenomenon was so unusual that it attracted the attention of many scientists of the day, including Galileo. Barium sulfide was thus the first synthetic phosphor, a property associated with its being a wide band-gap (3.9 eV) semiconductor. Cerium-doped BaS remains of interest as a phosphor for electroluminescent devices (Braithwaite and Weaver 1990).

$$BaSO_4 + 4H_2(950°C) \rightarrow BaS(s) + 4H_2O(g) \quad (11)$$

With Barium-Halogen Bonds

All four halides of barium are known; BaF$_2$ has the CaF$_2$ lattice type, BaBr$_2$ and BaI$_2$ display the PbCl$_2$ lattice, and BaCl$_2$ can crystallize in either type. They are not formed commercially from the metal and halogens, but rather from displacement reactions. BaF$_2$, for example, is formed from barium carbonate; the insolubility of the fluoride ($K_{sp} = 1.8 \times 10^{-7}$) drives the reaction (12). Barium chloride is made by the neutralization of barium hydroxide with HCl in water (13); the dihydrate product BaCl$_2$•2H$_2$O can be heated to leave the anhydrous BaCl$_2$. Industrially, BaCl$_2$ can be made from the high-temperature fusion of barium sulfide and calcium chloride (14); the barium chloride is removed with water, leaving the less soluble calcium sulfide behind. Barium bromide can be made from barium carbonate or barium sulfide (15, 16). Barium iodide can be made exactly as the bromide (17), but it is also formed by the reaction of BaH$_2$ with ammonium iodide in pyridine (18).

$$BaCO_3(aq) + 2NH_4F(aq) \rightarrow BaF_2(s) + (NH_4)_2CO_3(aq) \quad (12)$$

$$Ba(OH)_2 + 2HCl(aq) \rightarrow BaCl_2(aq) + 2H_2O \quad (13)$$

$$BaS + CaCl_2 \rightarrow BaCl_2(s) + CaS(s) \quad (14)$$

$$BaCO_3 + 2HBr(aq) \rightarrow BaBr_2(aq) + H_2O + CO_2(g) \quad (15)$$

$$BaS + HBr \rightarrow BaBr_2 + H_2S(g) \quad (16)$$

$$BaCO_3 + 2HI(aq) \rightarrow BaI_2(aq) + H_2O + CO_2(g) \quad (17)$$

$$BaH_2 + 2NH_4I(aq) \rightarrow BaI_2(aq) + 2NH_3(g) + H_2(g) \quad (18)$$

Although brittle and rather soft (Mohs hardness = 3), crystalline barium fluoride is used to make optical lenses and windows for infrared spectroscopy, as it is suitably transparent from roughly 150 nm

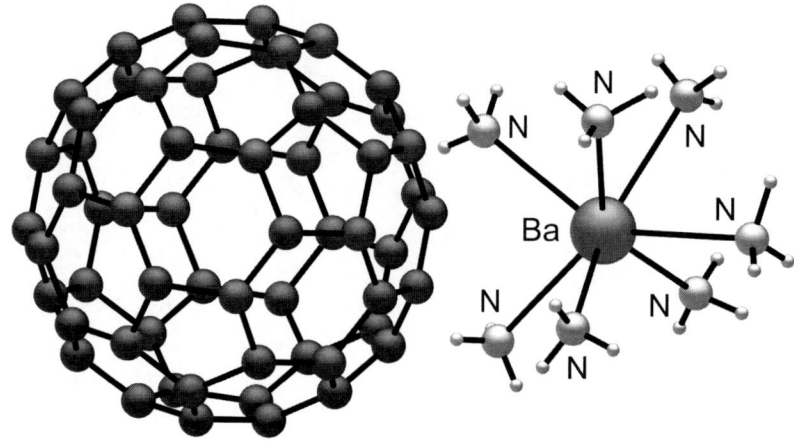

Barium, Physical and Chemical Properties, Fig. 4 Solid state structure of $[Ba(NH_3)_7]^{2+}[C_{60}]^{2-}$

(ultraviolet) to 12 μm (infrared). Barium fluoride is also a fast scintillator material for detecting X-rays, gamma rays, alpha and beta particles, and neutrons.

Barium hydride, which is made by the high temperature reaction of elemental barium and hydrogen (19), can be fused with the barium halides to produce the barium hydride halides BaHX (X = Cl, Br, I). An alternative method of preparation is to heat elemental barium and a barium halide in a hydrogen atmosphere at 950°C (20). The BaHX compounds have the PbClF crystal structure, and melt without decomposition at high temperatures (e.g., mp of BaHCl is 850°C).

$$Ba(s) + H_2(g) \rightarrow BaH_2(s) \quad (19)$$

$$Ba(s) + BaX_2(s) + H_2(g)$$
$$\rightarrow 2BaHX(s)\,(X = Cl, Br, I) \quad (20)$$

With Barium-Nitrogen Bonds

Barium dissolves in liquid ammonia to give a deep blue-black solution that yields a copper colored ammoniate, $Ba(NH_3)_6$, on evaporation. With time, the ammoniate will decompose to form the amide (21). In the presence of the carbon fullerenes C_{60} and C_{70}, however, the $[Ba(NH_3)_n]^{2+}$ cations (n = 7, 9) are generated from barium in liquid ammonia. The X-ray crystal structure of $[Ba(NH_3)_7]C_{60} \cdot NH_3$ reveals a monocapped trigonal antiprism around the metal (Ba–N = 2.85–2.94 Å), with an ordered C_{60} dianion (Fig. 4) (Himmel and Jansen 1998).

$$Ba(NH_3)_6(s) \rightarrow Ba(NH_2)_2(s) + 4NH_3(s) + H_2(g) \quad (21)$$

Whereas the parent amide $Ba(NH_2)_2$ possesses an ionic lattice, replacement of the hydrogen atoms with groups of increasing size leads to molecular complexes. Such amido complexes are versatile reagents, and can be used to prepare other main-group and transition metal complexes. The chemistry of $Ba[N(SiMe_3)_2]_2$ (Fig. 2) has been reviewed (Westerhausen 1998).

With Barium-Carbon Bonds

Barium compounds with bonds to carbon (organobarium complexes) are relatively rare species. Owing to the largely ionic character of the bonding in barium compounds, all are reactive species, and decompose in the presence of air and moisture. They are formed by a variety of methods, often involving exchange reactions (e.g., (22)) (Hanusa 2007). Some organobarium species can function as initiators for polymerization reactions and serve as precursors to oxides under CVD conditions. The bis(cyclopentadicnyl) metallocenes usually display "bent" (nonlinear) geometries (Fig. 5), similar to the bent gas-phase dihalides; related issues in their metal-ligand bonding (e.g., reverse polarization effects, d-orbital participation) may be involved.

$$2K[C_5R_5] + BaI_2(s) \rightarrow (C_5R_5)_2Ba + 2KI(s) \quad (22)$$

Biological Features of Barium Chemistry

The soluble salts of barium are toxic in mammalian systems. They are absorbed rapidly from the gastrointestinal tract and are deposited in the muscles, lungs, and bone. At low doses, barium acts as a muscle

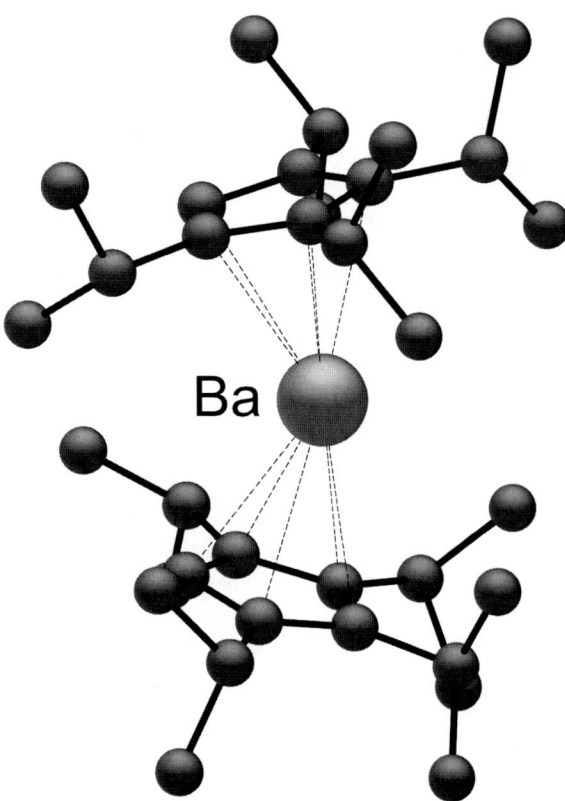

Barium, Physical and Chemical Properties, Fig. 5 Solid state structure of [C$_5$(i-Pr)$_4$H]$_2$Ba. The molecule is "bent," with an angle between the cyclopentadienyl ring planes of 153°

stimulant; at higher doses, barium affects the nervous system, eventually leading to paralysis. Oral doses of barium cause vomiting and diarrhea, followed by decreased heart rate and elevated blood pressure. Higher doses result in cardiac irregularities, tremors, and difficult or labored breathing. Death can occur from cardiac and respiratory failure. A single 0.8 g dose of barium chloride can be fatal to a 70 kg human. The similar size of the barium and potassium ions (1.35 Å and 1.38 Å, respectively), is thought to be the major source of the toxicity. Barium may interfere in the functioning of ▶ potassium channels, and the symptoms of barium poisoning are characteristic of the effects observed with a drop in serum potassium levels.

Industrial workers exposed to barium dust, usually in the form of barium sulfate or carbonate, may develop a benign pneumoconiosis referred to as "baritosis." Although it results in incidences of hypertension, baritosis is unusually reversible once the source of exposure is removed.

An important medical application of barium is the use of a slurry of the sulfate known as a "barium meal." When swallowed, the material coats the lining of the gastrointestinal tract, allowing for improved imaging by X-ray techniques. The minimal toxicity of barium sulfate is a consequence of its low solubility, and the ingested material is rapidly excreted.

Cross-References

▶ Barium(II) Transport in Potassium(I) and Calcium (II) Membrane Channels
▶ Potassium Channels, Structure and Function

References

Braithwaite N, Weaver G (1990) Electronic materials. Butterworth, London

Diefenbach M, Schwarz H (2005) High-electron-density C$_6$H$_6$ units: stable ten-electron benzene complexes. Chem Eur J 11:3058–3063

Drake SR et al (1993) Group IIA metal β-diketonate complexes; the crystal structure of [Sr$_3$(tmhd)$_6$(Htmhd)]•C$_6$H$_5$Me•C$_5$H$_{12}$ and [Ba$_4$(tmhd)$_8$] (Htmhd = 2,2,6,6-tetramethylheptane-3,5-dione). J Chem Soc, Dalton Trans 2883–2890

Hanusa TP (2007) Alkaline-earth metals: beryllium, magnesium, calcium, strontium, and barium. In: Crabtree RH, Mingos DMP (eds) Comprehensive organometallic chemistry-III, vol 2. Elsevier, Oxford, pp 67–152

Hargittai M (2000) Molecular structure of metal halides. Chem Rev 100:2233–2301

Himmel K, Jansen M (1998) Synthesis and single-crystal structure analysis of [Ba(NH$_3$)$_7$]C$_{60}$•NH$_3$. Inorg Chem 37:3437–3439

Kaupp M, Schleyer PVR (1992) The structural variations of monomeric alkaline earth MX$_2$ compounds (M = Ca, Sr, Ba; X = Li, BeH, BH$_2$, CH$_3$, NH$_2$, OH, F). An ab initio pseudopotential study. J Am Chem Soc 114:491–497

Murugavel R et al (2001) Reactions of 2-mercaptobenzoic acid with divalent alkaline earth metal ions: synthesis, spectral studies, and single-crystal X-ray structures of calcium, strontium, and barium complexes of 2,2´-dithiobis(benzoic acid). Inorg Chem 40:6870–6878

Pierson HO (1999) Handbook of chemical vapor deposition: principles, technology, and applications. Noyes, Norwich

Westerhausen M (1998) Synthesis, properties, and reactivity of alkaline earth metal bis[bis(trialkylsilyl)amides]. Coord Chem Rev 176:157–210

Behavior of Aluminum in Biological Systems

▶ Aluminum, Biological Effects

Berylliosis

▶ Beryllium as Antigen

Beryllium (Be) Exposure in the Workplace

▶ Beryllium as Antigen

Beryllium as Antigen

Vladimir N. Uversky
Department of Molecular Medicine, University of South Florida, College of Medicine, Tampa, FL, USA

Synonyms

Berylliosis; Beryllium (Be) exposure in the workplace; Beryllium sensitization (BeS); Beryllium-induced disease; Chronic beryllium disease (CBD); Contact allergy to beryllium

Definition

Beryllium exposure can have adverse health effects, starting from the contact allergy to Be and ending with the chronic beryllium disease (CBD) that leads to physiologic impairment and need for immunosuppressive medications, or even with the lung cancer.

Some Physicochemical Properties and Commercial Use

The chemical element beryllium is the fourth element in the chemical periodic table with the symbol of Be, atomic number of 4, and atomic weight of the 9.0122. Beryllium is a typical steel-gray metal, which in its free form exists as a strong, brittle, and lightweight alkaline earth metal. Beryllium is a relatively rare element in both the universe and in the crust of the Earth (it has a concentration of 2–6 parts per million (ppm) in the Earth's crust), being naturally found only in combination with other elements in minerals. It is almost always divalent in its compounds. Among the notable gemstones that contain beryllium are beryl (the precious forms of which are aquamarine, bixbite, and emerald) and chrysoberyl.

Its lightweight and unique chemical and physical properties make beryllium a highly desirable component with a wide use in high-technology industries, such as aerospace, ceramics, electronics, and defense. For example, when alloyed to aluminum, cobalt, copper, iron, and nickel, Be increases hardness and resistance to corrosion of the corresponding alloys. Since beryllium is nonmagnetic, tools made of this metal are used by naval or military explosive ordnance disposal teams for work on or near naval mines, since these mines commonly have magnetic fuzes. Beryllium-based instruments are also used to tune the highly magnetic klystrons, magnetrons, traveling wave tubes, and other gears, that are used for generating high levels of microwave power in the transmitters. In magnetic resonance imaging (MRI) machines, many maintenance and construction materials contain beryllium. Also, due to its high thermal stability, thermal conductivity, flexural rigidity, and low density (1.85 times that of water), beryllium has multiple structural applications and is considered as a unique aerospace material utilized in construction of high-speed aircrafts, missiles, various space vehicles, and communication satellites. Furthermore, due to its low density and atomic mass, beryllium is relatively transparent to X-rays and other forms of ionizing radiation. These properties make Be the most common window material for the X-ray equipment and in particle physics experiments. Finally, the high thermal conductivities of beryllium and beryllium oxide have led to their use in heat transport and heat-sinking applications (http://en.wikipedia.org/wiki/Beryllium).

Beryllium Exposure and Health Risks

Beryllium is not known to be necessary or useful for either plant or animal life. The commercial use of beryllium is challenged due to the toxicity of this metal. Since beryllium is chemically similar to magnesium, it can replace magnesium in some proteins, causing their malfunction (Emsley 2001). Although small amounts (approximately 35 μm) of beryllium are found in the human body, this content is not considered to be harmful (Emsley 2001).

Beryllium exposure primarily occurs through inhalation of the beryllium-containing dusts by workers involved in manufacturing the Be-containing products. It is estimated that approximately 200,000 current and at least one million total workers have been exposed to beryllium in the United States alone (Henneberger et al. 2004). In some people, beryllium exposure can cause a chronic life-threatening allergic disease, the chronic beryllium disease (CBD) (also known as berylliosis). Beryllium and beryllium compounds are also considered as category I carcinogens (http://en.wikipedia.org/wiki/Beryllium). Some of the consequences of the beryllium exposure are considered below.

Contact Allergy to Beryllium and Acute Beryllium Disease

Acute inhalation of large amounts of dust or fumes contaminated with beryllium over a short time or short and heavy exposure to beryllium might produce acute beryllium disease, which results from an irritant response at high exposure levels and which may take several forms, such as contact dermatitis, conjunctivitis, gingivitis, stomatitis, nasopharyngitis, tracheobronchitis, and pneumonitis. Severities of damage and recovery times are different for different forms of the acute beryllium disease and depend on the exposure level. Information on the acute beryllium disease is mostly based on the materials available at the Canadian Centre for Occupational Health and Safety web site (http://www.ccohs.ca/oshanswers/diseases/beryllium.html).

Contact dermatitis is an inflammation of the skin that is accompanied by itching, redness, rashes, swelling, and blisters. These symptoms appear on the exposed areas of the body, especially on the face, neck, arms, and hands. Skin effects (lesions, ulcerations, wart-like bumps) can also develop if beryllium penetrates into cuts or scratches. The membrane that covers the front of the eye can become inflamed in association with dermatitis. Beryllium may also cause conjunctivitis in occupationally exposed workers. Splashes of beryllium solutions may also burn the eyes causing fluid accumulation and reddening around the eyes. Conditions reflecting contact dermatitis usually improve a few weeks after the exposure ends.

Besides skin, beryllium may act on mucous membranes of oral cavities. As a result, persons with dental implants made from alloys including beryllium may develop gingivitis, stomatitis, and asymptomatic local sensitivity to beryllium.

Nasopharyngitis is an inflammation of the nose and throat. Symptoms include pain, swelling, and bleeding of the nose. This condition clears up 3–6 weeks after exposure ends.

Tracheobronchitis is an inflammation of the windpipe and the airways beyond it. Symptoms are coughing and discomfort, and tightness of the chest. Recovery takes about 1 month.

Acute beryllium inhalation can cause pneumonitis, which is an inflammation of the lungs confined to the walls of the air sacs. Pneumonitis is the most serious of the acute effects from the beryllium exposure. It varies in severity and can result in death. However, fatal cases are rare, and recovery is usually complete in about 6 months. Symptoms of acute beryllium pneumonitis are coughing, breathing difficulties, tightness of the chest, appetite and weight loss, and general weakness and tiredness.

Allergic reactions to beryllium in patients subjected to the skin-patch tests were considered as rather rare. However, recent systematic study on the patients with patch test reactions to beryllium chloride revealed that contact allergy to beryllium chloride may not be as unusual as the literature suggested (Toledo et al. 2011).

Beryllium Sensitization

Beryllium sensitization is an allergic reaction to beryllium that can develop after a person breathes beryllium dust or fumes. Here, beryllium acts as an antigen that induces a hypersensitive cellular immune response. Beryllium sensitization is not a disease, and sensitized people may not have symptoms. However, beryllium sensitization represents a potential foundation for both skin effect and chronic beryllium disease (CBD) associated with the secondary beryllium exposures. In fact, beryllium sensitization and CBD are both caused by the beryllium exposure and can be found based on the results of the beryllium lymphocyte proliferation test (BeLPT) (Newman et al. 2005). Since approximately 50% of individuals with beryllium sensitization have chronic beryllium disease at the time of their initial clinical evaluation, it was hypothesized that most beryllium-sensitized workers will eventually develop CBD (Newman et al. 2005). The hypothesis was supported by a longitudinal cohort study of a group of individuals with beryllium sensitization but without the initial evidence of CBD. The study showed that CBD developed in 31% of sensitized individuals within an average follow period of 3.8 years (range, 1.0–9.5 years), whereas 69% remained beryllium sensitized without disease after an average follow-up time of 4.8 years (range, 1.7–11.6 years) (Newman et al. 2005). This high probability of CBD development in originally asymptotic beryllium-exposed workers and the fact that beryllium is retained in lungs long after the actual exposure clearly suggested that beryllium sensitization is an adverse health effect that merits medical follow-up (Newman et al. 2005).

Chronic Beryllium Disease (CBD)

Inhalation of small amounts of beryllium-containing dusts or fumes contaminated with beryllium over a long time can lead to the development of berylliosis or chronic beryllium disease (CBD), which is an immunologically mediated granulomatous lung disease that could result from the inhalation of airborne beryllium particles. CBD is definitely a product of industrialization, since for the first time, this disease in the form of chemical pneumonitis was reported in Europe in 1933 and in the United States in 1943 (http://en.wikipedia.org/wiki/Beryllium). Symptoms of the disease can take up to 5 years to develop. It is estimated that about a third of the CBD patients die, whereas the survivors are left disabled (Emsley 2001).

CBD is characterized by the presence of noncaseating granulomatous inflammation primarily affecting the lung, although other organs may be involved (Fontenot and Kotzin 2003). There are several hallmarks of the CBD, such as the accumulation of Be-specific $CD4^+$ T cells in the lung (the healthy human lung contains few lymphocytes, whereas the lungs of patients with CBD are characterized by the accumulation of a remarkably large number of Be-responsive $CD4^+$ T cells; that is, by the $CD4^+$ T cell alveolitis (Saltini et al. 1989)), persistent lung inflammation, and the development of lung fibrosis (Falta et al. 2010). Because the histopathologic features of CBD resemble those of lung sarcoidosis, a disease in which abnormal collections of chronic inflammatory cells (granulomas) also form nodules in lungs, the CBD diagnosis is complicated and depends on the detection of a Be-specific immune response in blood and/or lung and the presence of noncaseating granulomas and/or mononuclear cell inflammation on a biopsy specimen (Falta et al. 2010). It was pointed out that the frequency of antigen-specific T cells in blood can serve as a noninvasive biomarker to predict disease development and severity of the Be-specific $CD4^+$ T-cell alveolitis (Martin et al. 2011).

Since the susceptibility to this disease depends on the nature of the exposure and the genetic predisposition of the individual (see below), this ailment develops in 2–16% of subjects exposed to the beryllium at the workplace (Falta et al. 2010). It is recognized now that the activation and accumulation of Be-responsive $CD4^+$ T cells in the lungs of the CBD patients is linked to the genetic alterations in the major histocompatibility complex class II (MHCII) molecules, which generally are implicated in susceptibility to various immune-mediated diseases. MHCII proteins are the protein/peptide receptors (T-cell receptors, TCRs) found only on antigen-presenting cells. The human MHC class II molecules (human leukocyte antigen complexes, HLA) are encoded by three different isotypes, HLA-DR, -DQ, and -DP, each being highly polymorphic. CBP-related MHCII molecules are from the HLA-DP isotype. These receptors are composed of two subunits, DPα and DPβ, each having two domains, $α_1$ and $α_2$ and $β_1$ and $β_2$, with the domains $α_1$ and $β1$ forming a recognition module, heterodimer that contains the peptide-binding groove, and with the domains $α_2$ and $β_2$ acting as transmembrane domains that anchor the MHC class II molecule to the cell membrane (Abbas and Lichtman 2009). MHCII subunits are encoded by the *HLA-DPA1* and *HLA-DPB1* loci found in the *MHCII* (or *HLA-D*) region located within the human chromosome 6 (http://en.wikipedia.org/wiki/HLA-DP). CBD predisposition is linked to *HLA-DP* alleles that contain a glutamic acid at position 69 of the β-chain (βGlu69) (Falta et al. 2010; Richeldi et al. 1993).

Here is a brief description of the development of the beryllium-induced granulomatous response. As it was already mentioned, CBD patients are characterized by the presence of the large amounts of Be-responsive $CD4^+$ T cells that contain Be-specific TCRs. Besides these Be-specific TCRs, the Be-responsive $CD4^+$

T lymphocytes express markers of previous activation and exhibit an effector memory T cell phenotype (Saltini et al. 1990; Fontenot et al. 2002). Recognition of beryllium by such memory CD4$^+$ T cells promotes their clonal proliferation and secretion of the T helper 1 (Th1)-type cytokines, such as interleukin-2 (IL-2), interferon-γ (IFN-γ), and tumor necrosis factor-α (TNF-α) (Fontenot et al. 2002). Then, IFN-γ and TNF-α promote macrophage accumulation, activation, and aggregation, resulting in the initiation of the granulomatous response (Falta et al. 2010).

Obviously, the existence of Be-specific TCRs (i.e., some mutant forms of MHCII) is crucial for the initiation of this process. In agreement with this hypothesis, *DPB1* alleles with βGlu69 mutation were shown to be strongly associated with disease susceptibility (Falta et al. 2010; Richeldi et al. 1993), since approximately 80% of patients with CBD were characterized by the presence of βGlu69-containing *DPB1* alleles (Falta et al. 2010), and since βGlu69-containing *DPB1* alleles were shown to be a serious risk factor for the development of Be sensitization and not simply a marker of progression from sensitization to disease (Falta et al. 2010). Furthermore, since these same alleles were capable of presenting beryllium to pathogenic CD4$^+$ T cells, and since beryllium recognition required the presence of βGlu69, it was proposed that a single polymorphic amino acid might dictate beryllium presentation and, more importantly, disease susceptibility (Falta et al. 2010). However, the fact that ~20% CBD patients do not possess a βGlu69-containing *HLA-DPB1* allele suggests that some other mutations in the MHCII molecules can be important for the genetic susceptibility to CBD (Falta et al. 2010).

The existing structural data on the HLA-DP2 in complex with a peptide derived from the HLA-DR α-chain (pDRA) revealed that the binding groove of the HLA-DP2 βGlu69 variant is characterized by the presence of the unique acidic pocket flanked by leucine residues at the positions 4 and 7 of the pDRA and the β-chain α-helix and was composed of three DP2 β-chain amino acids that contribute to the net negative surface charge of this pocket: βGlu68 and βGlu69 from the β-chain α-helix and βGlu26 from the floor of the peptide-binding groove (Dai et al. 2010). These findings clearly indicated that the acidic pocket induced by the βGlu69 mutation should be considered as the beryllium binding site (Falta et al. 2010). This hypothesis was supported by the fact that the mutated HLA-DP2 molecules that contained point substitutions at the positions βGlu26 and βGlu68 did not activate beryllium-specific T cells (Dai et al. 2010).

Therefore, existing data strongly suggest that the acidic pocket of the HLA-DP2 binding groove is the beryllium binding site. Formation of this pocket as a result of the βGlu69 mutation in the HLA-DP2 protein provides an explanation for the genetic linkage of HLA-DP2 to the development of granulomatous inflammation in the Be-exposed worker and therefore represents a molecular basis for better understanding of the beryllium-induced disease immunopathogenesis.

Beryllium Exposure and Lung Cancer

The topic of the potential carcinogenicity of beryllium is full of controversies. Early epidemiological and animal studies suggested that the occupational exposure to beryllium might cause lung cancer (Schubauer-Berigan et al. 2011). However, a recent systematic review of epidemiologic studies on cancer among workers exposed to beryllium showed that the available evidence does not support a conclusion that a causal association has been established between occupational exposure to beryllium and the risk of cancer (Boffetta et al. 2012). One of the major reasons for this discrepancy is the lack of adequate analysis of confounding factors in early studies. For example, some excess mortality from lung cancer was detected in the large cohort of patients with the beryllium exposure, which was partially explained by confounding by tobacco smoking and urban residence (Boffetta et al. 2012). Similarly, a weight-of-evidence analysis of the 33 animal studies and 17 epidemiologic studies showed that the evidence for carcinogenicity of beryllium is not as clear as suggested by previous evaluations, because of the inadequacy of the available smoking history information, the lack of well-characterized historical occupational exposures, and shortcomings in the animal studies (Hollins et al. 2009). Therefore, it was concluded that the studies of beryllium disease patients did not provide independent evidence for the carcinogenicity of beryllium and the results from other studies did not support the hypothesis of an increased risk of lung cancer or any other cancer associated with the occupational exposure to beryllium (Boffetta et al. 2012).

Cross-References

▶ Beryllium, Physical and Chemical Properties

References

Abbas AK, Lichtman AH (2009) Basic Immunology. Functions and disorders of the immune system, 3rd edn. Saunders/Elsevier, Philadelphia

Boffetta P, Fryzek JP, Mandel JS (2012) Occupational exposure to beryllium and cancer risk: a review of the epidemiologic evidence. Crit Rev Toxicol 42:107–118

Dai S, Murphy GA, Crawford F, Mack DG, Falta MT, Marrack P, Kappler JW, Fontenot AP (2010) Crystal structure of HLA-DP2 and implications for chronic beryllium disease. Proc Natl Acad Sci USA 107:7425–7430

Emsley J (2001) Nature's Building blocks: an A-Z guide to the elements. Oxford University Press, Oxford

Falta MT, Bowerman NA, Dai S, Kappler JW, Fontenot AP (2010) Linking genetic susceptibility and T cell activation in beryllium-induced disease. Proc Am Thorac Soc 7:126–129

Fontenot AP, Canavera SJ, Gharavi L, Newman LS, Kotzin BL (2002) Target organ localization of memory CD4(+) T cells in patients with chronic beryllium disease. J Clin Invest 110:1473–1482

Fontenot AP, Kotzin BL (2003) Chronic beryllium disease: immune-mediated destruction with implications for organ-specific autoimmunity. Tissue Antigens 62:449–458

Henneberger PK, Goe SK, Miller WE, Doney B, Groce DW (2004) Industries in the United States with airborne beryllium exposure and estimates of the number of current workers potentially exposed. J Occup Environ Hyg 1:648–659

Hollins DM, McKinley MA, Williams C, Wiman A, Fillos D, Chapman PS, Madl AK (2009) Beryllium and lung cancer: a weight of evidence evaluation of the toxicological and epidemiological literature. Crit Rev Toxicol 39(Suppl 1):1–32

Martin AK, Mack DG, Falta MT, Mroz MM, Newman LS, Maier LA, Fontenot AP (2011) Beryllium-specific CD4+ T cells in blood as a biomarker of disease progression. J Allergy Clin Immunol 128(1100–1106):e1101–e1105

Newman LS, Mroz MM, Balkissoon R, Maier LA (2005) Beryllium sensitization progresses to chronic beryllium disease: a longitudinal study of disease risk. Am J Respir Crit Care Med 171:54–60

Richeldi L, Sorrentino R, Saltini C (1993) HLA-DPB1 glutamate 69: a genetic marker of beryllium disease. Science 262:242–244

Saltini C, Kirby M, Trapnell BC, Tamura N, Crystal RG (1990) Biased accumulation of T lymphocytes with "memory"-type CD45 leukocyte common antigen gene expression on the epithelial surface of the human lung. J Exp Med 171:1123–1140

Saltini C, Winestock K, Kirby M, Pinkston P, Crystal RG (1989) Maintenance of alveolitis in patients with chronic beryllium disease by beryllium-specific helper T cells. N Engl J Med 320:1103–1109

Schubauer-Berigan MK, Deddens JA, Couch JR, Petersen MR (2011) Risk of lung cancer associated with quantitative beryllium exposure metrics within an occupational cohort. Occup Environ Med 68:354–360

Toledo F, Silvestre JF, Cuesta L, Latorre N, Monteagudo A (2011) Contact allergy to beryllium chloride: report of 12 cases. Contact Dermatitis 64:104–109

Beryllium Sensitization (BeS)

▶ Beryllium as Antigen

Beryllium, Physical and Chemical Properties

Fathi Habashi
Department of Mining, Metallurgical, and Materials Engineering, Laval University, Quebec City, Canada

Beryllium is the first member of the alkaline earths but its chemical properties are more similar to aluminum – a property described as "diagonal similarities" (Fig. 1) as discussed below. Beryllium is an expensive metal used in small and specialized industries. Its dust and fumes as well as vapours of its compounds are poisonous to inhale. Its compounds have sweet taste that is why it was initially called "glucinium." It is fabricated by powder metallurgy techniques because coarse grains tend to develop in the castings causing brittleness and low tensile strength. About 10 % of the metal is used in the metallic form, 80 % in form of beryllium–copper alloys (containing about 2 % Be), or other master alloys, and the remaining 10 % is used as a refractory oxide. In the metallic form it is used as a moderator to slow down fast neutrons in nuclear reactors because of its low atomic weight and low neutron

Beryllium, Physical and Chemical Properties, Fig. 1 Diagonal similarities in the periodic table

cross section. As an alloy with copper, it is particularly important in springs because such alloys possess high elasticity and great endurance.

Physical Properties

Atomic number	4
Atomic weight	9.0122
Relative abundance in the Earth's crust, %	6×10^{-4}
Atomic radius, pm	112.50
Atomic volume at 298 K, cm^3/mol	4.877
Crystal structure	
At 293 K	Hexagonal closest packed
At 1,523 K	Body-centered cubic
Melting point, °C	1,287
Boiling point, °C	2,472
Transformation point, °C	1,254
Density	
At 298 K, g/cm^3	1.8477
At 1,773 K, g/cm^3	1.42
Heat of fusion, J/g	1,357
Heat of transformation, J/g	837
Vapor pressure	
At 500 K	5.7×10^{-29}
At 1,000 K	4.73×10^{-12}
At 1,560 K	4.84×10^{-6}
Specific heat	
At 298 K, J g^{-1} K^{-1}	1.830
At 700 K, J g^{-1} K^{-1}	2.740
Thermal conductivity at 298 K, W m^{-1} K^{-1}	165 ± 15
Linear coefficient of thermal expansion, 298–373 K, K^{-1}	11.5×10^{-6}
Electrical resistivity at 298 K, Ω m	4.31×10^{-8}
Volume contraction on solidification	3%

Because of its low density beryllium is considered a light metal. It transmits X-rays well because of its low atomic number. Relatively long-wave gamma rays cause beryllium to emit neutrons, on account of a (InlinemediaObject γ, n) reaction.

Chemical Properties

Beryllium is a typical metal; it has the electronic structure 2, 2 and hence when it loses its two outer electrons it will have the electronic structure of the inert gas helium. Although beryllium is almost always divalent in its compounds, it is more similar to aluminum than to magnesium or calcium. For example, like aluminum, it is amphoteric: it dissolves in dilute nonoxidizing mineral acids, accompanied by hydrogen evolution and salt formation and is also attacked by aqueous hydroxide accompanied by hydrogen evolution and beryllate formation. Beryllium ion precipitates by alkali to form beryllium hydroxide, $Be(OH)_2$, which dissolves in excess alkali. Beryllium chloride, $BeCl_2$, is volatile and has a covalent bond like aluminum chloride, $AlCl_3$.

In humid air and water vapor, beryllium forms a strongly adhering surface layer of oxide that prevents further oxidation up to about 600°C. Above 600°C oxidation takes place. Beryllium is an excellent reducing agent because of its great affinity for oxygen:

$$Be + 1/2 O_2 \rightarrow BeO$$

At temperatures >900°C, it reacts violently with nitrogen or ammonia to form beryllium nitride, Be_3N_2. However, it does not react with hydrogen even at high temperatures. Below 500–600°C beryllium is not attacked by dry carbon dioxide and is attacked only very slowly by moist carbon dioxide. Beryllium powder reacts with fluorine at room temperature, and at elevated temperatures it reacts with chlorine, bromine, and iodine and with sulfur, selenium, and tellurium vapor; in each case, beryllium burns with a flame.

The adhering oxide film protects beryllium from attack by both cold and hot water. It also protects cold beryllium from attack by oxidizing acids. Since most beryllium compounds have a highly exothermic heat of formation, beryllium reduces the salts and the borates and silicates of many metals. For example, the only halides that are stable toward beryllium are those of the alkali metals and magnesium; all others are reduced by beryllium. Molten alkali-metal hydroxides react explosively with beryllium.

Beryllium is very reactive in the liquid state, reacting with most oxides, nitrides, sulfides, and carbides, including those of magnesium, calcium, aluminum, titanium, and zirconium.

References

Habashi F (2003) Metals from Ores. An Introduction to Extractive Metallurgy, Métallurgie Extractive Québec, Québec City, Canada. Distributed b Laval University Bookstore "Zone". www.zone.ul.ca

Petzow G et al (1997) In: Habashi F (ed) Handbook of extractive metallurgy. Wiley, Weinheim, pp 955–980

Beryllium-Induced Disease

▶ Beryllium as Antigen

Bestatin, N-[(2 S, 3R)-3-Amino-2-hydroxy-4-phenylbutyryl]-L-leucin

▶ Zinc Aminopeptidases, Aminopeptidase from Vibrio Proteolyticus (Aeromonas proteolytica) as Prototypical Enzyme

Biarsenical Fluorescent Probes

Artur Krężel
Department of Protein Engineering, Faculty of Biotechnology, University of Wrocław, Wrocław, Poland

Synonyms

Bisarsenical probes; FlAsH-EDT$_2$; Fluorescent arsenical helix/hairpin binder; ReAsH-EDT$_2$

Definition

Fluorescent biarsenical probes refer to chemical agents that are built on fluorescent platforms with two proton to arsenic substitutions at certain positions that are capped with two molecules of 1,2-ethanedithiol. The probes remain nonluminescent until protein conjugation when they became highly fluorescent. Due to stability of arsenic-sulfur bond, biarsenical probes have very high affinity to four cysteine residues, which may be genetically introduced to proteins as a short sequence appropriately spaced (e.g., CCPGCC) or by incorporation into secondary structure elements.

Principles of the Biarsenical Labeling Strategy

Many of the chemistry-driven strategies used for monitoring, controlling, and modifying protein functions require small fluorescent probes that are selectively attached to the protein of interest. The probe conjugation is mostly achieved by single or multiple amino acid residue chemical labeling. From all functional groups being modified with probes such as carboxylic, hydroxyl, amine, and thiol group (R-SH), the last one is the most applicable due to its low natural abundance and relatively high reactivity (Hermanson 2008). Additionally, successful usage of the probe in vivo requires unique amino acid sequence or unique structural motif in the target protein. In 1998 R.Y. Tsien and coworkers presented a new fluorescent protien labeling strategy that use the biarsenical fluorescein derivative, FLASH-EDT$_2$ (fluorescein arsenical helix/hairpin binder). Later, they introduced a new probe based in biarsenical modification of resorufin (ReAsH-EDT$_2$) as well as other less known probes (Tsien 2005). These probes, having two arsenic atoms at position 4$'$ and 5$'$ capped with small protectants – 1,2-ethanedithiol (EDT) or 2,3-dimercaptopropanol (BAL) – possess ultrahigh affinity for short tetracysteine sequence appropriately spaced with two amino acids (CCXXCC: TC tag), uncommon in naturally occurring proteins (Jakobs et al. 2008). TC tag has significantly higher affinity to biarsenical platforms than the two protecting molecules (Fig. 1). New genetically encoded biarsenical probe recognition tag has been optimized several times to date to obtain the most efficient sequence in terms of conjugate stability and fluorescent properties. The most important advantage of biarsenical labeling strategy is that EDT-protected probes are usually cell permeable (require less than 1 h incubation) and practically

Biarsenical Fluorescent Probes, Fig. 1 Structures of the most frequently used biarsenical probes for fluorescent labeling of protein with tetracysteine tag (TC tag); *green* FlAsH-EDT$_2$ and *red* ReAsH-EDT$_2$

nonfluorescent, which significantly decreases the background when they are used for fluorescent microscopy imaging, compared to other fluorescent probes. Biarsenical EDT-protected probes became highly fluorescent when they bind to TC tags both in vitro and in vivo, and its fluorescence is more than 50,000 times higher than that of protected probe. Moreover, fluorescent conjugation in the cell is relatively fast (typically a couple of minutes) which makes this strategy one of the most important on the market. However, there are a number of proteins with vicinal thiols in the cell that may nonspecifically coordinate biarsenical probes with lower affinity than TC tags and increase the background causing lower signal to noise ratio. The administration of EDT-protected biarsenicals to TC-tagged proteins in the presence of the protectant prevents poisoning of other cellular proteins with vicinal or surface-exposed thiols. In vitro addition of excess of protectants (typically 0.25–1 mM) to biarsenical complex with TC-tagged protein results in immediate decrease of the fluorescence due to reverse reaction.

Besides some drawbacks of the fluorescent labeling strategy, one of the most attractive feature of this methodology is low molecular weight of the fluorescent moiety and affinity tag (typically less than 2 kDa), which is much lower compared to fluorescent proteins family (FPs), luciferase, β-galactosidase, β-lactamase, and many other protein reporters (Crivat and Tarasaka 2012).

Design of New Biarsenical Probes

Currently, there are many known biarsenical probes based on various fluorescent dyes with different chemical and spectral properties. New ones are still developing. Synthesis of the most popular biarsenical probes relies on two plain steps. The first one is mercuration reaction (proton substitution with mercury) of fluorescein or resorufin at positions 4' and 5'. Biarsenical product is received by transmetalation reaction with arsenic(III) chloride followed by addition of dithiol protectant (Adams and Tsien 2008). Green (FlAsH) and red (ReAsH) moieties differ not only in spectral properties but also in reactivity. The red probe represents higher reactivity which is explained by lower molecular weight. This probe gives much faster response; however, its quantum yield is significantly lower when compared to the green probe.

FlAsH-EDT$_2$ and ReAsH-EDT$_2$ probes are significantly different and do not cover research requirements for new probes with novel or intermediate spectral or chemical properties. Simple proton substitution with

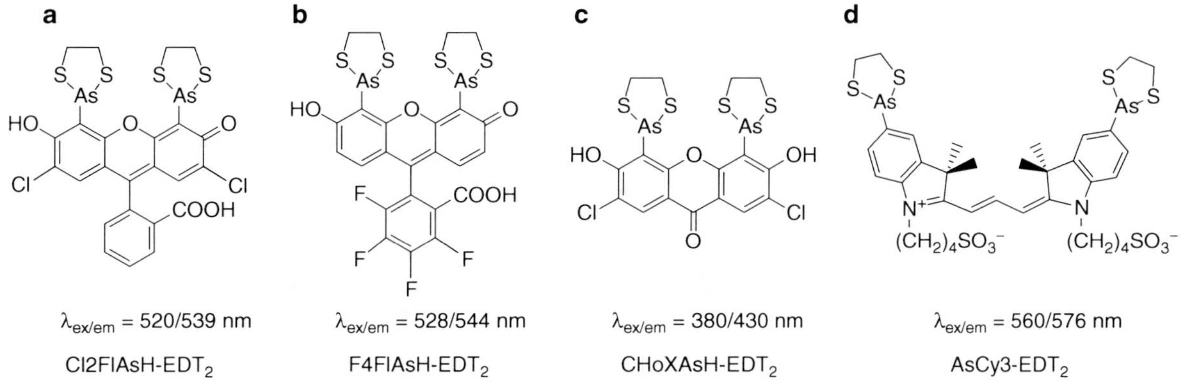

Biarsenical Fluorescent Probes, Fig. 2 Examples of multicolor EDT-protected biarsenical probes built on various types of fluorescent platforms: (**a**) 2′7′-dichlorofluorescein – Cl2FlAsH-EDT$_2$; (**b**) 3,4,5,6-tetrafluorofluorescein – F4FlAsH-EDT$_2$; (**c**) 2,7-dichloro-3,6-dihroxyxanthone – ChoXAsH-EDT$_2$; (**d**) Cy3 platform

chlorine or fluorine atoms at position 2′ and 7′ of fluorescein molecule resulted in new derivatives F2FlAsH-EDT$_2$ and Cl2FlAsH-EDT$_2$, respectively (Fig. 2). Halogenated biarsenical probes demonstrate higher absorbance, quantum yield, and Stokes shift, higher photostability, reduced pH dependence, and significantly different excitation and emission energies. Similarly, halogenation at positions 3–6 results in a new set of biarsenical probes: F4FlAsH-EDT$_2$, Cl4FlAsH-EDT$_2$. Blue and red shifts of new probes make them ideal tool for FRET pair constructions when attached to certain protein positions (Pomorski and Krężel 2011). Usage of 3,6-dihydroxy-xanthone and its 2,7-dichloro derivative allowed to construct blue biarsenical probes HoXAs-EDT$_2$ and CHoXAs-EDT$_2$, respectively. Although, they have been used for FRET pair construction with fluorescent proteins, their stability is significantly lower compared to FlAsH-EDT$_2$ and ReAsH-EDT$_2$ probes. The design of new biarsenicals is not only limited to fluorescein and resorufin derivatives that posses conserved interarsenic distance ∼ 4.8 Å. Usage of Cy3 platform for preparation of biarsenical derivative (AsCy3-EDT$_2$) resulted in significantly increased interarsenic distance of ∼ 14.5 Å (Fig. 2). The elongated distance requires different TC tag with longer spacer, CCKAEAACC. This probe exhibits extremely rapid response (<15 s) toward affinity tag and has a very high extinction coefficient (1.8×10^5 M^{-1} cm^{-1}), which makes the probe a very attractive tool in terms of the level of brightness ($\varepsilon \times \Phi = 5 \times 10^4$ M^{-1} cm^{-1}).

Some of the biarsenical products, such as rhodamines, are nonfluorescent even when bound to TC tags. Another example is biarsenical fluorescein spirolactams SplAsH (Spirolactam Arsenical Hairpin binder). These chemicals can be used as convenient site-specific handles for attaching other chemicals or other fluorescent probes.

The biarsenical probes built on 5(6)-carboxyfluorescein (CrAsH-EDT$_2$) or 5(6)-aminofluorescein (aminoFlAsH-EDT$_2$) bring enormous possibilities in utilizing the biarsenical chemistry. Usage of relatively highly reactive carboxylic or amine group allows to design another generation of dual or multifunctional biarsenical probes. Classic application was done by immobilization of CrAsH-EDT$_2$ onto amine-agarose resin for use in protein purification. Another successful applications were based on coupling of CrAsH-EDT$_2$ moiety with Ca(II) chelator (Calcium Green FlAsH-EDT$_2$, CaGF) and photocrosslinker (TRAP) for measuring available Ca(II) signals in neurons and protein-protein interactions linkage, respectively. Multifunctional derivatives were also achieved by conjugation of biotin (biotinFlAsH-EDT$_2$) and AlexaFluor 568 (AF568-FlAsH-EDT$_2$) to biarsenical platforms (Pomorski and Krężel 2011).

Tetracysteine Tags

The biarsenical affinity tag (TC tag) was originally designed in such a way to maintain coordination properties of two As(III) with highest possible affinity. To build a rigid structure, an α-helical motif based on

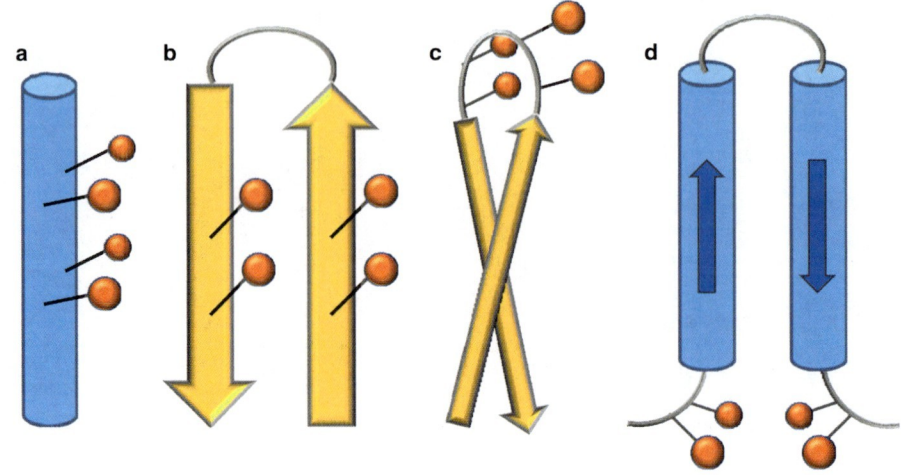

Biarsenical Fluorescent Probes, Fig. 3 The comparison of TC motifs successfully incorporated into protein structures. (**a**) α-Helix, (**b**) β-strand, (**c**) protein loop, (**d**) protein split/bipartite motif. *Yellow spheres* demonstrate exposed cysteine residues essential for efficient biarsenical conjugation

W(EAAAR)$_n$ sequence was applied in the case of the first biarsenical-protein conjugate. Four alanine to cysteine residue substitutions in α-helix formed a convenient coordination environment for biarsenical moiety in a characteristic parallelogram. Further studies showed that truncation of the original 18 amino acid peptide to shorter α-helix and substitution of ER to other helical amino acid did not significantly affect complex properties. However, use of helical-breaking proline residue in PG sequence resulted in formation of a much more stable complex. CCPGCC sequence reflects in high rate formation of the complex and has high affinity for biarsenicals ($K_d = 4$ pM). Additional optimization of the TC tag relied on optimization of flanking amino acids performed using high-throughput genetic screens and fluorescence-activated cell sorting. The optimized sequence FLNCCPGCCMEP indicates higher quantum yield and dithiol resistance in case of FlAsH-EDT$_2$ and ReAsH-EDT$_2$ probes. Improved fluorescent properties were explained by NMR studies that showed strong interaction of the N-terminal part of the TC tag with biarsenical moiety of the probe. FLNCCPGCCMEP sequence is so far the most frequently used TC tag for protein conjugation with biarsenical fluorescent probes (Adams and Tsien 2008).

The biarsenical labeling of proteins is significantly limited due to the fact that one TC tag may bind various biarsenical probes at the same moment if they are available. This fact limited multicolor use of the methodology. However, in a few studies, it has been shown that diversification of TC tags reflects in different probe affinities. Green FlAsH-EDT$_2$ was shown to selectively label CCPGCC sequence, while ReAsH-EDT$_2$ tends to label CCKACC sequence. The orthogonal labeling in vivo may be achieved by usage of CCPGCC sequence and its optimized version, FLNCCPGCCMEP, together with proper dithiol washing. The latter sequence tends to bind with ReAsH-EDT$_2$, while short tag is selectively labeled with FlAsH-EDT$_2$. In some cases, TC tandem sequence (TC$_n$ tag) may improve fluorescent brightness; however, it reflects in higher cysteine residues' oxidation tendency and significant increase in molecular weight of the affinity tag (Pomorski and Krężel 2011).

Essential requirement for biarsenical labeling technology is not to disturb protein structure, activity, and other functions of the protein, unless so desired. The easiest way to incorporate TC tag to the protein is an addition of the TC sequence to protein terminus that form N- or C-terminal fusion. This approach allows protein labeling, even when protein structure is unknown, and its immediate illumination. Since TC tag was originally designed to be a part of α-helix, there is a possibility to place TC motif into this structure as continuous sequence (Fig. 3). α-Helix may be also be elongated to introduce TC motif in such a manner to not disturb protein structure. The placement of TC motif to β-structure requires usually knowledge about 3D structure of the protein of interest. In this case, the most efficient fluorescence or complex affinity is usually achieved in several trials with cysteine residue placement. The most frequent structural motif of the protein used for TC environment design is protein loop. It may be based on a loop already present in protein or may be added into protein chain turns or domain connections. This approach is frequently used together with bioinformatic prediction of protein structure. One should remember that TC motif based on protein loop may differ from other loops and strongly depends on surrounding amino acids. Another type of tetracysteine coordination environment is built on protein splits and bipartite motifs (Scheck and Schepartz 2011). There are several ways to bind a probe using such coordination environment (Fig. 3). One includes the placement of four cysteine residues into two different a-helices in 2×2 Cys mode. Similarly, protein loops and β-strands may be structurally arranged in such a manner to form TC environment. Bipartite architecture of TC motif is very important in protein science, allowing one to study protein-protein interactions, oligomerization, activity, protein structure recognition, etc.

Although CCXXCC sequence and generally speaking TC motif is uncommon in naturally occurring proteins, there are some studies showing high affinity to nonrecombinant proteins, such as bacterial protein family SlyD or RBSS domain of human PML-PARα oncoprotein.

Protein Imaging with Biarsenical Fluorescent Probes

The main application of the biarsenical probes is visualization of the specific protein, that have an appropriate TC tag. Since the tag and probe are small sized, they have been used for imaging of proteins whose function or location is disrupted when fused with fluorescent proteins. Some of the early works indicated that unspecific binding to naturally occurring thiols prevented successful usage of the probes; however, recently published protocol shows that insufficient washing with dithiol protectant can lead to poor results (Hoffmann et al. 2010). However, one should keep in mind that different sequence-probe pairs have limited dithiol resistance and very high concentration can lead to probe release from sequence. Also high protectant concentration can also lead to cell detachment. The extracellular proteins as well as those located in endoplasmic reticulum and Golgi apparatus are oxidized, so in order for the probe to attach to TC tag, the cysteines have to be reduced first. In the first case, use of membrane-impermeant biarsenical probe such as CrAsH-EDT$_2$ or sFlAsH-EDT$_2$ is recommended. Although it has been shown that biarsenical probes can cause some toxicity in mitochondria, it was numerously stated that staining did not affect cell viability or cause cell death. It was previously mentioned that there is a strategy for orthogonal labeling of different TC sequences in the same cell, but it is also possible to perform pulse-chase experiments, where current population of TC-tagged proteins is stained with one probe, e.g., FlAsH-EDT$_2$, and then after additional incubation, the new pool of tagged protein is visualized using second biarsenical probe, e.g., ReAsH-EDT$_2$. This approach allows tracking of the fate of the protein and shows how new pool of the protein is incorporated into cellular structures. Additionally, ReAsH/DAB/OsO$_4$ strategy allows to gain ultrastructural information on the protein location. TC tag was added to GFP in order to develop multi-resolution tag, which combines GFP's excellent contrast with ReAsH's ability to generate singlet oxygen. The examples of biarsenical probes application include monitoring of mitochondrial proteins during apoptosis, study of protein translocation in response to the stimuli, or receptor trafficking. Additionally, since the probes attach very fast once TC tag is present, they can be used for visualization of the mRNA translation in vivo. The application of the probes for protein imaging is not only limited to mammalian cells, but can be also applied in bacteria and yeast. In this case, due to different nature of the cell's outer barrier, the biarsenical probes do not enter the cell so readily as with mammalian cells; however, there are published methods that allow their convenient use. It was shown that the biarsenical probes have negative effect on the doubling time of *S. cerevisiae* but there are many successful examples of their usage. Examples of application include study of spatial and temporal dynamics of flagellar filament protein in *E. coli* multicellular communities or labeling of secretion effectors of pathogenic bacteria during invasion of mammalian cells. The utilization of biarsenical probes was most successful in case of virus proteins, since the fluorescent proteins tend to disrupt their function. The probes allowed tracking of absorption, entry, and uncoating of vesicular stomatis virus. HIV-1 integrase was tracked in 4D. Also HIV Gag protein was thoroughly studied using biarsenical probes. There are also examples of successful imaging of proteins in plant cells and *Dictyostelium* (Pomorski and Krężel 2011).

Other Applications of Biarsenical Probes

Advantages of protein labeling with biarsenical probes are mostly based on their fluorescence switch feature, low molecular weight, and almost unlimited possibilities to use them to site-controlled protein conjugation. The last advantage opens enormous application possibilities in protein science both in vitro and in vivo. In early study FlAsH-EDT$_2$ was used as a reporter of folding state of the cellular CRABP I protein in vivo. Incorporation of TC motif in appropriate place of the protein and its subsequent conjugation with biarsenical probe functions as protein denaturation indicator. By measuring fluorescence changes, one may control the protein denaturation progress, mechanism, as well its kinetics. Based on similar conformational changes, aggregation of α-synuclein was studied in living cells. FlAsH-based fluorescence was also used to recognize different β-amyloid oligomeric states. The usage of biarsenical probes is limited not only to dramatic conformational changes during protein denaturation or aggregation. In most examples studied so far, conformational changes were related to protein activity and its function upon various stimuli. The first structural studies were performed on calmodulin, where Ca(II)-induced structural changes monitored by FlAsH moiety led to increase fluorescence ~40%. In later study, more environmental sensitive ReAsH-EDT$_2$ with fluorescence anisotropy showed that distinct conformations of helix A are formed in response to Ca(II) binding to N- and C-terminal sites. Studies performed on calmodulin example proved that FlAsH moiety can be applied in single molecule anisotropy measurement to probe protein dynamics on a nanosecond scale.

Since TC tags allow precise location of small fluorophores like no other method, they have been widely used for FRET applications, where protein conformational changes are measured. Usage of FlAsH moiety for FRET pair constructions gave better resolution than previously often used FITC. Biarsenical probes were mostly used to replace fluorescent proteins as donors or acceptors in FRET. This approach has several advantages over FP-FP pair, such as reduced size, no FRET control before probe addition, and variety of colors of the probes. Use of the biarsenical probes overcomes the probel of obligatory dimerization of some FPs. There are several examples where FP-biarsenical pair was used to observe conformational changes of G-protein coupled receptors (GPCRs) such as changes induced by activation of human A$_{2A}$ adenosine receptor and mouse α$_{2A}$ adrenergic receptor. FRET between FlAsH and CFP or FlAsH and YFP was shown to give better signal amplitude compared to previously used CFP-YFP pair. Moreover, FlAsH moiety was utilized together with conventional fluorescent probe, such as Alexa Fluor 488, to investigate ligand-induced movement of C-terminal part of β$_2$-adrenoceptor relative to the cytoplasmic end of transmembrane helix 6 of this protein.

One of the very attractive and powerful applications of biarsenical probes is the control of protein activity. Since biarsenical probes can be placed at desired protein place, they can function as molecular switch of protein functions, such as allosteric inhibitor or activator. Tests on mutants with differently placed TC tag revealed that tag location can have various effect on protein activity (e.g., V_{max}, K_m) upon biarsenical conjugation when compared to WT protein. In one of the first examples, TC environment was introduced into loops of protein tyrosine phosphatase (PTP) to find the most efficient regulatory site. One of the investigated mutants showed 12 × reduced catalytic efficiency. Alternatively, FlAsH-EDT$_2$ was applied to mimic the inhibitory effect of Cd(II) on renal Na$^+$/glucose co-transporter and its mutants, providing evidence that endogenous CXXC is responsible for transporter sensitivity toward Cd(II). Next application related to protein control activity was achieved by chromophore/fluorescence-assisted light inactivation (CALI/FALI). Biarsenical probes were found to be better inactivators than FPs. Study on molecular mechanism of calmodulin inactivation by FlAsH CALI revealed that singlet oxygen causes methionine oxidation that leads to destruction of protein structure and histidine oxidation which can form cross-links within calmodulin-MLCK peptide complex. In latter study ReAsH moiety was proved to generate more singlet oxygen species under strong irradiation. TC-tagged connexin43 in gap junctions was 95% inactivated after only 25 s light exposure when conjugated with ReAsH.

The affinity of biarsenical probes to its affinity tag is so efficient that protein complexes may be observed in SDS-PAGE gels. The detection limit depends on protein but reaches even as little as 1 ng. Naturally occurring bacterial chaperone SlyD has the ability to bind biarsenical probes and may be used as internal mass marker of ~27 kDa. Biarsenical labeling was also used in capillary electrophoresis techniques with detection limit of as little as 10^{-20} mol of FlAsH-TC complex. The tight binding of TC motif to biarsenical probe has led to the invention of affinity purification systems with immobilized biarsenical platforms. Due to reversible complex formation, TC-tagged proteins are eluted from beads with a solution of EDT or less odorous DMPS.

Since biarsenical fluorescent probes bind to TC tag with 1:1 stoichiometry, it can be used in fluorescence anisotropy (polarization) experiments. It allows, e.g., time-resolved monitoring of proteolytic reaction that can be used in high-throughput systems. Biarsenical probes can also be used as reporter for gene therapy. In the first study that compared usage of different genetically encoded reporters, TC-tagged proteins were expressed as well as WT and did not affect cell growth. Similarly, FlAsH-EDT$_2$ can be used to monitor real-time protein synthesis or gene expression.

Cross-References

▶ Arsenic-Induced Stress Proteins
▶ As
▶ Cadmium, Effect on Transport Across Cell Membranes
▶ Calcium in Nervous System
▶ Calcium Signaling
▶ Calmodulin
▶ Mercury, Physical and Chemical Properties
▶ Osmium, Physical and Chemical Properties

References

Adams SR, Tsien RY (2008) Preparation of the membrane-permeant biarsenicals FlAsH-EDT$_2$ and ReAsH-EDT$_2$ for fluorescent labeling of tetracysteine-tagged proteins. Nat Protoc 3:1527–1534
Crivat G, Tarasaka JW (2012) Imaging proteins inside cells with fluorescent tags. Trends Biotechnol 30:8–16
Hermanson GT (2008) Bioconjugate techniques. Academic, Amsterdam
Hoffmann C, Gaietta G, Zürn A, Adams SR, Terrillon S, Ellisman MH, Tsien RY, Lohse MJ (2010) Fluorescent labeling of tetracysteine-tagged proteins in intact cells. Nat Protoc 5:1666–1677
Jakobs S, Andresen M, Wurm CA (2008) "FlAsH" protein labeling. In: Miller LW (ed) Probes and tags to study biomolecular function: for proteins, RNA, and membranes. Wiley-VCH, Weinheim, pp 73–88
Pomorski A, Kręzel A (2011) Exploration of biarsenical chemistry – challenges in protein research. Chembiochem 12:1152–1167
Scheck RA, Schepartz A (2011) Surveying protein structure and function using bis-arsenical small molecules. Acc Chem Res 44:654–665
Tsien RY (2005) Building and breeding molecules to spy on cells and tumors. FEBS Lett 579:927–932

Bicarbonate

▶ Zinc and Iron, Gamma and Beta Class, Carbonic Anhydrases of Domain Archaea

Bilayer

▶ Chromium and Membrane Cholesterol

Bind

▶ Calcium-Binding Protein Site Types

Binding of Fluorescent Proteins at Gold Nanoparticles

▶ Gold Nanoparticles and Fluorescent Proteins, Optically Coupled Hybrid Architectures

Binding of Platinum to Metalloproteins

▶ Platinum Interaction with Copper Proteins
▶ Platinum-Containing Anticancer Drugs and proteins, interaction

Binding of Platinum to Proteins

▶ Platinum (IV) Complexes, Inhibition of Porcine Pancreatic Phospholipase A2

Binding Site: Binding Motif

▶ Magnesium Binding Sites in Proteins

Binuclear Mixed-Valent Electron Transfer Copper Center

▶ Cytochrome c Oxidase, CuA Center

Biodistribution of Gold Nanomaterials

▶ Gold Nanomaterials as Prospective Metal-based Delivery Systems for Cancer Treatment

Bioinformatics

▶ Zinc-Binding Proteins, Abundance

Bioinspired Materials

▶ Silicateins

Biological Activity

▶ Lanthanides, Toxicity

Biological Copper Acquisition

▶ Biological Copper Transport

Biological Copper Transport

David L. Huffman and Alia V. H. Hinz
Department of Chemistry, Western Michigan University, Kalamazoo, MI, USA

Synonyms

Biological copper acquisition; Copper trafficking in eukaryotic cells

Definition

Biological copper transport describes the acquisition and delivery of Cu(I) to cellular destinations, as well as the removal of excess Cu(I). Eukaryotic cells transport Cu(I) via permeases, metallochaperones, Cu(I)-specific pumps, and accessory factors responsible for *holo*protein maturation. Key cellular challenges include the prevention of deleterious adventitious reactions and the facile (i.e., rapid) movement of Cu(I) in transit. Inborn errors of certain genes in these pathways are linked to Wilson's disease, Menkes disease, ALS (Lou Gehrig's disease), and cytochrome c oxidase deficiency.

A Copper Requirement

Copper has been refined and processed by humans since the dawn of civilization but its role in living processes was only appreciated in the twentieth century. The vital processes of respiration, iron uptake, neurotransmitter synthesis, and free radical detoxification all include enzymes with copper cofactors. Copper deficiency is rare since we typically acquire sufficient copper in our diets.

Copper: A Double-Edged Sword

While copper is a necessary cofactor for certain enzymes, excess copper is toxic. Copper toxicity occurs by several different mechanisms. Cu(I) can disproportionate into Cu(0) and Cu(II) in aqueous solution. Additionally, Cu(I) is also able to react with

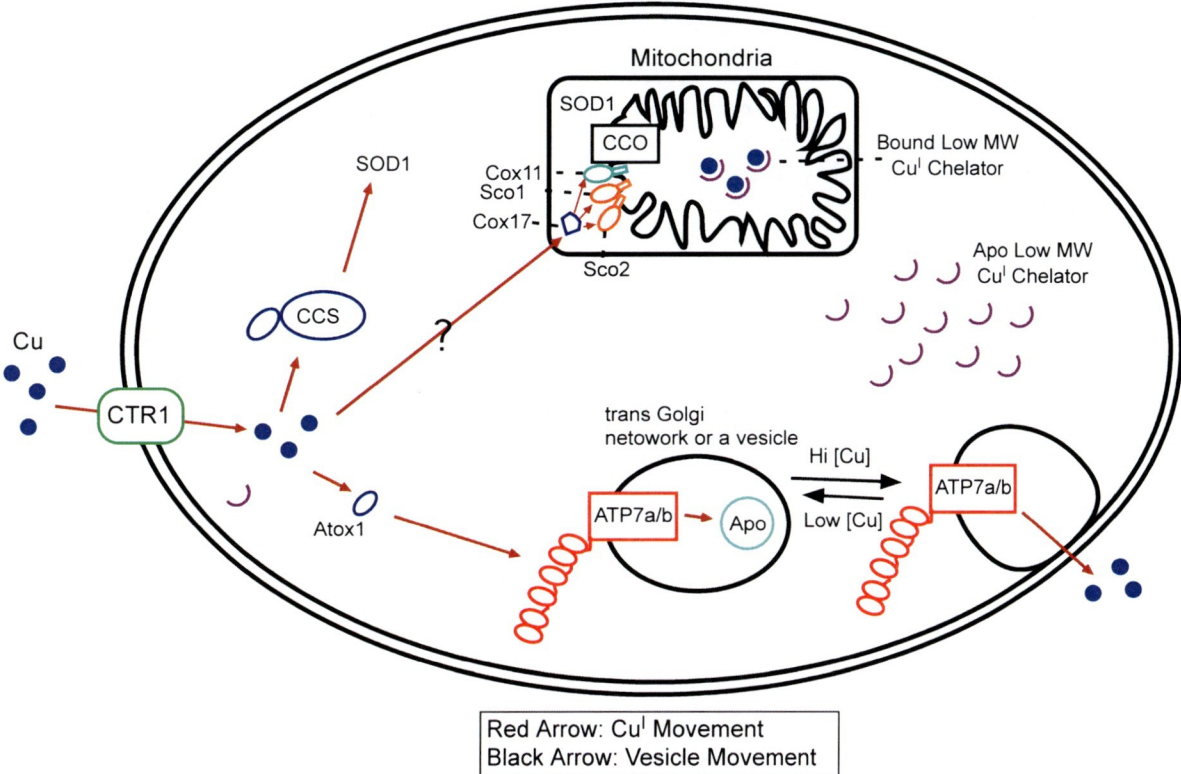

Biological Copper Transport, Fig. 1 Entry of Cu(I) (●) into the cell is mediated by Ctr1, a Cu permease. The chaperone for superoxide dismutase (CCS) delivers Cu to SOD. The metallochaperone Atox1 delivers Cu to ATP7a (Menkes) and ATP7b (Wilson) proteins, then Cu is pumped into the trans-Golgi network for incorporation into apoprotein. Under high copper conditions, ATP7a/b traffics toward the cell surface for expulsion of Cu. The metallochaperone Cox17 delivers Cu to Sco1, Cox11, and Sco2, for metallation of cytochrome c oxidase (CCO) from a matrix pool of Cu that is bound to a low molecular weight ligand

molecular oxygen to produce the superoxide anion or with hydrogen peroxide to produce the hydroxyl radical via Fenton-like chemistry. Both the superoxide anion and the hydroxyl radical are highly reactive and can damage nucleic acids, lipids, and proteins. Lastly, excess Cu(I) or Cu(II) can coordinate with the sites that are meant for other metal ions, thus altering the metal allocation of the system. Therefore, the cell must meticulously maintain proper copper ion levels (termed "homeostasis") to ensure that enough copper is available for the proteins that require it while preventing excess copper from accumulating (Huffman and O'Halloran 2001). This is achieved by carefully controlling the movement of copper from the moment it enters the cell until it reaches its ultimate destination (Fig. 1).

Copper(I) Coordination Chemistry

Biologically utilized oxidation states of copper include both Cu(I) and Cu(II), and the coordination preferences differ for these two oxidation states; however, Cu(I) is the form that is encountered in the copper delivery pathways. Cu(I) typically prefers softer ligands, such as the sulfur in cysteine or methionine, and can be found in digonal, trigonal, and tetrahedral environments in mononuclear coordination complexes. From an energetic standpoint, Cu(I) is less expensive to transport than Cu(II), but the chemical reactivity of Cu(I) poses a significant problem. The cell solves this issue by providing ligands that satisfy the coordination preferences of Cu(I), thus stabilizing the monovalent oxidation state of copper.

Mononuclear Cu(I) complexes within copper trafficking proteins are usually digonal or trigonal (Boal and Rosenzweig 2009), provided by thiols on the surface of the protein spatially arranged to provide facile transfer to a copper site of similar or greater affinity for Cu(I) (Xiao et al. 2011). Typically a digonal site is housed in a conserved MXCXXC motif, but the coordination sphere can be expanded by an exogenous ligand or another protein-based ligand. Interestingly, Cu(I) clusters with a nuclearity of 2–4 have been detected as in vitro metallochaperone crystal structures, usually at a dimer or trimer interface (Banci et al. 2010), and their in vivo relevance is not known. The repertoire of copper trafficking proteins has expanded with other cysteine-containing motifs (Robinson and Winge 2010), especially with the assembly factors for cytochrome c oxidase (CcO) found in the mitochondria, including Cox17, Sco1, Sco2, and Cox11, demonstrating the versatility and wealth of Cu(I) coordination.

Copper(I) Permeases

Ctr1, a membrane protein that is conserved among many organisms and part of the Ctr family, plays a significant role in the import of Cu(I) into the cell (Kim et al. 2008). Ctr1 has several distinct domains, including three transmembrane domains, a MX_3M element in the second transmembrane domain, a cluster of cysteine and histidine residues in the C terminus, and multiple methionine residues in the N terminus. The MX_3M element has been shown to be vital for copper uptake, with mutation of the methionine to alanine or serine resulting in decreased copper uptake. Electron microscopy and two-dimensional crystallography studies have indicated that Ctr1 is functionally active as a homotrimer in the membrane, with a putative pore forming between the interfaces of the subunits to allow Cu(I) to traverse the membrane and enter the cell (De Feo et al. 2007). A multiple Met-containing motif is becoming a common theme in extracytosolic Cu(I) acquisition, including prokaryotic metallo-oxidases, a term coined by Daniel J. Kosman. The FRE family of metalloreductases and Steap proteins are candidates for the reduction of Cu(II) to Cu(I) on the cell surface, facilitating uptake with the Cu(I) permeases. Through a process of endocytosis, Ctr1 can also relocalize to intracellular vesicles to decrease Cu(I) uptake (Kim et al. 2008).

Metallochaperones

Prior to the mid-1990s, metalloproteins were believed to acquire their metal cofactors directly by encountering metal ions that had diffused into the cell. Since then it has been found that proteins, termed "metallochaperones," bind and deliver specific metal ions, including copper, to target proteins. Given the potentially devastating effects of excess copper, cells have developed mechanisms for preventing the accumulation of unbound, excess copper. The amount of free copper in the cell at any one point in time is very low, with estimates of less than one free atom per cell (Huffman and O'Halloran 2001).

Metallochaperones That Target P_{IB}-Type ATPases

The single domain metallochaperones are small proteins, approximately 7–8 kD. The most studied of these proteins, yeast Atx1, contains a ferrodoxin fold ($\beta\alpha\beta\beta\alpha\beta$) that binds Cu(I) in a conserved MXCXXC motif (Fig. 2) between the first beta sheet and alpha helix of the protein (Boal and Rosenzweig 2009). Atx1 delivers copper to Ccc2a, an N-terminal cytosolic domain of a P_{1B}-type ATPase, named Ccc2 that possesses a ferrodoxin fold as well. The interaction between Atx1 and Ccc2a is copper-dependent and also depends upon a complex interaction interface, consisting of basic residues in Atx1 and acidic residues in Ccc2a as well as a surface complementarity (Banci et al. 2010). During the copper transfer process between Atx1 and Ccc2a a trigonal copper complex is formed as observed by NMR studies, provided by two cysteine thiolates from one protein and one from its partner; mutagenesis studies followed by NMR titrations indicate that the most predominant intermediate possesses one thiolate from Atx1 and two from Ccc2a. The human homolog of Atx1, Atox1 (or Hah1), functions in a similar way. In the trans-Golgi, Atox1 delivers copper to the N-terminal metal-binding domains of ATP7a or ATP7b (also P_{1B}-type ATPases), which in turn pumps copper across a membrane concomitant with ATP hydrolysis.

Atox1 forms NMR-observable complexes with copper-binding domains 1, 2, and 4 (e.g., Fig. 3) of ATP7b and domains 1 and 4 of ATP7a (Banci et al. 2010). Insights into the nature of this complex is

Biological Copper Transport, Fig. 2 Solution NMR structure of the yeast metallochaperone Cu^I-Atx1, PDB ID 1FD8, with ferrodoxin fold (*green*), cysteine thiolates (*yellow*), and digonal Cu (*orange sphere*) site

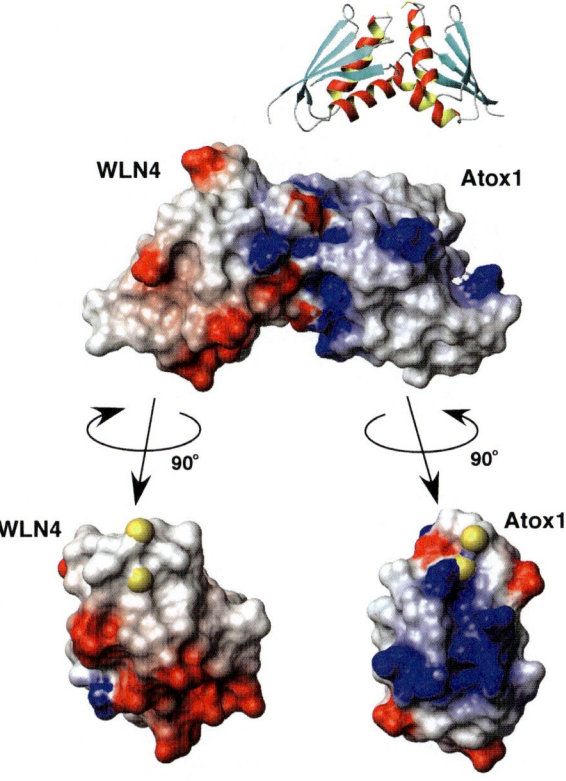

Biological Copper Transport, Fig. 3 Model of Cu-dependent complex formation between HMBD 4 of ATP7b, WLN4, and the human metallochaperone Atox1, based on the structure of Atx1/Cu/Ccc2a (PDB ID 2GGP) and NMR titrations. Potential surface of WLN4 shows array of acidic residues that interact with complementary basic surface of Atox1

provided by the X-ray structures of the metal-bridged homodimers Hg(II)-[Atox1]$_2$ (PDB ID 1FE4), Cd(II)-[Atox1]$_2$ (PDB ID 1FE0), Cu(I)-[Atox1]$_2$ (PDB ID 1FEE), the X-ray and NMR structures of the Cd-bridged heterodimer Atox1-Cd(II)-MNK1 (PDB ID 3CJK and 2K1R), and the NMR structure of the Cu(I)-bridged Atx1-Ccc2a heterodimer (PDB ID 2GGP). This review focuses on eukaryotes, but many other structures exist for homologues of these proteins, in archea, bacteria, and cyanobacteria (Boal and Rosenzweig 2009; Banci et al. 2010).

Metallochaperones That Target SOD1

The only identified protein of this class to date is CCS. Under copper limiting conditions, this protein transports copper to Cu, Zn superoxide dismutase (SOD1), an enzyme that detoxifies the reactive superoxide anion and protects the cell from oxidative damage (Leitch et al. 2009). CCS (*c*opper *c*haperone for *S*OD) is found in yeast and other eukaryotes and located primarily in the cytoplasm, though a small amount is found in the inner membrane space of the mitochondria where SOD is believed to neutralize any superoxide inadvertently produced by the electron transport chain.

At 30–32 kD and with three functionally distinct domains, the CCS proteins are larger and more complex than their single domain counterparts. Domain I is found at the amino terminus and has a ferrodoxin fold like yeast Atx1, including the MXCXXC motif used for copper binding. Displaying a similar sequence homology to SOD, Domain II binds to SOD to properly orientate the protein (Fig. 4) and facilitate the copper transfer, but is not believed to actually bind copper. Domain III, located at the carboxy terminal and possessing a conserved CXC

Biological Copper Transport, Fig. 4 Crystal structure of the complex between yeast CCS and yeast Cu,Zn SOD1 (*yellow*), PDB ID 1JK9. Domain I of CCS is *green*, SOD-like Domain II is *blue*, and Domain III containing the CXC motif is *pink*

motif, acts in conjunction with Domain I to transfer the copper to SOD (Leitch et al. 2009). The similarity of specific domains to other proteins with which they interact is a common theme in copper trafficking proteins.

Copper(I)-Pumping Enzymes

The copper transporting P_{IB}-type ATPases are commonly referred to as Cu(I) pumps, powered by ATP hydrolysis (Lutsenko et al. 2007). Though they are found in prokaryotes as well as eukaryotes, the human homologues differ by the presence of six high-affinity copper-binding sites (HMBD) at the N-terminus of ATP7a and ATP7b (Fig. 5). This provides an entropic driving force for copper capture and thereby increases the local concentration of copper in the vicinity of the transmembrane copper-binding sites. The apparent trafficking of ATP7b and ATP7a as a function of intracellular copper concentration and the interaction of the N-terminus of the protein with the actuator (A) and the ATP-binding domain point to a sophisticated role for the N-terminus (Lutsenko et al. 2007). The HMBD functions to capture copper and transduce conformational changes to other cytosolic facing domains. Each HMBD possesses the conserved MXCXXC motif and ligates one atom of Cu(I) in a distorted linear arrangement via the sulfur atoms in the cysteines. These proteins, in response to elevated cytosolic copper concentrations, translocate from the trans-Golgi network to vesicles that ferry copper to the plasma membrane for extracellular release. The 650 amino acid N-terminal domain interacts with other cytosolic facing domains, including the actuator (A) domain and the ATP-binding domain (comprised of the nucleotide binding N- and phosphorylation P-subdomains) (Lutsenko et al. 2007). Copper-binding domains 5 and 6, WLN5-6, function as a unit in solution and rotate little with respect to one another (Banci et al. 2010), whereas domains 3 and 4 are mobile with respect to one another.

The N-terminal copper-binding sites of ATP7a and ATP7b can obtain Cu(I) from the metallochaperone Atox1 or from low molecular weight chelators (Banci et al. 2010). The exchange of copper between sites has been determined empirically and a low barrier for copper transfer exists, as in the Atx1-Ccc2a paradigm. One question that has not been addressed is the rate of copper movement between sites in the N-terminal of ATP7b. The three-dimensional structure has not yet been solved for full-length ATP7a or ATP7b, but NMR structures exist for a number of the metal-binding domains as well as the A domain and the N domain. The way in which the cytosolic facing domains of these ATPases interact is not yet known, but clues have been garnered from cryoEM structures (PDB ID: 2VOY, 3J08, 3J09) of *Archaeoglobus fulgidus* CopA and from the X-ray structure of *Legionella pneumophila* CopA (PDB ID: 3RFU, Fig. 6), gram negative bacterial P_{IB}-type ATPases. Neither of these structures has pinpointed the exact location of the N-terminal copper-binding domain (HMBD), but it coincides with the position of the N-terminal portion of the A-domain in the SERCA1 structure, just prior to the M1 helix.

The six HMBD of ATP7a and ATP7b are spatially separated, suggesting that different portions of the N-terminus interact with discrete portions of the other cytosolic-facing domains. For example in ATP7b, domain 6 of WLN5-6 would occupy the same position as the N-terminal HMBD of CopA, but WLN1-4 is separated by 57 disordered residues from WLN5-6, and therefore WLN1-4, interacts at different positions on the cytosolic surface *or* it functions only to acquire copper from the Atox1 metallochaperone. The exact

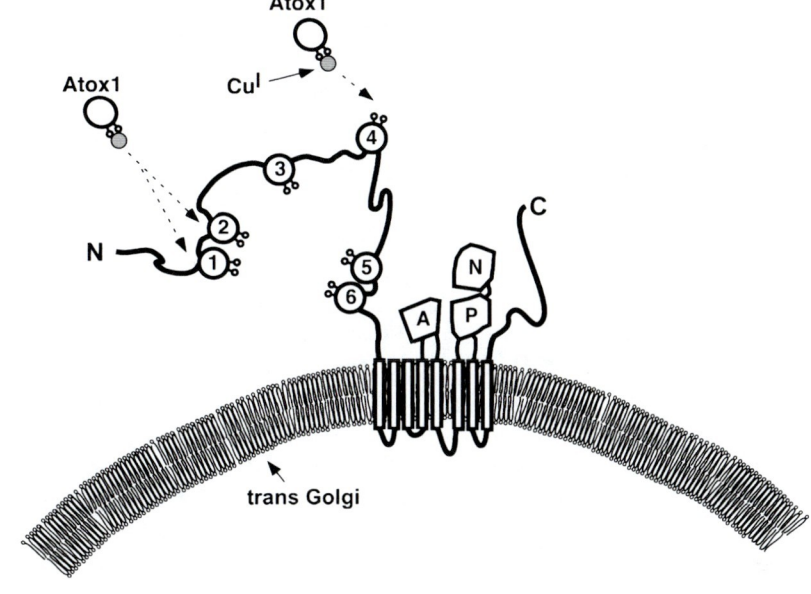

Biological Copper Transport, Fig. 5 Cartoon of ATP7b (ATP7a has similar topology), showing the interaction, observed by NMR, of Cu^I-Atox1 with domains one, two, and four. Cytosolic domains include the 650 residue N-terminus with six HMBD, the A (actuator), N (nucleotide binding), P (phosphorylation) domains, and the 90 residue C terminus. ATP7b has eight transmembrane helices

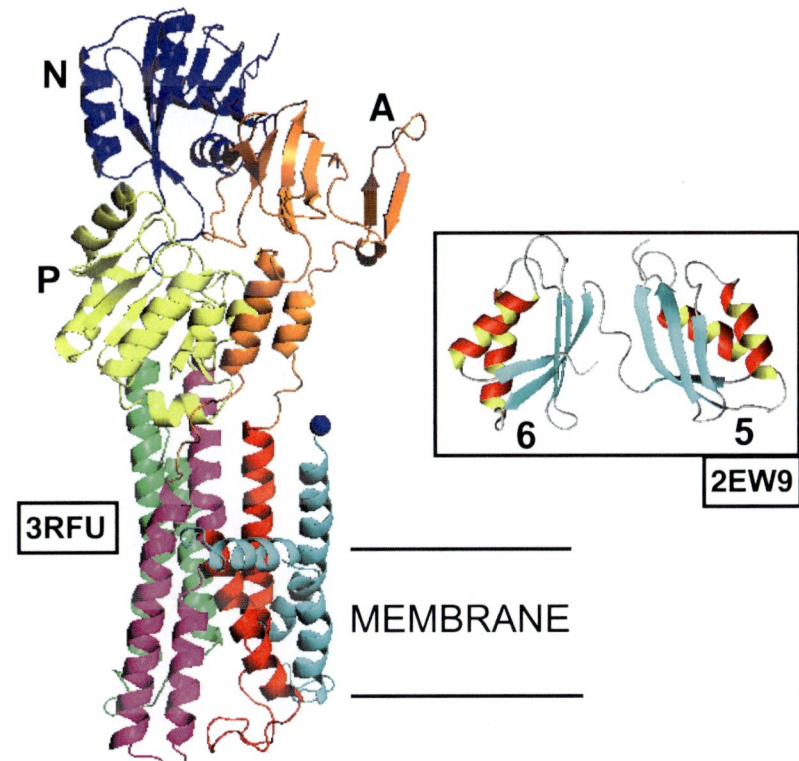

Biological Copper Transport, Fig. 6 Crystal structure of *Legionella pneumophila* CopA (PDB ID 3RFU) copper-transporting P$_{IB}$-type ATPase. The most N-terminal resolved feature (HMBD is disordered) is Val74 (*blue sphere*). The first two transmembrane (TM) helices MA and MB are shown in *blue*, TM helices M1 and M2 in *red*, A (actuator) domain in *orange*, TM helices M3 and M4 in *magenta*, P (phosphorylation) in *yellow*, N (nucleotide binding) in *blue*, and TM helices M5 and M6 in *green*. Membrane spanning portion is indicated by *bars*. The solution structure of WLN5-6 (PDB ID 2EW9) consisting of two HMDB is shown for visual comparison. The human homolog, ATP7b (or ATP7a), has six HMBD, whereas CopA has one

function of the N-terminal domains has been a matter of discussion, and like the structures of the bacterial homologues, they are likely more mobile than the rest of the structure. Do they simply regulate the function of the enzyme *or* do they have a role in passing Cu(I) to the binding sites at and within the membrane? Perhaps the HMBDs function *both* to capture Cu(I) and also possess a regulatory role. The low free concentration of copper in the cytosol suggests a role of the HMBD in copper acquisition. Does the Atox1 metallochaperone directly pass copper to binding sites in the membrane portion of the enzyme? The trafficking of ATP7a and ATP7b and the role of the N- and C-termini in this process facilitate removal of copper from the cytosol, and the possession of six HMBDs likely aids this process. Several groups have shown that ATP7a and ATP7b undergo kinase-mediated phosphorylation at specific sites within the N-terminus and C-terminus, and this event has an effect on the trafficking of these proteins (Inesi 2011). Cisplatin, the most widely prescribed chemotherapeutic agent, binds to the copper-binding site of Atox1 (PDB ID 3IWL), and also interacts with the N-terminus of ATP7b; detoxification of cisplatin by proteins of the copper trafficking pathway is an area of active investigation.

Mitochondrial Copper Handling Proteins

The only mitochondrial cuproenzymes identified to date are Cu,Zn SOD1 and cytochrome c oxidase (CcO). Both utilize distinct metallochaperones to acquire their copper. Cu,Zn SOD1 is supplied with copper from CCS (Leitch et al. 2009), a metallochaperone discussed above. Several proteins are involved in transporting copper to CcO, a large, transmembrane complex that is the final enzyme in the mitochondrial electron transport chain (Robinson and Winge 2010). CcO synthesis and metallation occurs in the mitochondria, though how exactly the copper enters the mitochondria after it has entered the cell is still being elucidated. From a mitochondrial matrix pool, Cu(I) is acquired by Cox17, a small, hydrophilic protein in the intermembrane space that can bind between one and four Cu(I) atoms. Cox17 then transfers Cu(I) to two intermembrane associated proteins, Cox11 and Sco1 (Robinson and Winge 2010). Cox11 transfers Cu(I) to the mononuclear Cu_B site on Cox1 of CcO, while Sco1 transfers Cu(I) to the dinuclear Cu_A site on Cox2 of CcO. How Cox17 acquires Cu(I) is an area of much interest. It was originally thought that since Cox17 is present in both the cytosol and the intermembrane space of the mitochondria, Cox17 was shuttling Cu(I) from the cytosol to the mitochondria. However, an experiment in which Cox17 was tethered revealed that the protein did not need to leave the mitochondria in order to supply Cu(I) to CcO. Perhaps Cox17 is supplied Cu(I) by a small, copper-binding ligand that is present in both the cytosol and the mitochondrial matrix (Robinson and Winge 2010).

Conclusion

The cellular requirement for copper dictates the need for uptake, distribution, and export pathways, carefully handled by task-specific proteins at each checkpoint. Details about the mechanism and structures of these proteins are continuing to emerge, and the way in which copper trafficking proteins interact with each other. Insights are being gained by the development of exciting new methods to track copper ions in situ. The binding affinity of copper to these proteins has been under intense investigation, but unanswered questions remain about the rates of copper transfer. Areas of future study include (1) copper delivery to, and regulation of the P_{IB}-type ATPases, (2) and the maintenance of mitochondrial copper pools.

Acknowledgments DLH gratefully acknowledges support from the National Science Foundation (CAREER 0645518) and AVHH acknowledges Western Michigan University Graduate Student Research Fund.

Cross-References

▶ Copper-Binding Proteins
▶ Copper, Biological Functions
▶ Copper, Physical and Chemical Properties
▶ Copper-Zinc Superoxide Dismutase and Lou Gehrig's Disease
▶ Metallothioneins and Copper

References

Banci L, Bertini I, McGreevy KS, Rosato A (2010) Molecular recognition in copper trafficking. Nat Prod Rep 27:695–710, http://www.ncbi.nlm.nih.gov/entrez/query.fcgi?cmd=Retrieve&db=PubMed&dopt=Citation&list_uids=20442960

Boal AK, Rosenzweig AC (2009) Structural biology of copper trafficking. Chem Rev 109:4760–4779, http://www.ncbi.nlm.nih.gov/entrez/query.fcgi?cmd=Retrieve&db=PubMed&dopt=Citation&list_uids=19824702

De Feo CJ, Aller SG, Unger VM (2007) A structural perspective on copper uptake in eukaryotes. Biometals 20:705–716, http://www.ncbi.nlm.nih.gov/entrez/query.fcgi?cmd=Retrieve&db=PubMed&dopt=Citation&list_uids=17211682

Huffman DL, O'Halloran TV (2001) Function, structure, and mechanism of intracellular copper trafficking proteins. Annu Rev Biochem 70:677–701, http://www.ncbi.nlm.nih.gov/entrez/query.fcgi?cmd=Retrieve&db=PubMed&dopt=Citation&list_uids=11395420

Inesi G (2011) Calcium and copper transport ATPases: analogies and diversities in transduction and signaling mechanisms. J Cell Commun Signal 5:227–237. http://www.ncbi.nlm.nih.gov/entrez/query.fcgi?cmd=Retrieve&db=PubMed&dopt=Citation&list_uids=21656155

Kim BE, Nevitt T, Thiele DJ (2008) Mechanisms for copper acquisition, distribution and regulation. Nat Chem Biol 4:176–185, http://www.ncbi.nlm.nih.gov/entrez/query.fcgi?cmd=Retrieve&db=PubMed&dopt=Citation&list_uids=18277979

Leitch JM, Yick PJ, Culotta VC (2009) The right to choose: multiple pathways for activating copper, zinc superoxide dismutase. J Biol Chem 284:24679–24683, http://www.ncbi.nlm.nih.gov/entrez/query.fcgi?cmd=Retrieve&db=PubMed&dopt=Citation&list_uids=19586921

Lutsenko S, Barnes NL, Bartee MY, Dmitriev OY (2007) Function and regulation of human copper-transporting ATPases. Physiol Rev 87:1011–1046, http://www.ncbi.nlm.nih.gov/entrez/query.fcgi?cmd=Retrieve&db=PubMed&dopt=Citation&list_uids=17615395

Robinson NJ, Winge DR (2010) Copper metallochaperones. Annu Rev Biochem 79:537–562. http://www.ncbi.nlm.nih.gov/entrez/query.fcgi?cmd=Retrieve&db=PubMed&dopt=Citation&list_uids=20205585

Xiao Z, Brose J, Schimo S, Ackland SM, La Fontaine S, Wedd AG (2011) Unification of the copper(I) binding affinities of the metallo-chaperones Atx1, Atox1, and related proteins: detection probes and affinity standards. J Biol Chem 286:11047–11055. http://www.ncbi.nlm.nih.gov/entrez/query.fcgi?cmd=Retrieve&db=PubMed&dopt=Citation&list_uids=21258123

Biological Effect

▶ Lanthanides, Toxicity

Biological Effect of Vanadium

▶ Vanadium in Live Organisms

Biological Functions of Ca2+ in Mitochondria

▶ Calcium and Mitochondrion

Biological Implications: Cell Death

▶ Hexavalent chromium and DNA, biological implications of interaction

Biological Implications: Chromosomal Aberrations

▶ Hexavalent chromium and DNA, biological implications of interaction

Biological Implications: Genomic Instability

▶ Hexavalent chromium and DNA, biological implications of interaction

Biological Implications: Mutagenesis

▶ Hexavalent chromium and DNA, biological implications of interaction

Biological Methane Oxidation Catalyst

▶ Particulate Methane Monooxygenase

Biology

▶ Tungsten in Biological Systems

Biomacromolecular Coordination of Germanium

▶ Rubredoxin, interaction with germanium

Biomarker

▸ Lanthanides, Luminescent Complexes as Labels

Biomarkers for Cadmium

Walter C. Prozialeck
Department of Pharmacology, Midwestern University, Downers Grove, IL, USA

Synonyms

Cadmium: Cd; Cd^{2+}

Definitions

Cadmium (Cd): Is an important industrial agent and environment pollutant that is a major cause of kidney disease in many regions of the world.

Nephrotoxicity/proximal tubule: The proximal tubule is the primary target of Cd toxicity in the kidney. Injury of proximal tubule epithelial results in increases in urine volume and excretion of low-molecular-weight proteins, amino acids, glucose, and electrolytes. These effects of Cd may result from even low levels of exposure and are often irreversible.

Biomarkers: The United States National Institutes of Health have broadly defined the term "biomarker" as a "characteristic that is objectively measured and evaluated as an indicator of normal biological processes, pathogenic processes, or pharmacologic responses to a therapeutic intervention." In this entry, the topic of Cd biomarkers will be considered from a more narrow perspective, as "any substance or molecule that can serve as an indicator of the functional state or level of toxic injury in the kidney."

Biological implications: Biomarkers have proven to be useful tools for evaluating Cd exposure and nephrotoxicity in human populations as well in laboratory studies on animals. However, many fundamental issues regarding the selection of markers and definition of their critical levels have yet to be fully resolved. Recent studies have provided hope that new and even more sensitive biomarkers of Cd-induced kidney injury can be developed.

Cadmium Overview

Cd as an Environmental Health Problem

Cd is an important industrial agent and a widespread environmental pollutant that currently ranks seventh on the United States Environmental Protection Agency's priority list of hazardous substances. Cd is normally found at low concentrations throughout the lithosphere but has become increasingly concentrated in the biosphere through smelting, mining, agriculture, and industrial activities of humans. As a stable, divalent cation, Cd is not biodegradable and persists in the environment for long periods of time. Despite efforts by many agencies to reduce the usage of Cd, Cd pollution continues to be a major public health problem in many regions of the world (for reviews see ATSDR 2008; Jarup and Akesson 2009).

Workers in smelting industries or industries that utilize Cd and its compounds can be exposed in the workplace by inhaling Cd oxide fumes or Cd-contaminated dust. The general population is more likely to be exposed by the ingestion of Cd-contaminated food or water (ATSDR 2008; Jarup and Akesson 2009). In addition, tobacco contains large amounts of Cd and is a major source of exposure among smokers (ATSDR 2008).

Depending on the dose, route, and duration of exposure, Cd can damage various organs including the lung, liver, kidney, and bone (ATSDR 2008; Jarup and Akesson 2009), and it is carcinogenic (ATSDR 2008). With the chronic, low-level patterns of exposure that are common in human populations, the kidney is the primary target of toxicity, where Cd accumulates in the epithelial cells of the proximal tubule, resulting in a generalized reabsorptive dysfunction that is characterized by polyuria, glucosuria, and low-molecular-weight proteinuria (Jarup and Akesson 2009; Prozialeck and Edwards 2010).

The nephrotoxic actions of Cd are, to a great extent, a consequence of the unique toxicokinetics of Cd in the body (for reviews see ATSDR 2008; Prozialeck and Edwards 2010). Following respiratory exposure, Cd is efficiently absorbed from the lung; up to 40–60% of inhaled Cd reaches the systemic circulation. With oral exposure, the absorption of Cd from the gastrointestinal tract is considerably lower (only 5–10%). However, with long-term exposure, even this low level of absorption from the gastrointestinal tract can lead to systemic accumulation of Cd and subsequent toxicities.

Cd is initially transported to the liver where it is taken up by hepatocytes. In the hepatocytes, Cd induces the synthesis of metallothionein, which binds Cd. However, over time, the Cd-metallothionein complex can be released into the bloodstream. Even though the complex is nontoxic to most organs, it can be filtered at the glomerulus and taken up by the epithelial cells of the proximal tubule. Thus, Cd-metallothionein can have the paradoxical effect of facilitating the delivery of Cd from the liver to the kidney. In addition, Cd in plasma binds to a variety of other proteins as well as low-molecular-weight thiols, such as cysteine and glutathione. The Cd that is associated with the low-molecular-weight compounds is filtered at the glomerulus and be taken up by epithelial cells of the proximal tubule. In addition, there is some evidence for the uptake of ionic Cd^{2+} through metal ion transporters in the proximal tubule. Regardless of the form or speciation of Cd that is present in the bloodstream, Cd eventually accumulates in the epithelial cells of the proximal tubule.

Monitoring of Human Populations

The monitoring of human populations for early signs of Cd exposure and toxicity has posed a major challenge (Bernard 2004; Prozialeck and Edwards 2010). Intuitively, it might seem that the most direct way to monitor levels of Cd exposure would be to simply measure blood or urinary levels of Cd. However, this issue is greatly complicated by the tendency of Cd to be sequestered in organs such as liver and kidney. While blood levels of Cd can yield information regarding recent exposures, they often do not provide information regarding the total body burden of Cd or the severity of injury in specific target organs. Likewise, the monitoring and interpretation of data on urinary levels of Cd are not as straightforward as one might expect. With low, or even moderate, levels of exposure, any Cd that is filtered at the glomerulus is almost completely reabsorbed by epithelial cells of the proximal tubule; little or no Cd is excreted in the urine. It is only when the body burden of Cd is fairly large and/or kidney injury begins to appear that urinary excretion of Cd increases significantly. As a result of these limitations in interpreting data on blood and urinary levels of Cd, investigators have utilized various biomarkers to assess levels of Cd exposure and toxicity. As a result of Cds tendency to accumulate in epithelial cells of the proximal tubule, the kidney is, in effect, a sentinel of Cd exposure. Consequently, the most useful biomarkers of Cd exposure and toxicity have been markers of the various effects of Cd in the kidney.

This entry will highlight some of the urinary biomarkers that have proven to be most useful in monitoring Cd exposure and toxicity in human populations and in experimental animals. In addition, novel markers that appear to offer increased sensitivity will be described. All of the biomarkers that will be described represent different events in the pathophysiology of Cd-induced kidney injury. With continuous exposure, the levels of Cd in the proximal tubule cells continue to increase until a critical threshold concentration of about 150–200 μg/g of tissue is reached. The classic view is that as this threshold concentration is approached, the cells undergo oxidative stress that leads to injury and either necrotic or apoptotic cell death (reviewed by Prozialeck and Edwards 2010). The cellular injury causes alterations in proximal tubule function as well as the shedding of injured cells and cytosolic contents into the urine. The shedding of dead or injured cells triggers a repair process in which neighboring noninjured cells dedifferentiate in a process known as epithelial-mesenchymal transformation. The dedifferentiated cells migrate to the denuded area of the basement membrane and replace the injured cells.

Biomarkers of Cd Nephrotoxicity

As noted previously, the traditional urinary biomarkers that have been used to monitor Cd toxicity reflect various steps in this sequence of pathologic events. These urinary markers can be classified into four broad categories: (1) Cd-binding proteins such as metallothionein; (2) low-molecular-weight proteins; (3) proteins and enzymes derived from the brush border, intracellular organelles, or the cytosol of proximal tubule epithelial cells; and (4) proteins expressed in response to cellular injury. Figure 1 shows the typical patterns for the urinary excretion of each of these classes of markers along with a timeline describing

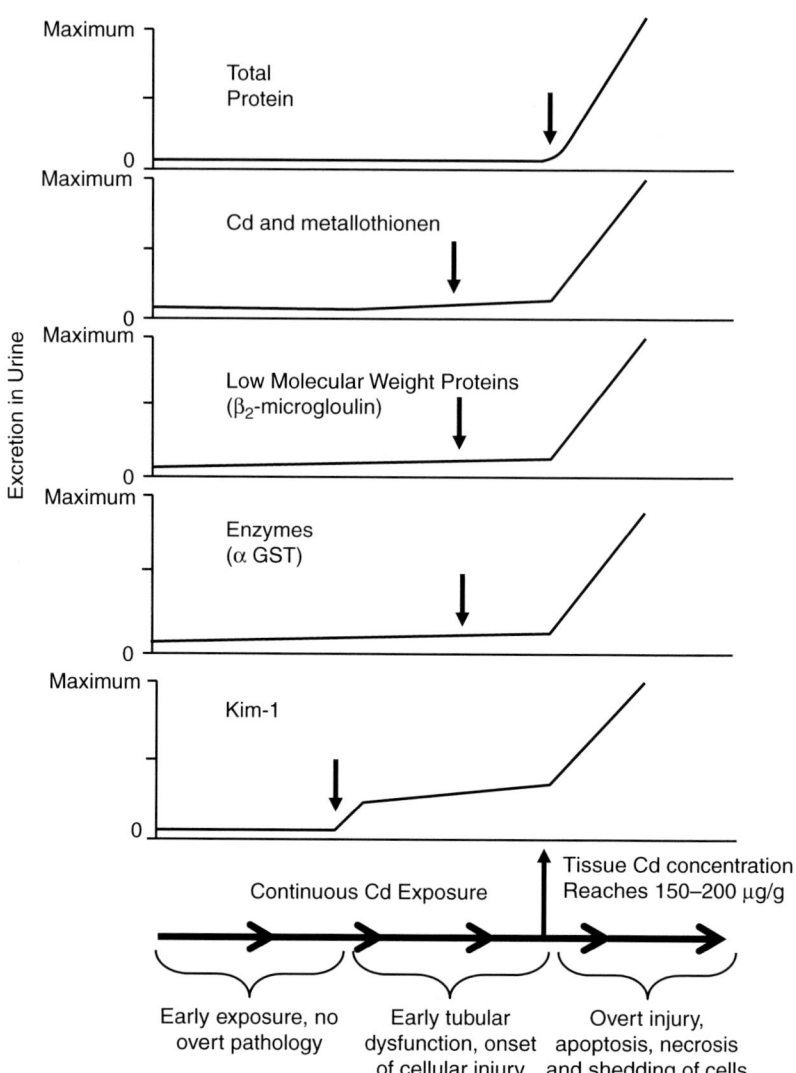

Biomarkers for Cadmium, Fig. 1 Patterns of urinary excretion of representative markers of Cd exposure and toxicity. The graphs show the general patterns for the urinary excretion of various classes of biomarkers of Cd-induced kidney injury. These are not the results of single experiments but rather depict the general results of a wide variety of studies (reviewed in Prozialeck and Edwards 2010 and Shaikh and Smith 1986). The y-axis shows the relative urinary excretion of the various markers (0 to maximum), and the x-axis indicates the relative duration of Cd exposure. The actual time frame for the various events depends on the level of Cd exposure. In humans who are exposed to low levels of Cd, the time frame can involve years of exposure. With higher levels of exposure that are used in experimental animals, the same events can occur in weeks. The *bracketed comments* at the *bottom* describe the stages of exposure and proximal tubular injury. The *downward arrows* (↓) indicate the point at which excretion of each marker usually becomes significantly elevated

specific pathophysiologic events in the proximal tubule. Reference will be made to this figure as each of the classes of markers is described below.

Cd and Metallothionein

The urinary excretion of Cd and metallothionein can serve as markers of both Cd exposure and of Cd-induced proximal tubule injury (Bernard 2004; Prozialeck and Edwards 2010; Shaikh and Smith 1986). During early stages of exposure, circulating Cd which is bound to low-molecular-weight molecules such as metallothionein, cysteine, or glutathione in the plasma is filtered at the glomerulus and efficiently taken up by the epithelial cells of the proximal tubule. Only extremely small amounts are excreted in the urine. During this stage of exposure (labeled as "Early exposure" in the timeline in the figure), the presence of Cd or metallothionein in the urine most likely results from the normal turnover and shedding of epithelial cells and is a reflection of the level of Cd exposure and the body burden of Cd (Prozialeck and Edwards 2010). However, over time, the concentration of Cd in the epithelial cells increases to the point that Cd injures the cell and/or disrupts tubular reabsorptive processes. At this stage (labeled "Intermediate exposure" in the figure), the excretion of Cd and metallothionein begins to increase in a linear manner. However, as the intracellular levels of Cd increase further, more of the epithelial cells begin to die and slough off. At this point, the urinary excretion of Cd and metallothionein increases markedly (Prozialeck and Edwards 2010; Shaikh and Smith 1986). This surge in the urinary excretion of Cd and metallothionein coincides with the onset of polyuria and proteinuria. Thus, the early, linear phases of Cd and metallothionein excretion are a reflection of Cd exposure, whereas the later increases in excretion are a reflection of Cd-induced tubular injury.

The World Health Organization, United States Environmental Protection Agency and other agencies have established guidelines for the monitoring of populations for Cd exposure and for Cd exposure limits (ATSDR 2008; Huang 2004; World Health Organization (WHO) 2000). Even though there are variations among the standards from these different agencies, some generalizations can be made. The blood levels of Cd in nonexposed populations are typically less than 0.5 μg/L. Blood levels higher than 1.0 μg/L are generally indicative of Cd exposure; levels higher than 5 μg/L are considered hazardous. Urinary levels of Cd in nonexposed populations are usually below 0.5 μg/g creatinine; values above 1–2 μg/g are indicative of exposure or elevated body burden. The critical urinary Cd concentration that is associated with the onset of renal injury is usually about 2–10 μg/g creatinine, which corresponds to a renal cortical Cd concentration of about 150–200 μg/g tissue (Prozialeck and Edwards 2010). It should be emphasized that these generalizations are derived from consensus-based standards from various regulatory agencies. However, there is evidence that even lower urinary levels of Cd may be associated with adverse effects (for review see Prozialeck and Edwards 2010). With regard to metallothionein, the critical urinary level that is associated with the onset of overt kidney injury is about approximately 300 μg/g creatinine (Chen et al. 2006; Shaikh and Smith 1986).

It should be noted that the majority of studies on which these standards/recommendations are based involved the measurement of total urinary metallothionein; they did not differentiate/identify specific metallothionein isoforms. There are currently four known isoforms of metallothionein. Although the relationships between urinary excretion of Cd and these different metallothionein isoforms have not been established, there are a few reports indicating possible differential effects of Cd on the expression of the different isoforms. The application of using specific isoforms of metallothionein as biomarkers of Cd exposure remains to be fully explored.

Low-Molecular-Weight Proteins

The second category of Cd urinary biomarkers includes a variety of low-molecular-weight proteins such as β_2-microglobulin, Clara cell protein (CC-16), α_1-microglobulin, retinol-binding protein, and vitamin D–binding protein. These low-molecular-weight proteins are present in plasma and are small enough to be easily filtered at the glomerulus. Under normal circumstances, these filtered proteins are efficiently reabsorbed by the proximal tubule and are not excreted to any great extent in the urine (Bernard 2004; Prozialeck and Edwards 2010). However, as Cd accumulates in the proximal tubule, absorption of these proteins becomes impaired, and the proteins begin to appear in the urine. Of these proteins, β_2-microglobulin has been most widely employed as a standard marker for monitoring the early stages of Cd exposure and toxicity in humans. Urinary levels of β_2-microglobulin of 1,000 μg/g creatinine (or greater) are considered to indicate specific renal injury. This level is typically associated with urinary Cd of greater than 5 μg/g creatinine. For population monitoring, a cutoff value of 300 μg/g β_2-microgobulin/creatinine has been used (Huang 2004). However, other investigators have recommended lower critical exposure levels (Uno et al. 2005). Even though β_2-microglobulin has proven to be a very useful biomarker, its lack of stability in acidic urine can be problematic.

As with β_2-microglobulin, increased levels of retinol-binding protein are suggestive of impairment of tubular reabsorptive function. Unlike β_2-microglobulin, however, retinol-binding protein is stable in acidic urine, and no special preservative or alkaline treatment is required (for review see Prozialeck and Edwards 2010).

Proximal Tubule–Derived Enzymes

Some of the most extensively used markers of Cd-induced proximal tubule injury have been enzymes that are expressed in proximal tubule epithelial cells. A variety of enzymes including: N-acetyl-β-D-glucosaminidase (NAG), lactate dehydrogenase (LDH), alkaline phosphatase, and more recently, alpha-glutathione-S-transferase (α-GST) have been studied in this context. The appearance of these enzymes in urine is classically thought to result from the leakage of intracellular contents when necrotic proximal tubule epithelial cells lose their membrane integrity and/or slough off into the urine (Vaidya et al. 2008).

NAG has proven to be especially useful in the monitoring of human populations. NAG is a lysosomal enzyme that exists as multiple isoforms. Both forms A and B are expressed in kidney. However, the B form, which is more abundant in the proximal tubule, is regarded as the more sensitive and reliable marker of Cd-induced injury. However, assays that do not differentiate between the two isoforms can also yield useful results. Several epidemiologic studies have shown that NAG outperforms other traditional markers (Jin et al. 1999; Moriguchi et al. 2009; Noonan et al. 2002; Suwazono et al. 2006). However, it is also noteworthy that it does not perform as well in animal (rat) models of Cd nephrotoxicity (Prozialeck and Edwards 2010). One major advantage of NAG for large-scale population studies is that it is relatively stable in nonpreserved urine.

Recent studies suggest that α-GST may also be especially useful early marker of Cd-induced kidney injury. Garcon et al. (2007) reported that α-GST was a sensitive indicator of kidney injury in workers who had been exposed to Pb and Cd. Results of studies from our laboratories showed that α-GST was a more sensitive marker of kidney injury than NAG in a rat model of Cd-induced kidney injury (Prozialeck and Edwards 2010).

Miscellaneous Markers of Proximal Tubule Dysfunction, Amino Acids, Glucose, Na$^+$, K$^+$, and Ca^{2+}

In addition to its effects on these protein biomarkers, Cd causes a generalized proximal tubule dysfunction that results in an increase in the urinary excretion of amino acids, Na$^+$, K$^+$, PO$_4^-$, and Ca^{2+} (Shaikh and Smith 1986). Even though these effects are characteristic of Cd nephrotoxicity, the urinary excretion of these substances can be influenced by many factors other than Cd exposure (Shaikh and Smith 1986). In general, these substances have not been widely applied in the monitoring of human populations for early signs of Cd exposure. One notable exception, however, is Ca^{2+}, which has been shown to be a reliable indicator of Cd-induced proximal dysfunction in exposed human subjects (Wu et al. 2001).

Injury Response Proteins

Most of the traditional biomarkers of Cd nephrotoxicity are based on the assumption that Cd causes necrotic or apoptotic cell death of proximal tubule epithelial cells. However, an increasing volume of evidence indicates that the early stages of Cd toxicity involve changes in proximal tubule cell adhesion and function that occur before the onset of cell death (Prozialeck et al. 2009a; Prozialeck and Edwards 2010). In addition, several recent studies indicate that the onset of Cd-induced kidney injury may be preceded by changes in specific markers of metallothionein expression, immune function, and glucose metabolism (for review see Prozialeck and Edwards 2010). Together, these recent findings raise the possibility of identifying more specific and earlier biomarkers of Cd exposure and toxicity. One of the more promising urinary markers that have been described recently is kidney injury molecule-1 (Kim-1).

Kim-1 is a transmembrane protein that is not detectable in normal kidney but is expressed at high levels in the proximal tubule after ischemic or toxic injury (Vaidya et al. 2008). Kim-1 acts as a regulator of cell adhesion and endocytosis in regenerating cells of the injured tubule as they reform a functional epithelial barrier. This process is associated with the proteolytic cleavage of the ectodomain of Kim-1 into the urine. The ectodomain is stable in urine and has been shown to be a sensitive marker of renal injury induced by a variety of agents (Vaidya et al. 2008).

In studies utilizing a rat model of Cd-induced kidney injury, Kim-1 outperformed traditional urinary markers (Prozialeck et al. 2007; Prozialeck et al. 2009a, 2009b). Kim-1 was detected in the urine 4–5 weeks before the onset of proteinuria and 2–5 weeks before the appearance of other markers such as metallothionein and CC-16. Other studies showed that the Cd-induced increase in Kim-1 expression occurred at a time when there was little or no evidence of either necrosis or apoptosis of proximal tubule epithelial cells (Prozialeck et al. 2009b). The fact that Kim-1 can be detected at a time before lethal injury to proximal tubule epithelial cells has occurred may be especially significant. Perhaps, with earlier detection via Kim-1, it may be possible to reverse, or at least more effectively treat, Cd-induced kidney injury. In light of this possibility, studies on the utility of Kim-1 as marker of Cd toxicity in humans are certainly warranted.

Markers of Glomerular Injury

While the proximal tubule is the primary target of Cd-induced kidney injury, there is evidence that Cd can also affect the glomeruli. Changes in classic markers of glomerular dysfunction such as BUN and serum or urinary creatinine are generally not seen during the early or mild stages of Cd-induced kidney injury (Prozialeck and Edwards 2010). However, other investigators have reported associations between Cd exposure and alterations (in glomerular function). For example, Weaver et al. (2011) have recently reported significant increases in creatinine clearance in subjects exposed to Cd and Pb. At present, the relative contributions and relationship of glomerular injury and proximal tubule injury to these finding remain unclear.

Summary and Perspective

Biomarkers have proven to be useful tools for evaluating Cd exposure and nephrotoxicity in human populations as well in laboratory studies on animals. While many fundamental issues regarding the selection of markers and definition of their critical levels have yet to be fully resolved, recent studies have provided hope that new and even more sensitive Cd biomarkers can be developed.

Cross-References

- ▶ Cadmium Absorption
- ▶ Cadmium and Metallothionein
- ▶ Cadmium and Oxidative Stress
- ▶ Cadmium and Stress Response
- ▶ Cadmium Exposure, Cellular and Molecular Adaptations
- ▶ Cadmium, Effect on Transport Across Cell Membranes
- ▶ Cadmium Transport
- ▶ Cadmium, Physical and Chemical Properties

References

ATSDR (2008) Toxicological profile for cadmium. http://www.atsdr.cdc.gov/cercla/toxprofiles/tp5.html

Bernard A (2004) Renal dysfunction induced by cadmium: biomarkers of critical effects. Biometals 17:519–523

Chen L, Jin T, Huang B et al (2006) Critical exposure level of cadmium for elevated urinary metallothionein–an occupational population study in China. Toxicol Appl Pharmacol 215:93–99

Garcon G, Leleu B, Marez T et al (2007) Biomonitoring of the adverse effects induced by the chronic exposure to lead and cadmium on kidney function: usefulness of alpha-glutathione S-transferase. Sci Total Environ 377:165–172

Huang J (2004) Chinese National health standards for occupational exposure to cadmium and diagnostic criteria of occupational chronic cadmium poisoning. Biometals 17(5):511. doi:10.1023/B:BIOM.0000045833.30776.84

Jarup L, Akesson A (2009) Current status of cadmium as an environmental health problem. Toxicol Appl Pharmacol 238:201–208

Jin T, Nordberg G, Wu X et al (1999) Urinary N-acetyl-beta-D-glucosaminidase isoenzymes as biomarker of renal dysfunction caused by cadmium in a general population. Environ Res 81:167–173

Moriguchi J, Inoue Y, Kamiyama S et al (2009) N-acetyl-beta-D-glucosaminidase (NAG) as the most sensitive marker of tubular dysfunction for monitoring residents in non-polluted areas. Toxicol Lett 190:1–8

Noonan CW, Sarasua SM, Campagna D et al (2002) Effects of exposure to low levels of environmental cadmium on renal biomarkers. Environ Health Perspect 110:151–155

Prozialeck WC, Edwards JR (2010) Early biomarkers of cadmium exposure and nephrotoxicity. Biometals 23:793–809

Prozialeck WC, Vaidya VS, Liu J et al (2007) Kidney injury molecule-1 is an early biomarker of cadmium nephrotoxicity. Kidney Int 72:985–993

Prozialeck WC, Edwards JR, Lamar PC et al (2009a) Expression of kidney injury molecule-1 (Kim-1) in relation to necrosis and apoptosis during the early stages of Cd-induced proximal tubule injury. Toxicol Appl Pharmacol 238:306–314

Prozialeck WC, Edwards JR, Vaidya VS et al (2009b) Preclinical evaluation of novel urinary biomarkers of cadmium nephrotoxicity. Toxicol Appl Pharmacol 238:301–305

Shaikh ZA, Smith LM (1986) Biological indicators of cadmium exposure and toxicity. Experientia Suppl 50:124–130

Suwazono Y, Sand S, Vahter M et al (2006) Benchmark dose for cadmium-induced renal effects in humans. Environ Health Perspect 114:1072–1076

Uno T, Kobayashi E, Suwazono Y et al (2005) Health effects of cadmium exposure in the general environment in Japan with special reference to the lower limit of the benchmark dose as the threshold level of urinary cadmium. Scand J Work Environ Health 31:307–315

Vaidya VS, Ferguson MA, Bonventre JV (2008) Biomarkers of acute kidney injury. Annu Rev Pharmacol Toxicol 48:463–493

Weaver VM, Kim NS, Jaar BG et al (2011) Associations of low-level urine cadmium with kidney function in lead workers. Occup Environ Med 68(4):250–256

World Health Organization (WHO) (2000) Cadmium. http://www.euro.who.int/document/aiq/6_3cadmium.pdf

Wu X, Jin T, Wang Z et al (2001) Urinary calcium as a biomarker of renal dysfunction in a general population exposed to cadmium. J Occup Environ Med 43:898–904

Biomedical Application

▶ Gold Nanomaterials as Prospective Metal-based Delivery Systems for Cancer Treatment

Biomedical Plausibility of Exposure and Mechanism of Action

▶ Mercury and Alzheimer's Disease

Biomimicry

▶ Silicateins

Biomineralization

▶ Silicateins

Biomineralization of Gold Nanoparticles from Gold Complexes in *Cupriavidus Metallidurans* CH34

▶ Gold Biomineralization in Bacterium Cupriavidus metallidurans

Biomonitoring

▶ Arsenic in Tissues, Organs, and Cells

Biosensors

▶ Silicon Nanowires

Biosilica

▶ Silicateins

Biosynthesis of Selenoproteins

▶ Selenoproteins and the Biosynthesis and Activity of Thyroid Hormones

Bisarsenical Probes

▶ Biarsenical Fluorescent Probes

Bismuth in Brain

▶ Bismuth in Brain, Distribution

Bismuth in Brain, Distribution

Agnete Larsen
Department of Biomedicine/Pharmacology Health, Aarhus University, Aarhus, Denmark

Synonyms

Bismuth in brain

Definitions

Encephalopathy: This term involves a large group of transient or permanent brain disorders with very diverse etiologies. Among other causes it can be mentioned metabolic alterations, bacterial, food poisoning, trauma, hypoxia, pharmacological, etc. When unresolved, encephalopathies lead to neurodegeneration. The hallmark of encephalopathy is mental alteration, impairment of cognitive functions, lethargy, confusion, tremors, seizures, convulsions, and others.

Autometallography: This method is used for the imaging of metal-containing clusters in tissues sections. This methodology involves the reduction of silver ions from a silver donor e.g., silver lactate (Ag^+) to metallic silver (Ag^0) by metals such as bismuth in the presence of a developer. In consequence, Ag^0 is deposited in the site where the metal of interest is located and those deposits can be photographed by both light and electron microscopies.

In contrast to many other heavy metals, bismuth is generally considered a safe or even "green" heavy metal. Due to the chemical and physical properties of the metal, bismuth is often used as a replacement of lead and mercury in many industrial settings, and in 2004, the industrial use of bismuth reached a yearly consumption of as much as 5,000 t (Palmieri 2004). Possible source of human bismuth exposure include the widespread use of bismuth in cosmetics and as the soft metal has excellent ballistic qualities, bismuth shotgun pellets are a known source of direct bismuth exposure to wild life especially wounded game animals which might live on for prolonged periods exposed to fragments from shotgun pellets (Pamphlett et al. 2000; Stoltenberg 2004).

Medical use of bismuth dates back more than a 100 years (The Lancet 1857) but even in this millennium, bismuth compounds are still in use for gastrointestinal disorders such as the well-known US over-the-counter remedy Pepto-Bismol. Bismuth salts have also been shown to have a bacteriostatic effect on the ulcer-generating bacteria *Heliocobacter pylori* (Rokkas and Sladen 1988; Stoltenberg et al. 2001a). Modern medical use also includes bismuth adjuvants for chemotherapy (Tiekink 2002). Reported side effects to bismuth pharmaceuticals include gingival discoloration, erythema, and osteopathy. More dangerous side effects are nephropathy and a potentially fatal encephalopathic condition (Slikkerveer and de Wolff 1989). Bismuth encephalopathy is characterized by a series of neuropsychiatric changes such as emotional changes, insomnia, and changes in eating habits followed by confusion and motor dysfunctions stretching from tremors and ataxia to myoclonus and overt convulsions (Bes et al. 1976). Though sometimes fatal, this encephalopathic state is most often reversible within a few weeks after cessation of bismuth exposure. Most available knowledge regarding the pathogenesis of bismuth encephalopathy dates back to an epidemic incident taking place in France and neighboring countries in the 1970s. During this episode, approximately 1,000 cases of bismuth-induced encephalopathies were reported. The epidemic occurred at a time and place at which the use of bismuth compounds for various gastrointestinal disorders was quite high and with great individual variation however despite the high number of patients evaluations of the incidence expect that only 0.1% of all bismuth consumers were affected. There was no apparent relationship between disease susceptibility, bismuth blood values, or the amount or period of bismuth ingestion, and the pathogenesis of the disease is still far from unraveled and it thus remains unclear if it was a random clustering or if unknown environmental factors contributed (Martin-Bouyer 1978; Martin-bouyer et al. 1981).

Following the epidemic, Ross and coworkers (1988) set out to create an animal (mouse) model for bismuth encephalopathy and much of our knowledge of the distribution of bismuth in brain tissue is derived from this model. The model is based on intraperitoneal injections of bismuth subnitrate (2,500 mg/kg) functioning as a permanent deposit for bismuth liberation. Using this model, many mice display neurotoxic features similar to the human condition, reaching a brain Bi level (∼8 ppm) similar to the one reported in deceased patients. In a later bismuth distribution study, the model was modified to a single injection, giving raise to brain Bi levels of 1–5 ppm within a 1–4-week period of exposure (Ross et al. 1994).

Applying atomic absorption spectrometry (AAS), Ross and coworkers could show that given blood-borne bismuth exposure, there is a clear pattern of regional difference in bismuth content throughout the brain. In these studies, it was evident that the highest accumulation of bismuth is found in the olfactory bulb and subsequently the hypothalamus. The olfactory bulb is seldom examined but a high degree of hypothalamic accumulation has been reported also in more qualitative studies (Pamphlett et al. 2000; Larsen et al. 2005).

Other areas with relatively high quantitative bismuth concentration were the septum region and the brain stem containing approximately half the concentration of the olfactory bulb. Outside these regions, the amount of bismuth accumulation was relatively small compared to the olfactory bulb and the hypothalamus. About a third of the maximal olfactory bulb concentration was found in tissue samples from the spinal cord and a little less in the thalamus and cerebellum. The by far lowest bismuth concentrations were seen in cortical areas and the striatum, e.g., in the temporal cortex the bismuth concentration was about one seventh of concentration seen in the olfactory bulb (1 ppm to 7.1 ppm respectively). It should be noted that the regional distributions seen here have also been reported on intraperitoneal exposure to water-soluble bismuth compounds (Ross et al. 1994).

In their 1994 and 1996 articles (Ross et al. 1994; Larsen et al. 2005), the Ross group also purposed a technique for in situ bismuth visualization, the autometallographic silver enhancement technique (AMG), later to be refined by Danscher and Stoltenberg (Ross et al. 1996; Stoltenberg and Danscher; Stoltenberg et al. 2007). With AMG around 96% of injected bismuth can be visualized (Stoltenberg et al. 2007). Using AMG, Ross could confirm distinct regional differences in bismuth distribution with olfactory bulb and hypothalamus exhibiting the highest amount of AMG-detectable bismuth accumulation. Overall, the regional staining pattern showed little inter-animal variation in the Ross studies; however, some inter-animal variation in AMG-staining intensity was evident and this phenomenon is supported in other animal studies (Pamphlett et al. 2000; Larsen et al. 2005). The AMG study made by Ross depicted a regional distribution of bismuth centered around and gradually spreading from the circumventricular organs (CVOs), i.e., it was evident that the accumulations of bismuth present decreased as the distance to the CVO increased (Ross et al. 1994, 1996). The lack of the blood–brain barrier allowed the way in for the blood-borne bismuth exposure, e.g., the high amount of bismuth in the hypothalamus seems associated to the near presence to fenestrated blood vessels. In accordance with this, a very high amount of bismuth is seen in the vessels and the underlying lamina and in areas with only minor bismuth accumulations, most metals seem to be captured within the vasculature (Ross et al. 1994, 1996). However, throughout the brain, a high amount of bismuth accumulations was also seen in lysosomes in both neurons and glia (Pamphlett et al. 2000; Stoltenberg 2004; Ross et al. 1994, 1996). The choroid plexus was also characterized by a very high amount of bismuth in the epithelia, corresponding to morphological findings of alteration in the ventricular system with hydrocephalus as a striking feature in the murine model. The neuronal staining was most evident in larger neurons especially in the brain stem. Other animal studies also reported that large motor neuron is a center for bismuth accumulation (Pamphlett et al. 2000; Ross et al. 1994). The cerebellum, especially the Purkinje cells (especially those close to the fourth ventricle), is also described as among the neurons most heavily loaded with AMG-detectable bismuth.

The only human study (Stoltenberg et al.) employing AMG on brain slices from six victims of bismuth encephalopathies also confirmed metal accumulations in both neuron and glia with cerebellum as one of the "hot spots." Unlike the mouse model, this study did however find equally large mainly neuronal accumulations in thalamus and neocortex, areas of lesser bismuth content in the rodent model of encephalopathy. In one human sample, a widely distributed bismuth pattern was seen involving also areas like the hippocampus, indicating that bismuth accumulation takes place in any brain region depending on the amount of metal present. Retrograde axonal transport of bismuth occurs both within the CNS and from the periphery into spinal cord (Ross et al. 1994; Stoltenberg et al. 2001b) and both such active transport and the mode of entry will also affect the distribution patterns as seen, e.g., following direct intercranial exposure in which the highest amount of bismuth accumulation is seen in the vicinity of a bismuth implant (Stoltenberg et al. 2003).

To sum up, the presently available information on bismuth distribution indicates that bismuth uptake is affected by the mode of entry, i.e., blood-borne intra via fenestrated blood vessels contra intra-muscularly or intra-cranially, with CVO-near accumulation of especially the hypothalamus as a heavily loaded area. Olfactory bulb with its projection to the periphery is another central area of accumulation. Additionally, cerebellar Purkinje cells and large sensory and especially motor neurons in the brain stem seem prone to bismuth accumulation. Bismuth can be found not only in vasculature but also in glia cells, many of them astrocytes in the white matter as well as in neurons. The ependyma of the choroid plexus also contains a high amount of bismuth. On the ultrastructural level, cellular bismuth inclusions are seen in lysosomal-like structures of both glia and neurons as well as extracellular in basement membranes. Blood concentration of bismuth shows a high degree of intra-personal variation, and even in the laboratory setting, there are differences in the degree of neuronal accumulation between individuals, although the regional pattern of distribution remains the same for many types of bismuth exposure. Especially with regard to the concentration of bismuth in larger, mainly motor neurons and the choroid plexus, the distribution of bismuth shows similarity to other heavy metals such as silver and mercury (Møller-Madsen 1993; Cassano et al. 1969; Rungby and Danscher 1983).

Cross-References

▶ Germanium, Toxicity
▶ Germanium, Physical and Chemical Properties
▶ Germanium-Containing Compounds, Current Knowledge and Applications

References

Bes A et al (1976) Toxic encephalopathy due to bismuth salts. Rev Med Touluse 12:801–813

Cassano GB et al (1969) The distribution of inhaled mercury (Hg^{203}) vapors in the brain of rats and mice. J Neuropathol Exp Neurol 28:214–255

Larsen A et al (2005) In vivo distribution of bismuth in the mouse brain; influence of long-term survival and intracranial placement on the uptake and transport of bismuth in neuronal tissue. Basic Clin Pharmacol Toxicol 97(3):188–196

Martin-Bouyer B (1978) Poisoning by orally administred bismuth salts. Gastrointest Clin Biol 2(4):349–356

Martin-bouyer G et al (1981) Epidemiological study of encephalopathies following bismuth administration per os. Characteristics of intoxicated subjects: comparison with control group. Clin Toxicol 11:1277–1283

Møller-Madsen B (1993) Localization of mercury in CNS of the ratII interperitoneal injection of methylmercuric chloride (CH_3HgCl) and mercuric chloride ($HgCl_2$). Toxicol Appl Pharmacol 103:303–323

Palmieri Y (2004) About bismuth. htpp://www.bismuth.be/bismuthpdf

Pamphlett R et al (2000) Tissue Uptake of bismuth from shotgun pellets. Envion Res 82(3):258–262

Rokkas T, Sladen GE (1988) Bismuth: effects on gastritis and peptic ulcer. Scand J Gastroentral Suppl 142:82–86

Ross JF et al (1988) Characterization of a murine model for human bismuth encephaloapathy. Neurotoxicoly 9:581–586

Ross JF et al (1994) Highest brain bismuth levels and neuropathology are adjeacent to fenestrated blood vessels in mouse brain after intraperitonal dosing of bismuth subnitrate. Toxicol Appl Pharmacol 124:191–200

Ross JF et al (1996) Distribution of bismuth in the brain: intraperitoneal dosing of bismuth subnitratein mice: implications for the route of entry of xenobiotic metals into the brain. Brain Res 725:137–154

Rungby J, Danscher G (1983) Localization of exogenous silver in brain and spinal cord of silver exposed rats. Acta Neuropathol 60:92–98

Slikkerveer A, de Wolff FA (1989) Pharmacokinetices and toxicity of bismuth compounds. Toxicol Manag Rev 4:303–323

Stoltenberg M (2004) Bismuth. Some aspects of localization, transport and pathological effects of metalic bismuth and bismuth salts with special emphasis o nits neurotoxicity to man and experimental animals. Thesis, University of Aarhus, Denmark

Stoltenberg M, Danscher G. Histochemical differentiation of autometallographically traceable metals (Au, Ag, Bi, Zn): Protocols for chemical removal of separate autometallographic metal clusters in Epon Sections. Histochem J 32:645–652

Stoltenberg M et al Autometallographic tracing of bismuth in human brain autopsies. J Neuropathol Exp Neurol 60:705–710

Stoltenberg M et al (2001) Histochemical tracing of bismuth in *Heliocobacter pylori* after in vitro exposure to bismuth citrate. Scand J Gastroenterol 36(2):144–148

Stoltenberg M et al (2001) Retrograde axonal transport of bismuth. An autometallograhic study. Acta Neuropathol 101:123–128

Stoltenberg M et al (2003) In vivo cellular uptake of bismuth ions from shotgun pellets. Histol Histopathol 18(3):781–785

Stoltenberg M, Juhl S, Danscher G (2007) Bismuth ions are metabolized into autometallographic traceable bismuth-sulphur quantum dots. Eur J Histochem 51(1):53–57

The Lancet (1857). 2:185–187

Tiekink ER (2002) Antimony and bismuth compounds in oncology. Crti Rev Oncol 42(3):217–224

Bismuth, Interaction with Gastrin

Graham S. Baldwin
Department of Surgery, Austin Health, The University of Melbourne, Heidelberg, VIC, Australia

Definition

The peptide hormone gastrin was originally identified as a stimulant of acid secretion, but is now known to also act as a growth factor in the gastrointestinal tract (Dockray et al. 2001). Gastrin is synthesized as a precursor of 101 amino acids (*preprogastrin*) which, on removal of the signal peptide of 21 amino acids, yields *progastrin* (80 amino acids). Proteolytic processing in antral G cells in the stomach generates a number of intermediate peptides, including *glycine-extended gastrin$_{17}$* (Ggly), which has the sequence ZGPWLEEEEEAYGWMDFG. Transamidation of the C-terminal glycine yields the C-terminal amidated phenylalanine characteristic of *amidated gastrin* (Gamide). Both Ggly and Gamide are independently active, via different receptors, in the gastrointestinal tract.

Basic Characteristics

Iron: Interaction with Gastrin

Fluorescence experiments have revealed that both Ggly and Gamide bind two ferric ions with high affinity ($K_d = 0.6$ μM, pH 4.0) in aqueous solution (Baldwin et al. 2001), via the carboxylate groups in the side chains of Glutamates 7, 8, and 9 (Pannequin et al. 2002).

Progastrin also binds two ferric ions, and the ferric ion–progastrin complex is very stable, with a half-life of 117 ± 8 days at pH 7.6 and 25°C (Baldwin 2004).

Binding of ferric ions is essential for the biological activities of Ggly in vitro. Mutation of Glutamate 7 of Ggly to Alanine reduced the stoichiometry of ferric ion binding from 2 to 1, and completely abolished biological activity in cell proliferation and migration assays (Pannequin et al. 2002). The iron chelator desferrioxamine also completely blocked Ggly activity in cell proliferation and migration assays in vitro (Pannequin et al. 2002) and in the colorectal mucosa in vivo (Ferrand et al. 2010). The minimum biologically active Ggly fragments are the heptapeptides LEEEEEA and EEEEEAY, and their activity is still dependent on ferric ions (He et al. 2004). Interestingly, mutation of Glutamate 7 of Gamide to Alanine had no effect on the biological activity of Gamide, even though the stoichiometry of ferric ion binding was again reduced from 2 to 1, presumably because the receptors for Ggly and Gamide are distinct (Pannequin et al. 2004a).

Bismuth: Interaction with Gastrin

Bi^{3+} ions also bind to Ggly, although with lower affinity than Fe^{3+} ions ($K_d = 5.8 \pm 1.4$ μM, pH 4.0) (Pannequin et al. 2004b). NMR spectroscopy indicated that, as with Fe^{3+} ions, binding was via the carboxylate groups in the side chains of Glutamates 7, 8 and 9. Because the Bi^{3+}–Ggly complex is not recognized by the Ggly receptor, Bi^{3+} ions act as competitive inhibitors in both cell proliferation and migration assays in vitro (Pannequin et al. 2004b). In contrast Bi^{3+} ions did not reduce the binding of Gamide to its receptor, or inhibit the biological activity of Gamide.

Physiological Significance

Bismuth salts have been used for many years for the treatment of gastrointestinal disorders such as ulcers and diarrhea (Gorbach 1990). The direct anti-bacterial effect of Bi^{3+} ions on the gastric bacterium *Helicobacter pylori* provided the rationale for the use of bismuth salts, in combination with inhibitors of gastric acid production and with antibiotics, in the treatment of *H. pylori*–induced ulcers (Houben et al. 1999). The recognition of the ability of Bi^{3+} ions to inhibit the biological activity of Ggly suggests that bismuth salts may also interfere with the potentiation by Ggly of the Gamide-induced secretion of gastric acid (Chen et al. 2000). Since non-amidated gastrins like Ggly appear to act as growth factors for colorectal cancer (Aly et al. 2004), inhibition of Ggly activity by Bi^{3+} ions may offer a novel and selective therapy for this all too common disease.

References

Aly A, Shulkes A, Baldwin GS (2004) Gastrins, cholecystokinins and gastrointestinal cancer. Biochim Biophys Acta 1704:1–10

Baldwin GS, Curtain CC, Sawyer WH (2001) Selective, high-affinity binding of ferric ions by glycine-extended gastrin(17). Biochemistry 40:10741–10746

Baldwin GS (2004) Properties of the complex between recombinant human progastrin and ferric ions. The Protein Journal 23:65–70

Chen D, Zhao CM, Dockray GJ et al (2000) Glycine-extended gastrin synergizes with gastrin 17 to stimulate acid secretion in gastrin-deficient mice. Gastroenterology 119:756–765

Dockray GJ, Varro A, Dimaline R et al (2001) The gastrins: their production and biological activities. Annu Rev Physiol 63:119–139

Ferrand A, Lachal S, Bramante G et al (2010) Stimulation of proliferation in the colorectal mucosa by gastrin precursors is blocked by desferrioxamine. Am J Physiol Gastrointestinal and Liver Physiology 299:G220–G227

Gorbach SL (1990) Bismuth therapy in gastrointestinal diseases. Gastroenterology 99:863–875

He H, Shehan BP, Barnham KJ et al (2004) Biological activity and ferric ion binding of fragments of glycine-extended gastrin. Biochemistry 43:11853–11861

Houben MH, van de Beek D, Hensen EF et al (1999) A systematic review of *Helicobacter pylori* eradication therapy–the impact of antimicrobial resistance on eradication rates. Aliment Pharmacol Ther 13:1047–1055

Pannequin J, Barnham KJ, Hollande F et al (2002) Ferric ions are essential for the biological activity of the hormone glycine-extended gastrin. J Biol Chem 277:48602–48609

Pannequin J, Tantiongco JP, Kovac S et al (2004a) Divergent roles for ferric ions in the biological activity of amidated and non-amidated gastrins. J Endocrinol 181:315–325

Pannequin J, Kovac S, Tantiongco JP et al (2004b) A novel effect of bismuth ions: selective inhibition of the biological activity of glycine-extended gastrin. J Biol Chem 279:2453–2460

Bismuth, Interaction with Transferrin

Graham S. Baldwin
Department of Surgery, Austin Health, The University of Melbourne, Heidelberg, VIC, Australia

Definition

Transferrin is an abundant serum glycoprotein of molecular mass 80 kDa (kilodaltons) (Wally and

Buchanan 2007). Sequence similarity between the N- and C-terminal lobes of the molecule indicates that transferrin has evolved by duplication of an ancestral gene. Each lobe binds a single ferric ion with high affinity, via the side chains of two tyrosines, a histidine, and an aspartic acid. The remaining two ligands in the distorted octahedral binding site are provided by the synergistic anion bicarbonate. Transferrin is responsible for the transport of iron around the body, and delivers iron to target cells via interaction with the transferrin receptor TfR1. A related protein, lactoferrin, is found in milk and other secreted fluids, where its ability to bind iron tightly limits bacterial growth.

Basic Characteristics

Binding of Bismuth by Transferrins

In the presence of bicarbonate, bismuth ions bind to both ferric ion–binding sites of transferrin (Li et al. 1996; Sun et al. 1999, 2001). Binding to the C-terminal lobe is tighter than to the N-terminal lobe (Li et al. 1996; Sun et al. 1999), and the affinity of the C-terminal site for bismuth ions (K_d(Dissociation constant) = 1.5×10^{-17} M) is greater than for the binding of ferric ions ($K_d = 1 \times 10^{-16}$ M) at I = 0.2 M, 298 K (Miquel et al. 2004). Lactoferrin also binds two bismuth ions, but in this case the affinity is weaker than for ferric ions (Zhang et al. 2001).

Interaction with the Transferrin Receptor

Bismuth-loaded transferrin is recognized by the transferrin receptor TfR1, although the interaction is considerably weaker ($K_d = 4 \times 10^{-6}$ M) than for iron-loaded transferrin ($K_d = 2.3 \times 10^{-9}$ M) at I = 0.2 M, 310 K (Miquel et al. 2004). Further studies with human cell lines will be required to establish whether or not the interaction with the transferrin receptor TfR1 is responsible for the cellular uptake of bismuth ions.

References

Li H, Sadler PJ, Sun H (1996) Unexpectedly strong binding of a large metal ion (Bi^{3+}) to human serum transferrin. J Biol Chem 271:9483–9489

Miquel G, Nekaa T, Kahn PH et al (2004) Mechanism of formation of the complex between transferrin and bismuth, and interaction with transferrin receptor 1. Biochemistry 43:14722–14731

Sun H, Li H, Mason AB et al (1999) N-lobe versus C-lobe complexation of bismuth by human transferrin. Biochem J 337(Pt 1):105–111

Sun H, Li H, Mason AB et al (2001) Competitive binding of bismuth to transferrin and albumin in aqueous solution and in blood plasma. J Biol Chem 276:8829–8835

Wally J, Buchanan SK (2007) A structural comparison of human serum transferrin and human lactoferrin. Biometals 20:249–262

Zhang L, Szeto KY, Wong WB et al (2001) Interactions of bismuth with human lactoferrin and recognition of the Bi (III)-lactoferrin complex by intestinal cells. Biochemistry 40:13281–13287

Bismuth, Physical and Chemical Properties

Fathi Habashi
Department of Mining, Metallurgical, and Materials Engineering, Laval University, Quebec City, Canada

Bismuth is a metalloid of no useful mechanical properties. It is mainly used as an alloying component in fusible alloys. Only one stable isotope, ^{209}Bi, is known, but there are several unstable isotopes (^{199}Bi–^{215}Bi). Isotopes with mass number > 210 are found in the natural decay chains of radioactive elements. Isotopes with mass number < 208 have been formed in nuclear transformations. The volume of the molten metal increases by about 3% on solidification.

Physical Properties

Atomic number	83
Atomic weight	208.98
Relative abundance in Earth's crust,%	2×10^{-5}
Atomic radius, nm	0.18
Density at 20°C, g/cm^3	9,790
Melting point, °C	271.40
Boiling point, °C	1,564
Crystal system	Rhombohedral
Lattice constant, nm	$a = 0.47457$
	$\alpha = 57.24°$
Latent heat of fusion, J/mol	11,280
Latent heat of vaporization, J/mol	178,632
Coefficient of linear expansion, K^{-1}	13.5×10^{-6}

(continued)

Electrical resistivity, μΩ cm	
At 0°C	106.8
At 1,000°C	160.2
Specific heat at 25°C, J mol^{-1} K^{-1}	25.5
Thermal conductivity, J s^{-1} m^{-1} K^{-1}	
At 0°C	8.2
At 300°C	11.3
At 400°C	12.3
Vapor pressure, bar	
At 893°C	1.013×10^{-3}
At 1,053°C	1.013×10^{-2}
At 1,266°C	1.013×10^{-1}
Surface tension, mN/m	
At 300°C	376
At 400°C	370
At 500°C	363
Absorption cross section for thermal neutrons, m^2/atom	$(3.4 \pm 0.2) \times 10^{-30}$
Hardness, Brinell, N/mm^2	184
Hardness, Mohs	2.5
Poisson's ratio	0.33
Shear modulus, MPa	12,400
Modulus of elasticity, GPa	338
Viscosity, MPa·s	
At 300°C	1.65
At 350°C	1.49
At 400°C	1.37
At 500°C	1.19
At 600°C	1.06

Chemical Properties

Bismuth does not oxidize in dry air. Liquid bismuth is covered by an oxide film of Bi_2O_3 that protects it from further oxidation. In most compounds, bismuth has a +3 oxidation state. Bismuth is precipitated by hydrolysis on dilution or partial neutralization:

$$BiCl_3 + H_2O \rightarrow Bi(OH)Cl_2 + HCl$$

$$Bi(OH)Cl_2 \rightarrow BiOCl + HCl$$

or by cementation with iron turnings:

$$2BiCl_3 + 3Fe \rightarrow 3FeCl_2 + 2Bi$$

Bismuthine, BiH_3, is a colorless gas unstable at room temperature decomposing to bismuth and hydrogen. Bismuth is precipitated as Bi_2S_3 from slightly acidic solution with H_2S. The precipitate is brown black and soluble in strong acids and hot dilute nitric acid. Yellow-green Bi_2S_3 precipitates from alkaline solution on addition of Na_2S.

References

Habashi F (2001) Arsenic, antimony, and bismuth production. In: Encyclopedia of materials: science & technology, pp 332–336

Krüger J et al (1997) In: Habashi F (ed) Handbook of extractive metallurgy. Wiley, Weinheim, pp 845–871

Blood Clotting

▶ Calcium-Binding Proteins, Overview

Blood Clotting Impairments as Result of Zinc Disorders

▶ Zinc in Hemostasis

Blood Clotting Proteins

Sheryl R. Bowley, Mingdong Huang, Barbara C. Furie and Bruce Furie
Division of Hemostasis and Thrombosis, Beth Israel Deaconess Medical Center, Harvard Medical School, Boston, MA, USA

Synonyms

EGF-like domain: epidermal growth factor-like domain; *Gla*: Γ-carboxyglutamic acid; *Procofactors*: procoagulant cofactors

Definitions

Coagulation: blood clotting
Gla: γ-carboxyglutamic acid is formed by post-translational addition of a carboxyl group to the γ-carbon of glutamic acid

Thrombosis: formation of a blood clot inside a blood vessel, obstructing the flow of blood through the circulatory system

Procofactors: proteins whose binding to another coagulation protein requires proteolytic cleavage for full activity of the complex

Tenase complex: a complex formed by the assembly of factors IXa and VIIIa on a cell membrane that converts factor X to active factor Xa

Prothrombinase complex: a complex formed by the association of factors Va and Xa on a cell membrane that converts prothrombin to thrombin

EGF-like domain: a protein domain conserved through evolution that includes an N-terminal two-stranded beta-sheet followed by a loop and a short C-terminal two-stranded beta-sheet

Blood coagulation is an important host defense mechanism that maintains the integrity of the vascular system in response to injury. To prevent excessive bleeding after vascular injury, platelets rapidly bind to the subendothelium to form the initial platelet plug. This is followed by a series of reactions involving the conversion of soluble blood coagulation proteins from their proenzyme forms to serine proteases and the proteolytic activation of plasma-derived procofactors to active cofactors. This culminates in the production of thrombin which catalyzes the formation of the fibrin clot that stabilizes the initial platelet plug. Many of these enzymatic reactions are calcium dependent and occur on negatively charged cellular surfaces.

The extrinsic pathway, responsible for rapid and efficient coagulation in vivo, includes vitamin K–dependent proteins, plasma and cellular cofactors and enzymes. The vitamin K–dependent proteins (prothrombin, factor VII, factor IX, and factor X) contain γ-carboxyglutamic acid residues essential for calcium-dependent membrane binding. The posttranslational carboxylation of glutamic acid residues is mediated by the γ-carboxylase in the presence of vitamin K, oxygen, and carbon dioxide during protein biosynthesis. Indeed, vitamin K deficiency results in impaired blood clotting. Plasma procofactors, factor V and factor VIII, serve as cofactors for two vitamin K–dependent serine proteases, factor Xa and factor IXa, respectively. Both procofactors are activated proteolytically by thrombin and require calcium to maintain productive association between different domains.

The key to initiation of the extrinsic pathway is the cellular cofactor tissue factor (TF), an integral membrane protein that is constitutively expressed by certain cells within the vessel wall and cells surrounding blood vessels (Furie and Furie 2008). Upon exposure to plasma, TF binds factor VII and its active form factor VIIa. The TF/VIIa complex activates both factor IX and factor X to factor IXa and factor Xa, respectively. Factor IXa assembles on membrane surfaces with its active cofactor, factor VIIIa in the presence of calcium. The factor IXa/factor VIIIa complex, referred to as the "tenase" complex, converts factor X to active factor Xa. Similarly, factor Xa binds to its cofactor active factor Va bound on membrane surfaces in the presence of calcium to form the "prothrombinase" complex that converts the enzymatically inactive prothrombin to its enzyme form thrombin. Thrombin mediates the conversion of the soluble plasma protein fibrinogen into monomeric fibrin molecules that rapidly polymerize to form the insoluble fibrin clot. The fibrin clot is further reinforced by covalent cross-linking between fibrin molecules by factor XIIIa. As with other proenzymes, factor XIII is activated to factor XIIIa by thrombin and requires calcium as a cofactor.

Vitamin K–Dependent Blood Coagulation Proteins

Prothrombin

Prothrombin is a plasma glycoprotein with a molecular weight of 72,000 Da. The most abundant of the vitamin K–dependent blood coagulation proteins, it circulates as a single polypeptide chain zymogen. Ten γ-carboxyglutamic acid residues are located in the Gla domain, a region that anchors the protein to membranes in the presence of calcium ions. γ-Carboxyglutamic acid is generated as a vitamin K–dependent post-translational modification of glutamic acid during protein biosynthesis. Adjacent to the Gla domain are two kringle domains, important for protein complex formation with factor Va. The C-terminal region of prothrombin is the catalytic domain that includes the proteolytic site involved in its conversion from a zymogen to an active enzyme, thrombin. Prothrombin is converted to thrombin by the prothrombinase complex, factor Xa, and factor Va bound on a cell membrane exhibiting exposed phosphatidylserine.

The calcium ions in the Gla domain are mostly bound internally to stabilize the omega loop that characterizes all Gla domains in the vitamin K–dependent

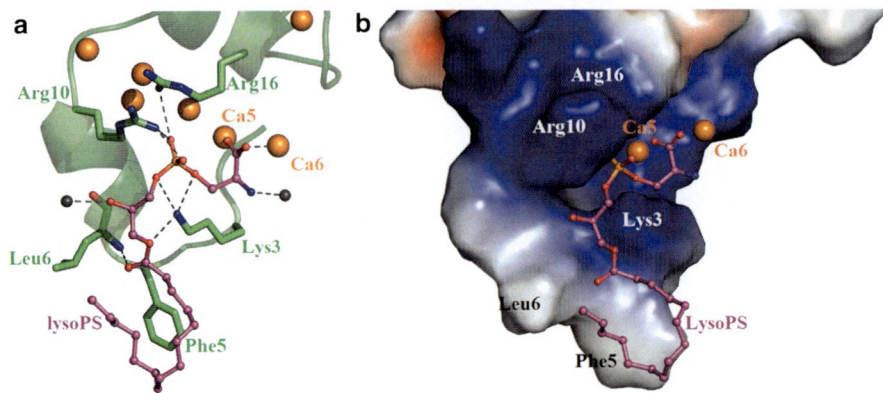

Blood Clotting Proteins, Fig. 1 *Gla domain of prothrombin.* (**a**) Calcium ions (*orange spheres*) in the Gla domain of prothrombin form electrostatic interactions with the serine head group of lysophosphoserine, lysoPS (*magenta*). The phosphate group interacts with positively charged lysine and arginine side chains (Huang et al. 2003). (**b**) The calcium-dependent conformational transition via the Gla domain of prothrombin leads to exposure of hydrophobic residues forming an external patch (*gray area*) that align with fatty acid tails in the membrane

proteins (Soriano-Garcia et al. 1992). In the presence of calcium ions, the serine head group of the acidic phospholipid, phosphatidylserine, of an activated cell membrane, forms electrostatic interactions with the γ-carboxyglutamic acid-calcium network in the Gla domain of prothrombin, allowing the substrate to bind the membrane and the prothrombinase complex (Fig. 1a) (Huang et al. 2003). A calcium-dependent conformational transition in the prothrombin Gla domain leads to exposure of a hydrophobic patch (e.g., amino acids Phe5, Leu6, and Val9 in bovine prothrombin) that aligns with fatty acid tails in the membrane (Fig. 1b) (Huang et al. 2003).

Factor VII

Factor VII, a protein with a molecular weight of 50,000 Da, circulates in plasma in two forms: the dominant albeit inactive zymogen factor VII and a trace amount of enzymatically active factor VIIa. Typical of vitamin K–dependent blood coagulation proteins, factor VII contains a Gla domain with ten γ-carboxyglutamic acid residues. Adjacent to the Gla domain are two epidermal growth factor (EGF)-like domains, followed by the serine protease domain with an active site homologous to trypsin and chymotrypsin. Tissue factor, an integral membrane protein expressed on the plasma membrane of many cells, can bind either factor VII or factor VIIa in a calcium-dependent manner. The tissue factor/factor VIIa complex initiates blood coagulation via activation of factor IX and factor X to their active enzyme forms.

The structure of the complex of the active site–inhibited factor VIIa and the extracellular domain of tissue factor has been determined by X-ray crystallography (Fig. 2) (Banner et al. 1996). Salient features of this complex include extensive "embracing" of extended conformations of one protein by the other. The structure also shows the Gla domain with seven bound calcium ions. Factor VII binds to negatively charged membranes via the Gla domain in a reversible and calcium-dependent manner. The structure of the calcium-stabilized Gla domain is homologous to that of prothrombin, factor X, and factor IX. Of the two EGF domains of factor VII, only EGF1 binds calcium ions. The EGF1, together with the protease domain, serves as the main site of contact with tissue factor. The serine protease domain also binds a single calcium ion. Zinc binding to the protease domain inhibits the enzymatic activity of factor VIIa.

Factor IX

Factor IX has a molecular weight of 56,000 Da and consists of a Gla domain, two adjacent EGF domains, and a serine protease domain. The Gla domain contains 12 γ-carboxyglutamic acid residues that bind calcium ions crucial for interaction with membranes. Factor IX is activated to its serine protease form, factor IXa, by either the tissue factor/factor VIIa complex or factor XIa, in the presence of calcium ions. Factor IXa and factor VIIIa assemble to form a membrane surface–bound tenase complex that activates factor X to factor

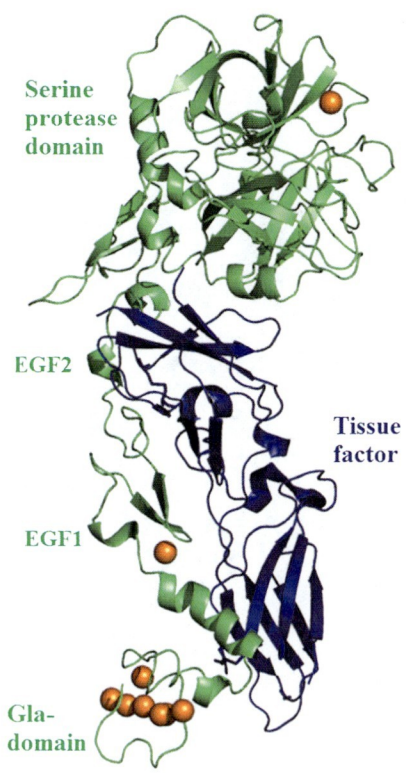

Blood Clotting Proteins, Fig. 2 *Domain organization of vitamin K–dependent proteins.* The structure of the complex of active site–inhibited factor VIIa and the extracellular domain of tissue factor shows the domain organization of factor VIIa (Banner 2006). Interaction with tissue factor involves mostly the EGF1 domain and the serine protease domain of factor VIIa. Factor VII binds to negatively charged phospholipid membranes via the Gla domain. Calcium ions are shown as *orange spheres*

site that includes a β-hydroxyaspartic acid. The post-translational modification of this aspartic acid, however, is not required for factor IX function.

Factor X

Factor X, with a molecular weight of 56,000 Da, contains a Gla domain, two EGF domain, and a serine protease domain. The Gla domain of factor X includes 11 γ-carboxyglutamic acid residues and has marked structural homology with other Gla domains. The Gla domain binds Ca^{2+}, exposing a membrane binding site which is a prerequisite for anchoring the vitamin K–dependent blood clotting proteins to membranes. The N-terminus of the Gla domain shows little stable structure in the absence of Ca^{2+}. The two EGF domains are important for binding factor Va.

Factor X is activated to factor Xa by either the tissue factor/factor VIIa complex or by the factor IXa/factor VIIIa complex in the presence of phospholipid membranes and calcium ions. Factor Xa then associates with factor Va on the membrane surface to form the prothrombinase complex. This complex activates prothrombin to thrombin in the presence of calcium ions.

Procofactors

Factor VIII

Factor VIII is a plasma glycoprotein that is synthesized as a single polypeptide precursor with domains organized as A1–A2–B–A3–C1–C2. Intracellular proteolytic cleavage at the B–A3 junction or within the B-domain produces forms of factor VIII that circulate in the blood: a light chain (A3–C1–C2, 80,000 Da) and a heterogeneous heavy chain with a molecular weight varying between 90,000 Da (A1–A2) and 200,000 Da (A1–A2–B). Further proteolysis between the A1 and A2 domains by thrombin results in the generation of active factor VIII. The activated cofactor forms a complex with factor IXa on membrane surfaces to activate factor X during blood coagulation. Defects in the factor VIII gene that lead to diminished or absent factor VIII activity cause a bleeding disorder, hemophilia A.

The three-dimensional structures of the human factor VIII C2 domain (Pratt et al. 1999) and the B-domain-deleted human factor VIII (Ngo et al. 2008, Shen et al. 2008) have been determined by

Xa in the presence of calcium ions. Defective activity or deficiency of factor IX due to mutation of the gene is the cause of hemophilia B.

The Gla domain of factor IX binds both internal calcium ions as well as calcium ions on the protein surface (Huang et al. 2004). The crystal structures of the Gla domain confirm the similarity of this domain in all of the vitamin K–dependent proteins involved in blood coagulation. The Gla domain of factor IX undergoes two metal-dependent conformational transitions. The calcium-stabilized conformer expresses a phospholipid-binding site that is located in the Gla domain at the amino terminus of the protein. Factor IXa, but not factor IX, binds to factor VIIIa on the surface of activated platelets via the second EGF domain and the serine protease domain. The first EGF-like domain has a high-affinity calcium-binding

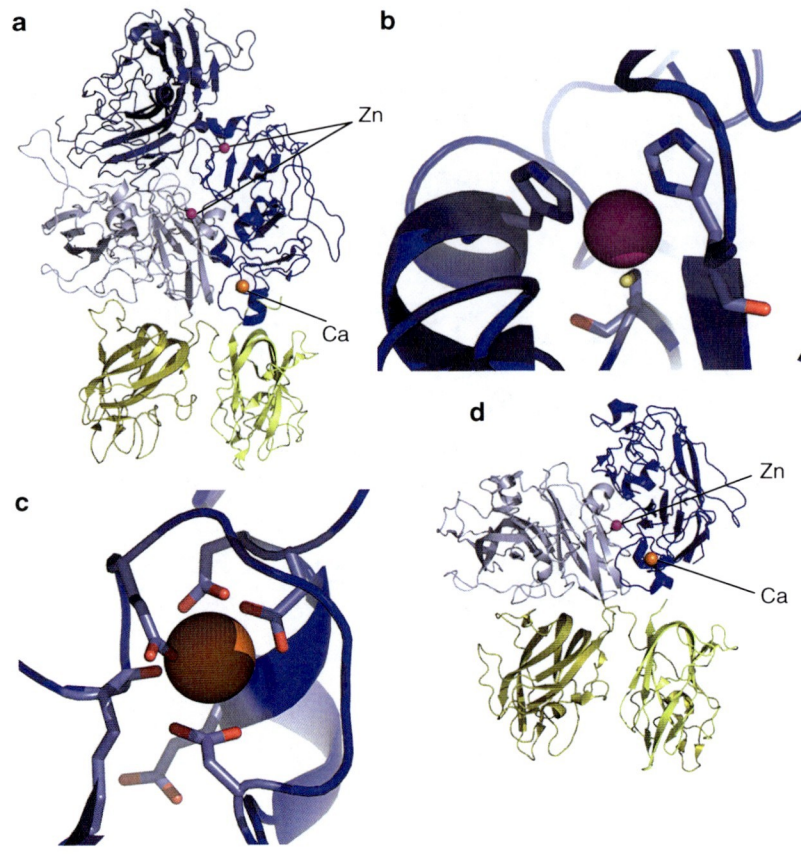

Blood Clotting Proteins, Fig. 3 *Structures of procofactors.* (**a**) The three-dimensional structure of B-domain-deleted human factor VIII showing a triangular heterotrimer of the A domains (*blue*) stacked on two smaller globular C domains (*yellow*)(Ngo et al. 2008). Factor VIII proteins bind copper ions (*magenta spheres*) and calcium ions (*orange spheres*). (**b**) The copper ion in the A1 domain of factor VIII is coordinated by two His residues and a single Cys. (**c**) The A1 domain of factor VIII contains a single calcium-binding site coordinated by Glu and Asp side chains as well as the backbone carbonyl of a Lys and a Glu. (**d**) Bovine-activated protein C–inactivated factor Va has a similar domain structure as factor VIII (Adams et al. 2004)

X-ray crystallography (Fig. 3a). The B-domain has no apparent function with regard to the procoagulant activity of factor VIII. The overall structure of the B-domain-deleted factor VIII has been described as a triangular heterotrimer of the A domains stacked on two smaller globular C domains. Each C domain projects loops containing hydrophobic and basic residues that likely contribute to the interaction of factor VIII with the membrane. The A1, A2, and A3 domains resemble a typical cupredoxin fold. Two copper ions are present within the A1 and A3 domains that are homologous to the copper-binding motif in ceruloplasmin, the major copper-carrying protein in the blood. Both of the copper ions in the A1 and A3 domains are coordinated by two His residues and a single Cys (Fig. 3b). The strategic locations of these copper ions likely maintain the structural integrity of the interdomain interface and enhance subunit affinity. The A1 domain also contains a single calcium-binding site defined by Glu and Asp side chains as well as the backbone carbonyl of a Lys and a Glu (Fig. 3c). This calcium ion may be important in the maintenance of the C2-A1 domain interface, thus promoting the active conformation of factor VIII.

Factor V

Plasma-derived factor V is synthesized in the liver and circulates as a single 330,000 Da polypeptide chain. Factor V and factor VIII share the same A1–A2–B–A3–C1–C2 multidomain architecture, with the A and C domains in the two proteins sharing approximately 40% homology. Proteolytic removal of the B-domain produces factor Va composed of a 105,000 Da heavy chain (A1–A2 domains) and a 74,000 or 71,000 Da light chain (A3–C1–C2 domains) that are held together by a single calcium ion and hydrophobic interactions. In a calcium-dependent manner, factor Va assembles with factor Xa on membrane surfaces and enhances the rate at which factor Xa cleaves prothrombin to thrombin. Factor Va is inactivated by activated protein C through cleavage adjacent to Arg 506, leading to the release of the A2 domain. The Arg506Gln

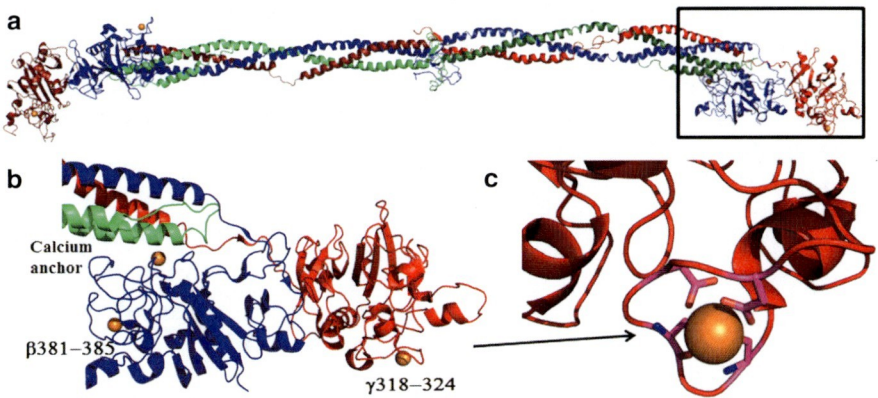

Blood Clotting Proteins, Fig. 4 *Calcium-binding sites in fibrinogen.* (**a**) Fibrinogen has a trinodular structure with calcium-binding sites in the distal globular region of the molecule (Kollman et al. 2009). (**b**) The β-chain (*blue*) and γ-chain (*red*) contain homologous calcium (*orange spheres*) binding sites. Another calcium-binding site is located at the junction where all three chains meet. This calcium ion appears to anchor the globular region of the molecule to the coiled-coil region. (**c**) The high-affinity calcium-binding site in the γ-chain has an EF-hand binding motif

mutation, known as factor V_{Leiden}, is associated with thrombosis due to inefficient inactivation of factor Va.

Only the C2 domain crystal structure has been determined for human factor V. While the multidomain structure of human factor V is yet to be determined, the crystal structure of activated protein C–inactivated bovine factor Va (Factor Vai) has revealed important details regarding factor V function (Fig. 3d). The bovine factor Vai structure reveals a domain arrangement similar to the human factor VIII structure, albeit missing the A2 domain due to APC-mediated inactivation. In contrast to factor VIII, the bovine factor Vai has a single copper ion coordinated by two His residues and an Asp residue. This copper ion is located within the buried surface between the A1 and A3 domains and likely provides additional structural stabilization of the interface but no ascertained functional role. The A1 domain also contains a high-affinity calcium-binding site. This calcium site may be critical for the packing of the A3 domain against the A1 domain (Adams et al. 2004).

Fibrinogen and Factor XIII

Fibrinogen, the most abundant plasma blood coagulation protein, has a molecular weight of 340,000 Da and consists of three pairs of nonidentical polypeptide chains, $(A\alpha,B\beta,\gamma)_2$. These chains are folded into three distinct structural regions: two distal globular regions linked by coiled-coil connectors to one central region forming a trinodular structure (Fig. 4a). Thrombin catalyzes the conversion of fibrinogen into fibrin monomer. Fibrin monomers self-assemble, polymerizing in an orderly sequence leading to the formation of fibrin strands.

Calcium ions enhance fibrin polymerization to produce thicker fibers and modulate fibrinogen susceptibility to proteolytic enzymes. Calcium-binding sites in fibrinogen have been identified in the crystal structure (Fig. 4b). The γ-chain has a high-affinity calcium-binding site located in the γ318–324 loop with amino acid sequence homology to the EF-hand in calmodulin (Fig. 4c) (Kollman et al. 2009). This calcium-binding site influences fibrinogen-mediated platelet aggregation and imparts protection to fibrinogen from proteolytic degradation. Since the γ- and β-chain polypeptides are homologous, there is a corresponding calcium-binding site located in the β381–385 loop for which function has yet to be ascertained. Another calcium ion–binding site involves two β-chain Asp residues and a γ-chain Glu residue. This site serves as an anchor between the β-globular region and the coiled-coil connector. This calcium may also impact access to tissue plasminogen activator–binding sites, a protein involved in the breakdown of clots (Weisel 2005).

Factor XIII, a plasma glycoprotein transglutaminase of 320,000 Da molecular weight, is a tetramer consisting of two catalytic A subunits and two carrier B subunits (A_2B_2). The activated form, factor XIIIa, stabilizes the fibrin clot by catalyzing covalent cross-linking between Glu and Lys residues in the fibrin molecule. In the presence of calcium, factor XIII is activated by thrombin-mediated cleavage of the activation peptides from the A subunits. The carrier subunits then dissociate from the catalytic subunits, leading to a conformational change that unmasks the catalytic site. The X-ray crystal structure of factor XIII A_2 showed a single major Ca^{2+}-binding site in each A subunit (Komaromi et al. 2011).

Cross-References

- Calcium-Binding Protein Site Types
- Calcium-Binding Proteins, Overview
- Calcium in Biological Systems
- EF-Hand Proteins
- Zinc in Hemostasis

References

Adams TE, Hockin MF, Mann KG, Everse SJ (2004) The crystal structure of activated protein C-inactivated bovine factor Va: implications for cofactor function. Proc Natl Acad Sci USA 101:8918–8923

Banner DW, D'Arcy A, Chene C, Winkler FK, Guha A, Konigsberg WH, Nemerson Y, Kirchhofer D (1996) The crystal structure of the complex of blood coagulation factor VIIa with soluble tissue factor. Nature 380:41–46

Furie B, Furie BC (2008) Mechanisms of thrombus formation. N Engl J Med 359:938–949

Huang M, Rigby AC, Morelli X, Grant MA, Huang G, Furie B, Seaton B, Furie BC (2003) Structural basis of membrane binding by Gla domains of vitamin K-dependent proteins. Nat Struct Biol 10:751–756

Huang M, Furie BC, Furie B (2004) Crystal structure of the calcium-stabilized human factor IX Gla domain bound to a conformation-specific anti-factor IX antibody. J Biol Chem 279:14338–14346

Kollman JM, Pandi L, Sawaya MR, Riley M, Doolittle RF (2009) Crystal structure of human fibrinogen. Biochemistry 48:3877–3886

Komaromi I, Bagoly Z, Muszbek L (2011) Factor XIII: novel structural and functional aspects. J Thromb Haemost 9(1):9–20

Ngo JC, Huang M, Roth DA, Furie BC, Furie B (2008) Crystal structure of human factor VIII: implications for the formation of the factor IXa-factor VIIIa complex. Structure 16:597–606

Pratt KP, Shen BW, Takeshima K, Davie EW, Fujikawa K, Stoddard BL (1999) Structure of the C2 domain of human factor VIII at 1.5 Å resolution. Nature 402:439–442

Shen BW, Spiegel PC, Chang CH, Huh JW, Lee JS, Kim J, Kim YH, Stoddard BL (2008) The tertiary structure and domain organization of coagulation factor VIII. Blood 111:1240–1247

Soriano-Garcia M, Padmanabhan K, de Vos AM, Tulinsky A (1992) The Ca^{2+} ion and membrane binding structure of the Gla domain of Ca-prothrombin fragment 1. Biochemistry 31:2554–2566

Weisel JW (2005) Fibrinogen and fibrin. Adv Protein Chem 70:247–299

Blood Platelets

▶ Nanosilver, Next-Generation Antithrombotic Agent

Blue Copper Protein

▶ Plastocyanin

Blue Copper Proteins

▶ Monocopper Blue Proteins

BOR1 – Borate Transporter Protein

▶ Boron, Biologically Active Compounds

Borane

▶ Amine-Boranes

Borano-Amine

▶ Amine-Boranes

Boric Acid Transport

▶ Atr1, Boron Exporter in Yeast

Boron and Aquaporins

Gerd Patrick Bienert and François Chaumont
Institut des Sciences de la Vie, Universite catholique de Louvain, Louvain-la-Neuve, Belgium

Synonyms

Aquaporin-mediated boron transport; Boron channel; Major intrinsic proteins and boron transport

Definition

Aquaporins or major intrinsic proteins (MIPs) are transmembrane channel proteins, which facilitate the passive and bidirectional diffusion of water and a variety of small and noncharged compounds across biological membranes. Aquaporins are found in organisms of all kingdoms of life and are present in all main subcellular membrane systems. The substrate specificity/spectra of aquaporins are highly isoform-dependent. Some plant isoforms were shown to facilitate the transmembrane diffusion of boric acid, the most frequent chemical boron species that organisms have to deal with. All organisms face the challenge to handle considerable variations in the concentration of metalloids they are exposed to, in terms of either the demand to acquire sufficient amounts for their metabolism or, conversely, the necessity to extrude them to prevent toxicity. This is achieved through homeostatic processes that require, among others, aquaporin-mediated transport across membranes at the cellular level.

Essentiality and Roles of Boron in Living Organisms

In the past years, a large number of studies have demonstrated that boron (B) is an essential or at least beneficial element for animals and humans (reviewed in Hunt 2007). Boron is crucial for vertebrate development as boron deprivation resulted in abnormal or necrotic embryonic phenotypes. Molecular mechanisms and functions underlying boron demand are unknown. Interestingly, although the molecular targets are likely to be different in animal and plant metabolism, boron seems to be highly needed at early stages of tissue differentiation (Goldbach et al. 2007).

Plant biology can be seen as the pioneering research field driving advances in the understanding of biological boron handling and roles. This is due to the fact that the essential role of boron in plants was already established more than 80 years ago (Warington 1923). An adequate nutritional boron status is essential for plants in general and for high yield and crop quality in particular, as boron deficiency causes many disadvantageous morphological, physiological, and biochemical alterations (reviewed in Goldbach et al. 2007; Marschner 2011). These defects mainly result in rapid cessation of root elongation and reduced vegetative growth and fertility.

Underlying molecular mechanisms for most of the negative effects are still a matter of speculation. The only known molecular function of boron in plants is its role in cross-linking apiose residues from rhamnogalacturonan II moieties of pectin and, therefore, sustaining cell-wall integrity and functioning (reviewed in Miwa et al. 2010).

Despite its complex chemistry and a large range of binding forms, boron is exclusively found as borates in most soils. Bioavailable boron species are solely boric acid ($B(OH)_3$) and the borate anion ($B(OH)_4^-$) which stay in chemical equilibrium ($B(OH)_3 + H_2O = B(OH)_4^- + H^+$) at a pKa of 9.25. In most soils and at physiological pH ranges in organisms, the equilibrium is greatly shifted toward boric acid. Therefore, the chemistry of boron in soils and organisms is very simple. For plants, boron is the only essential microelement which was shown to be mainly taken up as an uncharged boric acid molecule but not in an ionic form. This represents a peculiarity of plant boron management in plant nutrition (Marschner 2011).

Membrane Permeability to Boric Acid

The undissociated boric acid molecule was initially thought to cross the membrane by simple diffusion because it is among the smallest of all metalloid species. Furthermore, the amount of boron species taken up by plants in physiological uptake assays was closely correlated to boric acid concentrations in the medium (Miwa and Fujiwara 2010). However, permeability coefficients of plant membranes for boric acid were lower than predicted. For instance, the permeability coefficient of plasma membrane from squash roots was one to two orders of magnitudes smaller than the values obtained in experiments using liposomes made of different artificial lipid compositions (Dordas et al. 2000). The low permeability coefficient values for boric acid could not meet plant boron demand and uptake rates, respectively. In addition, physiological studies have suggested the existence of channel-mediated facilitated diffusion as well as energy-dependent active transport systems for boron (Miwa and Fujiwara 2010). It was hypothesized that boric acid transport occurred via aquaporins (see below), as the application of aquaporin inhibitors (mercuric chloride and phloretin) to plasma membrane vesicles of squash roots partially inhibited boron uptake (Dordas and Brown 2001). Accordingly, the expression of a PIP (plasma membrane intrinsic protein) aquaporin from maize in *Xenopus laevis* oocytes resulted in an increased boron permeability of the plasma membrane further suggesting the involvement of aquaporins in transmembrane boric acid transport in plants.

Aquaporins Are Important Players in the Uptake and Distribution of Boron in Plants

In a complete set of studies including direct uptake assays *in planta*, functional tests in heterologous expression systems, and physiological and molecular genetic studies in *Arabidopsis thaliana*, Fujiwara and coworkers solved the long-standing puzzle of boron transport in plants by independently identifying two types of transport proteins regulating the uptake and redistribution of boron: the aquaporin nodulin-26-like intrinsic proteins (NIPs) and the BOR transport proteins (Takano et al. 2002, 2006). Members of the BOR transporter family have similarities to anion exchange proteins and represent active efflux transporters for borate anions. NIPs belong to the aquaporin or MIP family (▶ Aquaporins and Transport of Metalloids). Under boron-limited conditions, the spatially coordinated action of AtNIP5;1 conducting boric acid uptake and AtBOR1 ensuring borate efflux into the xylem was shown to be essential for the flux of boron from the soil to the shoot in *A. thaliana* (Miwa and Fujiwara 2010). AtNIP5;1 mediates the diffusion of boric acid through root endodermis cells and provides the substrate for AtBOR1, which is localized in proximal plasma membrane domains of pericycle cells. Nutrient export from pericycle cells that surround the vascular tissues is part of the xylem-loading process for subsequent long-distance transport and distribution of nutrients to the shoot. Such a spatial cooperation between an import (NIP-mediated) and an export (BOR-mediated) function of boron transporters may also be expected in other tissues or cell types since the transport of this uncharged molecule cannot benefit from the plasma membrane electrochemical gradient (Miwa and Fujiwara 2010). Similar cooperative work of a passive NIP channel and an active efflux transporter (Lsi2) was also shown to be crucial to regulate the uptake and allocation of the beneficial metalloid species silicic acid in rice (*Oryza sativa*) (▶ Silicon and Aquaporins). Characterization of NIP-mediated boron transport processes has shown that other NIP isoforms including OsNIP3;1, AtNIP5;1, and AtNIP6;1 constitute passive boric acid channels playing key roles in boron transport under boron-limiting conditions (Miwa and Fujiwara 2010; Tanaka et al. 2008). Interestingly, quantitative PCR analyses showed that *OsNIP3;1*, *AtNIP5;1*, and *AtNIP6;1* are consistently upregulated under low boron supply suggesting that plants (1) perceive their boron status and (2) actively regulate NIP-mediated transport processes at the transcriptional level to ensure boron homeostasis (Miwa et al. 2010). Altogether, these data clearly support the idea that NIP aquaporins are crucial for boron uptake and allocation in both monocot and eudicot plant species.

However, boron uptake under boron-deficient conditions is not the only challenge for plants. Bioavailable soil boron concentrations are very important for plant fitness, as the range between boron deficiency and toxicity is very narrow compared to many other nutrients. Excess of boron supply, which can occur during dry periods, in irrigated fields of arid areas or in (over-) fertilized fields, was shown to rapidly lead to leaf necrosis and growth inhibition. Plants have therefore to carefully regulate boron transport processes and cope with both boron deficiency and toxicity during the growth period.

Plant tolerance mechanisms against an excess of boron in soils also seem to involve NIP aquaporins. *HvNIP2;1* gene from barley (*Hordeum vulgare*) is associated to a quantitative trait locus (QTL) conferring boron tolerance. A significantly lowered expression of *HvNIP2;1* under high-boron conditions was shown to prevent toxic boron influx into the root of a boron-tolerant barley cultivar compared to a nontolerant one (Schnurbusch et al. 2010).

NIPs can be classified into three subgroups: NIPIs, NIPIIs, and NIPIIIs based on the amino acid residue composition of the aromatic/Arginine selectivity filter of aquaporins which represents the narrowest part of the channel path (▶ Aquaporins and Transport of Metalloids). The physicochemical properties of these residues are involved in the determination of the substrate specificity of the MIP pore. Until now, all boric acid-permeable NIP aquaporins belong to either the NIPII or NIPIII subgroup (Miwa and Fujiwara 2010).

Other Aquaporins and Their Role in Plant Boron Transport

In addition to NIPs, other members of the aquaporin family might also be involved in boron transport. For instance, expression of maize ZmPIP1;1 in *X. laevis* oocytes resulted in an increased boron permeability (Dordas et al. 2000) and heterologous expression of *Solanaceae* XIPs and barley PIPs in *Saccharomyces cerevisiae* yeast mutants resulted in an increased sensitivity of yeast toward boron (Bienert et al. 2011; Fitzpatrick and Reid 2009). XIPs possess striking similarities in their sequence and selectivity with members of the boron-permeable NIPII subgroup (Bienert et al. 2011). Interestingly, XIPs are absent from genomes of boron-deficiency-sensitive eudicot plants (e.g., species belonging to the genus of *Brassica*) and the low boron demanding and boron-deficiency-insensitive gramineous species such as wheat, barley rice, and maize. When boron efficiency was studied in Arabidopsis, the identified QTLs encompassed two *TIP* aquaporin genes, namely, *AtTIP3;1* and *AtTIP4;1* (Zeng et al. 2008). Additionally, the ectopic overexpression of the tonoplast and/or mitochondrial-localized aquaporin AtTIP5;1 in Arabidopsis conferred tolerance to boron toxicity, suggesting its capacity to transport boron and/or its involvement in boron regulation in physiological conditions (Pang et al. 2010). As the expression of *AtTIP5;1* is pollen-specific, its function in boron transport has a potentially high interest as pollen germination, pollen viability, and pollen tube growth of most plants are highly sensitive to boron deficiency (Marschner 2011). Nevertheless, physiologically relevant roles of TIPs, PIPs, and XIPs in boron transport remain to be resolved.

Aquaporins and BOR Transporters from Organisms Other than Plants Are Involved in Boron Transport Processes

Knowledge on boron transport processes in organisms other than plants is scarce. Nevertheless, the discovery of AtBOR1 and AtNIP5;1 encouraged the search and characterization of homologous transport proteins in other organisms. In *S. cerevisiae* ScBor1p (a BOR transport protein) and ScFps1p (an aquaglyceroporin) seem to have similar roles as AtBOR1 and AtNIP5;1 in borate efflux and boric acid influx, respectively (Nozawa et al. 2006). HsNaBC1 in humans (a Na^+-coupled borate cotransporter) which is the mammalian homolog of AtBOR1 catalyzes the uptake of borate across the plasma membrane, coupled to the Na^+ gradient. The importance of the boron transporter HsNaBC1 is supported by the fact that *NaBC1* knocked-out cell lines stop to develop and proliferate (Park et al. 2004). It is expected that mammals possess, besides NaBC1 transporter, a high-capacity efflux system for boron as a surplus of boron in the diet leads to massive excretion of this compound via the urine. It is tempting to speculate that such massive flux is mediated by aquaglyceroporins releasing boric acid out of the cells.

Concluding Remarks

NIPs of several plant species have been demonstrated to be involved in boric acid uptake and allocation. Boron demand varies a lot between plant species. Therefore, a detailed molecular and physiological characterization of different NIP isoforms will allow a more complete understanding of the mechanisms regulating boron transport in different plant species. Furthermore, it has to be resolved whether aquaporins of organisms other than plants play a similar important role in boron transport processes as NIPs do.

Funding

This work was supported by grants from the Belgian National Fund for Scientific Research (FNRS), the Interuniversity Attraction Poles Programme–Belgian Science Policy, and the "Communauté française de Belgique–Actions de Recherches Concertées." GPB was supported by a grant from the FNRS.

Cross-References

- Arsenic and Aquaporins
- Aquaporins and Transport of Metalloids
- Selenium and Aquaporins
- Silicon and Aquaporins

References

Bienert GP, Bienert MD, Jahn TP et al (2011) Solanaceae XIPs are plasma membrane aquaporins that facilitate the transport of many uncharged substrates. Plant J 66:306–317

Dordas C, Brown PH (2001) Evidence for channel mediated transport of boric acid in squash (*Cucurbita pepo*). Plant Soil 235:95–103

Dordas C, Chrispeels MJ, Brown PH (2000) Permeability and channel-mediated transport of boric acid across membrane vesicles isolated from squash roots. Plant Physiol 124:1349–1362

Fitzpatrick KL, Reid RJ (2009) The involvement of aquaglyceroporins in transport of boron in barley roots. Plant Cell Environ 32:1357–1365

Goldbach HE, Huang L, Wimmer MA (2007) Boron functions in plants and animals: recent advances in boron research and open questions. Springer, Dordrecht

Hunt CD (2007) Dietary boron: evidence for essentiality and homeostatic control in humans and animals. Springer, Dordrecht

Marschner H (2011) Marschner's Mineral Nutrition of Higher Plants. Academic, London

Miwa K, Fujiwara T (2010) Boron transport in plants: co-ordinated regulation of transporters. Ann Bot 105:1103–1108

Miwa K, Tanaka M, Kamiya T et al (2010) Molecular mechanisms of boron transport in plants: involvement of Arabidopsis NIP5;1 and NIP6;1. Landes Bioscience-Springer, New York

Nozawa A, Takano J, Kobayashi M et al (2006) Roles of BOR1, DUR3, and FPS1 in boron transport and tolerance in *Saccharomyces cerevisiae*. FEMS Microbiol Lett 262:216–222

Pang Y, Li L, Ren F et al (2010) Overexpression of the tonoplast aquaporin AtTIP5;1 conferred tolerance to boron toxicity in Arabidopsis. J Genet Genomics 37:389–397

Park M, Li Q, Shcheynikov N et al (2004) NaBC1 is a ubiquitous electrogenic Na+ −coupled borate transporter essential for cellular boron homeostasis and cell growth and proliferation. Mol Cell 16:331–341

Schnurbusch T, Hayes J, Hrmova M et al (2010) Boron toxicity tolerance in barley through reduced expression of the multifunctional aquaporin HvNIP2;1. Plant Physiol 153:1706–1715

Takano J, Noguchi K, Yasumori M et al (2002) Arabidopsis boron transporter for xylem loading. Nature 420:337–340

Takano J, Wada M, Ludewig U et al (2006) The Arabidopsis major intrinsic protein NIP5;1 is essential for efficient boron uptake and plant development under boron limitation. Plant Cell 18:1498–1509

Tanaka M, Wallace IS, Takano J et al (2008) NIP6;1 is a boric acid channel for preferential transport of boron to growing shoot tissues in Arabidopsis. Plant Cell 20:2860–2875

Warington K (1923) The effect of boric acid and borax on the broad bean and certain other plants. Ann Bot 37:629–672

Zeng C, Han Y, Shi L et al (2008) Genetic analysis of the physiological responses to low boron stress in Arabidopsis thaliana. Plant Cell Environ 31:112–122

Boron Channel

- Boron and Aquaporins

Boron Efflux Pump

- Atr1, Boron Exporter in Yeast

Boron Stress Tolerance, YMR279c and YOR378w

Ahmet Koc and İrem Uluisik
Department of Molecular Biology and Genetics,
Izmir Institute of Technology,
Urla, İzmir, Turkey

Synonyms

ATR1 paralogs in yeast; The role of ATR1 paralogs in boron stress response

Definition

The essentiality of boron for plants has been known for about 90 years. And it is also necessary for the growth and development of various species. Boron deficiency and toxicity is associated with several defects in these organisms. Thus, regulation of the boron intake and efflux is essential. Transporters constitute an important place in this sense.

There are several genes that play a role in boron transport and tolerance. The first identified boron transporter was *Arabidopsis thaliana BOR1* and homologs of

BOR1 were found in many organisms (Takano et al. 2002). The yeast homolog of *BOR1 (YNL275W)* was the first identified gene related with boron metabolism. It is localized to the plasma membrane and has an efflux function (Takano et al. 2007). Dur3 and Fps1 were suggested to have roles in boron tolerance in yeast, but the mechanism in which these genes function is not clear (Nozawa et al. 2006). Atr1 was characterized in 2009 as the main boron exporter in yeast *Saccharomyces cerevisiae*. It is regulated strictly by the boron treatment. Both yeast Bor1 and Atr1 were shown to reduce the intracellular boron concentrations. The study also revealed that *S. cerevisiae* genome contains two *ATR1* paralogs. Bioinformatic analysis showed that the paralogs *YMR279C*, *YOR378W*, and *ATR1* have 70% similarity in their sequences (Kaya et al. 2009). Both genes belong to DHA2 family of drug:H + antiporters like Atr1 (Sa-Correia et al. 2009). Thus it is thought that these genes may play a role in boron tolerance mechanisms as well.

Deletion of either of these genes did not result in sensitivity to boron. The N-terminal regions of these genes were found to be heterogeneous. They are not controlled by the same transcription factors suggesting they probably function in response to different stresses. In addition, wild-type cells carrying *YMR279C* gene can grow in the presence of toxic amount of (150 mM) boric acid; however, the overexpression of *YOR378W* had no effect on the growth of cells. To examine the role of *YMR279C* in boron stress response, intracellular boron levels were measured and found out that *YMR279C* expressing wild-type and *ymr279cΔ* cells has 60% and 40% intracellular boron, respectively. Thus, Ymr279c has a potential to be a boron transporter that can pump out boron from the cell; however, Yor378w does not have a role in boron stress (Bozdag et al. 2011).

Cross-References

▶ Atr1, Boron Exporter in Yeast

References

Bozdag GO, Uluisik I et al (2011) Roles of ATR1 paralogs YMR279c and YOR378w in boron stress tolerance. Biochem Biophys Res Commun 409(4):748–751

Kaya A, Karakaya HC et al (2009) Identification of a novel system for boron transport: atr1 is a main boron exporter in yeast. Mol Cell Biol 29(13):3665–3674

Nozawa A, Takano J et al (2006) Roles of BOR1, DUR3, and FPS1 in boron transport and tolerance in *Saccharomyces cerevisiae*. FEMS Microbiol Lett 262(2):216–222

Sa-Correia I, dos Santos SC et al (2009) Drug:H(+) antiporters in chemical stress response in yeast. Trends Microbiol 17(1):22–31

Takano J, Noguchi K et al (2002) Arabidopsis boron transporter for xylem loading. Nature 420(6913):337–340

Takano J, Kobayashi M et al (2007) *Saccharomyces cerevisiae* Bor1p is a boron exporter and a key determinant of boron tolerance. FEMS Microbiol Lett 267(2):230–235

Boron Toxicity

▶ Atr1, Boron Exporter in Yeast

Boron Transport in Yeast

▶ Atr1, Boron Exporter in Yeast

Boron, Biologically Active Compounds

Leonid Breydo
Department of Molecular Medicine, Morsani College of Medicine, University of South Florida, Tampa, FL, USA

Synonyms

AI-2 – autoinducer 2; BOR1 – borate transporter protein; RGII – rhamnogalacturonan II

Definition

Boron-containing compounds are primarily utilized by bacteria as quorum autoinducers and by plants as components of the cell wall. Many synthetic organoboron compounds are used as enzyme inhibitors.

Boron, Biologically Active Compounds,
Fig. 1 Structure of RGII, a boron-containing polysaccharide present in the plant cell walls (**1**)

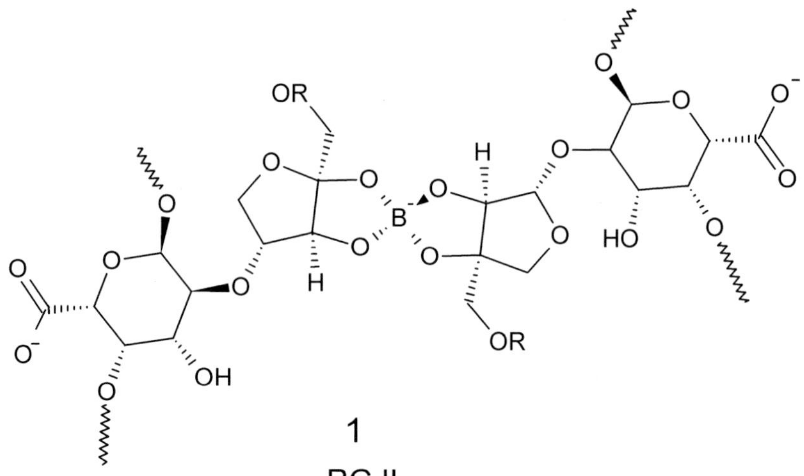

1
RG II

Introduction

Boron is a ubiquitous element in rocks, soil, and water; its average concentration ranging from 1 mg/kg in water to 100 mg/kg in rocks. Boron is electron-poor and prefers to form tetracoordinate complexes with "hard" nucleophiles. In physiological environment, it is present in +3 oxidation state, usually in the form of borate anion or borate esters. Borate anions form stable complexes with organic acids, polysaccharides, and other biopolymers. Usually borate complexes two hydroxyl groups (either a diol or a hydroxycarboxylic acid) to form a borate diester. Since boron can bind four ligands, in many cases, borate esters cross-link two organic molecules together.

Boron in Plants

Boron is a required nutrient for plants. Its main role is structural: a borate-cross-linked polysaccharide rhamnogalacturonan II (RGII) is an important component of cell walls (O'Neill et al. 2004). Rhamnogalacturonan II (RGII) is a structurally complex pectic polysaccharide. It is a 5–10 kDa polysaccharide composed primarily of D-galactose along with 11 other glycosyl residues (O'Neill et al. 2004). It was later discovered that it exists as a dimer covalently cross-linked by a borate diester (Fig. 1). Cross-linking of RGII is required for the formation of a three-dimensional pectic network in the cell walls and thus for normal plant development. Cross-linking is accelerated by divalent metal ions that are believed to bind to polysaccharides bringing them closer together. There is considerable evidence that RGII is present in cell walls of all gymnosperms and angiosperms. Interestingly, boron-cross-linked RGII is a major component of red wine (about 150 mg/l) where it complexes the majority of heavy metal ions present in wine (Pellerin and O'Neill 1998).

Boron uptake in plants is controlled by BOR1, a borate transporter (Takano et al. 2002). It is analogous to bicarbonate transporter, and its overexpression allows the plant to survive in low boron environment. Boron deficiency leads to significant problems with plant growth as the tissues of boron-deficient plants are often brittle and contain cells that do not expand normally.

Boron in Bacteria

Boron-containing compound also plays a role in quorum sensing in bacteria (Federle 2009). Quorum sensing is the process of cell-to-cell communication in bacteria. It allows bacterial populations to coordinate gene expression and increases the effectiveness of biofilm formation, antibiotic production, and other communal responses to the environment (Miller and Bassler 2001). It is accomplished via exchange of small molecules called autoinducers. These molecules are ligands for cytoplasmic or membrane-bound receptors that act as transcriptional regulators upon ligand binding. If ligand concentration is high enough, its

Boron, Biologically Active Compounds, Fig. 2 Boron-containing natural products: AI-2, an autoinducer of bacterial quorum sensing (**2**); antibiotics borophycin (**3**), boromycin (**4**), aplasmomycin (**5**), and tartrolon B (**6**)

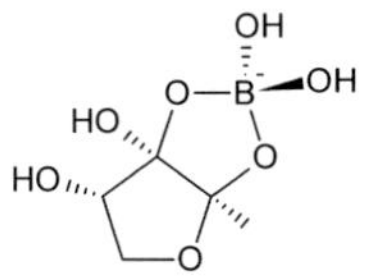

2
AI-2

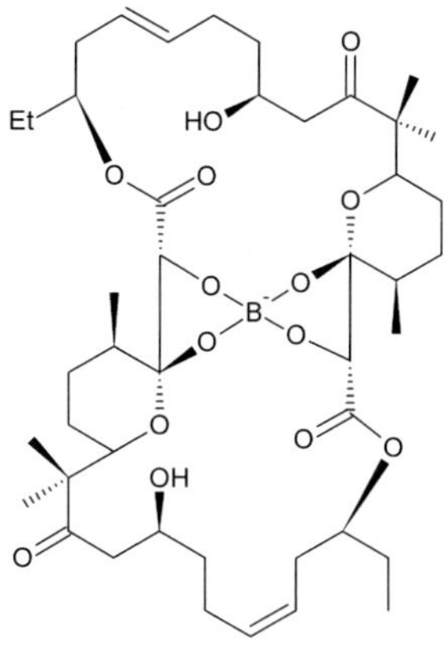

3
borophycin

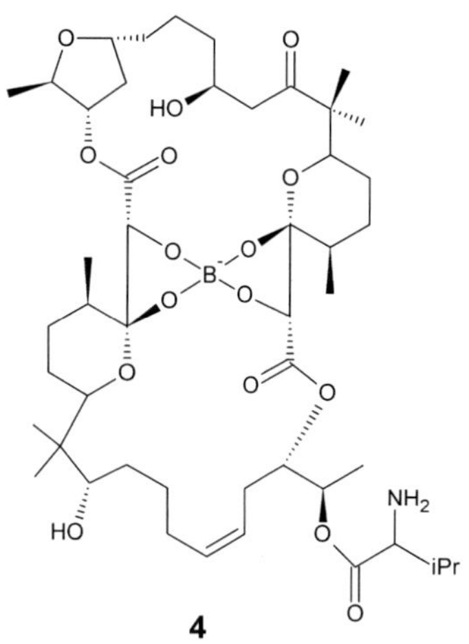

4
boromycin

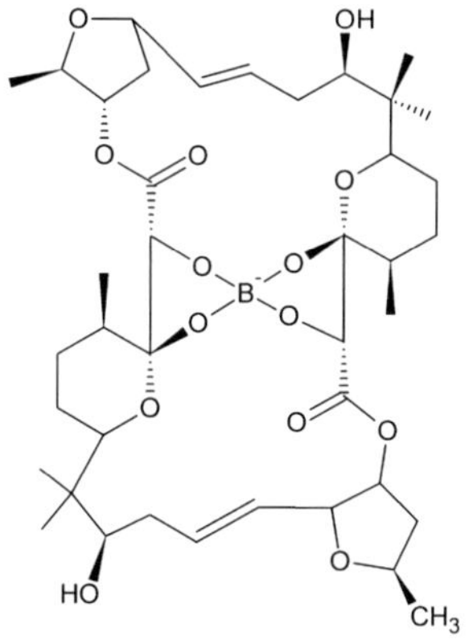

5
aplasmomycin

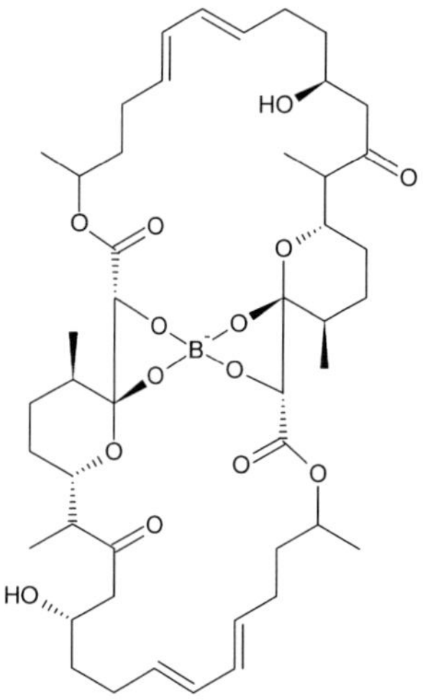

6
tartrolon B

receptor activates transcription of several genes including the one for the enzyme producing the autoinducer (Hodgkinson et al. 2007). Most autoinducers are specific to a single bacterial species while some are produced by a large number of species. AI-2 is one of those multi-species autoinducers. It actually corresponds to either of two diastereomers of 2-methyl-2,3,3,4-tetrahydroxytetrahydrofuran (THMF) that are formed by cyclization of a linear precursor. S-diastereomer of this compound exists as a borate diester while R-diastereomer is not boronated because its hydroxyl groups are in trans-configuration (Federle 2009). Borate diester is synthesized in a reaction of S-THMF with boric acid. Usage of S- versus R-diastereomer of AI-2 depends on the bacterial species. In the bacteria that use S-enantiomer of AI-2, signaling is significantly accelerated by addition of 10–100 μM boric acid (Chen et al. 2002). Upon uptake, AI-2 is phosphorylated and binds to a transcriptional repressor LsrR, releasing it from DNA and activating transcription of several genes (Xavier and Bassler 2005). Boronate ester likely stabilizes this autoinducer since in its absence, THMF is prone to isomerization, reductive elimination, and oligomerization via acetal formation (Dembitsky et al. 2011). Synthetic boron-containing analogs of AI-2 have been prepared, and some of them inhibit biofilm formation at micromolar concentrations (Dembitsky et al. 2011).

Boron, Biologically Active Compounds, Fig. 3 Boronic acids (**7, 8**) and other boron-containing enzyme inhibitors (**9**)

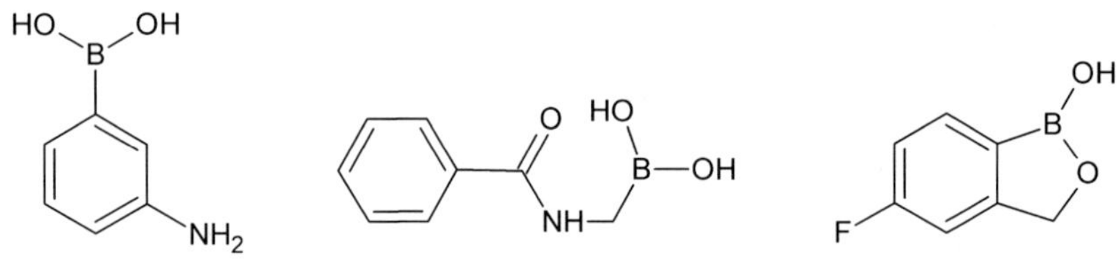

Bacteria and fungi synthesize a variety of antibiotics to compete with other microorganisms. These natural products have been the source of many commercial antibiotics and drugs against other diseases such as cancer. Since boric acid is abundant in nature, boron is incorporated in some of these antibacterial natural products (Dembitsky et al. 2011). All known boron-containing antibiotics are polyketide antibiotics that contain a boronate diester bound to four hydroxyls (Fig. 2). These antibiotics include borophycin, boromycin, aplasmomycin, tartrolon B, and several of their derivatives. These antibiotics act as ionophores increasing permeability of the cell membranes for cations. Removal of boron from aplasmomycin resulted in loss of its ionophore activity, and this activity could be restored by addition of boric acid (Chen et al. 1980). It is likely that the role of boronate diester in these antibiotics is to maintain vicinal cis-diols in a specific conformation and increase their stability.

Boron in Animals

Boric acid and most organoboron compounds are nontoxic to humans with LD_{50} at 6 g/kg. In fact, boron is an essential or at least beneficial element in animals. Presence of specific borate transporter (a homolog of BOR1 in plants) in the human proteome indicates the importance of this element. Congenital endothelial dystrophy type 2, a rare form of corneal dystrophy, is associated with mutations in borate transporter, causing loss of function of this protein (Vithana et al. 2006).

Several physiological roles for boron in animals have been proposed although its exact function is unknown. Boron is important for calcium and magnesium metabolism. For example, low boron diet leads to poor absorption of calcium and magnesium and bone abnormalities (Armstrong et al. 2000). In addition, studies have shown the importance of boron-containing compounds in embryonic development in animals with low levels of borate leading to high percentage of necrotic embryos (Fort et al. 2002). It is also possible that borate diesters of galactose derivatives perform similar functions in animal membranes as RGII performs in the plant cell walls. Specifically, it has been proposed that borate diester–cross-linked polysaccharides stabilize lipid rafts and enhance membrane binding by GPI anchors of membrane proteins (Brown et al. 2002).

Biological Activity of Synthetic Organoboron Compounds

Structure of boronic acid resembles a transition state for the ester and amide hydrolysis. Thus, boronic acids such as **7** and **8** (Fig. 3) are effective inhibitors of hydrolytic enzymes (serine proteases, β-lactamases, and others) acting as transition state analogs (Dembitsky et al. 2011). In addition, high affinity of boron for oxygen ligands enables organoboron compounds such as boroxazole AN2690 (Fig. 3) to inhibit other enzymes. Boroxazoles similar to AN2690 inhibit leucyl t-RNA synthetase by forming stable adducts with oxygen atoms in the enzyme's active site (Rock et al. 2007).

Conclusions

Chemical properties of boron (stable (+3) oxidation state and ability to form strong covalent complexes with cis-diols and other oxygen ligands) drive the biological activity of this element. Stabilization of cis-diols is crucial to boron's role in both plants and bacteria. Its role in animals is not well understood but may involve the same chemistry. Many synthetic boron-containing compounds function as effective enzyme inhibitors.

Cross-References

▶ Atr1, Boron Exporter in Yeast
▶ Boron-containing Compounds, Regulation of Therapeutic Potential
▶ Boron Stress Tolerance, YMR279c and YOR378w

References

Armstrong TA, Spears JW, Crenshaw TD, Nielsen FH (2000) Boron supplementation of a semipurified diet for weanling pigs improves feed efficiency and bone strength characteristics and alters plasma lipid metabolites. J Nutr 130:2575–2581

Brown PH, Bellaloui N, Wimmer MA, Bassil ES, Ruiz JH, Pfeffer H, Dannel F, Romheld V (2002) Boron in plant biology. Plant Biol 4:205–223

Chen TS, Chang CJ, Floss HG (1980) Biosynthesis of the boron-containing antibiotic aplasmomycin. Nuclear magnetic resonance analysis of aplasmomycin and desboroaplasmomycin. J Antibiot (Tokyo) 33:1316–1322

Chen X, Schauder S, Potier N, Van Dorsselaer A, Pelczer I, Bassler BL, Hughson FM (2002) Structural identification of a bacterial quorum-sensing signal containing boron. Nature 415:545–549

Dembitsky VM, Al Quntar AA, Srebnik M (2011) Natural and synthetic small boron-containing molecules as potential inhibitors of bacterial and fungal quorum sensing. Chem Rev 111:209–237

Federle MJ (2009) Autoinducer-2-based chemical communication in bacteria: complexities of interspecies signaling. Contrib Microbiol 16:18–32

Fort DJ, Rogers RL, McLaughlin DW, Sellers CM, Schlekat CL (2002) Impact of boron deficiency on *Xenopus laevis*: a summary of biological effects and potential biochemical roles. Biol Trace Elem Res 90:117–142

Hodgkinson JT, Welch M, Spring DR (2007) Learning the language of bacteria. ACS Chem Biol 2:715–717

Miller MB, Bassler BL (2001) Quorum sensing in bacteria. Annu Rev Microbiol 55:165–199

O'Neill MA, Ishii T, Albersheim P, Darvill AG (2004) Rhamnogalacturonan II: structure and function of a borate cross-linked cell wall pectic polysaccharide. Annu Rev Plant Biol 55:109–139

Pellerin P, O'Neill MA (1998) The interaction of the pectic polysaccharide Rhamnogalacturonan II with heavy metals and lanthanides in wines and fruit juices. Analusis 26:32–36

Rock FL, Mao W, Yaremchuk A, Tukalo M, Crepin T, Zhou H, Zhang YK, Hernandez V, Akama T, Baker SJ et al (2007) An antifungal agent inhibits an aminoacyl-tRNA synthetase by trapping tRNA in the editing site. Science 316:1759–1761

Takano J, Noguchi K, Yasumori M, Kobayashi M, Gajdos Z, Miwa K, Hayashi H, Yoneyama T, Fujiwara T (2002) Arabidopsis boron transporter for xylem loading. Nature 420:337–340

Vithana EN, Morgan P, Sundaresan P, Ebenezer ND, Tan DT, Mohamed MD, Anand S, Khine KO, Venkataraman D, Yong VH et al (2006) Mutations in sodium-borate cotransporter SLC4A11 cause recessive congenital hereditary endothelial dystrophy (CHED2). Nat Genet 38:755–757

Xavier KB, Bassler BL (2005) Regulation of uptake and processing of the quorum-sensing autoinducer AI-2 in *Escherichia coli*. J Bacteriol 187:238–248

Boron: Physical and Chemical Properties

Fathi Habashi
Department of Mining, Metallurgical, and Materials Engineering, Laval University, Quebec City, Canada

Boron is a metalloid of no useful mechanical properties but used as an alloying element in steel. It is the second hardest element after diamond and is an essential plant nutrient. Relatively large quantities of amorphous boron are used as additives in pyrotechnic mixtures, solid rocket propellant fuels, and explosives. High-purity boron (>99.99%) is used in electronics. It is used as a ppm additive for germanium and silicon to make *p*-type semiconductors. Crystalline high-purity boron is used in thermistors. Boron filaments have been developed as reinforcing material for light-weight, stiff composites for use in commercial and military aircraft recently replaced by graphite filaments. In nuclear technology thin films of boron are used in neutron counters. Boron powder dispersed in polyethylene castings is used for shielding against thermal neutrons. The isotope boron-10 has a large neutron absorption cross section and is used as a control for nuclear reactors, as a shield for nuclear radiation, and in instruments used for detecting neutrons.

Boron compounds are extensively used in the manufacture of borosilicate glasses which have low coefficient of expansion hence can resist thermal shock, known under the trade name Pyrex. The borax bead test is a historical method of qualitative analysis for metals invented by Berzelius in 1812 and is based on the solubility of metal oxides in borax glass to give a distinctive color.

Physical Properties

There are several allotropic forms of boron. The β-rhombohedral form is the thermodynamically stable

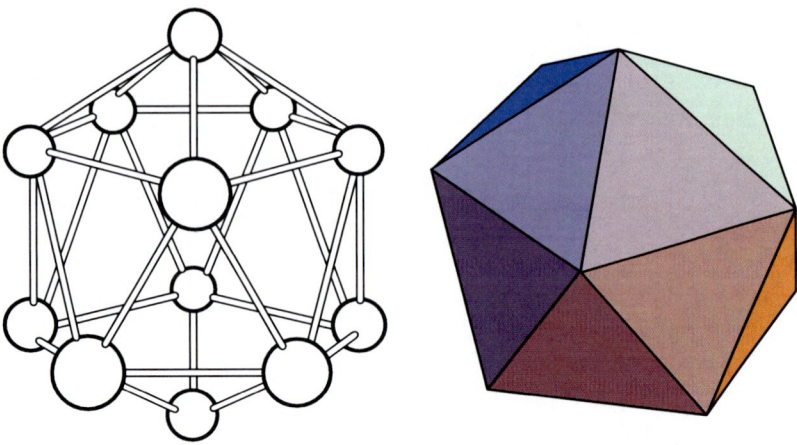

Boron: Physical and Chemical Properties, **Fig. 1** Elemental boron, B_{12}, and the icosahedron

modification at all temperatures. Amorphous boron slowly converts to the β-rhombohedral form at ≈1,200°C and to α-rhombohedral boron above 1,500°C. Any type of boron recrystallizes in β-rhombohedral structure when heated above the melting point and cooled. In the stable modification, boron forms an icosahedron, B_{12}, in which the atoms are connected together by a covalent bond (Fig. 1). An icosahedron is a polyhedron having 12 vertices, 30 edges, and 20 equivalent equilateral triangle faces.

Atomic number	5
Atomic weight	10.811
Melting point, °C	2,050 ± 50
Sublimation point, °C	2,550
Density, g/cm³	
Amorphous, at 20°C	2.3
β-rhombohedral, at 20°C	2.35
α-rhombohedral, at 20°C	2.46
Liquid at melting point	2.99
Solid at melting point	2.13
Color	
Amorphous	Brown to gray
α-rhombohedral	Red to brown
β-rhombohedral	Lustrous gray black
Hardness (Knoop), kg/mm²	
Crystallized from melt	2,390
Vapor deposited	2,690
Electrical resistivity at 300 K, Ω cm	
Amorphous	7.5×10^2
β-rhombohedral, single crystal	7×10^5
β-rhombohedral, polycrystalline	10^6–10^7
Heat capacity C_p, J K⁻¹ mol⁻¹	
Amorphous at 300 K	12.054
β-rhombohedral at 300 K	11.166
(continued)	
Solid at melting point	33.955
Liquid at melting point	39.063
Entropy S at 298 K, J K⁻¹ mol⁻¹	
Amorphous	6.548
β-rhombohedral	5.875
Enthalpy of fusion, ΔH_f, kJ/mol	50.2
Enthalpy of sublimation, ΔH_s, kJ/mol	572.7

Chemical Properties

Boric acid, $B(OH)_3$ is a weak acid. Boron oxide, B_2O_3, like SiO_2 has the ability to dissolve metallic oxides to form borates and glasses which are difficult to crystallize. Boron hydrides are gases of the composition B_nH_{n+4} and B_nH_{n+6}, are known as boranes. They are easily oxidized with considerable energy liberation and have been studied for use as rocket fuels. Boron nitride, BN, a soft white powder, has a structure like graphite and similar lubricating properties at high temperature. When heated at high temperature and under pressure it becomes very hard like diamond.

References

Baudis U et al (1997) Boron. In: Habashi F (ed) Handbook of extractive metallurgy. Wiley-VCH, Weinheim, pp 1985–2063

Habashi F (2009) Boron. Its history and its position in the periodic table. In: Konuk A et al. (eds) Fourth international boron symposium, Eskişehir, Turkey, pp 355–363

Boronated Amine

▶ Amine-Boranes

Boron-containing Compounds, Regulation of Therapeutic Potential

Ion Romulus Scorei
Department of Biochemistry, University of Craiova, Craiova, DJ, Romania

Synonyms

Cellular control mechanism for therapeutic potential of boron-containing compounds; The molecular mechanisms of the therapeutic action of boron-containing compounds

Definition

Boron is a micronutrient element necessary for the growth and development of vascular plants, marine algae and algal flagellates, diatoms, and cyanobacteria. Although boron has not been yet shown to be an essential nutrient in animal cells, more data will probably support this role in the future.

Boron compounds with therapeutic potential are the inorganic or organic compounds that contain boron as an integral part of the molecule and possess interesting pharmacological properties like hypolipidemic, anti-inflammatory, anti-osteoporosis, and antineoplastic.

Introduction

Boron (B) has been shown to possess the following characteristics: (1) It is a cell signaling molecule. (2) It is a co-factor of the enzymes it regulates. (3) It is a nonenzymatic cofactor. (4) It plays both structural and functional roles, including electron transfer, redox sensing, and structural modules. (5) It plays a role in the cytoskeleton structure (Scorei 2012).

Various types of B-containing molecules already exist and have been investigated as therapeutic agents. These molecules include B-containing analogues of natural biomolecules; the antibacterial and antimalarial agent diazaborine; antibacterial oxazaborolidines; antibacterial diphenyl borinic esters; the antifungal benzoxaboroles; and a B-N bond containing an estrogen receptor modulator (Scorei and Popa 2010).

Boron contains an empty p-orbital which makes it a strong electrophilic compound and a Lewis acid. It can readily form dative bonds with nucleophiles and thus, it transforms from an uncharged, trigonal-planar structure to an anionic, tetrahedral structure. This feature allows it to form dative bonds with nucleophiles in enzyme-active sites, providing additional binding affinity. The attractive characteristics of boron have not been let unnoticed by the pharmaceutical industry and the use of boron in pharmaceuticals has been expanding. Except for the drug Bortezomib, the majority of B-containing compounds are currently used in the cancer treatment and they belong to the Boron Neutron Capture Therapy (BNCT) class. Identification of the bacterial quorum sensor autoinducer 2 (AI2) as a B-containing stable complex (Park et al. 2005), of the transporters responsible for efficient B uptake in animal cells, and of the borate ability to inhibit many enzyme systems represents the discoveries of the boron chemistry and constitutes the basis of new drugs with boron atom included.

The Use of Boron-Containing Compounds as Potential Therapeutics

Natural Boron Compounds
Boric Acid and Borate Esters
Boric acid (BA) is an astringent, mild disinfectant, and is good for eye wash. Sodium borate is used in cold creams, eye washes, and mouth rinses.

BA is an inhibitor of peptidases, proteases, proteasomes, arginase, nitric oxide synthase, and transpeptidases (Hunt 1996). The inhibition of serine protease and dehydrogenase activities can be explained by the BA capacity to bind OH groups from NAD and serine. It has been demonstrated that BA controls the proliferation of some cancer cell types (Barranco and Eckhert 2004). The Prostatic Serum Antigen (PSA) is a serine protease and a putative target for BA (Scorei and Popa 2010). Based on the PSA inhibition, the use of BA in the chemical therapy of prostate carcinoma has been proposed (Devirian and Volpe 2003).

A high dose of BA (1–50 mM) slows down the cell replication and induces apoptosis in both melanoma cells and MDA. Thus, the inhibition of cancer cells by BA involves a diversity of cellular targets, such as direct enzymatic inhibition, apoptosis, receptor binding, and mRNA splicing. Recently, it has been experimentally demonstrated that 1 mM of BA inhibits the ZR-75-1 breast cancer cell line, but not the MCF-7 cell line (Meacham et al. 2010). The lack of BA-mediated inhibition of MCF-7 cellular growth could be caused by the presence of the "sodium-boron co-transporter (NaBC1)." This co-transporter exists on the cell surface and is able to pump out boron molecules from the cell, in exchange for Na^+ ions. This co-transporter is not present in the ZR-75-1 cells. ZR-75-1 is a nonmetastatic epithelial breast cancer cell line, which is estrogen/progesterone receptor–positive. MCF-7 is a metastatic epithelial cell line of breast cancer. These cells are positive for estrogen and progesterone receptors. If BA becomes an anticancer agent for breast cancer, these data will encourage women with increased cancer risk factors to raise their boron intake, in order to diminish the evolution of this disease (Scorei 2011).

Calcium fructoborate (CF) is a commercially marketed borate ester naturally found in fresh and dried fruits, vegetables, and herbs, as well as in wine or produced by chemical synthesis (Scorei and Popa 2010). CF is efficient in the treatment (as adjuvant) of osteoporosis and osteoarthritis (Scorei and Rotaru 2011). In addition, CF has shown inhibitory effects on MDA-MB-231 breast cancer cells. CF enters most likely the cell through a co-transport mechanism, via a sugar transporter. MDA-MB-231 is a metastatic cancer cell line and it is negative for the estrogen receptor expression. Inside cells, CF acts as an antioxidant and induces the over-expression of apoptosis-related proteins, and eventually apoptosis (Scorei and Rotaru 2011).

Boron Polyketides
Generally speaking, B-containing macrolides such as boromycin, borophycin, tartrolon B, and aplasmomycin are antibiotic borodiesters and act as ionophores (Dembitsky et al. 2002).

Boromycin is a natural bacteriocidal polyether-macrolide produced by *Streptomyces antibioticus*, with antibiotic activity against Gram-positive bacteria. It acts at the cell membrane level and affects cells by losing the intracellular potassium. It also disrupts selectively the cell cycle in some cancer cell types, making them sensitive to specific anticancer agents, albeit the mechanism of action against eukaryotic cells remains little understood. Boromycin was recently discovered to be a potent antihuman immune-deficiency virus (HIV) antibiotic. It strongly inhibits the replication of the clinically isolated HIV-1 strain and apparently blocks the release of infectious HIV particles from the cells chronically infected with HIV-1, by unknown mechanisms. Synthesis, biosynthesis, and biological activities of boromycin derivatives have been described and reviewed (Dembitsky et al. 2011).

Borophycin is a polyketide and it is extracted from the species of *Nostoc*. It was shown to have inhibitory effects on several cancer cell lines. The compound exhibits potent cytotoxicity against human epidermoid carcinoma (LoVo) and human colorectal adenocarcinoma (KB) cell lines (Scorei and Popa 2010).

Tartrolons are macrolides with a chemical structure related to boromycin and aplasmomycin. They have antiviral and antineoplastic chemotherapeutic properties. Tartrolon B acts against Gram-positive bacteria and notably, it strongly inhibits the growth of mammalian cells (mouse fibroblasts) in culture.

Aplasmomycin, secreted by *Streptomyces griseus*, has known inhibitory effects against Gram-positive bacteria and *Plasmodium berghei*. It has not yet been verified against cancer cells, but is similar in structure with tartrolons which have anticancer properties.

Synthetic Boron Compounds

Boranes are a large class of B-containing derivatives, relevant in cancer treatments. Amine-carboxyboranes are efficient antineoplastic and cytotoxic agents, with selective action against unicellular tumors and leukemia-derived solid tumors, lymphoma, sarcoma, and carcinoma. Amine-cyanoboranes and amine-carboxyboranes have previously been shown to inhibit the induced inflammation in rodents. Dicarba-closo-dodecaborane (carborane) is a novel class of androgen receptor antagonists, with a hydrophobic skeletal structure and possible antitumor activity. Amine-boranes have cytotoxic activity and are of potential use in BNCT. Trimethylamine cyanoborane (TACB) inhibits DNA and protein synthesis in Ehrlich ascite cells, gene regulation via chromatin phosphorylation and methylation.

Borinic esters are a class of boron-containing compounds that showed broad-spectrum antibacterial activity with minimum inhibitory concentrations (MIC). They have been designed and synthesized with low μ/mL range.

Boronic acids are potent and selective inhibitors for the migration and viability of the cancer cells.

From the structural point of view, boronic acids are trivalent boron-containing organic compounds that possess one alkyl substituent (i.e., a C–B bond) and two hydroxyl groups, to fill the remaining valences on the boron atom. Due to the easy interconversion of boronic acids between the neutral sp2 (trigonal planar substituted) and the anionic sp3 (tetrahedral substituted) hybridization states, the B-OH unit replaces the $C = O$ bond, at a site where an acyl group transfer takes place (Groziak 2001).

Phenylboronic acid (PBA) and diphenylboronic esters (DPBE) are the most efficient types of boronic acid derivatives, which act as serine protease inhibitors. PBA is more efficient than BA and decreases the cancer cell viability in 8 days. Non-tumorigenic cells are at least five times less sensitive to PBA, at the effective dose for cancer cells. These data suggest that PBA could be a promising cancer treatment and could be used prophylactically. PBA shows a selective inhibition of breast and prostate cancer migration in vivo and of tumor metastasis in mice (McAuley et al. 2012).

The drug Bortezomib (PS-341) is a boronic acid derivative and a proteasome inhibitor, in other words a novel target in cancer therapy. It disrupts the regulation of cell cycle and induces apoptosis. Bortezomib has been approved by the US Food and Drug Administration for the treatment of patients with chemorefractory multiple myeloma and for some forms of non-Hodgkin's lymphoma (Paramore and Frantz 2003). In cell cultures, Bortezomib induces apoptosis in both hematologic and solid tumor malignancies, including myeloma, mantle cell lymphoma, cell lung cancer, ovarian cancer, pancreatic cancer, prostate cancer, and head and neck cancers. Good correlation was seen between the Bortezomib dose, proteasome inhibition, and positive modulation of serum PSA. This indicates that Bortezomib could be used efficiently in combination with radiation or chemotherapy, for controlling androgen-independent prostate cancer. Until now, clinical experiments with Bortezomib have demonstrated only a limited activity against solid tumors when it is used as a single agent. However, Bortezomib has demonstrated its activity either as a single agent or in combination with several other cytotoxic agents, such as 5-fluorouracil, irinotecan, gemcitabine, doxorubicin, and docetaxel, or with radiation, enhancing both chemotherapy- and radiation therapy (RT)-induced apoptosis.

Benzoxaboroles – derivatives of boronic acids – were first described over 50 years ago. However, most of them have recently been investigated due to their exceptional properties and wide applications. For instance, it has been recently discovered that 5-fluoro-1,3-dihydro-1-hydroxy-2,1-benzoxaborole (AN2690) is an efficient broad-spectrum antifungal agent. The potency of molecules is believed to arise from the ability of boron atoms to form a stable adduct with the oxygen atoms of the leucyl-tRNA synthetase, effectively inhibiting the enzyme. After the discovery of the excellent antifungal activity of the 5-fluoro-substituted benzoxaborole (AN2690) against onychomycosis, a systematic investigation of the medical applications of benzoxaboroles is conducted. Some of them are currently in preclinical and clinical trials (Baker et al. 2011).

Oxazaborolidines are compounds that possess a B-N bond, and are readily obtained from an amino alcohol and a boronic acid. Nevertheless, despite their ubiquity in organic synthesis, the effect of oxazaborolidines on bacterial adhesion, biofilm formation, or any other pharmacological activity has been recently known. Several representative oxazaborolidines have been synthesized and evaluated for their antibacterial activity against *S. mutans*, which is one of the most predominant bacteria in the etiology of dental caries (Dembitsky et al. 2011).

The Molecular Mechanisms of the Boron-Containing Compounds as Potential Therapeutics

Boron compounds like BA, borates, boranes, boronic esters, boronic acids, and borate esters influence many cellular processes. A short list includes B transport, cell growth, mitogen-activated protein kinase pathway, proteasomes, and apoptosis. Regarding the identification of the genes involved, most of the research work was realized on BA and borate esters versus B transport, and also on boronic-esters versus proteasomes. The mechanisms involved are largely unknown. The ratio between borate and BA and the pH are important in controlling the borate transport (Park et al. 2005). In yeasts, it is likely that the signal transduction pathways, with roles in B exchange, B resistance, and amino acid biosynthesis, share a common activator. Such work has not been realized

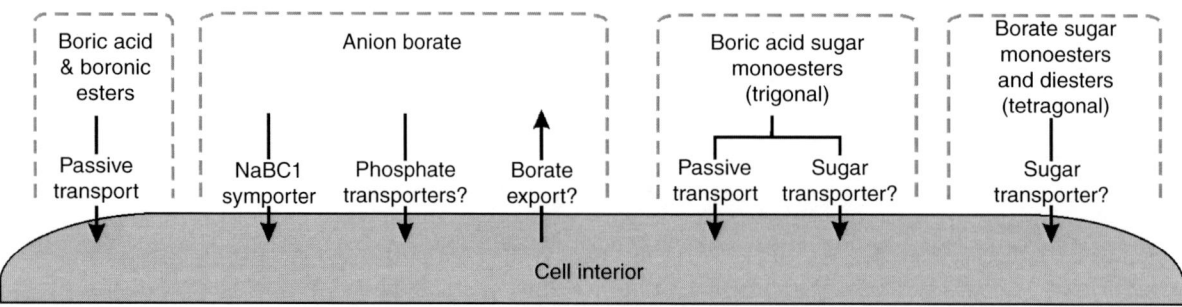

Boron-containing Compounds, Regulation of Therapeutic Potential, Fig. 1 Known and putative mechanisms for the transport of boron compounds (boric acid, borate, borate sugar esters, and boronic esters) into cells

yet on animal cells. If so, it would help decipher the interaction between B and cellular processes. Hence, the regulation of borates exporters may not be related to a redox process, but to a BA/borate speciation.

Boron as Signaling Molecule

One of the most exciting discoveries of the past years was the identification of the bacterial quorum sensor autoinducer 2 (AI2) as a B-containing stable complex (Dembitsky et al. 2011). Quorum sensing is a bacterial intercellular communication process that coordinates gene expression across cells, to initiate events such as bioluminescence, biofilm formation, production of virulence factors, etc. The B-containing autoinducer AI-2, a hydrated boric acid complex with a tetrahydroxy-dihydrofuran, was identified as ligand in an X-ray structure of the auto-sensor protein LuxP (Baker et al. 2011). Regulation of the quorum sensing as an approach to antimicrobial therapy was the subject of a recent review (Lowery et al. 2010).

Some main paradigms have been explored in developing the QS modulators as potential therapeutics: (1) interference with the signal synthase, (2) sequestration of the autoinducer, (3) antagonism of the receptor, with receptor antagonism having received the most attention to date for the discovery of QS modulators, (4) prevention of the signal secretion, and (5) inhibition of the downstream signaling events. Several arylboronic acids were found to inhibit AI-2-induced bioluminescence, with IC50 values of low or sub-micromolar concentrations.

Regulation of the Boron Transport in Cell

B is needed by cells in small amounts. In excess, BA and borate are toxic. Living cells regulate the internal concentration of borate, using specialized transporters, though the mechanism of regulation remains little understood.

Electrically neutral boron compounds can cross passively the membrane, while the borate anion is transported by aquaporins, such as APQ9, or by the specialized borate transporter, NABC1. Borate ions activate the mitogen-activated protein kinases pathway and stimulate the growth and the proliferation of human embryonic kidney 293 cells (Park et al. 2005). The B-transporter NaBCl controls the plasma borate levels in human kidney cells. The finding that low concentrations of borate activate the MAPK pathway and the knockdown of NaBC1 halted the cell growth and proliferation provide further evidence for the essential functional role of B in animal metabolism. Potential paths for the transport of B compounds in animal cells are summarized in Fig. 1.

B transporters practically limit the use of BA in cancer treatments. Some cancer cells have impaired ability to eliminate the excess of borate and are about five times more boron/apoptosis sensitive than the normal cells (Park et al. 2005).

Boron and the Enzymatic Activity

The mechanisms involving boron activity on human cells are based on the inhibition of a variety of enzymatic activities including serine proteases, cyt b5 reductase, dehydrogenase of xanthine oxidase, glutamyl transpeptidase, alkaline phosphatase, alcohol NAD-dehydrogenases, mRNA splicing and cell division, but also on receptor binding mimicry and on the induction of apoptosis. Borate forms complex between amino and hydroxy groups in proteins and targeting residues of lysine, glutamine, serine, hystidine, and proline. They also bind polyhydroxy compounds including sugars, such as mannitol, xylitol, sorbitol, glucose, fructose, and ribose from NAD and FAD; phenols, such as catechol and pyrogallol; and α-hydroxy acids, such as 2-hydroxyisobutyric acid, salicylic acid, and cis-2-hydroxycyclopentanecarboxyric acid. The effect of borate as phosphate substituent might be also common, but it was less studied. Borate stabilizes the architecture and activity of alkaline phosphatase, protecting it from the oxidative stress. Boron was also shown to increase the resistance of heme-containing proteins, such as cyt c and metmyoglobin, to thermal stress, by increasing thermal stability to inactivation and resistance to denaturizing. These effects might be also due to boron protection of the 3D structure. Interestingly, cyt b is regulated by the boron concentration and inhibited when boron is above or below the optimal concentrations.

This fact has therapeutic implications, as boron can be used to stabilize the peroxide-sensitive heme proteins.

Boron and Proteasome Inhibition

Proteasomes are large protein complexes from eukaryotes, responsible for recycling ubiquitinated proteins. Proteasome alterations were associated to several diseases, such as cardiac dysfunction, cataract, neurodegenerative disorders, cachexia, and rheumatoid diseases. They have not been linked yet with cancer development. Several anti-apoptotic and proliferative signaling pathways require proteasomal activity. In the pro-oncogenic NF-κB pathway, which is activated and prevalent in the regulation of many tumor types, NF-κB proteins are kept in the inactive state by IκBs inhibitors. In order to remove this inhibition, IκBs have to be phosphorylated, polyubiquitylated, and then recycled by proteasome. The proper functioning of the proteasome complex seems to require boron within a specific concentration range. Less or more boron will inhibit the activity of proteasomes (Scorei 2012). In high concentration, boric acid and boronic acid slow down the carcinogenic progression, because they block the IκBs degradation, which in turn downregulates the NF-κB signaling. Bortezomib inhibits proteasomes through the formation of complexes with their active site. The relationship between apoptosis and the Bortezomib effects on cells is rather complex. The over-expression of the anti-apoptotic protein Bcl-2 in H460 cells does not affect the proteasomal activity, but decreases the effect of Bortezomib on apoptosis.

Boron-Containing Compounds as Inhibitors of Serine Proteases

Serine proteases, a large and functionally diverse class of proteolytic enzymes, are prominent therapeutic targets, due to their involvement in a host of physiological processes. They catalyze the peptide bond cleavage by acylation and deacylation of the active site serine residue, in a sequence that involves two tetrahedral intermediates. Peptide derivatives with electron-deficient ketones, aldehydes, boronic acids, and phosphonylating agents have been devised as analogues of the second tetrahedral intermediate, with their selectivity among the various proteases related to the substrate specificity. The expression of many of these proteases is thought to be linked with the pathogenicity of Gram-negative bacteria. This requires further studies to obtain more profound concepts. It has been shown that these enzymes facilitate the bacterial colonization of the skin and mucous membranes. Boronic acids are a very appealing class of serine proteases inhibitors.

Boron and Vitamin D3

Epidemiological studies have shown an inverse correlation between the exposure to solar radiation and breast cancer incidence and mortality (Scorei 2011). This was suggested to be linked with Vitamin D2 production. Vitamin D3 (calcitriol), a derivative of D2, has an important role in regulating the prostatic cell growth, with demonstrated effects on the prostate cancer cell line LNCaP31. Vitamin D3 arrests the cell cycle, induces apoptosis, and inhibits metastases. It also inhibits the proliferation of the prostate cancer cells. Its activity of tumor inhibition might be due to the induction of cyclin-dependent kinase inhibitor p21 and G1-G0 cell cycle arrest. This may explain the regression of the cancer cell growth in vitamin D3–treated rats. Vitamin D3 also initiates its own inactivation because it is able to induce the expression of a protein CYP24 that initiates the vitamin D3 catabolism, which is a mechanism present in the cells. It was shown that boron, in general, and fructoborate, in particular, increase the intracellular concentration of vitamin D3 (Scorei 2011). This effect might occur due to the fact that boron up-regulates the 25-hydroxylation step or suppresses the vitamin D3 catabolic pathway. Boron readily forms covalent complexes with cis-vicinal dihydroxy compounds and it is reasonable to assume that boron forms complexes with

24,25-dihydroxyvitamin D, the final product of the 25-OH-D reaction with 24-hydroxylase. This complex acts as a competitive inhibitor for the 24-hydroxylase reaction or as a downregulator for this enzyme. Boron may be an inhibitor for microsomal enzymes (24-hydroxylase and estradiol hydroxylases) that catalyze the insertion of the hydroxyl group vicinal to the existing hydroxyl groups in steroids. Consequently, the combinations between vitamin D3 and boron may become widespread in prostate cancer therapies.

Borate: Phosphate Similarities

A negative correlation has been reported to exist between the phosphate concentration and borate, in various types of cells from plants and in normal versus osteoporotic bones (Nielsen 2008). Borate was shown to enhance the phosphorylation. In human cells, it affects living cells via a mediator, putatively TNF-alpha, in which transduction signal involves a cascade of phosphorylations. Although many similarities exist in structure and activity between borate and phosphate, the borate: phosphate substitution and its effect was relatively little studied. Phosphate esters are important in cellular-energetic, biochemical activation, signal transduction, and conformational switching. The borate:phosphate similarity, combined with borate's ability to spontaneously esterify the hydroxyl groups, suggests that phosphate ester recognition sites on proteins might exhibit significant affinity for nonenzymatically formed borate esters. In normal cells, the borate:phosphate competition is little because phosphate is thousands of times more abundant and in some complexes, such as with cytidine-$2',3'$ and RNAse, the affinity of phosphate is higher than that of borate. Will borate increase the effects of borate: phosphate similarity in phosphate-starved treatment? These results are consistent with the recent reports suggesting that in situ formation of borate esters, which mimic the corresponding phosphate esters, supports the enzyme catalysis (Gabel and London 2008).

Boron Esters as Anti-inflammatory Agents

The relationship between inflammation and cancer is realized through the elaboration of cytokines and growth factors, favoring the cancer cell growth, the induction of the COX-2, a protein that controls the synthesis of prostaglandins linked with tumor proliferation, and the generation of mutagenic reactive chemical species of oxygen and nitrogen (Nielsen and Meacham 2011). Cytokines produced during inflammation increase the expression of 5-lipoxygenase (5-LOX), leading to metabolites driving cancer development. Inhibiting 5-LOX metabolites triggers apoptosis in prostate cancer cells. Boron's anti-inflammatory activity may occur via the suppression of serine proteases released by inflammation-activated white blood cells, inhibition of leukotriene synthesis, reduction of reactive oxygen species generated during neutrophils respiratory burst, and suppression of T-cell activity and antibody concentrations. Boron inhibits the synthesis of arachidonic acid (AA)–derived eicosanoids, a class of pro-inflammatory prostaglandins. High boron intake leads to boron incorporation in membrane phospholipids, partially substituting AA-derived eicosanoids and increasing the abundance of the omega-3-derived eicosapentaenoic acid. The treatment with calcium fructoborate of the LPS-stimulated RAW264.7 macrophages inhibited the synthesis of the cytokines IL-1β and IL-6 and released the nitric oxide (NO). IL-1b, IL-6, and NO are inflammation mediators and researchers have proposed the use of CF as an anti-inflammatory agent (Scorei and Rotaru 2011).

Future Directions

Normal and Cancer Cell Boron Toxicity: A Novel Avenue in the Fight Against Cancer

The recommended daily dose of boron in humans is between 0.14 and 0.28 mg per day Kg body weight (bw). The 50% lethal dose of B as boric acid is of 2,660 mg per Kg bw. It is relatively close to that of the table salt (3,000 mg per Kg bw). For these reasons, B is not considered to be a toxic element (Baker et al. 2011). The daily doses of borate between 2.5 and 24.8 mg per kg bw, used in the past to treat epilepsy, were shown to have nonlethal side effects such as alopecia, and reversible side effects such as dermatitis, anorexia, and indigestion. In cell cultures, all cancer cells died at B concentrations varying from 1 to 50 mM boric acid. Non-cancer cells were shown to be about five times more resistant to B. No bibliographic evidence for a correlation between B exposure and carcinogenesis has been found, and no mutagenic effects have yet been reported for BA.

Based on the apoptosis measurements, some cancer cells were shown to be more sensitive to B than normal cells (Scorei and Popa 2010). The BA sensitivity varies among different cancer cell lines and has shown that the sensitivity of the cultured cancer cells was negative correlated with the expression of the NaBC1 borate transporter. This indicates that apart from its borate import function, NaBC1 might be also used to regulate the level of the intracellular borate through export. These works were done in the range of 1–50 mM BA. Based on the fact that about 90% of boron is eliminated from the human body within 22–24 h, the above presented exposure treatments led to cancer cell apoptosis, corresponding to an approximate daily intake of B between 4.4 and 51.4 mg per Kg bw. This exposure is below the B toxic levels (see above) and is near the B exposure levels used in the past to treat epilepsy. For these reasons, B is a promising avenue for inducing apoptosis in some forms of cancer. B treatments are proposed to be the most efficient against cancer cell lines, showing an under-expression of NaBC1, when B is administered as nontoxic compounds such as boron sugar esters (BSE) or as BA. The BSE would be particularly more efficient against cancer cells that, unlike normal cells, also show an over-expression of the sugar transporters, glucose and fructose channels. Thus, cancer cells are exposed to an increased risk in the presence of B that enters masked, as a sugar derivative (Scorei and Rotaru 2011).

Boron and the Fight Against Osteoporosis

The trace element boron, previously thought to have no nutritional value, might play a key role in the fight against osteoporosis, the bone-thinning disease. B impacts steroid hormone metabolism in humans, affecting the levels of estrogens and testosterone. It has been hypothesized that B interacts with steroid hormones by facilitating the hydroxylation reactions and possibly by acting in some manner to protect steroid hormones from rapid degradation. Nielsen et al. reported that boron supplementation of 3 mg per day decreases urinary Ca and P excretion in humans. A significant increase in the concentration of plasma steroid hormones has also been demonstrated in rats and humans. The synthesis of vitamin D seems necessary, as well. It is believed that optimum boron supplementation regulates the catabolic enzymatic hydroxylation. Given all these effects of boron, it is not surprising that it is beneficial for the optimum calcium metabolism. This is very useful in the prevention of bone loss and generally in some disorders of unknown etiology such as osteoporosis. The latter exhibits disturbed major mineral metabolism and an impact on osteoarthritis by increasing the synthesis of corticosteroids. Recent results have shown that additional boron has a positive effect on the ostrich tibia growth and development, but a high dosage of boron has a negative effect. According to the present data, a boron supplementation of 200 mg/L in water represents the optimal dosage for bone development in ostrich chicks. The dietary boron supplements can increase the serum content of boron in osteoporotic rats to stimulate the bone formation and to inhibit the bone resorption, producing an obviously therapeutic effect against osteoporosis (Hosmane 2011).

Thus, boron supplements might be useful in the treatment of osteoporosis and in the maintenance of healthy women in the future.

Cross-References

▶ Amine-Boranes
▶ Atr1, Boron Exporter in Yeast
▶ Boron and Aquaporins
▶ Boron Stress Tolerance, YMR279c and YOR378w
▶ Boron, Biologically Active Compounds

References

Baker SJ, Tomshob JW, Benkovicc SJ (2011) Boron-containing inhibitors of synthetases. Chem Soc Rev 40:4279–4285
Barranco WT, Eckhert CD (2004) Boric acid inhibits human prostate cancer cell proliferation. Cancer Lett 216(1):21–29
Dembitsky VM, Smoum R, Al-Quntar AAA, Ali HA, Pergament I, Srebnik M (2002) Natural occurrence of boron-containing compounds in plants, algae and microorganisms. Plant Sci 163:931–942
Dembitsky VM, Al-Quntar AAA, Srebnik M (2011) Natural and synthetic small boron-containing molecules as potential inhibitors of bacterial and fungal quorum sensing. Chem Rev 111:209–237
Devirian T, Volpe S (2003) The physiological effects of dietary boron. Crit Rev Food Sci Nutr 43:219–223
Gabel SA, London RE (2008) Ternary borate-nucleoside complex stabilization by Ribonuclease A demonstrates phosphate mimicry. J Biol Inorg Chem 13(2):207–217
Groziak MP (2001) Boron therapeutics on the horizon. Am J Ther 8:321–328
Hosmane NS (ed) (2011) Boron science: new technologies and applications. CRC Press/Northern Illinois University, Dekalb
Hunt CD (1996) Biochemical effects of physiological amounts of dietary boron. J Trace Elem Exp Med 9(4):185–213

Lowery CA, Salzameda NT, Sawada D, Kaufmann GF, Janda KD (2010) Medicinal chemistry as a conduit for the modulation of quorum sensing. J Med Chem 53(21):7467–7489

McAuley EM, Bradke TA, Plopper GE (2012) Phenylboronic acid is a more potent inhibitor than boric acid of key signaling networks involved in cancer cell migration. Cell Adh Migr 5:382–386

Meacham S, Karakas S, Wallace A, Altun F (2010) Boron in human health: evidence for dietary recommendations and public policies. Open Miner Process J 3:36–53

Nielsen FH (2008) Is boron nutritionally relevant? Nutr Rev 66:183–191

Nielsen F, Meacham S (2011) Growing evidence for human health benefits of boron. J Evid Based Complement Alternat Med. doi:10.1177/2156587211407638

Paramore A, Frantz S (2003) Bortezomib. Nat Rev Drug Discov 2:611–612

Park M, Li Q, Shcheynikov N, Muallen S, Zeng W (2005) Borate transport and cell growth and proliferation: not only in plants. Cell Cycle 4(1):24–26

Scorei R (2011) Boron compounds in the breast cancer cells chemoprevention and chemotherapy. In: Gunduz E, Gunduz M (eds) Breast cancer – current and alternative therapeutic modalities. InTech, Rijeca, pp 91–114

Scorei R (2012) Is boron a prebiotic element? A mini-review of the essentiality of boron for the appearance of life on Earth. Orig Life Evol Biosph 42(1):3–17

Scorei R, Popa R (2010) Boron-containing compounds as preventive and chemotherapeutic agents for cancer. Anti-Cancer Agents Med Chem 10:346–351

Scorei RI, Rotaru P (2011) Calcium fructoborate – potential anti-inflammatory agent. Biol Trace Elem Res 143(3):1223–1238

Botano-Remediation

▶ Lead and Phytoremediation

BSA-Coated AgNPs

▶ Colloidal Silver Nanoparticles and Bovine Serum Albumin

Burn Dressing

▶ Silver, Burn Wound Sepsis and Healing

Burn Wound Infection

▶ Silver, Burn Wound Sepsis and Healing

Burn Wound Reepithelialization

▶ Silver, Burn Wound Sepsis and Healing

Burn Wound Treatment

▶ Silver, Burn Wound Sepsis and Healing

C

C2 Domain

▶ C2 Domain Proteins
▶ Calcium-Binding Proteins, Overview

C2 Domain Proteins

Thomas A. Leonard
Max F. Perutz Laboratories, Vienna, Austria

Synonyms

2nd conserved domain of protein kinase C; C2 domain; Calcium-binding region; CBR; Endocytosis; Exocytosis; Intracellular signaling; LH2 domain; Lipoxygenase homology domain; Membrane trafficking; Second messenger signaling; Signal transduction; Vesicle fusion

Definition

C2 domain proteins are proteins containing the second conserved domain of protein kinase C. The C2 domain is an all-beta domain, consisting of a beta-sandwich of two antiparallel four-stranded beta sheets. C2 domains are found in a wide range of both enzymatic and non-enzymatic proteins involved in signal transduction and membrane trafficking.

Introduction

Comprising approximately 130 amino acids, the C2 domain was first identified as a conserved domain of the calcium-dependent protein kinases C (PKC). 233 C2 domains are predicted in 127 human proteins, making the C2 domain, after the pleckstrin homology (PH) domain, the second most abundant lipid-binding domain found in eukaryotic proteins. Since the original identification of the C2 domain as the calcium sensor in PKC, both Ca^{2+}-binding and non-Ca^{2+}-binding C2 domains have been characterized in a wide variety of lipid signal transduction enzymes and membrane-trafficking proteins. By sensing the lipid microenvironments within cells, C2 domains, coupled to a variety of effector domains, spatially and temporally regulate both lipid second messenger signaling and membrane fusion events.

To date, there are structures available for 31 C2 domains from 24 proteins; the C2 fold is a β-sandwich of two four-stranded antiparallel β-sheets (Fig. 1). Three loops at the top of the domain and four at the bottom connect the β-strands. Topologically, the C2 domain comes in two flavors that differ in the connectivity of their β-strands: By fusing the N- and C-termini and cutting the loop between strands β1 and β2, type I topology can be converted into type II. A third C2 topology, called the LH2 domain, will be discussed later in more detail. Despite adopting two different topologies, the C2 domains of synaptotagmin (C2A; type I) and phospholipase C-δ1 (PLC-δ1) (type II) differ by a root mean square deviation of only 1.4 Å. It is, as yet, unclear why C2 domains exist in two

V.N. Uversky et al. (eds.), *Encyclopedia of Metalloproteins*, DOI 10.1007/978-1-4614-1533-6,
© Springer Science+Business Media New York 2013

C2 Domain Proteins,

Fig. 1 *C2 domain structure and topology.* The structures of type I (I, synaptotagmin), type II (II, PKCε), and type III (III, *Clostridium perfringens* alpha toxin LH2 domain) topology C2 domains are illustrated in the *upper panels*. The beta sandwich fold is highlighted by the coloring of the two antiparallel β-sheets in *blue* and *orange*. Strand β1 of the type I topology C2 domain and the equivalent β-strands in type II and type III C2 domains are colored in *red* to orient the viewer. The Ca^{2+}-binding regions (CBRs) are shown in *green*, together with the bound Ca^{2+} ions, represented as *spheres*. The corresponding topology diagrams (*middle panels*) illustrate the connectivity of the β-strands. The strand numbers, colors, and annotation of the topology diagrams correspond to the structural representations shown in the *upper panels*. Numbers in *circles* reflect the CBRs. The topologically distinct CBR3 of the LH2 domain is highlighted with *yellow circles*. Below each topology diagram and structure is a list of representative proteins containing the respective topology C2 domain

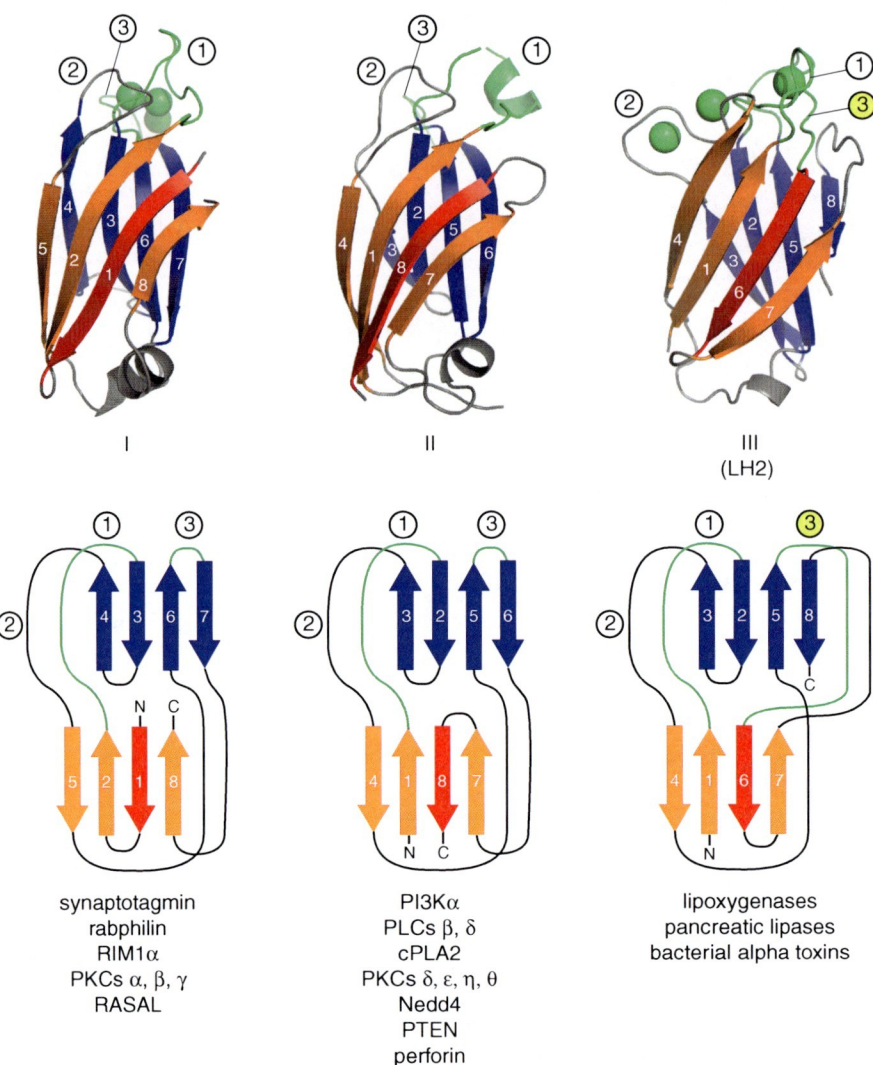

distinct topologies, but it has been speculated that the topology could influence the relative orientation of a C2 domain with respect to its neighboring domains; in type I C2 domains (synaptotagmin, PKCβII), the N- and C-termini are on the top surface of the domain, whereas in type II C2 domains (PLC-δ1, cPLA2), the termini are at the bottom. Despite a high degree of structural conservation in the core β-sandwich of the C2 domain, there is low sequence conservation between the C2 domains of different proteins. The type I C2 domains of synaptotagmin and PKCβII exhibit only 30% sequence identity over 126 equivalent residues, while the type II C2 domains of PLC-δ1 and PLA2 are only 18% sequence identical (Nalefski and Falke 1996; Rizo and Sudhof 1998).

Ca^{2+}-Binding C2 Domains

Intracellular signaling elicited by external stimuli is often mediated by an increase in cytosolic Ca^{2+} concentration. Release of intracellular stores of Ca^{2+} in response to agonist activation of G protein–coupled receptors or growth factor activation of receptor tyrosine kinases recruits C2 domain–containing proteins to the membrane. The recruitment of effectors to the membrane by small molecules, which diffuse freely in the cytosol, enables signals originating at the membrane to be propagated in a rapid and synchronous manner. The involvement of C2 domains in metal ion binding was first postulated on the basis that the calcium-independent PKCs apparently lacked this domain (in fact, they do contain a non-Ca^{2+}-binding C2 domain with an alternate topology, but this was not discovered until later). Subsequent studies on PKCs and other C2 domain–containing proteins such as synaptotagmins and cytosolic phospholipase A2 (cPLA2) identified the C2 domain as being responsible for Ca^{2+}-dependent membrane binding. Ca^{2+} binding is mediated exclusively by residues in the loops connecting the β-strands at the top of the domain (e.g., PKCβII, Fig. 1 I); these loops were thus designated the Ca^{2+}-binding regions (CBRs), although this designation is somewhat misleading as a number of C2 domains do not bind calcium (e.g., PKCε, Fig. 1 II). Of those that do, however, both C2 domain topologies mediate binding to calcium in identical ways, since the circular permutation that converts the two topologies does not affect the CBRs. The Ca^{2+} binding sites are primarily formed by aspartate side chains that serve as bidentate ligands for two or three Ca^{2+} ions (Nalefski and Falke 1996; Rizo and Sudhof 1998). The intrinsic affinities for Ca^{2+} at each of the binding sites vary dramatically between C2 domains and are up to three orders of magnitude higher in the presence of phospholipids. Half-maximal binding of C2 domains to lipid vesicles typically occurs at 5-50 μM; Ca^{2+} has been shown to simultaneously increase the association rate constant (k_a) of C2 domains with membranes, and decrease the dissociation rate constant, (k_d). Electrostatic interactions mediated by Ca^{2+} primarily accelerate the rate of association with anionic membranes, while penetration of the hydrophobic core of the membrane has been proposed as the primary mechanism by which dissociation from the membrane is slowed. Ca^{2+} ions have been observed in four distinct positions in isolated C2 domain structures, but in the absence of structural information on lipid binding for the majority of these domains, it is unclear whether all four sites are capable of playing functionally equivalent roles. However, the structural arrangement of the Ca^{2+}-binding motifs provides an unambiguous explanation for the cooperativity that is observed in Ca^{2+} binding. Hill coefficients of 2 have been reported for the isolated C2A domains of the DOC2β (double C2 domain) C2A domain and the C2 domain of PKCβ, while synaptotagmin 1, which binds 3 Ca^{2+} ions, exhibits a Hill coefficient of 3. The affinity for Ca^{2+} at site III in synaptotagmin 1 is considerably lower than the respective affinities of sites I and II, and calcium is not observed at this position in the C2 domains of PKCs α and βII, cPLA2, and PLCδ1. In synaptotagmin I, the coordination sphere of the bound Ca^{2+} ion is incomplete, explaining the low affinity of this site. It has been speculated that anionic phospholipids may fill unsatisfied coordination sites at site III, increasing the affinity for Ca^{2+}. Ca^{2+} is observed at site I in cPLA2 and PLCδ1, but not at sites II and III. Instead, a second binding site, site IV, is created by the juxtaposition of two conserved aspartates in CBR1. While equivalent aspartates exist in CBR1 of synaptotagmin and PKCs α and βII, Ca^{2+} is not observed at this site. Structurally, there is no impedance to Ca^{2+} binding at site IV, and variations in Ca^{2+} occupancy may be due to its observation in the absence of membranes. Only lanthanum (La^{3+}) was observed at site II of PLCδ1, despite an identical arrangement of the Ca^{2+}-binding residues to that found in synaptotagmin 1. It seems likely, however, that Ca^{2+} is bound at site II in the presence of phospholipids. In cPLA2, conservative substitutions of calcium-chelating aspartates in CBR3 probably reduce the affinity for Ca^{2+} at site III, though since the binding of Ca^{2+} is cooperative and since site II is not occupied in the absence of phospholipids, it is conceivable that all of these C2 domains contain a total of four Ca^{2+}-binding sites. The structures of Ca^{2+}-free and Ca^{2+}-bound C2 domains show that Ca^{2+} does not induce a substantial conformational change, suggesting that the C2 domain acts as a calcium sensor in the cell. NMR studies have shown that Ca^{2+} induces a stabilization and reduction in conformational flexibility in the CBRs. In this way, Ca^{2+} binding by the C2 domain differs markedly from another class of Ca^{2+} effector domains, EF-hands, in which binding of a single Ca^{2+} ion to a contiguous helix-turn-helix motif induces a conformational change that exposes hydrophobic surfaces. Ca^{2+} causes an increase in the thermal denaturation temperature and confers resistance to proteolytic degradation of the C2A domain of synaptotagmin I. Ca^{2+} binding contributes to membrane binding of C2 domains in two distinct ways: first, Ca^{2+} dramatically alters the electrostatic landscape of the C2 domain surface that favors interaction with anionic membranes; secondly, Ca^{2+} provides a stereospecific bridge between the C2 domain and anionic phospholipids. The bridge model is supported by the crystal structure of the C2 domain of PKCα in complex with Ca^{2+} and a short chain phosphatidylserine (PS) molecule. Ca^{2+} binding has been postulated to induce local conformational changes that could perturb inter-domain interactions in C2 domain–containing proteins, but such conformational changes have only been observed in a splice variant of the C2A domain of the active zone protein piccolo, which contains a unique nine-amino acid insertion in the loop opposite the Ca^{2+}-binding loops. It remains to be seen whether this phenomenon is a property of C2 domains in general or a characteristic peculiar to piccolo (Cho and Stahelin 2006).

C2 domains exhibit a diversity of membrane-recognition motifs, which has drawn comparison with the complementarity determining regions (CDRs) of immunoglobulins. The similarity does not end there, either, since the immunoglobulin fold is the same β-sandwich of two antiparallel β-sheets. Three variable loops between β-strands in the immunoglobulin domain are responsible for antigen recognition in a manner analogous to the Ca^{2+}- and lipid-binding CBRs of C2 domains. Indeed, the β-sandwich scaffold appears to have been utilized for the evolution of an enormous diversity of recognition modules with capacities for lipid binding (C2 domains), antigen recognition (immunoglobulins), and protein-protein interaction.

Lipid Selectivity of C2 Domains

Lipid selectivity in the presence of Ca^{2+} is determined by the residues of the CBRs; C2 domains containing cationic residues in the CBRs prefer anionic lipids over zwitterionic ones, while those containing aliphatic and aromatic residues in their CBRs (cPLA2 and 5-lipoxygenase) strongly favor neutral phosphatidylcholine (PC) membranes over anionic ones. Penetration of hydrophobic side chains into PC membranes may be favored because the desolvation penalty is less than that for penetration into anionic membranes. Stereospecific binding of L-α-PS by PKCα and PLC-δ1 is mediated by two Ca^{2+} ions and depends on residues in the CBRs that specifically recognize the serine head group of PS. The C2A domain of JFC1, a membrane-trafficking protein, has been reported to bind 3′-phosphoinositides both in vivo and in vitro (Cho and Stahelin 2006).

In addition to Ca^{2+}-dependent lipid binding, Ca^{2+}-independent phospholipid binding is highlighted by the C2B domains of synaptotagmins II and IV, which bind soluble inositol polyphosphates. Multiple structures of C2 domains have contained ordered phosphate or sulfate ions in the vicinity of a patch of basic residues, suggesting that the domain might be able to associate electrostatically, but nonspecifically, with anionic membranes. A large number of C2 domains contain this cationic patch on the concave surface of the β-sandwich, which has been designated as the β-groove. Although the size and electrostatic potential of this patch varies widely among C2 domains, its functional significance was highlighted by the recent observation that this constitutes a second phospholipid-binding site in the C2 domain of PKCα; the phosphates of the phosphatidylinositol-4,5-bisphosphate (PIP_2) head group are coordinated by a cluster of conserved lysine residues. However, despite structural validation of PIP_2 binding to the β-groove of PKCα, there are conflicting reports as to the specificity of this interaction. In the synaptotagmins, it is believed that the β-groove is key to their fusion activity. In addition to the synaptotagmins and PKCα, Ca^{2+}-independent lipid binding of C2 domains through their β-grooves has been reported for cPLA2, class II PI3Ks, and PKCs θ and ε. It is likely that these C2 domains engage multiple membrane components simultaneously, and that their membrane-bound orientations are a product of these interactions. The diversity in membrane and lipid recognition, both Ca^{2+} dependent and independent, of C2 domains suggests a complexity of membrane targeting that likely has well-defined signaling consequences (Lemmon 2008). Modulating the ability of C2 domains to bind phospholipids by posttranslational modification can also regulate the membrane targeting of C2 domains. WNK1 phosphorylates synaptotagmin 2 within its C2 domain, increasing the amount of Ca^{2+} required for membrane binding. Mutations in WNK1 have been linked to a heritable form of hypertension, pseudohypoaldosteronism. Whether posttranslational modification of C2 domains is a universal mechanism employed by cells is yet to be determined.

Membrane-Bound Orientation of C2 Domains

The orientation of C2 domains with respect to the membrane is dependent on a number of factors, not least the nature and specificity of the interaction between the CBRs and phospholipid. Dual phospholipid-binding sites in some C2 domains and C2-protein interactions are also expected to play significant roles. Monolayer penetration studies of PKCα and cPLA2 C2 domains show that electrostatic interactions between cationic residues in the CBRs of PKCα and anionic phospholipid head groups maintain PKCα at the membrane surface, whereas hydrophobic residues in the CBRs of cPLA2 actually penetrate PC membranes. Electron paramagnetic resonance (EPR) and X-ray reflectivity analyses have also confirmed that the depth of membrane penetration from the phosphate group varies between the C2 domains of cPLA2, PKCα, and synaptotagmin I. The orientation of the PKCα C2 domain takes into account its ability to interact simultaneously with PS and PIP_2, but, noting that the β-groove is found in many other C2 domains, this orientation is not expected to be unique to PKCα (Cho and Stahelin 2006).

C2 Domain–Containing Proteins

C2 domains contain neither catalytic activity nor the capacity to undergo gross conformational rearrangements in response to ligand binding. Furthermore, C2 domains have never been found to constitute individual proteins, consistent with the notion that they are sensor domains. The flow of information in a system dictates that sensors must talk to effectors such that a physiological response can be elicited. C2 domains, therefore, are of much greater significance within the context of their full-length proteins. C2 domains are found in enzymes generally involved in signal transduction (PLC, PKC, lipoxygenase, PI3K, phosphatase and tensin homolog (PTEN), ubiquitin ligases) as well as in membrane-trafficking proteins without catalytic activity (synaptotagmins, rabphilin, regulating synaptic membrane exocytosis protein (RIM), Munc-13) (Fig. 2). Despite a wealth of structural data on isolated C2 domains, there are relatively few structures of full-length, C2 domain–containing proteins.

Catalytic C2 Domain–Containing Proteins
C2 domains are found in proteins with phospholipase, ubiquitin ligase, lipid kinase and phosphatase, protein kinase and phosphatase, and lipid peroxidizing activities. Structures of a number of these holoenzymes provide a framework for a discussion of the role of their C2 domains. The structure of the original C2 domain–containing protein, PKC, was recently published, but the C2 domain adopts two positions in the crystal lattice that are the result of lattice packing interactions and are not deemed physiologically relevant. The Ca^{2+}-dependent C2 domains of the conventional PKCs bind PS with high specificity and affinity, but do not measurably penetrate the hydrophobic layer. Conventional and novel PKCs contain type I and II C2 domains, respectively, perhaps with consequences for the orientation of their catalytic domains with respect to the membrane.

The type II C2 domain of PI3Kα makes a number of intramolecular contacts with both the p110α and p85α subunits. The putative membrane-binding CBRs contact the helical domain of p85α, while a 22-amino acid insertion in the loop connecting strands β7 and β8 mediates extensive intramolecular contacts with the p110α kinase domain. The PI3Kα C2 domain does not contain a cationic β-groove. Removal of the p85α nSH2 domain from inhibitory constraints with the kinase domain is required for activation, but it is unclear whether this is also coupled to displacement of the C2 domain from its intramolecular contacts. However, somatic mutations in the interface between the p110α C2 domain and the helical domain of p85α, as well as between the C2 domain and the nSH2 domain of p85α, have been detected in a number of diverse tumor types; PI3KCA, the gene encoding the catalytic subunit of PI3Kα, is one of the most frequently mutated oncogenes in human cancer (Huang et al. 2007).

PLCβ and δ1 contain type II C2 domains that are sandwiched between the lipase and EF-hand domains in the full-length proteins. The interface buries 1106 Å, or 40%, of C2 domain surface area, partially occluding the β-groove. However, neither of the PLC C2 domains contain a cationic patch in the β-groove and the CBRs are free to engage the membrane. While the architecture of the two holoenzymes is highly conserved, the principal differences are found between the C2 domains. PLCδ-1 C2 binds PS in a Ca^{2+}-dependent manner, whereas the C2 domain of PLCβ interacts with activated $Gα_q$ subunits of heterotrimeric G proteins with nanomolar affinity. Binding of GTP-bound $Gα_q$ results in PLCβ activation at the membrane. Membrane association of the $Gα_q$ subunit via

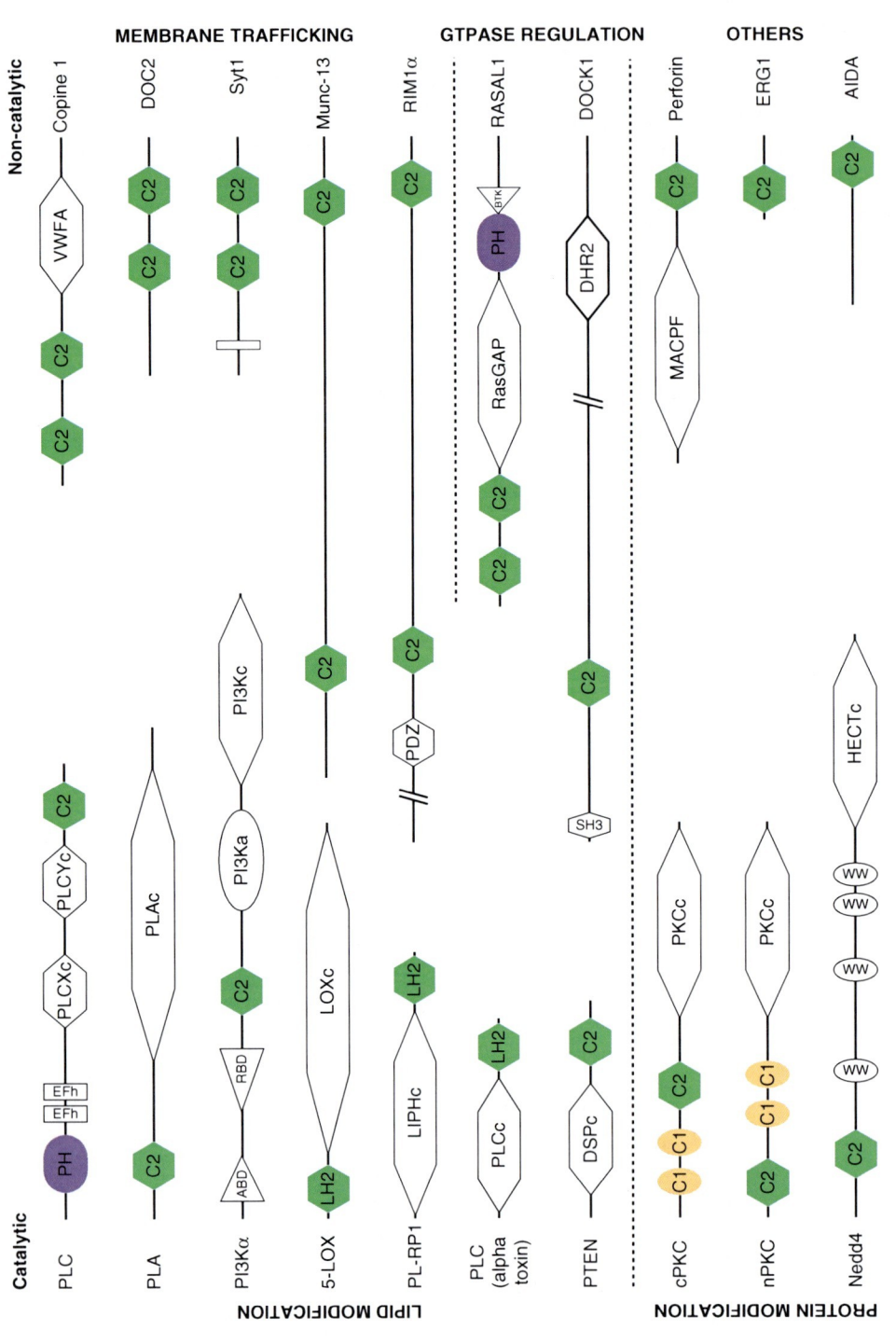

C2 Domain Proteins, Fig. 2 *C2 domain-containing proteins.* On the *left* of the figure are C2 domain-containing proteins containing effector domains with catalytic activity, grouped into those that catalyze lipid modifications and those that catalyze protein modifications. (PLC = phospholipase C, PLA = phospholipase A, PI3Kα = phosphoinositide 3-kinase alpha, 5-LOX = 5-lipoxygenase, PL-RP1 = pancreatic lipase-related protein 1, PTEN = phosphatase and tensin homolog, cPKC = conventional protein kinase C, nPKC = novel protein kinase C, Nedd4 = neural precursor cell expressed developmentally downregulated protein 4). On the *right* of the figure are C2 domain-containing proteins with non-catalytic effector domains, grouped into those that regulate membrane trafficking, those involved in GTPase regulation, and others. (DOC2 = double C2 domain, Syt1 = synaptotagmin 1, Munc-13 = mammalian unc-13 homolog, RIM1α = regulating synaptic membrane exocytosis protein 1 alpha, RASAL1 = RasGAP-activating-like protein 1, DOCK1 = dedicator of cytokinesis 1, ERG1 = elicitor responsive gene 1, AIDA = axin interaction dorsal-associated protein). C2 domains are colored *green;* other membrane-targeting domains are colored in *purple* (PH domains) and *yellow* (C1 domains)

prenylation of its carboxy terminus may compensate for the apparent lack of membrane binding by the C2 domain of PLCβ. The C2 domain of PLCs has been proposed to help orient the catalytic domain with respect to the membrane in order to facilitate hydrolysis of PIP$_2$. Like the C2 domain of PI3Kα, a loop between strands β7 and β8 mediates a number of crucial intramolecular contacts with both the lipase and PH domains. The type I C2 domains of PKCα and βII superimpose well with that of PLCδ1, suggesting that perhaps the C-terminal α-helix that follows strand β8 (which is topologically equivalent to the β7-β8 loop in type II C2 domains) plays a similar role in mediating intramolecular contacts in conventional PKCs. Despite diverse modes of activation, including by protein-protein interactions, the basal activity of all PLCs is low relative to their maximal activation; it is likely, therefore, that inter-domain conformational changes, which may or may not involve dissociation of the C2 domain from intramolecular contacts, play an important role in PLC activation (Rhee 2001).

In the dual specificity phosphatase, PTEN, the C2 domain and phosphatase domain interact across an extensive interface. Both the C2 domain and the phosphatase domain are necessary for membrane recruitment. The C2 domain has been proposed to play a dual role in targeting PTEN to the plasma membrane and assisting a productive orientation of the phosphatase domain at the membrane surface. As such, truncation of the C-terminus of the C2 domain results in a complete loss of phosphatase activity. The type II C2 domain can be superimposed on the C2 domain of cPLA2 with a r.m.s.d. of just 1.9 Å over 85 Cα atoms. However, the ligands for Ca^{2+} binding are absent in PTEN, in which membrane binding appears to be mediated by electrostatic interactions between the cationic surfaces of both the C2 and phosphatase domains and hydrophobic residues in the CBR3 loop; equivalent hydrophobic residues in the CBR3 loop are known to insert into the lipid bilayer in the C2 domain of synaptotagmin. The basic surface of the C2 domain and the CBR3 loop are on the same face as the phosphatase active site, consistent with a role in orienting the phosphatase domain with respect to the membrane and its lipid substrate, phosphatidylinositol-3,4,5-phosphate (PIP$_3$). Mutation of PTEN is a common event in about 50% of gliobastomas, endometrial and prostate carcinomas, and melanomas. Tumor mutations map evenly to both the phosphatase and C2 domains and reduce or eliminate PTEN's growth suppression activity (Leslie and Downes 2002).

The Nedd4 E3 ubiquitin ligases contain an N-terminal Ca^{2+}-binding type II topology C2 domain, superimposable on the C2 domains of Munc-13, PKCs ε and η, and synaptotagmin. Ca^{2+} binding to the C2 domain of Nedd4 promotes its interaction with acidic lipids. In addition to its role in membrane binding, the Nedd4 C2 domain interacts with the SH2 domain of Grb10, which recruits Nedd4 to IGFR1 (insulin-like growth factor receptor 1) for ubiquitin-mediated degradation. Ca^{2+} binding activates Nedd4 by displacing the C2 domain from autoinhibitory interactions.

The lipoxygenase family of lipid-peroxidizing enzymes metabolizes arachidonic acid to produce both pro-inflammatory leukotrienes and anti-inflammatory lipoxins. Lipoxygenases comprise a C2-like N-terminal LH2 (lipoxygenase homology) domain (Fig. 1, III) and a C-terminal, predominantly α-helical, catalytic domain containing a non-heme iron (Allard and Brock 2005). The LH2 domains of 5-, 8-, and 12-lipoxygenases, like many C2 domains, mediate Ca^{2+}-dependent association with the nuclear membrane. Three Ca^{2+} ions are observed bound to the LH2 domain of 8-lipoxygenase, chelated by conserved aspartate residues; however, due to the topological change in CBR3 that distinguishes the LH2 domain from the type I and type II topology C2 domains (Fig. 1, yellow circles), the sites are structurally distinct from those in synaptotagmin, PKCs, and PLCs. Mutation of the Ca^{2+}-ligating residues results in both impaired membrane binding and Ca^{2+}-dependent enzyme activity. Mutation of surface-exposed hydrophobic residues in the CBRs of 5- and 8-lipoxygenase also impairs membrane binding. Since the Ca^{2+}-binding loops are found on the same side as the putative entrance into the substrate-binding pocket, the LH2 domain likely orients the catalytic domain to facilitate fatty acid acquisition from the membrane. The LH2 domain is a third permutation of the canonical C2 domain that is generated from C2 type II topology by connecting strand β5 to strand β8 and β8 to β7 (Fig. 1, II and III). The C2 β-sandwich fold of two four-stranded antiparallel β-sheets is maintained by breaking and rejoining strands on the same side of the domain, but in contrast to type I and II C2 domains, the LH2 domain starts and ends in different β-sheets.

This has topological consequences for the connection of LH2 domains to their various effector domains. There are 19 human proteins predicted to contain LH2 domains, of which only lipoxygenase homology domain 1 contains multiple LH2 domains (seven). Lipoxygenases and pancreatic lipases are the principal members of this family (Chahinian et al. 2000; Allard and Brock 2005). However, the LH2 domain is also found in bacterial alpha toxins, critical virulence factors that have phospholipase activity. The *Clostridium perfringens* alpha toxin LH2 domain has significant structural homology with the LH2 domains of nonbacterial lipoxygenases and pancreatic lipases, as well as with the C2 domain of synaptotagmin. Moreover, eukaryotic phospholipases also contain membrane-targeting C2 domains. Membrane targeting of bacterial alpha toxins, like that of the lipoxygenases and pancreatic lipases, is mediated by Ca^{2+}-dependent phospholipid binding. The Ca^{2+}-binding loops of the LH2 domain are oriented on the same face of the alpha toxin as the catalytic site, supporting the notion that LH2 domains, just like their related C2 domains, assist the productive orientation of the catalytic domain at the interface with the membrane. Indeed, the LH2 domain of the bacterial alpha toxins is required for their hemolytic activities (Popoff and Bouvet 2009). The LH2 domain of pancreatic lipase is also oriented with respect to the catalytic domain such that the putative membrane-binding CBRs are on the same face of the molecule as the active site (Chahinian et al. 2000). While lipoxygenases, pancreatic lipases, and bacterial alpha toxin C2 domains all appear to facilitate a productive orientation of their catalytic domains with respect to their membrane-bound substrates, the inter-domain contacts are different in each family. The LH2 domains of the three proteins superimpose with r.m.s.d.s of less than 2.8 Å over 110 matching $C\alpha$ atoms, but the effector domains of each protein contact three distinct surfaces of the LH2 domain, of which two only partially overlap.

Non-Catalytic C2 Domain Containing Proteins

Of the non-enzymatic C2 domain–containing proteins, the remainder are primarily involved in membrane trafficking. The prototypical C2 domain–containing proteins in trafficking are the approximately 400-amino acid synaptotagmins, which contain tandem type I C2A and C2B domains C-terminal of a transmembrane α-helix. There are 15 members of the mammalian synaptotagmin family, of which eight bind Ca^{2+}. Ca^{2+} binding to synaptotagmin 1 triggers the displacement of complexin from "primed" SNARE complexes, resulting in the high speed and synchronicity of Ca^{2+}-mediated neurotransmitter release. In addition to Ca^{2+}-dependent membrane binding to PS, the C2 domains of synaptotagmins bind phosphoinositides via their cationic β-grooves. Synaptotagmins have been shown to lower the energy needed for membrane fusion; membrane insertion of residues in the CBRs of the two C2 domains has been postulated to induce positive membrane curvature under the SNARE complex ring, causing buckling of the plasma membrane towards the vesicle, therefore reducing the distance between the two membranes and reducing the energy barrier for hemifusion (Sudhof 2004). Other multiple C2-containing monotopic transmembrane proteins involved in membrane trafficking include the ferlins, tricalbins, and MCTPs.

RIM, Munc-13, rabphilin, and DOC2 are non-transmembrane soluble C2-containing proteins that regulate membrane fusion. The active zone protein RIM1α is a 190 kDa protein containing two type I C2 domains widely separated in primary sequence. Munc-13 is a multidomain protein found in presynaptic active zones where it mediates the priming of synaptic vesicle fusion, although the mechanism is as yet unclear. Munc-13 contains three C2 domains also widely separated in primary sequence. The type II C2B domain of Munc-13 contains an unusual amphipathic helix in one of the Ca^{2+}-binding loops that confers on it a preference for phosphoinositides. Mutation of Ca^{2+} binding to this domain does not block neurotransmitter release in response to isolated action potentials, but depresses the release evoked by action potential trains. Rabphilin is a regulator of synaptic vesicle recovery from use-dependent depression. Like the synaptotagmins, rabphilin contains tandem type I C2A and C2B domains C-terminal of the rabphilin effector domain, a binding domain for the small G protein Rab3A (Nalefski and Falke 1996; Sudhof 2004). The DOC2 family of synaptic proteins also contains tandem C2A and C2B domains C-terminal of the DOC2 effector MID (Munc-13 interacting domain). The C2A domain of DOC2B binds membranes containing PS in a Ca^{2+}-dependent manner. The copine family of proteins, found to be associated with secretory vesicles, also contains tandem Ca^{2+}-binding C2 domains in front of a C-terminal von Willebrand factor type A (VWFA) domain.

It is not fully understood why many of the proteins involved in membrane trafficking contain multiple C2 domains; in cell signaling proteins, multiple C2 domains are the exception rather than the rule. Multiple C2 domains may increase the avidity of the host protein for the membrane. However, the tandem C2 domains of synaptotagmin, rabphilin, and DOC2 are nonequivalent sensor modules, suggesting that they have evolved functional specialization. As such, the C2 domains of synaptotagmin have different structural features and exhibit different lipid selectivities. Since the membrane fusion reaction is dependent on bringing the vesicle and target membranes into close juxtaposition, an attractive hypothesis is that multiple C2-containing proteins evolved to bridge two separate membranes. Ca^{2+}-mediated synchronous exocytosis of neurotransmitter presumably requires simultaneous local remodeling of both the vesicle and presynaptic membranes, which would be consistent with distinct roles for the C2A and C2B domains of synaptotagmin in membrane binding.

Natural killer cells and cytotoxic T lymphocytes are critical for the elimination of virus-infected and neoplastic cells. The pore-forming protein perforin is necessary for the delivery of pro-apoptotic granzymes into the cytosol of the target cell. Perforin contains a C-terminal Ca^{2+}-binding type II topology C2 domain that targets perforin to membranes; the C2 domain is important for the activity of perforin: at low Ca^{2+} concentrations, perforin is not activated, but upon granule exocytosis, elevated Ca^{2+} and neutral pH promote membrane binding. The recently determined structure of the perforin monomer shows a limited inter-domain interface between the bottom of the C2 domain and the C-terminal and EGF domains, but, significantly, a number of mutations linked to familial hemophagocytic lymphohistiocytosis (FHL) map to this region (Pipkin and Lieberman 2007).

Other non-catalytic C2-containing proteins include the DOCK (dedicator of cytokinesis) family of Rho family guanine nucleotide exchange factors (GEFs) and the RASAL (RasGAP-activating-like) family of Ras GTPase–activating proteins. The surface loops (CBRs) of the DOCK type II C2 domain create a basic pocket for the recognition of the PIP_3 head group. A 40 amino acid segment between strands β7 and β8 forms a helical scaffold that, by comparison to the β7–β8 loop insertions of PI3Kα and PLCβ/δ C2 domains, could mediate intramolecular or protein-protein interactions.

The tandem C2 domains of the RASAL proteins are homologous to the C2 domains of synaptotagmin and PKCβII; the C2 domains bind to PS- and PC-containing membranes in a Ca^{2+}-dependent manner in vitro. RASAL responds to oscillations in intracellular Ca^{2+} concentration via its type I C2 domains to regulate the activation state of Ras. The AIDA (axin interaction dorsal-associated) family of cytoskeleton-interacting proteins contains a C2 domain of type II topology. Putative C2 domains have additionally been detected bioinformatically in ciliary basal body–associated proteins (annotated in the PFAM database as B9 domains), as well as in several microfilament and endocytosis-related proteins. Finally, C2 domains are also found in plant proteins; rice ERG1 (elicitor-responsive gene) contains a single type II topology C2 domain that translocates to the plasma membrane in response to fungal pathogen elicitors. ERG1 C2 is structurally most similar to the C2B domain of Munc-13 and binds to PS-containing vesicles in a Ca^{2+}-dependent manner (Zhang and Aravind 2010).

Subcellular Targeting of C2 Domain–Containing Proteins

Subcellular targeting studies of GFP-tagged C2 domains have shown that their behaviors in vivo largely mimic their in vitro binding properties. The C2 domains of PKCα and PLCδ1, which preferentially bind anionic membranes containing PS, translocate to the plasma membrane in response to Ca^{2+} while the C2 domains of cPLA2 and 5-lipoxygenase, which bind neutral and PC-containing membranes, translocate preferentially to the PC-rich perinuclear region. The calcium concentrations reported for the half-maximal binding of many C2 domains to membranes in vitro are in the micromolar to millimolar range, yet cytosolic calcium concentrations are typically submicromolar in both resting and stimulated cells. The C2 domains of conventional PKCs, cPLA2, and RASAL all respond linearly to cellular calcium oscillations, suggesting that translocation is a function of calcium concentration in the cell. However, membrane residence is not simply a function of Ca^{2+} concentration considering that all C2 domain–containing proteins are coupled to effector domains, many of which exert their catalytic function on membrane-bound substrates (PI3K, PLC, cPLA2, lipoxygenase) and potentially interact with other proteins at the membrane. Furthermore, many proteins contain membrane-binding modules in addition to C2 domains, including PKC and Munc-13 (C1 domains), PLC (PH domain), and rabphilin (FYVE domain) (Fig. 2), which extend the duration of their residence at cellular membranes. It is likely, therefore, that apparent in vitro affinities are artificially lower due to a higher k_d than is physiologically the case in vivo, rather than any change in k_a.

Conclusions

Since the discovery, 23 years ago, that the second conserved domain (C2) of PKC was a calcium sensor, biochemical, cell biological, and structural studies of these domains and their host proteins have contributed enormously to efforts aimed at understanding the processes of signal transduction and membrane trafficking. The C2 domain has been widely described as having one of two topologies, though the LH2 domain can be classified as a third topology of the C2 domain. Indeed, the LH2 domains of the lipoxygenases and pancreatic lipases have been widely discussed in the context of C2 domain properties. The C2 domain is regarded as an exclusively eukaryotic domain, and a SMART search of bacterial genomes fails to find C2 domain–containing proteins; in fact, there is at least one example of a bacterial C2 domain–containing protein in the alpha toxins. C2 domains, like their immunoglobulin counterparts, have an enormous repertoire of ligand-binding capabilities, whether those ligands are calcium, phospholipids, or other proteins. Within the context of signal transduction enzymes, C2 domains recruit the proteins to the appropriate cellular membrane while simultaneously ensuring the productive orientation of the catalytic domain. In at least a subset of these enzymes, calcium and phospholipid binding is also coupled to the relief of autoinhibitory intramolecular interactions between the C2 domain and the remainder of the protein. Multiple C2 domain–containing proteins, on the other hand, appear to have employed the C2 domain to regulate membrane fusion. One of the most fundamental processes in life, membrane fusion is critical to intracellular trafficking pathways, synaptic transmission, and fertilization. Inherently complex and very tightly regulated, the mechanisms governing membrane fusion are still unclear. While it is evident that the C2 domain plays a significant role, much more work is needed to define the contributions of individual C2 domain–containing proteins and, within that framework, the contributions of each C2 domain.

Cross-References

▶ Bacterial Calcium Binding Proteins
▶ Calcium in Biological Systems
▶ Calcium Ion Selectivity in Biological Systems
▶ Calcium, Neuronal Sensor Proteins
▶ Calcium-Binding Proteins, Overview
▶ Calcium-Binding Protein Site Types
▶ Lipases
▶ Phosphatidylinositol 3-kinases

References

Allard JB, Brock TG (2005) Structural organization of the regulatory domain of human 5-lipoxygenase. Curr Protein Pept Sci 6(2):125–131

Chahinian H, Sias B et al (2000) The C-terminal domain of pancreatic lipase: functional and structural analogies with c2 domains. Curr Protein Pept Sci 1(1):91–103

Cho W, Stahelin RV (2006) Membrane binding and subcellular targeting of C2 domains. Biochim Biophys Acta 1761(8):838–849

Huang CH, Mandelker D et al (2007) The structure of a human p110alpha/p85alpha complex elucidates the effects of oncogenic PI3Kalpha mutations. Science 318(5857):1744–1748

Lemmon MA (2008) Membrane recognition by phospholipid-binding domains. Nat Rev Mol Cell Biol 9(2):99–111

Leslie NR, Downes CP (2002) PTEN: the down side of PI 3-kinase signalling. Cell Signal 14(4):285–295

Nalefski EA, Falke JJ (1996) The C2 domain calcium-binding motif: structural and functional diversity. Protein Sci 5(12):2375–2390

Pipkin ME, Lieberman J (2007) Delivering the kiss of death: progress on understanding how perforin works. Curr Opin Immunol 19(3):301–308

Popoff MR, Bouvet P (2009) Clostridial toxins. Future Microbiol 4(8):1021–1064

Rhee SG (2001) Regulation of phosphoinositide-specific phospholipase C. Annu Rev Biochem 70:281–312

Rizo J, Sudhof TC (1998) C2-domains, structure and function of a universal Ca^{2+} –binding domain. J Biol Chem 273(26):15879–15882

Sudhof TC (2004) The synaptic vesicle cycle. Annu Rev Neurosci 27:509–547

Zhang D, Aravind L (2010) Identification of novel families and classification of the C2 domain superfamily elucidate the origin and evolution of membrane targeting activities in eukaryotes. Gene 469(1–2):18–30

Ca(II)

▶ Calcium-Binding Protein Site Types

Ca^{2+}

▶ Calcium and Mitochondrion
▶ Calcium-Binding Protein Site Types

Ca^{2+} Pump

▶ Calcium ATPase and Beryllium Fluoride

Ca^{2+}-/Calmodulin-Dependent Protein Serine/Threonine Phosphatase

▶ Calcineurin

Ca^{2+}-Activated K^+ Channels

▶ Barium Binding to EF-Hand Proteins and Potassium Channels

Ca^{2+}-Binding Protein

▶ EF-hand Proteins and Magnesium

Ca^{2+}-Pump

▶ Calcium ATPases

Ca^{2+}-Translocating ATPase

▶ Calcium ATPases

Cadherin Family Members in Embryonic Development, Morphogenesis, and Disease

▶ Cadherins

Cadherins

Paco Hulpiau[1] and Frans van Roy[1,2]
[1]Department for Molecular Biomedical Research, VIB, Ghent, Belgium
[2]Department of Biomedical Molecular Biology, Ghent University, Ghent, Belgium

Synonyms

Cadherin family members in embryonic development, morphogenesis, and disease; Cadherins, protocadherins, and cadherin-related proteins; Calcium-dependent cell–cell adhesion proteins; Signaling by cadherins and cadherin-associated proteins; Structures and functions of extracellular calcium-binding cadherin repeats

Definitions

Genuine *cadherins* are metazoa-specific, calcium-dependent transmembrane (TM) proteins that generally mediate cell–cell adhesion or cell–cell recognition and are characterized by the presence of at least two consecutive extracellular cadherin-specific motifs in their extracellular domains. Each motif, called a cadherin repeat (EC), is about 110 amino acid residues (AA) long and comprises highly conserved calcium-binding AA.

All members of the major cadherin (CDH) gene family encode proteins with an ectodomain comprising at least five consecutive calcium-binding ECs. There is strong sequence conservation within the family. The family includes the paradigmatic type-I "classical" cadherin genes, encoding proteins with five ECs and a single TM domain. Comprehensive phylogenetic analysis of related cadherins (Hulpiau and van Roy 2009) identified several more branches in this major family, including the closely related type-II cadherins, also called atypical cadherins; the type-III and type-IV cadherins, which have elongated ectodomains with about ten ECs and at least one LamG domain (a 180-AA long globular domain originally found at the C-terminus of laminin-α chains), and which are not anymore present in mammals; the desmogleins and desmocollins, present in mammalian desmosomes; and the CELSR/Flamingo cadherins having, besides an ectodomain of nine ECs, a seven-transmembrane domain, which is exceptional for cadherins. Cadherins of types I to IV all have a cytoplasmic domain containing two conserved motifs for binding to members of the armadillo protein family.

The *protocadherin* (PCDH) gene family encodes proteins with an ectodomain comprising six or seven calcium-binding ECs with high sequence conservation within the family but weaker homology to the ECs of members of the CDH family. Further, protocadherins have a single TM domain and a distinct, protocadherin-specific cytoplasmic domain. Protocadherins can be further subdivided into clustered and non-clustered protocadherins on the basis of particular genomic organizations.

The *cadherin-related* (CDHR) genes do not fit into the above two families because they are clearly separated phylogenetically. The encoded proteins comprise at least two typical, consecutive, calcium-binding ECs, but often many more than the typical members of the CDH and PCDH families. Their overall domain organization is more diverse and also includes unique cytoplasmic domains.

Armadillo proteins have a central armadillo domain composed of tandemly arranged armadillo repeats. These repeats are about 40 AA long and were originally identified in the Armadillo protein, which is the β-catenin ortholog in *Drosophila melanogaster*. The armadillo domain has a typical curved structure and functions for multiple specific protein–protein interactions.

Orthologs are genes in different species that are closely homologous to each other because they represent the divergent copies of a single gene in the last common ancestor. During evolution, orthologs were separated by a speciation event, in contrast to paralogs, which were separated by a gene duplication event.

PDZ is a protein interaction domain of about 70–90 AA, named after a common structure found in the Post-Synaptic Density protein PSD-95, in Discs Large and in Zona Occludens 1 proteins. A PDZ domain in one protein generally interacts with a PDZ recognition motif at the C-terminus of another protein. Such sequence-specific interactions contribute to the formation of protein scaffolds leading to signaling networks.

Further Classification and Phylogenetics of Cadherin Superfamily Members

In mammals, the cadherin superfamily comprises more than 100 different types of cadherins, but ancestral proteins with similar structures were present in the most ancestral metazoans (Hulpiau and van Roy 2009). As the complexity of newly emerging organisms increased, the cadherin superfamily expanded and diverged. As defined above, the cadherin superfamily can be phylogenetically divided into three major families (CDH, PCDH, and CDHR) merely on the basis of ectodomain homologies (Fig. 1) (Hulpiau and van Roy 2009). These families are subdivided mainly on the basis of the domain composition and sequence homology of the ectodomains, TM, and cytoplasmic regions.

The cadherin family consists of classical cadherins (CDH) and flamingo cadherins (CELSR). Each type had an ancient ancestor in early metazoan life, which evolved into about 31 members in mammals by two rounds of whole genome duplications and several individual gene duplications throughout evolution. In addition to the type-I "classical" cadherins, the CDH branch includes type-II or atypical cadherins, the desmosomal cadherins, and several cadherins with longer ectodomains containing additional motifs besides ECs. The latter cadherins are the type-III and type-IV classical cadherins and have additional non-EC calcium-binding motifs in their ectodomain. These motifs consist of alternating epidermal-growth-factor-like (EGF-like) and laminin G (LamG) domains localized between the N-terminal EC repeats and the single TM domain (Fig. 1). The cytoplasmic domains of classical cadherins contain two conserved motifs for binding to proteins with a central armadillo domain, that is, p120ctn and β-catenin. Three other cadherins

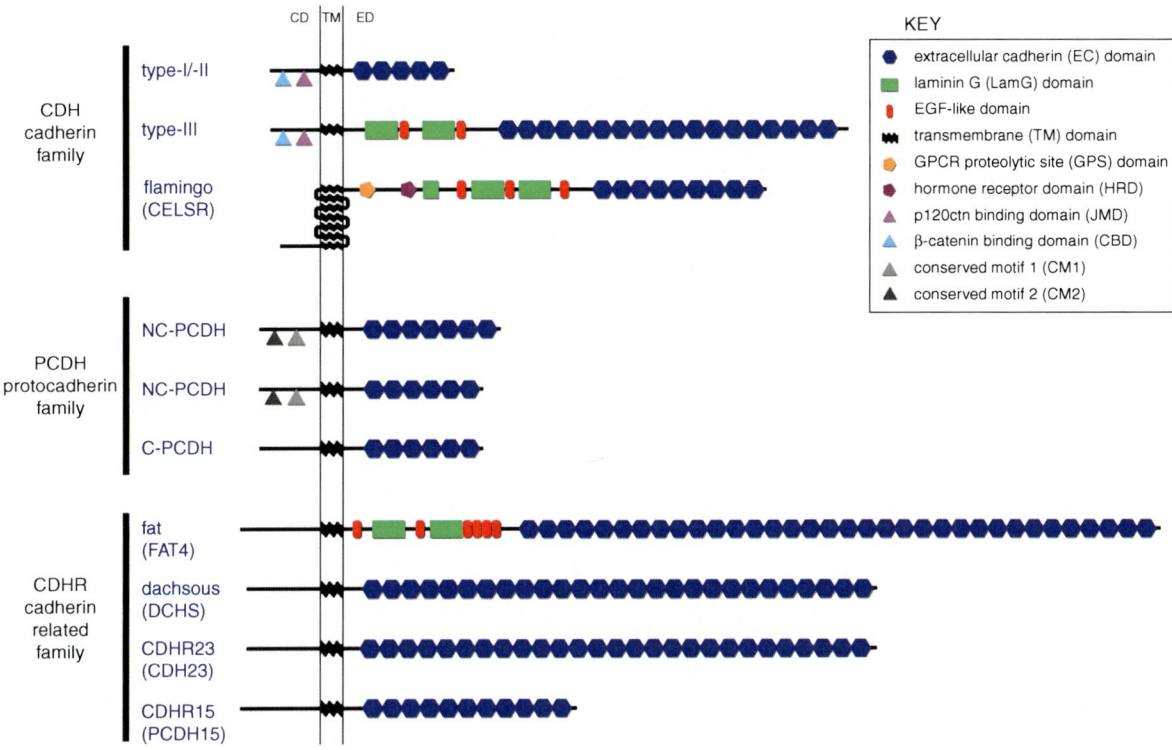

Cadherins, Fig. 1 Classification and domain composition of representative members of the cadherin superfamily. The molecules are depicted as mature mammalian proteins with their N-terminal ends on the right. Their sizes range between about 800 and 5000 amino acid residues. On *top*: the cadherin family (CDH) represented by a type-I/-II classical cadherin with five EC domains, a type-III classical cadherin with at least 13 EC domains, and a flamingo (CELSR) cadherin with 9 EC domains. Other typical domains are indicated and explained in the key. In the *middle*: the protocadherin family with non-clustered protocadherins (NC-PCDH) with seven or six ECs, and clustered protocadherins (C-PCDH) with six ECs. At the *bottom*: major members of the cadherin-related (CDHR) family. *CD* cytoplasmic domain, *ED* extracellular domain or ectodomain, *TM* transmembrane domain, *JMD* conserved cadherin-specific juxtamembrane domain

belonging to the CDH family are exceptional. Kidney-specific (Ksp) cadherin-16 and liver intestine (LI) cadherin-17 have seven ECs as a result of duplication of the first two ECs. Cadherin-13, also called T-cadherin (truncated) or H-cadherin (heart) stands out because it is the only known cadherin lacking a TM domain and is instead linked to the cell surface by a glycosylphosphatidylinositol (GPI) anchor (reviewed by Berx and van Roy in Nelson and Fuchs 2010). Cadherin-16, -17, and -13 also lack in their N-terminal region the conserved Trp residue(s) important for the classical adhesion mechanism described below. Finally, the family includes Flamingo/CELSR proteins, which typically have a longer ectodomain with nine ECs, a seven-pass TM domain, and a cytoplasmic domain not related to those of other cadherins.

Most non-vertebrate metazoans have only one type-III classical cadherin with 13 or more ECs, and one flamingo/CELSR cadherin with nine ECs (Hulpiau and van Roy 2009). Vertebrates have many more classical cadherins of type-I and type-II, and they have only five ECs due to the evolutionary loss of N-terminal EC repeats and internal loss of the EGF-like and LamG domains (Hulpiau and van Roy 2009). The desmosomal cadherin genes, comprising clustered desmoglein (DSG) and desmocollin (DSC) genes, evolved from classical cadherin genes in gnathostomes fairly recently.

The largest cadherin family, the protocadherins (PCDH), represents more than half of the cadherin repertoires. They emerged independently and expanded greatly in vertebrates. This expansion is consistent with their important role in the nervous system. Based on their genomic structure, they can be divided into clustered and non-clustered protocadherins. Humans have over 50 clustered protocadherin genes pooled in neighboring α, β, and γ clusters. Whereas the β cluster consists of repeats of single-exon genes only, mature transcripts of α- and γ-protocadherins are generated from one of several variable exons spliced to a set of three constant exons (Morishita and Yagi 2007). This generates a large number of unique protocadherins with homologous but different N-terminal domains comprising six ECs, one TM, and a short cytoplasmic domain. The latter is elongated in α- and γ-protocadherins by a shared sequence encoded by the constant exons. The non-clustered protocadherins have been called δ-protocadherins and comprise ten members in vertebrates (Hulpiau and van Roy 2009). They have ectodomains with six or seven ECs and are generally expressed as either short or long isoforms, the latter having an elongated cytoplasmic domain with conserved sequence motifs (CM-1 to -3).

Of the remaining cadherin-related (CDHR) proteins in the superfamily, the best studied are Dachsous (DCHS) and fat/FAT4, which were present in the most ancient multicellular animals (Hulpiau and van Roy 2009). Mammals express four proteins, named FAT1 to FAT4, but only FAT4 is the genuine ortholog of the fruit fly protein fat and a heterophilic interaction partner of DCHS. Indeed, although all mammalian FAT proteins share a long ectodomain with 34 ECs, FAT1 to FAT3 are in fact homologs of *Drosophila* fat-like (ftl or fat2). In mammalian auditory hair cells, cadherin-related-23 (often called cadherin-23 or CDH23) has 27 ECs in its ectodomain and makes heterophilic contact with cadherin-related-15 (often called protocadherin-15 or PCDH15), which has 11 ECs in its ectodomain. This interaction is very important for correct functioning of the inner ear (see below). The cadherin-related proteins with the smallest number of ECs are the proto-oncogene Ret (four ECs) and the calsyntenins (two ECs).

Molecular Structures of Cadherin Domains

Basic structure of the classical cadherin domain (EC). Structural studies on the adhesion mechanisms of cadherins have focused mainly on the vertebrate classical cadherins ([reviewed by Shapiro and Weis in Nelson and Fuchs 2010]; Hulpiau and van Roy 2009; van Roy and Berx 2008). One cadherin repeat (EC domain) folds into a structural unit consisting of seven β-strands forming an immunoglobulin-like Greek key topology. Two consecutive EC domains are stabilized by three calcium ions binding to a cluster of specific AA situated in the interdomain linker. In the primary structure of each EC, these calcium-binding AA are arranged as four conserved motifs: a Glu (E) residue around position 11, a DRE motif in the center, a DxNDxxPxF motif near the C-terminal of the EC, and a DxD motif further downstream in the next EC (Fig. 2a). However, in each cadherin ectodomain, the membrane-proximal last EC typically lacks the C-terminal motif because there is no subsequent EC domain to support binding of a calcium ion at the end of this last one. For that reason, this atypical sequence was also named MPED (membrane-proximal extracellular domain). The *Xenopus* C-cadherin ectodomain, which has five EC domains, has three calcium ions bound in each interdomain region, which fully agrees with the presence of 12 calcium ions in classical cadherin proteins.

Homophilic adhesion mode. The ectodomains of cadherins form a molecular layer composed of two interfaces involved in the assembly of junctions: *trans* adhesive interfaces between cadherins on apposed cell surfaces and *cis* interactions between cadherins in the same cell surface (Harrison et al. 2011).

It is widely accepted that there are two dimeric homophilic binding modes in *trans*. Both of them are based on free N-terminal ends of opposing EC1 domains; this occurs when a signal peptide and a prodomain are removed from the preproprotein. In the first binding configuration, the EC1 structures of two cadherin molecules on apposed cells exchange their N-terminal β-strands. For type-I cadherins, a Trp residue docks into a hydrophobic pocket generated by a conserved His-Ala-Val sequence in the apposed EC. For type-II cadherins, this anchoring occurs by two Trp residues. This type of adhesive *trans* interface is called the strand swapping dimer (Figs. 2b and 3a). The second *trans* binding mode is called the X-dimer mode and is considered an intermediate form in type-I and type-II cadherins. Here, the apposed cadherin molecules contact each other at the interdomain of the first and second EC domains to form an X-shaped structure (Figs. 2c and 3a). T-cadherin (CDH13) cannot form a strand swap dimer because it

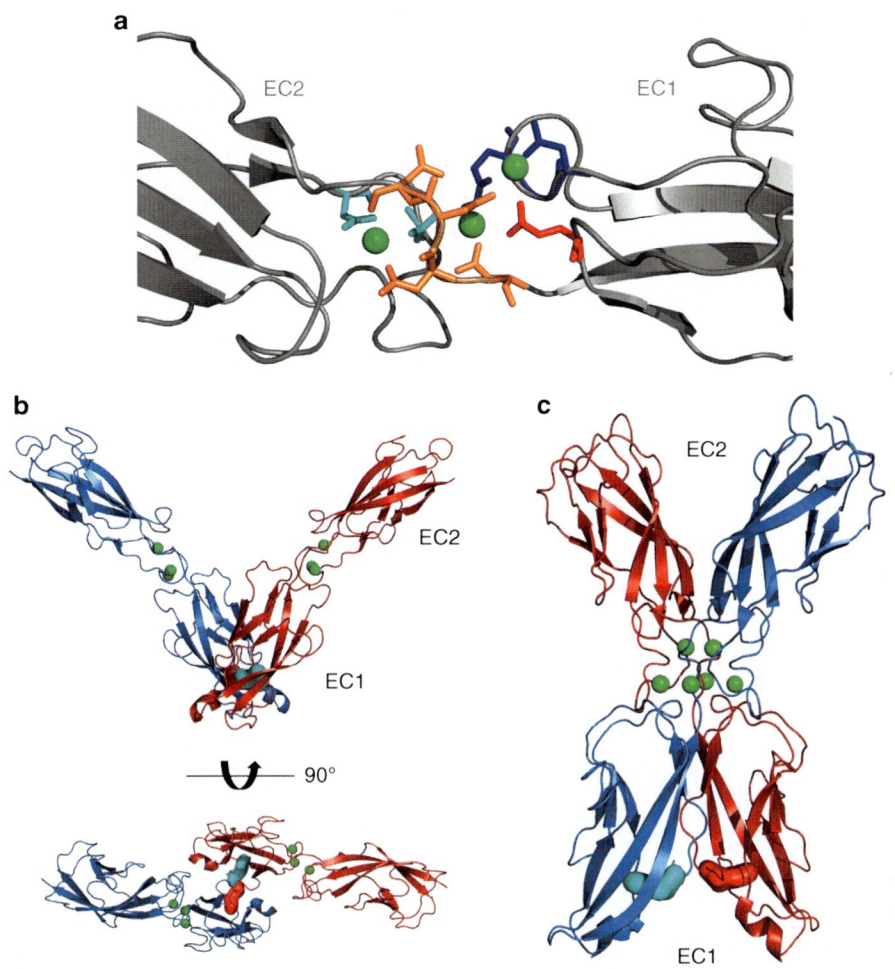

Cadherins, Fig. 2 Detailed view of calcium-binding motifs and adhesion interfaces in cadherins. (**a**) Binding of three Ca^{2+} ions in the linker region between successive EC domains (http://www.rcsb.org/; PDB (protein data bank) code: 1FF5). Four conserved motifs are involved: in the first EC domain, Glu11 in the PEN (Pro-Glu-Asn) motif is indicated as a *red stick*; Asp67 and Glu69 in the LDRE (Leu-Asp-Arg-Glu) motif are indicated in *blue*; Asp100 and the subsequent residues in the DxNDN (Asp-x-Asn-Asp-Asn) motif are indicated in *orange*; in the DxD motif of the second EC, Asp134 and Asp136 are shown in *cyan*. Ca^{2+} ions are shown as *green spheres*. (**b**) The strand swapping adhesion interface. The tryptophan (Trp) residues of the N-terminal adhesion arm in each EC1 are highlighted in *red* and *blue* surface representation, and they anchor into a hydrophobic pocket in the opposite EC1. (**c**) The X-dimer adhesion interface. Interaction occurs at the EC1-EC2 interdomain near the Ca^{2+} ions. The Trp residues in the adhesion arms of EC1 dock into their own protomer

has no Trp residue near its N-terminus. Instead, the X-dimer mode is used as a final adhesive interface by this GPI-anchored cadherin. Structural studies on an E-cadherin EC1–EC2 dimer have shown that the X-dimer is formed at intermediate Ca^{2+} concentrations of 500 μM to 1 mM when the Trp residue docks into its own protomer (reviewed by Hulpiau and van Roy 2009; van Roy and Berx 2008). At Ca^{2+} concentrations above 1 mM, strand swapping interactions start to predominate. Below 50 μM Ca^{2+}, an ectodomain structure of consecutive ECs is not stable. A third model, the so-called interdigitation model, involves a much more protrusive interaction of the apposed ectodomains, with the N-terminal EC1 binding to internal ECs up to EC4 (reviewed by van Roy and Berx 2008).

Further, the curved shape of the full size C-cadherin ectodomain has been fitted onto cryoelectron tomography images of mouse desmosomes (reviewed by Shapiro and Weis in Nelson and Fuchs 2010).

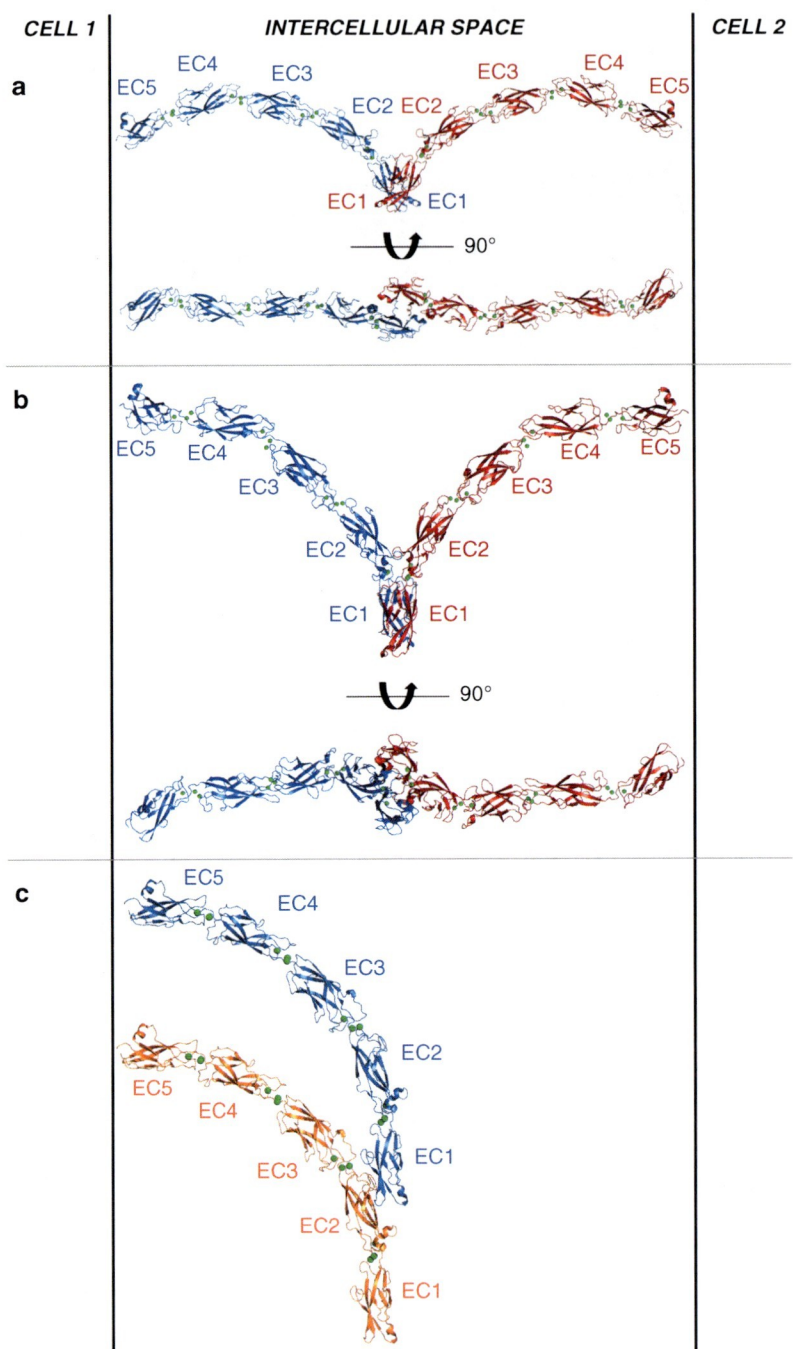

Cadherins, Fig. 3 Structural overview of the different adhesion mechanisms used by classical cadherins. Ectodomains of cadherins are represented by five EC domains in the intercellular space, with the EC5 proximal to the surface of the cell in which the respective cadherins reside. The 12 Ca^{2+} ions are represented by *green spheres*: four sets of three ions in between EC domains. (**a**) Structure of the strand swap dimer in *trans*. Adhesion occurs near the N-terminus of EC1. There is no *trans* contact between the EC1–EC2 interdomain regions. (**b**) Structure of the X-dimer in *trans*. The apposed cadherins contact each other at the EC1–EC2 interdomain region. (**c**) Structure of the *cis* interface. The concave side of the EC1 domain in the cadherin ectodomain (*blue*) interacts with the EC2 convex side of the neighboring cadherin (*orange*)

The visualized interactions resemble W, S, and λ shapes that are consistent with Trp swapping models. An interaction network of three molecules has been proposed for the λ shape. In this model, the Trp of the first molecule is near the hydrophobic pocket of the second one. The Trp of the second molecule is then near the hydrophobic pocket of a third molecule and the Trp of the third molecule is free. Furthermore, other molecular networks with even more cadherin molecules are conceivable, for example, a λ shape with an additional molecule in *cis* forming a desmosomal knot.

The *cis* interface found in E-, N-, and C-cadherin molecules consists of an asymmetric interaction between the EC1 domain of one molecule and the EC2 of a neighboring molecule on the same cell surface (Fig. 3c). The concave face of EC1 opposite the *trans* dimer interface interacts with the convex side of EC2 of the next protomer. This means that each cadherin ectodomain can form two *cis* interfaces by using both its EC1 concave side and its EC2 convex face. These *cis* interactions can then be further combined with the strand-swapping *trans* interaction to complete the junctional structure. Whereas mutations that interfere with the *trans* interface abolish cell–cell adhesion completely, *cis* interface mutations make junctions extremely mobile and unstable (Harrison et al. 2011).

Much less is known about the homophilic adhesion mechanisms of other types of cadherins. Sequence analyses of protocadherins, cadherin-related proteins, and invertebrate classical cadherins point to a mechanism other than the strand swapping binding mode. Future studies might reveal whether their interfaces resemble that of the X-dimer or of another unknown adhesion mechanism. Initial structures have already been determined for the EC1 of one protocadherin (PCDH-α4) and one cadherin-related protein (CDH23), but not in homophilic adhesion mode. EC1 of PCDH-α4 lacks a Trp near its N-terminus and consequently the hydrophobic pocket in which Trp can dock is also absent.

Heterophilic adhesion mode. Until recently, little was known about cadherin-mediated heterophilic adhesions. N-cadherin (CDH2) and R-cadherin (CDH4), both of which are classical type-I cadherins, can form *cis* heterodimers, presumably by adapting a strand swapping conformation using their single Trp residue nearby the N-terminal end (reviewed by Shapiro and Weis in Nelson and Fuchs 2010). A specific *trans* interaction has been reported between E-cadherin (CDH1) and LI-cadherin (CDH17), which are prominently present in intestinal epithelial cells (reviewed by Hulpiau and van Roy 2009). The following two important heterophilic interactions between cadherin-related molecules have been recognized, but the mechanisms remain unclear. The first interaction occurs between mammalian fat (FAT4) and Dachsous (DCHS1). The other is the *trans* interaction between a cadherin-related-23 (CDH23) homodimer and a cadherin-related-15 (CDHR15/PCDH15) homodimer. Current structural knowledge is limited to the CDHR23 EC1–EC2 domains (Elledge et al. 2010): unexpectedly, there is an extra calcium (Ca0) at the N-terminus. Mutating Asn3 and Arg4 residues nearly abolishes the heterotypic interaction; this means that they are crucial in this novel adhesion mechanism. Finally, protocadherins interact with classical cadherins, but the mechanism is not known. N-cadherin was shown to bind to Pcdh8 (reviewed by Nishimura and Takeichi 2009). In *Xenopus*, paraxial protocadherin (PAPC) is one of the three homologs of mammalian protocadherin-8 (Hulpiau and van Roy 2009), and it can form a functional complex with fibronectin leucine-rich domain transmembrane protein-3 (FLRT3) and with C-cadherin. Many of the structures mentioned above need further structural and experimental validation to elucidate the mechanisms involved. There is not even a single adhesion structure reported for several types of cadherins and cadherin-related molecules, such as the longer type-III cadherins, flamingo (CELSR) cadherins, Fat, Fat-like, and Dachsous cadherins.

Heterotypic adhesion modes (between different cell types) might be fully mediated by cadherins, but they might also involve other types of adhesion molecules. For instance, the localization of intraepithelial lymphocytes in the intestine is based on a heterophilic, heterotypic interaction between E-cadherin on keratinocytes and αEβ7 integrins on lymphocytes (reviewed by van Roy and Berx 2008). The Glu31 residue at the tip of the EC1 structure is essential for this binding as well as, surprisingly, the MPED (or EC5) domain.

Adhesion Characteristics of Cadherins

Type I classical cadherins have been shown to mediate strong and selective cell–cell adhesion that is generally both homophilic (like molecules bind each other on apposed cell surfaces; Figs. 3 and 4) and homotypic

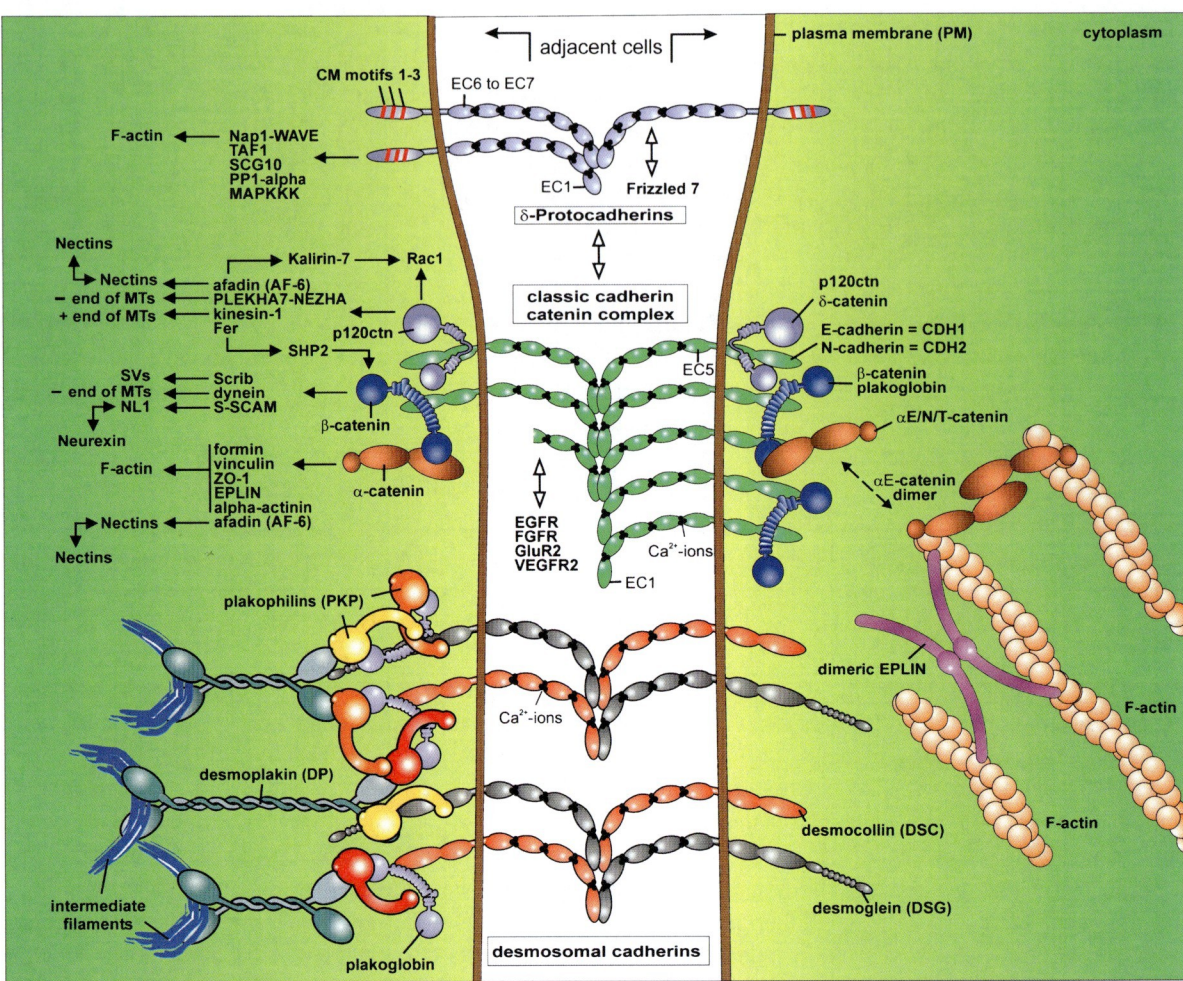

Cadherins, Fig. 4 Schematic overview of representative cadherin superfamily members making intercellular contacts. The scheme represents a "universal" cell type. Natural cell types, such as epithelial cells or neurons, generally express only some of the molecules depicted, while other cadherin superfamily members and associated molecules not shown here might also be expressed. *On top*: homophilic interaction between δ-protocadherins having six to seven extracellular cadherin repeats (EC1 to EC6 or -7). Their cytoplasmic domains may contain conserved motifs CM1–CM3, and they may interact with the intracellular proteins listed on the *left*. In the *middle*: the classical cadherin–catenin complex (CCC), with five ECs in their ectodomain, binding to each other in both *cis* and *trans* adhesion modes (see text). The cytoplasmic domains of classic E- and N-cadherin share two conserved domains for interaction with armadillo proteins, including p120ctn binding to a juxtamembranous domain (JMD), and β-catenin binding to a more C-terminal catenin-binding domain (CBD). At the *bottom*: composition of typical desmosomes; various desmosomal cadherins (DSG and DSC isoforms) contribute to this junction. Their cytoplasmic domains associate with a variety of armadillo proteins, including plakoglobin and plakophilins (PKP), and possibly also with the aminoterminal ends of desmoplakin (DP) dimers. Alternatively, the main role of plakoglobin and PKP may be to link DSG and DSC to DP. The drawing also summarizes several other molecular interactions, either in the extracellular space (indicated by *double open arrows*), or in the cytoplasm. Briefly, monomeric α-catenin binds to the CCC via β-catenin. It can occur as three isoforms: αE-, αN-, and αT-catenin. Dimeric αE-catenin cannot bind β-catenin but instead binds and cross-links filamentous actin (F-actin). The *broken double arrow* indicates a possible dynamic conversion between the monomeric and dimeric states. A selection of additional cytoplasmic interaction partners of p120ctn, β-catenin and α-catenin is listed on the *left* and described in the text. *Single arrows* indicate molecular interactions or activating effects; *double bent arrows* indicate additional systems of intercellular junctions. *SVs*: synaptic vesicles (Modified after Brigidi and Bamji (2011); van Roy and Berx (2008))

(like cells bind each other). A good example of homophilic, heterotypic interaction mediated by E-cadherin is the binding between keratinocytes and melanocytes. Historically, the founding member E-cadherin, a type-I epithelial cadherin, was discovered as a mammalian transmembrane protein of about 120 kDa that was protected by Ca^{2+} against iodination and trypsinization, and mediates Ca^{2+}-dependent cell–cell adhesion (reviewed by van Roy and Berx 2008). Antibodies reactive with its ectodomain revealed that different tissues express related but not identical cadherin types. Distinct spatiotemporal expression patterns have been consistently observed for the large cadherin superfamily. Further, use of neutralizing antibodies confirmed the essential contribution of classic cadherins to cell–cell adhesion. Once the corresponding cDNAs were cloned and used in dual transfection experiments, individual cells in mixed aggregation cultures were found to sort out from each other on the basis of homophilic cadherin-mediated adhesion activity (reviewed by van Roy and Berx 2008). The protein motif that is mainly responsible for this homophilic recognition is localized in the N-terminal EC1 repeat. A stretched and stable structure of the whole ectodomain is essential and possible only upon Ca^{2+} binding.

The cadherin-dependent cell adhesion strength is much enforced by clustering due to interactions with a cortically organized actin cytoskeleton. In the case of type-II classic cadherins, evidence for strong and homophilic cell–cell adhesion is often lacking, and the binding pattern might be rather promiscuous. For protocadherins, the evidence regarding cell–cell adhesion is still sparse but points to weak interactions, but nevertheless, these proteins might have important functions in sorting out different cell types (see below).

The GPI anchor of cadherin-13 forms the basis for its unusual location at apical instead of basolateral membranes in polarized epithelial cells (reviewed by Berx and van Roy in Nelson and Fuchs 2010). The cadherin-13 adhesion mode is homophilic but relatively weak, possibly due to its structural peculiarities summarized above. This weakness, in combination with the absence of direct coupling to intracellular structures, implies that cadherin-13 functions mainly in cell recognition and signaling rather than in adhesion (see also below). Phylogenetically, cadherin-16 and cadherin-17 also branch separately from the typical cadherins (Hulpiau and van Roy 2009). They show a duplication of the first two N-terminal ECs generating a seven-EC ectodomain, in combination with a short cytoplasmic domain lacking both conserved motifs for interaction with armadillo proteins. These two elongated cadherins were found to interact homophilically but with lower adhesion strength than classical cadherins. Evidence was recently provided for a heterophilic interaction between cadherin-17 and E-cadherin.

Expression Patterns of Cadherin Family Members

E-cadherin is widely expressed in most epithelial tissues and cell lines. In simple monolayered epithelia with well-organized intercellular contacts, the junctions are concentrated in the lateral cell–cell contacts. E-cadherin is expressed along the whole lateral plasma membrane but can be enriched in more compact adherens junctions (AJ), also named zonulae adherentes (ZA). These junctions are associated with a subcortical concentration of actin filaments, and are generally localized near the apical surface but below the tight junctions (zonulae occludentes). Tight junctions serve as a selective barrier against the "outside world" by controlling the paracellular transport of molecules and small particles. In the lateral membranes, there are also gap junctions for chemical and direct electrical intercellular communication, and the spot-like desmosomes (maculae adherentes), which form shear-resistant intercellular junctions distributed rather randomly on the lateral surfaces of simple epithelia (reviewed in Nelson and Fuchs 2010). Due to their multimolecular cytoplasmic plaque structure, desmosomes are rigidly connected by the linker protein desmoplakin to intermediate filaments, which are cytokeratins in epithelial cells. Interestingly, in both desmosomes and AJ, the cells are connected by members of the cadherin superfamily, but the packing density of desmosomal cadherins is much higher (Harris and Tepass 2010). Desmosomal cadherins occur as several isoforms of either desmogleins (DSG) or desmocollins (DSC), all of which have a single ectodomain with four typical calcium-binding ECs and a membrane-proximal anchor region (reviewed by Delva et al. in Nelson and Fuchs 2010). Both DSG and DSC seem to be required for strong cell–cell interactions, although it is unclear whether this occurs in a homophilic way (for instance, DSG1 binding to DSG1) or a heterophilic way (for instance, DSG1 binding to DSG2 or DSC1). The cytoplasmic domains of these desmosomal cadherins are longer than those of classical cadherins because they contain more motifs. Interaction occurs with plakoglobin, plakophilins, and desmoplakin (Fig. 4).

In multilayered stratified epithelia, such as the epidermis, only the basal cells contacting the basal lamina show a polarity comparable to that of simple epithelia. The suprabasal layers express E-cadherin and form desmosomes all over their perimeter, and in that way the layers are bound together tightly in the tissue. The expression of particular DSG, DSC, and PKP isoforms is influenced by the differentiation status of the suprabasal cells (reviewed by Delva et al. in Nelson and Fuchs 2010). In other tissues, the situation is varied. For instance, N-cadherin in neurons is localized at tiny synaptic contacts (puncta adhaerentia) and contributes to synaptogenesis by circumscribing the central neurotransmitter release zone (reviewed by Giagtzoglou et al. in Nelson and Fuchs 2010). In the heart, N-cadherin, a type-I classic cadherin, is localized at so-called intermediate junctions (fasciae adherentes) in the intercalated discs connecting individual cardiomyocytes (reviewed by Franke in Nelson and Fuchs 2010). These junctions also contain peculiar molecular complexes that comprise both classical desmosomal proteins (plakophilin-2, desmoplakin) and adherens junctional (AJ) proteins (beta-catenin, alpha-E-, and alpha-T-catenin). In endothelial cell junctions, N-cadherin is strongly expressed but not at homotypic junctions (between endothelial cells). Indeed, the main cell–cell adhesion function is taken over by a class-II classic cadherin, VE-cadherin or cadherin-5. In muscle cells, cadherin-15 (M-cadherin) is the major class-II cadherin. In mesenchymal cells, including osteoblasts, cadherin-11 (OB-cadherin) is an important classic cadherin. Cadherin-16 and -17 are expressed predominantly in renal and intestinal epithelia, respectively. GPI-anchored cadherin-13 is expressed by endothelial cells, mostly at the leading edge of migrating cells, and this has a pro-survival and pro-angiogenic effect (reviewed by Berx and van Roy in Nelson and Fuchs 2010).

Much less is known about the expression patterns of other cadherin superfamily members. Vertebrate CELSR2 and CELSR3 are expressed by several types of neurons. FAT1, the vertebrate homolog of *Drosophila* fat-like, is expressed at lateral cell junctions of epithelial cells but it is more abundant in basal regions than in AJs. Nonetheless, loss of FAT1 indirectly affects AJ integrity by influencing F-actin organization (Nishimura and Takeichi 2009). Mammalian FAT4, the ortholog of *Drosophila* fat, is expressed in various cell types, where it interacts heterophilically with Dachsous-1 (DCHS1) in neighboring cells of identical or different types. Protocadherins are differentially expressed at high levels in various neural tissues, but the delta-protocadherins are also widely expressed by other cell types. As mentioned above, the cadherin-related proteins CDHR15 (PCDH15) and CDHR23 interact with each other in auditory hair cells of the inner ear, both as transient links between adjacent stereocilia in developing hair cells and as tip links in mature hair cells (reviewed by El-Amraoui and Petit in Nelson and Fuchs 2010).

Molecular Interaction Partners and Functions of Cadherin Superfamily Members

Agreement on the biological relevance of the cadherin superfamily members is evidenced by numerous functional studies. Cadherins play key roles during early embryonic development and in morphogenesis (reviewed by Stepniak et al. in Nelson and Fuchs 2010; reviewed by Nishimura and Takeichi 2009). Not surprisingly, they are affected in many genetic diseases and other pathologies, and particularly in cancer (see below). Mechanistically, one can divide the functional parts of cadherins into two domains: the extracellular one and the cytoplasmic one. First, the number of EC repeats in the N-terminal extracellular domain is variable, and there are additional EGF-like and LamG repeats in several family members (Fig. 1). This extracellular domain is mainly involved in *cis* and *trans* binding with other cadherins, but it can also associate with other transmembrane proteins, as pointed out below. Second, the cytoplasmic domains exhibit more structural variability among family members but often have conserved domains within subfamilies. These domains are either known to bind specific cytoplasmic interaction partners for structural or signaling purposes or are expected to do so. In the case of classic cadherins, these interaction partners are armadillo proteins, which have been studied extensively. More recently, cytoplasmic interaction partners for protocadherins and cadherin-related proteins have also been identified, but other binding proteins will surely be discovered.

Cytoplasmic interaction partners of classical cadherins. The cytoplasmic domains of classic cadherins (types I to IV) bind two types of armadillo proteins: p120ctn and related protein family members, and beta-catenin and the homologous plakoglobin (gamma-catenin). More specifically, p120ctn binds to a conserved juxtamembrane domain (JMD), whereas β-catenin binds to a conserved C-terminal domain (Fig. 4). Association of p120ctn protects cadherins from premature endocytosis and hence stabilizes the junctions (reviewed by van Roy and Berx 2008). On the other hand, β-catenin forms a molecular bridge between the cadherins and α-catenins, which are F-actin-binding vinculin-related proteins. In a classical cadherin–catenin complex, also referred to as CCC, classic cadherins are linked to the F-actin cytoskeleton via a β-catenin/α-catenin molecular bridge, which is sensitive to regulation, for instance by reversible phosphorylation or irreversible ubiquitination (reviewed by van Roy and Berx 2008). However, recent data reported by the Nelson and Weis groups have cast doubt on this model: other actin-binding linker molecules, such as α-actinin and Eplin, might be involved in the cytoskeletal anchoring of cadherin-dependent junctions (Fig. 4). Interestingly, both p120ctn and β-catenin have important signaling functions as cytoplasmic or nuclear proteins outside the CCC, but these functions fall beyond the scope of this essay. Consequently, sequestering these armadillo proteins by cadherins at the junctions inhibits their alternative functions. This generates a more quiescent phenotype that is less proliferative and less migratory. Even more molecular complexity is reached by the binding of the protein afadin (AF-6) to either p120ctn or α-catenin in the CCC (reviewed by Nishimura and Takeichi 2009; van Roy and Berx 2008) (Fig. 4). Indeed, the central PDZ domain of afadin can bind at the same time to the C-termini of nectins. The latter are immunoglobulin-like transmembrane adhesion molecules binding either homo- or heterophilically. In this way, nectin–nectin interactions facilitate, via cytoplasmic bridging, cadherin-mediated AJ formation.

It is important to realize that the catenins in the classic CCC serve not only in mechanistic assembly but also in extensive focal signaling. This concept has become progressively clear in recent years and has been studied most extensively at excitatory synapses (reviewed by Brigidi and Bamji 2011; Harris and Tepass 2010). On one end, the C-terminus of β-catenin has a PDZ-binding motif through which it acts as a scaffold for recruitment of various proteins to the CCC, including the presynaptic scaffolding protein Scribble and the postsynaptic scaffolding protein S-SCAM (Magi2) (Fig. 4). Synaptic vesicles (SVs) are recruited via Scribble to the presynaptic membrane. Moreover, cadherin-associated p120ctn recruits the cytoplasmic tyrosine kinase Fer to the CCC. Fer phosphorylates and activates the tyrosine phosphatase SHP2, which in turn dephosphorylates β-catenin and thereby promotes the stability of the CCC and the local clustering of SVs. At the postsynaptic side, CCC-associated S-SCAM recruits the cell adhesion protein neuroligin-1 (NL1) to the synapse, where it engages in a *trans*-synaptic binding with neurexin. This has a cooperative effect on the clustering of SVs in the presynaptic compartment. Likewise, δ-catenin binds several PDZ-domain-containing proteins, including postsynaptic receptor scaffolding proteins. δ-catenin is a p120ctn homolog exclusively expressed in the brain. By modulating small GTPases, both CCC-associated and cytoplasmic p120ctn and δ-catenin contribute to the regulation of postsynaptic spine morphology (spine head width, length, and density) (reviewed by Brigidi and Bamji 2011). Moreover, spine head width is increased by the clustering of postsynaptic N-cadherin. The associated α-catenin recruits the adaptor molecule afadin (AF-6), which in turn recruits the Rac-GEF Kalirin-7, promoting Rac1-dependent spine maturation. Functional interaction of catenins, be it β-catenin, p120ctn, or α-catenin, with small GTPases and actin remodeling proteins is a recurrent theme that has evident implications for correct tissue morphogenesis (reviewed by Harris and Tepass 2010).

More recently, it has been demonstrated that AJs also interact with microtubules, either the plus ends of microtubules via β-catenin, or the minus ends of microtubules via p120ctn (reviewed by Harris and Tepass 2010) (Fig. 4). Microtubular minus ends are associated with the centrosomes in single cells, but in polarized epithelial cells lateral microtubules are typically oriented with their minus ends directed toward the AJ in the apical part of the lateral membranes. Association of the AJ with microtubules facilitates its assembly, and in addition it allows the AJ to influence intracellular structuring. For instance, the orientation of both symmetric and asymmetric cell divisions turned out to be influenced by AJs (reviewed by Harris and Tepass 2010).

Cytoplasmic interaction partners of non-classic cadherins. Knowledge about molecular interaction partners of non-classic cadherin types is sparser. The cytoplasmic sequences of non-clustered δ-protocadherins are rather diverse and so are their interaction partners (reviewed by Redies et al. 2005) (Fig. 4): the phosphatase PP1α binds to the conserved motif CM3 (sequence RRVTF), which is present in a subclass of long isoforms; the transcriptional regulator TAF1/Set binds to the cytoplasmic domain of protocadherin-7; the adaptor protein Dab-1 binds to the cytoplasmic domain of protocadherin 18. For the clustered protocadherins, which have within each cluster a largely shared cytoplasmic domain (see above), the tyrosine kinase Fyn has been shown to associate with α-protocadherins, whereas the microtubule-destabilizing protein SCG10 associates with at least one member of the γ-protocadherins.

The functional consequences of these specific interactions are generally poorly understood. A recent report describes a specific complex between the actin-organizing complex Nap1/WAVE and the cytoplasmic domain of protocadherin-10 (OL-protocadherin) (reviewed by Nishimura and Takeichi 2009). Evidence shows that this complex stimulates migration: at sites of cell–cell contacts, protocadherin-10 becomes enriched and it recruits Nap1/WAVE, which in turn leads to weakening of the classic CCC at the junctions. These molecular interactions might explain the aberrant migration of particular neurons in mice in which protocadherin-10 is knocked out. Also the cytoplasmic tail of the cadherin-related protein FAT1 interacts with an actin organizer, Ena/VASP, and this regulates actin dynamics such as polymerization of stress fibers (reviewed by Nishimura and Takeichi 2009). It is noteworthy that the cytoplasmic domain of FAT4 deviates completely from those of FAT1 to FAT3 (Hulpiau and van Roy 2009). Upon binding of FAT4 to DCHS1, FAT4 signals to the Hippo-YAP pathway, which has tumor suppressor activity in humans, but the underlying mechanism is unresolved (reviewed by Berx and van Roy in Nelson and Fuchs 2010). Both CDHR23 and CDHR15 (PCDH15) occur as alternative splice forms differing in sequence only in their cytoplasmic domains (reviewed by El-Amraoui and Petit in Nelson and Fuchs 2010). Particular isoforms interact with the PDZ domain protein harmonin. The cytoplasmic domain of one CDHR15/PCDH15 isoform interacts also with myosin VIIa.

Transmembrane interaction partners. On the other hand, also the ectodomains of classic cadherins are involved in interactions with other protein types. For instance, the E-cadherin ectodomain interacts physically with the ectodomains of receptor tyrosine kinases, such as EGFR and the HGF-receptor c-Met. These interactions have either an inhibitory effect on the growth stimulating effects of the ligand, or a stimulatory effect ascribed to co-endocytosis of E-cadherin with the receptor (reviewed by van Roy and Berx 2008). It has been demonstrated that the ectodomain of N-cadherin interacts molecularly and functionally with the FGF-receptor ectodomain to prevent FGFR internalization and allow sustained receptor activation and downstream signaling (reviewed by Berx and van Roy in Nelson and Fuchs 2010). In neural cells, the ectodomain of N-cadherin also associates with the AMPA receptor GluR2, and this interaction appears to be essential for GluR2-mediated spine maturation (reviewed by Brigidi and Bamji 2011). Interestingly, the ectodomain of the *Xenopus* δ-protocadherin PAPC (Hulpiau and van Roy 2009) interacts with the ectodomain of Frizzled-7, and both of these transmembrane proteins turned out to be essential for planar cell polarity (PCP) phenomena, such as convergent extension movement of mesoderm during gastrulation (reviewed by Nishimura and Takeichi 2009; Redies et al. 2005). GPI-anchored cadherin-13 shows a peculiar type of binding (reviewed by Berx and van Roy in Nelson and Fuchs 2010). Upon homophilic ligation on endothelial cell surfaces, this atypical cadherin becomes linked to Grp78/BiP. The latter is normally ER-retained but can be secreted under particular conditions. The formation of the complex of Grp78 with cadherin-13 triggers the anti-apoptotic Akt kinase pathway.

In view of these findings on extracellular interactions, it is somewhat surprising that several transmembrane proteins seem to interact with the cytoplasmic domains of classic cadherins. For instance, in endothelial cells, VEGFR2 associates with the cytoplasmic domain of VE-cadherin (the type-II cadherin-5). This interaction also leads to prevention of receptor internalization and signaling, which ultimately inhibits cell growth. Another interesting case is the cytoplasmic interaction in synaptic junctions of N-cadherin with arcadlin, which corresponds to the rat protocadherin-8, a non-clustered δ-protocadherin (reviewed by Nishimura and Takeichi 2009). Excitation of hippocampal neurons upregulates arcadlin, and this promotes N-cadherin internalization, a process accelerated by the homophilic interaction of arcadlin ectodomains and involving an associated MAPKKK. A similar observation of cross-inhibition involves the induction by activin of PAPC expression in *Xenopus*, which is linked to decreased adhesion activity of C-cadherin, a class-I cadherin in frog.

Cadherin-Associated Pathologies in Mouse Models and Human Patients

Several zygotic (total) and conditional knockout mice have been generated for members of the cadherin superfamily and for the cytoplasmic catenins associated with them (reviewed by Stepniak et al. in Nelson and Fuchs 2010). The cadherins analyzed in this way are mainly from the type-I and type-II families. A zyogotic knockout of E-cadherin is early lethal (at E4) due to defects in trophectoderm. Interestingly, these defects could not be rescued by expressing an N-cadherin cDNA from the E-cadherin locus in a gene replacement approach. Zygotic N-cadherin ablation leads to death at E10 due to heart defects. Various mice with tissue-specific knockout of E-cadherin, N-cadherin, or both of them often develop lethal defects associated with impaired differentiation and induced apoptosis. The zygotic knockout of cadherin-5 (VE-cadherin) is lethal at E9.5 due to severe vascular defects. The zygotic knockout of protocadherin-10 (OL-protocadherin), which interacts with the actin-regulating complex Nap1-WAVE, results in abnormal migration of the growth cones of striatal neurons (reviewed by Nishimura and Takeichi 2009). The importance of a few other δ-protocadherins has been studied by gene silencing in *Xenopus* (reviewed by Redies et al. 2005). Zygotic deletion of the entire cluster of γ-protocadherins revealed a role for these proteins in spinal synaptic development and activity (reviewed by Morishita and Yagi 2007). Further, it is noteworthy that the zygotic knockout of cadherin-related protein 23 (CDHR23) leads to degeneration of inner ear structures, vestibular defects, and deafness. Also other mutants of mouse and zebrafish demonstrate the importance of CDHR23–CDHR15 interactions for correct auditory hair organization and associated mechano-electrical transduction. Most importantly, Usher syndrome 1 (USH1), which is the most severe form of hereditary deaf-blindness in humans, is a monogenic disorder caused by mutations in any one of five USH1 genes; four of these genes encode the above-mentioned proteins: CDHR23, CDHR15 (PCDH15), harmonin, and myosin VIIa (reviewed by El-Amraoui and Petit in Nelson and Fuchs 2010).

Human mutations in P-cadherin (class-I cadherin-3) have been linked to the HJMD and EEM syndromes, which have common features such as hair loss and progressive blindness. Also mutations in the human desmoglein-4 (DSG4) gene are associated with an inherited hair disorder. The importance of appropriate desmosome formation is also demonstrated by the following findings (reviewed by Stepniak et al. in Nelson and Fuchs 2010). Autoantibodies against either DSG1 or DSG3 cause, respectively, the blistering diseases pemphigus foliaceus and pemphigus vulgaris. Heterozygous mutations in the human genes encoding the desmosomal components DSG2, DSC2, plakoglobin, plakophilin-1 and -2, and desmoplakin are at the basis of about 50% of arrhythmogenic right ventricular cardiomyopathy (ARVC). One more example of a cadherin superfamily member associated with disease in humans is protocadherin-19, a δ-protocadherin encoded by the X-chromosome. Mutations in PCDH19 have been identified in a female-restricted epilepsy and cognitive impairment syndrome and also in the Dravet syndrome, which is featured by epileptic encephalopathy mainly in females. It has been proposed that protocadherin-11Y (PCDH11Y), encoded by the human Y-chromosome, might compensate for the PCDH19 defects in males. Finally, the small CLSTN1, which has only 2 ECs, has been related to Alzheimer's disease.

Importantly, several cadherin family members are involved in the major human pathology of cancer (reviewed by Berx and van Roy in Nelson and Fuchs 2010). The inactivation of E-cadherin in malignant epithelial cancers has been a paradigm for a cell–cell adhesion molecule functioning as both tumor suppressor and invasion suppressor. E-cadherin can become inactivated in cancer by different events: inactivation mutations in combination with allelic loss, as frequently observed in invasive lobular breast cancers, germline mutations in families with hereditary diffuse gastric cancer (HDGC), promoter methylation, and transcriptional silencing by members of the Snail/Slug or ZEB families. Also posttranslational modifications, such as specific phosphorylation of cadherins or catenins, and increased endocytosis of the CCC might play a role in increasing the malignancy of tumor cells. Regularly seen in tumor cells at the invasive front is the process of epithelial-mesenchymal transition (EMT), which is concomitant with cadherin switching: epithelial E-cadherin is downregulated and replaced by so-called mesenchymal cadherins, such as N-cadherin, cadherin-6, or cadherin-11. It is most interesting that similar cadherin switches do occur during dynamic morphogenetic processes. Impressive examples are the processes of neurulation, somitogenesis, and neural crest morphogenesis and directed migration (reviewed by Stepniak et al. in Nelson and Fuchs 2010).

There is mounting evidence that other cadherin superfamily members can serve as tumor suppressors, including several δ-protocadherins, such as protocadherin-8, -10, and -20. In contrast, it has been proposed that cytoplasmic expression of a protocadherin-11Y isoform truncated at its N-terminus contributes to the androgen resistance of advanced prostate cancers. Overexpression of cadherin-5 (VE-cadherin) in melanomas causes vascular mimicry and promotes malignancy. Also, constitutive activation of the atypical RET protein, with EC motifs in the short ectodomain and a cytoplasmic tyrosine kinase domain, causes thyroid cancer and multiple endocrine neoplasias. A special case is the GPI-anchored cadherin-13. Its expression by cancer cells inhibits their growth but the underlying mechanism is unclear. On the other hand, its expression by the microvasculature in the tumor environment promotes angiogenesis and cancer metastasis. Further analysis of the cadherin superfamily members in various advanced culture systems and in different organs in informative model organisms and in human pathologies will doubtlessly unveil even more about the important role of cadherins in metazoan life.

Cross-References

▶ Calcium and Extracellular Matrix
▶ Calcium-Binding Proteins
▶ Calcium-Binding Proteins, Overview
▶ Calcium in Biological Systems

References

Brigidi GS, Bamji SX (2011) Cadherin-catenin adhesion complexes at the synapse. Curr Opin Neurobiol 21(2):208–214

Elledge HM, Kazmierczak P, Clark P et al (2010) Structure of the N terminus of cadherin 23 reveals a new adhesion mechanism for a subset of cadherin superfamily members. Proc Natl Acad Sci USA 107:10708–10712

Harris TJ, Tepass U (2010) Adherens junctions: from molecules to morphogenesis. Nat Rev Mol Cell Biol 11:502–514

Harrison OJ, Jin X, Hong S et al (2011) The extracellular architecture of adherens junctions revealed by crystal structures of type I cadherins. Structure 19:244–256

Hulpiau P, van Roy F (2009) Molecular evolution of the cadherin superfamily. Int J Biochem Cell Biol 41:343–369

Morishita H, Yagi T (2007) Protocadherin family: diversity, structure, and function. Curr Opin Cell Biol 19:584–592

Nelson WJ, Fuchs E (eds) (2010) Cell-cell junctions, Cold Spring Harbor perspectives in biology. Cold Spring Harbor Laboratory Press, Woodbury

Nishimura T, Takeichi M (2009) Remodeling of the adherens junctions during morphogenesis. Curr Top Dev Biol 89:33–54

Redies C, Vanhalst K, van Roy F (2005) Delta-protocadherins: unique structures and functions. Cell Mol Life Sci 62:2840–2852

van Roy F, Berx G (2008) The cell-cell adhesion molecule E-cadherin. Cell Mol Life Sci 65:3756–3788

Cadherins, Protocadherins, and Cadherin-Related Proteins

▶ Cadherins

Cadmium Absorption

Samuel Ogheneovo Asagba
Department of Biochemistry, Delta State University, Abraka, Delta State, Nigeria

Synonyms

Assimilation of cadmium; Uptake of cadmium

Definition

Cadmium (Cd) is a soft, ductile, silver-white electropositive metal, which is widely distributed in the environment. Environmental contamination by Cd is due to human and natural activities. Food and drinking water are the main routes of exposure to Cd for nonsmoking general population. The absorption of Cd after dietary exposure ranges from 0.5% to 3.0% in majority of animal species, while in man a range of 3.0–8.0% has been reported. The metal accumulates mainly in the liver and kidney where it has multiple cytotoxic and metabolic effects. Cd bioavailability, retention, and consequent toxicity are influenced by several factors such as diet composition, chemical forms of Cd, nutritional status, among others.

Introduction

Contamination of the environment by Cd may occur through anthropogenic and natural sources (WHO 1992). The main routes of Cd exposure from the environment are inhalation and ingestion. Although the efficiency of Cd absorption through inhalation (25–50%) is much higher than that through ingestion (1–10%) (Zalups and Ahmad 2003; WHO 1992), concerns about airborne exposure are limited to special populations, including smokers, people living near smelters, and metal-processing workers. In contrast, dietary Cd intake is an important public health issue for the general world population despite the lower bioavailability of Cd through the gastrointestinal tract (GIT). This is because the element is readily distributed to tissues after oral exposure, where it inhibits antioxidant enzymes (Asagba and Eriyamremu 2007). This inhibition can lead to increased oxidative stress which may result in membrane damage and loss of membrane-bound enzymes such as the ATPases (Asagba et al. 2004; Asagba and Obi 2005). Tissues in which these effects have been reported are the liver and kidney which are considered the main target organs in acute and chronic cadmium exposure (Zalups and Ahmad 2003; Asagba and Obi 2000). Other tissues involved in Cd toxicity include the testis, heart, bone, eye, and brain (WHO 1992; Asagba et al. 2002; Eriyamremu et al. 2005; Asagba 2007).

Since the major route of human contamination is oral, understanding the mechanisms by which Cd crosses the intestinal barrier, and what regulates its absorption is of prime interest. Following oral exposure, Cd is absorbed in mammals preferentially in the duodenum and proximal jejunum (Zalups and Ahmad 2003). After uptake into the organism, Cd is transported via the blood to various tissues, particularly to the liver where it induces the synthesis of metallothionein (MT), an ubiquitous metal-binding protein of low molecular mass and high cysteine content (Zalups and Ahmad 2003). This protein may be required in the gastrointestinal absorption of Cd (WHO 1992).

Although knowledge of the absorption of Cd from experiments with human subjects is very limited, a considerable body of animal experiments is available in literature (WHO 1992; Andersen et al. 1992). These experiments indicate that different factors seem to affect the degree of absorption and toxicity of cadmium. Besides age, intestinal absorption of cadmium is influenced by a variety of factors including its chemical form, dose and route of exposure, intestinal content, diet composition, nutritional status, and interactions of cadmium with other nutrients (WHO 1996; Zalups and Ahmad 2003).

According to Andersen et al. (1992), the two aspects of intestinal cadmium uptake that have been studied mainly in experimental animals are (1) the effects of dietary components on the intestinal uptake of ionic cadmium administered after mixing with the diet, in drinking water, or given as a single oral dose and (2) the bioavailability for intestinal uptake of cadmium incorporated into various foodstuffs. Besides these two aspects, this entry will also be focused on the findings, assertions, and hypothesis pertaining to the molecular mechanisms involved in absorption of oral cadmium.

Absorption of Cadmium

This entry is focused mainly on gastrointestinal absorption of cadmium because the body burden of Cd is derived primarily from ingestion of food and drinking water contaminated with Cd. Besides, a significant fraction of inhaled Cd ends up in the gastrointestinal tract as a result of mucociliary clearance and subsequent ingestion. In fact, studies have demonstrated that as much as 60% of the inhaled dose of Cd ended up being translocated to the gastrointestinal tract in rats exposed acutely to aerosols containing Cd carbonate (Zalups and Ahmad 2003). It has been shown that the efficiency of gastrointestinal absorption of Cd is only about 1–2% in mice and rats, 0.5–3% in monkeys, 2% in goats, and 5% in pigs and lambs.

In humans, the efficiency of gastrointestinal absorption of Cd has been reported to be approximately 3–8% of the ingested load (Zalups and Ahmad 2003).

The major site of cadmium uptake in the intestine is not certain; however, a low pH of the gastric content emptied into the duodenum is thought to contribute to improve uptake (Andersen et al. 1992). Beyond the duodenum, the pH increases, and cadmium will rapidly be chelated by various dietary components, and therefore, its bioavailability will be reduced.

The chemical form of Cd is an important factor on its gastrointestinal absorption. It has been reported that Cd in foods such as meat, seafoods, and vegetable exists mainly as Cd–MT or MT-like Cd-binding proteins (Ohta et al. 1993). As is already well known, MT is induced in the intestinal tissue by oral Cd administration, and discussion on the mechanism of gastrointestinal absorption of Cd has been based on Cd ion mainly (Kikuchi et al. 2003). However, the chemical form of Cd such as Cd–MT in foods or Cd bound by MT induced in the intestinal tissue has also been considered in literature (Zalups and Ahmad 2003; Groten et al. 1991; Groten et al. 1994).

It has been shown that during dietary Cd exposure, Cd may be absorbed as complexes with MT or other dietary constituents which may be soluble or insoluble. However, there is conflicting information in literature on the availability of Cd–MT for intestinal uptake in relation to that of ionic Cd. Some reports indicate lower Cd absorption in rats fed Cd–MT than in that fed with ionic Cd (Andersen et al. 1992; Zalups and Ahmad 2003). These reports also indicate that the ratio of the concentration of Cd in the kidney to the concentration of Cd in the liver was higher after oral administration of Cd–MT than after oral administration of $CdCl_2$. On the other hand, similar uptake of cadmium acetate and Cd–MT has been reported when these different forms of Cd^{2+} were exposed to rats by gavage (Andersen et al. 1992). Other animal studies also indicate no difference in the bioavailability of Cd incorporated into dietary components such as wheat during growth or oysters in relation to inorganic Cd (Andersen et al. 1992). Similarly, the feeding of cadmium incorporated in pigs' livers resulted in about half the accumulation of cadmium in the rats' livers that took place after intake of a diet containing cadmium chloride (Groten et al. 1990). Conversely, a similar comparative study (Asagba 2010) revealed that more Cd was accumulated in the tissues of rats fed with fish-incorporated Cd in diet relative to those exposed to the metal in drinking water after 3 months of exposure. This disparity in the bioavailability of Cd from these foods may be due to differences in the solubility of proteins or ligands associated with Cd, as postulated by Lind et al. (1995).

Diet and Absorption of Oral Cadmium

There are several studies in literature which indicate that the absorption of Cd is influenced by the type or composition of the diet of a population. Numerous studies have shown that diets rich in fibers such as unrefined whole diets (Andersen et al. 1992) and a Nigerian-like diet (Asagba et al. 2004; Asagba and Eriyamremu 2007) decreased accumulation of Cd in rats.

Little is known about the influence of dietary fibers on metal absorption in human subjects. However, studies have revealed that absorption of cadmium in nonsmoking women 20–50 years of age depends on intake of dietary fibers (Zalups and Ahmad 2003). The findings indicate a tendency toward higher blood Cd (BCd) and urinary Cd (UCd) concentrations with increasing fiber intake; however, the concentrations were not statistically significant at the 5% level, indicating an inhibitory effect of fiber on the gastrointestinal absorption of cadmium.

The role of proteins in the absorption of cadmium has also received attention in biological literature. Studies on the effect of glycinin, ovalbumin, and gelatin on the absorption of cadmium indicate that all three proteins reduced intestinal Cd uptake which suggests that these proteins can protect against Cd toxicity (Andersen et al. 1992). Low protein diets generally enhance Cd toxicity, while high protein diets reduce the toxicity. This is not surprising since low protein diets generally enhance the uptake of Cd, while the reverse effect is observed with a high protein diet (Andersen et al. 1992).

It is well known that many toxic effects of Cd arise from interactions with essential elements, such as zinc (Zn). These interactions can take place at different stages of absorption, distribution in the organism, and excretion of both metals and at the stage of Zn biological functions. Numerous studies show that enhanced Zn consumption may reduce Cd absorption and accumulation and prevent or reduce the adverse actions of Cd, whereas Zn deficiency can intensify Cd accumulation and toxicity (Brzóska et al. 2008; Roqalska et al. 2009). Studies have shown that the nutritional status of animals or humans with regard to zinc (Zn), iron (Fe), and/or calcium (Ca) can have a profound effect on the rate of Cd absorption from the gut. If the long-term intake of one or more of these minerals is low, the nutritional status is reduced, and Cd absorption increases; by contrast, if long-term intake is high, nutritional status is enhanced, and Cd absorption is decreased (Reeves and Chaney 2001; Brzóska and Moniuszko-Jakoniuk 1998).

A study on the effect of marginal nutritional status of Zn, Fe, and Ca on the bioavailability of Cd in sunflower kernels (SFK) demonstrated a much higher rate of absorption and organ retention of Cd in rats given a marginal supply of these mineral nutrients than in those receiving an adequate supply (Reeves and Chaney 2001). In addition, it was shown that the intrinsic, natural concentration of Zn, but not Ca and Fe, was enough to reduce the absorption and organ retention of dietary Cd supplied by the SFK. Foods such as rice, on the other hand, contain a very low intrinsic amount of Zn, Fe, or Ca (Reeves and Chaney 2001). Previous studies conducted to assess the risk of food-chain Cd indicated that rice, because of its poor supply of Zn, Fe, and Ca, could have caused populations of subsisting rice consumers to suffer a high incidence of Cd-induced renal tubular dysfunction. These individuals consumed rice raised in soils that were contaminated by a mixture of ore wastes of Cd and Zn in a ratio of 0.5–1:100 μg. These populations seem to be more susceptible to Cd poisoning than those who consume more nutritious diets but with similar intakes of Cd (Reeves and Chaney 2002 and Reeves et al. 2005). It has therefore been hypothesized that the low nutritional status of rice consumers, which results from an inadequate supply of these minerals from rice, could contribute significantly to a higher apparent susceptibility to soil Cd contamination from rice than the higher nutritional status of those who consume other grains with higher mineral content.

To test this hypothesis, a study was conducted in which rats were fed diets with adequate or marginal amounts of dietary Zn, Fe, or Ca (Reeves and Chaney 2002 and Reeves et al. 2005). The results obtained from this study support the hypothesis that populations exposed to dietary sources of Cd and subsisting on marginal mineral intakes could be at greater risk than well-nourished populations exposed to similar amounts of dietary Cd. The studies by Reeves et al. (2005) also indicate that MT induction is not involved in duodenal Cd accumulation in animals with marginal dietary status of Fe, Zn, and Ca. In addition, these studies support the hypothesis that marginal deficiencies of Fe, Zn, and Ca, commonly found in certain human populations subsisting on rice-based diets, play an important role in increasing the risk of dietary Cd exposure.

The effect of calcium supplementation on absorption and retention of cadmium in the suckling period was evaluated in Wistar rat pups of both sexes by Sarić et al. (2002). Results showed that after oral exposure, cadmium concentrations in all calcium-supplemented groups were significantly decreased in the organs and carcass and that the effect was dose related. No such effect of calcium was found after parenteral cadmium exposure. The authors concluded that calcium supplementation during the suckling period could be an efficient way of reducing oral cadmium absorption and retention without affecting tissue essential trace element concentrations.

Studies by Grosicki (2004) indicate that vitamin C supplement decreased the carcass Cd burden and the Cd content in the liver, kidneys, testicles, and muscles; the highest decreases were found in the testicles, the lowest ones in the muscles. Similarly, Sauer et al. (1997) observed a significant less cadmium accumulation in the lung, kidney, and testis in retinol-pretreated rats. Another study by Prasad et al. (1982) indicates that vitamin B_6 and B_1 deficiencies in rats resulted in a nonspecific increase in Cd ions.

Mechanisms of Oral Cadmium Absorption

Although food intake is among the most important routes of Cd exposure, not many details are known about the intestinal absorption mechanisms of Cd.

Available research evidence (Foulkes et al. 1981; Foulkes 1989) indicates that the mechanism of cellular Cd uptake in the rat jejunum consists of nonspecific binding to anionic sites on the membrane, followed by a temperature-dependent and rate-limiting internalization step which is probably related to membrane fluidity. Completion of the absorptive process is by transport across the basolateral membrane into serosal fluid. This step proceeds at only 1–2% of the rate of uptake from the lumen. Transport of cadmium from the small intestine is also thought to be facilitated by other possible mechanisms, including metal transport proteins such as divalent metal transporter 1 (DMT1), calcium ion channels, amino acid transporters (as cysteine–cadmium conjugates), and by endocytosis of cadmium–metallothionein (Cd–MT) complexes (Reeves and Chaney 2005).

The gastrointestinal tract produces metallothionein (MT) which can sequester cadmium. MTs are a family of low-molecular-weight heavy-metal-binding proteins, unique in their high cysteine content (Chang et al. 2009). The role of MT in the absorption of Cd has not been fully elucidated. Earlier studies have shown that the ability of the intestine to produce MT is limited but increase from the proximal to the distal small intestine (Elsenhans et al. 1994, 1999). This would improve the ability of the distal small intestine to handle Cd and thus make the metal less bioavailable in this region (Eriyamremu et al. 2005). Similarly, the findings of Min et al. (1992) suggest that mucosal MT in the small intestine might trap Cd absorbed from the intestinal lumen. However, the report by Lind and Wicklund (1997) does not support the hypothesis that intestinal Cd absorption is increased when the Cd-binding capacity of intestinal MT is saturated. It is also noteworthy to point out that the findings of Liu et al. (2001) support the hypothesis that endogenous MT does not function as a protective barrier against Cd absorption or alter its tissue distribution.

Summary and Conclusion

Humans are generally exposed to Cd by two main routes, inhalation and ingestion. The body burden of Cd is derived primarily from ingestion of food and drinking water contaminated with Cd. Most of the absorption of Cd appears to occur primarily in the duodenum, and this process is aided by the low pH in this region.

Intestinal Cd absorption is influenced by diet, nutritional status, the chemical form of Cd, among other factors. The mechanism of Cd absorption consists of nonspecific binding to anionic sites on the membrane, followed by a rate-limiting internalization which is temperature dependent. Absorption of cadmium from the small intestine is also facilitated by metal transport proteins such as divalent metal transporter 1 (DMT1), calcium ion channels, amino acid transporters (as cysteine–cadmium conjugates), and by endocytosis of cadmium–metallothionein (Cd–MT) complexes. However, it is noteworthy that the role of intestinal MT in Cd absorption and subsequent tissue distribution is not fully agreed upon.

The lack of conclusive information on the availability Cd in Cd–MT for intestinal uptake in relation to that of ionic Cd is noticeable, and there is a great need for further studies in this area. Also, since humans are usually exposed to Cd–MT in foods and rarely to inorganic Cd, the toxicity of food-incorporated Cd deserve further investigation, in view of the observed difference in tissue accumulation from these forms of Cd.

Cross-References

▶ Cadmium and Metallothionein
▶ Cadmium, Physical and Chemical Properties

References

Andersen O, Nielsen JB, Nordberg GF (1992) Factors affecting the intestinal uptake of cadmium from the diet. In: Nordberg GF et al (eds) Cadmium and the environment. International Agency for Research on Cancer, Lyon, pp 173–187

Asagba SO (2007) Alterations in activities of tissue enzymes in oral cadmium toxicity. Nig J Sci Environ 6:91–102

Asagba SO (2010) Comparative effect of water and food-chain mediated cadmium exposure in rats. Biometals 23(6): 961–971

Asagba SO, Eriyamremu GE (2007) Oral cadmium exposure and haematological and liver function parameters of rats fed a Nigerian-like diet. J Nutr Environ Med 16(3–4):267–274

Asagba SO, Eriyamremu GE, Adaikpoh MA, Ezeoma A (2004) Levels of lipid peroxidation, superoxide dismutase and Na^+/K^+-ATPase in some tissues of rats exposed to a Nigerian diet and cadmium. BiolTrace Elem Res 100(1):75–86

Asagba SO, Obi FO (2000) Effect of cadmium on kidney and liver cell membrane integrity and antioxidant enzyme status: implications for Warri River cadmium level. Trop J Environ Sci Health 3(1):33–39

Asagba SO, Obi FO (2005) A comparative evaluation of the biological effects of environmental cadmium contaminated control diet and laboratory cadmium supplemented test diet. Biometals 18:155–161

Asagba SO, Isamah GK, Ossai EK, Ekakitie AO (2002) Effect of oral exposure to cadmium on the levels of vitamin A and lipid peroxidation in the eye. Bull Environ Contam Toxicol 68:18–21

Brzóska MM, Moniuszko-Jakoniuk J (1998) The influence of calcium content in diet on cumulation and toxicity of cadmium in the organism. Arch Toxicol 72:63–73

Brzóska MM, Galazyn-Sidorczuk M, Roszczenko A, Jurczuk M, Majewska K, Moniuszko-Jaconiuk J (2008) Beneficial effect of zinc supplementation on biomechanical properties of femoral distal end and femoral diaphysis of male rats chronically exposed to cadmium. Chem Biol Interact 171(3):312–324

Chang X, Jin T, Chen L, Nordberg M, Lei L (2009) Metallothionein 1 isoform mRNA expression in peripheral lymphocytes as a biomarker for occupational cadmium exposure. Exp Biol Med (Maywood) 234(6):666–672

Elsenhans B, Strugala G, Schmann K (1999) Longitudinal pattern of enzymatic and absorptive functions in the small intestine of rats after short term exposure to dietary cadmium chloride. Arch Environ Contam Toxicol 36:341–346

Elsenhans B, Schüller N, Schümann K, Forth W (1994) Oral and subcutaneous administration of cadmium chloride and the distribution of metallothionein and cadmium along the villus-crypt axis in rat jejunum. Biol Trace Elem Res 42(3):179–190

Eriyamremu GE, Asagba SO, Onyeneke EC, Adaikpoh MA (2005) Changes in carboxypeptidase A, dipeptidase and Na^+/K^+-ATPase activities in the intestine of rats orally exposed to different doses of cadmium. Biometals 18:1–6

Foulkes EC (1989) On the mechanism of cellular cadmium uptake. Biol Trace Elem Res 21:195–200

Foulkes EC, Johnson DH, Sugawara N, Bonewitz RF, Voner C (1981) Mechanism of cadmium absorption in rats. NTIS, Springfield, VA

Grosicki A (2004) Influence of vitamin C on cadmium absorption and distribution in rats. J Trace Elem Med Biol 18(2):183–187

Groten JP, Sinkeldam EJ, Luken JB, van Bladeren PJ (1990) Comparison of the toxicity of inorganic and liver incorporated cadmium: a 4-wk feeding study in rats. Food Chem Toxicol 28(6):435–441

Groten JP, Sinkeldam EJ, Luten JB, van Bladeren PJ (1991) Cadmium accumulation and metallothionein concentrations after 4-week dietary exposure to cadmium chloride or cadmium-metallothionein in rats. Toxicol Appl Pharmacol 111:504–513

Groten JP, Koeman JH, Van Nesselrooij JHJ, Luten JB, Fentener Van Vlissingen JM, Stenhuis WS, Van Bladeren PJ (1994) Comparison of renal toxicity after long-term oral administration of cadmium chloride and cadmium-metallothionein in rats. Toxicol Sci 23(4):544–552

Kikuchi Y, Nomiyama T, Kumagai N, Dekio F, Uemura T, Takebayashi T, Matsumoto Y, Sano Y, Hosoda K, Watanebe S, Sakurai H, Omae K (2003) Uptake of cadmium in meals from the digestive tract of young non-smoking Japanese female volunteers. J Occup Health 45:43–57

Lind Y, Wicklund GA (1997) The involvement of metallothionein in the intestinal absorption of cadmium in mice. Toxicol Letts 91(3):179–187

Lind Y, Wicklund GA, Engman J, Jorhem L (1995) Bioavailability of cadmium from crab hepatopancreas and mushroom in relation to inorganic cadmium: a 9-week feeding study in mice. Food Chem Toxicol 38(8):667–673

Liu Y, Liu J, Klaassen CD (2001) Metallothionein-null and wild-type mice show similar cadmium absorption and tissue distribution following oral cadmium administration. Toxicol Appl Pharmacol 175:253–259

Min KS, Nakatsubo T, Kawamura S, Fujita Y, Onosaka S, Tanaka K (1992) Effects of mucosal metallothionein in small intestine on tissue distribution of cadmium after oral administration of cadmium compounds. Toxicol Appl Pharmacol 113(2):306–310

Ohta H, Seki Y, Imamiva S (1993) Possible role of metallothionein on the gastrointestinal absorption and distribution of cadmium. Kitasato Arch Exp Med 65:137–145

Prasad R, Lyall V, Nath R (1982) Effect of vitamin B6 and B1 deficiencies on the intestinal uptake of calcium, zinc and cadmium. Ann Nutr Metab 26(5):324–330

Reeves PG, Chaney RL (2001) Mineral status of female rats affects the absorption and organ distribution of dietary cadmium derived from edible sunflower kernel (*Helianthus annuus* L.). Environ Res 85(3):215–225

Reeves PG, Chaney RL (2002) Nutrient status affects the absorption and whole body and organ retention of cadmium in rats fed rice based diets. Environ Sci Technol 36:2684–2692

Reeves PG, Chaney RL, Simmons RW, Cherian MG (2005) Metallothionein induction is not involved in cadmium accumulation in the duodenum of mice and rats fed diets containing high-cadmium rice or sunflower kernels and a marginal supply of zinc, iron and calcium. J Nutr 135:99–108

Roqalska J, Brzóska MM, Roszczenko A, Moniuszko-Jakonuik J (2009) Enhanced zinc consumption prevents cadmium induced alterations in lipid metabolism in male rats. Chem Biol Interact 177(2):142–152

Sarić MM, Blanusa M, Piasek M, Varnai VM, Juresa D, Kostial K (2002) Effect of dietary calcium on cadmium absorption and retention in suckling rats. Biometals 15:175–182

Sauer JM, Waalkes MP, Hooser SB, Baines AT, Kuester RK, Sipes IG (1997) Tolerance induced by all-trans-retinol to the hepatotoxic effects of cadmium in rats: role of metallothionein expression. Toxicol Appl Pharmacol 143:110–119

WHO (1992) Environmental health criteria, 134, Cadmium. World Health Organization, Geneva

WHO (1996) Trace elements in human nutrition and health. World Health Organisation, Geneva

Zalups RK, Ahmad S (2003) Molecular handling of cadmium in transporting epithelia. Toxicol Appl Pharmacol 186:163–188

Cadmium and Health Risks

Tim S. Nawrot[1], Jan A. Staessen[2,3], Harry A. Roels[4], Ann Cuypers[1], Karen Smeets[1] and Jaco Vangronsveld[1]
[1]Centre for Environmental Sciences, Hasselt University, Diepenbeek, Belgium
[2]Study Coordinating Centre, Department of Cardiovascular Diseases, KU Leuven, Leuven, Belgium
[3]Unit of Epidemiology, Maastricht University, Maastricht, The Netherlands
[4]Louvain Centre for Toxicology and Applied Pharmacology, Université catholique de Louvain, Brussels, Belgium

Synonyms

Epidemiology; Public health

Definition

Cadmium (Cd) is a natural element in the Earth's crust. All rocks, soils, and waters contain some Cd. It is a global environmental pollutant potentially with multiple health consequences.

Introduction

Populations worldwide are exposed to a low-level intake of this toxic element through their food, causing an age-related cumulative increase in the body burden (Järup et al. 1998).

People living in the vicinity of industrial emissions and other point sources of Cd release can be exposed to an increased level of Cd other than food (▶ Cadmium Absorption). Prevention strategies have been proposed and discussed: (1) reduction of the transfer of Cd from soil to plants (passage of Cd into the human food chain) by maintaining the pH of agricultural and garden soils close to neutral and (2) reduction of the intestinal Cd absorption by preserving a balanced iron status (Nawrot et al. 2010a). Here an overview of the recently gained evidence that contributed to our understanding about the effects of Cd on human health is given.

Cadmium-Related Morbidity

Keynote:
- The bone and kidney are the critical target organs for Cd toxicity in humans.
- Both direct and indirect effects on bone.
- Cadmium is a human carcinogen.

Kidney

Microproteins in urine are sensitive biomarkers of Cd-induced renal damage reflecting a tubulotoxic effect. Among them is β_2 microglobulin, a small plasma protein, which passes the glomerular filter and subsequently almost completely reabsorbed in the absence of Cd-induced tubular dysfunction. Depending on the biomarker of nephrotoxicity, thresholds of urinary Cd can range from about 2 µg/g creatinine for the onset of early biochemical alterations (e.g., hypercalciuria) to 10 µg/g creatinine for the development of the classic tubular microproteinuria (Buchet et al. 1990; Roels 2003). A cross-sectional analysis of 14,778 subjects (NHANES) showed that subjects in the highest quartile of blood Cd (>0.6 µg/L) were almost two times more likely to exhibit albuminuria (≥30 mg/g creatinine) and 32% more likely to have reduced glomerular filtration rate (<60 mL/min per 1.73 m^2) (Navas-Acien et al. 2009).

Epidemiological evidence shows higher susceptibility for persons with diabetes to develop Cd-induced renal dysfunction (Nawrot et al. 2010a). Although there is strong evidence that elevated levels of tubular biomarkers of renal dysfunction are associated with urinary Cd, a surrogate of the Cd body burden, there is less agreement about the clinical significance and predictability of these changes. Prospective epidemiological evidence from Belgium (Nawrot et al. 2008) and the USA (Menke et al. 2009) suggests that the increased Cd-related mortality was directly related to the toxic effects of Cd, rather than being mediated by renal dysfunction.

Osteoporosis

Osteoporosis is usually an age-related bone disorder. Evidence accumulates that besides the kidney the bone is a primary target organ of Cd toxicity as well. Clinical features associated with osteoporosis include increased risk of new fractures and increased mortality. Studies among populations from Belgium, China, Japan, and Sweden showed associations between osteoporosis and low-level environmental Cd exposure. The commonly accepted explanation is that Cd-induced renal tubular damage reduces the calcium reabsorption in the nephron, resulting in hypercalciuria and decreased bone mineral density, and hence increased fracture risk (Järup et al. 1998; Staessen et al. 1994), particularly in postmenopausal women and older men. However, a recent study also discovered a dose-response association between the odds of osteoporosis in young men (mean age 45) and urinary Cd (Nawrot et al. 2010b).

Hypercalciuria (a urinary concentration of more than 200 mg of calcium per liter) should be considered an early tubulotoxic effect of Cd, because it may exacerbate the development of osteoporosis, especially in the elderly. In the studies performed in northeast Belgium, bone mineral density was negatively associated with urinary Cd in postmenopausal women. A twofold increase in urinary Cd excretion at baseline was associated with a 73% increased risk of fractures in women (95% confidence interval [95%CI], 1.16–2.57). The corresponding results for men were 1.20 (0.75–1.93) (Staessen et al. 1999). Swedish investigations showed a doubling of the risk for osteoporosis for urinary Cd levels of 0.5–3 µg Cd/g creatinine (middle tertile) compared with the lowest tertile (<0.5 µg Cd/g creatinine) (Alfven et al. 2000).

Recent data (as reviewed in Nawrot et al. 2010a) provide more insight into the mechanisms supporting a direct osteotoxic effect of Cd independent of the status of kidney function, in that urinary excretion of pyridinium cross-links from bone collagen is increased. The shape of this association was linear with effects observed at low levels. The Cd-induced effect on bone is not mediated via impaired activation of vitamin D.

Cancer

Keynote: Four lines of evidence explain why Cd is classified as a human carcinogen:

1. First, as reviewed by Verougstraete and colleagues (2003), several studies in workers revealed a positive association between the risk of lung cancer and occupational exposure to Cd. The combined estimate showed an increased risk of 20% in workers exposed to Cd compared with those not exposed (Verougstraete et al. 2003).
2. Second, data obtained from rat studies showed that the pulmonary system is a target site for carcinogenesis after Cd inhalation. However, exposure to toxic metals in animal studies has usually been much higher than that reported in humans environmentally exposed to toxic metals (Nawrot et al. 2010a).
3. Third, several in vitro studies have indicated plausible toxicodynamic pathways, such as increased oxidative stress, modified activity of transcription factors, and inhibition of DNA repair (Jin et al. 2003). Most errors that arise during DNA replication can be corrected by DNA polymerase proof reading or by post-replication mismatch repair. Inactivation of the DNA repair machinery is an important primary effect of Cd toxicity, because repair systems are required to deal with the constant DNA damage associated with normal cell function. The latter mechanism might indeed be relevant for environmental exposure because it has been shown that chronic exposure of yeast to environmentally relevant concentrations of Cd can result in extreme hypermutability (Jin et al. 2003). In this study, the DNA-mismatch repair system was already inhibited by 28% at Cd concentrations as low as 5 µM. For example, the prostate of healthy unexposed humans contained Cd concentrations of 12–28 µM and human lungs of nonsmokers contained Cd concentrations of 0.9–6 µM (Jin et al. 2003).
4. Further, in vitro studies provide evidence that Cd may act like an estrogen, forming high-affinity complexes with estrogen receptors, suggesting a positive role in breast cancer carcinogenesis.

Along with this experimental evidence, recent epidemiological studies (reviewed in: Nawrot et al. 2010a), summarized in Table 1, provided new insights on the role of exposure to Cd in the development of cancer in humans. First, the results of a population-based case-control study noticed a significant twofold increased risk of breast cancer in women in the highest quartile of Cd exposure compared with those in the lowest quartile (McElroy et al. 2006). In a population-based prospective cohort study with a median follow-up of 17.2 years in an area close to three zinc smelters, the association between incident lung cancer and urinary Cd was assessed (Nawrot et al. 2006). Cd concentration in soil ranged from 0.8 to 17.0 mg/kg. At baseline, geometric mean urinary Cd excretion was 12.3 nmol/day (1.78 µg/day) for people in the high-exposure area, compared

Cadmium and Health Risks, Table 1 Studies on cancer in association with environmental cadmium exposure

Site	Reference	Population	Effect size	Shape of the association
Breast	McElroy et al. (2006)	Case-control study n = 254 cases n = 246 controls USA, based on NHANES sample	Odds ratio: 2.29 (95% CI: 1.3–4.2) comparing the highest quartile of urinary Cd ($\geq$0.58 µg/g crt) to the lowest (<0.26 µg/g crt)	Continuous linear increase in risk
Endometrium	Åkesson et al. (2008)	Cohort study n = 30,210 postmenopausal women 16 years follow-up, Sweden	Relative risk: 1.39 (95% CI: 1.04–1.86) for highest tertile of intake of cadmium $\geq$16 µg Cd/day versus <13.7 µg Cd/day (lowest tertile)	Third tertile significantly different from first. Shape linear
Lung caner	Nawrot et al. (2006)	Cohort study n = 994 15 years follow-up Belgium	Relative risk 1.31 (95% CI: 1.03–1.65) for doubling in urinary cadmium	Continuous linear increase in risk
Pancreas	Kriegel et al. (2006)	Case-control study n = 31 cases n = 52 controls Egypt	Odds ratio 1.12 (95% CI: 1.04–1.23) per µg/L serum Cd	Continuous increase risk
Prostate	Vinceti et al. (2007)	Case-control study n = 45 cases n = 58 controls Italy	Odds ratio: 4.7 (95% CI: 1.3–17.5) for highest quintile toenail Cd ($\geq$0.031 µg/g) versus <0.007	Threshold observed $\sim$0.015 µg/g toenail Cd
	Van Wijngaarden et al. (2008)	Cross-sectional 1,320 men NHANES population sample, United States	Significant cadmium–zinc interaction Men with zinc intake <12.7 mg/day a urinary cadmium increase of 1 µg/g crt is associated with a 35% increase in serum PSA	Effect size depends on zinc intake
Urinary bladder	Kellen et al. (2007)	Case-control study n = 172 bladder cases n = 359 controls Belgium	Odds ratio: 5.7 (95% CI: 5.0–13.8) comparing the highest ($\geq$1 µg/L) to the lowest tertile (<0.2 µg/L) of blood cadmium	Continuous linear increase in risk

with 7.7 nmol day^{-1} (0.87 µg/day) for those in the reference (low exposure) area. The risk of lung cancer was 3.58 higher in the high-exposure area compared to the area with low exposure. The 24-h urinary excretion is a biomarker of lifetime exposure to Cd. The risk for lung cancer was increased by 70% for a doubling of 24-h urinary Cd excretion. Confounding by co-exposure to arsenic was unlikely.

Epidemiological studies did not convincingly imply Cd as a cause of prostate cancer. Of 11 cohort studies, only 3 found a positive association (Verougstraete et al. 2003). However, a recent case-control study (Vinceti et al. 2007) with 40 cases, and 58 controls showed an excess cancer risk in subjects in the third and fourth (highest) quartiles (above 0.0145 µg Cd/g) of toenail Cd concentration [OR = 1.3 (95% CI, 0.3–4.9)] and 4.7 µg/g [(95% CI, 1.3–17.5), respectively, p-trend = 0.004] compared with subjects in the bottom quartile. In the NHANES population, which included a sample of 1,320 men (Van Wijngaarden et al. 2008), an effect modification by zinc on the blood prostate-specific antigen levels (PSA) and urinary Cd has been reported (Table 1). An increase in urinary Cd by 1 μg/g creatinine was associated with a 35% increase in PSA levels, in subjects with a zinc intake below the median (12.7 mg/day). In a case-control study, pancreatic cancer was associated with serum Cd levels (Table 1) (Kriegel et al. 2006). For each 1 μg Cd/L serum increase, the odds for pancreatic cancer increased with 12%. In a study of bladder cancer Kellen et al. (2007) showed a 5.7-fold increase in risk between subjects with blood Cd at the lowest tertile (<0.2 μg/L) versus the highest tertile (≥1 μg/L) (Table 1).

Prospective Mortality Studies

- *Keynote*: Environmental exposure to Cd increases total mortality continuously without evidence of a threshold, independently of kidney function and other classical factors associated with mortality including sex, age, smoking, and socioeconomic status.

Recently, two population-based cohort studies showed an increased risk for premature death in association with Cd exposure. The average urinary Cd concentration at baseline was about three times higher in the Belgian cohort (~1 μg/g creatinine) (Nawrot et al. 2008) compared with the US cohort (Menke et al. 2009). The hazard ratios for all-cause mortality, associated with a twofold higher urinary Cd were 1.28 (95% CI, 1.15–1.43) in men and 1.06 (95% CI: 0.96–1.16) for women in the US cohort (NHANES III), and 1.20 (95% CI, 1.04–1.39) in the Belgian cohort men and women combined. In the Belgian cohort, the hazard rates were not different between men and women (no urinary Cd by gender interaction in relation to mortality observed). The cause-related mortality pattern differed between the two cohorts. In the Belgian cohort, deaths from non-cardiovascular but not cardiovascular causes increased with higher 24-h urinary Cd excretion (Nawrot et al. 2008). In the NHANES study, both non-cardiovascular and cardiovascular disease increased with higher urinary Cd concentrations in men whereas in women non-cardiovascular diseases were borderline significantly associated but not cardiovascular mortality.

Conclusion

There has been substantial progress in the evaluation of the health effects of Cd and the exploration of the shape of the concentration-response function at different organ systems. These results have important scientific, medical, and public health implications. Indeed, the mean exposure for adults across Europe is close to the tolerable weekly intake of 2.5 μg/kg body weight. Subgroups such as vegetarians, children, smokers, and people living in highly contaminated areas may exceed the tolerable weekly intake about twofold.

To reduce the transfer of Cd from soil to plants, the vegetable bioavailability of Cd in soils should be diminished by maintaining pH of agricultural and garden soils close to neutral. A balanced iron intake is effective in reducing the bioavailability of Cd present in the intestine, by reducing its absorption. Along with the recent knowledge concerning low-dose Cd exposure, the current exposure to Cd at the population level should be kept as low as possible so that the urinary Cd concentration is kept below 0.66 μg/g creatinine (margin of safety = 3) as proposed by the EU (European Community report, EUR 23424 EN).

Cross-References

▶ Cadmium Absorption

References

Åkesson A, Julin B, Wolk A (2008) Long-term dietary cadmium intake and postmenopausal endometrial cancer incidence: a population-based prospective cohort study. Cancer Res 68:6435–6441

Alfven T, Elinder CG, Carlsson MD, Grubb A, Hellstrom L, Persson B et al (2000) Low-level cadmium exposure and osteoporosis. J Bone Miner Res 15:1579–1586

Buchet JP, Lauwerys R, Roels H, Bernard A, Bruaux P, Claeys F et al (1990) Renal effects of cadmium body burden of the general population. Lancet 336:699–702

Järup L, Berglund M, Elinder CG, Nordberg G, Vahter M (1998) Health effects of cadmium exposure – a review of the literature and a risk estimate. Scand J Work Environ Health 24(Suppl 1):1–51

Jin YH, Clark AB, Slebos RJ, Al-Refai H, Taylor JA, Kunkel TA et al (2003) Cadmium is a mutagen that acts by inhibiting mismatch repair. Nat Genet 34:326–329

Kellen E, Zeegers MP, Hond ED, Buntinx F (2007) Blood cadmium may be associated with bladder carcinogenesis: the Belgian case-control study on bladder cancer. Cancer Detect Prev 31:77–82

Kriegel AM, Soliman AS, Zhang Q, El-Ghawalby N, Ezzat F, Soultan A, Abdel-Wahab M, Fathy O, Ebidi G, Bassiouni N, Hamilton SR, Abbruzzese JL, Lacey MR, Blake DA (2006) Serum cadmium levels in pancreatic cancer patients from the East Nile Delta region of Egypt. Environ Health Perspect 114:113–119

McElroy JA, Shafer MM, Trentham-Dietz A, Hampton JM, Newcomb PA (2006) Cadmium exposure and breast cancer risk. J Natl Cancer Inst 98:869–873

Menke A, Muntner P, Silbergeld EK, Platz EA, Guallar E (2009) Cadmium levels in urine and mortality among U.S. adults. Environ Health Perspect 117:190–196

Navas-Acien A, Tellez-Plaza M, Guallar E, Muntner P, Silbergeld E, Jaar B et al (2009) Blood cadmium and lead and chronic kidney disease in US adults: a joint analysis. Am J Epidemiol 170:1156–1164

Nawrot T, Plusquin M, Hogervorst J, Roels HA, Celis H, Thijs L et al (2006) Environmental exposure to cadmium and risk of cancer: a prospective population-based study. Lancet Oncol 7:119–126

Nawrot TS, Van HE, Thijs L, Richart T, Kuznetsova T, Jin Y et al (2008) Cadmium-related mortality and long-term secular trends in the cadmium body burden of an environmentally exposed population. Environ Health Perspect 116:1620–1628

Nawrot TS, Staessen JA, Roels HA, Munters E, Cuypers A, Richart T et al (2010a) Cadmium exposure in the population: from health risks to strategies of prevention. Biometals 23:769–782

Nawrot T, Geusens P, Nulens T, Nemery B (2010b) Occupational cadmium exposure, calcium excretion, bone density and risk for osteoporosis. J Bone Miner Res 25(6):1441–1445

Roels H (2003) The Belgian cadmium experience in the industrial setting and the general population. Arbeidsgezondheidszorg Ergon 40(3):101108

Staessen JA, Lauwerys RR, Ide G, Roels HA, Vyncke G, Amery A (1994) Renal function and historical environmental cadmium pollution from zinc smelters. Lancet 343:1523–1527

Staessen JA, Roels HA, Emelianov D, Kuznetsova T, Thijs L, Vangronsveld J et al (1999) Environmental exposure to cadmium, forearm bone density, and risk of fractures: prospective population study. Public health and environmental exposure to cadmium (PheeCad) study group. Lancet 353:1140–1144

van Wijngaarden E, Singer EA, Palapattu GS (2008) Prostate-specific antigen levels in relation to cadmium exposure and zinc intake: results from the 2001–2002 national health and nutrition examination survey. Prostate 68:122–128

Verougstraete V, Lison D, Hotz P (2003) Cadmium, lung and prostate cancer: a systematic review of recent epidemiological data. J Toxicol Environ Health B Crit Rev 6:227–255

Vinceti M, Venturelli M, Sighinolfi C, Trerotoli P, Bonvicini F, Ferrari A et al (2007) Case-control study of toenail cadmium and prostate cancer risk in Italy. Sci Total Environ 373:77–81

Cadmium and Metallothionein

Ivan Sabolić, Davorka Breljak, Carol M. Herak-Kramberger and Marija Ljubojević
Unit of Molecular Toxicology, Institute for Medical Research and Occupational Health, Zagreb, Croatia

Synonyms

Interaction of cadmium and metallothionein; Metallothioneins as cadmium-binding proteins; Metallothioneins in cadmium-induced toxicity; Role of metallothioneins in cadmium-induced hepatotoxicity and nephrotoxicity

Definition

Cadmium (Cd) is a nonessential metal which is an important environmental pollutant, heavily toxic to the living world. In mammals, the major target organs of Cd are liver and kidneys, in which the toxicity is closely connected with the expression of metallothioneins (MTs), a family of small, cysteine-rich metal-binding proteins with Cd detoxification, antioxidative, and antiapoptotic functions. In the liver, Cd binds to the endogenous and/or de novo synthesized MTs and accumulates as an inert CdMT complex. This complex is released into circulation from the intoxicated cells and reaches the kidney, where it is filtered, endocytosed by the proximal tubule epithelium, and degraded. The liberated Cd binds to the endogenous and newly synthesized MTs and becomes sequestered as CdMT in the cell cytoplasm. When the binding capacity of MTs is surpassed, free Cd ions become cytotoxic, damaging the structure and function of proximal tubule cells. MTs in the urine can be used as an indicator of kidney damage in the general population exposed to environmental Cd.

Cadmium

Cadmium (Cd) is a chemical element from the group of transition elements/metals. Some of these metals (Fe, Co, Cu, Zn) represent essential elements that play important roles in cell biology and physiology, whereas some others, like Cd, Pb, and Hg, are highly toxic for humans, animals, and plants. Cd occurs widely in nature, and in the Earth's crust and oceans it is closely associated with Zn. According to physicochemical properties, Cd is in many respects similar to Zn, and numerous Cd-induced toxic effects in the mammalian cells reflect the interactions of Cd with Zn-related functions.

In the environment, volcanic activity is the major natural source of Cd, whereas various anthropogenic activities (combustion of fossil fuels, mining, metal smelting and refinery, production of steel, cement, batteries, paints, plastics and phosphate fertilizers, waste disposal, tobacco smoking) contaminate soil, water, air, and food with this environmental health hazard. In the general population, tobacco smoking represents the most important single source of chronic Cd exposure. Absorption of Cd from the lungs is much more effective than that from the intestine; up to 50% of the Cd inhaled via cigarette smoke may be absorbed, whereas the intestinal absorption of Cd accounts for 5–8% of the load (Foulkes 1986; Järup et al. 1998). In humans, an acute intoxication with Cd is also possible; it occurs as a result of occupational exposure to high Cd concentrations via inhalation of fumes or dust or via ingestion of contaminated food and/or water, causing severe injuries in lungs and gastrointestinal tract. However, the Cd-polluted living environment represents the base for continuous, long-term, and wide-scale exposure of humans and animals to small doses of this toxic metal. With time, Cd accumulates in various mammalian organs and (a) causes damage to the liver, kidneys, and gastrointestinal and reproductive tract, (b) interferes with embryonal development, (c) affects functions of neuronal and immune system, (d) causes high blood pressure, (e) inhibits bone mineralization, (f) potentiates diabetes, (g) contributes to mutagenicity and metaplastic transformation of cells, and (h) acts as an endocrine disruptor. Various aspects of Cd biology and toxicology in plants, humans, and experimental animals were described in a book edited by Foulkes 1986, a detailed status of Cd as an environmental and health problem relevant to humans was reviewed by Järup et al. 1998, and two excellent up-to-date collections of reviews and original articles dedicated to various epidemiological, clinical, and molecular aspects of Cd toxicity in mammalian cells appeared recently as special issues of relevant scientific journals (Prozialeck 2009; Moulis and Thevenod 2010).

Metallothioneins

Various aspects of biology of metallothioneins (MTs) and their roles in mammalian and nonmammalian cells were reviewed and collected in a recent book by Sigel et al. (2009), and a few extensive reviews describing the proposed roles of MTs in the mammalian cells were published recently in scientific journals (Davis and Cousins 2000; Miles et al. 2000; Prozialeck 2009; Moulis and Thevenod 2010).

Numerous studies have shown that actions of Cd in the mammalian cells are closely connected with the expression and function of MTs, a family of small (6–7 kilodaltons (kDa)) metal-binding proteins. MTs represent evolutionary conserved, heat-stable single-chain polypeptides, with 61–68 amino acids, among them are 20 cysteines. In the metal-free form of proteins (thioneins, apo-MTs), cysteines are assembled in a characteristic pattern, and the tertiary-structured protein is arranged in two domains (α- and β-clusters) with different metal-binding capacity. These structural and Cd-binding characteristics are schematically shown in Fig. 1 for MT1, an MT isoform common to most mammalian cells. In Western blot of the rat kidney and liver cytoplasm, this protein usually exhibits several bands, a strong one at $\sim$7 kDa, reflecting the monomeric forms, and one or more weaker bands of higher molecular mass that reflect the presence of oligomers, such as dimers and/or trimers (Fig. 1c).

The thiol groups in apo-MTs are highly reactive, and the proteins exhibit high-affinity/high-capacity binding properties with various transition metal ions. One molecule of apo-MT can bind seven or more Zn, Cd (shown in Fig. 1b), Cu, or Hg ions. Although in physiological conditions, Zn is the usual content of the metal-MT complex, the common apo-MTs in the mammalian cells have higher affinity for Cd, Hg, Pb, or Cu than for Zn, and the stability constant of the CdMT complex is up to 1,000-fold higher than that of ZnMT. These phenomena have two important consequences for the onset and development of toxicity induced by Cd or other toxic metals: (a) higher-affinity metals can displace Zn from ZnMT, and (b) higher-affinity metal-MT complexes exhibit higher resistance to degradation by proteases. However, some studies

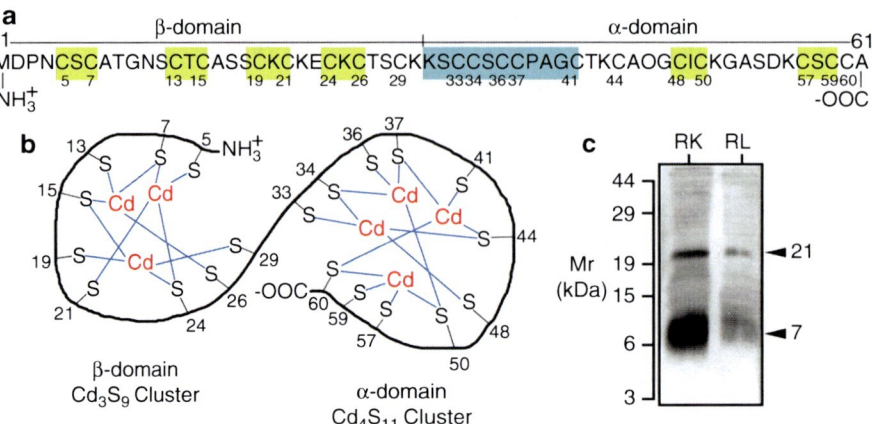

Cadmium and Metallothionein, Fig. 1 (a) Amino acid sequence of the rabbit MT1 contains 61 residues, 20 of them being cysteines (C; the numbers under the Cs denote their place within the sequence). Six C-X-C repeats (C is cysteine and X can be any other amino acid) are labeled yellow, whereas the blue-labeled sequence K-X-X-C-C-X-C-C-P-X-X-C denotes a characteristic pattern common to all mammalian MTs. (b) Schematic representation of the MT1 tertiary structure. The protein exhibits two functionally independent Cd-binding domains (clusters); the α-domain binds four Cd ions (Cd_4S_{11} cluster), whereas the β-domain binds three Cd ions (Cd_3S_9 cluster). Blue lines represent interactions of Cd ions with the thiol (S) groups of specific cysteines indicated by numbers. The scheme was generated using the information in Klaassen et al. (1999) and Sigel et al. (2009). In physiological conditions, the molecule can bind an equal number of Zn ions. Recent studies have shown that in various tumor cells, established cell lines, and in fresh tissues from various rat organs, only a part of MTs is complexed with Zn; a large proportion of the total MT (up to 90% in some cell lines and tumor cells and up to 54% in the rat tissues) exists as the metal-free form (apo-MT) (Petering et al. 2006; Sigel et al. 2009). (c) Western blot of MT in cytosolic fractions from the rat kidney cortex (RK) and liver (RL) homogenates. In these samples, a monoclonal antibody, which recognizes a highly conserved domain common to MT1 and MT2, labeled a strong ~7 kDa and a weak ~21 kDa protein band related to monomeric and oligomeric (trimeric) forms of the proteins, respectively. *Mr* relative molecular mass

have shown that the metal-MT complex is less sensitive to proteolysis than the apo-MT in the acidic lysosomes (pH ~ 5.5), but in the neutral cytoplasmic conditions (pH ~ 7), apo-MT is more stable than the metal-saturated MT (Foulkes 1986; Miles et al. 2000; Sigel et al. 2009).

Regulation of Metallothionein Expression: Effect of Cadmium

Various effectors of MT gene activity and MTs expression at the level of mRNA and protein have been reviewed in more detail elsewhere (Andrews 2000; Davis and Cousins 2000; Miles et al. 2000; Sigel et al. 2009; Sabolic et al. 2010). MTs in rats and mice are controlled by four genes which code for apoproteins MT1, MT2, MT3, and MT4, whereas human MTs are controlled by 17 genes, 10 of which are functional. The common four MTs (MT1–MT4), present in most mammalian organs/tissues and studied most extensively, exhibit significant polymorphism in their amino acid sequences, and their expression in some organs shows species and sex differences.

MT3 is predominantly expressed in the brain, while MT4 is expressed exclusively in stratified squamous epithelia, where they may function as intracellular regulators of Zn and Cu and cell protectors by scavenging reactive oxygen species (ROS). MT3 and MT4 are poorly responsive to the usual inducers of MT expression, such as Zn and Cd. On the other hand, MT1 and MT2 represent the most prevalent thioneins expressed in most mammalian tissues, and their roles in homeostasis of essential metals (Zn, Cu) and Cd toxicity have been extensively documented in various experimental models in vivo and in vitro. A number of important information regarding the possible roles of MTs in the cell physiology and Cd-induced toxicology has been collected from the experiments in genetically modified mice with inactivated MT1 or MT2 (or both) genes and nominal absence of the respective proteins (knockout (KO) mice) and in mice with multiple identical MT genes and overproduction of the respective proteins (transgenic (MT-TG) mice).

A variety of factors are known to activate the MT1 and MT2 genes and upregulate the production

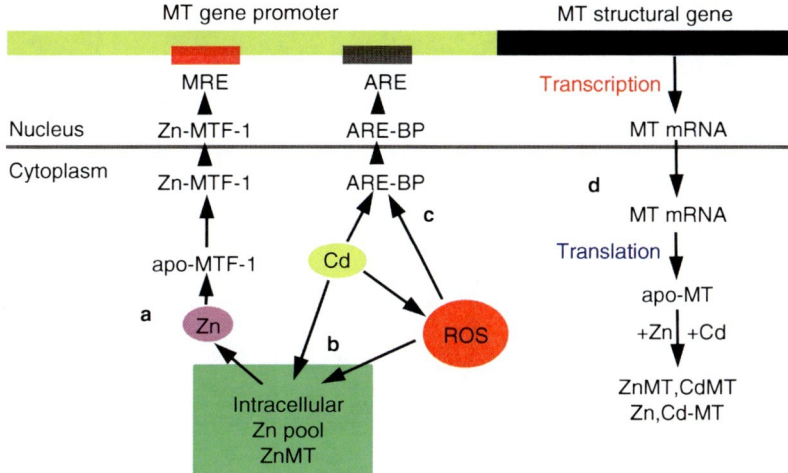

Cadmium and Metallothionein, Fig. 2 A simplified scheme showing regulation of the mouse *MT1* gene activity with Zn or Cd. More complicated interactions, with other activators and inhibitors of the gene activity, were elaborated in detail previously (Davis and Cousins 2000; Miles et al. 2000; Sigel et al. 2009; Sabolic et al. 2010). MRE (metal response element) and ARE (antioxidant response element) are specific locations in the gene promoter. ARE-BP, ARE-binding protein; MTF-1, metal-regulated transcription factor 1; ROS, reactive oxygen species (free radicals, such as H_2O_2). (**a**) The apo-MTF-1 is an inactive cytoplasmic protein, which can be activated by Zn-induced phosphorylation and then migrates as Zn-MTF-1 into the nucleus to associate with MRE, a step which activates the MT transcription. In case of Cd intoxication, (**b**) Cd ions and ROS can activate MTF-1 indirectly, by mobilizing Zn from the intracellular pool (including from ZnMT) and, (**c**) independently on Zn, by activating the ARE-BP, which translocates into the nucleus, binds to ARE, and activates the *MT* gene. (**d**) The final product of the *MT* gene activation is apo-MT, which can bind Zn (ZnMT) or Cd (CdMT) or both ions (Zn, CdMT) in any stoichiometric combination. A synergistic activation of the MT gene via MRE and ARE may be important in Cd-induced hepatotoxicity and nephrotoxicity, when a high intracellular concentration of Cd is associated with an increased production of free radicals (Modified from Sabolic et al. 2010, with kind permission from Springer Science+Business Media B.V.)

of apo-MTs. Cd and various stress-related conditions are particularly strong inducers, largely via generating free radicals, e.g., ROS and reactive nitrogen species (RNS) in the affected cells. Zn and ROS/RNS, generated during normal metabolism, may be the major players in regulating MT gene expression in physiological conditions. As shown schematically in Fig. 2, the induction of mouse liver MT1 gene results in an enhanced transcription following interaction of inducers with specific response elements in the gene promoter.

Roles of Metallothioneins in Physiology and Cadmium-Induced Toxicity

MTs are predominantly intracellular proteins detected largely in the cell cytoplasm but also in lysosomes, mitochondria, and nuclei. The presence of MTs in a variety of mammalian cells and their organelles suggests possible important intracellular functions of these metalloproteins. However, the MT-null (KO for MT1 and MT2) and MT-TG mice have no significant phenotypic or reproductive problems, indicating that these functions are not vital for animal survival and reproduction. The following roles for MTs in the mammalian cells have been proposed: (a) intracellular storage, transport, and homeostasis of the essential metals Zn and Cu; (b) provision of Zn and Cu for metalloproteins, enzymes, and transcription factors in various intracellular compartments during prenatal, perinatal, and postnatal periods; (c) binding and neutralizing the highly reactive ionic forms of toxic metals, such as Cd, thus providing protection against their cytotoxicity, genotoxicity, and carcinogenicity; (d) scavenging free radicals (ROS/RNS) generated in normal metabolism and in oxidative stress induced by toxic metals and other factors; and (e) protection of cell vitality in neurodegenerative and other diseases (Foulkes 1986; Klaassen et al. 1999; Miles et al. 2000; Kang 2006; Sabolic 2006; Prozialeck 2009; Sigel et al. 2009; Moulis and Thevenod 2010).

The scheme of intracellular MT roles in the absence and presence of Cd is shown in Fig. 3. The redox status in the cell seems to be the major regulator of interaction of Zn, apo-MT, and other apo-proteins.

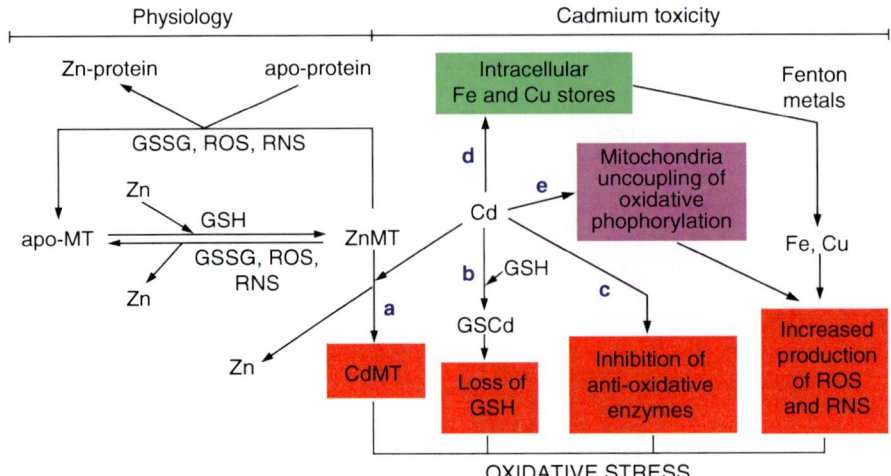

Cadmium and Metallothionein, Fig. 3 Schematic presentation of the role of intracellular MTs in Zn homeostasis in physiological conditions and in the presence of toxic concentrations of Cd. In physiological conditions, a relative abundance of apo-MT and ZnMT depends on the intracellular redox state; the reduced conditions, with higher GSH/GSSG ratio, favor formation of ZnMT, whereas the oxidized conditions, with lower GSH/GSSG ratio and a limited abundance of reactive species (ROS/RNS), favor the release of Zn and its transfer to various Zn-dependent metalloproteins. (**a**) In Cd-induced toxicity, due to higher affinity of SH groups for Cd, Zn is released and replaced by Cd, thus forming the nontoxic CdMT complex. When the Cd-binding capacity of MTs is surpassed, (**b**) the free Cd ions bind to and deplete GSH and other intracellular antioxidants, thus decreasing the GSH/GSSG ratio, (**c**) inhibit the activity of antioxidative enzymes, (**d**) displace and release the Fenton metals Fe and Cu from intracellular stores, which then promote generation of free radicals, and (**e**) uncouple the oxidative phosphorylation in mitochondria, causing a massive production of free radicals. An overall manifestation of this toxic condition represents oxidative stress. As discussed in Fig. 2, the liberated Zn and ROS enhance the activity of MT gene and production of apo-MT. Being itself redox active, the newly synthesized apo-MT can fight oxidative stress by binding and neutralizing free radicals (ROS/RNS scavenger). However, ROS and RNS can also act as signaling molecules and activate a cascade of reactions that can lead to cell death in form of apoptosis or necrosis. All these processes have been elaborated previously (Andrews 2000; Miles et al. 2000; Kang 2006; Sabolic 2006; Prozialeck 2009; Sigel et al. 2009; Moulis and Thevenod 2010)

In physiological conditions, when the metabolic production of ROS/RNS is limited, and the ratio of reduced (GSH) versus oxidized (GSSG) glutathione is high, the reactive thiol groups of apo-MT exhibit high affinity for Zn, forming the ZnMT complex. The prooxidative molecules (GSSG, ROS, RNS) interact with ZnMT, oxidize cysteines into cystines, and enable release of Zn and its transfer to various intracellular apo-proteins. The presence of Cd ions shifts these reactions into prooxidative direction. At low Cd concentration, due to higher affinity of MT thiol groups for Cd, Zn becomes released and replaced, and the toxic metal becomes entrapped and inactivated/detoxicated as the CdMT complex. In case of overloading with Cd, when the capacity of MTs to buffer toxic metals is lost, free Cd ions directly target various intracellular structures and functions, and the redox status becomes heavily compromised due to loss of antioxidants, inhibition of antioxidative enzymes, and enhanced production of prooxidants, resulting in the state named "oxidative stress" (Fig. 3).

Role of Metallothioneins in Cadmium Entry into the Organism

Various aspects of Cd handling in epithelia, as well as the role of MTs and other proteins in Cd transport, have been reviewed by Zalups and Ahmad 2003 and Bridges and Zalups 2005. The major routes of Cd entry into the mammalian body are lungs and intestine; absorption via the skin and/or eyes plays only a minor role in this respect. In the lung cells, MTs may protect from oxidative injury induced by Cd or other factors. Following inhalation of Cd-contaminated aerosols, the lungs react with inflammation, edema, and hemorrhage. These symptoms are strongly diminished in experimental animals with higher content of MTs in their lung tissue and intensify in the MT-null animals. A possible role of MTs in Cd transport across the alveolar epithelium into the blood has not been documented. In the gastrointestinal tract, Cd in the contaminated drink and food is absorbed largely by enterocytes in the proximal small intestine. The endogenous MTs are expressed in the intestinal epithelium, but their primary function may be storage and regulation of the intracellular Zn concentration and neutralization of the locally generated free radicals. Cd stimulates the expression of MTs in the intestinal cells and is sequestered as CdMT, but a possible role of MTs in the intestinal Cd absorption is unclear.

Following absorption in the lungs or intestine, Cd is distributed by systemic circulation to various organs largely bound to albumin and other thiol-containing reactive biomolecules in the plasma, and less as CdMT. In humans, a small concentration of MTs is always present in the plasma, which in a healthy organism may distribute Zn and Cu among organs. Much higher (>tenfold) concentration of plasma MTs was measured in workers occupationally exposed to Cd and in experimental animals treated with $CdCl_2$ (Foulkes 1986; Prozialeck 2009; Sigel et al. 2009).

Role of Metallothioneins in Cd-Induced Hepatotoxicity

Although Cd is heavily toxic for all mammalian organs, liver and kidneys represent the major targets where interactions of Cd and MTs in acute and chronic Cd intoxication have been studied most extensively. In these organs, MTs are important for prevention and protection against Cd-induced toxic effects, but in the kidneys, MTs can also be mediators of Cd toxicity.

Cd circulating in the blood enters the liver cells by poorly defined mechanisms; while free Cd ions may cross the hepatocyte plasma membrane by various transporters, Cd-albumin and similar protein complexes may be internalized by endocytosis (Zalups and Ahmad 2003; Bridges and Zalups 2005). However, this may not be valid for the CdMT complex. As found in our immunocytochemical studies in cryosections of the liver tissue from control rats (Fig. 4), individual hepatocytes exhibited a variable expression of endogenous MTs in their cytoplasm, whereas reticuloendothelial (Kupffer) cells showed no significant staining for MTs (Fig. 4a and inset). However, 15 min following intravenous (i.v.) injection of CdMT, a massive, endocytosis-mediated accumulation of this complex was observed in intracellular vesicles of Kupffer cells (Fig. 4b and inset), whereas hepatocytes exhibited no visible internalization of this complex. On the other hand, the process of endocytosis in the same liver cells was highly active, as shown in Fig. 4c, already 5 min after i.v. injection of the fluorescent marker FITC-dextran in rats both Kupffer cells and hepatocytes exhibited a vigorous, endocytosis-mediated accumulation of this marker in numerous intracellular vesicles. This indicates that the CdMT complex is probably not a substrate for endocytosis in the rat hepatocytes in vivo, possibly because megalin, a scavenger receptor for MTs, is missing in the hepatocyte plasma membrane (Sabolic et al. 2010). Rather, the complex may be endocytosed by macrophages, e.g., the Kupffer cells, and degraded in their lysosomes, and the liberated Cd ions may be released into the pericellular space and/or blood (Fig. 4d) and taken up by hepatocytes via the poorly defined ionic transports (Zalups and Ahmad 2003). This further means that in chronic Cd intoxication, when hepatocytes produce and release into circulation an increased amount of CdMT (see later), an amount of Cd from the complex could continuously recycle between the Kupffer cells and hepatocytes and cause toxicity for a long time, while the bulk of CdMT in circulation ends in the kidneys.

In various experimental animals, Cd acts as a potent hepatotoxin after acute or chronic exposure. Inside the hepatocytes, Cd binds to cytoplasmic proteins, largely to the existing (endogenous) MTs, and becomes trapped inside the cytoplasm, inactivated, and detoxicated. During a short-term poisoning with high Cd doses, it is assumed that Cd ions initially bind to the existing MTs until saturation, and thereafter, they become toxic, causing an extensive oxidative stress and ROS/RNS-mediated damage of the cell structure and function, manifested by inflammation, apoptosis, and/or necrosis of hepatocytes (Foulkes 1986). Various studies have shown that in acute Cd-induced hepatotoxicity, MTs have strong protective and antioxidative functions. Acute hepatotoxicity is (a) weak or absent in immature rats, which in the liver exhibit much higher concentrations of MTs than the adult animals; (b) prevented by pretreating animals with small doses of Cd or Zn, which in the liver and other organs induce synthesis of new apo-MTs; (c) absent in TG mice, which in the liver have elevated levels of MT1; and (d) increased in MT-null mice (Miles et al. 2000; Prozialeck 2009; Sigel et al. 2009; Moulis and Thevenod 2010).

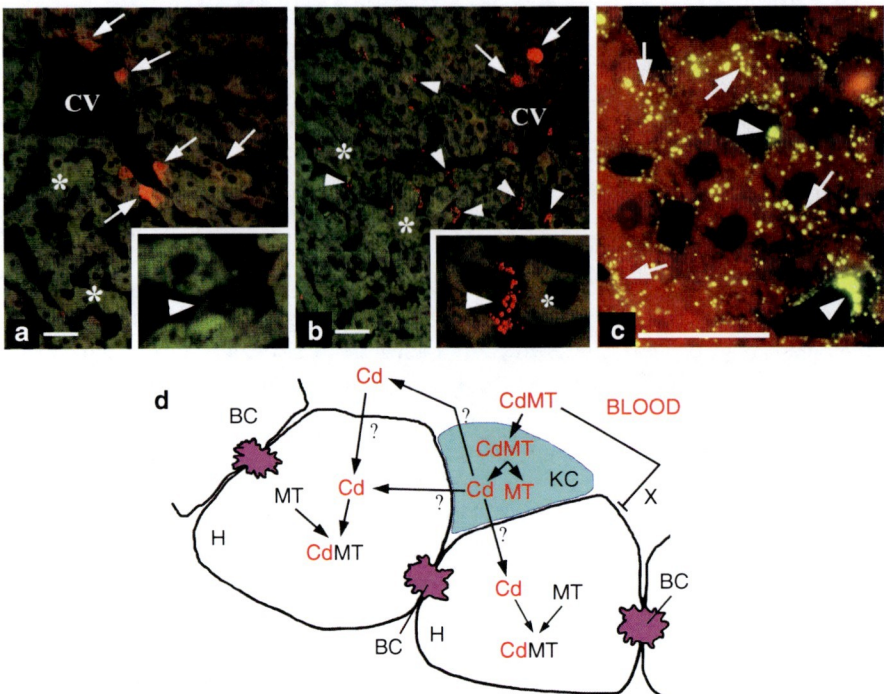

Cadmium and Metallothionein, Fig. 4 Fate of CdMT in the rat liver. (**a**) Immunolocalization of MTs in the liver of rats that had been injected i.v. with saline 15 min before sacrifice. Individual hepatocytes, largely those around the central vein (CV), exhibited a variable cytoplasmic staining of endogenous MT (*red* fluorescence, *arrows*), whereas most hepatocytes (*asterisks*) and the Kupffer cells (*inset, arrowhead*) did not show a significant MT staining. (**b**) In rats injected i.v. with CdMT (0.4 mg Cd/kg body mass) 15 min after the injection, intracellular organelles with the red-stained material indicated the presence of endocytosed CdMT in the Kupffer cells (*arrowheads* and *inset*). Some hepatocytes exhibited a variable content of endogenous MT (*arrows*), but most of them were negative for endogenous MT (*asterisks*), and none of them showed any endocytosis-mediated internalization of CdMT. However, (**c**) both the Kupffer cells (*arrowheads*) and hepatocytes (*arrows*) exhibited a vigorous endocytosis of the fluorescent marker FITC-dextran; already 5 min following the i.v. injection of this marker, the fluorescence accumulated in numerous yellow-stained intracellular vesicles randomly scattered in the cell cytoplasm (*arrows*). Bars, 20 μm. (**d**) Scheme of the fate of CdMT in the rat liver. CdMT is internalized by endocytosis in the Kupffer cells (KC) and degraded in lysosomes. The liberated Cd ions are released into the pericellular space and/or blood and taken up by the surrounding hepatocytes (H) via the poorly defined transport mechanisms (?). The endocytosis-mediated internalization of CdMT in hepatocytes seems to be very low, if present at all (X). In hepatocytes, the CdMT complex is reformed by binding of Cd ions to endogenous MTs. *BC* bile canaliculi

Chronic Cd hepatotoxicity in humans can result from a long-term exposure to small Cd concentrations, whereas in experimental animals, this condition can be mimicked with repeated small doses of $CdCl_2$ from a few weeks to a few months injected intraperitoneally or subcutaneously (s.c.) or by oral exposure for several months to years. Such treatments result in a time- and dose-dependent accumulation of Cd and increase in MT content in the liver due to continuous stimulation of MT synthesis. With time, the cytoplasmic CdMT accumulates and can reach a very high concentration. A time-dependent upregulation of MTs, GSH, and antioxidative enzymes and amelioration of the toxicity by various antioxidants indicate oxidative stress as the underlying mechanism in this condition (Moulis and Thevenod 2010). An example of Cd and MT accumulation in a subchronic model of Cd hepatotoxicity in rats is demonstrated in Fig. 5.

The final result of either acute or chronic Cd-induced hepatotoxicity is a release of CdMT from the damaged (apoptotic and/or necrotic) hepatocytes into the blood circulation, by which this complex reaches the kidneys.

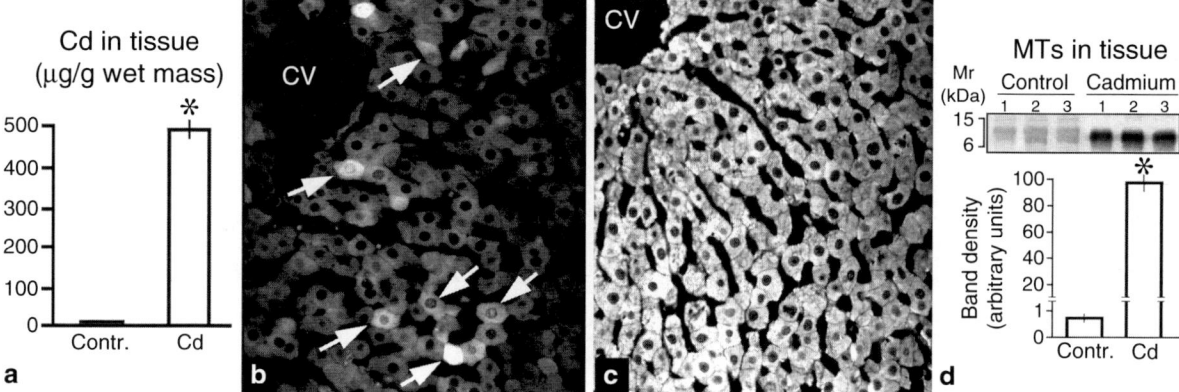

Cadmium and Metallothionein, Fig. 5 Cd and MTs in the liver of rats ($n = 3$) treated s.c. with saline (0.5 mL/kg body mass/day; control) or $CdCl_2$ (2 mg Cd/kg body mass/day) for 14 days. (**a**) After 14 days of treatment, the tissue content of Cd increased from nearly zero in controls to ~500 μg/g wet mass in Cd-treated rats. (**b**) Immunocytochemical localization of MTs in the liver of control rats; the individual hepatocytes expressed a variable staining intensity, reflecting a variable content of endogenous MTs in their cytoplasm (*arrows*). (**c**) In $CdCl_2$-treated rats, all hepatocytes exhibited a very strong staining for MTs. (**d**) Western blot of proteins in the liver tissue extract revealed a weak MT band (~7 kDa) in control rats and a strongly upregulated band in $CdCl_2$-treated rats. By densitometry, the band density in Cd-treated rats was ~135-fold stronger than in control animals. *CV* central vein. *Mr* relative molecular mass; *, Vs. control, $P < 0.05$ (Modified from Sabolic et al. 2010, with kind permission from Springer Science+Business Media B.V.)

Role of Metallothioneins in Cd-Induced Nephrotoxicity

In humans and experimental animals, kidneys represent the major target in the long-term environmental and/or occupational exposure to Cd. Cd primarily targets the structure and function of proximal tubule (PT) cells, resulting in reabsorptive and secretory malfunctions with urinary symptoms resembling the acquired Fanconi syndrome (Foulkes 1986; Järup et al. 1998; Prozialeck 2009). The interaction of Cd and MTs in Cd-induced nephrotoxicity has been extensively studied in various animal models and mammalian cell cultures treated with $CdCl_2$ or CdMT. To induce Cd-induced nephrotoxicity in rodents, the animals can be treated in different ways with $CdCl_2$ from a few days to a few years, and such treatments result in kidney damage, which is always associated with the accumulation of Cd and upregulation of MTs in the tissue (Foulkes 1986; Järup et al. 1998; Sabolic et al. 2010; Prozialeck 2009; Sigel et al. 2009). An example of these phenomena is shown in Fig. 6, which contains the data obtained in the rat model of subchronic Cd-induced nephrotoxicity, where the accumulation of Cd in the kidney cortex was correlated with the expression of MTs by immunocytochemistry in tissue cryosections and by Western blotting of the tissue extract.

Although Cd may enter renal cells by different mechanisms (Zalups and Ahmad 2003; Bridges and Zalups 2005), there is a general opinion that chronic Cd-induced nephrotoxicity is primarily induced by the CdMT complex that arrives by circulation from the Cd-injured organs, largely from the liver. Such a pattern was proven (a) in rats following transplantation of the Cd-intoxicated rat liver into a healthy animal, where CdMT from the donor liver was released into circulation and caused toxic injury in the acceptor's kidneys, and (b) in rodents following s.c. or i.v. application of CdMT (Foulkes 1986; Sabolic 2006; Sabolic et al. 2010). Indeed, as shown in Fig. 7, in rats that had been injected i.v. with CdMT, the complex was filtered in the glomeruli and endocytosed by the PT cells; 15 min following i.v. injection, CdMT was localized in numerous intracellular organelles (endocytic vesicles and/or lysosomes) randomly scattered in the PT cell cytoplasm (Fig. 7b). This finding fits the current model of chronic Cd-induced nephrotoxicity, as depicted in Fig. 7c.

The existent and the newly synthesized apo-MTs in the cell cytoplasm during chronic Cd-induced nephrotoxicity are important for sequestration and accumulation of the ionic Cd, and in this way, they protect from and ameliorate the toxic actions of Cd ions as the ROS/RNS scavengers. This is supported by the

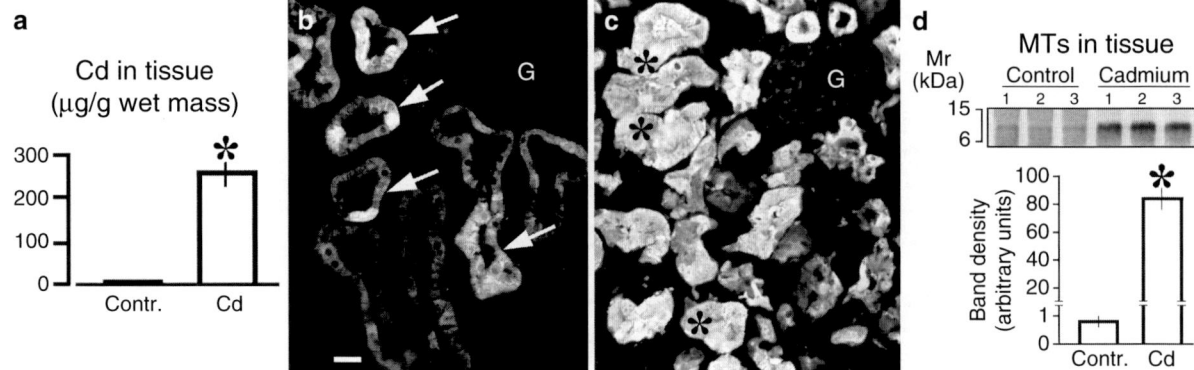

Cadmium and Metallothionein, Fig. 6 Cd and MTs in the kidney cortex of rats ($n = 3$) treated s.c. with saline (0.5 mL/kg body mass/day; control) or $CdCl_2$ (2 mg Cd/kg body mass/day) for 14 days. (**a**) After 14 days of treatment, the tissue content of Cd increased from nearly zero in controls to ~280 μg/g wet mass in Cd-treated rats. (**b**) Immunocytochemical localization of MTs in the cortical tubules of control rats; the cells in PT expressed a variable staining intensity, reflecting a variable content of endogenous MTs in their cytoplasm (*arrows*). (**c**) In $CdCl_2$-treated rats, the cells in all types of cortical tubules were strongly positive for MTs. In PT, the cells were edematous, and the tubule lumen was largely clogged with the debris from the injured epithelium (*asterisks*). (**d**) Western blot of proteins in the kidney cortex tissue extract revealed a weak MT band (~7 kDa) in control rats and a strongly upregulated band in $CdCl_2$-treated rats. By densitometry, the band density in Cd-treated rats was ~95-fold stronger than in control animals. *G* glomerulus; Bar, 20 μm, *Mr* relative molecular mass. *, Vs. control, $P < 0.05$ (Modified from Sabolic et al. 2010, with kind permission from Springer Science+Business Media B.V.)

following observations: (a) in MT-null mice treated with $CdCl_2$, the initial accumulation of Cd is not affected, but the rate of Cd elimination is faster, indicating that the cytosolic MTs are necessary for binding and sequestering Cd ions; (b) MT-null mice are more sensitive to chronic $CdCl_2$-induced nephrotoxicity than the normal (wild-type) mice; and (c) pretreatment of rodents with small doses of $ZnCl_2$ and $CdCl_2$ upregulates MTs and protects from acute nephrotoxicity induced by CdMT. However, in case of overloading conditions with Cd ions, it is assumed that the binding capacity of cytoplasmic MTs eventually becomes insufficient and that Cd ions attack other molecules and induce a full-scale oxidative stress and toxic condition that can end in apoptosis or necrosis of the tubular epithelium (Miles et al. 2000; Thevenod 2003; Sabolic 2006; Prozialeck 2009; Sigel et al. 2009; Moulis and Thevenod 2010).

Cadmium-Related Expression of Metallothioneins in Other Mammalian Organs

In rodents, acute or chronic treatment with $CdCl_2$ induced various toxic effects in reproductive organs, such as vascular damage, testicular necrosis, degenerative changes in ovaries, and loss of reproductive potency. These effects were not associated with significant changes in the expression of MTs at the level of mRNA and/or protein so that the role of MTs in these organs remains unknown (Foulkes 1986; Järup et al. 1998; Prozialeck 2009; Moulis and Thevenod 2010). However, MTs may play a protective role in Cd-induced osteotoxicity; a long-term exposure to Cd in humans and experimental animals is associated with a loss of calcium from the bones and its excretion in the urine, thus increasing the risk of kidney stones, osteomalacia, and osteoporosis. Following Cd treatment in rats, MT1 and MT2 proteins and their mRNA in bone cells are upregulated, playing a protective and antiosteoporotic role (Järup et al. 1998; Prozialeck 2009). Furthermore, a chronic environmental or experimental exposure to Cd in humans and animals can induce anemia and impaired immunity. Platelets and various blood cells contain a limited amount of Cd-inducible MTs, which seem to protect from Cd-related oxidative stress (Järup et al. 1998; Prozialeck 2009). MTs were also found to be protective against the ROS/RNS-mediated cell damage in ischemia-reperfusion injury of the heart. Finally, recent epidemiological and experimental data have indicated that chronic exposure to Cd in humans and animals can be associated with cancerogenesis in various organs. The animals with higher content of MTs in the specific organs (lungs, liver) had a lower incidence of Cd-induced carcinoma, thus suggesting that

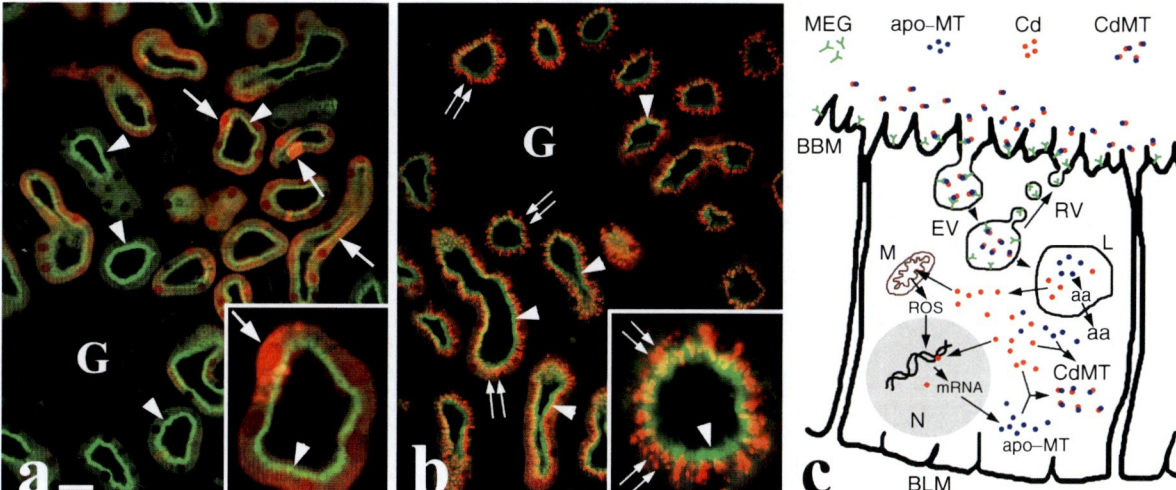

Cadmium and Metallothionein, Fig. 7 Fate of the i.v. injected CdMT and localization of megalin in the rat kidney, as shown by immunofluorescence staining with specific antibodies. (**a** and *inset*) In control rats injected i.v. with saline (0.5 mL/kg body mass) 15 min before sacrifice, many cells of the cortical PT exhibited a variable content of endogenous MTs in their cytoplasm (red fluorescence; arrows), whereas megalin was localized largely in the cell subapical domain (*green* fluorescence; *arrowheads*). (**b** and *inset*) In rats injected i.v. with CdMT (0.4 mg Cd/kg body mass), 15 min later, the injected complex was detected in numerous intracellular organelles (endosomes and lysosomes) randomly scattered in the cell cytoplasm (*red*-stained organelles; *thin double arrows*), indicating that the complex had been filtered in the glomeruli and endocytosed by the epithelial cells. Megalin was labeled with green fluorescence (*arrowheads*). *G* glomerulus. Bar, 20 μm. (**c**) Schematic presentation of the current model of nephrotoxicity induced by CdMT in chronic Cd exposure. Being a small molecule, the circulating CdMT is freely filtered in the glomeruli, it binds to the receptor protein megalin at the bottom of the epithelial cell BBM, and this complex (CdMT-megalin) is internalized by endocytosis and delivered to the endolysosomal compartment for degradation. Following dissociation of this complex in the acidic medium, megalin recycles back to the BBM via recycling vesicles, whereas CdMT is further dissociated into ionic Cd and apo-MT. The latter is degraded, whereas the ionic Cd is transported into the cell cytoplasm, where it primarily binds to endogenous cytoplasmic MTs, thus becoming detoxicated, but it also indirectly induces production of free radicals (ROS/RNS) and stimulates the machinery for synthesis of apo-MTs. All these phenomena have been described elsewhere in more detail (Christensen et al. 2009; Prozialeck 2009; Sigel et al. 2009; Moulis and Thevenod 2010). BBM brush-border membrane, BLM basolateral membrane, MEG megalin, EV endocytic vesicles, RV, recycling vesicles, L lysosome, M mitochondrion, aa amino acids

sequestering Cd and scavenging ROS/RNS during the oxidative stress may protect from activation of oncogenes and uncontrolled cell proliferation (Miles et al. 2000; Prozialeck 2009; Moulis and Thevenod 2010).

Cadmium-Related Expression of Metallothioneins as a Diagnostic Tool

As reviewed in more detail previously (Prozialeck 2009; Moulis and Thevenod 2010), MTs in blood and urine can be used in the human population for diagnostic purposes in conditions related to Cd-induced toxicity. Thus, the expression of MT1/MT2 mRNA in blood and peripheral lymphocytes can be used as a sensitive biomarker of environmental and occupational exposure to Cd, whereas recent epidemiological studies revealed that Cd-exposed humans exhibit anti-MT antibodies in their blood plasma and that these antibodies could be used as a biomarker for severity of this toxic condition. In the urine, MTs were found to be closely linked to urine Cd and may represent a perspective indicator of environmental and/or occupational Cd exposure and associated kidney damage.

Acknowledgments This work was supported by grant 022-0222148-2146 from Ministry for Science, Education and Sports, Republic of Croatia (I.S.).

References

Andrews GK (2000) Regulation of metallothionein gene expression by oxidative stress and metal ions. Biochem Pharmacol 59:95–104

Bridges CC, Zalups RK (2005) Molecular and ionic mimicry and the transport of toxic metals. Toxicol Appl Pharmacol 204:274–308

Christensen EI, Verroust PJ, Nielsen R (2009) Receptor-mediated endocytosis in renal proximal tubule. Pflugers Arch Eur J Physiol 458:1039–1048

Davis SR, Cousins RJ (2000) Metallothionein expression in animals: a physiological perspective on function. J Nutr 130:1085–1088

Foulkes EC (ed) (1986) Cadmium, vol 80, Handbook of experimental pharmacology. Springer, Berlin/Heidelberg, pp 1–400

Järup L, Berglund M, Elinder CG, Nordberg G, Vahter M (1998) Health effects of cadmium exposure – a review of the literature and a risk estimate. Scand J Work Environ Health 24(suppl 1):1–51

Kang YJ (2006) Metallothionein redox cycle and function. Exp Biol Med 231:1459–1467

Klaassen CD, Liu J, Choudhuri S (1999) Metallothionein: an intracellular protein to protect against cadmium toxicity. Annu Rev Pharmacol Toxicol 39:267–294

Miles AT, Hawksworth GM, Beattie JH, Rodilla V (2000) Induction, regulation, degradation, and biological significance of mammalian metallothioneins. Crit Rev Biochem Mol Biol 35:35–70

Moulis J-M, Thevenod F (Guest eds) (2010) New perspectives in cadmium toxicity. BioMetals 23(5):763–960

Petering DH, Zhu J, Krezoski S, Meeusen J, Kiekenbush C, Krull S, Specher T, Dughish M (2006) Apo-metallothionein emerging as a major player in the cellular activities of metallothionein. Exp Biol Med 231:1528–1534

Prozialeck W (Guest ed) (2009) New insights into the mechanisms of cadmium toxicity – advances in cadmium research. Toxicol Appl Pharmacol 238(3):1–326

Sabolic I (2006) Common mechanisms in nephropathy induced by toxic metals. Nephron Physiol 104:107–114

Sabolic I, Breljak D, Skarica M, Herak-Kramberger CM (2010) Role of metallothionein in cadmium traffic and toxicity in kidneys and other mammalian organs. Biometals 23:897–926

Sigel A, Sigel H, Sigel RKO (eds) (2009) Metallothioneins and related chelators, vol 5, Metal ions in life sciences. Royal Society of Chemistry, Cambridge, UK, pp 1–514

Thevenod F (2003) Nephrotoxicity and the proximal tubule. Insights from cadmium. Nephron Physiol 93:87–93

Zalups RK, Ahmad S (2003) Molecular handling of cadmium in transporting epithelia. Toxicol Appl Pharmacol 186:163–188

Cadmium and Oxidative Stress

Ann Cuypers, Tony Remans, Vangronsveld Jaco and Karen Smeets
Centre for Environmental Sciences, Hasselt University, Diepenbeek, Belgium

Synonyms

Cellular redox balance; Oxidative challenge

Definition

Oxidative stress is a process in which the cellular redox balance between pro- and antioxidants is disturbed in favor of the former.

Introduction

Cadmium (Cd) contamination is a widespread complication of industrial's reliance on metals and the intensive use of agrochemicals containing Cd. Cadmium is not biodegradable, and therefore, the global environmental risk is constantly increasing due to its accumulation via the food chain. Furthermore, as Cd is classified as a type I carcinogenic element, it poses a serious threat to humans and animals but also other organisms, such as plants, fungi, and bacteria, present in all compartments of different ecosystems, are negatively affected by Cd.

Cadmium has a very high affinity for sulfhydryl (thiol) groups, and through binding with these functional groups of, for example, enzymes or structural proteins, it causes metabolic disruptions (Sharma and Dietz 2009; Cuypers et al. 2010). Many micronutrients are cations and are essential components of metalloproteins, as cofactor in the enzymatic catalysis and manifold of other cellular processes (Cuypers et al. 2009). Under physiological conditions, Cd is present as a bivalent cation that can either replace essential elements from their complexes or interfere with their uptake and, in this way, exert its toxic action. Another important underlying mechanism of Cd-induced morphological and physiological damage is oxidative stress, a process in which the cellular redox balance between pro- and antioxidants is disturbed in favor of the former. Although Cd is not redox active and therefore unable to directly induce reactive oxygen species (ROS) production, multiple studies report elevated ROS levels when organisms are exposed to Cd (Cuypers et al. 2009, 2010). The resulting Cd-induced oxidative challenge is important for downstream induction of both damage and signaling processes (Fig. 1). The interest in this apparent paradox is rapidly increasing in current research and is focused on in this entry. When comparing the results of different studies, one should bear in mind to discuss them in relation to the experimental setup used,

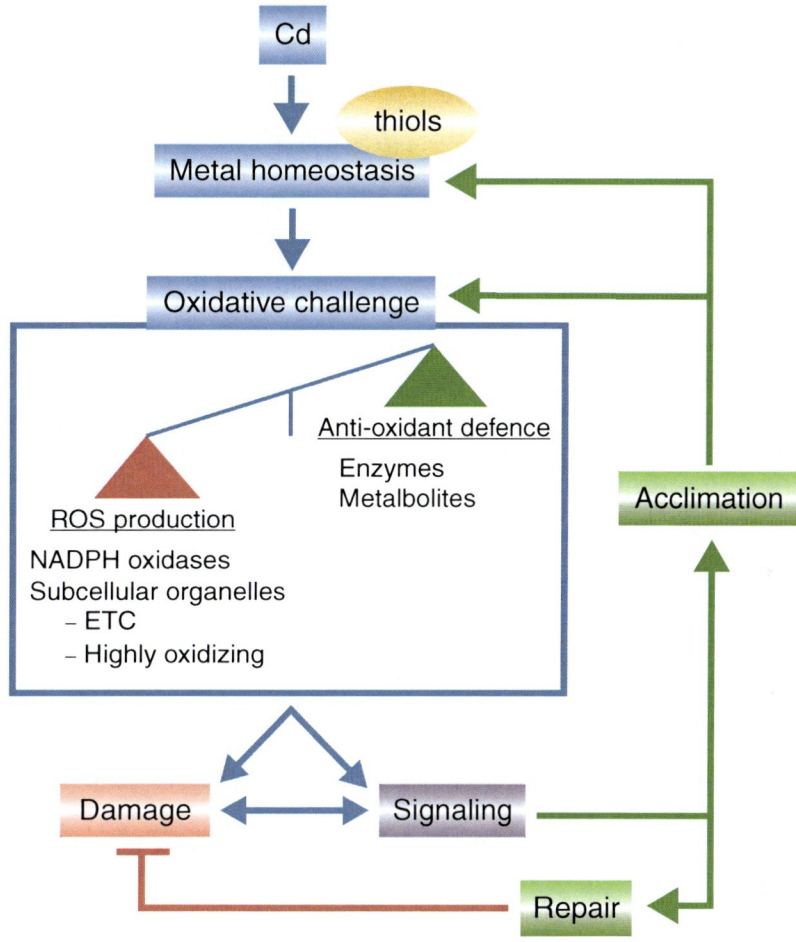

Cadmium and Oxidative Stress, Fig. 1 Generalized model for the Cd-induced cellular oxidative challenge

that is, (1) exposure time (acute or chronic), (2) the Cd concentration used, (3) the chemical form in which the metal is applied (e.g., $CdCl_2$ or $CdSO_4$), (4) the way of administration (e.g., food, drinking water, intraperitoneal in case of animals; growth media in case of plants), (5) in vitro or in vivo experiments, and (6) organism specific characteristics such as species, age, and developmental stage. The mechanisms of Cd-induced responses clearly depend on all these variables.

A disturbed redox homeostasis is one of the main reported outcomes of all Cd-based studies, and fluctuations in the oxidative challenge, determined by pro- and antioxidants leading to damage or signaling, are outlined in the next sections.

Cadmium and Thiols: Toxicity and Detoxification

In general, almost all metals strongly bind to thiol groups and thus to cysteine-rich proteins. As mentioned earlier, this is an important toxicity mechanism of Cd, but it also makes up one of the most important detoxification systems. Metabolites and peptides containing thiol groups are of primary importance in Cd detoxification through chelation, that is, formation of cellular Cd–thiol complexes. In human and animal systems, cysteine-rich metallothioneins bind and detoxify Cd, while phytochelatins form important Cd-chelating agents in plants (Valko et al. 2006; Sharma and Dietz 2009; Cuypers et al. 2010).

Metallothioneins are gene-encoded cysteine-rich polypeptides, while phytochelatins are short cysteine-rich metal-chelating peptides in plants that are enzymatically formed. Glutathione (GSH; γ-Glu-Cys-Gly) is a highly abundant tripeptide, containing one cysteine molecule that forms the core of its properties as a chelating agent, but also as an antioxidant metabolite. As such, diverse detoxification pathways are combined in GSH.

$$Fe^{2+} + H_2O_2 \longrightarrow Fe^{3+} + OH^- + OH^\circ \text{ (Fenton)}$$
$$O_2^{\bullet-} + H_2O_2 \xrightarrow{Fe(III)/Cu(II)} O_2 + OH^- + OH^\circ \text{ (Haber Weiss)}$$

Cadmium and Oxidative Stress, Fig. 2 Fenton and Haber–Weiss reaction

Cadmium-Induced ROS Production

All aerobic organisms need oxygen for energy production in mitochondria. As a consequence, electrons can leak from the electron transport chain to oxygen (O_2), and during this monovalent reduction, ROS can be formed such as superoxide ($O_2^{\bullet-}$), hydrogen peroxide (H_2O_2), and hydroxyl radical ($^\bullet OH$) (Halliwell 2006).

Redox-active metals such as iron (Fe) can directly induce ROS production via Fenton-like reactions (Fig. 2), whereas Cd, as a bivalent cation, is unable to perform these reactions. It can merely alter the cellular redox status via targeting thiol, carboxyl, and other functional groups present in the electron transport chain or antioxidative defense enzymes. Also, it activates prooxidative enzymatic mechanisms such as NADPH oxidases, which results in a disturbed redox balance in favor of the prooxidants. This status is currently known as oxidative stress, but the term "stress" has a negative connotation as ROS also exert beneficial functions within a cell, alternatively the term "oxidative challenge" is used.

Replacement of Redox-Active Elements
When freely available in the cytoplasm, redox-active elements directly enhance the production of $^\bullet OH$ through the Fenton reaction (Fig. 2). By replacing metals in various proteins, Cd increases the amount of free redox-active metals in the cell. Reduction of the oxidized metal ion can be achieved by the Haber–Weiss reaction with $O_2^{\bullet-}$ as a substrate (Fig. 2) (Cuypers et al. 2010).

Mitochondrial ROS Production
Due to the presence of the electron transport chain and its associated ROS production, mitochondria are a major target for different environmental stressors including Cd (Cannino et al. 2009). Different studies point toward the generation of ROS and mitochondrial dysfunction during Cd stress caused by morphological alterations and physiological disturbances. The thiol groups of respiratory complexes I to V are critical targets for Cd, and activities of complexes II and III appear the most inhibited in liver, brain, and heart of Cd-exposed guinea pigs. Production of ROS is maximal at the level of complex III and is related to the opening of the mitochondrial permeability transition (MPT) pore. A disrupted mitochondrial electron transport and corresponding mitochondrial permeability lead to an increased ROS production and/or leakage as compared to healthy tissues (Cannino et al. 2009; Cuypers et al. 2010).

Mitochondrial ROS production might lead to the oxidation of membrane phospholipids, mitochondrial DNA cleavage, and impaired ATP generation with resulting mitochondrial damage and induction of apoptosis, effects that resemble aging-related processes. It is noteworthy to mention that Cd also affects the regulation of mitochondrial gene expression and is involved in mitochondrial retrograde signaling to the nucleus (Cannino et al. 2009). In this way, not only apoptosis but also cell survival and/or (malignant) cell proliferation could be linked to mitochondrial ROS production with Cd as a transcriptional regulator of cell proliferation processes (Liu et al. 2009).

Induction of NADPH Oxidases
NADPH oxidases (NOX) are multicomponent enzymes that use electrons derived from intracellular NADPH sources, to reduce molecular O_2 generating $O_2^{\bullet-}$ in the extracellular space. Their role as key components in human innate host defense is clearly demonstrated in multiple studies. As they also generate ROS in their metabolic reaction without external stress factors, the NOX family contributes to cellular signaling, regulation of gene expression, posttranslational protein processing that can lead to cell differentiation, and apoptosis (Thévenod 2009).

The gene expression of *NOX4* was upregulated in mice kidneys following chronic exposure to low levels of Cd. Unlike other NOX proteins, NOX4 does not depend on cytoplasmic cofactors and is regulated only at the level of transcription. Increased *NOX4* gene expression, thus could have led to the observed increase in NOX activity in these kidneys. The exact role NOX4 plays in Cd toxicity is not yet described but may be to produce ROS as a signal to activate the antioxidative defense system leading to acclimation (Fig. 1). In Cd-exposed hepatocytes, NADPH-dependent ROS production is also described and triggers signal transduction leading to protection mechanisms. Alternatively, excess ROS production may also cause Cd-induced damage in, for example, mouse neuronal cells. These authors describe a cyclooxygenase-2 upregulation induced by NADPH oxidase–dependent ROS, culminating in cell death after Cd exposure (Cuypers et al. 2010 and references therein).

In summary, cellular NOX activity can be affected by Cd, resulting in either signaling leading to the onset of cellular protection mechanisms or, alternatively, in cellular toxicity via the induction of cell death or (malignant) cell proliferation. Therefore, controlled levels of ROS production are crucial to ensure correct levels for signaling or defense responses. A large network of antioxidative mechanisms is described and will be discussed in the next paragraphs.

Cadmium-Induced Antioxidative Defense

During evolution, cells have developed protection mechanisms to minimize oxidative damage. Oxidants such as ROS are balanced against a well-developed antioxidative defense system consisting of both enzymes and metabolites present in all subcellular compartments and aiming at maximal protection (Halliwell 2006). In stress conditions, an increased ROS production triggers cells to activate and expand their antioxidative network. As such, Cd exerts a dual role on antioxidative capacity: it inhibits antioxidants (cfr. *supra*) but also activates several antioxidative components by disturbing the redox balance and, consequently, inducing a signal transduction cascade (Fig. 1).

Antioxidative Enzymes

Superoxide dismutases (SOD) are metalloenzymes catalyzing the dismutation of $O_2^{\bullet-}$ to the less reactive H_2O_2 with remarkably high efficiency. Multiple SOD isoforms exist, each containing a redox-active metal in their catalytic site to perform the reaction: CuZnSOD and MnSOD are expressed in different subcellular locations (Halliwell 2006). Activation of SOD during Cd stress is studied at transcriptional and metabolic levels. It is, however, difficult to generalize the SOD response due to the different exposure conditions and organs studied. In rats exposed to 50 mg Cd per liter of drinking water, during 12 weeks, an increased versus decreased total SOD activity in kidney and liver, respectively, was noticed. Acute exposure (24 h), however, resulted in a decrease of both kidney and liver SOD activities after an intraperitoneal administration of a single dose of 5 mg/kg. An in vitro study where neuronal cortical cells were exposed for 24 h to different Cd concentrations showed an increase in total SOD activity (Cuypers et al. 2010 and references therein).

Different SOD isoforms, according to their metallic cofactor, show specific subcellular distribution as well as response rates. Their involvement also has to be taken into account when comparing oxidative signatures described in different studies. Fluctuations in specific SOD isoforms can result in different subcellular ROS levels and species, but do they eventually lead to different physiological outcomes? Although there are clear indications for this, further research is needed to elucidate whether and how subcellular ROS levels are involved in downstream signaling (Valko et al. 2006; Thévenod 2009).

Both catalases (CAT) and peroxidases (Px) are involved in H_2O_2 quenching, and different roles during Cd exposure are suggested depending on the intensity of the stressor. Whereas CATs are active in severe stress situations, peroxidases (Px) are suggested to protect the cell against low levels of oxidative stress, possibly indicating a role for these enzymes in the fine-tuning of ROS levels important in signal transduction. In most organisms, CAT activity is mainly located in the peroxisomes and, to a lesser extent, in the cytoplasm of erythrocytes, the nucleus, and mitochondria, and it converts H_2O_2 to O_2 and H_2O. In contrast to CAT, Px-mediated H_2O_2 detoxification occurs via the oxidation of other organic substrates.

Glutathione peroxidase (GSH–Px), with GSH as a substrate, appears in five isoforms in most mammals and has a selenocysteine in its active site. Also in this case, as mentioned for SOD, the activation of CAT and Px is probably differentially regulated and dependent on tissue and organism studied, exposure conditions, etc. (Valko et al. 2006; Cuypers et al. 2010).

In mice, an increased CAT activity in liver was detected after 6 days of intraperitoneal Cd exposure that was similar to outcomes of experiments using rats exposed for 5 days to Cd via gastric gavage, where an increased blood CAT activity was observed. On the other hand, reduced CAT activities were observed in both kidney and liver of rats after acute intraperitoneal administration (24 h) as well as via Cd exposure through drinking water (10–30 days). Several possible underlying mechanisms for decreased CAT activities are hypothesized, for example, an interaction between Cd and the catalytic subunit of CAT; an observed Fe deficiency in kidney and liver of rats chronically exposed to Cd possibly decreased the CAT activity, since Fe is an essential element in the active center of CAT (Cuypers et al. 2010 and references therein). As for CAT, both increases and decreases in GSH–Px were observed in different studies. In rats, GSH–Px activity was found in blood cells after acute (24 h) intraperitoneal exposure to Cd, whereas decreased GSH–Px activities were reported after chronic Cd exposure in liver and kidney of mice and rats. As selenium (Se) forms the core of the active site of the GSH–Px, a possible underlying mechanism for a decreased activity is via Se depletion through Cd–Se–Cys complex formation. Furthermore, competition between GSH–Px and metallothioneins for S–amino acids could be a potential cause for a reduction in GSH–Px activity during Cd stress (Cuypers et al. 2010 and references therein).

Antioxidative Metabolites

Whereas antioxidative enzymes are specifically involved in ROS scavenging, several metabolites are – next to their antioxidative properties – also essential in diverse metabolic processes. These metabolites can be classified into two groups: (1) water-soluble or hydrophilic metabolites such as glutathione (GSH) and ascorbic acid (AsA) reacting with Cd-induced prooxidants in cells and blood plasma and (2) lipid-soluble or hydrophobic metabolites such as vitamin E that protect cell membranes from Cd-induced lipid peroxidation (Valko et al. 2006; Cuypers et al. 2010).

The widely distributed tripeptide GSH (cfr. *supra*) is one of the most important metabolites involved in defense against Cd-induced oxidative stress. Its antioxidant properties are attributed to the thiol (SH) group on the cysteine residue (Halliwell 2006), which enables GSH to transfer its reducing equivalents to GSH–Px, glutathione S-transferases, glutaredoxins, and AsA. In the cell, GSH is maintained in its reduced form by glutathione reductase (GR), and it has a dual role during Cd stress as it neutralizes ROS but also detoxifies Cd directly through chelation (cfr. *supra*). The latter lowers the cellular amount of damaging free Cd and can therefore reduce Cd-induced oxidative stress.

Ascorbic acid (AsA) also directly neutralizes Cd-induced ROS production. However, humans, primates, and some other species require dietary AsA uptake since they lost the ability for the complete AsA biosynthesis during evolution. Reduced AsA is diminished under Cd stress in mouse testes, although no effects on renal AsA content were detected. Several animal studies indicate that AsA supplementation reverses the adverse effects of Cd due to its antioxidative properties, although it also influences Cd absorption and distribution. Vitamin E (tocotrienol) is part of the tocochromanol family and is only synthesized by photosynthetic cells, again requiring dietary uptake in animals. α-Tocopherol is the most abundant and active isoform in human and animal tissues, where it incorporates into lipophilic environments. Vitamin E significantly decreases Cd-induced lipid peroxidation in different organs and body fluids of rats, but the exact mechanisms of uptake, distribution, and cellular effects have to be elucidated (Valko et al. 2006; Cuypers et al. 2010).

Cadmium Stress: A Real Challenge for Plants

Plants, crop plants in particular, form a link between the Cd contamination in the soil and the accumulation of this nondegradable contaminant in the food chain. As they are sessile organisms, plants cannot escape stress situations; consequently, uptake of the nonessential element Cd on contaminated soil is unavoidable as it enters the plant via uptake systems for essential elements like Zn, Fe, Mg, Cu, and Ca. This is due to the chemical similarity of Cd^{2+} ions with the ions of these essential elements, which also causes deregulation of their homeostasis (DalCorso et al. 2008; Kucera et al. 2008; Sharma and Dietz 2009). Once taken up by the plant, Cd causes phytotoxic reactions, mainly through (a) displacement of essential cations from their functional site in biomolecules – again due to chemical similarity; (b) direct interaction with thiol, histidyl, and carboxyl groups of proteins, leading to loss of protein function; and (c) excessive generation of ROS (Sharma and Dietz 2009; Cuypers et al. 2009).

Cadmium-Induced ROS Production in Plants

Similar to animals, ROS generation can be due to the induction of enzymatic ROS production, for example, through plant NADPH oxidases (RBOHs; respiratory burst oxidase homologues) or lipoxygenases. In contrast to animals, plant NADPH oxidases are integral plasma membrane enzymes, but they also produce extracellular $O_2^{\bullet -}$. Lipoxygenase activity causes the formation of lipid peroxides, which in controlled levels are precursors for oxylipin signaling molecules, but Cd exposure can also lead to excessive lipid peroxidation, causing membrane damage and loss of membrane functionality (DalCorso et al. 2008; Kucera et al. 2008; Cuypers et al. 2009; Sharma and Dietz 2009). Cadmium also stimulates peroxisome biogenesis that contributes to excessive ROS production: xanthine oxidase and NADPH-dependent oxidase generate $O_2^{\bullet -}$, and glycolate oxidase, flavin oxidase, and β-oxidation of fatty acids yield H_2O_2, which is metabolized by CAT activity (Sharma and Dietz 2009).

Like in animals, Cd influences the electron transport chain in mitochondria, but this is only under recent investigation because a lot of attention has been given to an extra major source of excessive ROS in plants, more specifically the chloroplast. Cadmium hampers the photosynthetic activities of the chloroplast by damaging the light harvesting complex II and photosystem (PS) II. Cadmium (Cd^{2+}) is a competitive inhibitor of the Ca^{2+} site in the catalytic center of PSII in *Chlamydomonas reinhardtii*, inhibiting PSII photoactivation. Blockage of the electron flow in PSII leads to the formation of excited triplet chlorophyll that reacts with molecular oxygen ($^3O_2$) to form the highly reactive singlet oxygen ($^1O_2$) (Kucera et al. 2008; Sharma and Dietz 2009).

The electron transport is also impaired under Cd stress because of the peroxidation and loss of thylakoid membrane integrity. Furthermore, Cd inhibits chlorophyll synthesis leading to decreased total chlorophyll content, and Cd can replace Mg in chlorophyll, together resulting in decreased photosynthesis capacity. Cadmium negatively influences the carbon fixation, since Cd exposure induces stomatal closure and inhibits enzymes involved in CO_2 fixation (Kucera et al. 2008; DalCorso et al. 2008).

Antioxidative Defense in Plants Under Cd Stress

As plants are sessile organisms, they cannot escape from adverse mineral conditions, and in order to cope with sublethal stress conditions, they have mechanisms to adjust their metabolism. Comparison of redox-related components between tolerant/hyperaccumulator phenotypes and nontolerant relatives that were exposed to Cd revealed a relationship between metal sensitivity and redox imbalance: it was apparent that oxidative stress is a major component of Cd phytotoxicity as all tolerant/hyperaccumulator plants showed elevated antioxidative capacities, especially constitutive high levels of H_2O_2 and $O_2^{\bullet -}$ decomposing enzymes like ascorbate peroxidase (APx), CAT, and SOD (Sharma and Dietz 2009).

The antioxidative defense system of plants consists of enzymes and metabolites, similar to animals. Superoxide is scavenged by superoxide dismutases (SODs) and H_2O_2 by peroxidases and catalases (CATs). As is demonstrated for CAT and GSH–Px in animals, also APx has a strong affinity for H_2O_2 and therefore can result in a tight regulation of H_2O_2 levels for signaling purposes (Sharma and Dietz 2009). Antioxidative enzyme activities can be increased or inhibited by Cd, depending on several factors, such as exposure concentration, exposure time, and plant species. For example, SOD activity in plants exposed to Cd was increased in wheat but decreased in pea plants (DalCorso et al. 2008). SOD isoforms in plants exist in the cytoplasm, chloroplast, and mitochondria, and these may be differentially influenced. For example, Cd inhibited gene expression of the chloroplastic isoforms of SOD (Cu/ZnSOD2 and FeSODs), whereas cytoplasmic (Cu/ZnSOD1) and mitochondrial (MnSOD) SODs were not influenced (Cuypers et al. 2009).

Two important antioxidative metabolites that are special to plants are ascorbic acid and tocopherols in comparison to humans. Ascorbic acid takes part in the ascorbate–glutathione pathway for ROS detoxification. Initially, APx oxidizes ascorbate when it reduces H_2O_2 to H_2O. Consecutively, the ascorbate–glutathione cycle uses the reducing power of NADPH to recycle AsA back to a reduced state, after they have been oxidized in the removal of excess ROS. Enzymes of this cycle are localized in chloroplast, mitochondria, peroxisome, and at the plasma membrane (Kucera et al. 2008). The AsA–GSH cycle is stimulated under Cd stress, and the expression and activities of APX were higher in the metal tolerant *Arabidopsis halleri* than in *Arabidopsis thaliana* (DalCorso et al. 2008; Cuypers et al. 2009).

Also, tocopherols are important plant compounds, which can reduce damage to lipids and protect cell membranes. Tocopherols react with singlet oxygen ($^1O_2$), and they can terminate the chain reactions that are occurring during lipid peroxidation. α-Tocopherol levels increased strongly immediately after Cd exposure along with an increased expression of hydroxypyruvate dioxygenase, an enzyme involved in tocopherol biosynthesis, and the tocopherol deficient *vte1* mutant was hypersensitive to Cd (Sharma and Dietz 2009).

A Disturbed Redox Balance: Two Different Outcomes?

The oxidative changes that arise during Cd exposure are often linked to damaging processes, hence the term oxidative stress. However, fluctuations in redox homeostasis also function as a regulator of diverse signaling processes. Signaling cascades, among which the MAPK (mitogen-activated protein kinase) pathway, efficiently respond to small changes in ROS content or composition, aiming at an acclimation of the cellular state. All of these kinases are described in literature as being influenced under Cd stress. Likewise are Ca-dependent kinases, calmodulins, and redox-related transcription factors (Thévenod 2009). A clear sequence of events, however, is not obvious to describe. Neither is a link between exposure condition and outcome. Most studies associate long-dose chronic exposure with the induction of signaling processes that lead to acclimation and repair mechanisms (Fig. 1). Signal transduction pathways activated under acute exposure to higher levels of Cd, on the other hand, often lead to the induction of programmed/apoptotic cell death (Cuypers et al. 2010; Templeton and Liu 2010 and references herein). Whereas, an increased accumulation of ROS and lipid peroxidation induces programmed cell death or apoptosis; also, necrotic events and uncontrolled cell proliferation are associated with (severe) redox-related damage (Liu et al. 2009).

In conclusion, cross talk and spatiotemporal interactions between signaling pathways depend on the strength and duration of the ROS levels as well as on the influence of the surrounding environment. ROS are produced at different subcellular locations, within different time intervals, and in a different composition and concentration. The main challenge remains how to link a disturbed redox balance with the induction of either damaging or signaling processes.

References

Cannino G, Ferruggia E, Luparello C, Rinaldi AP (2009) Cadmium and mitochondria. Mitochondria 9:377–384

Cuypers A, Smeets K, Vangronsveld J (2009) Heavy metal stress in plants. In: Hirt H (ed) Plant stress biology: from genomics to systems biology. Wiley-VCH, Weinheim, pp 161–178

Cuypers A, Plusquin M, Remans T, Jozefczak M, Keunen E, Gielen H, Opdenakker K, Nair AR, Munters E, Nawrot T, Vangronsveld J, Smeets K (2010) Cd stress: an oxidative challenge. Biometals 23:927–940

DalCorso G, Farinati S, Maistri S, Furini A (2008) How plants cope with cadmium: staking all on metabolism and gene expression. J Integr Plant Biol 50:1268–1280

Halliwell B (2006) Reactive species and antioxidants. Redox biology is a fundamental theme of aerobic life. Plant Physiol 141:312–322.

Kucera T, Horakova H, Sonska A (2008) Toxic metal ions in photoautotrophic organisms. Photosynthetica 46:481–489

Liu J, Qu W, Kadiiska MB (2009) Role of oxidative stress in cadmium toxicity and carcinogenesis. Toxicol Appl Pharmacol 238:209–214

Sharma SS, Dietz K-J (2009) The relationship between metal toxicity and cellular redox imbalance. Trends Plant Sci 14:43–50

Templeton DM, Liu Y (2010) Multiple roles of cadmium in cell death and survival. Chem Biol Interact 188:267–275

Thévenod F (2009) Cadmium and cellular signaling cascades: to be or not to be? Toxicol Appl Pharmacol 238:221–239

Valko M, Rhodes CJ, Moncol J, Izakovic M, Mazur M (2006) Free radicals metals and antioxidants in oxidative stress-induced cancer. Chem Biol Interact 160:1–40

Cadmium and Stress Response

Masanori Kitamura
Department of Molecular Signaling, Interdisciplinary Graduate School of Medicine and Engineering, University of Yamanashi, Chuo, Yamanashi, Japan

Synonyms

Endoplasmic reticulum stress response; Oxidative stress response; Unfolded protein response

Definition

Exposure to cadmium results in cellular stress and evokes responses that involve injurious signaling and protective reactions. The most typical examples are oxidative stress and endoplasmic reticulum (ER) stress. Oxidative stress represents an imbalance between the production of reactive oxygen species (ROS) and the potential to detoxify ROS. The oxidative stress response comprises (1) generation of ROS and consequent modulation of cellular components including proteins, lipids, and DNA, leading to cellular activation or apoptosis and (2) induction of antioxidant systems that protect cells from ROS-mediated activation and cell injury. ER stress is defined as accumulation of unfolded proteins in the ER. It triggers an adaptive program, namely the unfolded protein response (UPR). The UPR alleviates ER stress by suppression of protein synthesis, facilitation of protein folding, and reinforced degradation of unfolded proteins. The UPR may activate an array of kinases and transcription factors, leading to expression of target genes. However, when stress is beyond the capacity of the cytoprotective machinery, cells undergo apoptosis. Like the pro-survival process, the pro-apoptotic process is also mediated by the UPR.

Introduction

Cadmium is a toxic metal and a potent environmental pollutant. In humans, cadmium intoxication is caused by its intake through contaminated water, food, and air – especially cigarette smoke. Because of its low excretion rate, cadmium accumulates in different organs over time and causes various negative effects on human's health, including renal dysfunction, osteoporosis/bone fractures, and development of cancers. Cadmium accumulates in some target organs, especially in the kidney where its highest concentrations are observed. Human studies indicated that 7% of the general population have renal dysfunction caused by cadmium exposure. Cadmium nephropathy is characterized by proteinuria, aminoaciduria, glucosuria, phosphaturia, and reduction in glomerular filtration rate. In the kidney, accumulation of cadmium occurs mainly in the proximal tubules. Following prolonged and/or high levels of exposure to cadmium, apoptosis is induced in the proximal tubules (Thevenod 2003).

Several underlying mechanisms have been postulated, and currently, ROS are considered as crucial mediators for the cadmium-triggered tissue injuries (Cuypers et al. 2010). However, recent investigation disclosed that, in addition to oxidative stress, cadmium causes ER stress in vitro and in vivo, which also plays a crucial role in the induction of cell injury (Kitamura and Hiramatsu 2010).

Principles

Oxidative Stress Response

Under normal conditions, ROS are generated by the leakage of activated oxygen from mitochondria during oxidative phosphorylation. Under pathological situations, mitochondria are also the major source of ROS that leads to activation of specific signaling pathways and consequent regulation of cell behavior including proliferation, migration, differentiation, and apoptosis. In particular, excessive ROS cause damage of cellular proteins, lipids, and DNA, leading to apoptotic and necrotic cell death. On the other hand, in response to generation of ROS, cells are able to exert defense mechanisms to control cellular homeostasis. Xanthine oxidase, NADPH oxidases, and cytochromes P450 contribute to the production of ROS, and cellular antioxidant enzymes such as superoxide dismutase (SOD), catalase, and glutathione peroxidase are involved in scavenging of ROS.

In biological systems, sequential reduction of oxygen leads to the generation of superoxide anion ($O_2^{\bullet-}$)

and hydrogen peroxide (H_2O_2). SOD scavenges $O_2^{\cdot-}$ by catalyzing conversion of $O_2^{\cdot-}$ to H_2O_2. $O_2^{\cdot-}$ also rapidly reacts with nitric oxide, yielding another reactive species, peroxynitrite ($ONOO^-$). All of these ROS are potential triggers for apoptosis.

ER Stress Response

ER stress is induced under a variety of pathological situations. Regardless of types of triggers, ER stress induces a coordinated adaptive program, the UPR. Three major transducers for sensing ER stress are present at the membrane of the ER, that is, protein kinase-like ER kinase (PERK), activating transcription factor 6 (ATF6), and inositol-requiring enzyme 1 (IRE1). Activation of PERK leads to phosphorylation of eukaryotic translation initiation factor 2α (eIF2α), which causes general inhibition of protein synthesis. In response to ER stress, 90 kDa ATF6 (p90ATF6) transits to the Golgi where it is cleaved by proteases, yielding an active transcription factor, 50 kDa ATF6 (p50ATF6). Similarly, activated IRE1 catalyzes removal of a small intron from the mRNA of X-box binding protein 1 (XBP1). This splicing event creates a translational frameshift in *XBP1* mRNA to produce an active transcription factor. Active p50ATF6 and XBP1 subsequently bind to the ER stress response element (ERSE) and the UPR element (UPRE), leading to expression of target genes including ER chaperone 78 kDa glucose-regulated protein (GRP78) (also called BiP) and ER-associated degradation factors involved in elimination of unfolded proteins. Primarily, these pathways contribute to attenuation of ER stress (Kitamura 2008).

During the UPR, however, death signals may also be transduced from the ER (Kim et al. 2006). For example, activation of the PERK-eIF2α pathway causes selective induction of transcription factor ATF4. Subsequently, ATF4 triggers expression of pro-apoptotic CCAAT/enhancer-binding protein-homologous protein (CHOP) through binding to the amino acid response element. ER stress activates caspase-12 localized at the ER membrane through an interaction with IRE1 and TNF receptor-associated factor 2 (TRAF2), leading cells to undergo apoptosis. The IRE1–TRAF2 interaction also allows for recruitment and activation of apoptosis signal-regulating kinase 1 and downstream c-Jun N-terminal kinase (JNK), both of which are involved in a variety of pro-apoptotic signaling.

Evidence

Induction of Oxidative Stress by Cadmium

Metals such as iron, copper, chromium, vanadium, and cobalt are capable of redox cycling in which a single electron may be accepted or donated by the metal. This action catalyzes reactions that produce reactive radicals and generate ROS. However, cadmium is a bivalent cation and unable to generate free radicals directly, although production of ROS in cadmium-exposed cells has been documented in many studies (Liu et al. 2009). Because cadmium possesses a high affinity for thiols, the major thiol antioxidant glutathione (GSH) that is abundant in cells is considered as a primary target for cadmium. That is, depletion of the reduced GSH pool results in a disturbance of the antioxidant defense mechanism. Cadmium depletes glutathione and protein-bound sulfhydryl groups, resulting in enhanced production of ROS such as $O_2^{\cdot-}$, H_2O_2, and hydroxyl radicals. The spin-trapping technique in conjunction with electron spin resonance also provided direct evidence for cadmium-triggered generation of these ROS in vitro and in vivo (Liu et al. 2009).

Induction of ER Stress by Cadmium

Cadmium has the potential to induce ER stress in various cell types, as reviewed recently (Kitamura and Hiramatsu 2010). Cadmium triggers expression of GRP94 in thymocytes and GRP78 in lung epithelial cells and renal tubular cells. Cadmium causes expression of CHOP and activation of caspase-12 in mesangial cells and hepatoma cells. It also induces splicing of *XBP1* mRNA in fibroblasts. We showed that, in renal tubular cells, cadmium caused phosphorylation of PERK and eIF2α, activation of ATF6, and splicing of *XBP1* mRNA, and it was associated with induction of GRP78, GRP94, and CHOP. Furthermore, administration of cadmium in mice also triggered expression of GRP78 and CHOP in the kidney and liver. Thus, cadmium is able to activate the three major branches of the UPR.

Molecular mechanisms underlying the induction of ER stress by cadmium are not fully understood, but ROS are possible candidates to mediate induction of ER stress. This issue is described in detail in the section "Cross Talk Between Oxidative Stress and ER Stress." In addition to ROS, several recent reports suggested release of calcium from the ER mediates cadmium-induced apoptosis of several cell types. Depletion of calcium store in the ER is a well-known trigger to induce ER stress, and induction of ER stress by cadmium may, in part, be mediated by mobilization of intracellular calcium.

Role of Oxidative Stress in Cadmium-Induced Cell Death

Several previous studies showed involvement of ROS in cadmium-induced cell injury. As described, exposure of cells to cadmium causes generation of ROS, which is associated with a decrease in glutathione levels and consequent cellular death. Another line of evidence shows that cadmium-triggered apoptosis is inhibited by treatments with antioxidants, suggesting crucial roles of ROS.

Several mechanisms have been postulated to the cytotoxic effect of ROS in cadmium-exposed cells. First and foremost, excessive ROS cause damage of cellular components including proteins, lipids, and DNA, leading to apoptotic and necrotic cell death. When applied in low to moderate concentrations, cadmium mainly causes apoptosis. However, exposure of cells to its high concentrations leads to necrotic cell death. Both apoptosis and necrosis are induced by increased accumulation of ROS and associated with facilitation of lipid peroxidation. Cadmium-induced ROS also interact with the cellular defense machinery via activation of mitogen-activated protein (MAP) kinases and other signaling pathways. For example, cadmium-induced ROS triggers activation of p38 MAP kinase, leading to both pro- and anti-apoptotic events, for example, activation of caspase-3 and induction of 70 kDa heat shock proteins. Cadmium also increases phosphorylation of JNK via changes in cytosolic Ca^{2+} fluxes and/or changes in the cellular redox balance, leading to apoptotic cell death (Cuypers et al. 2010).

Role of ER Stress in Cadmium-Induced Cell Death

Cadmium induces both ER stress and apoptosis in vitro and in vivo. Cells transfected with ER chaperones such as GRP78 and 150 kDa oxygen-regulated protein are resistant to cadmium-induced apoptosis (Yokouchi et al. 2007), suggesting a crucial role of ER stress.

There are three proximal transducers in the ER for sensing of ER stress, and all the transducers are activated by cadmium. The first transducer is ATF6. Cadmium induces activation of the ATF6 pathway, and treatment of cells with an inhibitor of ATF6 attenuates cadmium-induced apoptosis. Similarly, dominant-negative inhibition of ATF6 suppresses cadmium-induced apoptosis. The second proximal transducer for ER stress is IRE1. Activated IRE1 catalyzes the removal of a small intron from *XBP1* mRNA, leading to production of the active transcription factor. Treatment with cadmium rapidly induces splicing of *XBP1* mRNA. Transfection with a dominant-negative mutant of XBP1 leads to cellular resistance against cadmium-induced apoptosis. Similarly to the pro-apoptotic effect of XBP1, JNK, another molecule downstream of IRE1, is also phosphorylated by cadmium, leading to induction of apoptosis. Thus, the ATF6 and IRE1 pathways participate in the induction of apoptosis by cadmium (Yokouchi et al. 2007).

The third transducer for ER stress is PERK, the activation of which leads to phosphorylation of eIF2α and blockade of protein synthesis. Cadmium rapidly induces phosphorylation of PERK and downstream eIF2α. Salubrinal, a selective activator of eIF2α, inhibits cadmium-induced apoptosis. Furthermore, dominant-negative inhibition of eIF2α significantly enhances cadmium-induced apoptotic cell death. In contrast to the ATF6 and IRE1 pathways, the PERK-eIF2α pathway is anti-apoptotic in cadmium-exposed cells (Yokouchi et al. 2007).

Taken together, these results suggest differential, bidirectional regulation of apoptosis by the UPR in cadmium-exposed cells.

Cross Talk Between Oxidative Stress and ER Stress

Oxidative stress and ER stress occur in cells under the exposure to cadmium. The oxidative stress response and the UPR may occur independently, but cross talk is also present between these stress responses under the exposure to cadmium. Accumulating evidence suggests that protein folding and generation of ROS are closely linked with each other (Malhotra and Kaufman 2007; Kitamura and Hiramatsu 2010). Prolonged activation of the UPR may result in oxidative stress and consequent cellular death. Accumulation of ROS by the UPR is caused through two mechanisms; the ER-dependent and the mitochondria-dependent ROS generation. On the other hand, another line of evidence suggests that induction of the UPR occurs under oxidative stress, which is an adaptive mechanism to preserve cell function. We found that cadmium-induced ER stress was attenuated by antioxidants in renal tubular cells. Exposure of cells to ROS donors caused ER stress, whereas suppression of ER stress did not attenuate cadmium-triggered oxidative stress, suggesting that ER stress is the event downstream of oxidative stress (Yokouchi et al. 2008).

In the induction of ER stress by cadmium, particular ROS may play dominant roles. We found that exposure to $O_2^{\bullet-}$, H_2O_2, or $ONOO^-$ induced apoptosis of renal tubular cells, whereas ER stress was caused only by $O_2^{\bullet-}$ and $ONOO^-$. Scavenging of $O_2^{\bullet-}$ attenuated cadmium-induced ER stress and apoptosis, whereas inhibition of $ONOO^-$ was ineffective. Furthermore, $O_2^{\bullet-}$ was involved in the activation of the ATF6 and IRE1 pro-apoptotic pathways in cadmium-exposed cells (Yokouchi et al. 2008). These results provide evidence that $O_2^{\bullet-}$ is preferentially involved in cadmium-triggered, ER stress-mediated apoptosis.

Oxidative Stress, ER Stress, and Other Stress Signaling

Cadmium triggers an array of signaling pathways. For example, cadmium has the potential to activate stress-related signaling including MAP kinase cascades, the Akt pathway, and NF-κB signaling, all of which are involved in the regulation of cell survival and death (Thévenod 2009). It is well known that ROS trigger activation of MAP kinases, Akt, and NF-κB (Kamata and Hirata 1999). ER stress also has the potential to activate Akt and NF-κB, at least in part, through the ATF6 pathway (Kitamura 2009). Recently, we reported that ER stress induces activation of extracellular signal-regulated kinase, p38 MAP kinase and JNK through activation of the PERK-eIF2α pathway and the IRE1 pathway (Zhao et al. 2011). The activation of Akt, MAP kinases, and NF-κB by cadmium seems to be nonselective and possibly the events downstream of oxidative stress and ER stress. Oxidative stress and ER stress may be at the top of the chain of most signaling pathways activated by cadmium.

Therapeutic Implications

Previous reports suggested that antioxidants are useful for the prevention of cadmium-related tissue injury. Currently, treatments against cadmium toxicity include antioxidant therapy with melatonin and vitamin E. In addition to antioxidants, some reagents that attenuate ER stress may also be useful for this purpose. For example, chemical chaperone 4-phenylbutyric acid (4-PBA) is an agent that stabilizes protein conformation, improves ER folding capacity, and facilitates trafficking of mutant proteins. Oral administration of 4-PBA to a murine model of type 2 diabetes alleviates ER stress, reduces diabetic symptoms, and lowers systemic inflammation. Based on the fact that ER stress is involved in cadmium-induced toxicity, chemical chaperoning may be another possible strategy for the treatment of cadmium intoxication. However, it is worthwhile to note that some reports indicated caution in the use of antioxidants against cadmium toxicity. For example, studies in cancer cells have shown that use of vitamin C (ascorbate) in the presence of some metal (cadmium or nickel) caused damage of double-stranded DNA and apoptotic cell death. Toward therapeutic intervention in cadmium-related pathologies, the complex interactions among metals, antioxidants, and chemical chaperones need further investigation.

Conclusion

Exposure to cadmium causes oxidative stress and ER stress, both of which evoke injurious signaling as well as cytoprotective reactions. The oxidative stress is located upstream (or downstream) of ER stress, and cross talk of these stress responses forms proximal events that regulate activation of downstream signaling. Appropriate control of oxidative stress and ER stress is a potential strategy for therapeutic intervention in cadmium-related disorders.

Cross-References

▶ Cadmium Absorption
▶ Cadmium and Metallothionein
▶ Cadmium and Oxidative Stress
▶ Cadmium and Stress Response
▶ Cadmium Carbonic Anhydrase
▶ Cadmium, Effect on Transport Across Cell Membranes
▶ Cadmium Exposure, Cellular and Molecular Adaptations
▶ Cadmium, Physical and Chemical Properties
▶ Cadmium Transport
▶ Chromium(VI), Oxidative Cell Damage

References

Cuypers A, Plusquin M, Remans T, Jozefczak M, Keunen E, Gielen H, Opdenakker K, Nair AR, Munters E, Artois TJ, Nawrot T, Vangronsveld J, Smeets K (2010) Cadmium stress: an oxidative challenge. Biometals 23:927–940

Kamata H, Hirata H (1999) Redox regulation of cellular signalling. Cell Signal 11:1–14

Kim R, Emi M, Tanabe K, Murakami S (2006) Role of the unfolded protein response in cell death. Apoptosis 11:5–13

Kitamura M (2008) Endoplasmic reticulum stress and unfolded protein response in renal pathophysiology: Janus faces. Am J Physiol – Renal 295:F323–F342

Kitamura M (2009) Biphasic, bidirectional regulation of NF-κB by endoplasmic reticulum stress. Antioxid Redox Signal 11:2353–2564

Kitamura M, Hiramatsu N (2010) The oxidative stress – endoplasmic reticulum stress axis in cadmium toxicity. Biometals 23:941–950

Liu J, Qu W, Kadiiska MB (2009) Role of oxidative stress in cadmium toxicity and carcinogenesis. Toxicol Appl Pharmacol 238:209–214

Malhotra JD, Kaufman RJ (2007) Endoplasmic reticulum stress and oxidative stress: a vicious cycle or a double-edged sword? Antioxid Redox Signal 9:2277–22793

Thevenod F (2003) Nephrotoxicity and the proximal tubule. Insights from cadmium. Nephron Physiol 93:87–93

Thévenod F (2009) Cadmium and cellular signaling cascades: to be or not to be? Toxicol Appl Pharmacol 238:221–239

Yokouchi M, Hiramatsu N, Hayakawa K, Kasai A, Takano Y, Yao J, Kitamura M (2007) Atypical, bidirectional regulation of cadmium-induced apoptosis via distinct signaling of unfolded protein response. Cell Death Differ 14:1467–1474

Yokouchi M, Hiramatsu N, Hayakawa K, Okamura M, Du S, Kasai A, Takano Y, Shitamura A, Shimada T, Yao J, Kitamura M (2008) Involvement of selective reactive oxygen species upstream of proapoptotic branches of unfolded protein response. J Biol Chem 283:4252–4260

Zhao Y, Tian T, Huang T, Nakajima S, Saito Y, Takahashi S, Yao J, Paton AW, Paton JC, Kitamura M (2011) Subtilase cytotoxin activates MAP kinases through PERK and IRE1 branches of the unfolded protein response. Toxicol Sci 120:79–89

Cadmium Carbonic Anhydrase

Claudiu T. Supuran
Department of Chemistry, University of Florence, Sesto Fiorentino (Florence), Italy

Synonyms

Cadmium(II); Cd(II); Zinc carbonic anhydrases

Definition

Cadmium: Cadmium is a metallic element with the oxidation state of +2, occurring rarely in metalloproteins, except some carbonic anhydrases. It is currently considered highly toxic, but in Cd-CAs, this metal ion has a catalytic role.

Carbonic anhydrase: Superfamily of metalloenzymes catalyzing CO_2 hydration to bicarbonate and protons. Ubiquitous in all life kingdoms.

Biological implications: Cd-CAs are involved in CO_2 fixations and photosynthesis in many diatoms.

Carbonic Anhydrase Families

The carbonic anhydrases (CAs, EC 4.2.1.1.) are a superfamily of metalloenzyme which evolved independently several times, with five genetically distinct enzyme classes known to date: the α-, β-, γ-, δ-, and ζ-CAs (Supuran 2008). The α-, β-, and δ-CAs use Zn(II) ions at the active site (Supuran 2008, 2010); the γ-CAs are probably Fe(II) enzymes (but they are active also with bound Zn(II) or Co(II) ions) (Ferry 2010), whereas the ζ-class uses Cd(II) or Zn(II) to perform the physiologic reaction catalysis (Xu et al. 2008). All these enzymes, including the Cd-CA catalyze a very simple but essential reaction, hydration of carbon dioxide to bicarbonate and protons (Supuran 2008). The 3D fold of the five enzyme classes are very different from each other, as it is their oligomerization state: α-CAs are normally monomers and rarely dimers; β-CAs are dimers, tetramers, or octamers; γ-CAs are trimers, whereas the δ- and ζ-CAs are probably monomers, but in the case of the last family, three slightly different active sites are present on the same protein backbone which is in fact a pseudotrimer, at least for the best investigated member of the class, the enzyme from the marine diatom *Thalassiosira weissflogii* (Xu et al. 2008). Many representatives of all these enzyme classes have been crystallized and characterized in detail, except the δ-CAs. Cd-CAs were described so far only in diatoms (Xu et al. 2008).

Biological Role of Cd-CAs

CAs are key enzymes involved in the acquisition of inorganic carbon for photosynthesis in phytoplankton (Xu et al. 2008). Most of the phytoplankton operate a carbon-concentrating mechanism (CCM) to increase the CO_2 concentration at the site of fixation by RuBisCO several folds over its external concentration, allowing the enzyme to function efficiently. Marine diatoms possess both external and internal CAs. It has been hypothesized in the model diatom *T. weissflogii* that the external CA catalyzes the dehydration of HCO_3^- to CO_2 to increase the gradient of the CO_2 diffusion from the external medium to the cytoplasm, and the internal CA in the cytoplasm catalyzes the rehydration of CO_2 to HCO_3^- to prevent the leakage of CO_2 to the external medium again (Xu et al. 2008). One of the most remarkable findings regarding the ζ-CA from *T. weissflogii* was that this is a Cd(II)-containing enzyme, which can also work with Zn(II) bound at the active site, and that there is a rather rapid metal exchange between zinc and cadmium, depending on the availability of metal ions in the marine environment (Xu et al. 2008). It is not known yet where this enzyme is localized in *T. weissflogii* or other diatoms.

Role of Cd(II) in the Catalytic Mechanism

The Cd(II) ion is essential for catalysis being coordinated by two Cys and one His residues, with a water molecule completing the coordination sphere. Cd(II) is thus in a distorted tetrahedral geometry, but a second water molecule is nearby, so that it is not clear whether Cd(II) is indeed tetrahedral or trigonal bipyramidal in this cadmium enzyme (Xu et al. 2008; Viparelli et al. 2010).

It is generally accepted that in Cd-CAs, a cadmium hydroxide moiety acts as nucleophile in the catalytic cycle, similar with the zinc hydroxide one in other classes of CAs (Supuran 2010). The enzyme can be inhibited by inorganic, metal-complexing anions or sulfonamides, similar to the Zn(II)-containing CAs. The *T. weissflogii* Cd-CA is one of the most effective enzymes known in nature, with a turnover number for the hydration of CO_2 to bicarbonate close to the limit of the diffusion-controlled processes (Xu et al. 2008; Viparelli et al. 2010). Their function in diatoms is extremely important for the CO_2 fixation process, and with the climate change in fact, a better study of such enzymes may prove critical for understanding these intricate phenomena. It is also interesting to note that a metal ion, such as cadmium, normally associated with high toxicity for mammals, plays a crucial role in this enzyme found in diatoms, being essential to their life cycle.

Cross-References

▶ Cadmium Exposure, Cellular and Molecular Adaptations
▶ Cadmium, Physical and Chemical Properties
▶ Zinc and Iron, Gamma and Beta Class, Carbonic Anhydrases of Domain Archaea
▶ Zinc and Zinc Ions in Biological Systems
▶ Zinc Carbonic Anhydrases

References

Ferry JG (2010) The gamma class of carbonic anhydrases. Biochim Biophys Acta 1804:374–381
Supuran CT (2008) Carbonic anhydrases: novel therapeutic applications for inhibitors and activators. Nat Rev Drug Discov 7:168–181
Supuran CT (2010) Carbonic anhydrase inhibitors. Bioorg Med Chem Lett 20:3467–3474
Viparelli F, Monti SM, De Simone G, Innocenti A, Scozzafava A, Xu Y, Morel FM, Supuran CT (2010) Inhibition of the R1 fragment of the cadmium- containg ζ class carbonic anhydrase from the diatom *Thalassiosira weissflogii* with anions. Bioorg Med Chem Lett 20:4745–4748
Xu Y, Feng L, Jeffrey PD, Shi Y, Morel FM (2008) Structure and metal exchange in the cadmium carbonic anhydrase of marine diatoms. Nature 452:56–61

Cadmium Exposure, Cellular and Molecular Adaptations

Jean-Marc Moulis
Institut de Recherches en Sciences et Technologies du Vivant, Laboratoire Chimie et Biologie des Métaux (IRTSV/LCBM), CEA–Grenoble, Grenoble, France
CNRS, UMR5249, Grenoble, France
Université Joseph Fourier-Grenoble I, UMR5249, Grenoble, France

Synonyms

Adaptation: Acclimatization; Refitting
Resistance: Insusceptibility; Obstruction; Opposition; Resistivity

Definition

Adaptation: The fact for a living species to become better suited to its environment.

Resistance: The process of opposing the harmful threat of a toxic compound.

Speciation: Describes the associated forms in which elements, e.g., metals, occur in complex systems, such as a biological medium.

Redox: Acronym for oxidation-reduction. Here, refers to the set of reactions in which electrons can be exchanged within a cell.

Reactive oxygen species (ROS): Designates three oxygen-containing compounds, the superoxide radical anion, hydrogen peroxide, and hydroxyl radical, which all derive from dioxygen by successive reduction steps. By extension, ROS may refer to strongly oxidative (i.e., electron withdrawing) oxygen-containing molecules such as the hypochlorous anion or peroxynitrite.

Introduction

Cadmium compounds are overwhelmingly toxic in their interactions with living bodies (Satarug et al. 2010), even though apparently safe and functional replacement of zinc in a carbonic anhydrase isoform from diatoms is documented (Morel 2008). Depending on the means of exposure, on the considered living species, and on speciation, cadmium has different molecular targets: cadmium jeopardizes the exposed organism by inducing cell death or at least malfunction.

A characteristic feature of cadmium toxicity is the range of molecules being able to interact with the ionic form of the metal (Cd^{2+}, the only one of significant biological importance). The affinity of divalent cadmium for thiolates is often put forward, but binding to nitrogen-containing ligands (histidine imidazole, amines, amides, etc.) and carboxylic acids also occurs (Moulis 2010). Regarding this chemical versatility, the major challenge of cadmium toxicity is to identify the most relevant biological pathways that are perturbed by the toxic metal. Changing the cellular environment by introduction of a toxic compound such as cadmium triggers a cellular response that can be described in two phases, a homeostatic one aiming at maintaining the cellular functions as they were prior to the insult and an adaptive one helping cells to cope with the new situation in the long run. Schematically, the first phase is preeminent upon acute exposure, and it may be sufficient to keep viability if the exposure does not last and is not too large. The second phase is expected to gain more importance upon chronic, often low concentration, exposure to the toxic metal. The fundamental molecular mechanisms underlying these two sets of situations are sketched in this essay to provide a general view of cadmium toxicity, with emphasis on the changes allowing cells to sustain long-term cadmium exposure.

Cadmium-Induced Cell Death

As most of the antioxidant systems available to cells rely on thiol-containing molecules, divalent cadmium is expected to displace the cellular balance between oxidants and reducing molecules (i.e., *redox* equilibrium). The intracellular abundant cysteine-containing tripeptide glutathione (GSH) chelates cadmium, and the function of prominent redox-active proteins such as thioredoxins, glutaredoxins, and peroxiredoxins is perturbed by the presence of cadmium ions. The imbalance in the availability and production of antioxidant molecules triggers the upraising of radicals and other reactive molecules, including the now well-known reactive oxygen species (ROS), which modify and inactivate scores of biological components and lead to necrosis and death at high concentrations (Liu et al. 2009; Cuypers et al. 2010).

In a generally less brutal way, the challenge generated by moderate levels of cadmium to the redox balance and the ensuing increase of the concentration of ROS and other oxidizing molecules impact several pathways without immediate damage. The cellular response aims at correcting the changes by inducing various activities known as the antioxidant response. But the drastic changes cadmium induces may not allow cells to cope with the unbalanced situation, and apoptosis may occur.

Although cell death is a likely outcome upon cadmium exposure, some cells may remain viable. They originate from a few, or even a single, clones of the initial population, and they play a prominent role in cadmium-induced carcinogenesis in mammals when they proliferate. This process requires an adaptive response which efficiently handles the cadmium threat and in which many distinct mechanisms may be at play.

Resistance to Cadmium Toxicity

Chronologically, cells have to be able to sustain the insult of toxic compounds before being able to recover or in a position to proliferate. Means enhancing the activities that are immediately challenged by cadmium allow cells to resist, at least transiently.

Resistance by counteraction against oxidants. The molecules involved in the antioxidant response are naturally at the forefront of the resistance mechanisms. Enzymes participating to the synthesis and the turnover of GSH, such as γ-glutamate cysteine ligase (GCL) or GSH synthetase and GSH reductase (GR), respectively, are upregulated in some cadmium resistant cells. In mammalian cells, the enhanced antioxidant response is mainly mediated by the nuclear respiratory (transcription) factor Nrf2. Nrf2 binds to antioxidant response elements found in the promoter sequences of genes triggering the antioxidant response, including the ones encoding heme oxygenase 1, GR, or GCL for instance.

But not all resistance mechanisms against cadmium are directly dependent on Nrf2. Reallocation of cellular resources can help fighting the cadmium challenge. For instance, increase of the cytosolic $NADP^+$-dependent isocitrate dehydrogenase diverts cells from apoptosis, most likely by increasing cytosolic NADPH, one of the few electron donors (i.e., reductant) in this cellular subcompartment.

Resistance by enhancing cadmium-inactivated activities. More generally, any metabolic activity which is inhibited by cadmium has to be reestablished to a sufficient level to allow cells to resist to the toxic metal. Various ways to this aim have been identified, including overproduction of potent antioxidant activities mediated by Nrf2-induced transcription. Transcriptional upregulation of cadmium-inactivated activities is often observed to compensate or substitute jeopardized pathways and targets. An interesting mechanism has been evidenced in yeast: the increased demand for GSH called for by increased cadmium concentrations is partly met by reallocation of the cell sulfur resources to GSH synthesis by switching isozymes production (pyruvate decarboxylase, aldehyde dehydrogenase, enolase) from cysteine-rich to cysteine-poor forms, with the likely additional advantage of decreased sensitivity to cadmium inhibition (Fauchon et al. 2002). In plants, cadmium exposure increases synthesis of participants to major metabolic pathways, such as glycolysis and the pentose phosphate pathway, with the apparent purpose of maintaining energy-rich metabolites (ATP, NADH, NADPH) and of building molecules involved in cadmium binding (Villiers et al. 2011).

Resistance by cadmium trapping. Among them, phytochelatins (PC) are strongly produced as a response to cadmium. GSH is the precursor of these sulfur-rich molecules which participate in cadmium traffic and vacuolar sequestration (see "▶ Accumulation"). This upregulation occurs at different levels, and phytochelatin synthase activity is strongly stimulated by cadmium bound to GSH. In a similar tone, metallothioneins (MT) increased synthesis is a common observation resulting from cadmium exposure in other cells. These cysteine-rich proteins display a strong affinity for cadmium, and they also participate in cadmium traffic (Klaassen et al. 2009; Maret 2011). For instance, ingested cadmium in mammals rapidly reaches the liver where induced MT synthesis protects the organ from cadmium damage. The chelated form of the metal is excreted from the initial target organs (duodenum, liver, etc.), and it ends up in the renal glomerulus via the circulation. The Cd-MT complex enters proximal tubule cells via endocytosis, and the metal accumulates in kidneys as the result of Cd-MT processing. This series of events illustrates that different cell types handle cadmium in different ways, and that the physiological metal speciation in multicellular organisms has a major influence on the toxicological consequences of cellular cadmium exposure. MT induction is mainly a transcriptional mechanism involving a specialized transcription factor, MTF-1, which binds metals such as zinc and cadmium and interacts with metal-responsive elements in the promoter regions of dependent genes, including MT ones. Despite the importance of MT in the first line of defense against cadmium, the very long half-life of the metal (several decades in humans) once absorbed implies that MTs and PCs cannot be considered as single mediators of long-term cadmium detoxification, and additional mechanisms have to be implemented to allow cells to overcome the cadmium insult.

Resistance by competition from other metals. Many processes involved in cadmium distribution within the body correspond to situations in which the toxic metal hijacks the molecular devices devoted to the traffic of essential metals (Clemens 2006; Moulis 2010;

Thévenod 2010). Competition between (some of) these metals and cadmium may occur for transporters, chelators (e.g., GSH, MT, or PC), and other molecules. Hence, increased availability of selected essential (nontoxic) metals may oppose cadmium toxicity. The most obvious example is zinc, an abundant and hardly toxic biological metal, which strongly resembles cadmium. Zinc supplementation contributes to the antioxidant response, by increasing the Nrf2-triggered transcription of the catalytic subunit (heavy chain) of glutamate-cysteine ligase for instance, and it often protects from cadmium-induced cell death. Other indirect mechanisms may mediate the cellular protection afforded by essential metals against cadmium. The viability of a human epithelial cell line that was developed in high zinc concentrations was far larger upon exposure to high cadmium concentrations as compared to naïve cells. But instead of reflecting a mere competition between zinc and cadmium, adaptation to zinc inhibited cadmium entry into the cells via a transporter which was responsible for manganese and cadmium uptake but not directly involved in zinc traffic.

This single example shows that the permeation of the plasma membrane to cadmium is a major player in the sensitivity/resistance to the toxic effects of the metal. Many molecular components of this membrane (transporters of various kinds and ionic channels) can potentially move cadmium in and out in different cell types. In a given case, one specific molecule often supersedes other possible contributors. Consequently, inhibiting, in the case of inward movement, or inducing, in that of extrusion, this transporter protects cells from toxicity. This often occurs upon sustained exposure to cadmium and cellular selection for viability. For instance, mouse testicular cadmium toxicity is associated with SLC39A8 (Zip8) activity, and resistance to necrosis results from inactivation of the gene (He et al. 2009).

Resistance via signaling. As a stressful agent for cells, cadmium impacts a range of signaling cascades, and resistance mechanisms have been associated with changes in the efficiency of signal transduction or escape of targets from the regulation. In the presence of cadmium, resistance mechanisms associated with alterations of signaling pathways must enhance proliferative signals or dampen apoptotic ones (Thévenod 2009). Biochemically, most of these effects can be traced back to the interaction of cadmium ions with reactive sites in signal sensing molecules or the modifications of these sites induced by cadmium-triggered redox imbalance. Examples include extracellular cadmium interaction with E-cadherin with subsequent effects on the epithelial-mesenchymal transition favoring metastasis spreading and decreased activation of MAPK (mitogen-activated protein kinases) lowering the efficiency of cadmium-triggered apoptosis in resistant cells.

The above variety of mechanisms which have been evidenced in different cells under different conditions does not imply that conditions circumventing cell death may last for long periods of time. In most reports, the information about the length of the resistance effect is lacking, as is that of its reversibility (both aspects are generally connected). But in the cases in which viability is maintained with or without the continuous presence of cadmium, the selected cells have turned the resistance process into an adaptive, most often malignant, change. Many of the resistance mechanisms described above are relevant to acute exposure of naïve cells to cadmium, and they aim to maintain viability upon a sudden chemical threat (Fig. 1). Yet, sustained exposure to the poison requires other or adjusted mechanisms. For instance, it is not always clear that low cadmium doses need to be counteracted by a massive antioxidant response. The continuous exposure of the cells is accompanied by a shift of the homeostatic status, and adaptation aims at affording the necessary corrections in a permanent way.

Adaptation and DNA Lesions

A well-established mechanism to permanently affect any cell behavior is through genetic imprint. Solid evidence for cadmium as a mutagenic chemical is virtually nonexistent, but it is a likely comutagen with other genotoxic compounds or upon physicochemical stress, e.g., irradiation. Direct DNA oxidative damage induced by cadmium is probably a feature to be associated with acute poisoning. But cadmium does not need to enhance the formation of DNA lesions to be toxic. Various DNA modifications are continuously produced at low levels through side reactions of metabolism, and they need to be repaired to maintain genetic integrity. Accumulated evidence shows that all DNA repair systems (nucleotide excision, base excision, and mismatch repair) are particularly

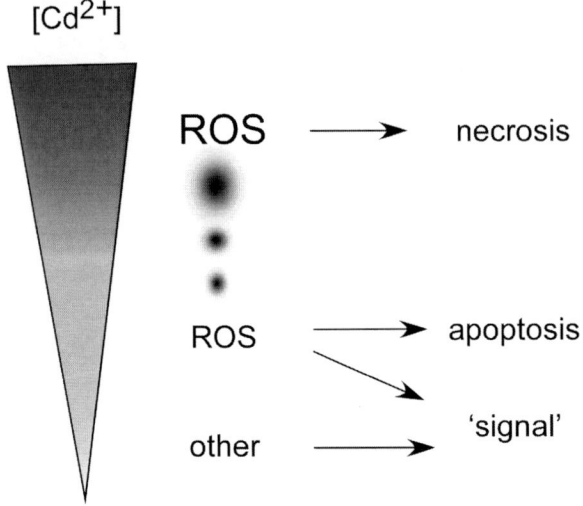

Cadmium Exposure, Cellular and Molecular Adaptations, Fig. 1 Cellular fate as a function of the cadmium concentration. Cells may be exposed to a range of cadmium concentrations (*left*) which mainly leads to a proportional oxidative stress (*ROS* reactive oxygen species) and to additional effects ("signal" discussed in the text) at low concentrations. The outcome for the cells (*right*) ranges from death, mainly by necrosis at high concentrations, to resistance and adaptation, when the cadmium threat can be overcome

sensitive to cadmium. Several components and steps (damage recognition, removal, and efficiency of the repairing activities) involved in these correcting processes may be impaired by cadmium (Hartwig 2010). The defects in repairing DNA damage have a significant influence on the balance between cellular proliferation and death, and they most probably should take the blame for cadmium-induced transformation, hence adaptation and carcinogenesis (Waalkes 2003; Hartwig 2010).

Furthermore, modifications of the chromosomal integrity and of the fidelity of the genetic information may not be the only way through which cadmium impacts the cellular fate. Epigenetic changes, e.g., the methylation status of DNA, have been evidenced upon cadmium exposure, with particular consequences on the expression of tumor suppressors such as *RASSF1A* or *CDKN2A* and induced regulatory defects of the cell cycle and of DNA repair. It is likely that the impact of cadmium on epigenetics will be more significantly documented in the near future.

Beyond methylation and histone modifications, noncoding DNA is also susceptible to the presence of cadmium. MicroRNA molecules (miRNA) regulate different steps of the turnover of specific genes, such as those carried out by transcription factors and degradation labeling activities (E3 ubiquitin ligases). Examples include miR393 and miR171 in legumes and miR146a in leukocytes, which regulate plant development and cancer (pancreas and glial) cells invasion and metastasis, respectively.

Despite accumulated data, it is usually difficult to discriminate between permanent epigenetic changes that can be transmitted to daughter cells and alterations of signal transduction pathways adjusting to the continuous presence of cadmium.

Adaptation by Transcription/Signaling

A range of signaling events accompanies the exposure of mammalian cells to cadmium with consequences on apoptosis and differentiation, not to mention development. Cell survival and proliferation, decreased DNA repair capacity, and genomic instability all contribute to adaptation to permanent cadmium resistance in transformed cells (Beyersmann and Hechtenberg 1997; Thévenod 2009).

Proliferation occurs via enhancing growth-promoting factors or by inhibiting apoptosis, by enhancing proapoptotic or decreasing antiapoptotic signals. All detailed situations cannot be recapitulated here, but a variety of signaling pathways is generally involved. Kinases and phosphatases, such as mitogen-activated protein kinases, protein kinases C and A, Wnt (wingless-type mammary tumor virus integration site family), nuclear factor-kappa B, and protein phosphatases 2A and 5 to name but a few, appear to be mediators of these effects. For instance, the proapoptotic activation of c-Jun N-terminal kinases (JNK), a member of the MAPK family, by cadmium may be impaired under chronic exposure. Subsequent failure to direct cells to apoptosis provides a strong selective advantage to subpopulations in which the inactivation of this pathway overrides proapoptotic activities. Efficiency of Nrf2-driven transcription is among the multifarious consequences of MAPK enhanced activity.

The presence of cadmium in the cellular environment can also activate several surface receptors without entering cells. Such mechanisms are expected to lead to transient variations of second messengers (e.g., Ca^{2+}), and chronic exposure to cadmium may permanently change the steady-state concentrations of these messengers, or it may blunt the dynamics of the cellular response via the involved receptors. Candidate molecules which are sensitive to cadmium include Ca^{2+}-sensing G protein–coupled receptors, receptor tyrosine kinases, and steroid receptors, all targeting many downstream events. For instance, cadmium interacts with the ligand-binding domain of estrogen receptor α, and it may interfere with the conformational change of the receptor needed for activation. Although not surface receptors per se, intercellular junction molecules, such as E-cadherin and connexins, can interact with cadmium with both changes of the adhesion properties and activation of signaling pathways by translocation of β-catenin or secretion of messengers (e.g., prostaglandins, insulin, or glutamate) for instance.

The molecular adaptations to cadmium via signaling pathways seem almost countless when data obtained in many experimental systems are cumulated, but some of the practical consequences can be organized into the following few.

Adaptation by Impaired Import

In many examples of persistent resistance to cadmium, cells are protected by the strongly diminished ability of the toxic metal to enter into cells. The physiological function of the membrane components responsible for cadmium uptake may be the transport of metal ions. Instances include calcium channels of different types (voltage or ligand dependent, store operated), metal cotransporters for iron or zinc, and receptors triggering the input of cadmium complexes by endocytosis, such as with megalin and cubilin or the neutrophil gelatinase–associated lipocalin-2 receptor. Adaptation mechanisms via modifications of these different transporters remain ill-documented with the exception of ZIP 8 (Zrt-/Irt-like protein 8, SLC39A8). This divalent cation-bicarbonate cotransporter is probably responsible for high-affinity manganese influx in many cells, and it can transport other cations including cadmium. The molecular changes inactivating it are not completely elucidated, but they may involve epigenetic silencing of expression or inactivation by (protein kinase C) phosphorylation.

Adaptation by Accumulation

A constant feature displayed by cadmium is the strong induction of metallothioneins occurring upon exposure to the metal. In mammals, transcriptional upregulation of MT provides a higher cadmium-binding capacity to targeted cells, yet the cadmium-metallothionein complexes are subjected to distribution and turnover among different cell types, with limited excretion of the noxious metal from the body. Thus, MT upregulation cannot be considered as an efficient and lasting adaptation mechanism in mammalian cells, and there is no evidence that accumulation occurs safely in mammals.

The situation is different in microorganisms and plants. The cells can be protected by extracellular binding to the cell wall, intracellular safe binding (e.g., to polyphosphates) and sequestration into vacuoles, and active export. These different processes can be coupled with, for instance, cadmium binding to strong chelators such as PCs, transport to suitable locations (e.g., plant shoots), and loading into specific organelles such as vacuoles. The movements of cadmium-containing species through membranes are carried out by an extensive range of transporters, including many ATP-binding cassette (ABC) transporters and (P1)-ATPase pumps.

The ability of these proteins to handle toxic compounds such as cadmium is enhanced by the possible adjustment of their regulation in exposed organisms. For example, cadmium accumulator plants, such as *Arabidopsis halleri*, *Thlaspi caerulescens*, or rice to a smaller degree, appear to display a distinct transcriptional response when exposed to cadmium as compared to sensitive plants such as *Arabidopsis thaliana*. The range of contributing genes is large including many encoding transporters, stress (i.e., heat-shock) proteins, and other activities supporting basic metabolic and structural functions. In a similar way, it appears that different yeast strains may adjust the sulfur use of some of their enzymatic isoforms according to their natural habitat, including with respect to its potential cadmium content.

Adaptation by Efficient Export

A last possibility to escape cadmium toxicity is to get rid of the poisonous metal once present inside cells.

As internal ligands such as GSH or PCs are readily mobilized upon cellular cadmium uptake, transporters of these complexes contribute to the resistance to the metal. Cadmium can thus be displaced either to safer places, such as in the vacuole of plants and in the model fission yeast *S. pombe* via ABC transporters, or back to the outside as through the cystic fibrosis transmembrane conductance regulator (CFTR) or the multidrug resistance-associated protein 1 (MRP1) in mammalian epithelial cells. Other transporters, such as those involved in the removal of zinc (the Znt family) or iron (ferroportin, which has been recently shown to be regulated by MTF-1), may participate in cadmium efflux, but the only well-established example of adaptation by sustained enhancement of export in mammalian cells is that of MRP1 upregulation. This adaptive mechanism is mediated, at least in part, by phosphorylation by MAPK (p38) which may shift to a basal higher activity promoting the transport of GSH adducts, including with cadmium. Yet, p38 activation has parallel effects such as counteracting cadmium-driven endoplasmic reticulum stress and activating autophagy, aiming at preserving cells from the poison. As already mentioned above, decreased apoptotic signals can promote cell survival, and cellular efflux proteins can also contribute to remove them (e.g., ceramide metabolites) instead of directly transporting cadmium.

Conclusion

The above description of the cellular response to cadmium exposure draws a complex picture with elusive features depending on the cell type, the cadmium dose, and many other parameters. Yet, a tentative general description of the successive events allowing cells to get through cadmium threat and maintain viability may be proposed (Fig. 2). It should be noticed that the following scenario is likely to apply to many situations in which cells are exposed to toxic compounds; only the specific molecular adjustments may change depending on the properties of the poison. First, the sudden change in the cellular environment must be overcome by cells with means that are quite similar between acute and the initial stages of chronic exposure. Counterbalancing the cadmium-induced oxidative stress and expending various resistance

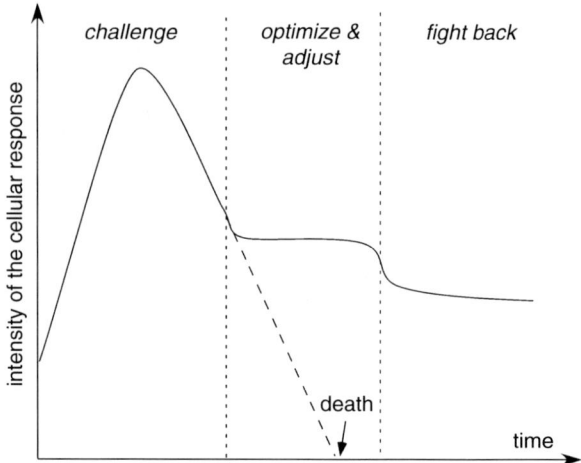

Cadmium Exposure, Cellular and Molecular Adaptations, Fig. 2 Schematic view of the different phases leading to cellular adaptation in the presence of cadmium. Naïve cells initially mobilize various means to respond in proportion to the cadmium insult (the "challenge" phase, *left*). They then optimize their resources to adjust to the continuous threat (the "optimize and adjust" phase, *middle*). These two phases may be considered as reversible for the most part. If the cadmium insult exceeds the homeostatic potential of the cells, death occurs (*dashed line*). Last, adaptation mechanisms described in the text are set in place (the "fight back" phase, *right*) to allow cells to permanently and irreversibly thrive in the presence of cadmium

mechanisms contribute to maintain viability. The efficiency of the cellular response depends on many parameters. For instance, challenging by a mixture of toxic compounds is a deleterious situation in most cases, whereas successive low-level doses of cadmium usually trigger hormosis and help cells to survive, by a mechanism resembling actual adaptation (see below). But this phase is unlikely to last for long as it puts a strong pressure on the cell resources which cannot be exclusively summoned for the single purpose of canceling out the noxious cadmium effects. Therefore, a second step is needed to optimize ways of adjusting to the new situation. This homeostatic response can be considered as a preparation phase to handle the permanently altered conditions in the case of chronic exposure. Most, if not all, processes implemented up to this point, such as those keeping the redox balance or activating specific signaling cascades, are probably reversible. But, if the presence of cadmium persists, a last phase should come, which is the "adaptation" phase discussed herein. At this stage, cells are irreversibly modified to thrive.

This permanent adjustment may involve epigenetic modifications and definitive metabolic and regulatory changes. Consequently, the cellular status may also be permanently affected with possible drastic changes in the cellular fate: for instance, exposed quiescent cells may be tuned to proliferate, hence developing as neoplasms. In such terminal phases of adaptation, presently available evidence indicates that a complex interaction between genetic, metabolic, and regulatory networks occurs, rather than the up- or downregulation of a single pathway. Henceforth, the present scientific challenge is to decipher the interplay between the involved networks to better understand cadmium toxicology and to relieve the deleterious effects of this lingering pollutant.

Summary

The divalent form of cadmium is a persistent pollutant in the environment to which cells of all kinds may be exposed. The toxicity of this metal is largely mediated by disequilibrium between the pro- and antioxidant cellular components with many molecular consequences ending with cell death. However, cells can cope with the cadmium threat in many ways, and some may even thrive in the continuous presence of moderate concentrations of the metal. This adaptive process optimizes the resistance mechanisms by anchoring them into the genetic, metabolic, and regulatory properties of cells. Cadmium handling can thus result into impeded uptake, safe accumulation, or efficient removal, all maintaining viability. Once naïve cells resist to the initial cadmium exposure, adaptation is a succession of events, the main principles of which are outlined in this essay.

Cross-References

- ▶ Biomarkers for Cadmium
- ▶ Cadherins
- ▶ Cadmium Absorption
- ▶ Cadmium and Health Risks
- ▶ Cadmium and Metallothionein
- ▶ Cadmium and Oxidative Stress
- ▶ Cadmium and Stress Response
- ▶ Cadmium Carbonic Anhydrase
- ▶ Cadmium, Effect on Transport Across Cell Membranes
- ▶ Cadmium, Physical and Chemical Properties
- ▶ Cadmium Transport
- ▶ Calcium-Binding Proteins, Overview
- ▶ Zinc and Zinc Ions in Biological Systems
- ▶ Zinc Cellular Homeostasis
- ▶ Zinc-Binding Proteins, Abundance
- ▶ Zinc-Binding Sites in Proteins

References

Beyersmann D, Hechtenberg S (1997) Cadmium, gene regulation, and cellular signalling in mammalian cells. Toxicol Appl Pharmacol 144:247–261

Clemens S (2006) Toxic metal accumulation, responses to exposure and mechanisms of tolerance in plants. Biochimie 88:1707–1719

Cuypers A, Plusquin M, Remans T, Jozefczak M, Keunen E, Gielen H, Opdenakker K, Nair AR, Munters E, Artois TJ, Nawrot T, Vangronsveld J, Smeets K (2010) Cadmium stress: an oxidative challenge. Biometals 23:927–940

Fauchon M, Lagniel G, Aude JC, Lombardia L, Soularue P, Petat C, Marguerie G, Sentenac A, Werner M, Labarre J (2002) Sulfur sparing in the yeast proteome in response to sulfur demand. Mol Cell 9:713–723

Hartwig A (2010) Mechanisms in cadmium-induced carcinogenicity: recent insights. Biometals 23:951–960

He L, Wang B, Hay EB, Nebert DW (2009) Discovery of ZIP transporters that participate in cadmium damage to testis and kidney. Toxicol Appl Pharmacol 238:250–257

Klaassen CD, Liu J, Diwan BA (2009) Metallothionein protection of cadmium toxicity. Toxicol Appl Pharmacol 238:215–220

Liu J, Qu W, Kadiiska MB (2009) Role of oxidative stress in cadmium toxicity and carcinogenesis. Toxicol Appl Pharmacol 238:209–214

Maret W (2011) Redox biochemistry of mammalian metallothioneins. J Biol Inorg Chem 16(7):1079–1086

Morel FM (2008) The co-evolution of phytoplankton and trace element cycles in the oceans. Geobiology 6:318–324

Moulis JM (2010) Cellular mechanisms of cadmium toxicity related to the homeostasis of essential metals. Biometals 23:877–896

Satarug S, Garrett SH, Sens MA, Sens DA (2010) Cadmium, environmental exposure, and health outcomes. Environ Health Perspect 118:182–190

Thévenod F (2009) Cadmium and cellular signaling cascades: to be or not to be? Toxicol Appl Pharmacol 238:221–239

Thévenod F (2010) Catch me if you can! Novel aspects of cadmium transport in mammalian cells. Biometals 23:857–875

Villiers F, Ducruix C, Hugouvieux V, Jarno N, Ezan E, Garin J, Junot C, Bourguignon J (2011) Investigating the plant response to cadmium exposure by proteomic and metabolomic approaches. Proteomics 11:1650–1663

Waalkes MP (2003) Cadmium carcinogenesis. Mutat Res 533:107–120

Cadmium Flux

▶ Cadmium Transport

Cadmium Permeation

▶ Cadmium Transport

Cadmium Transport

Frank Thévenod
Faculty of Health, School of Medicine, Centre for Biomedical Training and Research (ZBAF), Institute of Physiology & Pathophysiology, University of Witten/Herdecke, Witten, Germany

Synonyms

Cadmium flux; Cadmium permeation

Definition

Transport of cadmium ion across cellular membranes of eukaryotic cells through receptors, channels, solute carriers, or ATPases.

Introduction

Cadmium is a toxic transition metal and has no biological function in eukaryotic organisms (with the exception of certain forms of marine phytoplankton). The toxicologically relevant chemical form of cadmium is the divalent cadmium ion (Cd^{2+}). Cd^{2+} taken up by cells and organisms accumulates because it is eliminated very slowly (e.g., in the human kidney Cd^{2+} has a half-time of 10–30 years). Cd^{2+} binds to many cellular and/or extracellular proteins which store the metal and also partly contribute to its detoxification. Binding may be nonspecific and is characterized by a low affinity between Cd^{2+} and the protein, like serum albumin with an equilibrium dissociation constant (K_D) of $\sim 10^{-4}$ M, or the interaction may be more specific by binding to certain amino acid residues, in particular cysteine sulfurs. Examples of the latter are the antioxidative tripeptide glutathione ($K_D \sim 10^{-10}$ M) or the metal-binding proteins metallothioneins ($K_D \sim 10^{-17}$ M) where Cd^{2+} may displace essential metals, such as Zn^{2+}. Metal exchange evoked by Cd^{2+} binding, as well as disruption of the biological function of these proteins, is just an example of the manifold toxic effects elicited by Cd^{2+} in cells (Waisberg et al. 2003; Thévenod 2009; Moulis 2010).

To accumulate in cells and interfere with cellular functions, hydrophilic Cd^{2+} that is present in extracellular fluids as a free ion or complexed to proteins or peptides must permeate lipophilic cellular membranes. Cd^{2+} can only cross membranes through intrinsic proteinous pathways, and this requires either passive (facilitated diffusion) or active (energy dependence) transport mechanisms. Passive transport of Cd^{2+} may involve channels or solute carriers, whereas active mechanisms comprise primary active pumping of Cd^{2+} out of cells and receptor-mediated endocytosis of Cd^{2+} complexes. The similar chemical properties of Cd^{2+} and essential metals ("ionic mimicry") as well as of Cd^{2+} complexes and endogenous substrates ("molecular mimicry") determine the ability of eukaryotic cells to transport Cd^{2+} (Clarkson 1993; Bridges and Zalups 2005). Transport (and toxicity) can only occur if cells possess transport pathways for essential metals or biological molecules which interact with Cd^{2+} or Cd^{2+} complexes in a specific manner. After entering the intracellular compartment, higher-affinity complex formation of Cd^{2+} to other proteins may take place by ligand exchange, thereby also preventing back diffusion and forming a kinetic trap for Cd^{2+}.

Candidates for Cadmium Transport Pathways

This entry summarizes evidence solely for eukaryotic Cd^{2+} transport pathways that has been obtained using stringent experimental approaches, such as radiotracer, fluorescent dye, and/or electrophysiological transport assays combined with molecular biology techniques, because proof of Cd^{2+} fluxes can only be given by direct demonstration of Cd^{2+} transport across biological membranes. Using this "conservative" approach, several likely candidates have been identified.

SLC39 Transporters

The SLC39 transporters are members of the ZIP family of metal-ion transporters (Eide 2004). This designation stands for Zrt-, Irt-like proteins and reflects the first members of this transporter family to be identified. The products of some of these genes, for example, SLC39A1, SLC39A2, and SLC39A4, are involved in Zn^{2+} uptake across the plasma membrane. Other members may function in the uptake of other metal ions (e.g., Fe^{2+} and Mn^{2+}) or may transport metal ion substrates across membranes of intracellular organelles. Various levels of experimental evidence have been obtained for SLC39A8 (ZIP8) and SLC39A14 (ZIP14) involvement in Cd^{2+} transport in tissues such as kidney, intestine, and testis (He et al. 2009). These two transporters function physiologically as Zn^{2+}/HCO_3^- or Mn^{2+}/HCO_3^- symporters. In mice, ZIP8 expression is highest in lung, testis, and kidney. In ZIP8 cRNA–injected Xenopus oocyte cultures, it was shown that ZIP8-mediated $^{109}Cd^{2+}$ and $^{65}Zn^{2+}$ uptake exhibits very low Michaelis constant (K_m) values of ~0.48 and ~0.26 μM, respectively, and studies of electrogenicity to determine whether ZIP8 moves an electrical charge (i.e., carries current) showed an influx of two HCO_3^- anions per one Cd^{2+} (or one Zn^{2+}) cation, that is, electroneutral complexes. SLC39A14 has two exon 4s giving rise to ZIP14A and ZIP14B alternatively spliced products. Mouse ZIP14A expression is highest in liver, duodenum, kidney, and testis; ZIP14B expression is highest in liver, duodenum, brain, and testis. ZIP14B has a higher affinity than ZIP14A toward Cd^{2+} (K_m values of 0.14 vs. 1.1 μM). Given the high affinity for ZIP14, Cd^{2+} is likely to displace Zn^{2+} or Mn^{2+} and enter the body to cause toxicity. Hence, both ZIP8 and ZIP14 are candidates for high-affinity Cd^{2+} transporters mediating toxicity in cells and tissues expressing these solute transporters (He et al. 2009).

SLCA11A2 Transporter

The solute carrier SLCA11A2 or divalent metal-ion transporter-1 (DMT1 also abbreviated to DCT1 for divalent cation transporter-1 or NRAMP2 for natural-resistance-associated macrophage protein-2) is the second member of the *SLC11* gene family of metal-ion transporters that are energized by the H^+ electrochemical gradient (Mackenzie and Hediger 2004). DMT1 transports ferrous ions (Fe^{2+}) but not ferric ions (Fe^{3+}) as a symport/cotransport with H^+ and displays optimal transport activity at acidic pH. DMT1 plays a crucial role in Fe^{2+} homeostasis. DMT1 is ubiquitously expressed, but most notably in the proximal duodenum (where it is the major transporter for iron entry), in red blood cell precursors (where it is essential for full hemoglobinization), in macrophages, but also in the kidney and brain. DMT1 expression is strongest on the brush border of mature villous enterocytes where the expression of DMT1 is tightly regulated by body iron status and is consistent with a role for DMT1 in luminal Fe^{2+} uptake. In macrophages, DMT1 is restricted at the membrane of phagosomes where red cells are engulfed. In the kidney, the majority of DMT1 resides in late endosomal and lysosomal membranes of proximal tubule (PT) cells in the renal cortex, suggesting that DMT1 is involved in movement of Fe^{2+} across late endosomal and lysosomal membranes. DMT1 is not very selective for Fe^{2+}. In fact, a broad range of transition metals (Cd^{2+}, Zn^{2+}, Mn^{2+}, Cu^{2+}, Co^{2+}, Ni^{2+}, and Pb^{2+}) can also evoke inward currents in Xenopus oocytes expressing DMT1. Furthermore, radiotracer assays in Xenopus oocytes and transfected HEK293 cells have established that DMT1 is capable of transporting Cd^{2+}, Mn^{2+}, and Zn^{2+}. The K_m for Cd^{2+} is ~1.04 μM, and Cd^{2+} appears to be transported even more effectively by DMT1 than Fe^{2+} (K_m ~2 μM). A link between Cd^{2+} uptake (and toxicity) and *DMT1* expression has been proposed under conditions where body iron homeostasis is disrupted, resulting in increased DMT1 expression in duodenum, kidneys, and liver. These conditions with increased demand for iron or pathologically elevated intestinal Fe^{2+} absorption appear to be associated with increased Cd^{2+} uptake from the gastrointestinal tract and an increased Cd^{2+} burden of those tissues, but also testes, brain, and blood. These studies indicate that the disruption of iron homeostasis increases the sensitivity to Cd^{2+} because of increased expression of DMT1. Consequently, DMT1 appears to be a key transporter involved in Cd^{2+} toxicity (Thévenod 2010).

Megalin and Cubilin Receptors

Megalin and cubilin are receptors involved in endocytosis which bind multiple ligands and are expressed primarily not only in luminal (apical) plasma membranes of polarized epithelial cells but also in neurons (Christensen et al. 2009). Megalin and cubilin are highly expressed in the early parts of the endocytotic apparatus of the renal PT where they are largely responsible for the tubular clearance (i.e., reabsorption) of most proteins filtered in the glomeruli. Megalin and cubilin are structurally very different, and each binds distinct ligands with varying affinities. Megalin is a 600-kDa transmembrane protein belonging to the low-density lipoprotein (LDL) receptor family. Cubilin is a 460-kDa peripheral membrane protein identical to the intrinsic factor–vitamin B12 receptor from the small intestine. Cubilin physically associates with megalin, and it seems that megalin is responsible for internalization of cubilin and its ligands, in addition to internalizing its own ligands. Cubilin also binds amnionless, a 50-kDa transmembrane protein that is required for its membrane expression and internalization. In addition to direct receptor-receptor interaction, megalin and cubilin share several ligands, including vitamin D–binding protein, albumin, hemoglobin, myoglobin, immunoglobulin light chains, and receptor-associated protein (RAP). Following binding to these receptors at the apical membrane, ligands are internalized into coated vesicles and delivered to early and late endosomes. Whereas the receptors are recycled to the apical membrane, the ligands are transferred to lysosomes for protein degradation. Though it has been commonly assumed that the iron-binding protein transferrin (Tf) with a molecular weight of 78 kDa is above the filtration cutoff for glomerular sieving, filtration of Tf at the glomerulus and its subsequent reabsorption by megalin-cubilin-mediated endocytosis has been convincingly demonstrated. This observation is of utmost importance for the understanding of uptake of Cd^{2+} complexed to metallothioneins (CdMT), a major form of Cd^{2+} present in the blood circulation and a major source of Cd^{2+} in nephrotoxicity induced by chronic oral or pulmonary exposure to the metal. CdMT is easily filtered by the glomerular sieve because of its molecular weight of $\sim$7 kDa and is reabsorbed by the PT by a mechanism analogous to receptor-mediated endocytosis (RME) of proteins. Indeed, metallothionein binds to megalin with a K_D of $\sim$100 μM, and endocytosis of CdMT by the megalin-cubilin complex significantly contributes to CdMT toxicity in cultured PT cells. Additional experimental evidence indicates that following degradation of metallothionein moiety in late endosomes and lysosomes Cd^{2+} is extruded by DMT1 expressed in the membrane of these acidic compartments (see above) into the cytosol where ionic Cd^{2+} triggers cell death via apoptosis of PT cells (Thévenod 2010).

Lipocalin-2 Receptor

Because metallothionein uptake is only partially antagonized by typical megalin-cubilin ligands, such as RAP, additional receptors have been postulated to mediate CdMT toxicity in epithelia. A cell-surface receptor for lipocalin-2 (24p3, NGAL, $K_D \sim$100 pM), a secreted eukaryotic protein which delivers iron to cells by binding to iron-containing siderophores, high-affinity iron-chelating compounds secreted by microorganisms such as bacteria and fungi, has been recently cloned and is expressed in a variety of tissues, including kidney, intestine, and liver. This receptor (Lip2R) modulates iron uptake and apoptosis in cancer cells. In the kidney, Lip2R is mainly expressed in the distal nephron. In cultured distal tubule cells lipocalin-2 reduces CdMT toxicity, suggesting that Lip2R contributes to RME of CdMT and other protein-metal complexes. Transiently transfected CHO cells (which do not express megalin-cubilin) mainly express Lip2R in their plasma membranes. When transfected cells are exposed to fluorescently labeled metallothionein, Tf, or the plant Cd^{2+}-detoxifying phytochelatin (PC), all three proteins are selectively taken up by Lip2R-overexpressing cells. The K_D of binding of all three ligands to Lip2R is $\sim$1 μM. CdMT induces apoptotic cell death in Lip2R, but not in control vector-transfected cells and saturates at 1.4 μM MT/10 μM Cd^{2+} after 24 h exposure. Hence, MT, Tf, and PC are ligands of the Lip2R which, in addition to megalin-cubilin, mediates RME of CdMT and cell death (Langelueddecke et al. 2012). The Lip2R expressed in the distal nephron could represent a second high-affinity pathway for clearance of CdMT (and other metal-protein complexes) that has not been endocytosed by the low-affinity system of megalin-cubilin in the initial portion of the nephron.

ABC Transporters

ATP-binding cassette transporters (ABC transporters) are members of a large protein superfamily expressed in many living cells from prokaryotes to humans. They are transmembrane proteins that in their majority utilize the energy of adenosine triphosphate (ATP) hydrolysis to pump various large hydrophobic, anionic, or cationic substrates across extra- and intracellular membranes, including metabolic products, lipids, sterols, and drugs. ABC transporters are involved in tumor resistance, cystic fibrosis, bacterial multidrug resistance, and a range of other inherited human diseases.

ABCB1

Multidrug resistance P-glycoprotein (ABCB1) pumps a broad range of structurally unrelated, hydrophobic, amphiphilic, and cationic xenobiotics out of cells. ABCB1 is expressed in the apical membrane of epithelia, including the kidney proximal tubule (PT). ABCB1 is highly upregulated in cancer cells, resulting in resistance to chemotherapeutic agents. The expression of ABCB1 can be regulated by a number of stress responses, including nuclear factor kappa B (NF-κB), AP-1, and also Wnt signaling. Pertinently, it was discovered almost 20 years ago that Cd^{2+} causes upregulation of ABCB1 which was found to be associated with decreased Cd^{2+} toxicity in renal PT cells. The simplest mechanism to account for ABCB1-mediated abrogation of Cd^{2+} toxicity is direct Cd^{2+} efflux by ABCB1, as has been proposed by several studies. However, recent experiments indicate that ABCB1 protects PT cells from apoptosis independently of Cd^{2+} efflux. The sphingolipid glucosylceramide may be the proapoptotic substrate extruded by ABCB1 in PT cells, leading to cell survival and possibly propagating Cd^{2+} carcinogenesis (Lee et al. 2011).

ABCC1

The multidrug resistance–associated protein 1 (MRP1; ABCC1) is a high-affinity transporter of the physiological substrate cysteinyl leukotriene C(4) and releases this cytokine from leukocytes into the extracellular space during inflammatory processes. But ABCC1 also functions as a multispecific organic anion transporter, for instance, with oxidized glutathione (GSH) and activated aflatoxin B1 as substrates. In addition, ABCC1 transports glucuronides and sulfate conjugates of steroid hormones and bile salts. ABCC1 and other members of this branch of lipophilic anion transporters are expressed in various epithelial tissues, such as liver and kidney, as well as in endothelia of the blood-brain barrier and in neurons. It is likely that ABCC family members are efflux pumps for Cd^{2+} in the form of Cd^{2+}-GSH complexes (similarly to the yeast vacuolar glutathione S-conjugate transporter YCF1, which has a role in detoxifying metals in yeast and resembles ABCC1), but though GSH efflux by ABCC1 has been demonstrated, Cd^{2+}-GSH transport has not been proven so far.

ABCC7

The cystic fibrosis transmembrane conductance regulator (CFTR; ABCC7) is a cAMP-dependent Cl^- channel expressed in the apical membrane of salt-transporting tissues, such as secretory epithelia and exocrine glands, where it controls ion and fluid homeostasis on the epithelial surfaces, and mutations in the ABCC7 gene cause cystic fibrosis. ABCC7 is also expressed in the apical pole of kidney proximal and distal tubule cells. Recent studies indicate that ABCC7 also mediates GSH export from cells. However, evidence for Cd^{2+} transport by ABCC7 is scarce. Only in one recent study with mouse PT cells expressing ABCC7, low micromolar Cd^{2+} concentrations induced ABCC7-like Cl^- channel gating indirectly by activation of the extracellular signal-activated protein kinase (ERK1/2). Moreover, Cd^{2+}-induced activation of Cl^- currents was associated with ABCC7-mediated extrusion of GSH and Cd^{2+}, possibly as a Cd^{2+}-GSH complex (L'hoste et al. 2009).

Ca^{2+} Channels

Ca^{2+} and Cd^{2+} have similar ionic radii suggesting permeation of Ca^{2+} channels by Cd^{2+} as a mechanism of entry into cells. Many reports have postulated Cd^{2+} uptake by L- and N-type voltage-dependent calcium channels (VDCCs), which are opened by changes in the electrical membrane potential difference, or store-operated calcium channels (SOCs), which are activated by intracellular calcium stores depletion (particularly the endoplasmic reticulum) by hormones or neurotransmitters. However, studies supporting Cd^{2+} flux through these channels using electrophysiological and/or radiotracer techniques are rare. Moreover, Cd^{2+} enters VDCCs and SOCs but potently blocks the pore at submicromolar concentrations and can therefore only permeate the channel if strong nonphysiological potentials are applied. In addition, Cd^{2+} influx through VDCCs or SOCs is likely to be insignificant. The open probability of VDCCs is strongly voltage dependent. Most VDCCs are open over a very narrow voltage range and closed at or near resting membrane potential. The same line of reasoning applies to SOCs which open only upon Ca^{2+} store depletion and are closed under nonstimulated conditions. Nevertheless, several candidate Ca^{2+} permeable ion channels have been described that are likely to be permeated by Cd^{2+} as well.

TRP Channels

Some members of the transient receptor potential (TRP) superfamily of ion channels have been shown to carry Cd^{2+}. TRPs mediate various sensory functions, such as temperature, touch, pain, osmolarity, or taste. Other proposed functions include repletion of intracellular Ca^{2+} stores, receptor-mediated excitation, and cell cycle regulation. A nonselective cation channel that is stimulated by the drug maitotoxin and differs from SOCs has been characterized in renal MDCK cells which carries $^{109}Cd^{2+}$ (K_m ~1.2 μM) and is blocked by Ni^{2+}, Mn^{2+}, and the specific inhibitory drugs SK525a and loperamide (Olivi and Bressler 2000). Though this channel has been postulated to be associated with TRPC1 (transient receptor potential cation channel 1), a member of the subfamily of "canonical" TRPs that is thought to mediate Ca^{2+} store repletion, its molecular identity has remained elusive so far. Transient receptor potential cation channel M7 (TRPM7) is another member of the TRP channel superfamily which belongs to the "melastatin-related" subfamily. It is blocked by the specific blocker 2-aminoethoxydiphenyl borate (2-APB) and was found to be responsible for Cd^{2+} uptake in human and murine osteoblast-like cells where it could play a role in bone damage induced by Cd^{2+} (Levesque et al. 2008). It is ubiquitously expressed and permeable to Ca^{2+} and Mg^{2+}, but also conducts divalent metals such as Zn^{2+}, Mn^{2+}, Co^{2+}, Ni^{2+}, Cd^{2+}, Ba^{2+}, and Sr^{2+}. TRPM7 is currently believed to regulate Ca^{2+} and Mg^{2+} fluxes to affect cell adhesion, cell growth, proliferation, and cell death. As a caveat, the ability of Cd^{2+} to permeate TRPM7 is the lowest of all divalent metal ions tested. Finally, recently TRPV6, a member of the subfamily of "vanilloid-related" TRP channels (also known as CaT1), which is a highly Ca^{2+} selective channel expressed in duodenum, kidney, and placenta and is involved in vitamin D–dependent Ca^{2+} transport in epithelia, has been shown to transport Cd^{2+} at concentrations ≥ 10 μM using the fluorescent indicator for divalent metals Mag-Fura-2 and to be permeable to 2 mM Cd^{2+} using the electrophysiological patch-clamp technique to study single ion channels in cells (Kovacs et al. 2011).

CACNA1G-I Channels

Among the voltage-gated Ca^{2+} channels, $Ca_V3.1$–3 (CACNA1G-I) T-type calcium channels may be the best candidates for Ca^{2+} channels permeated by Cd^{2+} and other divalent metals under physiological conditions (Perez-Reyes 2003). Under physiological conditions, T-type calcium channels have a pacemaker function in the sinoatrial node of the heart and contribute to tonic bursting activation patterns of neurons in the thalamus. The threshold for T-type channel activation in physiological conditions is between −75 and −60 mV. Moreover, based on the characteristics of the voltage dependence of activation and steady-state inactivation curves, these channels are predicted to exhibit considerable "window" currents close to resting membrane potential, so they could mediate significant tonic Cd^{2+} entry. In preliminary experiments, CACNA1G ($Ca_V3.1$, α1G) T-type Ca^{2+} channels have been shown to carry inward Cd^{2+} (2–10 mM) currents at voltages near resting membrane potential. CACNA1G is also permeable to Fe^{2+} and Mn^{2+}, and though transport rates are low by channel standards, calculations have shown that they are similar to those of Fe^{2+} transporters, such as DMT1. CACNA1G is expressed in excitable cells, such as heart and neurons, but it has also been detected in smooth muscle cells of vessels and organs as well as in the distal nephron of the kidney where its function remains unclear.

Mitochondrial Ca^{2+} Uniporter

The mitochondrial Ca^{2+} uniporter (MCU) is located in the inner mitochondrial membrane (MCU) where it is blocked by nanomolar concentrations of ruthenium red or Ru360. Mitochondrial Ca^{2+} uptake via the MCU controls the rate of energy production, shapes the amplitude and spatiotemporal patterns of intracellular Ca^{2+} signals, and is instrumental to cell death. Patch-clamp experiments have convincingly demonstrated that the MCU is a Ca^{2+} channel (Ryu et al. 2010): A highly Ca^{2+} selective (K_D for Ca^{2+} <2 nM) inwardly rectifying current with a single-channel conductance of 2.6–5.2 pS named MiCa has been demonstrated in mitoplasts from COS-7 cells and is also blocked by nanomolar concentrations of the specific inhibitors ruthenium red and Ru360. In addition, two voltage-dependent Ca^{2+} channels in human heart mitoplasts, mCa1 and mCa2, have been characterized and exhibit high Ca^{2+} selectivity. Like MiCa, mCa1 is inhibited by nanomolar Ru360 but has a higher mean single-channel unitary conductance (13.7 pS). mCa2 shares the same voltage dependence with mCa1, but exhibits a smaller unitary conductance (7.67 pS) and is relatively insensitive to Ru360. The molecular structure of the MCU Ca^{2+} channel has been recently elucidated (De Stefani et al. 2011). With regard to Cd^{2+} transport, kidney cortex mitoplasts have been shown to take up Cd^{2+} using the Cd^{2+}-sensitive fluorescent indicator FluoZin-1, and Cd^{2+} uptake was blocked by MCU inhibitors (such as micromolar concentrations of La^{3+} or nanomolar concentrations of ruthenium red and Ru360) (Lee and Thévenod 2006). Hence, Cd^{2+} may permeate the MCU directly but more direct evidence, for example, by patch-clamp techniques and/or $^{109}Cd^{2+}$ radiotracer experiments, would substantiate these observations.

Outlook and Future Directions

In recent years, several transport proteins have been identified that carry the toxicant Cd^{2+} across biological membranes, and it is likely that more candidate transporters will be discovered in the future. These proteins transport free Cd^{2+} ions respective Cd^{2+}-protein or -peptide complexes at the low concentrations that are found in biological fluids of organisms chronically exposed to Cd^{2+} in the environment (food, cigarette smoke, etc.). Hence, various putative entry pathways are available to mediate Cd^{2+} uptake into cells which could explain the observation that a variety of tissues and organs are affected by Cd^{2+}. Interestingly, so far, only one group of transporters has been found to mediate Cd^{2+} efflux out of cells with the potential to protect against toxicity, namely, the ABC transporters. However, a drawback of transport by ABC transporters is that they consume energy by hydrolyzing ATP. Moreover, Cd^{2+} efflux also occurs at the expense of cellular loss of the radical scavenger GSH which needs to chelate Cd^{2+} to allow efflux of the metal. Hence, Cd^{2+} efflux is costly and potentially damaging to the cell and may also ultimately cause death or mutagenesis and malignant transformation of Cd^{2+} affected cells (Waisberg et al. 2003; Thévenod 2009). Consequently, strategies to prevent and circumvent the harmful effects of Cd^{2+} should aim at targeting the uptake pathways rather than focusing on efflux mechanisms mediated by ABC transporters which promote Cd^{2+} toxicity of cells.

Acknowledgments Funded by the Deutsche Forschungsgemeinschaft (TH 345/8-1, 10-1, and 11-1), ZBAF, and Stiftung Westermann-Westdorp, Essen, Germany

References

Bridges CC, Zalups RK (2005) Molecular and ionic mimicry and the transport of toxic metals. Toxicol Appl Pharmacol 204:274–308

Christensen EI, Verroust PJ, Nielsen R (2009) Receptor-mediated endocytosis in renal proximal tubule. Pflugers Arch 458:1039–1048

Clarkson TW (1993) Molecular and ionic mimicry of toxic metals. Annu Rev Pharmacol Toxicol 33:545–571

De Stefani D, Raffaello A, Teardo E et al (2011) A forty-kilodalton protein of the inner membrane is the mitochondrial calcium uniporter. Nature 476:336–340

Eide DJ (2004) The SLC39 family of metal ion transporters. Pflugers Arch 447:796–800

He L, Wang B, Hay EB et al (2009) Discovery of ZIP transporters that participate in cadmium damage to testis and kidney. Toxicol Appl Pharmacol 238:250–257

Kovacs G, Danko T, Bergeron MJ et al. (2011) Heavy metal cations permeate the TRPV6 epithelial cation channel. Cell Calcium 49:43–55

L'hoste S, Chargui A, Belfodil R et al (2009) CFTR mediates cadmium-induced apoptosis through modulation of ROS level in mouse proximal tubule cells. Free Radic Biol Med 46:1017–1031

Langelueddecke C, Roussa E, Fenton RA et al (2012) The lipocalin-2 (24p3/NGAL) receptor is expressed in the distal nephron and mediates protein endocytosis. J Biol Chem 287:159–169

Lee WK, Torchalski B, Kohistani N et al (2011) ABCB1 protects kidney proximal tubule cells against cadmium-induced apoptosis: Roles of cadmium and ceramide transport Toxicol Sci 121:343–356

Lee WK, Thévenod F (2006) A role for mitochondrial aquaporins in cellular life-and-death decisions? Am J Physiol Cell Physiol 291:C195–C202

Levesque M, Martineau C, Jumarie C et al (2008) Characterization of cadmium uptake and cytotoxicity in human osteoblast-like MG-63 cells. Toxicol Appl Pharmacol 231:308–317

Mackenzie B, Hediger MA (2004) SLC11 family of H^+–coupled metal-ion transporters NRAMP1 and DMT1. Pflugers Arch 447:571–579

Moulis J-M (2010) Cellular mechanisms of cadmium toxicity related to the homeostasis of essential metals. Biometals 23:877–896

Olivi L, Bressler J (2000) Maitotoxin stimulates Cd influx in Madin-Darby kidney cells by activating Ca-permeable cation channels. Cell Calcium 27:187–193

Perez-Reyes E (2003) Molecular physiology of low-voltage-activated t-type calcium channels. Physiol Rev 83:117–161

Ryu SY, Beutner G, Dirksen RT et al (2010) Mitochondrial ryanodine receptors and other mitochondrial Ca^{2+} permeable channels. FEBS Lett 584:1948–1955

Thévenod F (2009) Cadmium and cellular signaling cascades: to be or not to be? Toxicol Appl Pharmacol 238:221–239

Thévenod F (2010) Catch me if you can! Novel aspects of cadmium transport in mammalian cells. Biometals 23:857–875

Waisberg M, Joseph P, Hale B et al (2003) Molecular and cellular mechanisms of cadmium carcinogenesis. Toxicology 192:95–117

Cadmium(II)

▸ Cadmium Carbonic Anhydrase

Cadmium, Effect on Transport Across Cell Membranes

E. Van Kerkhove[1], V. Pennemans[2] and Q. Swennen[2]
[1]Department of Physiology, Centre for Environmental Sciences, Hasselt University, Diepenbeek, Belgium
[2]Biomedical Institute, Hasselt University, Diepenbeek, Belgium

Introduction

Cadmium (Cd^{2+}) is an ever-present and global environmental pollutant. Current industrial Cd^{2+} emission has been drastically reduced, but Cd^{2+} continues to be a health hazard (Nawrot et al. 2010). Historically, accumulated Cd^{2+} cannot be degraded, and its half-life in the body is long (10–30 years). Several organs are affected by this heavy metal. In general, long-term exposure studies show skeletal damage, lung disease, and lung cancer. Since 1993, Cd^{2+} has been classified as a group I carcinogen by the International Agency for Research on Cancer. The main target for chronic, low-level Cd^{2+} exposure, however, is the kidney. Cd^{2+} intoxication leads to proximal tubule dysfunction. The specific mechanisms, however, by which it produces adverse effects on the kidney, have yet to be fully unraveled. Next to reactive oxygen species (ROS) production (Cuypers et al. 2010), interference with gene expression, and repair of DNA (Hartwig 2010), Cd^{2+} also interacts with transport across cell membranes and epithelia. This can lead to a distortion of the cell's homeostasis and function, with serious consequences on general health. In the renal tubules, it has been observed that Cd^{2+} leads to a diminished reabsorption of sodium (Na^+), potassium (K^+), magnesium (Mg^{2+}), calcium (Ca^{2+}), chloride (Cl^-) and phosphate (P_i), low-molecular-weight proteins (LMWP), amino acids (AA), and glucose, resulting in an increased ejection fraction of these molecules. Clinically this can be observed as polyuria, hypercalciuria, glucosuria, and hyperphosphaturia without a change in glomerular filtration rate (GFR). Although a lot of research has been done to elucidate the mechanisms by which Cd^{2+} influences membrane transport in the kidney, it is not always clear whether Cd^{2+} has primary or secondary effects on cell membrane transport. Also, the Cd^{2+} doses applied in vitro do not always correspond to the in vivo situation. A general overview of, and critical comments on, the research concerning the effect of Cd^{2+} on epithelial transport mechanisms can be found in the review of Van Kerkhove et al. (2010).

This entry gives a short overview of the transport mechanisms that are affected by Cd^{2+}. The main focus will be on the transport in renal proximal tubule cells, since these have been more widely investigated.

Transport Mechanisms in Epithelial Cells

Epithelial cells form a barrier between the internal and external milieu. In case of the kidney for instance, urinary fluid in the lumen of the tubule (external) is separated from the rest of the body by the epithelial cells of the nephron. Epithelial cells are closely connected by tight junctions, through which free diffusion of solutes and fluids is possible, depending on the permeability. The tight junctions around the cells also delimit the two parts of the epithelial cell membrane: the apical membrane at the luminal side and the basolateral membrane at the blood side. In most epithelia, the apical membrane faces the external milieu, while the basolateral membrane is in contact with the extracellular body fluid compartment. Transport of solutes over the epithelial layer can occur in two ways. First, solutes can sequentially pass across the basolateral and the apical membrane (or vice versa) in an either passive (through electrochemical driving forces) or active way (against the electrochemical gradient). Second, solutes can diffuse through the tight junctions; this is called paracellular transport. The Na^+K^+ pump, which is situated at the basolateral membrane, is a key membrane transporter in cell homeostasis and the prime mover in most of the active transepithelial transport. It couples the hydrolysis of ATP to the vectorial transport of Na$^+$ and K$^+$ across the plasma membrane, maintaining electrochemical gradients for Na$^+$ and K$^+$. Several secondary and tertiary active transport systems of inorganic ions and small organic molecules are driven by this Na$^+$ electrochemical gradient. Moreover, an extra driving force for electrogenic transport is provided by the intracellular negative membrane potential, created by the K$^+$ electrochemical gradient.

Effect of Cadmium on the Na$^+$K$^+$-ATPase

In several animal models, administration of Cd^{2+} leads to a loss of basolateral invaginations in the cortical proximal tubule and a decrease in Na$^+$K$^+$-ATPase. Although a direct effect of Cd^{2+} on the ATPase has been investigated, the group of Thévenod suggests a more indirect causal mechanism (for references see the review by Van Kerkhove et al. (2010)). They state that Cd^{2+} increases the production of ROS intermediates, which decrease the stability of the α1 subunit of the Na$^+$K$^+$-ATPase which is subsequently degraded by both endolysosomal proteases and the ubiquitin protease complex.

It seems that after an initial decline in the activity of the ATPase, intoxicated animals show a defensive pattern. The cells try to cope with oxidative stress and upregulate the Na$^+$K$^+$-ATPase, while if doses are too high and/or time of exposure becomes too long, the toxic effect overrules.

Besides the effect of Cd^{2+} on Na$^+$K$^+$-ATPase in kidney tubules, reduced activity of the Na$^+$K$^+$-ATPase by Cd^{2+} was seen in hepatic microsomes and isolated microsomes of the rat brain.

Cadmium and Ion Transport Across Membranes and Epithelia

Next to the inhibitory effect of Cd^{2+} on the Na$^+$K$^+$-ATPase, a large range of other channels and pumps are affected by this heavy metal. Since it is impossible to overview all of the ion transport mechanisms that are influenced by Cd^{2+}, a short selection of some important ions is given.

Zinc Transport

Zinc is a trace element that is involved in a wide variety of biological functions, such as gene expression, cellular proliferation and differentiation, growth and development, apoptosis, and immune response. Zinc reabsorption, which occurs in the renal tubules of the kidney, is inhibited by Cd^{2+}. A clear overview of the influence of Cd^{2+} on the homeostasis of other essential metals is given in a review by Moulis (2010).

Chloride Transport

The kidney is of major importance for the salt and water homeostasis of the body. One of the mechanisms responsible for this is the Na$^+$Cl$^-$ reabsorption. Cd^{2+} intoxication leads to a mobilization of intracellular Ca^{2+} which activates chloride channels, leading to chloride secretion (see Table 7 in Van Kerkhove et al. (2010)). This, together with some other affected Cl$^-$ transport mechanisms results in a decrease in Cl$^-$ reabsorption, which influences the water homeostasis of the body.

Potassium Transport

K$^+$ channels exist in many forms and have diverse functions. Cd^{2+} has been shown to affect these channels (see Table 7 in Van Kerkhove et al. (2010)). The K$^+$ conductance is important in stabilizing the membrane potential of excitable cells or in creating a negative cell potential in transporting epithelia. Apical K$^+$ channels also play a role in K$^+$ reabsorption or secretion in the kidney. Interference of Cd^{2+} with these channels and with the membrane potential may have an impact on the transepithelial electrogenic transport and deregulate normal transport rates.

The Effect of Cadmium on Membrane Transport of Glucose and Amino Acids

Under normal circumstances, glucose and AAs are completely reabsorbed in the kidney. The Na$^+$-dependent transport processes take place in the proximal tubule. As mentioned earlier, clinical signs of Cd^{2+} intoxication can be glucosuria and aminoaciduria, suggesting an impairment of these reabsorption mechanisms. In several animal models, Na$^+$-dependent glucose uptake, as well as several Na$^+$-dependent amino acid transport systems, is reduced in the proximal tubule in the kidney. In many cases, Cd^{2+} seems to affect the V_{max} and therefore the amount and/or expression of the transporter, but not the K_m, i.e., the properties of the transporters (Table 9 in Van Kerkhove et al. (2010)).

The Effect of Cadmium on Calcium and Phosphate and its Role in Bone Demineralization

It is known that Cd^{2+} exposure causes loss of bone (Bhattacharyya 2009). Whether this bone loss is a direct effect on bone of Cd^{2+} intoxication or is secondary to the impairment of Ca^{2+} and P_i reabsorption in the kidney remains to be elucidated.

Normally, more than 99% of the filtered Ca^{2+} is reabsorbed by the nephron. The Ca^{2+} transport mechanisms that could be involved are apical Ca^{2+} channels (all segments), solvent drag, the basolateral Ca^{2+} ATPase and the Na/Ca^{2+} exchanger (all segments). In the proximal tubule of the kidney, Cd^{2+} intoxication will cause a decrease in fluid reabsorption that in turn will have a negative influence on solvent drag, and paracellular Ca^{2+} uptake. Moreover, transcellular Ca^{2+} uptake is also affected by Cd^{2+}. It was found that epithelial Ca^{2+} channels (ECAC) were blocked by Cd^{2+}. However, not only the uptake of Ca^{2+} at the luminal site of the epithelial cell will be impaired. Due to a dose-dependent reduction of the Ca^{2+}-ATPase, an even larger inhibition of Ca^{2+} extrusion across the basolateral membrane will occur. The net effect will be a decreased Ca^{2+} reabsorption with an increased Ca^{2+} content in the kidney cortex. The influence of Cd^{2+} on Ca^{2+}-ATPase was confirmed in several other animal models such as the gill of the rainbow trout, in permeabilized red blood cells and in intestinal epithelium of the rat. In *Tilapia* intestine, an inhibition of the Na^+/Ca^{2+} exchange was observed. Finally, Cd^{2+} may use Ca^{2+} channels to enter cells.

As P_i plays a key role in the possible bone loss in Cd^{2+} intoxication, it is of interest to study the influence of Cd^{2+} on the P_i reabsorption in the kidney. Normally, 90% of the filtered P_i is reabsorbed. In the proximal tubule, P_i reabsorption occurs mainly through secondary active transport. Key player in this process is the Na^+-P_i cotransporter which has been identified in the brush border membrane (BBM) of the proximal tubule. A direct dose-dependent inhibitory effect of Cd^{2+} on this cotransporter is observed in several animal models. This inhibition leads to phosphaturia (see Table 8 in Van Kerkhove et al. (2010)).

The Effect of Cadmium on Membrane Transport of Other Organic Substances and Metals

Metal transporters play a role in the uptake of essential metals into cells, the uptake via the intestine or the extrusion via the liver and the kidney. Cd^{2+} may use these transporters to enter cells and/or impair the transport of essential metals (Thevenod 2010; Moulis 2010).

The membranes of kidney and liver cells also contain several proteins involved in the uptake and/or excretion of xenobiotics and endogenous organic compounds, all that is not reclaimed is excreted.

Organic Anions

Kidney tubules transport a variety of organic substances other than glucose and AA.

The proximal tubule cells secrete many organic anions that need to be removed from the body. This process depends at least in part on the activity of the Na^+K^+-ATPase and the K^+ and Na^+ gradients it creates. As discussed before, Cd^{2+} influences the Na^+K^+-ATPase activity, and therefore, it might also have an effect on the transport of organic anions. Indeed, animals intoxicated with Cd^{2+} showed a reduced excretion of para-aminohippuric acid (PAH) in vivo. Similarly, several in vitro models demonstrated a reduction of the PAH uptake in vitro after exposure to Cd^{2+}. Interestingly, incubation of microdissected proximal tubule segments of the rabbit with Cd^{2+} induced a bell-shaped curve with a twofold increase of PAH transport at low Cd^{2+} levels (1 μM Cd^{2+}). Interestingly, these concentrations seemed to stimulate cell growth. It might be that Cd^{2+}, although damaging cellular processes and transport systems in the end at these low concentrations, observed in the general and the exposed population, also has a signaling function and induces cell dedifferentiation, cell growth/proliferation in an attempt of the cells to defend themselves. This aspect of Cd^{2+} has not really been well studied up to now.

Endogenous metabolites such as the monocarboxylates (lactate, pyruvate), di- and tricarboxylates (malate, citrate, succinate), and bile salts are organic anions that need to be recovered.

Experiments suggest that transport of endogenous organic anions was reduced by exposure to Cd^{2+}, thus interfering with the cell's function. Again, the V_{max} was primarily affected and therefore the amount and/or expression of the transporter, but not the K_m, i.e., the properties of the transporters (see Table 10 in Van Kerkhove et al. (2010)).

Not much is known about Cd^{2+} effects on anion transporters in the intestine or in liver cells.

Organic Cations

Many cell membranes possess transporters for metals, i.e., Mn, Zn, Fe, Ca, Ni, and Co. It has been shown that Cd^{2+} may use these transporters to enter the cells and therefore interfere with the uptake of the essential metals, which in turn may interact with the normal functioning of the cell. ZIP8, ZIP14, DMT1, and Ca^{2+} channels and transporters are listed as candidates for Cd^{2+} transport (Moulis 2010). As the expression of these transporters is high in the intestine, it seems likely that dietary intake of Fe, Ca, Zn, and Mn may influence the intestinal absorption of Cd^{2+} and vice versa.

The Effect of Cadmium on Water Channels and Proton Pathways

Voltage sensitive proton channels and epithelial proton and water channels are sensitive to Cd^{2+} and other metals. Cd^{2+} is known to interfere with the renal vacuolar H^+-ATPase, inducing a strong decrease in the activity and amount of V-ATPase in the proximal tubule brush border membrane. In doing so, Cd^{2+} intoxication might interfere with protein reabsorption, as will be discussed in the next section. Further interaction between Cd^{2+}, proton pathways (see Table 12 in Van Kerkhove et al. (2010)), and cell pH and its role in Cd^{2+} toxicity need to be explored.

Cd^{2+} intoxication can cause polyuria, due to either impaired solute uptake or by blockage of antidiuretic hormone–sensitive water channels. Effects on water channels and their role in fluid loss are questions that need further research.

The Effect of Cadmium on Endocytosis

Endocytosis is a process used by many cells to absorb large molecules by engulfing them with their cell membrane. Cd^{2+} has been shown to interfere with this process in a wide range of organisms (see Table 13 in Van Kerkhove et al. (2010)).

In unicellular and invertebrate organisms, phagocytosis and endocytosis are used in acquisition of food, immune responses, and are crucial for the survival of these lower life forms. Exposure to very low concentrations of Cd^{2+} (ranging from 10 to 100 nM $CdCl_2$) had no effect or slightly stimulated phagocytosis in bivalve species. Somewhat higher doses (10–100 μM $CdCl_2$) on the other hand suppressed phagocytosis in worm as well as bivalve species.

In the kidney of vertebrates, receptor-mediated endocytosis is essential in the reabsorption of plasma proteins from the renal ultrafiltrate. Chronic exposure to Cd^{2+} leads to proteinuria. In vitro as well as in vivo models demonstrated that exposure to Cd^{2+} diminished endocytosis in kidney cells. This effect was coupled to a decreased activity and/or amount of the endosomal V-ATPase. Thus, the acidification of the endosomes was impaired, causing a fall in receptor-ligand dissociation and a reduced recycling of the receptor, which reduced the overall efficiency of endocytosis. The impaired endocytosis would then lead to a decreased reabsorption of filtered proteins and thus proteinuria. In addition, it might also derange intracellular vesicle trafficking causing loss of specific transporters from the brush border membrane of the kidney.

The Effect of Cadmium on Integrity of the Transporting Epithelium

Cells of transporting epithelia and vascular endothelia are attached to each other by specialized junctional complexes which determine the transepithelial permeability and regulate the transport of substrates across the epithelium. These junctional complexes are composed of specific junction-associated proteins such as integrins, cadherins, connexins, etc., and are closely associated with the cytoskeleton.

In in vitro models, Cd^{2+} causes loosening of the intercellular junctions, followed by a rapid decline of the transepithelial electrical resistance. Electron microscopy showed that the electron density of the intracellular plaques that are associated with the adhering junctions decreased markedly. This suggested that Cd^{2+} might cause proteins, involved in linking membrane-associated cell adhesion molecules to the cytoskeleton, to dissociate and possibly diffuse into the cytosol. Thus, adhesion would disappear at this site, leading to a breakdown in the linkage between the junctional complexes and the cytoskeleton. This mechanism could explain the temporal relationship between the Cd^{2+}-induced breakdown of cell-cell junctions and the dramatic change in cell shape from a flat to a round appearance. A possible candidate site to be affected by Cd^{2+} is the Ca^{2+}-dependent cell adhesion molecule E-cadherin, which plays a role in the Ca^{2+}-dependent cell-cell adhesion. In vitro models of exposure to Cd^{2+} showed that the amount of E-cadherin was decreased, and this coincided with the disruptions of cell-cell junctions. In addition, Cd^{2+} has been shown to have similar effects on several other cadherins as well as on the actin cytoskeleton in various experimental in vitro models. In addition, the effect of Cd^{2+} on E-cadherin distribution is similar to that caused by the removal of extracellular Ca^{2+}. This suggests a direct effect on E-cadherin, displacing Ca^{2+} from its binding site due to the higher affinity of Cd^{2+} for these sites, changing the adhesive properties of the molecule and possibly its interaction with the actin.

In rats exposed to $CdCl_2$, the patterns of N-cadherin, E-cadherin, and β-cadherin localization in the epithelium were profoundly changed. This alteration in cadherin localization was not secondary to cell death, and in addition, Cd^{2+} only induced very low levels of oxidative stress. Taken together, this suggests that the cadherin/catenin complex might be a very early target of Cd^{2+}-induced toxicity in the proximal tubule in vivo, which confirms the data collected using in vitro models as described above.

The causal mechanisms need to be studied further, but given the importance of cadherins as regulators of epithelial function, the disruption of cell-cell junctions in the epithelium might help to explain some of the Cd^{2+}-induced changes in epithelial function (Prozialeck et al. 2003; Prozialeck and Edwards 2010).

Concluding Remarks

When discussing the literature and comparing the in vivo and in vitro situations, we must keep in mind the experimental model and the range of Cd^{2+} concentrations as well as the Cd^{2+} species used.

Indeed, most studies use the free, ionic form of Cd^{2+} to study its effect on transport or epithelial integrity. However, in vivo, practically all Cd^{2+} that reaches the systemic circulation is bound to proteins and other materials in the blood, e.g., (Sabolic et al. 2010), which makes it difficult to extrapolate the results of in vitro studies to the in vivo situation.

Acute in vitro experiments trying to discover direct effects of Cd^{2+} on cell membrane transporters need to be considered with caution. With a few exceptions, the extrapolation to the in vivo situation may not be entirely justified. The doses of Cd^{2+} applied in the in vitro experiments are often a few orders of magnitude higher than those encountered in vivo. It is possible that Cd^{2+} does not affect the transport proteins and transport directly in vivo but rather indirectly via oxidative stress, destruction of the cytoskeleton and/or disruption of cell-cell contacts, suppression or stimulation of expression of transporters, cell death, or still other indirect pathways. To unravel the molecular mechanisms of changes in membrane transport functions due to Cd^{2+}, experimental work is necessary in conditions that are much closer to the in vivo situation (Van Kerkhove et al. 2010).

References

Bhattacharyya MH (2009) Cadmium osteotoxicity in experimental animals: mechanisms and relationship to human exposures. Toxicol Appl Pharmacol 238:258–265

Cuypers A, Plusquin M, Remans T et al (2010) Cadmium stress: an oxidative challenge. Biometals 23:927–940

Hartwig A (2010) Mechanisms in cadmium-induced carcinogenicity: recent insights. Biometals 23:951–960

Moulis JM (2010) Cellular mechanisms of cadmium toxicity related to the homeostasis of essential metals. Biometals 23:877–896

Nawrot TS, Staessen JA, Roels HA et al (2010) Cadmium exposure in the population: from health risks to strategies of prevention. Biometals 23:769–782

Prozialeck WC, Edwards JR (2010) Early biomarkers of cadmium exposure and nephrotoxicity. Biometals 23:793–809

Prozialeck WC, Lamar PC, Lynch SM (2003) Cadmium alters the localization of N-cadherin, E-cadherin, and beta-catenin in the proximal tubule epithelium. Toxicol Appl Pharmacol 189:180–195

Sabolic I, Breljak D, Skarica M et al (2010) Role of metallothionein in cadmium traffic and toxicity in kidneys and other mammalian organs. Biometals 23:897–926

Thevenod F (2010) Catch me if you can! Novel aspects of cadmium transport in mammalian cells. Biometals 23:857–875

Van Kerkhove E, Pennemans V, Swennen Q (2010) Cadmium and transport of ions and substances across cell membranes and epithelia. Biometals 23:823–855

Cadmium, Physical and Chemical Properties

J. David Van Horn
Department of Chemistry, University of Missouri-Kansas City, Kansas City, MO, USA

Synonyms

Cd, Element 48, [7440-43-9]

Definition

Cadmium is a metallic element in the second transition series, Group IIB (12).

General Cadmium Chemistry

Cadmium (Cd, atomic weight, 112.41; atomic number 48) is a nonabundant, toxic metallic element usually found associated with minerals such as ZnS and other zinc ores. One mineral form, greenockite (CdS), is significant, but the usual Cd source is the oxide in industrial slag associated with the refining of Zn or Pb. The industrial uses of Cd include anticorrosion coatings, nickel-cadmium rechargeable batteries, and in neutron moderating alloys. Cadmium sulfide is used as a bright yellow pigment, and all cadmium chalcogenides (CdS, CdSe, CdTe) are used in semiconductor, photonic, and nanoparticle materials and for research.

Cd is in the "zinc group" of the periodic table and exhibits the common +2 oxidation state of this group arising from the loss of two s-orbital electrons; specifically, Cd(II) has the stable [Kr] $3d^{10}$ electronic configuration. The +2 oxidation state dominates the inorganic chemistry of Cd; its chemistry is similar to that of zinc and other M(II) cations, even though Cd(II) exhibits significant ionic radius differences (Shannon 1976) compared with essential M(II) cations, except for Ca(II) (Table 1). Many Cd(II) compounds are isomorphous with Mg(II). The Cd(II) cation is soluble in a wide range of aqueous solutions, but a number of its compounds (e.g., CdS, $Cd_3(PO_4)_2$) are quite *insoluble*. Cd(II) is a borderline acid in the hard/soft sense; thus, it can interact with a range of biological ligands that include oxygen, sulfur, and nitrogen donor atoms. Like Zn(II), the Cd(II) ion prefers a four-coordinate, tetrahedral environment, but its biological coordination chemistry is not limited to this arrangement. Placement of the ion in a wide variety of proteins leads to Cd(II) sites with coordination numbers ranging from three to six and geometries that include trigonal, square planar, square pyramid, trigonal bipyramid, and octahedral. Some distorted geometrical coordination environments and sites lacking one ligand are also observed in the protein crystal data (Henkel and Krebs 2004). The hydrolysis of Cd(II) is comparable to other transition metal ions; (Lide 2009) Cd(II) is slightly less susceptible to hydrolysis than Zn(II) or Ni(II) but significantly more susceptible to hydrolysis than the alkaline earth metal cations, Ca(II) and Mg(II) (Table 2). The reduction potential of Cd(II) to the neutral metal is -0.402 V, about half that of zinc. The $Cd/Cd(OH)_2$ redox couple is utilized in rechargeable batteries.

The full $3d^{10}$ electronic configuration of Cd(II) renders this transition metal cation diamagnetic and spectroscopically silent to paramagnetic resonance or electronic spectroscopy. Analysis of Cd may be conducted by standard methods of metal analysis, including various forms of titrimetry for concentrated samples, and with atomic absorption spectroscopy or neutron activation analysis for dilute samples. Bulk material or consumer products containing Cd may be analyzed by X-ray fluorescence. Finally, ^{111}Cd and ^{113}Cd nuclei are NMR active and are used for evaluating the biochemistry of Cd compounds and characterizing solid- or solution-state Cd compounds. Cadmium NMR is also used to evaluate Zn-binding sites in proteins by replacing Zn with Cd in in vitro experiments.

Cadmium, Physical and Chemical Properties, Table 1 Comparison of M(II) ionic radii (Å) (Shannon 1976)

Coordination number	4	5	6	7	8	12
Cd(II)	0.78	0.87	0.95	1.03	1.10	1.31
Ca(II)	–	–	1.0	1.06	1.12	1.34
Mg(II)	0.57	0.66	0.72	–	0.89	–
Mn(II) (high spin)	0.66	0.75	0.83*	0.90	0.96	–
Zn(II)	0.60	0.68	0.74	–	0.90	–

*Mn(II) CN = 6 has a low-spin value of 0.67

Cadmium, Physical and Chemical Properties, Table 2 Comparison of M(II) hydroxide solubility product; a larger pK_{sp} indicates a more favorable reaction (Lide 2009)

Hydroxide	K_{sp}	pK_{sp}
$Cd(OH)_2$	7.2×10^{-15}	14.1
$Zn(OH)_2$	3×10^{-17}	16.5
$Ni(OH)_2$	5.47×10^{-16}	15.3
$Mn(OH)_2$	2×10^{-13}	12.7
$Mg(OH)_2$	5.61×10^{-12}	11.3
$Ca(OH)_2$	5.02×10^{-6}	5.3

Ingestion and inhalation are the major routes of contamination from Cd metal, alloy, or compound sources in foods, agricultural chemicals, dust, and primary cigarette smoke. Cd is cleared slowly from the body and presents a chronic health hazard. The element is considered a carcinogen, probably via indirect mechanisms, the most effected organs being the lungs, testes, and prostate. Cd causes other disease, especially in the lung and in bone, notably affecting bone mineralization processes. Acute doses of Cd (>200 ppm bloodstream concentration) are nephrotoxic and also target the liver. Cd(II) may affect various Zn(II)- and Ca(II)-dependent biochemical pathways, protein-binding sites, and related processes. Targeted therapy for Cd(II) poisoning in the form of chelation agents is problematic due to selectivity issues; Zn(II) and Ca(II) supplementation may provide some alleviation of the effects of Cd-induced disease and biological overload.

Cross-References

- ▶ Cadmium Absorption
- ▶ Cadmium and Health Risks
- ▶ Cadmium and Metallothionein
- ▶ Cadmium and Oxidative Stress
- ▶ Cadmium and Stress Response
- ▶ Cadmium Carbonic Anhydrase
- ▶ Cadmium, Effect on Transport Across Cell Membranes
- ▶ Cadmium Exposure, Cellular and Molecular Adaptations
- ▶ Cadmium Transport
- ▶ Metallothioneins and Mercury
- ▶ Oxidative Stress

References

Henkel G, Krebs B (2004) Metallothioneins: zinc, cadmium, mercury, and copper thiolates and selenolates mimicking protein active site features - structural aspects and biological implications. Chem Rev 104:801–824

Lide DR (ed) (2009) CRC handbook of chemistry and physics, 90th edn. The Chemical Rubber Co, Cleveland, OH

Shannon RD (1976) Revised effective ionic radii and systematic studies of interatomic distances in halides and chalcogenides. Acta Cryst A32:751–767

Calbindin D_{28k}

Karin Åkerfeldt[1], Tommy Cedervall[2], Mikael Bauer[2] and Sara Linse[2]
[1]Department of Chemistry, Haverford College, Haverford, PA, USA
[2]Department of Biochemistry and Structural Biology, Lund University, Chemical Centre, Lund, Sweden

Synonyms

Calbindin D_{28k}=calbindin=calbindin D28=D-28K=Vitamin D–dependent calcium-binding protein, avian-type; EF-hand superfamily = calmodulin superfamily = troponin superfamily; Gene name: CALB1=CAB27

Definitions

EF-hand: helix-loop-helix motif obeying a 29-residue consensus sequence with hydrophobic and Ca^{2+}-ligating residues in defined positions, typically forming a pentagonal bipyramidal coordination sphere.

Hexa EF-hand (HEF) proteins: Ca^{2+}-binding proteins containing six EF-hand motifs.

Domain: Independent folding unit.

Subdomain: Super-secondary motif which is part of a domain.

Deamidation: Spontaneous hydrolysis of a carboxamide to a carboxylic acid via a cyclic intermediate. Deamidation of asparagine typically leads to iso-aspartate and aspartate in a 2:1 molar ratio.

Chromophoric chelator: Metal-binding dye which changes its optical spectrum upon chelation.

Discovery and Prevalence

Calbindin D_{28k} was originally purified from chicken intestine (Wasserman and Taylor 1996) and has since been found in many species and tissues. In mammals, calbindin D_{28k} is noted for its abundance in brain, kidney, and sensory neurons. In brain, the expression pattern of calbindin D_{28k} in particular neuronal subtypes is distinctly different from those of the homologous proteins calretinin and secretagogin. In birds, calbindin D_{28k} is expressed in the intestine, whereas in mammals, the intestinal role appears to be carried out by the smaller protein calbindin D_{9k}.

Sequence

The amino acid sequence of calbindin D_{28k} is highly conserved; human and rat calbindin D_{28k} share 257 identities over 261 residues (98.5% identity), whereas the human and chicken sequences are only 80% identical. The sequence is dominated by six EF-hand motifs and is characterized by a high prevalence of acidic residues. The connecting linkers are considerably longer than in many other EF-hand proteins, including calmodulin and calbindin D_{9k}. The human sequence is shown below with the four canonical and two variant EF-hand loops highlighted with solid and dotted underlining. Differences in the rat and the chicken sequences are displayed above and below the human version, respectively, indicating the positions that differ.

```
R
H    MAESHLQSSLITAS  QFFEIWLHFDADGSGYLEGKELQNLIQEL43   EF1
C       T  GVE S A          H Y S   N  MD     F
R    L
H    QQARKKAGLELSP   EMKTFVDQYGQRDDGKIGIVELAHVLPTE85   EF2
C             D T        A      KAT          Q
R
H    ENFLLLFRCQQLKSCE EFMKTWRKYDTDHSGFIETEELKNFLKDL130  EF3
C       F        S D  Q        S     DS   S
R
H    LEKANKTVDDTKLAE YTDLMLKLFDSNNDGKLELTEMARLLPVQ174   EF4
C       Q    QIE S T      EI  RM    A        L
R
H    ENFLLKFQGIKMCGK EFNKAFELYDQDGNGYIDENELDALLKDL218   EF5
C          I   V  A       S        M                S
R       E                 S                          S
H       CEKNKQDLDINN  ITTYKKNIMALSDGGKLYRTDLALILCAGDN261 EF6
C          K         LA   S          AE    EE
```

Posttranslational Modifications

Different types of chemical modifications of specific residues affect the structure and function of calbindin D_{28k}. There are five cysteine residues in human calbindin D_{28k}, located at positions 94, 100, 187, 219, and 257, while chicken and rat contain four each. All cysteine residues are located in the linker regions connecting the EF-hand subdomains, except for cysteine 257, which is located at the end of the very last EF-hand helix, near the C-terminal residue 261. The cysteines are sensitive to changes in redox potential. An intramolecular disulfide bond between residues 94 and 100 has been observed in the Ca^{2+}-bound form of human and rat calbindin D_{28k} (Vanbelle et al. 2005; Hobbs et al. 2009). The three C-terminal cysteine residues can also be modified with one glutathione each (Cedervall et al. 2005a, see further below). As shown by mass spectrometry, an extra modification, consistent with the conversion of a cysteine thiol to a sulfenic acid or a disulfide to a disulfide-S-monoxide, can form in the human protein. The cysteine residues can also be S-nitrosylated with S-nitrosoglutathione, which leads to structural changes in the protein (Tao et al. 2002). Calbindin D_{28k} is additionally susceptible to deamidation at residue N203 at neutral and higher pH. The deamidation reaction goes through a cyclic intermediate and converts N203 into a 1:2 mixture of aspartate and iso-aspartate (Vanbelle et al. 2005). The deamidation is slower when the protein is Ca^{2+} loaded, and is avoided by purifying the protein at slightly acidic pH (Thulin and Linse 1999). Mutating N203 to D, and Q182 to E eliminates the possibility to deamidate and provides a homogeneous calbindin D28k sample with a uniquely defined NMR spectrum (Helgstrand et al. 2004).

The EF-Hand

The EF-hand represents one of the most abundant Ca^{2+}-binding motifs. This 29 helix-loop-helix structure ligates one Ca^{2+} ion, although some EF-hands lack Ca^{2+}-binding functionality. The motif was discovered in parvalbumin by Kretsinger and coworkers, in which helices E and F are folded in a manner reminiscent of a hand, with helices E and F corresponding to the index finger and thumb, respectively.

Calbindin D$_{28k}$,

Fig. 1 Structure. Space filling models of the calbindin D$_{28k}$ NMR structure (2G9B.pdb), calbindin D$_{28k}$ modeled with bound IMPase peptide (coordinates kindly provided by Drs. Bobay and Cavanagh), secretagogin crystal structure (2BE4.pdb), and a crystal structure of the calmodulin sMLCK-peptide complex (1CDL.pdb). The color coding is as follows: EF1 *blue*, EF2 *light blue*, EF3 *green*, EF4 *light green*, EF5 *red*, and EF6 *pink*. Bound peptide is shown in *yellow*

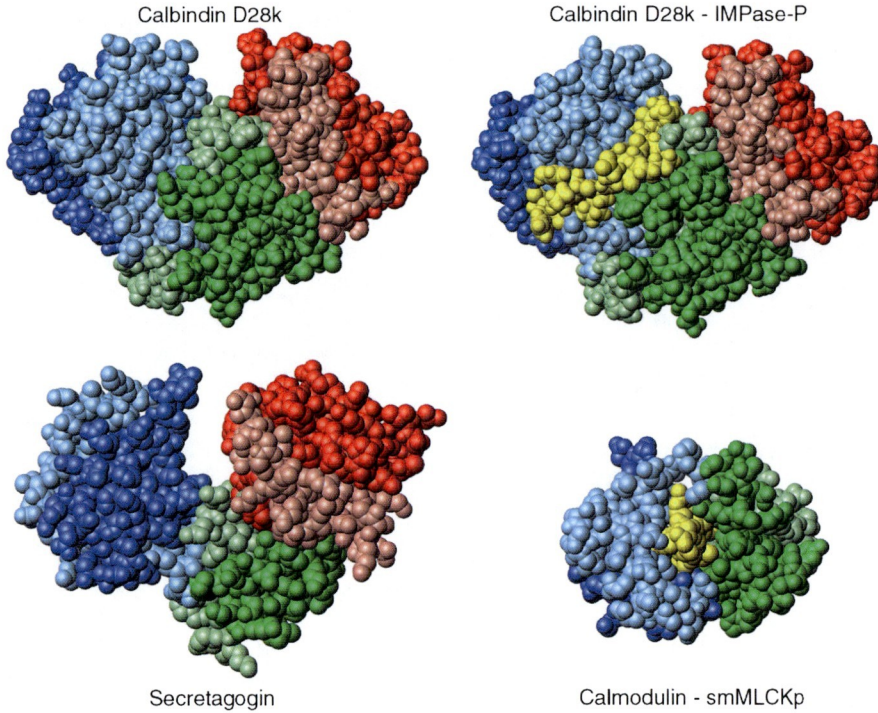

The canonical EF-hand loop (underlined with a solid line in the sequence above) consists of 12 residues and includes all the Ca^{2+}-coordinating residues, although residues 1 and 12 are part of the flanking helices. The loop displays a high degree of conservation, commonly including glutamate, aspartate, glutamine, and serine residues. Another hallmark is the highly conserved glycine at position 6. In each site, the central Ca^{2+} is coordinated in a pentagonal bipyramidal geometry to carboxyl oxygens of residues 1, 3, 5, and 12 (bidentate), a backbone amide carbonyl oxygen of residue 7, and an oxygen derived from a water molecule, which is hydrogen bonded to the side chain of residue 9. However, the Ca^{2+} ion is somewhat flexible in its coordination demands, both in terms of geometry and number of coordinating oxygen atoms.

In general, EF-hand proteins contain a large number of surface charges and display high solubility and high thermal stability, particularly in the Ca^{2+}-loaded state. Eukaryotic proteins have thus far been shown to contain between 2 and 12 EF-hands (Permyakov and Kretsinger 2011) in which the motif is found almost exclusively in pairs, as exemplified by a single pair in calbindin D$_{9k}$ and two pairs in calmodulin.

Domain Organization and the HEF Family

Single EF-hands, derived from their constituent proteins, contain significant hydrophobic patches and have a tendency to oligomerize (Cedervall et al. 2005b). The interactions between naturally occurring EF-hand partners can be very strong and selective and have been used to determine the domain organization for calbindin D$_{28k}$ (Linse et al. 1997; Berggård et al. 2000a). Remarkably, all six EF-hands of this protein interact in a single globular domain. Among the hexa EF-hand (HEF)-proteins, calretinin and secretagogin show a high degree of similarity to calbindin D$_{28k}$, but with some distinct structural and functional features. Studies by Palczewska et al. indicate that calretinin contains two globular domains, one consisting of EF1 and EF2 and the other of EF3-EF6. Secretagogin is not fully separated into two independent domains and its crystal structure reveals a deep cleft in the structure between EF1-EF2 and EF3-EF6 (Fig. 1). The three proteins appear to share an evolutionary origin, with four moderate or high affinity Ca^{2+} sites, although their expression patterns within neuronal subtypes appear distinct. Other, less studied HEF proteins are Eps15 homology domain, *Plasmodium falciparum* surface protein, calsymin and CREC proteins, including reticulocalbin (Permyakov and Kretsinger 2011).

Structure

A high-resolution structure has been solved by NMR spectroscopy for Ca^{2+}-loaded rat calbindin D_{28k} in the reduced form (Kojetin et al. 2006; Fig. 1). This confirms the single globular domain (Linse et al. 1997) and provides detailed information on the three-dimensional structure of calbindin D_{28k}. Within the domain, the EF-hands are organized in pairs, as they occur in the sequence (EF1 with EF2, EF3 with EF4, and EF5 with EF6) with significant contacts between the pairs. For example, the EF1-EF2 pair is proximal to EF3-EF4 and the latter pair packs against EF6. Within each pair, structured as a four helix bundle, are seen the typical contacts, for example, the loop-loop interactions between the short anti-parallel β-strands positioned at the same end of the bundle, the hydrophobic contacts between the residues at position 8 of each loop, which dock into the protein core, and various contacts between the helices. The relatively long linkers connecting the EF-hands display extensive close interactions with one another. The linker between EF2 and EF3 contacts the EF3 to EF4 linker, and the EF4-EF5 linker contacts the sequence connecting EF5-EF6.

Based on the high-resolution structure (Kojetin et al. 2006; Figs. 1 and 2) and interaction data (Åkerfeldt et al. 1996; Linse et al. 1997; Cedervall et al. 2005b; Kordys et al. 2007), the following properties of each individual EF-hand are known.

EF1 (residues 15–43) is a canonical EF-hand containing one of the four high affinity sites and is found in a semi-open conformation in the Ca^{2+}-bound state of calbindin D_{28k}. It pairs with EF2 but also contacts EF4 (e.g., Phe17 of EF1 packs against Leu153 of EF4). It has a hydrophilic side facing the solution, but a significant number of hydrophobic groups are seen both on the opposite side and on the edges of this EF-hand. As an individual synthetic peptide, EF1 folds upon Ca^{2+} binding and its dimerization is Ca^{2+} dependent.

EF2 (residues 57–85) does not have a standard loop sequence and does not bind Ca^{2+}. As a separate peptide, it displays random coil structure, but in the intact protein, it is folded in a helix-loop-helix conformation with a more extended loop than the other EF-hands. It pairs with EF1 and engages in hydrophobic contacts with EF4. EF2 provides a large fraction of the target-binding surface of calbindin D_{28k}.

EF3 (residues 102–130) is a canonical EF-hand containing one of the four high affinity sites and in the Ca^{2+}-bound state of calbindin D_{28k} is found in a relatively closed conformation. It pairs with EF4 and contacts both EF2 and EF6. As an individual sequence, EF3 has significant secondary structure even in the absence of Ca^{2+} and forms dimers both with and without Ca^{2+} present. Ca^{2+} binding promotes further association into higher oligomers.

EF4 (residues 146–174) is a canonical EF-hand containing one of the four high affinity sites. EF4 forms a pair with EF3 and also contacts EF1, EF2, and EF6. Thus, EF4 appears to play the most central role in the generation of a single globular domain with the other EF-hands as it displays short-range interactions with all EF-hands except EF5. This is reflected in the distribution of hydrophobic side chains, which decorate all sides of the EF4 surface. Its fold is more flat than the other EF-hands and it is deeply inserted in the intact protein. As a separate peptide, EF4 folds upon Ca^{2+} binding; however, dimerization and Ca^{2+} binding are not energetically coupled for this fragment.

EF5 (residues 190–218) is a canonical EF-hand containing one of the four high affinity sites. This is the most peripheral EF-hand in calbindin D_{28k}, which exhibits no short-range interactions beyond the pairing with EF6. As a separate peptide, EF5 displays some secondary structure in the absence of Ca^{2+}, but the helical content increases upon Ca^{2+} binding and its dimerization is Ca^{2+} dependent. EF5 has one hydrophilic side facing the surface and the hydrophobic side chains are buried in the interface with EF6.

EF6 (residues 231–259) does not have a standard EF-hand loop sequence: EF6 binds Ca^{2+} very weakly and is Ca^{2+}-free at physiologically relevant Ca^{2+} concentrations. Still its fold is very similar to a Ca^{2+}-loaded EF-hand. EF6 pairs with EF5 and displays short-range interactions also with EF3 and EF4.

Calcium-Binding Properties

Calbindin D_{28k} contains four Ca^{2+} binding sites with physiologically relevant affinity (Berggård et al. 2002a; Venters et al. 2003). At physiological Kcl

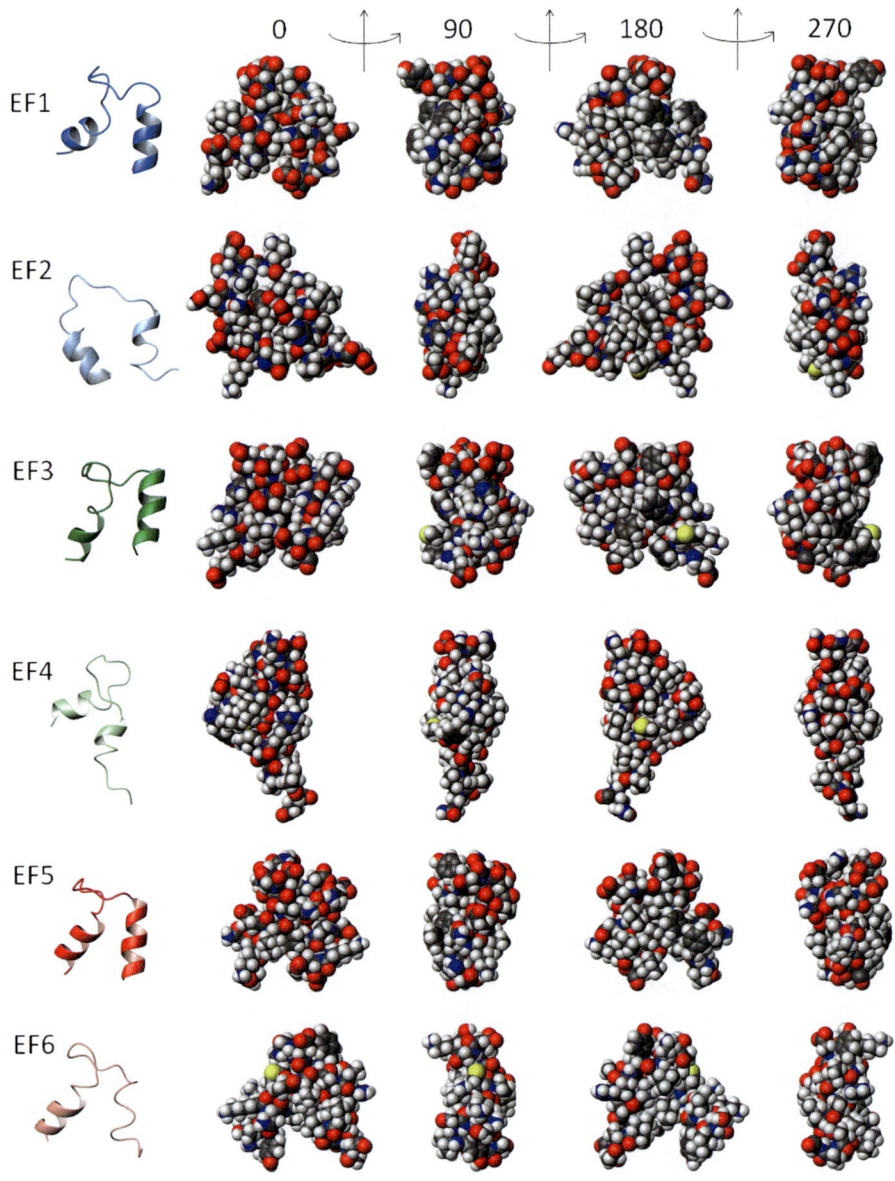

Calbindin D$_{28k}$, Fig. 2 The six EF-hands of calbindin D$_{28k}$. Space filling models with hydrogen atoms in *white*, aliphatic carbons in *gray*, aromatic carbons in *dark gray*, oxygens in *red*, nitrogens in *blue*, and sulfur in *yellow*. Each EF-hand is shown in four orientations related by 90° rotations as indicated, and the *left-most* orientation is also shown in *ribbon* representation

concentration, the average Ca^{2+} affinity for human calbindin D$_{28k}$ is K=2.4 10^6 M^{-1} (K$_D$=0.41 μM). Dissecting the chicken protein into single EF-hand peptides helped assign the high affinity Ca^{2+} binding functionality to EF-hands 1, 3, 4, and 5, while EF-hand 6 binds Ca^{2+} with very low affinity and EF-hand 2 does not appear to bind Ca^{2+} even at very high concentration (Åkerfeldt et al. 1996).

Two modes of Ca^{2+} binding have been reported for intact calbindin D$_{28k}$: parallel with a low degree of positive cooperativity for human and chicken calbindin D$_{28k}$ (Fig. 3a, Berggård et al. 2002a; Leathers et al. 1990), and sequential for the rat protein with at least two separate binding events (Venters et al. 2003).

Rat and human calbindin D$_{28k}$ have very similar sequences (98.5% identical); however, the Ca^{2+} binding properties of rat and human protein have not been studied under the same conditions. The different binding modes reported therefore most likely reflect the paradoxical combination of high thermodynamic stability toward denaturation and high sensitivity to solution conditions and chemical modification, including deamidation. Moreover, the apo form seems to

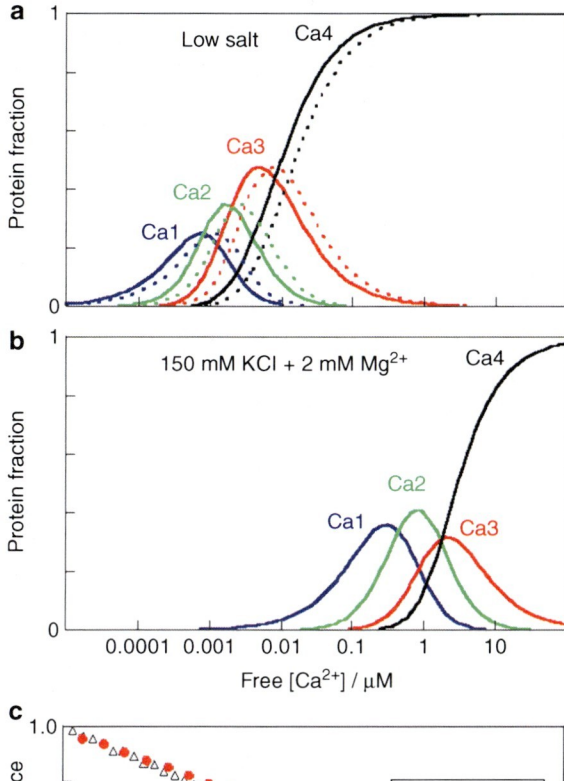

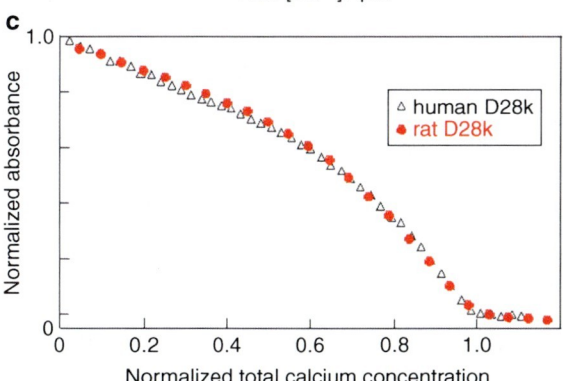

Calbindin D$_{28k}$, Fig. 3 Ca^{2+} binding to calbindin D$_{28k}$. (**a–b**) Fraction of protein in complex with 1, 2, 3, or 4 Ca^{2+} ions bound as a function of free Ca^{2+} concentration, calculated from the macroscopic Ca^{2+}-binding constants determined at low ionic strength in the presence of Quin2 for human calbindin D$_{28k}$ under oxidizing (*solid lines*) and reducing conditions (*dotted lines*). (**c**) Raw data (absorbance at 263 nm versus total Ca^{2+} concentration) at low ionic strength for reduced rat (*red filled circles*) and human (*open black triangles*) calbindin D$_{28k}$ in the presence of Quin2. The absorbance at 263 nm reflects the Ca^{2+} loading level of Quin2 with higher absorbance in its Ca^{2+}-free state. The data were fitted using CaLigator (André and Linse 2002)

exist in different folding states depending on its environment, and it is suggested that this will affect the Ca^{2+} binding mechanism (Hobbs et al. 2009; Kojetin et al. 2006). To find out whether rat and human proteins behave in a similar manner under identical conditions, the rat protein was cloned in Pet3a plasmid, expressed, and purified at pH 6 as described for human protein (Thulin and Linse 1999), a procedure that is known to provide a homogenous protein devoid of deamidation. Figure 3d includes the new Ca^{2+} binding data for reduced rat calbindin D$_{28k}$ in comparison with reduced human protein in the presence of the chromophoric chelator Quin2. Under these conditions, the two proteins produce highly similar data.

At neutral or higher pH, the human protein deamidates at residue N203, and two major isoforms arise, one with Asp203 and one containing the rearranged residue, iso-Asp203. The Asp203 form binds four Ca^{2+} ions with high affinity and positive cooperativity, similarly to the wild type, while the iso-Asp203 form exhibits sequential Ca^{2+} binding (Vanbelle et al. 2005; Fig. 3c).

Structural Transitions upon Ca^{2+} Binding

Calbindin D$_{28k}$ is a highly hydrophobic protein both in the absence and presence of Ca^{2+}; however, the exposure of hydrophobic surfaces changes in response to Ca^{2+} binding as seen by changes in the ANS fluorescence spectrum (Berggård et al. 2000b). There is also a change in tertiary structure involving aromatic side chains as evidenced by near-UV CD spectroscopy. Ca^{2+} binding leads to a conformational change which is very different from that seen for calmodulin (Berggård et al. 2002a; Kojetin et al. 2006). In calmodulin, there is a major change from closed to open domains, revealing target-binding surfaces, while the change in calbindin D$_{28k}$ appears to involve more subtle changes in the packing of EF-hands relative to one another.

Binding of Other Metal Ions

The EF-hand is sufficiently flexible to adapt to the coordination of Mg^{2+}. Calbindin D$_{28k}$ thus binds Mg^{2+} to the same four sites as Ca^{2+}. At physiological salt concentration (0.15 M KCl), the average Mg^{2+} affinity for calbindin D$_{28k}$ is $1.4·10^3$ M^{-1} (K$_D$=0.71 mM), that is, 1,700-fold lower than for Ca^{2+}. Considering that the Ca^{2+} concentration changes from ca. 100 nM in the resting state to 1–10 µM upon cellular activation, and the intracellular concentration of Mg^{2+} is relatively constant at 1–2 mM, Mg^{2+} is still a relevant physiological competitor. The apparent Ca^{2+} affinity is reduced by a factor of 2 in the presence of 2 mM Mg^{2+} but the positive cooperativity of Ca^{2+} binding is more pronounced (Berggård et al. 2002a). Renal calbindin D_{28k} appears to play a role in both Ca^{2+} and Mg^{2+} transport.

A number of EF-hand proteins have been shown to bind other divalent metal ions with rather high affinity, for instance, calretinin, which is closely related to calbindin D_{28k} and has been shown to bind both Zn^{2+} and Cu^{2+}. Similarly, calbindin D_{28k} was recently shown to bind three Zn^{2+} ions to sites different from the Ca^{2+}-binding sites and involving at least one of the four histidine residues (His80; Bauer et al. 2008). The Zn^{2+}-loaded state is structurally different from the Ca^{2+}-loaded state and binding of the two metal ions shows negative allostery. Given that the Zn^{2+} site with the highest affinity ($K_D=1.2$ μM at low ionic strength) is about two orders of magnitude weaker than the Ca^{2+}-sites and the very low free Zn^{2+}-concentration in the cell, calbindin D_{28k} will under normal conditions not be ligated to Zn^{2+}. However, there could be a role for Zn^{2+} binding in cell types with high concentrations of Zn^{2+}, as in pancreatic cells, or temporarily, when the concentration of Zn^{2+} is elevated, for instance during conditions triggering apoptosis.

Target Binding

Several target protein have been reported to bind to calbindin D_{28k}, for example, IMPase (discovered by Berggård et al. 2002b and confirmed by several investigators, for example, Schmidt et al. 2005; Kordys et al. 2007), RAN-binding protein (Kordys et al. 2007), TRPV5 (Lambers et al. 2006), and caspase-3 (Bellido et al. 2000). The target-binding surface of calbindin D_{28k} has been mapped using high-resolution NMR spectroscopy and peptides from target proteins (Kojetin et al. 2006; Kordys et al. 2007). The binding site seems to involve EF2, EF4, and the linker between EF2 and EF3 (Fig. 1). In the case of IMPase, the interaction with calbindin D_{28k} leads to enhanced enzymatic activity to regenerate *myo*-inositol from *myo*-inositol monophosphate. The effect of calbindin D_{28k} is more pronounced under conditions approaching apoptosis, for example, at mild acidosis (Berggård et al. 2002b). TRPV5 is a plasma membrane Ca^{2+} entry channel, and calbindin D_{28k} associates with the channel and tightly buffers the incoming Ca^{2+} to prevent channel inactivation (Lambers et al. 2006).

Redox Regulation

Each cysteine exerts a unique influence on the structure and function of calbindin D_{28k} as revealed by a series of cysteine to serine mutations (Cedervall et al. 2005a). The interplay between the cysteine residues is complex. Cysteines 94 and 100 form a disulfide bridge under nonreducing conditions (Vanbelle et al. 2005), and are necessary for a redox-driven structural change to take place. This change occurs within physiologically relevant redox potentials (between –250 and –175 mV) and results in an altered exposure of hydrophobic surfaces (Fig. 4). Both Ca^{2+} binding and IMPase target activation are affected by the redox potential. The redox-regulated structural change occurs on the tertiary level with no observable differences in the secondary structure content, as judged by far UV CD spectroscopy. The reduced form of calbindin D_{28k}, in which residues 94 and 100 are present as free thiols, displays a slightly higher affinity for Ca^{2+} than the oxidized state (Fig. 3a, b; Vanbelle et al. 2005; Cedervall et al. 2005a). Under reducing conditions, IMPase is less activated by calbindin D_{28k}. Upon oxidation and disulfide bond formation, the hydrophobic surface of calbindin D_{28k} increases, as evidenced by increased ANS binding. The redox potential in cells is mainly controlled by the ratio of reduced and oxidized glutathione. At physiologically relevant glutathione concentrations, cysteine residues 187, 219, and 257 can each react with one glutathione (Cedervall et al. 2005a). Glutathione modifications of residues 187 and 219 further increase the exposure of hydrophobic surfaces in calbindin D_{28k}. These results suggest an intricate interplay between the cysteine residues present in calbindin D_{28k}, which affects how it regulates target proteins, including IMPase.

Stability Toward Denaturation

Calbindin D_{28k} is highly soluble and also highly stable toward thermal denaturation. The Ca^{2+} form does not denature below 100°C and boiling can be used as an

Calbindin D$_{28k}$

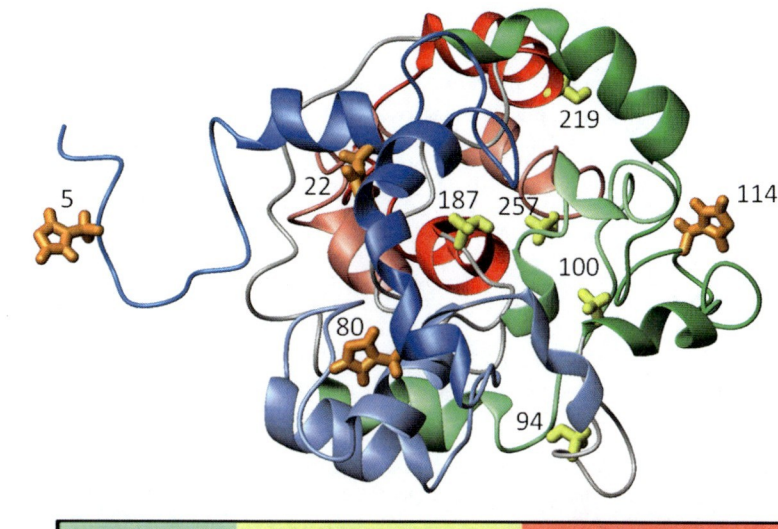
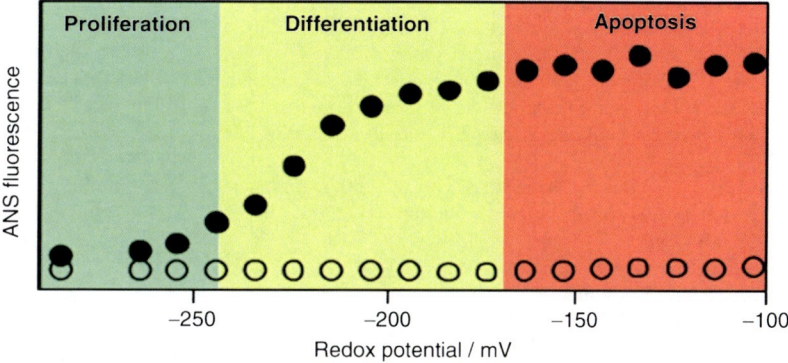

Fig. 4 Redox regulation. *Top*: *Ribbon* representation of rat calbindin D$_{28k}$ (2G9B.pdb) with Cys 94, 100, 187, 219, and Ser 257 (Cys in the human protein) in yellow, His 5, 22, 80, and 114 in *orange*. The EF-hands colored as in Fig. 1. *Bottom*: Structural transition as a function of redox potential (higher ANS fluorescence reflects more extensive exposure of hydrophobic surfaces on the protein) for wild-type human calbindin D$_{28k}$ (*filled circles*) and a mutant where all five cysteines have been mutated to serine (*open circles*)

efficient purification step that precipitates many contaminants. Release of Ca^{2+} reduces the stability, and unfolding of the apo form can be observed in the presence of urea. At pH 7, the urea concentration at the transition midpoint is 4.5 M for the reduced protein. The stability is pH dependent between pH 6 and 8, with higher stability at pH 6. The oxidized protein displays a more cooperative denaturation process (Cedervall et al. 2005a). Glutathione modification further modulates the urea unfolding process.

pH Sensing

The structure and function of apo-calbindin D$_{28k}$ is very sensitive to small changes in the proton concentration near or close to neutral pH. A change from a pH of 6.5–7.8 results in distinct tertiary structural changes, as indicated by a variety of spectroscopic methods (near-UV CD, ANS fluorescence, tryptophan fluorescence, and UV absorbance spectroscopy); however, the secondary structure remains unaffected, as deduced by far UV CD spectroscopy. The Ca^{2+}-loaded protein is less sensitive to pH changes. The structural changes that take place in response to a variation in both pH and Ca^{2+} concentration near neutral pH suggest that these represent important means physiologically by which calbindin D$_{28k}$ modulates its interactions with target proteins. This has, for example, been shown to be the case with IMPase, which displays an increased activation upon calbindin D$_{28k}$ binding at slightly acidic conditions (Berggård et al. 2002b).

Cross-References

▶ Calcium in Biological Systems
▶ Calcium in Health and Disease
▶ Calcium in Nervous System
▶ Calcium-Binding Proteins, Overview
▶ Calcium-Binding Protein Site Types
▶ Calmodulin

- Copper-Binding Proteins
- EF-Hand Proteins
- EF-hand Proteins and Magnesium
- Magnesium Binding Sites in Proteins
- Parvalbumin
- Zinc-Binding Sites in Proteins

References

Åkerfeldt KS, Coyne AN, Wilk RR et al (1996) Ca^{2+}-binding stoichiometry of calbindin D_{28k} as assessed by spectroscopic analyses of synthetic peptide fragments. Biochemistry 35:3662–3669

André I, Linse S (2002) Measurement of Ca^{2+}-binding constants of proteins and presentation of the CaLigator software. Anal Biochem 305:195–205

Bauer MC, Frohm B, Malm J et al (2008) S. Zn^{2+} binding to human calbindin D_{28k}. Protein Sci 17:760–767

Bellido T, Huening M, Raval-Pandya M et al (2000) Calbindin-D28k is expressed in osteoblastic cells and suppresses their apoptosis by inhibiting caspase-3 activity. J Biol Chem 275:26328–26332

Berggård T, Thulin E, Åkerfeldt KS et al (2000a) Fragment complementation of human calbindin D_{28k}. Protein Sci 9:2094–2108

Berggård T, Silow M, Thulin E et al (2000b) Ca^{2+} and H^+ dependent conformational changes of calbindin D_{28k}. Biochemistry 39:6864–6873

Berggård T, Miron S, Önnerfjord P et al (2002a) Calbindin D_{28k} exhibits properties characteristic of a Ca^{2+} sensor. J Biol Chem 277:16662–16672

Berggård T, Szczepankiewicz O, Thulin E et al (2002b) myo-inositol monophosphatase is an activated target of calbindin D_{28k}. J Biol Chem 277:41954–41959

Cedervall T, Berggård T, Borek V et al (2005a) Redox sensitive cysteine residues in Calbindin D_{28k} are structurally and functionally important. Biochemistry 44:684–693

Cedervall T, André I, Selah C et al (2005b) Calbindin D_{28k} EF-hand ligand binding and oligomerization – four high affinity sites, three modes of action. Biochemistry 44:13522–13532

Helgstrand M, Vanbelle C, Thulin E, Linse S, Akke M (2004) Sequential 1H, 15N and 13C NMR assignment of human calbindin D28k. J Biomol NMR 28:305–306

Hobbs CA, Deterding LJ, Perera L et al (2009) Structural characterization of the conformational change in calbindin-D28k upon calcium binding using differential surface modification analyzed by mass spectrometry. Biochemistry 48:8603–8614

Kojetin DJ, Venters RA, Kordys DR et al (2006) Structure, binding interface and hydrophobic transitions of Ca^{2+}-loaded calbindin-D_{28K}. Nat Struct Mol Biol 13:641–647

Kordys DR, Bobay BG, Thompson RJ et al (2007) Peptide binding proclivities of calcium loaded calbindin-D28k. FEBS Lett 581:4778–4778

Lambers TT, Mahieu F, Oancea E et al (2006) Calbindin-D28K dynamically controls TRPV5-mediated Ca^{2+} transport. EMBO J 25:2978–2988

Leathers VL, Linse S, Forsen S et al (1990) Calbindin-D28K, a 1 alpha, 25 dihydroxyvitamin D3-induced calcium-binding protein, binds five or six Ca^{2+} ions with high affinity. J Biol Chem 265:9838–9841

Linse S, Thulin E, Gifford LK et al (1997) Domain organization of calbindin D_{28k} as determined from the association of six synthetic EF-hand fragments. Protein Sci 6:2385–2396

Permyakov EA, Kretsinger RH (2011) EF-hand proteins. In: Uversky VN (ed) Calcium binding proteins, Chapter 11, Section 11.5, Proteins with Six EF-Hands. Wiley, Singapore

Schmidt H, Schwaller B, Eilers J (2005) Calbindin D28k targets myo-inositol monophosphatase in spines and dendrites of cerebellar purkinje neurons. Proc Natl Acad Sci USA 102:5850–5855

Tao L, Murphy ME, English AM (2002) S-nitrosation of Ca(2+)-loaded and Ca(2+)-free recombinant calbindin D(28K) from human brain. Biochemistry 41:6185–6192

Thulin E, Linse S (1999) Expression and purification of human calbindin D_{28k}. Protein Expr Purif 15:271–281

Vanbelle C, Halgand F, Cedervall T et al (2005) Deamidation and disulfide bridge formation in human calbindin D28k with effects on calcium binding. Protein Sci 14:968–979

Venters RA, Benson LM, Craig TA (2003) The effects of Ca(2+) binding on the conformation of calbindin D(28K): a nuclear magnetic resonance and microelectrospray mass spectrometry study. Anal Biochem 317:59–66

Wasserman RH, Taylor AN (1996) Vitamin D_3 induced calcium-binding protein in chick intestinal mucosa. Science 152:791–793

Calbindin D_{28k}=Calbindin=Calbindin D28=D-28K=Vitamin D–Dependent Calcium-Binding Protein, Avian-Type

- Calbindin D_{28k}

Calcineurin

Jagoree Roy and Martha S. Cyert
Department of Biology, Stanford University, Stanford, USA

Synonyms

Calcineurin: Ca^{2+}-/calmodulin-dependent protein serine/threonine phosphatase; PP2B; PP3

Immunosuppressant: Cyclosporin A (CsA); FK506; Tacrolimus; Calcineurin inhibitor

Docking site: PxIxIT; LxVP; Docking motif

Definition

Calcineurin: Calcineurin is a phosphatase that reverses the modification of proteins by kinase-mediated phosphorylation on either serine or threonine residues. Calcineurin requires Ca^{2+} and Ca^{2+}-bound calmodulin for full activity. In different classification systems, calcineurin has also been labeled as PP2B or PP3.

Immunosuppressant: Drugs that inhibit the function of the immune system are termed immunosuppressants. Two commonly used immunosuppressants are cyclosporin A and FK506 (also called tacrolimus). Both of these immunosuppressants are also inhibitors of calcineurin.

Docking site: Docking sites, also termed docking motifs, are short linear sequences, 4–8 amino acids in length, which are contained within a substrate protein and mediate the interaction of that substrate with its cognate enzyme. In substrates of calcineurin, two such docking sites have been identified; these are termed PxIxIT and LxVP, based on the sequence of these sites in a particularly well-studied calcineurin substrate, i.e., the NFAT family of transcription factors. Docking sites do not interact with the active site of CN and are located in regions of the substrate that are distinct from residues that are dephosphorylated by calcineurin.

Introduction

Calcineurin (CN) is a highly conserved, protein serine/threonine phosphatase that is regulated by Ca^{2+} and calmodulin. Originally identified as a calmodulin-binding protein, CN received its name due to its Ca^{2+}-binding properties and abundance in neuronal tissue. Subsequent discovery of its phosphatase activity in 1982 led to its classification as protein phosphatase 2B (PP2B). It is now termed PP3 and is the third member of the phosphoprotein phosphatase (PPP) family of serine/threonine phosphatases (PP1-PP7), which also includes the major protein phosphatases, PP1 and PP2A (Shi 2009).

In 1991, identification of CN as the target of the immunosuppressive agents cyclosporin A (CsA) and tacrolimus (FK506) established the enzyme's critical role in T-cell activation (Aramburu et al. 2000; Rusnak and Mertz 2000). Because of its clinical importance, CN has received considerable attention, and wide use of CsA and FK506 as CN inhibitors has revealed its many functions in organisms as diverse as yeast and humans.

Structure of CN

CN is a heterodimer, composed of a catalytic (CN-A) and regulatory subunit (CN-B), and in its active form, binds one molecule of Ca^{2+}/calmodulin (Aramburu et al. 2000; Rusnak and Mertz 2000), see Fig. 1. CN-A contains four regions: a globular catalytic domain, followed by an α-helical region that binds CN-B, a calmodulin-binding domain, and a C-terminal autoinhibitory (AI) domain, which blocks the catalytic site and keeps CN inactive under basal conditions (Aramburu et al. 2000; Rusnak and Mertz 2000). Proteolytic removal of this domain results in constitutive enzyme activity, and a peptide encoding the AI sequence inhibits CN in vivo and in vitro.

The regulatory subunit of CN, CN-B, is a highly conserved Ca^{2+}-binding protein consisting of two lobes, each of which contains a pair of Ca^{2+}-binding EF-hand domains. Sites in one lobe have tenfold lower affinity for Ca^{2+} than those in the other lobe (Aramburu et al. 2000; Rusnak and Mertz 2000). The N-terminal glycine of CN-B is myristoylated in all species, promoting thermal stability of the enzyme.

Mechanism of Activation

Ca^{2+} activates CN through two distinct mechanisms: First, via Ca^{2+} binding to CN-B and subsequently by the binding of activated calmodulin to CN-A. At low cellular concentrations of Ca^{2+} ($<10^{-7}$ M), only the high-affinity Ca^{2+}-binding sites on CN-B are occupied and the enzyme is inactive. A rise in $[Ca^{2+}]$ to 1 μM results in occupation of the low-affinity Ca^{2+}-binding sites on CN-B and limited activation of CN. This basal activity is greatly stimulated upon further binding of Ca^{2+}-bound calmodulin to CN-A, which results in a conformational change that displaces the AI domain from the catalytic active site (Aramburu et al. 2000; Rusnak and Mertz 2000). This cooperative mode of activation enables CN to respond to small and localized changes in cellular Ca^{2+}.

Metal-Dependent Catalysis

The catalytic core of CN is similar to other phosphoesterases and consists of a dinuclear metal active center, composed of one Zn^{2+} and one Fe^{2+}

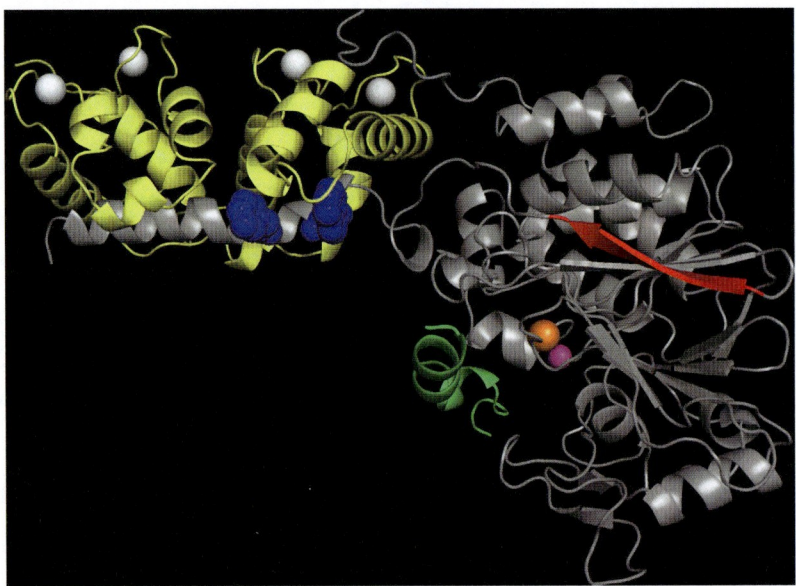

Calcineurin, Fig. 1 X-ray structure of calcineurin (based on Protein Data Bank (PDB) accession code 1AUI (Kissinger et.al. 1995)). CN-B is shown in *yellow* and its bound Ca^{2+} ions are shown in *white*. CN-A, residues 14–373, shown in *silver*. The β-14 sheet that binds the "PxIxIT" docking motif is shown in *red* and residues that interact with the "YLxVP" motif are shown in *blue*. CN-A residues 469–486, encoding the AID, shown in *green*. The active site is marked by Zn^{2+} (*purple*) and Fe^{2+} (*orange*) ions. Note that residues 374–468 of CN-A, which include the calmodulin-binding domain, are not visible in the crystal structure

cofactor (see Fig. 1). The phosphoesterase motif, comprised of three β sheets and two α helices in a β–α–β–α–β format, provides a scaffold for interaction with the two metal ions, which are 3–4 Å units apart. Highly conserved residues coordinate these ions and form a bridge between them (Aramburu et al. 2000; Rusnak and Mertz 2000). Conservation of this active-site structure between CN and other phosphoesterases such as λ–phosphatase suggests a common catalytic mechanism involving direct transfer of the phosphoryl group to a metal-coordinated water molecule (Shi 2009). The identity of the metal cofactors, however, varies in different members of the family.

For CN, Fe^{2+} rather than Fe^{3+} is likely required for catalysis, and the redox state of the active site iron ion may regulate enzyme activity. Purified CN is sensitive to oxidizing agents such as H_2O_2, and anaerobic conditions are required to retain full activity of the purified enzyme (Ghosh et al. 2003). Superoxide dismutase protects CN from inactivation both in vitro and in vivo (Aramburu et al. 2000), and association of SOD1 (Cu-Zn superoxide dismutase) with CN is observed in vivo (Agbas et al. 2007).

CN Interaction with Substrates and Regulators

CN has restricted substrate specificity; however, the phosphosites it regulates share little or no sequence similarity. In fact, unlike serine/threonine protein kinases, CN and related phosphatases act poorly on short peptide substrates, and dephosphorylation is only moderately influenced by residues flanking the phosphosite. Instead, CN recognizes short stretches of amino acids in its substrates, termed docking motifs. These docking motifs are distinct from the substrate's phosphosites and interact at low affinity with surfaces of CN that are spatially distinct from its catalytic site (Li et al. 2011; Roy and Cyert 2009).

The first docking motif for CN was identified in NFAT (Nuclear Factor of Activated T cells), a family of four closely related transcription factors that mediate CN-regulated gene expression in T cells and many other tissues (see below). A region of NFAT that bound directly to CN was identified, and its consensus within the NFAT family is defined as a six amino acid "PxIxIT" motif. The PxIxIT motif, whose actual sequence consensus is considerably broader than the name indicates, has now been identified in CN-regulated proteins from many different organisms, and is present in most, but not all, substrates (Li et al. 2011; Roy and Cyert 2009).

PxIxIT interacts both with the active and inactive forms of CN and mediates substrate binding to a region of CN that is removed from its active site. PVIVIT, a nonnative PxIxIT peptide that was selected in vitro for increased affinity for CN, interacts with a hydrophobic channel on the surface of CN-A (Li et al. 2011; Roy and Cyert 2009) (see Fig. 1). This mode of enzyme-substrate interaction is evolutionarily conserved; the structurally analogous region on PP1 interacts with the RVxF docking motif that is shared by most PP1-interacting proteins. Mutation of conserved residues in this hydrophobic surface abrogates substrate binding in vitro and severely compromises CN function in vivo. Conversely, mutations in the PxIxIT motif in substrates that prevent binding to CN also disrupt dephosphorylation (Li et al. 2011; Roy and Cyert 2009).

Native PxIxIT sites vary considerably in sequence and affinity for CN, and this motif is a major determinant of overall substrate affinity for CN. Mutating the PxIxIT sequence of Crz1, the CN-activated transcription factor from yeast, to increase or decrease its affinity for CN, results in corresponding changes in Crz1-CN-binding affinity, the extent of Crz1 dephosphorylation, and the amount of Crz1-dependent transcription generated by a defined Ca^{2+} signal (Li et al. 2011; Roy and Cyert 2009). Similarly, replacement of the PxIxIT motif in the mammalian CN substrate, NFAT, with the high-affinity PVIVIT sequence, results in increased NFAT dephosphorylation and activity in vivo (Li et al. 2011; Roy and Cyert 2009). The PxIxIT motif, therefore, influences substrate recognition and also fine-tunes the Ca^{2+} dependence of substrate regulation in vivo. Furthermore, PxIxIT sequences are a general mode of protein interaction with CN and are found in CN regulators and inhibitors, such as RCANs, CAIN/Cabin-1, and the A238L viral inhibitor, and in scaffold proteins such as AKAP-79/150 (discussed below).

A second docking motif in CN substrates, with the four amino acid consensus, "LxVP," interacts only with the active, Ca^{2+}-/calmodulin-bound form of the enzyme and may position residues for dephosphorylation at the active site. The LxVP motif, first defined as a conserved sequence in NFAT family members is required for their dephosphorylation by CN, and has now been identified in several other mammalian and yeast substrates. Although the structure of LxVP bound to CN has yet to be determined experimentally, computer modeling and mutational studies predict that it recognizes a hydrophobic groove formed by residues in both CN-A and CN-B, which is also the binding site for immunosuppressants (Roy and Cyert 2009) (see Fig. 1). This surface is accessible only after Ca^{2+} and Ca^{2+}/calmodulin bind to and activate CN. Thus, rather than mediating independent modes of substrate binding to CN, PxIxIT and LxVP may interact sequentially with the enzyme to coordinate substrate recognition and subsequent presentation of phosphosites to the active site. Additional docking motifs in CN substrates may yet be identified – PP1 contains three protein-interaction surfaces and each is conserved in CN (Roy and Cyert 2009). Additional docking sites could act in concert with PxIxIT and LxVP or define distinct mechanisms of substrate recognition by CN.

Immunosuppressant-Binding Sites Overlap with Substrate Docking Regions in CN

Characterization of the LxVP motif provided new insights into the mechanism by which CsA and FK506 inhibit CN. These immunosuppressants (ISs) inhibit CN when complexed with the immunophilin proteins (IPs), cyclophilin, and FKBP, respectively. The IS-IP complexes competitively inhibit CN dephosphorylation of peptide and protein substrates. However they do not block the active site and do not inhibit dephosphorylation of para-nitrophenol phosphate (p-NPP), a small molecule that can readily diffuse into the CN active site and does not require docking (Aramburu et al. 2000; Rusnak and Mertz 2000). X-ray crystallography revealed that both IS-IP complexes bind to the same region of CN, i.e., the hydrophobic groove formed by residues in CN-A and CN-B that is the proposed binding site for LxVP (Aramburu et al. 2000; Rusnak and Mertz 2000; Roy and Cyert 2009). Mutations in this region of CN reduce binding of both IS-IPs and the LxVP motif, and the ISs compete with LxVP for CN binding

(Roy and Cyert 2009). Thus, IS-IP apparently inhibits CN by preventing LxVP sites in substrates from interacting with CN.

Physiological Roles for CN

Regulation of Gene Expression

In mammals, CN regulates gene expression through dephosphorylation of several transcription factors (Creb, Mef2, CRTC-1, Elk-1, etc.), but is best known for its regulation of the Nuclear Factor of Activated T cells (NFAT) family of transcription factors (NFAT1-4) (Aramburu et al. 2000; Crabtree and Schreiber 2009; Rusnak and Mertz 2000). NFATs were originally identified as key factors for T lymphocyte activation, and CsA and FK506, suppress the immune response by inhibiting CN-mediated dephosphorylation of NFAT. In resting T cells, NFAT proteins are phosphorylated and cytoplasmic. Antigen binding increases intracellular $[Ca^{2+}]$ and activates CN, which dephosphorylates NFAT and triggers its rapid accumulation in the nucleus. NFAT cooperates with other transcription factors to induce expression of genes, such as IL-2, whose products are required for T-cell activation and the cellular immune response (Crabtree and Schreiber 2009). This pathway couples Ca^{2+} signals to gene expression in a rapid and reversible manner and mediates key Ca^{2+}-dependent processes in many tissues. In skeletal muscle, CN/NFAT contributes to muscle fiber-type specification because in slow-twitch, but not fast-twitch muscles, Ca^{2+} signals generated in response to neuron firing allow sufficient accumulation of NFAT in the nucleus to trigger gene expression (Schiaffino 2010). CN/NFAT signaling also plays critical roles in heart, lungs, vasculature, and neuronal tissue development, and misregulation of CN/NFAT signaling is a major factor underlying several diseases (Crabtree and Schreiber 2009).

CN Environmental Stress Response Pathway in Yeast and Fungi

NFAT arose late in evolution and is only present in vertebrates. However, CN similarly regulates gene expression in simpler eukaryotes such as yeast and other fungi. In *Saccharomyces cerevisiae*, cytosolic $[Ca^{2+}]$ increases during environmental stress, i.e., exposure to osmotic or heat shock, metal ions, and alkaline or cell wall stress, and CN is essential under these conditions (Cyert 2003). Like NFAT, the yeast transcription factor, Crz1 (CN-regulated zinc-finger protein), rapidly accumulates in the nucleus upon dephosphorylation by CN, where it activates a transcriptional response that promotes cell survival. Crz1 and NFAT both contain PxIxIT sites (Roy and Cyert 2009), but are otherwise evolutionarily unrelated. The CN/Crz1 signaling pathway is conserved in diverse fungi including *Schizosaccharomyces pombe* and the pathogens such as *Candida albicans*, *Aspergillus fumigatus*, *Magnaporthe grisea*, *Ustilago hordei*, and *Cryptococcus neoformans*, where it is required for virulence (Cervantes-Chavez et al. 2011; Stie and Fox 2008; Zhang et al. 2009). Thus, this Ca^{2+}-/CN-dependent signaling pathway is a favored solution to a common microbial problem – the need to adapt to a constantly changing environment. In *S. cerevisiae*, CN's role in promoting survival during environmental stress extends beyond its regulation of gene expression. Additional CN substrates include Slm1 and Slm2, related PH domain-containing proteins that are required for stress-induced endocytosis, and Hph1, an ER-localized protein that associates with the posttranslational translocation machinery and may stimulate the translocation of proteins required for survival into the ER during stress (Roy and Cyert 2009; Piña et al. 2011).

Regulation of Protein Trafficking

CN regulates protein trafficking in cells as diverse as neurons and yeast. During synaptic transmission, Ca^{2+} influx triggers neurotransmitter release through exocytosis of synaptic vesicles and also promotes the retrieval and recycling of synaptic vesicles by their subsequent endocytosis (synaptic vesicle endocytosis or SVE). CN coordinately dephosphorylates a group of nerve terminal proteins, called the dephosphins, that are essential for SVE and include dynamin I, syndapin, amphiphysin, synaptojanin, epsin 1, eps 15, and AP180 (Cousin and Robinson 2001). Dephosphorylation of the dephosphins regulates their interactions with other proteins or, in the case of dynamin I, its enzymatic activity. Subsequent rephosphorylation of these proteins enables them to stimulate further rounds of SVE. Thus CN coordinates the Ca^{2+} signal generated by neuron membrane depolarization with SVE. Regulation of endocytosis by CN is evolutionarily conserved and occurs in the nerve cord and intestinal epithelium of *C. elegans* (Song and Ahnn 2011). In the mammalian brain, regulation of SVE by CN is key for neuronal communication and the response to chronic stimulation such as in drug addiction.

In the hippocampus, CN also promotes endocytosis of the AMPA receptor in postsynaptic densities of excitatory synapses by dephosphorylating the GluR1 subunit. This dephosphorylation requires CN targeting by AKAP79/150, a scaffolding protein that colocalizes CN and several kinases with the AMPA receptor. This NMDA-dependent downregulation of the AMPA receptor is critical for long-term depression (LTD), a key component of synaptic plasticity, and is one of the molecular processes that mediates many aspects of learning and memory. At the behavioral level, forebrain-specific CN knockout mice display reduced levels of LTD and impaired working memory (Li et al. 2011; Kvajo et al. 2010).

Regulation of the Cytoskeleton

In many cells, Ca^{2+} signals act locally through CN- and Ca^{2+}-dependent kinases to reorganize the cytoskeleton during motility. In cultured *Xenopus* neurons, Ca^{2+} inhibits neurite outgrowth through CN-dependent remodeling of the actin cytoskeleton (Lautermilch and Spitzer 2000). Paradoxically, Ca^{2+} mediates opposing responses in migrating neurons, i.e., − attraction to or repulsion from an extracellular guidance cue, with the outcome determined by the relative activities of CN and Ca^{2+}-/calmodulin-dependent kinases (Wen et al. 2007). The magnitude of the Ca^{2+} signal is critical for this decision, as CN is activated at low Ca^{2+} concentrations that fail to activate CAM kinase. CN substrates that regulate actin dynamics include GAP-43, an actin-binding protein that is highly concentrated in growth cones and promotes actin polymerization when phosphorylated, and Slingshot, a phosphatase that dephosphorylates and activates ADF (actin-depolymerizing factor)/cofilin, a major regulator of actin polymerization (Lautermilch and Spitzer 2000; Wen et al. 2007). Cofilin is phosphorylated by LIM kinase, and the balance of LIM kinase vs. Slingshot activity determines whether neurons are attracted or repulsed by BMP-7 gradients (Wen et al. 2007). Ca^{2+} signals activate CN, causing dephosphorylation of Slingshot, which locally activates cofilin to promote growth cone repulsion.

Regulation of the Cell Cycle

Ca^{2+} and calmodulin regulate the cell cycle in many eukaryotes by controlling protein kinases and CN (Kahl and Means 2003). In *Xenopus laevis*, unfertilized eggs are arrested at metaphase of the second meiotic division and a Ca^{2+} spike evoked by fertilization releases arrest by activating CAM kinase II and CN. Fzy/Cdc20, a regulator of the APC (anaphase-promoting complex), which degrades cyclin, may be a critical CN target in this process (Li et al. 2011). During meiosis in *Drosophila*, CN releases cells from metaphase I arrest and is regulated by the RCAN, Sarah (Sra) (see below) (Takeo et al. 2010). CN also regulates G1; CsA induces G1 arrest in several mammalian cell types, and CN promotes the cyclin D1 synthesis in fibroblasts (Kahl and Means 2003). In *S. cerevisiae*, CN is reported to modulate the G2 to M transition by regulating the Swe1 kinase, which inhibits Cdk1 and mediates cell cycle arrest in response to stress (Miyakawa and Mizunuma 2007).

Regulation of Apoptosis

Apoptosis, or programmed cell death, occurs in some tissues during normal development and, in many cell types, during stress and injury. Apoptosis is triggered by a combination of extrinsic signals and intrinsic factors, which lead to mitochondrial dysfunction, cytochrome c release, and caspase activation. Ca^{2+} is a critical control factor for apoptosis, especially in cardiac cell death triggered by heart attack and failure and, for neuronal death, during stroke and neurodegenerative disease (Mukherjee and Soto 2011). Activation of CN and calpain, the Ca^{2+}-regulated protease, is critical for induction of Ca^{2+}-dependent apoptosis. CN dephosphorylates Bad, a member of the Bcl2 protein family, causing it to translocate to mitochondria where it stimulates cytochrome c release (Aramburu et al. 2000). Activation of calpain during apoptosis further enhances CN signaling through degradation of Cabin1/Cain, an endogenous inhibitor of CN (see below).

Regulation of Ion Channels, Pumps, Exchangers

CN regulates ion channels in many tissues, particularly in neurons and cardiomyocytes. In neurons, CN directly regulates activity of Ca^{2+} channels, including the NMDA receptor and the L-type voltage-gated Ca^{2+}, $Ca_v1.2$ (Li et al. 2011). The scaffolding protein

AKAP-79/150 (see below) interacts with $Ca_v1.2$, CN, and PKA and targets both modulators to the channel. PKA and CN have opposing roles in channel regulation, with PKA increasing current amplitude and CN reversing this effect. Thus, channel opening initiates a negative feedback loop by locally activating CN and promoting its own dephosphorylation; in addition, CN is thought to activate NFAT that localizes to the mouth of the channel, thus directly tying channel activity to transcriptional activation (Li et al. 2011). In cardiomyocytes, CN also modulates $Ca_v1.2$ activity; however, in this tissue, CN interacts directly with the channel and does not require AKAP79/150. Recent evidence suggests that CN positively regulates $Ca_v1.2$ in this tissue and acts independently of PKA; however, its mechanism and function are not yet fully elucidated (Tandan et al. 2009). Other ion channels regulated by CN include TRESK, a two-pore domain K^+ channel expressed widely in the immune system and neuronal tissues that may play a role in nociception (Li et al. 2011). CN-dependent activation of TRESK negatively regulates neuronal excitability by promoting the background or leak K^+ conductance that maintains negative membrane potential. CN also regulates a critical ion pump, i.e., the Na^+/K^+ ATPase. Activation of the Na^+/K^+ ATPase by CN promotes critical extrusion of Na^+ in the kidney, explaining in part the nephrotoxic effects of CsA and FK506, whereas in neurons, glutamate binding to NMDA receptor activates influx of both Ca^{2+} and Na^+, and CN-mediated activation of Na^+/K^+ ATPase activity protects cells from glutamate toxicity (Aramburu et al. 2000). CN also controls ion balance by regulating ion exchangers, including negative regulation of the Na^+/Ca^{2+} exchanger, NCX1, in cardiomyocytes and the vacuolar Ca^{2+}/H^+ exchanger, Vcx1, in *S. cerevisiae* (Cyert 2003; Katanosaka et al. 2005).

CN-Dependent Regulation of Other Signaling Pathways

Several CN substrates, including KSR2 and DARPP-32, mediate Ca^{2+}-dependent regulation of other major signaling pathways to allow coordination of Ca^{2+} signals with other second messengers in the cell and promote cross talk between signaling pathways.

DARPP-32 is a key regulator of dopamine-induced signaling in the brain. Dopamine is the primary neurotransmitter involved in reward pathways, and changes in dopaminergic neurotransmission are implicated in many disorders including Parkinson's, schizophrenia, and drug and alcohol abuse. Dopamine elicits an increase in cAMP resulting in PKA-dependent phosphorylation of DARPP-32 on thr-34. This form of DARPP-32 inhibits PP-1, which reverses PKA phosphorylation of many proteins, and thereby amplifies PKA-dependent signaling. In contrast, glutamate generates Ca^{2+} signals through the NMDA receptor, stimulating CN to dephosphorylate DARPP-32, which activates PP-1 and antagonizes PKA-dependent signaling (Kvajo et al. 2010). This modulation of DARPP-32 allows signals from different neurotransmitters to be integrated to control behavior. A closely related PP-1 inhibitor, Inhibitor-1 (I-1), is similarly regulated by CN and PKA and coordinates hormonal signaling in the heart, liver, and other tissues (Aramburu et al. 2000; Rusnak and Mertz 2000).

In pancreatic β-cells and neurons, Ca^{2+} signals stimulate ERK/MAP kinase signaling through CN-dependent dephosphorylation of KSR2, a scaffold for the ERK/MAP kinase signaling pathway. In pancreatic β-cells, glucose stimulation causes Ca^{2+}-dependent ERK activation, which results in insulin secretion. Dephosphorylation of KSR2 by CN abrogates its binding to 14-3-3 proteins, resulting in KSR2 translocation to the cell surface where it binds to ERK and its upstream kinases and promotes their activation by Ras. An LxVP motif in KSR2 is required for CN binding and this Ca^{2+}-dependent activation. Similarly, KSR2 mediates Ca^{2+} activation of ERK/MAP kinase signaling in neurons, evoked by K^+-induced depolarization (Li et al. 2011).

Modulation of CN Activity by Endogenous Protein Regulators

CN catalytic activity is regulated in vivo by several classes of regulators. The conserved RCAN family has been linked to both Down's syndrome and cancer and is comprised of three genes in humans, each with multiple splice isoforms. Regulation of CN by RCANs is complex; these proteins are regulators and substrates of CN and have both positive and negative effects on signaling (Roy and Cyert 2009). In the case of the single yeast RCAN, Rcn1 and human RCAN1-4, the proteins are present at low level under basal conditions and are transcriptionally induced under CN signaling conditions, i.e., by Crz1 and NFAT, respectively. When present at high levels, the proteins bind to and inhibit CN, thus forming a negative feedback loop. RCANs utilize both PxIxIT and LxVP docking motifs to bind CN and may inhibit CN function by preventing substrate access to the enzyme. However, a peptide encoding the last exon of human RCAN1 inhibits CN activity toward p-NPP and peptide substrates in vitro, suggesting that catalytic activity is also directly inhibited (Li et al. 2011; Roy and Cyert 2009).

RCANs can also positively regulate CN. Yeast mutants lacking Rcn1 are deficient for CN signaling, and mice deleted for RCAN1 and RCAN2 display properties characteristic of CN-deficient mice, including reduced expression of NFAT-dependent genes and a decreased propensity to develop cardiac hypertrophy (Li et al. 2011). RCAN phosphorylation is required for this positive effect, which occurs through an undetermined mechanism. RCAN phosphorylation also triggers its ubiquitin/proteosome-mediated degradation and can be reversed by CN-dependent dephosphorylation producing yet another layer of negative feedback on CN signaling. Thus, unraveling each of RCAN's effects on CN signaling remains an exciting if somewhat daunting challenge for the future.

Cabin1/Cain also inhibits CN through an undetermined mechanism. Cabin1/Cain is highly expressed in the brain in a pattern that mirrors that of CN, and it binds to CN-A via a C-terminal proline-rich domain that likely contains a PxIxIT motif. Cabin1/Cain binds to amphiphysin, one of the dephosphins through which CN regulates SVE. Overexpression of Cabin1/Cain in HEK293 cells decreases endocytosis, and during SVE, Cabin1/Cain inhibition of CN is thought to promote rephosphorylation of the dephosphins to terminate endocytosis. Cabin1/Cain also inhibits CN signaling during T-cell activation. In activated T lymphocytes, PKC phosphorylates Cabin1/Cain, increasing its affinity for CN and providing negative feedback on signaling (Liu 2003).

Because of CN's key role in T-cell activation, it is targeted by viral proteins to dampen the host immune response. A238L, made by African swine fever virus, contains two domains: the C-terminal 80 amino acids bind to and inhibit CN and the rest of the protein inhibits NFκB, which activates additional immune responses. A238L contains a PxIxIT motif; however, the mechanism by which it inhibits CN is unknown (Roy and Cyert 2009).

CN Targeting Proteins

In vivo, CN can be directed to specific substrates by scaffolding proteins. AKAP79 in humans and its rodent homologue AKAP150 contains defined binding sites for CN, PKA, and PKC (Li et al. 2011). AKAP79/150 also interacts with substrates for these enzymes and is required for regulation of these substrates in vivo. As discussed above, in hippocampal neurons, AKAP79/150 facilitates modification of the $Ca_v1.2$ L-type Ca^{2+} channel by PKA and CN. In cells that express mutant AKAP79/150 lacking the CN-binding site, the channel fails to be regulated by the phosphatase, and the PKA interaction motif of AKAP79/150 must be present for the channel to be phosphorylated by this kinase. Similarly, during LTD, AKAP79/150 localizes CN and PKA for regulation of AMPA receptor endocytosis by cobinding these regulators and adaptors, such as PSD-95, that interact with the AMPA receptor (Li et al. 2011). AKAP79/150 contains a PxIxIT docking motif and was initially characterized as an inhibitor because CN bound to AKAP is inactive in vitro. However, in vivo, CN is thought to interact dynamically with AKAP79/150 and partition between AKAP-bound and substrate-bound states, with the AKAP7/150 maintaining a high local concentration of the enzyme.

In cadiomyocytes, the Ca^{2+}-and integrin-binding protein 1 (CIB1) specifically targets CN to the cell membrane, where it is activated by Ca^{2+} influx through the L-type channel and activates NFAT (Heineke et al. 2010). CIB1 is specifically upregulated during cardiac hypertrophy (see below). Interestingly, CIB1 interacts with CN via CN-B rather than CN-A, showing that both subunits can mediate specific interactions with other proteins.

Impact of CN on Mammalian Diseases

CN signaling is altered in a growing list of pathologies, some which are associated with chronic activation of CN, such as neurodegenerative diseases and cardiac hypertrophy, while in others, such as diabetes and Down's syndrome, CN signaling is compromised. CN also contributes to cancer progression, where its role varies by cell type.

CN is Chronically Activated in Neurodegenerative Disease

Hyperactivation of CN is strongly associated with neurodegenerative disorders such as Alzheimer's (AD) and Huntington's (HD) disease (Mukherjee and Soto 2011). During these and other neurodegenerative diseases, misfolded, aggregated proteins accumulate in cells, causing ER stress and provoking Ca^{2+} release from the ER. The resulting elevation in CN activity can be further enhanced by proteolytic cleavage of the CN-A AID by caspases and/or inactivation of Cabin1/Cain by calpain. Brain samples from AD patients are enriched in truncated, constitutively active forms of CN-A, and chronic activation of CN promotes neuronal death through dephosphorylation of BAD and induction of apoptosis. In HD, CN further contributes to neuronal death by dephosphorylating huntingtin, which reduces huntingtin's ability to promote secretion of neurotrophic factors such as BDNF (brain-derived neuronal growth factor). CN also causes synaptic dysfunction by inactivating CREB, increasing endocytosis of AMPA receptors and misregulating SVE. CN inhibitors reduce the abnormal focal swelling observed in dendrites and axons during neurodegenerative disease and, as indicated by studies with mouse models, have significant therapeutic potential (Mukherjee and Soto 2011).

CN Signaling in Heart Disease

Activation of CN signaling accompanies heart disease and results in cardiac hypertrophy. CN is both necessary and sufficient to induce cardiac hypertrophy in rodent models and does so primarily by activating NFAT and MEF2, both of which induce expression of genes required for the pathology (Heineke and Molkentin 2006). Inhibition of CN with FK506, CsA, or transgenic expression of CN inhibitor proteins alleviates hypertrophy. Cib1, which is upregulated during pathological but not physiological cardiac hypertrophy, promotes stress-induced, Ca^{2+}-dependent activation of CN in this process by positioning the enzyme near the L-type Ca^{2+} channel (Heineke et al. 2010).

CN in Diabetes

Pathophysiological responses that result from abrogation of CN signaling are sometimes revealed as complications associated with clinical use of FK506 or CsA. These CN inhibitors impair insulin production by pancreatic β cells, and patients receiving them often develop diabetes mellitus. CN/NFAT signaling is required for insulin transcription and for other aspects of β cell function. CN may also directly regulate insulin secretion (Heit 2007).

CN in Skin Cancer

Treatment with CsA or FK506 also results in significantly increased risk for squamous cell carcinoma (Wu et al. 2010). In keratinocytes, inhibition of CN/NFAT signaling leads to increased expression of ATF3, an AP-1 family protein that suppresses p53-mediated senescence. Thus, in these cells, the CN/NFAT pathway is required for p53 to function effectively as a tumor suppressor.

CN/NFAT Signaling Is Perturbed in Down's Syndrome

The portion of chromosome 21 that is present in three copies in Down's syndrome patients contains two genes that combine to decrease CN/NFAT signaling: DSCR1 encodes the Rcan1 CN regulator, whose increased dose inhibits CN, and DYRK1a encodes a protein kinase that phosphorylates NFAT and promotes its export from the nucleus. Mice overexpressing Rcan1 and Dyrk1a exhibit many of the same defects associated with Down's syndrome (Li et al. 2011).

Multiple Roles for CN/NFAT in Cancer

Interestingly, individuals with Down's syndrome also exhibit significantly reduced incidence of leukemia and solid tumors as a result of decreased NFAT signaling. Thus, in contrast, the tumor-suppressive role described for keratinocytes, in many cancer cells, the CN/NFAT signaling promotes tumor progression by stimulating proliferation, migration, invasion, and angiogenesis (Mancini and Toker 2009) and may therefore be an attractive therapeutic target.

CN and Schizophrenia

The molecular causes of schizophrenia, one of the most common mental disorders, are poorly understood. CN-deficient mice display behaviors common to this disorder, and genetic studies show that some alleles of PP3CC, which encodes CN-A, are significantly associated with schizophrenia. Thus, identifying CN-regulated processes that are perturbed in schizophrenic patients may provide insight into the causes of the disorder (Kvajo et al. 2010).

Future Perspectives: Development of Novel Strategies to Modulate CN Signaling

CN's many associations with pathology suggest a huge therapeutic potential for drugs that target this phosphatase. However, CN is ubiquitously expressed, and the multiple side effects associated with immunosuppressive therapy illustrate the need for drugs that are either tissue specific or disrupt CN-dependent regulation of particular substrates. Inhibitors of substrate binding, like PVIVIT and LxVP, could potentially be tailored for specificity. Also, several small molecule inhibitors of CN have been discovered: CN585, a noncompetitive, reversible inhibitor (Erdmann et al. 2010), PD144795, which also has anti-HIV and anti-inflammatory properties (Rusnak and Mertz 2000), and the INCA compounds, which are allosteric inhibitors of CN/NFAT (Roy and Cyert 2009). Elucidating how CN recognizes and dephosphorylates its substrates and identifying the mechanisms by which its endogenous protein regulators act may yield novel strategies to modify the activity and function of this critical enzyme in vivo.

Cross-References

- ▸ Calcium and Mitochondria
- ▸ Calcium in Biological Systems
- ▸ Calcium in Health and Disease
- ▸ Calcium in Heart Function and Diseases
- ▸ Calcium in Nervous System
- ▸ Calcium-Binding Protein Site Types
- ▸ Calcium-Binding Proteins, Overview
- ▸ Calmodulin
- ▸ EF-Hand Proteins

References

Agbas A, Hui D, Wang X, Tek V, Zaidi A, Michaelis EK (2007) Activation of brain calcineurin (Cn) by Cu-Zn superoxide dismutase (SOD1) depends on direct SOD1-Cn protein interactions occurring in vitro and in vivo. Biochem J 405:51–59

Aramburu J, Rao A, Klee CB (2000) Calcineurin: from structure to function. Curr Top Cell Regul 36:237–295

Cervantes-Chavez JA, Ali S, Bakkeren G (2011) Response to environmental stresses, cell-wall integrity, and virulence are orchestrated through the calcineurin pathway in Ustilago hordei. Mol Plant Microbe Interact 24:219–232

Cousin MA, Robinson PJ (2001) The dephosphins: dephosphorylation by calcineurin triggers synaptic vesicle endocytosis. Trends Neurosci 24:659–665

Crabtree GR, Schreiber SL (2009) SnapShot: Ca2+ −calcineurin-NFAT signaling. Cell 138:210

Cyert MS (2003) Calcineurin signaling in Saccharomyces cerevisiae: how yeast go crazy in response to stress. Biochem Biophys Res Commun 311:1143–1150

Erdmann F, Weiwad M, Kilka S, Karanik M, Patzel M, Baumgrass R, Liebscher J, Fischer G (2010) The novel calcineurin inhibitor CN585 has potent immunosuppressive properties in stimulated human T cells. J Biol Chem 285:1888–1898

Ghosh MC, Wang X, Li S, Klee C (2003) Regulation of calcineurin by oxidative stress. Methods Enzymol 366:289–304

Heineke J, Molkentin JD (2006) Regulation of cardiac hypertrophy by intracellular signalling pathways. Nat Rev Mol Cell Biol 7:589–600

Heineke J, Auger-Messier M, Correll RN, Xu J, Benard MJ, Yuan W, Drexler H, Parise LV, Molkentin JD (2010) CIB1 is a regulator of pathological cardiac hypertrophy. Nat Med 16:872–879

Heit JJ (2007) Calcineurin/NFAT signaling in the beta-cell: From diabetes to new therapeutics. Bioessays 29:1011–1021

Kahl CR, Means AR (2003) Regulation of cell cycle progression by calcium/calmodulin-dependent pathways. Endocr Rev 24:719–736

Katanosaka Y, Iwata Y, Kobayashi Y, Shibasaki F, Wakabayashi S, Shigekawa M (2005) Calcineurin inhibits Na+/Ca2+ exchange in phenylephrine-treated hypertrophic cardiomyocytes. J Biol Chem 280:5764–5772

Kvajo M, McKellar H, Gogos JA (2010) Molecules, signaling, and schizophrenia. Curr Top Behav Neurosci 4:629–656

Lautermilch NJ, Spitzer NC (2000) Regulation of calcineurin by growth cone calcium waves controls neurite extension. J Neurosci 20:315–325

Li H, Rao A, Hogan PG (2011) Interaction of calcineurin with substrates and targeting proteins. Trends Cell Biol 21:91–103

Liu JO (2003) Endogenous protein inhibitors of calcineurin. Biochem Biophys Res Commun 311:1103–1109

Mancini M, Toker A (2009) NFAT proteins: emerging roles in cancer progression. Nat Rev Cancer 9:810–820

Miyakawa T, Mizunuma M (2007) Physiological roles of calcineurin in Saccharomyces cerevisiae with special emphasis on its roles in G2/M cell-cycle regulation. Biosci Biotechnol Biochem 71:633–645

Mukherjee A, Soto C (2011) Role of calcineurin in neurodegeneration produced by misfolded proteins and endoplasmic reticulum stress. Curr Opin Cell Biol 23:223–230

Piña FJ, O'Donnell AF, Pagant S, Piao HL, Miller JP, Fields S, Miller EA, Cyert MS (2011) Hph1 and Hph2 are novel components of the Sec63/Sec62 posttranslational translocation complex that aid in vacuolar proton ATPase biogenesis. Eukaryot Cell 10:63–71

Roy J, Cyert MS (2009) Cracking the phosphatase code: docking interactions determine substrate specificity. Sci Signal 2:re9

Rusnak F, Mertz P (2000) Calcineurin: form and function. Physiol Rev 80:1483–1521

Schiaffino S (2010) Fibre types in skeletal muscle: a personal account. Acta Physiol (Oxf) 199:451–463

Shi Y (2009) Serine/threonine phosphatases: mechanism through structure. Cell 139:468–484

Song HO, Ahnn J (2011) Calcineurin may regulate multiple endocytic processes in *C. elegans*. BMB Rep 44:96–101

Stie J, Fox D (2008) Calcineurin regulation in fungi and beyond. Eukaryot Cell 7:177–186

Takeo S, Hawley RS, Aigaki T (2010) Calcineurin and its regulation by Sra/RCAN is required for completion of meiosis in Drosophila. Dev Biol 344:957–967

Tandan S, Wang Y, Wang TT, Jiang N, Hall DD, Hell JW, Luo X, Rothermel BA, Hill JA (2009) Physical and functional interaction between calcineurin and the cardiac L-type Ca2+ channel. Circ Res 105:51–60

Wen Z, Han L, Bamburg JR, Shim S, Ming GL, Zheng JQ (2007) BMP gradients steer nerve growth cones by a balancing act of LIM kinase and Slingshot phosphatase on ADF/cofilin. J Cell Biol 178:107–119

Wu X, Nguyen BC, Dziunycz P, Chang S, Brooks Y, Lefort K, Hofbauer GF, Dotto GP (2010) Opposing roles for calcineurin and ATF3 in squamous skin cancer. Nature 465:368–372

Zhang H, Zhao Q, Liu K, Zhang Z, Wang Y, Zheng X (2009) MgCRZ1, a transcription factor of *Magnaporthe grisea*, controls growth, development and is involved in full virulence. FEMS Microbiol Lett 293:160–169

Calcineurin Inhibitor

▶ Calcineurin

Calcium

▶ Calcium and Mitochondrion
▶ Magnesium and Inflammation

Calcium (Ca^{2+})

▶ Magnesium and Vessels

Calcium and Apoptosis

▶ Mercury and Lead, Effects on Voltage-Gated Calcium Channel Function

Calcium and Death

▶ Calcium in Health and Disease

Calcium and Extracellular Matrix

Erhard Hohenester
Department of Life Sciences, Imperial College London, London, UK

Synonyms

Connective tissue

Definition

Extracellular matrix is the network of polysaccharides and glycoproteins that occupies the extracellular space of invertebrate and vertebrate tissues. Many matrix proteins bind calcium ions, which serve to stabilize individual domains or the interface between domains, or to mediate interactions between different components of the extracellular matrix. Mutations of calcium-binding residues in matrix proteins cause protein misfolding and human diseases.

The extracellular matrix (ECM) is defined as the aggregate of all secreted molecules that are immobilized in the extracellular space of animal bodies. ECMs can be highly ordered, with distinct morphological features (e.g., aligned fibers) visible in electron micrographs. Other ECMs can have an amorphous appearance, but they nevertheless contain typical ECM proteins (e.g., in the brain). The functions of ECM are manifold, but may be summarized as either mechanical or instructive. In vertebrates, the various forms of ECM lend mechanical stability to tissues and organs, such as bone, tendons, cartilage, blood vessels, skin, and many others. Far from being a passive scaffold for tissue formation, however, the ECM regulates the development and functions of almost all cell types in animals. Cell-ECM interactions are mediated by a variety of cellular receptors and are crucial for cell survival, proliferation, polarity, differentiation, and migration. Furthermore, by acting as a medium for the diffusion and delivery of morphogens and growth factors, the ECM plays a major role in the establishment of instructive gradients during development. The essential nature of ECM molecules for animal development and homeostasis is underscored by their conservation in even the simplest invertebrates. Furthermore, a number of human diseases are caused by genetic or acquired defects in ECM structure or cell-ECM communication (Reichardt 1999; Hynes 2009).

In contrast to the cytosol of cells, the extracellular compartment of vertebrate tissues and organs is characterized by a high and constant concentration of calcium. This concentration is commonly assumed to be the same as in blood plasma, that is, 1.2 mM, but it is important to realize that experimental verification for this assumption is often lacking. Local fluctuations in calcium concentration may exist in certain tissues and conceivably have a regulatory role. All ECM molecules fold in the lumen of the endoplasmic reticulum, where calcium is abundant, and are partially assembled into supramolecular structures during their passage through the secretory pathway. To what extent calcium ions are involved in these pre-secretion processes is largely unknown. As will be discussed below, there are several ECM proteins whose functions depend critically on the presence of calcium. It is commonly assumed that the role of calcium in these proteins is structural, and the harsh conditions of many extracellular compartments certainly seem to favor domains with robust folds strengthened by disulfide bridges and/or metal ion–binding sites.

The major classes of ECM molecules are collagens, non-collagenous glycoproteins, and proteoglycans. There are ∼30 different collagen types in humans. Type I, II, III, V, and XI collagens assemble into supramolecular fibrils, which are a prominent feature of many vertebrate ECMs. Collagen fibrils are a major constituent of bone and teeth, the two mineralized tissues that contain a large amount of inorganic calcium in the form of hydroxyapatite. Several non-collagenous phosphorylated proteins, for example osteopontin, are important for bone mineralization, but recent results suggest that the collagen fibrils themselves may have a role in this process. Proteins are assumed to promote biomineralization by nucleating hydroxyapatite crystals, but the precise mechanisms have remained obscure due to a lack of structural information. The non-fibrillar collagen types form irregular sheets (type IV collagen), polygonal three-dimensional networks (types VIII and X collagen), or other assemblies. The crystal structure of the C-terminal NC1 domain of type X collagen revealed a tight homotrimer with a cluster of four calcium ions near the trimer axis, but the closely related type VIII collagen NC1 trimer is stabilized in a different manner, without the need for calcium ions.

Proteoglycans consist of a core protein, to which are attached one or more glycosaminoglycan chains. These GAG chains are linear polysaccharides that are modified by sulfation. Many of the proteoglycan functions are carried out by the GAG chains, but the core proteins have important functions as well. Calcium is not believed to play a major role in proteogylcan biochemistry. One notable exception is the calcium-dependent interaction of the C-type lectin domain of the aggrecan core protein with tenascin.

Among the non-collagenous ECM proteins, calcium binding is critical for the structure and function of fibrillins, thrombospondins, and laminins, as well as for collagen binding by the EF-hand protein SPARC. It is worth noting that all of these proteins are evolutionarily ancient and found in even the simplest animals. The remainder of this brief survey will describe these important proteins in more detail.

Elastic fibers, which are abundant in elastic tissues, such as the aorta, arteries, skin, lung, and ligaments, are composed of a core of covalently cross-linked elastin surrounded by a mantle of microfibrils. The major component of the microfibrils are two large (∼350 kD) glycoproteins, fibrillin-1 and fibrillin-2 (Kielty 2006). Apart from their functions in elastic fibers, the fibrillins also play an important role in the extracellular control of transforming growth factor β and bone morphogenetic protein signaling. The two fibrillins have a modular architecture, consisting of 47 epidermal growth factor (EGF)-like domains interspersed with eight transforming growth factor β-binding protein-like (TB) domains or eight-cysteine domains (Fig. 1). Forty-three of the 47 EGF domains have a consensus sequence for calcium binding, which was first identified in the calcium-binding EGF (cbEGF) domains of blood coagulation factor IX. Extensive structural and biophysical studies of fibrillin cbEGF pairs have revealed that the calcium ions are bound near the domain interfaces and serve to stabilize the extended conformation of cbEGF tandems

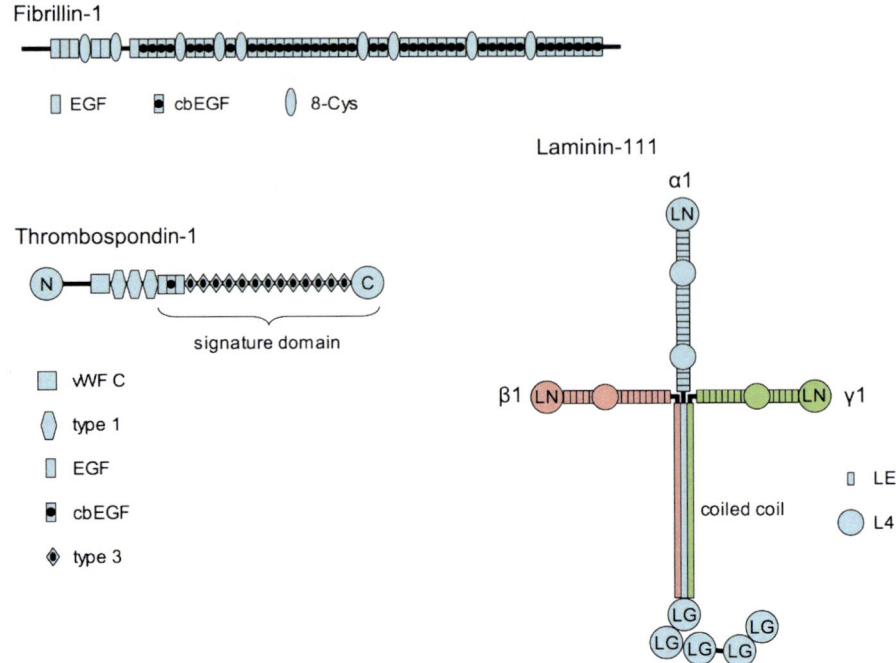

Calcium and Extracellular Matrix, Fig. 1 Domain organization of fibrillin-1, thrombospondin-1, and laminin-111. Fibrillin-1 consists of multiple epidermal growth factor (EGF)-like domains, most of which are of the calcium-binding type (cbEGF), interspersed with eight-cysteine (8-Cys) domains. Thrombospondin-1 is a heterotrimer consisting of an N-terminal domain; an α-helical trimerization domain; a von Willebrand factor type C (vWF C) domain; multiple thrombospondin type 1, type 2 (EGF), and type 3 domains; and a C-terminal domain. The three laminin-111 chains consist of a laminin N-terminal (LN) domain and multiple laminin-type EGF (LE) domains interspersed with laminin 4 (L4) domains, followed by an α-helical region that forms a coiled coil in the laminin-111 heterotrimer. The laminin α chain additionally contains five laminin G (LG)-like domains at its C-terminus

(Handford et al. 2000). The dissociation constants of the calcium ions depend on the context: While individual fibrillin cbEGF domains bind calcium with low affinity, cbEGF domains flanked by their neighboring domains bind calcium with dissociation constants of ≤ 300 μM. The cbEGF domains of fibrillin are, therefore, likely to be largely saturated with calcium at the physiological calcium concentration.

How the fibrillin molecules associate to form the microfibrils of ~10 nm diameter has been the subject of much research and debate, and a consensus on the microfibril structure has yet to be reached. All the current microfibril models feature lateral packing of fibrillin molecules, but differ in the stagger between adjacent fibrillin molecules and in the extent to which the fibrillin molecules are folded back on themselves in the microfibril. Microfibrils extracted from tissues appear as beads on a string in rotary shadowing electron micrographs, with bead-to-bead distances varying from ~50 to ~150 nm depending on the tissue source and extraction procedure used. It is likely that essential microfibril-associated proteins are lost upon extraction. Indeed, the filamentous inter-bead regions are not seen in quick-freeze deep-etch micrographs of ciliary zonule microfibrils, suggesting that there is still much left to learn about the microfibril structure in situ. Mutations in fibrillin genes cause two genetic disorders in humans: Marfan syndrome (fibrillin-1) and congenital contractural arachnodactyly (fibrillin-2). Missense mutations causing Marfan syndrome affect all regions of the fibrillin-1 molecule, including cbEGF residues involved in calcium binding. It has been shown that the latter mutations weaken the domain interfaces in a manner that is consistent with the structural models.

Thrombospondins (TSPs) are a family of calcium-binding glycoproteins that are characterized by their oligomeric nature and a typical modular architecture containing calcium-binding motifs not found in any other metazoan protein. Mammalian TSP-1 and TSP-2 are trimers and consist of an N-terminal domain, followed by an α-helical oligomerization domain, a von Willebrand factor type C domain, three thrombospondin type I domains, three EGF domains (the second of which is of the cbEGF type), a stretch of 13 calcium-binding type 3 repeats (to be described), and a C-terminal lectin-like domain (Fig. 1) (Carlson et al. 2008). Mammalian TSP-3, TSP-4, and TSP-5 are pentamers and lack the type I repeats. TSP-5, also called cartilage oligomeric matrix protein (COMP), additionally lacks the N-terminal domain. The best studied member, TSP-1, is not primarily a structural ECM protein, but is involved in regulating cellular processes, such as angiogenesis, synaptogenesis, and inflammation. TSP-5/COMP, on the other hand, appears to serve an important structural role in cartilage and may be regarded as a bona fide ECM protein. Mutations in the COMP gene cause two skeletal disorders in humans: pseudoachondroplasia and multiple epiphyseal dysplasia (Briggs and Chapman 2002).

TSP-1 was initially isolated from human platelets, and its biochemical properties and ultrastructure were found to be very sensitive to calcium removal. When the sequence of TSP-1 was elucidated in 1986, it was noticed that the type 3 repeats contain aspartic acid-rich motifs that resemble the calcium-binding EF-hand motif. The structural details of calcium binding by TSPs were eventually revealed by two crystal structures nearly two decades later. The first crystal structure established the novel mode of calcium binding by the type 3 repeats (Kvansakul et al. 2004), and the second structure showed how the EGF domains, the type 3 repeats, and the C-terminal domain are assembled into an intricate superstructure, which was termed the "TSP signature domain" (Carlson et al. 2008). In the signature domain, the type 3 repeats surround a core of 26 calcium ions, creating an unprecedented structure termed "the wire" (Fig. 2). There are two variants of the type 3 repeat, N-type and C-type, which differ in length and in some details of the calcium coordination. The full wire has the sequence 1C-2N-3C-4C-5N-6C-7C-8N-9C-10N-11C-12N-13C; it lacks a conventional hydrophobic core and is stabilized by the 26 calcium ions and 8 disulfide bridges. Each repeat contains a D-X-D-X-D/N-G-X-X-D/N-X-X-D/N motif that typically binds two calcium ions (the few deviations from the consensus are of no importance here). The first ion is coordinated by the side chains at positions 1, 3, 5, and 12 of the consensus motif; the second calcium ion is coordinated by the side chains at positions 3, 5, and 12. The coordination spheres of the calcium ions are completed by an asparagine side chain at position 14, 15, or 19, main chain carbonyl oxygens, and water molecules. The wire is wrapped around the globular C-terminal domain, and the third EGF domain also contributes prominently to the superstructure of the TSP signature domain. The lectin-like C-terminal domain of TSPs binds 3–4 calcium ions, and there is some evidence that these ions mediate interactions with other ECM proteins. Detailed biophysical measurements support the view that the TSP signature domain behaves essentially as a single unit. A total of 22–27 calcium ions were detected by atomic absorption spectroscopy and equilibrium dialysis, in good agreement with the structural results. Interestingly, only about ten of these ions are exchangeable, suggesting that most of the calcium ions in the type 3 repeats may be bound very tightly. Missense mutations in COMP that cause pseudoachondroplasia and multiple epiphyseal dysplasia frequently affect residues involved in calcium coordination, suggesting that they may destabilize the entire signature domain and thereby lead to the intracellular aggregation of COMP that is characteristic of these skeletal disorders. The disease mechanism is likely to be complex, but endoplasmic reticulum stress due to misfolded COMP protein is now widely believed to be an important contributor.

The laminins constitute a family of adhesive glycoproteins that are characteristic components of basement membranes (BMs), a sheet-like type of ECM that underlies all epithelia and surrounds muscle, peripheral nerve, and fat cells. Laminins are essential for embryo development, as well as for tissue function in adult animals. Laminins are large, cross-shaped molecules consisting of three polypeptide chains (α, β, and γ). The three short arms of the cross are composed of one chain each, while the long arm is an α-helical coiled coil of all three chains, terminating in a tandem of five laminin G (LG)-like domains provided by the α chain (Fig. 1). The human genome encodes five α, three β, and three γ chains, which are assembled into at least 15 laminin heterotrimers (Aumailley et al. 2005). Laminin-111 (α1β1γ1) has long been known to be a calcium-binding protein: In the presence of ≥50 μM calcium at 37°C,

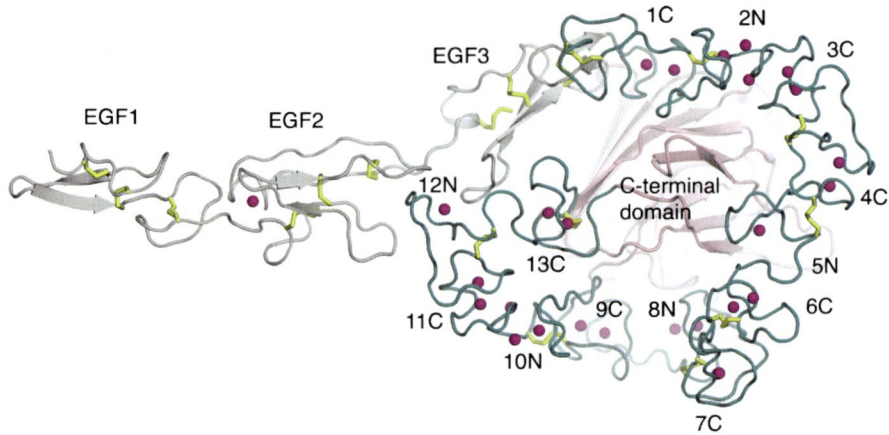

Calcium and Extracellular Matrix, Fig. 2 Crystal structure of the thrombospondin-2 signature domain (pdb 1yo8; Carlson et al. 2008). Disulfide bridges are in *yellow*, and calcium ions are shown as *magenta spheres*. The signature domain consists of three epidermal growth factor (EGF)-like domains; 13 calcium-binding type 3 repeats, grouped into either N- or C-type repeats (1N, 2C, 3N, and so on); and a C-terminal lectin-like domain

laminin-111 aggregates in solution, a process that is believed to mimic the biologically relevant polymerization of laminin at the cell surface. Laminin aggregation in solution is reversible, and the polymer can be dissociated by calcium chelation or cooling to 4°C. Biochemical experiments and genetic evidence from mouse mutants have established that laminin polymerization is mediated by the laminin N-terminal (LN) domains at the tips of the three short arms. However, there is currently no structural information to explain how calcium mediates laminin short arm interactions.

The five LG domains at the C-terminus of laminin α chains contain important binding sites for cellular receptors, including integrins, α-dystroglycan (α-DG), heparan sulfate proteoglycans, and sulfated glycolipids. Of greatest interest here is the interaction of the LG4-LG5 pair with α-DG, which is critical for muscle function and has long been known to require calcium. The LG domain is a β-sandwich, the core of which adopts the jelly roll topology found in numerous lectins and other proteins (Timpl et al. 2000). Both LG4 and LG5 have a calcium ion bound to one edge of the β-sandwich. The coordination of these ions is identical in the two LG domains, with the calcium ion bound by two aspartic acids and two main chain carbonyl oxygens. In all crystal structures of laminin LG domains, the coordination shell of the calcium ion is completed by an acidic buffer molecule or an acidic side chain from a neighboring molecule in the crystal lattice. This observation suggests that an acidic α-DG moiety might be recognized by the LG calcium site(s). Mutagenesis of the calcium-binding aspartic acid residues in the laminin α2 LG4-LG5 pair has shown that the calcium site in LG4, but not the one in LG5, is essential for α-DG binding. Consistent with this result, LG4 has been shown to contain the α-DG-binding site in the laminin α1 chain. LG domains also occur in the BM proteins agrin and perlecan, and α-DG binding to these proteins is likely to follow the principle established for laminin. The α-DG moiety recognized by laminins, agrin and perlecan has remained elusive. α-DG is a heavily glycosylated protein and it was recently shown that a rare carbohydrate modification involving a phosphodiester bond is essential for laminin binding.

SPARC (secreted protein acidic and rich in cysteine), also called osteonectin or BM-40, is a small (~43 kDa) glycoprotein that is not a structural ECM component itself, but interacts with collagens and may play a role in collagen secretion and fibril formation (Bradshaw 2009). SPARC consists of an acidic N-terminal segment, a follistatin (FS)-like domain, and a characteristic α-helical domain containing two EF-hand calcium-binding motifs (EC domain). The second of the EF-hands has a canonical motif; the first EF-hand has an unusual one-residue insertion into the canonical motif. Despite this deviation from the consensus, the EF-hand pair of SPARC binds two calcium ions with high affinity (dissociation constants of ~50 and ~500 nM, respectively), ruling out a regulatory role of the EF-hands in the extracellular milieu. The collagen-binding site of SPARC has been located by mutagenesis: it resides in the EC domain and is composed of the loop connecting the two EF-hands and a long α-helix that is cradled by the EF-hand pair. The collagen-binding site is only partially accessible in intact SPARC, and proteolytic cleavage of an obstructing loop increases the affinity for collagen tenfold. Recent biochemical and structural studies have revealed the mode of collagen binding. SPARC binds to a unique Gly-Val-Met-Gly-Phe-Hyp (Hyp is hydroxyproline) motif in fibrillar collagens. The phenylalanine residue appears to be particularly important for binding, as it is inserted into a deep pocket in the EC domain of SPARC that forms only upon collagen binding. Calcium is not directly involved in collagen recognition, but the intimate involvement of the loop connecting the EF-hands explains why calcium chelation abrogates collagen binding.

Cross-References

▶ Calcium in Biological Systems
▶ Calcium-Binding Proteins, Overview
▶ EF-Hand Proteins

References

Aumailley M, Bruckner-Tuderman L, Carter WG et al (2005) A simplified laminin nomenclature. Matrix Biol 24:326–332
Bradshaw AD (2009) The role of SPARC in extracellular matrix assembly. J Cell Commun Signal 3:239–246
Briggs MD, Chapman KL (2002) Pseudoachondroplasia and multiple epiphyseal dysplasia: mutation review, molecular interactions, and genotype to phenotype correlations. Hum Mutat 19:465–478
Carlson CB, Lawler J, Mosher DF (2008) Structures of thrombospondins. Cell Mol Life Sci 65:672–686
Handford PA, Downing AK, Reinhardt DP et al (2000) Fibrillin: from domain structure to supramolecular assembly. Matrix Biol 19:457–470
Hynes RO (2009) The extracellular matrix: not just pretty fibrils. Science 326:1216–1219
Kielty CM (2006) Elastic fibres in health and disease. Expert Rev Mol Med 8:1–23
Kvansakul M, Adams JC, Hohenester E (2004) Structure of a thrombospondin C-terminal fragment reveals a novel calcium core in the type 3 repeats. EMBO J 23:1223–1233
Reichardt L (1999) Extracellular matrix molecules and their receptors: introduction. In: Kreis T, Vale R (eds) Guidebook to the extracellular matrix, anchor and adhesion proteins, 2nd edn. Oxford University Press, Oxford
Timpl R, Tisi D, Talts JF et al (2000) Structure and function of laminin LG modules. Matrix Biol 19:309–317

Calcium and Mitochondria

▶ Calcium and Mitochondrion

Calcium and Mitochondrion

Thomas E. Gunter
Department of Biochemistry and Biophysics,
University of Rochester School of Medicine and
Dentistry, Rochester, NY, USA

Synonyms

Calcium and mitochondria

Definition

Calcium: An alkaline earth metal having atomic number 20, typically found in the 2+ oxidation state.

Mitochondrion (plural: mitochondria): An intracellular organelle noted for being the site of oxidative phosphorylation and for producing over 90% of the ATP in a typical animal cell.

Importance of Mitochondrial Ca^{2+} Transport

Mitochondrial Ca^{2+} transport is known to perform three functions: (1) to activate steps in the metabolic pathways and increase the rate of ATP production, (2) to modify the shape and distribution of cytosolic Ca^{2+} transients, and (3) to induce the mitochondrial permeability transition in mitochondrially induced apoptosis or programmed cell death. These functions are extremely important. Intramitochondrial free Ca^{2+} concentration ($[Ca^{2+}]_m$) activates important steps in the metabolic pathways, such as pyruvate dehydrogenase, α-ketoglutarate dehydrogenase, isocitrate dehydrogenase, and the F_1F_0 ATP synthase, which taken together can increase the rate of ATP production by oxidative phosphorylation by a factor of around 3 (Balaban 2009). Many types of cells, particularly energetically active neurons like those of the basal ganglia, could not survive and function without this increased

ATP production. In energetically active tissue like the heart under high pumping load, where Ca^{2+} transients control muscle contraction as well as ATP supply, Ca^{2+}'s role in matching the supply of ATP with demand is critically important to avoiding heart failure (Balaban 2009). Furthermore, recent studies have shown other signaling mechanisms which function through mitochondrial Ca^{2+} uptake such as the suppression of prosurvival microautophagy by the release of Ca^{2+} through inositol trisphosphate receptors (IP3R) and its subsequent uptake by mitochondria (Cardenas 2010). It has been known for some time that mitochondria change the shape and distribution of cytosolic Ca^{2+} transients. In an extreme example of this, mitochondria set up a "firewall" completely shielding regions of the pancreatic acinar cells from Ca^{2+} transients (Gunter 2004). The mitochondrial role in apoptosis or programmed cell death has been recognized since the 1990s. Often, excessive mitochondrial Ca^{2+} uptake leads to induction of the mitochondrial permeability transition (MPT) by opening the permeability transition pore (PTP), and this can be central to the mitochondrion's role in controlling apoptosis or programmed cell death. The MPT will be discussed below.

Known Mechanisms of Mitochondrial Ca^{2+} Transport

The Uniporter

The mitochondria of all vertebrates and many other forms of animal life have been found to sequester large amounts of Ca^{2+}. The mechanism through which this uptake occurs was found through several different types of experiments to be a Ca^{2+} uniporter (Gunter 1990) or a mechanism which transports Ca^{2+} down its electrochemical gradient without coupling that transport to the transport of any other ion. The essential evidence was: (1) When Ca^{2+} uptake was energized by substrate oxidation, it could be inhibited by metabolic inhibitors but not by oligomycin, an inhibitor of the F_1F_0 ATP synthase, while if it was energized by ATP hydrolysis, it could be inhibited by oligomycin but not by metabolic inhibitors. This showed that uptake was being driven by the common denominator of these two energization schemes, the membrane potential ($\Delta\psi$). (2) Passive swelling experiments showed that the uptake was not mediated by exchange of Ca^{2+} for 2 K^+ or for 1 or 2 H^+ but seemed independent of the transport of other ions. (3) The membrane potential dependance of uptake showed that the thermodynamic driving force for this transport was the electrochemical Ca^{2+} gradient, as would be expected of a uniporter (Gunter 1990).

The Ca^{2+} uniporter transports Ca^{2+} and other ions in the selectivity series $Ca^{2+} > Sr^{2+} >> Mn^{2+} > Ba^{2+} >> La^{3+}$. It will also transport other lanthanides but very, very slowly. The uniporter shows a second-order concentration dependance and has a Hill coefficient close to 2 (Gunter 1990). The form of the $\Delta\psi$ dependance is consistent with the equations for diffusion down the electrochemical Ca^{2+} gradient. The combined concentration and membrane potential dependance is given by:

$$v = V_{max}\left(\frac{[Ca^{2+}]^2}{K_{0.5}^2 + [Ca^{2+}]^2}\right)\left(\frac{e^{\Delta\phi/2}(\Delta\phi/2)}{\sinh(\Delta\phi/2)}\right), \quad (1)$$

where $\Delta\phi = 2bF(\Delta\psi - \Delta\psi_0)/RT$. In this expression, V_{max}, $K_{0.5}$, b, and $\Delta\psi_0$ are fitting parameters while v is the velocity of transport, $\Delta\psi$ is the membrane potential, and F, R, and T are the Faraday constant, the gas constant, and the Kelvin temperature, respectively. The values of b and $\Delta\psi_0$ found for rat liver mitochondria were 1 and 91 mV, respectively (Gunter 1990). The values of $K_{0.5}$ found in the literature vary over a wide range undoubtedly because very rapid Ca^{2+} uptake decreases $\Delta\psi$. For that reason, it is likely that the actual value of $K_{0.5}$ is near 10 μM (Gunter 1990). The highest value for transport velocity reported in the literature is 1,750 nmol/mg/min, and even that value was probably decreased because of effects of Ca^{2+} uptake on $\Delta\psi$ (Gunter 1990). The uniporter's second-order dependance has been shown to be due to the existence of an activation site as well as a transport site. Ca^{2+} as well as other ions such as Pr^{3+} have been shown to activate the transport of the ions in the selectivity series. The most commonly used inhibitors of the uniporter are ruthenium red, Ru 360, a component of ruthenium red, which has about tenfold more potency, and lanthanides (Gunter 2009). Many drugs and other factors have been found to inhibit the uniporter (see Gunter 2009). Polyamines such as spermine and spermidine have been shown to increase Ca^{2+} transport via the uniporter at low Ca^{2+} concentrations (Gunter 1990).

The Ca^{2+} conductance of the mitochondrial inner membrane has also been studied using electrophysiological techniques, and specific characteristics of conductance associated with the mitochondrial Ca^{2+} uniporter have been identified. It has been identified as a specific high conductance channel with more than one conductance state.

Recently, Perocchi et al. set up a clever strategy utilizing RNA interference (RNAi) to identify a component of the uniporter, which they call MICU1 (Perocchi 2010). The MitoCarta proteomic inventory of mouse mitochondrial proteins was used to identify 18 proteins which were found in the inner membrane and were expressed in vertebrates and kinetoplasmids, as is the uniporter but not in yeast. Of these, 13 proteins had RNAi agents already available. They quantified mitochondrial Ca^{2+} uptake using a Hela cell line expressing the bioluminescent Ca^{2+} indicator mitochondrial aequorin. By following the effects on Ca^{2+} uptake of the RNAi agents for these 13 mitochondrial peptides, they identified one of these proteins, MICU1, whose knockdown also knocked down Ca^{2+} uptake independently of effects on $\Delta\psi$, cell number, and oxidation rate. This work and a number of further controls identified this protein as being a component of the mitochondrial Ca^{2+} uniporter.

The RaM

Another mechanism of mitochondrial Ca^{2+} uptake was discovered using a system designed to generate transients of free Ca^{2+} concentration ($[Ca^{2+}]$) inside a fluorescence cuvette, which had the same appearance as transients seen in the cytosol of cells (Gunter 2004). The apparatus consisted of a computer-controlled pipetter used in conjunction with a fluorescence spectrometer. Buffered Ca^{2+} was injected into a cylindrical cuvette by the pipetter to initiate the transient. After an appropriate time, the transient was ended by injection of a strong Ca^{2+} chelator. This system was designed to explore how mitochondrial uptake of Ca^{2+} from transients like those seen in the cell cytosol might differ from that seen when Ca^{2+} was added to a suspension of isolated mitochondria. The results provided evidence both for uptake via the Ca^{2+} uniporter and also via another mechanism which sequestered a limited amount of Ca^{2+} very rapidly (Gunter 2004). Because of the speed of Ca^{2+} uptake, this second uptake mechanism was called the "rapid mode of uptake" or RaM. The initial evidence showed that fits to uniporter velocity at different $[Ca^{2+}]$ did not converge at zero uptake, as was expected, but at a significant uptake, indicating that some form of uptake had occurred before the first time point in each line of fit (see Fig. 1). Strong evidence for this mechanism was found in isolated rat liver and heart mitochondria. More limited data also suggested its existence in rat brain and avian heart (Gunter 2004). A different type of experiment using an even more rapid Ca^{2+} transient was set up to determine the uptake time of this RaM mechanism. This experiment used "caged Ca^{2+}," a UV labile Ca^{2+} chelator, which breaks up when exposed to UV and releases the Ca, to generate the rapid Ca^{2+} transients. In this set of experiments, rat liver mitochondria were loaded with the fluorescent Ca^{2+} indicator Flo 4 and bound to a cover slip. They were covered with medium containing the "caged Ca^{2+}" and then exposed to brief flashes of UV. By following Ca^{2+} uptake into individual mitochondria using these techniques, it was shown that complete RaM uptake of Ca^{2+} from the transients only required 25–30 ms (see Fig. 2) (Gunter 2004).

The RaM mechanisms in both liver and heart mitochondria sequester Ca^{2+} from a Ca^{2+} transient very rapidly. The RaM will also sequester Ca^{2+} from a $[Ca^{2+}]$ significantly below 200 nM, a much lower concentration than that necessary for uniporter uptake. RaM uptake into both types of mitochondria is inhibited by exposure of the mitochondria to $[Ca^{2+}]$ above 140–160 nM just prior to exposure to the Ca^{2+} transient. Ruthenium red (RR) inhibits Ca^{2+} uptake via the RaM in both liver and heart mitochondria but inhibits more strongly in liver mitochondria. RR inhibition of the RaM mechanism is not as strong as RR inhibition of the uniporter. Ca^{2+} uptake via the RaM is stopped by Ca^{2+} from the transient binding to an external inhibition site. This greatly limits the amount of Ca^{2+} that can be sequestered via the RaM mechanism. Ca^{2+} uptake via the RaM in both liver and heart is increased by the presence of physiological levels of spermine in the medium; however, it is increased much more in liver (by a factor of 6–10) than in heart (by a factor of 2). Around 1 mM, ATP or GTP increases RaM uptake in liver mitochondria by about a factor of 2 but not in heart. Ca^{2+} uptake via the RaM into heart mitochondria is significantly increased by [ADP] in the range around 20 μM and inhibited by 1 mM AMP,

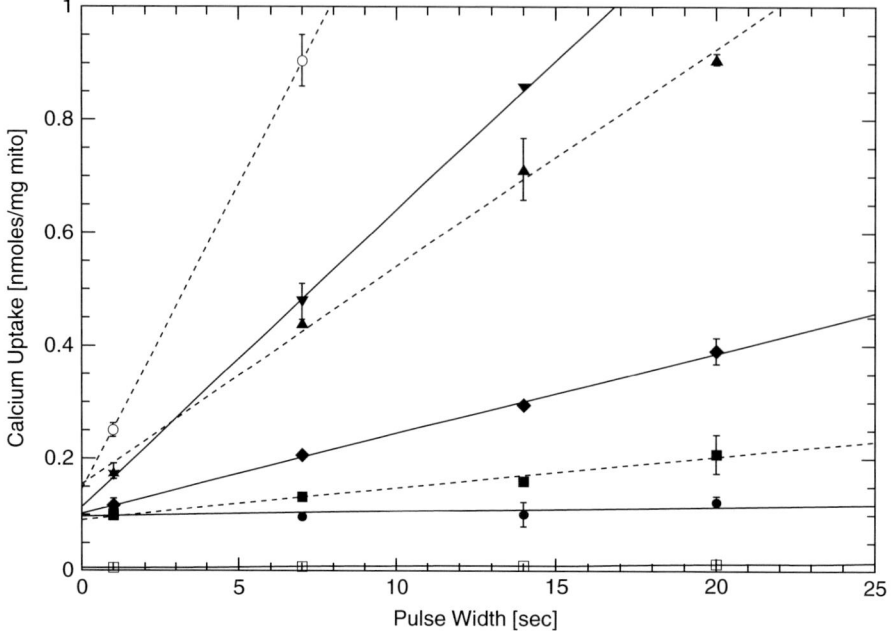

Calcium and Mitochondrion, Fig. 1 Ca uptake vs pulse width as a function of pulse height for rat liver mitochondria. Pulses were made using 2.4 mM EGTA and 2.7 mM ^{45}CaCl$_2$-buffered with HEDTA to a free calcium in the pulse medium of 1 μM. The experiment was run in HEPES-buffered 150 mM KCl (pH 7.2) with 5 mM K-succinate present with mitochondria at a concentration of 0.5 mg/ml. The pulse heights are as follows: □, 474 ± 14 nM with ruthenium red; ●, 165 ± 12 nM; ■, 307 ± 7 nM; ♦, 408 ± 8 nM; ▲, 567 ± 8 nM; ▼, 719 ± 2 nM; ○, 877 ± 9 nM. *Error bars* are 1 standard deviation (Reprinted from Gunter, T. E., Buntinas, L.,Sparagna, G. C., and Gunter, K. K. (1998) The Ca^{2+} Transport Mechanisms of Mitochondria and Ca^{2+} uptake from Physiological Type Ca^{2+} Transients. Biochim. Biophys. Acta 1366, 5–15 with permission from Elsevier Science)

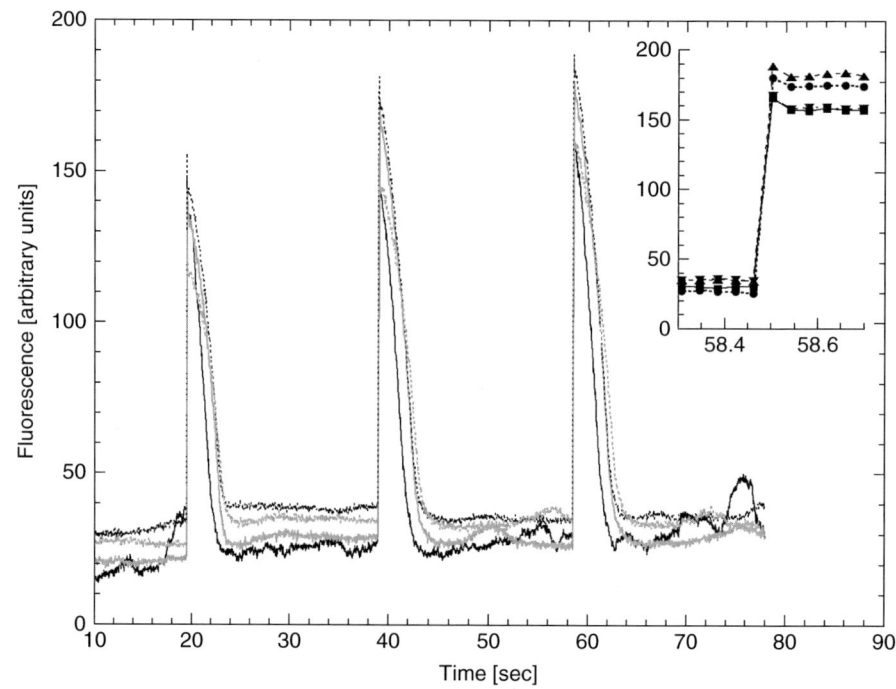

Calcium and Mitochondrion, Fig. 2 Fluorescence response (Fluo 4) indicating intramitochondrial free Ca^{2+} concentration in four individual mitochondria during 3 UV pulses of 0.5 ms duration, which release Ca^{2+} from "caged Ca^{2+}." Ram uptake is complete in less than 40 ms, however, [Ca^{2+}]$_m$ remains high in the mitochondrial matrix for seconds. The inset shows a time-expanded view of uptake during the third pulse (Reprinted from Gunter, T. E., Yule, D. I., Gunter, K. K., Eliseev, R. A., and Salter, J. D. (2004) Calcium and Mitochndria. FEBS Lett. 567, 96–102 with permission from Elsevier Science)

however, not in liver mitochondria. Because the length of time that RaM transport is active prior to Ca^{2+} binding to the external inhibition site seems to decrease as transient $[Ca^{2+}]$ increases, the total amount of uptake per transient does not increase greatly as the transient amplitudes increase. After Ca^{2+} binding to the external Ca^{2+} inhibition site and cessation of Ca^{2+} uptake via the RaM, RaM Ca^{2+} uptake can be activated again by exposing the mitochondria to $[Ca^{2+}]$ below 100 nM. In liver mitochondria, RaM activity is completely restored in less than 1 s. In heart mitochondria, complete restoration takes over 1 min. However, this does not mean that there is no RaM uptake into heart mitochondria in much shorter times. RaM activity may be half recovered in 10–12 s but takes a much longer time to completely recover. Maximum Ca^{2+} uptake via the RaM mechanism from a single transient into liver mitochondria can be as high as 8 nmol/mg protein, but it is less in heart mitochondria (Gunter 2004). 8 nmol/mg protein is about twice the amount estimated by McCormack to be necessary to activate the intramitochondrial Ca^{2+}-dependent metabolic reactions that activate ATP production (Gunter 2004). Uptake per pulse via the RaM into heart mitochondria is less than that into liver mitochondria, but because the number of Ca^{2+} transients per min is so much larger in heart than in liver, the amount of uptake via the RaM per unit time may be similar.

Interesting results, probably related to those discussed on the RaM, were also reported using pig heart mitochondria. In these studies, Ca^{2+} uptake was measured by the intramitochondrial Ca^{2+} indicator Rhod-2, NADH by NADH fluorescence, and fast oxygen use by calibrating hemoglobin as an O_2 indicator (Gunter 2004). In this work, the time resolution of the transient generating system was around 100 ms. The results showed Ca^{2+} uptake in around 100 ms, increased NADH production within 200 ms, and rapid use of O_2 in oxidative phosphorylation within 270 ms. The maximum responses in terms of ATP production did not come from very large Ca^{2+} transients but from Ca^{2+} transients around of 560 nM, decreasing at both higher and lower concentrations (Gunter 2004). 560 nM is close to the average size for Ca^{2+} transients in the cytosol. This shows that rapid uptake of Ca^{2+} from transients of relatively low $[Ca^{2+}]$ can activate oxidative phosphorylation in heart mitochondria.

Ryanodine Receptor

A ryanodine receptor (RYR) has been reported to exist in the mitochondrial inner membrane of rat and mouse cardiac muscle cells. Like the RYRs present in the endoplasmic or sarcoplasmic reticulum (ER or SR) of many types of cells, these could represent another rapid mechanism of Ca^{2+} uptake (Gunter 2009). Unlike the RYR of the SR of cardiac muscle cells (RYR2), the one present in the mitochondrial membrane was of the RYR1 type. This receptor has not been found in the mitochondria of liver cells.

Evidence for a Ca^{2+}/H^+ Exchanger

A genome-wide high-throughput RNA interference screen was conducted by Jiang et al. to identify genes that control Ca^{2+} flux in drosophila S2 cells (Jiang 2009). Mitochondrially targeted ratiometric pericam was used at different wavelengths to monitor both Ca^{2+} (measuring the fluorescence change at 405 nm) and H^+ (measuring the fluorescence change at 488 nm) flux in these cells. Through this methodology, CG4589, the drosophila analog of the human gene Letm1, was identified as a gene strongly affecting both $[Ca^{2+}]_m$ and $[H^+]_m$. Prior to this time, Letm1 had been shown to be a K^+/H^+ exchanger involved in mitochondrial volume control (Jiang 2009) and to be a candidate gene for seizures in Wolf-Hirschhorn syndrome (Jiang 2009). At low $[Ca^{2+}]$ uptake, this mechanism appears to be faster than the uniporter (Jiang 2009). Following work in S2 cells, Hela cells, and reconstituted Letm1 in liposomes, the authors concluded that Letm1 is a $Ca^{2+}/1\ H^+$ exchanger, which takes up Ca^{2+} down to 50 nM but whose effects are not observed above 1 µM where only evidence for the uniporter is seen. The authors found that uptake via this mechanism increases the alkalinity of the mitochondrial matrix similar to uptake via the uniporter, and that knockdown of Letm1 increases the membrane potential significantly. They suggested that this mechanism generally functions as an uptake mechanism, as would be expected of a $Ca^{2+}/1\ H^+$ exchanger.

Efflux Mechanisms

Since Ca^{2+} can move passively down its electrochemical gradient when moving from the external space into the matrix space, it must be energetically uphill for it to move from the matrix to the external space.

Assuming that $[Ca^{2+}]_o = 100$ nM, $[Ca^{2+}]_i = 300$ nM, $T = 25\,°C$ or $298\,°K$ and $\Delta\psi = 180$ mV, a mole of Ca^{2+} ions would lose

$$2.303RT \log\left([Ca^{2+}]_o / [Ca^{2+}]_i\right) + 2F\Delta\psi$$
$$= 2.303RT \log(0.333) + 2F(0.18)$$
$$= 32.01 \text{ kJ}, \qquad (2)$$

in moving from the external space into the matrix, and this same amount of energy must be supplied from some source before the Ca^{2+} can be transported back to the external space. Energy for Ca^{2+} efflux could be supplied either from the energy of cations moving inward via a passive exchanger, anions moving outward via a passive cotransporter, or from the mitochondrial ETC or ATP hydrolysis for an active mechanism. Experiments have identified two separate transport mechanisms which mediate efflux of Ca^{2+} from mitochondria. Both appear to be present in all vertebrate mitochondria in which this has been carefully studied; however, the Na^+-dependent mechanism is dominant in heart, brain, and many other types of mitochondria, while the Na^+-independent mechanism is dominant in liver and kidney mitochondria (Gunter 2009).

Na^+-Dependent Efflux

The work on the nature of this mechanism has suggested that it is a passive exchanger of Na^+ ions for Ca^{2+} (Gunter 1990). Li^+ may substitute for Na^+ on the mechanism and Sr^{2+} for Ca^{2+}; however, the mechanism does not transport Mn^{2+} (Gunter 1990). The kinetics for efflux via this mechanism have been fit to the form:

$$v = V_{max}\left(\frac{[Na^+]^2}{K_{0.5}^2 + [Na^+]^2}\right)\left(\frac{[Ca^{2+}]}{K_{Ca} + [Ca^{2+}]}\right), \qquad (3)$$

for both heart and liver mitochondria (Gunter 1990). For heart mitochondria, V_{max} has been found to be 18 nmol/mg/min, K_{Ca} to be near 10 nmol/mg protein, and K_{Na} to be 7–12 mM (Gunter 1990). For liver mitochondria, V_{max} has been found to be 2.6 ± 0.5 nmol/mg/min, K_{Ca} to be 8.1 ± 1.4 nmol/mg protein, and K_{Na} to be 9.4 ± 0.6 mM (Gunter 1990). The V_{max} for this mechanism has been found to be around 30 nmole/mg/min for brain mitochondria making it even faster than in heart mitochondria. The Na^+ electrochemical gradient closely follows the proton electrochemical gradient because of a very fast Na^+/H^+ exchanger in the mitochondrial inner membrane (Gunter 2009). Therefore, the Na^+ electrochemical gradient is about the same as the proton electrochemical gradient ($2.303RT\Delta pH + F\Delta\psi$), and twice the energy of the Na^+ electrochemical gradient, at the values of $[Ca^{2+}]_o$, $[Ca^{2+}]_i$, $\Delta\psi$, and T used in the section above yields 46.15 kJ/mole. This suggests that this efflux mechanism could function as a $Ca^{2+}/2Na^+$ exchanger. However, it is not enough to show that it could function as this type of exchanger under normal conditions; it must be shown to function as this type of exchanger under the most extreme conditions, known as "null point conditions." In other words, the mechanism must be able to receive enough energy from two times the energy of the Na^+ gradient at the point where the Ca^{2+} gradient is as high as the mechanism will pump against. This is the point where the velocity of the efflux becomes zero because the energy of the Ca^{2+} concentration gradient has grown too large for it to pump outward at a higher Ca^{2+} gradient. This type of analysis was carried out by Baysal et al. (Gunter 2009) for this Na^+-dependent efflux mechanism, and it was found that there was not enough energy in twice the energy of the Na^+ electrochemical gradient to account for this efflux up to the null point. At the time the original work was published, it was believed that the Na^+-dependent efflux mechanism was nonelectrogenic; however, further work showed that the mechanism was actually electrogenic (Gunter 2009), and the simplest mechanism consistent with all of the data is a $Ca^{2+}/3Na^+$ exchanger.

CGP37157 and tetraphenylphosphonium (TTP) (Gunter 2009) are the most commonly used inhibitors of this mechanism; however, many drugs inhibit it (see Gunter 2009).

A protein previously identified as an nNa^+/Ca^{2+} exchanger (NCLX) has recently been associated with the Na^+-dependent Ca^{2+} efflux mechanism (Palty 2010). NCLX had already been shown to exchange Ca^{2+} for Li^+ as the mitochondrial Na^+-dependent Ca^{2+} efflux mechanism has been shown to do. Gold-labeled NCLX antibodies allowed the observation of gold particles on the inner membrane within the cristae of NCLX-overexpressing SHSY-5Y cells. Mitochondrial Na^+-dependent Ca^{2+} efflux was enhanced in NCLX-overexpressing cells, reduced by expression of siRNA effective in knockdown of

NCLX, and NCLX was inhibited by CGP-37157, which also inhibits mitochondrial Na$^+$-dependent efflux. Mitochondrial Na$^+$/Ca^{2+} exchange was blocked in cells expressing an inactive mutant form of NCLX.

Na$^+$-Independent Efflux

The Na$^+$-independent Ca^{2+} efflux mechanism transports not only Ca^{2+} but also Sr^{2+}, Mn^{2+}, and Ba^{2+} from the mitochondrial matrix to the external space (Gunter 1990). It is believed to be nonelectrogenic (Gunter 2009). Since it is very difficult to quantify transport of H$^+$ because its concentration is dominated by the water equilibrium (H$^+$ + OH$^-$ = H$_2$O) and binding, the mechanism was believed by many to be a passive Ca^{2+}/2 H$^+$ exchanger. However, the "null point" method, discussed in the section above, was developed to test whether or not this mechanism had enough energy from the inward transport of two protons to transport 1 Ca^{2+} ion outward. The results showed that enough energy was not available from the transport of two protons inward to transport a Ca^{2+} outward at the null point; therefore, the mechanism cannot be a passive Ca^{2+}/2 H$^+$ exchanger (Gunter 2009)}, and there must be an additional source of energy available to this mechanism. There have been several publications suggesting that a passive Ca^{2+}/2H$^+$ exchanger could not explain the properties of this transport mechanism (see Gunter 1990). It has been found, for example, that that inhibitors of electron transport inhibit this mechanism independently of their effect on pH gradient (Gunter 1990). This suggests that it may receive some energy from the mitochondrial ETC and therefore be an active Ca^{2+}/2H$^+$ exchanger (Gunter 2009). This mechanism is inhibited by CN$^-$, low levels of uncouplers, and very high levels of both ruthenium red and tetraphenyl phosphonium (Gunter 1990).

The kinetics for this mechanism in liver mitochondria take the mathematical form:

$$v = V_{max}\left(\frac{[Ca^{2+}]^2 + a[Ca^{2+}]}{K_m^2 + [Ca^{2+}]^2 + 2a[Ca^{2+}]}\right), \quad (4)$$

where V_{max}, K_m, and a are fitting parameters. The values of these parameters for liver mitochondria are V_{max} = 1.2 ± 0.1 nmol/mg/min, K_m = 8.4 ± 0.6 nmol/mg, and a = 0.9 ± 0.2 nmol/mg (Gunter 1990).

The Mitochondrial Ca^{2+}-Induced Permeability Transition

When mitochondria are exposed to high [Ca^{2+}]$_m$ often in the presence of other inducers, a large pore can open in the mitochondrial inner membrane and dissipate the membrane potential and pH gradient. This is called the mitochondrial permeability transition (MPT), and the pore is called the permeability transition pore (PTP). This MPT is believed to occur in almost all types of mitochondria. Somewhat similar behavior has even been seen in yeast mitochondria. Pore opening causes rapid mitochondrial swelling. The PTP is sometimes classified as a Ca^{2+} efflux mechanism because it does cause Ca^{2+} to leave the matrix space; however, it has no specificity and rapidly transports all small ions and molecules with a molecular weight below around 1,500 Da. Cyclosporin A is a very potent inhibitor of the MPT, and it has been shown to function by binding to a mitochondrial protein, cyclophillin D. Induction of the MPT has been associated with binding of cyclophillin D to the pore protein, and knockout or knockdown of cyclophillin D decreases the probability of inducing the MPT. Pore opening can usually also be reversed by chelating the Ca^{2+}. While long-term opening of the PTP leads to cell death, transient opening of the pore or flickering has been reported and may have a physiological role. Pore opening is one of the ways in which mitochondria can control induction of apoptosis. Many workers believed that the adenine nucleotide translocase (ANT) was the core PTP protein; however, MPT induction has now also been seen when this protein is knocked out. Therefore, it has been suggested that the ANT only represents one of a set of proteins which can function as core PTP proteins. It is necessary to understand the primary characteristics of the MPT in order to understand the way mitochondria handle Ca^{2+}. Many good review articles are available on the MPT (see for example Halestrap 2010).

Questions and Issues

Fluorescent Ca^{2+} indicators led to the discovery that mitochondria were usually exposed to [Ca^{2+}]$_c$ of 85–100 nM except during Ca^{2+} transients when the average cytosolic concentration might temporarily go above 1 µM. Calculations based on the known activity of the mitochondrial Ca^{2+} uniporter at that time suggested that the uniporter was not able to sequester sufficient Ca^{2+} from the observed Ca^{2+} transients to activate the process of oxidative phosphorylation. This dilemma led to the discovery of two independent pathways by which sufficient Ca^{2+} might be sequestered by mitochondria. One of these pathways led to the discovery of mitochondrial mechanisms which can sequester Ca^{2+} from lower $[Ca^{2+}]$ sources such as the RaM, the ryanodine receptor, and the $Ca^{2+}/1\ H^+$ exchanger, mechanisms with higher Ca^{2+} affinity than the uniporter. These were discussed above. There are numerous reports in the literature of Ca^{2+} uptake into mitochondria both very rapidly and also from $[Ca^{2+}]_c$ much lower than that which can be sequestered via the uniporter. Many questions still exist about the function of these mechanisms. The other pathway led to the discovery of high Ca^{2+} microdomains. Currently, it seems likely that mitochondria sequester Ca^{2+} through one or more of these higher affinity mechanisms and also through the uniporter acting within high $[Ca^{2+}]$ microdomains.

The heart of the microdomain hypothesis was the early recognition that mitochondria are often found close to the Ca^{2+} release sites on sarcoplasmic or endoplasmic reticulum (SR or ER). Immediately following release of Ca^{2+} from stores within the SR or ER, the concentration of Ca^{2+} in the regions between the SR or ER and mitochondria can be high enough (over 50 µM) for a significant amount of uptake to occur into the mitochondria, and activation of ATP production can occur. Most of the work showing that this view is realistic was carried out on cultured cells using increasingly sophisticated fluorescent probes. Today, this generally means probes that are expressed genetically and are often ratiometric and targeted to the site of interest in the cell. Excellent reviews of this work can be found, and recent work has directly measured the $[Ca^{2+}]$ in the microdomains using sophisticated fluorescence techniques (Shin 2010).

An obvious question is why are there redundant mechanisms providing intramitochondrial Ca^{2+} for activation of ATP production? First, having the necessary ATP supply is simply too important not to have some kind of backup (Balaban 2009). Second, in some cell types, there are plenty of mitochondria which are not exposed to microdomains, so the mechanisms are not necessarily redundant. One clear case is the mitochondria moving inside the neurites of nerve cells, which must produce considerable ATP.

Another important question about the mitochondrial Ca^{2+} transport system relates to the observation that the velocity of the uptake mechanisms seems much larger than that of the efflux mechanisms. This is not so much an issue with uptake via the RaM where the uptake per pulse is limited by Ca^{2+} binding to the external inhibition site, but it is an issue for uniporter uptake within high Ca^{2+} microdomains, where the amount of uptake could be considerable. The problem is that over the long term, total influx and total efflux must be about the same, otherwise, if influx dominates, as the kinetics seems to suggest, the mitochondria sequester more and more Ca^{2+} and probably undergo the MPT. Taking heart as a case in point, the largest transport velocity for the uniporter measured for heart mitochondria was 1,750 nmol/mg/min (Gunter 1990), and even this rate was probably decreased by decreases in $\Delta\psi$ caused by the rapid Ca^{2+} uptake. The portion of the heart beat cycle covered by the transient is significant, and the V_{max} for the Na^+-dependent efflux mechanism is only 18 nmol/mg/min (Gunter 1990). Furthermore, the only faster efflux mechanism that has been suggested is transient flickering of the PTP. Are the uptake velocities measured in isolated mitochondria faster than physiological?

Another question is whether the RaM is actually a high-conductance conformation of the uniporter complex, as has been suggested (Gunter 2004; Gunter 2009), or could it possible be the $Ca^{2+}/1\ H^+$ exchanger? It does share with the exchanger the characteristic of being able to sequester Ca^{2+} from low $[Ca^{2+}]$. The former seems more likely, since the exchanger does not seem as fast as the RaM; however, it should also be kept in mind that the very fast transport per site calculations for the RaM (Gunter 2004), suggesting that it functioned as a simple pore, was estimated on the assumption that the number of RaM mechanisms was the same as that for the uniporter. It could also be a slower mechanism present in much higher numbers. More data is necessary to settle this question.

While we know quite a bit about how mitochondrial Ca^{2+} transporters function in isolated mitochondria, and while we are learning much more about how they function in cells, many questions still remain unanswered. The advent of the use of molecular techniques in this important field should allow many of these questions to be answered.

Cross-References

▶ Calcium ATPases

References

Balaban RS (2009) The role of Ca^{2+} signaling in the coordination of mitochondrial ATP production with cardiac work. Biochim Biophys Acta 1787:1334–1341

Cardenas C, Miller RA, Smith I et al (2010) Essential regulation of cell bioenergetics by constitutive insp3 receptor Ca^{2+} transfer to mitochondria. Cell 142:270–283

Gunter TE, Pfeiffer DR (1990) Mechanisms by which mitochondria transport calcium. Am J Physiol 258:C755–C786

Gunter TE, Sheu S-S (2009) Characteristics and possible functions of mitochondrial Ca^{2+} transport mechanisms. Biochim Biophys Acta 1787:1291–1308

Gunter TE, Buntinas L, Sparagna GC, Gunter KK (1998) The Ca^{2+} transport mechanisms of mitochondria and Ca^{2+} uptake from physiological type Ca^{2+} transients. Biochim Biophys Acta 1366:5–15

Gunter TE, Yule DI, Gunter KK et al (2004) Calcium and mitochondria. FEBS Lett 567:96–102

Halestrap AP (2010) A pore way to die: the role of mitochondria in reperfusion injury and cardioprotection. Biochem Soc Trans 38:841–860

Jiang D, Zhao L, Clapham DE (2009) Genome-wide RNAi screen identifies LETm1 as a mitochondrial Ca^{2+}/H^+ antiporter. Science 326:144–147

Palty R, Silverman WF, Hershfinkel M et al (2010) Nclx is an essential component of mitochondrial Na^+/Ca^{2+} exchange. Proc Natl Acad Sci USA 107:436–441

Perocchi F, Gohil VM, Girgis HS et al (2010) Micu1 encodes a mitochondrial ef hand protein required for Ca^{2+} uptake. Nature 467:291–296

Shin DF, Muallem S (2010) What the mitochondria see. Mol Cell 39:6–7

Calcium and Neurotransmitter

▶ Mercury and Lead, Effects on Voltage-Gated Calcium Channel Function

Calcium and Viruses

Yubin Zhou[1,2], Shenghui Xue[3] and Jenny J. Yang[1,4]
[1]Department of Chemistry, Georgia State University, Atlanta, GA, USA
[2]Division of Signaling and Gene Expression, La Jolla Institute for Allergy and Immunology, La Jolla, CA, USA
[3]Department of Biology, Georgia State University, Atlanta, GA, USA
[4]Natural Science Center, Atlanta, GA, USA

Synonyms

Apoptosis, programmed cell death; Calcium homeostasis: Calcium metabolism; Calcium: Ca(II)

Definition

Calcium: A metallic element with atomic number of 20. Calcium is one of the most abundant elements on earth. It usually exists as oxidation form with an oxidation state of +2. Ca^{2+} plays essential roles to regulate a variety of biological events.

Apoptosis: A type of cell death featured by chromatin condensation, chromosomal DNA fragmentation, blebbing, and cell shrinkage. Unlike necrosis, apoptosis is a programmed cellular event to eliminate the cells without releasing the harmful materials.

Calcium homeostasis: A mechanism by which the cell or body regulates calcium concentration at the correct level.

Introduction

Ca^{2+}, a universal "signal for life and death," acts as an important intracellular messenger inside eukaryotic cells to regulate a wide range of cellular processes, including cell motility, secretion, muscle contraction, gene transcription, cell proliferation, and apoptosis (▶ Calcium in Biological Systems). The diversity of Ca^{2+} signals is achieved by the exquisite choreography of a repertoire of signaling components, including

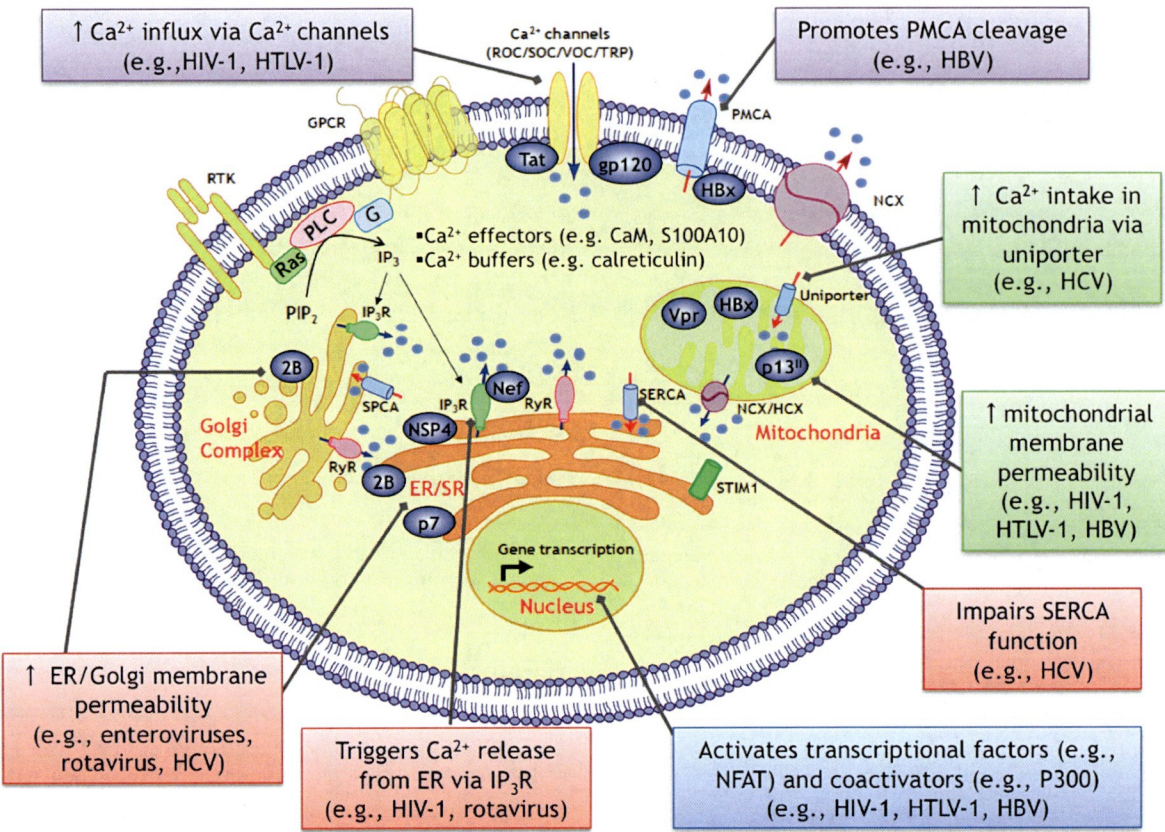

Calcium and Viruses, Fig. 1 The choreography of Ca^{2+} signaling and examples of virus-induced disruption of Ca^{2+} homeostasis (Adapted with permission from reference Zhou et al. 2009). See text for details on the Ca^{2+} signaling components. Viral proteins are capable of perturbing the intracellular Ca^{2+} homeostasis by (1) modulating Ca^{2+} pumps and/or channels on the plasma membrane (e.g., Tat and gp120 of HIV-1, HBx of HBV), (2) triggering Ca^{2+} release from internal stores via IP_3R (e.g., NSP4 of rotavirus and Nef of HIV) or altering membrane permeability and pump activity of internal stores (e.g., 2B of coxsackievirus, p7 of HCV, and core protein of HCV), (3) disrupting mitochondrial membrane permeabilization or potential (e.g., Vpr of HIV-1, p13^{II} of HTLV-1, core protein of HCV and HBx of HBV) and (4) activating Ca^{2+}-responsive transcriptional factors or coactivators, such as p300 (e.g., p12^{I} of HTLV-1 and Vpr of HIV-1) in the nucleus. Additionally, a variety of viral proteins interact with important cellular CaBPs, such as CaM, S100A10, and calreticulin, to remodel the Ca^{2+} signaling network. Calcium ions are shown as *blue dots*

receptors, ion channels, pumps, exchangers, Ca^{2+} buffers, Ca^{2+} effectors, Ca^{2+}-sensitive enzymes, and transcriptional factors that reside in distinct cellular compartments (Berridge et al. 2003). The Ca^{2+} signaling system undergoes constant remodeling to meet the specific spatiotemporal requirements in a flexible yet precise manner. This flexibility, on one hand, allows the host cell to adjust to various stimuli, such as viral infection. On the other hand, viruses may take advantage of the universal Ca^{2+} signal to create a tailored cellular environment to meet their own demands (Fig. 1).

Viruses choose Ca^{2+}, instead of other metal ions (e.g., Mg^{2+}, K^+, Na^+), to benefit their own life cycles because of the irreplaceable and important physiochemical and physiological nature of Ca^{2+}: First, Ca^{2+} has been chosen by nature through evolution as a versatile second messenger to regulate almost all cellular events. Evolving as intimate intracellular parasites that are adept at hijacking the host cell machinery, viruses can conveniently target Ca^{2+} signals to affect a diverse range of downstream effectors and pathways to maximize virus replication, while still achieving their coexistence with host cells. Second, a >10,000-fold gradient of Ca^{2+} is maintained across the plasma membrane, which is substantially larger than the dynamic range of monovalent K^+ and Na^+ (<100-fold) and Mg^{2+} (<10-fold) in mammalian cells. This enables the viruses to easily manipulate Ca^{2+} gradients between membranous compartments to transduce information encoded by any particular spatiotemporal Ca^{2+} pattern. Third, acute change of K^+ and Na^+ at the millimolar range is more likely to cause osmotic shock and/or to abruptly disrupt membrane potential than the change of Ca^{2+} (at nM or μM level), thereby circumventing these detrimental effects on host cells.

Viruses appropriate or hijack the Ca^{2+} signaling network in various ways that favor virus entry, virus replication, virion assembly, maturation, and/or release (Zhou et al. 2009; Chami et al. 2006). The common scenarios encountered during virus – Ca^{2+} interplays, as summarized and briefly exemplified in the following paragraphs, include the following:
- Viral proteins disrupt Ca^{2+} homeostasis by altering membrane permeability and/or manipulating key components of the Ca^{2+}-signaling toolkit.
- Ca^{2+} directly binds to viral proteins to maintain structural integrity or functionality.
- Critical virus-host interactions depend on cellular Ca^{2+}-regulated proteins or pathways.

Disruption of Calcium Homeostasis During Viral Infection

The Ca^{2+} concentration in different compartments of the cell is strictly maintained (Fig. 1) (Putney 1998). The cellular ionized Ca^{2+} gradient approximately follows the order of extracellular space ($[Ca^{2+}]_o$: $\sim 10^{-3}$ M) > endoplasmic reticulum/sarcoplasmic reticulum (ER/SR) ($[Ca^{2+}]_{ER}$: $\sim 10^{-4}$ M) > cytosol ($[Ca^{2+}]_{CYT}$: $\sim 10^{-7}$ M to $\sim 10^{-5}$ M). Following signal stimulation or alteration in the membrane potential, $[Ca^{2+}]_{CYT}$ may be elevated by 100-fold from 10^{-7} M to 10^{-5} M. This is made possible by Ca^{2+} from two major sources: the extracellular space and the internal Ca^{2+} stores (mainly ER or SR). Extracellular Ca^{2+}, sensed by the extracellular Ca^{2+}-sensing receptor, is believed to maintain the long-term Ca^{2+} homeostasis by replenishing the internal calcium stores, whereas the internal calcium stores are directly responsible for the changes in $[Ca^{2+}]_{CYT}$ through the activity of two major ER-resident Ca^{2+} release channels, e.g., the ryanodine (RyR) receptor and the inositol 1,4,5-triphosphate (IP3) receptor. At the resting state, $[Ca^{2+}]_{CYT}$ is maintained at submicromolar range by extruding excessive Ca^{2+} outside of the plasma membrane via plasma membrane Ca^{2+}-ATPase (PMCA; ▶ Calcium ATPases) and Na^+/Ca^{2+} exchanger (NCX) or by pumping Ca^{2+} back into internal stores through sarcoplasmic/endoplasmic reticulum Ca^{2+}-ATPase (SERCA) (▶ Calcium ATPases), secretory pathway Ca^{2+}-ATPase, and the mitochondrial uniporter. The Ca^{2+} signals are delivered by affecting the activity of Ca^{2+} buffers, Ca^{2+} effectors, and Ca^{2+}-regulated enzymes (▶ Calcium-Binding Proteins, Overview). The signals can also have "long-term" effects by modulating the activity of several Ca^{2+}-responsive transcriptional factors including nuclear factor of activated T cells (NFAT), cyclic AMP response element-binding proteins (CREB), and downstream regulatory element modulator (DREAM).

How Does the Altered Ca^{2+} Signaling Benefit the Life Cycle of a Virus?

Most often, albeit not always, viral infection tends to cause an increase in $[Ca^{2+}]_{CYT}$ by using strategies outlined below. The modest increase of $[Ca^{2+}]_{CYT}$ may benefit the life cycle of viruses in the following ways:

First, a modestly elevated $[Ca^{2+}]_{CYT}$ would activate or accelerate a number of Ca^{2+}-dependent enzymatic processes in the cytosol, as well as Ca^{2+}-sensitive transcriptional factors (e.g., NFAT), to promote virus replication or to establish persistent infection. The Ca^{2+} released from ER can be readily uptaken by mitochondria. A modest increase in mitochondrial matrix Ca^{2+} may activate Ca^{2+}-dependent Krebs cycle dehydrogenases and increase production of ATP, thereby meeting a higher demand for energy to aid virus replication.

Second, the decrease of Ca^{2+} in ER and Golgi complex disrupt the protein trafficking and sorting pathways, which effectively perturb the host defense mechanism against viral infection by alleviating host antiviral immune responses and escaping premature clearance by the host. Intracellular accumulation of ER or Golgi-derived secretory vesicles, where viral RNA replication takes place for some RNA viruses (e.g., enteroviruses as discussed below), creates a microcosmic environment favoring viral replication.

Third, Ca^{2+}-flux between ER and mitochondria plays a critical role in determining the fate of host cells when exposed to apoptotic stimuli, such as viral infections. Modulation of ER-mitochondria Ca^{2+} coupling may either prevent apoptosis or induce apoptosis, depending on the stages of the viral life cycle and types of viruses. Apoptosis is usually elicited as an innate defense mechanism to counteract viral infection and control virus production. In general, an anti-apoptotic strategy is employed by the viral to prevent host immune clearance and promote virus replication in the early or middle stage of infection. Meanwhile, a viral infection may induce apoptosis to aid egress of virions to the outside and dissemination of progeny at a later stage.

The dysregulated Ca^{2+} signaling scenario observed during enterovirus infection can probably best illustrate some of the points outlined above. A typical enterovirus contains a single-stranded, plus-sense RNA genome encoded by four structural proteins (VP1–4) and ten nonstructural proteins ($2A^{pro}$, 2B, 2C, 3A, 3B, $3C^{pro}$, $3D^{pol}$, 2BC, 3AB, and $3CD^{pro}$) that are produced by a proteolytic processing cascade from a single translational precursor. Among those, the 2B protein can induce substantial decrease (~40%) of Ca^{2+} in ER and Golgi complex by direct permeabilization of the ER/Golgi membrane. Such dramatic change in intracellular Ca^{2+} serves two purposes (van Kuppeveld et al. 2005): (a) It inhibits intracellular protein trafficking pathways, thus favoring virus replication and downregulating host antiviral immune response. A reduced intravesicular Ca^{2+} level would result in the inhibition of vesicular protein transport and accumulation of ER/Golgi-derived vesicles, where the replication complex forms and viral RNA replication takes place. (b) It exerts an anti-apoptotic activity on infected cells. Upon enterovirus entry, the cellular apoptotic process is immediately triggered as an innate defense mechanism in response to infection, but is abruptly suppressed during the middle stage of infection. The apoptotic process resumes at late stage after viral replication. The perturbation of Ca^{2+} homeostasis at 2–4 hours post infection (hpi) coincides with the inhibition of the apoptotic cell response triggered at about 2 hpi. Thus, the middle-stage interruption of apoptosis is linked to downregulated ER-mitochondrial Ca^{2+} fluxes that prevent cytotoxic Ca^{2+} overloading in mitochondria (▶ Calcium and Mitochondria). Overall, 2B-induced reduction of Ca^{2+} in ER/Golgi provides the virus a favorable cellular environment and optimal time window for viral RNA and protein synthesis. The exquisite synchronization of rise in $[Ca^{2+}]_{CYT}$ and anti-apoptotic activity allows virus particles to complete assembly before cell lysis.

What Strategies Do Viruses Devise to Alter Intracellular Ca^{2+} Signaling?

Although different viruses use a wide range of ways to perturb the cellular Ca^{2+}-signaling network, one still can discern mechanistic similarity and find common themes shared among viruses (Fig. 1) (Zhou et al. 2009).

Modulation of Ca^{2+} Channels or Pumps in the Plasma Membrane

The plasma membrane provides a solid boundary between the inside and outside of the cell. Ionic concentration gradients are maintained across the plasma membrane by a number of ion channels, pumps, and exchangers.

In excitable cells (e.g., muscle and neuronal cells), any external stimulus that depolarizes the plasma membrane is capable of activating voltage-gated or receptor-operated Ca^{2+} channels (VOC/ROC) and eliciting swift Ca^{2+} flux into cytoplasm. During a viral infection, these Ca^{2+} entry components may become the immediate targets of attack. For instance, two viral proteins encoded by the genome of human immunodeficiency virus 1 (HIV-1), the glycoprotein gp120 and the transcriptional transactivator Tat, are able to elicit the elevation of $[Ca^{2+}]_{CYT}$ in mammalian cells. The increase in $[Ca^{2+}]_{CYT}$ is attributable to the activation of L-type voltage-gated Ca^{2+} channels and N-methyl-D-aspartate (NMDA) receptors. External application of recombinant gp120 to fetal neurons and astrocytes causes a dose-dependent rise of $[Ca^{2+}]_{CYT}$ up to ~2 μM and induces neurotoxicity in these cells. Similarly, Tat-induced dysregulation of Ca^{2+} homeostasis leads to neurotoxicity and contributes to HIV-related dementia.

In non-excitable cells (e.g., epithelial cells and lymphocytes) that generally do not possess VOC/ROC, the depletion of Ca^{2+} stores often triggers sustained Ca^{2+} entry across the plasma membrane via store-operated Ca^{2+} channels (SOC). $p12^{I}$ encoded by the human

T-lymphotropic virus 1 (HTLV-1) and the core protein encoded by hepatitis C virus (HCV) are among a handful of viral proteins capable of activating SOC channels in T lymphocytes and/or Jurkat T cells. A sustained elevation of $[Ca^{2+}]_{CYT}$ and accompanying high-frequency cytosolic Ca^{2+} oscillation specifically activate NFAT, a transcriptional factor that initiates a highly coordinated choreography of gene expression. Activation of NFAT exerts long-term effects on cells of the immune system by inducing gene expression (e.g., IL-2 and Bcl-2) and further promoting lymphocyte activation and survival to support viral infection.

PMCA and NCX, as the major Ca^{2+} extrusion apparatus, can transport the Ca^{2+} outside of the cell to maintain Ca^{2+} concentration gradient across the plasma membrane. Viruses can directly or indirectly modulate the activity of these components to induce changes in cytosolic Ca^{2+} concentration. For example, overexpression of the hepatitis B virus X protein (HBx) has been shown to stimulate caspase-3-dependent cleavage of PMCA. The decrease in PMCA activity, along with the HBx-mediated release of Ca^{2+} from mitochondria (as discussed below), leads to a net increase in $[Ca^{2+}]_{CYT}$ that enhances HBV DNA replication and increases HBV core assembly. In addition, it activates or promotes Ca^{2+}-responsive pathways that regulate cell survival, apoptosis, and proliferation.

Modulation of ER-Resident Ca^{2+} Release Channels and Pumps

Ca^{2+} immobilization from internal stores (mainly ER) is another important contributor to the intracellular Ca^{2+} increase. IP_3R and RyR are two of the central players in switching on/off Ca^{2+} release from internal stores. The engagement of receptors on lymphocytes or agonist binding to cell surface receptors activates phospholipase C-γ (PLC-γ), which hydrolyzes phosphatidylinositol-4,5-biphosphate (PIP_2) to produce inositol-1,4,5-triphosphate (IP_3). IP_3 activates IP_3R and triggers Ca^{2+} release from ER. The replenishing of Ca^{2+} stores is achieved by Ca^{2+} influx across plasma membrane through SOC channels and pumping of cytosolic Ca^{2+} back to the ER lumen via the SERCA pump. By virtue of ER's roles as a hub in coordinating the ebb and flow of intracellular Ca^{2+}, viruses containing either RNA or DNA genomes have evolved astute means to manipulate IP_3R-mediated Ca^{2+} release (e.g., HIV-1 and rotavirus) or to modulate the activity of SERCA (e.g., HCV).

Nef is an accessory protein encoded by HIV-1 that plays important roles in the pathogenesis of acquired immunodeficiency syndrome (AIDS). Overexpression of Nef could induce a cytosolic Ca^{2+} increase in Jurkat T cells possibly through the interaction with IP_3R and subsequently promote the T cell receptor-independent activation of the NFAT pathway, without notable increase in PLCγ-catalyzed production of IP_3. The activated NFAT further promotes viral gene transcription and replication. Given the observation that a Src-like protein-tyrosine kinase (PTK) coimmunoprecipitates with both Nef and IP_3R, Nef might modulate the IP_3R activity via its interaction with the Src-like PTK.

The double-stranded RNA virus, rotavirus, is the major etiological agent of viral diarrhea in young children. The rotavirus nonstructural glycoprotein NSP4, a multifunctional enterotoxin, has been shown to induce Ca^{2+} release from ER in infected human epithelial cells through a PLC-dependent pathway. Exogenously applied NSP4 can cause diarrhea in rodent pups and increases cytosolic Ca^{2+} concentration via the activation of PLC and the resultant ER Ca^{2+} depletion through IP_3R. In addition, endogenous NSP4 can also be secreted from the apical surface of polarized epithelial cells or released outside after cell lysis, thus exerting exogenous action on neighboring noninfected cells. Several lines of evidence suggest that the NSP4-induced Ca^{2+} release from ER is linked to the stimulation of chloride secretion and diarrhea.

HCV is the major pathogen responsible for non-A non-B hepatitis in humans. It belongs to a family of positive-polarity, single-strand RNA viruses. Transient and stable expression of the HCV core protein in Huh7 cells, a human liver carcinoma line used to study HCV replication, induces ER Ca^{2+} depletion by impairing the function of the SERCA pump. The inhibition of SERCA activity is possibly caused by overexpression of inducible nitric oxide synthase and calreticulin (▶ Calnexin and Calreticulin) in response to HCV core protein–induced ER stress. HCV core protein–induced ER stress further activates the proapoptotic Bcl2-associated X protein, which induces opening of the mitochondrial voltage-dependent anion channel, cytochrome c release, apoptosis, and eventually liver damage.

ER/Golgi Membrane Permeabilization Induced by Pore-Forming Viral Proteins

Aside from altering the ER Ca^{2+} store level via ER Ca^{2+} release channels as described above, some viruses exploit alternative ways to adjust Ca^{2+} store filling status by directly forming pores on membrane or disrupting membrane permeability of Ca^{2+} stores. This approach is best employed by enteroviruses, HCV, and rotavirus.

The enterovirus 2B protein is characteristic of viroporin, a group of integral membrane proteins containing amphipathic α-helices and capable of forming pores on membrane to aid virion production and dissemination. 2B has been demonstrated to multimerize and form hydrophilic pore when incorporated into liposomes with an estimated pore size of 6 Å. The aqueous pores on membrane allow solutes with molecular mass <1 kDa to pass through freely. The 2B protein of enteroviruses has been shown to directly cause decrease of Ca^{2+} concentrations in subcellular compartments, such as Golgi complex and ER by forming pores on the membranes of these organelles and subsequently causing Ca^{2+} efflux from the lumen of these organelles.

The HCV viroporin p7, a 63-residue hydrophobic protein found in the ER membrane, forms hexamers on artificial lipid membranes and functions like a Ca^{2+} ion channel. These findings suggest that p7 might be responsible for the flow of Ca^{2+} from ER to the cytoplasm in HCV-infected cells, which is crucial for the assembly and release of infectious HCV virions.

The rotavirus NSP4 is primarily embedded in the ER membrane of rotavirus-infected cells and exhibits viroporin activity. Endogenously expressed NSP4 in Sf9 insect cells and HEK293T cells can alter the ER membrane permeability and cause a sustained increase of cytosolic Ca^{2+} concentration that is independent of the aforementioned PLC pathway. The membrane destabilization activity of NSP4 is mapped to residues 47 to 90, which share a structural similarity to those of the enterovirus 2B protein. Elevation of $[Ca^{2+}]_{CYT}$ may facilitate rotavirus replication and assembly of infectious virions (Ruiz et al. 2000).

Disruption of Mitochondrial Membrane Permeabilization and/or Potential

The multifunctional mitochondrion is an integral part of the internal Ca^{2+} pool and functions as a hub of energy production and apoptosis (▶ Calcium and Mitochondria). In mitochondria, Ca^{2+} can easily pass through outer mitochondrial membrane (OMM) pores and cross the inner mitochondrial membrane (IMM) through membrane-embedded channels and transporters. The resting inner membrane potential of the mitochondrion ($\Delta\psi_m$) is maintained at -150 to -180 mV by actively pumping proton across the inner membrane. Disruption of membrane permeability and dissipation of $\Delta\psi_m$ may lead to ATP depletion and cell death. Excess Ca^{2+} in mitochondrial matrix is pro-apoptotic since it activates the opening of PTP and results in loss of $\Delta\psi_m$ and release of cytochrome c. In contrast, actions that reduce matrix Ca^{2+} (e.g., downregulation of ER/mitochondria Ca^{2+} flux) may protect host cells from apoptosis (Rizzuto et al. 2004). During viral infection, a number of viral proteins can target mitochondria and exert either pro-apoptotic or anti-apoptotic action by altering mitochondrial Ca^{2+} signaling in host cells, depending on the stages of the viral life cycle.

Mitochondrial Ca^{2+} uptake is mediated by the mitochondrial voltage-dependent anion channel (VDAC) across OMM and the Ca^{2+} uniporter of IMM. The HBx protein encoded by HBV may exert pro-apoptotic effects on infected cells by modulating the activity of VDAC, whereas the HCV core protein does so by enhancing the activity of Ca^{2+} uniporter.

Ca^{2+} exits mitochondria through the opening of a nonselective high-conductance channel permeability transition pore (PTP) in IMM and the Na^+/Ca^{2+} exchanger. The IMM protein adenine nucleotide translocator (ANT) also contributes to the permeability transition. The well-characterized pro-apoptotic reagent, HIV-1 viral protein R (Vpr), may tightly interact with ANT and induce mitochondrial membrane depolarization, Ca^{2+} leakage, and cytochrome c release from mitochondria. The C-terminal fragment of Vpr (residues 52–96), along with ANT, has been shown to form ion channels in synthetic membrane and control mitochondrial membrane permeability. Another viral protein, the HTLV-1 accessory protein p13^{II}, is a mitochondrial protein primarily located at IMM. p13^{II} exhibits viroporin activity and may cause welling and depolarization of mitochondria by increasing inner membrane permeability to cations, such as Ca^{2+}, Na^+, and K^+. Such changes are responsible for promoting ceramide-induced or Fas ligand–induced apoptosis in infected T lymphocytes.

In summary, different viral proteins could disrupt intracellular Ca^{2+} homeostasis by targeting the same Ca^{2+} signaling components. Alternatively, to make full use of the limited viral gene products, a single viral protein may sometimes manipulate multiple Ca^{2+} signaling mechanisms by targeting different Ca^{2+}-modulated apparatus.

Ca^{2+}-Modulated Viral Proteins

Ca^{2+}-binding sites in viral proteins can be divided, although oversimplified, into two types: discontinuous and continuous ones (▶ Calcium-Binding Protein Site Types). While both types of Ca^{2+}-binding sites have been found in the proteins of diverse virus families, more is known about virus proteins with discontinuous binding sites. The majority of known viral Ca^{2+}-binding proteins (CaBPs) are structural proteins, including both coat and envelope proteins. In addition, discontinuous sites are found in the envelope-associated neuraminidase protein of influenza B virus and the nonstructural NSP4 protein of rotaviruses.

Coat Proteins

Ca^{2+} binding to coat proteins is required to maintain the structural integrity and/or the proper assembly and disassembly of virions. Viruses with both helical and icosahedral symmetry are represented in this category. Tobacco mosaic virus, a typical helical virus, has been shown to bind Ca^{2+} with apparent affinities <100 μM. With the extracellular Ca^{2+} concentration at mM range, such an affinity would ensure a tight binding of Ca^{2+} to capsid proteins.

Among viruses with icosahedral symmetry, the numbers of bound Ca^{2+} ions and coordinating geometry of the Ca^{2+}-binding sites differ in that the Ca^{2+} ions may be situated between the interacting interfaces of capsid subunits (Fig. 2a) or sit on the symmetric threefold or fivefold axis (Fig. 2b). For example, the virion of cocksfoot mottle virus (CfMV), a plant virus with a single-stranded, positive-sense RNA genome, has an icosahedral capsid composed of 180 copies of the coat protein monomer assembled in $T = 3$ quasi-equivalent symmetry. Each monomer has a jelly-roll β-sandwich topology and can assume one of three slightly different conformations, denoted as quasi-equivalent A, B, and C subunits. These subunits are assembled into asymmetric units which then coalesce to form the icosahedral capsid. Three Ca^{2+} ions, each incorporated between the interacting surfaces of subunits (A–B, B–C, and A–C), function as reusable "glue" to stick adjacent subunits together and stabilize the capsid. Each Ca^{2+} ion is coordinated by two residues from one subunit and three residues from the other interacting subunit (Fig. 2a). Such Ca^{2+}-binding sites seem to adopt an octahedral geometry with only five ligands. A similar scenario is seen in the trimeric VP7 of rotaviruses, but it contains two Ca^{2+} ions at each subunit interface (Aoki et al. 2009). The withdrawal of Ca^{2+} during virus entry is speculated to trigger the uncoating of rotavirus in endosome.

Another example of the incorporation of Ca^{2+} into an icosahedral virus particle is the human rhinoviruses. In the virions of human rhinovirus 1A (HRV1A), HRV3, and HRV14, Ca^{2+} ions are located at the fivefold axis of symmetry (Fig. 2b). With two additional oxygen atoms from water molecules above or below the metal ion as coordinating ligands, the Ca^{2+}-binding pocket forms a pentagonal bipyramidal geometry.

Envelope Proteins

An example of Ca^{2+} binding by a virion envelope protein is the neuraminidase of ortho- and paramyxoviruses, both families of negative-polarity, single-stranded RNA viruses of animals. The neuraminidase catalyzes the cleavage of the glycosidic linkages between the terminal siliac acid residues and the body of carbohydrate moieties on the surface of infected cells. This activity is required for the release of virus from the cell surface. Two distinct Ca^{2+}-binding sites are found in the neuraminidase. One high-affinity site (Fig. 2c) is close to the active site and adopts an octahedral geometry with five coordinating protein ligands and one molecule of water. This Ca^{2+}-binding site is conserved among influenza A virus, influenza B virus, and the parainfluenza viruses. The other site is a relatively low-affinity site, located on the fourfold axis of the tetrameric neuraminidase of the influenza B viruses. The high-affinity Ca^{2+}-binding site is needed for the thermostability and optimal activity of the enzyme, whereas the low-affinity site has been postulated to hold the tetramer together.

Nonstructural Proteins

The rotavirus NSP4 is required for the budding of immature viral particle into the ER lumen and plays

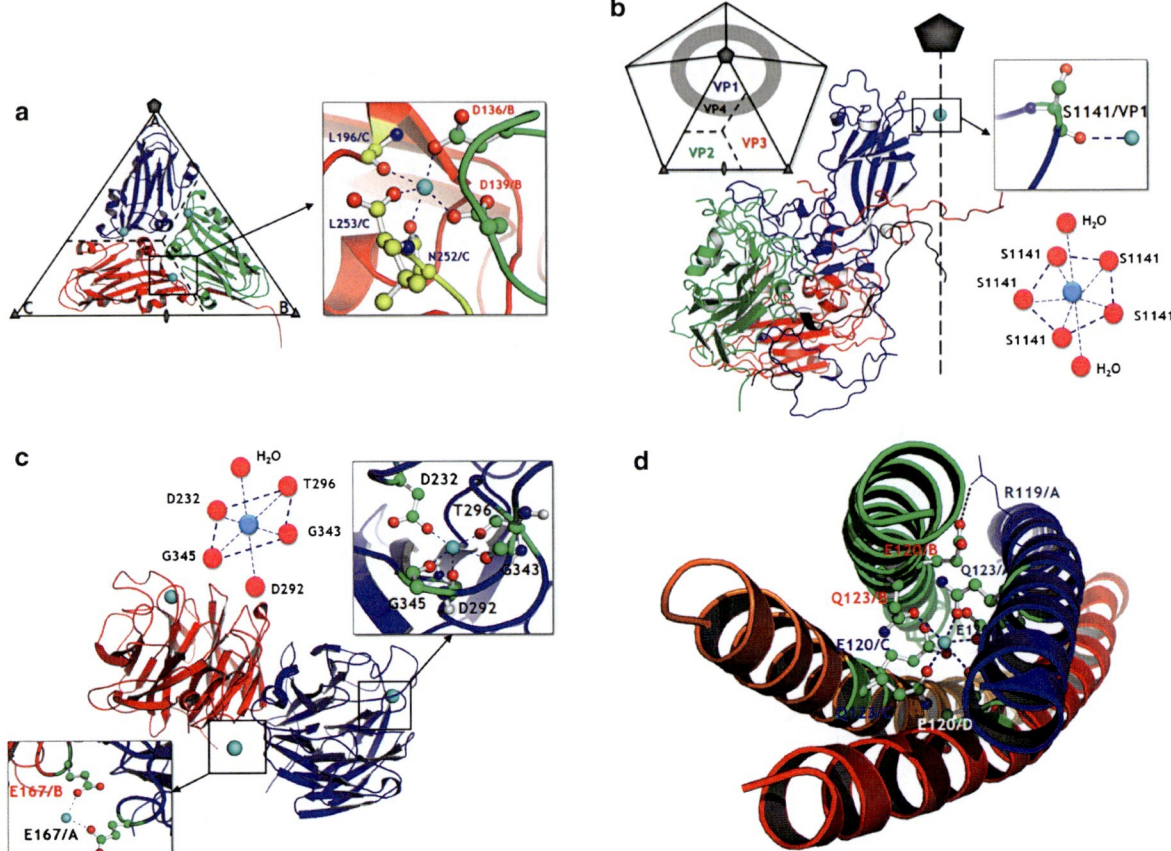

Calcium and Viruses, Fig. 2 Examples of Ca^{2+}-modulated viral proteins (Adapted with permission from reference Zhou et al. 2009). (**a**) 3D representation of the icosahedral asymmetric unit of the cocksfoot mottle virus capsid and the location of the incorporated Ca^{2+} ions (PDB entry: 1ng0). The assembling unit is formed by three subunits, A (*blue*), B (*green*), and C (*red*), that are chemically identical but slightly different in conformational arrangement. Ca^{2+}, situated between the interfaces of neighboring subunits (A–B, A–C, or B–C), is coordinated by oxygen atoms from the side chains of D136 and D139 in one subunit and oxygen atoms from the main chain of L196, the side chain of N252, and the C-terminal carboxyl group of L253 in the other neighboring subunit (*enlarged area*). The *solid pentagon*, *triangle*, and *oval* represent five-, three-, and two-fold axes of the icosahedron. (**b**) Ca^{2+} ion located on the fivefold axis of the capsid of human rhinovirus 3 (HRV3) (PDB entry: 1rhi). The icosahedral capsid of HRV3 is composed of 60 copies of each of the four capsid proteins VP1 (*blue*), VP2 (*green*), VP3 (*red*), and VP4 (*black*). VP1, VP2, and VP3 are exposed to the external surface of the viral particle, whereas VP4 lies in the internal surface. A Ca^{2+} ion is found situated on the fivefold axis of the capsid and coordinated by five oxygen atoms from the main chain carbonyl group of the fivefold symmetry-related S1141 on VP1 (*enlarged area*). With two additional oxygen atoms from water molecules above or below the metal ion as coordinating ligands, the Ca^{2+}-binding pocket forms a pentagonal bipyramidal geometry. (**c**) The 3D structure of neuraminidase of influenza B virus (PDB entry: 1nsb). The cartoon only represents half of the tetrameric form of this enzyme. Three Ca^{2+}-binding sites are found in two identical subunits, A (*blue*) and B (*red*). Each subunit contains one octahedral Ca^{2+}-binding site (*upper panel*). Another site (*lower panel*), coordinated by a fourfold symmetry-related residue, holds the oligomer together. (**d**) The core Ca^{2+}-binding pocket in the oligomerization domain (aa. 95–137) of NSP4 from rotavirus (PDB entry: 2o1j). The domain self-assembles into a paralleled tetrameric coiled coil. Chains A, B, C, and D are shown in *blue*, *green*, *orange*, and *red*, respectively. The Ca^{2+} ion is coordinated by six oxygen atoms from the side chains of Q123 on chains A–D, as well as the side chains of E120 on chains B and D. Calcium ions are shown as *cyan spheres*

a central role in the morphogenesis of rotaviruses, even though it is not incorporated into the virus particle. The rotavirus NSP4 contains a core Ca^{2+}-binding site when it oligomerizes into a functional homo-tetramer. The Ca^{2+} ion is coordinated by the side chains of several residues from the four identical polypeptides within the tetramer (Fig. 2d). Ca^{2+}-binding at this site appears to stabilize the NSP4 tetramer.

In contrast to the relative abundance of discontinuous Ca^{2+} sites, the literature reports on continuous viral Ca^{2+}-binding sites are scarce. The prototype continuous Ca^{2+}-binding site is the EF-hand and EF-hand-like motifs (Zhou et al. 2006). The EF-hand Ca^{2+}-binding motif contains a helix-loop-helix topology, much like the spread thumb and forefinger of the human hand, in which the Ca^{2+} ions are coordinated ligands within the loop. Only four such cases can be found, namely, the transmembrane protein gp 41 of HIV-1, VP7 of rotavirus, VP1 of polyomavirus, and the protease domain of rubella virus (RUBV). Among these, the EF-hand motif within the RUBV nonstructural protease is the most well characterized (Zhou et al. 2007). The Ca^{2+}-binding affinity of this putative EF-hand motif was determined to be ~200–300 μM, which agrees with the Ca^{2+} concentration of late endosomes and lysosomes (400–600 μM), where the RUBV replication complex forms and replication occurs. After removal of the Ca^{2+}-binding site, the protease activity is decreased and rendered temperature sensitive, indicating that the function of Ca^{2+} binding is to stabilize the protease and maintain optimal virus infectivity.

Given the widespread occurrence of EF-hand motifs in cellular proteins and the importance of Ca^{2+} as an intracellular messenger, it is surprising that only three of these motifs were reported in viral proteins. Indeed, a comprehensive search for potential viral EF-hand motifs with bioinformatic approaches resulted in the detection of over 90 putative EF-hand and EF-hand-like motifs in proteins encoded by the genomes of almost 80 different viruses, covering the majority of virus families (Zhou et al. 2009). In contrast to EF-hand motifs in cellular proteins, almost all of these predictions were single EF-hand motifs. These putative EF-hand-motif-containing proteins are involved in a wide range of viral processes. Notably, the functions of almost 20% of the proteins with predicted Ca^{2+}-binding motifs remain uncharacterized.

Calcium and Viruses, Table 1 Interactions between cellular Ca^{2+} binding proteins (CaBPs) and viral proteins

Cellular CaBP	Viral molecular identity	Virus	Consequences of interaction
Annexin II	p55^{GAG}	HIV-1	Facilitates virus entry and fusion in macrophages
	Glycoprotein B	CMV	Enhances binding and fusion to membranes
Annexin V	Small HBsAg	HBV	Participates in initial steps of HBV infection
Calmodulin	Nef	HIV-1	Alters T lymphocyte signaling pathway
	gp 160/gp41	HIV-1	Disrupts CaM signaling pathway
	gp 41	SIV	
	p17^{GAG}	HIV-1	
Calreticulin/calnexin	E1 and E2	RUB	Regulates viral glycoprotein maturation
	Viral RNA	RUB	
	MP	TMV	Regulates cell-to-cell virus movement
	F, HN	SeV	Mediates maturation of glycoproteins
	gp 160	HIV-1	Facilitates protein maturation
	P12^I	HTLV-1	
	Tax	HTLV-1	Facilitates viral protein folding; possibly mediates the interaction with MHCI
ERC-55	E6	HPV	
Fibulin-1	E6	HPV	Regulates cell migration and invasion
S100A10 (p11)	pol	HBV	Inhibits viral replication
	NS3	BTV	Mediates nonlytic virus release

HIV human immunodeficiency virus, *CMV* cytomegalovirus, *HBV* hepatitis B virus, *SIV* simian immunodeficiency virus, *RUB* rubella virus, *SeV* Sendai virus, *TMV* tobacco mosaic virus, *HTLV* human T cell lymphotropic virus, *HPV* human papillomavirus, *EBV* Epstein-Barr virus, *BTV* bluetongue virus (Adapted with permission from Zhou et al. 2009)

Ca^{2+}-Dependent Virus-Host Interactions

Compared to the paucity of reported viral CaBP's, host cells contain abundant CaBPs. Accordingly, viral proteins utilize a number of important cellular CaBPs, including proteins in the cytoplasm (e.g., annexin, calmodulin, and S100), endoplasmic reticulum (e.g., ERC-55, calreticulin, and calnexin), and extracellular matrix (e.g., fibulin-1) in their replication cycles (listed in Table 1).

Take the ubiquitously expressed calmodulin (CaM) as an example (▶ Calmodulin). CaM consists of two globular and autonomous domains, each of which contains two EF-hand motifs. Through its binding to Ca^{2+} and the concomitant conformational changes that result, CaM is capable of transducing the intracellular Ca^{2+} signal changes into divergent cellular events by binding to an array of cellular proteins. Two HIV proteins, Nef, an accessory protein, and gp160, the glycoprotein precursor, have been shown to interact with CaM in a Ca^{2+}-dependent fashion. Nef is a myristoylated protein expressed early in infection. Nef has been shown to downregulate both CD4 and MHC-Class I cell surface receptors, both important in the cell-mediated response, and to alter T lymphocyte signaling pathways. The latter effect is partially associated with its ability to strongly interact with CaM. In addition to Nef, the precursor gp160/gp41 has been shown to interact with CaM. Such interaction is postulated to disrupt the anti-apoptotic CaM signaling pathway by either reducing the amount of free cytosolic CaM or changing its subcellular localization. A similar CaM-targeting sequence is also detected in simian immunodeficiency virus gp41. In view of the diverse roles of Ca^{2+}/CaM-dependent signaling pathways, the interaction between all these HIV proteins with CaM is expected to play multiple roles to fit the HIV life cycle in response to altered Ca^{2+} signals.

Summary

As briefly described in this entry, viruses constantly disrupt or remodel the intracellular Ca^{2+} signaling network, from direct binding of Ca^{2+} by viral proteins to binding of virus proteins to cellular CaBPs to alteration of Ca^{2+} homeostasis. The remodeled Ca^{2+} network affects multiple steps of virus replication as well as cellular outcomes. The prediction of a large number of poorly characterized, putative EF-hand or EF-hand Ca^{2+}-binding motifs in a diverse collection of virus proteins further expands the repertoire of virus-Ca^{2+} interactions. The Ca^{2+} signaling field has been constantly reinvigorated with the continuous discovery of new components and expansion of the Ca^{2+}-signaling toolkits. It is anticipated that, with the continuing development of new research techniques and tools, more and more virus-Ca^{2+} interplays will be elucidated at molecular levels.

Cross-References

▶ Calcium and Mitochondria
▶ Calcium ATPase
▶ Calcium in Biological Systems
▶ Calcium-Binding Protein Site Types
▶ Calcium-Binding Proteins, Overview
▶ Calmodulin
▶ Calnexin and Calreticulin

References

Aoki ST, Settembre C, Trask SD et al (2009) Structure of rotavirus outer-layer protein VP7 bound with a neutralizing Fab. Science 324:1444–1447

Berridge MJ, Bootman MD, Roderick HL (2003) Calcium signalling: dynamics, homeostasis and remodeling. Nat Rev Mol Cell Biol 4:517–29

Chami M, Oules B, Paterlini-Brechot P (2006) Cytobiological consequences of calcium-signaling alterations induced by human viral proteins. Biochim Biophys Acta 1763: 1344–1362

Putney JW Jr (1998) Calcium signaling: up, down, up, down...what's the point? Science 279:191–2

Rizzuto R, Duchen MR, Pozzan T (2004) Flirting in little space: the ER/mitochondria Ca^{2+} liaison. Sci STKE 2004:re1

Ruiz MC, Cohen J, Michelangeli F (2000) Role of Ca^{2+} in the replication and pathogenesis of rotavirus and other viral infections. Cell Calcium 28:137–49

van Kuppeveld FJ, de jong AS, Melchers WJ et al (2005) Enterovirus protein 2B po(u)res out the calcium: a viral strategy to survive?. Trends Microbiol 13:41–44

Zhou Y, Frey TK, Yang JJ (2009) Viral calciomics: interplays between Ca^{2+} and virus. Cell Calcium 46:1–17

Zhou Y, Tzeng WP, Yang W et al (2007) Identification of a Ca^{2+}-binding domain in the rubella virus nonstructural protease. J Virol 81:7517–7528

Zhou Y, Yang W, Kirberger M et al (2006) Prediction of EF-hand calcium-binding proteins and analysis of bacterial EF-hand proteins. Proteins 65:643–655

Calcium as a Secondary Messenger

▶ Calcium in Biological Systems
▶ Calcium-Binding Proteins, Overview

Calcium as an Intracellular Messenger

▶ Calcium, Local and Global Cell Messenger

Calcium ATPase

▶ Calcium-Binding Proteins, Overview

Calcium ATPase and Beryllium Fluoride

Hiroshi Suzuki
Department of Biochemistry, Asahikawa Medical University, Asahikawa, Hokkaido, Japan

Synonyms

Ca^{2+} pump

Definition

Sarco(endo)plasmic reticulum Ca^{2+}-ATPase, calcium pump, catalyzes Ca^{2+} transport coupled with ATP hydrolysis into the lumen against ~10,000 time concentration gradient. The ATPase is activated by high-affinity binding of two cytoplasmic Ca^{2+} ions at the transport sites in the transmembrane region and forms an autophosphorylated intermediate by transferring ATP γ-phosphate to a catalytic aspartate (Asp351) in the cytoplasmic region. The subsequent large conformational change disrupts the Ca^{2+} binding sites and releases Ca^{2+} into the lumen and produces the catalytic site for hydrolysis of the Asp351-phosphate bond. Beryllium in beryllium fluoride (BeF_3^-) compound is directly ligated with the catalytic aspartyl oxygen, producing a very stable analog of the covalently bound phosphate at Asp351 with an equivalent tetrahedral structure and bond lengths. The biochemical studies of the Ca^{2+}-ATPase/BeF_3^- complexes with and without bound Ca^{2+} at the transport sites revealed characteristic properties of the phosphorylated intermediates with and without bound Ca^{2+}. The Ca^{2+}-free complex was successfully crystallized and provided an atomic model of the Ca^{2+}-released phosphorylated intermediate. Furthermore, in comparison with the complexes of Ca^{2+}-ATPase with AlF_x (transition-state analogs of the phosphorylation and hydrolysis) and with MgF_4^{2-} (an analog of hydrolysis product complex, Ca^{2+}-ATPase with non-covalently bound inorganic phosphate P_i (HPO_4^{2-})), the structural mechanism of energy coupling in the ATP-driven Ca^{2+} transport has been deeply understood.

Ca^{2+}-ATPase: Function and Reaction Sequence

Sarco(endo)plasmic reticulum Ca^{2+}-ATPase (SERCA) catalyzes ATP-driven Ca^{2+} transport from the cytoplasm into the lumen (Fig. 1). It is a representative member of P-type ion transporting ATPases, which include plasma membrane Ca^{2+}-ATPase; Na^+,K^+-ATPase; H^+,K^+-ATPase (responsible for gastric acid secretion); Cu-ATPases; and Golgi Ca^{2+},Mn^{2+}-ATPase. The ATPases form an autophosphorylated intermediate (EP) in the ion transport cycle thereby named "P-type." The family members possess three large cytoplasmic domains: nucleotide-binding (N), phosphorylation (P, which contains the autophosphorylation catalytic aspartate), and actuator (A) domains and ten transmembrane helices (M1–M10) (see Fig. 2 in Cross-Reference by C. Toyoshima). The ion transport sites (high-affinity ion binding sites) consist of residues on M4, M5, M6, and M8. Essential residues in the catalytic site for the ATP hydrolysis and in the ion transport (binding) sites are well conserved among the members, and the ATPases work in the common mechanism for the specific ion transport.

The three members of SERCA (1, 2, and 3) encoded by three different genes (ATP2A1, ATP2A2, ATP2A3) are highly homologous and further form alternative splice variants. SERCA1a and 1b are expressed in skeletal fast-twitch muscle as an adult type and a fetal type, respectively. SERCA2a is expressed in heart muscle, and 2b is in smooth muscle and non-muscle cells as a housekeeping isoform. SERCA3 isoforms are expressed in specific tissues such as in nerve tissues and immune cells. Mutations of SERCA genes cause diseases, such as Brody disease (autosomal recessive disease characterized by exercise-induced impairment in muscle relaxation) and Darier disease (autosomal dominant genetic skin disease with abnormal keratinization). Disruption of SERCA function and expression is related with cancer and diabetes.

In the Ca^{2+} transport cycle by SERCA (Fig. 1), the enzyme is first activated by cooperative binding of two

Calcium ATPase and Beryllium Fluoride,
Fig. 1 Reaction scheme of SERCA

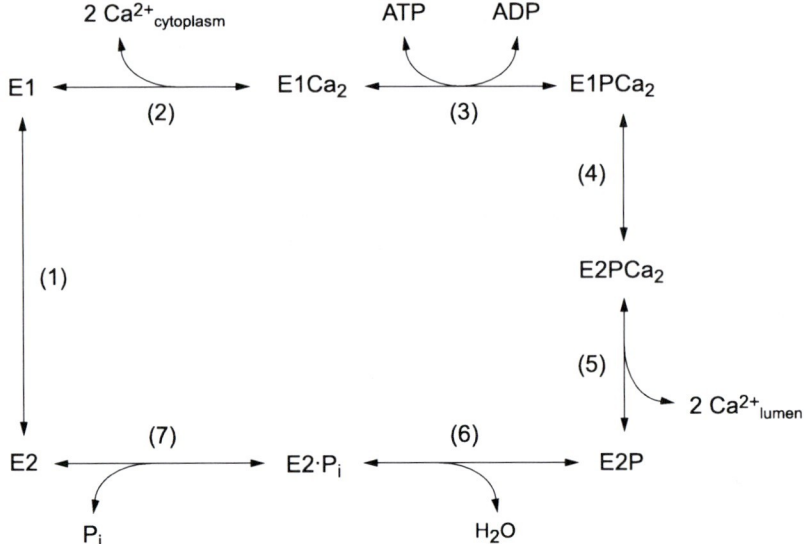

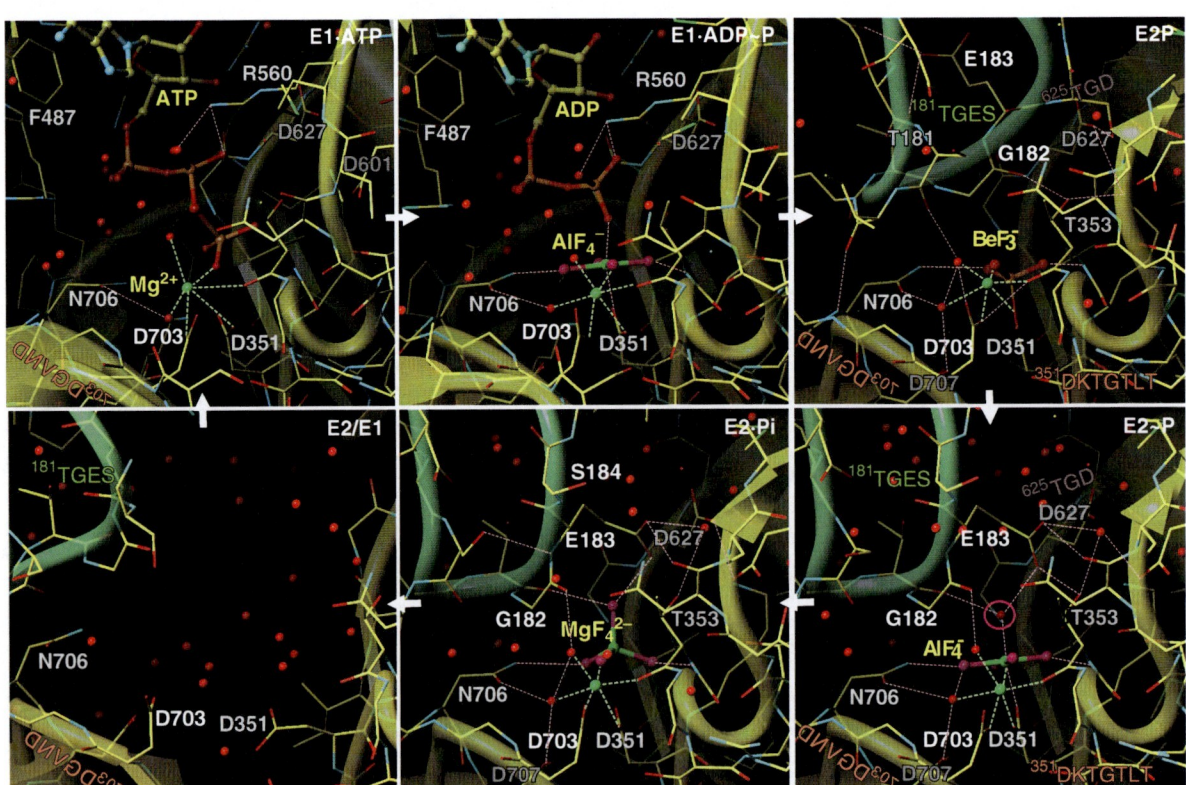

Calcium ATPase and Beryllium Fluoride, Fig. 2 Structural events that occur around the phosphorylation site (Kindly provided by Prof. Chikashi Toyoshima, University of Tokyo, see also review article by Toyoshima (2008)). Crystal structures of $E1Ca_2 \cdot AMPPCP$, $E1Ca_2 \cdot ATP$ analog; $E1Ca_2 \cdot AlF_4^- \cdot ADP$, $E1PCa_2 \cdot ADP^\ddagger$ transition-state analog; $E2 \cdot BeF_3^-$ (TG), E2P ground-state analog; $E2 \cdot AlF_4^-$ (TG), $E2 \sim P^\ddagger$ transition-state analog; $E2 \cdot MgF_4^{2-}$ (TG), $E2 \cdot P_i$ product-state analog; and E2 fixed with TG and BHQ. Broken lines in *pink* show likely hydrogen bonds, and those in *light green* show Mg^{2+} coordination. Small *red spheres* represent water molecules. Note in $E2 \sim P^\ddagger$, an attacking water molecule for the hydrolysis is coordinated by aluminum (phosphorus) atom and Glu183-oxygen and Gly182-carbonyl oxygen on the TGES184 loop of the A domain. Conserved sequence motifs are labeled. *TG* thapsigargin, *BHQ* tert-butylhydroquinone

cytoplasmic Ca^{2+} ions at the high-affinity transport sites ($E2$ to $E1Ca_2$, steps 1–2) and autophosphorylated at a catalytic aspartate (Asp^{351}) with MgATP to produce an ADP-sensitive EP ($E1P$, which reacts rapidly with ADP to regenerate ATP in the reverse reaction, step 3). Upon the $E1P$ formation, the bound two Ca^{2+} ions become inaccessible from both cytoplasmic and lumenal sides, thus occluded in the transport sites ($E1PCa_2$). The subsequent isomeric transition (a large structural change) to the ADP-insensitive form ($E2P$) results in rearrangements of the transmembrane helices and consequent disruption of the Ca^{2+} binding sites to deocclude Ca^{2+}, reduce largely its affinity, and open the lumenal gate, thus releasing Ca^{2+} into the lumen (steps 4–5). This Ca^{2+}-release process is very rapid; therefore, the $E2PCa_2$ transient state does not accumulate in the wild-type enzyme. Upon the structural change for the Ca^{2+} release, the catalytic site is largely rearranged so as to gain hydrolytic activity; thereby the Asp^{351}-acylphosphate in the Ca^{2+}-released $E2P$ is hydrolyzed to form the Ca^{2+}-unbound inactive $E2$ state (steps 6–7). The Ca^{2+} ligands in the transport sites of the Ca^{2+}-released $E2P$ bind protons (totally $2 \sim 3$) from the lumenal side, and these protons are released into the cytoplasm upon the Ca^{2+}-binding $E2 \rightarrow E1Ca_2$. The Ca^{2+} pump is therefore actually Ca^{2+},H^+ pump and electrogenic. Mg^{2+} bound at the catalytic Mg^{2+} subsite is required as a catalytic cofactor in the phosphorylation and dephosphorylation for the Ca^{2+} transport. The transport cycle by SERCA1a turns maximally $\sim$50 times in 1 s, and this rapid turnover is responsible for rapid muscle relaxation. The cycle is totally reversible, e.g., $E2P$ can be formed from P_i in the presence of Mg^{2+} and absence of Ca^{2+}, and the subsequent lumenal Ca^{2+} binding at lumenally oriented low-affinity transport sites of $E2P$ reverses the Ca^{2+}-releasing step and produces $E1PCa_2$, which is then dephosphorylated to $E1Ca_2$ by ADP. Both the phosphoryl transfer to Asp^{351} from ATP and the Asp^{351}-acylphosphate hydrolysis take place with in-line associative mechanism, and therefore, a trigonal bipyramidal structure of the pentacoordinated phosphorus is formed in the transition states.

The affinity of SERCAs for the transporting Ca^{2+} ($K_d \approx$ sub-μM) is highest among other Ca^{2+} removal systems such as Na^+,Ca^{2+} exchanger. Therefore, SERCAs are responsible for setting the submicromolar cytoplasmic Ca^{2+} level in resting cells. Also, the lumenal $\sim$ mM Ca^{2+} level, which is critical for the functions of endoplasmic reticulum, is appropriately set by the SERCA's functional property, i.e., by the lumenally oriented low-affinity Ca^{2+} sites in $E2P$ to which lumenal Ca^{2+} binds when its level becomes $\sim$mM level high, thereby causing the lumenal Ca^{2+}-induced feedback inhibition of the pump function.

The Use of Metal Fluoride Compounds Provided Essential Information for Understanding Ca^{2+} Transport Mechanism

The major question on the Ca^{2+}-ATPase is how the catalytic site and transport sites communicate each other to accomplish the energy coupling for the Ca^{2+} transport despite their long-distance separation by $\sim$50 Å. The mechanism has been thought to be brought about by mutual communication via structural changes between the catalytic site in the cytoplasmic domains and the transport sites in the transmembrane domain. It is now actually revealed to be achieved by large motions of cytoplasmic domains coupled with rearrangements of the transmembrane helices. Note that the phosphorylated intermediates decay within a few tens of milliseconds; therefore, exploring and characterizing their structural states have been extremely difficult. One major breakthrough was the development of stable structural analogs of the phosphorylated intermediates with metal fluoride compounds: beryllium fluoride (BeF_3^-), aluminum fluoride (AlF_x), and magnesium fluoride (MgF_4^{2-}), which are analogs of the phosphate group (Danko et al. 2004, 2009, Daiho et al. 2010). The beryllium in beryllium fluoride can be directly coordinated with the catalytic aspartyl oxygen, therefore mimics the covalent acylphosphate bond.

To produce the complexes of Ca^{2+}-ATPase with metal fluoride, the sarcoplasmic reticulum Ca^{2+}-ATPase (SERCA1a) was incubated with these compounds typically in a solution at pH $6 \sim 7$ containing $1 \sim 5$ mM F^-, $10 \sim 50$ μM Be^{2+} or Al^{3+} (for BeF_x or AlF_x), and $1 \sim 15$ mM Mg^{2+} (for its binding at the catalytic Mg^{2+} subsite as required for the metal fluoride binding at the catalytic site as well as for the phosphorylation). For the MgF_x binding, the incubation was made without Be^{2+} and Al^{3+}, otherwise as above. The incubation was performed either in the presence of a saturating concentration of Ca^{2+}

Calcium ATPase and Beryllium Fluoride, Table 1 Ca^{2+}-ATPase complexes with metal fluoride and their assignments to intermediate states in the transport cycle

Intermediate states in Ca^{2+} transport cycle									
	E2 (no ligands)	E1Ca$_2$	E1Ca$_2$·ATP	E1PCa$_2$·ADP‡ E1PCa$_2$·ADP	E1PCa$_2$	E2PCa$_2$	E2P	E2P‡	E2·P$_i$
Stable analog	E2 or E2(TG)	E1Ca$_2$	E1Ca$_2$·AMPPCP	E1Ca$_2$·AlF$_4^-$·ADP E1PCa$_2$·AMPPN	E1Ca$_2$·BeF$_3^-$	E2Ca$_2$·BeF$_3^-$	E2·BeF$_3^-$	E2·AlF$_4^-$	E2·MgF$_4^{2-}$

TG thapsigargin

(10 ~ 100 μM) or in its absence (by removing free Ca^{2+} with EGTA) to obtain the complexes of the Ca^{2+}-ATPase with and without bound Ca^{2+} at the transport sites. The formation of the Ca^{2+}-ATPase complexes with the metal fluoride, i.e., its binding at the catalytic site, was verified by the functional analysis: the loss of the ATPase activity and of phosphorylation from ATP and/or P$_i$. By these experiments, the complexes of the Ca^{2+}-ATPase with the phosphate analogs were successfully developed, and appropriate conditions for the formation of complexes were established.

The Ca^{2+}-ATPase complexes thus produced were remarkably stable and not decomposed even for ~a month. With MgF$_x$ binding, the metal fluoride binding was found to be quasi-irreversible, which is probably brought about by chemical nature of fluoride. Namely, it possesses a significantly higher electronegativity than oxygen (actually the highest among all atoms) and a small size, therefore it produces the stronger coordination of the metal fluoride than the phosphate group and fixes the structures analogous to the phosphorylated intermediates. As summarized in Table 1 and as described in detail in the following sections, each of the complexes were successfully ascribed to the intermediate states on the basis of the nature in ligation chemistry of the metal fluoride compounds, and the characteristic properties of the Ca^{2+}-ATPase complexes developed. Detailed biochemical and crystallographic studies of the complexes were enabled by their remarkable stability and revealed hitherto unknown features of the intermediate states and provided the atomic structural models. The structural mechanism of Ca^{2+} transport thus revealed is depicted very nicely as the cartoon model by Toyoshima (see Fig. 2 in Cross-Reference by C. Toyoshima, 2008, 2009).

Ligation Chemistry of Beryllium Fluoride as Compared with Aluminum Fluoride and Magnesium Fluoride

Beryllium ion, Be^{2+}, possesses four coordinate bonds, and the beryllium fluoride compounds are known to adopt the tetrahedral geometry with the Be-F bond length 1.55 Å, thereby making them strictly isomorphous to the tetrahedral phosphate group. Moreover, because of the high charge density due to its small size, the beryllium in beryllium fluoride is able to attract an aspartate-oxygen atom (Asp^{351}-oxygen in the case of Ca^{2+}-ATPase) to coordinate in addition to three F^- and thus generate –O-BeF$_3^-$ adduct with the fourth coordination from the aspartate-oxygen. The -O-BeF$_3^-$ formed with the catalytic aspartate in fact was shown to possess the tetrahedral geometry superimposable with the covalently bound phosphate at the aspartate in the crystallographic studies of phosphotransferases (including the members of the haloacid dehalogenase (HAD) superfamily to which P-type ATPases belong, the bacterial response regulators, and later by the crystal structure of Ca^{2+}-free Ca^{2+}-ATPase complexed with BeF$_3^-$, *E*2·BeF$_3^-$, which is the *E*2P ground-state analog (for more details, see reference by Danko et al. (2004) and the references cited in).

Aluminum ion, Al^{3+}, ligates fluoride and forms either AlF$_3$ or AlF$_4^-$, depending on pH in the incubation medium, and binds to enzymes. As seen in their crystallographic structural studies, the bound AlF$_3$ and AlF$_4^-$ possess the planar (trigonal (AlF$_3$)) and square (AlF$_4^-$)) geometry, in which two oxygen atoms, e.g., from the specific aspartate and from the ADP β-phosphate or the hydrolytic water molecule, coordinate to the aluminum at apical positions. The AlF$_x$ thus coordinated is superimposable (AlF$_3$) or analogous (AlF$_4^-$) to the trigonal bipyramidal structure of the pentacoordinated phosphorus in the transition states of the in-line associative phosphoryl transfer and acylphosphate hydrolysis, which are the case for the Ca^{2+}-ATPase.

As Be^{2+}, Mg^{2+} possesses four coordinate bonds and produces the tetrahedral geometry MgF_4^{2-}, which mimics the phosphate group. However, being distinct from the beryllium, the magnesium in MgF_4^{2-} is not able to (or having much less ability to) attract aspartate-oxygen to coordinate directly, and thus the oxygen atom cannot substitute the fluoride in MgF_4^{2-}. Therefore, MgF_4^{2-} bound to the catalytic site mimics a non-covalently bound P_i (HPO_4^{2-}). In fact, in the crystal structure of the Ca^{2+}-free Ca^{2+}-ATPase ($E2$ state), the magnesium fluoride bound at the catalytic site was shown to be MgF_4^{2-} of which tetrahedral geometry is superimposable to the non-covalently bound P_i. The $E2 \cdot MgF_4^{2-}$ complex therefore represents the product complex of the $E2P$ hydrolysis, $E2 \cdot P_i$.

Analogs of E2P Without Bound Ca^{2+} During Asp351-Phosphate Hydrolysis

For understanding the Ca^{2+}-free structure and properties of EP ($E2P$) and its hydrolysis mechanism, the Ca^{2+}-free Ca^{2+}-ATPase ($E2$ state) was complexed with beryllium fluoride, aluminum fluoride, and magnesium fluoride (Danko et al. 2004). The EP formation from ATP and P_i, and ATP hydrolysis were completely inhibited by these compounds, indicating that the metal fluoride compounds were bound to the catalytic site. The complexes thus produced are all very stable and not degraded even after extensive washing for removing unbound metal fluoride. The detailed analyses of the complexes were then performed to reveal and compare their characteristic properties. In the complex formed with beryllium fluoride, lumenal Ca^{2+} at ~mM high concentration is accessible to the transport sites with the low affinity comparable to that in $E2P$ (the Ca^{2+}-released state of EP formed from P_i). Thus the complex with beryllium fluoride possesses the lumenally oriented transport sites, i.e., lumenally opened Ca^{2+}-release pathway, as $E2P$. The nature of the catalytic site was explored by a hydrophobicity-sensitive fluorescence development of trinitrophenyl (TNP)-AMP, which binds to the ATP binding site with a high affinity. The complex with beryllium fluoride as well as $E2P$ developed an extremely high fluorescence ("superfluorescence") of bound TNP-AMP, showing that the catalytic site is strongly hydrophobic and therefore a closed structure excluding nonspecific water molecules (but a specific water molecule attacking the Asp351-phosphate bond). Thus the $E2 \cdot BeF_x$ complex possesses characteristic properties known for the $E2P$ intermediate. The results showed that $E2 \cdot BeF_x$ is the analog of the $E2P$ ground state, which is the state immediately after the Ca^{2+} release with the lumenally opened release pathway and before (ready for) the acylphosphate hydrolysis at the catalytic site. The conclusion agrees with the chemical nature of beryllium fluoride (BeF_3^-), i.e., Be^{2+} in BeF_3^- can be directly coordinated by Asp351-oxygen mimicking the covalent acylphosphate bond. Detailed biochemical analyses further revealed that the complex is completely resistant against (i.e., sterically protected from) protease attacks by trypsin, proteinase K, and V8 protease at their specific cleavage sites on the surface of the cytoplasmic domains. The results indicated that the three cytoplasmic domains are tightly associated with each other producing their compactly organized headpiece structure. The atomic structure solved later clearly showed that the complex is actually $E2 \cdot BeF_3^-$ and possesses all these biochemically predicted properties, i.e., the lumenally opened Ca^{2+}-release pathway and the closed and hydrophobic catalytic site with direct ligation between Asp351-oxygen and BeF_3^- (Toyoshima et al. 2007, Olesen et al. 2007, Toyoshima 2008, 2009, Møller et al. 2010).

Other two complexes developed, $E2 \cdot AlF_x$ and $E2 \cdot MgF_4^{2-}$, were also found to be completely resistant against the proteases as $E2 \cdot BeF_3^-$; therefore, they possess the compactly organized cytoplasmic domains. However, being distinct from $E2 \cdot BeF_3^-$, the catalytic site in the two complexes is hydrophilic (no fluorescence development of bound TNP-AMP); therefore, they possess a more opened conformation of the catalytic site to which nonspecific water molecules have entered. Furthermore, in the two complexes, the lumenal Ca^{2+} is inaccessible to the transport sites, and therefore, the lumenal Ca^{2+}-release pathway (gate) is closed. These Ca^{2+}-free complexes probably represent the structural states during and after the $E2P$ hydrolysis. In agreement, the ligation chemistry of AlF_x and MgF_4^{2-} and the atomic structures of the Ca^{2+}-ATPase complexes (see review articles by Toyoshima 2008, 2009) clearly showed that the complexes are the analogs of the in-line hydrolysis transition-state $E2P^{\ddagger}$ ($E2 \cdot AlF_4^-$) and the product complex of the E2P hydrolysis $E2·P_i$ ($E2·MgF_4^{2-}$). Thus the three complexes of the Ca^{2+}-free Ca^{2+}-ATPase, $E2·BeF_3^-$, $E2·AlF_4^-$, and $E2·MgF_4^{2-}$, were successfully developed, characterized, and assigned to the structural states in the E2P hydrolysis. The distinct nature of the three complexes revealed that, after Ca^{2+} release into the lumen from E2PCa$_2$, the release pathway structure is changed from its opened state to closed state upon the E2P hydrolysis ($E2P \rightarrow E2P^\ddagger$ and $E2·P_i$), thereby prevents possible Ca^{2+} leakage from the lumen to cytoplasm. The gate closure takes place by and is coupled with the rearrangement of the catalytic site configuration upon the Asp351-phosphate hydrolysis, i.e., in E2P ground state $\rightarrow$ $E2P^\ddagger$ transition state.

In the atomic structures of the three complexes, the biochemically predicted characteristic properties are all seen. Furthermore, the ligations of the metal fluoride compounds (phosphate group), attacking H$_2$O molecule, and the catalytic Mg^{2+} are all seen in the atomic level with the critical residues involved in ligations and catalysis, e.g., the TGES184 outermost loop of the A domain and essential residues of the P domain around Asp351. As predicted, the beryllium in BeF_3^- is directly ligated with the Asp351-oxygen in $E2·BeF_3^-$, the planar AlF_4^- in $E2·AlF_4$ is coordinated with Asp351 and with the attacking water molecule from the apical positions, and MgF_4^{2-} bound in $E2·MgF_4^{2-}$ is mimicking the non-covalently bound P_i. The atomic structures demonstrated how the configuration in the catalytic site structure changes (Fig. 2, see also Toyoshima 2009) and consequently how the Ca^{2+}-release gate is closed during $E2P \rightarrow E2P^\ddagger$ in the hydrolysis, i.e., the A domain slightly inclines and this motion results in the gate closure.

Analogs of Ca^{2+}-Bound Intermediate States During the Phosphorylation

The complexes of the Ca^{2+}-bound Ca^{2+}-ATPase (E1Ca$_2$) with AlF$_x$ and BeF$_x$ were also developed. The complexes first produced and crystallized are $E1Ca_2·AlF_4^-·ADP$, the transition-state analog of the phosphorylation ($E1PCa_2·ADP^\ddagger$) and $E1Ca_2·AMPPCP$, the enzyme-substrate complex analog ($E1Ca_2·ATP$) (Fig. 2, see review articles by Toyoshima 2008, 2009). Their atomic structures revealed that the N and P domains largely move, thereby being cross-linked by ATP binding for the phosphoryl transfer to Asp351 and that the two Ca^{2+} ions at transport sites become occluded upon this structural change. The planar AlF_4^- is coordinated from the apical positions by ADP β-phosphate and Asp351 and thus revealed the transition-state atomic structure in the in-line phosphorylation mechanism.

Then the genuine $E1PCa_2$ analog complex was developed with BeF$_x$, most probably BeF_3^- (Danko et al. 2009). Its formation required $\sim$mM level Mg^{2+}, i.e., the Mg^{2+} binding at the catalytic Mg^{2+} subsite, as required for the phosphorylation with ATP. The $E1Ca_2·BeF_3^-$ complex possessed two Ca^{2+} occluded in the high-affinity binding (transport) sites and was very stable in the presence of $\sim$mM lumenal Ca^{2+} at least for 2 weeks. When the lumenal Ca^{2+} was removed, the complex was autoconverted very slowly in $\sim$10 h at 25°C to $E2·BeF_3^-$ releasing Ca^{2+}, as mimicking the EP isomerization and Ca^{2+}-release $E1PCa_2 \rightarrow E2P + 2Ca^{2+}$. The $E1Ca_2·BeF_3^-$ complex was produced also from $E2·BeF_3^-$ by low-affinity lumenal Ca^{2+} binding at the transport sites, as mimicking the lumenal Ca^{2+}-induced reverse conversion $E2P + 2Ca^{2+} \rightarrow E1PCa_2$. The addition of ADP to the $E1Ca_2·BeF_3^-$ complex destroyed this complex to $E1Ca_2$ state, as mimicking the reverse decomposition of the ADP-sensitive EP ($E1PCa_2$ + ADP $\rightarrow$ $E1Ca_2$ + ATP). All these properties of $E1Ca_2·BeF_3^-$ met requirements as a stable analog of the genuine $E1PCa_2$, the product of the phosphorylation reaction. MgF_4^{2-} was not able to produce a complex with $E1Ca_2$, in agreement with the fact that there is no state with non-covalently bound P_i in the in-line phosphorylation mechanism.

The detailed biochemical analyses of the stable $E1Ca_2·BeF_3^-$ complex revealed characteristic properties of $E1PCa_2$ (hitherto unexplored due to its rapid decay) for further understanding of the successive structural events, the EP formation, its isomerization, and Ca^{2+} release (Danko et al. 2009). Namely, in the change from the transition-state $E1PCa_2·ADP^\ddagger$ to the product-state $E1PCa_2$, the A domain slightly rotates parallel to the membrane plane and thereby comes close to the P domain (but not yet completely as in the Ca^{2+}-released E2P). Upon this motion, the transmembrane helices are also rearranged although keeping the Ca^{2+}-occluded state. The structure of $E1PCa_2$ thus produced is now ready for the subsequent large structural changes, i.e., the large rotation of the

A domain and the resulting tight association with the P domain to cause the loss of ADP sensitivity and Ca^{2+} deocclusion and release into the lumen ($E1PCa_2 \rightarrow E2PCa_2 \rightarrow E2P + 2Ca^{2+}$).

E2PCa$_2$ Transient State and Its Analog E2Ca$_2$·BeF$_3^-$

Finally, the transient $E2PCa_2$ state in the EP processing and Ca^{2+}-release $E1PCa_2 \rightarrow E2PCa_2 \rightarrow E2P + 2Ca^{2+}$ was successfully trapped, and its analog was developed with BeF_3^-. This transient state has been postulated but never been identified because of the very rapid Ca^{2+} release after the EP isomerization ($E1PCa_2 \rightarrow E2PCa_2$).

However, a mutation study (Daiho et al. 2007) successfully identified a crucial structural element, i.e., the length of the A/M1-linker (Glu40-Ser48, connecting the A domain and M1) that critically functions for these structural events, and thereby the $E2PCa_2$ transient state was successfully trapped by its mutation. When the linker is elongated by insertion of two or more amino acids, the EP isomerization is markedly accelerated and, surprisingly, the subsequent Ca^{2+}-release $E2PCa_2 \rightarrow E2P + 2Ca^{2+}$ is almost completely blocked. Thus the transient-state $E2PCa_2$ was successfully trapped and identified for the first time by the A/M1-linker elongation. When the linker is shortened by deletion of any single residues within the linker, the EP isomerization $E1PCa_2 \rightarrow E2PCa_2$ is almost completely blocked. Thus the proper length of the linker critically functions for the structural changes $E1PCa_2 \rightarrow E2PCa_2 \rightarrow E2P + 2Ca^{2+}$.

The kinetic and structural properties of the $E2PCa_2$ state trapped by the linker elongation were analyzed in detail. In this state, the bound Ca^{2+} ions are occluded in the transport sites, and the A domain has been largely rotated from its position in $E1PCa_2$ ($E1Ca_2$·BeF_3^-) and associated with the P domain at the two positions (at the Val200 loop and at TGES184 loop of the A domain). However, in this state, the hydrophobic interactions among the A and P domains and the top part of M2 at Tyr122, i.e., Tyr122-hydrophobic cluster, are not yet produced properly. Actually, the formation of this hydrophobic cluster has been demonstrated by the substitution mutations of this cluster to be critical for producing the Ca^{2+}-released $E2P$ structure (Yamasaki et al. 2008).

The finding demonstrated that the length of A/M1-linker should be properly long for the $E1PCa_2 \rightarrow E2PCa_2$ isomerization and then appropriately short for the subsequent Ca^{2+}-release $E2PCa_2 \rightarrow E2P + 2Ca^{2+}$. The results dissected and predicted the critical structural changes to occur in these processes with a critical function of the A/M1-linker as follows. In the $E1PCa_2 \rightarrow E2PCa_2$ isomerization, the A domain largely rotates parallel to the membrane plane, thus comes above and docks onto the P domain. Because of the A domain's positioning above the P domain, the A/M1-linker is strained in $E2PCa_2$ state, and the strain thus imposed will cause structural change for the Ca^{2+} release from $E2PCa_2$, i.e., inclination of the A and P domains and the connected helices (M2 and M4/M5), which results in rearrangements of the transmembrane helices so as to disrupt Ca^{2+} sites and deocclude Ca^{2+}, opening the gate, thus releasing Ca^{2+} into the lumen. Upon these motions, the A and P domains and top part of M2 associate and produce the tight hydrophobic interactions, Tyr122-hydrophobic cluster, which stabilizes the inclined and Ca^{2+}-released structure of $E2P$. Also upon these changes, the $E2P$ catalytic site for the acylphosphate hydrolysis is produced. Thus the $E2P$ hydrolysis takes place after the Ca^{2+} release, accomplishing the Ca^{2+} transport and energy coupling.

Then with the elongated A/M1-linker mutant, the analog of $E2PCa_2$, $E2Ca_2$·BeF_3^-, was successfully developed by incubating the $E1Ca_2$ state with beryllium fluoride. The complex is remarkably stable, not decomposing at least for 2 weeks. The complex $E2$·BeF_3^- thus produced possesses all the characteristic properties revealed with the trapped $E2PCa_2$. The complex was produced also from $E2$·BeF_3^- by the binding of lumenal Ca^{2+} (at $\sim$ mM level) to the lumenally opened low-affinity transport sites in $E2$·BeF_3^-. Thus the $E2Ca_2$·BeF_3^- formation mimicked the $E2PCa_2$ formation in both the forward and reverse reactions, i.e., from $E1Ca_2$ by ATP via the $E1PCa_2$ isomerization and from $E2P$ by the lumenal Ca^{2+} binding to the lumenally opened transport sites. Importantly, AlF_x and MgF_x are not able to produce this $E2PCa_2$ analogous complex either from $E1Ca_2$ or from $E2$·AlF_4^- and $E2$·MgF_4^{2-}. BeF_x is thus unique in this regard. The observation agrees with the fact that the transient $E2PCa_2$ state possesses the covalent Asp351-phosphate bond, i.e., the $E2PCa_2$ state is produced only from the states with the covalent acylphosphate bond, $E1PCa_2$ ($E1Ca_2$·BeF_3^-) and $E2P$

($E2 \cdot BeF_3^-$), but not from the transition states (mimicked by $E1Ca_2 \cdot AlF_x$ and $E2 \cdot AlF_x$) and the non-covalently bound P_i state (mimicked by $E2 \cdot MgF_4^{2-}$). The finding shows that the large structural changes for the EP isomerization and Ca^{2+} deocclusion/release in the forward and reverse reactions are strictly coupled with the formation of the particular configuration of the acylphosphate covalent bond within the catalytic site. Namely, upon formation of the covalent Asp351-phosphate bond, the Ca^{2+}-ATPase structure becomes ready for its structural isomerization and Ca^{2+}-release processes, but the transition-state structures are not yet prepared.

Concluding Remarks

As described above, the use of beryllium fluoride as the phosphate analog provided essential knowledge for understanding the structural mechanism of energy coupling in the Ca^{2+}-ATPase, the representative member of P-type ion transporting ATPase family. Because of the chemical nature of the beryllium fluoride as well as aluminum fluoride and magnesium fluoride and because of the remarkable stability of the enzymes complexed with these metal fluoride compounds, detailed biochemical and structural studies of many ATPase systems are feasible and being performed.

Cross-References

▶ Calcium ATPase

References

Daiho T, Yamasaki K, Danko S, Suzuki H (2007) Critical role of Glu^{40}-Ser^{48} loop linking actuator domain and 1st transmembrane helix of Ca^{2+}-ATPase in Ca^{2+} deocclusion and release from ADP-insensitive phosphoenzyme. J Biol Chem 282:34429–34447

Daiho T, Danko S, Yamasaki K, Suzuki H (2010) Stable structural analog of Ca^{2+}-ATPase ADP-insensitive phosphoenzyme with occluded Ca^{2+} formed by elongation of A-domain/M1′-linker and beryllium fluoride binding. J Biol Chem 285:24538–24547

Danko S, Yamasaki K, Daiho T, Suzuki H (2004) Distinct natures of beryllium fluoride-bound, aluminum fluoride-bound, and magnesium fluoride-bound stable analogues of an ADP-insensitive phosphoenzyme intermediate of sarcoplasmic reticulum Ca^{2+}-ATPase CHANGES IN CATALYTIC AND TRANSPORT SITES DURING PHOSPHOENZYME HYDROLYSIS. J Biol Chem 279:14991–14998

Danko S, Daiho T, Yamasaki K, Liu X, Suzuki H (2009) Formation of stable structural analog of ADP-sensitive phosphoenzyme of Ca^{2+}-ATPase with occluded Ca^{2+} by beryllium fluoride STRUCTURAL CHANGES DURING PHOSPHORYLATION AND ISOMERIZATION. J Biol Chem 284:22722–22735

Møller JV, Olesen C, Winther AM, Nissen P (2010) The sarcoplasmic Ca^{2+}-ATPase: design of a perfect chemi-osmotic pump. Q Rev Biophys 43:501–56

Olesen C, Picard M, Winther AM, Gyrup C, Morth JP, Oxvig C, Møller JV, Nissen P (2007) The structural basis of calcium transport by the calcium pump. Nature 450:1036–42

Toyoshima C (2008) Structural aspects of ion pumping by Ca^{2+}-ATPase of sarcoplasmic reticulum. Arch Biochem Biophys 476:3–11

Toyoshima C (2009) How Ca^{2+}-ATPase pumps ions across the sarcoplasmic reticulum membrane. Biochim Biophys Acta 1793:941–946

Toyoshima C, Norimatsu Y, Iwasawa S, Tsuda T, Ogawa H (2007) How processing of aspartylphosphate is coupled to lumenal gating of the ion pathway in the calcium pump. Proc Natl Acad Sci USA 104:19831–19836

Yamasaki K, Wang G, Daiho T, Danko S, Suzuki H (2008) Roles of Tyr^{122}-hydrophobic cluster and K$^+$ binding in Ca^{2+}-releasing process of ADP-insensitive phosphoenzyme of sarcoplasmic reticulum Ca^{2+}-ATPase. J Biol Chem 283:29144–29155

Calcium ATPases

Chikashi Toyoshima
Institute of Molecular and Cellular Biosciences,
The University of Tokyo, Tokyo, Japan

Synonyms

Proteins: sarco(endo)plasmic reticulum Ca^{2+}-ATPase, SERCA; Plasma membrane Ca^{2+}-ATPase (PMCA).
DNA: ATP2A1-3 (human genes for SERCA1-3); ATP2B1-4 (PMCA1-4)
Biological functions: Ca^{2+}-pump; Ca^{2+}-translocating ATPase

Definition

Enzyme, EC 3.6.3.8; an ATP-powered ion pump that transports Ca^{2+} across the membrane against a large (up to 15,000-fold) concentration gradient. A representative member of the P-type ATPase family, which is in turn a member of the haloacid dehalogenase superfamily (Burroughs et al. 2006). EC 3.6.3.8 includes sarco(endo)plasmic reticulum calcium ATPase, commonly abbreviated as SERCA, and plasma membrane Ca^{2+}-ATPase, abbreviated as PMCA. As SERCA transports 2 Ca^{2+} ions per ATP hydrolyzed from the cytoplasm into the lumen of sarcoplasmic reticulum and 2 or 3 H^+ in the opposite direction, it may be termed "Ca^{2+}, H^+-ATPase," similarly to Na^+, K^+-ATPase. PMCA transports one Ca^{2+} per ATP.

Biological Function and Amino Acid Sequence Information

In this entry, description is limited almost exclusively to SERCA pumps, as they are much better studied than PMCA. SERCA pumps exist from some bacteria (e.g., *Bacillus subtilis*), fungi (e.g., *Neurospora crassa*), plants to human, although their functional meaning is unclear in some bacterial species (Vangheluwe et al. 2009). In human, three genes (ATP2A1-3) generate multiple SERCA isoforms (SERCA1a,b, SERCA2a–c, and SERCA3a–f) by alternative splicing. SERCA1a is the adult form of the fast twitch skeletal muscle sarcoplasmic reticulum (SR) Ca^{2+}-ATPase and consists of 994 amino acid residues; SERCA1b is the neonatal form having additional seven residues at the C-terminus. In muscle contraction, Ca^{2+} is released from SR into muscle cells through Ca^{2+} release channels and pumped back into SR by SERCA1 to relax muscles cells. The concentration of Ca^{2+} within the muscle cell is maintained at less than micrometer in the relaxed state. SERCA2 is by far the most widespread of all SERCA isoforms and phylogenetically the oldest. SERCA2 has three splice variants and SERCA2a (997 amino acid residues) is the main isoform in cardiac muscle and slow twitch skeletal muscle. SERCA2b is considered to be the housekeeping ATPase and has an extra transmembrane segment at the C-terminus (Brini and Carafoli 2009).

Signature sequence of the P-type ATPase family is written as D-K-T-G-T-[LIVM]-[TI]. The starting Asp is the phosphorylated residue, corresponding to Asp351 in SERCA1a. This signature sequence constitutes motif PS00154 in the PROSITE database and makes P-type ATPase a member of haloacid dehalogenase superfamily (Burroughs et al. 2006). The atomic structure around the phosphorylation site is similar to that of bacterial two-component regulators, which have different supersecondary structures. A database on P-type ATPases is available at http://traplabs.dk/patbase/. Various types of SERCA are reviewed by (Vangheluwe et al. 2009). A complete compilation of all mutagenesis is provided by Andersen (http://sercamutation.au.dk/).

Reaction Cycle

The mechanism of active ion transport by Ca^{2+}-ATPase is conventionally described by E1/E2 theory (Fig. 1) (Inesi 1985). In the E1 state, the Ca^{2+}-binding sites have high affinity and face the cytoplasm in the E1 state, and have low affinity and face the lumenal (or extracellular) side in E2. Hence, Ca^{2+}-ATPase is said to be a member of E1/E2-type ATPase that includes Na^+, K^+-ATPase and gastric H^+, K^+-ATPase among others. This type of ATPases is also called P-type ATPase, because the enzyme is auto-phosphorylated (denoted by "P") at invariant Asp during the reaction cycle, distinct from F_1F_0-type ATPase, for example.

As SERCA1a transfers two Ca^{2+} ions from the cytoplasm and 2 (or 3) H^+ into the cytoplasm, the reaction cycle is electrogenic (Inesi 1985). However, since SR membrane is leaky to all monovalent ions, pumped protons are dissipated away (presumably through H^+ channel). Therefore, SR membrane has no significant membrane potential, and Ca^{2+}-ATPase is in reality a Ca^{2+} pump. SERCA1a can also transport Mn^{2+} but at a slower rate with unknown stoichiometry. H^+-countertransport (and dissipation) is necessary partly for stabilizing empty Ca^{2+}-binding sites in the E2 states and for changing the affinity of the Ca^{2+}-binding sites.

The SERCA pump itself is not regulated by phosphorylation. As long as Ca^{2+} and ATP are present, the pump runs continuously. In cardiac muscle, however, there are regulatory proteins phospholamban (52 amino acid residues) and sarcolipin (31 residues). They are related short transmembrane peptides and regulated by phosphorylation (MacLennan and Kranias 2003). PMCA is regulated by calmodulin and protein kinases. Their binding sites are located in

Calcium ATPases, Fig. 1 A simplified reaction scheme according to the E1/E2 model. Only forward direction is shown

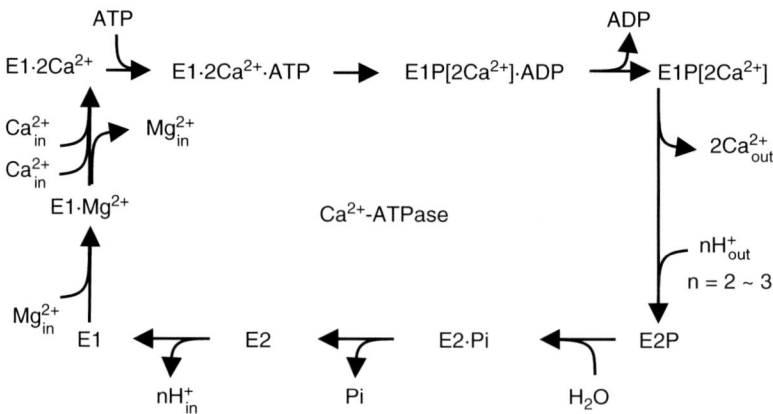

the cytoplasmic extension after the last transmembrane helix (Brini and Carafoli 2009).

Energetics

The free energy required for transferring 2 mol of Ca^{2+} against a 15,000-fold concentration gradient requires 12.6 kcal, even when no membrane potential exists. This number clearly exceeds the standard free energy of ATP (-7.3 kcal/mol). The amount of free energy required is obtained, in muscle cells, by reducing the concentration of ADP by converting it to ATP with creatine kinase. In fact, the whole reaction cycle can be reversed and ATP can be synthesized under certain conditions (Inesi 1985). In particular, the ATPase in the E1P state (but not in the E2P state) readily synthesizes ATP, as the standard free energy liberated by hydrolysis of aspartylphosphate (11.7 kcal/mol) is substantially higher than that of ATP. Therefore, E1P and E2P are also called ADP-sensitive and -insensitive phosphorylated forms, respectively. For PMCA, because of the membrane potential, transfer of only one Ca^{2+} per ATP hydrolyzed is possible.

Inhibitors

There are three well-known inhibitors of high affinity, namely, thapsigargin (TG), a sesquiterpene lactone from a plant, cyclopiazonic acid, an indole tetramic acid fungal toxin, and 2,5-di-*tert*-butyl-1,4-dihyhydroxybenzene (BHQ). Of these three, TG has the highest affinity with a subnanomolar dissociation constant (K_d). All of them fix the enzyme in the E2 state. No strong inhibitor is known to stabilize the ATPase in the E1 states. The binding sites of BHQ and cyclopiazonic acid are located on the cytoplasmic surface of the transmembrane region and partially overlap; TG binds between the M3 and M7 transmembrane helices.

Protein Production and Purification

The most common source of SERCA is rabbit hind leg white muscle, in which only SERCA1a is present. Approximately 500 mg of SR membrane ("light" SR fraction devoid of T-tubule containing Ca^{2+}-release channels) can be obtained from 150 g of muscle by differential centrifugation. This preparation contains usually 30 mg/ml proteins, ~60% of which is SERCA1a, and can be stored for a few months at -80°C with 0.3 M sucrose. SERCA1a can be affinity purified with Reactive Red 120 after solubilization with 2% octaethyleneglycol dodecylether ($C_{12}E_8$) and hydrolyzes ~20 ATP molecules/sec at 25°C (in 123 mM KCl, 6.15 mM $MgCl_2$, 0.12 mM $CaCl_2$, 2 mM ATP, 0.1% $C_{12}E_8$, 61.5 mM MOPS, pH 7.0).

Recombinant SERCA can be expressed in HEK293, COS1 and COS7. Large-scale productions have been achieved using adenovirus vectors, yielding the expression level of SERCA1a as high as 20% of total microsomal proteins. For a large-scale production, an yeast (*Saccharomyces cerevisiae*) system has also been used and yielded crystals of mutants.

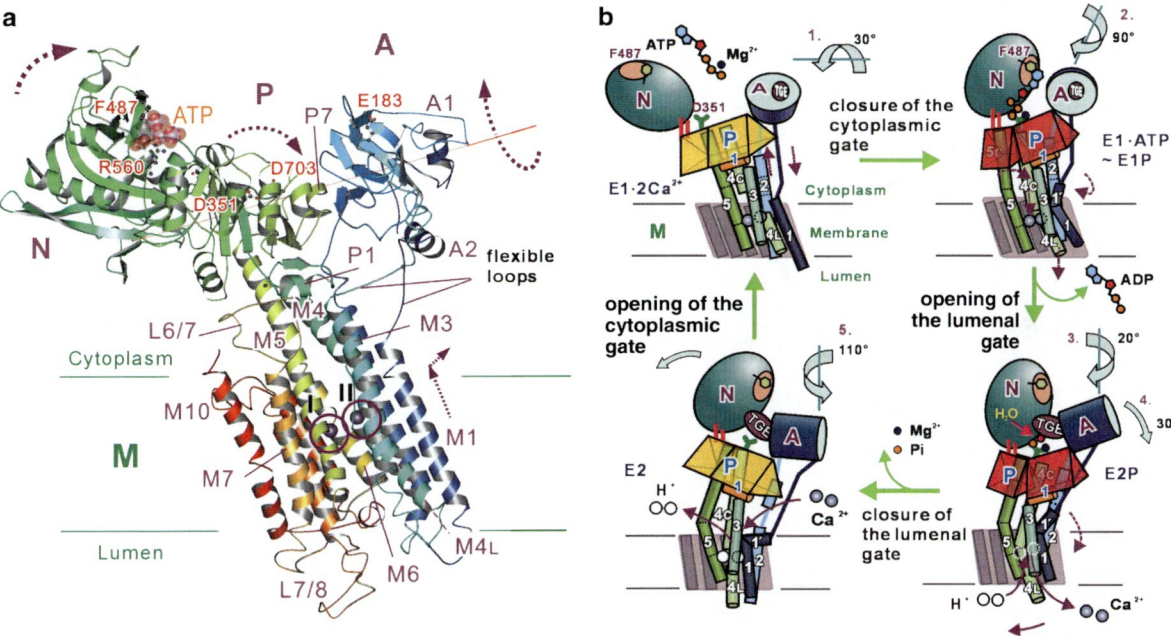

Calcium ATPases, Fig. 2 Architecture of Ca^{2+}-ATPase and its ion pumping mechanism. (**a**) A ribbon representation of Ca^{2+}-ATPase in the $E1 \cdot 2Ca^{2+}$ state, viewed parallel to the membrane plane. Colors change gradually from the amino terminus (*blue*) to the carboxy terminus (*red*). Two *purple spheres* (numbered and circled) represent bound Ca^{2+}. Three cytoplasmic domains (A, N, and P), some of the α-helices in the A-domain (A1–A2), P-domain (P1 and P7), and those in the transmembrane domain (M1–M10) are indicated. Docked ATP is shown in transparent *space fill*. Several key residues – E183 (A, activation of attacking water), F487 and R560 (N, ATP binding), D351 (P, phosphorylation site), and D703 (magnesium binding) are shown in *ball-and-stick*. Axis of rotation (or tilt) of the A-domain is indicated with *thin orange line*. PDB accession code is 1SU4 ($E1 \cdot 2Ca^{2+}$). (**b**) A cartoon illustrating the structural changes of Ca^{2+}-ATPase during the reaction cycle, based on the crystal structures in nine different states

Architecture of the Molecule

Ca^{2+}-ATPase is a tall (~150 Å high) integral membrane protein, comprising three large cytoplasmic domains designated as A (actuator), N (nucleotide binding), and P (phosphorylation), 10 (M1–M10) transmembrane α-helices and short (except for L7/8 that connects M7 and M8) lumenal loops (Fig. 2a) (Toyoshima 2008). The M4 and M5 helices are very long (~60 Å), extending from the lumenal surface to the top of the P-domain. The cytoplasmic extensions of the transmembrane helices may appear to form a "stalk (S)" segment. The distance between the transmembrane Ca^{2+}-binding sites and the phosphorylation site is >50 Å. Since 12 out of 13 Trp residues reside in the transmembrane region, intrinsic fluorescence from Trp is widely used to monitor the movements of transmembrane helices. Limited proleolysis with proteinase K and trypsin provides useful information on the arrangement of cytoplasmic domains.

Structure of the A-domain: The A-domain is the smallest (~160 residues) of the three cytoplasmic domains and acts as the "actuator" of the gates that control the binding and release of Ca^{2+}. It is connected to the M1 and M3 helices with flexible links and more directly to the M2 helix with a swivel (Fig. 2). The A-domain contains the [183]TGES loop, one of the signature sequences of the P-type ATPase, which plays a critical role in activating the water molecule that attacks the aspartylphosphate (Fig. 2b).

Structure of the P-domain: The P-domain is the catalytic core domain and contains the phosphorylated residue Asp351, Mg^{2+}-binding residue Asp703 (Fig. 3), and other critical residues that classify Ca^{2+}-ATPase as a member of the haloacid dehalogenase superfamily (Burroughs et al. 2006). The P-domain has a Rossmann fold, commonly found in nucleotide-binding proteins, consisting of a parallel β-sheet and associated α-helices. A key feature is that the cytoplasmic extension of M5 is integrated into the P-domain as one of the associated

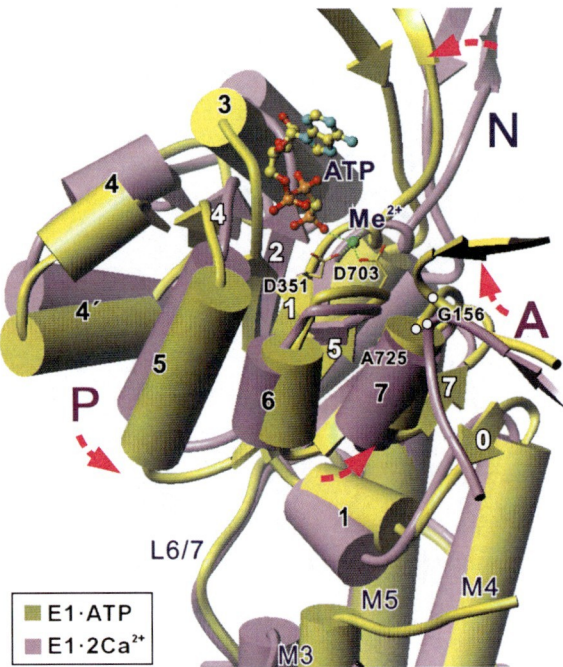

Calcium ATPases, Fig. 3 Structural changes caused by the binding of ATP and Mg^{2+} to the P-domain. The bound metal (Me^{2+}, small *green sphere*) is most likely Ca^{2+} rather than Mg^{2+} in the crystal structure. Note that the P-domain is bent (*arrows in red broken lines*) by coordination of the metal by the γ-phosphate, D351, and D703 (shown in stick model) and that the A-domain tilts because of the inclination of the P7 helix

accelerates the hydrolysis of aspartylphosphate and stabilizes the E2 state.

Structure of the N-domain: The N-domain is the largest (~240 residues) of the three cytoplasmic domains and the least conserved domain in P-type ATPases. The N-domain is a long insertion between the two parts that constitute the P-domain and connected to them with two strands having a β-sheet like hydrogen bonding (Fig. 3). This is a primary hinge and allows a 60° change in inclination to allow phosphoryl transfer from ATP. The N-domain contains the binding site for ATP (Fig. 2). The side chain of Phe487 stacks with the adenine ring and Arg560 fixes the β-phosphate. Lys515, a critical residue located at the depth of the binding cavity, can be labeled specifically at alkaline pH with FITC, which is used for many spectroscopic studies.

Organization of the transmembrane domain: SERCA1a has 10 (M1–M10) transmembrane α-helices (Fig. 2), two of which (M4 and M6) are partly unwound throughout the reaction cycle. M1–M2 form a rigid V-shaped structure and follow the movements of the A-domain, and thereby play many roles in opening and closing the transmembrane gates. They show very large movements in both within and perpendicular (up to two turns of α-helix) to the membrane. M1, connected to the N-terminal half of the A-domain, consists of two parts and form a kinked helix to accommodate different positions and orientations of the A-domain. The N-terminal part (M1′) is amphipathic and lie on the membrane surface when M1 is kinked. This kink is important in making a pivot for the movements of M3 and M4. The length of the linker that connects the A-domain and M1′ is critical. M2 is a long contiguous helix from E1 to E2P, but becomes disrupted in transition to E2·Pi. M3–M4 also form a helix pair and show both lateral and vertical movements. M5 is the spine of the molecule but can bend at two Glys. The lumenal part (M5L) stays unchanged. M6 and M7 are located far apart and connected by a loop of ~20 residues (L6/7; Fig. 2a). This loop, running along the bottom of the P-domain and connected to M5 with critical hydrogen bonds, is rigid and serves as a limiter for movements of M5 and the socket of M3. M7–M10 in SERCA1a appear to be an anchor to the membrane and do not undergo a large rearrangement or conformational changes. They presumably have specialized functions in each subfamily

helices, and the cytoplasmic end of M4 is clamped to M5 by a short antiparallel β-strand; the M3 helix is connected to the P-domain by critical hydrogen bonds and also moved by the P1 helix that runs along the bottom of the P-domain (Fig. 3). Therefore, structural events that occur in the P-domain are readily transmitted to the transmembrane domain. The central β-sheet consists of two parts (strands 1–4 and 5–7), which allow bending of the P-domain when phosphate and Mg^{2+} bind (arrows in Fig. 3). In this sense, the P-domain has an integrated hinge that alters the orientation of the P-domain with respect to the M5 helix and also the inclination of the N-domain with respect to the P-domain depending on the phosphorylation (Fig. 3). The Mg^{2+} coordinated by Asp351 and Asp703 is essential for phosphoryl transfer. However, almost any divalent cations can occupy this site and interferes with the reaction. This is the reason why high concentration of Ca^{2+} is inhibitory. There is a K^+ (or Na^+)-binding site at the bottom of the P-domain. This monovalent ion

Calcium ATPases, Fig. 4 Structure of the transmembrane Ca^{2+}-binding sites. (**a**) atomic model in the $E1 \cdot 2Ca^{2+}$ state. (**b**) Cartoons illustrating the binding sites in $E1 \cdot 2Ca^{2+}$ and E2 viewed approximately normal to the membrane. Two Ca^{2+} appear as *cyan spheres* and water molecules as *red spheres*. Note that site II Ca^{2+} is exchangeable by conformation change of the E309 side chain in $E1 \cdot 2Ca^{2+}$ (*double-headed arrow* in **a**). The *arrows* in (**b**) indicate the movements of the helices in the transition $E2 \rightarrow E1 \cdot 2Ca^{2+}$. *Red circles* represent protonation of the oxygen atoms. *Dotted lines* indicate hydrogen bonds (*pink*) and coordination of Ca^{2+} (*light green*)

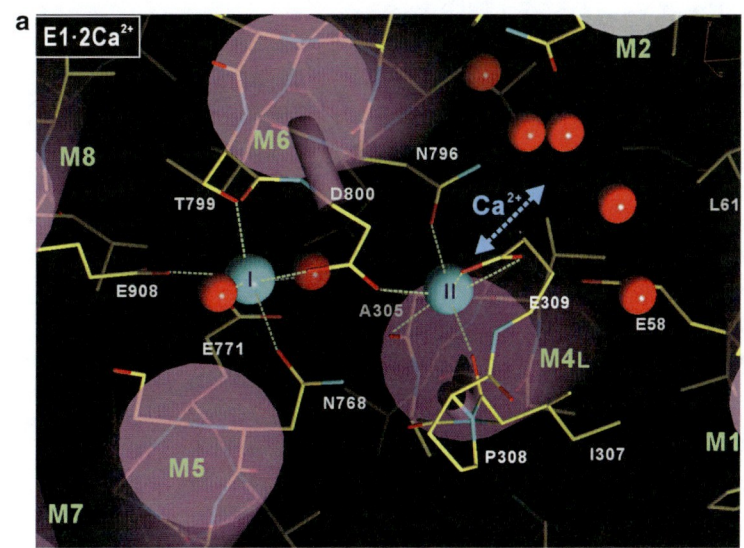

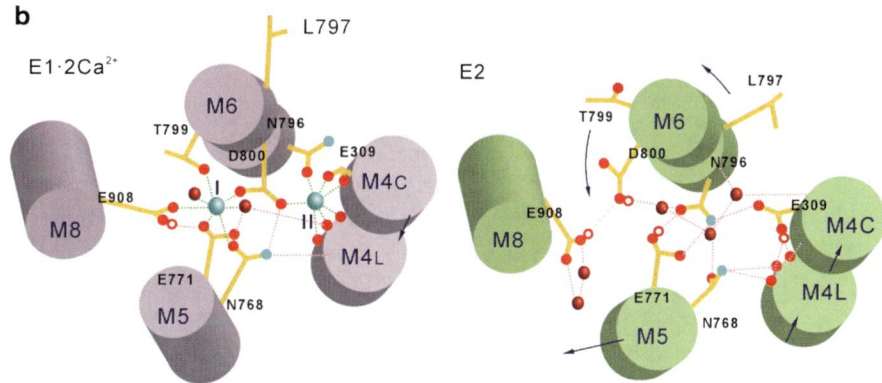

and are in fact lacking in type I P-type ATPases (heavy metal pumps). The amino acid sequence is well conserved for M4–M6 but not for M8 even within the members of closely related P-type ATPases, such as Na^+, K^+-, and H^+, K^+-ATPases.

Details of the Ca^{2+}-Binding Sites

The two Ca^{2+}-binding sites (I and II) are located side by side near the cytoplasmic surface of the lipid bilayer (Fig. 4), with the site II ~3 Å closer to the surface (Fig. 2a). Yet, the binding of two Ca^{2+} is sequential and cooperative. The binding sites have K_ds of 1.7 and 0.18 μM and the Hill coefficient of 1.7–1.8 (Inesi 1985). They show a strong pH dependence ($pK_a = 6.4$). The Glu309Gln mutant shows K_d of 0.23 μM (Hill coefficient: 0.94) and pK_a of 7.8.

Site I, the binding site for the first Ca^{2+}, is located at the center of the transmembrane domain in a space surrounded by the M5, M6, and M8 helices, and formed by the side chain oxygen atoms and two water molecules (Fig. 4). M8 is located rather distally and the contribution of Glu908 is not essential in that Gln can substitute Glu908 to a large extent. Electrostatic calculation shows that the carboxyl group of Glu908 is protonated throughout the reaction cycle.

Site II is formed "on" the end of the lumenal half of the M4 helix (M4L) with the contribution of four side chain and three main chain oxygen atoms (Fig. 4). For providing main chain oxygen atoms, the M4 helix is partly unwound (between Ile307-Gly310) and Glu309, the gating residue, caps the bound Ca^{2+}. This arrangement of oxygen atoms is reminiscent of the EF-hand motif (see "EF-Hand Proteins").

Synopsis of Ion Pumping

Since the first crystal structure of SERCA1a in E1·2Ca^{2+} published in 2000, more than 20 crystal structures for nine different states that roughly cover the entire reaction cycle have been deposited in the Protein Data Bank (Toyoshima 2008). Rapid advancement was made by the use of metal fluorides adducts (BeF$_3^-$ for the ground state, AlF$_4^-$ for the transition state, and MgF$_4^{2-}$ for the product state) as stable phosphate analogs. Here presented is a brief scenario of ion pumping by SERCA1a. More detailed account is found in (Toyoshima 2008). Movies illustrating these movements and aligned coordinates of the crystal structures can be downloaded from the author's Web site (http://www.iam.u-tokyo.ac.jp/StrBiol/).

E1 → E1·2Ca^{2+}: binding of two Ca^{2+}. In the absence of Ca^{2+} and at pH 7, most of SERCA1a is in the E1 state, in which Ca^{2+}-binding sites have high affinity and the cytoplasmic headpiece is open. The first Ca^{2+} will enter the binding cavity through improperly formed site II (otherwise, it will be trapped there) and binds to site I. This will rotate M6 and position Asp800 properly. The M5 helix will straighten (Fig. 1b) and alter the arrangement of oxygen atoms in site II to form a higher affinity binding site. Finally the carboxyl of Glu309 side chain will cap site II Ca^{2+}, and the binding signal is transmitted to the phosphorylation site some 50 Å away. Because there is enough space around Glu309, the cytoplasmic gate remains unlocked, and site II Ca^{2+} is exchangeable with those in the cytoplasm (Fig. 4a).

E1·2Ca^{2+} → E1P: occlusion of bound Ca^{2+}. ATP cross-links the P- and N-domains, so that the γ-phosphate of ATP and a Mg^{2+} bind to the P-domain to bend it in two directions (Fig. 3). This bending of the P-domain tilts the A-domain sitting on the P7 helix by ~30° (1 in Fig. 2b), and thereby places strain on the link between the A-domain and the M3 helix. This strain appears to be the driving force for the A-domain rotation in the next step. At the same time, the M1 helix is pulled up (small arrows in broken lines; Fig. 2a) to fix the side chain conformation of Glu309 by occupying the space around it. Thus, the cytoplasmic gate is locked and two Ca^{2+} are occluded in the transmembrane binding sites.

E1P → E2P: release of Ca^{2+} into the lumen of SR. Phosphoryl transfer to Asp351 allows the dissociation of ADP, which triggers the opening of the N- and P-domain interface. The A-domain rotates 90° (2 in Fig. 2b) and brings the ^{181}TGES loop of the A-domain deep into the gap between the N- and P-domains above the aspartylphosphate so that it occupies the space where ADP was in E1P·ADP to prevent rebinding of ADP. At the same time, it shields the aspartylphosphate from bulk water. The A-domain rotation causes a 30° change in inclination of the P-domain toward M1, which in turn causes a drastic rearrangement of the transmembrane helices M1–M6, including a large (~5.5 Å) downward movement of M4, sharp bending of M5 toward M1 (Fig. 2b), and rotation of M6 (Fig. 4b), which destroy the Ca^{2+}-binding sites. The V-shaped structure formed by the M1 and M2 helices pushes against M4L, opening the lumenal gate and releasing the bound Ca^{2+} into the lumen (Fig. 2b). This will allow protons and water molecules to enter and stabilize the empty Ca^{2+}-binding sites.

E2P → E2: hydrolysis of aspartylphosphate and closing of the lumenal gate. The A-domain rotates further by 25° around a different axis (3 in Fig. 2b). This rotation places further strain on the M2 helix, which, as a result, partly unwinds and releases the V-shaped structure formed by the M1 and M2 helices so that it takes a ~5 Å lower position toward the lumenal side. This movement imposes more upright position on M4L and closes the lumenal gate. At the same time, the rotation of the A-domain introduces one water molecule in the phosphorylation site. Glu183 in the TGES loop activates the water molecule to attack aspartylphosphate. Then, the resultant inorganic phosphate and Mg^{2+} are released and the P-domain becomes relaxed (4 in Fig. 2b). This in turn releases the M1 and M2 helices to lock the lumenal gate and places the ATPase into the E2 state.

E2 → E1: release of protons and change in affinity of the Ca^{2+}-binding sites. The E2 state, or more rigorously, E2·nH$^+$ (n = 2 or 3) state is unstable at pH 7, and protons are released spontaneously into the cytoplasm to confer high affinity on the Ca^{2+}-binding sites. ATP can bind to the ATPase in the E2 state and accelerate this transition presumably by opening the cytoplasmic headpiece. One Mg^{2+} may bind to the Ca^{2+}-binding sites with mM K_d and accelerate the reaction cycle presumably by achieving most of structural changes that occur on the binding of first Ca^{2+}. Thus, the ground state under physiological conditions is E1·Mg^{2+}. There is confusion on this point in the literature.

Cross-References

▶ Calcium ATPase
▶ Calcium in Health and Disease
▶ Calcium in Heart Function and Diseases
▶ Calcium Ion Selectivity in Biological Systems
▶ Calcium, Physical and Chemical Properties
▶ Calcium-Binding Protein Site Types
▶ Calcium-Binding Proteins
▶ EF-Hand Proteins
▶ Magnesium in Biological Systems
▶ Na(+)/K(+)-Exchanging ATPase

References

Brini M, Carafoli E (2009) Calcium pumps in health and disease. Physiol Rev 89:1341–1378

Burroughs AM, Allen KN, Dunaway-Mariano D, Aravind L (2006) Evolutionary genomics of the HAD superfamily: understanding the structural adaptations and catalytic diversity in a superfamily of phosphoesterases and allied enzymes. J Mol Biol 361:1003–1034

Inesi G (1985) Mechanism of calcium transport. Annu Rev Physiol 47:573–601

MacLennan DH, Kranias EG (2003) Phospholamban: a crucial regulator of cardiac contractility. Nat Rev Mol Cell Biol 4:566–577

Toyoshima C (2008) Structural aspects of ion pumping by Ca^{2+}-ATPase of sarcoplasmic reticulum. Arch Biochem Biophys 476:3–11

Vangheluwe P, Sepulveda MR, Missiaen L, Raeymaekers L, Wuytack F, Vanoevelen J (2009) Intracellular Ca^{2+}- and Mn^{2+}-transport ATPases. Chem Rev 109:4733–4759

Calcium Body Balance

▶ Calcium in Health and Disease

Calcium Channel Protein

▶ Calcium-Binding Proteins, Overview

Calcium Channels

▶ Calcium in Nervous System

Calcium Concentration [Ca^{2+}]

▶ Calcium Sparklets and Waves

Calcium Cytotoxicity

▶ Calcium in Health and Disease

Calcium Diffusion

▶ Calcium, Local and Global Cell Messenger

Calcium Homeostasis: Calcium Metabolism

▶ Calcium and Viruses

Calcium in Biological Systems

Robert H. Kretsinger
Department of Biology, University of Virginia, Charlottesville, VA, USA

Synonyms

Calcium as a secondary messenger; Calmodulin; Cytosolic signaling; EF-hand proteins; Myosin light chains; Troponin C

Definition

The variation in the concentration of free Ca^{2+} ion within the cytosol over time is a form of information. This information can be transduced into a change in protein conformation and/or enzyme activity by interaction with a calcium-modulated protein(s).

Overview

Calcium is unique in biological systems. Ca^{2+} is the only metal cation demonstrated to function as a secondary messenger in the cytosol of eukaryotes. The information in this pulse of Ca^{2+} ions (Berridge 2006) is transduced into a change of conformation of a calcium-modulated protein(s). Many of these calcium-modulated proteins contain two to twelve tandem EF-hand domains. Others contain one or two C2 domains or four annexin domains. Numerous calcium channels and calcium pumps in the membranes of the cell and its various organelles, especially the endoplasmic reticulum, are involved in tuning the calcium concentration in the cytosol (Permyakov and Kretsinger 2010).

This standard model requires several refinements. Some EF-hand proteins are involved in temporal buffering of waves or pulses of calcium in the cytosol; others appear to be involved in facilitated diffusion. Many extracellular proteins bind calcium; it is essential for their stabilities and functions. The concentration of free Ca^{2+} ions is usually held constant in the extracellular fluid, or plasma, of multicellular organisms. However, there is some evidence that $[Ca^{2+}]_{out}$ may vary in a controlled manner over restricted volumes and/or brief times. Calcium may be involved in extracellular signaling (Hofer 2005); these target proteins are considered to be calcium modulated. Some prokaryotes have calcium-binding proteins; however, there is no evidence of calcium functioning as a second messenger (Permyakov and Kretsinger 2009).

Over 30 distinct biominerals have been identified; most contain calcium (Lowenstam and Weiner 1989). These calcium carbonates and phosphates contain proteins that are inferred to be essential in determining their rates of formation and domain structures as well as the mechanical properties of shells, teeth, and bones. Many diseases are associated with mutations and/or malfunctions of these diverse calcium-binding proteins including pumps and channels.

Calcium Coordination

Many of these unique properties of calcium in biology can be related to its basic chemistry. Calcium (Latin calc, "lime") is the fifth most abundant element and the third most abundant metal in the earth's crust. The stable ion Ca^{2+} has the electron configuration of argon – $1s^2\ 2s^2\ 2p^6\ 3s^2\ 3p^6$ – plus two additional protons; it is less reactive than the alkaline metals. Calcium has four stable isotopes (^{40}Ca, ^{42}Ca, ^{43}Ca, and ^{44}Ca), plus two more isotopes (^{46}Ca and ^{48}Ca) that have such long half-lives that for all practical purposes, they can be considered stable. It also has a radioactive isotope ^{41}Ca, which has a half-life of 103,000 years. Ninety-seven percent of naturally occurring calcium is ^{40}Ca.

Seawater contains approximately 400 ppm calcium, nearly all free Ca^{2+} ion (~10 mM). Calcium makes up 3.5% of the earth's crust and occurs almost solely in inorganic compounds. It is obtained from carbonate minerals like chalk, limestone, dolomite, and marble. Six metals are present in the human body in high concentrations: calcium (1,700 g per 70 kg of body mass), potassium (250 g), sodium (70 g), magnesium (42 g), iron (5 g), and zinc (3 g). Other metals are present less than 1 g per 70 kg of body mass (Permyakov and Kretsinger 2010).

Fifteen metals are essential for at least some organisms. They are considered in two groups: non-transition (Na, K, Mg, Ca, Zn, Cd) and transition elements (V, Cr, Mn, Fe, Co, Ni, Cu, Mo, W). Atoms of alkali elements, sodium and potassium, possess one s-electron in the outer shell besides the electron structure of rare gas atoms; therefore, they are characterized by low ionization potentials (5.138 and 4.339 eV, respectively). Their ions have relatively large radii (1.02 and 1.38 Å). The ionization potential of the calcium atom ($1s^2\ 2s^2\ 2p^6\ 3s^2\ 3p^6\ 4s^2$) for the first electron is 6.1132 eV; the ionization potential for the second electron is 11.871 eV. In normal geological and biological conditions, the Ca^{1+} ion does not exist. The completely filled octets of the Mg^{2+}, Ca^{2+}, and Sr^{2+} ions ($2s^2\ 2p^6$, $3s^2\ 3p^6$, and $4s^2\ 4p^6$, respectively) have no preferences with respect to the direction of bond formation and can be modeled by spheres with increasing radii (0.72, 0.99 and 1.35 Å) and decreasing charge density. The similar sizes of the ions of calcium and of the lanthanides (0.99 and 1.06–0.85 Å) allow the lanthanides to replace calcium in many binding sites; this is useful for spectroscopic studies. The alkali and alkaline earth metals have very weak tendencies to form covalent bonds.

The transition elements involved in biological processes – V, Cr, Mn, Fe, Co, Ni, Cu, Mo, W – have incompletely filled d-orbitals; Cu is actually $4s^1 3d^{10}$. They, including copper, are characterized by variable valences and formation of colored complexes and paramagnetic substances due to unpaired electrons. The transition elements are often involved in redox reactions.

Zinc and cadmium are not transition elements since they have no empty d-orbitals. Zn^{2+} ions are different from the other non-transition metal ions. While the radius of Zn^{2+} ion (0.74 Å) is close to the radius of the Mg^{2+} ion (0.72 Å), its ionization potentials are higher than those of calcium and magnesium. The high values of zinc ionization potentials (9.394 and 17.964 eV) are reflected in its stronger tendency to form covalent bonds. It is bound by many proteins.

Metal cations in aqueous solution are surrounded by water molecules oriented by the electric field of the ion; this creates charge–dipole interactions. The smaller the radius of an ion, the greater the charge density and the stronger is its interaction with the dipole moment of water. Some ions possess a rather rigid and stable first hydration shell that can have tetrahedral (Li^+) or octahedral (Mg^{2+}, Co^{2+}, Ni^{2+}) geometry. The Zn^{2+} ion can have both tetrahedral and octahedral coordination in the first hydration shell. The Mg^{2+} ion, with its +2 charge and small radius, tightly orders six water molecules in an octahedral arrangement in the first hydration shell. The second, and perhaps third, layers of water are also organized by the charge of the ion and contribute to the overall hydration free energy of -455 kcal mol^{-1}. The K^+ ion is larger and has only a +1 charge; as a result, eight or nine water molecules pack around the ion in a less well-ordered manner; the hydration energy is -80 kcal mol^{-1}. Because the Mg^{2+} ion interacts strongly with six water molecules in $[Mg(H_2O)_6]^{2+}$, larger anions do not readily replace the water to give precipitates. The larger Ca^{2+} ion binds more strongly to those larger anions relative to water; hence, these anions displace water from calcium more readily. Large anions, CO_3^{2-} and PO_4^{3-}, precipitate with calcium at lower metal ion concentrations than with magnesium. As a consequence, calcium has a lower concentration in sea water ($\sim$10 mM for [Ca^{2+}] versus $\sim$50 mM for [Mg^{2+}]). One can find deposits, both geochemical and biochemical, of $CaCO_3$, $CaSO_4$, and $Ca_3(PO_4)_2$, but not of the corresponding magnesium salts except mixed in the calcium salts, for example, dolomite ($CaMg(CO_3)_2$). Calcium oxalate is insoluble; its crystals are found in plant tissue, but magnesium oxalate is soluble. Moreover, calcium tends to precipitate many polyanions, such as DNA, RNA, and some acidic proteins (Lowenstam and Weiner 1989; Mann 2001).

Calcium and monovalent metal cations, except Li^+, have a wide variety of geometries of the first hydration shell. The number of water molecules associated with the cation and the distance between them and the central ion vary and increase with increasing ion size. Because the six oxygens of hexa-aquo-magnesium are in optimal van der Waals contact and leave slowly, magnesium association rate constant, $k_{on}(Mg)$, for proteins is relatively low. In contrast, the waters of hepta-aquo-calcium have more lateral mobility; thus, $k_{on}(Ca)$ is much faster; this accounts for most of the difference in affinity of calcium-binding proteins for calcium versus magnesium. The Ca^{2+} ion is hepta-hydrate and has a diffusion coefficient of 1.335 10^{-9} m^2 s^{-1} in water at 25°C and a water exchange rate of 10^8–10^9 s^{-1}.

The favored coordination for the Mg^{2+} ion is sixfold octahedral. Ca^{2+} shows a greater diversity of coordination numbers, with seven- and eightfold coordination the most common in crystal structures of small organic molecules complexed with calcium. Bond distances between the Ca^{2+} ion and its ligands vary more than do those of Mg^{2+}. The radius of the coordination sphere is significantly larger for calcium than for magnesium ions: bond distances to oxygen donor atoms typically range from 2.0 to 2.1 Å for Mg^{2+} and 2.2 to 2.5 Å for Ca^{2+}. Compared to Mg^{2+} ions, Ca^{2+} forms looser complexes of higher and variable coordination number, without directionality, and with more variable bond lengths (Permyakov and Kretsinger 2010).

Calcium-Modulated Proteins

Those calcium-binding proteins in the cytosol or bound to membranes facing the cytosol are inferred to be calcium modulated. That is, when the cell is quiescent, the concentration of the free Ca^{2+} ion is less than 10^{-7} M (pCa > 7) and the calcium-modulated protein is in the apo- or magnesi-form. Following stimulus, the concentration of calcium rises (pCa < 5.5), and the protein binds calcium (Whitiker 2010). The attendant change in structure is involved in the transduction of the information of a pulse or wave of

Ca^{2+} ions to an ultimate target enzyme or structure. Most of these calcium-modulated proteins contain from two to twelve copies of the EF-hand domain (Nakayama et al. 2000). There are other proteins in the cytosol that bind calcium and also appear to be modulated by calcium; these include proteins that contain one or several C2 domains, such as protein kinases C or synaptotagmin, the annexins, as well as calcium pumps and channels.

Many of the functional characteristics of calcium and of calcium-modulated proteins can be rationalized from the geometry of calcium coordination. In proteins, the Ca^{2+} ion (atomic radius 0.99 Å) is usually coordinated by seven oxygen atoms in an approximate pentagonal bipyramidal conformation at average Ca-O distance 2.3 ± 0.2 Å; the oxygen atoms have some lateral flexibility. The Mg^{2+} ion (atomic radius 0.65 Å) is usually bound by six oxygen atoms at the vertices of an octahedron with Mg-O distance 2.0 Å; these oxygens are in tight van der Waals contact with one another. Although many small molecules bind magnesium with greater affinity than they bind calcium, most intra- and extracellular proteins bind calcium with much greater affinity than they bind magnesium.

The dissociation constant is the ratio of the off rate to the on rate: K_d (M) = k_{off} (s^{-1})/k_{on} (M^{-1} s^{-1}). The rate-limiting dehydration of Ca(H$_2$O)$_7^{2+}$ is fast, ~10$^{8.0}$ s^{-1}, while that of Mg(H$_2$O)$_6^{2+}$ is slow, 10$^{4.6}$ s^{-1}. This difference reflects the loose pentagonal bipyramidal versus the tight octahedral packing of the oxygen ligands. The cation must be (partially) dehydrated before it can bind to the protein. These rates are important for modeling the flux of Ca^{2+} ions through the cytosol and the attendant binding of proteins. The increase in affinity of most proteins for calcium relative to magnesium derives primarily from this difference in k_{on} (Permyakov and Kretsinger 2010).

Calcium-Binding Proteins

Certainly, calcium-modulated proteins bind calcium; the term "calcium-binding proteins" may be short hand for those that are not demonstrated to be calcium modulated and/or are not intrinsic membrane proteins. Calcium may be essential to maintain the structure, stability, and functions of these calcium-binding proteins. However, they are not directly involved in information transduction pathways. The apo-forms of calcium-modulated proteins usually have well-defined structures. They undergo a change in conformation upon binding calcium as an essential step in information transduction, as illustrated for the N-terminal EF-lobe of calmodulin (Fig. 1). In contrast, the apo-forms of (non-calcium-modulated) calcium-binding proteins are often disordered and are not usually encountered under physiological conditions.

Biomineralization

Cells, especially bacteria and protists, can alter the ion concentrations in their immediate environment, thereby inducing precipitation of (usually) amorphous biominerals, so-called "*b*iologically *i*nduced *m*ineralization" (Mann 2001).

Biomineralization can occur intracellularly in vesicles derived from the Golgi apparatus or endoplasmic reticulum. These membrane-enclosed vesicles, for instance, those containing amorphous calcium phosphate in bone, may be retained within the cell. Or, following exocytosis, or polarized budding of the cell membrane, the vesicles, so-called exosomes, may fuse and/or dissolve to release their minerals into the extracellular environment, for example, mollusk shells or chiton teeth. Packets of amorphous calcium phosphate attached to collagen fibrils are the precursors of plates of hydroxyapatite attached to theses fibrils.

Animals can create an extracellular space (partially) enclosed by cells and/or an impermeant matrix. They change its composition by ion pumps in the cell or vesicle membranes. And they synthesize and/or secrete polysaccharides and proteins in(to) this space – "*b*iologically *c*ontrolled *m*ineralization." The resultant minerals are (usually) amorphous precipitates. These usually, but not always, undergo one or several transitions to the lowest free energy crystal form. The resultant crystals may be randomly oriented, in terms of their crystallographic axes, or aligned relative to the organic matrix. In either case, they will have incorporated some of the protein or polysaccharide into the macroscopic crystals observed in the matured or maturing organism. These organics may be incorporated into the macroscopic crystal in fault lines between smaller, perfect crystal domains. Or, the organic(s) may form a distinct layer with its fiber axes aligned and thereby directing the orientation of the

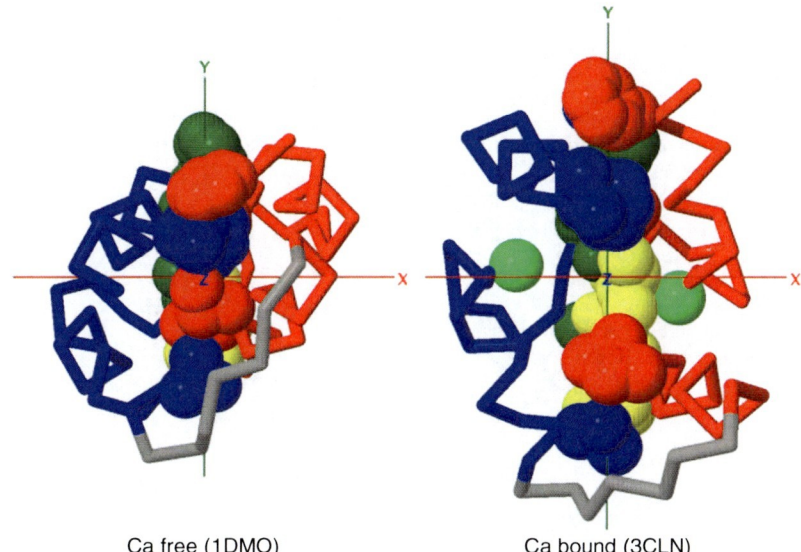

Ca free (1DMO)   Ca bound (3CLN)

Calcium in Biological Systems, Fig. 1 The change in conformation of the first (N-terminal) EF-lobe of calmodulin. EF-hand domains, 1 (odd) and 2 (even), are related by an approximate twofold axis, z. The apo-form (*Protein Data Bank*, 1DMO) is on the *left*; the calci-form (PDB 3CLN) is on the *right*. Both EF-lobes are represented as α-carbon traces. The atoms (both side chain and main chain) within ± 2.0 Å of the yz-plane are shown as van der Waals spheres. These are mostly side chain atoms at the interface. The close contacts between the side chains of residues Met36, Phe19 (odd) and Met72, Phe68 (even) in the apo-form are relieved upon binding calcium. *Red*, odd domain; *blue*, even domain; *gray*, loop connecting EF-hand odd and EF-hand even; *dark green*, loop connecting helix E and helix F; *yellow*, side chains of the two loops EF; *green*, Ca^{2+} ions (Figure courtesy of Hiroshi Kawasaki, Department of Supramolecular Biology, Graduate School of Nanobioscience, Yokohama City University, Suehirocho, 1-7-29, Tsurumi-ku, Yokohama 230–0045, Japan: kawasaki@yokohama-cu.ac.jp. Generated using http://calcium.sci.yokohama-cu.ac.jp/efhand.html)

mineral crystallites. This implies some sort of (quasi) epitaxis, a subject of much speculation but limited experimental evidence.

The overall shapes of these crystallites usually differ from those of their inorganic counterparts. A plausible mechanism, a sort of anti-epitaxis, posits that certain inhibitory proteins, often containing γ-carboxyglutamic acid, selectively bind to specific faces, thereby inhibiting growth in that direction but permitting it in other directions (Mann 2001). The details of these calcium–proteins interactions in the solid state are poorly understood.

Summary

The chapters in this volume focus on the soluble calcium-binding and calcium-modulated proteins because more is known of their structures and modes of actions. They provide a reference or comparison for the intrinsic membrane proteins and those associated with biominerals.

Cross-References

▶ Annexins
▶ Bacterial Calcium Binding Proteins
▶ C2 Domain Proteins
▶ Calcium and Mitochondria
▶ Calcium and Viruses
▶ Calcium in Health and Disease
▶ Calcium in Nervous System
▶ Calcium Sparklets and Waves
▶ Calcium-Binding Protein Site Types
▶ Calcium-Binding Proteins, Overview
▶ Calmodulin
▶ EF-Hand Proteins

References

Berridge MJ (2006) Calcium microdomains: organization and function. Cell Calcium 40:405–412
Hofer AM (2005) Another dimension to calcium signaling: a look at extracellular calcium. J Cell Sci 118:855–862

Lowenstam HA, Weiner S (1989) On biomineralization. Oxford University Press, New York

Mann S (2001) Biomineralization: principles and concepts in bioinorganic materials chemistry. Oxford University Press, New York

Nakayama S, Kawasaki H, Kretsinger RH (2000) Evolution of EF-hand proteins. Top Biol Inorg Chem 3:29–58

Permyakov EA, Kretsinger RH (2009) Cell signaling, beyond cytosolic calcium in eukaryotes. J Inorg Biochem 103:77–86

Permyakov EA, Kretsinger RH (2010) Calcium binding proteins. John Wiley, Hoboken

Whitiker M (ed) (2010) Calcium in living cells methods in cell biology, vol 99. Elsevier/Academic, Burlington

Calcium in Health and Disease

Marcus C. Schaub[1] and Claus W. Heizmann[2]
[1]Institute of Pharmacology and Toxicology, University of Zurich, Zurich, Switzerland
[2]Department of Pediatrics, Division of Clinical Chemistry, University of Zurich, Zurich, Switzerland

Synonyms

Calcium and death; Calcium body balance; Calcium cytotoxicity; Calcium signaling in disease; Calcium signaling in health

Definition of the Subject

Calcium is involved in almost every cellular activity. As a second messenger it is a critical component of the signaling network. It, furthermore, links mitochondrial energy production by oxidative phosphorylation to the actual energy demand. As complex animals left the oceanic calcium-rich environment to adopt terrestrial life they carried with them a huge calcium reservoir in their bones that guarantees the supply of the hundred times smaller amount of calcium vital for cell survival by maintaining calcium homeostasis. In contrast, too much calcium in the cell under pathologic condition or physical cell lesion presents a death signal (cardiac infarction, brain stroke). The human genome contains over 200 genes coding for calcium binding proteins (primarily EF-hand proteins). Their functions and relations to disease are discussed. The final remarks refer to the genetic makeup which may affect many of the regulatory proteins responsible for calcium homeostasis. Since longevity has significantly increased in recent times medicine is confronted with late-onset diseases caused by genetic defects dating from ancient times, hundreds of thousands to millions of years ago.

Scope of This Brief Essay on Calcium in Health and Disease

How and why has Nature come to rely on calcium as a ubiquitous dynamic signaling (free Ca^{2+} ions) as well as a structural (in organic and inorganic complexes) component. Calcium is the most abundant mineral in the body, bones and teeth, accounting for >99% of the total of around 1 kg (Williams and Fausto da Silva 2006). For Ca^{2+} to act as intracellular messenger receptor proteins have evolved that recognize the Ca^{2+} signal (Berridge 2005; Bradshaw and Dennis 2009; Haiech et al. 2009, 2011). These calcium sensors contain highly specific Ca^{2+} binding sites, so-called EF-hands (a characteristic helix-loop-helix structure with one Ca^{2+} binding to the loop), which will be described in detail in several other contributions in this Encyclopedia. Suffice it to mention that starting with the first Ca^{2+} sensing protein, troponin-C in striated muscles, followed by brain calmodulin in the 1960s, over 66 subfamilies with Ca^{2+} binding EF-proteins stemming from over 200 human genes have been described so far (Schaub and Heizmann 2008; Nakayama et al. 2000; Heizmann et al. 2002; Krebs and Michalak 2007; Leclerc and Heizmann 2011). Many of these proteins proved to be related to various diseases and a selection of them have been compiled in Table 2 toward the end of the article. If not tightly controlled Ca^{2+} is a highly cytotoxic substance. Any drastic Ca^{2+} overload in a living cell leads to its death. The routes of the cellular Ca^{2+} death converge in the mitochondria where they will bring the energy production to a halt (Carafoli and Brini 2008). The mitochondrial structures on which the Ca^{2+} load imprints its kiss of death will briefly be discussed. Several aspects of calcium in health and disease will be highlighted by selected examples rather than treated in general terms. Above all, however, only detailed knowledge of the normal functioning of Ca^{2+} physiology allows one to better understand disease pathways. Because of space limitation, only few overviews and books could be cited from where the primary literature can be retrieved. However, references specifically pertaining to the clinical relevance of individual EF-hand proteins for various diseases as listed in Table 2, are also incorporated in the reference list.

Is Calcium More Toxic than Botulinum Toxin?

The answer is not as easy as it may look. True, the toxin (a disulfide-linked dimer of two polypeptides of 50 and 100 kDa) produced by the anaerobic, Gram-positive bacterium *Clostridium botulinum* (in Latin botulus means "sausage") is the most acutely toxic substance known. Ninety to 200 ng of botulinum toxin applied intravenously or by inhalation suffices to kill an adult person, and 4 kg would be more than enough to eliminate the entire human population of the world. The heavy chain of the toxin targets the poison to the nerve axon terminals (Brunger et al. 2008). The light chain of the toxin is a proteinase attacking some of the fusion proteins (SNARE proteins) in the neuromuscular junction, thereby preventing it from anchoring nerve vesicles to the presynaptic membrane, thus inhibiting release of acetylcholine. By interference with the nerve impulses it causes flaccid muscle paralysis. Fortunately, however, one seldom encounters this lethal toxin, and if canned meat is under suspicion, heating destroys the toxin. It is nevertheless feared as a potential bioterror weapon.

How does the cytotoxicity of Ca^{2+} compare to botulinum toxin? Ca^{2+} as ubiquitous intracellular second messenger presents a vital signaling component involved in virtually all cell activities including regulation of metabolic enzymes, membrane-linked processes (excitation-contraction, -secretion, and -transcription coupling, nerve impulse transmission), hormonal regulations, control of contractile and motile systems, cell cycle, fertilization, and vision. In cardiomyocytes the cytoplasmic Ca^{2+} concentration is very low at rest $\sim 10^{-7}$ M which is almost 20,000-fold lower than in the extracellular space (Table 1). The Ca^{2+} signal for a heart beat (heart muscle contraction) transiently increases the very low cytoplasmic Ca^{2+} concentration about 100-fold to 10^{-5} M. This signal results from Ca^{2+} entry into the cell through the voltage-gated L-type Ca^{2+}-channel (LCC) upon surface membrane depolarization (Fig. 1), and subsequently further Ca^{2+} is released from the intracellular

Calcium in Health and Disease, Table 1 Approximate extracellular and intracellular ion concentrations of a cardiomyocyte

Ion	Extracellular concentration (mmol/L)	Intracellular concentration (mmol/L)	Ratio of extracellular to intracellular ion concentrations
Na^+	145	15	10
K^+	4	150	0.03 (40-fold difference)
Cl^-	120	5–30	4–25
Mg^{2+}	1	1	1
Ca^{2+}	2	0.0001 (cytoplasm)	20,000

sarcoplasmatic reticulum (SR) via the ryanodine Ca^{2+} release channel (RYR). After a fraction of a second the activator Ca^{2+} will be removed again by the Ca^{2+}-pumps and the Na^+/Ca^{2+}-exchanger (NCX) for muscle relaxation (more detailed description in Sect. Regulation of Cardiac Contraction by Calcium Signaling). Such highly precise regulation of cytoplasmic Ca^{2+} requires extremely fast (operating in the millisecond time range) mechanisms to shift Ca^{2+} between extracellular and intracellular compartments involving various pumps and channels (Fig. 1). For instance, a voluntary skeletal muscle that goes from rest into mechanical activity increases its energy requirement 2,000-fold within a fraction of a second. The signaling Ca^{2+} for contraction at the same time must also assure the necessary metabolic activation for sufficient ATP production. Should anything go off course with the Ca^{2+} control, in particular, too much Ca^{2+} in the cytoplasm would severely damage if not kill the cell. Since all cells of the body are bathedin this highly toxic environment containing ~ 2 mM Ca^{2+} it is most likely that an excess of Ca^{2+} may enter the cytoplasm following any cell lesion. Any cell death results from Ca^{2+} overload in the cytoplasm including myocardial infarction, brain stroke, degenerative nerve cell diseases, inflammatory cell death, ischemia-reperfusion injury etc.

Thus, in contrast to botulinum toxin, our body cells are constantly exposed to a massive burden of deadly toxic Ca^{2+}. Each living cell must have developed robust devices to rid it from excessive Ca^{2+}. Early in the sixteenth century the Swiss-born Philippus Aureolus Theophrastus Bombastus von Hohenheim, called Paracelsus, introduced some rationale in the patient treatment, in particular, he stated: "the dose

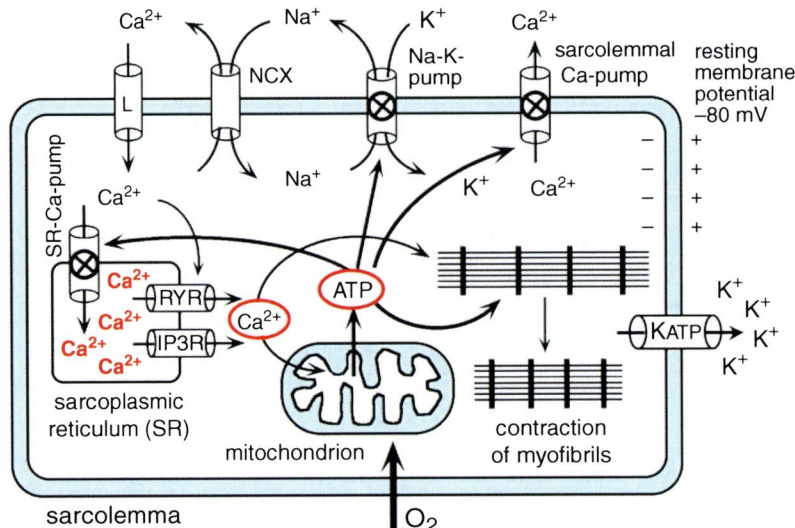

Calcium in Health and Disease, Fig. 1 Simplified schematic of cardiomyocyte subcellular arrangement of energy (ATP) consuming ion pumps (*marked by crosses*) and ion channels. L = voltage-gated L-type Ca^{2+}-channel (LCC); NCX = Na^+/Ca^{2+}-exchanger; RYR = ryanodine receptor, Ca^{2+} release channel of the SR; IP3R = inositol trisphosphate receptor, Ca^{2+} release channel of the SR; KATP = ATP-dependent K^+-channel. The resulting imbalances of ion distribution across the cell surface membrane are given in Table 1. Collectively, the various ion gradients give rise to the membrane resting potential of around −80 mV (positive outside and negative inside). For further explanations see text

makes the poison." This statement holds true until today: any medicine with an impact becomes poisonous when overdosed. It also applies to both Ca^{2+} and botulinum toxin. Minute amounts of this latter toxin injected into specific muscles will weaken them for a period of several months and is used to prevent formation of wrinkles by paralyzing facial muscles. In fact, it is the most common cosmetic operation today. On the other hand, in view of the large amount of calcium in the bones and the complicated hormonal control of its partition over different pools it is almost impossible to influence the body calcium content. Only the small and dynamic intracellular fraction may be affected by the so-called calcium channel blockers that bind to LCC and reduce extracellular Ca^{2+} entry during the action potential. As the dihydropyridine drugs presenting the major class of calcium channel blockers bind to LCC, these channels are also known as dihydropyridine receptors (DHPRs). The ion gradients between extra- and intracellular compartments of a cardiomyocyte are provided by energy (ATP) consuming ion pumps in functional interaction with ion channels and exchangers (see Fig. 1 and Table 1).

Bodily Calcium Balance

Calcium is the fifth most abundant element in the human body. It is an essential element that is available only through dietary sources. Its metabolism depends on (i) intestinal absorption, (ii) renal reabsorption, and (iii) bone turnover. These processes are in turn regulated by a set of interacting hormones including parathyroid hormone (PTH), 1,25-dihydroxyvitamin-D (1,25D), ionized calcium itself, and their corresponding receptors in the gut, kidney, and bone. Up to 50% of bone is made up of the inorganic calcium hydroxyapatite (known as bone mineral) interwoven with the collagen fiber network providing structural strength. The non-bone calcium (10–15 g in an adult) representing <1% of total body calcium is in constant and rapid exchange among the various calcium pools. Serum calcium ranges from 2.2 to 2.6 mM in healthy subjects. It comprises free Ca^{2+} ions (∼51%), protein bound complexes (∼40%) mainly with albumin and globulin, and ionic complexes (∼9%) mainly Ca^{2+}-phosphate, Ca^{2+}-carbonate, and Ca^{2+}-oxalate. Only free Ca^{2+} ions are in equilibrium between the different calcium pools.

To avoid calcium toxicity, the concentration of serum ionized Ca^{2+} is most tightly maintained within a physiological range of 1.10–1.35 mM. This concentration range extends to the extracellular fluid compartment (EFC), which provides the backup for delivery to and reuptake of Ca^{2+} from the cells. Serum Ca^{2+} homeostasis is regulated by a rapid negative feedback hormonal pathway. A fall of serum Ca^{2+} inactivates the Ca^{2+}-receptor (CaR) in the parathyroid cells and increases PTH secretion, which restores serum Ca^{2+} by activating the parathyroid receptor (PTHR) in bone, to increase Ca^{2+} resorption, and in kidney, to increase tubular Ca^{2+} reabsorption. In kidney, the increased PTH secretion augments its Ca^{2+} restorative effect by increasing secretion of 1,25D which through the vitamin-D receptor (VDR) in the gut, increases active Ca^{2+} absorption and increases Ca^{2+} resorption in bone. Calcitonin produced by the parafollicular C-cells in the thyroid gland inhibits the activity of osteoclasts which digest bone matrix, thereby, releasing Ca^{2+} and phosphorus into the blood. In the kidney, calcitonin inhibits tubular reabsorption of both Ca^{2+} and phosphorus promoting their loss in the urine. The contribution of calcitonin in serum Ca^{2+} regulation varies from species to species and is almost negligible in humans.

During youth the calcium-bone balance is positive for bone formation (formation > resorption), in young adults it is neutral (formation = resorption), while with advancing age it becomes negative (formation < resorption). The threatening occurrence of osteoporosis at higher age is often therapeutically encountered by higher calcium intake combined with biphosphonate drugs. However, the efficiency of this treatment is questionable as almost the entire excess calcium offered over the normal daily nutritional intake is excreted via kidney, feces, and sweat. What the experts agree on is that this treatment activates bone metabolic turnover. The bone structure is, however, remodeling throughout life, renewing the entire skeletal tissue on average every 10–12 years. In order to fend off osteoporosis it is important to combine calcium supply with the correct amount of vitamin-D. In addition, estradiol levels may have to be redressed in women with osteoporosis. In elderly patients (>70 years of age) with cardiovascular risks calcium supplementation should be limited to the dietary reference intake of around 1,200 mg in order to avoid ectopic calcification of the vascular system.

In summary, the tight regulation of Ca^{2+} in serum and EFC is a prerequisite for its safe use as intracellular signaling component. About 1 g of calcium is in the plasma and EFC, while all tissue cells together comprise 6–8 g of calcium mostly stored in intracellular vesicles such as sarcoplasmatic (SR) and endoplasmatic (ER) reticulum (Fig. 1). This dynamic calcium-system is safely backed up by the massive 1 kg calcium-store in the bones. Approximately 400 Ma ago, the ancestors of the bony fish (osteichthyes) and cartilaginous fish (chondrichthyes) diverged in the Silurian period. The structural strength of a calcified skeleton was critical for complex animals at the end of the Devonian period when they left the calcium-rich ocean to resume life on land after plants had already conquered dry land. So the calcium-rich bones made up for the loss of calcium in the oceanic environment supporting the calcium-dependent regulation of cells and nervous system.

Calcium as Intracellular Second Messenger

How can calcium function as highly specific intracellular signal transmitter? Divalent cations such as Ca^{2+} and Mg^{2+} can form more stable and more specific complexes with organic substances like proteins than the other abundant monovalent cations Na^+ and K^+. They do, however, not form covalent bonds in biology. Energy usage in all living cells relies on phosphoryl bond breakdown and Ca^{2+} more readily than Mg^{2+} forms insoluble salts with the originating inorganic phosphate and pyrophosphate; this limits its solubility to around 10^{-3} M. This lower solubility of calcium salts required Ca^{2+} to be ejected from the cytoplasm while Mg^{2+} could be kept in the cells in millimolar concentrations where it serves as cofactor in enzyme reactions involving the MgATP complex. Consequently, all living cells have developed robust devices to rid the intracellular space from Ca^{2+} by energy consuming ion pumps and ion channels (Fig. 1). As mentioned, Ca^{2+} in the EFC and in the cells (outside the bony skeleton) amounts to 10–15 g while the total bodily Mg^{2+} content is about 25 g. One third of the Mg^{2+} is in the cells (~9 g, corresponding to a concentration of ~18 mM^{-1}) of which only a small fraction is ionized (~1 mM^{-1}) with almost the same concentration inside and outside of the cells (Table 1). Calcium as member of the 4th row in the periodic table of elements

(4 electron shells) has a more complex electron configuration than magnesium (in the 3rd row of the periodic table with 3 electron shells) allowing for greater flexibility of Ca^{2+} in coordination with protein ligands (usually with 6–8, but up to 12 coordination points are possible). In contrast, Mg^{2+} (ionic radius = 0.64 Å) is much smaller than Ca^{2+} (ionic radius = 0.99 Å) and requires a strictly fixed geometry with 6 coordinating oxygen atoms in binding with proteins. Importantly, ligand binding by Ca^{2+} to the protein sensor sites is ~1,000 times faster than with Mg^{2+}. What kind of animal life would that be with Mg^{2+} as signal transmitter operating at a 1,000-fold slower time scale? Slow thinking and slow moving, a concert pianist could not play a trill with 10 key strokes per second. Finally, the enormous Ca^{2+} gradient between cell inside and outside presents a large electrochemical potential that can be used to create signaling Ca^{2+} spikes by letting Ca^{2+} ions quickly enter via the LCC in the cell membrane.

Canonical EF-Hand Calcium Sensor Proteins

In the 1960s, troponin-C (TNC, ~18 kDa mass), which is involved in regulation of contraction in striated muscle (skeletal and heart muscle), was the first characterized intracellular Ca^{2+} sensor protein, being joined soon after by parvalbumin (PARV, 12 kDa) isolated from fish and frog muscles and calmodulin (CAM, ~16.8 kDa) obtained from the brain. TNC is firmly incorporated in the actin filament structure of striated muscle and in cooperation with troponin-I (TNI, ~24 kDa), troponin-T (TNT, ~36 kDa), and tropomyosin (TM, 28.5 kDa) presents the Ca^{2+}-dependent switch for muscle activity. There are two TNC isoforms, cardiac TNC (cTNC, exclusively expressed in cardiac and slow contracting aerobic muscle fibers) and fast skeletal muscle sTNC, which are the products of two different genes, TNNC1 (cTNC) and TNNC2 (sTNC). Mutations in all troponin components as well as in tropomyosin, collectively known as striated muscle regulatory proteins, have been found that may cause different forms of hereditary heart disease including hypertrophic (HCM), dilated (DCM), restrictive (RCM), and non-compaction cardiomyopathies. Most of these hereditary cadiomyopathies are either caused or accompanied by changes in Ca^{2+} signaling quality, i.e., increased or decreased Ca^{2+}-sensitivity of the contractile machinery.

CAM is a ubiquitous cytoplasmic Ca^{2+}-sensor that by interaction with over 100 different proteins regulates the activity of a myriad of soluble and structural components in different cell compartments. CAM is the prototypical intracellular Ca^{2+}-sensor containing four canonical Ca^{2+} binding EF-hands. It contains two EF-hands in the N-terminal and two EF-hands in the C-terminal portions of the dumbbell-shaped molecule. It is assumed that the compact structure of CAM with four EF-hands may have arisen in prokaryotes by two subsequent gene duplications from a single EF-hand precursor protein. This prokaryote Ca^{2+} binding CAM seems to have been so successful that it was carried over into the eukaryotic lineage where it was preserved in the kingdoms of protista, fungi, plantae, and animalia. CAM is one of the most conserved proteins with 100% amino acid (AA) sequence identity among all vertebrates. It consists of 149 AA including the N-terminal methionine which may be removed in the mature protein. Around 500 Ma ago the genome of all vertebrates contained three separate CAM genes. Mammalians including humans still contain three different genes, all coding for the same AA sequence. Why are there three genes for the same protein? It seems that the 5'- and 3'- untranslated regions (UTR) may vary to some degree between the three genes possibly permitting different protein expression levels at discrete cellular sites during differentiation and in highly specialized cell types such as neurons or striated muscle cells. No structural mutations have been reported so far for CAM exons indicating they probably would not be compatible with life. On the other hand, several point mutations in the introns flanking exons 3 and 4, whose significance remains unknown, have been described.

The evolutionary relation based on sequences indicates that CAM and cTNC together with sTNC, myosin essential light chain (ELC), myosin regulatory light chain (RLC), and parvalbumin (PARV) belong to the so-called CTER family of EF-hand proteins. CTER stands for CAM, TNC, ELC, and RLC. All these proteins contain 4 EF-hand motifs with the exception of PARV which has lost one of them in the distant past. In addition, some of the EF-hands in these proteins have also lost the Ca^{2+} binding capacity due to evolutionary sub-functionalization. CAM does not only bind to targets when it is saturated with Ca^{2+}, but it can also be firmly associated with some proteins in its apo-form (no Ca^{2+} bound) where it may function as a Ca^{2+}-sensor subunit in a heteromeric protein complex. This is the case with proteins relevant for Ca^{2+} regulation such as RYR, the pore forming subunit of the LCC, the inositol 1,4,5-trisphosphate (IP3) -operated Ca^{2+} release channel (IP3R = IP3 receptor) of the SR or ER, and phosphorylase kinase. In other Ca^{2+} regulated proteins a single EF-hand motif is found integrated in the primary structure as, for instance, in calcineurin-B, dystrophin, calpain proteases, vertebrate phosphodiesterases and in RYR.

The S100 Protein Family

The first S100 protein was isolated from brain also in the 60s of the last century, which is now known as S100B. It is a small (~10.7 kDa) acidic protein soluble in 100% saturated ammonium sulfate solution and hence named "S100." The S100 proteins are, compared to CAM, phylogenetically late-comers exclusively expressed in vertebrates, fishes, amphibians, reptiles, birds, and mammalians. They represent the largest subgroup within the superfamily of Ca^{2+} binding EF-hand proteins. At least 20 S100 protein coding genes have been located in the human genome 16 of which are clustered in a region of chromosome 1q21. This region is prone to molecular rearrangement linking S100 proteins and cancer. The S100 proteins derived from the 1q21 gene cluster are designated as S100A followed by Arabic numbers (S100A1, S100A2, S100A3 etc.). In contrast, proteins from other chromosomal regions just bear a letter (S100B, S100C, S100D etc.).

The S100 proteins (molecular weight 9–13 kDa) contain two Ca^{2+} binding motifs, one classical C-terminal EF-hand (as the four EF-hands in CAM with 12 AA in the Ca^{2+} binding loop) and one S100-specific N-terminal "pseudo EF-hand" (with 14 AA in the binding loop). Moreover, the Ca^{2+} binding loop in the S100 pseudo EF-hands varies considerably between the different proteins allowing for functional differentiation. Usually two S100 proteins form a functional dimer with the two protein monomers running side-by-side in opposite direction (antiparallel), though also heterodimers and few multimers can be found. Upon Ca^{2+} binding the C-terminal EF-hand undergoes a large conformational change resulting in the exposure of a hydrophobic surface responsible for target binding. The expression of the S100 members is often tissue and cell type specific. A unique feature of these proteins is that some members are secreted from cells upon stimulation. In the high extracellular Ca^{2+} concentration larger complexes are formed which act as pro-inflammatory signaling components through activation of RAGE (receptor for advanced glycation end-products) and Toll-like receptor-4 (Leclerc and Heizmann 2011).

Ca^{2+} binding to S100 proteins induces structural changes that allow interaction with target proteins and modulation of their activity. These protein-protein interactions also significantly affect the Ca^{2+} binding properties (affinity and kinetics). Furthermore, the regulatory properties of specific S100 proteins can be altered by various post-translational modifications including phosphorylation, nitrosylation, citrullination, carboxymethylation, glutathionylation, transamidation, or sumoylation.

In contrast to CAM and TNC, the S100 proteins are not directly involved in switching on and off key cell functions such as muscle contraction or nerve transmission, but rather operate as modulators. This is reflected in the relatively mild phenotypes in mice in response to overexpression or knockout of individual S100 proteins. Alternatively, functional redundancy may be operating among different S100 proteins. Nevertheless, specific S100 proteins are associated with serious diseases including cancer, autoimmune and inflammatory diseases, neurodegeneration, and cardiomyopathy. Due to their specific disease association several of the S100 proteins have proven their value as diagnostic and prognostic markers (see Sect. Calcium Binding Proteins in Disease).

Magnesium Acting as Modulator of Calcium Signaling

It is important that fast and selective binding of Ca^{2+} to its sensor protein occurs in the presence of ~1 mM Mg^{2+} (Table 1). That is 10,000 times higher than cytoplasmic Ca^{2+} at rest and still at least 100 times higher when the activator Ca^{2+} is at its peak. Thus Ca^{2+} signaling is possible only if Mg^{2+} has significantly lower affinities to the regulatory protein sites.

Beside different affinities also the binding kinetics play an important role. The two C-terminal EF-hands of cTNC and sTNC seem to confer stability to the protein in the filamentous structure and have a high Ca^{2+} affinity ($K_d \sim 10^{-8}$ M). However, in the presence of physiological concentrations of Mg^{2+} the apparent affinity for Ca^{2+} is lowered by two orders of magnitude. It may thus be assumed that these C-terminal "structural" sites may always be occupied be it by Ca^{2+} or by Mg^{2+} at rest as well as during activation. Only the N-terminal sites, two in sTNC and one in cTNC, are involved in regulation of contraction. These "regulatory" sites have an affinity for Ca^{2+} of $K_d \sim 10^{-6}$ M which is virtually not affected by Mg^{2+} because the affinity for Mg^{2+} is as low as $K_d \sim 10^{-2}$ M (10 mM Mg^{2+} would be required for 50% binding saturation even in the absence of Ca^{2+}). This situation ensures that in striated muscle (heart and skeletal) signaling depends solely on Ca^{2+} and is not disturbed by Mg^{2+}.

In CAM the affinity of the two N-terminal EF-hands for Ca^{2+} is about one order of magnitude lower ($K_d \sim 10^{-5}$ M) than that of the two C-terminal EF-hands ($K_d \sim 10^{-6}$ M). This allows CAM to sense Ca^{2+} transients in the cytoplasm over a relatively wide concentration range. For full activation of CAM, Ca^{2+} binds with positive cooperativity to all four CAM EF-hands comparable in a way to binding of oxygen to the hemoglobin tetramer. Beginning with the first Ca^{2+}, conformational changes are stepwise induced as the saturation of CAM with Ca^{2+} proceeds. At low cytoplasmic Ca^{2+} in the resting state, most of the CAM molecules will be in the apo-form (no bound divalent cation). The affinity for Mg^{2+} is again so much lower that it hardly interferes with the Ca^{2+} signal. However, during rest some sites of CAM may become occupied by Mg^{2+}. Since the binding kinetics of Mg^{2+} are so much slower than those of Ca^{2+}, the slow exchange rate at the sites occupied by Mg^{2+} could slightly delay the spread of the Ca^{2+}-CAM signal.

The question remains, however, can relatively small variations of Mg^{2+} levels affect intracellular Ca^{2+} signaling in a physiologically relevant manner? Hypermagnesemia is uncommon except in kidney insufficiency and may be counteracted by intravenous application of Ca^{2+} and loop diuretics. High blood Mg^{2+} antagonizes Ca^{2+} function by attenuating nerve transmission. The more common hypomagnesiemia reduces the Ca^{2+} antagonism of Mg^{2+} thus unmasking the effects of Ca^{2+}. It can be cured by Mg^{2+} supplementation. Low Mg^{2+} typically accompanies chronic diseases including hypertension, diabetes, cardiovascular disease, hyperthyreosis, diarrhea, and alcohol abuse. Its symptoms are increased excitability of nerves and muscle, paresthesia, muscle cramps, cardiac insufficiency and arrhythmias. In hypertension with hypomagnesemia administration of Mg^{2+} alone can lower the blood pressure significantly. Changes in intracellular Mg^{2+} obviously affect Ca^{2+} function in various signaling pathways at every point where Ca^{2+} is involved. This leads to significant cumulative effects on Ca^{2+} function particularly in the central nervous system (CNS), in muscles, and blood vessels.

Parvalbumin: A Delayed Calcium Buffer and Food Allergen

In contrast to the Ca^{2+} sensing signaling proteins, PARV has unique properties that render it an intracellular "delayed Ca^{2+} buffer." It contains three EF-hands of which the first N-terminal motif is incomplete, and therefore, unable to bind Ca^{2+} (Schaub and Heizmann 2008). The two functional EF-hands have a high affinity for Ca^{2+} ($K_d = 10^{-7}$–10^{-9} M) and a moderate affinity for Mg^{2+} ($K_d = 10^{-3}$–10^{-5} M). Ca^{2+} and Mg^{2+} compete for the same binding sites in PARV, therefore, the actual binding is determined by the relative concentrations of the two cations. Given the 10,000 times higher Mg^{2+} concentration ($\sim$1 mM) than that of Ca^{2+} at rest (0.0001 mM) PARV has bound Mg^{2+} in the resting myocyte. Ca^{2+} can bind to PARV only when Mg^{2+} is dissociated. The off-rate of Mg^{2+} from PARV measured in vitro is about 3 per second. Even if this off-rate should be considerably faster in the living cell, it still remains the rate-limiting step for Ca^{2+} binding to PARV. Muscle physiology has shown that PARV is present in fast twitching muscles of small mammalians and frogs at submillimolar concentrations. PARV is not found in slow contracting muscles nor in the hearts of larger animals. In contrast it is present in especially high concentrations in the muscles of fishes performing fast burst swimming with high amplitude undulations. PARV is designed to bind Ca^{2+} with a certain delay after Mg^{2+} has dissociated from the binding sites so that Ca^{2+} binding to PARV does not compete with the initial binding of Ca^{2+} to the regulatory sites of cTNC for triggering contraction.

In this way PARV accelerates relaxation in fast moving muscles by facilitating the removal of Ca^{2+} from cTNC and its reuptake into the SR. For these properties PARV has been successfully used in transgenic animal experiments to alleviate diastolic heart dysfunction, when cytoplasmic Ca^{2+} is not sufficiently removed between two heart beats.

Excitatory GABAergic (GABA = gamma-aminobutyric acid) neurons contain high concentrations of PARV in the soma, the axons as well as in the presynaptic terminals. For its Mg^{2+} and Ca^{2+} binding properties PARV accelerates the initial decay phase of the Ca^{2+} transients without affecting the Ca^{2+} peak height thus allowing fast firing rates. These characteristics have been shown to contribute to short-term synaptic plasticity.

Food allergy induced by gastrointestinal sensitization provokes the most common life-threatening IgE-mediated anaphylaxis. PARV with bound Ca^{2+} is remarkably resistant to heat and denaturing chemicals; this might contribute to its allergenic properties (Schaub and Heizmann 2008). It shares this allergenic potential with many other Ca^{2+} binding proteins isolated from pollen (grass, trees, and weeds), parasites, and fish. Interestingly, PARV with mutations in both EF-hands preventing Ca^{2+} binding results in almost complete loss of antigenicity.

Regulation of Cardiac Contraction by Calcium Signaling

Regulation of cardiac and skeletal muscle contraction is one of the best studied regulatory systems (Fig. 1). One reason may be that in this system Ca^{2+} induced alterations in protein-protein interactions directly translate into mechanical output by the sarcomeric contractile actomyosin machine. Contractile parameters reflect the complex regulatory finesses and can easily be measured. Figure 1 displays the topical repartition of the devices (ion channels and energy using ion pumps) that guide the Ca^{2+} signaling pathways in cardiac muscle. Ca^{2+} entering through the LCCs in response to the depolarizing action potential (AP) approaches the closely positioned RYR of the SR terminal cisternae where it induces a much larger Ca^{2+} release from the SR (Ca^{2+}-induced Ca^{2+} release = CICR, indicated in Fig. 1 by the bent arrow pointing to RYR). This Ca^{2+} serves at the same time its two major purposes (i) induction of contraction by binding to the N-terminal regulatory site in cTNC, and (ii) stimulation of ATP synthesis in the mitochondria. Akin to the close apposition of LCCs and terminal cisternae of the SR, also the mitochondria are located in proximity to the SR Ca^{2+} release channels so that Ca^{2+} from the SR directly attains the mitochondrial matrix through the Ca^{2+} uniporter of the mitochondrial inner membrane (MIM). Consequently, the Ca^{2+} concentration in the mitochondrial matrix (MM) closely follows the Ca^{2+} outflux from the SR. This allows for coordination of energy production and demand.

For termination of contraction Ca^{2+} is quickly sequestered from the cytoplasm into the SR by the high affinity Ca^{2+}-pump of the SR (SERCA) which operates against a steep gradient. In the SR calcium can reach as high a concentration as several mM, mostly bound to the SR calcium storage protein calsequestrin (CSQ) which is complexed to triadin, junctin, and the luminal side of the RYR. The sudden drop in cytoplasmic Ca^{2+} causes the activator Ca^{2+} to dissociate from the cTNC regulatory EF-hand inducing relaxation. While in larger animals, including humans, around 70% of the systolic Ca^{2+} is taken up by the SERCA, the remainder is extruded from the cell by the sarcolemmal NCX. The sarcolemmal Ca^{2+}-pump (PMCA; PMCA4 in cardiomyocytes) is located in the caveolae and hardly contributes to Ca^{2+} extrusion, but rather functions as a regulator of Ca^{2+} signal transmission to target molecules including nNOS (neural nitric oxide (NO) synthase), calcineurin (PP2B, Ser/Thr protein phosphatase) and others.

The Na^+/K^+-pump (NKA) plays a pivotal role by establishing and maintaining the Na^+ and K^+ gradients between in- and outside the cell (Fig. 1 and Table 1). The electrochemical potentials of Na^+ and K^+ are linked to each other by the pump, while the Na^+ gradient provides the driving potential for Ca^{2+} extrusion via the NCX uphill against the high Ca^{2+} concentration outside (the required energy for this uphill Ca^{2+} extrusion is thus provided indirectly by the Na^+/K^+-pump). The sarcolemma is impenetrable for the positively charged Na^+ ion while a series of different K^+-channels allow for K^+ escape down its gradient to the outside of the cell. In fact, the inside negative resting potential of -80 mV results primarily from the constant outflow of K^+ ions.

The resting potential is additionally modulated by the ATP-dependent K^+-channel (KATP channel) in the sarcolemma (sarcKATP channel). This channel is a hetero-octamer consisting of four Kir6.2 subunits forming the pore and four sulfonylurea regulatory subunits (SUR; SURA2 in cardiomyocytes). Different metabolites and drugs can bind to the SUR subunits while nucleotides may bind to both types of subunits. Generally, binding of ATP closes the channel while ADP activates it. Since usually ATP is far in excess over ADP the channels are not very active. However, when the cells are under metabolic stress (for instance during ischemia or in acute heart failure) the cytoplasmic Ca^{2+} may increase because it is no more efficiently extruded from the cell due to energy shortage of the pumps for maintaining the ion gradients (Zaugg and Schaub 2003). In such situations the fall in the ATP/ADP ratio stimulates the sarcKATP channels through which K^+ ions escape strengthening the inside negative membrane potential. The larger the potential negativity (hyperpolarization) the shorter the AP and the lessCa^{2+} enters the cell, while at lower negativity (hypopolarization) more Ca^{2+} enters the cell during a longer AP. If during increased heart work the beat frequency and the number of APs also increases, more Ca^{2+} accumulates per time interval in the myocyte; this, in turn, translates into positive inotropism (stronger contractions). Since contractility and metabolic activity are coupled by Ca^{2+} the sarcKATP channels directly link cellular activity to the actual metabolic capacity. High activity of the sarcKATP channels under metabolic stress causes hyperpolarization (less Ca^{2+} enters the cell) which, as a protective mechanism, attenuates myocyte activity. The subunit composition of the sarcKATP channels varies from tissue to tissue according to function. KATP channels with varying subunit composition are also found in the cell nucleus and in the inner mitochondrial membrane.

Orchestration of Calcium Signaling in the Cardiomyocyte

The regulation of contraction and energy production in the cardiomyocyte is in reality even more complex than discussed above and depicted in Fig. 1. The cardiomyocyte is subject to a three-tiered control system, (i) immediate and fast feedback in response to mechanical load on a beat-to-beat basis (Frank-Starling relation depending on changes in Ca^{2+}-sensitivity of the contractile machinery), (ii) more sustained regulation involving transmitter and hormones as primary messengers triggering various intracellular signaling cascades, and (iii) long-term adaptation by changes in the gene expression profile. All three stages are critically dependent on location and time-specific deployment of Ca^{2+} (Schaub et al. 2006).

First, the ion pumps and the Ca^{2+} release channels of the SR have additional regulatory subunits which may be fixed in the protein complex or may reversibly interact with them. Second, most of the polypeptides involved in regulation can be reversibly modified, primarily by phosphorylation. Such covalent protein modification requires at least two enzyme systems, mostly different types of protein kinases for phosphorylation and protein phosphatases to reverse the phosphorylation effects. Many of these modifiers of the Ca^{2+} handling components are themselves dependent on Ca^{2+} for activity. Interestingly, a number of cardiac proteins involved in sarcomere regulation can be phosphorylated in their N-terminal region at Ser or Thr that are missing in their skeletal muscle counterparts; these include cTNI, phospholamban (PLN, the regulatory subunit of SERCA in the SR), SERCA2a (the cardiac-specific isoform of SERCA), cardiac myosin binding protein-C (cMyBPC), and cardiac titin. In the cases of cTNI, PLN, cMyBPC, and cardiac titin, these phosphorylatable sites reside on an additional peptide stretch contained in the primary sequence of the cardiac species. cTNI contains a 31 AA insert with two Ser's in addition to four more phosphorylatable Ser and Thr sites in the remainder of the protein. The six sites in cTNI can be reversibly phosphorylated by at least seven different protein kinases (PKA, PKC, PKG, Rho kinase, CaMKII, PAK1, and PAK3; some of these kinases are themselves Ca2+−dependent) partly overlapping and partly specific for particular sites. The different kinases are under control of different signaling cascades, but all affect the Ca^{2+} binding properties of cTNC by its interaction with the modified cTNI.

Third, adding to the complexity of intracellular Ca^{2+} handling almost all regulatory proteins come in different isoforms displaying different properties, which are specific for cardiac and other tissues. For instance, at least 11 isoforms of SERCA encoded by three separate genes have been discovered so far. Alternative splicing of the gene ATP2A2 produces three isoforms, SERCA2a, SERCA2b, and SERCA2c. SERCA2a is typical for cardiac and slow muscle SR, SERCA2b is the major isoform expressed in the skin epidermis, and SERCA2c is a ubiquitous minor partner. Over 130 mutations in the ATP2A2 gene have been reported mostly affecting all three isoforms. However, mutations causing the rare Darier's skin disease reside in exon 21 which is specific for SERCA2b in the epidermis and not present in the other isoforms. Consequently, the mutated protein is associated with severe disruption of Ca^{2+} transport function in the skin but does not affect the heart.

The major players in cardiovascular regulation comprise the catecholamines, which stimulate the beta-adrenergic receptors as well as angiotensin and endothelin, which stimulate their respective receptor systems. All the receptors involved in these regulatory systems are G-protein-coupled receptors (comprising seven transmembrane segments, GPCRs) whose stimulation affects different signaling cascades resulting in an increase of cellular Ca^{2+}. Ca^{2+}-dependent, reversible phosphorylation of their intracellular C-terminal domain governs their activities, switch of G-protein coupling, and internalization.

Calcium as Regulator of Mitochondrial Energy Metabolism

Over 95% of a vertebrate's energy requirement is produced in the mitochondria by oxidative phosphorylation in the respiratory chain. The human heart generates over 30 kg ATP during a day. Eighty to ninety percent of the required ATP is met by the catabolism of free fatty acids (FFAs) via beta-oxidation. The entire ATP (concentration $\sim$10 mM) pool of the heart is turned over within $\sim$1 min at body rest (basal heart activity) and within less than $\sim$10 s at maximum workload. Thus even a small mismatch between ATP production and utilization would rapidly lead to acute heart failure.

The heart is an omnivore, able to oxidize different substrates to support ATP synthesis: FFAs, glucose, lactate, and even AAs. The mitochondrial outer membrane (MOM) allows, via porins (voltage-dependent anion channel, VDAC, letting through molecules up to $\sim$5,000 kD), free solute exchange between the cytoplasm and the mitochondrial intermembrane space (MIMS), but the MIM remains impenetrable. There are several active mitochondrial transport systems for substrate transfer to the matrix; these are Ca^{2+} regulated. As the energy production relies on the availability of the substrates, cytoplasmic Ca^{2+} could regulate mitochondrial metabolism without entering the matrix via the uniporter.

Nevertheless, direct regulation of mitochondrial metabolism by Ca^{2+} has been established. Under normal conditions glucose covers $\sim$10% of the energy costs. This fraction can be increased several fold during high mechanical loads or during pathologic conditions. Even with maximal stimulation of glycolysis the FFA oxidation remains dominant. First, Ca^{2+} activates glycogen phosphorylase kinase in the cytoplasm; this subsequently phosphorylates and activates glycogen phosphorylase thus increasing the delivery of glucose for glycolysis yielding pyruvate. After energy consuming transport into the mitochondrial matrix pyruvate is decarboxylated and transformed to acetyl-CoA by the pyruvate dehydrogenase complex, which is then fed into the tricarboxylic acid cycle (TCAC). This enzyme complex comprises over 90 subunits with three functional activities: pyruvate dehydrogenase (PDH), dihydrolipoyl transacetylase, and dihydrolipoyl dehydrogenase. The activity of the PDH is dependent on Ca^{2+}, so are also isocitrate dehydrogenase and alpha-ketoglutarate dehydrogenase of the TCAC. The protons of NADH (nicotinamide adenine dinucleotide), which originate from these reactions, will be transferred to the respiratory chain. Finally, also the ATP synthase (FoF1, complex-V assembled from about 16 polypeptide subunits) is stimulated when Ca^{2+} is increased in the matrix probably by a post-translational modification not yet understood in detail, or alternatively, by association of FoF1 with the Ca^{2+}-sensor protein S100A1. The metabolic link to increased energy demand by cytoplasmic and mitochondrial Ca^{2+} is given by the following scheme:

$$\begin{aligned}
&\text{Energy demand} \uparrow \rightarrow \text{cytoplasmic } Ca^{2+} \uparrow \\
&\rightarrow \text{mitochondrial } Ca^{2+} \uparrow \\
&\rightarrow \text{activities of dehydrogenases} \uparrow \\
&\rightarrow NADH/H^+ \uparrow \rightarrow ATP \uparrow \\
&\rightarrow \text{energy supply} \uparrow
\end{aligned}$$

In fact, oxidative phosphorylation is stimulated more rapidly by Ca^{2+} than by the dephosphorylated phosphoryl carriers creatine or ADP. Thus intramitochondrial

Ca^{2+} functions as metabolic mediator by matching energy supply to actual demand. What is the effect of excess Ca^{2+} in the cytosol and especially in the mitochondria?

Mitochondrial Calcium Death

When Ca^{2+} inundates a cell because of physical cell membrane lesion or under a pathologic condition this cell is doomed to die. The site which determines over life or death is the mitochondrion that cannot fend off being swamped by the high cytoplasmic Ca^{2+}. Besides stimulating most cell activities in vain, the excess cytoplasmic Ca^{2+} dissipates energy by producing heat. In the mitochondrion excess Ca^{2+} precipitates as Ca-phosphate; this brings the respiratory chain to a halt. Normally Ca^{2+} exits the mitochondrion via the Na^+/Ca^{2+}-exchanger. However, when the cytoplasmic Ca^{2+} is very high this exchange becomes insufficient. Under such strenuous conditions when the respiratory chain can no longer sustain the proton gradient across the MIM (with a 150–180 mV potential inside negative) the FoF1 ATP synthase starts operating in reverse, that is it hydrolyzes ATP, as it does when isolated in vitro. Consumption of ATP by the FoF1 ATPase contributes to maintaining the membrane potential at suboptimal levels by pumping protons out of the matrix in order to gain time for potential recovery of the cell. During this period the adenine nucleotide translocase (ANT, ~33 kDa) also operates in reverse by letting ATP produced by anaerobic glycolysis in the cytoplasm flow into the matrix of the mitochondrion. Ca^{2+} overload, depletion of ATP and oxidative stress (reactive oxygen species, ROS) resulting from uncoupling of the respiratory chain may lead to opening of a non-specific pore in the MIM, known as the mitochondrial permeability transition pore (mPTP). Which proteins are integral constituents of this pore is still debated, but the ANT and the mitochondrial phosphate carrier (PiC, ~40 kDa) appear to be involved in the pore formation with cyclophilin-D (CyP-D, a matrix peptidyl-prolyl cis-trans isomerase, ~19 kDa) as regulator. This pore has a diameter of 1.2–1.5 nm and allows solutes passing <1,500 Da. With this free exchange of solutes between mitochondrial matrix and cell cytoplasm the residual proton gradient finally collapses and the cell death processis initiated.

There are three types of cell death pathways: apoptosis, autophagy, and necrosis. Much attention has been paid to apoptosis and autophagy because of their genetically determined or "programmed" pathways that mediate both these processes. They occur under specified conditions during development, maintenance and repair of tissues in response to extracellular or intracellular signals. Necrosis presents a more general death reaction elicited by too much Ca^{2+} in the cytoplasm and the mitochondria. The Ca^{2+}-dependent cysteine proteinases, calpains, in the cytoplasm significantly contribute to the destruction of the cell. The caspase (cysteine proteinases) cascade activation in apoptosis may be triggered by either extracellular inducers via Fas ligand and death receptor, or intrinsic inducers (stress, DNA damage, chemotherapy, irradiation). The intrinsic inducers lead to mitochondrial damage resulting in the release of pro-apoptotic proteins from the Blc2 family as well as the pro-apoptotic cytochrome-c. Both extrinsic and intrinsic apoptotic pathways merge on the Ca^{2+}-activated effector caspase3. Furthermore, characteristic pro-apoptotic proteins have been shown also to be involved in necrotic cell death. In cardiac infarction, brain stroke, atherosclerotic and other pathologic lesions excess Ca^{2+} is the decisive factor which drives the living matter to death by necrosis.

Calcium Binding Proteins in Disease

A selection of human EF-hand proteins associated with various diseases is given in Table 2. The field has exploded during the present decade; the processes are extremely complex with many of the S100 proteins specifically involved in particular diseases. To help the reader finding the primary literature references are given at the bottom of Table 2. The different specificities reflect sub-functionalization within the S100 protein family and may allow for potentially new therapeutic approaches. As mentioned in Sect. The S100 Protein Family, these proteins modify the Ca^{2+} signals and may thus not be as critical for survival as TNC and CAM. Instead, derangement of individual S100 proteins (under- or overexpression, non-lethal mutations) within certain limits may be compatible with life, though disease may develop. Many of these diseases associated with neurodegenerative processes, cardiomyopathy, or mood disorders occur later in life.

Calcium in Health and Disease, Table 2 Selected human EF-hand proteins of clinical relevance

Proteins		References
	Diagnostic biomarkers	
Cardiac troponin (TN-I; TN-T)	Myocardial infarction	Labugger et al. (2000), Eriksson et al. (2005)
S100 B	Brain damage; melanoma; mood disorders	Donato and Heizmann (2010)
S100 A2/A4/P	Cancer	Ismail et al. (2010); Sack and Stein (2009)
Calbindin-D28K	Classification of tumors;	Pelc et al. (2002), Camby et al. (1999)
S100 proteins	Improvement of prognosis	
Calretinin	Malignant mesothelioma	Raiko et al. (2010)
S100A7	Psoriasis	Wolf et al. (2010)
S100A8/A9/A12	Inflammation	Hofmann Bowman et al. (2010), Loser et al. (2010)
S100 proteins	Monitoring the development of fetal/newborn oral structures	Castagnola et al. (2011)
	Cellular markers for CNS neurons	
Parvalbumin	Subpopulation of GABAergic neurons	Heizmann and Braun (1995)
Calbindin D-28K	Neurons	Mikoshiba (2009)
Calretinin	Neurons	
Caldendrin	Neurons, postsynaptic densities	Mikhaylova et al. (2006); Rogers (1989)
	Potential gene therapy	
Parvalbumin	Heart failure	Davis et al. (2008)
S100A1	Heart failure	Rohde et al. (2011)
S100A10(p11)	Depression	Alexander et al. (2010)
	Involved in disease mechanisms	
Calpain, sorcin, calcineurin, KChIp4, S100B	Alzheimer's disease	Meydyouf and Ghysdael (2008), Zatz and Starling (2005), Trinchese et al. (2008), Sturchler et al. (2008), Leclerc et al. (2009)
CIB1/calmyrin	Cardiac hypertrophy	White et al. (2006), Heineke et al. (2010)
Parvalbumin	Fish allergy	Swoboda et al. (2007)
Procalcins (Bet V4, Phlp7)	Pollen allergens	Magler et al. (2010)
Calflagin	Parasitemia in trypanosoma brucei infection	Emmer et al. (2010)
Polycystin-2 (Ca^{2+}-channel)	Polycystic kidney disease	Petri et al. (2010)
Swiprosins	Immune and brain functions	Dutting et al. (2011)
Ca^{2+}-binding virus proteins	Viral calciomics	Zhou et al. (2009)
FH 8	Fasciola hepatica	Fraga et al. (2010)
Troponin (mutations)	Cardiomyopathy	Pinto et al. (2009)
Nesfatin/nucleobindins	Anti-obesity treatment	Shimuzu et al. (2009)

This presumably reflects cumulative perturbations each time the organism approaches stress situations until it transgresses into overt disease.

Cardiac troponin components (cTNI and cTNT) present two established biomarkers (Table 2). cTNI and cTNT together with cTNC form the Ca^{2+} binding muscle troponin complex. Several isoforms of cTNT are expressed in heart and skeletal muscle during development, while cTNI exists only in the heart. The new highly sensitive automated cTNI and cTNT tests (using polyclonal antibodies) are the "gold standard" for cardiac infarct detection with a cutoff limit at ~0.014 ng/ml in the blood. Only a hundred dead myocytes can give detectable increases in cTNI and cTNT. Two hours after onset of pain symptoms the cTN-tests already read significantly higher than in healthy controls. Serial determinations with 2–6 h intervals allow one to estimate infarct size and prognosis for the future. The cTNT or cTNI test is superior to all other clinically available biomarkers including myoglobin, MB fraction of creatine kinase (CK-MB), myeloperoxidase, and heart fatty acid binding protein. The ECG (electrocardiogram) itself is often insufficient to diagnose acute myocardial infarction since

ST-segment elevation in the ECG may occur under other conditions as well.

Gene therapy with several Ca^{2+}-dependent proteins to rescue experimental models of heart failure has been performed with mice and rats. In human severe heart failure systolic dysfunction is accompanied by downregulation of SERCA2a. Less Ca^{2+} is available in the SR and an increase of heart rate cannot compensate for the lower Ca^{2+} content. Transgenic expression of SERCA2a improves cardiac contraction in a mouse heart failure model. In isolated cardiomyocytes from rats with cardiac insufficiency transfection of S100A1 restores contractile parameters. Cardiac targeted transgenic expression of PARV in mice improves relaxation thus improving diastolic dysfunction (Sect. Parvalbumin a Delayed Calcium Buffer and Food Allergen). Although these examples provide proof of principle for gene therapy with Ca^{2+} handling proteins to restore Ca^{2+} homeostasis or to delay disease onset. It is, however, not known at present how many gene copies will be incorporated into the host genome nor where these copies will be inserted with regard to genomic regulation. A steady expression at a defined level in the target cells without affecting the endogenous gene activity profile would be required. These conditions are far from being realized in the near future.

Final Remarks

Disturbance of Ca^{2+} homeostasis can be lethal. There are many Ca^{2+} binding and Ca^{2+} regulating proteins in the cell that are interrelated in complex signaling networks; these provide a robust framework for cell function. However, under significant pathological stress or even with physical cell lesions the control over the Ca^{2+} homeostasis may get out of hand. Ca^{2+} then accumulates in the cell and invades the mitochondria, the gate to death. As Ca^{2+} is involved in almost all physiological cell functions, so it plays a key role in most, if not all, diseases. A much larger proportion of diseases than one might be inclined to believe, ultimately depends on the individual genetic makeup. The most common DNA sequence deviations are single nucleotide polymorphisms (SNPs). About 15–20 million SNPs are unevenly spread over the entire human genome, on an average one in every 100–200 nucleotides. Added to these are several thousand structural variants, for example, shorter and longer deletions, insertion, inversions, copy number variants, and segmental duplications. The various Ca^{2+}-related protein genes are estimated to harbor several thousand sequence variants giving rise to major and minor functional alterations (gain-of-function or loss-of-function). The vast majority of these variants were established thousands to millions of years ago. Most of the lethal defects have been eliminated; while those variants compatible with life have survived and become fixed in the extant genomes.

Why then do fatal diseases still develop during aging in the modern human populations? In the industrialized countries life span has almost doubled from ~40 to ~80 within little more than 100 years. This is mainly due to the introduction of improved hygiene and nutrition, as well as the prevention and treatment of infectious diseases. Despite these benefits, the longer life is burdened with ailments including chronic degenerative disease, cancer, cardiovascular disease, and neuropathy. This may be due to the accumulation of mutations for late onset diseases that have not been subjected to evolutionary pressure. Natural selection does not select for healthy octagenarians, but only for survival through the reproductive and child rearing phase in order to assure species preservation. Today's medicine is thus confronted with an elderly and disease-prone population that may be referred to as "evolutionary" or "Darwinian" medicine.

References

Alexander B et al (2010) Reversal of depressed behaviors in mice by p11 gene therapy in the nucleus accumbens. Sci Transl Med 2:54ra76

Berridge MJ (2005) Unlocking the secrets of cell signalling. Annu Rev Physiol 67:1–21

Bradshaw RA, Dennis EA (2009) Handbook of cell signaling, 2nd edn. Academic, Oxford

Brunger AT, Jin R, Breidenbach MA (2008) Highly specific interactions between botulinum neurotoxins and synaptic vesicle proteins. Cell Mol Life Sci 65:2296–2306

Camby I et al (1999) Supratentorial pilocytic astrocytomas, astrocytomas, anaplastic expression of S100 proteins. Brain Pathol 9:1–19

Carafoli E, Brini M (2008) Calcium Signalling and Disease (Subcell Biochem, vol 45. Springer, Berlin)

Castagnola P et al (2011) The surprising composition of the salivary proteome of preterm human newborn. Mol Cell Proteomics 10(1):M110.003467

Davis J, Westfall MV, Townsend D et al (2008) Designing heart performance by gene transfer. Physiol Rev 88:1567–1651

Donato R, Heizmann CW (2010) S100B protein in the nervous system and cardiovasc. Apparatus in normal and pathological conditions. Cardiovasc Psych Neurol 2010:929712

Dutting S et al (2011) Fraternal twins: Swiprosin-1/EFhd2 and Swiprosin-2/EFhd1, two homologous EF-hand containing calcium binding adaptor proteins with distinct functions. Cell Commun Signal 9:2

Emmer BT et al (2010) Calflagin inhibition prolongs host survival and suppresses parasitemia in trypanosoma brucei infection. Eukaryot Cell 9:934–942

Eriksson S et al (2005) Negative interference in cardiac troponin I immunoassays by circulating troponin autoantibodies. Clin Chem 51:839–847

Fraga H et al (2010) FH8 – a small EF-hand protein from Fasciola hepatica. FEBS J 277:5072–5085

Haiech J, Heizmann CW, Krebs J (eds) Biochim Biophys Acta Mol Cell Res (Special Issue) vol. 1793 (2009) and vol. 1813 (2011)

Heineke J et al (2010) CIB1 is a regulator of pathological cardiac hypertrophy. Nat Med 16:872–879

Heizmann CW, Braun K (1995) Calcium Regulation by Calcium-Binding Proteins in Neurodegenerative Disorders (Neurosci Intelligence Unit). Springer/Landes Company, Heidelberg/Austin

Heizmann CW, Fritz G, Schäfer BW (2002) S100 proteins: structure, functions and pathology. Front Biosci 7: d1356–d1368

Hofmann Bowman MA et al (2010) S100A12 mediates aortic wall remodeling and aortic aneurysm. Circ Res 106:145–154

Ismail TM et al (2010) Self-association of calcium-binding protein S100A4 and metastasis. J Biol Chem 285:914–922

Krebs J, Michalak M (eds) (2007) Calcium: a Matter of Life or Death (New Comprehensive Biochemistry, vol 41. Elsevier, Amsterdam

Labugger R et al (2000) Extensive troponin I and T modification detected in serum from patients with acute myocardial infarction. Circulation 102:1221–1226

Leclerc E, Heizmann CW (2011) The importance of Ca^{2+}/Zn^{2+} signaling S100 proteins and RAGE in translational medicine. Front Biosci (Schol Ed) 3:1232–1262

Leclerc E et al (2009) Crosstalk between calcium, amyloid beta and the receptor for advanced glycation endproducts in Alzheimer's disease. Rev Neurosci 20:95–110

Loser K et al (2010) The toll-like receptor 4 ligands Mrp8 and Mrp14 are crucial in the development of autoreactive CD8+ T cells. Nat Med 16:713–718

Magler I et al (2010) Molecular metamorphosis in polcalcin allergens by EF-hand rearrangements and domain swapping. FEBS J 277:2598–2619

Meydyouf H, Ghysdael J (2008) The calcineurin/NFAT signaling pathway: a novel therapeutic target in leukemia and solid tumors. Cell Cycle 7:297–303

Mikhaylova M et al (2006) Neuronal Ca2+ signaling in the brain via caldendrin and calneurons. Biochim Biophys Acta 1763:1229–1237

Mikoshiba K (ed) (2009) Handbook of Neurochemistry and Molecular Neurobiology. Neural Signaling Mechanisms, 3rd edn. Springer, Berlin

Nakayama S, Kawasaki H, Kretsinger RH (2000) Evolution of EF-hand proteins. In: Carafoli E, Krebs J (eds) Calcium Homeostasis, Topics in Biological Inorganic Chemistry. Springer, Berlin, pp 29–58

Pelc K et al (2002) Calbindin-D28k: a marker of recurrence for medulloblastomas. Cancer 95:410–419

Petri ET et al (2010) Structure of the EF-hand domain of polycystin-2 suggests a mechanism for Ca2+-dependent regulation of polycystin-2 channel activity. Proc Natl Acad Sci 107:9176–9181

Pinto JR et al (2009) A functional and structural study of Troponin C mutations related to hypertrophic cardiomyopathy. J Biol Chem 284:19090–19100. doi:10.1074/jbc.M109.007021

Raiko I et al (2010) Development of an enzyme-linked immunosorbent assay for the detection of human calretinin in plasma and serum of mesothelioma patients. BMC Cancer 10:242

Rohde D, Brinks H, Ritterhoff J et al (2011) S100A1 gene therapy for heart failure: a novel strategy on the verge of clinical trials. J Mol Cell Cardiol 50:777–784

Rogers H (1989) Immunoreactivity for calretinin and other calcium binding proteins in cerebellum. Neuroscience 31:711–721

Sack U, Stein U (2009) Wnt up your mind—intervention strategies for S100A4-induced metastasis in colon cancer. Gen Physiol Biophys 28:F55–F64

Schaub MC, Heizmann CW (2008) Calcium, troponin, calmodulin, S100 proteins: from myocardial basics to new therapeutic strategies. Biochem Biophys Res Commun 369:247–264

Schaub MC, Hefti MA, Zaugg M (2006) Integration of calcium with the signaling network in cardiac myocytes. J Mol Cell Cardiol 41:183–214

Shimuzu H et al (2009) Nesfatin-1: an overview and future clinical application. Endocr J 56:537–543

Sturchler E et al (2008) Site-specific blockade of RAGE-Vd prevents amyloid-β oligomer neurotoxicity. J Neurosci 28:5149–5158. doi:10.1523/JNEUROSCI.4878-07.2008

Swoboda I et al (2007) A recombinant hypoallergenic parvalbumin mutant for immunotherapy of IgE-mediated fish allergy. J Immunol 178:6290–6296

Trinchese F et al (2008) Inhibition of calpains improves memory and synaptic transmission in a mouse model of Alzheimer disease. J Clin Invest 118:2796–2807

White C et al (2006) CIB1, a ubiquitously expressed Ca2+-binding protein ligand of the InsP3 receptor Ca2+ release channel. J Biol Chem 281:20825–20833

Williams RJP, Frausto da Silva JJR (2006) The chemistry of evolution. The development of our ecosystem. Elsevier, Amsterdam

Wolf R et al (2010) Gene from a psoriasis susceptibility locus primes the skin for inflammation. Sci Trans Med 2:61ra90

Zatz M, Starling A (2005) Calpains and disease. N Engl J Med 352:2413–2423

Zaugg M, Schaub MC (2003) Signaling and cellular mechanisms in cardiac protection by ischemic and pharmacological preconditioning. J Muscle Res Cell Motil 24:219–249

Zhou Y et al (2009) Viral calciomics: interplays between Ca^{2+} and virus. Cell Calcium 46:1–17

Calcium in Heart Function and Diseases

Giuseppe Inesi
California Pacific Medical Center Research Institute,
San Francisco, CA, USA

Synonyms

Calcium ion; Cardiac excitation/contraction coupling; Cardiac hypertrophy; Relaxation of cardiac muscle

Definition

Ca^{2+} signaling refers to coupling of membrane excitation to contractile activation of heart muscle. Mechanisms permitting cyclic rise and reduction of Ca^{2+} concentration within cardiac muscle cells are required for optimal heart beating. Alterations of these mechanisms are followed by deficient contraction and relaxation.

Introduction

Ca^{2+} signaling serves as a common mechanism to couple membrane excitation to intracellular functions in most biological tissues (Clapham 2007). This mechanism is based on: (a) a high gradient between extracellular (or intracellular stores) Ca^{2+} concentration (mM) and intracellular (cytosolic) Ca^{2+} concentration (0.1 μM); (b) passive fluxes of signaling Ca^{2+} through selective channels from extracellular fluids (or intracellular stores) into the cytosol; (c) the presence of intracellular proteins, such as calmodulin and troponin, mediating activation of specific enzymes and/or functions upon Ca^{2+} binding; and (d) active transport of Ca^{2+} from the cytosol to extracellular fluids (or intracellular stores) to reduce the cytosolic Ca^{2+} concentration and terminate the Ca^{2+} signal.

Ca^{2+} Signaling in Cardiac Muscle

In cardiac muscle, variations of cytosolic Ca^{2+} are involved in several signaling functions including activation of transcription and contraction (Bers 2008).

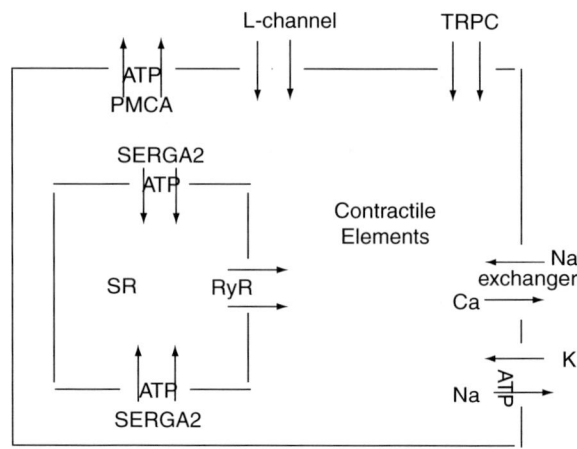

Calcium in Heart Function and Diseases, Fig. 1 *Diagram of Ca^{2+} movements that couple membrane excitation to contraction and induce subsequent relaxation in cardiac myocytes. Arrows* indicate direction of ion movements. Initial influx of trigger Ca^{2+} occurs through the voltage-dependent L-channel upon membrane depolarization. This initial Ca^{2+} influx then triggers release of a larger amount of Ca^{2+} from the sarcoplasmic reticulum (SR) through the ryanodine receptor channel (RyR). This Ca^{2+} is sufficient to raise the cytosolic Ca^{2+} concentration to the μM level and to induce contraction of myofilaments. Following repolarization of the membrane, the excess cytosolic Ca^{2+} is actively transported back into the lumen of the sarcoplasmic reticulum (SR) by SERCA2. Furthermore, a smaller amount of cytosolic Ca^{2+} is extruded from the cell by the plasma membrane ATPase (PMCA) and the Na^+/Ca^{2+} exchanger. The cytosolic Na^+ concentration is kept low by active extrusion through the Na^+/K^+ ATPase. Under certain circumstance, Ca^{2+} influx may occur though the transient receptor potential channel (TRPC)

Variation of cytosolic Ca^{2+} may occur as transient and spatially diffuse events triggered by electrical depolarization of the plasma membrane or waves and "sparks" confined to limited subcellular areas. Initiation and decay of signals is dependent on various proteins that render possible passive Ca^{2+} fluxes down a gradient, active Ca^{2+} transport against a gradient, and Ca^{2+}/Na^+ exchange sustained by respective gradients (Fig. 1).

The sequence of Ca^{2+} passive fluxes and Ca^{2+} active transport involved in contractile activation and subsequent relaxation in cardiac myocytes is shown in the diagram of Fig. 1. Initially, influx of low quantities of extracellular Ca^{2+} occurs through voltage-gated L-type Ca^{2+} channels that reside on the plasma membrane. The L-type channels are normally closed at resting membrane potential but are activated (i.e., opened) at depolarized membrane potentials. The functional importance of this channel has led to the development of pharmacological "Ca^{2+} blockers," which are currently used in treatment of cardiovascular diseases. Influx of extracellular Ca^{2+} may also occur through transient receptor potential channels (TRPC), whose importance was recognized in cardiac myocytes subjected to experimental hypertrophy. In turn, a prominent contribution to the rise of cytosolic Ca^{2+} is attributed to release from internal stores of sarcoplasmic reticulum (SR), which is elicited by the initial entry of extracellular Ca^{2+} through the L-type channels. Ca^{2+} release from the sarcoplasmic reticulum occurs through the ryanodine channel, which is the major mediator of "calcium-induced calcium release" (CICR) in myocytes.

When a Ca^{2+} signal reaches its peak, decay is due in part to return of cytosolic Ca^{2+} to the extracellular fluids, mediated by the plasma membrane Ca^{2+} ATPase (PMCA) which utilizes ATP for active transport of Ca^{2+} and by the Na^+/Ca^{2+} exchanger which allows efflux of cytosolic by utilization of the Na^+ gradient originally produced by the Na^+ pump across the plasma membrane. On the other hand, the major contribution to signal decay is due to return of cytosolic Ca^{2+} to intracellular stores, which is accomplished through active transport by the sarcoplasmic reticulum Ca^{2+} ATPase. In fact, the Ca^{2+} transport ATPase (SERCA2) of cardiac sarcoplasmic reticulum plays an important role as it fills intracellular stores with Ca^{2+} to be released to initiate contractile activation and, in turn, sequesters cytosolic Ca^{2+} to allow relaxation. It is noteworthy that SERCA2 protein is present in cardiac muscle at much higher stoichiometric levels than the PMCA or the Na^+/Ca^{2+} exchanger.

The Ca^{2+} Transport ATPase of Sarcoplasmic Reticulum

Vesicular fragments of sarcoplasmic reticulum (SR) membrane originally prepared from skeletal and cardiac muscle were referred to as "relaxing factor" since they prevented "superprecipitation" (i.e., contraction analog) of native actomyosin (containing myosin, actin, and the troponin complex) upon addition of ATP. It was soon established that this effect was produced by ATP-dependent sequestration of Ca^{2+} by the vesicles from the reaction medium. In fact, the same relaxing effect could be produced simply by Ca^{2+} chelation with EGTA added to the medium. ATP-dependent sequestration of Ca^{2+} by the SR vesicles is operated by the Ca^{2+}-activated sarco-/endoplasmic reticulum ATPase (SERCA) which is the prominent protein component of the SR membrane. The SERCA protein is encoded by three highly conserved genes (SERCA1, 2, and 3) localized on different chromosomes and undergoes alternative splicing following transcription (Periasamy and Kalyanasundaram 2007). The cardiac SR ATPase (SERCA2a) is encoded by the (human nomenclature) ATP2A2 gene, which also yields the SERCA2b isoform found in the endoplasmic reticulum of most cells. The SERCA2 gene (ATP2A2) encodes a transcript that can be alternatively spliced into three different isoforms of the SERCA protein: SERCA2a expressed predominantly in cardiac and slow twitch skeletal muscle; SERCA2b expressed in all tissues at low level including muscle and nonmuscle cells; and SERCA2c detected in epithelial, mesenchymal, and hematopoietic cell lines, primary human monocytes, and recently even in cardiac muscle.

Mechanism of ATP Utilization and Coupled Ca^{2+} Transport

The SERCA enzyme is a prototype ion-motive ATPase of the P-type family. The SERCA1 isoform, but also the SERCA2 isoform, has been studied in detail, demonstrating that the catalytic and transport cycle includes Ca^{2+} binding, ATP utilization by formation of a phosphorylated enzyme intermediate, translocation of bound Ca^{2+} across the SR membrane against a concentration gradient, and final hydrolytic cleavage of Pi from the phosphoenzyme intermediate (Inesi et al. 1990; Toyoshima and Inesi 2004). The amino acid sequence (Fig. 2) comprises ten transmembrane, mostly helical segments, where the fourth, fifth, sixth, and eighth segments contribute residues to the binding of two Ca^{2+} (TMBS) that are required for enzyme activation and undergo active transport. The SERCA headpiece includes an N domain with the ATP binding site, a P domain with the aspartyl residue undergoing phosphorylation as a catalytic intermediate, and an A domain containing the conserved TGE motif required for catalytic assistance of the final hydrolytic reaction. A high-resolution three-dimensional structure has been obtained (Fig. 2) under conditions

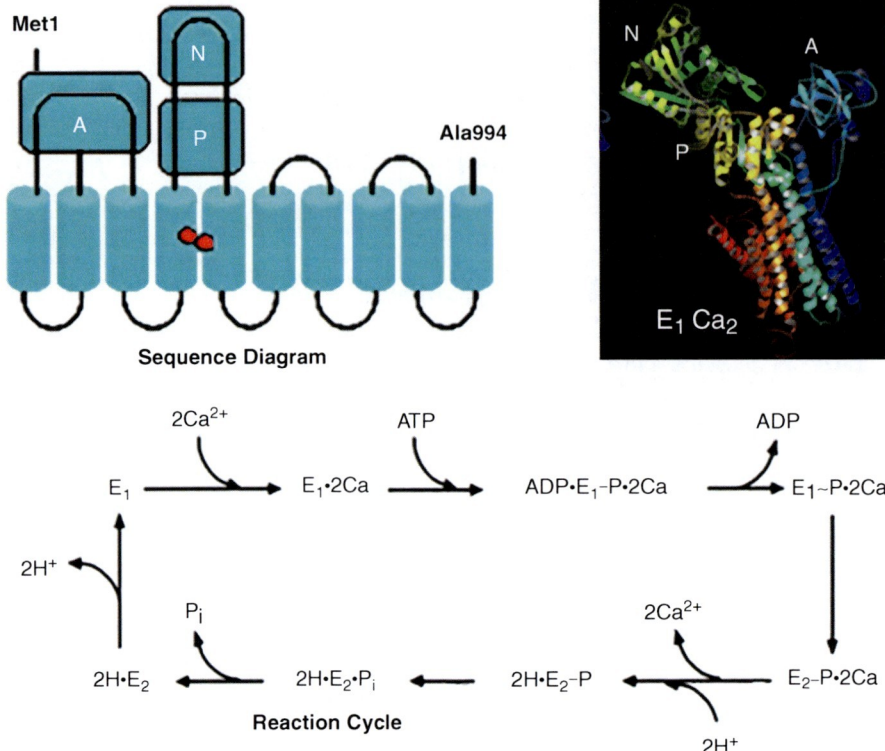

Calcium in Heart Function and Diseases, Fig. 2 Two-dimensional SERCA sequence diagram, crystal structure of the $E_1 \cdot Ca_2$ state (Protein Data Bank code 1su4), and sequential reactions comprising an ATPase catalytic and Ca^{2+} transport cycle. The sequence diagram shows ten transmembrane segments, with two *red dots* indicating two Ca^{2+}-bound. It also indicates the sequence distribution into N (ATP binding), P (phosphorylation), and A (catalytic actuator), which are then assembled in the cytosolic headpiece of the protein, as shown in the crystal structure. The membrane-bound segments containing the Ca^{2+} binding sites are then clustered in the assembled structure. Note that in the catalytic cycle diagram, two Ca^{2+} are exchanged for two H^+ (at pH 7), and the enzyme undergoes sequential reactions including phosphorylation and conformational changes. Ca^{2+} binding, phosphorylation, isomeric transition of the phosphoenzyme, and H^+ dissociation are all involved in triggering conformational changes as the cycle proceeds in the forward direction

corresponding to various steps of the reaction cycle, demonstrating that the ATPase protein undergoes a series of long-range conformational changes as the catalytic and transport cycle proceeds with sequential reactions. Diverse conformational states correspond to the enzyme with high affinity and outward orientation of the Ca^{2+} sites with respect to the lumen of SR (E1), the enzyme following binding of two Ca^{2+} ($E_1 \cdot Ca_2$), the initial phosphoenzyme intermediate obtained by utilization of ATP ($E_1 \cdot Ca_2$-P), the phosphoenzyme intermediate after a conformational change producing lower affinity and lumenal orientation of the calcium sites ($E_2 \cdot Ca_2$-P), the phosphoenzyme after luminal dissociation of (E_2-P), and the enzyme following hydrolytic cleavage of Pi that still retains low affinity and lumenal orientation of the calcium sites (E_2). At neutral pH, binding and dissociation of two Ca^{2+} is accompanied by exchange with two H^+ that derive from acidic chains of residues participating in binding (Fig. 2). It is apparent that, overall, the free energy of ATP is utilized to reduce the Ca^{2+} binding affinity of SERCA. In fact, if the free energy required for a catalytic and transport cycle is considered to be

$$\Delta G = nRT \ln\left(K_a^{CaE-P}/K_a^{CaE}\right)$$

where $n =$ two calcium ions transported per cycle, and the equilibrium constants correspond to the association constants of the enzyme for Ca^{2+} in the ground state

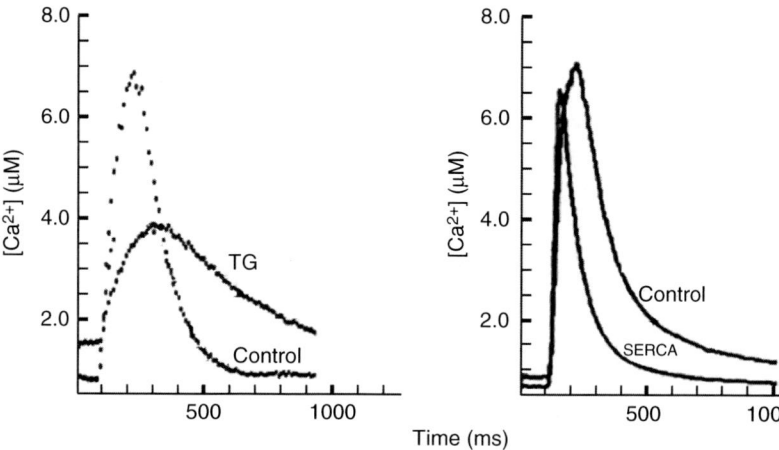

Calcium in Heart Function and Diseases, Fig. 3 *Effects of thapsigargin (TG) and heterologous SERCA expression (SERA) on cytosolic Ca^{2+} transients of cardiac myocytes subjected to electrical field stimulation.* Myocytes treated with TG (10 nM) or overexpressing exogenous SERCA following gene transfer by adenovirus vectors are compared with control myocytes. Note the reduction in peak height and decay rate following TG treatment and the increased decay rate following SERA overexpression (Adapted from Prasad and Inesi 2009; and Cavagna et al. 2000)

and following activation by ATP. The derived value corresponds approximately to the standard free energy deriving from hydrolytic cleavage of the terminal phosphate of ATP (Inesi et al. 1990).

The Sarcoplasmic Reticulum ATPase in Cardiac Muscle

From the cell physiology point of view, the important role played by SERCA in Ca^{2+} signaling is well apparent in primary cultures of cardiac myocytes, as they provide a rather simple system to study complementary features of SERCA transport activity and cytosolic Ca^{2+} signaling. In fact, myocytes subjected to electric stimuli develop cytosolic Ca^{2+} transients that include a rapid rise followed by a slower decay. The rapid Ca^{2+} rise does not show significant dependence on temperature, as expected of passive flux through a channel. On the contrary, the rate of decay exhibits strong temperature dependence, similar to that exhibited by the transport activity of isolated SR vesicles. This behavior is consistent with passive flux through a channel giving rise to the Ca^{2+} transient and dependence of the transient decay on active transport by the SERCA enzyme. The dependence of Ca^{2+} signaling on transport by SERCA can be demonstrated by inhibiting its catalytic and transport activity with thapsigargin (TG), a plant-derived sesquiterpene lactone which (when used at nanomolar concentrations) inhibits specifically SERCA activity without effecting cell growth or other cell characteristics. In cardiac myocytes subjected to electrical stimuli, it can then be shown that severe alterations of Ca^{2+} signaling are produced following specific inhibition of SERCA2 transport activity with TG (Fig. 3) or SERCA2 gene silencing with short interference RNA. In more complex models of cardiac muscle, severe alterations of Ca^{2+} signaling as well as contractile function have been demonstrated following specific inhibition of SERCA2 transport activity with TG, or reduction of expression by a SERCA2 gene null mutation. On the other hand, increased expression of endogenous SERCA, or expression of heterologous SERCA in addition to endogenous SERCA, allows faster decay of Ca^{2+} signals (Fig. 3).

A specific feature of SERCA2 in cardiac muscle is its regulation by phospholamban (MacLennan and Kranias 2003). Phospholamban is a small protein that interacts with the transmembrane and stalk regions of the SERCA protein. The functional effect resulting from this interaction is a displacement of the ATPase activation curve to a higher Ca^{2+} concentration range, therefore yielding lower activity at relevant cytosolic Ca^{2+} concentrations. This effect is overcome by phospholamban phosphorylation following adrenergic activation of protein kinase, thereby providing a mechanism for increased cardiac contraction and more efficient relaxation upon sympathetic discharge. Numerous and elegant studies with transgenic animals have shown that phospholamban is a key determinant of cardiac function and dysfunction. In addition, specific mutations in the human phospholamban gene may result in severe cardiomyopathies.

Defects of Ca^{2+} Signaling in Diseases of Cardiac Muscle

A most important point to consider regarding the involvement of Ca^{2+} ATPase in cardiac diseases is SERCA2 downregulation in cardiac hypertrophy, leading to deficient Ca^{2+} signaling and failure (Prasad and Inesi 2010). It is apparent that downregulation occurs since SERCA2 is not included in the transcriptional program leading to hypertrophy of cardiac muscle. Even though the mechanism of SERCA2 downregulation in hypertrophy is not yet clear, altered regulation of transcription and expression is undoubtedly an important factor. In addition to SERCA, other Ca^{2+}-handling proteins such as Na^+/Ca^{2+} exchanger and the TRPC proteins may be involved. Calcineurin, a Ca^{2+}/calmodulin-activated phosphatase and transcriptional activator that plays an important role in heart development and remodeling, is likely to be involved in adaptive responses to alterations of cytosolic Ca^{2+} homeostasis. However, the consequences of its activation are likely to be interwoven with additional mechanisms of transcriptional regulation that may enhance or counteract the overall effect, directing it to expression of specific proteins and excluding others. Presently, clarification of adaptive mechanisms for up- and downregulation of endogenous SERCA and other Ca^{2+}-handling proteins is a most promising endeavor in order to gain understanding and hopefully improving treatment of cardiac hypertrophy and failure.

The evidence indicating that defective expression, function, and regulation of Ca^{2+} cycling proteins is a significant factor in the pathogenesis of cardiac hypertrophy and failure has led to attempts to relieve related shortcomings by overexpression through introduction of exogenous cDNA, targeting calcium cycling proteins in heart failure through gene transfer (Del Monte and Hajjar 2003). On the other hand, development of pharmacological agents aimed at specific transcriptional pathways would be most helpful, in order to control the expression levels of calcium signaling proteins and the hypertrophy program. The present knowledge on activators and inhibitors of various enzyme involved in transcription is likely to provide a guide for evaluation of chemical structures and synthesis of new compounds. The advantage of pharmacological agents, as compared to exogenous gene transfer, is related to their easier administration and wider population targeting.

Acknowledgments The author is partially supported by NHLBI Grant RO1-69830

Cross-References

▶ Biological Copper Transport
▶ Calcium ATPase
▶ Calcium Signaling
▶ Cardiac Excitation/Contraction Coupling
▶ Cardiac Hypertrophy
▶ Na^+/Ca^{2+}-K^+ Exchanger
▶ Ryanodine Receptors (RyRs)

References

Bers DM (2008) Calcium cycling and signaling in cardiac myocytes. Annu Rev Physiol 70:23–49
Cavagna M, O'Donnell JM, Sumbilla C, Inesi G, Klein MG (2000) Exogenous Ca2+-ATPase isoform effects on Ca2+ transients of embryonic chicken and neonatal rat cardiac myocytes. J Physiol 528:53–63
Clapham DE (2007) Calcium signaling. Cell 131(6):1047–1058
Del Monte F, Hajjar RJ (2003) Targeting calcium cycling proteins in heart failure through gene transfer. J Physiol 546:49–61
Inesi G, Sumbilla C, Kirtley ME (1990) Relationships of molecular structure and function in Ca^{2+} transport ATPase. Physiol Rev 70(3):749–760
MacLennan DH, Kranias EG (2003) Phospholamban: a crucial regulator of cardiac contractility. Nat Rev Mol Cell Biol 4:566–577
Periasamy M, Kalyanasundaram A (2007) SERCA pump isoforms: their role in calcium transport and disease. Muscle Nerve 35(4):430–442
Prasad AM, Inesi G (2009) Effects of thapsigargin and phenylephrine on calcineurin and protein kinase C signaling functions in cardiac myocytes. Am J Physiol Cell Physiol 296(5): C992–C1002
Prasad AM, Inesi G (2010) Downregulation of Ca^{2+} signalling proteins in cardiac hypertrophy. Minerva Cardioangiol 58(2):193–204
Toyoshima C, Inesi G (2004) Structural basis of ion pumping by Ca2± ATPase of the sarcoplasmic reticulum. Annu Rev Biochem 73:269–292

Calcium in Nervous System

R. Scott Duncan and Peter Koulen
Vision Research Center and Departments of Basic Medical Science and Ophthalmology, School of Medicine, University of Missouri, Kansas City, MO, USA

Synonyms

Calcium channels; Calcium signaling; Neuronal calcium; Second messenger

Definition

Calcium contained or stored within neurons or extracellular calcium readily available for neurons to utilize for receptor binding, influx through channels, flux through transporters or pumps or for intracellular second messenger signaling is calcium in the nervous system.

Ca^{2+} and Ca^{2+} Signaling

The calcium ion (Ca^{2+}) is a ubiquitous ion and second messenger found in all living organisms. Ca^{2+} facilitates cellular responses to external and internal stimuli and is critical for normal cellular function and survival. It regulates the activity of a large number of enzymes involved in a variety of physiological processes. In neurons, Ca^{2+} controls many cellular processes such as synaptic plasticity, neurotransmitter release, gene expression, differentiation, and cell fate. Ca^{2+} signals activate specific cellular processes at different rates. For example, some processes such as synaptic vesicle release occurs within microseconds after Ca^{2+} signals, while neuronal gene expression can be activated hours after Ca^{2+} signals occur (reviewed by Berridge et al. 2003).

Ca^{2+} concentration gradients must be maintained within neurons to allow Ca^{2+} to be an effective second messenger. The extracellular space and the lumen of intracellular organelles such as the endoplasmic reticulum (ER), nuclear envelope (NE), and the mitochondrial matrix contain high concentrations of Ca^{2+} (low millimolar concentrations) and thus function as Ca^{2+} stores. The cytoplasm of the cell contains low Ca^{2+} concentration (high nanomolar concentrations) relative to Ca^{2+} maintained in stores thereby providing an adequate driving force for Ca^{2+} flux. In neurons, excessive accumulation of intracellular Ca^{2+} can initiate apoptosis. Ca^{2+} released from intracellular stores into the cytosol is quickly taken back up into stores or extruded out of the cell by Ca^{2+} pumps that consume energy in the form of ATP or by ion exchangers that are regulated by ion concentration gradients. Mitochondria take up Ca^{2+} from the cytosol through the Ca^{2+} uniporter located in the inner mitochondrial membrane into the mitochondrial matrix. Uptake of Ca^{2+} into the mitochondria increases mitochondrial metabolism and energy generation. Excessive Ca^{2+} uptake, however, can lead to the opening of the mitochondrial permeability transition pore facilitating apoptosis (reviewed by Rimessi et al. 2008).

While active transport keeps the concentration of Ca^{2+} in the cytoplasm low, a variety of Ca^{2+} channels generate an increase in cytoplasmic Ca^{2+} through influx of Ca^{2+} from the extracellular space or through release of Ca^{2+} from intracellular stores. In neurons, the main types of Ca^{2+} channels include voltage-gated Ca^{2+} channels (VGCCs), ligand-gated Ca^{2+} channels (LGCCs), and intracellular Ca^{2+} channels (ICCs) that might have functional characteristics similar to VGCCs and LGCCs. These channel types are activated by different stimuli and control different neuronal processes. In the neuronal postsynapse, Ca^{2+} influx occurs primarily via either glutamate receptors, VGCCs, or ICCs (reviewed by Higley and Sabatini 2008). In many neuronal types in the brain, NMDA receptors and VGCCs elicit a significant portion of Ca^{2+} influx.

Ca^{2+} Channels in the Nervous System

VGCCs are widely expressed in the nervous system and they are activated by membrane depolarization leading to influx of Ca^{2+} into the cell from the extracellular space. There are two types of VGCCs: high-voltage-activated (Ca_v1 and Ca_v2) and low voltage-activated (Ca_v3) VGCCs (reviewed in Duncan et al. 2010). The involvement of VGCCs in synaptic transmission, neuronal development, and gene expression is well documented. VGCC activity is influenced by subunit composition, phosphorylation, G-protein signaling and changes in channel subunit expression, splicing, or targeting (Koulen et al. 2005a; Martín-Montañez et al. 2010, reviewed in Duncan et al. 2010).

Ligand-gated Ca^{2+} channels are activated by a variety of small molecules and neurotransmitters to facilitate Ca^{2+} influx into neurons. One well-characterized example is the N-methyl-D-aspartate (NMDA) receptor which is activated by the excitatory neurotransmitter, L-glutamate. The NMDA receptor exhibits widespread expression in the brain and plays a crucial role in long-term potentiation (LTP) and synaptic plasticity (reviewed in Duncan et al. 2010). The NMDA receptor is regulated, in part, by magnesium (Mg^{2+}), accessory proteins, and subunit phosphorylation. Other ionotropic glutamate receptors, kainate and α-amino-3-hydroxy-5-methylisoxazole-4-propionic acid (AMPA) receptors, as well as some serotonin, acetylcholine, and purinergic receptors, are nonselective cation channels that conduct mainly sodium (Na^+) and depolarize the plasma membrane. Membrane depolarization by AMPA receptor activation relieves the Mg^{2+} block of NMDA receptors allowing them to open and conduct Ca^{2+} upon glutamate binding (reviewed in Duncan et al. 2010). Additionally, the LGCCs above can depolarize the plasma membrane thereby activating VGCCs.

Nicotinic acetylcholine receptors (nAchRs) are LGCCs widely expressed in the nervous system and are activated by the neurotransmitter, acetylcholine. Nicotinic AChRs are involved in neuronal development, synaptic plasticity, and gene expression (reviewed in McKay et al. 2007). One particular nAchR, composed of α7 subunits, is much more selective for Ca^{2+} than other nAChRs. Ca^{2+}-conducting nAChRs can activate intracellular Ca^{2+} channels such as ryanodine receptors (RyRs) via Ca^{2+}-induced Ca^{2+} release (CICR) (McKay et al. 2007).

Transient receptor potential (TRP) channels are nonselective cation channels and consist of a diverse family of receptors containing multiple subgroups (reviewed by Berridge et al. 2003). The TRP channel family is activated by a very wide range on endogenous agonists. TRP channels typically exhibit low conductances and can function over more prolonged periods of time without overwhelming the cell with Ca^{2+} (Berridge et al. 2003). TRPC plays a role in capacitive Ca^{2+} entry to refill intracellular Ca^{2+} stores (Berridge et al. 2003). TRPC channels are linked to IP_3Rs via Homer proteins stimulating opening of channels (Duncan et al. 2005).

In addition to the plasma membrane LGCCs and VGCCs, neurons also express intracellular Ca^{2+} channels (ICCs) predominantly localized to intracellular membranes such as the endoplasmic reticulum (ER) (Stevens et al. 2008). The ICCs are activated to release Ca^{2+} by second messengers such as inositol 1, 4, 5-trisphosphate (IP_3). IP_3 receptors (IP_3Rs) and RyRs are the major ICCs releasing Ca^{2+} from intracellular stores and they are widely expressed in the nervous system (Koulen et al. 2005a, b, 2008; Mafe et al. 2006; Duncan et al. 2007). Some TRP proteins of the polycystin family are also ICCs and contribute to CICR (Koulen et al. 2002; 2005c). Ca^{2+} release from and reuptake into intracellular stores control cytoplasmic Ca^{2+} concentrations as well as the spatial and temporal characteristics of cytoplasmic Ca^{2+} signals. IP_3Rs and RyRs are essential for a variety of cellular processes influenced by Ca^{2+} such as cell proliferation, development, gene expression, and neurotransmitter release.

Neurons utilize distinct signaling cascades to elicit different physiological responses from Ca^{2+} (reviewed in Berridge et al. 2003, Berridge 2006). For example, some G-protein-coupled receptors (GPCRs) that couple to the IP_3 signaling cascade elicit rapid Ca^{2+} transients, whereas others elicit Ca^{2+} responses that arise slowly but continue over a longer duration (reviewed in Berridge et al. 2003). In addition, activation of the metabotropic glutamate receptor (mGluR) mGluR1 generates single Ca^{2+} transients, while mGluR5 produces Ca^{2+} oscillations (Berridge et al. 2003).

RyRs are activated by the same ion that they conduct – Ca^{2+}. They are CICR channels as they augment Ca^{2+} signals generated from other sources (Hayrapetyan et al. 2008; Rybalchenko et al. 2008). In neurons, VGCCs emit Ca^{2+} which, in turn, activates RyRs releasing additional Ca^{2+} via CICR (Berridge et al. 2003).

RyR accessory proteins also regulate RyR activity. Homer proteins bind RyR1 thereby increasing channel open probability and Ca^{2+} release rate (reviewed in Duncan et al. 2005). Homer dimerization is important for relevant regulation of RyR1 activity (Duncan et al. 2005). The short dominant-negative form of Homer, Homer 1a, competes with the long Homer 1c form for RyR1 binding (Hwang et al. 2003, reviewed in Duncan et al. 2005). Homer 1c reduces RyR2 activity by reducing channel open probability and Homer 1a decreases this inhibition (Westhoff et al. 2003, reviewed in

Duncan et al. 2005). In addition, Homer-1 has been shown to interact with polycystin-1 in hippocampal neurons, suggesting that Homer proteins are important functional regulators of ICCs and other ion channels (Stokely et al. 2006).

In addition to Ca^{2+} channels, Ca^{2+}-binding proteins including Ca^{2+} buffers and Ca^{2+} sensors also regulate processes critical for proper neuronal function (reviewed in Amici et al. 2009). Ca^{2+} sensors such as calmodulin (CaM), hippocalcin, and neuronal Ca^{2+} sensor-1 (NCS-1) play a major role in regulating synaptic plasticity in neurons (Amici et al. 2009). In addition, Ca^{2+} buffers such as calbindin D-28, calretinin, and parvalbumin influence the spatial and temporal characteristics of a Ca^{2+} response (reviewed in Berridge et al. 2003).

Ca^{2+} and Neurotransmission

Neurotransmitter release is controlled by cytosolic Ca^{2+}. Ca^{2+} influx into synaptic terminal leads to vesicle release and priming of new synaptic vesicles (reviewed in Berridge et al. 2003). Ca^{2+}-mediated synaptic transmission occurs through various sources of Ca^{2+} including NMDA receptors and VGCCs (Berridge et al. 2003). Activation of VGCCs at synaptic terminal depolarizes the membrane and relieves the Mg^{2+} block of NMDA receptors thus allowing Ca^{2+} influx upon glutamate binding. This phenomenon is known as coincidence detection as the activation of NMDA receptors sometimes requires sufficient membrane depolarization (Berridge et al. 2003).

Action potentials along the neuronal axon generate synchronous release of neurotransmitter followed by asynchronous release which occurs at much lower rates (Neher and Sakaba 2008). Synchronous release occurs via brief Ca^{2+} microdomains that develop and quickly subside near VGCCs (Neher and Sakaba 2008). In many of the synapses studied, vesicle releases rate are affected by a narrow local Ca^{2+} concentration range (Neher and Sakaba 2008). In some synapses, short-term depression during prolonged activity is, in part, due to exhaustion of the pool of "readily releasable" vesicles (Neher and Sakaba 2008). The Ca^{2+} concentration required for vesicle recruitment is lower than that of the Ca^{2+} concentration required for vesicle release (Neher and Sakaba 2008).

Ca^{2+} release from IP_3Rs and RyRs regulates neuronal and astrocyte exocytosis (reviewed in Berridge 2006). For example, astrocytes utilize Ca^{2+} release from IP_3Rs to initiate exocytosis of glutamate-containing vesicles (Berridge 2006). Peptides released from terminals of hypothalamic neurons also require localized Ca^{2+} release from RyRs (Berridge 2006).

In the hippocampus, spontaneous transmitter release from mossy fiber boutons generates miniature excitatory postsynaptic currents (mEPSCs) in hippocampal CA3 neurons. The mEPSC frequency is normally low during periods of relative inactivity but increases considerably after stimulation of presynaptic nAChRs introducing Ca^{2+} into the cell (Berridge 2006). This Ca^{2+} influx does not initiate transmitter release, but it enhances Ca^{2+} release from RyRs to elicit spontaneous Ca^{2+} sparks that initiate exocytosis (Berridge 2006).

Ca^{2+} and Neuronal Differentiation and Development

Ca^{2+} signaling regulates the neuronal migration, differentiation, and development in the central nervous system. Neuronal precursor cells in the brain migrate from the ventricular zone (VZ) to future cortical areas during early development. Ca^{2+} regulates neuronal motility, axonal elongation and guidance, and dendritic development (reviewed in Gomez and Zheng 2006). VGCCs are a key player in regulating neuronal motility. For example, activation of N-type VGCCs influences the migration of cerebellar granule cells (reviewed in Gomez and Zheng 2006). Directional cues are, in part, mediated by intracellular Ca^{2+} gradients. The Ca^{2+} concentration within cellular microdomains, particularly near the leading neuritic processes, underlies the translocation of neuronal soma (reviewed in Gomez and Zheng 2006).

In addition, the Ca^{2+} concentration gradient range can control growth cone repulsion or attraction (reviewed in Gomez and Zheng 2006). Extension of growth cones is mediated by an optimal Ca^{2+} concentration range, and the frequency of Ca^{2+} waves influences the rate of axonal elongation (reviewed in Gomez and Zheng 2006).

The Ca^{2+}-dependent kinase, Ca^{2+}/calmodulin-dependent protein kinase (CaMK) II isoform is involved in axonal extension and guidance (reviewed in Gomez and Zheng 2006). The α-CaMKII isoform requires relatively large Ca^{2+} transients to influence axonal branching, while the β-CaMK II isoform requires lower Ca^{2+} concentration to regulate neurite extension (reviewed in Gomez and Zheng 2006). Additionally, Ca^{2+} can regulate the cGMP/cGMP-dependent kinase (PKG) as well as the Rho GTPase pathways of neurite growth and extension (reviewed in Gomez and Zheng 2006). The Ca^{2+}-dependent phosphatase, calcineurin, controls neurite extension differentially depending upon the nature of the Ca^{2+} signals (reviewed in Gomez and Zheng 2006).

Numerous neuronal functions, such as synapse development, are controlled by spatially restricted Ca^{2+} transients (reviewed in Berridge 2006). Spatially restricted Ca^{2+} signals arise in each stage of synapse development and these Ca^{2+} transients perform essential functions in the establishment and turnover of synapses (reviewed in Michaelsen and Lohmann 2010). The initial stage of synapse formation is the establishment of a contact between axons and dendrites, which is normally initiated by dendritic filipodia (Michaelsen and Lohmann 2010). Creation of these synaptic contacts generates spatiotemporally regulated Ca^{2+} transients in dendrites which, in turn, regulate this interaction (Michaelsen and Lohmann 2010). These contact-generated Ca^{2+} responses likely start in the filopodium, and spread into the dendritic shaft, where they terminate rapidly (Michaelsen and Lohmann 2010).

In developing hippocampal neurons, a rise in spontaneous Ca^{2+} signaling is associated with a reduction in filopodial motility (Michaelsen and Lohmann 2010). Elevating intracellular Ca^{2+} concentrations reduces filopodial motility, while inhibiting spontaneous dendritic Ca^{2+} responses initiates filopodial growth (Michaelsen and Lohmann 2010). Likewise, axon-dendrite contacts showing strong contact-generated dendritic Ca^{2+} signals become stabilized while dendrites exhibiting insignificant Ca^{2+} signaling become unstable (Michaelsen and Lohmann 2010).

To accomplish restricted Ca^{2+} signals in dendrites of aspiny interneurons, Ca^{2+} entry must be spatially confined and efficiently eliminated via buffering or extrusion (Michaelsen and Lohmann 2010). Synaptic Ca^{2+} transients are spatially restricted to microdomains in aspiny interneurons due in large part to the presence of Ca^{2+}-permeable AMPARs which exhibit rapid inactivation kinetics, thus limiting the overall Ca^{2+} influx (reviewed in Berridge 2006; Michaelsen and Lohmann 2010). In addition, synaptic clustering of these Ca^{2+}-permeable AMPARs thereby generate more spatially restricted Ca^{2+} transients compared to VGCCs. These Ca^{2+}-permeable AMPARs are transiently expressed in immature, developing pyramidal cells, but are generally absent later in development when spines mature (Michaelsen and Lohmann 2010).

In aspiny neurons, endogenous Ca^{2+} buffers also help to confine synaptic Ca^{2+} responses allowing for highly localized Ca^{2+} transients (Michaelsen and Lohmann 2010). Aspiny interneurons abundantly express the Ca^{2+}-binding protein parvalbumin, while young pyramidal neurons transiently express calbindin or calretinin providing adequate intracellular Ca^{2+}-buffering capability (Michaelsen and Lohmann 2010).

The confinement of Ca^{2+} transients within individual synapses may be critical for certain forms of synaptic plasticity throughout this stage of development. Young hippocampal neurons can undergo LTP prior to the formation of dendritic spines (Michaelsen and Lohmann 2010). Mature neurons require CaMKII activation through elevations in synaptic Ca^{2+} to generate LTP. Unlike in mature neurons, LTP in immature neurons requires cAMP-dependent protein kinase A instead of CaMKII (Michaelsen and Lohmann 2010).

In later developmental stages, spines appear on pyramidal neuron dendrites. Spine formation is regulated by CaMKs, including CaMKI, CaMKII, and CaMKIV, indicating that localized Ca^{2+} signals may elicit growth of individual spines (Michaelsen and Lohmann 2010). Activation of the CaMKIα isoform initiates axonal growth, while CaMKIγ activation supports dendritic outgrowth (Neal et al. 2010).

Ca^{2+} signaling from intracellular stores can regulate dendritic branching. For example, overexpression of the long Homer-1 isoform, Homer-1 L, in cerebellar Purkinje neurons results in elevated IP_3R-mediated Ca^{2+} signaling that leads to the reduction of dendritic branching (Tanaka et al. 2006).

Ca^{2+} and Neuronal Gene Expression

Postsynaptic Ca^{2+} influx generates signaling events ultimately leading to the alteration or regulation of the expression of genes involved in dendritic development and synaptic plasticity (Greer and Greenberg 2008).

Differential gene expression through Ca^{2+} signals largely depends upon which Ca^{2+} channels generate the Ca^{2+} response (Duncan et al. 2007; Greer and Greenberg 2008).

The L-type VGCCs exhibit specific biophysical properties leading to high Ca^{2+} conductance with a relatively long duration, which is necessary for increases in nuclear Ca^{2+} and likely critical for changes in gene expression (Greer and Greenberg 2008). Furthermore, the somatic localization of L-type VGCCs places the receptors in close proximity to the nucleus compared to other channels exhibiting synaptic localization thereby allowing them to generate nuclear Ca^{2+} signals important for changes in gene expression (Greer and Greenberg 2008). In addition, the expression of differentially distributed IP_3R subtypes within subcellular compartments, such as the nuclear envelope, can generate nuclear Ca^{2+} transients which likely control processes important in nuclear function such as gene expression (Duncan et al. 2007).

The Ca^{2+}-binding protein, calmodulin (CaM), is a major mediator of Ca^{2+}-induced signals. CaM can activate CamKII, which can activate the transcription factors CREB and NeuroD in the nucleus (Greer and Greenberg 2008). In addition, the MEF2 transcription factor is activated by calcineurin (Greer and Greenberg 2008).

NFkB- and NFAT-mediated gene transcription can be initiated by fast Ca^{2+} oscillations while slower oscillations stimulate NFkB (Mellström et al. 2008). This effect demonstrates how Ca^{2+} can discretely activate transcription factors. The transcription factor CREB is important in neuronal function and Ca^{2+} influx through synaptic NMDA receptors or VGCCs can activate CREB to initiate BDNF gene expression (Mellström et al. 2008). Interestingly, activation of extrasynaptic NMDA receptors counteracts this CREB-mediated activation of BDNF expression (Mellström et al. 2008). Furthermore, different Ca^{2+} signaling machinery in neurons underlies different phases of neuroprotection. Synaptic Ca^{2+} influx through NMDA receptors leads to Akt activation and short-term neuronal survival (Mellström et al. 2008). On the other hand, sustained neuroprotection involves CREB activation followed by nuclear Ca^{2+}/calmodulin activity (Mellström et al. 2008).

Some CamKII and CaMKIV forms are predominantly nuclear. Mice not expressing CaMKIV are defective in hippocampal LTP and LTD in Purkinje neurons of the cerebellum, both attributable to CREB phosphorylation status (Mellström et al. 2008).

Ca^{2+} and Neurodegeneration

In a variety of neurodegenerative diseases, normal neuronal function and viability is influenced by intracellular Ca^{2+} signaling. Excessive increases in intracellular Ca^{2+} can initiate apoptosis or necrosis depending upon the concentration, location, and duration of released Ca^{2+}. Accordingly, strategies for regulating intracellular Ca^{2+} concentrations and Ca^{2+} signaling may be useful therapeutic interventions against neurodegenerative diseases (Duncan et al. 2010; Hwang et al. 2009; Rybalchenko et al. 2009).

Age-related changes in neuronal Ca^{2+} regulation include increased intracellular Ca^{2+} concentrations, increased Ca^{2+} influx through VGCCs, and impaired mitochondrial Ca^{2+} buffering together with dysregulated intracellular Ca^{2+} stores (Duncan et al. 2010). Oxidative stress, which occurs in numerous neurodegenerative diseases, can significantly alter intracellular Ca^{2+} signaling (Kaja et al. 2010).

Excessive NMDA receptor activation by glutamate causes massive Ca^{2+} influx leading to subsequent excitotoxicity and neuronal death (Duncan et al. 2010). In retinal ganglion cells, glutamate administration stimulates NMDA receptors leading to delayed Ca^{2+} dysregulation preceeding neuronal death (reviewed in Duncan et al. 2010). Brain tissue obtained from Alzheimer's disease (AD) patients exhibits significant changes in the expression of neuronal Ca^{2+} signaling proteins (Reviewed in Duncan et al. 2010). Furthermore, NMDA receptor activation or glutamate receptor expression contributes to AD pathophysiology in experimental models (reviewed in Duncan et al. 2010). For example, Aβ reduces NMDA receptor activity and cell surface expression levels (Duncan et al. 2010).

In animal models of AD, mutations in PS1 and PS2 lead to changes in IP_3R function resulting in increased Ca^{2+} release from ER stores and decreased neuronal viability (reviewed in Duncan et al. 2010). Interestingly, IP_3 concentrations in brains of patients with AD are diminished (Duncan et al. 2010). Mutant PS-1 and triple transgenic (PS-1/PS-2/APP) mice exhibit elevated RyR levels, increased Ca^{2+} release from RyRs, and increased vulnerability to Aβ (Duncan et al. 2010).

Furthermore, exposure of cultured neurons to Aβ elevates RyR3 expression and activity resulting in neuronal death (Duncan et al. 2010). In accordance, AD patients exhibit alterations in RyR expression and ryanodine binding in affected brain regions (Duncan et al. 2010).

Recently, the role of PS1 and PS2 on intracellular Ca^{2+} signaling has been examined at the molecular level. Application of an 82 amino acid PS1 fragment to the cytoplasmic side of RyR increases single channel RyR current and open probability (Rybalchenko et al. 2008). Similarly, application of an 87 amino acid PS2 fragment significantly increases RyR single channel activity without affecting Ca^{2+} sensitivity of the receptor (Hayrapetyan et al. 2008). These effects of PS1 and PS2 may represent a role for presenilins in regulating RyR activity and Ca^{2+} homeostasis in general.

Acknowledgments Partial support by Grants EY014227 and EY022774 from NIH/NEI; RR027093 and RR022570 from NIH/NCRR; and AG010485, AG022550, and AG027956 from NIH/NIA as well as by a Research to Prevent Blindness Challenge Grant and by the Felix and Carmen Sabates Missouri Endowed Chair in Vision Research (P.K.) is gratefully acknowledged.

Cross-References

- ▶ Calcium and Extracellular Matrix
- ▶ Calcium and Mitochondrion
- ▶ Calcium and Viruses
- ▶ Calcium ATPase
- ▶ Calcium in Health and Disease
- ▶ Calcium in Heart Function and Diseases
- ▶ Calcium in Vision
- ▶ Calcium Ion Selectivity in Biological Systems
- ▶ Calcium, Neuronal Sensor Proteins
- ▶ Calcium, Physical and Chemical Properties
- ▶ Calcium Sparklets and Waves
- ▶ Calcium-Binding Protein Site Types
- ▶ Calcium-Binding Proteins, Overview
- ▶ Calmodulin
- ▶ Calnexin and Calreticulin
- ▶ Calsequestrin
- ▶ EF-Hand Proteins
- ▶ Penta-EF-Hand Calcium-Binding Proteins
- ▶ S100 Proteins
- ▶ Sarcoplasmic Calcium-Binding Protein Family: SCP, Calerythrin, Aequorin, and Calexcitin

References

Amici M, Doherty A, Jo J, Jane D, Cho K, Collingridge G, Dargan S (2009) Neuronal calcium sensors and synaptic plasticity. Biochem Soc Trans 37:1359–1363

Berridge MJ (2006) Calcium microdomains: organization and function. Cell Calcium 40:405–412

Berridge MJ, Bootman MD, Roderick HL (2003) Calcium signaling: dynamics, homeostasis and remodeling. Nat Rev Mol Cell Biol 4:517–529

Duncan RS, Hwang SY, Koulen P (2005) Effects of Vesl/Homer proteins on intracellular signaling. Exp Biol Med 230(8):527–535

Duncan RS, Hwang SY, Koulen P (2007) Differential inositol 1,4,5-trisphosphate receptor signaling in a neuronal cell line. Int J Biochem Cell Biol 39:1852–1862

Duncan RS, Goad DL, Grillo MA, Kaja S, Payne AJ, Koulen P (2010) Control of intracellular calcium signaling as a neuroprotective strategy. Molecules 15:1168–1195

Gomez TM, Zheng JQ (2006) The molecular basis for calcium-dependent axon pathfinding. Nat Rev Neurosci 7:115–125

Greer PL, Greenberg ME (2008) From synapse to nucleus: calcium-dependent gene transcription in the control of synapse development and function. Neuron 59:846–860

Hayrapetyan V, Rybalchenko V, Rybalchenko N, Koulen P (2008) The cytosolic N-terminus of presenilin-2 increases single channel activity of brain ryanodine receptors through direct protein-protein interaction. Cell Calcium 44:507–518

Higley MJ, Sabatini BL (2008) Calcium signaling in dendrites and spines: practical and functional considerations. Neuron 59:902–913

Hwang SY, Wei J, Westhoff JH, Duncan RS, Ozawa F, Volpe P, Inokuchi K, Koulen P (2003) Differential functional interaction of two Vesl/Homer protein isoforms with ryanodine receptor type 1: a novel mechanism for control of intracellular calcium signaling. Cell Calcium 34:177–184

Hwang JY, Duncan RS, Madry C, Singh M, Koulen P (2009) Progesterone potentiates calcium release through IP_3 receptor by an Akt-mediated mechanism in hippocampal neurons. Cell Calcium 45:233–242

Kaja S, Duncan RS, Longoria S, Hilgenberg JD, Payne AJ, Desai NM, Parikh R, Burroughs SL, Gregg EV, Goad DL, Koulen P (2010) Novel mechanism of increased Ca^{2+} release following oxidative stress in neuronal cells involves type 2 inositol-1,4,5-trisphosphate receptors. Neuroscience 175:281–291

Koulen P, Cai Y, Geng L, Maeda Y, Nishimura S, Witzgall R, Ehrlich BE, Somlo S (2002) Polycystin-2 is an intracellular calcium release channel. Nat Cell Biol 4:191–197

Koulen P, Duncan RS, Liu J, Cohen NE, Yannazzo JA, McClung N, Lockhart CL, Branden M, Buechner M (2005a) Polycystin-2 accelerates Ca^{2+} release from intracellular stores in C. elegans. Cell Calcium 37:593–601

Koulen P, Liu J, Nixon E, Madry C (2005b) Interaction between mGluR8 and calcium channels in photoreceptors is sensitive to pertussis toxin and occurs via G protein βγ subunit signaling. Invest Ophthalmol Vis Sci 46:287–291

Koulen P, Wei J, Madry C, Liu J, Nixon E (2005c) Differentially distributed IP$_3$ receptors and Ca^{2+} signaling in rod bipolar cells. Invest Ophthalmol Vis Sci 46:292–298

Koulen P, Madry C, Duncan RS, Hwang JY, Nixon E, McClung N, Gregg EV, Singh M (2008) Progesterone potentiates IP$_3$-mediated calcium signaling through Akt/PKB. Cell Physiol Biochem 21:161–172

Mafe O, Gregg E, Medina-Ortiz W, Koulen P (2006) Localization of inositol 1, 4, 5-trisphosphate receptors in mouse retinal ganglion cells. J Neurosci Res 84:1750–1758

Martín-Montañez E, Acevedo MJ, López-Téllez JF, Duncan RS, Mateos AG, Pavía J, Koulen P, Khan ZU (2010) Regulator of G-protein signaling 14 protein modulates Ca^{2+} influx through Cav1 channels. Neuroreport 21:1034–1039

McKay BE, Placzek AN, Dani JA (2007) Regulation of synaptic transmission and plasticity by neuronal nicotinic acetylcholine receptors. Biochem Pharmacol 74:1120–1133

Mellström B, Savignac M, Gomez-Villafuertes R, Naranjo JR (2008) Ca^{2+}-operated transcriptional networks: molecular mechanisms and in vivo models. Physiol Rev 88:421–449

Michaelsen K, Lohmann C (2010) Calcium dynamics at developing synapses: mechanisms and functions. Eur J Neurosci 32(2):218–223

Neal AP, Molina-Campos E, Marrero-Rosado B, Bradford AB, Fox SM, Kovalova N, Hannon HE (2010) CaMKK-CaMKI signaling pathways differentially control axon and dendrite elongation in cortical neurons. J Neurosci 30(8):2807–2809

Neher E, Sakaba T (2008) Multiple roles of calcium ions in the regulation of neurotransmitter release. Neuron 59:861–872

Rimessi A, Giorgi C, Pinton P, Rizzuto R (2008) The versatility of mitochondrial calcium signals: from stimulation of cell metabolism to induction of cell death. Biochim Biophys Acta 1777:808–816

Rybalchenko V, Hwang SY, Rybalchenko N, Koulen P (2008) The cytosolic N-terminus of Presenilin-1 potentiates mouse ryanodine receptor single channel activity. Int J Biochem Cell Biol 40:84–97

Rybalchenko V, Grillo MA, Gastinger MJ, Rybalchenko N, Payne AJ, Koulen P (2009) The unliganded long isoform of estrogen receptor beta stimulates brain ryanodine receptor single channel activity alongside with cytosolic Ca^{2+}. J Recept Signal Transduct Res 29(6):326–341

Stevens S, Duncan S, Koulen P, Prokai L (2008) Proteomic analysis of mouse brain microsomes: identification and bioinformatic characterization of endoplasmic reticulum proteins in the mammalian central nervous system. J Proteome Res 7:1046–1054

Stokely M, Hwang SY, Hwang JY, Fan B, King M, Inokuchi K, Koulen P (2006) Polycystin-1 can interact with Homer 1/Vesl-1 in postnatal hippocampal neurons. J Neurosci Res 84:1727–1737

Tanaka M, Duncan RS, McClung N, Yannazzo JS, Hwang SY, Marunouchi T, Inokuchi K, Koulen P (2006) Homer proteins control neuronal differentiation through IP$_3$ receptor signaling. FEBS Lett 580:6145–6150

Westhoff JH, Hwang SY, Duncan RS, Ozawa F, Volpe P, Inokuchi K, Koulen P (2003) Vesl/Homer proteins regulate ryanodine receptor type 2 function and intracellular calcium signaling. Cell Calcium 34:261–269

Calcium in Vision

Jason D. Kenealey[1], Arthur S. Polans[2] and Nansi Jo Colley[2]
[1]Department of Biomolecular Chemistry, University of Wisconsin, Madison, WI, USA
[2]Department of Ophthalmology and Visual Sciences, UW Eye Research Institute, University of Wisconsin, Madison, WI, USA

Synonyms

Calcium ions in eyesight; Calcium regulation of phototransduction and light adaptation

Definition

Phototransduction refers to the process of converting light into electrical signals underlying the neural activity of the retina. Photoreceptors, the site of transduction in both invertebrate and vertebrate vision, can modulate their cellular responses as a function of light intensity and background illumination in order to cope with the wide range of lighting in the environment. These adaptation and transduction mechanisms rely on calcium signaling. A variety of ocular pathologies, including different forms of retinal degeneration and the aberrant growth of retinal precursors or other cell types in the eye, often are due to mutations in gene products regulating calcium. The discovery of these events is now providing opportunities for intervention in the treatment of very disparate diseases of the eye.

Calcium in Vertebrate Photoreceptors

Vertebrate photoreceptor cells can detect the absorption of a single photon of light and continue to respond in a near linear fashion over four to six log units of light intensity before the response saturates. The intracellular concentration of calcium and the "toolbox" of ▶ calcium-binding proteins found in the photoreceptor cell contribute to this process, modulating the range of light intensity to which the cell can respond (Nakatani et al. 2002). Photoreceptor cells also are capable of adjusting their responses to compensate for changes in background illumination. This sort of adaptation allows one to see in a darkened theater and then to drive a car upon exiting from a matinee on a sunny day. While the intricacies of the molecular mechanisms underlying these events are not fully understood, the role of calcium and its targets has been partially elucidated. Of further significance, the binding proteins that mediate the effects of calcium also have been implicated in a variety of ocular diseases.

Phototransduction

Vertebrate photoreceptor cells are comprised of an inner and an outer segment. The inner segment contains the organelles characteristic of a eukaryotic cell and maintains synaptic contact with the next layer of retinal neurons. A thin modified cilium connects the inner segment to the highly specialized outer segment. Of the two types of photoreceptor cells, rods and cones, found in the vertebrate retina, rods are better studied, but the basics of phototransduction are believed to be the same in cones (Yau and Hardie 2009). The rod outer segment is comprised of a plasma membrane enclosing a separate stack of flattened saccules referred to as disks (Fig. 1a). Approximately 1,000–2,000 of these disks per rod outer segment house the roughly 10^7–10^9 visual pigment molecules, rhodopsin. Rhodopsin consists of a chromophore, the 11-*cis* aldehyde of vitamin A (11-*cis* retinal) bound in the dark to the protein opsin, an intrinsic membrane protein. Cone cells use the same chromophore but differ in their types of opsin, resulting in different spectral properties underlying the detection of color. The direct effect of light is to cause the isomerization of 11-*cis* retinal to its all-*trans* configuration. In vertebrate photoreceptor cells this results in the dissociation of retinal from opsin and its subsequent recycling through the pigmented cells lying adjacent to the outer segments. Isomerization also initiates the phototransduction enzymatic cascade (Fig. 1b).

Rhodopsin is perhaps the best characterized member of the large family of G-protein coupled receptors (GPCRs). Following photoisomerization of the chromophore, rhodopsin binds a G-protein, referred to in the earlier literature as transducin (G_t). Once G_t binds to photoactivated rhodopsin, GTP is exchanged for GDP causing the alpha submit ($G_{t\alpha}$) of the heterotrimeric G-protein to dissociate. $G_{t\alpha}$ subsequently interacts with the inhibitory gamma subunits of the disk-associated cGMP-phosphodiesterase (PDE), resulting in PDE activation.

cGMP acts as the internal transmitter mediating between the absorption of light by rhodopsin in the disk membranes and the change in conductance of the outer segment plasma membrane. Cation channels in the outer segment plasma membrane are directly gated by cGMP. These channels remain open in the dark when cGMP levels are high. Activation of PDE and its subsequent hydrolysis of cGMP cause these cation channels to close. A hyperpolarization of the photoreceptor cell ensues, ultimately leading to a decrease in the release of synaptic transmitter and conveyance of the light signal to the next order of retinal neurons.

To inactivate these light events and reset the system, several counter measures must be instigated. First, rhodopsin is phosphorylated by a specific kinase (GRK1), causing the binding of an additional protein, arrestin. These two events, phosphorylation and the binding of arrestin, block the activation of further G-proteins by photoactivated rhodopsin. Next, GTP is hydrolyzed to GDP by the intrinsic GTPase activity of $G_{t\alpha}$. The fidelity of this process is aided by a GTPase-activating protein (GAP) complex. Following the hydrolysis of GTP, the PDE-$G_{t\alpha}$ complex dissociates; PDE then binds once again with its inhibitory subunits and $G_{t\alpha}$ with its beta and gamma subunits. These steps ensure that no further cGMP is hydrolyzed. Guanylate cyclase (GC) in the outer segment then converts GTP to cGMP and the cation channels in the plasma membrane reopen. Newly delivered 11-*cis* retinal binds to opsin, arrestin uncouples, rhodopsin is dephosphorylated, and the system is essentially reset to receive the next photon.

While sodium ions comprise the majority of the cations conducted through the cGMP-gated channels of the outer segment plasma membrane, calcium ions make up roughly 15% of the current. Therefore, in the dark when the cGMP-gated channels are open, the intracellular concentration of calcium is high, approximately 500–700 nM. Calcium is then returned to the extracellular milieu by its extrusion from the outer segment through the activity of plasma membrane Na^+/Ca^{2+}-K^+ exchangers. The rate of the exchanger is unaffected by light, so the closure of the cGMP-gated channels upon illumination decreases the amount of calcium entering the cell, and the intracellular concentration of calcium drops as the exchanger continues to operate. This decline in intracellular calcium upon light exposure in turn activates GC, mediated by specific calcium-binding proteins referred to as GCAPs (guanylate cyclase–activating proteins). GCAPS are 23 kDa members of the ▶ EF-hand superfamily of calcium-binding proteins. Three mammalian isoforms of GCAPs (GCAP1-3) have been identified in the retina, each containing four EF-hand motifs, but only EF-hands 2–4 are capable of binding calcium. GCAPs, in their calcium-bound form, inhibit GC activity. Therefore, when calcium levels are highest in the dark GC activity is marginal owing to the inhibitory effect of calcium-bound GCAP. As calcium levels fall upon illumination, GCAP in its unbound form is no longer inhibitory and GC synthesizes cGMP to return the cell to its dark-adapted state. In this manner, calcium acts as a counter measure to light, thus allowing the photoreceptor cell to extend its responsive range of light intensity. In knock-out mice lacking GCAP, rod cells have a diminished capacity to return to dark-adapted conditions following light stimulation, thus supporting the contention that calcium-regulated GC activity mediated by GCAP is necessary for adaptation.

The decline in calcium concentration following photoactivation has at least two additional consequences for adaptation. GRK1 activity is thought to be regulated by another EF-hand calcium-binding protein, which is relatively specific for photoreceptor cells. It is known by several names in the literature: recoverin, S-modulin, and visinin. This protein also contains four EF-hand motifs. Like calmodulin, EF1 and EF2 in recoverin interact to form one domain, while EF3 and EF4 form a second domain. However, recoverin overall takes a globular shape with the two domains in close proximity, connected through a short U-shape linker, rather than the long central helix providing the classic dumbbell shape of calmodulin. In dim light, calcium-bound recoverin purportedly interacts with GRK1, slowing rhodopsin phosphorylation and the subsequent binding of arrestin. As the light intensity increases and calcium levels in the photoreceptor cell decline, the interaction with recoverin diminishes, allowing GRK1 to phosphorylate rhodopsin and arrestin to bind, thus reducing the lifetime of photoactivated rhodopsin. Again, calcium acts as a counter measure to light, in this instance by negatively regulating the activity of GRK1 through recoverin.

Finally, the affinity of the cGMP-gated channel is modulated by the calcium concentration. In the dark when calcium levels in the outer segment are elevated, the affinity of the plasma membrane cation channels for cGMP is lower. This is thought to be due to the negative interaction of calmodulin or a similar EF-hand calcium-binding protein with the channel. As the intracellular concentration of calcium declines upon illumination, the modulatory affect of calmodulin also decreases and the affinity of the channel to cGMP increases. This allows channels to reopen, thus accelerating the return to the dark-adapted state.

In summary, the light-dependent decline in intracellular calcium can affect the synthesis of cGMP through the action of GCAPs, regulate the lifetime of photoactivated rhodopsin by recoverin, and alter the sensitivity of the cation channel through calmodulin or a comparable calcium-binding protein. These processes then contribute to the photoreceptor's ability to adapt to different levels of illumination in the host's environment, a key feature of vision.

Ocular Pathologies in Humans

(a) Degenerative Cone Diseases – Patients afflicted with autosomal dominant cone dystrophies experience impaired central and color vision. Mutations underlying these dystrophies are heterogeneous but often involve genes expressed in photoreceptor cells. Some involve phototransduction-specific genes encoding, for example, visual pigment proteins, while others affect gene products expressed at the other end of the cell at the synaptic processes. The most extensively studied genes contributing to cone dystrophies, however, are associated with cGMP synthesis involving GC and GCAP1 (Jiang and Baehr 2010). GC mutations often involve its dimerization domain, while mutations in GCAP1 primarily involve residues flanking the EF-hands or residues within the EF-hand calcium-binding loops. GCAP1(Y99C) involves the replacement of a hydrophobic tyrosine residue with cysteine flanking EF-hand 3. This alters both the structure and the calcium binding of EF-hand 3. EF-hand 3 is critical in transforming GCAP1 from an activator to an inhibitor upon binding calcium. In contrast to a flanking residue, GCAP1(N104K) contains a mutation within the loop region of EF3 causing a significant reduction in the binding affinity for calcium. This renders the GCAP1 mutant

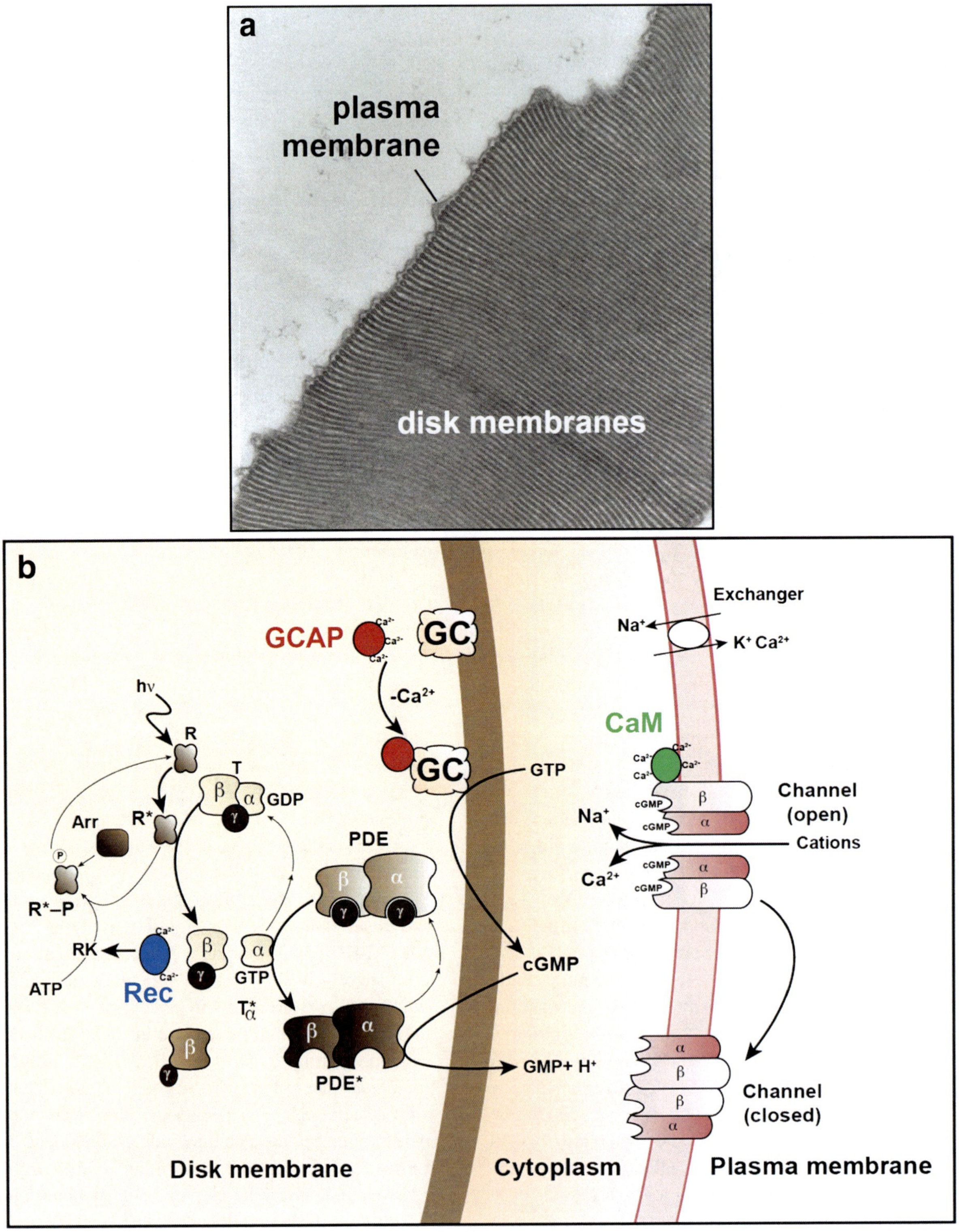

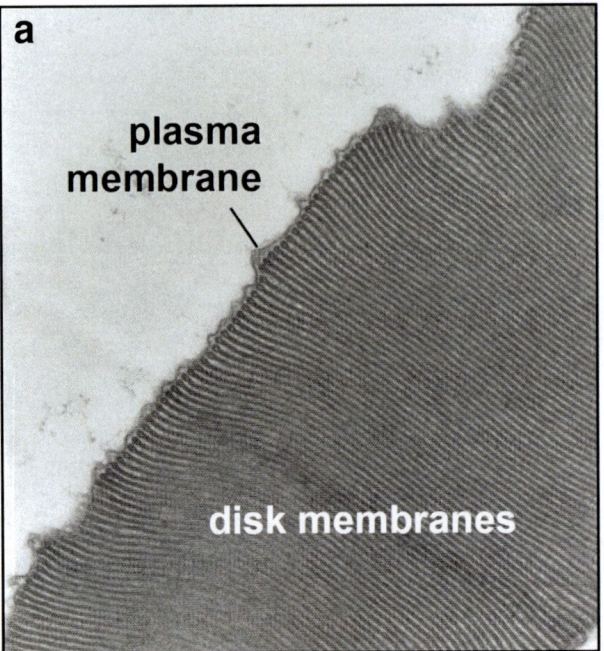

Calcium in Vision, Fig. 1 (continued)

unable to inhibit GC even at elevated levels of calcium. Since calcium-bound GCAP1 is necessary to inhibit GC, both mutations are predicted to result in elevated levels of cGMP and a greater frequency of open cation channels, thus allowing more persistent entrance of calcium into the photoreceptor outer segment. Calcium is a potent apoptotic signal, perhaps underlying the ensuing degeneration of the cone photoreceptors.

Similar to Y99C, a mutation in a flanking hydrophobic residue at EF-hand 4, GCAP1(I143NT), also has been identified in a patient population afflicted with cone dystrophy. The mutation results in the reorientation of the N-terminal helix, lowering the affinity for calcium binding at EF4. GCAP1(E155G) changes an invariant glutamate residue at position 12 of EF4, essential for calcium binding. Again, these mutations are expected to result in elevated GC activity in dark-adapted cone cells and aberrant levels of cGMP and channel conductance.

(b) Cancer-Associated Retinal Degenerations – Paraneoplastic syndromes encompass a spectrum of neurodegenerative diseases associated with the early onset of cancer in a portion of the body distinct from the degenerative site in the nervous system. These remote effects of cancer are thought to be mediated by an autoimmune reaction. *Can*cer-*A*ssociated *R*etinopathy (CAR) is one such disease, whereby a tumor, often a small cell carcinoma of the lung, initiates the loss of photoreceptor cells in the retina. The calcium-binding protein recoverin is directly involved in this effect (Ohguro and Nakazawa 2002).

Recoverin has been shown to be expressed in some primary tumors. Since recoverin is normally found in the eye, an immune privileged site, its presence in the tumor leads to an immune response in the patient. Both antibodies and activated T cells specific for recoverin have been found in CAR patients. The autoantibodies react to several different regions of the recoverin sequence, but an immunodominant region was identified around residues 64–70. When anti-recoverin antibodies are generated in animals, the same regions of recoverin are antigenic and residues 64–70 remain dominant. This region of recoverin corresponds to the first α-helix of EF-hand 2, consisting primarily of hydrophobic amino acids that normally would not be expected to be so antigenic. Recoverin is normally expressed at high levels in both rod and cone cells, localized throughout the cells, from the tips of their outer segments to their synaptic processes. Recoverin also has been found in a subset of bipolar and non-retinal cells but is considered essentially photoreceptor-specific. In rodents inoculated with recoverin, the resultant antibodies are taken up by the photoreceptor cells and then activate an apoptotic process leading to the degeneration of both rod and cone cells. A peptide corresponding to residues 64–70 also induces photoreceptor degeneration. Therefore, this unsuspecting region of recoverin is both immunodominant and immunopathogenic. (A second pathogenic site in recoverin corresponding to residues 136–167 also has been identified). Importantly, inoculation with recoverin or its immunopathogenic regions recapitulates what happens in humans with CAR, whereby widespread photoreceptor degeneration is observed. The remaining retinal layers are intact, but without photoreceptor cells to transduce light into the electrical activity of the nervous system the afflicted individuals are left visually impaired or blind depending upon the extent of photoreceptor cell damage. Patients are often treated with immunosuppressants such as prednisone in an attempt to

Calcium in Vision, Fig. 1 Phototransduction in the vertebrate rod photoreceptor cell. The vertebrate rod outer segment contains a stacked series of double-membranous disks surrounded by a plasma membrane, as seen in this electron micrograph. (**a**) Disk membranes are densely packed with the visual pigment molecule, rhodopsin, to optimize photon capture. Other components of the phototransduction enzymatic cascade involved in the regulation of cytoplasmic cGMP are associated with the cytoplasmic surface of the disks. (**b**) Channels in the plasma membrane are modulated by cGMP and thereby the entrance of sodium and calcium ions. Sodium ions determine to the larger extent the membrane potential and the subsequent signaling of synaptic transmitter release, while calcium ions regulate adaptation through a series of calcium-binding proteins (see text). Abbreviations: *hv* light, *R* rhodopsin, *R** photoactivated rhodopsin, *RK* rhodopsin kinase, *R*-P* phosphorylated rhodopsin, *Arr* arrestin, *T* transducin, *PDE* phosphodiesterase, *GC* guanylate cyclase, *GCAP* guanylate cyclase–activating protein, *Rec* recoverin, *CaM* calmodulin

dampen the immunological response to recoverin's aberrant expression in tumor cells. Potentially more relevant, the visual symptoms often precede the detection of cancer, so once autoantibodies to recoverin are detected in the patient's serum, that individual can be moved from the ophthalmology setting to an oncology clinic where the tumor can be treated. If the tumor at these early stages can be eradicated then the antigen, recoverin, is no longer a source for the autoimmune-mediated loss of photoreceptor cells. Therefore, the combined suppression of existing anti-recoverin antibodies and the elimination of the source of recoverin offer the best hope for salvaging vision.

While many diseases are thought to be mediated by autoimmunity, they often remain invalidated. CAR is one of the best examples of autoimmunity, since the antigen, recoverin, has been found in tumor cells of patients afflicted with the visual disorder, the immune response leads to specific anti-recoverin antibodies and specific activated T cells, the antigen is normally localized in photoreceptor cells, the site of degeneration, and the disease can be recapitulated in animal models using recoverin.

(c) Uveal Melanoma – The primary malignancy of the eye originates in the pigmented structures of the iris, choroid, and ciliary body, collectively referred to as the uvea. Each of these structures contains melanocytes that can become transformed into melanomas. One of the genes expressed at much lower levels in melanomas compared to their normal cell counterparts is ALG-2, *A*poptosis *L*inked *G*ene-2 or PCD6, *P*rogrammed *C*ell *D*eath Protein 6. ALG-2 is pro-apoptotic. Transient transfection studies to reduce the expression of ALG-2 protected cells from apoptosis. In contrast, its overexpression renders cells more susceptible to apoptotic stimuli. The conclusion from such experiments is that lowering the expression of ALG-2 might interfere with normal execution pathways associated with programmed cell death. It therefore might be advantageous to melanoma cells, as well as other cancer cell types, to downregulate ALG-2 and thereby improve their survival status.

ALG-2 forms dimers in cells independent of calcium concentration; however, calcium is essential for its interaction with such targets as Alix/AIP. ALG-2 is a member of the ▶ penta-EF-hand calcium-binding protein family, and mutations in either EF-hands 1 or 3 eliminate binding to Alix/AIP. Mutations in EF-hand 5 reduce but do not abrogate binding. These three EF-hands correspond to the two high affinity and one low affinity calcium-binding domains in ALG-2. It is thought that ALG-2 might function in the ER stress response. The downregulation of ALG-2 therefore could protect cancer cells by short-circuiting the connection between elevated levels of calcium associated with ER stress and the subsequent activation of cell death pathways.

Interestingly, calcium could play a significant role in the eventual treatment of uveal melanoma, as well as other types of cancer and neurodegenerative diseases (van Ginkel et al. 2008). Certain nontoxic natural products, for example, resveratrol, EGCG, and quercetin, display antiproliferative, anti-angiogenic, and pro-apoptotic features during the treatment of uveal melanoma and other cancers. These compounds are especially important since they can differentiate between tumor cells and normal cells to a greater extent than conventional chemotherapeutics. They activate the ER stress pathway as well as the intrinsic or mitochondrial apoptotic pathway in tumor cells. They also can inhibit endothelial cell migration and tube formation involved in new blood vessel formation – a requirement for tumor growth. In both tumor and endothelial cells, but not most normal cells, these nontoxic compounds induce calcium signals. The endoplasmic reticulum (ER) appears to be the initial source of calcium, whereby resveratrol and the other compounds cause a rapid rise in cytoplasmic calcium by releasing ER stores, likely through the activation of the IP_3 receptor and/or partial inhibition of SERCA, the pump necessary for sequestering calcium in the ER. Following the first rise in cytoplasmic calcium, mitochondria respond by taking up calcium, contributing to their depolarization and a calcium-induced calcium release, constituting a second, longer increase in cytoplasmic calcium. The rise in cytoplasmic calcium is then linked to many events, including the activation of the intrinsic apoptotic pathway ultimately involving caspase-3. The elevated levels of calcium also activate the calpains, a second family of cysteine proteases. Targets of calpain then include plasma membrane calcium pumps and exchangers, resulting in longer-term disruption of calcium homeostasis and tumor cell death.

Summary

The fundamentals of the phototransduction enzymatic cascade have been deciphered, and numerous gene mutations in components of the pathway leading to visual impairment and blindness have been identified. Since the retina is the most accessible portion of the central nervous system, lessons from its study may extend to the brain and other tissues. Not surprisingly, the retina has provided invaluable information about the normal physiology and pathology of calcium and calcium-binding proteins. This entry has focused on the calcium-binding proteins most readily associated with diseases of the human eye. Their further study represents an opportunity to bridge structural biology and key cellular pathways with aberrations leading to human disease, as well as new therapies based on the intricacies of calcium signaling. In addition, there is a host of other calcium-binding proteins that already have been identified in the retina, such as calretinin, the centrins, and certain ▶ neuronal calcium sensor proteins. These additional calcium-binding proteins await further examination to more definitively identify their function and possible connection with ocular pathologies.

Calcium in Invertebrate Photoreceptors

Phototransduction

The Drosophila compound eye is comprised of ~800 individual eye units called ommatidia. Each ommatidium contains eight photoreceptor cells. The R1-6 photoreceptor cells extend the length of the retina and express the major rhodopsin in the eye, Rh1 (rhodopsin1), which absorbs maximally in the blue-green region of the spectrum (λ_{max} 480 nm). Rh1, encoded by the *ninaE* gene, displays 22% amino acid identity with human rhodopsin. The R7 photoreceptor cells are located distally and express two ultraviolet-sensitive, Rh3 and Rh4, opsins. The R8 photoreceptors are proximal and express Rh5 (blue) and Rh6 (green) opsins. Like vertebrate cone photoreceptors, R7 and R8 cells function in color vision, while the R1-6 photoreceptor cells are specialized for several functions including brightness detection, orientation behavior, and motion detection (Vogt and Desplan 2007).

Each photoreceptor cell contains a photoreceptive rhabdomere, which is comprised of numerous tightly packed microvilli that contain the rhodopsin photopigments and the other components of phototransduction (Fig. 2). The rhabdomeres are functionally equivalent to the outer segments of the rod and cone photoreceptor cells in the vertebrate retina (Colley 2010; Fain et al. 2010; Yau and Hardie 2009).

Drosophila visual transduction is the fastest known G-protein-coupled signaling cascade. *Drosophila* Rh1 initiates the phototransduction cascade by interacting with a heterotrimeric Gq protein, which, in turn, activates phospholipase C (PLC-β) encoded by the *norpA* gene. Interestingly, PLC- β in *Drosophila* is closely related to PLC- β4 expressed in the vertebrate retina. In *Drosophila*, activation of PLC leads to the opening of two cation-selective, tetrameric, transient receptor channels (TRP and TRPL), resulting in a dramatic rise in intracellular Ca^{2+}. The photoreceptors depolarize and intracellular Ca^{2+} rises from resting, 100 nM levels, to tens of micromolar to as high as 1 mM in the microvilli. The downstream mechanisms for how PLC activity opens the TRP and TRPL channels remain elusive. Unlike other rhabdomeric photoreceptor cells, the IP_3-receptor is not involved, and hence, it is thought that either lipids generated via PLC activity or that PLC induced changes to the physical properties of the membrane bilayer may gate the channels. These lipid candidates include, DAG and/or downstream metabolites of DAG lipase, such PUFAs (polyunsaturated fatty acids), as well as a reduction in PIP_2 (Fig. 2). Indeed, PIP_2 has been recently shown to display both inhibitory and excitatory effects on TRP channels (Fain et al. 2010; Katz and Minke 2009; Wang and Montell 2007; Yau and Hardie 2009).

Following influx via the TRP channels, Ca^{2+} is removed from the rhabdomeres by extrusion from the cell via the sodium/Ca^{2+} exchanger (CalX), as is also the case in vertebrate rod photoreceptors. Ca^{2+} is also removed by diffusion into the cell body, where Ca^{2+} rises from approximately 100 nM resting levels to approximately 10 μM. These high levels of Ca^{2+} in the cell body are thought to be removed by

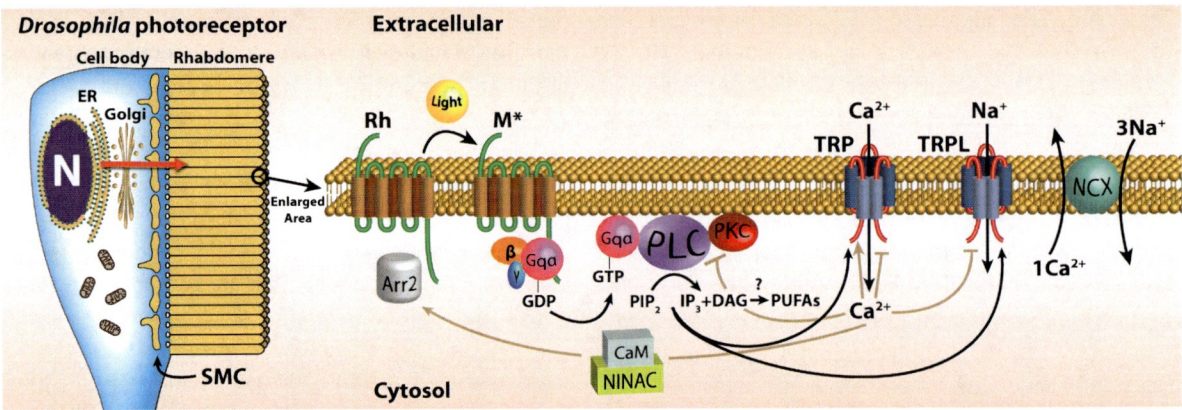

Calcium in Vision, Fig. 2 Phototransduction in *Drosophila*. Absorption of light by rhodopsin (Rh) converts it to active thermostable metarhodopsin (M*), which in turn stimulates the heterotrimeric Gq protein by GDP-GTP exchange. Activated Gqα-GTP is liberated from G_γ and G_β and stimulates PLC-β (phospholipase C). PLC hydrolyzes phosphatidylinositol 4,5-bisphosphate (PIP_2) to inositol triphosphate (IP_3) and diacylglycerol (DAG). DAG-kinase activity may subsequently produce polyunsaturated fatty acids (PUFAs). The two TRP and TRPL channels open due to events downstream of PLC, that are not yet known, and the photoreceptor cells depolarize. The light response is dominated by TRP, and TRP is chiefly Ca^{2+} permeable. Ca^{2+} influx acts sequentially for both positive and negative regulation of phototransduction (indicated by *brown lines* with *arrowheads* or *bars*, respectively). Initially, Ca^{2+} influx enhances TRP channel activation, but as Ca^{2+} rises to millimolar concentrations, Ca^{2+} inactivates both channels, and is thought to be the main mechanism of light adaptation. Rhodopsin inactivation by arrestin (Arr2) is Ca^{2+} dependent and requires calmodulin (CaM) and myosin III (NINAC). Ca^{2+} also terminates the light response by inactivating PLC via protein kinase C (PKC) (Yau and Hardie 2009). *Left*, diagram of the photoreceptor cell showing the nucleus (*N*), endoplasmic reticulum (*ER*), Golgi (*G*), submicrovillar cisternae (*SMC*), and rhabdomere with numerous tightly packed micovilli

sequestration via the sacro-endoplasmic reticulum ▸ Ca^{2+} ATPase (SERCA) and by buffering via Ca^{2+}-binding proteins, such as ▸ calnexin (Colley 2010; Fain et al. 2010; Katz and Minke 2009; Yau and Hardie 2009).

In *Drosophila* photoreceptor cells, Ca^{2+} is critical for excitation as well as for the termination of the light response. There are several proteins involved in Ca^{2+}-mediated termination including protein kinase C (PKC), calmodulin (CaM), and NINAC (myosin III) (Fig. 2). Failure in termination of the light response and loss of precise control of spatial and temporal profiles of Ca^{2+} are devastating to cells and can lead to cell death and retinal degeneration (Fain et al. 2010; Katz and Minke 2009; Wang and Montell 2007; Yau and Hardie 2009).

Retinal Degeneration in Drosophila

It is widely accepted that retinal degeneration can be caused by mutations in almost every constituent of the photoreceptor cells. These mutations fall into two classes. One class encompasses mutations leading to unregulated activities of phototransduction, including Ca^{2+} toxicity. These are termed light dependent. The second class involves defects in rhodopsin maturation and does not involve light activation of phototransduction. These are termed light independent (Colley 2010; Wang and Montell 2007).

Proper Ca^{2+} removal or sequestration is vital to photoreceptor cells. Ca^{2+} extrusion from photoreceptors via the ▸ Na^+/Ca^{2+} exchanger is critical to photoreceptor cell survival. Indeed, loss of the *Drosophila* Na^+/Ca^{2+} exchanger, CalX in the *calx* mutants causes high levels of sustained Ca^{2+} leading to retinal degeneration. In addition, overexpression of Calx is able to prevent the retinal degeneration resulting from constitutive activity of the TRP channels and Ca^{2+} overload. Therefore, Calx, indeed, plays a critical role in Ca^{2+} extrusion and cell viability (Wang and Montell 2007).

TRP channel function is critical to photoreceptor cell survival. Photoreceptor cells that lack TRP undergo light-dependent retinal degeneration. They are unable to sustain a steady-state current and are defective in Ca^{2+} influx. The mechanism for light-dependent retinal degeneration in *trp* null mutants is thought to result from the failure in a combination of the Ca^{2+} and PKC-dependent inhibition of PLC. This leads to the unregulated stimulation of PLC and subsequent depletion of microvillar PIP_2. Therefore, a lack of TRP protein causes a light-dependent retinal degeneration because of unregulated light-stimulated activities of PLC (Wang and Montell 2007; Yau and Hardie 2009).

In some cases, mutations in rhodopsin itself, or in other cases, mutations in the *arrestin* gene, lead to light-dependent retinal degeneration. Arrestin functions to deactivate rhodopsin; therefore, loss in arrestin leads to unregulated rhodopsin and uncontrolled light-stimulated activities of phototransduction. Additionally, it is thought that loss of arrestin leads to decreased endocytosis of Rh1 and all of these defects lead to retinal degeneration (Colley 2010).

Since the initial 1983 report that mutations in rhodopsin lead to retinal degeneration in *Drosophila*, in excess of 100 mutations in human rhodopsin have been identified in autosomal dominant Retinitis Pigmentosa (adRP). A large number of these mutations lead to misfolded rhodopsin that is incorrectly transported through the secretory pathway. However, the mechanism by which the mutant rhodopsins act dominantly to cause retinal degeneration was not known. Studies in *Drosophila*, in 1995, on dominant rhodopsin mutations revealed that the retinal degeneration is caused by interference in the maturation of normal rhodopsin by the mutant forms of rhodopsin. These studies in *Drosophila* provided a mechanistic explanation for the cause of certain forms of adRP (Colley 2010).

Rhodopsin is synthesized on membrane-bound ribosomes and undergoes translocation, posttranslational modifications, folding, and quality control in the endoplasmic reticulum (ER). In the face of these intricate and error-prone processes, the photoreceptor cells have evolved systems of molecular chaperones to promote the proper processing of newly synthesized rhodopsin. To become functionally active, rhodopsin must be correctly folded and is required to precisely traverse the secretory pathway to the rhabdomeres for its role in phototransduction. The mechanisms for regulating rhodopsin maturation including folding, chaperone interaction, glycosylation, chromophore attachment, and transport are crucial for photoreceptor cell survival as defects in these processes lead to retinal degeneration in both flies as well as in humans. These types of retinal degenerations are generally in the light-independent class (Colley 2010).

In *Drosophila*, as in humans, rhodopsin undergoes N-linked glycosylation during biosynthesis, and in flies, elimination of the glycosylation site (N20I) results in the retention of rhodopsin in the secretory pathway and retinal degeneration. In addition, like in humans, *Drosophila* rhodopsin must attain its vitamin-A-derived chromophore at a lysine residue in the seventh transmembrane domain. Defects in chromophore production in the *Drosophila* mutants *ninaB*, *ninaD*, *ninaG*, and *santa maria*, cause a defect in Rh1 transport to the rhabdomere, causing a severe reduction in Rh1 and retinal pathology (Colley 2010; Wang and Montell 2007).

In Drosophila, the successful transport of Rh1 from the ER to the rhabdomere requires the cyclophilin, NinaA. Cyclophilins are peptidyl-prolyl *cis–trans* isomerases and are thought to play a role in protein folding during biosynthesis. Consistent with a function in protein folding, NinaA resides in the ER. NinaA is also detected in transport vesicles together with Rh1 and forms a specific and stable complex with Rh1, consistent with a broad role for NinaA as a chaperone in the secretory pathway. Mutations in NinaA lead to severe retinal pathology in flies. Similarly, in mammals, a cyclophilin-like protein (RanBP2/Nup358) plays a role in protein biogenesis (Colley 2010).

In photoreceptor cells, successful maturation of rhodopsin and regulation of Ca^{2+} are essential for cell function and viability. There is a growing list of proteins that serve multifunctional roles. One example is calnexin, which serves as both a molecular chaperone for Rh1 and as a regulator of Ca^{2+} that enters photoreceptor cells during phototransduction. Mutations in *Drosophila calnexin* lead to severe defects in Rh1 maturation as well as defects in the control of cytosolic Ca^{2+} levels following activation of the light-sensitive TRP channels. As a result, the photoreceptor cells undergo a light-enhanced retinal degeneration that is due to the combined detrimental effects of defective Rh1 maturation and Ca^{2+} regulation (Colley 2010).

Summary

In *Drosophila* photoreceptor cells, the precise regulation of spatial and temporal profiles of Ca^{2+} is essential for cell function and prolonged elevation of Ca^{2+} can be toxic, leading to retinal degeneration. Hence, proper Ca^{2+} removal or sequestration following a transient rise in cytoplasmic Ca^{2+} during phototransduction is vital to photoreceptor cells. In addition to Ca^{2+} regulation, successful folding, maturation, and transport of rhodopsin and the other components of phototransduction is key to photoreceptor cell survival. Failure in Rh1 maturation and Ca^{2+} regulation leads to retinal degeneration in both *Drosophila* and vertebrates. Despite its perceived simplicity, *Drosophila* is surprisingly complex and contains a genetic makeup that is remarkably similar to humans. Therefore, mechanisms of Ca^{2+} regulation, rhodopsin maturation, and retinal degeneration identified in *Drosophila* will undoubtedly continue to provide insights that are clinically relevant to hereditary human retinal degeneration diseases.

Acknowledgments JK's and ASP's research is funded by the National Cancer Institute and the Retina Research Foundation/M.D. Matthews Research Chair. NJC's research is funded by the National Eye Institute, the Retina Research Foundation, and the Retina Research Foundation/Walter H. Helmerich Research Chair. We gratefully acknowledge C. Vang for assistance in preparing the figures.

Cross-References

▶ Calcium ATPase
▶ Calcium, Neuronal Sensor Proteins
▶ Calcium-Binding Proteins
▶ Calmodulin
▶ Calnexin and Calreticulin
▶ EF-Hand Proteins
▶ Na^+/Ca^{2+}-K^+ Exchanger
▶ Penta-EF-Hand Calcium-Binding Proteins

References

Colley NJ (2010) Retinal degeneration through the eye of the fly. In: Dartt DA (ed) Encyclopedia of the eye, vol 4. Academic, Oxford, pp 54–61
Fain GL, Hardie R, Laughlin SB (2010) Phototransduction and the evolution of photoreceptors. Curr Biol 20:R114–R124
Jiang L, Baehr W (2010) GCAP1 mutations associated with autosomal dominant cone dystrophy. In: Anderson R, Hollyfield J, LaVail M (eds) Retinal degenerative diseases, advances in experimental medicine and biology. Springer Science, New York
Katz B, Minke B (2009) Drosophila photoreceptors and signaling mechanisms. Front Cell Neurosci 3:2
Nakatani K, Chen C, Yau K-W, Koutalos Y (2002) Calcium and phototransduction. In: Baehr W, Palczewski K (eds) Photoreceptors and calcium. Kluwer/Plenum, New York
Ohguro H, Nakazawa M (2002) Pathological roles of recoverin in cancer-associated retinopathy. In: Baehr W, Palczewski K (eds) Photoreceptors and calcium, 2nd edn. Kluwer/Plenum, New York
van Ginkel P, Darjatmoko S, Sareen D et al (2008) Resveratrol inhibits uveal melanoma tumor growth via early mitochondrial dysfunction. Invest Ophthalmol Vis Sci 49:1299–1306
Vogt N, Desplan C (2007) The first steps in Drosophila motion detection. Neuron 56:5–7
Wang T, Montell C (2007) Phototransduction and retinal degeneration in Drosophila. Pflugers Arch 454:821–847
Yau KW, Hardie RC (2009) Phototransduction motifs and variations. Cell 139:246–264

Calcium Ion

▶ Calcium in Heart Function and Diseases

Calcium Ion Selectivity in Biological Systems

Todor Dudev[1] and Carmay Lim[1,2]
[1]Institute of Biomedical Sciences, Academia Sinica, Taipei, Taiwan
[2]Department of Chemistry, National Tsing Hua University, Hsinchu, Taiwan

Synonyms

Association/binding constant; Dication/divalent metal cation; Metal-binding site/pocket/cavity/cleft; Monocation/monovalent metal cation; Non-physiological/non-biogenic/"Alien" metal cation; Physiological/biogenic metal cation

Background/Definitions

Metal selectivity of a ▶ Ca^{2+}-binding site is an outcome of the competition between the bulk solvent and the protein ligands for the "native" Ca^{2+} and a "rival" cation M^{q+} (M = Na, K, Mg, Ln, etc.; q = 1, 2, 3) and can be assessed by the free energy of the $M^{q+} \to Ca^{2+}$ exchange reaction:

$$(M^{q+}-aq) + (Ca^{2+}-protein) \to \\ (M^{q+}-protein) + (Ca^{2+}-aq) \qquad (1)$$

In Equation (1), (M^{q+}/Ca^{2+}–aq) represents a hydrated metal cation outside the metal-binding cavity, and (M^{q+}/Ca^{2+}–protein) represents a metal cation bound in the protein cavity/ion channel pore. Equation (1) can be considered as the difference between two reactions describing the binding of the individual metal ions to the apoprotein, i.e.,

$$(M^{q+}-aq) + \text{protein} \rightarrow (M^{q+}-\text{protein}) \quad (2)$$

$$(Ca^{2+}-aq) + \text{protein} \rightarrow (Ca^{2+}-\text{protein}) \quad (3)$$

The association constant, $K_a(M^{q+})$ or $K_a(Ca^{2+})$, characterizing the affinity of the binding site for M^{q+} or Ca^{2+}, is given by

$$K_a(M^{q+}) = [M^{q+}-\text{protein}]/([M^{q+}-aq] \cdot [\text{protein}]) \quad (4)$$

$$K_a(Ca^{2+}) = [Ca^{2+}-\text{protein}]/([Ca^{2+}-aq] \cdot [\text{protein}]) \quad (5)$$

Since the reaction free energy is related to the association constant via

$$\Delta G = -RT \ln K_a, \quad (6)$$

where R is the gas constant and T, the temperature, the free energy for the $M^{q+} \rightarrow Ca^{2+}$ exchange reaction (1) can be expressed as

$$\begin{aligned}\Delta G(M^{q+} \rightarrow Ca^{2+}) &= -RT\,[\ln K_a(M^{q+}) - \ln K_a(Ca^{2+})] \\ &= -2.303\,RT \log\,[K_a(M^{q+})/ \\ &\qquad K_a(Ca^{2+})]\end{aligned} \quad (7)$$

Competition Between Ca^{2+} and Physiological Metal Cations for Ca-Binding Sites

Proteins Containing EF-Hand Motifs

▶ The EF-hand motif is found in a large group of Ca^{2+} signaling and buffering/transport proteins such as ▶ calmodulin, ▶ troponin C, ▶ parvalbumin, recoverin, ▶ calcineurin, ▶ calbindin D, and ▶ S100 protein. The canonical ▶ EF-hand motif consists of a 12-residue Ca^{2+}-binding loop flanked by two helices creating a signature helix–loop–helix motif (Gifford et al. 2007). The aspartate/glutamate (Asp/Glu) carboxylates and asparagine/glutamine (Asn/Gln) backbone carbonyls from the loop coordinate to Ca^{2+}, which often retains a bound water molecule, in pentagonal bipyramidal geometry (Fig. 1a). This seven-coordinate geometry is contributed by the carboxylate-binding mode of the conserved Glu at the last position of the EF-hand binding loop (Glu-12), which binds Ca^{2+} bidentately via both carboxylate oxygens, whereas the other Asp/Glu residues bind Ca^{2+} monodentately via one of the carboxylate oxygens.

The EF-hand binding site is characterized by an exceptional selectivity for Ca^{2+}. It is designed to bind preferentially the "native" Ca^{2+} against the background of up to 10^5 times higher concentrations of the competing cytosolic Mg^{2+} (by a factor of 10^3–10^4) and Na^+/K^+ (by a factor of 10^6, see Table 1). The finely tuned structural and electrostatic properties of the binding cavity make Ca^{2+} the cation of choice for the EF-hand proteins (Drake et al. 1996): (1) Interactions among the metal ligands rigidify the Ca-binding loop and constrain the metal cavity to an optimal size for Ca^{2+} that disfavors larger cations. (2) The relatively rigid EF-hand cavity provides an optimal level of negative charge density that favors dications over monocations. (3) The EF-hand binding site's pentagonal bipyramidal geometry and relatively large size prevent the smaller Mg^{2+}, which strongly prefers octahedral geometry, from binding.

Under physiological conditions, however, the competition between Ca^{2+} and Mg^{2+} for the EF-hand binding site depends not only on the EF-hand protein properties but also on the cytosolic Ca^{2+} and Mg^{2+} concentrations. In the *resting* cell, the concentration of Ca^{2+} (10^{-7}–10^{-8} M) is three- to fivefold lower than that of Mg^{2+} (10^{-3}–10^{-4} M) and does not favor Ca^{2+} binding. In this *resting* state, Mg^{2+} occupies (at least partially) the EF-hand binding sites and stabilizes the resting EF-hand domains. However, Mg^{2+} binding to EF-hand sites does not trigger the extensive conformational changes characteristic of the Ca^{2+}-activated proteins and, thus, no signaling response occurs (Gifford et al. 2007). Interestingly, Glu-12 is *mono*dentately rather than *bi*dentately bound to *hexa*coordinated Mg^{2+} (Fig. 1b), yielding a physiologically silent protein. In an *activated* cell, the intracellular

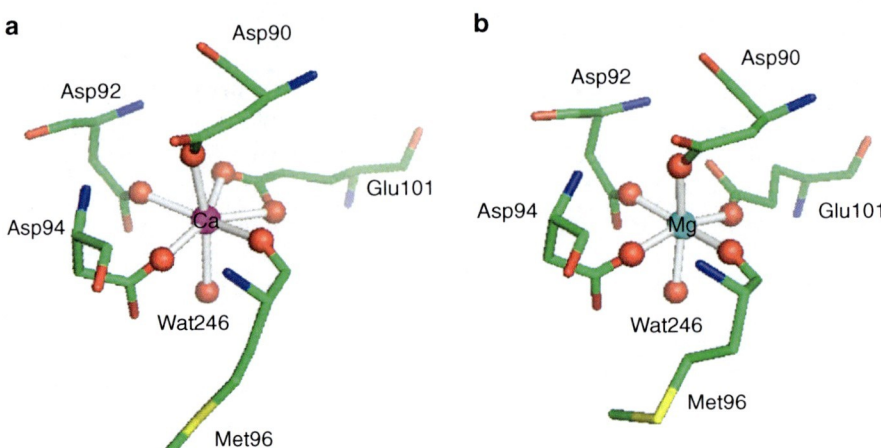

Calcium Ion Selectivity in Biological Systems, Fig. 1 Ball and stick diagram of pike 4.10 parvalbumin metal-binding site occupied by (**a**) Ca^{2+} in pentagonal bipyramidal geometry (1PAL, 1.68 Å) and (**b**) Mg^{2+} in octahedral geometry (4PAL, 1.80 Å)

Calcium Ion Selectivity in Biological Systems, Table 1 Experimental metal association constants, K_a in M^{-1}, and $M^{q+} \rightarrow Ca^{2+}$ exchange free energies, $\Delta G(M^{q+} \rightarrow Ca^{2+})$ in kcal/mol, in EF-hand proteins

Protein	Metal site	$K_a{}^a$		$\Delta G(M^{q+} \rightarrow Ca^{2+})$
		Ca^{2+}	Mg^{2+}	
Calmodulin	N-domain	3.5×10^6	2.7×10^3	4.2
	C-domain	2.0×10^7	5.8×10^2	6.2
Troponin C				
Skeletal	N-domain	3.3×10^5	2.0×10^2	4.4
	C-domain	2.0×10^7	5.0×10^3	4.9
Cardiac	N-domain	3.3×10^5	2.0×10^2	4.4
	C-domain	1.4×10^7	1.7×10^3	5.3
S100P	N-domain	8.9×10^3	$\sim 10^2$	~ 2.6
	C-domain	2.5×10^7		
Recombinant oncomodulin	CD-domain	1.2×10^6	6.6×10^2	4.4
	EF-domain	2.2×10^7	3.8×10^3	5.1
Parvalbumin		2.7×10^9	9.5×10^4	6.0
E. coli galactose-binding protein		7.1×10^5	8.3×10^1	5.3
			Na^+	
E. coli galactose-binding protein		7.1×10^5	$\leq 3.3 \times 10^{-1}$	≥ 8.6
			K^+	
E. coli galactose-binding protein		7.1×10^5	$\leq 2.5 \times 10^{-1}$	≥ 8.8
			Tm^{3+}	
E. coli galactose-binding protein		7.1×10^5	1.4×10^6	-0.4
			Lu^{3+}	
E. coli galactose-binding protein		7.1×10^5	1.8×10^6	-0.5
			Yb^{3+}	
E. coli galactose-binding protein		7.1×10^5	2.1×10^6	-0.6

^aBinding constants for *E. coli* galactose-binding protein are taken from Drake et al. (1996), while the rest are taken from Dudev and and Lim (2003)

concentration of Ca^{2+} increases to 10^{-5}–10^{-6} M in response to stimuli such as membrane depolarization or extracellular/intracellular messengers. This, in combination with the special intrinsic properties of the EF-hand binding site (see above), favors Ca^{2+} over Mg^{2+} binding to the protein. The Ca^{2+}-bound protein undergoes large conformational transformations, triggering a cascade of events along the signal transduction pathway. Thus, Ca^{2+} plays a regulatory role in the signal-transducing process: the metal-binding site does not bind Ca^{2+} until the respective signal occurs ("on reaction"), but it can debind the metal cation when the latter is no longer needed ("off reaction").

Proteins Containing C2 Domains

▶ C2 domains are found in membrane trafficking proteins (e.g., synaptotagmin, rabphilin-3A) and signal-transducing proteins (e.g., cytosolic phospholipase A_2, protein kinase C), which perform critical cellular functions such as lipid second messenger generation, protein phosphorylation, vesicular transport, GTPase regulation, and ubiquitin-mediated protein degradation (Nalefski and Falke 1996). Upon Ca^{2+} binding, C2 domains dock to cell membrane targets (e.g., phospholipids, inositol polyphosphates). At one rim of the C2 domain β-sandwich structure, two or three Ca^{2+} bind in a cleft lined by carboxylate and carbonyl groups from loops. In response to intracellular Ca^{2+} signaling, the C2 domains of cytosolic phospholipase A_2 and rabphilin-3A and the C2B domain of synaptotagmin I bind two Ca^{2+} ions with positive cooperativity, and the activated C2 domains subsequently dock to the respective target membrane. In contrast, the activation of the C2 domain of protein kinase C-β and the C2A domain of synaptotagmin I requires binding of three Ca^{2+} ions.

Although designed to specifically bind Ca^{2+}, the C2 domains, when free in solution, exhibit lower affinity for Ca^{2+} ($K_a \sim 10^2$–10^6 M^{-1}) than the EF-hand motifs ($K_a \sim 10^5$–10^7 M^{-1}). Generally, the Ca^{2+} affinity of the C2 modules increases in the presence of the target membrane. Yet, like the EF-hand proteins, the C2 domains are very efficient in discriminating between the native *micromolar* Ca^{2+} and rival *millimolar* Na^+, K^+, and Mg^{2+} cations in the cytoplasm. As in the EF-hand motif, monocations are excluded from the C2 metal-binding site since they are unable to overcome the electrostatic repulsion between the many carboxylate/carbonyl oxygens in the site and, thus, cannot immobilize the Ca-binding loops. On the other hand, Mg^{2+} dications are excluded because compression of the site to accommodate the shorter Mg^{2+}–O distances is countered by repulsion between the ligating oxygens. Thus, at *millimolar* physiological concentrations, Na^+, K^+, and Mg^{2+} fail to bind to the C2 domains and initiate membrane docking. However, at elevated supraphysiological concentrations, Mg^{2+} does bind to the C2 domains (Nalefski and Falke 2002). The Mg^{2+}-bound synaptotagmin III C2A domain is able to induce membrane docking, though less efficiently than the Ca^{2+}-loaded module, whereas the Mg^{2+}-bound C2 domain in cytosolic phospholipase A_2 fails to do so (Nalefski and Falke 2002).

Proteins Containing GLA Domains

GLA domains are essential structural motifs of several proteins involved in the ▶ blood coagulation process such as factor II (prothrombin), VII, IX, and X, as well as protein C and protein S. These domains contain a network of 10–13 γ-carboxyglutamic acid residues (Gla) coordinated to 7–8 Ca^{2+} and Mg^{2+} ions, which stabilize the structure and maintain the proper conformation for subsequent membrane docking. The rare Gla residue is produced by posttranslational carboxylation of specific Glu residues by a vitamin K–stimulated reaction. Gla has two carboxylate groups attached to the C^γ atom, resulting in a net charge of −2. It binds the metal ion predominantly in a chelation bidentate mode where both carboxylates are monodentately bound. Consistent with the presence of a second carboxylate group in Gla, Gla is involved mostly in binding divalent (rather than monovalent) cations during the blood coagulation process. Notably, mutating Gla back to its precursor, Glu, abolishes Ca^{2+} binding (Stenflo and Suttie 1977). Ca^{2+} ions are absolutely required for blood to coagulate: They are needed to stabilize an essential membrane-binding loop. This loop becomes disordered when only Mg^{2+} ions are present in the GLA domain, which fails to dock properly to the target membrane, thus disrupting the coagulation process. On the other hand, Mg^{2+} binding to a few of the metal-binding sites in the GLA domains, with Ca^{2+} occupying the other sites, accelerates the activation of the respective coagulation factors and enhances membrane binding (Shikamoto et al. 2003).

Since binding of both Ca^{2+} and Mg^{2+} (in a proper ratio) to the GLA domain promotes optimal binding to the target molecule, it is not surprising that GLA domains possess two types of metal-binding sites exhibiting different metal selectivity. Interior binding sites are selective for Ca^{2+} and are usually occupied by 4–5 Ca^{2+} ions. Peripheral binding sites, comprising usually three metal cations, exhibit poorer metal selectivity and can be occupied by Ca^{2+} or Mg^{2+} depending on their relative concentration. At physiological conditions where the blood plasma concentration of free Ca^{2+} is ~1.1–1.3 mM while that of Mg^{2+} is ~0.4–0.6 mM, Ca^{2+} occupies the interior binding sites, whereas three Mg^{2+} ions bind to the other nonselective binding sites (Shikamoto et al. 2003).

Notably, the affinity of Ca^{2+} for the GLA domain ($K_a \sim 10^2$–10^3 M^{-1}) is weaker than that for the EF-hand ($K_a \sim 10^5$–10^7 M^{-1}) and C2 ($K_a \sim 10^2$–10^6 M^{-1}) binding sites (see above). This difference can be attributed to the difference in the concentration ratio between Ca^{2+} and Mg^{2+} in the intracellular and extracellular compartments. Since the EF-hand and C2 domains have to sequester *micro*molar Ca^{2+} against much higher *milli*molar Mg^{2+}, Na^+, and K^+ in the cytosol, their affinity toward the "native" Ca^{2+} has to be precisely tuned. On the other hand, no such stringent requirements are needed for the GLA domains, as the concentration of Ca^{2+} in the blood plasma is quite high (millimolar range) and even surpasses that of its rival, Mg^{2+}.

Calcium Ion Channels

Calcium ion channels transport Ca^{2+} across the cell membrane. They play an essential role in a plethora of biological processes such as the skeletal, smooth, and cardiac muscle contraction; signal transduction; hormone and neurotransmitter secretion; and gene expression. The voltage-gated L-type Ca channel or homologs such as T-, N-, R-, and P-/Q-type Ca channels control selective passage of Ca^{2+} from the *extra*cellular to *intra*cellular compartments, while ryanodine receptor (RyR) or inositol trisphosphate receptor Ca^{2+} channel controls the release of Ca^{2+} from intracellular stores (sarcoplasmic or endoplasmic reticulum). Voltage-gated calcium channels possess remarkable ability to discriminate between the "native" Ca^{2+} and other competing cations, which are usually present in much higher concentrations in the respective biological compartments. Typically, they select Ca^{2+} over Na^+ and K^+ at a ratio of over 1,000:1 and do not conduct Mg^{2+} (Hille 1992). However, the RyR channel exhibits only moderate divalent/monovalent cation selectivity, not exceeding a factor of 7 and poorly discriminates between Ca^{2+} and Mg^{2+} (permeability ratio $P_{Ca2+}/P_{Mg2+} = 1.1$; Tinker et al. 1992).

The ion selectivity of the channel is controlled by its selectivity filter – the narrowest part of the pore comprising several protein residues that specifically interact with the passing metal ion. Although X-ray structures of Ca^{2+}-bound Ca^{2+} channels and their selectivity filters are still lacking, a series of site-directed mutagenesis and channel-blocker binding experiments have determined the composition of the selectivity filters of several Ca^{2+} ion channels. These filters are lined with negatively charged Asp/Glu residues whose side chains face the pore lumen. The selectivity filter of the high voltage-activated voltage-gated Ca^{2+} channels is comprised of four Glu residues (EEEE locus) donated by the pore-forming loops of the channel's four domains, while that of the low voltage-activated counterpart has a ring of two Glu and two Asp residues (EEDD locus), whereas that of the RyR channel consists of four aspartates (DDDD locus). Other loci of Asp/Glu residues at the pore entrance are implicated in fine-tuning the selectivity for Ca^{2+}.

The voltage-gated L-type and RyR channels have different permeation properties: The former exhibits high Ca^{2+} selectivity (see above) and affinity ($K_a \sim 10^6$ M^{-1}) and low conductance, while the latter exhibits lower Ca^{2+} selectivity (see above) and affinity ($K_a \sim 10^3$ M^{-1}) but higher conductance. These differences in the channels' selectivity/conductance are thought to stem from the difference in their physiological function(s) (see above). When the L-type channel allows Ca^{2+} in the cell, the RyR channel releases Ca^{2+} over several (>5) milliseconds, thus elevating the cytoplasmic Ca^{2+} concentration and triggering muscle contraction. Such long-lasting Ca^{2+} efflux is possible because the RyR channel can conduct Ca^{2+} out of the intracellular stores, while allowing passage of K^+/Mg^{2+} in the opposite direction, resulting in a constant driving force for Ca^{2+} release. Thus, the RyR channel's low selectivity and high conductance for both monovalent and divalent cations make it ideal for its physiological role of releasing Ca^{2+} over a long period (Gillespie and Fill 2008).

Competition Between Ca^{2+} and Alien Ln^{3+} for Ca-Binding Sites

Trivalent lanthanide cations, Ln^{3+}, have higher affinity toward oxygen-containing biological ligands (Asp/Glu, Asn/Gln/backbone peptide groups) than divalent Ca^{2+} and, therefore, can successfully compete and substitute for Ca^{2+} in protein-binding sites (Table 1). Although the competition between the "native" Ca^{2+} and nonbiogenic Ln^{3+} does not have immediate implications for biological processes in vivo, the ► unique physicochemical properties of lanthanide cations make them very useful in probing Ca^{2+} and other dication-binding sites in vitro. Lanthanides are extensively being used in both crystallographic and spectroscopic studies of metalloproteins. They are employed in determining the phases of the diffracted X-rays by multiple isomorphous replacement or multiwavelength anomalous dispersion. Furthermore, ► luminescent properties of lanthanide ions are utilized in bioanalytical assays to determine the interdomain distance in proteins. The large anisotropic magnetic susceptibility of paramagnetic lanthanides, which gives rise to large pseudocontact shifts that can be observed for residues as far as 40 Å from the metal center, is used to obtain long-distance restraints for protein NMR structure determination.

Determinants of Ca^{2+} Selectivity in Biological Systems

Both experimental and theoretical studies have revealed some key factors governing the metal selectivity process in Ca-binding sites:

1. *Binding Site Dielectric Constant.* Interactions of Na^+, K^+, Ca^{2+}, Mg^{2+}, and Ln^{3+} with oxygen-containing biological ligands are predominantly electrostatic (charge–charge/dipole) and are enhanced in a low dielectric environment. Therefore, the dielectric properties of the metal-binding site play an important role in the metal selectivity process. A buried/partially buried binding pocket with a low dielectric constant enhances the affinity (and competitiveness) of the more positively charged cation of the competing pair of metal ions for the ligating oxygens. Thus, decreasing the solvent exposure of the binding site/channel pore disfavors the substitution of Ca^{2+} by Na^+/K^+ but facilitates the displacement of Ca^{2+} by trivalent lanthanides.
2. *Binding Site Size/Flexibility.* Calcium proteins discriminate between the two divalent contenders, Ca^{2+} and Mg^{2+}, mainly by maintaining a proper size/shape/flexibility of the binding cavity, which is optimized to better suit the coordination requirements of Ca^{2+} than those of Mg^{2+}. Interactions among the many metal ligands and the protein matrix constrain the host metal cavity to an optimal size for Ca^{2+} and endow it with enough stiffness so as not to let a rival cation readjust its size/geometry upon binding. Such a relatively rigid and large binding cavity fits Ca^{2+}, which prefers sevenfold (pentagonal bipyramidal) coordination geometry and binds carboxylate and carbonyl oxygens with Ca–O distances ranging from 2.1 to 2.8 Å, but not the smaller rival Mg^{2+}, which strongly favors sixfold (octahedral) coordination geometry and shorter Mg–O bond distances (2.0–2.2 Å).
3. *Binding Site Overall Charge.* Ca-binding sites and ion channel selectivity filters, which contain constellations of negatively charged Asp, Glu, and Gla residues, effectively discriminate between the "native" Ca^{2+} and monovalent competitors such as Na^+ and K^+ (see above). The elevated negative charge density in a relatively rigid, low dielectric binding site/selectivity filter favors binding of Ca^{2+} over Na^+/K^+, as the monocations lack sufficient positive charge to alleviate the repulsion among the multiple ligating oxygens. On the other hand, trivalent "alien" Ln^{3+}, due mainly to strong charge–charge interactions with the carboxylates, have affinities higher than that of Ca^{2+} and can successfully displace Ca^{2+} from Ca-binding sites. In general, increasing the net negative charge of a relatively rigid and buried Ca-binding site increases its protection against monocation attack but, at the same time, increases its vulnerability to trivalent metal substitution.
4. *Competing Metal Concentrations.* Under physiological conditions, the competition between Ca^{2+} and a rival cation for the ► Ca^{2+}-binding site depends not only on the protein properties but also on the cation concentrations in the respective compartments, as exemplified by the EF-hand proteins and Gla domains.

Cross-References

► Blood Clotting Proteins
► C2 Domain Proteins
► Calbindin D_{28k}
► Calcineurin
► Calcium-Binding Protein Site Types
► Calmodulin
► EF-Hand Proteins
► Lanthanide Ions as Luminescent Probes
► Lanthanides, Physical and Chemical Characteristics

- ▶ Parvalbumin
- ▶ S100 Proteins
- ▶ Troponin

References

Drake SK, Lee KL, Falke JJ (1996) Tuning the equilibrium ion affinity and selectivity of the EF-hand calcium binding motif: substitutions at the gateway position. Biochemistry 35:6697–6705

Dudev T, Lim C (2003) Principles governing Mg, Ca, and Zn binding and selectivity in proteins. Chem Rev 103:773–787

Gifford JL, Walsh MP, Vogel HJ (2007) Structures and metal-ion-binding properties of Ca^{2+}-binding helix-loop-helix EF-hand motifs. Biochem J 405:199–221

Gillespie D, Fill M (2008) Intracellular calcium release channels mediate their own countercurrent: the ryanodine receptor case study. Biophys J 95:3706–3714

Hille B (1992) Ionic channels of excitable membranes, 2nd edn. Sinauer Associates, Sunderland

Nalefski EA, Falke JJ (1996) The C2 domain calcium-binding motif: structural and functional diversity. Protein Sci 5:2375–2390

Nalefski EA, Falke JJ (2002) Cation charge and size selectivity of the C2 domain of cytosolic phospholipase A_2. Biochemistry 41:1109–1122

Shikamoto Y, Morita T, Fujimoto Z, Mizuno H (2003) Crystal structure of Mg^{2+} and Ca^{2+}-bound Gla domain of factor IX complexed with binding protein. J Biol Chem 278:24090–24094

Stenflo J, Suttie JW (1977) Vitamin K – dependent formation of gamma-carboxyglutamic acid. Annu Rev Biochem 46:157–172

Tinker A, Lindsay ARG, Williams AJ (1992) A model for ionic conduction in the ryanodyne receptor channel of sheep cardiac muscle sarcoplasmic reticulum. J Gen Physiol 100:495–517

Calcium Ions in Eyesight

▶ Calcium in Vision

Calcium Pump Protein

▶ Calcium-Binding Proteins, Overview

Calcium Regulation of Phototransduction and Light Adaptation

▶ Calcium in Vision

Calcium Signaling

▶ Calcium in Nervous System

Calcium Signaling in Disease

▶ Calcium in Health and Disease

Calcium Signaling in Health

▶ Calcium in Health and Disease

Calcium Sparklets and Waves

Luis F. Santana and Manuel F. Navedo
Department of Physiology and Biophysics, University of Washington, Seattle, WA, USA

Synonyms

A kinase anchoring protein 150 (AKAP150); Calcium concentration $[Ca^{2+}]$; Calcium-induced calcium release (CICR); Calmodulin (CaM); Cyclic adenosine monophosphate (cAMP); Excitation-contraction coupling (EC coupling), excitation-transcription coupling (ET coupling); Förster resonance energy transfer (FRET); Inositol triphosphate (IP_3); Inositol triphosphate receptors (IP_3Rs); Myosin light chain (MLC); nuclear factor of activated T-cell c3 (NFATc3); Protein kinase A (PKA); Protein kinase Cα (PKCα); Protein phosphatase 2B or calcineurin (PP2B); Ryanodine receptors (RyRs); Sarcoplasmic reticulum (SR); Timothy syndrome (TS); Total internal reflection fluorescence (TIRF)

Definitions

1. Ca^{2+} sparklets are local Ca^{2+} signals produced by the opening of dihydropyridine-sensitive,

voltage-gated L-type Ca^{2+} channels in the plasma membrane of neurons and muscle.
2. Ca^{2+} waves are propagating cell-wide increases in cytosolic $[Ca^{2+}]$.
3. Ca^{2+} sparks are local Ca^{2+} release events from the sarcoplasmic reticulum via ryanodine receptors.
4. Ca^{2+} puffs are IP_3-activated, Ca^{2+} release events produced by the opening of clusters of IP_3Rs from the sarcoplasmic reticulum.
5. Optical clamping is the recording of local Ca^{2+} signals using fluorescence microscopy.

Introduction

In eukaryotic cells, Ca^{2+} flux into their cytoplasm via Ca^{2+} permeable channels in the plasma membrane and intracellular organelles triggers a plethora of signaling events that regulate numerous physiological processes, including excitability, secretion, contraction, and gene expression (Hille 2001). The development of bright fluorescent Ca^{2+} indicators and highly sensitive cameras and photomultipliers has allowed the recording of multiple Ca^{2+} signal modalities with high temporal and spatial resolution. In this chapter, we will discuss two of these signals: Ca^{2+} sparklets and Ca^{2+} waves.

Ca^{2+} sparklets are local Ca^{2+} signals produced by the opening of dihydropyridine-sensitive, voltage-gated L-type Ca^{2+} channels in the plasma membrane of neurons and muscle. Ca^{2+} release via small clusters of ryanodine (RYRs) and inositol triphosphate (IP_3Rs) receptors in the endoplasmic or sarcoplasmic reticulum (SR) produces Ca^{2+} sparks and Ca^{2+} puffs, respectively. Ca^{2+} waves that travel within and between cells are initiated and propagated by the sequential activation of Ca^{2+} sparks and Ca^{2+} puffs. The goal of this chapter is to describe the biophysical mechanisms underlying a Ca^{2+} signaling hierarchy ranging from sparklets to waves and to discuss their functional role in health and disease.

Ca^{2+} Sparklets

Using patch-clamp electrophysiological approaches, a detailed biophysical model of L-type Ca^{2+} channels has emerged. These channels have a threshold for activation of about -50 mV. Under physiological conditions, the conductance of single L-type Ca^{2+} channels is about 5 pS, producing currents of about 0.5 pA at -40 mV. L-type Ca^{2+} channels are inactivated by Ca^{2+} and voltage. Furthermore, they can operate in two functional modes. In mode 1, L-type Ca^{2+} channels open briefly (<1 ms) allowing a small amount of Ca^{2+} to enter the cell. The mean open time of L-type Ca^{2+} channels in mode 2 is about 10 ms (Hille 2001). Accordingly, the magnitude and time course of whole-cell L-type Ca^{2+} current depends on voltage, the number of channels activated, their elementary current, as well as gating mode.

Imaging Ca^{2+} Sparklets. Although powerful, patch-clamp electrophysiology could not easily answer two important questions about L-type Ca^{2+} channels. First, what is the spatial organization of functional L-type Ca^{2+} channels? Second, does L-type Ca^{2+} channel activity vary throughout the cell? One reason why patch-clamp electrophysiology cannot adequately address these questions is that in the whole-cell configuration, currents represent an ensemble of the activity of all channels throughout the surface membrane. In the single-channel mode, recordings are limited to a small patch of membrane per cell. Accordingly, these approaches offer limited information on the spatial organization of functional L-type Ca^{2+} channels. Recent advances in fluorescent probe development and imaging technology have helped circumvent these limitations and made it possible to optically record (called "optical clamping") Ca^{2+} sparklets resulting from Ca^{2+} influx via single L-type Ca^{2+} channels (Santana and Navedo 2009).

Ca^{2+} sparklets have been imaged using line-scan confocal microscopy and total internal reflection fluorescence (TIRF). Although line-scan imaging has the potential of recording sparklets with high temporal resolution (1 kHz), it comes at the expense of providing spatial information as one can only image a small volume at a time. TIRF microscopy offers a way around these issues for the following reasons. In TIRF, fluorescence is limited to a thin evanescent field of about 100 nm. Thus, the resolution in the Z-axis of a TIRF system is nearly 10 times higher than that of the typical confocal microscope (about 1 μm) (Santana and Navedo 2009)). This has the potential of dramatically reducing noise. Furthermore, if an electron multiplying CCD cameras capable of detecting photons from a single fluorescent molecule is used, one could image relatively large areas of the cell at relatively fast rates (e.g., 100–500 Hz) and with exquisite sensitivity. In combination, these factors give TIRF microscopy the potential of imaging, with a high signal to noise ratio, small fluorescence signals

Calcium Sparklets and Waves, Table 1 Biophysical properties of L-type Ca^{2+} channels

Sparklets' parameters	Calcium sparklets		Unitary calcium currents (I_{Ca})		Unitary I_{Ca} parameters
	Arterial myocytes	Cav1.2 (tsA-201)	Arterial myocytes	Cardiac myocytes	
Modal gating	Yes[a]	Yes[b]	Yes[d]	Yes[e]	Modal gating
Coupling coefficient (K) median	0.20[f]	0.20[f]	ND	0.22[f]	Coupling coefficient (K) median
Voltage-dependent changes in amplitude and activity	Yes	Yes	6.3[d]	9.7[e]	Conductance (pS)
Quantal level (nM)	38[a]	36[b]	0.55[d] (20 mM Ca^{2+}); 0.2 (2 mM Ca^{2+})	~0.6[e] (10 mM Ca^{2+})	Amplitude of unitary current at -40 mV (pA)
Duration (ms)	low nP_s $\tau = 24$ high nP_s $\tau_{fast} = 23$ $\tau_{slow} = 104$[g]	ND	$\tau = 14.3$[d]	Mode 1 $\tau = 2.34$ Mode 2 $\tau = 11$[g]	Open times (ms)
Minimal signal Mass (fC)	7[e] (-40 mV; 2 mM external Ca^{2+})	ND	7 (For a 0.2 pA unitary current of 35 ms duration)	ND	Charge movement (-40 mV; 2 mM external Ca^{2+}) (fC)

ND not determined
[a]Navedo et al. PNAS 102:11112–11117
[b]Navedo et al. JGP 127:221–223
[c]Amberg et al. J Physiol 579:187–201
[d]Rubart et al. JGP 107:459–472
[e]Yue and Marbal JGP 95:911–939
[f]Navedo et al. Circ Res 106:748–756
[g]Costantin et al. J Physiol 507:93–103

produced by Ca^{2+} influx via single Ca^{2+} channels within relatively large portion of the surface membrane. This makes this technique the most favorable approach for the study of the spatial organization of functional L-type Ca^{2+} channels in cardiac and smooth muscle (Santana and Navedo 2010).

Biophysical properties of Ca^{2+} sparklets. An important feature of Ca^{2+} sparklets is that their amplitude is variable, ranging from about 20 nM to several hundred nM Ca^{2+} depending on membrane voltage. Ca^{2+} entry through Ca^{2+} sparklet sites is quantal in nature. Accordingly, the amplitude of Ca^{2+} sparklets depends on the number of quanta activated. Large amplitude, multi-quantal Ca^{2+} sparklets are likely produced by random overlapping openings of adjacent L-type Ca^{2+} channels in the surface membrane of cells (Santana and Navedo 2009) (see Table 1 for description of the biophysical properties of Ca^{2+} sparklets).

L-type Ca^{2+} channels can readily be distinguished from other Ca^{2+}-permeable channels by their unique pharmacological, biophysical, and molecular biological properties. Consistent with this, Ca^{2+} sparklets are activated by the dihydropyridine agonist Bay-K 8644 and are eliminated by dihydropyridine antagonists nifedipine and nisoldipine. Simultaneous recordings of Ca^{2+} signals and L-type Ca^{2+} currents indicate that Ca^{2+} sparklets are associated with an inward Ca^{2+} current. Importantly, Ca^{2+} sparklets have similar voltage dependencies of activity and amplitude as L-type Ca^{2+} channels. Furthermore, Ca^{2+} sparklets in tsA-201 cells expressing L-type Ca^{2+} channels reproduce all the basic features of native Ca^{2+} sparklets including block by nifedipine, activation by Bay-K 8644, bimodal gating modalities, amplitude of quantal event, and voltage dependencies. Finally, arterial myocytes expressing a mutant L-type Ca^{2+} channel that is insensitive to inhibition by dihydropyridines produced dihydropyridine-insensitive low activity and persistent Ca^{2+} sparklets. This finding is important because it eliminates the possibility that Ca^{2+} sparklets are produced by a transient receptor potential or store-operated channel in cardiac or arterial myocytes (Santana and Navedo 2009). Thus, Ca^{2+} sparklets meet all the generally accepted pharmacological, biophysical, and molecular biological criteria used to identify L-type Ca^{2+} channels.

An intriguing feature of Ca^{2+} sparklets sites is that their activity (i.e., nP_s, where n is the number of sparklets and P_s is the probability that a sparklet is activated) and spatial distribution vary within the surface membrane of cardiac and arterial myocytes and tsA-201 cells expressing L-type Ca^{2+} channels. Using an analytical scheme similar to the one used for analysis of single-channel currents, it was found that Ca^{2+} sparklet activity was bimodal, with sites of low activity and sites with high, "persistent" activity. Based on this analysis, Ca^{2+} sparklets were grouped into three categories: silent (i.e., $nP_s = 0$), low (i.e., $0 > nP_s < 0.2$), and high activity (i.e., $nP_s > 0.2$), persistent Ca^{2+} sparklets (Santana and Navedo 2009). The fact that Ca^{2+} sparklets' spatial distribution is variable is interesting because L-type Ca^{2+} channels are broadly distributed throughout the surface membrane of neurons, and cardiac and smooth muscle.

Examination of the physiological role of Ca^{2+} sparklet in arterial myocytes demonstrated that Ca^{2+} sparklet activity regulates local and global $[Ca^{2+}]$ in a voltage-dependent manner. In arterial smooth muscle, persistent Ca^{2+} sparklet activity contributes to $\approx 50\%$ of the dihydropyridine-sensitive Ca^{2+} influx required for maintenance of steady-state cytosolic $[Ca^{2+}]$ under physiological conditions (i.e., 2 mM Ca^{2+} and -40 mV) (Santana and Navedo 2009). On the basis of these data, a new model for steady-state Ca^{2+} influx in arterial myocytes was proposed. In this model, membrane depolarization increases Ca^{2+} influx, at least in part, by promoting L-type Ca^{2+} channels to operate in a persistent gating mode and by increasing the duration of low and high activity, persistent Ca^{2+} sparklets.

Mechanisms for subcellular variations in Ca^{2+} sparklet activity. As mentioned above, Ca^{2+} sparklets activity varies along the surface membrane of cells. This behavior of Ca^{2+} sparklets is not likely due to subcellular variations in the molecular composition or clustering of L-type Ca^{2+} channels, as electrophysiological, immunofluorescence, and imaging data suggest that channels of similar subunit composition are broadly distributed throughout the surface membrane of cells. Rather, recent work indicates that membrane targeting of protein kinase Cα (PKCα), protein kinase A (PKA), and the phosphatase calcineurin by the scaffolding protein **A**-kinase **a**nchoring **p**rotein (AKAP) 150 (the rodent ortholog of human AKAP79) to specific regions of the surface membrane where they can differentially regulate L-type Ca^{2+} channel's function underlies heterogeneous Ca^{2+} sparklet activity in cells (Fig. 1) (Santana and Navedo 2010).

Data supporting this model are compelling. First, AKAP150 is specially suited to perform these tasks as it binds to PKCα, PKA, calcineurin, and L-type Ca^{2+} channels in neurons and muscle. Second, AKAP150 and PKCα colocalize to specific foci at or near the surface membrane of arterial myocytes. Third, loss of AKAP150 prevents PKCα targeting to the surface membrane of arterial myocytes. Importantly and consistent with the model, the loss of AKAP150 also abolishes persistent Ca^{2+} sparklet activity in these cells. Fourth, AKAP is required for PKA-dependent modulation of L-type Ca^{2+} channels in arterial myocytes. Fifth, the actions of PKCα and presumably PKA on the induction of persistent Ca^{2+} sparklet activity are opposed by calcineurin. Accordingly, the level of Ca^{2+} sparklet activity will vary regionally depending on the relative activities of PKCα, PKA, and calcineurin. Sixth, only a subpopulation of L-type Ca^{2+} channels interacts with AKAP150 and associated effector proteins in arterial myocytes (Santana and Navedo 2010). Collectively, these observations suggest that heterogeneous Ca^{2+} sparklet activity results from the concerted regulation of a specific set of signaling proteins on L-type Ca^{2+} channels broadly distributed throughout the surface membrane of cells.

As mentioned above, persistent Ca^{2+} sparklets result from frequent openings of L-type Ca^{2+} channels following PKCα and PKA activation. Therefore, what are the mechanisms underlying regional variations in Ca^{2+} sparklet activity and spatial distribution? Recent studies from several labs provide potential answers to this question. For instance, an increase in cAMP production could activate AKAP150-targeted PKA near a subset of L-type Ca^{2+} channels in the surface membrane, thus promoting persistent Ca^{2+} sparklet activity in those areas. Similarly, an increase in diacylglycerol or cytosolic $[Ca^{2+}]$ could activate PKCα near a subpopulation of L-type Ca^{2+} channel's recruit to the AKAP150-associated protein complex. In this scenario, Ca^{2+} entering the cell binds to calmodulin and promotes the release of PKCα from the AKAP150 complex. Once liberated from the AKAP150 complex, PKCα is free to phosphorylate nearby L-type Ca^{2+} channels, thus inducing persistent Ca^{2+} sparklet activity and contributing to heterogeneous spatial activation of these Ca^{2+} influx events (Fig. 1).

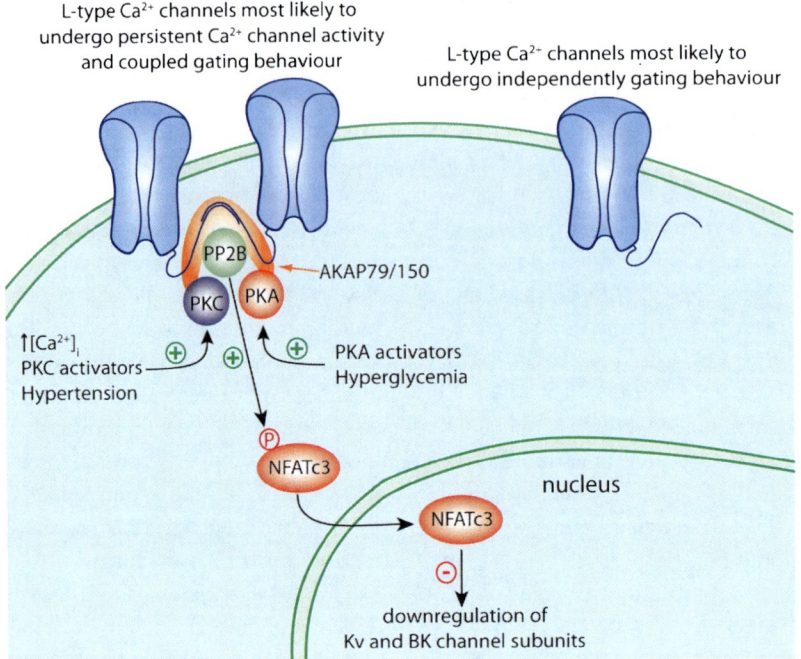

Calcium Sparklets and Waves, Fig. 1 *Proposed mechanisms for heterogeneous Ca^{2+} sparklet activity, coupled gating of L-type Ca^{2+} channels, and the activation of NFATc3 in cardiac and arterial myocytes.* In this model, L-type Ca^{2+} channels are broadly distributed throughout the sarcolemma of cardiac and arterial myocytes. Activation of these channels results in a subcellular increase in $[Ca^{2+}]_i$ called a "Ca^{2+} sparklet". A subpopulation of L-type Ca^{2+} channels is associated with a signaling tetrad composed of AKAP150, PKA, PKC, and PP2B. These L-type Ca^{2+} channels are rapidly modulated by AKAP150-associated PKC, PKA, and PP2B during physiological and pathological conditions such as hypertension and hyperglycemia. The association of AKAP150 with L-type Ca^{2+} channels also promotes coordinated openings and closings of these channels via transient interactions between variable numbers of L-type Ca^{2+} channels' C-termini. This signaling unit is thus able to modulate Ca^{2+} influx and by regulating the activity of the Ca^{2+}-activated phosphatase PP2B control NFATc3-dependent gene expression of Kv and BK channels in cardiac and arterial myocytes. Plus symbols indicate activation; negative symbols indicate inhibition/downregulation

Large amplitude Ca^{2+} sparklets could arise from the random overlapping opening of neighboring L-type Ca^{2+} channels with high open probability. However, not all L-type Ca^{2+} channels gate independently (Santana and Navedo 2010). It was found that small clusters of L-type Ca^{2+} channels could open and closed in a coordinated fashion (called "coupled gating"). Although the mechanisms inducing coupled gating of L-type Ca^{2+} channels are not entirely clear, data suggest this gating modality involves transient interactions between variable numbers of L-type Ca^{2+} channels' C-termini in an AKAP150-associated protein complex (Fig. 1). Multiple lines of evidence support this model. First, PKCα increases the probability of coupled gating between L-type Ca^{2+} channels. Second, displacement of calmodulin (CaM) away from its putative binding site (e.g., IQ domain) in the C-termini of L-type Ca^{2+} channels increases coupled gating activity. Third, AKAP150 is required for coupled gating between L-type Ca^{2+} channels. Accordingly, in the absence of AKAP150, L-type Ca^{2+} channel's gating is mostly stochastic even after the activation of PKCα. Fourth, an L-type Ca^{2+} channel construct lacking amino acids 1670–2171, which eliminates a large section of the C-termini of these channels that includes the AKAP150 binding region, showed no coupled gating activity. Consistent with this conclusion, Förster resonance energy transfer (FRET) analysis suggested that the C-termini of nearby L-type Ca^{2+} channels come into close proximity under conditions that favor coupled gating (Santana and Navedo 2010). Collectively, these data suggest that L-type Ca^{2+} channels are more likely to undergo coupled gating in regions of the cell where these channels interact with an AKAP150-associated protein complex. The mechanisms for coupled gating between L-type Ca^{2+} channels may involve a rearrangement of CaM within the IQ domain that induces transient interactions between a variable numbers of adjacent L-type Ca^{2+} channels via their C-termini. Future studies are required to address these issues.

Functional significance of Ca^{2+} sparklets during health and disease. In principle, functional coupling of L-type Ca^{2+} channels could have profound functional implications on excitation-contraction (EC) coupling and excitation-transcription (ET) coupling in cardiac and arterial myocytes. In cardiac myocytes, Ca^{2+} sparklets activate Ca^{2+} release via small clusters of RyRs located in nearby junctional SR via the mechanisms of Ca^{2+}-induced Ca^{2+} release (CICR). Tight, local control of Ca^{2+} spark activation by sparklets forms the basis for the generation of cell-wide increases in cytosolic [Ca^{2+}] that triggers contraction in cardiac muscle (Wang et al. 2004). In these cells, the coupling strength between sparklets and sparks is proportional to the amount of Ca^{2+} flux through L-type Ca^{2+} channels. Accordingly, at least in principle, the probability of Ca^{2+} spark activation could be higher in areas within the surface membrane in which L-type Ca^{2+} channels open coordinately.

Unlike cardiac myocytes, in arterial smooth muscle, RyRs and L-type Ca^{2+} channels do not form tight SR Ca^{2+} release units. In arterial myocytes, activation of L-type Ca^{2+} channels results in an increase in cytosolic [Ca^{2+}] that directly activates the contractile machinery. Thus, an increase in the probability of coupled L-type Ca^{2+} channels results in an increase in Ca^{2+} influx and global [Ca^{2+}] that activates contraction.

Arterial tone is elevated during pathological conditions such as diabetes and hypertension, which increases the probability of stroke, coronary artery disease, and cardiac hypertrophy. Multiple studies suggest that increased L-type Ca^{2+} channel activity is a major contributor to these pathological changes. Interestingly, increased L-type Ca^{2+} channel activity also underlies lethal cardiac arrhythmias in human with Timothy Syndrome (TS) caused by a single point mutation (G436R) in these channels. Consistent with this, Ca^{2+} sparklet activity is increased in arterial myocytes during hypertension and in type II diabetes, and in tsA-201 cells expressing TS-L-type Ca^{2+} channels (Santana and Navedo 2009, 2010). Accordingly, an increase in the number of low activity and persistent Ca^{2+} sparklet sites activated, and the frequency of coupled gating events underlie enhanced Ca^{2+} influx during these pathological conditions.

The mechanisms underlying increased Ca^{2+} sparklet activity during hypertension, type II diabetes, and TS vary depending on the pathological condition. Although AKAP is critical for local activation of persistent Ca^{2+} sparklets in all cases, enhanced Ca^{2+} sparklets' activity during hypertension results primarily from an increase in L-type Ca^{2+} channels' expression and PKCα activity, while activation of PKA contributes to higher persistent Ca^{2+} sparklets' activity during hyperglycemia and type II diabetes (Fig. 1). In the case of TS, an arginine for glycine substitution at position 436 in L-type Ca^{2+} channels increases the open time and P_o of these channels, thus enhancing Ca^{2+} sparklet activity (Santana and Navedo 2009, 2010). Regardless of the mechanism, an increase in Ca^{2+} sparklet activity and coupled gating could translate into enhanced Ca^{2+} influx that contributes to the development of type II diabetes, hypertension, and Ca^{2+}-dependent cardiac arrhythmias associated with TS.

The increase in persistent Ca^{2+} sparklet activity and coupled gating has important functional consequences in arterial myocytes during hypertension and possibly type II diabetes, as this Ca^{2+} signal activates the transcription factor NFATc3 via calcineurin (Fig. 1) (Santana and Navedo 2010). Activation of AKAP-targeted PKCα or AKAP-targeted PKA during hypertension or type II diabetes, respectively, induces persistent Ca^{2+} sparklet activity and enhances coupled gating behavior. This produces an increase in local intracellular Ca^{2+} influx that activates nearby calcineurin. Once activated, calcineurin is able to dephosphorylate the transcription factor NFATc3. Upon dephosphorylation, NFATc3 translocates to the nucleus of arterial myocytes where it can modify the expression of Kv and BK channel subunits. Downregulation of Kv and BK channel subunits, thus, decreases channel function, depolarizes arterial myocytes, and enhances Ca^{2+} sparklets activity, hence increasing Ca^{2+} influx, global [Ca^{2+}]$_i$, and myogenic tone. Importantly, and consistent with the model proposed above, loss of AKAP150 or PKCα protects against the development of experimental hypertension. Collectively, data support a model in which a signaling pentad composed by AKAP, PKCα, PKA, calcineurin, and L-type Ca^{2+} channels contributes to the regulation of gene expression, Ca^{2+} influx, and excitability during physiological and pathological conditions.

Ca^{2+} Waves

Though Ca^{2+} sparklets are highly localized Ca^{2+} signals produced by the opening of L-type Ca^{2+} channels in the surface membrane, Ca^{2+} waves are spontaneous transient increases in global cytosolic $[Ca^{2+}]$ produced by Ca^{2+} release via RyRs or IP_3Rs in the SR. This Ca^{2+} signaling modality could propagate from one end of a cell to the other. Ca^{2+} waves were first observed in the 1970s as elevations in cytosolic $[Ca^{2+}]$ in medaka fish eggs during the fertilizing process with the use of the chemiluminescent indicators (Jaffe 1991). Subsequently, the development of novel fluorescent indicators and optical imaging techniques helped expand our understanding of the mechanisms underlying the generation, propagation, and functional consequences of Ca^{2+} waves. To date, Ca^{2+} waves have been recorded from many different cell types including oocytes and eggs, as well as excitable (e.g., neuron, cardiac, and arterial myocytes) and non-excitable cells (e.g., hepatocytes, astrocytes, and endothelial cells) (Jaffe 1991). Ca^{2+} waves are initiated by different molecular entities and displayed different forms, velocities, and patterns, depending on the cell type and stimulus (Table 2). For the purpose of this entry, the discussion of Ca^{2+} waves will be centered on their generation, propagation, and functional role in cardiac and arterial myocytes.

Molecular mechanisms underlying the generation and propagation of Ca^{2+} waves in cardiac and smooth muscle. The SR in cardiac and arterial myocytes serves as an intracellular Ca^{2+} store that releases its content when RyRs and/or IP_3Rs is activated. The sequential activation of these channels generates Ca^{2+} waves that could propagate throughout the cytoplasm of these cells (Fig. 2) (Wier and Blatter 1991). In cardiac myocytes, Ca^{2+} waves are associated with a pathological state in which the SR is "overloaded" with Ca^{2+}. This increases the probability of spontaneous Ca^{2+} release via RyRs and the subsequent activation of nearby clusters of these channels by the mechanism of CICR (Fig. 2). Whether a Ca^{2+} wave develops would depend on the amplitude and duration of the triggering Ca^{2+} release event and the sensitivity of adjacent RyRs to this Ca^{2+} signal. Multiple studies indicate that increases in SR Ca^{2+} content enhance the sensitivity of RyRs to cytosolic Ca^{2+}. Thus, during SR Ca^{2+} overload, RyRs exist in a hypersensitive state that makes them susceptible to activation by the opening of nearby RyRs in the leading edge of the traveling wave (Keller et al. 2007). IP_3Rs are not involved in Ca^{2+} waves' generation and propagation in cardiac myocytes, but have been implicated in the regulation of cardiac hypertrophy via a calcineurin-dependent mechanism.

The mechanisms underlying the generation and propagation of Ca^{2+} waves in arterial myocytes are not clearly understood. In these cells, Ca^{2+} waves result from intracellular Ca^{2+} release via RyRs and/or IP_3Rs due to electrical, mechanical, or α-adrenergic stimulation (Fig. 2) (Wray and Burdyga 2010). Notably, several studies suggest that Ca^{2+} influx via L-type Ca^{2+} channels (i.e., Ca^{2+} sparklets) is important for the sustained generation of Ca^{2+} waves in arterial myocytes. However, it seems that the molecular entity or entities underlying the generation of Ca^{2+} waves will depend on the stimulus provided. For example, Ca^{2+} sparks at frequent discharge sites have been implicated in the generation of Ca^{2+} waves (Wray and Burdyga 2010). Consistent with this, caffeine, which increases the sensitivity of RyRs to cytosolic Ca^{2+}, has been shown to increase the frequency of Ca^{2+} waves in vascular myocytes. Meanwhile, ryanodine, an alkaloid that binds to RyRs and locks the receptor in a sub-conductance state that depletes SR Ca^{2+}, was shown to eliminate Ca^{2+} sparks and Ca^{2+} waves in pressurized cerebral arteries. These observations support the view that Ca^{2+} waves are produced and propagated by the gating of clusters of RyRs, at least in pressurized cerebral arteries that have not been exposed to agonists that elevate IP_3.

On the other hand, multiple studies have suggested that IP_3Rs are involved in agonist-induced Ca^{2+} waves, with possible involvement of RyRs via a CICR mechanism in arterial myocytes (Fig. 2). Indeed, vasoconstrictors, such as UTP, have been shown to promote Ca^{2+} waves through a mechanism that involved the activation of IP_3Rs. Whether Ca^{2+} release through IP_3Rs facilitates the recruitment of adjacent RyRs in order to amplify and propagate agonist-induced Ca^{2+} waves in arterial myocytes is not clear. A model in which stimulus-induced propagating Ca^{2+} waves will

Calcium Sparklets and Waves, Table 2 Properties of Ca^{2+} waves

Cell type	Velocity (μm/s)	Duration (s)	Frequency (Hz)	Stimulus
Cardiac myocytes	≈116 Range: 67–195[a]	289[a]	0.38[b]	SR Ca^{2+} overload
Arterial myocytes	≈47 Range: 7–121	ND	0.29	Pressure

ND not determined
[a]Ishide et al. Circ Res 57:844–855
[b]Kort et al. AJP-Heart 259:H940–H950; Jaggar AJP-Cell 281:C439–C448

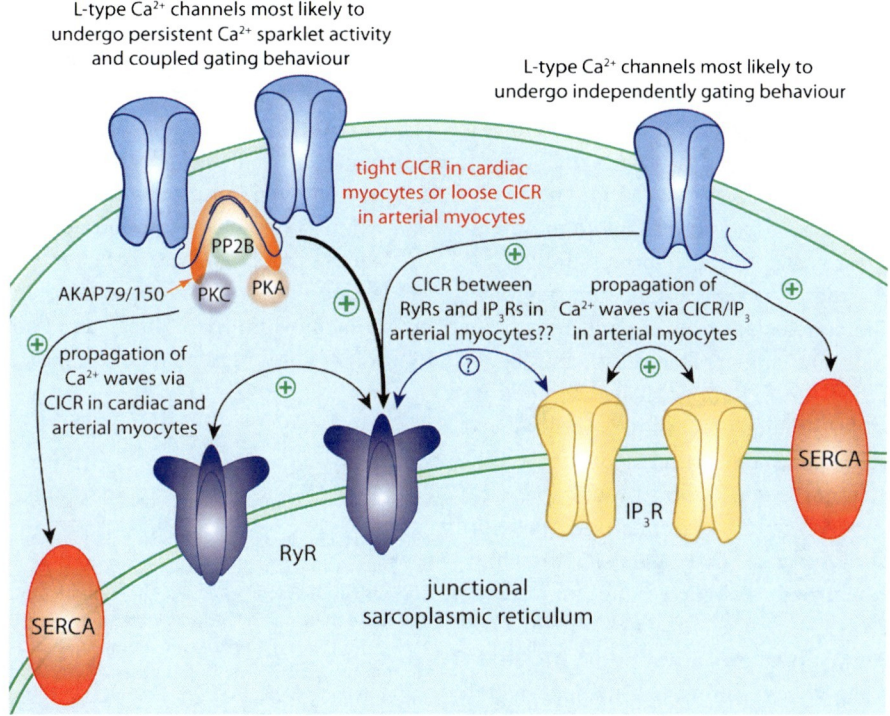

Calcium Sparklets and Waves, Fig. 2 *Proposed model for the generation of Ca^{2+} waves in cardiac and arterial myocytes.* Ca^{2+} waves are produced by the sequential activation of RyRs and/or IP$_3$Rs in the SR of cardiac and arterial myocytes. In cardiac myocytes, Ca^{2+} waves are a pathological phenomenon associated with SR Ca^{2+} overload. In these cells, Ca^{2+} waves are produced and propagated exclusively by spontaneous or triggered release of Ca^{2+} via RyRs by the mechanism of CICR. In arterial myocytes, Ca^{2+} waves result from SR Ca^{2+} release via RyRs and/or IP3Rs due to electrical, mechanical, or α-adrenergic stimulation. Whether CICR between RyRs and IP$_3$Rs occurs in these cells is not clear yet. In this model, independently gating or coupled L-type Ca^{2+} channels contribute to Ca^{2+} influx into a common global cytosolic Ca^{2+} pool from which SERCA pumps Ca^{2+} into the SR network of cardiac and arterial myocytes

depend on the level of expression, subcellular distribution, and activation of both RyRs and IP$_3$Rs in the SR of arterial myocytes could help reconcile these findings (Wray and Burdyga 2010). Alternatively, a mechanism similar to the one described in cardiac myocytes in which RyRs are sensitized to cytosolic Ca^{2+} during high SR Ca^{2+} load may play a pivotal role in the propagation of agonist-induced Ca^{2+} waves. Future studies should examine these issues.

Functional consequences of Ca^{2+} waves. In cardiac myocytes, Ca^{2+} waves do not seem to be physiologically relevant for either Ca^{2+}-dependent signal transduction or normal EC coupling. Rather, Ca^{2+} waves are believed to be a pathological manifestation of SR

Ca^{2+} overload that activates a depolarizing inward membrane current in cardiac myocytes responsible for delayed afterdepolarizations that could trigger ventricular arrhythmias (Cheng and Lederer 2008; Keller et al. 2007).

In smooth muscle, Ca^{2+} waves play a key physiological role. Ca^{2+} waves have been suggested to contribute to the development of myogenic tone during changes in response to activation of α-adrenergic receptor signaling (Wray and Burdyga 2010). Accordingly, activation of IP_3 signaling can evoke Ca^{2+} waves that elevate cytosolic $[Ca^{2+}]$, thus inducing the phosphorylation of the myosin light chain (MLC) by the myosin light chain kinase. Phosphorylation of MLC will induce cross-bridge cycling and cell shortening (Wray and Burdyga 2010).

The functional role of Ca^{2+} waves during the development of myogenic tone by arterial smooth muscle in response to increases in intravascular pressure is controversial. Two studies using cerebral arteries have reached opposite conclusions with regard to the role of Ca^{2+} waves on EC coupling. The Nelson and Jaggar groups (Jaggar et al. 1998) have published compelling data suggesting that the net effect of pressure-induced Ca^{2+} waves is to oppose contraction, and that, consistent with the current dogma, intravascular pressure elevates global cytosolic $[Ca^{2+}]$ primarily by promoting Ca^{2+} influx via L-type Ca^{2+} channels. Meanwhile, Donald Welsh's group recognizes the importance of Ca^{2+} influx via L-type Ca^{2+} channels on EC coupling, but argues that pressure increases the frequency of Ca^{2+} waves in a voltage-insensitive manner, which instead facilitates tone development by providing part of the Ca^{2+} necessary to the global Ca^{2+} pool that activates the contractile machinery in arterial myocytes. Although interesting, this model is at odds with work from many laboratories, indicating that global cytosolic $[Ca^{2+}]$ in arterial myocytes is most regulated by Ca^{2+} sparklets in a voltage-dependent manner (Santana and Navedo 2009; Wier and Blatter 1991; Wray and Burdyga 2010). Thus, while the reasons for these seemingly contradictory findings are unclear, it is possible that variations in experimental conditions (e.g., effectiveness of the inhibitors in a given experimental setting) may contribute to these differences. Future studies should further examine this issue and provide unambiguous answers about the functional role of Ca^{2+} waves on EC coupling in arterial myocytes.

Concluding Remarks

Advances in imaging technology have provided a unique opportunity to gain insight into the regulatory mechanisms and dynamics of Ca^{2+} signaling events and their functional role in cells. Such is the case with the optical recording of Ca^{2+} sparklets and Ca^{2+} waves in cardiac and arterial myocytes. Indeed, one of the most paradigm shifting observations made using optical approaches is that the open probability of L-type Ca^{2+} channels varies regionally within the surface membrane of cells. The recording of Ca^{2+} sparklets revealed the regulatory mechanisms underlying heterogeneous sparklet activity, uncovered a gating modality not described before for L-type Ca^{2+} channels (e.g., coupled gating), and provided insight into the functional role of this Ca^{2+} signal during physiological and pathological conditions. Similarly, the recording of Ca^{2+} waves and the groundbreaking discovery of Ca^{2+} sparks as the building block for the generation and propagation of Ca^{2+} waves provided an opportunity to determine how transient, local, and global increases in cytosolic $[Ca^{2+}]$ regulate cell excitability.

References

Cheng H, Lederer WJ (2008) Calcium sparks. Physiol Rev 88(4):1491–1545

Hille B (2001) Ionic channels of excitable membranes, 3rd edn. Sinauer Associates, Suderland

Jaffe LF (1991) The path of calcium in cytosolic calcium oscillations: a unifying hypothesis. Proc Natl Acad Sci USA 88(21):9883–9887

Jaggar JH, Wellman GC, Heppner TJ, Porter VA, Perez GJ, Gollasch M et al (1998) Ca^{2+} channels, ryanodine receptors and Ca^{2+}-activated K^+ channels: a functional unit for regulating arterial tone. Acta Physiol Scand 164(4):577–587

Keller M, Kao JP, Egger M, Niggli E (2007) Calcium waves driven by "sensitization" wave-fronts. Cardiovasc Res 74(1):39–45

Santana LF, Navedo MF (2009) Molecular and biophysical mechanisms of Ca^{2+} sparklets in smooth muscle. J Mol Cell Cardiol 47(4):436–444

Santana LF, Navedo MF (2010) Natural inequalities: why some L-type Ca^{2+} channels work harder than others. J Gen Physiol 136(2):143–147

Wang SQ, Wei C, Zhao G, Brochet DX, Shen J, Song LS et al (2004) Imaging microdomain Ca^{2+} in muscle cells. Circ Res 94(8):1011–1022

Wier WG, Blatter LA (1991) Ca^{2+}-oscillations and Ca^{2+}-waves in mammalian cardiac and vascular smooth muscle cells. Cell Calcium 12(2–3):241–254

Wray S, Burdyga T (2010) Sarcoplasmic reticulum function in smooth muscle. Physiol Rev 90(1):113–178

Calcium, Local and Global Cell Messenger

Martin D. Bootman[1] and Rüdiger Thul[2]
[1]Life, Health and Chemical Sciences, The Open University Walton Hall, Milton Keynes, UK
[2]School of Mathematical Sciences, University of Nottingham, Nottingham, UK

Synonyms

Calcium as an intracellular messenger; Calcium diffusion; Calmodulin; EF-hand proteins; Local signaling

Definition

All eukaryotic cells use ionized calcium (Ca^{2+}) as a messenger. An increase or decrease of Ca^{2+} concentration in the cytosol, or in organellar compartments, can lead to changes in cell function, structure, or viability. To encode information and control specific cellular processes, Ca^{2+} signals need to be temporally and spatially regulated. The diffusive movement of Ca^{2+} within cells is critical in turning local signals into whole-cell (global) events. Cells possess mechanisms to both aid and hinder the spread of Ca^{2+}, depending on their physiological context. Furthermore, Ca^{2+} channels themselves are regulated by Ca^{2+} to provide either positive or negative feedback control of Ca^{2+} fluxes.

Overview

Ca^{2+} is a versatile intracellular messenger. Cellular Ca^{2+} signals can span from brief, nanoscopic flickers, close to the mouth of channels, up to global events that spread regeneratively through cells and last for many tens of seconds (Berridge 2006). Furthermore, Ca^{2+} signals can spread between cells to coordinate the activities of tissues (Leybaert and Sanderson 2012). The spatial and temporal properties of Ca^{2+} signals are determined by the Ca^{2+} transport systems expressed within a particular cell type. There is a wide range of Ca^{2+} channels, pumps, buffers, and effector molecules encoded within the eukaryotic genome. Cells select from this "Ca^{2+} signaling toolkit" those components that generate Ca^{2+} signals to suit their physiology (Berridge et al. 2000). Specificity in Ca^{2+} signaling is encoded in the location, amplitude, and frequency of Ca^{2+} signals and, in some cases, by cross talk with other signal transduction cascades. In particular, specificity is achieved by expressing Ca^{2+} channels and effector molecules within privileged compartments where Ca^{2+} changes can arise with complete isolation from the rest of the cell. By compartmentalizing Ca^{2+} signals within specific domains, cells can use this simple divalent cation to control multiple, simultaneous processes. Aberrant changes in the distribution, amplitude, or kinetics of Ca^{2+} signals cause explicit pathological outcomes (Sammels et al. 2010).

In resting cells, the average cytosolic Ca^{2+} concentration is ~100 nM. Stimulation of cells via a variety of means (e.g., hormonal, electrical, mechanical) can elicit an increase of the cytosolic Ca^{2+} concentration. The amplitude of the Ca^{2+} rise depends on the cell type, the nature of the stimulus, and where it is measured. For example, hormonal stimulation of non-excitable cells typically evokes repetitive whole-cell Ca^{2+} oscillations. The frequency of such Ca^{2+} oscillations, or their cumulative Ca^{2+} signal, determines the extent of cellular response. Typically, Ca^{2+} oscillations in non-excitable cells arise via the repetitive release of Ca^{2+} from intracellular stores, principally the endoplasmic reticulum (ER). In contrast, electrical stimulation of excitable cells (e.g., myocytes, neurons) activates voltage-operated Ca^{2+} channels (VOCs) that allow the influx of Ca^{2+} from the extracellular space. This Ca^{2+} influx signal can remain localized around the mouth of the channel or trigger the release of Ca^{2+} from intracellular stores – a process known as Ca^{2+}-induced Ca^{2+} release (CICR). The self-amplification of Ca^{2+} signals by CICR is critical in turning local signals into whole-cell responses.

Ca^{2+} Channels

Ca^{2+} channels that mediate the influx of Ca^{2+} from the extracellular space are generally characterized by their activation mechanism. For example, receptor-operated Ca^{2+} channels (e.g., glutamate- and glycine-binding NMDA receptors) and second messenger-operated Ca^{2+} channels (e.g., cyclic nucleotide-gated ion channels and arachidonate-regulated Orai channels) are opened by the binding of an external or internal ligand, respectively. VOCs are widely expressed in excitable tissues and can be divided into three families – the Ca_V1 family of L-type channels; the Ca_V2 family of N-, P-/Q-, and R-type channels; and the Ca_V3 family of T-type channels – which have specific gating characteristics, pharmacologies, and functions (Catterall et al. 2005). The transient receptor potential (TRP) family includes a number of Ca^{2+}-permeable channels with distinct activation mechanisms (Gees et al. 2010). The release of Ca^{2+} from internal stores occurs through several different types of channel (Bootman et al. 2002), including inositol 1,4,5-trisphosphate receptors ($InsP_3Rs$), ryanodine receptors (RyRs), polycystin-2 (a member of the TRP family), and putative the NAADP-sensitive two-pore channels (TPCs).

Ca^{2+} Buffers

Cells express a number of Ca^{2+}-binding proteins that reversibly buffer Ca^{2+} changes within various cellular compartments. The involvement of Ca^{2+} buffers can be critical in shaping both the spatial and temporal properties of Ca^{2+} signals (Schwaller 2010). Two prominent cytosolic Ca^{2+} buffers are parvalbumin and calbindin D-28 k, which possess high-affinity (micromolar) Ca^{2+}-binding motifs known as EF-hands (see below) that allow them to adsorb Ca^{2+} ions as they diffuse away from channels. Within the ER, Ca^{2+}-binding proteins, such as calsequestrin and calreticulin, facilitate the accumulation of large amounts of Ca^{2+}, which is necessary for rapid cell signaling. Mitochondria also play a key role in buffering Ca^{2+} due to their ability to sequester substantial amounts of Ca^{2+} whenever the cytosolic Ca^{2+} concentration rises. The uptake of Ca^{2+} into the mitochondrial matrix stimulates the citric acid cycle to produce more ATP. When mitochondrial Ca^{2+} uptake is exaggerated, or occurs simultaneously with specific messengers (e.g., arachidonate, ceramide, reactive oxygen species), a substantial "permeability transition pore" (mPT) can be activated. mPT underlies the release of intramitochondrial proteins such as cytochrome C that trigger apoptosis (Duchen and Szabadkai 2010).

Proteins Mediating the Action of Ca^{2+}

The cellular actions of Ca^{2+} are almost entirely mediated by binding to specific proteins that then change their activity, or the activity of further binding partners. Some moieties, such as the Ca^{2+}-dependent transcriptional repressor DREAM, are active when the cytosolic/nuclear Ca^{2+} concentration is low and inhibited when the Ca^{2+} level rises. In the majority of cases, an elevation of Ca^{2+} concentration is an activation signal. Ca^{2+} binds to proteins via specific motifs. Two common motifs are EF-hands and C2 domains (Clapham 2007). The universal Ca^{2+} sensor, calmodulin (CaM), is plausibly the best-known example of an EF-hand-containing protein. CaM is ubiquitously expressed and mediates many effects of Ca^{2+} on cell metabolism and growth. In particular, CaM mediates the activation of Ca^{2+}-sensitive enzymes including Ca^{2+}/calmodulin-dependent protein kinases (CaMKs), the phosphatase calcineurin, myosin light-chain kinase (MLCK), and phosphorylase kinase. CaM has four EF-hands, which bind Ca^{2+} in a cooperative manner, thus allowing the Ca^{2+}/CaM interaction to provide a rapid switch in activity. Subtle changes in the amino acid sequences defining EF-hand motifs can alter the affinity of a protein for Ca^{2+}. A well-known C2 domain-containing protein is synaptotagmin, which facilitates neurotransmitter vesicle exocytosis in response to opening of nearby VOCs. The use of different Ca^{2+}-binding motifs with varying affinities allows Ca^{2+} to regulate cellular processes over a >100-fold range of concentrations. A pertinent example of this can be seen in neurons, which possess Ca^{2+}-binding proteins sensitive to submicromolar Ca^{2+} levels (e.g., neuronal Ca^{2+}-sensor proteins) up to synaptotagmin that is activated by Ca^{2+} in the ~ 100 micromolar range (Burgoyne and Weiss 2001).

Reversing Ca^{2+} Signals

Ca^{2+} signals are reversed by a range of pumps and exchanger molecules that are designed to reduce the resting cytosolic Ca^{2+} concentration to ~ 100 nM. Tissue-specific expression of different pump/exchanger combinations alters the kinetics of Ca^{2+} signal recovery. In tissues requiring rapid Ca^{2+}

transients, such as cardiac muscle cells, Na$^+$/Ca^{2+} exchangers provide a fast rate of Ca^{2+} transport critical for cells to reset between contractions. The plasma membrane Ca^{2+}-ATPase (PMCA) and sarco-/endoplasmic reticulum (SERCA) pumps have lower transport rates but higher affinities, which means that they operate at relatively low cytosolic Ca^{2+} concentrations.

Local Ca^{2+} Signaling

While whole-cell Ca^{2+} signals can regulate events such as contraction and gene transcription, it is becoming increasingly apparent that local Ca^{2+} signals are critically important in the regulation of cell function. Since cells have so many Ca^{2+}-sensitive processes, their simultaneous regulation can only be achieved by constraining Ca^{2+} within privileged compartments where Ca^{2+} source and effector proteins are in close proximity. A classic example of such a compartment is neuronal dendritic spines, where the folding of the plasma membrane delineates a microscopic signaling space. In other situations, Ca^{2+} signal compartmentalization relies on the close proximity of membranes from different structures or organelles. For example, privileged Ca^{2+} signaling compartments are formed by the close association of the ER to either the mitochondria to promote the flux of Ca^{2+} between these organelles or to the plasma membrane to regulate K$^+$ channels to control membrane excitability in neurons. Ca^{2+} can have sequential local and global signaling functions. For example, Orai1 channels on the plasma membrane are activated when ER stores are depleted of Ca^{2+} (Luik et al. 2006). The Ca^{2+} ions that enter cells through Orai1 channels can act locally to switch on mitogen-activated protein kinases and cAMP production before diffusing into cells to stimulate other events.

A particular local Ca^{2+} signaling compartment is found in cardiac myocytes of the heart. The plasma membrane of cardiac cells (sarcolemma) has deep invaginations known as "transverse tubules" (T-tubules). These thin (~100–200 nm in diameter) inward projections of the sarcolemma are like a system of tire spokes occurring at each of the Z-lines (longitudinal spacing ~1.8 μm). It is principally along the T-tubule membranes that VOCs are triggered to initiate cardiac contraction. Closely opposed to the T-tubules are elements of the sarcoplasmic reticulum (SR), bearing the Ca^{2+}-conducting RyRs. The VOCs on the T-tubules and the RyRs on the SR sit opposite each other across a small span (~15 nm) of cytoplasm. Together, the membrane with the VOCs and the membrane with the RyRs form a "dyadic junction." During electrical excitation of a cardiac myocyte, the action potential sweeps down the T-tubule and activates the VOCs to produce a small, local Ca^{2+} signal known as a "Ca^{2+} sparklet." The consequent Ca^{2+} influx signal is amplified by the RyRs via the CICR process described earlier. Ca^{2+} ions diffuse out of the dyadic junction, bind to the EF-hand-containing protein troponin C, and thereby trigger contraction. The simultaneous electrical activation of Ca^{2+} signals at many thousands of dyadic junctions leads to the global signal required for cardiac contraction. However, this whole-cell Ca^{2+} signal relies on the initial local communication between VOCs and RyRs.

The example of dyadic junctions is particularly interesting because it is one of the few Ca^{2+} signaling compartments where the locations and numbers of Ca^{2+} channels are reasonably well known. It has been estimated that ~10 L-type VOCs in the T-tubule membrane face ~100 RYR in the closely apposed SR junctional membrane. The Ca^{2+} sparklets produced by opening L-type VOCs represent the movement of ~300 Ca^{2+} ions from the extracellular space into the dyadic junction. Approximately 15 RyRs sense this Ca^{2+} influx and respond via CICR to produce a larger and long-lasting "Ca^{2+} spark." The transition from Ca^{2+} sparklet to Ca^{2+} spark and then to whole-cell response demonstrates how local events can feed global responses.

Another way of controlling the response of specific effector proteins is to alter their proximity to a particular Ca^{2+} source. For example, the protein synaptotagmin, mentioned above, has a relatively low affinity for Ca^{2+}, but it can be physically associated with a VOC so that it can rapidly respond to an elevation of Ca^{2+} concentration and trigger synaptic vesicle release. Other proteins may sense an elevated Ca^{2+} concentration and migrate toward it. An example of such a protein is the Ras GTPase-activating protein RASAL, which binds Ca^{2+} via a C2 domain and cycles on and off the plasma membrane in synchrony with Ca^{2+} oscillations. A further example is regulation of mitochondrial movement by Rho GTPases proteins called Miro. These proteins possess EF-hands to sense Ca^{2+} and also interact with other proteins involved in mitochondrial movement along microtubules. At resting cytosolic Ca^{2+} levels, Miro proteins facilitate mitochondrial movement by increasing the proportion of mitochondria associated with kinesin motor proteins, but elevation of Ca^{2+} concentration rapidly halts mitochondrial translocation. In this way, mitochondria can become trapped near the compartments with elevated Ca^{2+} concentration.

Ca^{2+} Diffusion

Diffusion is the main mechanism by which Ca^{2+} moves through the cell. Ca^{2+} moves from regions of relatively high concentration to areas of low concentration using two mechanisms: diffusion of free Ca^{2+} ions and diffusion of buffer-bound Ca^{2+}. Typically, less than 5% of the total Ca^{2+} concentration is free ions, so the majority of Ca^{2+} diffuses bound to mobile buffers. In addition to endogenous mobile buffers, exogenously introduced Ca^{2+} chelators such as EGTA, BAPTA, and fluorescent Ca^{2+} reporters (commonly used to measure cellular Ca^{2+}) affect the speed of intracellular Ca^{2+} transport. To characterize the combined effect of buffered and free diffusion on the spatiotemporal spread of Ca^{2+}, an effective Ca^{2+} diffusion D_{eff} coefficient is often considered. In a wider sense, D_{eff} corresponds to a coarse-grained description of the more detailed buffer kinetics, collapsing all binding and unbinding events to one number. A common approach to compute D_{eff} is the fast buffer approximation (Wagner and Keizer 1994; Smith et al. 2001). It is based on the assumption that buffer kinetics are fast, resulting in an instant equilibrium of bound and unbound buffer. As with all approximations, the fast buffer approximation is only valid under certain conditions. Importantly, D_{eff} is computed from the diffusion coefficients of the mobile buffers, the total buffer concentrations and the buffer dissociation constants. Therefore, D_{eff} varies between different cellular compartments such as the cytosol or the ER due to the expression of different buffers. Moreover, the local cellular geometry influences the speed of Ca^{2+} diffusion. In compartments like the ER, D_{eff} is further reduced since ER tortuosity acts as an obstacle for diffusion (Ölveczky and Verkman 1998).

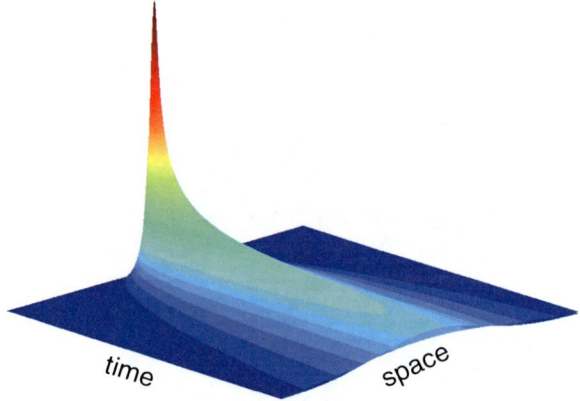

Calcium, Local and Global Cell Messenger, Fig. 1 Ca^{2+} diffusion profile from a point source located in the center of the spatial domain

As buffers generally diffuse more slowly than free Ca^{2+} ions, D_{eff} is smaller than the value of free Ca^{2+} diffusion. For example, cytosolic D_{eff} is in the range of 20–40 $\mu m^2/s$, while the diffusion coefficient for free cytosolic Ca^{2+} is 223 $\mu m^2/s$. Cells can hinder the diffusion of Ca^{2+} in order to keep Ca^{2+} signals localized to their site of origin. In particular, cells can express Ca^{2+} pumps or mitochondria to act as a firewall, preventing the diffusive spread of Ca^{2+} ions. Notable examples have been seen with apical-only Ca^{2+} signals in pancreatic acinar cells (Walsh et al. 2009), subsarcolemmal Ca^{2+} signals in atrial myocytes (Mackenzie et al. 2004), and the discontinuity between the cytosolic and subplasma membrane Ca^{2+} in smooth muscle (caused by the SR forming a "superficial buffer barrier").

A potential consequence of Ca^{2+} being transported by mobile buffers is facilitated diffusion. When Ca^{2+} diffusion is described by a single effective diffusion coefficient, D_{eff}, the spatial Ca^{2+} concentration profile around a point source evolves in time as shown in Fig. 1. The Ca^{2+} concentration decreases smoothly on either side of the point source, and the peak Ca^{2+} concentration decays steadily over time. The exact spatiotemporal profile of such a response depends on the conductance of the Ca^{2+} channel and the local buffering conditions. When buffers are considered explicitly, this monotonic evolution of the concentration profile may be disrupted. A localized peak in the Ca^{2+} concentration emerges at some distance from the Ca^{2+} source. This is the fingerprint of facilitated diffusion. The mechanism for this process resembles that of a "piggyback ride." Ca^{2+} binds to buffers at the source, and then Ca^{2+}-bound buffers diffuse away from it. After some time, which is determined by buffer off rates, Ca^{2+} dissociates from the buffers. In turn, this leads to a local increase in the Ca^{2+} concentration.

In addition to controlling cellular activities, local Ca^{2+} signals can regulate channel opening too. Many Ca^{2+} channels possess sites at which Ca^{2+} can bind and are positively or negatively regulated by the ions they liberate. Deciphering the profile of a Ca^{2+} signal around the mouth of an ion channel is key to understanding the Ca^{2+}-dependent gating of ion channels. As with all binding processes, the rate with which Ca^{2+} binds to a designated binding site depends on the number of Ca^{2+} ions in the vicinity of the binding site. The more Ca^{2+} ions around a particular binding site, the higher the probability that a Ca^{2+} ion binds in a given period of time. Therefore, averaged Ca^{2+} concentrations such as the cytosolic bulk Ca^{2+} concentration do not determine how Ca^{2+} signaling events proceed. All the information for channel gating is encoded in the local Ca^{2+} dynamics.

A prominent example of localized Ca^{2+} signaling is the "Ca^{2+} puff" that arises from activation of $InsP_3Rs$ on the ER (Bootman et al. 1997; Smith et al. 2009). Stimulation of cells with hormonal agonists evokes the production of $InsP_3$ from a membrane lipid precursor. $InsP_3$ is highly diffusible inside cells and thus rapidly makes its way from the plasma membrane to the ER, where it binds to, and activates, $InsP_3Rs$ to produce the Ca^{2+} oscillations mentioned above. Ca^{2+} puffs can be considered as an elementary event of Ca^{2+} signaling, rather like the Ca^{2+} sparks described earlier. An intriguing feature of Ca^{2+} signaling is that regular, periodic Ca^{2+} signals can rely on underlying stochastic activation of Ca^{2+} release channels. In the following section, how this random activation manifests itself for $InsP_3R$ to trigger a Ca^{2+} puff is illustrated.

Although $InsP_3Rs$ require the binding of $InsP_3$ to open, they are actually activated by Ca^{2+} itself. The activation of $InsP_3Rs$ by Ca^{2+} means that they have the ability to augment Ca^{2+} responses via CICR. $InsP_3Rs$ usually form clusters, which are expressed at spatially discrete locations on the membrane of the ER or SR. Although the number of $InsP_3Rs$ per cluster varies between cells and among cell types, it is generally accepted that a cluster comprises only tens of $InsP_3Rs$. This small number of ion channels per cluster has significant consequences for the cluster behavior: it is driven by randomness. Early experiments examining Ca^{2+} puff generation pointed toward the stochastic initiation of these events (Marchant et al. 1999; Bootman et al. 1997). After flash photolysis of $InsP_3$, it took a varying amount of time for the first Ca^{2+} puff to emerge. These experimental findings were corroborated by modeling studies, clearly demonstrating that stochastic effects dominate Ca^{2+} puff dynamics (Swillens et al. 1998; Falcke 2003) and see Thurley et al. (2012) for a recent review.

The randomness of Ca^{2+} puff generation originates from the small number of $InsP_3$- and Ca^{2+}-binding sites per cluster. Although the molecular details of Ca^{2+} and $InsP_3$ binding to the $InsP_3R$ are still being investigated, a large class of models agrees on $InsP_3$- and Ca^{2+}-dependent activation, as well as Ca^{2+}-dependent inactivation. One model that has received broad attention and ongoing modifications to accommodate new experimental results was originally proposed by De Young and Keizer (Shuai et al. 2007; De Young and Keizer 1992; Swaminathan et al. 2009). The De Young and Keizer model assumes three binding sites per $InsP_3R$ subunit (one for $InsP_3$, one for activating Ca^{2+}, and one for inhibitory Ca^{2+}), resulting in eight possible subunit states. One of these states corresponds to an activated subunit. Due to early measurements (Bezprozvanny et al. 1991), the tetrameric $InsP_3R$ is deemed open when at least three of the four subunits are in the activated state. Hence, there are four possible open states of an $InsP_3R$. Denoting the probability of an activated subunit as p_{ac}, the probability of an open $InsP_3R$ is given by $P_{op} = 4p_{ac}^3 - 3p_{ac}^4$.

The channel open probability, P_{op}, would be an adequate description at the cluster level if a cluster contained a large number of $InsP_3Rs$. However, the small $InsP_3R$ numbers per cluster necessitate a different approach. At the level of a single subunit, the continuous binding and unbinding of Ca^{2+} and $InsP_3$ translates into random transitions among the eight subunit states. Under the often-used assumption that the four subunits are independent, these transitions occur simultaneously at all four subunits. Therefore, a single binding event suffices to switch an $InsP_3R$ from inactivation to activation. It is this abrupt and stochastic state change that forms the basis for the fluctuating behavior of $InsP_3Rs$ and the randomness of Ca^{2+} puff initiation. Moving one level up in the kinetic hierarchy, the dynamics of an $InsP_3R$ as continuous jumps between different channel states may be conceptualized. While there is usually only

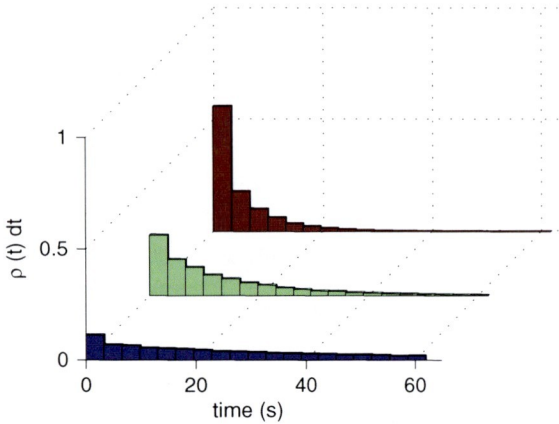

Calcium, Local and Global Cell Messenger, Fig. 2 Ca^{2+} puff initiation in a cluster of 8 DeYK channels for a Ca^{2+} concentration of 0.03 μM (*blue*), 0.05 μM (*green*), and 0.07 μM (*red*). The IP$_3$ concentration is 0.4 μM

one activated subunit state, the finding that at least three subunits have to be activated for channel opening leads to multiple activated channel states (see above). Hence, an InsP$_3$R opens when the channel arrives in one of these activated states for the first time. This idea led to the identification of Ca^{2+} puffs as a first passage time process, which in turn allowed for a thorough mathematical treatment of puff initiation times (Thul and Falcke 2006, 2007). On a conceptual level, this observation renders the initiation of a Ca^{2+} puff similar to the generation of a neuronal action potential. A neuron only fires an action potential once the membrane voltage is sufficiently depolarized for the first time, and reaching this threshold is completely random. It is important to note that the requirement of multiple subunits to simultaneously be in the activated state for the first time allows individual subunits to visit their respective activated state multiple times before channel opening. Therefore, if p_f denotes the probability of a single subunit to arrive at the activated subunit state for the first time, the probability P_f for an InsP$_3$R to open for the first time is *not* given by $4p_f^3 - 3p_f^4$. The computation of P_f requires more sophisticated mathematical tools (solution of a renewal equation, extensive combinatorics) as described in Thul and Falcke (2007).

In Fig. 2, the results from stochastic simulations of a cluster of eight tetrameric InsP$_3$Rs are plotted where each channel subunit is described by the De Young–Keizer (DK) model. The three histograms depict the probability to initiate a puff in a small time interval $[t, t + dt]$ for different values of the basal Ca^{2+} concentration. The parameter values of the original DK model are used, but adjust the binding constant for the activating Ca^{2+} binding site to reflect the large Ca^{2+} concentrations that occur at an open InsP$_3$R cluster. Upon increasing the basal Ca^{2+} concentration, Ca^{2+} puff generation is more likely to occur for shorter times. Moreover, the tail of the distribution shortens. This is consistent with the activating role of Ca^{2+} at low Ca^{2+} concentrations. A small increase in the basal Ca^{2+} concentration significantly raises the transition probability to the open channel state. The large peak at small times reflects the priming of InsP$_3$Rs by InsP$_3$. At a concentration of 0.4 μM, most subunits have activating InsP$_3$ bound. Hence, only activating Ca^{2+} has to bind to trigger a puff. This example shows how Ca^{2+} can control its own release within cells.

References

Berridge MJ (2006) Calcium microdomains: organization and function. Cell Calcium 40:405–412

Berridge MJ, Lipp P, Bootman MD (2000) The versatility and universality of calcium signalling. Nat Rev Mol Cell Biol 1:11–21

Bezprozvanny I, Watras J, Ehrlich BE (1991) Bell-shaped calcium-response curves of Ins(1,4,5)P$_3$-gated and calcium-gated channels from endoplasmic-reticulum of cerebellum. Nature 351:751–754

Bootman M, Niggli E, Berridge M, Lipp P (1997) Imaging the hierarchical Ca^{2+} signalling system in HeLa cells. J Physiol (Lond) 499(Pt 2):307–314

Bootman MD, Berridge MJ, Roderick HL (2002) Calcium signalling: more messengers, more channels, more complexity. Curr Biol 12:R563–R565

Burgoyne RD, Weiss JL (2001) The neuronal calcium sensor family of Ca^{2+}-binding proteins. Biochem J 353:1–12

Catterall WA, Perez-Reyes E, Snutch TP, Striessnig J (2005) International Union of Pharmacology. XLVIII. Nomenclature and structure-function relationships of voltage-gated calcium channels. Pharmacol Rev 57:411–425

Clapham DE (2007) Calcium signaling. Cell 131:1047–1058

De Young GW, Keizer J (1992) A single-pool inositol 1,4,5-trisphosphate-receptor-based model for agonist-stimulated oscillations in Ca^{2+} concentration. Proc Natl Acad Sci USA 89:9895–9899

Duchen MR, Szabadkai G (2010) Roles of mitochondria in human disease. Essays Biochem 47:115–137

Falcke M (2003) On the role of stochastic channel behavior in intracellular Ca^{2+} dynamics. Biophys J 84:42–56

Gees M, Colsoul B, Nilius B (2010) The role of transient receptor potential cation channels in Ca^{2+} signaling. Cold Spring Harb Perspect Biol 2:a003962

Leybaert L, Sanderson MJ (2012) Intercellular Ca^{2+} waves: mechanisms and function. Physiol Rev 92:1359–1392

Luik RM, Wu MM, Buchanan J, Lewis RS (2006) The elementary unit of store-operated Ca^{2+} entry: local activation of CRAC channels by STIM1 at ER-plasma membrane junctions. J Cell Biol 174:815–825

Mackenzie L, Roderick HL, Berridge MJ, Conway SJ, Bootman MD (2004) The spatial pattern of atrial cardiomyocyte calcium signalling modulates contraction. J Cell Sci 117:6327–6337

Marchant J, Callamaras N, Parker I (1999) Initiation of IP_3-mediated Ca^{2+} waves in Xenopus oocytes. EMBO J 18:5285–5299

Ölveczky BP, Verkman AS (1998) Monte Carlo analysis of obstructed diffusion in three dimensions: application to molecular diffusion in organelles. Biophys J 74:2722–2730

Sammels E, Parys JB, Missiaen L, De Smedt H, Bultynck G (2010) Intracellular Ca^{2+} storage in health and disease: a dynamic equilibrium. Cell Calcium 47:297–314

Schwaller B (2010) Cytosolic Ca2+ buffers. Cold Spring Harb Perspect Biol 2:a004051

Shuai J, Pearson JE, Foskett JK, Mak D-OD, Parker I (2007) A kinetic model of single and clustered IP_3 receptors in the absence of Ca^{2+} feedback. Biophys J 93:1151–1162

Smith G, Dai L, Miura R, Sherman A (2001) Asymptotic analysis of buffered calcium diffusion near a point source. Siam J Appl Math 61:1816–1838

Smith IF, Wiltgen SM, Shuai J, Parker I (2009) Ca^{2+} puffs originate from preestablished stable clusters of inositol trisphosphate receptors. Sci Signal 2:ra77

Swaminathan D, Ullah G, Jung P (2009) A simple sequential-binding model for calcium puffs. Chaos 19:037109

Swillens S, Champeil P, Combettes L, Dupont G (1998) Stochastic simulation of a single inositol 1,4,5-trisphosphate-sensitive Ca^{2+} channel reveals repetitive openings during "blip-like" Ca^{2+} transients. Cell Calcium 23:291–302

Thul R, Falcke M (2006) Frequency of elemental events of intracellular Ca^{2+} dynamics. Phys Rev E 73:061923

Thul R, Falcke M (2007) Waiting time distributions for clusters of complex molecules. Europhys Lett 79:38003

Thurley K, Skupin A, Thul R, Falcke M (2012) Fundamental properties of Ca^{2+} signals. Biochim Biophys Acta 1820:1185–1194

Wagner J, Keizer J (1994) Effects of rapid buffers on Ca^{2+} diffusion and Ca^{2+} oscillations. Biophys J 67:447–456

Calcium, Neuronal Sensor Proteins

James B. Ames
Department of Chemistry, University of California, Davis, CA, USA

Synonyms

CaM; DREAM; Frq1; GCAP; KChIP; NCS-1

Definitions

Calmodulin, Downstream response element antagonist modulator, Yeast frequenin, Guanylate cyclase activator protein, Potassium channel–interacting protein, Neuronal calcium sensor protein-1.

Background

Intracellular calcium (Ca^{2+}) regulates a variety of neuronal signal transduction processes in the brain and retina. The effects of changes in neuronal Ca^{2+} are mediated primarily by an emerging class of neuronal calcium sensor (NCS) proteins (Weiss et al. 2010) that belong to the EF-hand superfamily. The human genome encodes 14 members of the NCS family. The amino acid sequences of NCS proteins are highly conserved from yeast to humans (Fig. 1). Recoverin, the first NCS protein to be discovered (Dizhoor et al. 1991), and guanylate cyclase–activating proteins (GCAPs) (Dizhoor et al. 1994; Palczewski et al. 1994) are expressed exclusively in the retina where they serve as Ca^{2+} sensors in vision. Other NCS proteins are expressed in the brain and spinal cord such as neurocalcin (Hidaka and Okazaki 1993), frequenin (NCS1) (Pongs et al. 1993), visinin-like proteins (Braunewell and Klein-Szanto 2009), K^+ channel–interacting proteins (KChIPs) (An et al. 2000), DREAM/calsenilin (Carrion et al. 1999), and hippocalcin (Kobayashi et al. 1992). Frequenin is also expressed outside of the central nervous system as well as in invertebrates including flies, worms, and yeast (Frq1) (Hendricks et al. 1999). The common features of these proteins are an approximately 200-residue chain containing four EF-hand motifs, the sequence CPXG in the first EF-hand that markedly impairs its capacity to bind Ca^{2+}, and an amino-terminal myristoylation consensus sequence.

The structurally similar NCS proteins have remarkably different physiologic functions (Table 1). Perhaps the best characterized NCS protein is recoverin that serves as a calcium sensor in retinal rod cells. Recoverin prolongs the lifetime of light-excited rhodopsin by inhibiting rhodopsin kinase (RK) only at high Ca^{2+} levels. Hence, recoverin makes receptor desensitization Ca^{2+} dependent, and the resulting shortened lifetime of rhodopsin at low Ca^{2+} levels

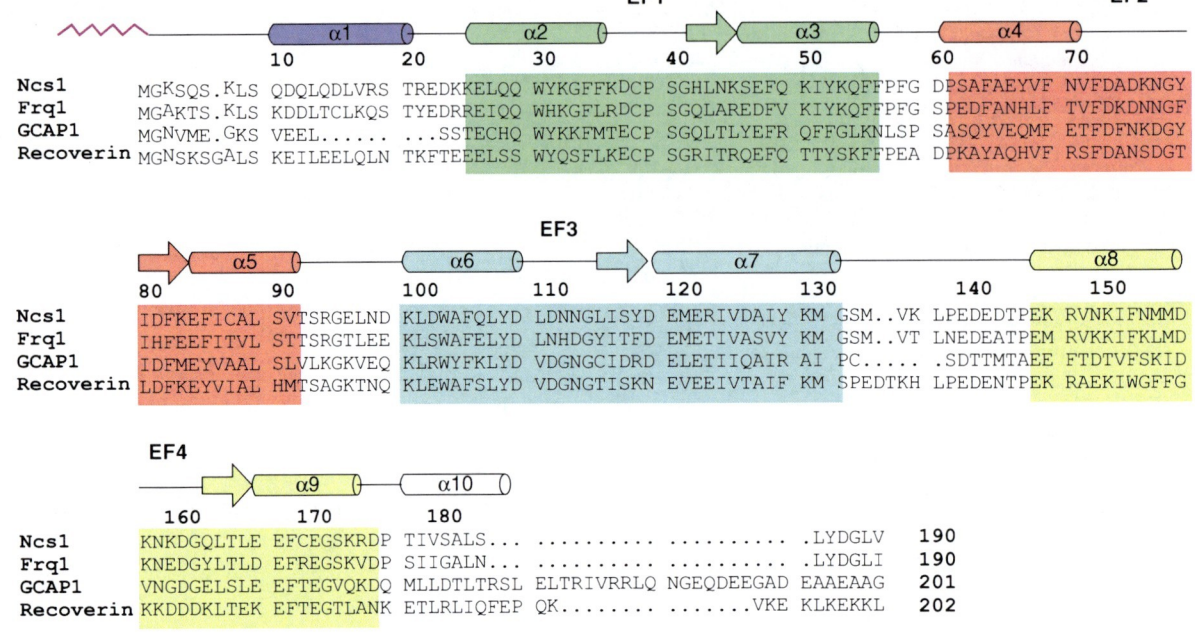

Calcium, Neuronal Sensor Proteins, Fig. 1 Amino acid sequence alignment of *S. pombe* NCS1 (Swiss-Prot accession no. Q09711), *S. cerevisiae* Frq1 (Q06389), bovine GCAP1 (P43080), and bovine recoverin (P21457). Secondary structure elements (helices and strands), EF-hand motifs (EF1 *green*, EF2 *red*, EF3 *cyan*, and EF4 *yellow*), N-terminal arm (*purple*), and residues that interact with the myristoyl group (*magenta*) are highlighted

Calcium, Neuronal Sensor Proteins, Table 1 Function of NCS proteins

NCS protein	Function
Recoverin	Inhibit rhodopsin kinase in retinal rods
GCAP1	Activate guanylate cyclase in retinal cones
GCAP2	Activate guanylate cyclase in retinal rods
GCIP	Inhibit guanylate cyclase in frog photoreceptors
KChIP1	Regulate K^+ channel gating kinetics in brain
KChIP2	Regulate K^+ channel gating kinetics in cardiac cells
Calsenilin/DREAM	Repress transcription of prodynorphin and c-fos genes
Frequenin (NCS1)	Activate PI(4) kinase and regulate Ca^{2+} and K^+ channels
Neurocalcin δ	Unknown
Hippocalcin	Activate phospholipase D and MAP kinase signaling
VILIP-1	Activate guanylate cyclase and traffic nicotinic receptors

may promote visual recovery and contribute to the adaptation to background light. Recoverin may also function in the rod inner segment and was identified as the antigen in cancer-associated retinopathy, an autoimmune disease of the retina caused by a primary tumor in another tissue. Other NCS proteins in retinal rods include the guanylate cyclase–activating proteins (GCAP1 and GCAP2) that activate retinal guanylate cyclase only at low Ca^{2+} levels and inhibit the cyclase at high Ca^{2+} (Dizhoor et al. 1994; Palczewski et al. 1994). GCAPs are important for regulating the recovery phase of visual excitation, and particular mutants are linked to various forms of retinal degeneration. Yeast and mammalian frequenins bind and activate a particular PtdIns 4-OH kinase isoform (Pik1 gene in yeast) (Hendricks et al. 1999; Strahl et al. 2007) required for vesicular trafficking in the late secretory pathway. Mammalian frequenin (NCS1) also regulates voltage-gated Ca^{2+} and K^+ channels. The KChIPs regulate the gating kinetics of voltage-gated, A-type K^+ channels (An et al. 2000). The DREAM/calsenilin/KChIP3 protein binds to specific DNA sequences in the prodynorphin and c-fos genes (Carrion et al. 1999) and serves as a calcium sensor and transcriptional repressor for pain modulation (Cheng et al. 2002). Hence, the functions of the NCS proteins appear to be quite diverse and nonoverlapping.

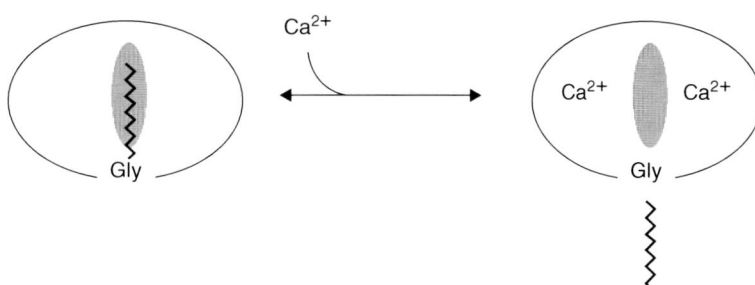

Calcium, Neuronal Sensor Proteins, Fig. 2 Schematic diagram of calcium-myristoyl switch in recoverin. The binding of two Ca^{2+} ions promotes the extrusion of the myristoyl group and exposure of other hydrophobic residues (marked by the *shaded oval*)

Mass spectrometric analysis of retinal recoverin and some of the other NCS proteins revealed that they are myristoylated at the amino terminus. Recoverin contains an N-terminal myristoyl (14:0) or related fatty acyl group (12:0, 14:1, 14:2). Retinal recoverin and myristoylated recombinant recoverin, but not unmyristoylated recoverin, bind to membranes in a Ca^{2+}-dependent manner. Likewise, bovine neurocalcin and hippocalcin contain an N-terminal myristoyl group, and both exhibit Ca^{2+}-induced membrane binding. These findings led to the proposal that NCS proteins possess a Ca^{2+}-myristoyl switch (Fig. 2). The covalently attached fatty acid is highly sequestered in recoverin in the calcium-free state. The binding of calcium to recoverin leads to the extrusion of the fatty acid, making it available to interact with lipid bilayer membranes or other hydrophobic sites. The Ca^{2+}-myristoyl switch function by recoverin also enables its light-dependent protein translocation in retinal rods.

In this entry, the atomic-level structures of various NCS proteins and their target complexes will be discussed and compared with that of calmodulin. The large effect of N-terminal myristoylation is examined on the structures of recoverin, GCAP1, and NCS1. Ca^{2+}-induced extrusion of the myristoyl group exposes unique hydrophobic binding sites in each protein that in turn interact with various target proteins. An emerging theme is that N-terminal myristoylation is critical for shaping each NCS family member into a unique structure, which upon Ca^{2+}-induced extrusion of the myristoyl group exposes a unique set of previously masked residues, thereby exposing a distinctive ensemble of hydrophobic residues to associate specifically with a particular physiological target.

Structure of Recoverin's CA^{2+}-Myristoyl Switch

The X-ray crystal structure of recombinant unmyristoylated recoverin (Flaherty et al. 1993) showed it to contain a compact array of EF-hand motifs, in contrast to the dumbbell shape of calmodulin and troponin C. The four EF-hands are organized into two domains: The first EF-hand, EF-1 (residues 27–56, colored green in Figs. 1 and 3), interacts with EF-2 (residues 63–92, red) to form the N-terminal domain, and EF-3 (residues 101–130, cyan) and EF-4 (residues 148–177, yellow) form the C-terminal domain. The linker between the two domains is U shaped rather than α-helical. Ca^{2+} is bound to EF-3, and Sm^{3+} (used to derive phases) is bound to EF-2. The other two EF-hands possess novel features that prevented ion binding. EF-1 is disabled by a Cys-Pro sequence in the binding loop. EF-4 contains an internal salt bridge in the binding loop that competes with Ca^{2+} binding. Myristoylated recoverin, the physiologically active form, has thus far eluded crystallization.

The structures of myristoylated recoverin in solution with 0, 1, and 2 Ca^{2+} bound have been determined by nuclear magnetic resonance (NMR) spectroscopy (Ames et al. 1997) (Fig. 3). In the Ca^{2+}-free state, the myristoyl group is sequestered in a deep hydrophobic cavity in the N-terminal domain. The cavity is formed by five α-helices. The two helices of EF-1 (residues 26–36 and 46–56), the exiting helix of EF-2 (residues 83–93), and entering helix of EF-3 (residues 100–109) lie perpendicular to the fatty acyl chain and form a boxlike arrangement that surrounds the myristoyl group laterally. A long, amphipathic α-helix near the N-terminus (residues 4–16) packs closely against and

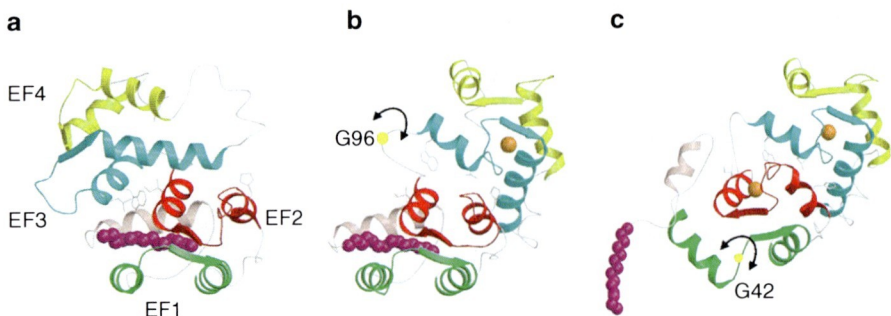

Calcium, Neuronal Sensor Proteins, Fig. 3 Three-dimensional structures of myristoylated recoverin with 0 Ca^{2+} bound (**a**), 1 Ca^{2+} bound (**b**), and 2 Ca^{2+} bound (**c**). The first step of the mechanism involves the binding of Ca^{2+} to EF3 that causes minor structural changes within the EF-hand that sterically promote a 45° swiveling of the two domains, resulting in a partial unclamping of the myristoyl group and a dramatic rearrangement at the domain interface. The resulting altered interaction between EF2 and EF3 facilitates the binding of a second Ca^{2+} to the protein at EF2 in the second step, which causes structural changes within the N-terminal domain that directly lead to the ejection of the fatty acyl group

runs antiparallel to the fatty acyl group and serves as a lid on top of the four-helix box. The N-terminal residues Gly 2 and Asn 3 form a tight hairpin turn that connects the myristoyl group to the N-terminal helix. This turn positions the myristoyl group inside the hydrophobic cavity and gives the impression of a cocked trigger. The bond angle strain stored in the tight hairpin turn may help eject the myristoyl group from the pocket once Ca^{2+} binds to the protein.

The structure of myristoylated recoverin with one Ca^{2+} bound at EF-3 (half-saturated recoverin, Fig. 3b) (Ames et al. 2002) represents a hybrid structure of the Ca^{2+}-free and Ca^{2+}-saturated states. The structure of the N-terminal domain (residues 2–92, green and red in Fig. 3) of half-saturated recoverin (Fig. 3b) resembles that of Ca^{2+}-free state (Fig. 3a) and is very different from that of the Ca^{2+}-saturated form (Fig. 3c). Conversely, the structure of the C-terminal domain (residues 102–202, cyan and yellow in Fig. 3) of half-saturated recoverin more closely resembles that of the Ca^{2+}-saturated state. Most striking in the structure of half-saturated recoverin is that the myristoyl group is flanked by a long N-terminal helix (residues 5–17) and is sequestered in a hydrophobic cavity containing many aromatic residues from EF-1 and EF-2 (F23, W31, Y53, F56, F83, and Y86). An important structural change induced by Ca^{2+} binding at EF-3 is that the carbonyl end of the fatty acyl group in the half-saturated species is displaced far away from hydrophobic residues of EF-3 (W104 and L108, Fig. 3a, b) and becomes somewhat solvent exposed. By contrast, the myristoyl group of Ca^{2+}-free recoverin is highly sequestered by residues of EF-3.

The structure of myristoylated recoverin with two Ca^{2+} bound shows the amino-terminal myristoyl group to be extruded (Ames et al. 1997) (Fig. 3c). The N-terminal eight residues are solvent exposed and highly flexible and thus serve as a mobile arm to position the myristoyl group outside the protein when Ca^{2+} is bound. The flexible arm is followed by a short α-helix (residues 9–17) that precedes the four EF-hand motifs, arranged in a tandem array as was seen in the X-ray structure. Calcium ions are bound to EF-2 and EF-3. EF-3 has the classic "open conformation" similar to the Ca^{2+}-occupied EF-hands in calmodulin and troponin C. EF-2 is somewhat unusual and the helix-packing angle of Ca^{2+}-bound EF-2 (120°) in recoverin more closely resembles that of the Ca^{2+}-free EF-hands (in the "closed conformation") found in calmodulin and troponin C. The overall topology of Ca^{2+}-bound myristoylated recoverin is similar to the X-ray structure of unmyristoylated recoverin described above. The RMS deviation of the main chain atoms in the EF-hand motifs is 1.5 Å in comparing Ca^{2+}-bound myristoylated recoverin to unmyristoylated recoverin. Hence, in Ca^{2+}-saturated recoverin, the N-terminal myristoyl group is solvent exposed and does not influence the interior protein structure.

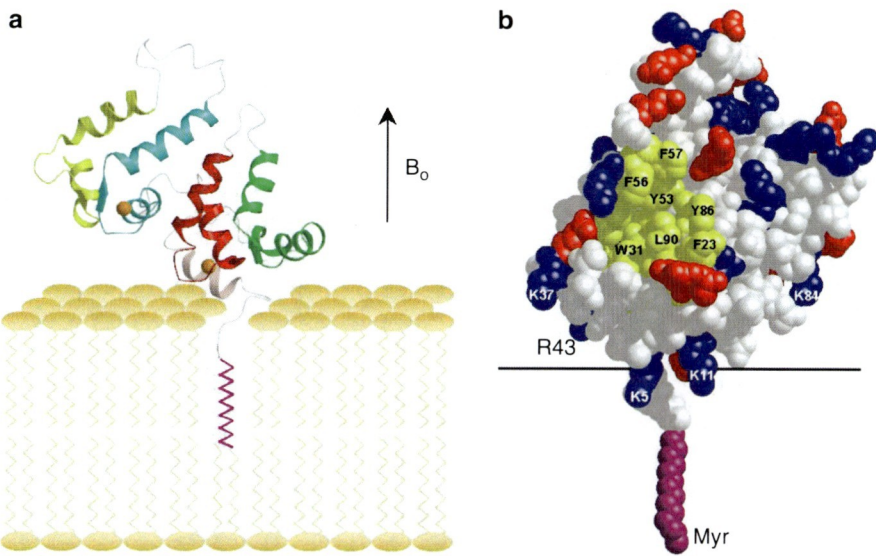

Calcium, Neuronal Sensor Proteins, Fig. 4 Main chain structure (**a**) and space-filling representation (**b**) of myristoylated recoverin bound to oriented lipid bilayers determined by solid-state NMR (Valentine et al. 2003). Hydrophobic residues are *yellow*, bound Ca^{2+} ions are *orange*, and charged residues are *red* and *blue*

The Ca^{2+}-induced exposure of the myristoyl group (Figs. 2 and 3) enables recoverin to bind to membranes only at high Ca^{2+} (Zozulya and Stryer 1992). Recent solid-state NMR studies have determined the structure of Ca^{2+}-bound myristoylated recoverin bound to oriented lipid bilayer membranes (Fig. 4) (Valentine et al. 2003). The protein is positioned on the membrane surface such that its long molecular axis is oriented 45° with respect to the membrane normal. The N-terminal region of recoverin points toward the membrane surface, with close contacts formed by basic residues K5, K11, K22, K37, R43, and K84. This orientation of membrane-bound recoverin allows an exposed hydrophobic crevice (lined primarily by residues F23, W31, F35, I52, Y53, F56, Y86, and L90) near the membrane surface that may serve as a potential binding site for the target protein, rhodopsin kinase (Fig. 4b).

Structural Diversity of NCS Proteins

Myristoylation Reshapes Structure of NCS Proteins. Three-dimensional structures have been determined recently for GCAP1 (Stephen et al. 2007) and NCS1 (Lim et al. 2011) that contain a sequestered myristoyl group (Fig. 5). Surprisingly, the myristoylated forms of GCAP1, NCS1, and recoverin all have very distinct three-dimensional folds (Fig. 5). The overall root-mean-squared deviations are 2.8 and 3.4 Å when comparing the main chain structures of Ca^{2+}-free NCS1 with recoverin and GCAP1, respectively. These very different structures reveal that the N-terminal myristoyl group is sequestered inside different protein cavities at different locations in each case. In NCS1, the N-terminal myristoyl group is sequestered inside a cavity near the C-terminus formed between the helices of EF3 and EF4 (Fig. 5a). The fatty acyl chain in NCS1 is nearly parallel to the helices of EF3 and EF4 that form walls that surround the myristoyl moiety (Fig. 5d). This arrangement in NCS1 is in stark contrast to recoverin where the myristoyl group is sequestered inside a protein cavity near the N-terminus (Fig. 5b). The myristate in recoverin is wedged perpendicularly between the helices of EF1 and EF2 (Fig. 5e) that contrasts with the parallel arrangement in NCS1 (Fig. 5d). For GCAP1 (Fig. 5c), the myristoyl group is located in between the N-terminal and C-terminal domains. In essence, the myristate bridges both domains of GCAP1 by interacting with helices at each end of the protein. The structural location and environment around the myristoyl group is very different in the various NCS proteins (Fig. 5). It is suggested that each NCS protein may adopt a distinct structure because its N-terminal myristoyl group associates with patches of hydrophobic residues that are unique to that protein.

Nonconserved residues of NCS proteins interact closely with the N-terminal myristoyl group and help stabilize the novel protein structure in each case. NCS1, recoverin, and GCAP1 all have nonconserved residues near the N-terminus (called an N-terminal arm

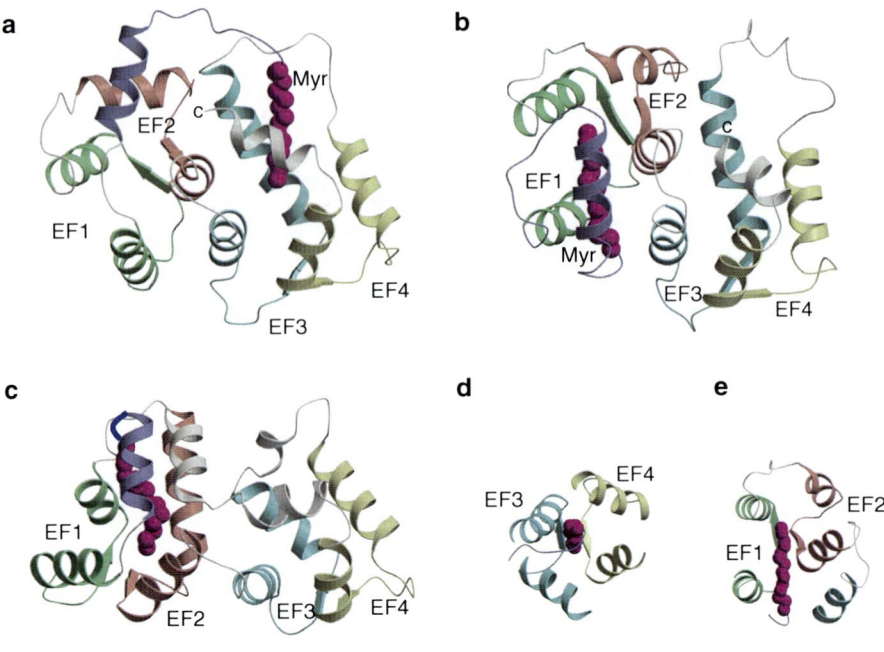

Calcium, Neuronal Sensor Proteins, Fig. 5 Main chain structures of Ca^{2+}-free myrisoylated NCS1 (2l2e) (a), recoverin (1iku) (b), and GCAP1 (2r2i) (c). Close-up views of the myristate-binding pocket in NCS1 (d) and recoverin (e). EF-hands and myristoyl group (*magenta*) are colored as defined in Fig. 1

highlighted purple in Fig. 5) that make specific contacts with the myristoyl moiety. GCAP1 also contains an extra helix at the C-terminus that contacts the N-terminal arm and myristoyl group (Fig. 5c). Thus, nonconserved residues at the N-terminus, C-terminus, and loop between EF3 and EF4 all play a role in creating a unique environment around the myristoyl group. In Ncs1, the long N-terminal arm and particular hydrophobic residues in the C-terminal helix are crucial for placing the C14 fatty acyl chain in a cavity between EF3 and EF4 (Fig. 5d). By contrast, the much shorter N-terminal arm in both recoverin and GCAP1 prevents the myristoyl group from reaching the C-terminal cavity and, instead, places the fatty acyl chain between EF1 and EF2 (Fig. 5e). It is proposed that nonconserved residues at the N-terminus, C-terminus, and/or loop between EF3 and EF4 may play a role in forming unique myristoyl-binding environments in other NCS proteins, such as VILIPs, neurocalcins, and hippocalcins that may help explain their capacity to associate with functionally diverse target proteins.

Structures of Ca^{2+}-bound NCS Proteins. Three-dimensional structures have been determined for unmyristoylated forms of Ca^{2+}-bound neurocalcin (Vijay-Kumar and Kumar 1999), frequenin (Bourne et al. 2001), KChIP1 (Zhou et al. 2004), and Frq1 (Strahl et al. 2007). The first eight residues from the N-terminus are unstructured and solvent exposed in each case, consistent with an extruded myristoyl group that causes Ca^{2+}-induced membrane localization of NCS proteins (Zozulya and Stryer 1992). The overall main chain structures of the Ca^{2+}-bound NCS proteins are very similar in each case, which is not too surprising given their sequence relatedness. However, if the main chain structures are so similar, then how can one explain their ability to bind unique target proteins? One distinguishing structural property is the number and location of bound Ca^{2+}. Recoverin has Ca^{2+} bound at EF-2 and EF-3; KChIP1 has Ca^{2+} bound at EF-3 and EF-4; and frequenin, neurocalcin, and GCAP2 have Ca^{2+} bound at EF2, EF3, and EF4. Another important structural property is the distribution of charged and hydrophobic residues on the protein surface. Surface representations of hydrophobicity and charge density of the various NCS structures are shown in Fig. 6. All NCS structures exhibit a similar exposed hydrophobic surface located on the N-terminal half of the protein, formed primarily by residues in EF-1 and EF-2 (F35, W31, F56, F57, Y86, and L90 for recoverin in Fig. 6a). The exposed hydrophobic residues in this region are highly conserved (labeled and colored yellow in Fig. 6) and correspond to residues of recoverin that interact with the myristoyl group in the Ca^{2+}-free state (Fig. 3a).

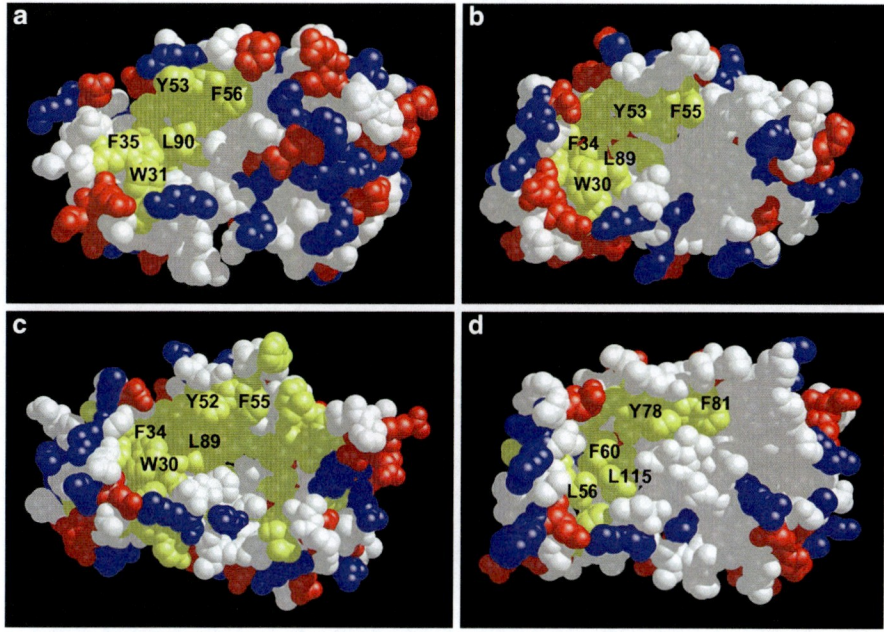

Calcium, Neuronal Sensor Proteins, Fig. 6 Space-filling representations of the Ca^{2+}-bound structures of recoverin (**a**), NCS1 (**b**), neurocalcin (**c**), and KChIP1 (**d**). Exposed hydrophobic residues are *yellow*, neutral residues are *white*, and charge residues are *red* and *blue*

A similar hydrophobic patch is also seen in membrane-bound recoverin (Fig. 4b). These exposed residues in the hydrophobic patch have been implicated in target recognition from mutagenesis studies, and these residues very likely form intermolecular contacts with target proteins as has been demonstrated in the recent crystal structure of KChIP1 (see below).

The distribution of charged (red and blue) and hydrophobic (yellow) residues on the surface of the C-terminal half of the NCS proteins is highly variable (Fig. 6). Frequenin exhibits exposed hydrophobic residues in the C-terminal domain that fuse together with the exposed hydrophobic crevice in the N-terminal domain, forming one continuous and elongated patch (Fig. 6b). By contrast, recoverin (Fig. 6a) has mostly charged residues on the surface of the C-terminal half, whereas neurocalcin (Fig. 6c) and KChIP1 (Fig. 6d) have mostly neutral residues shown in white. The different patterns of charge distribution on the C-terminal surface of NCS proteins might be important for conferring target specificity.

Ca^{2+}-sensitive dimerization of NCS proteins is another structural characteristic that could influence target recognition. Neurocalcin, recoverin, VILIP-1, and KChIP1 exist as dimers in their X-ray crystal structures. Hydrodynamic studies have confirmed that neurocalcin and DREAM form dimers in solution at high Ca^{2+} and are monomeric in the Ca^{2+}-free state. Indeed, the recent NMR structure of Ca^{2+}-bound DREAM forms a dimer in solution with intermolecular contacts involving Leu residues near the C-terminus. By contrast, GCAP2 forms a dimer only in the Ca^{2+}-free state and is monomeric at high Ca^{2+}. Ca^{2+}-sensitive dimerization of GCAP-2 has been demonstrated to control its ability to activate retinal guanylate cyclase. Ca^{2+}-sensitive protein oligomerization is also important physiologically for DREAM: The Ca^{2+}-free DREAM protein serves as a transcriptional repressor by binding to DNA response elements as a protein tetramer (Carrion et al. 1999). Ca^{2+}-induced dimerization of DREAM appears to disrupt DNA binding and may activate transcription of prodynorphin and c-fos genes (Carrion et al. 1999). In a related fashion, Ca^{2+}-bound KChIP1 forms a dimer in solution and in complex with an N-terminal fragment of the Kv4.2 K^+ channel (Zhou et al. 2004). By contrast, the full-length Kv4.2 channel tetramer binds to KChIP1 with a 4:4 stoichiometry, suggesting that KChIP1 dimers may assemble as a protein tetramer to recognize the channel. Such a protein tetramerization of KChIP1 may be Ca^{2+} sensitive like it is for DREAM. In short, the oligomerization properties of some NCS proteins appear to be Ca^{2+} sensitive, which may play a role in target recognition.

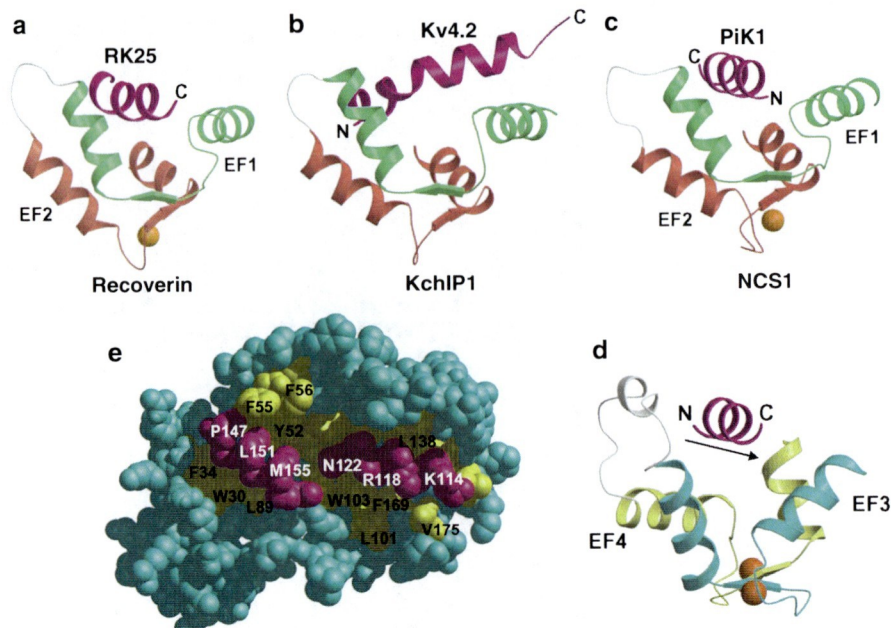

Calcium, Neuronal Sensor Proteins, Fig. 7 Ribbon diagrams illustrating intermolecular interactions for recoverin bound to RK25 (a), KChIP1 bound to Kv4.2 (b), NCS1 (N-domain) bound to Pik1(111–151) (c), NCS1 (C-domain) bound to Pik1(111–151) (d), and space-filling view of NCS1 bound to Pik1(111–151) (e). In each case, a target helix (*magenta*) is inserted in groove formed by the helices of the EF-hands. The intermolecular interactions are mostly hydrophobic as described in the text

Target Recognition by NCS Proteins

Recoverin Bound to Rhodopsin Kinase Fragment (RK25). The structure of Ca^{2+}-bound recoverin bound to a functional fragment of rhodopsin kinase (residues 1–25, hereafter referred to as RK25) was the first atomic-resolution structure of a Ca^{2+}-myristoyl switch protein bound to a functional target protein (Ames et al. 2006) (Fig. 7a). The structure of this complex revealed that RK25 forms a long amphipathic α-helix, whose hydrophobic surface interacts with the N-terminal hydrophobic groove of recoverin described above (Fig. 6). The structure of recoverin in the complex is quite similar to that of Ca^{2+}-bound recoverin alone in solution (root-mean-squared deviation = 1.8 Å). The structure of RK25 in the complex consists of an amphipathic α-helix (residues 4–16). The hydrophobic surface of the RK25 helix (L6, V9, V10, A11, F15) interacts with the exposed hydrophobic groove on recoverin (W31, F35, F49, I52, Y53, F56, F57, Y86, and L90). Mutagenesis studies on recoverin and RK have shown that many of the hydrophobic residues at the binding interface are essential for the high affinity interaction. These hydrophobic contacts are supplemented by a π-cation interaction involving F3 (RK25) and K192 from recoverin. Dipolar residues on the opposite face of the RK25 helix (S5, T8, N12, I16) are solvent exposed. The helical structure of RK25 in the complex is stabilized mostly by hydrophobic intermolecular interactions with recoverin, as free RK25 in solution is completely unstructured.

The Ca^{2+}-myristoyl switch mechanism of recoverin (i.e., Ca^{2+}-induced extrusion of the N-terminal myristoyl group, Fig. 2) is structurally coupled to Ca^{2+}-induced inhibition of RK. The exposed hydrophobic residues of recoverin that interact with rhodopsin kinase correspond to the same residues that contact the N-terminal myristoyl group in the structure of Ca^{2+}-free recoverin (Ames et al. 1997). The size of the myristoyl group is similar to the length and width of the RK25 helix in the complex, which explains why both effectively compete for binding to the exposed hydrophobic groove (Fig. 4b). The Ca^{2+}-induced exposure of the N-terminal hydrophobic groove therefore explains why recoverin binds to rhodopsin kinase only at high Ca^{2+} levels. In the Ca^{2+}-free state, the covalently attached myristoyl group sequesters the N-terminal hydrophobic groove and cover up the target-binding site. Ca^{2+}-induced extrusion of the myristoyl group of recoverin causes exposure of residues that bind and inhibit RK (Fig. 8b). This mechanism elegantly explains how recoverin controls both the localization and activity of RK in response to light. In the dark (high Ca^{2+}), Ca^{2+}-bound recoverin binds to RK (thereby inhibiting it) and delivers RK to the membrane via a Ca^{2+}-myristoyl switch that pre-positions RK near rhodopsin. Upon light activation (low Ca^{2+}), recoverin rapidly dissociates from both RK and the membrane, allowing RK to bind efficiently to its nearby substrate, rhodopsin, and cause rapid desensitization.

NCS1 Bound to Phosphatidylinositol 4-Kinase Fragment (Pik1). The structure of Ca^{2+}-bound NCS1 (or yeast Frq1 (Strahl et al. 2007)) bound to a functional fragment of Pik1 (residues 111–159, hereafter referred to as Pik1(111–159)) was determined by NMR (Lim et al. 2011) (Fig. 7c, d). The structure of NCS1 in the complex is very similar to the crystal structure in the absence of target (Bourne et al. 2001) with a concave solvent-exposed groove lined by two separate hydrophobic patches (highlighted yellow in Fig. 6). These two hydrophobic surfaces represent bipartite binding sites on NCS1 that interact with two helical segments in Pik1(111–159) (Fig. 7e). The structure of Pik1(111–159) in the complex adopts a conformation that contains two α-helices (residues 114–127 and 143–156) connected by a disordered loop. The N-terminal helix contains hydrophobic residues (I115, C116, L119, and I123) that contact C-terminal residues of NCS1 (L101, W103, V125, V128, L138, I152, L155, and F169). Interestingly, these same hydrophobic residues in Ca^{2+}-free NCS1 make close contacts with the myristoyl group. Therefore, Ca^{2+}-induced extrusion of the myristoyl group causes exposure of hydrophobic residues in Ncs1 that forms part of the Pik1-binding site (Figs. 7e and 8). The C-terminal helix of Pik1(111–159) contains many hydrophobic residues (V145, A148, I150, and I154) that contact the exposed N-terminal hydrophobic groove of NCS1 (W30, F34, F48, I51, Y52, F55, F85, and L89), very similar to the exposed hydrophobic groove seen in all NCS proteins (Fig. 6). The two helices of Pik1(111–159) do not interact with one another or with the unstructured connecting loop and are highly stabilized by interactions with NCS1.

Nonconserved residues in NCS1 at the C-terminus and immediately following EF3 may be structurally important for explaining target specificity. The nonconserved C-terminal region of NCS1 (residues, 180–190) is structurally disordered in the target complex, in contrast to a well-defined C-terminal helix seen in Ca^{2+}-free NCS1 in the absence of target (Lim et al. 2011). The C-terminal helix in Ca^{2+}-free NCS1 (target-free state) makes contact with the myristoyl group and residues in EF3 and EF4 (L101, A104, M121, I152, F169, and S173). These same residues in Ca^{2+}-bound NCS1 make contact with Pik1 in the complex (Fig. 7e). Therefore, the N-terminal Pik1 helix appears to substitute for and perhaps displace the C-terminal helix of Frq1, likely leading to the observed C-terminal destabilization in the complex (Fig. 8a). The corresponding C-terminal helix of KChIP1 is similarly displaced upon its binding to the Kv4.3 channel but not upon its binding to Kv4.2. The C-terminal helix in recoverin forms a stable interaction with EF3 and EF4, enabling the C-terminal helix to perhaps serve as a built-in competitive inhibitor that would presumably block its ability to bind to targets like Pik1 and Kv4.3. This role for the C-terminus may explain why the C-terminal sequences of NCS proteins are not well conserved (Fig. 1). Another nonconserved region of NCS1 implicated in target specificity is the stretch between EF3 and EF4 (residues 134–146). This region of NCS1 adopts a short α-helix in the complex that contacts the N-terminal helix of Pik1. By contrast, the region between EF3 and EF4 is unstructured in DREAM, Frq1, GCAP2, and KChIP1.

The structure of the Ncs1-Pik1 complex (Figs. 5a and 7c, d) suggest how a Ca^{2+}-myristoyl switch might promote activation of PtdIns 4-kinase (Fig. 8a). Under resting basal conditions, NCS1 exists in its Ca^{2+}-free state with a sequestered myristoyl group buried in the C-domain that covers part of its binding site for PtdIns 4-kinase (highlighted yellow in Figs. 7e and 8) and prevents binding of Ncs1 to Pik1. The fatty acyl chain has the same molecular dimensions (length and width) as the N-terminal helix of Pik1(111–159), which explains why the myristoyl group and Pik1 helix can effectively compete for the same binding site in Ncs1. A rise in cytosolic Ca^{2+} will cause Ca^{2+}-induced conformational changes in Ncs1, resulting in extrusion of the N-terminal myristoyl group. Ca^{2+}-induced extrusion of the myristoyl group exposes a hydrophobic crevice in the C-terminal domain of Ncs1, and concomitantly, Ca^{2+}-induced structural changes in its N-domain result in formation of a second exposed hydrophobic crevice, also seen in all known Ca^{2+}-bound NCS proteins. These two separate hydrophobic sites on the surface of Ca^{2+}-bound NCS1 are different from Ca^{2+}-bound recoverin that contains only one exposed hydrophobic patch (Fig. 8b) that interacts

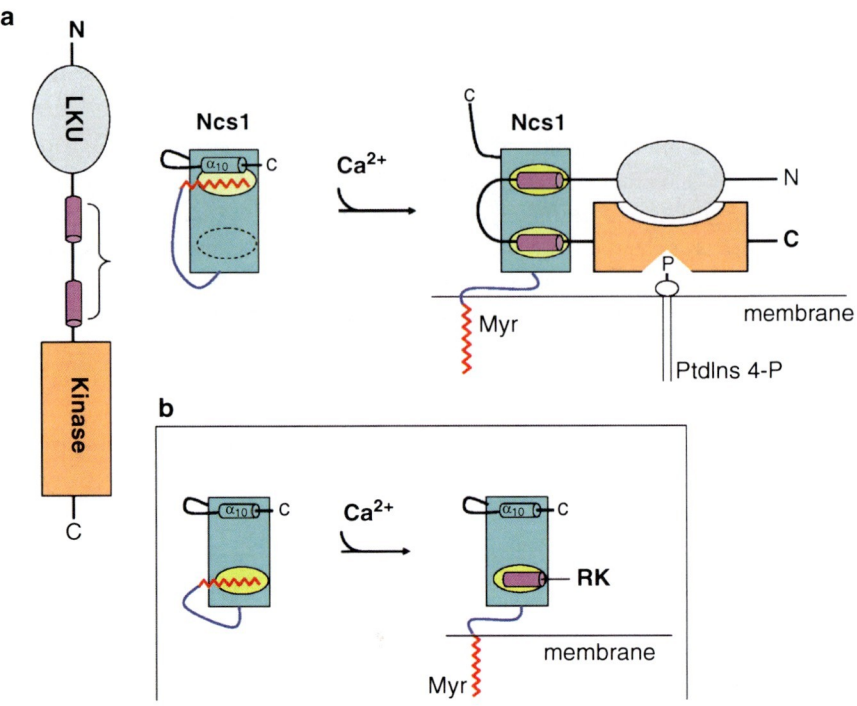

Calcium, Neuronal Sensor Proteins, Fig. 8 Schematic diagram of calcium-myristoyl switch coupled to target regulation illustrated for NCS1 (a) and recoverin (b)

with a single target helix in rhodopsin kinase (Ames et al. 2006). The two exposed hydrophobic sites on Ncs1 bind to the hydrophobic faces of the two antiparallel amphipathic α-helices in Pik1(111–159) (colored magenta in Fig. 8a). The Ca^{2+}-induced binding of NCS1 to PtdIns 4-kinase may promote structural changes that cause increased lipid kinase activity. Simultaneously, Ncs1 binding to PtdIns 4-kinase will also promote membrane localization of the lipid kinase because Ca^{2+}-bound NCS1 contains an extruded myristoyl group that serves as a membrane anchor. Thus, Ncs1 controls both delivery of PtdIns 4-kinase to the membrane where its substrates are located and formation of the optimally active state of the enzyme.

Mechanisms of Target Recognition. NCS proteins bind to helical target proteins analogous to the target binding seen for CaM (Fig. 7). Helical segments of target proteins bind to an exposed hydrophobic crevice formed by the two EF-hands in either the N-terminal or C-terminal domain in NCS proteins. In recoverin, the two N-terminal EF-hands form an exposed hydrophobic groove that interacts with a hydrophobic target helix from rhodopsin kinase (RK25) (Ames et al. 2006) (Fig. 7a). The N-terminal EF-hands of KChIP1 interact with a target helix derived from the T1 domain of Kv4.2 channels (Fig. 7b) (Zhou et al. 2004).

The orientation of the target helices bound to recoverin and KChIP1 is somewhat similar: The C-terminal end of the target helix is spatially close to the N-terminal helix of EF-1. By contrast, the Pik1 target helix binds to NCS1 in almost the exact opposite orientation (Fig. 7c). The N-terminal end of the Pik1 helix is closest to EF1 (green) in NCS1, whereas the C-terminal end of the RK25 target helix is closest to the corresponding region of recoverin. Nonconserved residues in NCS1 (G33 and D37) make important contacts with the Pik1 target helix and presumably assist in imposing the observed orientation of the helix. Thus, the requirement that the helix (in this case, from Pik1) must bind to NCS1 with a polarity opposite to that observed for the helices in other target-NCS family member complexes could clearly contribute to dictating the substrate specificity of frequenins, as compared to other NCS subtypes. Another important structural feature seen in the NCS1-Pik1 interaction is that two helical segments of the target are captured in the complex, whereas in the target complexes characterized for recoverin and KChIP1, only one helix is bound. Therefore, selective substrate recognition by NCS proteins may be explained by both by the orientation of the bound target helix and the number of target helices bound.

Summary

The molecular structures of NCS proteins were reviewed, and structural determinants important for target recognition were examined. N-terminal myristoylation has a profound effect on the structures of Ca^{2+}-free recoverin, GCAP1, and NCS1 (Fig. 5). Surprisingly, the sequestered myristoyl group interacts with quite different protein residues in each case and, therefore, is able to reshape these homologous NCS proteins into very different structures. The structures of the Ca^{2+}-bound NCS proteins all contain an extruded N-terminus with an exposed hydrophobic crevice implicated in target binding. It is proposed that N-terminal myristoylation is critical for shaping each NCS family member into a unique structure, which upon Ca^{2+}-induced extrusion of the myristoyl group exposes a unique set of previously masked residues, thereby exposing a distinctive ensemble of hydrophobic residues to associate specifically with a particular physiological target. Differences in their surface charge density and protein dimerization properties may also help to explain NCS target specificity and functional diversity. In the future, atomic-resolution structures of additional NCS proteins both with a sequestered myristoyl group and in their extruded forms bound to their respective target proteins are needed to improve our understanding of how this structurally conserved family of proteins can uniquely recognize their diverse biological targets.

Acknowledgments This work was supported by grants to J.B.A. from the NIH.

References

Ames JB, Ishima R, Tanaka T, Gordon JI, Stryer L, Ikura M (1997) Molecular mechanics of calcium-myristoyl switches. Nature 389:198–202

Ames JB, Hamasaki N, Molchanova T (2002) Structure and calcium-binding studies of a recoverin mutant (E85Q) in an allosteric intermediate state. Biochemistry 41:5776–5787

Ames JB, Levay K, Wingard JN, Lusin JD, Slepak VZ (2006) Structural basis for calcium-induced inhibition of rhodopsin kinase by recoverin. J Biol Chem 281:37237–37245

An WF, Bowlby MR, Betty M, Cao J, Ling HP, Mendoza G, Hinson JW, Mattsson KI, Strassle BW, Trimmer JS, Rhodes KJ (2000) Modulation of A-type potassium channels by a family of calcium sensors. Nature 403:553–556

Bourne Y, Dannenberg J, Pollmann VV, Marchot P, Pongs O (2001) Immunocytochemical localization and crystal structure of human frequenin (neuronal calcium sensor1). J Biol Chem 276:11949–11955

Braunewell KH, Klein-Szanto AJ (2009) Visinin-like proteins (VSNLs): interaction partners and emerging functions in signal transduction of a subfamily of neuronal Ca^{2+}-sensor proteins. Cell Tissue Res 335:301–316

Carrion AM, Link WA, Ledo F, Mellstrom B, Naranjo JR (1999) DREAM is a Ca^{2+}-regulated transcriptional repressor. Nature 398:80–84

Cheng HY, Pitcher GM, Laviolette SR, Whishaw IQ, Tong KI, Ikura M, Salter MW, Penninger JM (2002) DREAM is a critical transcriptional repressor for pain modulation. Cell 108:31–43

Dizhoor AM, Ray S, Kumar S, Niemi G, Spencer M, Rrolley D, Walsh KA, Philipov PP, Hurley JB, Stryer L (1991) Recoverin: a calcium sensitive activator of retinal rod guanylate cyclase. Science 251:915–918

Dizhoor AM, Lowe DG, Olsevskaya EV, Laura RP, Hurley JB (1994) The human photoreceptor membrane guanylyl cyclase, RetGC, is present in outer segments and is regulated by calcium and a soluble activator. Neuron 12:1345–1352

Flaherty KM, Zozulya S, Stryer L, McKay DB (1993) Three-dimensional structure of recoverin, a calcium sensor in vision. Cell 75:709–716

Hendricks KB, Wang BQ, Schnieders EA, Thorner J (1999) Yeast homologue of neuronal frequenin is a regulator of phosphatidylinositol-4-OH kinase. Nat Cell Biol 1:234–241

Hidaka H, Okazaki K (1993) Neurocalcin family: a novel calcium-binding protein abundant in bovine central nervous system. Neurosci Res 16:73–77

Kobayashi M, Takamatsu K, Saitoh S, Miura M, Noguchi T (1992) Molecular cloning of hippocalcin, a novel calcium-binding protein of the recoverin family exclusively expressed in hippocampus. Biochem Biophys Res Commun 189:511–517 (published erratum appears in Biochem Biophys Res Commun 1993 Oct 29;196(2):1017)

Lim S, Strahl T, Thorner J, Ames JB (2011) Structure of a Ca^{2+}–myristoyl switch protein that controls activation of a phosphatidylinositol 4-kinase in fission yeast. J Biol Chem 286:12565–12577

Palczewski K, Subbaraya I, Gorczyca WA, Helekar BS, Ruiz CC, Ohguro H, Huang J, Zhao X, Crabb JW, Johnson RS (1994) Molecular cloning and characterization of retinal photoreceptor guanylyl cyclase-activating protein. Neuron 13:395–404

Pongs O, Lindemeier J, Zhu XR, Theil T, Engelkamp D, Krah-Jentgens I, Lambrecht HG, Kock KW, Schwerner J, Rivosecchi R, Mallart A, Galceran J, Canal I, Barbas JA, Ferrus A (1993) Frequenin-a novel calcium-binding protein that modulates synaptic efficacy. Neuron 11:15–28

Stephen R, Bereta G, Golczak M, Palczewski K, Sousa MC (2007) Stabilizing function for myristoyl group revealed by the crystal structure of a neuronal calcium sensor, guanylate cyclase-activating protein 1. Structure 15:1392–1402

Strahl T, Huttner IG, Lusin JD, Osawa M, King D, Thorner J, Ames JB (2007) Structural insights into activation of phosphatidylinositol 4-kinase (Pik1) by yeast frequenin (Frq1). J Biol Chem 282:30949–30959

Valentine KG, Mesleh MF, Opella SJ, Ikura M, Ames JB (2003) Structure, topology, and dynamics of myristoylated recoverin bound to phospholipid bilayers. Biochemistry 42:6333–6340

Vijay-Kumar S, Kumar VD (1999) Crystal structure of recombinant bovine neurocalcin. Nat Struct Biol 6:80–88

Weiss JL, Hui H, Burgoyne RD (2010) Neuronal calcium sensor-1 regulation of calcium channels, secretion, and neuronal outgrowth. Cell Mol Neurobiol 30:1283–1292

Zhou W, Qian Y, Kunjilwar K, Pfaffinger PJ, Choe S (2004) Structural insights into the functional interaction of KChIP1 with Shal-type K(+) channels. Neuron 41:573–586

Zozulya S, Stryer L (1992) Calcium-myristoyl protein switch. Proc Natl Acad Sci USA 89:11569–11573

Calcium, Physical and Chemical Properties

Fathi Habashi
Department of Mining, Metallurgical, and Materials Engineering, Laval University, Quebec City, Canada

Calcium is the fifth most abundant element in the earth's crust. Some important, naturally occurring compounds are the carbonate (limestone), the sulfate, and complex silicates. It is a typical metal: it has the tendency to lose its outermost electrons and in doing so it achieves the inert gas electronic structure. Also its two outermost electrons are lost in one step. It is a ductile metal and can be formed by casting, extrusion, rolling, etc. The metal is used as a reducing agent, while CaO produced from limestone by calcination is used as a flux in iron and steel production, in gas desulfurization processes, and other metallurgical operations. Calcium hydroxide, obtained by dissolving CaO in water, is used as a pH control reagent in hydrometallurgy.

Both, gypsum, $CaSO_4 \cdot 2H_2O$, and anhydrite, $CaSO_4$, are widely distributed in the earth's crust and are used as materials of construction.

Physical Properties

Atomic number	20
Atomic weight	40.08
Relative abundance, %	3.63
Density at 20°C	1.55
Melting point, °C	838
Boiling point, °C	1,440
Specific heat (0–100°C), $J\ g^{-1}\ K^{-1}$	0.624
Heat of fusion, J/g	217.7
Heat of vaporization, J/g	4,187
Thermal expansion (0–400°C), K^{-1}	22.3×10^{-6}
Electrical resistivity at 0°C, Ω cm	3.91×10^{-6}
Thermal conductivity at 20°C, $W\ cm^{-1}\ K^{-1}$	1.26
Crystal structure at room temperature	Face-centered cubic
Lattice constant, nm	0.5582
Crystal structure above 448°C	Body-centered cubic

(continued)

Chemical Properties

Calcium is produced by the thermal reduction of lime with aluminum in a retort under vacuum at 1,200°C; the vapors of calcium are then collected in a condenser. The metal is unstable in moist air, rapidly forming a hydration coating. It can be stored in dry air at room temperature. It reacts spontaneously with water to form $Ca(OH)_2$ and hydrogen gas; when finely divided, it will ignite in air. Being a typical metal, all its compounds are colorless and exhibit only one valence state, +2, in all of its reactions.

Naturally occurring calcium phosphate is treated for the production of fertilizers. Calcium carbide is produced by reacting quicklime with coke in furnaces which are heated electrically to 2,000°C. It is used to produce acetylene by reaction with water. Calcium cyanamide is synthesized from calcium carbide and nitrogen at 900–1,000°C:

$$CaC_2 + N_2 \rightarrow CaCN_2 + C$$

It is used as a fertilizer since ammonia is generated when it comes in contact with water:

$$CaCN_2 + 3H_2O \rightarrow 2NH_3 + CaCO_3$$

References

Hluchan S et al (1997) Calcium. In: Habashi F (ed) Handbook of extractive metallurgy. Wiley-VCH, Weinheim, pp 2250–2341

Calcium: Ca(II)

► Calcium and Viruses

Calcium: Ca^{2+}, Ca^{2+} Ion

▶ Magnesium Binding Sites in Proteins

Calcium-Activated Photoproteins

▶ Calcium-Regulated Photoproteins

Calcium-Binding Constant

▶ α-Lactalbumin

Calcium-Binding Protein Site Types

Michael Kirberger[1] and Jenny J. Yang[1,2]
[1]Department of Chemistry, Georgia State University, Atlanta, GA, USA
[2]Natural Science Center, Atlanta, GA, USA

Synonyms

Calcium: Ca^{2+}; Ca(II); *Chelate*: Bind; Coordinate

Definitions

Calcium: A soft gray alkaline earth metal. Ionic calcium (Ca^{2+}) is essential for numerous cellular functions, as well as biomineralization.

Calcium-binding protein: Any protein or enzyme that requires the binding of a calcium ion to fulfill either a functional or structural role.

Binding site: Region in the protein where metal ions will bind as a result of electrostatic interactions with ligands of opposite charge.

Ligand: Ion, atom, or molecule that binds to a metal ion to form a coordination complex.

Tertiary structure: Three-dimensional shape of a protein, enzyme, or molecule.

Diverse Roles of Calcium-Binding Proteins

In the intracellular environment, calcium functions as a second messenger to facilitate a wide variety of functions related to muscle contraction, neurotransmitter release, and enzyme activation. Additionally, calcium and calcium-binding proteins (CaBPs) are involved in almost every aspect of the eukaryotic cell life cycle including cell differentiation and proliferation, membrane stability, apoptosis, and intracellular signaling. Control of these diverse functions is regulated by changes in cytosolic calcium levels, which increase from $\sim 10^{-7}$ M at rest to $\sim 10^{-5}$ M when activated. In response to increased cytosolic calcium, intracellular proteins such as calmodulin (CaM) and protein kinase C (PKC) bind calcium.

The cellular activities regulated by calcium, while numerous, are all dependent upon the abilities of different proteins to bind calcium selectively over other metals and to do so with affinities consistent with the concentration of free calcium available in any given environment. Therefore, the interface between calcium and biological activity can be localized to the protein calcium-binding sites; regions of the protein that have evolved to chelate calcium and translate the binding event into a conformational change capable of inducing activity not observed in the calcium-free state. Recent research efforts detailed in this entry demonstrate an increasingly diverse list of calcium-binding sites correlating with an equally diverse set of biological functions regulated by calcium-binding proteins. This entry will therefore focus on the description of common features and differences of different types of calcium-binding sites in proteins based in part on recent statistical analyses of structural data and identification of newly identified calcium-binding sites.

Sequential Classifications of Calcium-Binding Sites

Since the initial identification of EF-hand calcium-binding sites with a helix-loop-helix motif (Kretsinger and Nockolds 1973), calcium-binding sites have been divided into EF-hand and non-EF-hand types. Binding sites were also previously described as being either continuous or noncontinuous, based on the relative sequence positions of the binding residues.

Calcium-binding sites are further separated into three distinct classes. Class I sites are comprised of consecutive amino acids in the primary sequence which would include canonical EF-hand (e.g., calmodulin), pseudo EF-hand (e.g., S100 proteins), and other noncanonical EF-hand (e.g., calpain) motifs. Class II sites include a similar stretch of consecutive amino acids, but also include a coordinating ligand that is close to the other binding ligands in the three-dimensional structure, but distant in the sequence (e.g., HCV helicase, PDB ID 1hei). Class III sites, the least commonly observed, include multiple coordination ligands in close spatial proximity but still distant in the sequence (e.g., the C2 domain of the enzyme protein kinase C (PKC)).

Coordination Chemistry of a Calcium-Binding Site

Metal-binding sites can be described by a central shell of hydrophilic ligands that chelate the metal ion, with surrounding concentric shells of hydrophobic atoms, usually carbon, covalently bound to the ligand atoms and other carbon atoms (Fig. 1a). Binding of calcium is almost exclusively coordinated in proteins by oxygen ligands which originate in side-chain carboxyl (Asp, Glu), carboxamide (Asn, Gln), and hydroxyl (Ser, Thr) groups. Carbonyl oxygen from the main chain may also contribute to coordination, as well as oxygen from water molecules which are observed to form hydrogen bonds with Asp, Ser, and Asn residues. Nitrogen, which binds Zn^{2+} and may associate with calcium in small molecules, is rarely observed in the structures of calcium-binding sites.

The Protein Data Bank (PDB) currently includes structures for ~1,500 CaBPs not classified as EF-hand or EF-like motifs. These latter calcium-binding sites, some of which lack well-structured and recognizable geometric configurations and include binding ligands sequentially distant from each other in the structure, present interesting challenges for determining functional implications. A recent statistical evaluation of different calcium-binding sites from the PDB has demonstrated that EF-hand and non-EF-hand proteins are differentiable based on structural parameters, including distance between the ligand and the ion; distance between ligands; angles between the carbon covalently bound to the oxygen ligand, the oxygen ligand itself and the calcium ion; coordination number; ligand preference; and ligand charge (Fig. 1d–g) (Kirberger et al. 2008). Results of this analysis indicate that EF-hand sites typically utilize more formal charge (3 ± 1 vs. 1 ± 1), higher coordination numbers (7 ± 1 vs. 6 ± 2), and more protein ligands (6 ± 1 vs. 4 ± 2) than non-EF-hand sites. EF-hand sites also use fewer water ligands (1 ± 0 vs. 2 ± 2). Additionally, the order of ligand preference for EF-hand sites was found to be *sidechain Asp (29.7%) > sidechain Glu (26.6%) > mainchain carbonyl (21.4%) > H_2O (13.3%)*, while for non-EF-hand sites, the order was *H_2O (33.1%) > sidechain Asp (24.5%) > mainchain carbonyl (23.9%) > sidechain Glu (10.4%)*. The combination of ligand preference, decrease in charge utilization and increase in water coordination, account for fact that 20% of non-EF-hand sites exhibit a net change of zero.

Diversified Geometric Properties

Early work with calcium-binding proteins observed examples of highly organized coordination geometries, including either pentagonal-bipyramidal (Fig. 1b), where the ion is surrounded by a planar grouping of five oxygen atoms with additional oxygen atoms superior and inferior to the plane, or octahedral (Fig. 1c), with similar ligand coordination above and below a planar ring of only four oxygen atoms. Analyses of many of these binding sites indicate that the pentagonal-bipyramidal geometry may be constructed by ligands either continuous (e.g., CaM) or discontinuous (e.g., PKC) in the sequence, and this geometry is often observed with intracellular proteins that exhibit high affinity for calcium.

More recent analyses of calcium-binding sites in proteins exhibiting diversified functions in different cellular environment, however, reveal much greater variability in binding coordination geometries than was previously assumed. A simplified and more generalized set of coordination geometries may be applied based on a Hull property describing the spatial relationship of the calcium ion to the interior volume of the surrounding binding ligands. First, a holospheric binding geometry where oxygen ligands surround calcium on all sides is observed in both the pentagonal-bipyramidal and octahedral geometries (Fig. 2a).

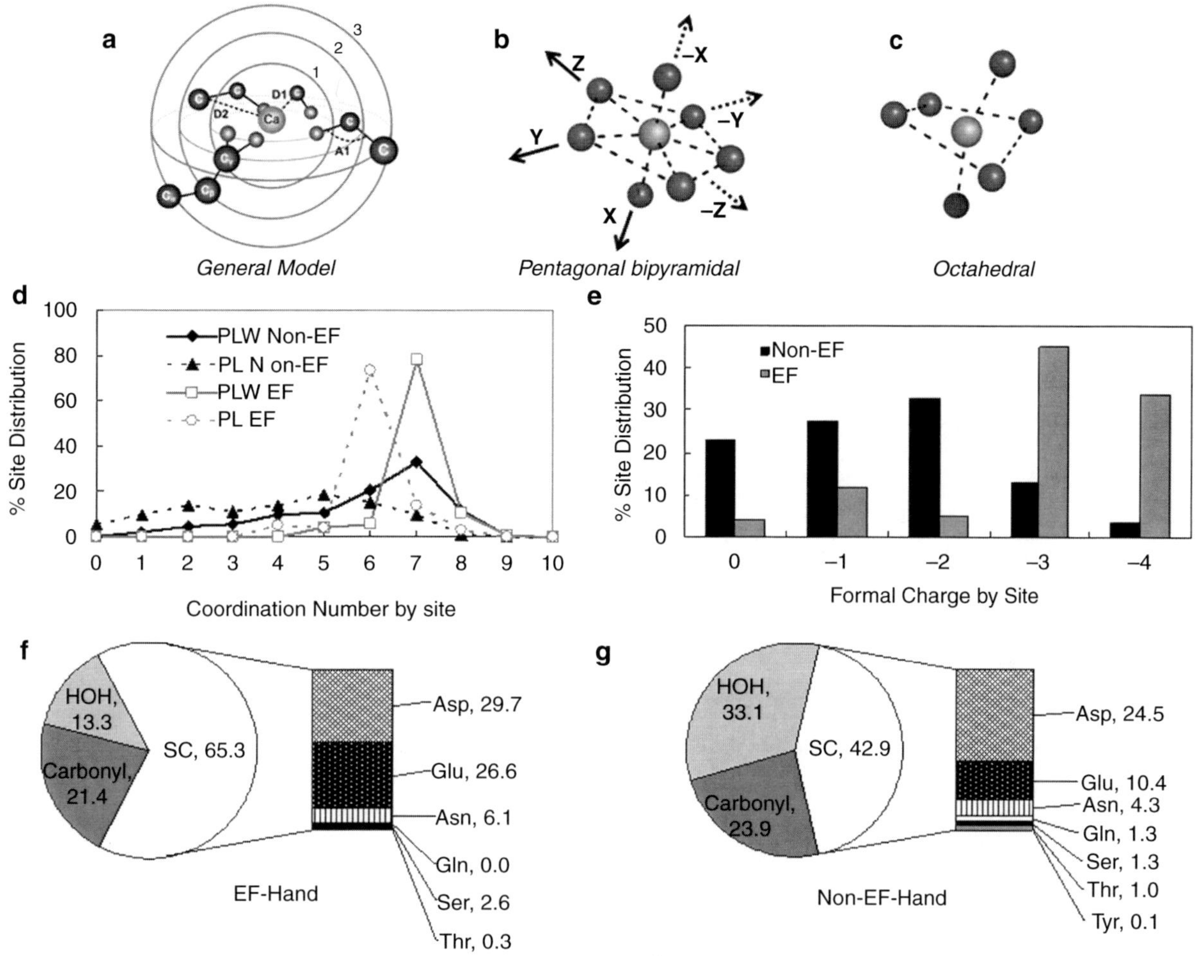

Calcium-Binding Protein Site Types, Fig. 1 (a) General binding model. Calcium is surrounded by a first shell of hydrophilic ligands (oxygen), which is in turn surrounded by concentric second and third shells of hydrophobic atoms; in this example, covalently bound carbon from side chains. (b) Pentagonal-bipyramidal geometry associated with EF-Hand binding motif. (c) Octahedral binding geometry for calcium. Also observed with binding of Mg^{2+}. Calcium-binding statistics for EF-hand and non-EF-hand motifs showing differences between (d) coordination number and (e) formal charge. The ligand distribution between (f) EF-hand and (g) non-EF-hand shows that non-EF-hand binding sites include fewer side-chain ligands, with increased substitution of water molecules. Also, side-chain Glu is reduced compared to EF-hand

This model is shared by the majority of calcium-binding sites with high coordination number for calcium (i.e., greater than four). When possible, proteins fold in such a way as to provide a pentagonal-bipyramidal geometry for binding.

Second, a hemispheric coordination scheme is observed with an open concavity (i.e., bowl structure) coordinating the ion such as calcium-binding sites in helicase and PKC C2 domain (Fig. 2b). A more irregular, planar binding site is described in Fig. 2c, which may be coordinated by as few as three binding ligands such as a calcium binding site in proteinase K. Completion of this geometry may involve increased utilization of water or other small molecules (e.g., sulfate ions) or cofactors in addition to ligand atoms contributed by the protein itself.

Third, a specific geometry is observed even in the C2 domain-binding site clusters, where binding ligand atoms are shared by more than one calcium ion (Fig. 2d). This type of calcium-binding site was first observed in thermolysin and later in the cadherin family. Additionally, when variations of this model

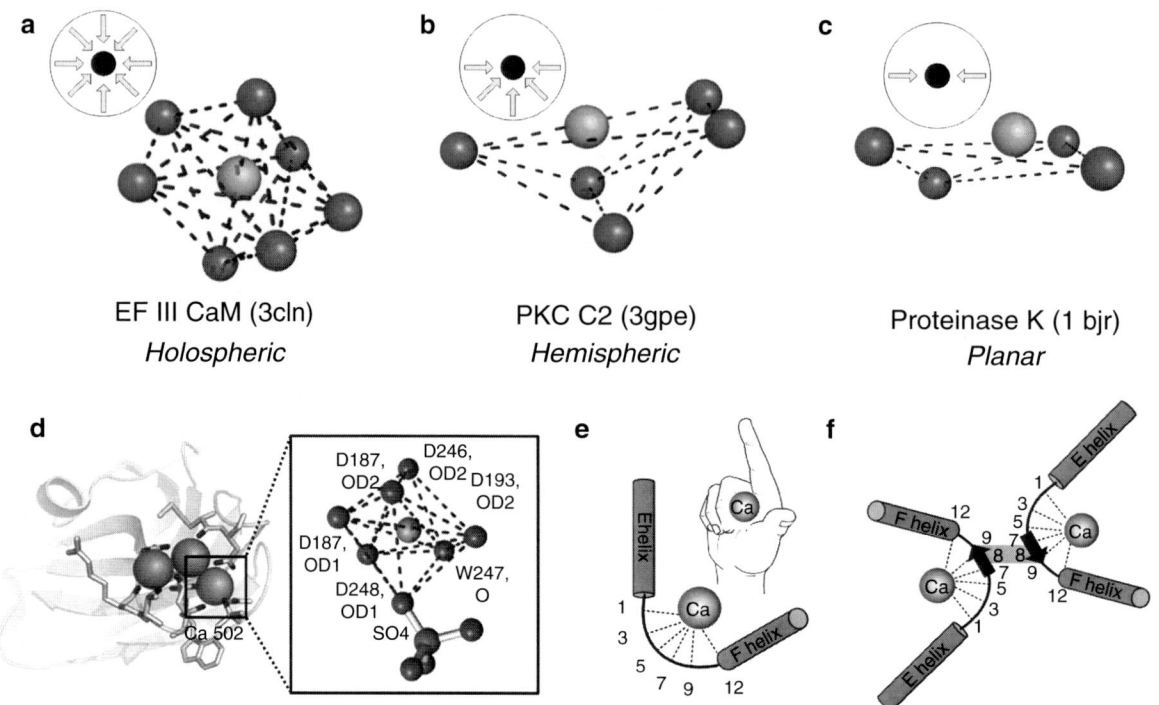

Calcium-Binding Protein Site Types, Fig. 2 (**a**) Holospheric binding where calcium is surrounded on all sides by oxygen ligands. This would include both pentagonal-bipyramidal and octahedral geometries. (**b**) Hemispheric binding where the calcium ion is exposed on one hemispheric surface. (**c**) Planar binding where the calcium ion is bound in a ring structure with exposure above and below the plane. (**d**) C2 domain from PKC. The coordination for the pentagonal-bipyramidal geometry for Ca-502 is completed with a sulfate ion. (**e**) EF-hand motif. (**f**) Paired EF-hand motifs with conserved hydrophobic residues in position 8 (Val, Leu, or Ile). Cooperative binding between the two EF-hand motifs is related to formation of hydrogen bonds between residues in loop position 8 which create two short antiparallel β-strands

are observed (e.g., octahedral, bowl, or planar geometry), these binding sites appear to be structurally incomplete pentagonal-bipyramidal geometries, rather than unique structures, or occur within monomeric units that form dimers which may then form a holospheric geometry. These observations suggest that the properties of calcium drive the evolution of CaBPs to provide binding sites unique to the chemistry of calcium.

Calcium-Binding Affinity and Selectivity

Corresponding to the cellular locations and concentration of calcium ions, calcium-binding proteins have affinity with K_d values ranging from nM to mM. Factors contributing to calcium-binding affinities of proteins are largely electrostatic interactions based on the number of charged ligand residues, residue types, and the binding site microenvironment. Since calcium-binding sites are often energetically coupled, the resulting cooperativity and calcium-induced conformational changes can also contribute to the measured affinity, and these factors make it difficult to experimentally determine affinity values for each of the sites as isolated binding events. Affinity values for paired binding sites have been reported, but are typically calculated based on upper and lower limits derived from the relationship between macroscopic and microscopic binding constants as described by Linse et al. (1991), or as relative affinities based on order of occupancy (i.e., higher affinity sites are populated first). Second, for higher affinity binding sites, analytical instruments lack sensitivity to analyze samples at concentrations comparable with their binding affinities, so experimental methods (e.g., high-resolution NMR) typically involve protein concentrations much higher than their metal binding affinity, which precludes accurate determination of binding affinity constants (Permyakov and Kretsinger 2010).

In addition, in many of the non-EF-hand binding sites, the lack of well-structured and recognizable geometric configurations, the fact that many sites include binding ligands sequentially distant from each other in the structure, and the lack of structures detailing proteins in both the *apo-* and *holo-*forms, present interesting challenges for determining functional implications and estimating contributions to affinity from calcium-induced conformational changes. This is why the literature remains relatively sparse with affinity data, or presents inconsistent data with respect to affinity.

Different approaches have been made to circumvent the issue of cooperativity. Early efforts involved analyses of individual EF-hand motifs isolated as peptides. Another approach involved spectrofluorometric analysis, where the macroscopic calcium affinities (expressed in dissociation constants K_d) for the N- and C-terminal domains of CaM each comprising a pair of cooperative binding sites (EF I and EF II in the N-terminal and EF III and EF IV in the C-terminal), was determined by monitoring Phe and Tyr residues which exhibit fluorescence changes upon binding of metal ions. Ye et al. determined site-specific K_d for CaM EF-loops I–IV (34, 245, 185, and 814 uM, respectively) by grafting the loops into a scaffold protein (Ye et al. 2005). This approach provides a new strategy to estimate cooperativity of coupled EF-hand proteins. The contribution of the cooperativity to the calcium binding affinity for the C-terminal domain is 40% greater than that of the N-terminal domain. Related studies have suggested optimal calcium binding requires four charged ligands, and affinity decreases with the addition or subtraction of charged residues from this number.

Proteins exhibit selectivity for different physiologically relevant metals depending on their environment and the nature of their functions, which is how calcium sensor proteins selectively bind intracellular calcium in an environment with fourfold higher levels of free magnesium. Binding of calcium is almost exclusively coordinated in proteins by oxygen ligands. Nitrogen, which binds Zn^{2+} and may associate with calcium in small molecules, is only infrequently observed in the structures of calcium-binding sites and, then, observed mostly in cases where the site has no net negative charge. Proteins that require metal cofactors, in general, are optimally activated only upon binding of their preferred target metal ions. However, CaBPs may bind magnesium ions at low affinity in a resting state, and the magnesium is then replaced by calcium resulting in a fully potentiated conformer. Similarly, the introduction of toxic metal ions can effectively occupy the native site, but can alter the overall conformation sufficiently enough to inhibit protein function. Typically, selectivity can be evaluated qualitatively, but not quantitatively, due to the previously noted issues with cooperativity and problems establishing accurate K_d values.

Characteristics of Different Types of Calcium-Binding Proteins

The increasingly populous superfamily of EF-hand proteins, comprising approximately 70 different genomic subfamilies, can be divided into two major groups based on calcium-binding sites: The canonical EF-hand motif which is the most common protein calcium binding structural domain and the more recently characterized noncanonical EF-loops which include the pseudo EF-hands observed in the N-termini of S100 and S100-like proteins.

Canonical EF-Hand Binding Motif

The canonical EF-hand motif (Fig. 2e) is highly conserved in both eukaryotes and prokaryotes. This sequential motif, described extensively in the literature, is 29 amino acids in length comprising a 12-residue loop surrounded by two flanking α-helices positioned in a relatively perpendicular orientation. Analyses of metadata from online databases (e.g., PFAM, ProSite) indicate that the length of the entering (E) and exiting (F) helices are typically nine and eight residues in length, respectively. Loop residues are assigned relative position numbers 1–12. Binding of calcium is coordinated by residues in loop positions 1(x-axis), 3(y), 5(z), 7($-y$), 9($-x$), and 12($-z$), forming a pentagonal-bipyramidal geometry (see section "Coordination Chemistry of a Calcium-Binding Site" and Fig. 1b).

Ligands observed within the EF-Loop are typically Asp at position 1; Asp or Asn at position 3; Asp, Ser, or Asn at position 5; a water molecule at position 9; and a bidentate Glu at position 12 which may initially anchor calcium and thus initiate rotation of loop residues to form the binding site. The coordinating ligand from position 7 is usually a carbonyl oxygen, while the noncoordinating residue in position 6 is frequently a flexible Gly.

Prior to 1997, structural analyses revealed that almost all EF-hand proteins included 2–6 paired motifs (e.g., calmodulin, parvalbumin, calcineurin, and troponin C). Interaction between the paired EF-loops may be related to the conserved hydrophobic residues in loop position 8 (Val, Leu, or Ile). Binding of Ca^{2+} in paired EF-hands is cooperative, and typically binding in one of the sites enhances the binding affinity of the second site (i.e., positive cooperativity). The hydrophobic residues in the canonical EF-loop position 8 between paired EF-hand sites form two short antiparallel β-strands (Fig. 2f), and it has been suggested that this EFβ-scaffold governs Ca^{2+} binding and the associated structural changes and represents the structural basis for positive cooperativity between the sites (Forsen et al. 1991).

Noncanonical EF-Hand Binding Motifs

Noncanonical EF-loops, including the pseudo EF-hand motif found in the S100 protein family, are characterized by binding geometries structurally similar to canonical sites, but exhibiting more variation in either the length of the binding loop and/or flanking helices, the absence of helices, or compositional changes within the binding loop.

The pseudo EF-hand motif is found in the S100 and S100-like proteins, including calbindin D_{9k} and calcyclin (S100A6). The S100 proteins generally are of lower molecular weight (~9–14 kDa). The full range of functions associated with S100 proteins remains unknown, but different S100 proteins have been identified with a substantial number of extracellular and intracellular activities, including regulatory activities related to phosphorylation, enzymes, and intracellular calcium release associated with ryanodine receptor function, as well as increased expression in inflammatory responses and cancer metastasis. In cells, these proteins may organize as covalently bound homodimers or heterodimers, with some exceptions including calbindin D_{9k} which is a monomer. Dimerization of S100 proteins appears to directly relate to their biological activities, and the structural basis for this self-assembly is driven by binding with calcium.

The pseudo EF-hand binding geometry is similar to the pentagonal-bipyramidal conformation observed with the canonical EF-hand, but significant differences are observed in the binding loop. Rather than a 12-residue loop, pseudo EF-hand extends to 14 residues where the calcium ion is coordinated predominantly with main-chain carbonyl oxygen atoms from residues occupying positions 1, 4, 6, and 9, with a water molecule coordinated by residue 11 and a bidentate (Asp or Glu) ligand in loop position 14. Because the majority of binding ligands originate from the backbone itself, the nature of the associated residue is less restricted than what is observed with canonical EF-hand binding sites. Additionally, where the canonical EF-loops typically have a formal charge between −2 and −4, less formal charge is observed in the pseudo EF-loops due to dominance of carbonyl oxygen binding ligands. An example of this can be seen with calprotectin (PDB ID 1×k4) which binds calcium with zero formal charge

This motif is usually observed to be paired with, and to sequentially precede, a canonical EF-hand which exhibits higher binding affinity for calcium. From the N-terminal, helices are labeled consecutively H1–H4. The pseudo EF-hand loop (L1) is flanked by H1 and H2, while the canonical EF-hand loop (L2) is surrounded by helices H3 and H4. The two motifs are separated by a flexible hinge region, while a short peptide extension appears at the C-terminal. Comparison of sequences indicates that the greatest homology is observed in the canonical EF-site, with the most variance observed in the hinge region and a C-terminal extension following the canonical EF-site.

The functional role of the pseudo EF-hand appears to be a more recent evolutionary feature producing lower affinity in the N-terminal domain and allowing significant calcium-induced changes in the canonical EF-hand, which in turn expresses a hydrophobic cleft necessary for target recognition and peptide binding.

Calpain, grancalcin, and ALG-2 are classified in the penta-EF-hand protein family. Penta-EF-hands have five binding sites. Not all sites are necessarily active and may be characterized as either EF-hand sites or incomplete EF-hands. The incomplete EF-hands typically exhibit the helix-loop-helix structure of canonical EF-hand, but a reduction in the number of residues in the loop sequence results in incomplete pentagonal-bipyramidal geometry. ALG-2 (apoptosis-linked gene 2, PDB ID 1hqv) includes four calcium-binding sites. The sites surrounding Ca-997 and Ca-998 are canonical EF-hand motifs, while the binding site chelating Ca-999 represents an incomplete pentagonal-bipyramidal geometry, comprising a short stretch of seven residues and water (Table 1). The final binding

Calcium-Binding Protein Site Types, Table 1 Examples of diverse calcium-binding sites

PDBID	Protein/enzyme/assembly	Binding site ligands	Seq. class	Struct class	Geometry
3gpe	PKC (Ca 501)	M186-O, D187-OD1, D246-OD2, D248-OD1, D248-OD2, D254-OD2, 1 H2O	III	Holospheric	Pentagonal-bipyramidal
3gpe	PKC (Ca 502)	D187-OD1, D187-OD2, D193-OD2, D246-OD1, W247-O, D248-OD1, PO$_4$	III	Holospheric	Pentagonal-bipyramidal
3gpe	PKC (Ca 503)	D248-OD2, D254-OD2, D254-OD2, R252-O, T251-OG1	I	Hemispheric	Bowl
1hml	α-lactalbumin	K79-O, D82-OD1, D84-O, D87-OD1, D88-OD1, 2 H2O	I	Holospheric	Pentagonal-bipyramidal
1aui	calaneurin (Ca 500)	D32-OD1, E41-OE2, D30-OD1, E41-OE2, S36-O, S34-OG	I	Holospheric	Pentagonal-bipyramidal
1aui	calaneurin (Ca 501)	E68-O, N66-OD1, D64-OD1, E73-OE1, E73-OE2, D62-OD1	I	Holospheric	Pentagonal-bipyramidal
1alv	calpain (Ca 4)	D135-OD1, N226-OD1, D225-OD2, D225-OD1, D223-OD1, D223-OD2, 2 H2O	III	Holospheric	–
1alv	calpain (Ca 1)	D110-OD1, E112-O, A107-O, E117-OE1, E117-OE2	I	Holospheric	Pentagonal-bipyramidal
1hqv	ALG-2 (apoptosis-linked gene-2, Ca 996)	N 106-OD1, 3 H2O	–	Planar	–
1hqv	ALG-2 (apoptosis-linked gene-2, Ca 997)	D38-OD2, D36-OD1, V42-O, E47-OE2, S40-OG, E47-OE1, D38-OD1 1 H2O	I	Holospheric	Pentagonal-bipyramidal
1hqv	ALG-2 (apoptosis-linked gene-2, Ca 999)	D171-OD1, D173-OD1, D169-OD1, W175-O, 2 H2O	I	Holospheric	Pentagonal-bipyramidal (incomplete)
1hz8	EGF (Ca 84)	N57-OD1, I42-O, E44-OE1, L58-O, D41-OD2, E44-OE2	III	Holospheric	Pentagonal-bipyramidal (incomplete)
3mi4	Trypsin	E70-OE1, N72-O, V75-O, EB0-OE2, 2H2O	II	Holospheric	Octahedral
2prk	Proteinase K (Ca 280A)	P175-O, V177-O, D200-OD1, D200-OD2, 4 H2O	II	Holospheric	Pentagonal-bipyramidal
2prk	Proteinase K (Ca 281A)	T16-O, D260-OD1, D260-OD2, 3 H2O	–	Hemispheric	–
1bci	Phospholipase A2 (cytosolic, C2 Domain, Ca 1950A)	T1041-O, N1065-OD1, D1043-Od1, D1043-OD2, D1040-OD1, MES4000-O3S, 1 H2O	III	Holospheric	Pentagonal-bipyramidal
1bci	Phospholipase A2 (cytosolic, C2 Domain, Ca 1951A)	D1093-OD1, D1093-OD2, D1040-OD1, D1040-OD2, D1043-OD2, A1094-O, N1095-OD1, 1 H2O	III	Holospheric	Pentagonal-bipyramidal
3qfy	Phospholipase A2 (extracellular)	Y28-O, G32-O, G30-O, D49-OD1, D49-OD2	II	Hemispheric	Bipyramidal (incomplete)
3l4m	MauG (Ca 400 Chain A)	N66-OD1, T275-O, P277-O, 4 H2O	–	Holospheric	Pentagonal-bipyramidal
1qmd	alpha-toxin (phospholipase C, Ca 403)	A337-O, D269-O, D336-OD1, G271-O	–	Hemispheric	Bowl
3mt5	Human BKK+ channel Ca2+ gated K+ channel in	D892-O, Q889-O, D897-OD2, D895-OD1	I	Hemispheric	Bowl
2aef	Methanobacterium autotrophicum (Ca602)	D184-OD1, D184-OD2, E210-OE2, E212-OE2, 3 H2O	II	Holospheric	Pentagonal-bipyramidal

PDB ID, File identifier in Protein Data bank. Seq Class, Sequential classification. Struct Class = General structural classification. Calcium ions listed in column 2 are identified by their sequence number in the PDB file (e.g., Ca 602). Binding site ligands are defined as [*Residue Type*][*Residue Sequence Number*][−][*Atom type*]. So, D184-OD1 describes the delta oxygen 1 for aspartic acid at sequence position 184

site for Ca-996 includes a single side-chain ligand atom and water molecules, resulting in a planar binding structure.

Interestingly, calcium-binding proteins with an odd number of EF-hand motifs such as calpain often form a dimer to ensure pairing of the odd EF-hand motifs. Additionally, some proteins originally identified with odd numbers of EF-hand motifs may in fact have "hidden" EF-hands in their structures. The N-terminal domain of STIM1, an ER protein that responds to depletion of luminal Ca^{2+} by activating store-operated Ca^{2+} (SOC) channels on the plasma membrane and thereby facilitating extracellular Ca^{2+} influx into the cytoplasm, contains several functionally important regions including an ER signal peptide and a canonical EF-hand domain. Preliminary investigations on STIM1 had indicated that the observed canonical EF-hand domain functioned as a solitary binding site for Ca^{2+}, but a second hidden EF-hand site, albeit lacking conserved calcium-binding ligand residues, was later revealed by X-ray crystallography and was found to stabilize the canonical EF-hand through hydrogen bonding between the paired loops and coupled flanking helices and to exhibit the cooperative binding effects associated with EF-hand pairs (Stathopulos et al. 2008). Single EF-hand motifs have also been identified in both bacteria and virus genomes, but it remains to be seen whether these single EF-hand motifs are stabilized with a hidden EF-hand site located elsewhere in the protein.

Non-EF-Hand Continuous Calcium-Binding Sites

A significant proportion of calcium-binding structures currently in the PDB includes structures not sequentially or structurally identifiable as canonical EF-hand motifs and may represent structural classes yet to be categorized. This group includes an increasing number of calcium-binding sites structurally similar to EF-hands but with increasing variability in the loop or helices. Table 1 summarizes many of the binding site examples discussed in this section.

Human low-density lipoprotein receptor contains two atypical calcium-binding sites found in extended loop regions. The NMR structure of this protein (PDB ID 1hz8) does not include water molecules, so the structures appear incomplete. This is apparent in the binding site surrounding Ca-84 which exhibits a distorted, incomplete pentagonal-bipyramidal geometry comprised of six binding ligand atoms. However, these ligands span an unusually long region of the loop (17 residues), and the loop itself is partially restricted by the formation of a disulfide bond. The second binding site coordinating Ca-83 is comprised of only four ligand atoms which fail to surround the ion, although this may be due to the absence of water molecules in the PDB file.

Another example is α-lactalbumin, an extracellular protein from the C-type lysozyme family that participates in the formation of lactose synthetase, a precursor enzyme involved in lactose synthesis. The calcium-binding site in α-lactalbumin consists of a short four-residue N-terminal side helix, a four-residue loop, and a longer (at least 12-residue) C-terminal side helix (PDB ID 1hml). Despite this significant variance from the canonical EF-hand loop, this site retains a pentagonal-bipyramidal binding geometry comprised of five protein ligand atoms and two water molecules. This site is also interesting because, with the exception of oxygen from the two water molecules and a single side-chain carboxyl oxygen from D82 in the loop region, the remaining binding ligand atoms are contributed by residues in α-helices.

Calcium and Enzymes

The potential for calcium binding to play a catalytic role in enzymes has been a subject of long-term debate. It is clear that calcium plays an important role in structural stability, folding, and regulation of enzymatic activity. The relationship between calcium and the generation of trypsin from trypsinogen was being investigated as early as 1913 (Mellanby and Woolley 1913), and research has shown that calcium can stabilize the structures of different enzymes, including thermolysin, trypsin, and proteinase K, which enables them to perform their catalytic activities. Many of the binding sites associated with these enzymes exhibit pentagonal-bipyramidal geometries, as observed with canonical EF-hand sites.

Trypsin contains a single high-affinity calcium-binding site with an octahedral binding geometry where the superior and inferior apices are both water molecules. Binding of calcium prevents autodegradation and is necessary for the structural integrity of the active enzyme. Proteinase K has two calcium-binding sites: a higher affinity site exhibiting pentagonal-bipyramidal geometry comprised of four ligand atoms from amino acid residues (two in close sequential proximity with a bidentate ligand more distant in the sequence) and four water molecules and a second calcium-binding site linked between residues in distant parts of the sequence (T16-O and bidentate D260) and bound with three water molecules. As with trypsin, binding of calcium in proteinase K is reported to stabilize the enzyme structure and facilitate structural changes necessary for catalytic activity.

Calpain is a cysteine protease that regulates cellular functions by cleaving a number of different substrate proteins (i.e., kinases, phosphatases, and transcription factors). Calpain binds three calcium ions: two in canonical EF-hand geometries and a third in a modified EF-hand involving two bidentate ligands Asp-225 and Asp-223.

Calcium-dependent phospholipase A2 (PLA2) enzymes include cytosolic and extracellular isoforms. Cytosolic PLA2 (cPLA2), which plays a role in production of lipid mediators of inflammation, contains a C2 domain with two calcium-binding sites. Both sites exhibit pentagonal-bipyramidal geometry despite sequential separation of binding ligand residues. The roles of extracellular phospholipase A2 enzymes differ significantly from PLA2 in the cytosol. Extracellular PLA2 in venom helps to immobilize prey, while pancreatic PLA2 plays an important role in the breakdown of phospholipids in dietary fat. Significant structural differences are also observed with extracellular enzyme PLA2 (PDB ID 3q4y), which incorporates a hemispheric calcium-binding site. Residues comprising the site are summarized in Table 1. This site includes a bidentate ligand D49 originating in an α-helix, similar to the bidentate anchoring ligand in the exiting helix of the canonical EF-loop. However, this ligand is sequentially distant from the other binding ligands. Additionally, binding ligands are not observed in the region in the pentagonal plane corresponding to the −y-axis (EF-loop position 7), which is normally occupied by carbonyl oxygen in the canonical EF-hand, or in the −x-axis space normally occupied by a water molecule. The resulting binding geometry in the crystal structure of phospholipase A2 therefore suggests an incomplete pentagonal-bipyramidal geometry.

Recently, calcium-binding sites were identified in a class of heme-containing peroxidases including canine myeloperoxidase, horseradish peroxidase, cationic peanut peroxidase, manganese peroxidase, lignin peroxidase, bacterial diheme cytochrome c peroxidase (BCCP), and the diheme enzyme MauG from *Paracoccus denitrificans*. The X-ray structure of MauG (PDB ID 3l4m) revealed that the calcium-binding site in MauG includes oxygen from the side-chain ligand atom (N66-OD1), two carbonyl oxygen atoms (T275 and P277), and four water molecules. Calcium "bridges" the two hemes via a hydrogen-bonded network of waters that connect to the propionate moiety of each heme. The presence of calcium was shown to be important in maintaining an appropriate structural environment for the hemes, which allows it to generate the *bis*-Fe(IV) intermediate and catalyze a complex six-electron oxidation of a protein substrate. This calcium-binding site is very unique since calcium is coordinated by oxygen atoms from four water molecules and three noncharged ligand residues. Despite the limited charge in the site, MauG apparently binds calcium with relatively high affinity (K_d 5.3 μM) (Shin et al. 2010). The reason for this unusually high affinity is not clear. It is possible that such strong affinity originates from the enhanced electrostatic force from the charged propionate side chain associated with the hydrophobic heme.

This example suggests that charge is not the sole factor influencing binding affinity. Moreover, the structural similarity of these calcium-binding sites in enzymes indicates calcium can be coordinated with much greater flexibility than what is suggested by the highly conserved and more densely charged canonical EF-hand loops.

C2 Domain with a Cluster of Calcium Ions

The C2 membrane-targeting domain is identified in cellular proteins that fulfill a role in signal transduction, including synaptotagmin I, phospholipase A, and the β-isoform of protein kinase C (PKC) (PDB ID 1a25). C2 domains contain a core calcium-binding region (CBR) where calcium binding, often accompanied by binding of additional cofactors as observed with certain isoforms of PKC, initiates conformational changes that allow the domain to identify membrane-attached targets, such as anionic phospholipids. C2 domain motifs diverge from the canonical EF-hand in several important ways. First, proteins with C2 domains exhibit β-sandwich architecture, compared to the predominantly α-helical nature of proteins with canonical EF-hands. Because of this architecture, a series of interstrand loops cluster at the end of the β-sandwich. In the C2 domain, a cleft is formed by these loops which are densely packed with aspartic acid residues. This cleft accommodates binding of multiple calcium ions as seen with PKC (Fig. 2d), which is presumed to be cooperative and necessary for stabilizing the structure in order for the domain to recognize its molecular target. Second, the binding ligands can originate from sequentially distant regions of the protein, as seen with Ca-502 in PKCα (PDB ID 3gpe) where the binding site is formed by ligands D187 (bidentate), D193, D246, W247 (carbonyl), D248, and SO_4. The overall geometry is holospheric (Fig. 2a), conforming to a pentagonal-bipyramidal geometry with oxygen from sulfate replacing the water molecule at the ($-x$) position. The crystal structure of PKCα shows binding of two additional calcium ions. One of these is coordinated in a pentagonal-bipyramidal geometry, while the other occupies a hemispheric bowl geometry (Fig. 2b). Unusually, several ligands appear to be shared between calcium ions in the structure (Table 1). Similar binding models in C2 domains are observed with PKCβ (PDB ID 1a25) and cadherins. Such clustering of calcium ions is not commonly seen for cytosolic calcium-binding proteins with strong calcium-binding affinities. Consistent with this observation, the reported lower calcium-binding affinities for this class of calcium-binding sites (~mM), correspond to the higher calcium content in the environment.

Calcium in Ion Channels

Ion channels describe transmembrane protein assemblies that allow the regulated movement of ions (Na^+, K^+, Ca^{2+}) across cellular compartments. Calcium channels, which facilitate the transfer of calcium across membranes, may be either ligand-gated or voltage-gated. Examples of ligand-gated calcium channels include IP_3 and ryanodine receptors. Voltage-gated calcium channels (VGCC) regulate the entry of calcium into the cell following changes in the membrane potential. This calcium influx in turn drives diverse cellular functions including cardiac muscle contraction and neurotransmitter release. Voltage-gated calcium channels, through their Ca_v protein subunits, may be regulated through an indirect calcium feedback mechanism mediated by binding of calcium to CaM which can interact with an isoleucine-glutamine (IQ) motif located in the N-terminal region of Ca_v.

Conversely, voltage-activated K^+ channels (e.g., BK or Slo1 channels) can be directly activated by increases in intracellular calcium which provides a feedback mechanism where opening of these channels hyperpolarizes the membrane and initiates closing of calcium channels, thereby reducing calcium influx. A calcium-binding site identified in BK K^+ channel is believed to contribute to this calcium regulatory mechanism. This binding site, identified as a "calcium bowl" (Schreiber and Salkoff 1997), is hemispheric, comprised of four binding residues (D892-O, Q889-O, D897-OD2, D895-OD1) originating in an Asp-rich sequence DQDDDDDPD, as seen in the PDB crystal structure for human BK (PDB ID 3mt5). Mutations or deletions of residues in this sequence have been shown to desensitize channel activity; however, further evidence suggests the existence of a second calcium-binding site that remains to be identified (Schreiber and Salkoff 1997). Similarly, the dimeric crystal structure of MthK calcium-gated K^+ channel in *Methanobacterium autotrophicum* (PDB ID 2aef) reveals two symmetrical calcium-binding sites believed to stabilize the RCK (regulate the conductance of K^+) domains. Unlike human BK, however, these binding sites conform to pentagonal-bipyramidal geometry with the addition of water molecules (Table 1), and the residues comprising the binding sites span a longer region of the sequence than the BK K^+ channel binding site.

Cross-References

- Bacterial Calcium Binding Proteins
- C2 Domain Proteins
- Cadherins
- Calcineurin
- Calcium and Viruses
- Calcium in Biological Systems
- Calcium in Health and Disease
- Calcium in Heart Function and Diseases
- Calcium ion Selectivity in Biological Systems
- Calcium-Binding Proteins, Overview
- Calmodulin
- Calsequestrin
- EF-Hand Proteins
- Parvalbumin
- Penta-EF-Hand Calcium-Binding Proteins
- Peroxidases
- S100 proteins
- Thermolysin
- Troponin

References

Forsen S, Linse S et al (1991) Ca2+ binding in proteins of the calmodulin superfamily: cooperativity, electrostatic contributions and molecular mechanisms. Ciba Found Symp 161:222–236

Kirberger M, Wang X et al (2008) Statistical analysis of structural characteristics of protein Ca(2+)-binding sites. J Biol Inorg Chem 13(7):1169–1181

Kretsinger RH, Nockolds CE (1973) Carp muscle calcium-binding protein. II. Structure determination and general description. J Biol Chem 248(9):3313–3326

Linse S, Helmersson A et al (1991) Calcium binding to calmodulin and its globular domains. J Biol Chem 266(13):8050–8054

Mellanby J, Woolley VJ (1913) The ferments of the pancreas: Part II. The action of calcium salts in the generation of trypsin from trypsinogen. J Physiol 46(2):159–172

Permyakov EA, Kretsinger RH (2010) Calcium binding proteins. Wiley, Hoboken

Schreiber M, Salkoff L (1997) A novel calcium-sensing domain in the BK channel. Biophys J 73(3):1355–1363

Shin S, Feng M et al (2010) The tightly bound calcium of MauG Is required for tryptophan tryptophylquinone cofactor biosynthesis. Biochemistry 50:144–150

Stathopulos PB, Zheng L et al (2008) Structural and mechanistic insights into STIM1-mediated initiation of store-operated calcium entry. Cell 135(1):110–122

Ye Y, Lee HW et al (2005) Probing site-specific calmodulin calcium and lanthanide affinity by grafting. J Am Chem Soc 127(11):3743–3750

Calcium-Binding Proteins

▶ Calnexin and Calreticulin
▶ EF-Hand Proteins

Calcium-Binding Proteins, Overview

Robert H. Kretsinger
Department of Biology, University of Virginia, Charlottesville, VA, USA

Synonyms

Annexin; Blood clotting; C2 domain; Calcium as a secondary messenger; Calcium ATPase; Calcium channel protein; Calcium pump protein; Calmodulin; Cytosolic signaling; EF-hand protein; Myosin light chain; Troponin C

Definitions

Metal ions play several major roles in the structures and functions of proteins. The bindings of some metal ions increase the stabilities of proteins or of protein domains. Some transition metal ions take part in catalysis in many enzymes. The oxygen, nitrogen, and sulfur atoms of several amino acid side chains, as well as the carbonyl oxygen of the peptide group and the carboxylate of the C-terminal residue may be involved in coordination of metal ions. Binding of the Ca^{2+} ion is limited to not only oxygen-containing ligands, primarily the carboxylates of Asp (aspartic acid) and of Glu (glutamic acid), but also the oxygens of Asn (asparagine), Gln (glutamine), Ser (serine), Thr (threonine), Tyr (tyrosine), and of the main chain peptides. A water oxygen may be in the primary coordination sphere of the Ca^{2+} ion; in turn, this water may be involved in a hydrogen bond with an oxygen from the protein. Sometimes, for example, in C2 domains or calsequestrin, several Ca^{2+} ions are bound near one another. However, they do not bind one another directly but are instead bridged by an oxygen. Most proteins that bind a metal cation, such as Ca^{2+}, bind it with reasonable specificity. The ligands involved might bind other metals; one infers that the selectivity of the protein reflects an optimization by natural selection.

Calcium-binding proteins can be considered in three categories – those in the cytosol whose activity or biological function is modulated by variation in $[Ca^{2+}]_{cyt}$; those, usually extracytosolic, whose binding of calcium, at constant $[Ca^{2+}]_{out}$, is essential to their functions; and those involved in calcium channels and pumps, where the concept of macroscopic $[Ca^{2+}]$ is problematic.

The variation in $[Ca^{2+}]_{cyt}$ over time and region is a form of information. This information can be transduced into a change in protein conformation and/or enzyme activity by interaction with a calcium-modulated protein(s) in the cytosol or associated with a membrane protein facing the cytosol.

Calcium binding is essential to the functions of many extracytosolic calcium-binding proteins; however, these functions are not (usually) modulated by calcium since $[Ca^{2+}]_{out}$ is (usually) constant.

Calcium pump proteins and calcium channels proteins are intrinsic to the membrane across which the calcium is transported.

All members of a protein domain have a similar tertiary structure, excepting those that are intrinsically disordered, have statistically significant similarity in amino acid sequence, and have evolved from a single precursor domain. A single protein domain belongs to only one domain family.

A chimeric protein is a single polypeptide chain that consists of two or more, nonhomologous domains. It has evolved by gene splicing.

Two proteins (domains) in different species are orthologs if they are related by speciation. If two proteins, in the same or in different species, are related by gene duplication they are paralogs.

Two, or more, subfamilies of proteins, each having two or more homologous domains are congruent if the common precursor to all of these subfamilies had the full complement of domains and all of the subfamilies are related by duplication of the entire, encoding gene. Congruent subfamilies are characterized by having all of their domains 1 more similar to one another than to other domains within the same proteins; correspondingly all domains 2, etc.

Horizontal gene transfer (HGT) is inferred to have occurred if two distantly related species share a protein domain with little evidence that the domain was present in an ancestor common to both species. HGT is well established to have occurred among bacteria and perhaps from eukaryotes to prokaryotes.

Overview

Calcium is unique in biological systems. Ca^{2+} is the only metal cation demonstrated to function as a secondary messenger in the cytosols of eukaryotes. The information in this pulse of Ca^{2+} ions (Berridge 2006) is transduced into a change of conformation of a calcium-modulated protein(s).

However, there are many extracytosolic proteins whose functions are dependent on the binding of calcium. Most intriguing, and least understood, are those intrinsic membrane proteins (complexes) involved in calcium pumps or channels. This article will summarize the characteristics of some representative, major families from each of these three groups – calcium-modulated proteins, extracytosolic calcium-binding proteins, and calcium pump proteins. It is not comprehensive but rather is a broad introduction to the more detailed articles in this volume.

Many of the functional characteristics of calcium-binding proteins (CaBPs) can be rationalized from the characteristics of calcium coordination. In proteins the Ca^{2+} ion (atomic radius 0.99 Å) is usually bound by seven oxygen atoms at the vertices of an approximate pentagonal bipyramid at average Ca-O distance 2.3 Å; the oxygen atoms have some lateral flexibility. In contrast the Mg^{2+} ion (ionic radius 0.65 Å) is usually coordinated by six oxygens with average Mg-O distance 2.0 Å at the vertices of an octahedron. These six oxygens are in close van der Waals contact with one another. Although many small molecules bind magnesium with higher affinity than they bind calcium, most proteins, both in the cytosol and extracellular, bind calcium with higher affinity.

The dissociation constant is the ratio of the off rate to on rate $[K_d (M) = k_{off} (s^{-1}) / k_{on} (s^{-1} M^{-1})]$. The rate limiting dehydration of calcium is fast, $Ca(H_2O)_7^{2+} \sim 10^{-8}$ s^{-1}; while that of $Mg(H_2O)_6^{2+}$ is slow, $\sim 10^{-4.6}$ s^{-1}. This difference reflects the loose pentagonal verses the tight octahedral packing of the oxygen ligands. The cation must be (partially) dehydrated before it can bind to the protein. These rates are important for modeling the flux of Ca^{2+} ions through the cytosol and their attendant binding by proteins. The increase in affinity of most proteins for calcium relative to magnesium derives primarily from the difference in k_{on}.

Most, CaBPs, in the cytosols of eukaryotic cells are involved in information transduction. These calcium-modulated proteins are in the apo- or magnesi-form in the so-called resting cell with calcium concentration $\sim 10^{-7}$ M, that is, pCa ~ 7.0. Following a stimulus to the cell, the concentration of cytosolic calcium, $[Ca^{2+}]_{cyt}$, rises to pCa ~ 4.0 and the calcium-modulated protein is then in the calci-form with a conformation and activity different from that in its apo-form. These calcium-modulated proteins might modify the activity of an enzyme or a structural protein or they themselves might be enzymes.

The second group consists of (usually) extracytosolic proteins. They require calcium for stability and/or activity but are not activated, or inhibited, by calcium since they reside in an environment of (near) constant calcium concentration.

Calcium pumps and channels are intrinsic membrane proteins or components of such complexes.

Calcium Modulated Proteins in the Cytosols of Eukaryotic Cells

Many of these calcium-modulated proteins contain 2–12 tandem EF-hand domains. Others contain one or two C2 domains or four annexin domains (Permyakov and Kretsinger 2010).

EF-Hand Proteins

Parvalbumin was the first EF-hand protein to have its crystal structure and its amino acid sequence determined. It has a canonical pair of EF-hands and it has been studied extensively (reviewed by Permyakov 2006, 2009; Permyakov and Kretsinger 2010). Many of the techniques used to study calcium-binding proteins in general, and EF-hands in particular, have been refined in investigations of parvalbumin. For these reasons, it has become a paradigm; even though, its full range of functions has yet to be determined.

The canonical EF-hand (Fig. 1) consists of α-helix E (forefinger, residues 1–10), a loop around the Ca^{2+} ion (clenched middle finger, 10–21), and α-helix F (thumb, 19–29). Residue 1 is often Glu; the insides of the helices (palmer surfaces) usually have hydrophobic residues that contact the insides of the other EF-hand of the pair. The side chains of the five residues, approximating the vertices of an octahedron (X, residue 10; Y, 12; Z, 14; −X, 18; and −Z, 21), provide oxygen atoms to coordinate the Ca^{2+} ion; residue 16 at -Y bonds to Ca^{2+} with its carbonyl oxygen. The positions of these ligands within the loop are often referred to as 1, 3, 5, 7, 9, and 12. The Ca^{2+} ion is actually seven coordinate in a pentagonal bipyramid with major axis: X, −X. There are five oxygens in the Y,Z plane since the −Z ligand (residue 21, usually Glu) coordinates Ca^{2+} with both oxygens of its carboxylate group. In magnesi-EF-hands the Mg^{2+} ion is six coordinate; this carboxylate coordinates with only one of its oxygens. Gly at 15 permits a tight bend; residue 17 has a hydrophobic side chain that attaches the loop to the hydrophobic core of the pair of EF-hands.

Several variations to this canonical calcium coordination scheme have been inferred from amino acid sequences and confirmed in crystal structures of other EF-hand proteins. Nearly one third of all known EF-hands do not bind calcium; those with no indels (insertions or deletions), have a non-oxygen containing side chain substituted at position 10, 12, 14, 18, or 21. Other EF-hands have indels, most notable is EF-hand 1 of the S-100 subfamily, in which several carbonyl oxygen atoms, instead of oxygens of side chains, coordinate the Ca^{2+} ion.

All members of a homolog family of proteins are inferred to have evolved from a common precursor domain in a single ancestral organism. The most parsimonious interpretation is that all of the EF-hand domains are homologs; however, the significances of sequence alignments are weak since the domains are only 30 residues long. Additional criteria, such as the fact that nearly all EF-hand proteins occur in pairs (Fig. 2) and that in most EF-hand proteins at least one of the two EF-hands binds calcium are included. Many (portions of) domains have been suggested to resemble the EF-hand based on their amino acid sequences. They are not recognized as EF-hands unless they have passed a Hidden Markov Model test based on unambiguous EF-hands of known structure. Most EF-hand containing proteins have been found in eukaryotes; however, there are several examples in eubacteria. EF-hand proteins, for example, calmodulin (Fig. 2), have been found in all eukaryotes subject to thorough investigation. This distribution might reflect an origin in the bacterium that gave rise to eukaryotic cells; other bacteria may have lost their EF-hands. Or, the precursor EF-hand may have arisen in an early eukaryote and its encoding gene have been transferred to a few bacteria by horizontal gene transfer.

Most EF-hand proteins contain 2–12 tandem copies of the EF-hand domain. EF-hands occur in pairs and are related by an approximate twofold axis of rotation, thereby forming an EF-lobe. Although about one third of all EF-hands are known or inferred not to bind calcium, usually at least one EF-hand domain in any protein does bind calcium with $pK_d(Ca^{2+})$ ~7.0. The protein, such as the ubiquitous and archetypical calmodulin, is in the apo- and/or magnesium-form prior to stimulation of the cell; following a rise in $[Ca^{2+}]_{cyt}$, the competent EF-hand(s) binds calcium with attendant change in conformation of itself and probably of the paired EF-hand of that EF-lobe. If the protein is heterochimeric with a non-EF-hand catalytic domain(s), the change in conformation of the EF-hand region activates the enzyme. If the EF-hand protein itself is not catalytic, the change in conformation causes the EF-hand protein to activate a target enzyme or structural protein.

Eighty distinct EF-hand proteins have been identified; however, the functions of only 25 are known. Ten of these are enzymes and have been demonstrated

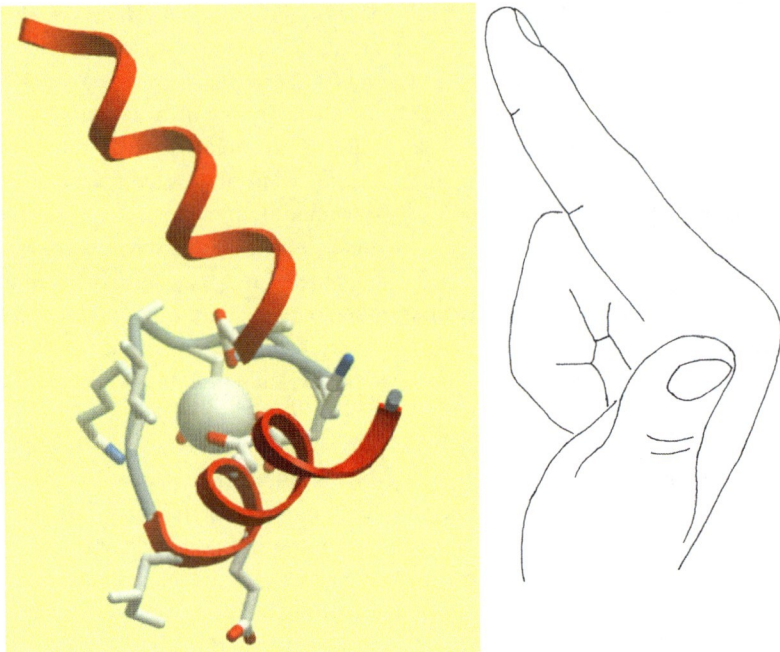

Calcium-Binding Proteins, Overview, Fig. 1 The EF-hand: The canonical EF-hand consists of α-helix E (forefinger, residues 1–10), a loop around the Ca^{2+} ion (clenched middle finger, 10–21), and α-helix F (thumb, 19–29). The side chains of the five residues, approximating the vertices of an octahedron (X, residue 10; Y, 12; Z, 14; −X, 18; and − Z, 21), provide oxygen atoms to coordinate Ca^{2+}; residue 16 at − Y bonds to Ca^{2+} with its carbonyl oxygen. The positions of these ligands within the loop are often referred to as 1, 3, 5, 7, 9, and 12. Nearly one third of all known EF-hands do not bind calcium; those with no indels (insertions or deletions), have a non-oxygen-containing side chain substituted at position 10, 12, 14, 18, or 21. Other EF-hands have indels, most notable is EF-hand 1 of the S-100 subfamily, in which several carbonyl oxygens, instead of oxygens of side chains, coordinate the Ca^{2+} ion

or inferred to be activated by the binding of calcium. Many, but certainly not all, of the remaining 55 function in information transduction pathways as summarized for a few calcium modulated proteins, such as calmodulin. However, others such as intestinal calcium-binding protein, in the S-100 subfamily, probably facilitate the diffusion of calcium through the cytosol; parvalbumin appears to function as a temporal buffer. Thirty, including the ten enzymes are heterochimeric. In addition to their EF-hands, they contain other domains of different evolutionary origin and different conformation. Members of the penta-EF-hand subfamily (calpain, sorcin, ALG-2, and peflin) have two canonical EF-lobes (domains 1 and 2 and 3 and 4). They form homo- or heterodimers by pairing their fifth EF-hands to form a canonical EF-lobe.

It is not unusual for a basic protein domain to find many uses, often spliced together with other domains. The EF-hand is one of the most widely distributed domains in eukaryotes, perhaps reflecting the range and subtlety of calcium signaling. For example, the downstream regulation element antagonist modulator (DREAM) upon binding calcium dissociates from a DNA-binding regulatory element that otherwise functions as a gene silencer. This might provide a precedent for long-term potentiation.

C2 Domain Proteins

The C2 domain was originally identified as the second of four domains (C1 though C4) in the α, β, and γ isoforms of mammalian calcium-dependent protein kinase C. It exists in both calcium-binding and nonbinding forms. Both interact with membranes and with multiple other proteins. The C2 domain is widely distributed in eukaryotes but rare or nonexistent in prokaryotes; it is, after the EF-hand, the most frequently occurring calcium sensor.

C2 domains are about 130 residues long; they fold autonomously and form a compact β-sandwich composed of two, four stranded β-sheets (Fig. 3).

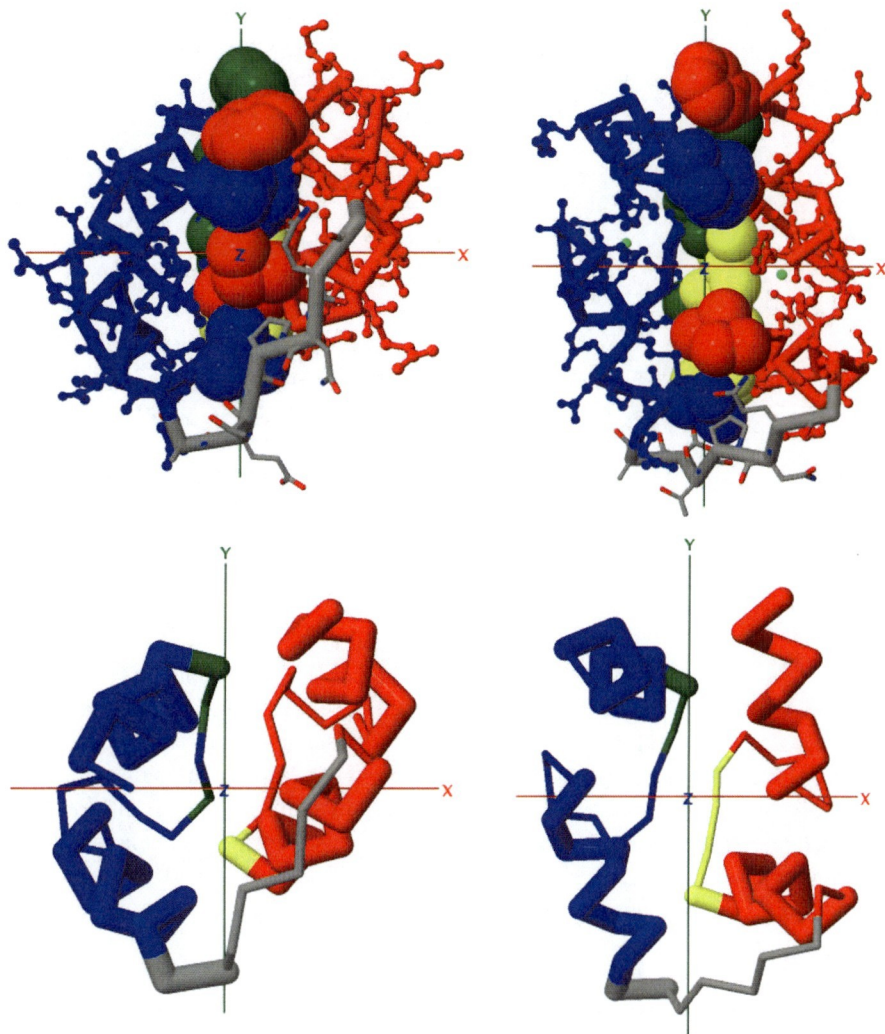

Calcium-Binding Proteins, Overview, Fig. 2 Calmodulin, N-lobe: EF-hand domains, 1 (Odd) and 2 (Even), are related by an approximate twofold axis, z, perpendicular to the plane of view. The apo-form (Protein Data Bank, 1DMO) is on the *left*; the calci-form (3CLN) is on the *right*. Both EF-lobes are represented as α-carbon traces, *lower pair*. The atoms (both side chain and main chain) within ± 2.0 Å of the yz-plane are shown as van der Waals spheres, *upper pair*. These are mostly side chain atoms at the interface of the two EF-hands in the EF-lobe. Upon binding calcium the E and the F helices tilt apart from one another and apart from the other EF-hand of the lobe. This "opening" relieves close contacts between the side chains of residues Met36, Phe19 (Odd) and Met72, Phe68 (Even) and permits binding of calmodulin to specific targets. *Red*, odd domain; *blue*, even domain; *green*, Ca^{2+} ions (Courtesy of Hiroshi Kawasaki, Department of Supramolecular Biology, Graduate School of Nanobioscience, Yokohama City University, Yokohama 230-0045, Japan)

Three loops at the top of the domain and four at the bottom connect the eight β-strands. Calcium binding occurs at the top three loops. The calcium-binding sites are formed primarily by Asp side chains that serve as bidentate ligands bridging two or three Ca^{2+} ions. Most, but not all, C2 domains are activated by calcium binding and then dock to a specific membrane.

Protein kinase Cα (PKCα) has two C2 domains that dock to different membrane surfaces during an intracellular calcium signal (Corbin et al. 2007). They discriminate two different mechanisms of C2 domain-directed intracellular targeting – messenger-activated target affinity (MATA) and target-activated messenger affinity (TAMA).

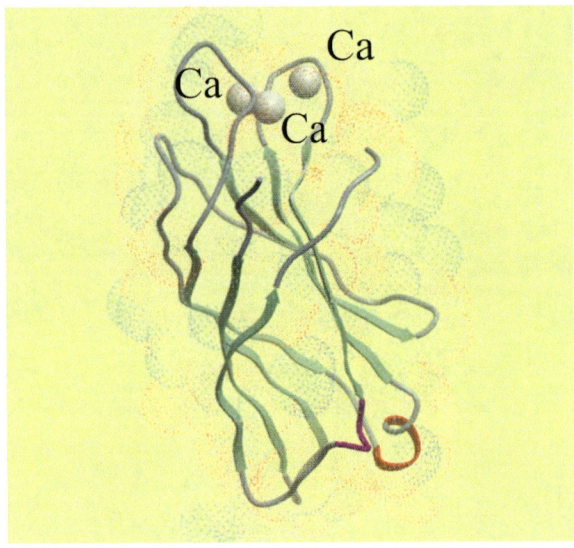

Calcium-Binding Proteins, Overview, Fig. 3 Structure of the C2A domain of synaptotagmin I (PDB file 1B7N): Three Ca^{2+} ions are bound at the top edge of the β-sandwich, which consists of two β-sheets, four strands each

Synaptotagmin has an established docking role in secretion and also appears to be a key component of the endocytosis machinery. It is a major calcium sensor that triggers release of neurotransmitters. The presence of distinct synaptotagmins on the membranes of synaptic vesicles and on active zones of the membranes that fuse during neurotransmitter release suggests an explanation for the existence of multiple synaptotagmins (Südhof 2002). The large cytosolic region of synaptotagmin contains two C2 domains. The first of them, C2A, may function in neurotransmitter release through its calcium-dependent interaction with syntaxin and phospholipids.

Phospholipase C (PLC) mediates the physiological effects of many extracellular stimuli by activation of inositol lipid-signaling pathways. In response to many extracellular stimuli, such as hormones, neurotransmitters, antigens, and growth factors, phospholipases C catalyze the hydrolysis of phosphatidylinositol (4,5)-bisphosphate, thereby generating two second messengers, inositol 1,4,5-trisphosphate and sn-1,2-diacylglycerol. Eleven phospholipase C isozymes encoded by different genes have been identified in mammals and, on the basis of their structure and sequence relationships, have been classified into five families designated PLCβ (1–4), PLCγ (1 and 2), PLCδ (1, 3 and 4), PLCε (1), and PLCζ (1).

The phospholipase A2 (PLA2) family consists of several nonhomologous groups of enzymes that catalyze the hydrolysis of the sn-2 ester bond in a variety of different phospholipids. The products of this reaction, a free fatty acid, and a lysophospholipid have many different physiological roles. It is assumed that the interaction of the C2 domain with membranes is analogous to the hydrophobic electrostatic switch that modulates reversible membrane binding of several myristoylated proteins. In this case, the electrostatic switch is calcium, which changes the electrostatic properties of the surface of the C2 domain, that is, neutralizes a cluster of negative charges and enables membrane binding. Full activation of phospholipase A2 requires calcium binding to the C2 domain and phosphorylation of several serines (Hirabayashi et al. 2004). The calcium binding induces translocation of phospholipase A2α from the cytosol to the perinuclear membranes.

Annexins

Annexins are ubiquitous calcium- and phospholipid-binding proteins whose many inferred functions have yet to be confirmed. They consist of a core of four homologous annexin domains and of a highly variable N-terminal tail. The core is inferred to have evolved by two cycles of gene duplication and fusion; domains 1 and 3 and domains 2 and 4 resemble one another more closely. No examples of proteins having only a single or only a pair of annexin domains are known. Annexin A6 (VI, old notation), and only A6, has eight domains; it evolved by another cycle of gene duplication and fusion, (Fig. 4). Each annexin domain consists of five α-helices; A, B, D, and E form a right-handed, four helix, bundle. The four domains pack into a flattened trapezoid; the axes of the four, four helix bundles are perpendicular to the surface of the trapezoid. Annexin domains 4 and 1 are related to domains 2 and 3 by a local, twofold axis, also perpendicular to the plane of the trapezoid. It has a slightly convex surface on which the amino and carboxyl termini of each domain come into close apposition and a concave surface on which the (4 × 2 =) eight potential calcium-binding loops – AB and DE – are located. Ca^{2+} ions that bind to this concave surface are inferred to cross-link carbonyl and carboxyl groups of (some of) the eight AB and CD loops with phosphoryl groups of membrane phospholipids. Annexins have lower affinity for calcium than do most EF-hands; therefore, they are inferred to be modulated by changes in calcium concentration within

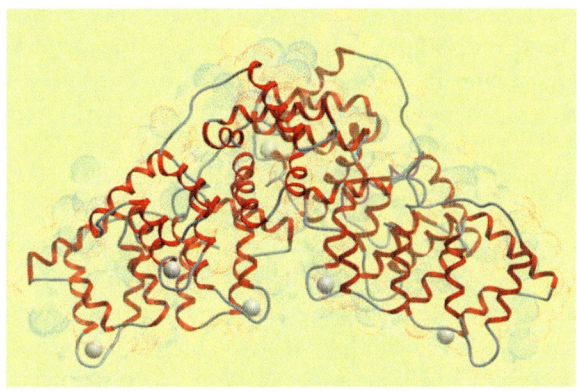

Calcium-Binding Proteins, Overview, Fig. 4 Structure of annexin A6 (PDB file 1AVC): Annexin A6 has eight domains; all other annexins have four. It evolved by gene duplication and fusion. The two flattened trapezoids (domains 1–4 and domains 5–8) are both seen approximately edge on; they are related by a twofold axis, vertical and in the plane of view. Domains 4 and 1 are related to domains 2 and 3 and domains 8 and 5 are related to domains 6 and 7 by twofold axes perpendicular to the planes of their respective trapezoids. The three Ca^{2+} ions in each trapezoid are indicated by *white spheres*; they are all on the inferred membrane binding, concave surface

limited regions of the cytosol. The convex side of annexin faces the cytoplasm and may interact with other proteins.

The N-terminal tails contain highly variable sequences 35–60 residues long; they are of low complexity and are inherently disordered; only portions of them are seen in crystal structures of annexins. The calcium-dependent membrane aggregation of annexins A1, A2, and A4, is strongly inhibited by phosphorylation of the N-tail; whereas, for annexin A7 aggregation is activated. The N-tail contains several putative Ser and Thr phosphorylation sites as well as consensus sequences for glycosylation and transglutamination. Annexin A1 also has several possible sites in the N-terminal domain for proteolysis; this profoundly modifies its physical and biological properties. The N-tails of the annexins sometimes fold back on this convex surface, reaching to the central pore in the trapezoid formed by packing of the four domains.

Many cytosolic proteins, including members of the EF-hand family, bind selectively to the N-terminal tails of annexins. Several S100 proteins and annexins interact in both calcium-dependent and calcium-independent manners, and form complexes that exhibit specific biological activities (Miwa et al. 2008).

Eukaryotes have ∼20 annexin genes; however, annexins are not found in fungi and prokaryotes. Most annexins are encoded by 12–15 exons, the variation depending in large part on the length of the N-terminal tails. For several annexins, particularly those with long N-tails, alternative splicing adds to the diversity of annexin isoforms; this may in turn amplify functional variability within the family as a whole. Any single cell type expresses a range of annexins, but no single annexin is expressed in all cells, implying that regulation of annexin gene expression is tightly controlled (Gerke et al 2005).

Possible Calcium-Modulated Proteins in Prokaryotes

Many prokaryotes have proteins that are inferred to bind calcium based on their amino acid sequences. However, the affinities or selectivities of these proteins for calcium have yet to be determined. There is only inferential evidence of calcium functioning as a second messenger in prokaryotes (Permyakov and Kretsinger 2009).

Extracytosolic Calcium-Binding Proteins

Since $[Ca^{2+}]_{out}$ is usually constant, calcium modulation of extracytosolic proteins is not (usually) possible. However, the functions of most of these proteins appears to be dependent their binding calcium.

Cell Matrix Proteins

Numerous proteins are associated with the very complex and varied extracellular matrices. The extracellular matrix is a network built up from a variety of proteins and proteoglycans. Its interaction with cells is mediated by cell adhesion molecules. Many matrix-matrix, cell-matrix, and cell-cell contacts involve calcium; however, the effective concentration and freedom of diffusion of calcium in these restricted volumes is not well understood. Most extracellular matrix proteins are multidomain, often chimeric proteins. The extracellular calcium-binding modules of known structure can be divided into two groups: the Ca^{2+} ion can either be bound to a single domain or can mediate interactions between independently folded domains in the same or in different proteins.

BM-40 (osteonectin or SPARC, Secreted Protein Acidic and Rich in Cysteine) is a small secreted glycoprotein, which is involved in the regulation of bone mineralization, tissue remodeling, and cell growth (Permyakov 2009). It is secreted by osteoblasts during bone formation, initiating mineralization and promoting crystal formation of the mineral. BM-40 also shows affinity for collagen in addition to bone mineral calcium. Its short N-terminal sequence, rich in glutamic acid residues, binds several Ca^{2+} ions with low affinity (dissociation constant 5–10 mM). The domain is followed by a domain homologous to follistatin and a C-terminal extracellular (EC) calcium-binding domain. The EC domain of BM-40 has a compact structure with high α-helical content and is stabilized by two disulfide bonds. The domain binds two Ca^{2+} ions through a canonical pair of EF-hands. One EF-hand of BM-40 is unusual: it has a one-residue insertion into the EF-hand consensus sequence; this is accommodated by a *cis*-peptide bond. Another unusual feature of the BM-40 EC domain is the presence of an amphiphilic α-helix that lies across the cleft formed by the helices of the EF-hand pair, as do target peptides in calmodulin.

Calcium-Binding Proteins Involved in Biomineralization

The main mineral components of bone and teeth are calcium and phosphate (of shells, calcium carbonate). Ninety per cent of the organic matrix of bone is collagenous and consists mainly of type I collagen, the remaining 10% consists of over 200 proteins of various sorts. The major bone matrix proteins, other than collagen, are osteocalcin, matrix GLA protein, osteonectin, proteoglycans, acidic glycoprotein 75, osteopontin, bone sialoprotein, fibronectin, vitronectin, and thrombospondin. These proteins bind calcium ions with $K_d(Ca^{2+})$ of about 3 mM and exhibit high affinity to microcrystals of hydroxyapatite, $Ca_5(PO_4)_3OH$, the basic bone mineral (Permyakov 2009).

Osteocalcin is the most abundant noncollagenous protein in bone. Its synthesis in osteoblasts is vitamin K dependent and its concentration in serum is closely linked to bone metabolism. Posttranslational modification by a vitamin K-dependent carboxylase produces three γ-carboxyglutamic acid (Gla) residues at positions 17, 21, and 24. Mature osteocalcin is largely unstructured in the absence of calcium and undergoes a transition to a folded, globular state at physiological concentrations of calcium. Three α-helices comprise a tightly packed core involving conserved hydrophobic residues Leu16, Leu32, Phe38, Ala41, Tyr42, Phe45, and Tyr46. All three Gla's implicated in hydroxyapatite binding are located on the same surface of helix 1 and, together with the conserved residue Asp30 from helix 2, coordinate five Ca^{2+} ions (Permyakov 2009; Permyakov and Kretsinger 2010).

Osteopontin is synthesized at elevated levels by not only osteoblasts, but also by kidney, vascular smooth muscle, and gall bladder cells. It is a potent inhibitor of hydroxyapatite formation. Osteopontin is phosphorylated; however, the role of its phosphate moieties in unclear. It might act as an inhibitor of crystal nucleation in physiological fluids at high super saturation of calcium phosphate. Osteopontin binds to integrins, which are expressed on osteoclasts and initiate bone resorption by mediating adhesion of the osteoclast to osteopontin in bone (Permyakov 2009).

Blood Clotting Proteins

Many enzymes involved in the blood clotting cascade have several Gla residues in an N-terminal Gla domain that specifically binds calcium. As with osteocalcin, their γ-carboxylation is vitamin K dependent. Prothrombin, factor X, factor IX, factor VII, protein C, and protein Z bind phospholipids in a calcium-dependent manner. The accumulation of these proteins on phospholipid surfaces requires a functional, calcium-loaded Gla domain. Factors VII, IX, and X, and protein C share the same domain architecture. The N-terminal Gla domain is followed by two domains homologous to the epidermal growth factor (EGF), and the C-terminal catalytic serine protease domain is homologous to trypsin. Factor X and protein C exist as heterodimers already in their zymogen forms, but factors VII and IX molecules dimerize on activation. The two chains are connected by a single disulfide bond. The Gla and EGF-like domains constitute the light chain; the protease domain is in the heavy chain. All Gla-containing plasma proteins are presumably involved in blood coagulation, some by promoting the process and others by attenuating it.

Thrombin is the central protease of the vertebrate blood coagulation cascade. It is derived from its inactive zymogen form, prothrombin, a 70-kD glycoprotein that is synthesized in the liver and secreted into the blood. Prothrombin, the most studied Gla containing protein, is composed of 579–582 residues, including ten Gla's, and three complex carbohydrate chains. Prothrombin has two Ca^{2+}-binding sites with a dissociation constant of 0.2 mM and several weaker sites. It has an N-terminal Gla-rich domain, two kringle domains, a short activation peptide, and a thrombin/catalytic/serine proteinase domain. During the activation of prothrombin to thrombin, the N-terminal peptide consisting of 156 residues is cleaved. This peptide, known as fragment-1, contains all ten Gla's and two of the carbohydrate chains of the intact prothrombin. Prothrombin binds to phospholipid dispersions in the presence of calcium and the activation of prothrombin to thrombin is accelerated in the protein-calcium-phospholipid complex. The phospholipid-binding site in prothrombin is located in its fragment-1 region (Permyakov 2009; Permyakov and Kretsinger 2010).

Coagulation factor VIIa consists of N- and C-terminal domains (residues 1–152 and 153–406). The N-terminal region is rich in Gla; it is followed by two EGF-like domains. The Gla and EGF-like domains are required for protein-protein and protein-membrane interactions responsible for complex formation between factor VII and tissue factor on the cell membrane. The C-domain includes the catalytic triad and shares the fold of chymotrypsin-like serine proteases. The function of the serine protease domain is to convert coagulation factors IX and X from zymogens to active enzymes (Permyakov 2009; Permyakov and Kretsinger 2010).

Salivary Protein A and Salivary Protein C

SPA consists of a single, proline-rich peptide chain of 106 residues; SPC contains the entire structure of SPA in its N-terminal part, but continues beyond residue 106 to the C-terminal residue 150. The proteins have a highly negatively charged N-terminus, which contains a total of 11 Gla's and two phosphoserines. Both salivary proteins A and C are found in parotid and submandibular saliva.

Epidermal Growth Factor (EGF)-like Domains

The EGF-like domain is one of the more widely distributed modules in extracellular proteins; it is characterized by six conserved cysteine residues organized in a characteristic pattern of disulfide bonds (1–3, 2–4, and 5–6) and two small β-sheets. EGF-like domains are found in a variety of proteins associated with various biological functions such as blood coagulation and fibrinolysis, activation of complement, cell adhesion and signaling, neurite outgrowth, formation of neuromuscular junctions, and involvement in basement membranes and connective tissue microfibrils. A subset of EGF-like domains contains a bipartite consensus sequence with a D/N-X-D/N-E/Q motif preceding the first cysteine and a X-D/N-X_n-Y/F-X motif between the third and fourth. This consensus contains a calcium-binding site. EGF-like domains play three functional roles: structural stabilization, protein-protein interactions, and spacing unit. High resolution structures of EGF-like domains have revealed a common fold, consisting of a major and a minor double-stranded β-hairpin, stabilized by the three consensus disulphide bonds. Each EGF-like domain contains a calcium-binding site, whose affinity is enhanced by N-terminal linkage to another domain.

Cadherins

Cadherins (calcium-dependent adhesion molecules) are cell adhesion proteins involved in establishing and maintaining intercellular connections and in controlling cell polarity and morphogenesis. They are transmembrane with large extracellular segments consisting of five repeated domains, each about 110 residues long; their adhesive properties reside in the N-terminal domain. Cadherin-mediated cell adhesion depends on the presence of extracellular calcium. Cadherins have conserved, repeat motifs in their extracellular domains, which are responsible for their calcium dependence. Classical cadherins, the first cadherins identified, are essential for cell-cell adhesion, serving to organize the adherens junction. Crystal structures of domains 1 and 2 of cadherin show that both domains are folded as independent modules made up entirely of β-sheets. The structure of the ten residue, interdomain linker is stabilized by three Ca^{2+} ions bound to a cluster of carboxylates contributed by both modules. The extracellular region of cadherins forms a rigid, rod-like structure only in the presence of calcium (Permyakov 2009).

Pentraxins

Pentraxins are a family of highly conserved, multimeric, pattern recognition proteins (Deban et al. 2009). They are divided into short and long classes. C reactive protein, the first pattern recognition receptor identified, and serum amyloid P component (SAP) are short pentraxins produced in the liver. Long pentraxins, including the prototype, PTX3, are expressed in a variety of cells and tissues, most notably dendritic cells and macrophages.

Through interaction with several ligands, including selected pathogens and apoptotic cells, pentraxins play a role in complement activation, pathogen recognition, and apoptotic cell clearance. In addition, PTX3 is involved in the deposition of extracellular matrix and in female fertility. SAP is the major DNA and chromatin-binding glycoprotein of plasma. The interaction is strictly calcium dependent. The physiological role of SAP is not yet defined. It is localized in the glomerular, basement membrane and in the peripheral, microfibrillar mantle of elastic fibers. Its binding to several substrates including oligosaccharides with terminal N-acetyl-galactosamine, 6-phosphate-mannose, glucuronic acid, and galactose is calcium dependent. In the crystal structure five identical subunits of SAP are arranged in a flat ring with a substantial hole in the center. The subunits are almost exclusively composed of β-sheets in a lectin-like fold. One face of the disk-shaped molecule contains the sites of interaction with ligands. An acidic functional group of large polyionic ligands bridges two Ca^{2+} ions; this accounts for the strict calcium dependence of its interactions.

Calcium-Binding Lectins

Lectins have no known enzymic activity but exhibit numerous biological activities that are related to their ability to bind carbohydrates in the presence of calcium. Lectins have specific binding sites for carbohydrates and thereby interact with specific cells, cell fractions, or glycoproteins. Proposed functions include promotion of symbiosis, involvement in cell recognition, and organization of supramolecular structures.

Concanavalin A

Con A is a tetramer composed of four identical polypeptide chains of 237 residues each. Each monomer can be schematically visualized as an ellipsoidal dome with a narrow, flat base. It has two calcium-binding sites and a carbohydrate-binding site; all near the apex of the dome. The tetramer structure appears to be required for its agglutination or precipitation activities. In the presence of calcium, Con A preferentially binds to α-D-mannopyranosyl, α-D-glucopyranosyl, and α-D-N-acetyl-glucosaminyl residues. The saccharide-binding site is located within 10–14 Å of the two calcium-binding sites. The residues that may participate in saccharide binding are found in close proximity in the three dimensional structure but are not clustered in the amino acid sequence. Most of the residues involved in calcium binding in Con A are clustered near its N-terminus. The two Ca^{2+} ions are bound in two adjoining octahedral sites that are 4.5 Å apart and have a common edge (Permyakov 2009; Permyakov and Kretsinger 2010).

Mannose-Binding Proteins

Mannose-binding proteins are members of the C-type lectin family; they are usually found in the serum and liver of mammals. Carbohydrate recognition domains of all C-type lectins have a similar core structure, which includes a cysteine-rich domain at the N-terminus followed by a collagenous domain, an oligomerization domain, and a C-terminal carbohydrate recognition domain. Monomers assemble into homotrimers, which further associate into larger oligomers. The mannose-binding proteins bind to a number of various monosaccharides containing vicinal, equatorial hydroxyl groups such as those found at the positions 3 and 4 of mannose, including N-acetylglucosamine and fucose. All C-type lectins have a calcium-binding site (site 2 in mannose-binding proteins) at which carbohydrates directly interact with the bound Ca^{2+} ion as well as with amino acids that serve as Ca^{2+} ligands. Mannose-binding proteins have three calcium-binding sites per molecule. The amino acid residues that form the binding site are highly conserved, especially a tripeptide sequence of two calcium ligands flanking a cis-proline residue (Permyakov 2009; Permyakov and Kretsinger 2010).

D-Galactose-Binding Protein

Sugar-binding proteins (molecular masses 25–45 kDa) are essential components of high affinity active transport systems for a large variety of carbohydrates, amino acids, and ions. They serve as initial receptors for the simple behavioral response of bacterial chemotaxis. There are remarkable similarities in both the tertiary structure and ligand-binding properties of the L-arabinose-, sulfate-, D-galactose-, Leu/Ile/Val- and leucine-specific binding proteins. All five proteins are ellipsoidal in shape (axial ratios about 2:1) and are composed of two similar yet distinct globular domains. The two domains in all of the structures are connected by three separate peptide segments. Although these interdomain connecting segments are widely separated in the primary structure, they are spatially nearby. The calcium-binding loop in D-galactose-binding protein adopts a conformation very similar to the loop in the

EF-hand, but it is not immediately preceded and followed by helices and is not homologous (Permyakov 2009; Permyakov and Kretsinger 2010).

Calcium-Binding Hydrolytic Enzymes

Many enzymes, which catalyze the hydrolysis of ester, phosphodiester, and peptide bonds, bind calcium. The binding of calcium to the enzymes might serve three functions. Ca^{2+} ions can stabilize an intermediate in the active site as in phospholipase A_2 and staphylococcal nuclease. Calcium stabilizes the enzyme at high temperatures as in thermolysin. It is also essential in the activation of zymogens as in trypsin, phospholipase A_2, and calpain.

Phospholipase A_2 (molecular mass 14 kDa) catalyzes the specific hydrolysis of the fatty acid ester bonds at the C2 position of 1,2-diacyl sn-phosphoglycerides. Both secreted and intracellular forms of phospholipase A_2 have been described. Phospholipases A_2 from mammalian pancreas as well as bee and snake venoms belong to the secreted forms; they are homologous monomers 118–129 residues long with particular conservation of the residues thought to be involved in calcium binding and in forming the active site. Calcium is an obligatory cofactor for interfacial catalysis by secreted phospholipase A_2 from virtually all sources. Calcium is coordinated near the active site by six ligands that form an octahedron. This relatively high affinity calcium-binding site is involved in the catalytic site of the enzyme. The binding of calcium affects His48, which play an essential role at the active site but is not directly involved in calcium binding. The mechanism for hydrolysis of phospholipids by phospholipase A_2 is similar to that for the hydrolysis of peptide bonds by serine proteases in which calcium stabilizes the tetrahedral intermediate. A low affinity calcium-binding site in phospholipase A_2 is located near its N-terminus and is postulated to be involved in micellar binding. The three-dimensional structure of phospholipase is characterized by a ring of amino acids surrounding the entrance to the active site; this ring structure has been proposed to be involved in micellar binding (Permyakov 2009; Permyakov and Kretsinger 2010).

Extracellular *Staphylococcus aureus* nuclease (molecular mass 16.8 kDa) requires millimolar concentrations of calcium for its activity. It is a monomer of 149 residues; it binds one Ca^{2+} ion. The calcium site is octahedral and consists of side chains and peptide carbonyl groups of Asp19, Asp21, Asp40, Glu43, and Thr41. Calcium coordination in Staphylococcal nuclease is similar to that in the high affinity phospholipase A_2 calcium-binding site. In this enzyme, calcium also takes part in stabilization of an intermediate during the reaction of hydrolysis of the phosphodiester bond according to the mechanism, which is similar to that proposed for phospholipase A_2. The active site of both enzymes is proximal to the calcium site (Permyakov 2009; Permyakov and Kretsinger 2010).

Thermolysin is the best characterized neutral protease produced by bacilli. Thermostable thermolysin like proteases are a group of metalloendopeptidases, which contain one catalytic Zn^{2+} ion and two to four Ca^{2+} ions that are important for stability. The N-terminal domain of thermolysin consists mainly of β-strands; the C-terminal domain is mainly α-helical. Ca^{2+} ions 1 and 2 (Ca1,2) are found in the C-terminal domain in the double calcium-binding site close to the active site zinc. Ca3 is located at the surface in the N-terminal domain, and Ca4 is bound by a surface-located ω-type loop in the C-terminal domain. Removal of calcium by chelators results in a partially unfolded, flexible molecule and in its rapid autolytic degradation. Ca3 and Ca4 are more important for thermal stability of thermolysin-like proteases than are C1,2. On the other hand, the double calcium site is so important that thermolysin cannot exist without this calcium site being occupied (Permyakov 2009; Permyakov and Kretsinger 2010).

Lipases

Lipases are water-soluble enzymes that hydrolyze ester bonds of water-insoluble substrates such as triglycerides, phospholipids, and cholesteryl esters. Lipases are versatile enzymes that have been isolated from a variety of eukaryotes and prokaryotes. They hydrolyze the ester bonds in long chain triacylglycerols. Although the overall similarity of lipases is low and molecular masses vary from 20 to 60 kDa, all lipases share a comparable three-dimensional fold. The amino terminal domain of pancreatic lipase consists of a series of nine β-sheets arranged in a fan-like pattern, termed an α/β hydrolase fold. The region of highest conservation is the active site, which contains a "classical" Ser-His-Asp catalytic triad. The amino terminal domain of pancreatic lipase has striking similarities to that found in bacterial and fungal lipases, which are single domain enzymes.

However, pancreatic lipase also has a separate, discrete carboxyl-terminal domain that is absent in the other lipases. A high affinity calcium-binding site with dissociation constant of 55 μM is found in lipase from *Staphylococcus hyicus*. The residual activity of the calcium-free enzyme compared to the activity of the calcium-loaded enzyme varies from 65% at 10°C to nearly zero at 40°C (Permyakov 2009; Permyakov and Kretsinger 2010).

Collagenase

Collagenase from the anaerobic spore-forming bacterium, *Histotoxic clostridia*, is responsible for the extensive tissue destruction of gas gangrene. The collagen-binding domain of *Clostridium histolyticum* class I collagenase also binds two Ca^{2+} ions between two loops and have limited solvent access. The coordination by oxygen atoms of Glu899, Glu901, Asp927, and Asp930 side chains, main chain carbonyl of Ser922, and one water is best described as a square antiprism. The water, Glu901, and Glu899 (bidentate) form one face while Ser922, Asp927, and Asp930 (bidentate) form the other face. The peptide bond between residues Glu901 and Asn902 adopts a *cis* conformation that is stabilized by calcium chelation. This dual ion structure is different from those in all other calcium-binding proteins. The apparent $K_d(Ca)$ of the domain is about 4 μM (Permyakov 2009; Permyakov and Kretsinger 2010).

α-Lactalbumin

α-Lactalbumin is homologous to lysozyme of eukaryotes. In spite of the fact that its calcium-binding domain is not an EF-hand, its structure, which includes a helix-loop-helix domain resembles an EF-hand domain. It is the modifier component of lactose synthase, which is synthesized in the lactating mammary gland. It complexes with galactosyl transferase, thereby altering the substrate specificity of the enzyme to favor glucose as the acceptor molecule. A single Ca^{2+} ion-binding site in α-lactalbumin consists of seven oxygens from carboxylates of Asp82, Asp87, and Asp88; carbonyls of Lys79 and Asp84; and two waters. The coordinating ligands form a slightly distorted pentagonal bipyramid. Ca-O distances are from 2.3 to 2.5 Å. The calcium-binding loop in α-lactalbumin is flanked by two helices. Conformation of the backbone chain in the calcium-binding region of α-lactalbumin is similar to the conformation of the corresponding region in hen egg lysozyme but the side chains in this region of α-lactalbumin and lysozyme are different. In contrast to the EF-hand, which has no disulfide bonds, bovine α-lactalbumin has four disulfide bonds, Cys6-Cys120, 28–111, 61–77, and 73–91. The removal of calcium has only minor effects on the structure of the metal-binding site and the largest structural change is observed in the cleft on the opposite side of the molecule. Tyr103 is shifted toward the interior of the cleft and water-mediated interactions with Gln54 and Asn56 replace the direct hydrogen bonds. These changes result in increased separation of the α and β domains, loss of a buried solvent molecule near the calcium-binding site, and the replacement of inter- and intra-lobe hydrogen bonds of Tyr103 by interactions with newly immobilized waters (Permyakov 2005).

Calcium-Buffering Proteins Within the Endoplasmic Reticulum (ER)

One of the primary functions of the ER is to store calcium either as a result of downregulation within the cytosol or in anticipation of upregulation. The ER is intracellular; however, the lumen of the ER is topologically extracytosolic. The relative advantage to the cell of using extracellular calcium versus ER calcium for cytosolic signaling is not obvious. Proteins that exit the endoplasmic reticulum are, for the most part, properly folded and assembled, owing to the coordinated activities of several folding enzymes, molecular chaperones, and a rigorous quality control system that retains and disposes of misfolded proteins. Some of the proteins that are involved in calcium sequestration are also involved in this protein-folding process.

Calsequestrin is found within the sarcoplasmic reticulum (SR) of skeletal and cardiac muscles; it contains no transmembrane segments and is therefore inferred to be located within the lumen of the SR. The total concentration of calcium in SR is as high as 50 mM, but a large portion of this calcium is bound to calsequestrin, which acts as a luminal buffering system; the concentration of free Ca^{2+} ion within the lumen can be maintained below the inhibitory level (~1.0 mM) of the calcium pump. For the fastest muscles, a limiting step in the contraction, relaxation cycle is pumping calcium into the SR. In this process, calsequestrin plays a key role through buffering the calcium levels within the lumen of the SR. It consists of three homologous domains, each with a thioredoxin fold, and a five strand β-sheet sandwiched by four α-helices. Each domain has a hydrophobic core with acidic residues on the exterior, forming electronegative surfaces. The connecting loops and the secondary structural elements that fill the interdomain spaces contain mostly acidic residues. Cations are required to stabilize the acidic center of calsequestrin. It is a glycoprotein with 30–50 calcium-binding sites with rather low affinity ($K_d(Ca^{2+}) \sim 10^{-3}$ M). Over 30% of its residues are Asp or Glu; its isoelectric point is 3.75. The binding of calcium makes calsequestrin insoluble. In the absence of calcium, calsequestrin is in near random coil conformation with α-helical content 11%. The binding of calcium increases its helical content up to 20% and changes its shape from elongated (Stokes radius, 45 Å) to much more compact (35 Å).

Calsequestrin has several functions in the lumen of the sarcoplasmic reticulum in addition to its well-recognized role as a calcium buffer. First, it is a luminal regulator of ryanodine receptor activity. In the presence of triadin and junctin, calsequestrin maximally inhibits the Ca^{2+} release channel when the free $[Ca^{2+}]$ in the lumen of the sarcoplasmic reticulum is 1 mM. This inhibition is relieved when the $[Ca^{2+}]$ changes, either because of small changes in the conformation of calsequestrin and/or its dissociation from the junctional face membrane. Calcium, but not magnesium, blocks the binding of calsequestrin to a 26 kDa protein of the junctional SR. It is inferred that calsequestrin is also involved in the regulation of calcium release from the ER (Beard et al. 2009).

Calreticulin and calnexin are the most intensively studied chaperones of the ER because of their unusual modes of substrate recognition, their intimate relationship with the Asn-linked glycosylation system, and the diversity of functions attributed to them. They are lectins that interact with newly synthesized glycoproteins that have undergone partial trimming of their core, N-linked oligosacharides. Simultaneously, they serve as molecular chaperones. Calreticulin, calnexin, and ERp57 (a glycoprotein-specific thiol-disulfide oxido-reductase) are components of the "calreticulin/calnexin cycle" that interacts with partially folded glycoproteins and determines whether the proteins are to be released from the endoplasmic reticulum or, alternatively, whether they are to be sent to the proteosome for degradation (Williams 2006). Accumulation of misfolded protein in the endoplasmic reticulum leads to activation of genes responsible for

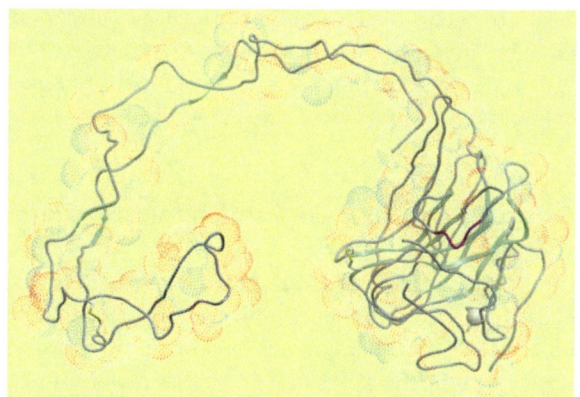

Calcium-Binding Proteins, Overview, Fig. 5 Structure of the luminal domain of calnexin (PDB file 1JHN): The 89 residue cytoplasmic tail of calnexin (65 kDa, 573 residues) (*left*) is phosphorylated and carries a C-terminal RKPRRE sequence that serves as a signal for endoplasmic reticulum localization. The luminal domain of calnexin binds substrates, (at least) four equivalents of calcium, and Mg-ATP

the expression of these chaperones. When accumulation of misfolded protein becomes toxic, apoptosis is triggered, possibly with membrane kinase, IRE1, involved in signaling via caspase-12 (Michalak et al. 2009). The C-domain of calreticulin is similar to the C-domain of calsequestrin. The C-domain terminates with the endoplasmic reticulum retrieval signal, KDEL.

Calnexin (65 kDa, 573 amino acid residues) is a non-glycosylated, transmembrane protein with an extracellular N-terminus and cytoplasmic C-terminus, so called type I. Its substrate-binding domain is located in the lumen of the ER. The luminal portion of calnexin has two domains: a globular β-sandwich that resembles legume lectins and an extended 140 Å arm consisting of two β-strands folded into a hairpin. Each β-strand is composed of four tandemly repeated, Pro-rich domains. The 89 residue cytoplasmic tail of calnexin is phosphorylated and carries a C-terminal RKPRRE sequence that serves as a signal for endoplasmic reticulum localization. The luminal domain of calnexin (Fig. 5) binds substrates, (at least) four equivalents of calcium, and Mg-ATP.

Possible Modulation of Extracellular Calcium-Binding Proteins

The concentration of free Ca^{2+} ions is usually held constant in the extracellular fluid, or plasma, of multicellular organisms. However, there is some evidence that $[Ca^{2+}]_{out}$ may vary in a controlled manner over restricted volumes and/or brief times. Calcium may be involved in extracellular signaling (Hofer 2005); these target proteins are considered to be calcium modulated.

Calcium Pump and Calcium Channel Proteins

Lipid bilayer membranes surround all cells and organelles, forming barriers that limit the free exchange of polar solutes. A wide variety of proteins responsible for controlling the diffusion or active transport of ions and nutrients are inserted into these membranes. Calcium is pumped out of the cytosols of both eukaryotic and prokaryotic cells and it is pumped out of the matrix of mitochondria and other plastids.

Calcium Pumps

The P-type ATPases, also known as E1-E2 ATPases, are a large group of evolutionarily related ion pumps that are found in bacteria, archaea, and eukaryotes. They are α-helical bundle, primary transporters; they all appear to interconvert between at least two different conformations, denoted by E1 and E2. They are encoded by five main gene families (I, II, III, IV, and V). Those that share specificity for Ca^{2+}, K^+, and Na^+ ions group together in a single clade and are designated as P-Type II ATPases. They include five subfamilies: A, B, C, D, and E; also known as SERCA (sarco/endoplasmic reticulum Ca^{2+}-ATPase), PMCA (plasma membrane Ca^{2+} ATPase), NK/HK (Na^+,K^+- and H^+, K^+-ATPase), ENA (P-type ATPases which are able to extrude Na^+, Li^+, and K^+; and ACU (P-type ATPases which mediate high affinity Na^+ and K^+ ion uptake and are encoded by ACU genes in fungi).

The central event in the activity of this family of P-type ATPases is the formation of an acid stable aspartyl phosphate intermediate (Inesi et al. 2008). This event is initiated by cooperative binding of two cytoplasmic Ca^{2+} ions to transport sites. The energy of this intermediate is used to induce a conformational change that closes the ion gate from the cytoplasm, reduces the affinity of these transport sites for Ca^{2+} ions, and opens the ion gate toward the lumenal, or extracellular, side of the membrane. After releasing calcium, protons bind to the transport sites and the aspartyl phosphate is hydrolyzed to complete the cycle.

The P-ATPase consists of four basic domains (Fig. 6). The transmembrane domain is almost entirely

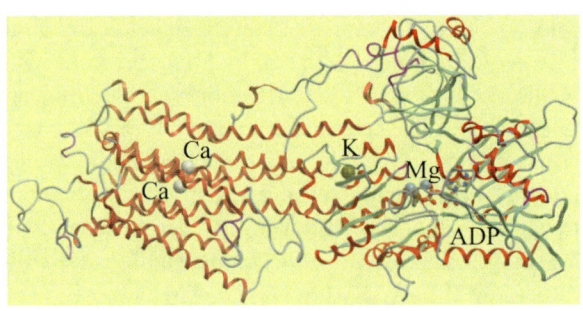

Calcium-Binding Proteins, Overview, Fig. 6 Structure of sarcoplasmic reticulum Ca^{2+}-ATPase in the Ca_2-E_1-ADP state (PDB file 1T5T): ADP is bound in the N domain, to the *right*. The transmembrane domain (helices M1–M10) is to the *left*; the Ca^{2+} ions indicate the channel. The transmembrane domain is almost entirely helical and has short loops on the luminal and cytoplasmic surfaces. Four of the transmembrane helices extend into the cytoplasm to form a stalk

helical (ten helices M1–M10) and has short loops on the lumenal and cytoplasmic surfaces. Four of the transmembrane helices extend into the cytoplasm to form a stalk. The three cytoplasmic domains are built from two large cytoplasmic loops between transmembrane helices M2/M3 and M4/M5. The M4/M5 loop forms the phosphorylation (P) domain and the nucleotide-binding (N) domain, which is inserted within the P domain. The third cytoplasmic loop, forming the transduction or anchor (A) domain, includes the smaller M2/M3 loop as well as the N terminus.

The Ca^{2+}-ATPase pump has to discriminate between Ca^{2+} and Na^+, K^+, and Mg^{2+} ions (Gouaux and Mackinnon 2005). An obvious difference between the Ca^{2+} ion-binding sites and the N^+ and K^+ sites is the greater importance of negatively charged oxygen atoms contributed by Glu and Asp side chains for calcium coordination. A higher charge density is apparently required to compensate for the dehydration of a divalent cation. Part of the selectivity for the Ca^{2+} ion derives from its being seven coordinate, as opposed to six for the Mg^{2+} ion.

Phospholamban, a 52-residue protein spanning the sarcoplasmic reticulum membrane, is an endogenous inhibitor of SERCA ATPase, lowering the apparent calcium affinity of the ATPase.

Calcium Channel Proteins

Calcium channels, like pumps, are located in both cell membranes and organelle membranes of eukaryotes.

Although there is strong evidence that calcium is extruded from the cytosols of bacteria, there have yet to be established channels permitting inflow of calcium from the surrounding medium. Calcium influx in many types of cells is regulated primarily through two types of plasma membrane channels: voltage-dependent calcium channels (VDCCs) and receptor-operated (ligand gated) channels (ROCs). VDCCs transduce changes in transmembrane potential into local cytosolic calcium transients that regulate enzyme activation, gene expression, neurotransmission, and neurite outgrowth or retraction. They are composed of an α1 (190 kDa) 1–10 subunit forming the Ca^{2+} ion-selective channel, and several accessory subunits, α2δ, β1–4, and γ with anchorage and regulatory functions. The α1 subunit contains four repeated domains (I–IV), each of which contains six transmembrane segments (S1–S6) and a membrane associated loop between transmembrane segments S5 and S6. Based on their electrophysiological and pharmacological properties and the type of α1 subunit, VDCCs are divided into five classes: Cav1.1–Cav1.4 (L-type), Cav2.1 (P/Q-type), Cav2.2 (N-type), Cav2.3 (R-type), and Cav3.1–3.3 (T-type). Low resolution structures have been developed from image reconstruction analysis of CaV1.1 channels purified from skeletal muscle membranes; however, no high resolution structures have been determined.

ROCs open in response to the binding of specific ligands, such as neurotransmitters, to the extracellular domain of the receptor. This interaction causes a change in the structure of the protein that leads to the opening of the channel pore and subsequent ion flux across the plasma membrane. Most ROCs are permeable to Ca^{2+} ions and represent an important mechanism for the generation of second messengers. Examples of ROCs include the glutamate (N-methyl-D-aspartate), α-amino-3-hydroxy-5-methylisoxazole-4-propionate acid (AMPA), kainite (KARs), nicotinicacetylcholine receptors (nACh), serotonin(5-HT3), and adenosine 5′-triphosphate (ATP) P2X receptors.

Summary

Magnesium and zinc, at physiological concentration, sometimes compete for calcium-binding sites. Given the precedents of known high affinity, high specificity calcium-binding sites, for instance, in the EF-hand proteins, one can safely assume that Nature could have evolved, in other proteins, high specificity calcium-binding sites. These observations pose two challenges: first to determine whether the competing cation is binding to the calcium-binding site or to a different site, second to determine whether the binding of the non-calcium cation exerts some sort of modulatory function on the calcium-binding protein. These possible physiological functions, cytosolic or extracellular, might occur under normal or pathologic conditions.

Cross-References

▶ Annexins
▶ Blood Clotting Proteins
▶ C2 Domain Proteins
▶ Cadherins
▶ Calcium ATPase
▶ Calmodulin
▶ Calnexin and Calreticulin
▶ Calsequestrin
▶ EF-Hand Proteins
▶ Lipases
▶ Sodium/Potassium-ATPase Structure and Function, Overview
▶ Thermolysin
▶ α-Lactalbumin

References

Beard NA, Wei L, Dulhunty AF (2009) Ca^{2+} signaling in striated muscle: the elusive roles of triadin, junctin, and calsequestrin. Eur Biophys J 39:27–36

Berridge MJ (2006) Calcium microdomains: organization and function. Cell Calcium 40:405–412

Corbin JA, Evans JH, Landgraf KE, Falke JJ (2007) Mechanism of specific membrane targeting by C2 domains: localized pools of target lipids enhance Ca^{2+} affinity. Biochemistry 46:4322–4336

Deban L, Bottazzi B, Garlanda C, de la Torre YM, Mantovani A (2009) Pentraxins: multifunctional proteins at the interface of innate immunity and inflammation. Biofactors 35:138–145

Gerke V, Creutz CE, Moss SE (2005) Annexins: linking Ca^{2+} signaling to membrane dynamics. Nat Rev Mol Cell Biol 6:449–461

Gouaux E, Mackinnon R (2005) Principles of selective ion transport in channels and pumps. Science 310(5753):1461–1465

Hirabayashi T, Murayama T, Shimizu T (2004) Regulatory mechanism and physiological role of cytosolic phospholipase A_2. Biol Pharm Bull 27(8):1168–1173

Hofer AM (2005) Another dimension to calcium signaling: a look at extracellular calcium. J Cell Sci 118:855–862

Inesi G, Prasad AM, Pilankatta R (2008) The Ca^{2+} ATPase of cardiac sarcoplasmic reticulum: physiological role and relevance to diseases. Biochem Biophys Res Commun 369(1):182–187

Michalak M, Groenendyk J, Szabo E, Gold LI, Opas M (2009) Calreticulin, a multi-process calcium-buffering chaperone of the endoplasmic reticulum. Biochem J 417(3):651–666

Miwa N, Uebi T, Kawamura S (2008) S100-annexin complexes – biology of conditional association. FEBS J 275(20): 4945–4955

Permyakov EA (2005) α-Lactalbumin. Nova Science, New York

Permyakov EA (2006) Parvalbumin. Nova Science, New York

Permyakov EA (2009) Metalloproteomics. Wiley, Hoboken

Permyakov EA, Kretsinger RH (2009) Cell signaling, beyond cytosolic calcium in eukaryotes. J Inorg Biochem 103:77–86

Permyakov EA, Kretsinger RH (2010) Calcium binding proteins. Wiley, Hoboken

Südhof TC (2002) Synaptotagmins: why so many? J Biol Chem 277:7629–7632

Williams DB (2006) Beyond lectins: the calnexin/calreticulin chaperone system of the endoplasmic reticulum. J Cell Sci 119:615–623

Calcium-Binding Region

▶ C2 Domain Proteins

Calcium-Dependent Cell–Cell Adhesion Proteins

▶ Cadherins

Calcium-Induced Calcium Release (CICR)

▶ Calcium Sparklets and Waves

Calcium-Modulated Potassium Channels

▶ Barium Binding to EF-Hand Proteins and Potassium Channels

Calcium-Regulated Photoproteins

Eugene S. Vysotski[1] and John Lee[2]
[1]Photobiology Laboratory, Institute of Biophysics Russian Academy of Sciences, Siberian Branch, Krasnoyarsk, Russia
[2]Department of Biochemistry and Molecular Biology, University of Georgia, Athens, GA, USA

Synonyms

Calcium-activated photoproteins

Definition

Ca^{2+}-regulated photoprotein is a "precharged" bioluminescent protein from which light emission is triggered by addition of calcium ions. The bioluminescence reaction does not require addition of molecular oxygen or any other cofactors, only the photoprotein and the triggering ion are necessary.

Background

Bioluminescence is widespread in the biosphere. Luminous organisms have been found among bacteria, fungi, protozoa, coelenterates, worms, mollusks, insects, and fish. Although these organisms occupy different places in the evolutionary ladder, the nature of their bioluminescence is always the same. In fact, bioluminescence is a chemiluminescent reaction whereby oxidation of a substrate, luciferin, is catalyzed by a specific enzyme, luciferase. Luciferins and luciferases of different organisms differ in structure; that is, the terms are generic and functional rather than structural and chemical. The many differences suggest that bioluminescence independently arose many times during evolution. Despite the more than 100-year history of studies on bioluminescence, the origin and functional advantage of bioluminescence in most organisms remain obscure.

Calcium-regulated photoproteins are "precharged" bioluminescent proteins that are triggered to emit light by binding Ca^{2+} or certain other inorganic ions. The reaction does not require the presence of molecular oxygen or of any other cofactors – the photoprotein and the triggering ion are the only components required for light emission. Since the energy emitted as light is derived from the "charged" photoprotein, that molecule can react only once, that is, it does not "turn over" as an enzyme does. In this respect, as well as in the lack of a requirement for molecular oxygen or any other cofactor, the reaction is strikingly different from that of classical bioluminescent systems in which an enzyme (luciferase) catalyzes the oxidation of a smaller organic substrate molecule (luciferin) with the creation of an excited state and the emission of light. This difference prompted Shimomura and Johnson (Shimomura 2006) to coin the term "photoprotein" to describe proteins that serve as the sole organic molecular species in bioluminescent reaction systems. Though other kinds of photoproteins have been described, the great majority of photoproteins now known to exist are stimulated to luminescence by calcium, and the term "calcium-activated photoproteins" was applied to them by Hastings and Morin (1969). Later, the term "calcium-regulated photoproteins" was suggested to refer to this group, first, because these proteins are members of the family of calcium-regulated effector proteins such as calmodulin and troponin C and, second, because calcium regulates the function of these proteins but is not essential for it.

Occurrence and General Features of Calcium-Regulated Photoproteins

All of the calcium-regulated photoproteins that have been discovered so far have been isolated from luminescent marine coelenterates. A great many luminescent coelenterates are known; calcium-triggered luminescence occurs in more than 25 of these and may eventually be found in all of them. Nevertheless, even at present, only a handful of photoproteins have been isolated and studied. These are aequorin, halistaurin (mitrocomin), and phialidin (clytin) from hydromedusae *Aequorea*, *Halistaura* (*Mitrocoma*), and *Phialidium* (*Clytia*), respectively, obelin from hydroids *Obelia longissima* and *Obelia geniculata*, and mnemiopsin and beroin from ctenophores *Mnemiopsis* and *Beroe* (Shimomura 2006).

The calcium-regulated photoprotein is a complex of a single-chain polypeptide with molecular mass ~22 kDa and a peroxy-substituted coelenterazine (2-hydroperoxycoelenterazine), which is tightly, though noncovalently, bound with the polypeptide (Fig. 1). Bioluminescence initiated by Ca^{2+} results from oxidative decarboxylation of the 2-hydroperoxycoelenterazine. The reaction yields an excited molecule of coelenteramide and CO_2. The transition of coelenteramide from the excited into the ground state is accompanied by light emission.

The bioluminescence spectral maxima of photoproteins are in the range 465–495 nm, the wavelength varying with source organism. For instance, the bioluminescence maximum is at 465 nm in the case of aequorin and 495 nm for obelin from *O. geniculata*. Photoproteins exhibit little fluorescence, but after the bioluminescence reaction, the bound product coelenteramide is brightly fluorescent (Shimomura 2006). With aequorin, for example, the fluorescence spectrum of the discharged protein coincides with the bioluminescence, whereas in case of obelin, the fluorescence is shifted to 25-nm longer wavelength than the bioluminescence (Vysotski and Lee 2004).

Structure of Calcium-Regulated Photoproteins and Mechanism of the Ca^{2+} Trigger

During the past 25 years, cloning and sequence analysis have been achieved for cDNAs coding for five Ca^{2+}-regulated photoproteins: aequorin, phialidin (clytin), halistaurin (mitrocomin), and two obelins from *O. longissima* and *O. geniculata* (Markova et al. 2002). The Ca^{2+}-regulated photoproteins show a high degree of sequence identity (76–63%) and contain three canonic EF-hand Ca^{2+}-binding sites. This places these photoproteins into the EF-hand calcium-binding protein family (Kawasaki et al. 1998), one containing the most numerous and extensively studied members of all the protein families. However, the degree of identity of photoproteins with other calcium-binding proteins, for example, sarcoplasmic calcium-binding protein, caltractin, calmodulin, troponin C, and even calcium-dependent coelenterazine-binding protein from *Renilla*, is significantly low, not more than 25%. The Ca^{2+}-regulated photoproteins are distinctive also in having a primary sequence with many tryptophan, cysteine, and histidine residues, which are not commonly found in other calcium-binding proteins.

Calcium-Regulated Photoproteins, Fig. 1 Coelenterazine (**a**), 2-hydroperoxycoelenterazine, oxygen preactivated coelenterazine bound within photoprotein molecule (**b**), oxidation product of coelenterazine, coelenteramide (**c**)

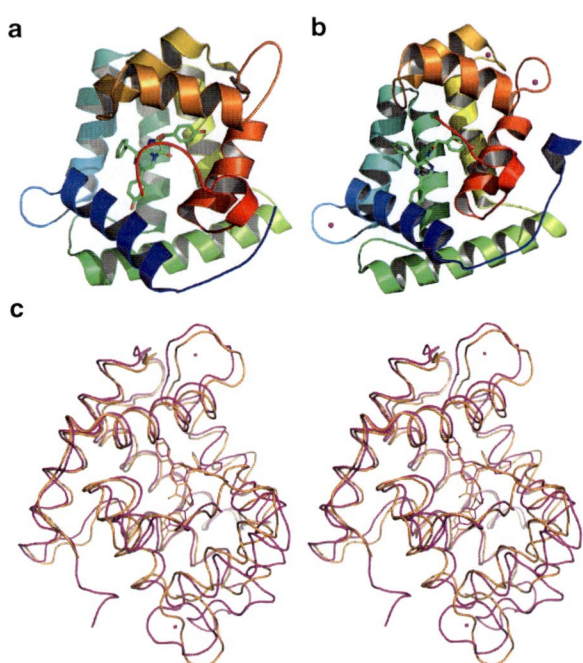

Calcium-Regulated Photoproteins, Fig. 2 The three-dimensional structures of obelin (**a**) (PDB code 1QV0) and Ca^{2+}-discharged obelin (**b**) (PDB code 2F8P). (**c**) Stereoview of the superimposition of obelin (*brown*) and Ca^{2+}-discharged obelin (*pink*). The 2-hydroperoxycoelenterazine and coelenteramide molecules are displayed by the stick models in the center of the protein structure; the calcium ions are shown as balls. The 2-hydroperoxycoelenterazine, coelenteramide, and calcium are colored according to the conformation state

The spatial structures of two photoproteins, obelin and aequorin, were determined in 2000 (Vysotski and Lee 2004; Shimomura 2006). As expected from the identity of their primary sequences, these and all subsequent photoproteins determined have the same compact globular structure (Fig. 2a). The spatial structure is formed by two sets of the four α-helices in the N- and C-terminal domains. Each domain can be thought of as a "cup" whose insides are lined with hydrophobic residues. The overall structure of photoprotein can then be considered as two "cups" joined at their rims. The 2-hydroperoxycoelenterazine resides in an internal cavity, which is surrounded by hydrophobic residues from the eight helices.

The three-dimensional structure of Ca^{2+}-discharged photoprotein (Fig. 2b) bound with the product of the bioluminescent reaction, coelenteramide, and calcium ions has been determined only for the case of obelin. The overall scaffold is retained and is the same as before bioluminescence discharge (Fig. 2). The RMS deviation from the C-α-atomic positions of Ca^{2+}-discharged obelin versus obelin is only 1.37 Å, which shows the well-conserved structural features between these protein states, the one primed with the 2-hydroperoxycoelenterazine and the other with the bound reaction product, coelenteramide. The coelenteramide is buried in a highly hydrophobic cavity situated at the center of the protein structure in the same place as its precursor 2-hydroperoxycoelenterazine, again surrounded by residues from each of the eight helices of the protein. This solvent-inaccessible cavity apparently provides the necessary environment for efficient generation of the product in the excited state and for its efficient fluorescence.

Since the photoprotein structure hardly changes following its reaction, it indicates that in the EF-hand protein family, the Ca^{2+}-regulated photoproteins belong to the category of Ca^{2+} signal modulators, such as a parvalbumin, rather than to the Ca^{2+} sensors, of which calmodulin is typical and the best-known representative (Nelson and Chazin 1998). This categorization probably has to do with function where, to be a sensor, protein-protein association is involved, whereas the photoproteins have to operate on the millisecond timescale, enabling just a subtle shift to disturb the hydrogen bond network that triggers bioluminescence.

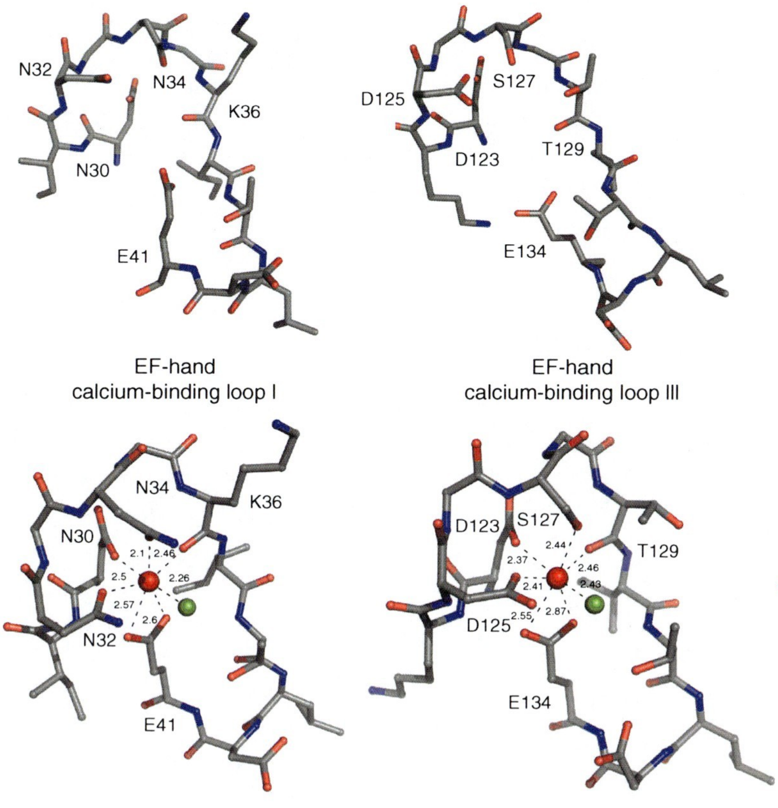

Calcium-Regulated Photoproteins, Fig. 3 Obelin EF-hand Ca^{2+}-binding loops I, III, and IV before bioluminescence reaction (*upper panel*) and after calcium binding (*bottom panel*). The calcium ion and the water molecule are shown as *red* and *green* balls, respectively

A calcium ion is found at each of the expected Ca^{2+}-binding sites, EF-hand loops I, III, and IV. For binding Ca^{2+} into the EF-hand, the 12 residues of the loop shift their positions to accommodate the calcium ion in its preferred configuration (Fig. 3). The typical geometrical arrangement of oxygen atoms in a pentagonal bipyramid is observed with the Ca^{2+} occupying the center of the pyramid. All three Ca^{2+}-binding sites of the photoprotein contribute six oxygen ligands to the metal ion, derived from the carboxylic side groups of Asp and Glu residues, the carbonyl groups of the peptide backbone, or the side chain of Asn, and the hydroxyl group of Ser, all with a coordination distance of ≈ 2.4 Å. The seventh ligand comes from the oxygen of a water molecule. Based on studies of the EF-hand Ca^{2+}-binding proteins from different sources, it has been observed that high-affinity Ca^{2+}-binding sites have either no water or at most one water ligand (Strynadka and James 1989). The three Ca^{2+}-binding sites of the Ca^{2+}-discharged obelin each contain only one water molecule as a ligand, suggesting that they all should have high affinity for calcium.

The residues comprising the binding cavity, that is, within 4 Å of the coelenteramide, after the bioluminescence discharge, essentially remain in place compared to their location in the obelin cavity, except Ile142, Ile111, Phe119, Ile142, Ile144, Trp135, and Tyr138, which are displaced, and Phe28, Gly143, and Thr172, which move into the cavity. As in obelin, there are two water molecules in the cavity of the discharged protein, but they are repositioned.

Aside from the loss of CO_2 from the chemical decomposition of coelenterazine, the biggest change in its molecular structure is in the reaction center around the C-2 position (C-2, O-33, and C-10) and C-8, resulting in an obvious deviation of the orientation of the phenol group at the C-10 position and the phenyl group at the C-8 (Fig. 4). Other parts of the

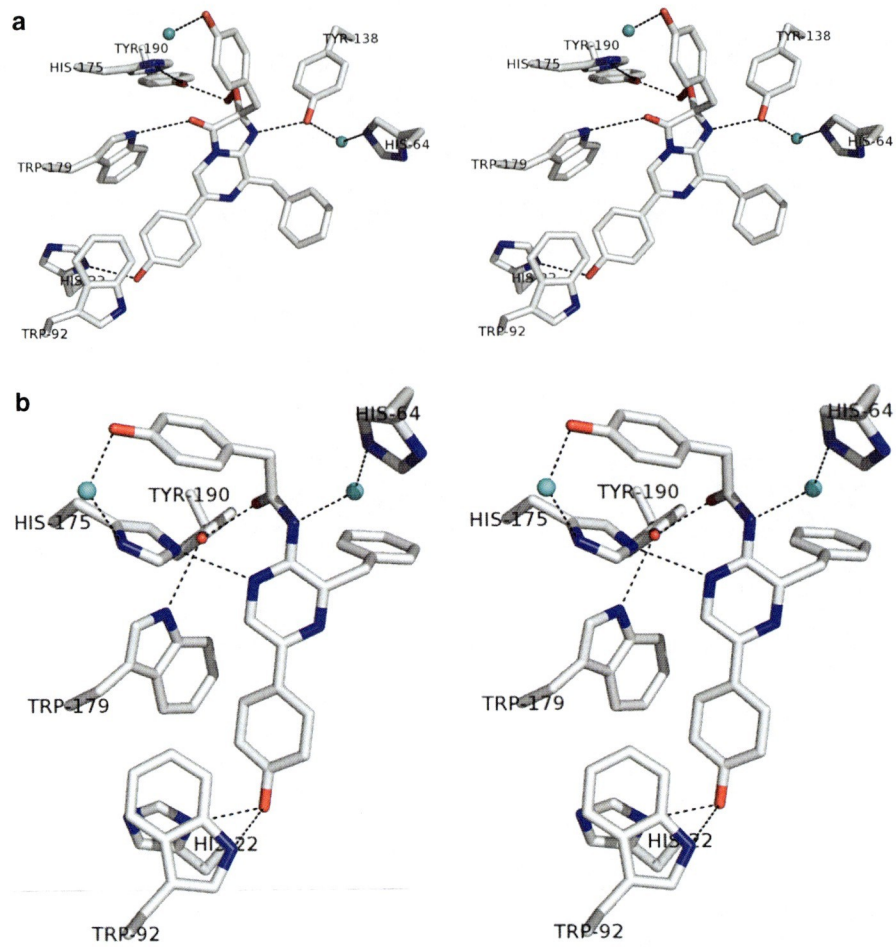

Calcium-Regulated Photoproteins, Fig. 4 Stereoview of the substrate-binding cavity with key residues of obelin before (**a**) (PDB code 1QV0) and after bioluminescent reaction (**b**) (PDB code 2F8P). Hydrogen bonds and the water molecules are shown with *dashed lines* and as *cyan balls*, respectively

molecule also adjust positions a little but not dramatically. His22 and Trp92, which were in hydrogen bond distances with the oxygen of the 6-(*p*-hydroxyphenyl) group of coelenterazine before reaction (Fig. 5a), are at practically the same distances to coelenteramide in the product cavity (Fig. 5b). In obelin, the Nε atom of Trp179 is 3.32 Å from the C3-carbonyl oxygen, but after reaction, this residue is moved in the direction of Tyr190 with formation of a new hydrogen bond to it. The Tyr190 OH group, which apparently stabilizes the 2-hydroperoxy group of coelenterazine in obelin by a hydrogen bond and is also hydrogen bonded to His175, is also slightly repositioned in the Ca^{2+}-discharged protein. The hydrogen bond to His175 is lost, but a new hydrogen bond develops to the carbonyl of coelenteramide. There is a major reorientation of the His175 with its imidazole ring now almost perpendicular to the original orientation (Vysotski and Lee 2007). His175 also forms new hydrogen bonds with the N1 atom of coelenteramide and the water molecule W_1 (Figs. 4b and 5b). Tyr138 is moved out of the binding cavity (Fig. 4a, b). The hydrogen bond from Tyr138 originally to the N1 of coelenterazine (Fig. 5a) now goes to Glu55 (Fig. 5b), and the Tyr138 appears to be replaced by that water molecule (W_2) originally connecting Tyr138 to His64 (Fig. 5a). As a result, the His64 is also slightly shifted toward coelenteramide. The second water molecule (W_1) is also moved, apparently after the change in position of the 2-(*p*-hydroxybenzyl) group. Thus, only His175 and Tyr138 undergo a noticeable repositioning in the binding cavity after the bioluminescence reaction.

In any protein crystal structure, hydrogen bonds are inferred if the separation of a putative H-donor and acceptor is less than about 3 Å, and in Fig. 5a, these inferred H-bonds are indicated by the dotted lines between the donor atom and the acceptor. The 2-hydroperoxycoelenterazine is not stable in free

Calcium-Regulated Photoproteins, Fig. 5 Two-dimensional representation to illustrate the hydrogen bond (*dashed lines*) network in the binding cavities of obelin (**a**) and Ca^{2+}-discharged obelin (**b**)

solution but in the hydrophobic environment of the binding site, the hydroperoxide group appears to be stabilized by the H-bond to Tyr190. In turn, there is an H-bond from Tyr190 to His175. This same arrangement is seen in other Ca^{2+}-regulated photoproteins: aequorin (PDB code 1EJ3), obelin from *O. geniculata* (PDB code 1JF0), and clytin (PDB code 3KPX).

The H-bond distance between Tyr190 and His175 is 2.72 Å (Fig. 5a), which is indicative of an H-bond with moderately strong electrostatic character. The hypothetical mechanism of the Ca^{2+} trigger (Vysotski and Lee 2004, 2007) is that as a direct result of Ca^{2+} binding, the H-bond between Tyr190 and His175 becomes stronger, increasing the electrostatic contribution, being equivalent to saying that the His175 becomes partially protonated. Because the tyrosine and hydroperoxide have similar pKs around 10, there will be a probability that the hydroperoxide will protonate the tyrosinate, and the peroxy anion then has another probability of irreversible nucleophilic addition to the C3-carbon of coelenterazine to form the committed dioxetanone intermediate. The exergonicity of this last step provides the thermodynamic feasibility of the overall process.

In the family of EF-hand calcium-binding proteins, the bound calcium ion is found associated specifically in a consensus sequence in the loop region of the helix-turn-helix motifs. In practice, the identification of the bound Ca^{2+} is by its strong electron density and, as well in almost all cases, by a bipyramidal pentagonal coordination with a bond length close to 2.4 Å, between the central atom and the coordination partner (Strynadka and James 1989; Nelson and Chazin 1998). The recent crystal structures of Ca^{2+}-discharged obelin, apoobelin, and apoaequorin bound with calcium, indicate adherence to these average bond length specifications.

In the published spatial structures of photoproteins, the loop structures are not prepositioned for calcium binding; that is, especially in the C-terminal loop III and IV, some movement of the residues must occur on Ca^{2+} binding to happily accommodate the coordinating atoms to the required 2.4 Å separation. In photoproteins, the first step in the generation of high-intensity bioluminescence must obviously be the binding of Ca^{2+} to the loops within the EF-hands. The Tyr138, His175, and Trp179 within the exiting helices of loops III and IV as well as Tyr190 have critical proximity to the substrate in the reactive center. Therefore, any conformational adjustment in the binding loops accompanying Ca^{2+} binding (or even spontaneous motions of the residues) can be expected to propagate into shifts of the hydrogen bond donor-acceptor separations around the coelenterazine, the ones apparently essential for the hydroperoxide stability, the networks O34, Tyr190, His175, O18, and N1 to Tyr138. Because the pKs of the tyrosine hydroxyl and the hydroperoxide are very close, as already noted, and the position of His175 is poised to act as a general base, the destabilization of the substrate is thereby triggered.

To initiate the shift of hydrogen donor-acceptor separations, the small spatial shift of the exiting α-helix of loop IV will be enough, since most of the residues mentioned above, including His175 which is the key residue, are found in this α-helix. The notion that a His residue in this position is very important for photoprotein activity is supported by substitution of this residue in aequorin to Ala, Phe, or Trp which leads to complete loss of activity. The suggestion that His175 can be the key residue for Ca^{2+} triggering of bioluminescence is also supported by the finding that in the crystal structure of Ca^{2+}-discharged obelin, the imidazole ring of this residue changes orientation to become almost perpendicular to its initial state. The process of hydrogen donor-acceptor separation may be fast and will be irreversible because it initiates the chemical reaction of coelenterazine decarboxylation, and its rate would then be independent of calcium concentration.

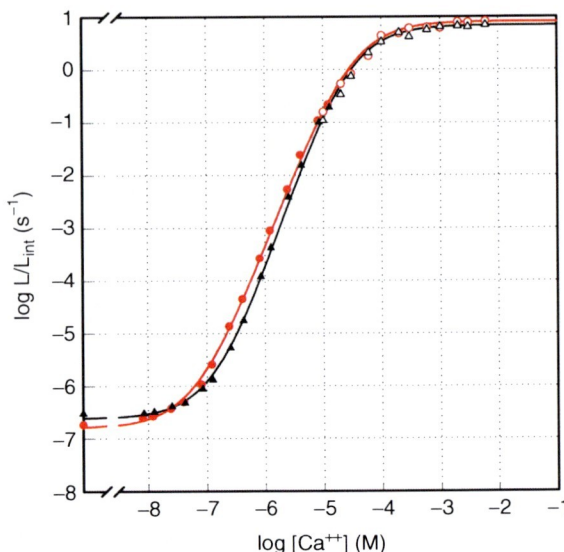

Calcium-Regulated Photoproteins, Fig. 6 Ca^{2+} concentration-effect curves for recombinant obelin from *O. geniculata* without (*red circles*) and with (*black triangles*) 1 mM Mg^{2+}, pH 7.0, 20°C. Symbols: *filled*, Ca-EGTA buffers; *open*, dilutions of $CaCl_2$

Applications of Calcium-Regulated Photoproteins

The mainstream applications of Ca^{2+}-regulated photoproteins take advantage of their inherent property to emit light on calcium binding. Owing to this property and because photoproteins are highly sensitive for detecting calcium and harmless when injected into living cells, they have been widely used as probes of cellular Ca^{2+}, both to estimate the intracellular Ca^{2+} concentration under steady-state conditions and to study the role of calcium transients in the regulation of cellular function. To estimate $[Ca^{2+}]$, purified natural photoproteins were delivered into the cell via labor-intensive procedures, such as microinjection or liposome-mediated transfer. Aequorin was more available and consequently has come to be commonly employed in measuring intracellular $[Ca^{2+}]$ notwithstanding its drawbacks.

Figure 6 shows Ca^{2+} concentration-effect curves for the recombinant photoprotein from *O. geniculata*, determined under conditions of pH and ionic strength likely to be encountered inside cells, in the absence and presence of 1 mM Mg^{2+}. The curves are log-log plots in which light intensities have been expressed in terms of a ratio that we have termed the fractional rate of discharge (L/L_{int}) (Markova et al. 2002). Under physiological conditions, the level of photoprotein luminescence rises rapidly as $[Ca^{2+}]$ is increased over the range 0.1–100 μM. The curve spans more than a million-fold range of light intensities. Therefore, in order to use photoprotein in the lower part of its range of sensitivity, one must pay close attention to maximizing the recorded signal. There is a natural limit to the extent to which one can reduce the threshold of detectability for $[Ca^{2+}]$ however, and that is imposed by the fact that the Ca^{2+} concentration-effect curve flattens out at very low $[Ca^{2+}]$. That is, there is a very low level of Ca^{2+}-independent luminescence. This fact has the obvious practical consequence that, under physiological conditions (Fig. 6), it is virtually impossible to detect calcium below about 10^{-8} M by means of wild-type photoprotein. From the theoretical point of view, it means that calcium ion should not be viewed as an indispensible ingredient in the chemical reaction leading to luminescence but as a factor that greatly accelerates the reaction (Blinks et al. 1982).

Another highly significant feature of the Ca^{2+} concentration-effect curve is the steepness of its midportion. The maximum slope has been found to be approximately 2.5 on log-log plots such as those of Fig. 6. This feature also has important implications from both theoretical and practical standpoints.

The most important theoretical implication is that more than two (i.e., at least three) calcium ions must interact with each photoprotein molecule in the control of the luminescent reaction (Blinks et al. 1982). This early proposition was subsequently confirmed by the spatial structure of Ca^{2+}-discharged photoprotein because calcium ion has been found in each Ca^{2+}-binding site of the photoprotein (Vysotski and Lee 2007). From the practical standpoint, the steepness of the curve means that changes in light intensity will give an exaggerated impression of the changes in $[Ca^{2+}]$ responsible for them. It should also be obvious from the Ca^{2+} concentration-effect curve that the photoproteins are not well suited to the measurement of free calcium concentrations in the range (mM) likely to be encountered in extracellular fluids or secretions, although their use for this purpose has been proposed.

The Ca^{2+}-regulated photoproteins are not uniquely sensitive to Ca^{2+}. A number of other di- and trivalent cations (e.g., Yb^{3+}, La^{3+}, Sr^{2+}) are capable of stimulating luminescence. For instance, all of the lanthanides that have been tested are more potent than Ca^{2+} in stimulating aequorin luminescence, and the maximum light intensity that they produce is either equal to (e.g., Yb^{3+}) or only slightly lower (e.g., La^{3+}) than that observed with Ca^{2+} (Blinks et al. 1982). However, calcium is the only ion likely to be found in living cells in sufficient quantities to trigger luminescence.

From the standpoint of intracellular measurements, free magnesium ion concentration is the most important factor known to influence the sensitivity of photoproteins to Ca^{2+}. In the case of aequorin, Mg^{2+} within the range of concentrations (in the vicinity of 1 mM) that likely might be encountered inside living cells reduces Ca^{2+}-independent luminescence and sensitivity to calcium, that is, shifts the Ca^{2+} concentration-effect curve to the right (Blinks et al. 1982). However, it may be not so critical for other photoproteins because for obelin, for example, the effect of Mg^{2+} is much less pronounced than on aequorin bioluminescence. A concentration of 1 mM Mg^{2+} has no effect on the Ca^{2+}-independent luminescence or the Ca^{2+} sensitivity (Fig. 6); only a nonphysiological concentration of magnesium ion of 10 mM produces a modest rightward shift of the Ca^{2+} concentration-effect curve. Another effect of Mg^{2+} on the luminescence of the obelin is more evident. Magnesium ions, even at 1 mM concentration, produce a decrease in maximum level of luminescence attainable under the influence of Ca^{2+}, and this effect cannot be surmounted by increases in $[Ca^{2+}]$ (Markova et al. 2002).

In fact, the Ca^{2+}-regulated photoproteins were the first intracellular calcium probes, and they were employed for measurement of calcium in cells for more than 40 years. However, the development of fluorescent dyes for detection of intracellular calcium, which have simplified the measurement procedure, essentially supplanted photoproteins. The successful cloning of cDNAs encoding apophotoproteins has opened new avenues for utilizing photoproteins, by expressing the recombinant apophotoprotein intracellularly, then adding coelenterazine externally which diffuses into the cell and forms the active photoprotein. Such cells and whole organisms have, in effect, a "built-in" calcium indicator. This technique is highly valuable because it does not require laborious procedures such as microinjection. To date, various types of cells expressing the apophotoprotein gene have been constructed, including mammalian and insect cells, bacteria, yeasts, plants, and fungi. They are widely employed in the studies of Ca^{2+} homeostasis in cells and the effect of various compounds. The application of such cells takes advantage of the fact that the light emission reports changes in intracellular Ca^{2+}, in particular those mediated by the interaction of various agonists and antagonists with cell receptors. Consequently, these mammalian cell lines are widely employed in high-throughput screening of new drugs, for instance, various chemical compounds affecting G-protein-coupled receptors, tyrosine kinase receptors, and ionic channels (Dupriez et al. 2002).

The most attractive and, probably, the most substantial idea was to selectively direct a photoprotein to individual cell compartments for local measurement of calcium. To direct a Ca^{2+}-regulated photoprotein to a necessary cell compartment, a chimeric DNA is constructed to fuse the apophotoprotein gene to a sequence encoding a molecular address, which might be either short amino acid sequences directing proteins to specific organelles or full-size proteins with known cell localization. For instance, for measurement of calcium in mitochondria, the N-terminus of aequorin has been fused with a cytochrome c oxidase fragment, consisting of the cleavable mitochondrial targeting signal and the first six residues of the enzyme. To monitor $[Ca^{2+}]$ in the intermembrane space of mitochondria, aequorin was linked with the

C-terminus of glycerophosphate dehydrogenase, which is located in the inner mitochondrial membrane with its C-terminus directed toward the intermembrane space. Similar approaches have been used to produce probes for measurement of [Ca^{2+}] in the nucleus, sarcoplasmic reticulum, and Golgi apparatus. Since Ca^{2+} concentration varies considerably among cell compartments and the role of local changes of [Ca^{2+}] in regulation of cell events is still not completely understood, the opportunity to measure local calcium concentrations gives unquestionable advantages for photoproteins with respect to fluorescent dyes, which are distributed throughout the cell, consequently allowing only an averaged estimation of [Ca^{2+}].

Assay of intracellular [Ca^{2+}] does not exhaust the applications of Ca^{2+}-regulated photoproteins; in many studies, it has been demonstrated that photoprotein is a very promising label for various in vitro assays. There are several factors accounting for that: (1) The sensitivity of analysis is comparable with that of radioisotope methods because photoprotein can be detected in attomole amounts using modern luminometers. (2) A background signal is practically absent because of the high selectivity of photoproteins to Ca^{2+}. (3) The light signal linearly depends on the photoprotein concentration throughout a virtually unlimited range. At a saturated [Ca^{2+}], the light intensity is in direct proportion to the protein amount because the photoprotein is the only molecule directly involved in the bioluminescent reaction. (4) The bioluminescent reaction is easy for triggering due to a lack of additional substrates and cofactors, and it is fast (reaction ceases within several seconds). (5) Photoproteins are not toxic. (6) Photoproteins are available because *E. coli* strains producing recombinant apophotoproteins have been constructed, and efficient procedures for their purification have been developed. (7) Photoproteins are stable upon storage in solution, lyophilization, and chemical or genetic modifications. These properties of photoproteins are very advantageous for their use in clinical diagnostics, for example, in ELISA and hybridization assays. To date, photoprotein-based reagents have been applied to the assay of numerous analytes of diagnostic value (hormones, proteins, etc.) and, conjugated to oligonucleotides, as a DNA probe for detection of specific nucleotide sequences. Direct comparison of hybridization assays utilizing different labels has shown that the use of a photoprotein increases the sensitivity of an analysis by 2–3 orders of magnitude. In turn, this makes the analysis more sensitive and less time-consuming.

Cross-References

▶ Calcium in Biological Systems
▶ Calcium ion Selectivity in Biological Systems
▶ Calcium-Binding Protein Site Types
▶ Calcium-Binding Proteins, Overview

References

Blinks JR, Wier WG, Hess P, Prendergast FG (1982) Measurement of Ca^{2+} concentrations in living cells. Prog Biophys Mol Biol 40:1–114

Dupriez VJ, Maes K, Le Poul E, Burgeon E, Detheux M (2002) Aequorin-based functional assays for G-protein-coupled receptors, ion channels, and tyrosine kinase receptors. Receptors Channels 8:319–330

Hastings JW, Morin JG (1969) Calcium-triggered light emission in *Renilla*. A unitary biochemical scheme for coelenterate bioluminescence. Biochem Biophys Res Commun 37:493–498

Kawasaki H, Nakayama S, Kretsinger RH (1998) Classification and evolution of EF-hand proteins. Biometals 11:277–295

Markova SV, Vysotski ES, Blinks JR, Burakova LP, Wang BC, Lee J (2002) Obelin from the bioluminescent marine hydroid *Obelia geniculata*: cloning, expression, and comparison of some properties with those of other Ca^{2+}-regulated photoproteins. Biochemistry 41:2227–2236

Nelson MR, Chazin WJ (1998) Structures of EF-hand Ca^{2+}-binding proteins: diversity in the organization, packing and response to Ca^{2+} binding. Biometals 11:297–318

Shimomura O (2006) Bioluminescence: chemical principles and methods. World Scientific, Singapore

Strynadka NCJ, James MNG (1989) Crystal structures of the helix-loop-helix calcium-binding proteins. Annu Rev Biochem 58:951–998

Vysotski ES, Lee J (2004) Ca^{2+}-regulated photoproteins: structural insight into the bioluminescence mechanism. Acc Chem Res 37:405–415

Vysotski ES, Lee J (2007) Bioluminescence mechanism of Ca^{2+}-regulated photoproteins from three-dimensional structures. In: Viviani VR, Ohmiya Y (eds) Luciferases and fluorescent proteins: principles and advances in biotechnology and bioimaging. Transworld Research Network, Kerala

Calerythrin

▶ Bacterial Calcium Binding Proteins
▶ Sarcoplasmic Calcium-Binding Protein Family: SCP, Calerythrin, Aequorin, and Calexcitin

Calexcitin

▶ Sarcoplasmic Calcium-Binding Protein Family: SCP, Calerythrin, Aequorin, and Calexcitin

Calmodulin

David O'Connell[1], Mikael Bauer[2], Christopher B. Marshall[3], Mitsu Ikura[3] and Sara Linse[2]
[1]University College Dublin, Conway Institute, Dublin, Ireland
[2]Department of Biochemistry and Structural Biology, Lund University, Chemical Centre, Lund, Sweden
[3]Ontario Cancer Institute and Department of Medical Biophysics, University of Toronto, Toronto, Ontario, Canada

Synonyms

EF-hand superfamily = Calmodulin superfamily = Troponin superfamily

Definitions

EF-hand: Helix-loop-helix motif obeying a 29-residue consensus sequence with hydrophobic and Ca^{2+}-ligating residues in defined positions, typically forming a pentagonal bipyramidal coordination sphere
Ca^{2+}-signaling proteins: Proteins for which Ca^{2+} binding preferentially stabilizes a form that activates other proteins
Domain: Independent folding unit
Chromophoric chelator: A metal-binding dye which changes its optical spectrum upon chelation

Discovery

The discovery of calmodulin (CaM) was reported in 1970 independently by Cheung and by Kakiuchi, Yamazaki, and Nakajima. They found that cyclic nucleotide phosphodiesterase was modulated by a Ca^{2+}-dependent activator protein, which was named calmodulin for *cal*cium-*modul*ated prote*in*.

Prevalence

CaM is a 17-kDa protein ubiquitously expressed in eukaryotic cells constituting at least 0.1% of total cellular protein, with higher abundance in brain and testes. The ubiquitous expression highlights the fundamental role of CaM in calcium homeostasis, further underscored by the observation that gene knockouts are lethal.

Sequence

The amino acid sequence of CaM is extremely conserved; the exact same protein sequence is found in all vertebrates, and the human genome contains three genes that encode an identical CaM sequence characterized by a high prevalence of acidic residues and dominated by four EF-hand motifs connected by linking segments. The vertebrate sequence is shown below with the four EF-hand loops underlined and other residues in consensus sequence positions in bold. The stars indicate positions responsible for the specificity of EF-hand pairing (see below).

```
              1  3 5   9  12
ADQLTEEQIAEFKEAFSLFDKDGDGTITTKELGTVMRSL₃₉    EF1
    GQNPTEAELQDMINEVDADGNGTIDFPEFLTMMARK₇₅    EF2
   MKDTDSEEEIREAFRVFDKDGNGYISAAELRHVMTNL₁₁₂   EF3
     GEKLTDEEVDEMIREADIDGDGQVNYEEFVQMMTAK₁₄₈   EF4
              *          *  *
```

Highly similar sequences are found also in yeast, drosophila, and plants. Point mutations have been studied in Drosophila but are lethal. Plants, in contrast to mammals, have a number of CaM genes coding for proteins of varying sequence (84–100% identity).

Posttranslational Modifications

Known posttranslational modifications include trimethylation of lysine 115 and oxidation of methionines. The loops of EF2 and EF3 contain classical deamidation sites (NG). There is evidence that a fraction of cellular CaM is constitutively phosphorylated in vivo. Phosphorylation in vitro by several kinases has been observed at Thr-26, Thr-29, Thr-79, Ser-81, Tyr-99, Ser-101, Thr-117, and Tyr-138.

The EF-Hand

The EF-hand and C2 motifs are the two most common motifs found in Ca^{2+}-binding proteins. The Ca^{2+} ion is rather flexible in its requirements for the geometry and number of coordinating oxygens. In the EF-hand, there are seven coordinating oxygens arranged as a pentagonal bipyramid. Receiving its moniker from the E and F helices in parvalbumin, the first EF-hand protein for which a high-resolution structure was solved by Kretsinger, the motif consists of a helix followed by a loop and then another helix. The EF-hand loop comprises 12 residues, although residues 1 and 12 are part of the entering and exiting helices. In the bound state, Ca^{2+} is coordinated by side-chain carboxyl or amide oxygens of residues 1, 3, 5, and 12 and the main-chain carbonyl oxygen of residue 7. Position 1 is most strictly conserved as an aspartate. Positions 3 and 5 are typically aspartate or asparagine but sometimes serine. The side-chain carboxylate in position 12 (glutamate or aspartate) acts as a bidentate Ca^{2+} ligand providing both its oxygens to Ca^{2+} chelation. The side-chain in position 9 may complete the Ca^{2+} coordination sphere, but more often, this residue is hydrogen-bonded to a water molecule which coordinates Ca^{2+}. The highly conserved glycine at position 6 allows for a turn to assemble the two halves of the loop in tight Ca^{2+} coordination. Proteins with EF hands constitute the EF-hand superfamily, also called the calmodulin superfamily or troponin superfamily. Whereas some proteins contain variant EF hands, CaM has four canonical EF hands that fulfill the consensus sequence at nearly all positions (see above).

EF-Hand Pairing

While some proteins contain a single or an odd number of EF hands, the most common structural feature is the EF-hand pair which may form a two EF-hand domain or be part of a larger domain with more EF hands. A significant hydrophobic surface area is buried in the interface between the two EF hands in each pair. A short two-stranded β-sheet connects the paired loops centered around their respective residue 8 with their bulky hydrophobic side chains interacting in the hydrophobic core. When excised from the protein, an EF hand may form homodimers, but with many orders of magnitude lower affinity (e.g., EF3-EF3 and EF4-EF4 of CaM versus EF3-EF4). The heterodimer over homodimer specificity is governed by electrostatic repulsion within one homodimer (EF4-EF4) and three hydrophobic positions which form one aromatic (F89, Y138, F141) and one aliphatic (A102, L105, I125) cluster (Linse et al. 2000; Fig. 3).

Structure

Since the late 1970s, a wealth of biochemical evidence demonstrated that CaM contains two independent folding units, each containing two EF hands. The first crystal structure of Ca^{2+}-loaded CaM revealed a dumbbell-shaped protein in which two globular domains are separated by a central helix (Babu et al. 1988; Fig. 1). The four EF hands are in a helix-loop-helix conformation, and a number of hydrophobic side chains are exposed at the base of each globular domain. The structure of vertebrate CaM was subsequently confirmed and refined by several investigators. NMR studies have revealed that in solution, the central helix is interrupted in the middle with a short flexible linker which allows the two globular domains to move relative to one another. In one crystal form, a compact conformation of CaM was selected by the crystallization procedure, indicating a conformational heterogeneity of the relative domain orientation in solution, supported by single-molecule Förster resonance energy transfer (FRET) studies. With the data available today, it seems most likely that the linker between the domains is flexible and samples multiple conformations including both extended and collapsed overall structures.

Stability Toward Denaturation

CaM has been optimized by evolution for stability and solubility to allow for high concentration in the cell. The highly negative net surface charge prevents aggregation and keeps the protein soluble at concentrations up to 5 mM (90 g/L). The Ca^{2+}-bound state is so stable that thermal denaturation occurs above 100°C; thus, boiling is used as an efficient purification step. The globular domains can be produced by proteolysis or cloning as separate fragments which are also very stable and do not denature upon boiling in the Ca^{2+} state. The stability toward thermal and chemical denaturation is substantially reduced upon Ca^{2+} release. In the apo state, the domains have similar midpoints for

Calmodulin 547

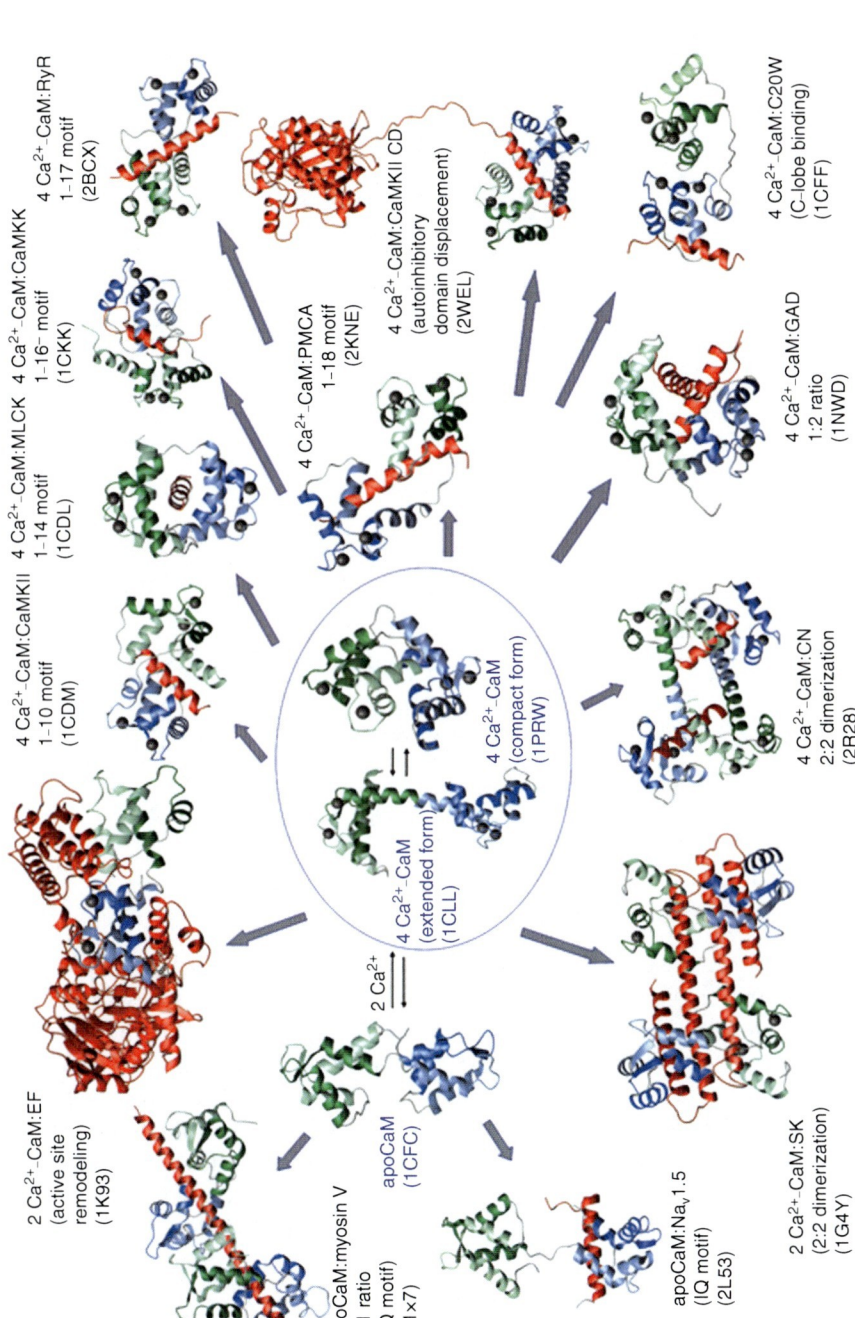

Calmodulin, Fig. 1 Structural models of calmodulin in its apo, calcium-bound, and target-complexed states. In each structure, the N-terminal domain of CaM is green (EF1 in *light green*), the C-terminal domain is blue (EF3 in *light blue*), Ca^{2+} ions are grey, and target peptides and proteins are *red*. The apo-CaM structure presented is one model from an ensemble of NMR structures, in which the two domains are found in variable relative orientations. Two forms of Ca^{2+}-bound CaM are shown, which likely exist in equilibrium. On the left are Ca^{2+}-independent complexes of apo-CaM with IQ motifs from myosin V and the voltage-gated sodium channel ($Na_V1.5$), followed by Ca^{2+}-dependent target complexes in which CaM is not fully saturated with Ca^{2+}, including *Bacillus anthracis* edema factor (EF), and a peptide from the small conductance Ca^{2+}-activated potassium channel (SK). Target complexes in which CaM is fully Ca^{2+}-bound include (clockwise from top) peptides from calmodulin-dependent kinase II (CaMKII), myosin light-chain kinase (MLCK), CaM-dependent kinase kinase (CaMKK), ryanodine receptor (RyR), plasma membrane Ca^{2+} ATPase pump (PCMA), plasma membrane calcium pump (C20W), petunia glutamate decarboxylase (GAD), and calcineurin (CN). The catalytic and regulatory domain of CaMKII in complex with Ca^{2+}/CaM is shown on the right. The Protein Data Bank (PDB) code for the structural coordinate file is given next to each structure, along with the type of motif and/or mechanism of target activation. More information on these structures and literature references can be obtained from www.rcsb.org/pdb

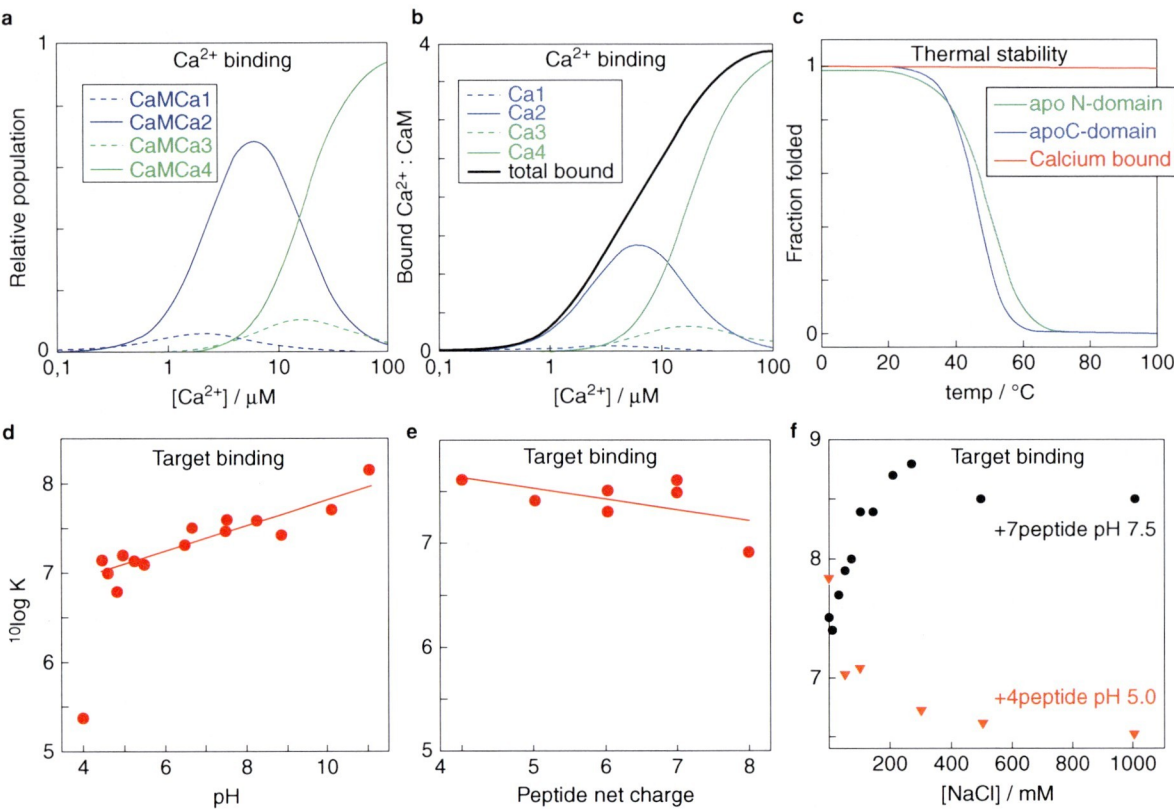

Calmodulin, Fig. 2 Biophysical properties of CaM and its globular domains. (**a**, **b**) Calcium binding as a function of free Ca^{2+} concentration. (**a**) Species distribution. (**b**) Concentration of bound Ca^{2+} in species with 1, 2, 3, and 4 Ca^{2+} ions bound, respectively. (**c**) Thermal stability of the globular domain fragments in the absence (*blue, green*) and presence (*red*) of calcium. (**d**) pH insensitivity of target peptide binding affinity. (**e**) Peptide charge insensitivity of target peptide binding affinity. (**f**) Salt dependence of target peptide binding affinity in a highly charged system (+7 peptide binding to CaM at pH 7.5) and in a moderately charged system (+4 peptide binding to CaM at pH 5.0)

thermal denaturation as the intact protein, with the N-terminal domain (thermal denaturation midpoint $T_m = 49°C$) slightly more stable than the C-terminal domain ($T_m = 46°C$) (Brzeska et al. 1983; Fig. 2).

Structural Transitions Due to Ca^{2+} Binding

Ca^{2+} binding leads to a large conformational change in each globular domain of CaM. This conformational change is key to the biological role of CaM and reveals the binding sites for target proteins (Zhang et al. 1995). The bidentate Ca^{2+}-coordinating carboxylate in position 12 of the EF-hand loop plays an important role in this transition from a closed apo state to a more open Ca^{2+}-bound state. The structural switch propagates from small alterations in the loop to a change in interhelical angles. In each domain, there is also a change in the angle between the two EF hands in the pair. The key feature is the exposure of a number of methionine and other hydrophobic residues at the base of each domain, providing an anchoring site for the large hydrophobic residues in target proteins. CaM is a dynamic system, and detailed NMR studies have revealed that the apo state conformational ensemble includes conformations similar to the Ca^{2+}-bound state and that Ca^{2+} binding shifts the ensemble toward those conformations (Evenäs et al. 2001).

Ca^{2+}-Binding Properties

CaM contains four Ca^{2+}-binding sites of physiological relevance. Each EF hand provides one of these sites. The Ca^{2+} affinity is ca. sixfold higher for the C-terminal compared to N-terminal domain (Linse et al. 1991; Fig. 2). At physiological KCl concentration, the average dissociation constant is $K_D = 3.5\ \mu M$

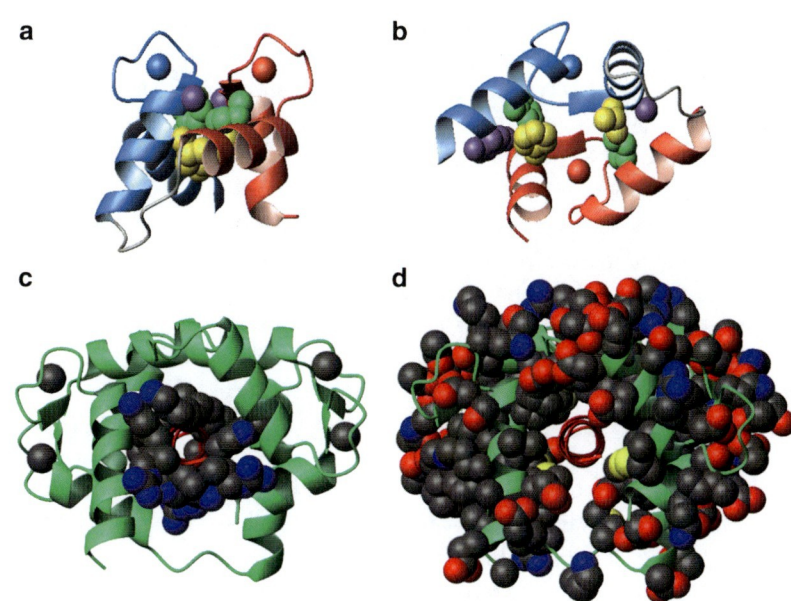

Calmodulin, Fig. 3 Structural models. (**a**, **b**) Hydrophobic residues responsible for specificity in EF-hand pairing are shown in *green* (F89, I125), *purple* (A102, Y138), and *yellow* (L105, F141). The backbone of EF3 is shown as a *blue ribbon* and EF4 as a red ribbon. (**a**) Side view and (**b**) bottom view of CaM C-terminal domain. (**c**, **d**) Calmodulin-MLCK peptide complex with CaM backbone in *green* and MLCKp backbone in *red*. Side chains are shown as space filling models with carbons in *black*, nitrogens in *blue*, and oxygens in *red* for MLCKp in (**d**) and for CaM in (**d**)

for the C-terminal sites and 22 μM for the N-terminal sites. Within each domain, the two Ca^{2+} ions are bound with positive cooperativity. Moreover, the Ca^{2+}-binding properties of each domain fragment is essentially the same as when the domain is part of the intact protein. In the presence of a bound target protein or peptide, the affinities are significantly increased, and the saturation curve is shifted toward lower Ca^{2+} concentration (Bayley et al. 1996).

CaM is optimized not for the highest possible Ca^{2+} affinity but for the most suitable affinity for its cellular functions. At physiological salt concentration, the enthalpy of Ca^{2+} binding is positive, thus unfavorable. Exposure of hydrophobic surface gives an unfavorable entropic contribution from ordering of water molecules. The free energy of Ca^{2+} binding is in fact dominated by the entropic gain from the release of water molecules that hydrate the Ca^{2+} ion in the unbound state. CaM is an intricate example of a protein that "uses" part of the available free energy for binding a small cofactor (metal ion) to drive a conformational change, which lowers the overall Ca^{2+} affinity but gives the system a regulatory capacity with huge biological impact. Proteins with higher Ca^{2+} affinity, for example, calbindin D_{9k}, parvalbumin, and the amylases, have a larger degree of preorganization of the protein, smaller conformational change, and/or fewer water molecules in the Ca^{2+} coordination sphere.

The rate of Ca^{2+} association is close to diffusion controlled, but the dissociation rates are relatively low, especially for the C-terminal domain with $k^{off} = 24\ s^{-1}$ while for the N-terminal domain $k^{off} = 240\ s^{-1}$, at 0.1 M KCl. (Martin et al. 1985). Because of the high positive cooperativity within each domain, dissociation of the first ion from each domain becomes rate limiting, and the two ions appear to dissociate together.

Mg^{2+} Binding

The EF hand is flexible enough to adapt to coordination of Mg^{2+}, but the affinity is 10^3–10^4-fold lower for Mg^{2+} binding compared to Ca^{2+} binding (Malmendal et al. 1989). Nevertheless, Mg^{2+} is still a relevant intracellular competitor because it is abundant. $[Mg^{2+}]$ is relatively constant around 1–2 mM, while $[Ca^{2+}]$ changes from ca. 100 nM in the resting state to 1–10 μM upon cellular activation. Indeed, in the resting cell, Mg^{2+} does bind to CaM, and the rate of Mg^{2+} dissociation may become rate limiting for Ca^{2+} binding to the regulatory sites.

Other Metal Ions

CaM has high affinity for many other divalent metal ions but also binds mono- and trivalent metal ions like Na^+, K^+, and the lanthanides. The fluorescent ions Tb^{3+}

and Eu^{3+} bind CaM with opposite domain preferences compared to Ca^{2+} and can be used to study ion competition, as well as structural transitions monitored by FRET to the metal ion sites from the tyrosine residues in the C-terminal domains. $^{113}Cd^{2+}$ is a sensitive NMR probe to study structural transitions in CaM. This spin-1/2 nucleus produces narrow NMR lines and superior spectroscopic properties compared to the quadrupolar $^{43}Ca^{2+}$ or $^{25}Mg^{2+}$ ions with very broad signals. ^{113}Cd NMR studies revealed cooperation of the globular domains in target binding (Linse et al. 1986), which was later confirmed by high-resolution structures (below).

Ca^{2+} Signaling Via Calmodulin

Ca^{2+} signaling is fundamentally important in regulating eukaryotic cellular homeostasis. The dynamics of Ca^{2+} signaling are governed by a very steep gradient between extracellular (mM) and intracellular ($\sim$100 nM) Ca^{2+} concentration. This gradient allows intracellular Ca^{2+} to fulfill a critical role as a second messenger. The cytosolic Ca^{2+} concentration is exquisitely regulated by the operation of transport systems responsible for its increase and decrease. Influx is through Ca^{2+} channels and $Na^+(H^+)/Ca^{2+}$ exchangers in the plasma membrane (PM), the endo(sarco)plasmic reticulum (ER), and/or the mitochondria. Decrease after a Ca^{2+} pulse is due to intracellular Ca^{2+} buffer proteins that bind with higher affinity but lower on rate compared to CaM and to extrusion transport systems represented by Ca^{2+} ATPases in PM and ER and Na^+/Ca^{2+} exchanger in PM. These transporter systems give rise to oscillations in the concentration of Ca^{2+} not only in the cytosol but also in the nucleus and intracellular organelles. These oscillations in $[Ca^{2+}]$ are recognized by CaM to mediate changes in gene expression, cell growth, development, cell survival, and cell death.

Target Repertoire

CaM binds to and regulates a very large number of proteins. Several hundred binding targets of CaM are currently known in a fantastic range of organisms spanning the natural kingdom from yeast, insects, plants to mammals based on classical biochemical experiments, affinity chromatography (Berggård et al. 2006; mouse), and more lately protein array screens (O'Connell et al. 2010). The target repertoire includes proteins from all compartments in the cell: cytosolic enzymes, proteins at the cell membrane, intracellular membrane proteins, organellar proteins, and proteins in the nucleus.

There is an entire family of cytosolic enzymes such as the CaM kinases whose activity is controlled by Ca^{2+} CaM. Upon increase in intracellular Ca^{2+}, Ca^{2+} CaM binds to and activates the kinases myosin light-chain kinase (MLCK), calmodulin-dependent kinases I, II, and IV (CaMKI, CaMKII, and CaMKIV), and CaM-dependent kinase kinase (CaMKK). Primary targets of CaMKK are CaMKI (cytosolic) and CaMKIV (nuclear), requiring binding of Ca^{2+} CaM to both CaMKK and CaMKI/CaMKIV (Wayman et al. 2008).

CaM has a large number of membrane protein targets, for example, subunits of the glutamate (NMDAR) receptor and voltage-gated potassium channel receptors (O'Connell et al. 2010; Pitt 2007). The role of these receptors in excitatory activity is fundamentally important in development of learning and memory and provides an excellent example of the regulation and propagation of a Ca^{2+} signal into the cell to achieve higher-order effects. NMDAR activity leads to an increase in intracellular $[Ca^{2+}]$, and binding of CaM to motifs in the cytoplasmic tails of the receptor regulates channel activity.

In several membrane proteins, including sodium and potassium channels, CaM has more than one interaction site in the cytoplasmic domain. One binding site is Ca^{2+} independent, serving to prelocalize CaM to the receptor, and the Ca^{2+} signal leads to binding of an additional receptor motif resulting in translocation and regulation. The voltage-gated potassium channel KCNQ1 is the pore-forming subunit regulating cardiac muscle function. CaM interacts with multiple sites in the cytoplasmic region of KCNQ1, including two IQ motifs (Pitt 2007), and is critical for the current through the channel.

CaM regulation of ion channels in the cell membrane underpins learning and memory, and nuclear CaM is critical in the formation of long-term memory in several animal species. The phosphorylation of Ser 133 of the transcription factor CREB is critical for its activation and is rapidly induced by brief synaptic activity in hippocampal neurons. A highly local rise in Ca^{2+}-ion concentration near the cell membrane causes a swift ($\sim$1 min) translocation of CaM from the cytoplasm to the nucleus, which culminates in activation of CaMKIV and phosphorylation of CREB. Translocation of CaM provides a form of cellular communication that combines the specificity of local Ca^{2+} signaling with the ability to produce action at a distance (Deisseroth et al. 1998).

Especially intriguing is the interaction of CaM with proteins that control cellular Ca^{2+} flux and refilling of the internal stores through which CaM regulates its own activity. In store-operated calcium entry (SOCE), ER luminal Ca^{2+} store depletion leads to the formation and opening of highly selective PM Ca^{2+} channels that facilitate a sustained increase in cellular $[Ca^{2+}]$ crucial for stimulating a cellular response (e.g., transcription activation) and for replenishing the ER stores. CaM is repeatedly identified to interact with the ER membrane proteins stromal interaction molecule 1 and 2 (STIM1 and 2). ER luminal Ca^{2+} depletion results in significant redistribution of STIM1 from homogeneous dispersion on the ER membrane to specific cluster sites at ER–PM junctions. The movement of STIM1 to the discrete ER–PM cluster sites facilitates the recruitment of the Orai1 component of the CRAC channel (Ca^{2+} release-activated Ca^{2+} channel) to the same junctions, a requisite for the sustained SOCE/CRAC entry. A lysine-rich region of STIM1 at the carboxyl terminal region may communicate with the Orai1 N-terminal pro-rich region, permitting the interactions which activate SOCE/CRAC entry while also targeting STIM1 to ER–PM junctions. This lysine-rich region of both STIM1 and STIM2 binds to CaM in a Ca^{2+}-dependent manner with dissociation constant in the nanomolar range (Bauer et al. 2008). This may implicate CaM in regulating STIM action in the ER–PM cooperation during SOCE.

Regulation and Competition

One of the most intriguing problems regarding CaM action is how one signal, the intracellular Ca^{2+} concentration, can be translated into the regulation of numerous targets in vivo. One possible mechanism relies on the frequencies of Ca^{2+} oscillations (Dolmetsch et al. 1998). Since Ca^{2+} oscillations occur on a rather fast time scale (seconds or faster), CaM binding of targets does not reach equilibrium during or between pulses. Thus, the binding kinetics and the local concentration of CaM and targets will determine the outcome of a Ca^{2+} oscillation train. To fine-tune the response to an oscillation, the concentration of target proteins is controlled spatially and temporally, while affinities and kinetics can be modified by phosphorylation of binding sequences. Another possible mechanism relies on "prelocalization" of CaM to Ca^{2+}-independent IQ motifs of certain targets such that they compete more effectively for CaM against the manifold of targets upon Ca^{2+} influx.

Electrostatic Interactions

CaM is a highly charged protein. At neutral pH, the apo protein has a net charge of -24, evenly distributed over the two globular domains, which changes to -16 when four Ca^{2+} ions are bound. The net negative charge makes the Ca^{2+} affinity and exchange rates strongly dependent on ionic strength and gives upshifted pKa values of ionizable side chains. Within each domain, there is marked charge separation between the two EF hands, as seen in several other proteins (Linse et al. 2000). In the Ca^{2+}-bound C-terminal domain, EF3 has zero net charge, whereas EF4 has net charge -7. This minimizes electrostatic repulsion between the EF hands, governing domain stability. Charge separation also governs the specificity of EF-hand pairing (EF3–EF4) due to electrostatic repulsion within the EF4–EF4 pair.

Anomalous Electrostatic Effects and Charge Regulation Mechanism

CaM–target interactions occur between two highly and oppositely charged species. CaM-binding segments typically carry a net positive charge, often as high as +6 to +8 over some 20 residues. High-resolution structures reveal a number of close contacts in which negatively charged side chains of CaM are in close proximity of positively charged residues in targets. However, addition of salt to screen electrostatic interactions actually increases the affinity between CaM and target peptide (André et al. 2006; Fig. 2), and disruption of ionic interactions by mutation or pH variation has very little effect on the affinity between CaM and peptide (André et al. 2004; Fig. 2). Peptides with +4 and +8 net charge have the same affinity, and a large change in the net charge of Ca^{2+}-loaded CaM is required for any measurable effect on peptide affinity. Thus, in this highly charged system, individual charges lose importance. This insensitivity is due to a charge regulation mechanism, which involves pKa-value modulation upon complex formation. The pKa values of ionizable groups in CaM shift closer to normal values upon target binding. Likewise, the downshifts of pKa values in the positive target are reduced upon CaM binding. In a more moderately charged system (e.g., CaM at pH 5 and a +4 peptide), electrostatic effects are normalized, and a monotonous decrease in affinity is observed upon increased ionic strength (Fig. 2).

Ca^{2+}-Dependent Target Motifs

Ca^{2+}-dependent targets can be classified according to the relative position of two hydrophobic residues that anchor the two lobes of CaM. The first high-resolution structures of CaM complexed with peptides derived from skeletal and smooth muscle MLCK revealed helical peptides with two hydrophobic side chains spaced by 14 residues bound by the N- and C-terminal lobes of CaM (Ikura et al. 1992, Meador et al. 1992). This CaM-binding motif called "1–14" (Rhoads and Freidberg 1997) has been found in several other CaM targets, including CaMKIV, calcineurin A, death-associated protein (DAP) kinase, and endothelial nitric oxide synthase (NOS), and may include additional hydrophobic residues in position 5 or 8. In complex with a CaMKII peptide, the two lobes of CaM are bound to more closely spaced hydrophobic residues (10-a.a. apart). This structure identified a class of 1–10 CaM targets, including synapsins and heat-shock proteins, which may have an additional hydrophobic residue in position 4 or 5. The structural plasticity of CaM is further illustrated by the complex with a CaMKK peptide which is bound in opposite orientation, is not entirely helical, and has more widely spaced hydrophobic anchors (1–16). The anchor for the N-lobe is a Trp residue on an 11-residue helix, whereas the C-lobe is anchored by a Phe residue found on a hairpin structure that folds back on the helix. This mode of 1–16 binding has only been observed for CaMKK homologues. The plasma membrane calcium pump has a 1–14 motif but is unusual in that it can be activated by an isolated C-lobe fragment of CaM. A peptide from the N-terminal portion of the CaM-binding region ("C20W") representing a splice variant of the pump binds to the C-terminal lobe of CaM while the linker and N-lobe remain flexible.

Two recent structures illustrate that CaM can extend considerably to accommodate long helical targets. A peptide derived from the ryanodine receptor is bound with hydrophobic anchor residues with "1–17" spacing, and a plasma membrane ATPase (PMCA) pump peptide interacts with CaM through a "1–18" motif. Because these targets are completely helical, the CaM lobes adopt similar relative orientations in the 1–18, 1–14, and 1–10 conformations, although the spacing between the lobes varies by 1–2 helical turns. As a result of the longer spacing in 1–17 and 1–18 motifs, there are few contacts between the two bound lobes of CaM, and much of the target sequence remains exposed to solvent.

The plasticity of CaM in binding to variably spaced anchors in single helices is due to the flexibility of the linker connecting the globular domains. CaM is also capable of simultaneous engagement of multiple helices. Ca^{2+}/CaM interacts with the C-terminal domain of plant glutamate decarboxylase (GAD) in a novel mode, simultaneously binding two GAD peptides leading to activation through dimerization. Each lobe of CaM binds a helical GAD peptide, and the two lobes are in close proximity making several contacts. The two GAD peptides are perpendicular to each other with two anchors separated by only two residues (1–4). Most CaM targets form helices that engage both lobes of CaM; however, the myristoylated alanine-rich C kinase substrate (MARCKS) is a notable exception. An elongated peptide has a single helical turn bearing a Phe residue that is grasped by the C-lobe hydrophobic pocket. The N-lobe forms a hydrophobic surface interacting with a Leu two residues away from the C-lobe anchor. Plasticity of CaM–target complexes is further highlighted by a 25-residue calcineurin (CN) peptide, which can form both a 1:1 complex with CaM and a domain swapped 2:2 complex in which each CN peptide interacts with the N-lobe of one CaM and the C-lobe of the other.

Ca^{2+}-Independent (Including IQ) Target Motifs

A large number of proteins interact with Ca^{2+}-free (apo) CaM, including myosin, nitric oxide synthase, cyclic nucleotide phosphodiesterase (PDE), phosphorylase kinase, neuromodulin, neurogranin, and IQGAPs. Many of these proteins contain a so-called "IQ motif" comprised of the sequence IQXXXRGXXXR. Apo-CaM binds some target proteins such as neuromodulin only in the absence of Ca^{2+}, whereas others remain associated with CaM in the presence and absence of Ca^{2+}. The CaM-binding region of PDE contains an "atypical" motif, which forms a helix that interacts with both lobes of Ca^{2+}/CaM. In the absence of Ca^{2+}, this peptide is less helical and binds with lower affinity to apo-CaM through the C-lobe only.

Two apo-CaM molecules bind to tandem IQ motifs from myosin V in the absence of Ca^{2+}, and the C-lobe is partially open with a shallow groove that binds the conserved N-terminal IQXXXR portion of the helical IQ motif. The N-lobe remains closed and forms weak interactions with the C-terminal portion of the IQ motif. By contrast, only the C-lobe of apo-CaM binds to an IQ motif from the $Na_V1.5$ channel in a semiopen conformation, and the N-lobe is fully closed and does not interact with target. Complexes of Ca^{2+}/CaM with voltage-activated calcium channel (Ca_V) IQ motifs highlight the remarkable plasticity of CaM, which binds similar IQ domains from Ca_V1 and Ca_V2 with opposite binding orientations.

Structural Plasticity

Several key features underlie the plasticity of CaM in target binding (Yamniuk and Vogel 2004, Ikura and Ames 2006). The C-lobe of apo-CaM samples both closed and open conformations in the absence of Ca^{2+} and is able to bind a subset of target motifs. In the Ca^{2+} state, each hydrophobic pocket is lined by four flexible and polarizable methionines and is structurally adaptable to accommodate a variety of hydrophobic side chains from diverse targets. The flexible interlobe linker allows the two lobes to simultaneously engage hydrophobic anchor residues with widely varied spacing. The remarkable intrinsic plasticity of CaM allows it to recognize a diverse set of target proteins through varied binding modes. This enables a number of fundamentally distinct mechanisms by which CaM can activate its targets, as exemplified below (Fig. 1).

Mechanisms of Target Protein Activation

How can the relatively small protein CaM regulate catalytic activity of a large enzyme by binding to a single helix? Many CaM-regulated enzymes are constitutively regulated by autoinhibitory domains, which are displaced upon CaM binding, activating enzymatic activity. The catalytic domains of CaM-regulated kinases are blocked by pseudosubstrate sequences followed closely by a CaM-binding site. In the resting state, the target motif of CaMKII is unstructured while the inhibitory domain forms a helix that occludes the substrate-binding pocket. Upon binding Ca^{2+}/CaM, the inhibitory domain becomes unstructured and dissociates from the substrate site while the target motif becomes helical. These changes expose the inhibitory region to autophosphorylation, which prevents it from rebinding and generates significant autonomous activity in neuronal cells. In the case of death-associated protein kinase 1(DAPK1), CaM binds its target helix and makes additional interactions with the catalytic domain and blocks the substrate binding site, suggesting that activation of this kinase may require more steps. Furthermore, CaM bound to the catalytic domain construct is significantly more extended than CaM bound to an isolated DAPK1 target peptide. The phosphatase CN also contains an autoinhibitory domain that is displaced by CaM binding, although the arrangement of domains differs from the kinases in that the autoinhibitory domain is C-terminal to the CaM-binding site. Some NOS isoforms are regulated by an autoinhibitory control element (ACE) that blocks the flow of electrons from flavins to heme. CaM binding activates the enzyme by displacing this inhibitory element.

In addition to relieving autoinhibition, CaM can activate target proteins by allosteric mechanisms. *Bacillus anthracis*, the cause of anthrax, secretes a toxic protein called edema factor (EdF), which is an adenylyl cyclase (AC) that converts the host cell's ATP to cAMP. EdF is activated by CaM, which is absent in the bacterium, thus ensuring no toxicity until EdF is released into an infected cell. EdF is activated by a novel mechanism termed "active site remodeling." The lobes of CaM recognize noncontiguous EdF sequences causing a large reorientation of the domains (Drum et al. 2002). The C-terminal lobe is Ca^{2+}-bound and grips a helix of EdF in the canonical manner, whereas the N-lobe is Ca^{2+} free and interacts with EdF in the closed conformation. CaM binds distal to the substrate-binding site of EdF and does not dramatically alter its structure but rather stabilizes the active conformation. *Bordetella pertussis*, the cause of whooping cough, also secretes an AC toxin, called CyaA, which has higher affinity than EdF for CaM and can be activated by the CaM C-lobe alone through a similar mechanism involving stabilization of the catalytic site.

CaM binding to ion channels modulates their gating by inducing dimerization. Small conductance Ca^{2+}-activated potassium (SK) channels are opened by intracellular Ca^{2+} via CaM associating with a cytoplasmic domain C-terminal to the pore. This triggers channel opening in a 2:2 complex. The 96-residue CaM-binding region of SK contains no recognizable CaM target motifs and forms two interacting helices connected by a loop. Two SK peptides interact through the longer helix to form an antiparallel dimer with a three-helix bundle at each end. In a previously unknown binding mode, the N- and C-lobes of an extended CaM contact different monomers, and each CaM interacts with a total of three helices. Each C-lobe binds one helix in a Ca^{2+}-independent manner while each N-lobe forms Ca^{2+}-dependent interactions with two helices, one from each subunit. It is proposed that through this N-lobe interaction, CaM stabilizes SK dimerization in response to Ca^{2+}, resulting in channel opening.

In contrast to SK channels, Ca_V channels respond to membrane potential, but their opening is modulated by CaM. A 77-residue peptide comprising "pre-IQ" and IQ motifs of the L-type $Ca_V1.2$ channel crystallized as a dimer in complex with four Ca^{2+}-bound CaM molecules. While Ca^{2+}/CaM bound the IQ motifs in a manner observed before, the pre-IQ regions formed a long coiled coil bridged by two Ca^{2+}/CaM molecules, each of which bound one subunit with its C-lobe and the other with its N-lobe, providing another example of channel dimerization. Other Ca^{2+} channels including the IP_3 and ryanodine receptors also bind CaM, yet the mechanisms by which CaM modulates their gating remain elusive.

Calmodulin in Protein Engineering and Biosensor Applications

CaM-GFP-based Ca^{2+} indicators have many advantages over synthetic fluorescent dyes in measuring free Ca^{2+} concentrations in living cells including the onset and termination of Ca^{2+} signaling in specific cellular compartments, such as cytoplasm, nucleus, or endoplasmic reticulum (Miyawaki et al. 1997). In the traditional yellow "cameleon" (YC), cyan fluorescent protein is fused to CaM, the CaM-binding peptide of MLCK and a yellow fluorescent protein. Upon increasing the free Ca^{2+} concentration, the CaM module binds Ca^{2+} and wraps around the fused peptide bringing the GFP variants closer together resulting in increased FRET. The dynamic range is improved by the cameleon YC6.1, in which a peptide from CaMKK is placed between the globular domains of CaM. The high affinity and Ca^{2+} dependence of CaM binding to kinase peptides is also exploited in affinity purification approaches using MLCK peptide as a fusion tag and calmodulin resin for purification.

Calmodulin Literature

Since its discovery in 1970, CaM has been extensively studied resulting in over 34,000 publications by several prominent groups. Only a small fraction is included in the reference list of this chapter.

Cross-References

▶ Bacterial Calcium Binding Proteins
▶ Barium Binding to EF-Hand Proteins and Potassium Channels
▶ C2 Domain Proteins
▶ Calcineurin
▶ Calcium in Biological Systems
▶ Calcium in Health and Disease
▶ Calcium in Nervous System
▶ Calcium, Local and Global Cell Messenger
▶ Calcium-Binding Protein Site Types
▶ Calcium-Binding Proteins, Overview
▶ EF-Hand Proteins
▶ EF-hand Proteins and Magnesium
▶ Magnesium Binding Sites in Proteins
▶ Parvalbumin

References

André I, Kesvatera T, Jönsson B et al (2004) The role of electrostatic interactions in calmodulin-peptide complex formation. Biophys J 87:1929–1938

André I, Kesvatera T, Jönsson B et al (2006) Salt enhances calmodulin-target interaction. Biophys J 90:2903–2910

Babu YS, Bugg CE, Cook WJ (1988) Structure of calmodulin refined at 2.2 Å resolution. J Mol Biol 204:191–204

Bauer MC, O'Connell D, Cahill DJ et al (2008) Calmodulin binding to the polybasic C-termini of STIM proteins involved in store operated calcium entry. Biochemistry 47:6089–6091

Bayley PM, Findlay WA, Martin SR (1996) Target recognition by calmodulin: dissecting the kinetics and affinity of interaction using short peptide sequences. Protein Sci 5: 1215–1228

Berggård T, Arrigoni G, Olsson O et al (2006) 140 mouse brain Ca^{2+}-calmodulin-binding proteins identified by affinity chromatography and tandem mass spectrometry. J Proteom Res 5:669–687

Brzeska H, Venyaminov SV, Grabarek Z et al (1983) Comparative studies on thermostability of calmodulin, skeletal muscle troponin C and their tryptic fragments. FEBS Lett 153: 169–173

Deisseroth K, Heist EK, Tsien RW (1998) Translocation of calmodulin to the nucleus supports CREB phosphorylation in hippocampal neurons. Nature 392:198–202

Dolmetsch RE, Xu K, Lewis RS (1998) Calcium oscillations increase the efficiency and specificity of gene expression. Nature 392:933–936

Drum CL, Yan SZ, Bard J, Shen YQ, Lu D, Soelaiman S, Grabarek Z, Bohm A, Tang WJ (2002) Structural basis for the activation of anthrax adenylyl cyclase exotoxin by calmodulin. Nature 415:396–402

Evenäs J, Malmendal A, Akke M (2001) Dynamics of the transition between open and closed conformations in a calmodulin C-terminal domain mutant. Structure 9:185–195

Ikura M, Clore GM, Gronenborn AM et al (1992) Solution structure of a calmodulin-target peptide complex by multidimensional NMR. Science 256:632–638

Ikura M, Ames JB (2006) Genetic polymorphism and protein conformational plasticity in the calmodulin superfamily: two ways to promote multifunctionality. Proc Natl Acad Sci USA 103:1159–1164

Linse S, Drakenberg T, Forsén S (1986) Mastoparan binding induces a structural change affecting both the N-terminal and C-terminal domains of Calmodulin. A ^{113}Cd-NNR Study. FEBS Lett 199:28–32

Linse S, Helmersson A, Forsén S (1991) Calcium binding to calmodulin and its globular domains. J Biol Chem 266:8050–8054

Linse S, Voorhies NE et al (2000) An EF-hand phage display study of calmodulin subdomain pairing. J Mol Biol 296:473–486

Malmendal A, Linse S, Evenäs J et al (1989) Battle for the EF-hands: magnesium-calcium interference in calmodulin. Biochemistry 38:11844–11850

Martin SR, Andersson Teleman A, Bayley PM, Drakenberg T, Forsen S (1985) Kinetics of calcium dissociation from calmodulin and its tryptic fragments. A stopped-flow fluorescence study using Quin 2 reveals a two-domain structure. Eur J Biochem 151:543–550

Meador WE, Means AR, Quiocho FA (1992) Target enzyme recognition by calmodulin: 2.4 a structure of a calmodulin-peptide complex. Science 257:1251–1255

Miyawaki A, Llopis J, Heim R et al (1997) Fluorescent indicators for Ca2+ based on green fluorescent proteins and calmodulin. Nature 388:882–887

O'Connell DJ, Bauer MC, O'Brien J et al (2010) Integrated protein array screening and high throughput validation of 70 novel neural calmodulin binding proteins. Mol Cell Proteomics 9:1118–1132

Pitt GS (2007) Calmodulin and CaMKII as molecular switches for cardiac ion channels. Cardiovasc Res 73:641–647

Rhoads AR, Friedberg F (1997) Sequence motifs for calmodulin recognition. FASEB J 11:331–340

Wayman GA, Lee YS, Tokumitsu H et al (2008) Calmodulin-kinases: modulators of neuronal development and plasticity. Neuron 59:914–931

Yamniuk AP, Vogel HJ (2004) Calmodulin's flexibility allows for promiscuity in its interactions with target proteins and peptides. Mol Biotechnol 27:33–57

Zhang M, Tanaka T, Ikura M (1995) Calcium-induced conformational transition revealed by the solution structure of apo calmodulin. Nat Struct Biol 2(9):758–767

Calmodulin (CaM)

▶ Calcium Sparklets and Waves

Calnexin and Calreticulin

Elzbieta Dudek[1] and Marek Michalak[1,2]
[1]Department of Biochemistry, University of Alberta, Edmonton, AB, Canada
[2]Faculty of Medicine and Dentistry, University of Alberta, Edmonton, AB, Canada

Synonyms

Calcium-binding proteins; Lectins; Molecular chaperones

Definition

Both of these proteins are present in endoplasmic reticulum that bind to misfolded proteins and assist in their folding and posttranslational modification. Calnexin is an integral membrane protein and endoplasmic reticulum–associated molecular chaperone. Calreticulin is known as a multifunctional Ca^{2+}-binding/buffering endoplasmic reticulum resident chaperone. The protein is responsible for buffering of over 50% of endoplasmic reticulum luminal Ca^{2+} and assisting in folding of newly synthesized glycoproteins.

Background

The endoplasmic reticulum (ER) is a multifunctional organelle responsible for many vital processes in the cell including the synthesis, intracellular transport, and quality control of membrane-associated and secreted proteins; lipid and steroid synthesis; Ca^{2+} signaling and homeostasis; communication with other intracellular organelles including the mitochondria and plasma membrane and ER stress responses. To perform these diverse functions, the ER contains a number of multifunctional integral and resident molecular chaperones and folding enzymes. Two important proteins in the ER are the key multifunctional lectin-like chaperones calnexin, an integral ER membrane protein, and calreticulin, a resident ER protein.

Calnexin

Calnexin was identified over 20 years ago as a Ca^{2+}-binding phosphoprotein and an ▶ ER lectin-like molecular chaperone. The amino acid sequence of calnexin is highly conserved between many species, for example, human, rat, mouse, and dog calnexin share 93–98% amino acid sequence identity (Jung et al. 2006). Calnexin homologs are also known in *Saccharomyces cerevisiae, Schisosaccharomyces pombe, Caenorhabditis elegans, Dictyostelium discoideum, Drosophila melanogaster, Ciona savignyi, Ciona intestinalis, and Arabidopsis thaliana* (Jung et al. 2006). The human gene of calnexin is localized on chromosome 5 (5p35) (Jung et al. 2006).

Structure

Calnexin is a 90-kDa type 1 integral membrane protein with a cytoplasmic C-terminal RKPRRE ER retention signal and an N-terminal signal sequence (Jung et al. 2006). The protein consists of four structural and functional domains: the N-globular domain, the P-arm domain, a short transmembrane domain anchors calnexin, and the C-terminal domain extends into the cytoplasm (Fig. 1). The crystal structure of the luminal portion of calnexin has been solved providing important insights into the function of the protein. The N-terminal globular domain of calnexin comprises two regions (residues 1–270 and 418–482) where a concave β-sheet with six strands and a convex β-sheet with seven strands are arranged antiparallel forming a hydrophobic core through interactions between hydrophobic residues (Fig. 1) (Schrag et al. 2001). There is also a disulfide bond within the luminal portion of calnexin formed between C^{161} and C^{195}. This region contains a glucose-binding site which is responsible for the lectin-like properties of calnexin. The carbohydrate-binding site has been mapped, and structural studies have suggested that amino acids M^{189}, Y^{165}, K^{167}, Y^{186}, E^{217}, C^{161}, and C^{195} are involved in substrate binding (Schrag et al. 2001). Recent mutagenesis studies showed that the W^{428}, within the N-terminal region protein, may also affect binding of the glycosylated substrate. The Ca^{2+}-binding site is also localized within the globular domain of calnexin. Ca^{2+} binding to the ER luminal portion of calnexin may play a structural role as the ion-binding site is located at a distance from the carbohydrate site and calnexin is known to undergo Ca^{2+}-dependent conformational changes (Williams 2006). There are also Zn^{2+}-binding sites localized to amino acid residues 1–270, and ion binding results in conformational changes thought to expose hydrophobic substrate binding amino acid residues of the protein. Calnexin also binds ATP which might be required for substrate-calnexin interactions although the binding site of this nucleotide has not yet been identified (Williams 2006). The P-arm domain of calnexin (residues 270–414) forms a structure that consists of an extended and flexible arm (Fig. 1). The P-domain contains four copies of two proline-rich motifs arranged in a linear pattern of 11112222, where motif 1 is PxxIxDPDAxKPEPWDE and motif 2 is GxWxPPxIxNPxYx (Fig. 1). The extended arm forms a long hairpin composed of head-to-tail interactions between a copy of motif 1 and a copy of motif 2. Each of the two repeat motifs has a three residue long β-strand (Schrag et al. 2001). The P-domain of calnexin might be involved in the interaction with other chaperones in the lumen of the ER. One of the best known binding partners of calnexin is oxidoreductase ERp57. Mutagenesis studies have shown that residues located at the tip of the P-domain (W^{343}, D^{344}, G^{349}, E^{352}, and E^{351}) are involved in binding to ERp57 (Coe and Michalak 2010). Although the P-arm domain of calnexin is not directly involved in substrate binding, the P-arm domain enhances the protein folding function of the globular domain. In addition, the P-arm domain may have a role in physically constraining the glycoprotein when bound to the N-domain (Williams 2006).

Amino acid residues 482–502 of calnexin form a transmembrane domain that anchors calnexin to the ER membrane and localizes the whole molecule to the immobile phase of the ER environment. It is unclear as to whether the transmembrane domain may have a role outside of its anchoring function. It has been suggested that the transmembrane domain may promote association of calnexin with membrane proteins by holding calnexin at the ER membrane, or the calnexin transmembrane domain may directly bind the transmembrane domains of substrates (Jung et al. 2006).

The C-terminal domain of calnexin (residues 502–593) forms an acid and charged tail oriented to the cytoplasm (Fig. 1). The function of the charged tail is not well established, but it is known that highly acidic sections of calnexin are high-capacity Ca^{2+}-binding sites. The C-domain of calnexin is phosphorylated at S^{552} and S^{562} by casein kinase II (CDK) and at S^{581} by protein kinase C (PKC) or proline-directed kinase (PDK) (Fig. 1) (Chevet et al. 2010). Phosphorylated calnexin is known to interact with major histocompatibility complex (MHC) class I molecules and affects the transport of MHC class I molecules through the ER (Chevet et al. 2010). The cytoplasmic tail has also been shown to interact with membrane-bound ribosomes, and it has been speculated that this binding may regulate the chaperone activity of calnexin (Jung et al. 2006).

Function

The best established role of calnexin is its chaperone function in ER. The protein, along with calreticulin, is a lectin-like molecular chaperone that recognizes glycosylated and nonglycosylated proteins and participates in the folding and quality control of newly synthesized glycoproteins (Maattanen et al. 2010). The addition of N-glycans to nascent proteins is a common posttranslational modification that occurs within the ER lumen. Both calnexin and calreticulin bind, in a Ca^{2+}-dependent manner, monoglucosylated carbohydrate intermediates (Williams 2006). Upon release of the terminal glucose by glucosidase II, calnexin-substrate interaction is broken, and if the protein is incorrectly folded, it is reglucosylated by UDP glucose: glycoprotein glucosytransferase (UGGT) to reassociate with calnexin for an additional folding cycle. Calnexin deficiency affects the maturation and cell surface targeting of some of its substrates and impacts quality control in the protein secretory pathway. Misfolded glycoproteins may also be substrates for ER-associated degradation (ERAD). In ERAD, a slow-acting enzyme, alpha-1,2-mannosidase, removes a mannose residue at Man_9, allowing the protein to reach Man_8. Exposure of Man_8 allows the misfolded protein to interact with ER degradation enhancing alpha-1,2-mannosidase-like protein (EDEM) (Maattanen et al. 2010).

The lumen of the ER is the main storage site of Ca^{2+} ions that are involved in a universal signaling role in the cell. The concentration of total (free and bound) Ca^{2+} within the ER is estimated to be about 2 mM, while the free ER Ca^{2+} varies from 50 to 500 µM (Williams 2006). Fluctuations of the ER luminal Ca^{2+} concentration result in impact on numerous cellular functions including motility; protein synthesis, modification and folding, and secretion; gene expression; cell-cycle progression; and apoptosis. Thus, changes in Ca^{2+} concentration may play a signaling role in both the lumen of the ER as well as in the cytosol. The enormous capacity of the ER for Ca^{2+} is mediated by the Ca^{2+} buffering capacity of the resident protein. Calnexin may also influence ER ▸ Ca^{2+} homeostasis. Calnexin interacts with SERCA2b (sarcoplasmic-endoplasmic reticulum Ca^{2+} ATPase) and inhibits SERCA function in Xenopus by interacting with D^{1036} in the COOH terminus of SERCA2b. Phosphorylated calnexin interacts with SERCA2b, but the interaction is lost upon dephosphorylation of the calnexin. In Drosophila, the calnexin homolog is required for rhodopsin maturation, and calnexin affects Ca^{2+} entry to photoreceptor cells during light stimulation. Deletion of the calnexin gene (calnexin 99A) leads to severe defects in rhodopsin expression, whereas other photoreceptor cell proteins are expressed normally (Rosenbaum et al. 2006). Mutations in calnexin also impair the ability of photoreceptor cells to control cytosolic Ca^{2+} levels following activation of the light-sensitive channels and finally lead to retinal degeneration that is enhanced by light (Rosenbaum et al. 2006). Altogether, the results suggest that calnexin deficiency may also affect visual pigments and illustrate a critical role for calnexin in rhodopsin maturation and Ca^{2+} regulation (Rosenbaum et al. 2006).

Through the interaction with integrins in the ER, calnexin is involved in cell adhesion. Heterodimers of α-β integrins mediate cell-cell adhesion and also adhesion of proteins to the cellular matrix and plasma membrane. The largest subgroup of the β-chain family, β1-integrins, are known to bind many adhesion proteins such as collagen, laminin, fibronectin, vitronectin, and VMCAM-1. Calnexin associates with the β1 subgroup of integrins and is involved in their assembly and retention of their immature form in the ER. Thus, calnexin provides a steady supply of β1-integrins to associate with α-chains and form functional heterodimers (Jung et al. 2006).

Additionally, using cell lines deficient in calnexin, it was possible to demonstrate that calnexin is a component of apoptotic pathways involving the ER.

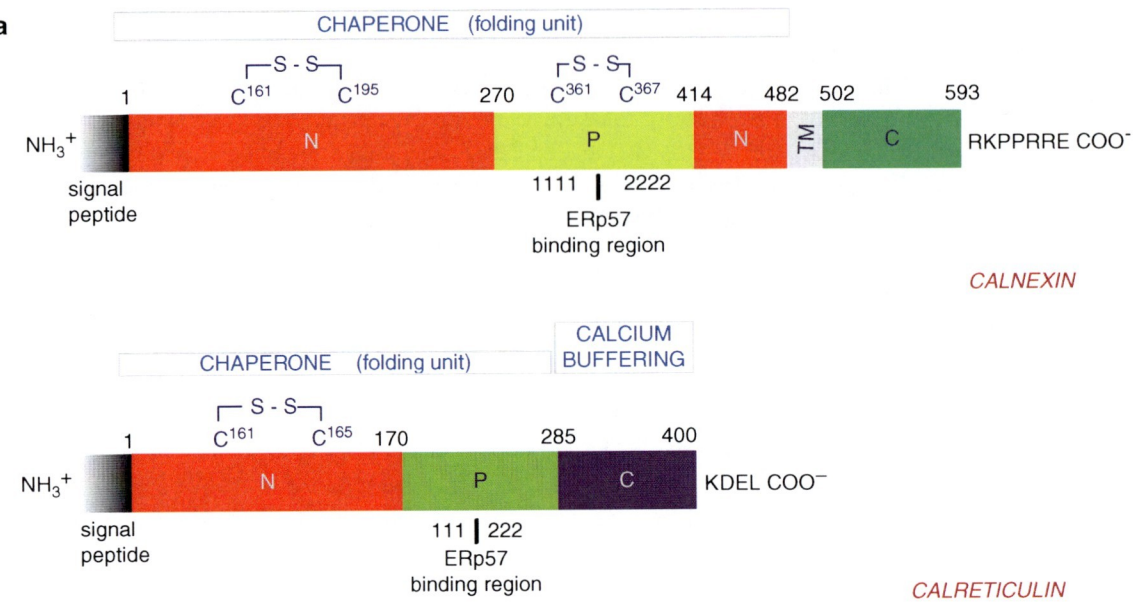

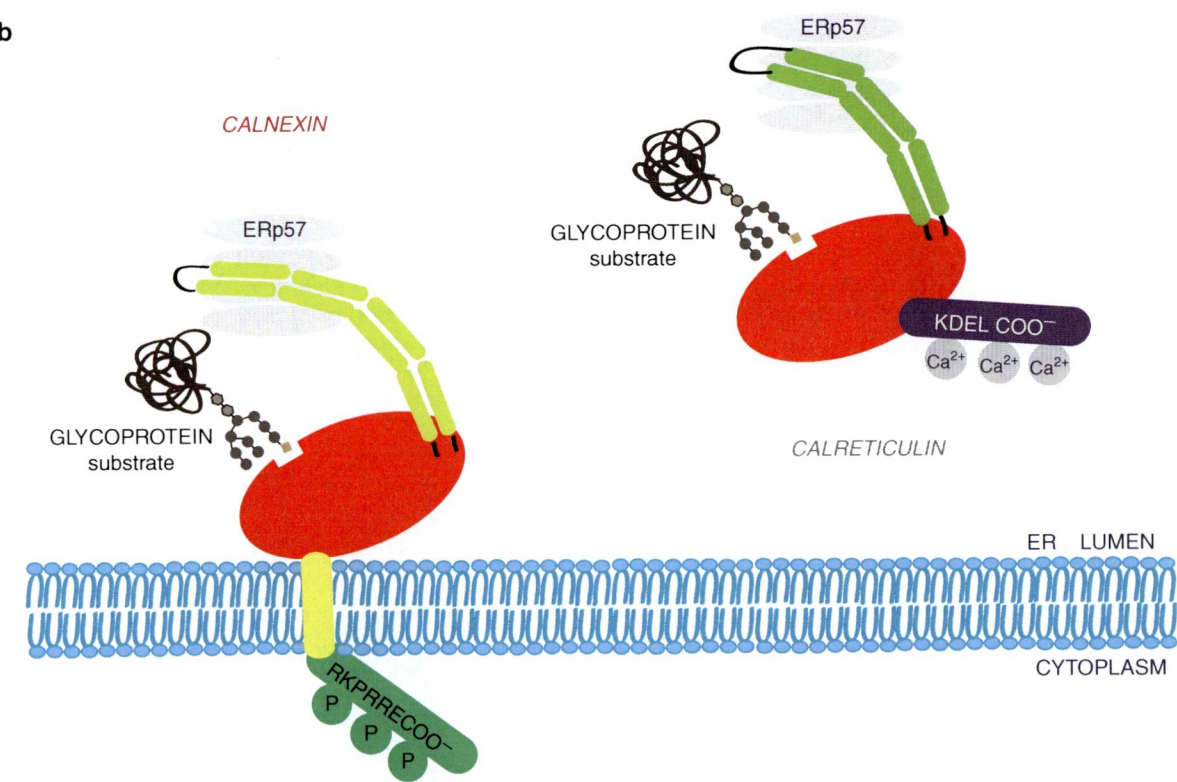

Calnexin and Calreticulin, Fig. 1 Linear models and schematic structures of calnexin and calreticulin. (**a**) In the upper panel the linear model of calnexin domains is shown. The N – terminal signal sequence (*grey box*), N domain (*red box*), P-domain (*light green box*) TM-transmembrane domain (*yellow box*) and a C-domain (*dark green box*) with the C-terminal ER retention sequence (RKPRRE). Repeats 1 (amino acids sequence PxxIxDPDAxKPEPWDE) and 2 (amino acids sequence GxWxPPxIxNPxYx) are indicated as 1111 and 2222 within the P-domain. The amino acids involved in the thiol

Although calnexin-deficient cells are resistant to ER stress-induced apoptosis, the absence of calnexin results in a decrease in caspase 12 expression as well as inhibition of Bap31 cleavage. Calnexin deficiency does not affect caspase 3, caspase 8, or cytochrome c release and may suggest that calnexin is not important in initiating apoptosis but rather has role in the induction of the latter stages of apoptosis. Moreover, calnexin itself is known to be cleaved by ER stress-inducing and noninducing conditions, and this cleavage is known to attenuate apoptosis.

Calnexin-Deficient Mouse Models

Two calnexin-deficient mouse models have been generated. Although calnexin-deficient mice are viable, they have a severe ataxic phenotype. Calnexin-deficient mice are 30–50% smaller than wild-type and show gait disturbance and splaying of the hindlimbs. Electron micrographs of the spinal cord and sciatic nerve indicate that there is severe dysmyelination in the peripheral nervous system (PNS). Thus, it is not surprising that reduced nerve conduction velocity is observed in calnexin-deficient mice (Kraus et al. 2010). A different strain of mouse with a disrupted calnexin gene expressing a truncated (15-kDa smaller) form of calnexin was also reported. Truncated calnexin possesses disruptions in amino acid residues within the carbohydrate pocket in the N-globular domain (C^{161}, C^{195}, Y^{165}, and K^{167}). This mutant mouse is phenotypically similar to the calnexin-deficient mouse suggesting that the chaperone function of calnexin is likely responsible for the observed neurological phenotype. Interestingly, there is no significant impact of calnexin deficiency on the development of immune system (Kraus et al. 2010).

Calreticulin

Calreticulin was discovered over 30 years ago as a Ca^{2+}-binding protein of the sarcoplasmic reticulum with high homology to calregulin. Calreticulin is differentially expressed under a variety of physiological and pathological conditions. A reduced level of calreticulin is observed in heart and brain in differentiated cells, while in highly differentiated cells or upon induction of ER stress, calreticulin is upregulated (Michalak et al. 2009).

Structure

Calreticulin is a 46-kDa ER resident protein involved in Ca^{2+} binding and is a well-known molecular chaperone. Calreticulin has an N-terminal cleavable signal sequence and an ER retention sequence KDEL at the C-terminus (Fig. 1). Similar to calnexin, calreticulin is composed of distinct structural and functional domains: the N-globular domain, the P-arm domain, and the C-domain (Maattanen et al. 2010). Based on the secondary structure, the N-terminal domain (residues 1–170) is predicted to be comprised of eight antiparallel β-strands (Fig. 1). The N-terminal domain of calreticulin contains the oligosaccharide- and polypeptide-binding sites. Although little is known about molecular features of substrate binding to calreticulin, a significant portion of the oligosaccharide domain has been mapped. Mutations to residues Y^{109} and D^{135} were identified to abolish interactions between calreticulin and oligosaccharides. Additionally, other amino acids were identified as responsible for sugar binding: K^{111}, Y^{128}, and D^{127} (Michalak et al. 2009). W^{302} and H^{153} in the N-domain have been shown to be critical for chaperone function and also affect the structure of calreticulin. Disruption of the

---

Calnexin and Calreticulin, Fig. 1 (continued) linkages are connected with an S-S, with the numbers delineating the amino acid residues at the transition between the various domains. In the lower panel the linear model of calreticulin domains is shown. The N-terminal signal sequence (*grey box*), N domain (*red box*), and P-domain (*green box*) and a C domain (*purple box*) with the C-terminal ER retention sequence (KDEL). Repeats 1 (amino acid sequences IxDPxA/DxKPEDWDx) and 2 (amino acid sequences GxWxPPxIxNPxYx) are indicated as 111 and 222 within the P-domain. The amino acids involved in the thiol linkage are connected with an S-S, with the numbers delineating the amino acid residues. (**b**) On the left, schematic structure of calnexin, transmembrane ER protein is shown. The P-domain (*light green*) contains residues involved in ERp57 binding (*grey*). The N-globular domain (*red*) contains residues involved in carbohydrate substrates binding. The TM domain (*yellow*) anchors calnexin to the ER membrane. The C-domain (*dark green*) is oriented to the cytoplasm and possesses three phosphorylation sites (*dark green spheres*). On the right schematic structure of calreticulin, soluble in ER lumen, is shown. The P-domain (*green*) contains residues involved in ERp57 binding (*grey*). The N-globular domain (*red*) contains residues involved in carbohydrate substrates binding. The C-domain (*purple*) has high capacity for binding Ca^{2+} (*grey spheres*)

disulfide bridge between C^{88} and C^{120} only partially affected the structure of calreticulin. The N-domain also binds Zn^{2+}, and this ion binding may have structural effects on the whole protein and thus affect protein function (Williams 2006). A proteolysis stable N-domain core is formed in the presence of Ca^{2+}, and development of this stable core may have specific pathophysiological implications. The P-domain is located in the middle of calreticulin's amino acid sequence (residues 170–285) and forms an extended, flexible arm (Fig. 1). The P-domain is a proline-rich region similar to the one found in calnexin and is composed of three copies of two repeat amino acid sequences (motif 1: IxDPxA/DxKPEDWDx and motif 2: GxWxPPxIxNPxYx). The repeat sequences are arranged in a 111222 pattern and are thought to be involved in oligosaccharide binding, together with N-domain. Based on NMR studies, it is known that the structure of the P-domain of calreticulin contains an extended region which is stabilized by three antiparallel β-sheets (Williams 2006). ERp57 can dock onto the tip of the P-domain, and mutational analysis showed that amino acids, E^{239}, D^{241}, D^{243}, and W^{244}, are involved in this interaction (Coe and Michalak 2010). The P-domain has also been found to bind Ca^{2+} with a high affinity ($K_d = 1$ μM) and low capacity (1 mol of Ca^{2+} per 1 mol of protein). The binding site appears to possess a potential ▶ EF-hand-like helix-loop-helix motif. The C-domain of calreticulin (residues 285–400) is mainly composed of negatively charged residues interrupted, at regular intervals, with one or more basic K or R residues (Fig. 1). It was found that disruption of these basic residues results in decreased in Ca^{2+}-binding capacity, and the binding can be directly attributed to the lysine amino group side chains. The C-domain is responsible for the Ca^{2+} buffering function of the protein and binding (Michalak et al. 2009). It is known that the C-domain is responsible for binding 50% of ER Ca^{2+} with a low affinity ($K_d = 2$ M) and high capacity (25 mol of Ca^{2+} per mol of protein), opposite to Ca^{2+} binding by the P-domain (Michalak et al. 2009).

Function

Many functions of calreticulin are related to the protein's ability to bind Ca^{2+} with high affinity through the C-domain. Thus, it is not surprising that calreticulin is known as a major ER Ca^{2+} buffer and similar to calnexin's function as a molecular chaperone (Michalak et al. 2009). Other functions within the cell are extensions of calreticulin's roles as a Ca^{2+} buffering and chaperone protein.

Being able to bind Ca^{2+} with high capacity and low affinity, calreticulin is classified as a class I Ca^{2+}-binding protein. In order to understand the Ca^{2+} binding and buffering properties of calreticulin, it is important to describe the cellular and animal models with calreticulin deficiency and overexpression. Calreticulin-deficient cells have reduced Ca^{2+} storage capacity in the ER, while overexpression of calreticulin leads to an increased amount of Ca^{2+} in the cell's intracellular stores (Williams 2006). For example, in calreticulin-deficient mouse embryonic fibroblasts (MEFs), a significant decrease in Ca^{2+} storage was observed. However, when the P + C-domains involved in Ca^{2+} binding were expressed in the calreticulin-deficient cells, there was a full recovery of ER Ca^{2+} storage capacity. Expression of the N + P-domains of calreticulin did not recover Ca^{2+} storage capacity in the ER of calreticulin-deficient cells. Taken together, these results both suggest that the main role of the C-domain is in Ca^{2+} binding. Moreover, the impact of Ca^{2+} fluctuations on the conformation and function of calreticulin was demonstrated by studies involving limited proteolysis of calreticulin by trypsin. These digestion studies showed that when the ER Ca^{2+} level was low (<100 μM) calreticulin was rapidly degraded by trypsin, while under high ER Ca^{2+} concentrations (500 μM–1 mM), the protein forms an N-domain protease-resistant core. The ability to regulate ER luminal Ca^{2+} concentrations affects other biological functions of calreticulin within the cell (Michalak et al. 2009). For example, cytotoxic T-lymphocytes release Ca^{2+}-activated perforin, granzymes/proteases, and calreticulin that, upon interaction with a target cell, cause lysis and apoptosis. Calreticulin regulates this process through Ca^{2+} binding. It has been found that increased levels of calreticulin result in increased amounts of chelated Ca^{2+} and a decrease in activated perforin involved in penetration of the plasma membrane of the targeted cell (Michalak et al. 2009). Calreticulin has also been found outside the ER lumen, in the nucleus, cytoplasm, and on the cell surface, where it may be involved in other biological functions within the cell. Cell surface calreticulin is involved in the immune system where it functions in antigen presentation and complement activation (Michalak et al. 2009). Calreticulin translocates to the cell surface with ERp57 allowing for presentation to T cells, initiation of the immune response, and apoptosis of the target cell. It is also an important protein for focal adhesion assembly and impacts phagocytosis and the proinflammatory response. Moreover, calreticulin induces the migration and motility of cells involved in wound healing such as keratinocytes, fibroblasts, monocytes, and macrophages.

Translocation of calreticulin to the cytoplasm is still controversial. There are three proposed mechanisms leading to the cytoplasmic localization of calreticulin: (1) the protein is not efficiently targetted to the ER, and this leads to accumulation of the calreticulin precursor in cytoplasm; (2) redistribution of calreticulin is a result of the protein leaking out of the ER; and (3) retrieval from the ER by reverse movement after processing (removal of the signal peptide) (Michalak et al. 2009). The functions of calreticulin in the cytoplasm are not well understood. It has been suggested that calreticulin is a cytoplasmic activator of integrins and is a signal transducer between integrins and Ca^{2+} channels in the plasma membrane. Other groups established that calreticulin could interact directly with the glucocorticoid and androgen hormone receptors and could inhibit steroid-sensitive gene transcription (Michalak et al. 2009).

There are only few studies that have shown calreticulin in the nuclear matrix in hepatocellular carcinomas and binding to core histones. It has also been suggested that calreticulin may act as a nuclear import protein (Michalak et al. 2009).

Calreticulin has been implicated in the cellular response to apoptosis. Modulation of ER Ca^{2+} stores impacts apoptosis as ER Ca^{2+} release is required for activation of transcriptional cascades. It was shown that an increase level of calreticulin results in a sensitivity to apoptosis, while calreticulin-deficient cells are resistant to apoptosis. Further studies showed that calreticulin, intraluminal Ca^{2+}, and disruption in Ca^{2+} regulation influence apoptosis events in cardiomyocytes. Cell-surface calreticulin seems to also play a role in apoptosis. It was shown that in RK3 cells, cell surface calreticulin mediates apoptosis through activating the tumor necrosis factor receptor type 1 (TNFR1). Many forms of cancer display cell surface calreticulin which invites immune cells to destroy them. Moreover, calreticulin was identified as a pro-phagocytic signal highly expressed on the surface of several human cancers, but minimally expressed on most normal cells. Increased calreticulin expression was an adverse prognostic factor in diverse tumors including neuroblastoma, bladder cancer, and non-Hodgkin's lymphoma. Preapoptotic translocation of calreticulin to the cell surface occurs with ERp57 and allows presentation to T cells, triggering the initiation of the immune response and subsequent apoptosis of the immunogenic cell, preventing organism damage. Disruption of the interaction of calreticulin with ERp57 as well as disruption of calreticulin not only prevents surface exposure of calreticulin but also renders the cell resistant to T-cell attack (Coe and Michalak 2010). Exogenous application of calreticulin is able to overcome this resistance. It was also shown that anthracyclines induce the rapid preapoptotic translocation of calreticulin to the cell surface. Blockade or knockdown of calreticulin suppresses the phagocytosis of anthracycline-treated tumor cells by dendritic cells and abolishes their immunogenicity in mice (Michalak et al. 2009).

Mouse Models

Unlike the calnexin-deficient mice described above, calreticulin deficiency is embryonic lethal. This indicates that calnexin and calreticulin, although having similar domain structure and sharing some common functions, may have fundamentally different roles in vivo. During development, calreticulin is expressed at a high level in central nervous system, liver, and heart, and embryonic lethality of calreticulin-deficient mice is due to impaired cardiac development, specifically a marked decrease in ventricular wall thickness. Embryonic fibroblasts derived from calreticulin-deficient mouse embryos show a significant decrease in ER Ca^{2+} capacity, but free ER Ca^{2+} remains unchanged (Williams 2006). Further studies on the embryonic stem cell demonstrated how critical the Ca^{2+} buffering function of calreticulin is to cardiac development as well as for mice survivability. Molecular studies indicate that calreticulin deficiency leads to impaired myofibrillogenesis. Moreover, deficiency of calreticulin leads to deficient intercalated disc formations in the heart, which are adherens-type junctions of cardiac muscle that contain vinculin, N-cadherin, and catenins (Michalak et al. 2009).

Studies examining the molecular level and functional consequences of overexpression of calreticulin have demonstrated a significant increase in Ca^{2+} capacity of the ER. Transgenic mice overexpressing calreticulin in the heart display bradycardia, complete heart block, and sudden death (Michalak et al. 2009).

Cross-References

▶ Calcium Homeostasis: Calcium Metabolism
▶ EF-Hand Proteins
▶ Molecular Chaperones

References

Chevet E, Smirle J, Cameron PH et al (2010) Calnexin phosphorylation: linking cytoplasmic signalling to endoplasmic reticulum luminal functions. Semin Cell Dev Biol 21:486–490
Coe H, Michalak M (2010) ERp57, a multifunctional endoplasmic reticulum resident oxidoreductase. Int J Biochem Cell Biol 42:796–799
Jung J, Coe H, Opas M et al (2006) Calnexin: an endoplasmic reticulum integral membrane chaperone. Calcium Bind Proteins 1:67–71
Kraus A, Groenendyk J, Bedard K et al (2010) Calnexin deficiency leads to dysmyelination. J Biol Chem 285:18928–18938
Maattanen P, Gehring K, Bergeron JJ et al (2010) Protein quality control in the ER: the recognition of misfolded proteins. Semin Cell Dev Biol 21:500–511
Michalak M, Groenendyk J, Szabo E et al (2009) Calreticulin, a multi-process calcium-buffering chaperone of the endoplasmic reticulum. Biochem J 417:651–666
Rosenbaum EE, Hardie RC, Colley NJ (2006) Calnexin is essential for rhodopsin maturation, Ca2+ regulation, and photoreceptor cell survival. Neuron 49:229–241
Schrag JD, Bergeron JJ, Li Y et al (2001) The structure of calnexin, an ER chaperone involved in quality control of protein folding. Mol Cell 8:633–644
Williams DB (2006) Beyond lectins: the calnexin/calreticulin chaperone system of the endoplasmic reticulum. J Cell Sci 119:615–623

Calpactins

▶ Annexins

Calsequestrin

ChulHee Kang
Washington State University, Pullman, WA, USA

Synonyms

CASQ; CSQ

Definition

Calsequestrin, a ~40 kD protein, binds Ca^{2+} with high capacity (40–50 mol Ca^{2+} mol^{-1} calsequestrin) but with moderate affinity ($K_d = 1$ mM) over a Ca^{2+} concentration range between 0.01 and 1 M and releases it with a high off-rate (10^6 s^{-1}). A high concentration of calsequestrin (up to 100 mg mL^{-1}) is present in the lumen of the junctional terminal cisternae of the sarcoplasmic reticulum (SR) (MacLennan and Chen 2009; Royer and Ríos 2009). In humans, there are two major isoforms; the skeletal muscle calsequestrin (Casq1) and the cardiac muscle calsequestrin (Casq2). They are encoded by *casq1* and *casq2* located on chromosomes 1q21 and 1p13.3-p11 and are the only isoforms in fast-twitch skeletal muscle and cardiac muscle, respectively. In slow-twitch skeletal muscle, Casq1 is the major isoform with Casq2 as minor. In general, the Ca^{2+}-binding capacity of Casq2 is higher than that of Casq1. Isoforms of calsequestrin are also present in the endoplasmic reticulum (ER) vacuolar domains of some neurons and smooth muscles (Dulhunty et al. 2009; MacLennan and Chen 2009; Knollmann 2009). In addition, the ER from plant cells contains very similar proteins that bind Ca^{2+} with a relatively high capacity and low affinity.

Biological Function

The regulation and ▶ transport of Ca^{2+} by the SR controls the state of the actin–myosin fibrils of muscle. Release of Ca^{2+} from the terminal cisternae (TC) of the SR brings about muscle contraction, and uptake of Ca^{2+} by the SR establishes relaxation. In this pump-storage release of Ca^{2+}, the ratio between total and free $[Ca^{2+}]$ in the TC is approximately between 100 and 500.

Thus, a Ca^{2+}-binding/storage motif of large capacity must exist inside the SR. At the same time, the same motif must have a relatively low affinity for Ca^{2+} to allow for its rapid release from the SR. Calsequestrin was suggested as this Ca^{2+} buffer inside the SR, lowering free Ca^{2+} concentrations and thereby facilitating further uptake by the ▶ Ca^{2+}-ATPase (SERCA). In addition, calsequestrin might participate in the Ca^{2+} release process by localizing Ca^{2+} at the release site, sensing the SR Ca^{2+} concentration and regulating the amount of Ca^{2+} released in response to an electrical signal from the motor nerve (Györke et al. 2009). A network of interacting proteins in the junctional face region of the SR such as junctin and triadin are implicated in the regulation of calsequestrin's interaction with the calcium channel known as the ryanodine receptor (RyR). Details of this association remain uncertain. Distribution of calsequestrin is uneven within the SR lumen with a markedly high concentration in close proximity to the RyR, possibly mediated by those calsequestrin-anchoring proteins. By localizing calsequestrin in the vicinity of the RyR during Ca^{2+} uptake and release, diffusion time for Ca^{2+} release could be drastically reduced.

Structure of Calsequestrin

The amino acid sequences of both Casq1 and Casq2 from several species have been characterized. In general, those isoforms exhibit 60–70% sequence identity (Fig. 1). Consistent with their high Ca^{2+}-binding capacity, the sequence of calsequestrin has a large proportion of negatively charged amino acids. Especially, the residues in the carboxyl terminus are the most acidic, which is also most variable area in terms of amino acid sequence among different isoforms. The sequence of calsequestrin does not show a typical KDEL tetrapeptide retrieval signal, thus the targeting mechanism for calsequestrin should be unique.

The crystal structures of both Casq1 and Casq2 show that it is made up of three, nearly identical tandem domains (Wang et al. 1998; Park et al. 2004; Kim et al. 2007). Each of those has the thioredoxin protein fold, although there is no apparent sequence similarity among the three domains (Fig. 2). Each thioredoxin-like domain is made up of a five stranded β-sheet with two α-helices on both sides and has a hydrophobic core with acidic residues on the exterior, generating highly electronegative potential surfaces (Fig. 3). The individual domains contain a high net negative charge, ranging from about -13 to -32, and also have high aromatic amino acid content, ranging from 9% to 13%. The highly negative C-terminal of both Casq1 and Casq2 are completely disordered.

Posttranslational Modification

The posttranslational modifications for calsequestrin have been reported, although their significance is not fully understood (Milstein et al. 2009; Sanchez et al. 2011). From the amino acid sequence, several potential phosphorylation and glycosylation sites were detected. Three serine residues (Ser378, 382, 386) in the canine cardiac isoform were shown to be phosphorylated, whereas the rabbit fast-twitch isoform is phosphorylated on Thr373. In addition, in vitro studies show that both Casq1 and Casq2 can be phosphorylated by casein kinase II. Posttranslational modifications of calsequestrin are associated with its complex cellular transport and physiological Ca^{2+} regulation (McFarland et al. 2010; Sanchez et al. 2011). By in vitro phosphoprylation of human recombinant Casq2, two phosphorylation sites, Ser^{385} and Ser^{393} were identified (Sanchez et al. 2011). Phosphorylation at those two positions produces a disorder-to-order transition of the C-terminus. This phosphorylation-induced extra α-helix and subsequent interaction with nearby electropositive patches lead to increased Ca^{2+}-binding capacity. Calsequestrin undergoes a unique degree of mannose trimming as it is trafficked from the proximal endoplasmic reticulum to the SR. The major glycoform of calsequestrin ($GlcNAc_2Man_9$) found in the proximal endoplasmic reticulum can severely hinder formation of the back-to-back interface, potentially preventing premature Ca^{2+}-dependent polymerization and ensuring its continuous mobility to the SR. Only trimmed glycans can stabilize both front-to-front and the back-to-back interfaces of calsequestrin through extensive hydrogen bonding and electrostatic interactions. Therefore, the mature glycoform of calsequestrin ($GlcNAc_2Man_{1-4}$) within the SR can be retained upon establishing a functional high-capacity Ca^{2+}-binding polymer (Sanchez et al. 2012).

Calsequestrin,

Fig. 1 *Multiple sequence alignment among calsequestrins from different species.* Homo sapien (hs), *Canis lupus familiaris* (cf), *Bos taurus* (bt), *Mus musculus* (mm), *Rattus norvegicus* (rn), *Gallus gallus* (gg), *Oryctolagus cuniculus* (oc), *Xenopus laevis* (xl), *Danio rerio* (dr), *Rana esculenta* (re), *Oncorhynchus mykiss* (om). Conserved residues are shown with a following amino acid color code: *red*, hydrophobic and aromatic amino acids, *blue*, acidic, *magenta*, basic, *green*, hydroxyl and amine containing as specified by the EMBL-EBI CLUSTALW 2.0.8 multiple sequence alignment program. Symbols used according to the ClustalW2 program: (*) denotes invariant amino acid positions, (:) denotes conserved substitutions, (.) denotes semi-conserved substitutions

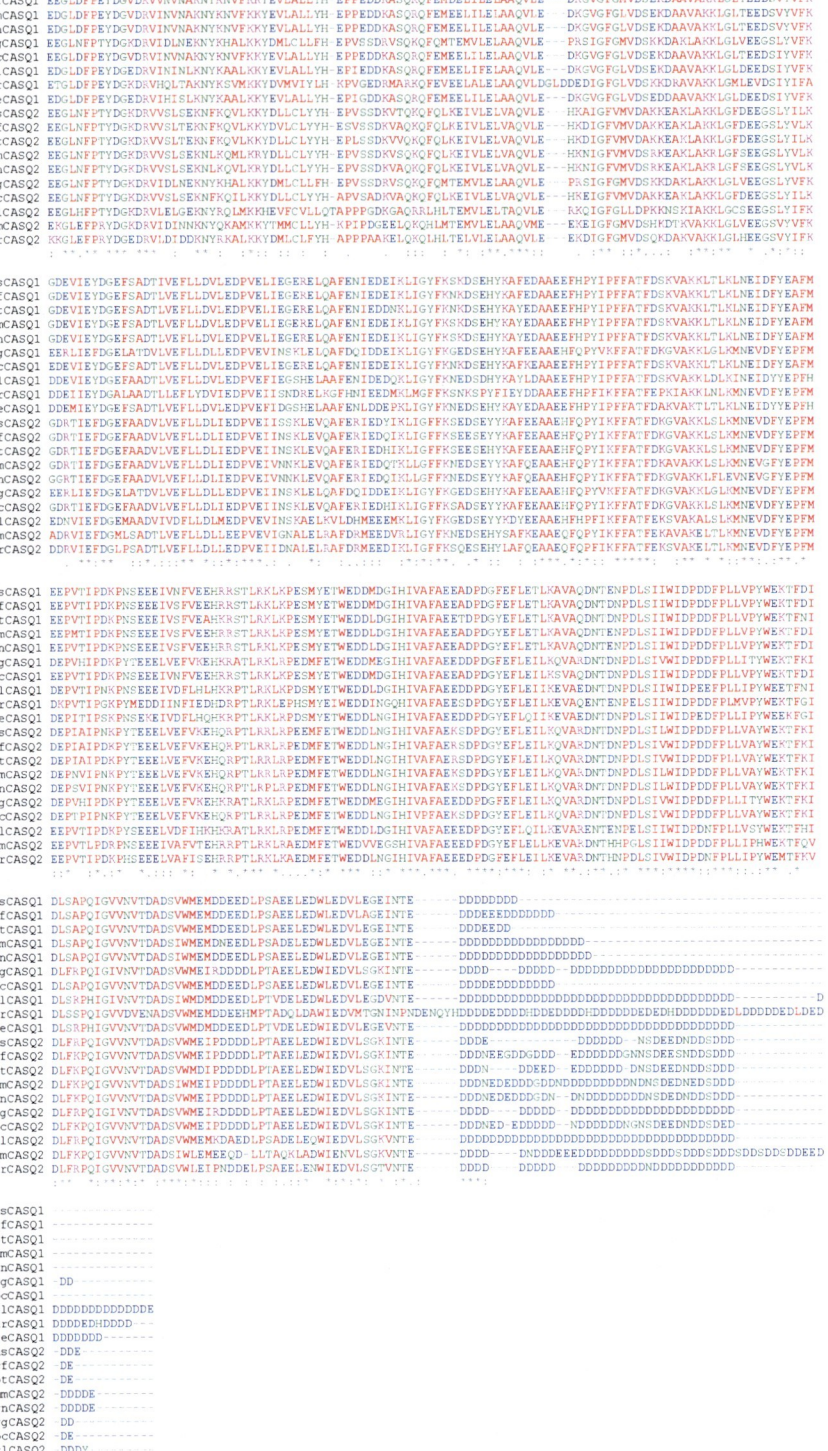

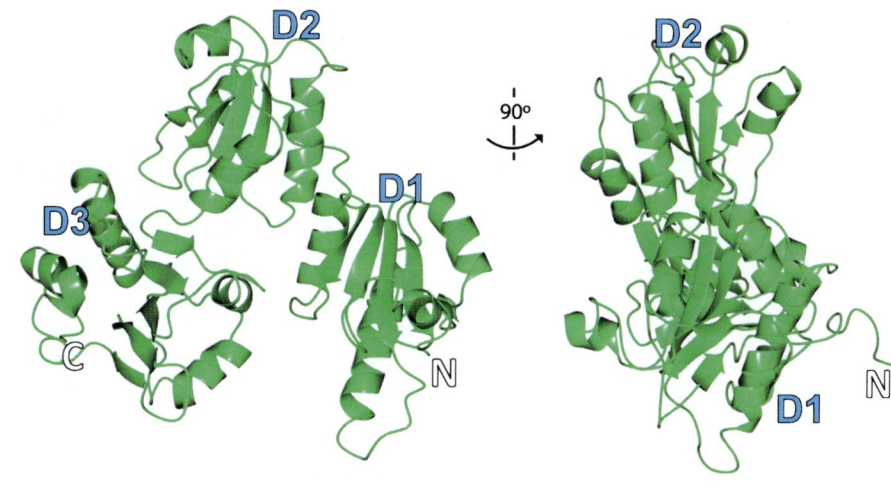

Calsequestrin, Fig. 2 Ribbon diagram showing the distribution of structural elements of calsequestrin. (**a**) The front and (**b**) the side of the calsequestrin. The N- and C-termini are labeled. Three domains are indicated by D1, D2, and D3

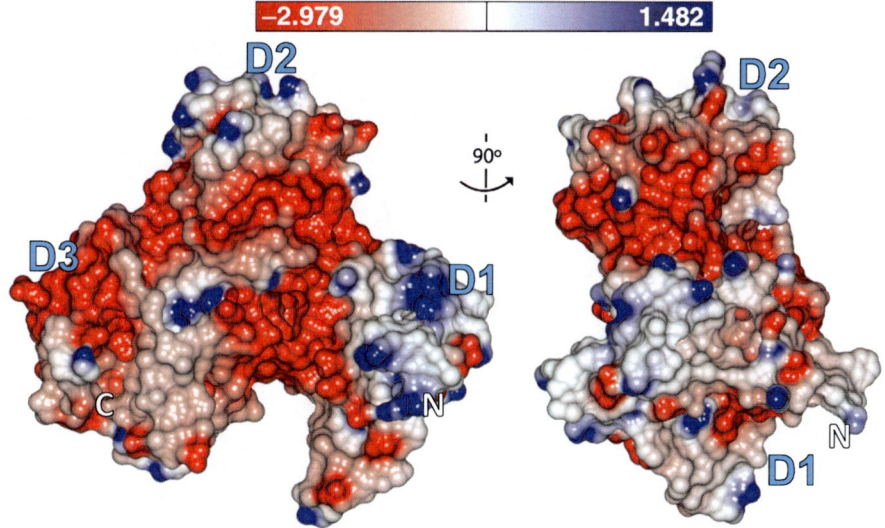

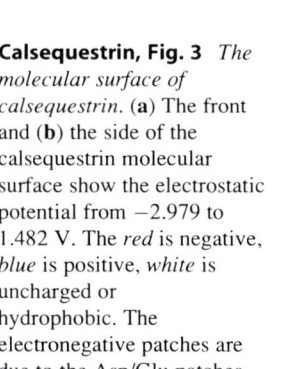

Calsequestrin, Fig. 3 The molecular surface of calsequestrin. (**a**) The front and (**b**) the side of the calsequestrin molecular surface show the electrostatic potential from −2.979 to 1.482 V. The *red* is negative, *blue* is positive, *white* is uncharged or hydrophobic. The electronegative patches are due to the Asp/Glu patches

High-Capacity and Low-Affinity Ca^{2+} binding of Calsequestrin

Crystal structures of Ca^{2+}-complexes for both human and rabbit skeletal calsequestrin are determined (Fig.4), clearly defining its Ca^{2+}-sequestration capabilities through resolution of high- and low-affinity Ca^{2+}-binding sites (Sanchez et al. 2012). Three high-affinity Ca^{2+} sites are in trigonal bipyramidal, octahedral, and pentagonal bipyramidal coordination geometries. However, instead of a typical ▸ EF-hand motif that is present in the high-affinity ▸ Ca^{2+}-binding proteins, most of the low-affinity sites, some of which are μ-carboxylate-bridged, are established by the pairs of acidic residues and solvent molecules on the surface through the net charge density. Major portion of those low affinity sites are established by the rotation of dimeric interfaces, indicating cooperative Ca^{2+}-binding that is consistent with the atomic absorption spectroscopic data (Sanchez et al. 2012). The low-affinity Ca^{2+} binding of calsequestrin is likely driven by the entropy gain from the liberation of many water molecules from the hydrated cations. Due to its large negative charge, calsequestrin is mostly unfolded random coil in low ionic strength but folds into a compact structure as the concentration of ions is increased. As folding progresses, cooperative interactions bring acidic surface residues together, literally creating new Ca^{2+}-binding sites. Cations such as Zn^{2+}, Sr^{2+}, and Tb^{2+} also bind to calsequestrin and cause changes

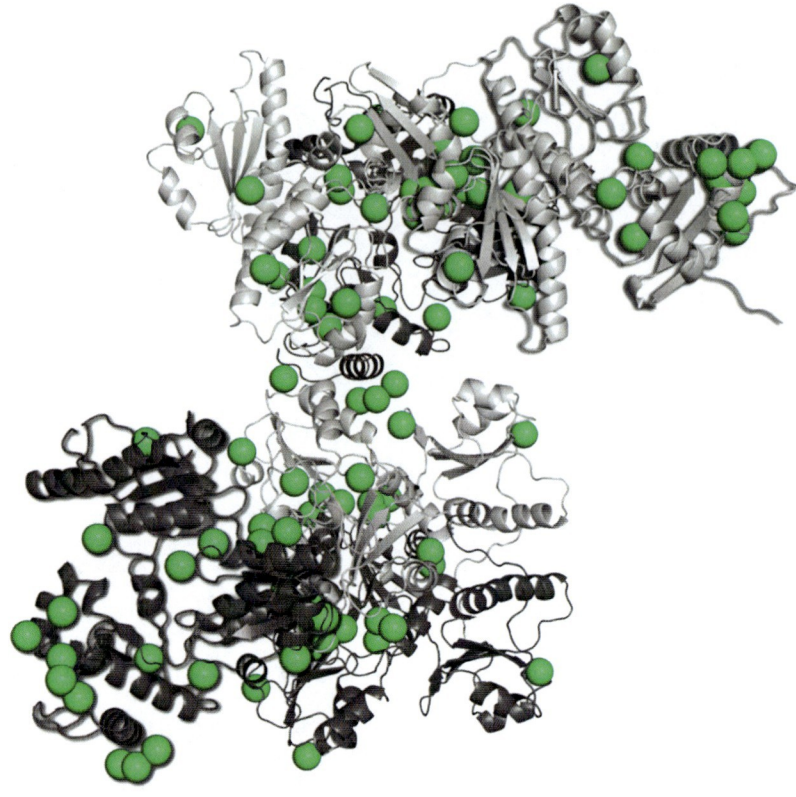

Calsequestrin, Fig. 4 Asymmetric unit of CASQ in P1 unit cell is shown in gray and black ribbon representations, with the Ca^{2+} ions shown as green spheres (Sanchez et al. 2012)

analogous to those caused by Ca^{2+}. Raising the concentration of Ca^{2+} leads to concomitant low-affinity binding of large numbers of Ca^{2+} and Ca^{2+}-induced calsequestrin aggregation. This strong cooperative Ca^{2+} binding accompanies the polymerization of calsequestrin, thus the Ca^{2+}-induced polymerization/ aggregation likely contributes to the function and localization of calsequestrin. On the contrary, monovalent ion-induced folding does not lead to aggregation and precipitation and Ca^{2+}/calsequestrin aggregates are even dissociated by K$^+$. It has been suggested that K$^+$ lowers the affinity of Ca^{2+} for calsequestrin from both cardiac and skeletal SR in vivo.

In the crystal lattices of both Casq1 and Casq2, individual molecules form a continuous, linear polymer of ~90 Å thick through two types of dimer interfaces (Wang et al. 1998). The first type of interface, the front-to-front interface, involves the fitting of the convex globule from domain II on one subunit into a concave depression on the other. This interface involves "arm exchange or domain swapping" between the extended amino terminal ends of the two adjacent molecules. Each extended arm comprises ten N-terminal residues from domain I. This extended arm binds along a groove between two β-strands of domain II in the neighbor. Also, the front-to-front interface contains a number of negatively charged groups. The second type of interface, the back-to-back interface, involves bringing together two jaw-like openings between domains I and III, thus creating a substantial and electronegative pocket within this interface (Park et al. 2004). Since both interfaces involve cross-bridging as acidic groups are brought into proximity, on-rates are likely to be close to the diffusion limit ($\sim 10^9$ M^{-1} s^{-1}), and off-rates to be approximately 10^6 s^{-1} for a binding constant of 10^3 M^{-1}. On the basis of the much greater negative charge of the back-to-back interface, the front-to-front type dimer forms before back-to-back type dimer. The residues involved in those interfaces are the most highly conserved residues in the entire sequence. This situation is reminiscent of the higher conservation of active site residues of other proteins, and thus strongly indicates that those two interfaces are the functional contacts involved in the coupled polymerization and low affinity Ca^{2+} binding (Park et al. 2003, 2004). Therefore, any disruption or interference of this critical intermolecular interaction by either mutation or small molecule binding could disrupt the functional integrity of the calsequestrin molecules (Kang et al. 2010).

The calsequestrin polymer, especially the highly charged disordered tail, provides an acidic and extended surface onto which Ca^{2+} can be adsorbed. The attractive forces exerted by such an extended surface would have a longer range than those from an isolated molecule. A sparingly soluble ion such as Ca^{2+} would tend to spread over the surface of this polymer forming a readily exchangeable film. An array of such polymers would create long, narrow, negatively charged channels leading to the calcium-release channel (RyR). One-dimensional diffusion of Ca^{2+} along the surface of the linear polymer would speed up the diffusion of Ca^{2+} to its release channel compared to random walk diffusion through liquid. Therefore, the calsequestrin polymer located near the RyR can act as a Ca^{2+} wire using a huge Ca^{2+} gradient ($\sim 10^4$) maintained between the cytosolic and the lumenal spaces. In addition, the lack of a fixed or stable structure of the Ca^{2+}-binding sites dictates that binding affinities would be low, and diffusion-limited on and off rates would be as fast as possible (MacLennan and Chen 2009). That is, the use of Ca^{2+} as a cross linker rather than as a tightly bound form speeds up its dissociation.

Related Human Disease

Both mutation and improper levels of calsequestrin have been implicated in several human diseases. Catecholamingeic polymorphic ventricular tachycardia (CPVT) is a familial arrhythmogenic cardiac disorder characterized by syncopal events, seizures, or sudden cardiac death at a young age (Knollmann 2009). The corresponding arrhythmogenic events are usually triggered in response to intensive exercise or emotional stress. Mutations of *casq2* have been found as the cause of autosomal recessive forms of CPVT. In addition, mutations in the cardiac RyR gene, *ryr2*, have been associated with the autosomal dominant forms of this CPVT. A typical CPVT phenotype caused by either *ryr2* or *casq2* is almost identical; thus, a similar causal mechanism has been suggested. So far, in vitro testing shows that all the mutations on *casq2* that cause CPVT either disrupt protein folding or diminish high capacity Ca^{2+} binding and polymerization (Kim et al. 2007; MacLennan and Chen 2009). Therefore, those mutations reduce the total Ca^{2+}-binding and Ca^{2+}-buffering capacity of the SR lumen by the amount that can be ascribed to calsequestrin. In the absence of calsequestrin or in the presence of calsequestrin with impaired polymerization, especially in a state where SERCA is fully activated by adrenergic stimulation, a reduced Ca^{2+}-buffering capacity in the SR will permit luminal Ca^{2+} concentrations to overshoot the normal threshold of RyR for a store overload-induced Ca2+ release (SOICR), causing arrhythmia (MacLennan and Chen 2009). The same pathophysiologic consequence can be ascribed to a malignant hyperthermia (MH) episode in the case of mutated skeletal calsequestrin, *casq1*. Calsequestrin was also identified as one of the major antigens in the thyroid-associated ophthalmopathy (TAO), a progressive orbital disorder associated with Grave's hyperthyroidism and, less often, with Hashimoto's thyroiditis.

Affinity to Small Molecules

Both Casq1 and Casq2 have shown significant affinity toward various types of pharmaceutical compounds such as tricyclic antidepressants and phenothiazine- and anthracycline-derivatives, and such binding results in a significant disruption of Ca^{2+}-binding capacity, polymerization, and its communication with other critical components of the channel complex (Kim et al. 2005; Park et al. 2005; Kang et al. 2010). Considering the concentration of calsequestrin in muscle tissue ($\sim$100 mg/ml) and the affinity between calsequestrin and these classes of drugs (1–100 μM), long-term or high-dose administration could lead to their accumulation in the SR, which substantially could reduce the calcium content of the SR (probably due to decreased buffering capacity and increased free Ca^{2+}) (Kang et al. 2010). Long-term exposure to such drugs could produce a chronic toxicity to both cardiac and skeletal muscles despite their moderate affinity.

All CPVT-related calsequestrin mutations result in disrupted behavior in both its Ca^{2+}-binding capacity and Ca^{2+}-dependent polymerization, which is similar to the interference caused by small molecule association. Therefore, those pharmaceutical compounds could seriously affect people who already have impaired calsequestrin, such as in the case of CPVT patients. The effects of some mutations or drug-association by themselves on the function of calsequestrin might be benign enough not to cause arrhythmias; however, there might exist an additive or synergetic interference effect on the normal function of calsequestrin by drug binding and hereditary mutation together. Therefore, specific caution has been posted for those people at high risk of serious cardiac complications (Kang et al. 2010).

Cross-References

▶ Biological Copper Transport
▶ Ca^{2+}-Binding Protein
▶ Calcium ATPase
▶ EF-Hand Proteins

References

Dulhunty A, Wei L, Beard N (2009) Junctin - the quiet achiever. J Physiol 587:3135–3137 (Review)

Györke S, Stevens S, Terentyev D (2009) Cardiac calsequestrin: quest inside the SR. J Physiol 587:3091–3094 (Review)

Kang C, Nissen M, Sachez E, Milting H (2010) Potential adverse interaction of human cardiac calsequestrin. Eur J Pharmacol 646:12–21

Kim E, Tam M, Siems WF, Kang C (2005) Effects of drugs with muscle-related side effects and affinity for calsequestrin on the calcium regulatory function of sarcoplasmic reticulum microsomes. Mol Pharmacol 68:1708–1715

Kim E, Youn B, Kemper L, Campbell C, Milting H, Varsanyi M, Kang C (2007) Biochemical analysis of human cardiac calsequestrin and its deleterious mutants. J Mol Biol 373:1047–1057

Knollmann B (2009) New roles of calsequestrin and triadin in cardiac muscle. J Physiol 587:3081–3087

MacLennan D, Chen S (2009) Store overload-induced Ca2+ release as a triggering mechanism for CPVT and MH episodes caused by mutations in RYR and CASQ genes. J Physiol 587:3113–3115 (Review)

McFarland T, Milstein M, Cala S (2010) Rough endoplasmic reticulum to junctional sarcoplasmic reticulum trafficking of calsequestrin in adult cardiomyocytes. J Mol Cell Cardiol 49:556–564

Milstein M, Houle T, Cala S (2009) Calsequestrin isoforms localize to different ER subcompartments: evidence for polymer and heteropolymer-dependent localization. Exp Cell Res 315:523–534

Park H, Wu A, Dunker A, Kang C (2003) Polymerization of Calsequestrin: Implications for Ca2+ Regulation. J Biol Chem 278:16176–16182

Park H, Park I-Y, Kim E, Youn B, Fields K, Dunker A, Kang C (2004) Comparing skeletal and cardiac calsequestrin structures and their calcium binding: a proposed mechanism for coupled calcium binding and protein polymerization. J Biol Chem 279:18026–18033

Park I, Kim E, Park H, Fields K, Dunker K, Kang C (2005) Interaction between cardiac calsequestrin and drugs with known cardiotoxicity; A potential source for the cardiac side effects. J Mol Pharmacol 67:95–105

Royer L, Ríos E (2009) Deconstructing calsequestrin. Complex buffering in the calcium store of skeletal muscle. J Physiol 587:3101–3111

Sanchez E, Munske G, Criswell A, Milting H, Dunker A, Kang C (2011) Phosphorylation of human calsequestrin: implications for calcium regulation. Mol Cell Biochem 353:195–204

Sanchez E, Lewis K, Danna B, Kang C (2012) High-capacity Ca^{2+}-binding of human skeletal calsequestrin. J Biol Chem 287:11592–11601

Sanchez E, Lewis K, Munske G, Nissen M, Kang C (2012) Glycosylation of skeletal calsequestrin, implications for its function. J Biol Chem 287:3042–3050

Wang S, Trumble W, Liao H, Wesson C, Dunker A, Kang C (1998) Crystal structure of calsequestrin from rabbit skeletal muscle sarcoplasmic reticulum. Nat Struct Biol 5:476–483

Calsymin

▶ Bacterial Calcium Binding Proteins

CaM

▶ Calcium, Neuronal Sensor Proteins

Cancer

▶ Lanthanides and Cancer
▶ Magnesium and Cell Cycle
▶ Selenium and Glutathione Peroxidases

Cancer Diagnosis and Treatment

▶ Gold Nanomaterials as Prospective Metal-based Delivery Systems for Cancer Treatment

Cancer: Carcinoma

▶ Selenium-Binding Protein 1 and Cancer

Carbon Dioxide

▶ Zinc and Iron, Gamma and Beta Class, Carbonic Anhydrases of Domain Archaea

Carbon Monoxide Dehydrogenase/Acetyl-CoA Synthase

▶ CO-Dehydrogenase/Acetyl-CoA Synthase

Carboplatin

▶ Platinum Anticancer Drugs

Carboxycathepsin

▶ Angiotensin I-Converting Enzyme

Carboxyethyl Germanium Sesquioxide

▶ Germanium-Containing Compounds, Current Knowledge and Applications

Cardiac Excitation/Contraction Coupling

▶ Calcium in Heart Function and Diseases

Cardiac Hypertrophy

▶ Calcium in Heart Function and Diseases

Cardiomyopathy

▶ Selenium and Muscle Function

Cardiovascular Disorders

▶ Selenium and Muscle Function

CASQ

▶ Calsequestrin

Catalases as NAD(P)H-Dependent Tellurite Reductases

Iván L. Calderón[1] and Claudio C. Vásquez[2]
[1]Laboratorio de Microbiología Molecular, Universidad Andrés Bello, Santiago, Chile
[2]Laboratorio de Microbiología Molecular, Departamento de Biología, Universidad de Santiago de Chile, Santiago, Chile

Synonyms

Bacterial tellurite resistance; Enzymatic tellurite reduction

Definition

Most aerobic organisms are exposed to oxidative stress, which results in the generation of free reactive oxygen species (superoxide, hydrogen peroxide, hydroxyl radical) that interfere with the cell's metabolism, cause oxidative damage of cellular macromolecules, and may eventually also cause cell death. Thus, eliminating these free oxygen radicals is absolutely mandatory for cell survival.

In this context, catalases are antioxidant enzymes that accelerate the rate of hydrogen peroxide decomposition to molecular oxygen and water with near kinetic perfection. Exhibiting one of the highest known turnover numbers, a catalase molecule can convert approximately 4×10^7 substrate molecules to the referred products each second. The catalytic efficiency (kcat/Km) of catalase (4.0×10^8 M^{-1} s^{-1}) is very high indeed. Because the efficiency is at the diffusion limit, catalase is said to have achieved "catalytic perfection." This activity is dependent on a heme cofactor with a bound iron atom, which is cycled between oxidation states. Many catalases also have been shown to be peroxidases, that is, they can oxidize short-chain alcohols including ethanol and other substrates in a two-step hydrogen peroxide-dependent reaction. In addition to their active-site heme groups, it has been described that many heme-containing catalases bind a second prosthetic group, NAD(P)H, which is not required for the peroxide dismutase activity. Bound NADPH can protect the enzyme against oxidative damage by the peroxide substrate by tunneling electrons toward the active-site heme group to regenerate its active oxidation state.

On the other hand, it has been found that some heavy metal/metalloid derivatives like the tellurium (Te) oxyanion tellurite (TeO_3^{2-}) act as natural substrates (electron acceptors) for the NAD(P) H-dependent oxidoreductase activity of bacterial and mammalian catalases. Particularly, tellurite is a strong oxidizing agent with high toxicity to both, prokaryotic and eukaryotic organisms.

Toxicant Reduction as a Mechanism of Resistance

Deciphering the origin of bacterial tellurite resistance has been the goal of many microbiologists for years. Several groups have reported the isolation and characterization of some genetic resistance determinants, and the biochemical mechanism underlying TeO_3^{2-} resistance suggests a multifactorial response that includes the reduction of tellurite to elemental tellurium.

The oxidant character of the tellurium oxyanions tellurite (TeO_3^{2-}) and tellurate (TeO_4^{2-}) makes them toxic to most microorganisms. Gram-negative bacteria are particularly sensitive to tellurite while some Gram-positive species including *Corynebacterium diphtheriae*, *Staphylococcus aureus*, *Streptococcus faecalis* and *Geobacillus stearothermophilus* V are naturally resistant to this salt.

Tellurite-resistant cells turn black when grown in liquid or solid media amended with this compound; the higher the tellurite concentration in the culture medium, the blacker the resulting cells (Fig. 1). However, this phenotype is also observed in sensitive bacteria growing at permissive tellurite concentrations. Electron microscopy studies have allowed the identification of the subcellular location of these black deposits both in Gram-positive and in Gram-negative bacteria. In turn, X-ray diffraction studies demonstrated that this insoluble, nondiffusible material corresponds to the less toxic Te form, elemental tellurium (Te^0).

The presence of tellurite-reducing activities that depend on an electron donor coenzyme such as NAD (P)H or $FADH_2$ has been observed in *Escherichia coli* (Cooper and Few 1952), *Mycobacterium avium* (Terai and Kamamura 1958), *Thermus spp.* (Chiong et al. 1988), *Rhodobacter sphaeroides* (Moore and Kaplan 1992), and *G. stearothermophilus* (Moscoso et al. 1998), among other bacteria.

Bacterial Tellurite Reductases

It has been proposed that several oxidoreductases, including the terminal oxidases of the bacterial respiratory chain contribute to tellurite reduction. An example of them is nitrate reductase, present in membrane fractions of *E. coli* that can mediate the reduction of tellurite (Avazéri et al. 1997). Tellurite-reducing activity was also identified in the E3 component of the pyruvate dehydrogenase complex from *E. coli*, *Zymomonas mobilis*, *Streptococcus pneumoniae*, and *G. stearothermophilus* (Castro et al. 2008). All these activities showed a NAD(P)H dependence, suggesting a wide distribution among microorganisms.

Catalases Exhibit Tellurite Reductase (TR) Activity

TR activity in catalases was first identified in a bacterial strain isolated from the effluent of a Chilean mining company, *Staphylococcus epidermidis* CH. This bacterium exhibits natural resistance to a series of quaternary ammonium compounds and also shows an important tellurite resistance phenotype. *S. epidermidis* CH is 5- and 100-fold more

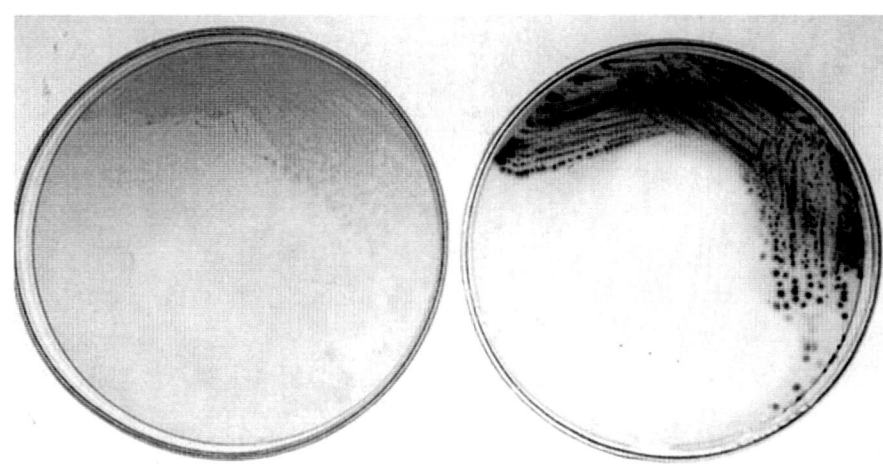

Catalases as NAD(P)H-Dependent Tellurite Reductases, Fig. 1 Bacterial reduction of K_2TeO_3. Bacteria growing on LB-agar plates containing (*right*) or not (*left*) sublethal tellurite concentrations

resistant to K_2TeO_3 than *G. stearothermophilus* V and *E. coli*, respectively. In order to gain some insight on the naturally occurring tellurite resistance in these Gram-positive bacteria, studies were focused on the nature of toxicant's bioreduction.

Crude extracts of *S. epidermidis* catalyze the in vitro, NADH-dependent reduction of K_2TeO_3, as shown by a standard spectrophotometric assay. An absorbance increase correlates directly with the blackening of the sample due to Te^0 production. Apart from being dialyzable, the enzymatic nature of the entity responsible for tellurite reduction was inferred by its sensitivity to heat, protein-denaturing agents, detergents, and proteases. The product of tellurite reduction by the *S. epidermidis* CH extract was identified as metallic tellurium by Induced Coupled Plasma-Optical Emission (ICP-OE) spectroscopy. The spectrum of NADH-dependent tellurite reduction obtained with the cell-free extract of *S. epidermidis* CH evidenced a major peak that matched with the Te^0 signal obtained by chemical reduction of potassium tellurite with 2-mercaptoethanol (Calderón et al. 2006).

A search for the enzyme responsible for TR activity was initiated. An enriched preparation coming from a series of chromatographic separations of the proteins present in cell-free extracts of *S. epidermidis* CH contained two major proteins with apparent M_r of ~60 and 75 kDa. Amino-terminal sequence analysis predicted that one of them was identical to catalase (predicted M_r ~58.2 kDa). Since in a previous work catalase was shown to be an NAD(P)H oxidase (Singh et al. 2004), the above evidence indicated that the NADH-dependent TR activity observed in this fraction would come likely from catalase (Fig. 2).

Heme-dependent monofunctional catalases differ in the strength with which they bind NAD(P)H and in substrate specificity of their accessory peroxidase activities. For example, bovine catalase binds both NADPH and NADH very tightly, whereas monofunctional bacterial catalases bind them more weakly. Bovine catalase and KatG, the bifunctional catalase/peroxidase from *E. coli*, display secondary peroxidase activities for a variety of substrates (Keilin and Hartree 1945; Oshino et al. 1973; Singh et al. 2004). Although somewhat diverse in their biochemical properties, monofunctional heme-containing catalases have core primary sequences conserved among prokaryotes and eukaryotes. Therefore, if the *S. epidermidis* CH monofunctional heme-containing catalase mediates tellurite reduction, the immediate assumption was that catalase from bovine liver was also able to reduce tellurite to Te^0 in vitro. In fact, purified bovine liver catalase showed a branch tellurite-reducing activity.

The in situ dismutation of hydrogen peroxide by purified bovine catalase in native polyacrylamide gels co-migrates with the TR activity. In a similar way, the *S. epidermidis* CH 60 kDa catalase also reduce tellurite in situ. On the other hand, the heterologous expression of *S. epidermidis* CH *katA* gene (encoding catalase) in *E. coli* (*E. coli*/pCAT), evidenced in situ activities for hydrogen peroxide dismutase and tellurite reduction. Experiments were carried out with extracts from exponentially growing *S. epidermidis* CH. A pBAD

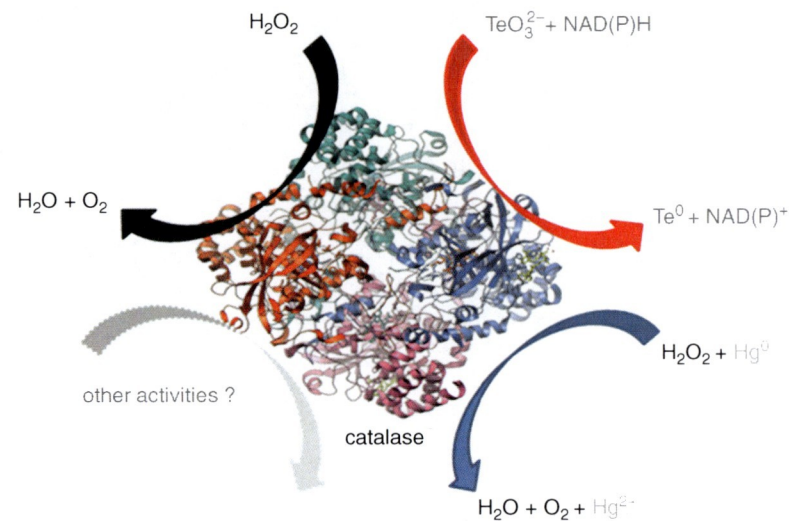

Catalases as NAD(P)H-Dependent Tellurite Reductases, Fig. 2 Catalase activities. A catalase molecule showing its various identified activities: peroxidase (*black arrow*), TR (*red*), mercury oxidase (*blue*), among others (*gray*)

vector-harboring *E. coli* (pBAD), which does not produce *S. epidermidis* catalase, was used as negative control. Also, the standard assay for tellurite-reducing activity was challenged with a known catalase inhibitor. The assay performed with cell-free extracts from *S. epidermidis* CH or *E. coli*/pCAT was inhibited by 50% in presence of increasing concentrations of sodium azide (Fig. 3). The same result was obtained when pure bovine liver catalase was used instead of the bacterial extract.

Catalases and Bacterial Resistance to Tellurite

E. coli defective in catalase is sensitive to tellurite. Expression of the *S. epidermidis* catalase gene confers increased resistance to tellurite and to hydrogen peroxide in this bacterium, arguing that catalase seems to provide a physiological defense line against these two strong oxidants. This recombinant strain also displays a higher level of resistance to hydrogen peroxide than its otherwise isogenic parent. Therefore, the expression of the *S. epidermidis* catalase gene complements the *E. coli* deficiency. Furthermore, these data indicate that the *S. epidermidis* CH *katA* gene is a tellurite resistance determinant that, at least in part, is responsible for the high resistance of this Gram-positive bacterium to tellurite in vivo.

The information collected to date strongly suggests that the phenomenon of bacterial tellurite resistance has its origin in the combined or synergic-action/cooperation of various different enzymatic activities that play defined roles in bacterial metabolism, such as catalases. Certainly that their participation in the reduction, and thus probably in tellurite resistance, could be merely an accident, but anyway beneficial to the microorganism.

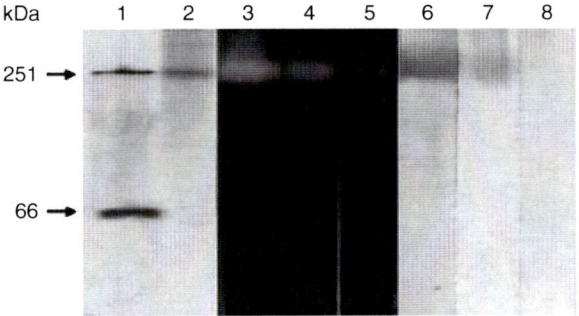

Catalases as NAD(P)H-Dependent Tellurite Reductases, Fig. 3 In situ determination of catalase and tellurite reductase activities of bovine liver catalase (BLC). The enzyme (255 kDa) was assayed to show catalase (lanes 3–5) and tellurite reductase (lanes 6–8) activities in a polyacrylamide gel under native conditions. Lane 1, MW standards [phosphoenol pyruvate carboxyquinase (PEPCK) from *Saccharomyces cerevissiae* (8 μg, 251 kDa) and bovine serum albumin (BSA, 12 μg, 66 kDa)]; lane 2, BLC with Coomasie blue staining; lane 3, BLC revealed for catalase activity; lane 4, as in 3, but previously incubated with 2 mM sodium azide for 3 min; lane 5, as in 3, but previously incubated with 10 mM sodium azide for 3 min; lane 6, BLC revealed for tellurite reductase activity; lane 7, as in 6, but previously incubated with 2 mM sodium azide for 3 min; lane 8, as in 6, but previously incubated with 10 mM sodium azide for 3 min

Kinetics Studies with Bovine Catalase

The reaction proposed for the catalase-mediated tellurite reduction is

$$TeO_3^{2-} + 3NAD(P)H + 3H^+ + 2O_2$$
$$\rightarrow Te^0 + 3NAD(P)^+ + 2O_2^- + 3H_2O$$

Thus, the reduction of tellurite by catalase requires molecular oxygen and produces superoxide as one of its products. This scheme is predicted to be favorable thermodynamically, involving as half-reactions the reduction of tellurite, the oxidation of NADH, and the formation of superoxide from oxygen.

The K_m of bovine liver catalase for tellurite was determined by assaying the rate of evolution of superoxide resulting from the reduction of tellurite by this enzyme (Calderón et al. 2006). Superoxide production rate is dependent on tellurite concentration and follows simple Michaelis-Menten kinetics. Under these experimental conditions, catalase has an apparent K_m for tellurite of 0.9 mM, a value that is comparable to that for peroxide (Nicholls and Schonbaum 1963), maybe indicating that tellurite could represent a natural substrate for the enzyme.

Catalase appears to be quite broad in its substrate range than originally thought, including two active sites that can participate in a variety of redox and condensation reactions. Tellurite is not the only metalloid derivative that behaves as catalase substrate. In fact, both eukaryotic and prokaryotic catalases carry out the heme-dependent oxidation of metallic mercury, a reaction that is stimulated by hydrogen peroxide (Magos et al. 1978; Du and Fang 1983; Ogata and Aikoh 1983; Smith et al. 1998). Thus, it is currently suspected that catalases would display a wider range of substrates than those of which we are currently aware, which could play multiple roles in the cell's defense against strong oxidizing agents encountered in nature, as a variety heavy metal ions.

Cross-References

▶ Heme Proteins, Heme Peroxidases
▶ Manganese and Catalases
▶ Nickel Superoxide Dismutase
▶ Peroxidases
▶ Selenium and Glutathione Peroxidases
▶ Tellurite-Detoxifying Protein TehB from *Escherichia coli*
▶ Tellurite-Resistance Protein TehA from *Escherichia coli*
▶ Tellurium in Nature
▶ Zinc in Superoxide Dismutase

References

Avazéri C, Turner R, Pommier J, Weiner J et al (1997) Tellurite reductase activity of nitrate reductase is responsible for the basal resistance of *Escherichia coli* to tellurite. Microbiology 143:1181–1189

Calderón IL, Arenas FA, Pérez JM, Fuentes DE et al (2006) Catalases are NAD(P)H-dependent tellurite reductases. PLoS One 20(1):e70

Castro ME, Molina R, Díaz W, Pichuantes SE et al (2008) The dihydrolipoamide dehydrogenase of *Aeromonas caviae* ST exhibits NADH-dependent telluritereductase activity. Biochem Biophys Res Commun 375:91–94

Chiong M, González E, Barra R, Vásquez C (1988) Purification and biochemical characterization of tellurite-reducing activities from *Thermus thermophilus* HB8. J Bacteriol 170:3269–3273

Cooper P, Few A (1952) Uptake of potassium tellurite by a sensitive strain of *Escherichia coli*. J Biochem (Tokyo) 51:552–557

Du S-H, Fang SC (1983) Catalase activity of C3 and C4 species and its relationship to mercury vapor uptake. Environ Exp Bot 23:347–353

Keilin D, Hartree EF (1945) Properties of catalase. Catalysis of coupled oxidation of alcohols. Biochem J 39:293–301

Magos L, Halbach S, Clarkson TW (1978) Role of catalase in the oxidation of mercury vapor. Biochem Pharmacol 27:1373–1377

Moore M, Kaplan S (1992) Identification of intrinsic high-level resistance to rare-earth oxides and oxyanions in members of the class *Proteobacteria*: characterization of tellurite, selenite, and rhodium sesquioxide reduction in *Rhodobacter sphaeroides*. J Bacteriol 174:1505–1514

Moscoso H, Saavedra C, Loyola C, Pichuantes S, Vásquez C (1998) Biochemical characterization of tellurite-reducing activities from *Bacillus stearothermophilus* V. Res Microbiol 49:389–397

Nicholls P, Schonbaum GR (1963) Catalases. In: Boyer P, Lardy H, Myrback K (eds) The enzymes. Academic, New York/London, pp 147–225

Ogata M, Aikoh H (1983) The oxidation mechanism of metallic mercury *in vitro* by catalase. Physiol Chem Phys Med NMR 15:89–91

Oshino N, Oshino R, Chance B (1973) The characteristics of the "peroxidatic" reaction of catalase in ethanol oxidation. Biochem J 131:555–563

Singh R, Wiseman B, Deemagarn T, Donald LJ, Duckworth HW et al (2004) Catalase-peroxidases (KatG) exhibit NADH oxidase activity. J Biol Chem 279:43098–43106

Smith T, Pitts K, McGarvey JA, Summers AO (1998) Bacterial oxidation of mercury metal vapor, Hg(0). Appl Environ Microbiol 64:1328–1332

Terai T, Kamamura Y (1958) Tellurite reductase from *Mycobacterium avium*. J Bacteriol 75:535–539

Catalysis

▶ Iron-Sulfur Cluster Proteins, Nitrogenases

Catalytic Inhibitor

▶ Gold(III), Cyclometalated Compound, Inhibition of Human DNA Topoisomerase IB

Catalytic Pathway: Reaction Pathway

▶ Monovalent Cations in Tryptophan Synthase Catalysis and Substrate Channeling Regulation

Catalytic Site Zinc

▶ Zinc Alcohol Dehydrogenases

Catalytic Zinc

▶ Zinc Alcohol Dehydrogenases

Catechol Oxidase and Tyrosinase

Catherine Belle
Département de Chimie Moléculaire, UMR-CNRS 5250, Université Joseph Fourier, ICMG FR-2607, Grenoble, France

Names and (Some) Synonyms

Catechol oxidase (CO), EC 1.10.3.1, 1,2-Benzenediol: oxygen oxidoreductase, Diphenoloxidase, *o*-Diphenolase, Catecholase

Tyrosinase (Ty), EC 1.14.18.1, Monophenol, L-dopa:dioxygen oxidoreductase, Monophenol monooxygenase, Monophenol oxidase, Cresolase

General Description and Properties

Catechol oxidase (CO) and tyrosinase (Ty) are oxidoreductases (EC1) that possess active sites with two copper atoms (type 3 copper centers). Reports on these enzymes' molecular weights are diverse and variable. Tertiary and quaternary structures of CO and Ty also display considerable variability in amino acid sequences. However, significant sequence homology between the active site regions is observed. The copper ions are surrounded by three nitrogen donor atoms from histidine residues. Ty and CO have been identified in three distinct forms in various stages of the catalytic cycle: a native *met* state (Cu^{II}-Cu^{II}), a reduced *deoxy* state (Cu^I Cu^I), and an *oxy* state with a dioxygen bound to the dicopper center (Cu^{II}-O_2^{2-}-Cu^{II}) (Fig. 1).

The copper(II) ions in the *met* state are bridged by aquo (hydroxo) ligands, which provide antiferromagnetic coupling between the copper ions, leading to EPR (electron paramagnetic resonance)-silent behavior.

This type of active site is analogous to hemocyanin active sites, which serve as oxygen carriers and are the third group of proteins discovered to have a type 3 copper active site. Ty enzymes mediate the hydroxylation of monophenols to *o*-diphenols (monophenolase or cresolase activity) and the subsequent oxidation to quinones. Catechol oxidases exhibit only diphenolase (or catecholase) activity (i.e., two-electron oxidation of *o*-diphenols to quinones) (Fig. 2). Ty catalyzes the oxygenation of a variety of phenols, while CO catalyzes oxidation of a variety of *o*-diphenol; both reduce molecular dioxygen to water during catalysis. A lag phase is present (in vitro) for monophenolase activity, but not for diphenolase activity.

CO and Ty activities are pH-dependent. The optimal pH values reported for each enzyme differ because of differences in the sources and substrates used for the activity measurements. These optima range from pH 5 to 7.5, but neither enzyme displays significant activity in basic conditions.

The *oxy* forms of the active sites can be obtained by treatment of the *met* forms with excess of dihydrogen

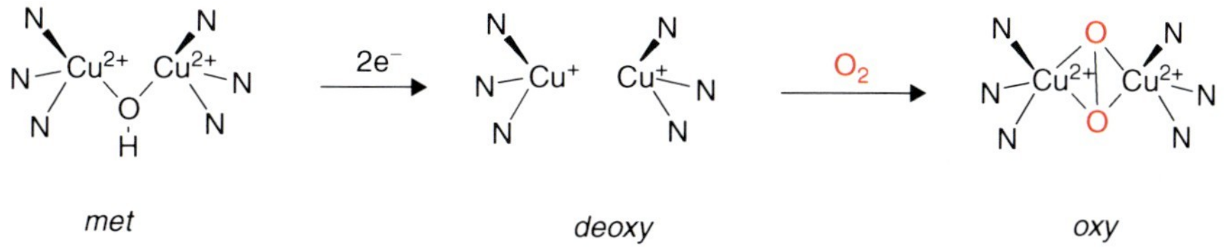

Catechol Oxidase and Tyrosinase, Fig. 1 Schematic representation of the three identified and structurally characterized forms for Ty and CO

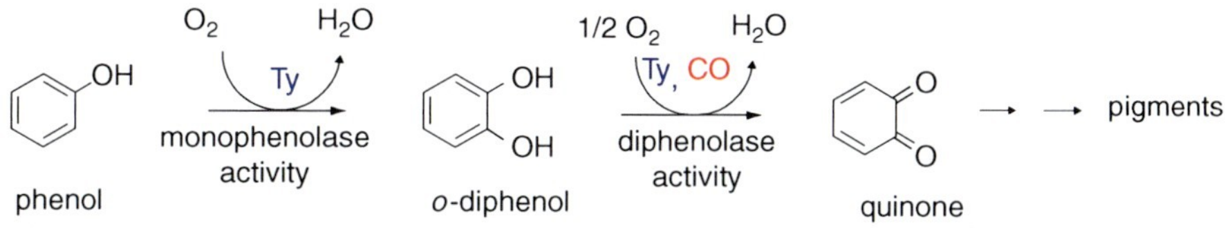

Catechol Oxidase and Tyrosinase, Fig. 2 Reaction pathway of the monophenolase and diphenolase activity catalyzed by Ty and CO, leading to the production of protective pigments

peroxide. The CO *oxy* form is less stable than that of Ty; CO cannot be fully converted to this form, and the *oxy* form spontaneously decays, with partial enzyme inactivation because the enzyme exhibits catalase activity (Gerdemann et al. 2002). The instability of the *oxy* form may contribute to CO's lack of monooxygenase reactivity toward monophenols, although this difference in activity may also be due to structural factors (see below).

Occurrence

CO is mainly found in and isolated from a large group of plants copper in plants, including potatoes and apples, and it is found in many different tissues of these plants such as leaves, fruit, and flowers. Ty is more widely distributed, and forms with slight variations are found in many organisms, including plants, fungi, bacteria, and mammals. These enzymes exist in complexes ranging from monomers to hexamers and are found freely dissolved (bacteria, plants) or bound within organelles or membrane (mammals) (Rolff et al. 2011).

Functions

The most evident proposed function of Ty and CO is in the biosynthesis of pigments of polyphenolic origin. Such pigments are widely distributed in plants (tannins), arthropods (cuticles), and mammals (melanins). These pigments are heterogeneous polyphenol-like biopolymers with complex structures. The colors of these pigments, which range from yellow to black, are a result of the autopolymerization of highly reactive quinones formed by the oxidization of phenol and catechol precursors. The browning of fruits, vegetables, and mushrooms after tissue damage induced by stress, pathogens or wounding is one of the most common outcomes of catechol oxydase- or tyrosinase-mediated melanization (Mayer 2006).

In mammals, the final products of Ty activity are melanin pigments responsible for skin, eye, and hair colors (Nordlund et al. 2006). The absence or inactivation of Ty leads to albinism. Melanin is produced in melanocytes, cells located in the basal layer of the dermis. Ty catalyzes the initial conversions of tyrosine leading to the formation of dihydroxyphenylalanine (DOPA) and dopaquinone (Fig. 3). It is also associated with the oxidation of 5,6-dihydroxyindole (DHI) in indole-5,6-quinone. This multistep process leads to the formation of eumelanin (the most protective melanin against photoinduced damage). Skin color depends principally on relative amounts of eumelanin (brown/black) and pheomelanin (red/yellow). In the human brain, tyrosinase is expressed at low levels, but Ty can perform the function of tyrosine hydroxylase (catalyzing the formation of tyrosine in DOPA in the brain) in its absence and may be linked to catecholamine neurotoxicity.

Inhibitors

Special attention has been devoted to CO inhibition to suppress the browning of plants and fruits during storage (Mayer 2006). Several inhibitors acting as chelating agents for copper or substrate analogues have been

Catechol Oxidase and Tyrosinase,
Fig. 3 Biosynthetic pathway of melanins and the involvement of tyrosinase enzyme. *Dotted line* is related to an additional pathway found in mammals

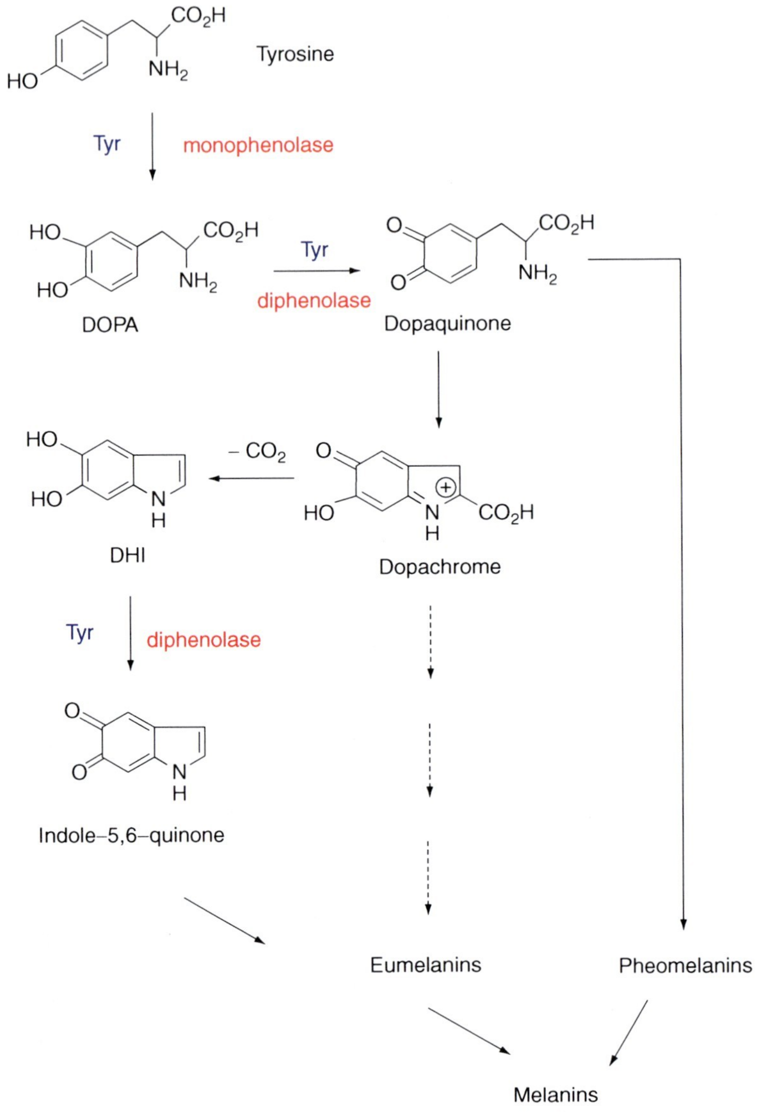

reported. In the food industry, sulfite is commonly used to inhibit enzymatic browning. In humans, high levels of melanin cause a variety of disorders (Nordlund et al. 2006), such as cutaneous hyperpigmentation (solar lentigo, melasma, naevi, freckles, age spots) and ocular retinitis pigmentosa. Moreover, Ty inhibitors are highly sought after as skin-lightening agents for cosmetic products. Human Ty is an important target in both cosmetics and medicine for the modulation of melanogenesis in different organs.

Spectroscopic Studies

Before the crystal structures of CO and Ty were reported, spectroscopic studies were the only tools for the investigation of the structures and mechanisms of these enzymes. A useful review describing the variety of spectroscopic techniques employed has been published (Tepper et al. 2010). Both type 3 copper centers have similar spectroscopic features. In the *met* state, they are characterized by an EPR-silent Cu(II) pair with an $S = 0$ ground state resulting from antiferromagnetic coupling between the two $S = 1/2$ electronic spins of each copper(II). UV/Vis spectra display a weak d-d transition from 600 to 700 nm, and an intensive UV/Vis absorption maximum around 345 nm is observed after binding of dioxygen because of a peroxo $O_2^{2-} \rightarrow$ Cu(II) charge transfer transition (Solomon et al. 1996; Gerdemann et al. 2002).

The Raman spectrum of the *oxy* form is characterized by an O-O stretching vibration around 750 cm^{-1}. Other spectroscopic techniques, such as EPR, X-ray absorption spectroscopy (XAS), and proton paramagnetic nucleus magnetic resonance (NMR), have yielded data supporting coordination by histidines for Ty or inhibitor binding (Tepper et al. 2010).

Structural Features

The nature of the related copper center has been discussed for over a century. X-ray crystal structures have recently been resolved for Ty and CO in various forms and from different sources. In 1998, Krebs and collaborators succeeded in the first structural determination of CO from *Ipomoea batatas* (*met*, *deoxy*, and an inhibited form with the bound inhibitor phenylthiourea) (Gerdemann et al. 2002). For Ty despite the large number of Ty sources, Ty have been recently solved from bacterial sources: (1) from recombinant *Streptomyces castanaeoglobisporus* (*met*, *deoxy*, and *oxy* forms) (Matoba et al. 2006), which contained an associated Cu "caddie" protein covering the entrance of the active site; (2) from *Bacillus megaterium* Ty (*met* form and a form with kojic acid) (Sendovski et al. 2011). The Cu...Cu pairs in the *met* forms are separated by distances in the range from 2.9 to 3.9 Å and are bridged by one or two water molecules. The observed Cu...Cu distances in the *deoxy* forms, generated by anaerobic reduction, increase to more than 4 Å, and bridging solvent molecules were not observed. The only structurally characterized *oxy* form was prepared by H$_2$O$_2$ addition to a *Streptomyces castanaeoglobisporus met* form. The Cu...Cu distance in this structure was 3.4 Å, similar to the *met* form, but the bridging density was modeled as a μ-η^2:η^2-peroxo peroxide ion. All of the reported structural data suggest that the dicopper active site is highly flexible along catalytic cycles. Although all the dicopper active sites structurally characterized can be well superimposed, interesting differences are observed and reveal structural reasons for the divergences in function and mechanisms as illustrated on Fig. 4 with superposition of *met* forms of *Streptomyces castanaeoglobisporus* Ty and *Ipomoea batatas* CO structures.

While the structure of one copper site (CuB) is very conserved in the analyzed enzymes, the other CuA site exhibits more pronounced differences. In CO, the

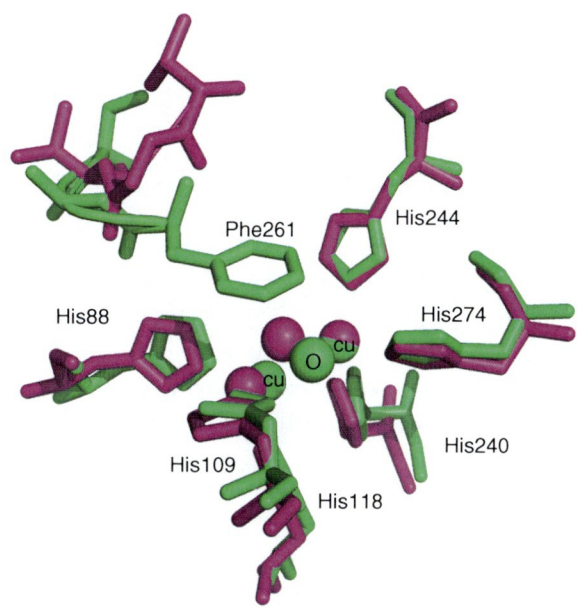

Catechol Oxidase and Tyrosinase, Fig. 4 Superposition of the copper active site in *met* forms for *Ipomoea batatas* CO (*green*, (PDB) Protein Data Bank code: 1BT3) and *Streptomyces castanaeoglobisporus* Ty (*pink*, PDB code: ZMX). The numbering is shown for CO (Graphic done by Dr. H. Jamet using PyMOL software (http://www.pymol.org))

partial obstruction of access to the dicopper center by a phenylalanine residue (Phe261) presumably creates substrate-binding specificity (Gerdemann et al. 2002). In the active sites of *Streptomyces castanaeoglobisporus* and *Bacillus megaterium* tyrosinases, the same location is occupied by a glycine (Gly204) or valine (Val218), respectively, both of which are less bulky than phenylalanine. The involvement of amino acid residues near the active site as proton acceptors has also been discussed (coordinated histidine or other noncoordinating residues located near the active site). Another interesting feature in the structure of *Ipomoea batatas* catechol oxidase (Gedermann et al.) is a covalent thioether bond between a carbon atom of a histidine ligand (His109) (one of the ligands of the CuA ion) and the cysteine sulfur atom of Cys 92. A similar type of bond is seen in *Vitis vinifera* PPO and tyrosinase from *Neurospora crassa*. The absence of such a Cys-His bridge, in human tyrosinase for instance, means that thioether modification does not play a direct role in the functional activity of the enzymes but instead imposes structural constraints around the copper center (CuA). In particular, such restraints may help to impose the trigonal pyramidal

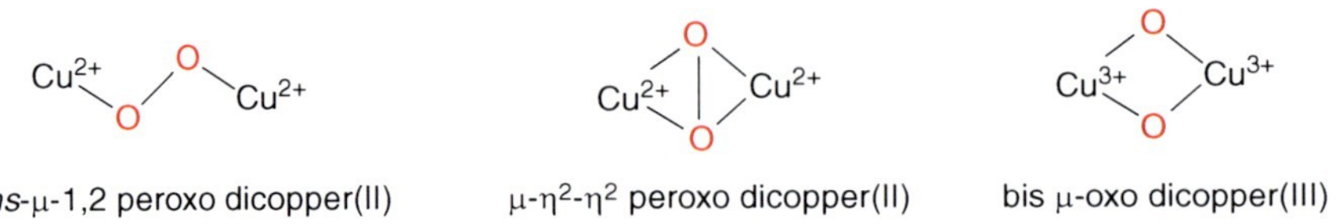

Catechol Oxidase and Tyrosinase, Fig. 5 Schematic representation of relevant dicopper-dioxygen cores

geometry (which can also be regarded as a distorted trigonal bipyramid with a vacant apical position) on the CuA ion in the +2 oxidation state of CO. This thioether bond may also prevent the displacement of His109 and a didentate binding mode of the substrate to a single Cu(II) ion. This feature may, in turn, optimize the redox potential of the metal for the oxidation of the catechol substrate and allow for rapid electron transfer in redox processes. Despite a clarification of the active site structures of different forms of CO and Ty by X-ray crystallography, the corresponding catalytic mechanisms are not yet fully characterized.

Synthetic Analogs

Because of the historical lack of structural data for catechol oxidase or tyrosinase, functional and structural analogs of the dicopper active site have been studied. Diverse models of Ty (Rolff et al. 2011 and Itoh and Fukuzumi 2007 for recent reviews) and CO (Koval et al. 2006) have been prepared and discussed. Model complexes have been especially useful in understanding the binding mode of molecular oxygen that in all cases, upon reaction with the Cu(I) Cu(I) *deoxy* form, results in a peroxide bound to a dicopper site. Various possible coordination modes of O_2 to dinuclear copper have been observed. Three of these motifs – bis(μ-oxo), μ-η^2:η^2-peroxo, and trans μ-1,2-peroxo (Fig. 5) – have proven to be dominant and can mediate monophenolase or diphenolase activity, although other examples have been identified or inferred. In the bis(μ-oxo) isomer, the O–O bond is broken, and the copper ions occupy a formal +3 oxidation state coordinated by two O_2^- ligands. The μ-η^2:η^2-peroxo and *trans* μ-1,2-peroxo isomers, in contrast, each features a peroxide dianion bound to two Cu(II) ions.

The different isomers exist in equilibrium with one another, with small activation energies for interconversion, and are alternative intermediates for substrate attack during catalysis although it has not yet been observed in the enzymes. Synthetic CO models have only achieved turnover numbers about 10,000-fold lower than those of the native enzymes (Gerdemann et al. 2002). Very few catalytic models of Ty have been synthesized (Rolff et al. 2011), despite the need in synthetic chemistry for friendly catalytic systems capable of performing *o*-hydroxylation of phenolic substrates using O_2.

Molecular Mechanism

The extensive structural, spectroscopic, and kinetic data available, in conjunction with synthetic analogs of the active site, have inspired proposals for the rational mechanisms of these enzymes (Solomon et al. 1996; Tepper et al. 2010; Rolff et al. 2011). A schematic scenario is described on Fig. 6, which includes the following steps:

- Peroxide is formed from the *deoxy* form in a μ-η^2:η^2 peroxo dicopper complex.
- The hydroxyl group of the phenol substrate binds to a copper.
- The peroxo electrophilically attacks the *ortho* position of the phenolic substrate (after rotation of the peroxo or substrate reorientation).
- The hydroxo group is protonated, the formed quinone is released, and the *deoxy* form is restored.

To explain the differences between catechol oxydase and tyrosinase, several factors are not yet known, including (1) substrate binding modes; (2) proton transfer, with the possibility that second-sphere hydrogen-bonding partners are involved in the mechanism; and (3) the exact step at which the copper-bound dioxygen is released.

Kinetic and spectroscopic studies of model systems have suggested other mechanistic possibilities, including (1) a one electron reduction of the dicopper(II)

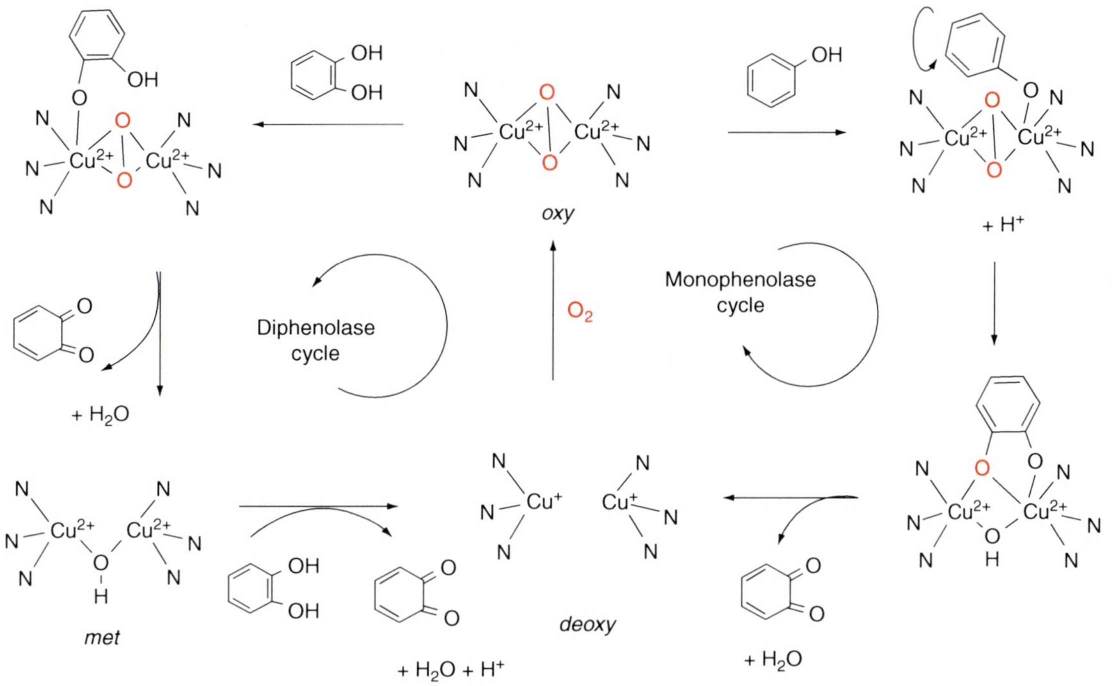

Catechol Oxidase and Tyrosinase, Fig. 6 Schematic illustration of the proposed mechanistic pathways for monophenolase and diphenolase activity

core, leading to the formation of a radical Cu(I)-semiquinone intermediate; (2) the involvement of other Cu_2O_2 adducts as intermediate species; and (3) the formation of hydrogen peroxide as a side product of the catalytic reaction. Although to the best of our knowledge, these species have been observed only under particular conditions during the substrate oxidation by Ty, CO, or model systems, the possibility of different oxidation pathways, either in the natural enzymes or associated with a malfunction, requires further study (Koval et al. 2006).

Cross-References

▶ Laccases

References

Gerdemann C, Eicken C, Krebs B (2002) The crystal structure of catechol oxidase: new insight into the function of type-3 copper proteins. Acc Chem Res 35:7019–7022

Itoh S, Fukuzumi S (2007) Monooxygenase activity of type 3 copper proteins. Acc Chem Res 40:592–600

Koval IA, Gamez P, Belle C, Selmeczi K, Reedijk J (2006) Synthetic models of the active site of catechol oxidase: mechanistic studies. Chem Soc Rev 35:814–840

Matoba Y, Kumagai T, Yamamoto A, Yoshitsu H, Sugiyama M (2006) Crystallographic evidence that the dinuclear copper center of tyrosinase is flexible during catalysis. J Biol Chem 281:8981–8990

Mayer AM (2006) Polyphenol oxidases in plant and fungi: going places? A review. Phytochemistry 67:2318–2331

Nordlund JJ, Boissy RE, Hearing VJ, King RA, Oetting WA, Ortonne JP (eds) (2006) The pigmentary system: physiology and pathophysiology, 2nd edn. Wiley/Blackwell, Oxford

Rolff M, Schottenheim J, Decker H, Tuczek F (2011) Copper-O_2 reactivity of tyrosinase models towards external monophenolic substrates: molecular mechanism and comparison with the enzyme. Chem Soc Rev 40:4077–4098

Sendovski M, Kanteev M, Shuster Ben-Yosef V, Adir N, Fishman A (2011) First structures of an active bacterial tyrosinase reveal copper plasticity. J Mol Biol 405:227–237

Solomon EI, Sudaram UM, Mackonkin TE (1996) Multicopper oxidases and oxygenases. Chem Rev 96:2563–2605

Tepper AWJW, Lonadi E, Bubacco L, Canters GW (2010) Structure, spectroscopy and function of tyrosinase; comparison with hemocyanin and catechol oxidase. In: Messerschmidt A (ed) Handbook of Metalloproteins. Wiley/Blackwell, Chichester

Cation Binding Sites and Structure-Function Relations

▶ Sodium-Binding Site Types in Proteins

CBR

▶ C2 Domain Proteins

Cd

▶ Biomarkers for Cadmium

Cd(II)

▶ Cadmium Carbonic Anhydrase

Cd, Element 48, [7440-43-9]

▶ Cadmium, Physical and Chemical Properties

Cd^{2+}

▶ Biomarkers for Cadmium

Cell Division

▶ Magnesium and Cell Cycle

Cell Labeling Techniques

▶ Labeling, Human Mesenchymal Stromal Cells with Indium-111, SPECT Imaging

Cell Signaling

▶ Nanosilver, Next-Generation Antithrombotic Agent

Cell Tracking

▶ Labeling, Human Mesenchymal Stromal Cells with Indium-111, SPECT Imaging

Cells that Exocytose Zinc – Cells that Exocytose Zn^{2+}

▶ Zinc Signal-Secreting Cells

Cellular Control Mechanism for Therapeutic Potential of Boron-Containing Compounds

▶ Boron-containing Compounds, Regulation of Therapeutic Potential

Cellular Electrolyte Metabolism

Olaf S. Andersen
Department of Physiology and Biophysics,
Weill Cornell Medical College, New York,
NY, USA

Definition

Electrolyte metabolism refers to the processes that regulate the electrolyte composition of the body fluid compartments, which in turn regulates the distribution of water among the different compartments – and thus the cell volume. The movement of ions and other polar solutes through cellular membranes is catalyzed by membrane proteins. Transmembrane differences in the concentrations of K^+, Na^+, and Cl^- are important for the generation of transmembrane potential

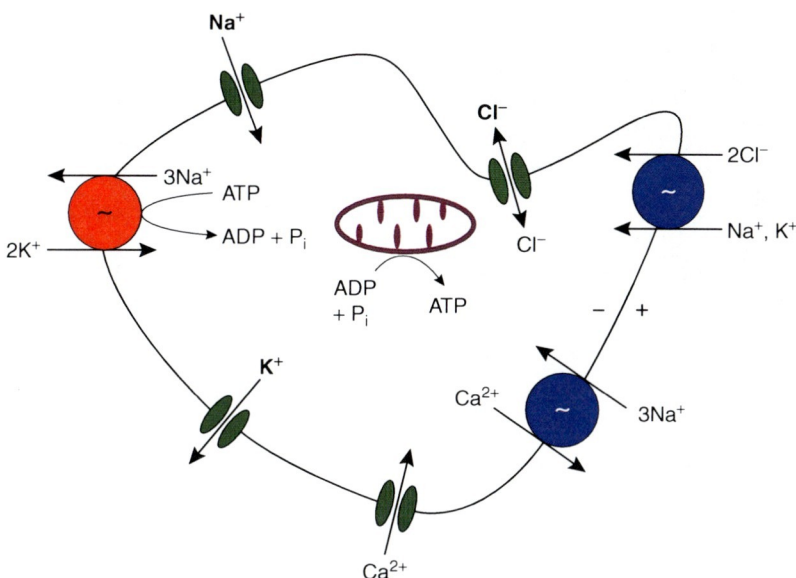

Cellular Electrolyte Metabolism, Fig. 1 Schematic representation of a cell, with channels (green) that are selective for K^+, Na^+, Ca^{2+} and Cl^- plus a primary active transporter, the Na^+,K^+-ATPase (red), which maintains the ion distribution between the extracellular and intracellular compartments, and two secondary active transporters (blue), which use the energy stored in the Na^+ concentration difference and the membrane potential difference to move Cl^- into the cell, by the Na^+,K^+,Cl^--cotransporter, and Ca^{2+} out of the cell, by the Na^+,Ca^{2+}-exchanger. The ATP is generated in the mitochondria by oxidative phosphorylation from ADP and inorganic phosphate (P_i). in resting cells, when there is no net charge movement across the plasma membrane, the intracellular fluid is electrically negative relative to the fluid compartment

differences. Transmembrane differences in the concentration of Na^+ are important for the maintenance of cell volume.

In addition to being structural elements in proteins (Yamashita et al. 1990), the alkali metal cations, together with the halide anions, are important for the generation of transmembrane potential differences (Sten-Knudsen 2002) and the regulation of cell volume (Hoffmann et al. 2009). Transmembrane potential differences (or membrane potentials) are generated when the ion concentrations in the extracellular and intracellular fluids differ. The cell volume changes when water moves across the membrane in response to a difference in the total solute concentrations (difference in water activity) in the two compartments. The major solutes in both compartments are the inorganic ions, which means that movement of water between the two compartments will cause the electrolyte concentrations (and thus the membrane potential) to change, and maintained changes in membrane potential will cause changes in the ion concentrations and the distribution of water.

The transmembrane movement of ions and water are catalyzed by integral membrane proteins that are embedded in the lipid bilayer that forms the barrier for nonselective solute movement. (Water also can move through the bilayer by dissolving into the bilayer hydrophobic core and diffuse across it, by the so-called solubility-diffusion mechanism (Finkelstein 1987)). Figure 1 shows a schematic cell with the four major ions (K^+, Na^+, Ca^{2+}, and Cl^-), which can cross the cell membrane by passive electrodiffusion through ion-selective channels, as well as by active (energy-dependent) transporters. The membrane proteins catalyze the transmembrane movement of ions and other polar solutes by providing a polar path by which the solutes can cross the membrane, thus avoiding the large desolvation penalty that would be incurred if the solutes were moving through bilayer hydrophobic core.

Fluid Compartments in Multicellular Organisms

In mammals and other multicellular animals, about 60% of the body mass is water, which is distributed

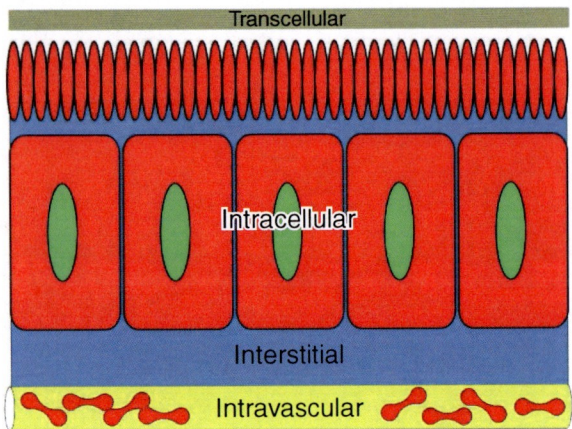

Cellular Electrolyte Metabolism, Fig. 2 The body fluid compartments (see text for details)

between two major fluid compartments (Fig. 2): the extracellular compartment (~20% of the body mass), the environment in which the body cells live, which is divided into the intravascular fluid compartment or plasma (~5% of the body mass) and interstitital fluid compartment (~15% of the body mass); and the intracellular compartment (~40% of the body mass).

The interstitial and intravascular fluid compartments have similar electrolyte compositions: high concentrations of Na^+, Cl^-, and HCO_3^-; low concentrations of K^+, Ca^{2+}, and Mg^{2+}; and a very low concentration of H^+ (Table 1). The two compartments are separated by the capillary wall, which is freely permeable to small electrolytes but imposes a barrier for the movement of proteins; the key difference thus is that the intravascular fluid contains the plasma proteins, some of which serve as carriers for trace metals, for example, serum albumin (Co^{2+}), ceruloplasmin (Cu^{2+}), and transferrin (Fe^{3+}). Though the intracellular fluid compartment is discontinuous, being distributed in all the body's cells, each of which being enveloped by a cell membrane, the intracellular fluid in all cells share common features that justify combining all the intracellular fluid into of a single compartment, namely, high concentrations of K^+, organic phosphates, and protein; low concentrations of Na^+, Mg^{2+}, Cl^-, and HCO_3^-; and very low concentrations of Ca^{2+} and H^+ (Table 1). The major variation among different cells is the intracellular $[Cl^-]$, which varies between ~5 mmoles/L in muscle cells and ~80 mmoles/L in red blood cells (intracellular water is ~2/3 of the cell mass, and the concentrations are per liter water). In addition to these two major compartments, the fluid in the gastrointestinal tract, the urine, and the cerebrospinal and intraocular fluids are lumped into the chemically transcellular fluid compartment, which have the common feature that they are lined by epithelia that regulate their composition.

The ion concentrations of the extracellular (intravascular) and intracellular fluids are listed in Table 1, together with the equilibrium potentials for the different ions (see Ion Permeation and Membrane Potentials, below). As noted above, the intracellular $[Cl^-]$ varies among cell types; the listed values pertain to a "typical" cell.

In Table 1, the values refer to the concentrations of the "free" ions, which may differ from the total concentration of the ion in plasma or cytoplasm. The total Ca^{2+} concentration in plasma, for example, is ~2.5 mM, but ~50% is bound to proteins, phosphate, and organic anions; the free $[Ca^{2+}]$ is only ~1.3 mM. Similarly, the free $[Mg^{2+}]$ is ~50% of the total Mg^{2+} concentration. This binding/buffering is even more pronounced in the intracellular compartment. For H^+, the buffer capacity β is defined as (Roos and Boron 1981)

$$\beta = \frac{\Delta B}{\Delta pH}, \quad (1)$$

where B is the amount of strong base that is needed to produce a given change in pH; $\beta \approx 50$ mmole/(L·pH). The intracellular fluid is well buffered.

In the case of Ca^{2+} (and Mg^{2+}), the buffering is described using the ion-binding ratio, κ_S, defined as (Zhou and Neher 1993)

$$\kappa_S = \frac{\Delta(\text{total Ion}^{2+})}{\Delta(\text{free }[\text{Ion}^{2+}])}; \quad (2)$$

for Ca^{+2}, $\kappa_S \approx 100$; for Mg^{2+}, $\kappa_S \approx 20$. The low intracellular $[Ca^{2+}]$, $[Ca^{2+}]_i$, in particular means that Ca^{2+} becomes an important intracellular messenger because the transmembrane Ca^{2+} fluxes are large enough to cause measurable changes in $[Ca^{2+}]_i$.

It is sometimes helpful to express the electrolyte concentrations in units of milliequivalents per liter (Eq/L or mEq/L). One converts from molar concentrations to Eq/L by multiplying the molar ion concentration by the absolute value of the ion's valence, such that the concentration in mEq/L denotes the total charge contributed by the ion in question. The ionic compositions (in mEq/L) of the major body fluid compartments are shown in Fig. 3.

Cellular Electrolyte Metabolism, Table 1 Extracellular and intracellular ion concentrations and equilibrium potentials

Ion	Extracellular concentration (mM)	Intracellular concentration (mM)	Equilibrium potential (mV)
Na^+	145	~12	+67
K^+	4.5	~150	−94
H^+	0.00004	~0.0001	−24
Ca^{2+}	~1.5	~0.0001	+129
Mg^{2+}	~0.5	~0.5	0
Cl^-	115	~10	−65
HCO_3^-	25	~10	−24

Ion concentrations in millimoles per liter water. Values for mammals, modified from (Andersen et al. 2009) Table 17–3
The $[Ca^{2+}]$ are the free ion concentrations; intracellular $[Cl^-]$ varies considerably among cell types, ranging from ~5 mM in skeletal muscle to ~80 mM in red blood cells
The equilibrium potentials (see Eq. 4) were calculated for $T = 37\ °C$ using the listed concentrations

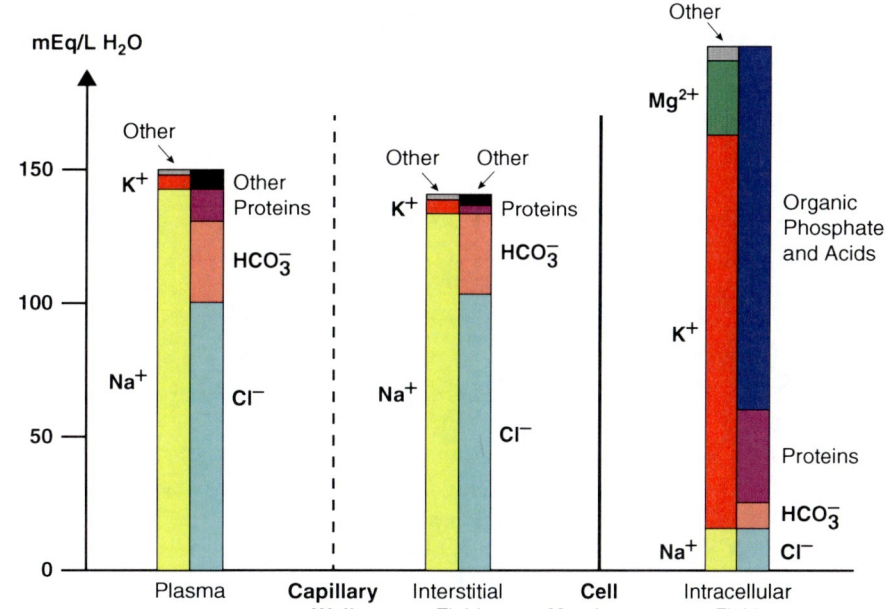

Cellular Electrolyte Metabolism, Fig. 3 Composition of plasma, interstitial fluid, and intracellular fluid. Note that the amount of positive and negative charge is the same in each compartment

In each fluid compartment, the total concentration of positive charge (the sum of the mEq/L for all cations) is equal to the total concentration of negative charge (the sum of the mEq/L for all anions). That is, there is electroneutrality in each compartment; this is important because even a very small charge imbalance, a difference between the amount of positive and negative charge present, would be associated with very large electrical potential differences between different fluid compartments (see Ion Permeation and Membrane Potentials, below).

Ion concentrations in the interstitial fluid differ slightly from those in plasma because the capillary wall allows ions, but not plasma proteins, to pass through. Plasma proteins carry a net negative charge, and the cation concentrations in plasma water will be higher than in interstitial water, whereas the inorganic anion concentrations will be less (the so-called Gibbs-Donnan effect).

The composition of the intracellular fluid differs from that of the extracellular fluid (Table 1). The different distributions of Na^+ and K^+ between the extracellular and intracellular solutions result from the action of ion transport systems in the plasma membrane (Fig. 1), which control the passive and active ion movement across the membrane. For the anions, the different distributions are due to both the ion transport systems and the large amounts of membrane impermeable multivalent protein anions and organic phosphates in the intracellular fluid.

Though not clear from the figure, the total free solute concentrations are the same in the intracellular and interstitial fluid compartment, the two compartments are in osmotic balance.

Considering the distribution of the body water between the intracellular and extracellular compartments and the chemical composition of the compartments, one can deduce that most of the body's Na^+ is in the extracellular fluid, whereas most of the K^+ is in the intracellular fluid. There are ~60 mmoles of Na^+ per kg body mass: ~10% in the intracellular fluid; 50–60% in the extracellular fluid; and the rest in bone. There are also ~60 mmoles of K^+ per kg body mass: ~95% in the intracellular fluid; ~2% in the extracellular fluid; and the rest in bone. Even small shifts of K^+ from the intracellular to the extracellular compartment thus may lead to large changes in the extracellular $[K^+]$ concentration, $[K^+]_e$, and vice versa for Na^+.

Ion Permeation and Membrane Potentials

Most cell membranes are endowed with different types of ion channels (Fig. 1) that differ in their ability to discriminate among different ions, their ion selectivity. Because the extracellular and intracellular ion concentrations differ (Table 1), the membrane potential ($V_m = V_i - V_e$, where V_i and V_e denote the electrical potential of the intracellular and extracellular compartment, respectively) will vary as a function of the number and type of conducting channels in the membrane.

The rate of ion movement through a channel, the single-channel current (i), varies as a function of V_m and the channel's reversal potential, V_{rev}, defined as the membrane potential where $i = 0$:

$$i = g \cdot (V_m - V_{rev}), \quad (3a)$$

where g is the single-channel conductance. g varies depending on the channel type, the permeant ion concentration(s) and V_m; most channels have conductances that range between 5 and 50 pS ((Hille 2001) Figure 12.8). The total membrane current (I) that is carried across by all channels of a given type will be:

$$I = N \cdot i = N \cdot g \cdot (V_m - V_{rev}) = G \cdot (V_m - V_{rev}), \quad (3b)$$

where N denotes the number of *conducting* channels in the membrane and G the total membrane conductance contributed by the channels in question.

In the case of highly selective channels that catalyze the transmembrane movement of only a single ion type, V_{rev} is equal to the ion's equilibrium (or Nernst) potential E:

$$E = \frac{-k_B T}{z \cdot e} \cdot \ln\left\{\frac{C_i}{C_e}\right\}, \quad (4)$$

k_B is Boltzmann's constant, T the temperature in Kelvin, z the ion valence, e the elementary charge, and C_i and C_e the intracellular and extracellular ion concentrations, respectively. (Membrane potentials, and equilibrium potentials are measured relative to the extracellular solution; membrane currents are defined to be positive when the current flow is from the intracellular to the extracellular solution).

Membrane potentials arise because ions diffuse from high to low concentrations; the diffusive flux, J_D, is given by Fick's first law

$$J_D = -A \cdot D \cdot \frac{dC}{dx}, \quad (5)$$

where A is the membrane area, D the diffusion coefficient, C the concentration and x distance (from the intracellular toward the extracellular solution). This diffusive ion movement will establish a *small* charge imbalance between the intracellular and extracellular fluids, which will give rise to a membrane potential difference, ΔV, given by

$$\Delta V = -\frac{\Delta Q}{C_m}, \quad (6)$$

where ΔQ is the net charge transfer across the membrane ($= \Delta n \cdot z \cdot e$, where Δn denotes the net transfer of ions across the membrane) and C_m the membrane capacitance ($= A \cdot C_{sp}$, where C_{sp} is the specific membrane capacitance, ~1 µF/cm^2). In response to the generation of ΔV, an electrical force will act on the ions causing them to migrate (in the opposite direction to the diffusive flux); the electromigrative flux, J_E, is given by

$$J_E = -A \cdot u \cdot C \cdot z \cdot e \cdot \frac{dV}{dx}, \quad (7)$$

where u is the ion mobility ($= D/k_B T$). The resting membrane potential then is the potential difference where diffusive and electromigrative fluxes exactly balance each other:

$$J_E + J_D = 0 \quad \text{or}$$
$$A \cdot D \left(\frac{dC}{dx} + C \cdot \frac{ze}{k_B T} \cdot \frac{dV}{dx} \right) = 0. \quad (8)$$

Equation 8 can then be integrated to yield the Nernst or equilibrium potential, E, cf. (4). (Strictly, E should be expressed in terms of the ion activities, but activity coefficients are neglected in (4) because the ionic strengths of the extracellular and intracellular solutions are similar).

If a cell membrane were endowed with only one type of highly selective channels that were closed (nonconductive) at time $t < 0$, for them to open at $t = 0$, V_m would move from 0 mV to E as

$$V_m(t) = E \cdot (1 - \exp\{-t/\tau\}), \quad (9)$$

where $\tau = C_m/G_m$; the net charge transfer would be, cf. (6), $E \cdot C_m$. G_m varies between 0.1 S/cm^2 (in nerve and muscle, where rapid potential changes are critical) and 10^{-8} S/cm^2 (red blood cells), meaning that τ varies between 10 μs and 100 s.

Membrane potential changes require charge transfer across the membrane, which means that electroneutrality cannot be exact – but the error is small, as illustrated by considering the net charge transfer required to move the membrane potential from 0 mV to the equilibrium potential for K$^+$, E_K. For a spherical cell of radius, r, equal to 10 μm and containing 150 mM K$^+$ (and an equal concentration of negative charge), there is $4 \cdot \pi \cdot r^3/3 \cdot [K^+]$ or $6.28 \cdot 10^{-13}$ moles of K$^+$ in the cell. The net charge transfer (in moles of charge) is $4 \cdot \pi \cdot r^3 \cdot C_m \cdot E_K/F$, where F is Faraday's constant, or $1.23 \cdot 10^{-17}$ moles. The net K$^+$ movement needed to establish the membrane potential is ~0.002% of the total amount of K$^+$ in the cell! Membrane potential changes results from very small net charge movements across the membrane; it is reasonable to invoke electroneutrality.

If there is only a single type of highly selective ion channels in the membrane, the resting membrane potential – the time-invariant potential of a cell "at rest" when the net current (charge transfer) across the membrane is 0 – will be equal to the equilibrium potential for the ion in question. If the membrane is endowed with several different types of highly selective ion channels – the major current-carrying ions are Na$^+$, K$^+$, Cl$^-$, and Ca^{2+} – then V_m becomes a weighted average of the equilibrium potentials for the different ions:

$$V_m = \frac{G_{Na} \cdot E_{Na} + G_K \cdot E_K + G_{Ca} \cdot E_{Ca} + G_{Cl} \cdot E_{Cl} - I_{active}}{G_{Na} + G_K + G_{Ca} + G_{Cl}}, \quad (10)$$

where the subscripts denote the values for the different ions and I_{active} is the current generated by the active (energy-dependent) transporters. In the resting cell, G_K usually is much larger than the membrane conductances for other ions and V_m will be close to E_K.

Cell Volume and Osmotic Balance

Cell membranes are, with few exceptions, very permeable to water (Finkelstein 1987). Water occupies volume, and a net water movement between the intracellular and extracellular compartments will be associated with changes in cell volume. Cell volume regulation thus requires osmotic balance across cell membranes – the chemical potential of water, μ_w, must be the same on both sides of the membrane such that there is no net water flow (Finkelstein 1987):

$$\mu_{W,i} = \mu_{W,e}, \quad (11)$$

where the subscripts "i" and "e" denotes the intracellular and extracellular compartments, respectively. For dilute solutions,

$$\mu_W = \mu_W^0 + RT \cdot \ln x_W + P \cdot \overline{V}_W, \quad (12)$$

where μ_W^0 denotes the standard chemical potential of water, x_W and $\overline{V}_W$ its mole-fraction and partial molar volume, and P the hydrostatic pressure. Combining (11 and 12),

$$\mu_W^0 + RT \cdot \ln x_{W,i} + P_i \cdot \overline{V}_W$$
$$= \mu_W^0 + RT \cdot \ln x_{W,e} + P_e \cdot \overline{V}_W \quad (13)$$

or,

$$\Delta \Pi = P_i - P_e = -\frac{RT}{\overline{V}_W} \cdot \ln \left\{ \frac{x_{W,i}}{x_{W,e}} \right\}, \quad (14)$$

where $\Delta\Pi$ is the osmotic pressure difference between the two solutions, and

$$\Pi_i = -\frac{RT}{\overline{V}_W} \cdot \ln\{x_{W,i}\} \quad \text{and}$$
$$\Pi_e = -\frac{RT}{\overline{V}_W} \cdot \ln\{x_{W,e}\} \tag{15}$$

are denoted the osmotic pressures of the intracellular and extracellular solution, respectively. x_W is, by definition, given by

$$x_W = \frac{n_W}{n_W + n_s} = 1 - \frac{n_s}{n_W + n_s}$$
$$= 1 - x_s, \text{ where } n_s = \sum n_i, \tag{16}$$

where n_W and n_i denote the number of moles of water and the different solute species, n_s is the total number of moles of solute, in the solution, and x_s the mole fraction of total solute. For dilute solutions, when $n_s << n_W$ (and $x_W \approx 1$), it becomes useful to express (15) in terms of the (total) solute mole fractions (concentrations), rather than x_W: and

$$\Pi_i = -\frac{RT}{\overline{V}_W} \cdot \ln\{1 - x_{s,i}\} \approx \frac{RT}{\overline{V}_W} \cdot x_{s,i}$$
$$\approx RT \cdot \frac{n_{s,i}}{\overline{V}_W \cdot x_{W,i}} \approx RT \cdot C_{s,i}$$

and $\tag{17}$

$$\Pi_e = -\frac{RT}{\overline{V}_W} \cdot \ln\{1 - x_{s,e}\} \approx \frac{RT}{\overline{V}_W} \cdot x_{s,e}$$
$$\approx RT \cdot \frac{n_{s,e}}{\overline{V}_W \cdot x_{W,e}} \approx RT \cdot C_{s,e}$$

where $C_{s,i}$ and $C_{s,e}$ denote the total solute concentrations in the two compartments. In the dilute solution limit, (14) thus can be rewritten as

$$\Delta\Pi = \Pi_i - \Pi_e = RT \cdot (C_{s,i} - C_{s,e})$$
$$= -RT \cdot (C_{W,i} - C_{W,e}) \tag{18}$$

Thus, if $x_{W,i}$ were different from $x_{W,e}$ (if $C_{s,i} \neq C_{s,e}$), there would be a hydrostatic pressure difference across the cell membrane. A difference in total solute concentration of 1 mM, would produce an osmotic pressure difference of 0.0246 atm or $2.49 \cdot 10^3$ Pa.

Any transmembrane pressure would be associated with a membrane tension, T_{mem}, given by Laplace's law (for a spherical cell of radius r):

$$T_{mem} = \frac{\Delta\Pi \cdot r}{2} \tag{19}$$

For $r = 10$ μm, a pressure difference of $2.5 \cdot 10^3$ Pa would produce a tension of 12.5 mN/m, which is sufficient to tear the membrane apart (Dai et al. 1998). Experimental determinations of T_{mem} in living cells show that $T_{mem} \approx 0.1$ mN/m (Dai et al. 1998), meaning that the total solute concentrations in the two compartments differ by only ~4 μM, or ~0.001%. It is reasonable to invoke osmotic balance.

The major intracellular solutes (the inorganic ions) are able to cross the membrane, and cell viability depends on exquisite control of the intracellular solute concentration, which is controlled by the combined action of the different membrane proteins that catalyze transmembrane solute movement (Fig. 1). One ion, Na^+, is effectively impermeant due to its relative low permeability and active extrusion by the Na^+,K^+-ATPase (the sodium pump), and cell volume is maintained by the Na^+,K^+-ATPase, which keeps $[Na^+]_i$ low. But it is not enough to keep $[Na^+]_i$ low; it is equally important to maintain $[Na^+]_e$ high, as Na^+ de facto becomes the major impermeant solute in the extracellular fluid. Extracellular Na^+ homeostasis is maintained by the combined actions of the gastrointestinal tract and the kidneys, where the Na^+,K^+-ATPase again plays a key role in the transepithelial ion movement, cf. (Palmer and Andersen 2008).

Cross-References

▶ Calcium in Biological Systems
▶ Magnesium in Biological Systems
▶ Potassium in Biological Systems
▶ Potassium in Health and Disease

References

Andersen OS, Ingólfsson HI, Lundbæk JA (2009) Ion channels. In: Bohr HG (ed) Handbook of molecular biophysics. Wiley-VCH, Weinheim, pp 557–592

Dai J, Sheetz MP, Wan X, Morris CE (1998) Membrane tension in swelling and shrinking molluscan neurons. J Neurosci 18:6681–6692

Finkelstein A (1987) Water movement through lipid bilayers, pores, and plasma membranes. Theory and reality. Wiley, New York

Hille B (2001) Ionic channels of excitable membranes, 3rd edn. Sinauer, Sunderland

Hoffmann EK, Lambert IH, Pedersen SF (2009) Physiology of cell volume regulation in vertebrates. Physiol Rev 89:193–277

Palmer LG, Andersen OS (2008) The two-membrane model of epithelial transport: Koefoed-Johnsen and Ussing (1958). J Gen Physiol 132:607–612

Roos A, Boron WF (1981) Intracellular pH. Physiol Rev 61:296–434

Sten-Knudsen O (2002) Biological membranes: theory of transport, potentials and electric impulses. Cambridge University Press, Cambridge

Yamashita MM, Wesson L, Eisenman G, Eisenberg D (1990) Where metal ions bind in proteins. Proc Natl Acad Sci USA 87:5648–5652

Zhou Z, Neher E (1993) Mobile and immobile calcium buffers in bovine adrenal chromaffin cells. J Physiol 469:245–273

Cellular Pharmacology of Platinum-Based Drugs

▶ Platinum-Containing Anticancer Drugs and proteins, interaction

Cellular Redox Balance

▶ Cadmium and Oxidative Stress

Cellular Respiration

▶ Heme Proteins, Cytochrome c Oxidase

Cerium

Takashiro Akitsu
Department of Chemistry, Tokyo University of Science, Shinjuku-ku, Tokyo, Japan

Definition

A lanthanoid element, the first element of the f-elements block, with the symbol Ce, atomic number 58, and atomic weight 140.115. Electron configuration [Xe]$4f^1 5d^1 6s^2$. Cerium is composed of three stable (^{140}Ce, 88.450%; ^{136}Ce, 0.185%; ^{138}Ce, 0.251%) isotopes and one radioactive (^{142}Ce, 11.114%) isotope, discovered (*Berzelius and Hisinge; Klaproth*) in 1803 and isolated by Mosander. Cerium exhibits oxidation states III and IV; atomic radii: 182 pm, covalent radii: 205 pm, redox potential (acidic solution) Ce^{3+}/Ce −2.34 V, Ce^{3+}/Ce^{2+} −3.2 V; electronegativity (Pauling) 1.12. Ground electronic state of Ce^{3+} is $^2F_{5/2}$ with S = 1/2, L = 3, J = 5/2 with $\lambda = 640$ cm^{-1}. Most stable technogenic radionuclide ^{144}Ce (half-life 284.9 days, $E_{max}(\beta) = 2.98$ MeV.) The most common compounds: CeO_2 (soluble in acids in the presence of reduction agent), $Ce(NO_3)_3 \cdot 6H_2O$, $CeCl_3 \cdot 6H_2O$, $CeSO_4 \cdot 4H_2O$. Biologically, cerium is of low to moderate toxicity, can cause itching, skin lesions, and death (at animals injected with large doses) due to cardiovascular collapse (Atkins et al. 2006; Cotton et al. 1999; Huheey et al. 1997; Oki et al. 1998; Rayner-Canham and Overton 2006).

Cross-References

▶ Cesium, Physical and Chemical Properties
▶ Lanthanide Ions as Luminescent Probes
▶ Lanthanide Metalloproteins
▶ Lanthanides and Cancer
▶ Lanthanides in Biological Labeling, Imaging, and Therapy
▶ Lanthanides in Nucleic Acid Analysis
▶ Lanthanides, Physical and Chemical Characteristics

References

Atkins P, Overton T, Rourke J, Weller M, Armstrong F (2006) Shriver and Atkins inorganic chemistry, 4th edn. Oxford University Press, Oxford/New York

Cotton FA, Wilkinson G, Murillo CA, Bochmann M (1999) Advanced inorganic chemistry, 6th edn. Wiley-Interscience, New York

Huheey JE, Keiter EA, Keiter RL (1997) Inorganic chemistry: principles of structure and reactivity, 4th edn. Prentice Hall, London

Oki M, Osawa T, Tanaka M, Chihara H (1998) Encyclopedic dictionary of chemistry. Tokyo Kagaku Dojin, Tokyo

Rayner-Canham G, Overton T (2006) Descriptive inorganic chemistry, 4th edn. W. H. Freeman, New York

Cerium, Physical and Chemical Properties

Fathi Habashi
Department of Mining, Metallurgical, and Materials Engineering, Laval University, Quebec City, Canada

Cerium is the most abundant rare-earth element and exceeds in abundance tin, cobalt, and lead. It is separated from other rare-earth elements by oxidation of solutions resulting from attack of bastnaesite or monazite. The most important uses of misch metal or cerium are their ability to react with oxygen, hydrogen, nitrogen, sulfur, arsenic, bismuth, and antimony, thus reducing the effects of these elements on the properties of the metals.

Misch metal or cerium-containing master alloys are added to cast iron to improve ductility, toughness, and the microstructure. Cerium allows graphite to form nodules, causing nucleation in spheroidal and vermicular cast iron, and neutralizes the harmful effect of the tramp elements. Addition of misch metal to copper alloys improves tensile strength and deep-drawing properties. The heat resistance and ductility of aluminum conductor cables are improved without any significant decrease in electrical conductivity. Lighter flint alloy consists basically of misch metal and iron. Some other metals are added in small amounts to modify the pyrophoric properties and to improve processing. The frictional pyrophoric properties of cerium-iron are based on a combination of microstructure and mechanical and chemical properties of the alloy. The typical crystal structure of commercial alloy consists of tough, brittle primary crystals of Ce_2Fe_{17} enclosed by a layer of $CeFe_2$ embedded in a soft matrix of $CeFe_2$ and Ce. The principal uses for cerium compounds are as polishing agents and as a component in glass.

Physical Properties

Atomic number	58
Atomic weight	140.12
Relative abundance in Earth's crust,%	2.5×10^{-3}
Density, g/cm^3	6.770
Crystal structure at room temperature, $a_0 = 0.51612$ nm	Face-centered cubic
Melting point, °C	798
Boiling point, °C	3,433
Heat of fusion ΔH kJ/mol	5.5
Specific heat at 25°C, J mol^{-1} K^{-1}	27.0
Coefficient of linear thermal expansion at 25°C, K^{-1}	27.0×10^{-6}
Thermal conductivity at 27°C, J s^{-1} cm^{-1} K^{-1}	0.114
Electrical resistivity at 0°C, μΩ cm	77

Chemical Properties

Cerium has not only the normal rare-earth +3 oxidation state but also the exceptional +4 state which simplifies its separation from the other rare-earth elements. Misch metal is a mixture of cerium, lanthanum, neodymium, and praseodymium with cerium as the major constituent. It is the lowest-priced rare-earth metal, because no expensive chemical separation is needed to produce it being produced by fused-salt electrolysis of rare-earth chlorides.

Misch metal is ductile. The freshly cut surface has a metallic gray appearance. In air, the surface oxidizes to form yellow to greenish-gray rare-earth hydroxide carbonates or oxide hydrates. Massive metal burns above 150°C in pure oxygen; however, chips, turnings, and powder burn at this temperature even in air. Misch metal dissolves in dilute mineral acids with evolution of hydrogen.

References

Habashi F (2003) Metals from ores. An introduction to extractive metallurgy. Métallurgie Extractive Québec, Québec City. Distributed by Laval University Bookstore, www.zone.ul.ca

Reinhardt K, Winkler H (1997) Cerium. In: Habashi F (ed) Handbook of extractive metallurgy. Wiley, Weinheim, pp 1743–1760

Cesium

▶ Cesium, Therapeutic Effects and Toxicity

Cesium Therapeutic Effects

▶ Cesium, Therapeutic Effects and Toxicity

Cesium Toxicity

▶ Cesium, Therapeutic Effects and Toxicity

Cesium, Physical and Chemical Properties

Fathi Habashi
Department of Mining, Metallurgical, and Materials Engineering, Laval University, Quebec City, Canada

Cesium is a very soft, ductile, silvery white metal. It is the last member of the alkali metals (if Francium, the metal following it in the group, is ignored since it is radioactive with very short half-life). Since it has the largest atomic size of the group then the outermost electron will be far away from the nucleus and therefore easily lost because of the weak electrostatic force of attraction. Consequently, cesium is the most reactive, not only the alkali metals but also of all metals. It also has the lowest hardness among all metals. Applications of nonradioactive cesium include photoelectric cells, optical character recognition devices, photomultiplier tubes, video camera tubes, optical components of infrared spectrophotometers, catalysts for several organic reactions, crystals for scintillation counters, and in magnetohydrodynamic power generators.

Physical Properties

Symbol	Cs
Atomic number	55
Atomic weight	132.91
Melting point, °C	28.7

(continued)

Symbol	Cs
Boiling point, °C	685
Atomic radius, nm	0.274
Ionic radius, nm	0.165
Density at 20°C, g/cm^3	1.873
Mohs hardness number	0.2
Specific heat, liquid, J g^{-1} K^{-1}	0.236
Heat of fusion, kJ/mol	2.13
Heat of vaporization at 0.1 MPa, J/mol	65.9
Ionization potential, eV	3.87
Standard electrode potential, V	−2.923
Electrical conductivity, Ω^{-1} cm^{-1}	
Solid, 25°C	4.9×10^4
Vapor, 1,250°C	2.0×10^2
Vapor pressure P, kPa	$\log P = -0.2185 \frac{A}{T} + B$

T, K	A	B
200–350	17543.0	6.0739
279–690	17070.7	5.8889

Chemical Properties

Cesium is a monovalent typical metal, with electronic structure 2, 8, 18, 18, 8, 1. When it loses its outer most electron, it will have the electronic structure of inert gases. Because of its extremely low ionization potential, cesium is usually more reactive than lithium, sodium, or potassium and pronouncedly more reactive than rubidium. When cesium is exposed to air, an explosion-like oxidation to form cesium superoxide, CsO_2, occurs; contact with water results in a vigorous reaction to form cesium hydroxide and hydrogen gas, which may ignite spontaneously.

Radioactive cesium isotope, ^{137}Cs (half-life 30.1 years), is a product of atomic bomb explosions and nuclear reactors disasters, together with ^{90}Sr, and both are widely present in the biosphere.

References

Bick M, Prinz H (1997) Chapter 54. In: Habashi F (ed) Handbook of extractive metallurgy. Wiley-VCH, Weinheim

Schreiter W (1961) Seltene metalle, vol 1. VEB Deutscher Velag für Grundstoffindustrie, Leipzig, pp 194–220

Cesium, Therapeutic Effects and Toxicity

Petr Melnikov[1] and Lourdes Zélia Zanoni[2]
[1]Department of Clinical Surgery, School of Medicine, Federal University of Mato Grosso do Sul, Campo Grande, MS, Brazil
[2]Department of Pediatrics, School of Medicine, Federal University of Mato Grosso do Sul, Campo Grande, MS, Brazil

Synonyms

Alkaline metals; Cesium; Cesium therapeutic effects; Cesium toxicity; Radioactive cesium

Definition

The aim of this entry is to resume and discuss the material currently available on therapeutic effects of cesium salts, their applications in medicine, and their toxicity.

Introduction

Cesium is naturally present as the stable ^{133}Cs in various ores and to a lesser extent in soil. The most important radioactive nuclides are ^{131}Cs and ^{137}Cs, with half-lives of 10 days and 30 years, respectively. Cesium enters easily into the plant and animal systems and is deposited in soft tissues. The total content of this intracellular form is very low, no more than 0.00131 g. Cesium salts are used as catalysts and for the production of special glasses and ceramics. In molecular biology, they are applied for decades for density gradient ultracentrifugation in order to isolate viral particles, subcellular organelles and fractions, and nucleic acids from biological samples.

An extensive literature including vast reviews arose concerning the distribution and residence times of cesium in the body and means of enhancing its excretion. The knowledge about cesium metabolism and toxicity is sparse, since acute and chronic toxicity is not sufficiently studied. However, cesium is by no means a novel topic in medicine. More than 120 years ago, Sidney Ringer, the scientist who is well known for his isotonic solution resembling blood serum in its salt constituents, engaged in a study of this rare alkaline element. For the first time, he postulated that cesium (as well as rubidium, another heavy element) was supposed to behave as physiological analog for potassium, and therefore employed for replacement therapy. Lately, this prognostic turned out to be confirmed only partially.

Once cesium enters the body, it is distributed through the system, with higher concentrations in the kidneys, skeletal muscle, liver, red blood cells, myocardium, placenta, and breast milk. Its biological half-life in humans is between 1 and 4 months. Naturally, it binds preferably to anionic intracellular components of erythrocytes and decreases their ability to give up oxygen in tissues. Absorption of cesium from the stomach to blood is assumed to be negligible, the same as for potassium. The absorption and distribution of cesium radioactive isotopes are identical to those of the stable ^{133}Cs.

Physiological Effects of Cesium

So far, no data has been located regarding gastrointestinal, hematological, musculoskeletal, hepatic, renal, endocrine, dermal, ocular, body weight, immunological, neurological, or development effects. No reports were located in which cancer in humans or animals could be associated with intermediate or chronic-duration oral exposure to stable cesium. Nevertheless, this element is known to produce remarkable and even dangerous metabolic effects regarding cardiovascular system.

It is worth reminding that cardiac K^+ channels are membrane-spanning proteins that allow the passive movement of K^+ ions across the cell membrane along its electrochemical gradient. Meanwhile, the pathophysiological mechanisms of the immediate Cs^+ effects on cardiac tissues consist in replacing K^+, leading to the blockade of inwardly rectifying K^+ current at ventricular level, and nodal hyperpolarization-activated cation current, both of which primarily affect the resting membrane potential. Furthermore, Cs^+ block of I_{k1} (inward rectifier current) involves interactions between cations at binding sites within the channel pore, and the inhibition of inward I_{k1} is likely to be explained by Cs^+ entry into I_{k1} channels, favored by hyperpolarization. In an animal model, the marked prolongation of the action potential duration (i.e., acquired long QTc syndrome) observed in man was also reproduced, but alterations were observed, such as ventricular tachycardia and torsade de pointes (Gmelin's 1973). These arrhythmogenic effects of Cs^+ ions have been linked to the inhibition of hyperpolarization-activated current and a reduction in cardiac K^+ currents, which may be blocked by Cs^+ ions from the intracellular or extracellular side. The investigation of such channelopathies continues to yield remarkable insights into the molecular basis of cardiac excitability.

Experimentally, results have been obtained in previous investigations in dogs and rabbits, when intravenous CsCl provoked an instant ventricular tachycardia associated with monophasic early afterdepolarizations. At tissue level, it was also shown that Cs^+ causes a voltage-dependent block of inward K^+ currents in resting skeletal muscle fibers. It is to suggest that the mechanisms involved are basically the same. However, it must be stressed that reaction rates are determined by molarity; so, in order to produce the same kinetic effect, cesium chloride might be taken in approximately double the amount in comparison to potassium chloride, and it is rarely achieved in practice.

The K^+ channels regulate the resting membrane potential, the frequency of pacemaker cells, and the shape and duration of the cardiac action potential. In mammalian cardiac cells, K^+ channels include, among others, rapid (I_{Kr}) and slow (I_{Ks}) components of the delayed rectifier current, as well as the inward rectifier current (I_{K1}). Changes in the expression of K^+ channels explain the regional variations in the morphology and duration of the cardiac action potential among different cardiac regions and are influenced by heart rate, intracellular signaling pathways, drugs, and cardiovascular disorders. A number of cardiac and noncardiac drugs are capable of blocking cardiac K^+ channels, and can cause an acquired long QTc syndrome as well as torsade de pointes. In cases of extreme gravity, the arrhythmias can deteriorate to ventricular fibrillation and cardiac arrest. It was concluded that cesium is a potent potassium blocker, and therefore it has been widely used to study the characteristics of these channels.

Recently, in animal studies, cesium salts have shown epileptogenic properties in respect to neocortical and hippocampal areas. This effect has been suggested to be related to an alteration in extracellular potassium regulation by glia. However, so far, no clinical equivalent is known with regard to human neurophysiology.

Clinical Effects of Cesium Intake

Oral intake of cesium chloride has been widely promoted on the basis of the hypothesis referred to as "high pH cancer therapy" advanced in 1984 by K.A. Brewer who did his Ph.D. in physics. Cesium chloride therapy named as "complementary alternative method" has never been approved, neither by the US Food and Drug Administration (FDA) nor by the European Agency for the Evaluation of Medicinal products. It is no surprise as this proposal plainly assumes that the usage of alkaline ions and, in particular, cesium might (or should) provoke substantial increase in pH within malignant cells. It assumes also that only tumor cells tend to incorporate cesium ions. This alone is enough to establish that the therapy is not based on good information. Nevertheless, the alkalinity is not a key point, since CsCl is not different from KCl, as dissociation degrees of their respective hydroxides are very close. Hence, the presence of cesium ions in the cell can, by no means, guarantee increased intracellular pH in human body, although, in principle, fluorescent pH probes could be useful at least for comparative studies of cesium action. So, there are neither theoretical nor experimental grounds for the idea of CsCl (or any other stable cesium salt) application in cancer treatment.

Meanwhile, desperate patients looking for an immediate miraculous relief from cancer would not go into these subtleties. So, the fact that this method is not currently officially endorsed led to adverse reaction, that is, blaming the "medical establishment" for their conservative and unsupportive standpoint. All this ever more stimulated widespread self-treatment without bearing in mind that it could lead to serious problems. Ordinary people with no medical knowledge and individuals with discovery delusions started to take and recommend large amounts of cesium chloride with no clinical evaluation of its possible hazardous consequences. Cesium chloride is sold as a routine dietary supplement in the USA. Unlike companies that produce drugs, the providers do not have to show evidence of safety or health benefits to the FDA before selling their products.

Finally, a paper has recently been published, bluntly stating that two patients suffering from terminal malignancies "were administered intravenous doses of an unapproved therapy," consisting of a solution containing cesium chloride. Both patients died. As the cases are unrelated, it seems that the procedures were carried out with experimental purposes. No "informed consent" has been issued. No data as to the dosage applied, clinical circumstances, or results of postmortem examination of these ethically inappropriate events were supplied.

Clinical Examples of Cesium Ingest

Herein, only representative clinical examples are discussed. In a report released in April 2004, US Agency for Toxic Substances and Disease Registry stated that no communications had been located in literature regarding death in humans following acute, intermediate, or chronic duration exposure to stable cesium. However, in 2003, a paper was published describing two deaths following acute exposure to this element. Case 1 was a 41-year-old male with kidney cancer, and case 2 was an 82-year-old male with lung cancer. As mentioned before, both patients were administered a solution containing cesium chloride. Forensic records collected on these two cases indicated that on one patient (case 1), two cesium therapies were administered on two consecutive days. Initially, the patient developed uncontrolled chills and seizures, and went into cardiac arrest while in the doctor's office. Case 2 apparently collapsed while he was receiving the intravenous injection containing the cesium chloride.

According to comparison of cesium levels in exposed and nonexposed tissues, in case 1, cesium content in liver tissue was 100,000 times higher than in control samples, and it was 10,000 times higher in brain than in controls. In case 2, cesium levels were substantially lower – 10,000 and 1,000 times, respectively. In both instances, cesium in whole blood was relatively lower, indicating its immediate migration into tissues, in accordance to the model of cesium distribution in the body. So, these cases should be qualified as acute poisoning with cesium chloride. It is worth reminding that even more innocuous potassium chloride is lethal, when injected intravenously in the form of concentrated solution.

In order to illustrate the safety of high pH therapy, another case report has described the effects of oral intake of cesium chloride. The author volunteered to experience on himself the effect of long-term oral administration of 6 g per day of cesium chloride. The drug was dissolved in water and consumed immediately after the morning and evening meals, which were diet-restricted to attain approximately 1% potassium intake. There was an initial general feeling of well-being and heightened sensory perception. A gradual decrease in appetite was noted initially before it was stabilized at a later date. Discontinuation of rich bread meals resulted in pre-nausea sensation which was followed by diarrhea. Almost immediately, a tingling sensation in the lip and cheek regions was experienced, but no harmful effects were noted in intellectual capacities or in driving skill.

Another self-treatment by alternate therapy is related to a woman presented to the emergency department following an episode of hypotension syncope. The patient was thirsty, disoriented, and hypotensive. An electrocardiogram indicated a sinus rhythm with a long QTc interval, with episodes of polymorphic ventricular tachycardia. The patient had a 2-year history of colon cancer with liver metastases and had received chemotherapy. At the same time, she had been self-treating with an alternative therapy of oral cesium salts for several weeks and a vegetarian diet. As in the previous case, the patient experienced numbness or tingling of the lips. She developed hypokalemia, but magnesium and calcium levels were unchanged. She was treated with saline solution supplemented with potassium, discharged herself 3 h late, but returned the next day following a second episode of syncope or possible seizure and still was hypokalemic. Electrocardiogram depicted sinus bradycardia, premature ventricular, and a more prolonged QTc interval. During the next 3 days, following discontinuance of cesium, the QTc interval gradually shortened to 390 ms, and potassium remained in the reference levels. This observation confirmed the history of cesium consumption but could not be directly related to the dose of cesium in the kinetic studies because those models involved intravenous administration and acute response rather than chronic exposure to oral cesium salts.

Another case of cesium chloride therapy was reported describing a man with recurrent syncope, who underwent a naturopathic treatment consisting of

2 g of cesium chloride four times a day intravenously for 2 weeks for prostate cancer. During treatment, he had his first episode of syncope. He continued to take cesium chloride three times a day. Two months later, he was hospitalized because of recurrent syncopes. The electrocardiogram showed a prolonged QTc interval and ventricular ectopic beats. Runs of torsade de pointes tachycardia were recorded on telemetry. The serum potassium level was 2.8 mEq/L. Analysis of a blood sample revealed a plasma cesium level of 830 μmol/L that is approximately 276,000 times higher than reference data and comparable to the values found in the case of acute poisoning discussed early but without a lethal outcome. The patient was treated with intravenous potassium and magnesium. The QTc interval remained prolonged and ventricular premature beats persisted after normalization of the serum potassium level. The patient agreed to stop taking cesium chloride. After 6 months of follow-up, he had not had any further episode of syncope and the corrected QTc interval had returned to normal.

Another case presentation, in this case with no cancer involvement, concerns a woman presented to a local hospital after experiencing three episodes of syncope during the past week. Prior to this, she had never had syncope or near-syncope. Her latest syncopal episode required cardiopulmonary resuscitation. She had no prior history of cardiovascular or neurological disease and was taking no prescription medications. However, she was taking an array of dietary supplements and natural products including cesium salt. She described this as part of a "detoxification program" for menorrhagia that entailed drinking 1–2 gal of water a day along with cesium salt. She had been doing this for the last 2 weeks. An electrocardiogram revealed normal rhythm and profound QTc prolongation. She had only mild hypokalemia and mild hypomagnesemia.

These were corrected with no significant change in the QTc. Although there was never electrocardiographic documentation of torsade de points, her physician recognized that cesium might have prolonged the QTc interval and induced arrhythmia. Hence, she was treated by prompt cessation of her "detoxification regimen" and correction of electrolyte abnormalities. A urine assay for cesium revealed a level of 750 mg/L, which is 65,000 times higher than the data for general population. Daily electrocardiograms showed gradual normalization of her resting QTc. The patient did well and returned to her previously asymptomatic syncope-free state.

Recently, a life-threatening torsade de pointes resulting from "nature" cancer treatment was described in a case report of a woman with recurrent syncope attacks. One of her naturopathic drugs was subsequently confirmed containing 89% CsCl by weight. Besides conventional treatment of QTc prolongation and torsade de pointes, the patient was given a 4-week course of oral Prussian blue to enhance gastrointestinal elimination of cesium. This is the first published case of a nonradioactive cesium poisoning treated with Prussian blue.

Practically no data are available for children, but recently a case was described in an adolescent 16-year-old girl with metastatic hepatocellular carcinoma. She had received courses of chemotherapy that resulted in minimal tumor regression. Against the advice of her oncologist, an alternative regimen was started that included cesium chloride supplements. Two weeks later, two brief syncopal episodes were observed. An electrocardiogram revealed occasional premature ventricular contractions, and QTc interval prolongation. After admitting to the hospital, she experienced monomorphic ventricular tachycardia. Her plasma cesium level was 2,400 mg/L. Two days later, the QTc interval on electrocardiogram had normalized.

Therapeutic Usage of Cesium

After the Chernobyl and recent Fukushima nuclear accidents, cesium was recommended as preventive therapy for radiation poisoning by the isotope ^{137}Cs. The principle of this treatment is quite clear: to saturate the body with the stable cesium (preferably in the form of cesium iodide) enhancing the clearance of radionuclide and effectively replacing it with a safer isotope. Actually, the data suggest that there is a threshold of maximum cesium saturation in the red blood cells and any additional exposure will not stimulate ^{137}Cs excretion. It is important that cesium salts are water soluble, which means that it is excreted from the body via urine and it happens very quickly. So, the most effective way of protection is to stay well hydrated.

As to the cancer treatment, ^{131}Cs and ^{137}Cs have been promoted to deal with various types of malignancies. The so-called cesium brachytherapy is a method of radiation usage in which sealed sources are employed to deliver a radiation doses at a distance of up to a few centimeters by surface, intracavitary, or intestinal applications. Radiation kills or arrests the growth of the cancer with minimal damage to healthy tissue. The isotope is usually incorporated into ^{137}Cs-Na borosilicate glass. The leach rates of the radioactive source are within permissible limits. Radioactive isotope 137 brachytherapy is used for carcinomas of cervix uteri, intestinal cancers, carcinoma of the tongue and floor of the mouth. Excellent results (with recurrence $< 3\%$) have been achieved in the treatment of the skin of nose epitheliomas. The radioactive isotope ^{131}Cs, which was approved in 2003 by the FDA for use in brachytherapy for prostate cancer and other malignancies, has the advantage of a shorter half-life. That means faster dose delivery, that cancer cells have less opportunity to repopulate, and less protracted radiation to normal healthy tissues, so side effects are minimal.

Summary

The knowledge about cesium metabolism and toxicity is sparse. Oral intake of cesium chloride has been widely promoted on the basis of the hypothesis referred to as "high pH cancer therapy," a complimentary alternative medicine method for cancer treatment. However, no properly confirmed tumor regression was reported so far in all probability because of neither theoretical nor experimental grounds for this proposal. The aim of this entry is to resume and discuss the material currently available on cesium salts and their applications in medicine. The presence of cesium in the cell does not guarantee high pH of its content, and there is no clinical evidence to support the claims that cancer cells are vulnerable to cesium. Cesium is relatively safe; signs of its mild toxicity are gastrointestinal distress, hypotension, syncope, numbness, or tingling of the lips. Nevertheless, total cesium intakes of 6 g/day have been found to produce severe hypokalemia, hypomagnesemia, prolonged QTc interval, episodes of polymorphic ventricular tachycardia, with or without torsade de pointes, and even acute heart arrest. However, full information on its acute and chronic toxicity is not sufficiently known. Health care providers should be aware of the cardiac complications, as a result of careless cesium usage as alternative medicine.

Radioactive cesium is successfully used for cancer treatment in the form of brachytherapy that is a sealed-source radiotherapy which is effective alternative for the standard external beam therapy.

Cross-References

▶ Potassium Channels, Structure and Function
▶ Rubidium, Physical and Chemical Properties

References

Gmelin's L (1973) Handbuch der AnorganischenChemie (System Number 20), 8th edn. Springer, Berlin/Heidelberg/New York

Melnikov P, Zanoni LZ (2010) Clinical effects of cesium intake. Biol Trace Elem Res 135(1–3):1–9

U.S. Department of Health and Human Services.Public Health Service (2004) Toxicological profile for cesium. Agency for Toxic Substances and Diseases Registry.Cesium- Health Effects, Atlanta

Chagas Disease

▶ Selenium and Muscle Function

Chalcogen Family

▶ Tellurite-Detoxifying Protein TehB from *Escherichia coli*

Chalcogen Resistance

▶ Bacterial Tellurite Processing Proteins
▶ Bacterial Tellurite Resistance

Channelopathies

▶ hERG (KCNH2) Potassium Channel, Function, Structure and Implications for Health and Disease

Chelation Therapy

▶ Mercury Toxicity

Chemoresistance to Platinum-Based Drugs

▶ Platinum-Resistant Cancer

Chromium

▶ Chromium and Allergic Reponses
▶ Chromium and Diabetes
▶ Chromium(III) and Immune System

Chromium [Cr (VI)] Compounds

▶ Hexavalent Chromium and Cancer

Chromium and Allergic Reponses

Peter Jensen[1] and Jacob P. Thyssen[2]
[1]Department of Dermato-Allergology, Copenhagen University Hospital Gentofte, Hellerup, Denmark
[2]National Allergy Research Centre, Department of Dermato-Allergology, Copenhagen University Hospital Gentofte, Hellerup, Denmark

Synonyms

Allergy; Chromium; Dermatitis; Epidemiology; Leather; Patch test; Sensitivity; Skin

Definitions

Contact dermatitis: A skin reaction (dermatitis, i.e., eczema) resulting from exposure to allergens (allergic contact dermatitis) or irritants (irritant contact dermatitis).

Introduction

Chromium (Cr) (Greek chroma, meaning color) is one of the most widely distributed metals and was discovered in 1797 by the French chemist Louis Nicolas Vauqelin who named it after its color characteristics. Cr is found in nature as Cr ironstone ($FeCr_2O_4$) and as red lead ore ($PbCrO_4$, chrocoite). Cr ironstone is used commercially and is mined in southern parts of Africa, the Philippines, and Kazakhstan. The majority is used for metallurgical purposes such as stainless steel and other alloys. It is also used for leather tanning, preservation of woods, color pigments in paints, and laboratory chemicals. In addition, Cr compounds are present in raw materials used for the production of cement. Skin exposure to Cr may cause contact allergy and allergic contact dermatitis (Lidén et al. 2010).

Allergic and Irritant Contact Dermatitis

Contact dermatitis is an eczematous skin reaction that results from exposure to either irritant substances (*irritant contact dermatitis*) or allergenic substances which are capable of eliciting an allergic reaction (*allergic contact dermatitis*). In both cases, an inflammatory reaction is elicited by immune cells present in the skin which culminates in an eczematous reaction (dermatitis) on exposed skin areas characterized in its acute phase by redness, swelling, scaling, itching, and vesicles containing a clear fluid (Fig. 1). Irritant contact dermatitis is often provoked by prolonged and frequent skin contact with water, chemical detergents, and solvents, often in combination and worsened by humid or dry environments. In other words, the inflammatory reaction and the resulting irritant contact dermatitis are provoked not by one specific substance but by a combination of several unspecific factors. Therein lies the key difference between irritant contact dermatitis and allergic contact dermatitis – the fact that allergic contact dermatitis is elicited by skin contact with a specific molecule from the environment which triggers specific memory immune T cells to initiate an allergic reaction which results in eczema (Rustemeyer et al. 2010).

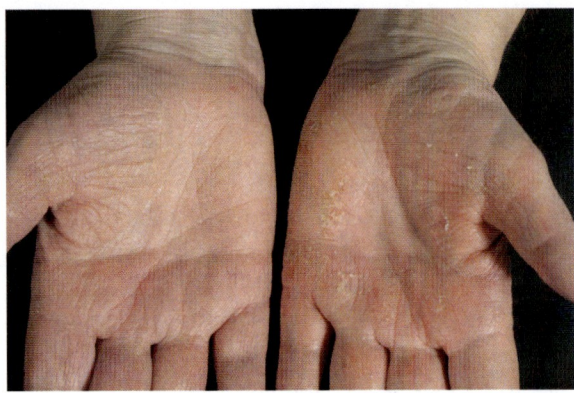

Chromium and Allergic Reponses, Fig. 1 Dermatitis (eczema) on the palms

Allergic contact dermatitis (i.e., allergic contact eczema) is the final stage of a series of events starting with contact sensitization. This first *sensitizing* stage is caused by a group of reactive chemicals referred to as allergens which are able to permanently change a subgroup of memory immune cells so that they will proliferate and target the skin component upon allergen reexposure. Mostly, these allergens are man-made, but naturally occurring contact-sensitizing chemicals are also known; however, they represent a very limited clinical problem. Contact sensitizers, also known as haptens, are small molecules with a weight below 500 Da. The hapten is able to penetrate the skin but is too small to trigger an immune response by itself. The sensitizing capacity is obtained by the hapten binding to proteins to form an antigenic hapten–protein complex in the upper layers of the skin. The next step involves the transport of the antigenic hapten–protein complex to the draining lymph nodes by Langerhans cells. Here, subsets of specific immune T cells recognize the antigenic complex, proliferate vigorously, and are then released in large numbers into the bloodstream. Upon renewed contact with the same allergen, a phase known as the *elicitation* or *effector* phase is initiated during which a range of chemical mediators are released culminating in an eczematous reaction which peaks within 18–72 h (Rustemeyer et al. 2010). Allergic contact dermatitis resulting from work-related exposure is known as *occupational allergic contact dermatitis*. Contact allergy is considered to be a chronic and potentially lifelong condition. Risk factors of contact allergy and allergic contact dermatitis include the sensitizing potential of the allergen, high allergen concentrations (dose per unit area), frequent exposure, occlusion, long duration of exposure, presence of penetration enhancing factors, and an altered skin barrier function.

Although occupational contact dermatitis is not life-threatening, it can have a considerable physical and psychosocial impact (Kadyk et al. 2003). Patients frequently complain of symptoms relating to their skin condition that also affect their mental state, and others describe interference with work and daily activities. Dermatitis on the hands is especially associated with significant impairment. An intact skin barrier on the hands is pivotal to protect the hands during everyday tasks. In the work setting, the hands are vital for both functionality and presentation, for instance, in occupations with direct interaction with other people – food service, child care, and sales. Even lesions that are noncontagious are often regarded as disgusting by others. Apart from the detrimental effect on quality of life of the dermatitis itself, it may also cause sick leave and occupational changes. The exact economical impact is unknown but may amount to several billion dollars annually in the USA alone (Kadyk et al. 2003).

Allergic Contact Dermatitis: Diagnosis and Treatment

Patients with a history and clinical picture of allergic contact dermatitis can be accurately diagnosed by a procedure known as *patch testing*. Basically, the patient is reexposed to the suspected allergen(s) under controlled conditions. This is done by applying appropriate amounts of allergen on to the skin, typically on the upper back. Here, it remains for 2 days, after which the patches containing the allergens in question are removed. In the case of contact allergy, the patient will develop an eczematous reaction at the skin site exposed to the allergen. The response can be graded according to severity of the reaction. Readings are typically done on days 2, 3, and 7, which ensures that even delayed reactions are recorded (Lindberg and Matura 2010). If a diagnosis of contact allergy is made, it is highly recommended that the patient avoids future contact with that particular allergen. Occupational changes can sometimes be necessary if avoidance of an allergen is impossible. Often, preventive measures, i.e., the use of protective gloves, aprons, or masks, are adequate to control the problem.

Most mildly-to-moderately affected patients can be treated effectively with topicals consisting of skin moisturizers combined with anti-inflammatory ointments while severely affected patients may require periodic phototherapy or systemic anti-inflammatory treatment. Interestingly, allergic Cr contact dermatitis is a particular challenge since it is associated with an often severe form of eczema which is difficult to treat. Cr may also elicit a systemic allergic dermatitis response (widespread eczema) when ingested, and this has been reported after oral intake of vitamin tablets containing only 150-µg Cr-chloride (Ozkaya et al. 2010).

Chromium as an Allergen

Cr is a transitional metal and exists in several oxidation states from 0 to +6, of which the ground states 0, +2, +3, and +6 are common. Only the trivalent Cr(III) and the hexavalent Cr(VI) oxidation states are stable enough to act as haptens. Metallic Cr, oxidation state 0, does not possess sensitizing potential since it tends to form an insoluble layer of oxide on the surface. Cr may fluctuate between the hexavalent and trivalent states depending on factors such as temperature and pH. Human as well as animal studies have shown that Cr(IV) may consistently be considered a stronger sensitizer and elicitor of allergic contact dermatitis when compared to Cr(III). This is explained by higher bioavailability of Cr(VI) since it is more water soluble than Cr(III), penetrates the skin more easily, and accumulates in the skin to a higher degree than Cr(III). Once inside the skin, Cr(VI) is converted to Cr(III), and the trivalent form binds more easily to proteins than Cr(VI), potentiating antigen presentation and elicitation of allergic dermatitis. Patch testing is routinely performed with 0.5% potassium dichromate, the hexavalent form of Cr.

Several studies have investigated how much Cr is required to elicit an allergic skin reaction. One such study showed that 1 of 14 Cr-allergic individuals reacted to 1 ppm occluded Cr(VI) (Allenby and Goodwin 1983). The patch test threshold was established at 10 ppm in 17 Cr-allergic patients who were patch tested on normal skin (Basketter et al. 2001), and furthermore, and in that same study, the presence of an irritant lowered the threshold from 2 to 1 ppm in 2 of 17 patients. A review on nine patch test studies on the elicitation threshold levels for Cr(VI) at different levels of pH found that the level was 15 ppm for 10% and 7.6 ppm for 5% of the sensitized population (Stern et al. 1993). A recent review on Cr allergy showed that exposure to occluded patch test concentrations of 7–45 ppm Cr(VI) resulted in allergic contact dermatitis in 10% of Cr-allergic patients (Hansen et al. 2002). Further data are needed to determine the exact eliciting capacity of Cr(III), but it is generally accepted that much higher concentrations of Cr(III) than Cr(VI) are needed to elicit dermatitis. This is supported by three studies; one included 54 patients and showed that one individual reacted to Cr(III) at a threshold concentration of 1,099 ppm (Nethercott et al. 1994); the second study estimated the minimal elicitation threshold $(MET)_{10\%}$ for Cr(VI) at a much lower 6 ppm (Hansen et al. 2003); the third study showed that 50 ppm Cr(III) was insufficient to elicit allergic Cr dermatitis in 18 Cr-sensitized patients (Iyer et al. 2002). All in all, Cr(VI) is a much more potent allergen than Cr(III).

Occupational Sources of Cr Exposure

Table 1 provides an overview of Cr-containing items and chemicals as well as work procedures that may be associated with occupational Cr exposure. As a metal, Cr can be found in various alloys such as stainless steel and on Cr-plated surfaces. Also, Cr is found in raw materials used for the production of cement. Paints, usually yellow, red, and orange often contain hexavalent Cr which can also be released by cutting or tooling of metals treated with anticorrosion agents. When zinc-galvanized sheet metal is cut or tooled, Cr may leak out and come into contact with the skin. Stainless and nonstainless steel may release Cr on welding, and the Cr-containing fumes may be distributed to the facial skin. Cr may also be released from metal products which are chromated in order to prevent rust or surface oxidation. Such products may include common household items such as screws, fittings, and other materials which are used both in an occupational and nonoccupational setting. Indeed, a recent study confirmed that the release of Cr(VI) from chromated metal products can be high enough to provoke allergic reactions in individuals already sensitized to Cr (Geier et al. 2009).

Chromium and Allergic Reponses, Table 1 Work procedures, chemicals, and metallic items which may release or cause occupational exposure to chromium

Stainless steel welding
Chromate production
Chrome plating
Ferrochrome industry
Chrome pigments
Leather tanning
Painters
Workers involved in the maintenance and servicing of copying machines and the disposal of some toner powders from copying machines
Battery makers
Candle makers
Dye makers
Printers
Rubber makers
Cement workers
Landfill sites with chromium-containing wastes
Cement-producing plants
Reagents used for analytic standards
Anticorrosion agents
Catalysts
Engraving
Metallic tools
Magnetic tapes
Wood preservatives

Chromium and Allergic Reponses, Table 2 The EU chromium directive (Directive 2003)

I	Cement and cement-containing preparations may not be used or placed on the market if they contain, when hydrated, more than 0.0002% soluble Cr(VI) of the total dry weight of the cement
II	If reducing agents are used, then without prejudice to the application of other Community provisions on the classification, packaging, and labeling of dangerous substances and preparations, the packaging of cement or cement-containing preparations shall be legibly and indelibly marked with information on the packing date, as well as on the storage conditions and the storage period appropriate to maintaining the activity of the reducing agent and to keeping the content of soluble Cr(VI) below the limit indicated in paragraph 1
III	By way of derogation, paragraphs 1 and 2 shall not apply to the placing on the market for, and use in, controlled closed and totally automated processes in which cement and cement-containing preparations are handled solely by machines and in which there is no possibility of contact with the skin

Chromium and Allergic Reponses, Table 3 Chemicals and metallic items which may cause nonoccupational exposure to chromium

Leather products, shoes, gloves, handbags, furniture
Tattoo pigments
Chromated metal products, i.e., screws, metal rings, tools, and fittings
Mobile phones
Eye shadows

As much as 90% of the global leather production is tanned using trivalent Cr(III) sulfate. This is performed to obtain softness, durability, and flexibility of the leather and make it usable for furniture, footwear, and clothing. During the tanning process, Cr(III) reacts with the leather, stabilizing certain proteins which make the finished product more resistant to degradation. Most of the Cr(III) is tightly bound in the leather, but some can still be released and come into contact with the skin (Lidén et al. 2010). One study showed that tannery workers who were exposed to Cr(III) developed allergic Cr dermatitis, emphasizing that the trivalent form of Cr may also sensitize (Estlander et al. 2000).

Historically, exposure to Cr and associated Cr-allergic contact dermatitis has mainly been caused by occupational contact with cement but also following exposure to leather, metal, paint, and plywood. In 1983, Danish legislation made the addition of ferrous sulfate mandatory in cement to reduce the water-soluble Cr content to not more than 2 ppm. Since then, the prevalence of Cr allergy in construction workers decreased, an observation that was confirmed in Germany. A similar decrease in countries without regulation could not be observed. As a direct result of the decreasing prevalence of Cr allergy observed in construction workers after the Danish legislation, the European Union (EU) passed a similar directive in 2005 (Directive 2003), Table 2.

Nonoccupational Sources of Cr Exposure

Table 3 provides an overview of Cr-containing items and chemicals that may be associated with nonoccupational Cr exposure. Leather is currently the most common cause of allergic Cr contact dermatitis and has lead to a significant increase in Cr allergy in Denmark between 1995 and 2007 (Thyssen et al. 2009).

Therefore, Cr-allergic contact dermatitis is now mainly a consumer problem which was underscored by a recent study (Thyssen et al. 2009) showing that 55% of 197 Cr-allergic patients with dermatitis had relevant leather exposure in their medical history. Women (39%) were more exposed to leather footwear compared to men (27.9%) and almost half the patients had dermatitis on the feet. A study measured Cr(VI) migration in a sample of leather goods and showed that Cr(VI) release from two shoes was 6 and 10 ppm, respectively (Hansen et al. 2002). An investigation by the German Risk Assessment Institute including more than 850 leather consumer items demonstrated that about 50% of these released more than 3 ppm Cr(VI) and that one tenth released amount in excess of 10 ppm Cr(VI) (http://www.bfr.bund.de/cd/9575). The Swedish Society for Nature and Conservation tested 21 pairs of leather shoes (http://www.naturskyddsforeningen.se/upload/press.badshoes.pdf) and found that levels of Cr(III) were very high ranging from 42 to 29,000 ppm, while no detectable levels of Cr(VI) were found. These values were well above the thresholds established by dose–response studies as described earlier. Other less frequent sources of Cr exposure in consumers include mobile phones, tattoo pigments, and eye shadows.

Acknowledgment Peter Jensen was supported by a grant from the Michaelsen Foundation.

Cross-References

▶ Chromium and Leather
▶ Chromium, Physical and Chemical Properties
▶ Chromium Toxicity, High-Valent Chromium
▶ Chromium(III) and Immune System
▶ Chromium(III) and Low Molecular Weight Peptides
▶ Chromium(III) and Transferrin
▶ Chromium(III), Cytokines, and Hormones
▶ Hexavalent Chromium and Cancer
▶ Hexavalent Chromium and DNA, Biological Implications of Interaction
▶ Trivalent Chromium

References

(2003) Directive 2003/53/EC of the European Parliament and of the Council of 18 June 2003 amending for the 26th time Council Directive 76/769/EEC relating to restrictions on the marketing and use of certain dangerous substances and preparations (nonylphenol, nonylphenol ethoxylate and cement). Official J Eur Commun 37:1–2

Allenby CF, Goodwin BF (1983) Influence of detergent washing powders on minimal eliciting patch test concentrations of nickel and chromium. Contact Dermatitis 9:491–499

Basketter D, Horev L, Slodovnik D et al (2001) Investigation of the threshold for allergic reactivity to chromium. Contact Dermatitis 44:70–74

Estlander T, Jolanki R, Kanerva L (2000) Occupational allergic contact dermatitis from trivalent chromium in leather tanning. Contact Dermatitis 43:114

Geier J, Lessmann H, Hellweg B et al (2009) Chromated metal products may be hazardous to patients with chromate allergy. Contact Dermatitis 60:199–202

Hansen MB, Rydin S, Menne T et al (2002) Quantitative aspects of contact allergy to chromium and exposure to chrome-tanned leather. Contact Dermatitis 47:127–134

Hansen MB, Johansen JD, Menne T (2003) Chromium allergy: significance of both Cr(III) and Cr(VI). Contact Dermatitis 49:206–212

Iyer VJ, Banerjee G, Govindram CB et al (2002) Role of different valence states of chromium in the elicitation of allergic contact dermatitis. Contact Dermatitis 47:357–360

Kadyk DL, McCarter K, Achen F et al (2003) Quality of life in patients with allergic contact dermatitis. J Am Acad Dermatol 49:1037–1048

Lidén C, Bruze M, Thyssen JP et al (2010) Metals. In: Johansen JD, Frosch PJ, Lepoittevin J-P (eds) Contact dermatitis, 5th edn. Springer, Berlin/Heidelberg, pp 644–671

Lindberg M, Matura M (2010) Patch testing. In: Johansen JD, Frosch PJ, Lepoittevin J-P (eds) Contact dermatitis, 5th edn. Springer, Berlin/Heidelberg, pp 439–464

Nethercott J, Paustenbach D, Adams R et al (1994) A study of chromium induced allergic contact dermatitis with 54 volunteers: implications for environmental risk assessment. Occup Environ Med 51:371–380

Ozkaya E, Topkarci Z, Ozarmagan G (2010) Systemic allergic dermatitis from chromium in a multivitamin/multimineral tablet. Contact Dermatitis 62:184

Rustemeyer T, van Hoogstraten IMW, von Blomberg BME et al (2010) Mechanisms of irritant and allergic contact dermatitis. In: Johansen JD, Frosch PJ, Lepoittevin J-P (eds) Contact dermatitis, 5th edn. Springer, Berlin/Heidelberg, pp 43–90

Stern AH, Bagdon RE, Hazen RE et al (1993) Risk assessment of the allergic dermatitis potential of environmental exposure to hexavalent chromium. J Toxicol Environ Health 40:613–641

Thyssen JP, Jensen P, Carlsen BC et al (2009) The prevalence of chromium allergy in Denmark is currently increasing as a result of leather exposure. Br J Dermatol 161:1288–1293

Chromium and Cancer

▶ Chromium Toxicity, High-Valent Chromium

Chromium and Diabetes

Nanne Kleefstra
Diabetes Centre, Isala clinics, Zwolle,
The Netherlands
Department of Internal Medicine, University Medical Center Groningen, Groningen, The Netherlands
Langerhans Medical Research Group, Zwolle,
The Netherlands

Synonyms

Chromium; Diabetes; Glucose tolerance factor; Trivalent chromium

Definition

Trivalent chromium plays an important role in potentiating the insulin response in cells sensitive to insulin. A deficiency in chromium could lead to an impaired response and thereby diabetes. Supplementing chromium in patients with diabetes with this deficiency could potentially improve glycemic control.

Introduction

The prevalence of diabetes mellitus worldwide was estimated to be 171,000,000 in the year 2000 and is expected to rise to 366,000,000 in the year 2030 (Wild et al. 2004). This rise is predominantly caused by the increase in patients having type 2 diabetes. Type 1 diabetes is characterized by an absolute shortage of endogenous insulin caused by an autoimmune reaction, which requires treatment with exogenous insulin to compensate for the absence of a patient's own insulin. Type 2 diabetes is to a large extent caused by an increase in insulin resistance, in other words a relative shortage of insulin. Patients are mainly treated by means of decreasing insulin resistance (by improving lifestyle or medication like metformine) and/or by decreasing the relative shortage of insulin (by medication like sulfonylurea compounds or exogenous insulin). In recent years, many additional modes of therapy have been introduced to treat patients, like bariatric surgery, incretines, and sodium-dependent glucose cotransporter 2 inhibitors, but are beyond the scope of this entry. In addition to the proportional rise in the aging population, overweight, in particular, plays an important role in this worrisome rise of patients with diabetes mellitus. The most important causes of being overweight are a decrease of physical activity and an increase of (unhealthy) food intake. Weight loss and an increase in physical exercise (even without weight loss) are effective self-care interventions and cost-effective, not only for the prevention of type 2 diabetes mellitus (T2DM), but also for tackling the cause for the majority of T2DM patients. Unfortunately, quite often these (general) self-care interventions are ineffective in daily practice. The majority of recently diagnosed T2DM patients, usually after a fruitless period of trying to improve their glycemic control by non-pharmacological means, will have to be treated with glucose-lowering drugs.

Various current guidelines for the treatment of T2DM advise the following non-pharmacological interventions (if applicable):

1. Stop smoking.
2. Maximize alcohol consumption to 2 units of alcohol a day.
3. Increase physical exercise to at least 30 min a day.
4. Reduce weight with 5–10%.
5. Restrict sodium to 50–100 mmol/day.
6. Improve food composition, with less fat and a normalized caloric intake.

In addition to these well-established interventions in regular medicine, patients are informed about other possibilities to improve their condition in many other ways. Patients are exposed to a multitude of suggested solutions outside the field of regular medicine, for example, in pharmacies, in supermarkets, on the Internet, in complementary medicine, etc. There is a very important difference between the six interventions mentioned above and the interventions like chromium supplementation that will be discussed below. The above-mentioned interventions require patients to actively change their habits, such as quitting their smoking habit (addiction) and/or reducing their alcohol consumption and/or increasing physical activity, etc. This means that patients really have to invest time and effort in changing their habits, and they may even suffer from withdrawal symptoms. Using supplements, like chromium, have a more passive character. The amount of effort and "pain" the patient has to invest influences a patients' motivation, and these factors could have positive and negative influences on the success of the intervention.

During the last decades, chromium has become the second most popular dietary supplement after calcium in the United States, with sales amounting to approximately 100 million dollars annually. Chromium supplements are mainly used for weight loss. However, patients with diabetes also use these supplements to improve their glycemic control by increasing insulin sensitivity.

Physiology (in Short)

Vincent and colleagues have done extensive research investigating chromium's mechanism of action at the cellular level (Sun et al. 2000) (Chromium and "Glucose tolerance Factor"; Chromium and Insulin Signaling). They discovered that the intracellular oligopeptide, Apo-low-molecular-weight chromium-binding substance (also known as Apo-chromoduline), plays an important role in potentiating the insulin response in cells sensitive to insulin. The degree of activation of the insulin receptor depends on the number of chromium ions bound to this peptide (with a minimum of 0 and a maximum of 4 ions), and this may lead to an eightfold difference in insulin receptor activation (when 4 ions are bound compared to 0). This could be one of the possible explanations for the insulin resistance seen with chromium deficiency.

Case Report

Several animal studies and a case report on humans have shown that absence of chromium in the diet leads to diabetes, and that the chromium deficit is associated with insulin resistance. In these animals, supplementation of chromium led to normoglycemia without the need of any glucose-lowering therapy. The case report, published in 1977, discussed a woman, aged 40, who had to undergo a total enterectomy as the result of a mesenterial thrombosis (Jeejeebhoy et al. 1977). Following the procedure, she received total parenteral nutrition through a subclavian catheter nightly. A little more than 3 years later, she lost more than 5 kg in a period of less than 3 months, and her plasma glucose concentration rose to values of manifest diabetes mellitus. To achieve a normoglycemic state, 45 units of zinc insulin were administered daily. Causes for the hyperglycemia were sought, because insulin resistance in a young woman who is not overweight is very rare. Chromium deficiency was considered as a possible cause after an article by Mertz from 1969 was discovered, in which the biological functions of chromium are discussed. The chromium concentration in her serum and hair was measured and found to be low (chromium in hair 154 ng/g (N > 500 ng/g), chromium in serum 0.55 ng/g (N 4.9–9.5 ng/g)). She was treated intravenously with 250 micrograms of chromium chloride daily for 2 weeks. This treatment resulted in a clear decrease in the amount of insulin required to treat her diabetes mellitus. After 4 months of chromium supplements, she no longer required insulin. She continued to receive 20 µg of chromium intravenously daily and remained normoglycemic after a period of 1 year.

Clinical Trials

Anderson et al. have performed a trial that can be considered as the Landmark trial of chromium-intervention studies in patients with T2DM (Anderson et al. 1997). In a randomized controlled trial performed in Chinese patients (a trial published in 1997), a first tentative conclusion about the effects of chromium on glycemic control was given. Chromium picolinate supplements were given over a period of 4 months to a group of 180 Chinese patients with T2DM. The patients were randomized into three groups: a group that received placebo, a group that received 200 µg chromium, and a group that received 1,000 µg of chromium daily as chromium picolinate. After 4 months, the HbA1c in the placebo group was unchanged (8.5%), while the HbA1c in the 200 µg group showed a significant decrease from 8.5% to 7.5%. In the group treated with 1,000 µg chromium a decrease in the HbA1c from 8.5% to 6.6% was seen (HbA1c is a measure of glycosylated hemoglobin that represents the level of glycemic control of predominantly the preceding 6–8 weeks.)

In 2007, Balk et al. performed a systematic review of randomized controlled trials concerning chromium supplementation in T2DM (Balk et al. 2007). They found 11 studies with 14 different chromium-based interventions measuring HbA1c as an endpoint: 11 out of 14 found no effect or non-significant effects of chromium supplementation. However, when performing a meta-analysis of these trials (including the study of Anderson et al.), the overall effect of chromium supplementation in people with diabetes was statistically significant in favor of chromium supplementation (−0.6% [95% confidence interval (CI): −0.9 to −0.2]). This mean −0.6% benefit, however, is largely due to the results of the study of Anderson et al. When excluding this study the effect of chromium on HbA1c is −0.3% (95% CI: −0.5 to −0.1). When stratifying the pooled results regarding methodological quality, sponsor involvement and western versus non-western studies, the effects of chromium supplementation turned out to be absent or non-relevant when studies were either of good methodological quality, or were without involvement of industry or were performed in western patients.

Since the review of Balk, two other double-blind, placebo-controlled randomized trials have been published. The first is a 6-month, double-blind study, performed in The Netherlands (Kleefstra et al. 2007). After 6 months the effect of chromium supplementation compared to placebo on HbA1c was +0.24 (95% CI: −0.06 to 0.54). The second study was a trial in which 30 Taiwanese patients with T2DM were enrolled (Albarracin et al. 2007). The patients were divided into three groups: a group which received placebo, a group that received 1,000 μg of chromium (chromium yeast), and a group that received 1,000 μg of chromium together with vitamin C and vitamin E. HbA1c levels dropped from 10.2% (standard deviation (SD) 0.5) to 9.5% (SD 0.2), which was a significant reduction. Unfortunately no between-group analyses (compared to placebo) were performed.

Discussion

There is, as discussed, a well-established role for trivalent chromium in glucose metabolism. However, especially in western patient with T2DM, this does not mean that supplementation of chromium in de forms and dosages as commonly used in studies leads to (relevant) effects on HbA1c. Questions that arise are among many others:

1. Are there patients, who are chromium deficient?
2. Can they somehow be selected, if chromium deficient patients do exist?
3. Is chromium supplementation beneficial in these patients?
4. Which forms of chromium in which dosages should be used?

It is of course of interest, that there are a few studies in non-western patients, which did find significant and clinically relevant effects of chromium supplementation on HbA1c. And of course it would be interesting to see these results replicated in methodologically sound trails. But, except for methodological issues, one could speculate about reasons for the differences in results between western and non-western patients, like genetic differences, but also the possible differences in intake of chromium with a low intake possibly leading to chromium deficiency, or the differences in dosages needed to be able to reach clinically meaningful effects in different patient populations.

Unfortunately, a reliable method for assessment of the chromium status is lacking as is sufficient information regarding the bioavailability of different forms of chromium. Future chromium research should focus on establishing a method for assessment of the chromium status. In addition, the bioavailability of different forms of chromium in Western patients (compared to non-western patients) should be investigated to be able to properly define a potentially effective dose.

Efficacy of chromium to improve glycemic control has not been established. Furthermore, there were some safety concerns with one form of chromium: chromium picolinate (Chromium Toxicity – Trivalent Chromium), and long-term safety has never been established. Last but not least, we should remember that patients with T2DM are not treated in order to lower HbA1c, but to improve the risk these patients have on microvascular and macrovascular complications (and to improve quality of life). At this moment, there are many non-pharmacological and pharmacological interventions possible in T2DM patients known, that reduce these risks. So, in the meantime, chromium should not be a treatment option in patients with T2DM, and non-pharmacological focus by diabetes health care providers together with patients should be on the items like quitting smoking, maximizing alcohol intake, increasing physical exercise, reducing weight, restricting sodium intake, and on improving food composition.

Cross-References

- Chromium and Glucose Tolerance Factor
- Chromium and Human Nutrition
- Chromium and Insulin Signaling
- Chromium and Membrane Cholesterol
- Chromium and Nutritional Supplement
- Chromium, Physical and Chemical Properties
- Chromium Toxicity, High-Valent Chromium
- Chromium(III) and Transferrin
- Trivalent Chromium

References

Albarracin C, Fuqua B, Geohas J, Juturu V, Finch MR, Komorowski JR (2007) Combination of chromium and biotin improves coronary risk factors in hypercholesterolemic type 2 diabetes mellitus: a placebo-controlled, double-blind randomized clinical trial. J Cardiometab Syndr 2:91–97

Anderson RA, Cheng N, Bryden NA, Polansky MM, Chi J, Feng J (1997) Elevated intakes of supplemental chromium improve glucose and insulin variables in individuals with type 2 diabetes. Diabetes 46:1786–1791

Balk EM, Tatsioni A, Lichtenstein AH, Lau J, Pittas AG (2007) Effect of chromium supplementation on glucose metabolism and lipids: a systematic review of randomized controlled trials. Diabetes Care 30:2154–2163

Jeejeebhoy KN, Chu RC, Marliss EB, Greenberg GR, Bruce-Robertson A (1977) Chromium deficiency, glucose intolerance, and neuropathy reversed by chromium supplementation, in a patient receiving long-term total parenteral nutrition. Am J Clin Nutr 30:531–538

Kleefstra N, Houweling ST, Bakker SJ, Verhoeven S, Gans RO, Meyboom-de Jong B, Bilo HJ (2007) Chromium treatment has no effect in patients with type 2 diabetes in a Western population: a randomized, double-blind, placebo-controlled trial. Diabetes Care 30:1092–1096

Sun Y, Ramirez J, Woski SA, Vincent JB (2000) The binding of trivalent chromium to low-molecular-weight chromium-binding substance (LMWCr) and the transfer of chromium from transferrin and chromium picolinate to LMWCr. J Biol Inorg Chem 5:129–136

Wild S, Roglic G, Green A, Sicree R, King H (2004) Global prevalence of diabetes: estimates for the year 2000 and projections for 2030. Diabetes Care 27:1047–1053

Chromium and Glucose Tolerance Factor

John B. Vincent
Department of Chemistry, The University of Alabama, Tuscaloosa, AL, USA

Synonyms

Glucose tolerance factor; GTF

Definition

As originally defined, glucose tolerance factor (GTF) is a material absent from *Torula* yeast that when fed to rats on a *Torula* yeast diet reverses apparent glucose intolerance in the rats. Early study on glucose tolerance factor equated the term with chromic ion. However, the term "glucose tolerance factor" was later used to refer to a material extracted and partially purified from brewer's yeast. The original studies have been shown to be methodologically flawed, and material from brewer's yeast has been shown to be an artifact. Given the considerable confusion over the name GTF and the history of GTF studies, the terms "glucose tolerance factor" and "GTF" should no longer be used use in chromium nutritional and biochemical research.

Initial Reports of a Glucose Tolerance Factor

The fields of Cr nutrition and Cr biochemistry had their beginnings in 1955 when Walter Mertz and Klaus Schwarz fed rats a semipurified *Torula* yeast–based diet (Mertz and Schwarz 1955); on the diet with Torula yeast as the sole protein source, the rats were reported to develop impaired glucose tolerance in response to an intravenous glucose load (1.25 g/kg body mass). Rats on the *Torula* yeast diet after an intravenous glucose challenge had a clearance rate of excess glucose (glucose above baseline at time zero), significantly greater than that of rats on a basal diet. The clearance rate was defined as the slope of a plot of ln (% excess glucose) versus time. The authors believed they had identified a new dietary requirement absent from the *Torula* yeast-based diet and whose absence was responsible for the glucose intolerance; they named this requirement "glucose tolerance factor" or GTF. Subsequently in 1959, these researchers claimed to identify the active ingredient of GTF as Cr^{3+} (Schwarz and Mertz 1959). Individually adding inorganic compounds of over 40 different elements to the Torula yeast diet could not restore glucose tolerance, while several inorganic Cr(III) complexes (200–500 μg Cr/kg body mass) did. Brewer's yeast and acid-hydrolyzed porcine kidney powder were identified as natural sources of "GTF," as extracts made from them could also restore proper glucose clearance in rats on the Torula yeast–based diet.

Unfortunately, while these studies are repeatedly cited and used as crucial evidence that Cr is an essential trace element for mammals, the methodology of these studies is flawed (Vincent 2001; Vincent and Stallings 2007). The Cr content of the diet was not determined, and the rats were maintained in wire mesh cages, allowing the rats to potentially obtain Cr by chewing on the metal. Thus, the actual Cr intake of the rats in these studies is impossible to gauge. The use of the large amounts of the metal ions is also of concern. Large doses of chromium may have pharmacological effect on subjects with impaired carbohydrate and lipid metabolism. The methods used to prepare the extracts were not described. Questions about the statistical significance of the effect in terms of the use of excess glucose rather than total glucose have also been raised.

Research prior to the 1970s was consistent with these early results of Schwarz and Mertz. Studies utilizing rats fed a variety of diets found that rats on some diets in addition to the *Torula* yeast–based diet apparently possessed low glucose removal rates; the glucose removal rates could be improved by addition of GTF concentrates. The concentrates were prepared from dried brewer's yeast, an enzymatic digest of brewer's yeast, or dried defatted porcine kidney powder. However, the methods of preparation of these materials were never described; while the amounts of the concentrates added to the diet where given, the amounts of Cr that these concentrates contained generally were not given. (The history of the early studies on the potential essentiality of chromium has been reviewed numerous times (Vincent 2001; Vincent and Stallings 2007)).

Attempts to Isolate Glucose Tolerance Factor

Nearly all of the attempts to isolate and characterize GTF have involved brewer's yeast, identified as a source of GTF in the 1959 study. Mertz and coworkers finally reported the details of the extraction and isolation of brewer's yeast GTF in 1977 (Toepfer et al. 1977), 20 years after the initial report. Brewer's yeast was extracted with boiling 50% ethanol; the ethanol was removed under vacuum, and the aqueous residual was applied to activated charcoal. Material active in bioassays was eluted from the charcoal with a 1:1 mixture of concentrated ammonia and diethyl ether. After removal of the ammonia and ether under vacuum, the resulting solution was hydrolyzed by refluxing for 18 h in 5 M HCl. Finally, the HCl was removed under vacuum, the solution was extracted with ether, and the pH of the solution was adjusted to 3. The cationic orange-red material was further purified by ion exchange chromatography; the material was passed through three (although no detail is provided about the chromatography including the identity of the three columns). Unfortunately, these incredibly harsh conditions would have destroyed any proteins, peptides, complex carbohydrates, or nucleic acids that initially could have been associated with the chromium. Thus, the possibility that the form of chromium recovered after the treatment resembles the form in the yeast is remote at best. The isolated GTF possessed a distinct feature at 262 nm in its ultraviolet spectrum, while mass spectral studies (no experimental details provided) suggested the presence of a pyridine moiety. Hence, nicotinic acid was proposed as a component of GTF as it apparently sublimed from the material and was identified by extraction with organic solvents (no experimental details provided). Amino acid analyses indicated the presence of glycine, glutamic acid, and cysteine as well as other amino acids, although the relative amounts were not reported. The isolated material was cationic and orange in color. The results were interpreted to indicate that GTF was a complex of Cr, nicotinate, glycine, cysteine, and glutamate (Toepfer et al. 1977). However, the orange color could suggest the formation of a chromium(III) ammonia complex (Vincent 2001). In paper chromatography experiments, the GTF gave several bands, only one of which from each material was active in the biological activity assays and migrated with the same R_f value; thus, the GTF characterized from the previous steps was impure. (Biological activity of a chromium-containing species referred to the ability of the species to potentiate the action of insulin to stimulate in vitro the metabolism of epididymal fat tissue from "chromium-deficient" rats, i.e., on the Torula yeast diet or another diet giving rise to similar effects when rats are given glucose tolerance tests.) The Cr in these active bands represented 6% of the total chromium from "purified" material (Toepfer et al. 1977). No chemical or physical characterization studies were performed on the most purified, active component.

Meaningful interpretation of these studies is not possible. Characterization of bulk materials of which only a tiny minority is active does not allow for deciphering the composition of the active component(s). Thus, one cannot assume that the amino acid analysis of the bulk product reflects the "active" component. Similarly, nicotinate detected in the bulk is not necessarily in the "active" component. Thus, nothing is known about the active component's composition except the apparent presence of some chromium. Because of the destructive isolation procedure used to obtain the Cr-containing material from brewer's yeast, the nature of the form of chromium in brewer's yeast cannot be deciphered from these studies. Based on this work, GTF has been proposed to be a Cr(III)-glutathione-nicotinate complex as glutathione is a tripeptide of glutamate, glycine, and cysteine, the three most abundant amino acids in the impure GTF extract. A three-dimensional structure has also been proposed for brewer's yeast GTF with two *trans* N-bound nicotinic acid ligands and amino acids occupying the remaining four sites of an octahedral around the chromic center (Mertz et al. 1974). This proposal, which actually appeared before the experimental report and has been reiterated numerous times in reviews and textbooks as the structure of the biologically active form of chromium, GTF, is without foundation. Yet the 1977 study dominated the understanding of the apparent nature of the biologically active form of chromium for the next two decades (Vincent 2001; Vincent and Stallings 2007).

Subsequent Studies

The results of the 1977 study have not been reproduced in several laboratories. Brewer's yeast has been found to contain amphoteric, anionic, and cationic complexes of Cr. However, many, if not all in one case, have been found to be artifacts formed between components of the growth media and chromium. The isolated species appear to have little if any similarity between research groups; thus, the isolated species may be very sensitivite to the conditions of the isolation, suggesting that artifacts are readily generated. Studies attempting to isolate and purify the active species from the yeast have repeatedly found that the active species could be separated from chromium-containing fractions. However, the identity of the active species is also controversial and may reflect that multiple species in the yeast (that no do contain Cr) may be able to give rise to effects in the bioassay. Gamma-aminobutyric acid was isolated and identified as the active species in one study. Similarly, studies using porcine kidney powder have not supported the original studies. A Cr-binding oligopeptide has been isolated porcine kidney powder; hydrolysis of this material produces species similar in properties to GTF from brewer's yeast and kidney powder. The studies since 1977 have been thoroughly reviewed (Vincent 2001; Vincent and Stallings 2007).

Recently, rats have been maintained on a purified diet (AIN-93G) without the added chromium in the mineral mix (16 μg Cr/kg food) in metal-free cages for 6 months. Other groups of rats received the diet with recommended 1,000-μg Cr/kg food in the mineral mix or the full diet with an additional 200-μg Cr/kg or 1,000-μg Cr/kg (Di Bona et al. 2011). No differences were noted in body mass or food intake between the groups. More significantly, the blood glucose of the rats on the diet with as low a Cr content as practically possible responded to an intravenous glucose challenge in a statistically identical fashion to that of the rats on the supplemented diets. The rates of glucose clearance and the areas under the glucose response curve were identical. When an intravenous glucose challenge (5 units insulin/kg body mass) was administered to the various groups of rats, the blood glucose levels of all the rats responded in an identical fashion. One notable difference was observed between the two groups of rats receiving the Cr supplementation beyond the Cr in the mineral mix and the rats on the diet without added chromium: The plasma insulin concentrations in response to the glucose challenge were lower for the supplemented rats that those concentrations for the rats on the diet without Cr supplementation (Di Bona et al. 2011). Thus, this study observed a pharmacological effect on high doses of Cr on insulin sensitivity (Vincent 2010). This study demonstrates that in the absence of other dietary stress (as could occur in the Torula yeast diet), no signs of Cr deficiency can be generated in rats provided a diet with as low a Cr concentration as practically possible, while high doses of Cr can have pharmacological effects. The authors proposed that since low-Cr diets do not generate any known deleterious effects and that effects previously attributed to low doses of Cr should be classified as pharmacological rather than nutritionally relevant effects, Cr should be removed from the list of essential trace elements for mammals (Di Bona et al. 2011).

Biological Activity Assays

The reproducibility and sensitivity of the assay for biological activity of chromium was improved when adipocytes isolated from epididymal fat tissue of "Cr-deficient" rats were utilized. Replicate assays at a series of insulin concentrations could now be readily performed, allowing for detailed kinetics experiments. These assays showed that Brewer's yeast extracts and the synthetic Cr-nicotinate complexes described above potentiated the ability of insulin to stimulate glucose oxidation at a variety of insulin concentrations. The degree of stimulation also depended on the chromium concentrations of either the extract or synthetic complexes, leading to the conclusion that these compounds contain biologically active forms of chromium. However, further analysis of these results has shown that the original interpretation is incorrect. Addition of the chromium sources without adding insulin stimulated the metabolism of glucose; the stimulation due to insulin above this increased background was actually decreased. The chromium sources in fact made insulin less effective, consistent with components of the extracts binding to insulin and not allowing the complexed insulin to bind to its receptor (Vincent 1994).

The use of a microbiological assay (yeast fermentation assay) has been proposed to determine biological activity of chromium-containing materials, but it is also problematic. The yeast fermentation assay reportedly yields results that parallel rat fat pad assays when using brewer's yeast "GTF" as a Cr source. However, as the active component of yeast extracts can be separated from chromium, what exactly this assay measures in regard to chromium is questionable (Vincent 2001).

Synthetic Models

The proposed identification of nicotinate (3-carboxypyridine) in "GTF" stimulated an interest in the synthesis of chromic-nicotinate or chromic-nicotinic acid ester complexes and chromic complexes of 2-carboxypyridine (picolinate acid) and 4-carboxypyridine (isonicotinic acid). However, the inability of the identity of GTF to be elucidated and the demonstration that the isolated species could be separated from Cr-containing species led to a rapid decline in inorganic chemistry studies after 1985 (Vincent 2001). "Chromium nicotinate" is the product of the reaction of two or three equivalents of nicotinic acid with chromic ions in aqueous solution at elevated temperatures. The structure and composition of chromium nicotinate have been poorly described. As a solid, chromium nicotinate is intractable, being insoluble or unstable in common solvents. Hence, studies on the solid have been limited even though it has gained substantial use as a nutritional supplement (under the trade name ChromeMate marketed by InterHealth Nutraceuticals), and studies of the solution from which the "compound" precipitates have additionally provided little additional data. The addition of a Cr(III) salt, CrX_3, to an aqueous solution of sodium nicotinate generates purple solutions from which gray-purple polymers of the general formula $[Cr(nic)_2(H_2O)_x(OH)]_n$ precipitate along with some $[nicH]X$, which can be removed by subsequent extraction. The addition of a Cr(III) salt to a hot aqueous solution of nicotinic acid yields blue solutions, which contain a mixture of species with 1:1 and 1:2 Cr-to-nicotinate ratios; increasing the pH results in the formation of a blue-gray polymer with a 1:1 Cr-to-nicotinate ratio. Dissolution of the polymers in mineral acid results in the formation of soluble species with an overall purple color and a 1:1 Cr-to-nicotinate ratio. None of the species have been crystallized, and NMR studies suggest the species are more complex than the proposed formulas would indicate (Rhodes et al. 2009). Chromium picolinate, $[Cr(picolinate)_3]$, is the most popular form of chromium in nutritional supplements and has been marketed primarily by Nutrition21. The complex has been well characterized by a variety of techniques including X-ray crystallography; the complex has limited solubility in water, ~500 μM (Vincent 2001). Unfortunately, the choice of these compounds as nutritional supplements stems primarily from the flawed studies on glucose tolerance factor. Their low solubility limits absorption of the compounds to about 1% efficiency (similar to dietary chromium). While the use of chromium compounds as nutritional supplements appears currently to be unfounded, chromium nicotinate and/or chromium picolinate are currently being studied for their potential as therapeutic agents in subjects with altered carbohydrate and lipid metabolism and associated conditions such as type 2 diabetes (Vincent 2010).

Summary and Conclusions

The term "glucose tolerance factor" has developed different meanings, leading to confusion. Studies often fail to distinguish the difference among the inorganic ion Cr^{3+}, Cr(III) complexes, and a purported biologically active form of chromium (i.e., the naturally occurring biomolecule(s) that has an inherent function when containing bound chromium(III)) when using the term "GTF." As originally proposed, GTF is a substance that is involved in maintaining normal glucose, prevents and cures impairment of glucose removal when given in the diet or by stomach tube, and results in impairment of intravenous glucose tolerance when it is deficient in the diet. As described above, this was soon thereafter equated with Cr^{3+}. As chromium must presumably interact with some organic biomolecule(s) to manifest any effect(s) in mammals, attempts were subsequently made to identify this (these) species. Unfortunately, the products of such attempts have also been termed GTF. The situation was not helped in the 1980s and early 1990s when nutritional supplements containing "glucose tolerance factor" were widely marketed; these products generally contained yeast extracts or synthetic complexes of Cr^{3+} and nicotinate, adding to the confusion (Vincent 2001; Vincent and Stallings 2007).

Over five decades ago, Schwarz and Mertz reported that Cr^{3+} (or GTF) was a nutrient for mammals; inorganic chromium(III) complexes apparently could restore the glucose tolerance of rats fed a *Torula* yeast–based (supposedly Cr-deficient) diet. Major portions of these studies, which were considered the pioneering work in the field of chromium biochemistry and nutrition, have been effectively refuted. The diet used by these workers has not been demonstrated to be chromium deficient; thus, any effects from supplementing the diet with quantities of Cr several fold larger than the normal dietary intake does not establish an essential requirement (Vincent 2010).

The biologically active form of chromium is not GTF, reportedly a Cr(III)-nicotinate-amino acid (or glutathione) complex. In fact, the composition of the artifact from yeast isolated by Mertz and coworkers was actually not established, and proposals for the three-dimensional structure of "GTF" are unfounded. The effects of materials isolated from brewer's yeast observed by Mertz and coworkers were probably serendipitous, as Cr^{3+} has been demonstrated repeatedly to be separable from agents in yeast responsible for in vitro stimulation of glucose metabolism in adipocytes. Yet even the detailed results by different laboratories of the most recent studies of yeast GTF still cannot be reconciled completely (Vincent 2001; Vincent and Stallings 2007). Given the considerable confusion over the name GTF and the history of GTF studies, the use of the term "glucose tolerance factor" or "GTF" should be terminated (Vincent 2001).

Cross-References

▶ Chromium and Human Nutrition
▶ Chromium and Nutritional Supplement

References

Di Bona KR, Love S, Rhodes NR et al (2011) Chromium is not an essential trace element for mammals: effects of a "low-chromium" diet. J Biol Inorg Chem 16(3):381–390. doi:10.1007/s00775-010-0734-y

Mertz W, Schwarz K (1955) Impaired intravenous glucose tolerance as an early sign of dietary necrotic liver degeneration. Arch Biochem Biophys 58:504–506

Mertz W, Toepfer EW, Rogionski EE et al (1974) Present knowledge on the role of chromium. Fed Proc 33:2275–2280

Rhodes NR, Konovalova T, Liang Q, Cassady CJ, Vincent JB (2009) Mass spectrometric and spectroscopic studies of the nutritional supplement chromium(III) nicotinate. Biol Trace Elem Res 130:114–130

Schwarz K, Mertz W (1959) Chromium(III) and the glucose tolerance factor. Arch Biochem Biophys 85:292–295

Toepfer EW, Mertz W, Polansky MM et al (1977) Preparation of chromium-containing material of glucose tolerance factor activity from brewer's yeast extracts and by synthesis. J Agric Food Chem 25:162–162

Vincent JB (1994) Relationship between glucose tolerance factor and low-molecular-weight chromium-binding substance. J Nutr 124:117–119

Vincent JB (2001) The bioinorganic chemistry of chromium. Polyhedron 20:1–26

Vincent JB (2010) Chromium: celebrating 50 years as an essential element? Dalton Trans 39:3787–3794

Vincent JB, Stallings D (2007) Introduction: a history of chromium studies (1955–1995). In: Vincent J (ed) The nutritional biochemistry of chromium(III). Elsevier, Amsterdam

Chromium and Human Nutrition

Barbara J. Stoecker
Department of Nutritional Sciences, Oklahoma State University, Stillwater, OK, USA

Synonyms

Chromium chloride; $3CrCl_3$; $CrCl_3 \cdot 6H_2O$

Definitions

Adequate intake (AI) – daily mean intake of a nutrient for apparently healthy people.

Introduction

Chromium can exist in multiple valence states but chromium (III) and chromium (VI) are of most practical importance (Gauglhofer and Bianchi 1991). Trivalent chromium is the most stable oxidation state and is likely to be the form in the diet because of the reducing substances in foods (Panel on Micronutrients 2001). Hexavalent chromium, a strong oxidizing agent which is rapidly reduced to trivalent chromium, is recognized as a pulmonary carcinogen (Gauglhofer and Bianchi 1991).

Determination of chromium in biological samples is very difficult because of the low concentrations of chromium in typical tissue samples. Urine, serum, and plasma are usually in the 0.1–0.2 ng/mL range which is close to the detection limit for most instruments. Contamination control and appropriate background correction systems are essential, and many chromium values reported before the 1980s were incorrect (Veillon 1989). An additional constraint is that chromium in serum or plasma may not be a satisfactory biomarker for chromium status (Offenbacher et al. 1997; Stoecker 2006). Consequently, many studies in the literature have tested effects of chromium supplementation without knowing initial chromium status of the participants. For chromium to have a nutritional effect, supplementation of chromium-deficient persons should be tested; a nutritional effect of chromium supplementation would not be expected in a clinical trial if participants already had adequate chromium status.

Food Sources

Chromium concentration in foods varies widely but is relatively low. Dairy products contained less than 1 µg chromium per serving and many meats contained less than 2 µg chromium per serving although some meats were higher perhaps reflecting chromium acquired during processing. Fruits and vegetables were variable, but consistently higher chromium concentrations were found in whole grain products, with some breakfast cereals being particularly rich in chromium (Anderson et al. 1992). Refined sugars and flour are generally lower in chromium than less refined products (Anderson et al. 1992). Acidic foods can take up chromium from storage and heating in stainless steel (Offenbacher et al. 1997). Chromium concentration of some beers was more than 10 ng/mL (Anderson 1987).

Chromium intake from self-selected diets has been investigated. Ten males and twenty-two females collected duplicate food and beverage samples for seven consecutive days. For men, the mean intake was 33 ± 3 µg/day with a 7 day mean range of 22–48 µg/day. For females, mean intake over 7 days was 25 ± 1 with a range of 13–36 µg/day. Mean chromium intake was approximately 15 µg /1,000 kcal (Anderson and Kozlovsky 1985). Other studies showing similar chromium intakes have been reviewed (Panel on Micronutrients 2001).

Absorption

Intestinal absorption of chromium from $^{51}CrCl_3$ is very low, demonstrating that most chromium from an oral dose remains unabsorbed in the intestine. Various studies which have reported the absorption of chromium to be between 0.5% and 2% have been reviewed (Anderson 1987; Offenbacher et al. 1997; Stoecker 2006).

A 12 day balance study was conducted on two men consuming 36.9 and 36.7 µg chromium/day. Their chromium balances were positive on these intakes and their average net absorption of chromium was 1.8% (Offenbacher et al. 1986). The chromium intake of 22 healthy elderly persons was 24.5 µg/day (Offenbacher et al. 1997). Using urinary chromium excretion as a proxy for absorption, Anderson and Kozlovsky reported that when dietary chromium intake was 10 µg, approximately 2% of the chromium was absorbed and when it was 40 µg, only 0.4–0.5% was absorbed (Anderson and Kozlovsky 1985). Adults receiving 200 µg of supplemental chromium daily as chromium chloride had urinary excretion of approximately 0.4% (Anderson 1987).

Absorption of chromium is primarily from the jejunum and is enhanced by the presence of amino acids, (Anderson 1987; Offenbacher et al. 1997;

Stoecker 2006). Simultaneous consumption of 1 mg chromium chloride with 100 mg ascorbic acid enhanced chromium absorption measured by area under an 8 h curve in three women to 1.4-, 2.7-, and 4.4-fold the levels obtained when chromium was consumed with water (Offenbacher et al. 1997). Several over-the-counter and prescription medications that reduce stomach acidity impeded chromium absorption in rats, but these studies have not been followed up in humans (Kamath et al. 1997).

Transport and Storage

Chromium competes with iron for a binding site on transferrin (Anderson 1987). Studies based on scans after dosing with $^{51}CrCl_3$ in control subjects and in patients with hemochromatosis showed that chromium disappeared rapidly from the blood, and organ deposition was highest in liver and spleen. Accumulation of ^{51}Cr was also seen in bone. Patients with hemochromatosis had only about half the retention of ^{51}Cr, and Lim and colleagues hypothesized that competition for binding sites on transferrin contributed to reduced uptake and whole body retention of ^{51}Cr (Lim et al. 1983; Sargent et al. 1979).

Functions

Chromium potentiates the action of insulin (Mertz 1969). Davis and Vincent demonstrated that a low molecular weight oligopeptide that binds four chromium ions markedly enhanced insulin receptor tyrosine kinase activity, thereby increasing the response of the receptor to insulin (Davis and Vincent 1997; Vincent 2000). However, there is not yet a biomarker that allows determination of baseline chromium status contributing to the equivocal results observed in the literature. Some studies show beneficial effects of chromium supplementation with various types of subjects while others show no significant effects of chromium (Offenbacher et al. 1997; Stoecker 2006).

In a randomized placebo-controlled trial with diabetic individuals in China, participants received, twice a day, placebo, 100 μg chromium or 500 μg chromium as chromium picolinate. Fasting blood glucose and glucose 2 h after a 75 g oral glucose load were both significantly reduced in the group receiving 1,000 μg chromium per day but not in the group receiving 200 μg chromium per day after both 2 and 4 months of supplementation. Fasting and 2 h insulin were significantly reduced in both groups at 2 and 4 months. Hemoglobin A_{1c} was significantly reduced by the higher chromium dose (7.4 ± 0.2% vs. 8.6 ± 0.2 for placebo). After 4 months of chromium supplementation, both supplemented groups had significantly reduced HbA_{1c} with the higher dose group being lower than the group dosed with 200 μg chromium per day. Unfortunately, data on dietary chromium intake was not available from this study (Anderson et al. 1997).

Seventeen nondiabetic subjects participated in a study in which they consumed diets containing 5 μg Cr/1,000 kcal for a total of 14 weeks. On the basis of an oral glucose tolerance test (1 g/kg body weight), they were divided into a control group (nine subjects with 90 min glucose values <5.56 mmol/L) and a hyperglycemic group (eight subjects with 90 min glucose values >5.56 mmol/L but <11.1 mmol glucose/L). The first 4 weeks of the study were for equilibration. At week 5, half of the subjects received 200 μg chromium as chromium chloride while the other half received a placebo. At week 10, supplement and placebo were reversed in a crossover design. There were no significant effects of chromium supplementation in the control subjects. However, in the hyperglycemic subjects, sums of glucose, insulin, and glucagon after a glucose tolerance were significantly lower with chromium supplementation than with placebo. Thus, it appeared that chromium intakes of only 5 μg/1,000 kcal were detrimental to persons with marginally elevated blood glucose (Anderson et al. 1991).

A carefully conducted study of chromium supplementation and resistance training in men (aged 56–69 years) showed significant effects of resistance training on skeletal muscle size, strength, and power. However, the double-blind daily supplementation of 924 μg chromium as chromium picolinate showed no significant effect on these parameters (Campbell et al. 1999).

Excretion

Urinary chromium losses have been related to the insulinogenic properties of carbohydrates

(Anderson 1987). Chromium excretion increases with exercise, but a study with a stable isotope of chromium also indicated that chromium absorption may be increased in response to acute exercise and strength training. Further study is required to determine if an increase in chromium absorption could be related to improvement in insulin response seen with exercise (Rubin et al. 1998).

Dietary Recommendations

An AI (adequate intake) for chromium was estimated as part of the process of developing the Recommended Dietary Intakes published by the Institute of Medicine (Panel on Micronutrients 2001). An AI is established when data are insufficient to identify dietary intakes at which half of healthy individuals would be expected to display a specified deficiency sign (Estimated Average Requirement (EAR)). Establishing an EAR is critical to the process of developing a recommended dietary allowance (RDA); when it is not possible to set an RDA for a nutrient, an AI is presented representing the mean intake of apparently healthy people. The AI for chromium for adults was based on an energy estimate for age multiplied by 13.4 μg chromium/1,000 kcal. The AI for women 19–50 years of age is 25 μg/day and for men of the same age is 35 μg/day. In older adults, average consumption of energy nutrients decreases, and the AI accordingly is set at 20 μg/day for women above the age of 50 and at 30 μg/day for men above the age of 50.

The AI for infants was derived from data showing that chromium concentration of breast milk is approximately 0.25 μg/L (Anderson, 1993). Based on the assumption of 0.78 L/day consumption of human breast milk in exclusively breast-fed infants, the AI for infants 0–6 months of age is set at 0.2 μg/day. The AI for infants 7–12 months of age is estimated to be 5.5 μg/day. This increase represents the increased intake of approximately 400 kcal per day of complementary food by infants of this age. The AI for chromium increases gradually throughout childhood based on metabolic weight (kg ¾) extrapolated from the adult AI.

Manifestations of Deficiency

Several reports have been reviewed that indicate a role for chromium in carbohydrate metabolism (Anderson 1987; Mertz 1993; Offenbacher et al. 1997). In 1977, a patient was reported who upon receiving total parenteral nutrition for 3.5 years developed unintentional weight loss, peripheral neuropathy, and impaired glucose tolerance (Jeejeebhoy et al. 1977). The high circulating levels of free fatty acids and the respiratory quotient of 0.66 indicated use of lipid as fuel. Increased infusion of glucose and 45 units of insulin daily did not control weight loss and other symptoms. Chromium concentrations in blood and hair were severalfold below normal values for that laboratory. A large infusion of intravenous chromium (250 μg daily for 2 weeks) normalized the glucose tolerance test, the respiratory quotient, and the peripheral neuropathy. Insulin infusion was stopped and glucose intake had to be reduced to avoid overweight. At the time of publication of this case report, infusion of 20 μg chromium as $CrCl_3$/day had maintained the patient for 18 months (Jeejeebhoy et al. 1977).

Cross-References

▶ Chromium and Glucose Tolerance Factor
▶ Chromium and Insulin Signaling
▶ Chromium and Nutritional Supplement
▶ Chromium(III) and Low Molecular Weight Peptides

References

Anderson RA (1987) Chromium. In: Mertz W (ed) Trace elements in human and animal nutrition, vol 1, 5th edn. Academic, San Diego/New York/Berkeley/Boston/London/Sydney/Tokyo/Toronto, pp 225–244

Anderson RA et al (1997) Elevated intakes of supplemental chromium improve glucose and insulin variables in individuals with type 2 diabetes. Diabetes 46(11):1786–1791

Anderson RA, Bryden NA, Polansky MM (1992) Dietary chromium intake. Freely chosen diets, institutional diet, and individual foods. Biol Trace Elem Res 32:117–121

Anderson RA, Kozlovsky AS (1985) Chromium intake, absorption and excretion of subjects consuming self-selected diets. Am J Clin Nutr 41(6):1177–1183

Anderson RA, Polansky MM, Bryden NA, Canary JJ (1991) Supplemental-chromium effects on glucose, insulin, glucagon, and urinary chromium losses in subjects consuming controlled low-chromium diets. Am J Clin Nutr 54(5):909–916

Campbell WW, Joseph LJ, Davey SL, Cyr-Campbell D, Anderson RA, Evans WJ (1999) Effects of resistance training and chromium picolinate on body composition and skeletal muscle in older men. J Appl Physiol 86(1):29–39 (Bethesda, MD: 1985)

Davis CM, Vincent JB (1997) Chromium oligopeptide activates insulin receptor tyrosine kinase activity. Biochemistry 36(15):4382–4385

Gauglhofer J, Bianchi V (1991) Chromium. In: Merian E (ed) Metals and their compounds in the environment: occurrence, analysis, and biological relevance. VCH, New York, pp 853–875

Jeejeebhoy KN, Chu RC, Marliss EB, Greenberg GR, Bruce-Robertson A (1977) Chromium deficiency, glucose intolerance, and neuropathy reversed by chromium supplementation, in a patient receiving long-term total parenteral nutrition. Am J Clin Nutr 30(4):531–538

Kamath SM, Stoecker BJ, Davis-Whitenack ML, Smith MM, Adeleye BO, Sangiah S (1997) Absorption, retention and urinary excretion of chromium-51 in rats pretreated with indomethacin and dosed with dimethylprostaglandin E2, misoprostol or prostacyclin. J Nutr 127(3):478–482

Lim TH, Sargent T 3rd, Kusubov N (1983) Kinetics of trace element chromium(III) in the human body. Am J Physiol 244(4):R445–R454

Mertz W (1969) Chromium occurrence and function in biological systems. Physiol Rev 49(2):163–239

Mertz W (1993) Chromium in human nutrition: a review. J Nutr 123(4):626–633

Offenbacher EG, Pi-Sunyer FX, Stoecker BJ (1997) Chromium. In: O'Dell BL, Sunde RA (eds) Handbook of nutritionally essential mineral elements. Marcel Dekker, New York, pp 389–411

Offenbacher EG, Spencer H, Dowling HJ, Pi-Sunyer FX (1986) Metabolic chromium balances in men. Am J Clin Nutr 44(1):77–82

Panel on Micronutrients (2001) Chromium. Dietary reference intakes for vitamin A, vitamin K, arsenic, boron, chromium, copper, iodine, iron, manganese, molybdenum, nickel, silicon, vanadium, and zinc. National Academy Press, Washington, DC, pp 197–223

Rubin MA et al (1998) Acute and chronic resistive exercise increase urinary chromium excretion in men as measured with an enriched chromium stable isotope. J Nutr 128(1):73–78

Sargent T 3rd, Lim TH, Jenson RL (1979) Reduced chromium retention in patients with hemochromatosis, a possible basis of hemochromatotic diabetes. Metab Clin Exper 28(1):70–79

Stoecker BJ (2006) Chromium. In: Bowman BA, Russell RM (eds) Present knowledge in nutrition, vol 2. International life sciences institute, Washington, DC, pp 498–505

Veillon C (1989) Analytical chemistry of chromium. Sci Total Environ 86(1–2):65–68

Vincent JB (2000) The biochemistry of chromium. J Nutr 130(4):715–718

Chromium and Insulin Signaling

Zbigniew Krejpcio
Division of Food Toxicology and Hygiene,
Department of Human Nutrition and Hygiene, The Poznan University of Life Sciences, Poznan, Poland
The College of Health, Beauty and Education in Poznan, Poznan, Poland

Synonyms

Insulin signal transduction; Insulin signaling; Insulin-signaling pathway

Definition

Insulin signaling refers to the cascade of processes triggered by insulin binding with its receptor, and as a result, cellular glucose uptake and utilization can take place. It has been extensively documented that trivalent chromium (Cr^{3+}) can improve insulin sensitivity by enhancing insulin signal transduction; however, the mechanisms of its action have not been fully understood. The contemporary theories explaining the possible mechanisms of Cr^{3+} action in insulin signaling are based on certain phenomena that include its role in the LMWCr, involvement in phosphorylation reactions, assisting insulin binding to its receptor, increasing number of insulin receptors, activation of Akt proteins, and interaction with cytokines and oxidation processes, as well as increasing cell membrane fluidity.

Insulin Signaling

Insulin (INS) is a hormone with pleiotropic functions that influences gene expressions and mitogenesis and regulate metabolism of glucose, lipids, and proteins. The insulin molecule contains 51 amino acids and is composed of two peptide chains (A and B) linked by disulfide bonds. In most species, the A chain consists of 21 amino acids, and the B chain, of 30 amino acids. These chains initially are contained within a single polypeptide chain (preproinsulin) that is converted to proinsulin and finally to insulin. Insulin is synthesized in the pancreas within the beta cells (β-cells) of the islets of Langerhans and can be released when stimulated by various factors. Increased blood glucose concentration is the primary stimulant to insulin secretion. After a meal, in response to elevated levels of the blood nutrients (glucose, lipids, acids), the β-cells of the pancreatic islets of Langerhans secrete insulin. Insulin regulates glucose homeostasis at many sites, reducing hepatic glucose output (via decreased gluconeogenesis and glycogenolysis) and increasing the rate of glucose uptake into striated muscle and adipose tissue. The insulin-signaling pathway therefore plays an essential role in glucose and energy homeostasis. Impairment of the insulin-signaling pathway (at various levels) leads to insulin resistance and development of type 2 diabetes mellitus (T2D) and the complications associated with this disease.

The first step of insulin action is its binding to insulin membrane receptor (IR) present in many cells (muscle, adipocytes, brain, liver). Insulin receptor belongs to the superfamily of receptor tyrosine kinases and is an integral membrane protein existing in dimeric form: α-subunit and β-subunit. These subunits are linked to each other by disulfide bonds present on the extracellular side of the plasma membrane. Each α-subunit contains a binding site for insulin.

Insulin binding to the α-subunit of IR induces a conformational change in the β-subunit, which triggers the cascade of signals inside target cells. This process is very complex, involving various interactions between intracellular biomolecules. The first step of this process, after binding of insulin with IR, is autophosphorylation of the IR β-subunit, which activates the tyrosine kinase domains. The insulin receptor tyrosine kinase (IRTK) is activated by two separate events, where three residues (Tyr_{1158}, Tyr_{1162}, and Tyr_{1163}), located in the regulatory loop, play a central role in this process. In the first phase, insulin binding increases the activity of IRTK and stimulates its β-subunit autophosphorylation. In the second phase, autophosphorylation of the insulin receptor fully activates the enzyme toward specific intracellular insulin receptor substrate proteins (IRS). IRS are related by functional properties and not sequence similarity. Four substrates belong to the IRS family (IRS1, IRS2, IRS3, IRS4). Other substrates include growth factor receptor–bound protein 2 (Grb-2), receptor-associated binder-1 (Gab1), $p60^{dok}$, the c-Cbl proto-oncogene (Cbl), adaptor protein with pleckstrin homology (PH) and Src homology 2 (SH2) domains (APS), and three isoforms of SH2 domain-containing alpha-2 collagen-related protein (Shc). IRS contain an NH_2-terminal PH domain and/or a phosphotyrosine-binding domain, COOH-terminal tyrosine residues that create SH2 protein-binding sites, proline-rich regions that engage Src homology 3 (SH3) domains, or WW domains (protein modules that bind proline-rich ligands) and serine-threonine-rich regions that bind other proteins. All substrates, except Shc, contain an SH2 domain that targets the substrate to the insulin receptor. Three main pathways propagate the signal generated through the insulin receptor: the IRS/phosphatidylinositol kinase (PI-3K) pathway, the retrovirus-associated DNA sequences (RAS)/mitogen-activated protein kinase (MAPK) pathway, and the Cbl-associated protein (CAP)/Cbl pathway. The RAS/MAPK cascade does not play a role in regulation of glucose transport but is largely involved in gene regulatory responses in insulin-sensitive tissues. The CAP/Cbl pathway is a PI-3K independent pathway participating in the insulin-mediated glucose transport through activation of TC10, a member of the Rho family of small guanosine triphosphate (GTP)–binding proteins. The most important role in insulin-dependent glucose transport is played by PI-3K. It consists of p10 catalytic subunits (three isoforms) and a p85 regulatory subunit (two isoforms) that have two SH2 domains that bind to phosphotyrosine motif on receptor tyrosine kinases or substrates. IRS-1 and IRS-2 dock with the p85 regulatory subunit, which in turn activates the p110 catalytic subunits. The PI-3K substrate phosphatidylinositol-4,5-bisphosphate (PIP_2) is phosphorylated on the 3-OH position of the inositol ring to produce phosphatidylinositol triphosphate (PIP_3), and formation of this lipid recruits PH domain-containing proteins, such as the serine-threonine kinases 3-phosphoinositide-dependent protein kinase-1 (PDK-1), protein kinase B (PKB)/Akt, and atypical kinases C zeta and lambda isoforms (aPKCs). PDK-1 phosphorylates PKB and aPKCs on a threonine residue located in the activation loop of the catalytic domain, causing their activation. Akt promotes the phosphorylation of p70S6

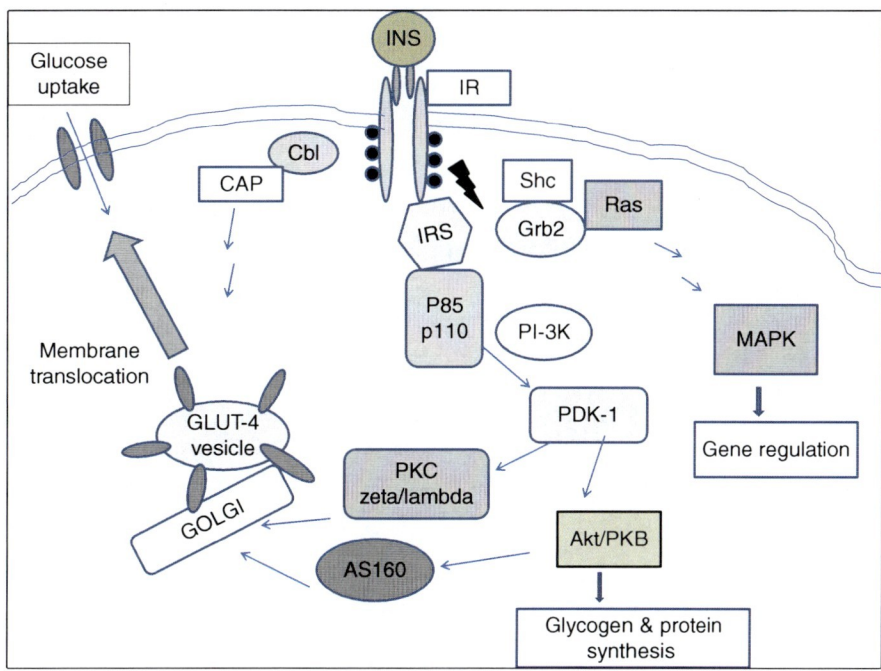

Chromium and Insulin Signaling, Fig. 1 Simplified model of insulin signaling pathways following insulin receptor activation in skeletal muscle cell

kinase and also plays a role in insulin signaling to glycogen synthesis by mediating the insulin-induced phosphorylation and inhibition of glycogen synthase kinase 3 (GSK-3) (Glund and Zierath 2005). Phosphorylation of Akt substrate (AS160) signals the Golgi to mobilize a family of glucose transporters (GLUT 1–4 proteins) for trafficking to the plasma membrane. Finally, activation of Akt results in the translocation of GLUT-4 vesicles (in muscle and adipose tissue) to the plasma membrane. These transporters then bind and fuse with the plasma membrane facilitating glucose transport via carrier-mediated-facilitated diffusion, initially enhanced by insulin binding to IR. The simplified schematic model of insulin-signaling events is presented in Fig. 1. Once the IR is inactivated, the transporter GLUT-4 returns to the Golgi. Generally, the efficacy of insulin signaling is determined by insulin binding to its receptor and activation of the kinases along the cascade.

In summary, insulin signaling can be enhanced by enhancing the binding of insulin to its receptor, the activation of kinases along the cascade, or inhibition of the dephosphorylation of insulin receptor.

Alterations in insulin signal transduction have serious consequences for glucose and lipid metabolism, leading to metabolic disorders such as insulin resistance and diabetes mellitus. Glucose homeostasis is impaired in patients with non-insulin-dependent diabetes mellitus (type 2 diabetes) as a result of defects in glucose transport in skeletal muscle. Although the primary causes of insulin resistance and type 2 diabetes have not been fully elucidated, genetic and environmental factors are thought to contribute to the development of the disease. A number of different altered metabolic states, such as persistent elevation of circulating glucose, insulin, fatty acids, and cytokines, can lead to peripheral insulin resistance. Insulin resistance may result, for example, from alterations in insulin receptor expression, binding, phosphorylation state, kinase activity, and/or the GLUT-4 vesicle budding, trafficking, docking, and fusion events. Reduction in the amounts of insulin receptor kinase itself, as well as a reduction in the extent of IRS protein tyrosine phosphorylation and PI-3K association/activation were found in patients with type 2 diabetes. However, whether these changes in insulin receptor function represent primary lesions that cause insulin resistance or whether they occur secondary to hyperinsulinemia or hyperglycemia is uncertain. There are also opinions that there may be no single or common defect that underlies peripheral insulin resistance. Most likely, insulin resistance is a complex phenomenon in which several genetic defects combine with environmental stresses to generate the phenotype (Pessin and Saltiel 2000).

The Role of Cr in Insulin Signaling

According to some researchers (Anderson 2003), some of the signs and symptoms of insulin resistance include glucose intolerance, hyperinsulinemia, increased LDL cholesterol, increased triacylglycerols, elevated total cholesterol, and decreased HDL cholesterol have been associated with decreased dietary intakes of chromium (Cr). For over 50 years, trivalent chromium (Cr) has been considered as an essential element for mammals, and for over 30 years for humans, plays an important role in carbohydrate and lipid metabolism and provides significant beneficial effects in the insulin-signaling system. In the presence of Cr, in its biologically active form, much lower amounts of insulin are required. However, the scientific base of this very concept was not fully explored and remains under question. Cr is believed to be required for optimal insulin activity and normal carbohydrate and lipid metabolism. A number of studies have considered the impact of Cr supplementation on blood glucose and insulin levels in animal models and humans. Several (but not all) randomized, placebo-controlled clinical trials have demonstrated statistically significant improvements in blood glucose and insulin levels following dietary supplementation of 200–1,000 μg Cr/day, mostly in the form of chromium tris(picolinate), Cr(pic)$_3$, by type 2 diabetic subjects for periods of 4 weeks to 4 months. Thus, Cr(pic)$_3$ has been considered as an insulin sensitizer. The bioavailability of Cr from Cr compounds depends on the type of ligand; however, it is generally low (0.1–2%). The mode of action of Cr on glucose metabolism involves several biochemical changes leading to increased insulin sensitivity. The mechanism of Cr action remains obscure despite multiple pathways of action being proposed, including a decrease in hepatic glucose production and an increase in peripheral glucose disposal. Early reports suggested that Cr enhances insulin binding, insulin receptor number expression, insulin internalization, and increases beta-cell sensitivity, but the exact mechanism whereby Cr participates in the functions of insulin has not been fully elucidated. Major theories explaining the mechanism of Cr action in insulin signaling are presented below.

The Role of LMWCr in Insulin Signaling

An interesting theory explaining the role of Cr^{3+} in the regulation of glucose metabolism is connected with a discovery of the oligopeptide LMWCr (low-molecular-weight-chromium-binding substance), also termed chromodulin. This peptide was isolated and purified from animal tissues (rabbit, bovine, dog, mouse, porcine liver or kidney) or fluids (bovine colostrum) exposed to potassium dichromate. Chromodulin has a molecular weight of approximately 1.5 kDa and is comprised of only four types of amino acid residues (glycine, cysteine, glutamate, and aspartate) and binds with high affinity to four chromic ions. The amino acid composition of the LMWCr isolated from bovine liver was identified to be approximately E:G:C:D::4:2:2:2 (Vincent 2007). In the absence of insulin, chromodulin does not affect the insulin receptor tyrosine kinase (IRTK) activity (and a membrane phosphotyrosine phosphatase in rat adipocyte) but stimulates its activity eightfold in the presence of insulin. Removal of Cr from the LMWCr results in the loss of kinase-potentiating activity (Davis and Vincent 1997). The proposed mechanism for the activation of IRTK activity by LMWCr has been described in detail by Vincent and Bennett (Vincent 2007) and is presented in Fig. 2. In short, this mechanism is explained on the assumption that the inactive form of LMWCr (as apoLMWCr without Cr) is stored in insulin-sensitive cells, while in response to insulin, Cr bound in transferrin is moved from the blood to these cells, loading of apoLMWCr with Cr to form the active form (holoLMWCr). The holoLMWCr then binds to the IR amplifying IRTK activity. A decrease of blood insulin level facilitates relaxation of the conformation of the IRS, excretion of the holoLMWCr from the cells into the blood, and finally brings about the termination of the insulin signaling. The site of activation appears to be located at or near the kinase activation site since addition of chromodulin to a fragment of the β-subunit that contains the active site and does not require insulin for activation resulted in a similar stimulation of kinase activity. Activation of insulin receptor kinase by the

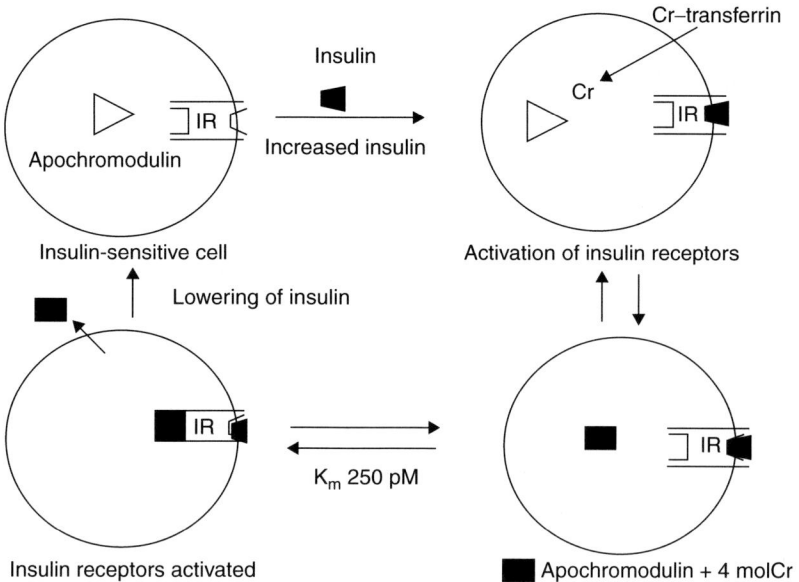

Chromium and Insulin Signaling, Fig. 2 A proposed mechanism for the activation of insulin receptor (IR) kinase activity by chromodulin in response to insulin. The inactive form of the IR is converted to the active form by binding insulin. This triggers the movement of Cr from transferrin into the insulin-dependent cells and the binding of Cr to apochromodulin. The holochromodulin containing 4 mol Cr/mol chromodulin then binds to the IR further activating IR kinase activity. Apochromodulin is unable to bind to the IR. When the concentration of insulin decreases, holochromodulin is released from the insulin-sensitive cells (Permission from Vincent 2000)

LMWCr requires 4 mol Cr/mol oligopeptide and is specific for Cr.

Cr Affects a Cascade of Phosphorylation Reactions

Cr is believed to activate the insulin receptor kinase and inhibit phosphotyrosine phosphatase (PTP-1) that inactivates the IR phosphatase, leading to increased phosphorylation of the IR, which is associated with increased insulin sensitivity. Studies conducted on Chinese hamster ovary cells showed that Cr activates insulin kinase activity at low doses of insulin.

Cr Binds Directly to Insulin Receptor

Cr in the form of multinuclear Cr assembly (like that occurring in LMWCr) has been proposed to bind to the extracellular α-subunits of the IR concomitant with insulin binding that in turn activates the IRTK activity. However, this mechanism has not been proven and remains speculative.

Cr Increases the Number of Insulin Receptors

Cr^{3+} has been suggested to increase IR number, as observed in red blood cells of hypoglycemic subjects after 6 weeks treatment. However, the mechanism of Cr action on IR was not explained, and other studies did not confirm such effects in animal models.

Cr Activates Akt

Some studies reported that supplemental Cr $(Cr(pic)_3)$ increases Akt phosphorylation in type 2 diabetic subjects. The same effect was observed in vitro in mouse 3T3-adipocytes treated with $Cr(phenylalanine)_3$; thus, this phosphorylation was suggested to be the mechanism of Cr action in insulin signaling (Yang et al. 2005).

Cr Interacts with Cytokines and Decreases Oxidation

Another possible mechanism of Cr action in the insulin-signaling pathway may involve interaction with cytokines (TNF-α, IL-6) and lipid peroxidation. In vitro studies showed that Cr (as $CrCl_3$) inhibits TNF-α secretion and oxidative stress induced by exposure of cultured U937 monocytes to high glucose medium. Cr has been proposed to enhance insulin sensitivity by lowering of TNF-α secretion, a cytokine known to inhibit the sensitivity and action of insulin. Cr was also shown to prevent lipid peroxidation (in vitro H_2O_2-treated cells); thus, its antioxidative effect might be essential in insulin signaling (Jain and Kannan 2001).

Cr Increases Cell Membrane Fluidity

Another possible mechanism of Cr action in insulin signaling involves changes in membrane lipids depots. Insulin-stimulated glucose transport has been observed to be decreased when membrane fluidity diminishes. Studies in vitro showed that in the presence of insulin, Cr (as $CrCl_3$ or $Cr(pic)_3$) increases membrane fluidity by decreasing plasma cholesterol level in 3T3-L1 adipocytes. The regulation of glucose transporter (GLUT-4) translocation by Cr did not involve known insulin-signaling proteins such as the insulin receptor, IRS-1, PI-3K, and Akt. Interestingly, cholesterol add-back to the plasma membrane prevented the beneficial effect of Cr on both GLUT-4 mobilization and insulin-stimulated glucose transport (Chen et al. 2009).

In skeletal and heart muscle cells, insulin directs the intracellular trafficking of the fatty acid translocase/CD36 to induce the uptake of cellular long-chain fatty acids (LCFA). Both insulin and Cr (as $Cr(pic)_3$) have been demonstrated to induce CD36 translocation to the plasma membrane in 3T3-L1 adipocytes. These data indicate that Cr improves glucose transport by modification of lipid depots and represses lipid-induced insulin resistance.

Despite over 50 years of intense studies on the role of Cr in glucose and lipid metabolism, the mechanisms of its action, as well as essentiality of Cr for animals and humans, remain under question. A recent well-controlled study on rats fed with a diet of as little Cr as reasonably possible failed to demonstrate any deleterious effects on body composition or glucose metabolism and insulin sensitivity compared to Cr "sufficient" diet, suggesting that Cr can no longer be considered as an essential element (Di Bona et al. 2011). The beneficial role of supplementary Cr may result from a pharmacological effect, increasing insulin sensitivity.

References

Anderson RA (2003) Chromium and insulin resistance. Nutr Res Rev 16:267–275

Chen G, Liu P, Pattar GR, Tackett L, Bhonagiri P, Strawbridge AB, Elemendorf JS (2009) Chromium activates glucose transport 4 trafficking and enhances insulin-stimulated glucose transport in 3 T3-L1 adipocytes via a cholesterol-dependent mechanism. Mol Endocrinol 20:857–870

Davis CM, Vincent JB (1997) Chromium oligopeptide activates insulin receptor kinase activity. Biochemistry 36:4382–4385

Di Bona KR, Love S, Rhodes NR, McAdory D, Sinha SH, Kern N, Kent J, Strickland J, Wilson A, Beaird J, Ramage J, Rasco JF, Vincent JB (2011) Chromium is not an essential trace element for mammals: effects of a "low-chromium" diet. J Biol Inorg Chem 16(3):381–390

Glund S, Zierath JR (2005) Tackling the insulin-signalling cascade. Can J Diabetes 29:239–245

Jain SK, Kannan K (2001) Chromium chloride inhibits oxidative stress and TNF-alpha secretion caused by exposure to high glucose in cultured U937 monocytes. Biochem Biophys Res Commun 289:687–691

Pessin JE, Saltiel AR (2000) Signaling pathways in insulin action: molecular targets of insulin resistance. J Clin Invest 106:165–169

Vincent JB (2000) The biochemistry of chromium. J Nutr 130:715–718

Vincent JB (ed) (2007) The nutritional biochemistry of chromium(III). Elsevier, Amsterdam

Yang X, Palanichamy K, Ontko AC, Rao MNA, Fanf CX, Ren J, Sreejayan N (2005) A newly synthetic chromium complex – chromium(phenylalanine)$_3$ improves insulin responsiveness and reduces whole body glucose tolerance. FEBS Lett 579:1458–1464

Chromium and Leather

Sreeram Kalarical Janardhanan, Raghava Rao Jonnalagadda and Balachandran Unni Nair
Chemical Laboratory, Central Leather Research Institute (Council of Scientific and Industrial Research), Chennai, Tamil Nadu, India

Synonyms

Chromium and tanning

Definition

The process of conversion of hides/skins which putrefy within hours of death of an animal to a commodity meeting the lifestyle, aesthetic, and fashion requirements of the user is described by the term "tanning." Chromium(III) ions based tanning owing to the kind of thermal stability and customer preferred properties that it confers is the most predominant methodology. It has been an intriguing factor to leather chemists as to why other metal ions or natural polyphenols are not able to match the properties conferred by chromium(III). Tanner, in the past, has been challenged by the poorer uptake and hence the release of chromium into the wastewaters. A reasonable understanding of the type of chromium(III) species found in tanning solutions, their interaction with collagen, the species present in the spent solutions, etc., has now been possible. Methodologies for overcoming the poorer uptake of chromium by the skin matrix, employing the principles of green chemistry have been developed and implemented at industrial levels. The challenge however is to enhance the rate of diffusion and fixation of chromium to the three-dimensional skin matrix.

Tanning: Moving Skin to a Fashionable Commodity

Viscoelasticity and a unique animal-dependent pore size distribution enables leather, a stabilized animal skin, to breathe and readjust to volume fluctuations such as that in foot. The process by which the skin protein, collagen, is stabilized against wet and dry heat, thermomechanical stress, and enzymatic attack is known as tanning. In this process, the tanning agent (inorganic metal ions like Cr^{3+}, Zr^{4+}, Ti^{4+}, Fe^{2+}/Fe^{3+}, etc.; organic products such as plant polyphenols, aldehydes, etc.) forms cross-links (H-bonds, covalent and coordinate covalent linkages) with active sites in the collagen (predominant skin protein). The consumption of leather and leather products is said to have increased by 55% over the past 30 years, with estimated trade value in USD being 120 billion per year. Leather, as a commodity, is traded based on lifestyle, aesthetic, and performance preferences of the consumer. Accordingly, the choice of the tanning agent is guided by the end properties achievable from each type. 1.6 billion square feet of leather (of the total 1.8 billion square feet processed annually) is tanned or retanned using chromium, thus signifying the role of this metal ion in conferring the consumer preferred properties. Chromium-tanned leathers tend to be softer and more pliable than their plant polyphenol counterparts. They also have the highest thermal stability, reported till date, and are very stable in water, and the production time is much shorter than other tanning agents.

Skin, after the operations prior to tanning, is predominantly composed of collagen. Collagen is distinctive, in that it is composed of a regular arrangement of amino acids in each of the three chains of collagen subunits (Ramachandran 1963). A Gly-Pro-X or Gly-X-Hyp, where X may be any of the various amino acids, is the sequence followed. Proline and hydroxyproline constitutes 1/4 of the total sequence and glycine accounts for 1/3 of the sequence. The tropocollagen or the collagen molecule, which is 300 nm long and 1.5 nm in diameter, is made of three left-handed polypeptide strands, twisted to form a right-handed coiled coil triple helix, a cooperative quaternary structure stabilized by numerous hydrogen bonds. Each triple helix associates into a right-handed microfibril, and they coil together to become the well-ordered crystalline fibril. Collagen fibers are bundles of fibrils, and these fibers are bundled up to provide fiber bundles, which are subsequently packed in a three-dimensional array, unique of the type of animal, to form the skin. Diffusion of the tanning material into the skin matrix, through the enormous number of pores, and then binding to the active sites is the key to tanning.

Chromium: The Most Preferred Tanning Agent

The aqueous chemistry of chromium is limited to di-, tri-, and hexavalent states. Chromium(III) forms $3d^2 4s 4p^3$ hybrid orbitals. This when coupled with the tripositive character of the metal ion enables the formation of thermodynamically stable coordinate covalent bonds with the side chain carboxyl sites of aspartic and glutamic acids in collagen. Historically, chromium-based tanning was performed using chrome alum applied in solution at pH of ~2, followed by a basification to pH 4 to fix the chromium to collagen. Subsequently, a two-bath tanning process where the skin was treated in a solution of chromic acid overnight, followed by its reduction to trivalent chromium at pH 4, in another bath had been employed. Cr(VI) being responsible for lung cancer, chrome ulcers, nasal septum perforation, and brain and kidney disorders, such two-bath processes have been replaced with one-bath processes employing chromium(III) salts,

Chromium and Leather, Fig. 1 Various chromium(III) species found in chromium(III) sulfate solution, 33% basic

$[Cr_3(OH)_4]^{5+}$	$[Cr(OH)_2Cr]^{4+}$	$[Cr_4(OH)_4(O)_2]^{4+}$
$[Cr]^{3+}$	$[Cr(OH)(SO_4)Cr]^{3+}$	$[Cr(OH)_2(fo)Cr]^{3+}$
$[Cr(O)(SO_4)Cr]^{2+}$	$[Cr(O)_2Cr]^{2+}$	$[Cr(OH)_2(SO_4)Cr]^{2+}$
$[Cr(OH)(SO_4)(fo)Cr]^{2+}$	$[CrSO_4]^{1+}$	$[Cr(ox)]^{1+}$
$[Cr(OH)(SO_4)_2Cr]^{1+}$	$[Cr(OH)(SO_4)]^{0}$	$[SO_4Cr(OH)_2CrSO_4]^{0}$
$[Cr(SO_4)_2]^{1-}$	$[Cr(ox)_2]^{1-}$	$[SO_4Cr(OH)_2(SO_4)CrSO_4]^{2-}$

* Coordinated H$_2$O is omitted for clarity

even though in earlier times, this change was prompted by a possible reduction in duration of tanning.

There are several factors that influence the fixation of chromium onto collagen and thus provide hydrothermal and enzymatic stability to collagen. These factors can in general be grouped as those relating to (a) the type of chromium salt employed for tanning, (b) the reactivity of collagen toward chromium, and (c) environmental factors (Chandrasekaran et al. 1999).

Collagen tanned with basic chromium(III) chloride or perchlorate did not provide for shrinkage temperatures as effective as that with coordinated sulfate. When the counterion employed was a chloride or perchlorate, the stability of the resultant chromium–collagen bond was only moderate. It is expected that chromium(III) along with its counterion must create a matrix with the supramolecular water structure. In this, the counterion can act as a water structure breaker or maker. While sulfate is a structure maker, the chloride is a structure breaker, indicating as to why chromium with its sulfate counterion is able to provide more stability to collagen (Covington et al. 2001). As the presence of coordinated sulfate is necessary for the efficient reaction of chromium(III), basic chromium(III) sulfate salt, which is a synergistic combination of Cr(III) molecular ions and their counterions, is the preferred tanning agent. Basic chromium(III) sulfate is prepared by the reduction of Cr(VI) using molasses or sulfur dioxide. The process variables such as temperature, concentration of reactants, nature of reductants, etc. influences the composition of the chromium tanning salt produced. Accordingly, 15 different types (Fig. 1) of chromium(III) species have been identified in basic chromium(III) sulfate solution. This diversity in chemical structure results in variable kinetic lability and thermodynamic affinity to the sites of collagen. Extended X-ray absorption fine structure (EXAFS) studies of chromium(III) bound to collagen showed that the dominating bound species are linear tetrachromium compounds (Covington 2001).

There is little direct evidence for the size and shape of chromium(III) species influencing their binding to collagen. While several authors have intuitively assumed the species to be linear, the solution chemistry does favor the formation of three-dimensional species such as the cyclic tetramer as well. While monomeric species such as the hexaaquachromium(III) ion is kinetically inert and hence not useful as a tanning agent, the dimeric and trimeric species are thermodynamically favorable in stabilizing the collagen matrix through complexation. It has been reported that irreversible binding of chromium is achieved only if the binding constants are greater than 15 M^{-1} for the (1):

$$[(H_2O)_4Cr(OH)_2Cr(H_2O)_4]^{4+} + P-COO^- \xrightleftharpoons{K}$$
$$[(H_2O)_3(P-COO)Cr(OH)_2Cr(H_2O)_4]^{3+} + H_2O \quad (1)$$

where P – COO$^-$ represents the carboxylic acid side chains of aspartic and glutamic acids in collagen (Rao et al. 1999).

It has been reported that Cr^{3+} and collagen segments form clusters of 20–40 nm between the fibrils, with Cr(III) acting as a cross-linker. Such external cross-links do not destroy the triple-helix conformation of collagen. Binding constants for collagen–chromium complexes are not available due to the heterogeneous character of collagen. One of the methods by which this has been circumvented is by studying the reaction of chromium(III) oligomers with thiocyanate. The trimeric species has been shown to react with thiocyanate with a bimolecular rate constant of $9.3 \times 10^{-4}\,M^{-1}\,s^{-1}$ at 30°C (Rao et al. 1999).

Another method by which the ability of chromium species to effectively cross-link with collagen is evaluated is the ease of collagen cleavage by CNBr. SDS-PAGE provides information on the number of collagen fragments formed on CNBr cleavage, and the trimeric chromium species was more effective toward collagen cross-linking than the dimeric species. Interaction studies carried out between rat tail tendon collagen and the chromium species followed by evaluation under an atomic force microscope provided evidence for organization of monomeric collagen into quarter staggered fibrils in the presence of Cr(III) dimer.

In the case of leather, the substrate, skin, is a three-dimensional matrix of finite thickness. This brings in an additional parameter of diffusion into the collagenous matrix. Diffusion into the skin matrix can be considered as those occurring through the macropores, micropores, or molecular pores. While diffusion through the macropores is limited only by the transport pathways, gel structures, swelling of collagenous matrix, and other mass transfer phenomena, that of micropores is predominantly governed by surface tension, solvent dielectrics, capillary phenomena, and electrolytes (Ramasami 2001). The molecular cavity size, polarity restrictions, charged sites, and range of distances for molecular interactions are expected to influence the diffusion through molecular process. In any event, the rate of diffusion can influence the rate of fixation as diffusion precedes fixation. In the case of chromium, the binding sites being the carboxyl groups of aspartic and amino acid residues, the pH and temperature play an important role in penetration and fixation. While penetration is achieved at a lower pH of 2.8–3.0, the fixation is achieved at pH 3.8–4.0. For this, the hide or skin is subjected to a process called as pickling, wherein the pH of the skin is reduced from near neutral conditions, employing slow but steady feeds of concentrated sulfuric acid. This process is carried out in the presence of neutral salts so as to avoid swelling of collagen in acidic condition.

After the penetration of chromium into the skin matrix is achieved, the pH of skin is raised to 3.8–4.0. During this process, the basicity of chromium is increased, and the chromium(III) molecules are polymerized by olation. With increase in molecular weight of the chromium(III) complex, hydrophobicity and hence the reactivity toward collagen increases. Raising the pH to 3.8–4.0 is considered as a high-risk process as any sudden jump of pH would result in the precipitation of chromium. This has resulted in a search for basification free processes.

One of the old methods to avoid basification was to leave enough residual alkali in the pelt, which later gets employed for basification. In the 1990s, a "ThroBlue process," where chromium tanning salts devoid of associated sodium sulfate, is applied at pH 7 along with a polyamide, thus ensuring a basification free process with higher uptake of chromium had been reported. In recent years, a polymeric compound containing an aromatic sulfonic acid was employed in the place of "pickle" and was followed by the use of basic chromium sulfate. This process also eliminated the need for basification (Sreeram and Ramasami 2003).

In spite of reducing the pH of the skin matrix to 2.8–3.0 to facilitate penetration of chromium(III) salts, some binding to the surface proteins can be expected. Mono- and dibasic carboxylic acids, referred to as masking agents, are generally added either to the pickle or to the tanning bath to control the rate of reaction and penetration of chromium(III) species into the hide protein. Coordination of chromium(III) to the masking agents reduces the cationic charge and then enhances the diffusion of chromium into the skin matrix. Masking increases the reaction rate, owing to an increase in the reactivity of collagen. The masking ratio, defined as the ratio of number of moles of carboxylic masking agent to the number of gram atoms of chromium, lies typically in the range of 0.5–1.0. In practical terms, only formate and phthalate are conventionally important masking agents (Covington 2001).

The second objective of reaction of chromium species with collagen is to bring about (to collagen) stability against enzymatic hydrolysis. Dimeric and trimer chromium(III) species was reported to bring about secondary and quaternary structural changes to collagenase, the enzyme which breaks the peptide bonds in collagen. The mode by which these species

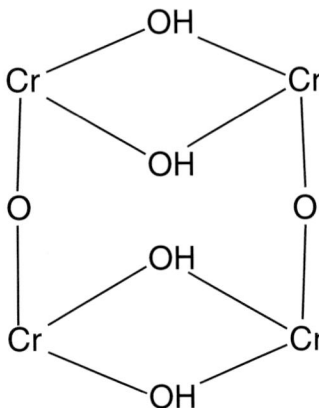

Chromium and Leather, Fig. 2 Structure of cyclic tetrameric chromium(III) species found in unspent solution of chromium (III) sulfate after tanning

inhibited the action of collagenase on collagen was reported to be competitive (Gayatri et al. 2000).

It has been estimated that only 10% of the bound chromium is involved in cross-linking, and therefore, only 1/40 of the carboxy groups are required for chromium–collagen cross-linking. Therefore, the earlier efforts of converting the amino groups into carboxyl terminated groups by way of pretreatment with glycine and formaldehyde or glyoxal and n-thioureidopyromellitamic acid and thus increase the uptake of chromium should be considered as one of providing more binding sites at the surface of the fiber bundles rather than improving the diffusion of chromium into the matrix (Wang and Zhou 2006).

Addressing Challenges in Chromium-Based Tanning

In spite of a greater level of understanding of how diffusion of chromium into the matrix as well as its fixation to the collagen can be improved, the uptake of chromium, in general, during the process of tanning has only reached levels of 70–80%, as against 40–60% achievable about two decades ago. Rao et al. reported that the spent chromium solution after the process of tanning predominantly contained a cyclic tetrameric species (Rao et al. 1998) (Fig. 2), which reacted with an anion like NCS^- with an equilibrium constant of 1.2 ± 0.1 M^{-1}, as against 15.7 ± 0.1 and 14.6 ± 0.1 M^{-1} for dimeric and trimeric species (Rao et al. 1999). The equilibrium constant values reflect the low thermodynamic affinity of the tetrameric species for the nucleophile. The reaction of collagenase with the cyclic tetrameric species was reported to be uncompetitive in character, indicating that the tetrameric species could not bring about stability to collagen. The molecular size of the tetrameric species (9.2 Å) is considered to be one of the causes of poorer uptake of this species during tanning. By introducing phthalates and employing appropriate molasses to acid ratios and temperature during the process of conversion of Cr(VI) to Cr(III), a modified basic chromium sulfate has been prepared, which exhibited over 85% uptake of chromium. There are also reports that tumbling a wet pickled pelt and chromium salt in a highly hydrophobic medium, such as paraffin wax, can result in the solute effectively partitioned between an inert solvent and water within the pelt. Consequently, a very rapid uptake of chromium is achievable. However, the subsequent complexation of the chromium(III) and its fixation are dependent on the surrounding water molecules, and therefore, the fixation needs to be carried out in the conventional manner. An alternative to this method is to control the solute–solvent interaction. Hydration reactions, being controlled by the dielectric constants, the addition of even small quantities of ethylene glycol, ethanolamine, or ethylenediamine (dielectric constants of 38, 34, and 14, respectively), markedly change the properties of water, reduce its solvating effect, and thus increase the rate of diffusion and thus fixation of chromium. It has been reported that the ethanolamine-based pretreatment of collagen results in an uptake of chromium to the levels of 90%. The mineral analogue of this reaction is the use of aluminum(III) salts prior to offering of chromium, which provides for about 90–95% uptake of chromium (Covington 2009).

Chromium(III) has been included among the essential trace elements and has considerable beneficial effects on glucose metabolism. However, the ability of chromium(III) to cross-link DNA and proteins, and also participate in nonenzymatic phosphorylation and influence the calcium transport channels, calls for caution while discharging chromium(III) into soil or wastewaters. Further, conversion of Cr(III) to Cr(VI) in the presence of manganese oxides present in the soil has also been reported. Based on these considerations, the discharge norms for industrial wastewaters specify the permissible levels of chromium to be below 2.0 ppm. This calls for recycling of chromium in the tanning process. Recycling of chromium can be either direct or indirect.

Direct recycling of spent chromium solution (from a conventional tanning) containing 0.15–0.5% chromium, 3.3–3.5% NaCl, and 2.8–3.3% Na_2SO_4 along with fresh chromium(III) salt for replenishing the concentration of chromium was found to have disadvantages of poorer uptake of chromium and quality of leather. This has been attributed to the structure and reactivity of the cyclic tetrameric species and adverse effect of accumulated neutral salts (Sreeram et al. 2005). Direct recycling of such chromium-bearing liquors into tanning requires membrane-based processes, which can reduce the neutral salts from 7% to 1%. In the absence of membrane-based processes, the recycling of spent chromium solutions into the pickle seems to be more advantageous. The direct recycling of spent chromium solutions from high exhaust tanning systems such as ethanolamine–chromium or aluminium–chromium directly without membrane separation processes into the pickle produces leathers devoid of grain harshness, possibly due to lower chromium content, or in turn lower higher-charged chromium(III) species. This is corroborated from the fact that the preacidification of spent chromium solution to a pH of 1 reduces the 3+ and above charged species to below 7% as against 73% in the case of pH 3.8.

Another way by which chromium(III) in the spent solutions can be completely recovered is the precipitative removal. Chromium(III) in solution can be easily precipitation as chromium(III) hydroxide by the addition of alkali. Though in principle, any alkali can be used, commercially adopted batch processes employ either magnesium oxide or magnesium oxide and lime as they afford lower-settled volumes, owing to higher bulk density of the chromium(III) hydroxide generated. However, the supernatant from such processes have a high total dissolved solids content (from the magnesium sulfate generated in the reaction). Discharge of such solutions to water bodies can adversely affect the salinity of the water. By suitable modifications to the conical reactors employed to precipitate chromium(III) hydroxide, hydrostatic pressure and turbulence within the reactor systems can be generated. Under such circumstances, a continuous mode of precipitation, employing sodium carbonate as precipitant, results in chromium(III) hydroxide precipitate volumes of around 20% as against 10% in the case of magnesium oxide (Sreeram and Ramasami 2003). Both batch and continuous methods are currently in commercial practice, and they result in discharge of supernatants having below 0.3% chromium. The precipitated chromium(III) hydroxide upon acidification results in basic chromium(III) sulfate solution, which is reemployed in tanning.

While direct or indirect recycling of spent solutions can overcome the discharge of chromium into wastewaters, the major challenge is the disposal of chromium(III) containing leather scraps as well as leather products as solid wastes. This waste amounts to 10% of the total raw hide/skin processed or approximately, at the current levels of production, 0.8 million tons, globally.

Leather wastes can be effectively used in the manufacture of boards for soling, heeling materials, general heavy components, light shoe inserts, and fancy goods, such as lamp shades, etc. Other alternate use for such protein rich waste is as a reducing agent in the manufacture of basic chromium sulfate. These chromium(III) salts thus generated have the added advantage of improved masking owing to the presence of oligopeptides. Other uses include that as adsorbents for dyes from wastewaters (Rao et al. 2004). The dye-bearing solid waste can subsequently be calcined under controlled conditions to generate chromium-based pigments.

Way Forward

The poorer uptake of chromium led to research initiatives toward replacing chromium. However, the higher thermal and enzymatic stability provided by chromium(III) remained unsurpassed. With increasing knowledge about the structure of collagen and newer analytical tools for the understanding of how chromium complexes react with collagen, the methodologies for improving the uptake and fixation of chromium are increasing. Some efforts toward speciation of chromium found in wastewaters led to the identification of species which are relatively unbound to collagen. This can lead to development of tanning salts devoid of such species. The chemistry of masking needs to improve so as to provide improved penetration of chromium at lower process times. The rate of reaction of chromium with collagen and the stability of chromium–collagen bonds need to be enhanced not only for increasing the fixation of chromium but also for reducing the discharge of chromium during subsequent posttanning operations. In essence, the profitable and sustainable employment of chromium as a tanning agent seems to be a reality in the future.

References

Chandrasekaran B, Rao JR, Sreeram KJ, Nair BU, Ramasami T (1999) Chrome tanning: state-of-art on the material composition and characterization. J Sci Ind Res 58(1):1–10
Covington AD (2001) Chrome tanning: exploding the perceived myths, preconceptions and received wisdom. J Am Leather Chem Assoc 96(12):467–480
Covington AD (2009) Tanning chemistry: the science of leather. RSC Publishing, Cambridge
Covington AD, Lampard GS, Menderes O, Chadwick AV, Rafeletos G, O'Brien P (2001) Extended X-ray absorption fine structure studies of the role of chromium in leather tanning. Polyhedron 20(5):461–466
Gayatri R, Rajaram R, Ramasami T (2000) Inhibition of collagenase by Cr(III): its relevance to stabilization of collagen. Biochim Biophys Acta Gen Subj 1524(2–3):228–237
Ramachandran GN (1963) Molecular structure of collagen. Int Rev Connect Tissue Res 1:127–182
Ramasami T (2001) Approach towards a unified theory for tanning: Wilson's dream. J Am Leather Chem Assoc 96(8):290–304
Rao JR, Chandrasekaran B, Subramanian V, Nair BU, Ramasami T (1998) Physico-chemical and structural studies on leathers tanned using high exhaust basic chromium sulphate salt. J Am Leather Chem Assoc 93(5):139–147
Rao JR, Gayatri R, Rajaram R, Nair BU, Ramasami T (1999) Chromium(III) hydrolytic oligomers: their relevance to protein binding. Biochim Biophys Acta Gen Subj 1472(3):595–602
Rao JR, Thanikaivelan P, Sreeram KJ, Nair BU (2004) Tanning studies with basic chromium sulfate prepared using chrome shavings as a reductant: a call for 'wealth from waste' approach to the tanning industry. J Am Leather Chem Assoc 99(4):170–176
Sreeram KJ, Ramasami T (2003) Sustaining tanning process through conservation, recovery and better utilization of chromium. Resour Conserv Recycl 38(3):185–212
Sreeram KJ, Ramesh R, Rao JR, Chandrababu NK, Nair BU, Ramasami T (2005) Direct chrome liquor recycling under Indian conditions: part 1. Role of chromium species on the quality of the leather. J Am Leather Chem Assoc 100(6):233–242
Wang HR, Zhou X (2006) A new pretannage with glyoxal and N-thioureidopyromellitamic acid for high exhaust chrome tannage. J Am Leather Chem Assoc 101(3):81–85

Chromium and Membrane Cholesterol

Jeffrey S. Elmendorf
Department of Cellular and Integrative Physiology and Department of Biochemistry and Molecular Biology, and Centers for Diabetes Research, Membrane Biosciences, and Vascular Biology and Medicine, Indiana University School of Medicine, Indianapolis, IN, USA

Synonyms

Bilayer; Glucose tolerance factor; Sterol

Definition

The surface of cells is defined by a membrane termed the plasma membrane. This membrane is composed of several classes of lipids including phospholipids, sphingomyelin, glycosphingolipids, and cholesterol. In this structure, the phospholipids are the primary component arranged into two parallel sheets or leaflets referred to as a bilayer. The other lipids, especially cholesterol, regulate the fluid consistency of the membrane. The rigid steroid ring of cholesterol, in particular, adds integrity to the membrane by partially immobilizing the fatty acid side chains of phospholipids. While this stiffening decreases plasma membrane fluidity, the high concentration of cholesterol in animal cell plasma membranes counters this firming effect by separating the phospholipid fatty acid chains and disrupting their crystallization. Cholesterol also plays an important role in membrane biology by forming specialized microdomains known as lipid rafts and caveolae, recognized to function in numerous cellular processes including regulation of glucose transport by insulin and chromium.

Basic Mechanisms of Insulin Action in Health and Disease

Glucose Transport Regulation by Insulin

Under normal insulin responsiveness, insulin promotes the removal of excess glucose from the circulation by stimulating the exocytic recruitment of intracellular

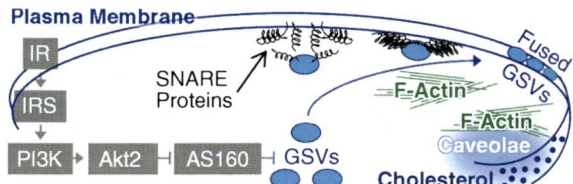

Chromium and Membrane Cholesterol, Fig. 1 Schematic illustration of insulin signals, cytoskeletal mechanisms, and plasma membrane parameters involved in insulin-stimulated GLUT4-storage vesicle (GSV) exocytosis. See text for abbreviations and details

glucose transporter GLUT4 storage vesicles (GSVs) to the plasma membrane of skeletal muscle and fat cells (Hoffman and Elmendorf 2011). This stimulated redistribution of intracellular GSVs results in plasma membrane GLUT4 accrual that facilitates cellular glucose uptake (Fig. 1). Activation of GSVs by insulin requires a phosphatidylinositol 3-kinase (PI3K) signal involving the upstream insulin receptor (IR) and insulin receptor substrate (IRS) activators and the downstream Akt2 and AS160 target enzymes (Fig. 1, shown in gray). Exocytosis of GSVs is mediated by interactions between specific GSV and plasma membrane protein complexes known as SNAREs (Fig. 1, shown in black). Vesicle SNAREs (v-SNARES, vesicle soluble N-ethylmaleimide-sensitive factor attachment protein receptors) bind target membrane SNAREs (t-SNAREs) in company with numerous accessory proteins. Syntaxin4 and SNAP23 (23 kDa synaptosomal-associated protein) are the t-SNARES, and VAMP2 is the v-SNARE involved in GSV fusion. At the plasma membrane, the enzyme phospholipase D1, together with its product phosphatidic acid, appear to be necessary for GSV and plasma membrane fusion. Mechanistically, phosphatidic acid has been suggested to act as a fusogenic lipid in biophysical modeling studies by lowering the activation energy for membrane bending (i.e., negative membrane curvature) during generation and expansion of fusion pores (Hoffman and Elmendorf 2011).

In addition to engaging signaling cascades to the GSV and plasma membrane, insulin also elicits a rapid, dynamic remodeling of actin filaments into a cortical mesh, and this mesh is necessary for GLUT4 translocation (Fig. 1, shown in green). Fluorescence confocal labeling of cortical filamentous actin (F-actin) shows actin filaments emanate from cholesterol-enriched plasma membrane caveolae microdomains (Hoffman and Elmendorf 2011). Caveolae, which are a special type of lipid raft, are small (50–100 nm) invaginations of the plasma membrane (Parton and Simons 2007). These flask-shaped structures are rich in proteins as well as lipids such as cholesterol and sphingolipids (Fig. 1, shown in light blue). Through the years, many functions for caveolae have been postulated in insulin and GLUT4 action. Although caveolae function needs to be cautiously interpreted because problems are associated with each of the numerous strategic approaches used to study these structures; substantial evidence supports a role for caveolae in cortical F-actin regulation (Hoffman and Elmendorf 2011). For example, biochemical disruption of this caveolin-associated F-actin, termed Cav-actin, structure does not affect the organization of clustered caveolae; however, disruption of the clustered caveolae disperses the Cav-actin structure. Quantitative electron microscopy and freeze-fracture analyses revealed that cytoskeletal components, including actin, are highly enriched in the membrane area underlying the neck part of caveolae. Together, these findings apparently assign caveolae a critical functionality in cortical F-actin organization. Given the unequivocal importance of cortical F-actin in insulin-regulated GLUT4 translocation, these findings also emphasize the importance of caveolae in GLUT4 regulation.

GLUT4 Dysregulation in Type 2 Diabetes

Despite the increase in the mechanistic knowledge of insulin action, the global prevalence of diabetes in 2010 was 284 million people worldwide, constituting around 6.4% of the world population. Projections for 2030 estimate the prevalence reaching 439 million individuals, comprising ~7.7% of the world population (Farag and Gaballa 2011). This is attributed in large part to the rising incidence of obesity worldwide, which makes it essential to focus attention on molecular mechanisms underlying insulin resistance that are fueled by obesity. In this regard, a large number of endocrine, inflammatory, neural, and cell-intrinsic pathways have been shown to be dysregulated in obesity and impair insulin signaling by increasing inhibitory serine phosphorylation of IRS1, which can have a direct impact on GLUT4 regulation (Qatanani and Lazar 2007).

However, several studies showing GLUT4 dysregulation without apparent defects in proximal insulin signaling point to the existence of other disabling factors of insulin-regulated GLUT4 exocytosis (Hoffman and Elmendorf 2011). Whether critical flaws in cytoskeletal F-actin organization, lipid bilayer composition, and/or GSV and plasma membrane fusion priming contribute to insulin resistance is not known, yet a clear association has been demonstrated between the diabetic milieu, cytoskeletal disorganization, and bilayer abnormalities (Hoffman and Elmendorf 2011). For example, isolated neutrophils from patients with type 2 diabetes display decreased actin polymerization compared to neutrophils from nondiabetic control subjects. This impairment is associated with persistent expression of the endothelial adhering beta 2-integrin CD11b/CD18, potentially exacerbating vascular dysfunction in diabetic patients. Also, diabetic rat retinal endothelial cells have a prominent reduction in F-actin integrity, a finding closely linked to vascular leakage. Loss of rat mesangial cell F-actin has also been reported after exposure to early diabetic-state-like conditions, possibly causing diabetic hyperfiltration. Further support has been derived from examination of erythrocytes from overweight, insulin-resistant individuals which show marked changes in the phospholipid composition of the plasma membrane.

With regard to insulin action and glucose transport, cholesterol complexing drug experiments have demonstrated that removal of cholesterol from the plasma membrane augments basal glucose uptake and metabolism (Hoffman and Elmendorf 2011). Consistent with cholesterol causing highly ordered gel-like states, moderate increases in PM fluidity increase glucose transport in adipocytes, and insulin-stimulated glucose transport declined when fluidity was diminished (Czech 1980). Mechanistically, it is now appreciated that nonphysiological plasma membrane cholesterol depletion greater than 50% reduces the rate of internalization of plasma membrane GLUT4 by more than 85% (Hoffman and Elmendorf 2011). Although this certainly explains the gain in plasma membrane GLUT4, it is an artificial, experimentally induced gain. In contrast, reductions of plasma membrane cholesterol less than 50% do not affect the rate of endocytosis but do increase plasma membrane GLUT4 content (Hoffman and Elmendorf 2011).

These later findings suggest that moderate plasma membrane cholesterol lowering augments GSV exocytosis. How these plasma membrane cholesterol-based aspects of GLUT4 regulation intermingle with insulin signaling and/or GSV regulation is unknown, yet coordinated signaling events and/or F-actin reorganization are areas of interest. In fact, plasma membrane cholesterol toxicity has been proposed to cause cortical F-actin disorganization and cellular insulin resistance (Fig. 2, shown in red) (Hoffman and Elmendorf 2011). In direct support of a more distal, perhaps membrane/cytoskeletal defect negatively impinging on GLUT4 translocation, various cell model systems of insulin resistance demonstrate intact proximal insulin signaling to Akt2/AS160.

Membrane Mechanisms of Chromium Action in Health and Disease

Membrane Fluidity Regulation by Chromium

New additions to the molecular details of GLUT4 regulation by insulin add important insight into the antidiabetic mechanism of chromium action. For example, research conducted in the early 1990s found that chromium picolinate increased membrane fluidity and the rate of insulin internalization in cultured rat skeletal muscle cells (Evans and Bowman 1992). Although endocytic internalization of insulin does not seem to enhance its regulation of GLUT4 translocation, studies suggest that this may compartmentalize and efficiently promote interaction of insulin with several of its intracellular substrates (Pattar and Elmendorf 2010). This is consistent with an enhancing effect of chromium on insulin signaling. However, with new data showing that only a small fraction ($\sim$5%) of insulin-stimulated Akt2 activity is necessary for a full GLUT4 response, another mechanism of chromium action is suggested to account for chromium-enhanced GLUT4 regulation by insulin (Pattar and Elmendorf 2010). Interestingly, treatment of insulin-resistant adipocytes or skeletal muscle cells with chromium lowers plasma membrane cholesterol, which positively influences insulin-stimulated GLUT4 translocation (Pattar and Elmendorf 2010). It was further demonstrated that reversing this action of chromium by adding back exogenous cholesterol completely rendered the enhancement of insulin action by chromium ineffective in these cells. These studies also showed that chromium action did not result from effects on known mediators of insulin action such as

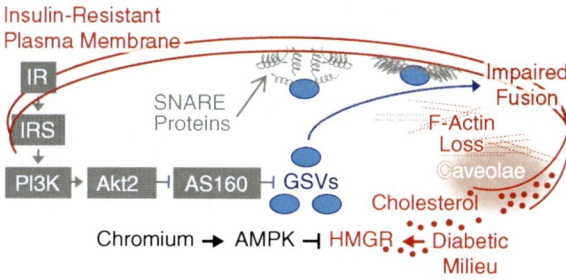

Chromium and Membrane Cholesterol, Fig. 2 Model of chromium action against plasma membrane cholesterol accrual that induces filamentous actin loss and impaired GSV regulation. See text for abbreviations and details

the IR, IRS-1, PI3K, and Akt2. Although this observed action of chromium suggests a mechanistic basis for chromium action being increased membrane fluidity, the effect on membrane fluidity, per se, may not underlie this nutrient's effect on insulin and GLU4 action. For example, as presented in the preceding section, cholesterol-enriched caveolae microdomains are now appreciated to critically influence cortical F-actin structure that is important in GLUT4 regulation. Therefore, chromium-induced modifications in plasma membrane cholesterol may enhance regulation of GLUT4 translocation by establishing an optimal cortical F-actin environment just beneath the plasma membrane.

Since chromium supplementation has the greatest benefit in overweight, insulin-resistant individuals (Balk et al. 2007), a presumption would also have to be that cells from these individuals possess a cholesterol-laden plasma membrane, whereas the plasma membrane cholesterol content of cells from nondiabetic subjects with normal insulin sensitivity and glucose tolerance would be lower and not affected by chromium. In line with this reasoning, chronic hyperglycemia has been shown to increase cholesterol ester accumulation concomitantly with reduced insulin-stimulated glucose uptake in cultured human skeletal muscle cells (Pattar and Elmendorf 2010). Moreover, skeletal muscle membranes from high-fat fed, insulin-resistant animal models are cholesterol-enriched and display a loss of F-actin. Similarly, human skeletal muscle biopsies reveal an inverse correlation between membrane cholesterol and whole-body glucose disposal (Pattar and Elmendorf 2010). Interestingly, in addition to hyperglycemia inducing cholesterol accrual as presented above, other key derangements of the diabetic milieu, namely, hyperinsulinemia and hyperlipidemia induce plasma membrane cholesterol accrual (Hoffman and Elmendorf 2011). Consistent with clinical data, the beneficial cholesterol-dependent action of chromium to mobilize GLUT4 to the plasma membrane is seen only in cells cultured in diabetic conditions, not in cells cultured in nondiabetic control conditions. These bench findings translate strikingly well to the observation that no significant alteration of chromium on glucose metabolism is seen in nondiabetic individuals, whereas there is a significant effect on chromium supplementation in diabetic patients (Balk et al. 2007). Taken together, chromium appears to display a unique ability to mitigate insulin resistance associated with plasma membrane cholesterol accrual.

Regulatory Cholesterol Metabolism Pathways

Mechanistically, it has been demonstrated that the activity of the AMP-activated protein kinase (AMPK) is enhanced by chromium (Pattar and Elmendorf 2010). This is of interest as a predicted result of AMPK activation is inhibition of energy-consuming biosynthetic pathways, such as fatty acid and cholesterol synthesis, and activation of ATP-producing catabolic pathways, such as fatty acid oxidation. Also, in muscle, AMPK activity can acutely and chronically affect basal and insulin-regulated glucose transport. New data suggest that these pathways that reduce fatty acids and cholesterol synthesis are likely engaged, as both acetyl CoA carboxylase (ACC) and 3-hydroxy-3-methyl-glutaryl coenzyme A reductase (HMGR) are phosphorylated in response to chromium (Pattar and Elmendorf 2010). Based on the observed effect of chromium on plasma membrane cholesterol, chromium-induced HMGR phosphorylation is of interest as this enzyme is the rate-limiting enzyme in the cholesterol biosynthesis pathway, and phosphorylation suppresses its activity (Fig. 2, shown in black). Also, of equal interest and importance is chromium's action on the AMPK/ACC pathway, as ACC catalyzes the conversion of acetyl CoA to malonyl coenzyme A (CoA), an inhibitor of translocation of long-chain fatty acyl (LCFA) groups from the cytosol to the mitochondrial matrix for fatty acid oxidation. Inactivation of ACC by AMPK in insulin-resistant skeletal muscle would be predicted to decrease excess fatty acids and several metabolites including acyl CoAs, ceramides, and diacylglycerol that activate kinases (e.g., *PKC*,

JNK, IKKβ) that have been demonstrated to increase the inhibitory serine phosphorylation of insulin-signaling cascades, perhaps providing an explanation to the beneficial effects of chromium on insulin signaling. Interestingly, new data show that chromium induces myocardial GLUT4 translocation to caveolar regions rich in caveolin-3 (expressed mainly in skeletal muscle) via activation of Akt, AMPK, and eNOS phosphorylation in streptozotocin-induced diabetic rats (Pattar and Elmendorf 2010). This finding is consistent with the caveolar-localized effect of chromium. Other work has recently shown an activation of AMPK with the chromium complex of D-phenylalanine (Cr(D-Phe)$_3$) in H9c2 myoblasts (Pattar and Elmendorf 2010). Activation of AMPK in this skeletal muscle cell line sufficiently stimulated glucose transport. Importantly, when AMPK activity was pharmacologically inhibited, the enhanced glucose uptake was blocked. It was suggested that Cr(D-Phe)$_3$ led to an increase in cellular AMP concentration and a decrease in mitochondrial membrane potential, both of which may account for the activation of AMPK. Finally, increased AMPK activity in chromium-treated 3T3-L1 adipocytes decreased secretion of resistin, an adipokine associated with impaired glucose transport (Pattar and Elmendorf 2010). Taken together, the proposed mechanism that AMPK activity is being increased by chromium may be a feasible explanation for its multiple actions in both relieving glucose and lipid disorders in metabolically challenged individuals.

Cholesterol Within and Beyond the Cell: Connecting Membrane and Circulating Cholesterol Levels

Although evidence relating chromium deficiency and cardiovascular disease is fragmentary, deficiency has been linked to reduced high-density lipoprotein cholesterol (HDL-C) (Simonoff 1984). Nevertheless, a link between chromium supplementation and benefits on raising high-density lipoprotein cholesterol (HDL-C) remains unclear (Balk et al. 2007). However, recent identification of a mechanism by which chromium may positively impact HDL-C generation stresses a plausible role of chromium in regulating circulating cholesterol (Sealls et al. 2011). This work stemmed from findings that membrane cholesterol accumulation in Niemann-Pick disease type C (NPC) cells impairs a rate-limiting step in HDL-C generation entailing cholesterol transporter ABCA1-mediated cholesterol efflux to lipid-poor apolipoprotein A1 (ApoA1). The HDL-C particle formed is preβ-1 HDL-C, a subclass which removes cholesterol from macrophages, representing a cardioprotective event. Thus, a growing appreciation is that this preβ-1 HDL-C particle likely represents the "functional" HDL-C subfraction. Examination of ABCA1 trafficking revealed that plasma membrane ABCA1 was diminished in hyperinsulinemia-induced insulin-resistant cells with plasma membrane cholesterol accrual, yet in the presence of chromium, this was prevented (Sealls et al. 2011). Endosomal membrane ABCA1 was elevated in the insulin-resistant cells and normalized by chromium. Mechanistically, ABCA1 is regulated by the endosomal-to-cytoplasm cycling of the GTPase Rab8. Whereas insulin resistance increased the Rab8 content in the endosomal membrane fraction and decreased this protein in the cytoplasm, these changes were normalized by chromium. This work further showed that similar to plasma membrane cholesterol accrual compromising GLUT4 function, endosomal membrane cholesterol accrual compromises Rab8/ABCA1/ApoA1 functionality, and that chromium, by increasing AMPK activity, diminishes endosomal membrane cholesterol and the cholesterol efflux deficiency (Fig. 3). Therefore, as the serum concentration of the preβ-1 HDL-C accounts for only a small fraction of total HDL-C, trials designed to assess the benefits of chromium on total HDL-C may have had an inherent flaw in understanding chromium's effect. Whether this cell-based model explains the benefits of chromium in humans with diabetes remains to be validated.

Summary and Clinical Perspective

In summary, while chromium is known to be an essential element for animals and humans, much work remains to elucidate the mechanism(s) by which chromium could have nutritional value. Nevertheless, a coherent picture is beginning to emerge from the present sum of experimental and clinical evidence. A model in which membrane cholesterol accrual may represent an unappreciated mechanism leading to insulin resistance and dyslipidemia is of interest, especially in the context of the putative mechanism of chromium action entailing the

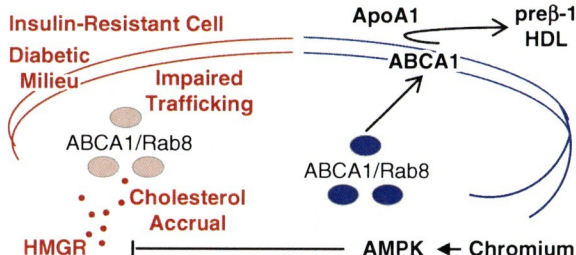

Chromium and Membrane Cholesterol, Fig. 3 Model of chromium protection against endosomal membrane cholesterol-associated impairment in ABCA1/Rab8 trafficking and ApoA1-mediated cholesterol efflux. See text for abbreviations and details

stimulation of AMPK that would suppress cellular cholesterol synthesis and offer subsequent improvements in cellular glucose uptake and cholesterol efflux, explaining longstanding health claims that chromium improves glycemic control in type 2 diabetes and that its deficiency may be a primary risk factor in cardiovascular disease. Finally, this membrane cholesterol model may provide an explanation for the existing controversies over the efficacy of chromium in the amelioration of the symptoms and complications of type 2 diabetes. For example, several agents that normalize blood glucose concentrations and/or improve insulin action, including metformin, phenformin, rosiglitazone, and troglitazone, have been shown to activate AMPK. Therefore, a prediction is that the nutritive value of chromium would be masked in patients receiving metformin or one of the other therapies listed above. Similarly, the action of AMPK to combat membrane cholesterol toxicity could explain why exercise, which also activates AMPK, can improve glycemia and dyslipidemia.

Cross-References

▶ Chromium and Allergic Reponses
▶ Chromium and Diabetes
▶ Chromium and Glucose Tolerance Factor
▶ Chromium and Human Nutrition
▶ Chromium and Insulin Signaling
▶ Chromium and Nutritional Supplement
▶ Chromium Binding to DNA
▶ Chromium, Physical and Chemical Properties
▶ Chromium Toxicity, High-Valent Chromium
▶ Chromium(III) and Immune System
▶ Chromium(III) and Low Molecular Weight Peptides
▶ Chromium(III) and Transferrin
▶ Chromium(III), Cytokines, and Hormones
▶ Hexavalent Chromium and Cancer
▶ Hexavalent Chromium and DNA, Biological Implications of Interaction
▶ Trivalent Chromium

References

Balk E, Tatsioni A, Lichtenstein A, Lau J, Pittas AG (2007) Effect of chromium supplementation on glucose metabolism and lipids: a systematic review of randomized controlled trials. Diabetes Care 30:2154–2163

Czech M (1980) Insulin action and the regulation of hexose transport. Diabetes 29:399–409

Evans G, Bowman T (1992) Chromium picolinate increases membrane fluidity and rate of insulin internalization. J Inorg Biochem 46:243–250

Farag Y, Gaballa M (2011) Diabesity: an overview of a rising epidemic. Nephrol Dial Transplant 26:28–35

Hoffman N, Elmendorf J (2011) Signaling, cytoskeletal and membrane mechanisms regulating GLUT4 exocytosis. Trends Endocrinol Metab 22(3):110–116

Parton R, Simons K (2007) The multiple faces of caveolae. Nat Rev Mol Cell Biol 8:185–194

Pattar G, Elmendorf J (2010) Dietary chromium supplementation and its role in carbohydrate and lipid metabolism. In: Avigliano L, Rossi L (eds) Biochemical aspects of human nutrition. Transworld Research Network, Trivandrum

Qatanani M, Lazar M (2007) Mechanisms of obesity-associated insulin resistance: many choices on the menu. Genes Dev 21:1443–1455

Sealls W, Penque B, Elmendorf J (2011) Evidence that chromium modulates cellular cholesterol homeostasis and ABCA1 functionality impaired by hyperinsulinemia. Arterioscler Thromb Vasc Biol 31:1139–1140

Simonoff M (1984) Chromium deficiency and cardiovascular risk. Cardiovasc Res 18:591–596

Chromium and Nutritional Supplement

Sreejayan Nair
University of Wyoming, School of Pharmacy, College of Health Sciences and the Center for Cardiovascular Research and Alternative Medicine, Laramie, WY, USA

Synonyms

Chromium complexes; Chromium dietary supplement; Inorganic chromium; Trivalent chromium

Definition

Chromium: Chromium is a common element that exists in the environment in several oxidation states. Trivalent chromium is a component of the natural diet which is thought to be essential for glucose and lipid homeostasis.

Obesity-associated diseases: Obesity is the abnormal or excessive fat accumulation that may impair health. Obesity increase the risk of cardiovascular disease includes insulin resistance, type-2 diabetes, dyslipidemia, stroke, atherosclerosis, and heart failure. Obesity is associated with increased risk of premature death.

Type-2 diabetes: Chronic disease characterized by elevated blood glucose levels caused by either a lack or the inability of the body to efficiently utilize insulin.

The use of nutritional supplements has increased during the past years and it is currently estimated that one half of all Americans used one or other form of nutritional supplements. Chromium is next only to calcium among the minerals sold as nutritional supplements and represents approximately 6% of the nutritional supplement market.

Chromium as an Essential Element

Chromium is an *essential* trace element that is believed to play an important role in carbohydrate and lipid metabolism (Vincent 2007). Way back in 1959, chromium was identified as the active component of "glucose tolerance factor" found in brewer's yeast. The *essentiality* of chromium in carbohydrate metabolism first came to light when chronically ill patients on total parenteral nutrition developed glucose intolerance which was reversed following supplementation of chromium (Anderson 1987). Several human and animal studies that followed demonstrated a beneficial effect of chromium in insulin resistant conditions and type-2 diabetes. However, this notion of the "essentiality" of chromium is now being questioned as recent studies have demonstrated that a "low chromium diet" does not alter glucose metabolism in the body (Di Bona et al. 2011).

Sources of Chromium and Recommended Daily Allowance

Dietary sources of chromium include meat, seafood, green vegetables, fruits, cheese, and wholegrain. A normal diet would therefore help meet the body's chromium requirement and chromium deficiency is therefore rare in normal healthy individuals. However, conditions such as diabetes and insulin resistance, pregnancy and ageing have been associated with depleted levels of chromium in the body. The recommended daily allowances for chromium as suggested by the Institute of Medicine of the National Academy of Sciences ranged from 50 to 200 µg/day for adults (National Research Council FaNB 1989). However, because of the difficulties in reliably assessing chromium levels in the food and in the human body, the Institute of Medicine has recently come up with adequate intakes of chromium rather than the traditional recommended daily allowance. The recommended adequate intake for chromium, which represents the level of chromium consumed by health individuals, is 23–39 µg/day for adult female and 35–50 µg/day for adult males (Institute of Medicine FaNB 2001). These numbers increase in pregnant or lactating women or under disease conditions.

Small-Molecule Organic Chromium Complexes

The poor oral bioavailability of inorganic chromium (0.4–2.5%) and the recognition that the "glucose tolerance factor" represents a chromium complex of amino acids and nicotinic acid prompted the synthesis and characterization of several organic chromium complexes as insulin-potentiating agents. Among them, chromium picolinate and chromium nicotinate have emerged as the most popular forms of organic chromium complexes that are available as nutritional supplements. Today, these chromium complexes are included in multivitamin tablets, breakfast cereals, and energy drinks. Studies have also shown that the absorption of chromium can be enhanced when taken along with vitamin B or C.

Health Benefits of Chromium Supplements

The primary purpose for which chromium supplementation is being used is for its potential beneficial effects to treat diabetes, lipid disorders, and overweight. There have been several clinical trials that have investigated the beneficial effects of nutritional supplementation with chromium in subjects with type-2-diabetes and insulin-resistant conditions. The outcome of these studies were however inconclusive, mainly owing to the small sample sizes, inadequate controls, varying doses used, different patient demographics, concurrent use of other antidiabetic agents, lack of robust methodologies and/or lack of randomization.

In a recently published meta-analysis, Balk and coworkers performed a systematic review of the major randomized controlled trials involving nutritional supplementation with chromium (Balk et al. 2007). For their analysis, these authors included 41 randomized controlled clinical trials that were performed in individuals with diabetes or glucose intolerance. Their end points were glycemic control and lipid outcomes. In addition, they only included the trials that had ten or more participants and those studies wherein chromium was supplement for at least 3 weeks. The meta-analysis revealed a modest, yet statistically significant beneficial effect of chromium supplementation in subjects with type-2 diabetes. Most of the studies included in this meta-analysis used chromium at doses ranging from 200 to 1,000 µg/day. The beneficial effects of chromium on hyperglycemia and insulin resistance were most prevalent in studies that used doses close to 1,000 µg/day. In type-2 diabetic patients, treatment with chromium caused a lowering of HbA_{1C} by 0.6% and fasting glucose levels by ~1 mmol/L. However, in contrast to glycemic parameters, the lipid levels were not altered following chromium supplementation, except for a significant raise in HDL-cholesterol observed in subjects who received brewer's yeast. In normal subjects (those without type-2 diabetes or insulin resistance), chromium supplementation did not have any effects on either the measures of glycemia or lipid parameters. Despite the smaller, albeit statistically significant favorable effects of chromium on glycemic parameters, the authors of the meta-analysis caution that the poor quality and heterogeneity of the clinical trials warrants further studies before definitive conclusions are made.

Due to the aforementioned conflicting reports, the American Diabetes Association's position statement concludes that there is inconclusive evidence for the benefit of chromium supplementation in diabetes (Bantle et al. 2008). The US Food and Drug Administration has previously issued a statement on similar lines stating that "chromium may reduce the risk of insulin resistance and therefore possibly reduce the risk of type-2 diabetes" and goes on to add, "the existence of such a relationship between chromium picolinate and either insulin resistance or type 2 diabetes is uncertain".

With a view to clear these contradictions, Cefalu and coworkers performed an exhaustive randomized, double-blind, placebo-controlled clinical trial in which they measured insulin-sensitivity, energy expenditure, and muscle and hepatic fat content in type-2 diabetic subjects who were not on any antidiabetic medications (Cefalu et al. 2010). The results of this study indicate that the diabetic population could be categorized as either "responders" and "non-responders" to chromium, and thus the beneficial effects of chromium supplementation may depend on the cohort of individuals tested. However, this claim is refuted by Kleefstra and coworkers who failed to observe any beneficial effect of chromium (at doses of 500 or 1,000 µg) compared to placebo on HbA_{1C} or insulin requirement in diabetic patients (Kleefstra et al. 2010).

In addition to the above human studies, chromium has been shown to reduce body fat, improve cardiovascular parameters, affect fetal-gene programming, regulate obesogenic genes, augment insulin signaling, mobilize glucose transporter, alter lipid metabolism, and affect cholesterol homeostasis in a variety of animal and cellular models suggesting the potential beneficial effects of chromium. However, the jury is till out in the debate over the effectiveness of chromium supplementation in type-2 diabetes and more large-scale, well-controlled trials are warranted to ascertain its benefits as a nutritional supplementation.

Cross-References

▶ Chromium and Diabetes
▶ Chromium and Glucose Tolerance Factor
▶ Chromium and Insulin Signaling
▶ Chromium(III) and Low Molecular Weight Peptides

References

Anderson RA (1987) Chromium. Academic, New York
Balk EM, Tatsioni A, Lichtenstein AH, Lau J, Pittas AG (2007) Effect of chromium supplementation on glucose metabolism and lipids: a systematic review of randomized controlled trials. Diabetes Care 30:2154–2163
Bantle JP, Wylie-Rosett J, Albright AL, Apovian CM, Clark NG, Franz MJ, Hoogwerf BJ, Lichtenstein AH, Mayer-Davis E, Mooradian AD, Wheeler ML (2008) Nutrition recommendations and interventions for diabetes: a position statement of the American Diabetes Association. Diabetes Care 31(Suppl 1):S61–S78

Cefalu WT, Rood J, Pinsonat P, Qin J, Sereda O, Levitan L, Anderson RA, Zhang XH, Martin JM, Martin CK, Wang ZQ, Newcomer B (2010) Characterization of the metabolic and physiologic response to chromium supplementation in subjects with type 2 diabetes mellitus. Metabolism 59:755–762

Di Bona KR, Love S, Rhodes NR, McAdory D, Sinha SH, Kern N, Kent J, Strickland J, Wilson A, Beaird J, Ramage J, Rasco JF, Vincent JB (2011) Chromium is not an essential trace element for mammals: effects of a "low-chromium" diet. J Biol Inorg Chem 16:381–390

Institute of Medicine FaNB (2001) Institute of Medicine, Food and Nutrition Board. Dietary reference intakes for vitamin A, vitamin K, arsenic, boron, chromium, copper, iodine, iron, manganese, molybdenum, nickel, silicon, vanadium, and zinc. National Academy Press, Washington, DC

Kleefstra N, Houweling ST, Groenier KH, Bilo HJ (2010) Characterization of the metabolic and physiologic response to chromium supplementation in subjects with type 2 diabetes mellitus. Metabolism 59:e17, author reply e18-9

National Research Council FaNB (1989) Recommended dietary allowances. National Academy Press, Washington, DC

Vincent JB (ed) (2007) The nutritional biochemistry of chromium (III). Elsevier, Oxford

Chromium and Tanning

▶ Chromium and Leather

Chromium Binding to DNA

Anatoly Zhitkovich
Department of Pathology and Laboratory Medicine, Brown University, Providence, RI, USA

Synonyms

Cr-DNA binding: Cr-DNA adduction; Inner-sphere Cr-DNA complexation

Ternary Cr-DNA adduct: Intermolecular Cr-DNA cross-link; Ligand-Cr-DNA cross-link

Definition

Stable Cr-DNA binding is inner-sphere coordination of Cr(III) atoms with DNA groups. This type of binding can also be described as covalent coordinate bonding.

Cr-DNA Binding

General Characteristics of Chromium-DNA Binding

Human tissues and other biological systems usually encounter Cr in one of its two stable oxidative states, namely, +3 or +6. Although both inorganic Cr(III) and Cr(VI) compounds can cause Cr-DNA binding, the conditions required for this reaction to take place are not identical. At neutral pH, Cr(VI) is completely unreactive toward DNA and requires reductive activation to elicit genetic damage and other forms of biological injury. The reduction process is associated with a transient formation of Cr(V) and Cr(IV) intermediates, organic radicals as by-products, and finally yields thermodynamically stable Cr(III) (Zhitkovich 2005). The most important biological reducers of Cr(VI) in human and other mammalian cells are ascorbate, glutathione, and cysteine (Fig. 1). Although its cellular concentration in vivo is similar to that of glutathione, ascorbate is the most important reducer, accounting for >90% of Cr(VI) metabolism in the main target tissues, such as the lung. This dominant role of ascorbate reflects its very fast rate of Cr(VI) reduction, which at physiological concentrations is up to 60 times faster than that of glutathione (Quievryn et al. 2003).

Initially, the genotoxic activity of carcinogenic Cr(VI) compounds had been largely attributed to oxidative stress and the resulting oxidative DNA damage. However, reduction of Cr(VI) in cultured cells or in vivo has also been known to result in stable Cr-DNA binding that persisted during rigorous, multistep purification procedures employing strong detergents and high-salt and organic solvent extractions. Early studies investigating Cr-DNA binding in the defined reactions in vitro with known biological reducers of Cr(VI) have yielded highly conflicting results, ranging from detecting large amounts of DNA-bound Cr to observing no binding at all. Careful examination of the buffer conditions and aqueous chemistry of Cr(III) led to the realization that the commonly used phosphate buffer was blocking Cr-DNA complexation through the formation of unreactive Cr(III)-phosphate chelates (Zhitkovich 2005). In the organic buffers containing the ionized sulfonate group (MES, MOPS, HEPES), reduction of Cr(VI) with the main biological reducers results in extensive Cr-DNA binding that is readily observed by a progressively decreased electrophoretic mobility

Chromium Binding to DNA, Fig. 1 *Chemical structures of the main biological reducers of Cr(VI). These reducers also form Cr(III)-mediated cross-links with DNA during Cr(VI) reduction in cells and in vitro*

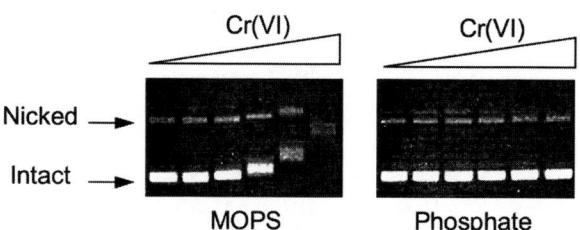

Chromium Binding to DNA, Fig. 2 *Altered electrophoretic mobility of plasmid DNA containing bound Cr. The ϕX174 plasmids were incubated with 0–200 μM chromate in the presence of 1 mM ascorbate for 30 min at 37°C. DNA bands in agarose gels were visualized by ethidium bromide fluorescence under UV illumination. In contrast to extensive Cr-DNA binding, reduction of Cr(VI) with its main biological reducer did not generate a detectable oxidative damage to the DNA sugar-phosphate backbone as evidenced by a lack of conversion of intact (supercoiled) plasmids into nicked (relaxed) conformation*

of the supercoiled plasmid DNA (Fig. 2). This phenomenon reflects unwinding of supercoiled DNA molecules, as the mobility of relaxed plasmids in the same reactions shows little or no changes. At high Cr(VI) concentrations, supercoiled plasmids could be completely unwound, resulting in their comigration with the relaxed DNA molecules. These conditions referred to as the coalescence point were used for the calculations of structural changes in DNA that gave an estimate of 1–2° unwinding by each DNA-bound Cr atom (Blankert et al. 2003). Although this unwinding angle is small in comparison to DNA adducts formed by platinum-based drugs, prolonged exposures of human and other mammalian cells to toxic chromate result in more than 100-fold cellular accumulation of Cr(VI) over its extracellular concentrations, leading to a very large production of Cr-DNA adducts at low doses (Zhitkovich 2005). Thus, a concentrated presence of even mildly duplex-distorting Cr-DNA complexes could elicit sufficiently large structural changes in chromatin loops with significant consequences for transcription and replication that are both known to be affected by the degree of DNA supercoiling.

Another readily detectable change in DNA molecules containing bound Cr atoms is their decreased staining with the commonly used DNA dye ethidium bromide (Fig. 2). The phenomenon of diminished ethidium bromide fluorescence was observed for all Cr(VI) or Cr(III) reactions permitting Cr-DNA binding. Ethidium bromide intercalates between DNA bases, which allows it to become fluorescent through energy transfer from UV radiation that is absorbed by DNA bases. While lower fluorescence of supercoiled DNA after indirect excitation by UV light may have a complex origin, fluorescence of linear DNA molecules after direct excitation of ethidium bromide reflects the amount of intercalated dye and can serve as a test for the degree of duplex distortions. Direct excitation experiments showed that Cr-DNA binding strongly inhibits ethidium bromide intercalation at a wide range of dye concentrations (Fig. 3), which is indicative of distorted DNA duplex that is unable to retain dye molecules. Unstacking of DNA bases is the most likely cause of diminished ethidium bromide intercalation due to the inability of dye molecules to establish strong interactions with top and bottom bases. Other structural changes elicited by Cr-DNA binding include increased stability of duplexes as evidenced by their higher resistance to heat- and alkali-induced denaturation. Positively charged Cr(III) complexes can also interact with DNA ionically, which makes up 40–60% of total DNA-bound Cr in reactions with low ionic strength (Zhitkovich 2005). The ionically bound Cr(III) is easily stripped from DNA by brief incubations with physiological concentrations of Na^+ or Mg^{2+} ions. Nonionic Cr-DNA complexes are largely resistant to dissociation even during prolonged incubations with

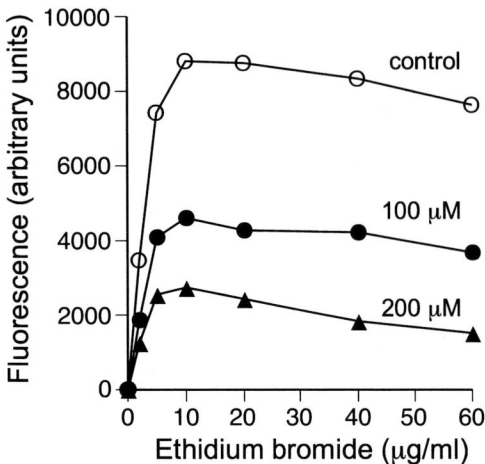

Chromium Binding to DNA, Fig. 3 *Diminished binding of ethidium bromide to Cr-adducted DNA.* Sheared chromosomal DNA was treated with Cr(VI) and 2 mM cysteine for 60 min at 37°C (25 mM MOPS, pH 7.0). Ethidium bromide fluorescence was recorded at 595 nm with excitation at 530 nm in the presence of 400 mM KCl and 5 mM EDTA

very high EDTA concentrations. However, smaller and less negatively charged phosphate ions can remove the majority of DNA-bound Cr(III) atoms. Binding of oligomeric Cr(III) products to DNA is more difficult to remove by phosphate and other chelators.

Interactions of Cr(III) with DNA

Cr(III) forms six-coordinate complexes with octahedral arrangement of ligands and displays a strong binding affinity for negatively charged oxygen groups. Direct coordination of Cr(III) to H_2O, SH group, and tertiary N atoms is also strong. Overall, Cr(III) complexes are usually described as kinetically inert that are slow to exchange their ligands, and consequently, are poorly reactive. Another factor impeding interactions of Cr(III) complexes with DNA is their hydrolysis. Dissolution of inorganic Cr(III) salts results in the initial formation of $Cr(H_2O)_6^{3+}$ as the main ionic species. However, these solutions undergo rapid hydrolysis resulting in the conversion of $Cr(H_2O)_6^{3+}$ into $Cr(OH)(H_2O)_5^{2+}$ and $Cr(OH)_2(H_2O)_5^{+}$. The formation of the Cr(III) hydroxo species initiates their polymerization, producing a mixture of low- and high-molecular-weight polymeric products. Monomeric and low oligomeric forms of Cr(III) remain soluble, whereas polymeric products form poorly soluble precipitates. Hydrolysis and polymerization reaction are inhibited at low pH but at neutral pH dissolution of inorganic Cr(III) salts leads to almost immediate formation of insoluble Cr(III) hydroxides and polymeric species. To avoid the solubility and oligomerization problems, the reactions of inorganic Cr(III) with DNA have been frequently performed near pH 6 or lower. Unlike $Cr(H_2O)_6^{3+}$, Cr(III) complexes with multidentate organic ligands are soluble at neutral pH and do not undergo oligomerization (Zhitkovich 2005). A rapid coordination of Cr(III) with the multidentate reducer molecules maintains its solubility in Cr(VI) reduction reactions carried out at neutral pH. Organic buffers containing negatively charged sulfonate group are capable of Cr(III) complexation, which inhibits the production of insoluble hydroxides and permits Cr-DNA binding. Cr^{3+} ions are unable to interact with neutral or positively charged TRIS molecules, and the solubility Cr(III) in diluted TRIS solutions with their very weak buffering capacity at neutral pH results from acidification of the reaction mixtures in the presence of strongly acidic $Cr(H_2O)_6^{3+}$.

Cr-DNA Binding Resulting from Cr(VI) Reduction

As mentioned above, reductive metabolism of Cr(VI) in biological buffers and inside the cells leads to extensive Cr-DNA binding. Cr-DNA complexes are a heterogeneous group including binary Cr-DNA adducts and various cross-links (Zhitkovich 2005). Binary Cr(III)-DNA adducts are usually the most abundant form of DNA-bound Cr produced during in vitro reduction of Cr(VI) with its main biological reducers. For Cr(VI) reduction reactions containing physiological concentrations of ascorbate, Cr-DNA adducts account for approximately 75% of total Cr-DNA binding. The exact amount of binary adducts formed in cells remains unknown due to the indirect methods of their quantitation as a fraction of DNA-bound Cr left after subtracting Cr-DNA cross-links. The purification of chromosomal DNA is associated with a significant loss of Cr-cross-linked ligands, which makes it difficult to get accurate estimates for binary adducts. Based on the limited recovery measurements for Cr-DNA cross-links, it was argued that in contrast to in vitro reactions, binary Cr-DNA adducts in cells probably constitute only a minor fraction of the total Cr-DNA binding (Zhitkovich 2005).

The most abundant form of Cr-DNA cross-links are ternary DNA adducts formed through Cr(III)-mediated DNA cross-linking of histidine and three main Cr(VI) reducers: ascorbate, cysteine, and glutathione (Zhitkovich 2005). Formation of stable complexes with O-, N-, and S-containing groups and the presence of six coordination sites are responsible for the ability of Cr(III) to link multidentate biological ligands to DNA, generating ligand-Cr(III)-DNA cross-links. The most common Cr-DNA modifications in Cr(VI)-treated cells were cysteine-Cr(III)-DNA and glutathione-Cr(III)-DNA cross-links, which in combination with histidine- and some other less frequent amino acid-Cr-DNA cross-links were estimated to constitute ~50% of total DNA-bound Cr even without corrections for losses during purification procedure. The production of ascorbate-Cr(III)-DNA cross-links is readily detectable during in vitro Cr(VI) reduction with physiological levels of ascorbate (1–3 mM in major human tissues) but is progressively decreased at lower concentrations. Formation of ascorbate-DNA cross-links in cells requires restoration of normal concentrations of vitamin C before Cr(VI) exposures. Under standard tissue culture conditions, all human and the vast majority, if not all, rodent cells contain barely detectable concentrations of ascorbate due to its low concentrations in fetal bovine serum and the absence in the commonly used synthetic media (Reynolds et al. 2007). Ternary Cr-DNA cross-links containing glutathione, cysteine, or histidine share the same mechanism of formation (Zhitkovich 2005):

1. Cr(VI) reduction to Cr(III)
2. Formation of binary ligand-Cr(III) complexes
3. Attachment of binary complexes to DNA forming ligand-Cr(III)-DNA cross-links

The last reaction is a rate-limiting step. For ascorbate-Cr(III)-DNA cross-links, the reaction mechanism is very similar with the exception of much higher rates at the final step of DNA conjugation.

In addition to small ligand-Cr-DNA cross-links, Cr(VI)-exposed cells also contain DNA-protein cross-links. These bulky lesions constitute only a small fraction of total DNA adducts in cultured cells, but the availability of a robust methodology for their measurements has led to their frequent use as a biomarker of genetic damage in Cr-exposed human populations and aquatic organisms. Although DNA-protein cross-links have a general structure of ternary adducts (protein-Cr(III)-DNA), their formation mechanism differs from that of small ascorbate/amino acid-Cr-DNA cross-links (Macfie et al. 2010). The three-step protein-DNA cross-linking proceeds as follows:

1. Cr(VI) reduction to Cr(III)
2. Cr(III)-DNA binding
3. Protein capture by DNA-bound Cr(III)

The rate-limiting step is protein conjugation by Cr(III)-DNA, which is particularly slow in vivo and explains a delayed buildup of DNA-protein cross-links in Cr(VI)-treated cells with limited DNA repair capacity.

Reduction of Cr(VI) in vitro also generates a small number of Cr(III)-mediated interstrand DNA cross-links (O'Brien et al. 2003). Interstrand cross-linking was promoted in reactions with low ratios of reducer to Cr(VI). At optimal conditions in cysteine-driven reactions, interstrand DNA cross-links made up approximately 1% of total Cr-DNA adducts, but their yield was much lower or even undetectable at more environmentally relevant Cr(VI) concentrations (Zhitkovich 2005). DNA interstrand cross-linking in Cr(VI)-ascorbate reactions was also highly nonlinear, with no cross-linking detected at physiologically relevant ratios of the main reactants. While every DNA nucleotide contains at least one point of attachment for Cr(III), potentially offering multiple opportunities for linkage of the opposite DNA strands, the formation of interstrand cross-links by Cr(III) is difficult to explain when steric factors are taken into consideration (Zhitkovich 2005). It has been therefore suggested that the most likely cause of interstrand DNA cross-linking under in vitro conditions with limited amounts of the Cr(III)-binding multidentate ligands are the oligomeric Cr(III) species, which are unlikely to be formed in cells. There is also strong genetic evidence arguing against the formation of interstrand DNA cross-links by Cr(VI) in mammalian cells (Salnikow and Zhitkovich 2008).

Nucleotide-Specificity of Cr-DNA Binding

As a hard Lewis acid, Cr(III) has a high affinity for the negatively charged oxygen, which predicts that DNA phosphates should act as preferred binding sites. In agreement with this prediction, studies with oligonucleotides of base-specific composition showed nearly identical Cr-DNA binding irrespective of the presence of a particular base and group of bases (Zhitkovich 2005).

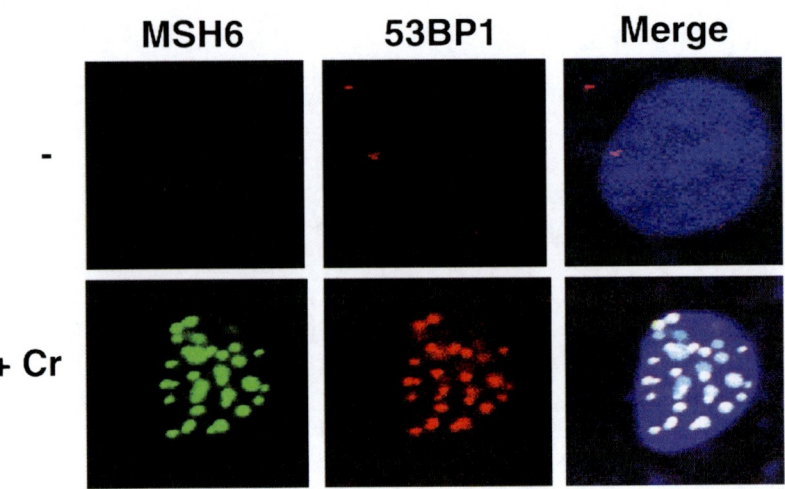

Chromium Binding to DNA, Fig. 4 *Presence of the mismatch repair protein MSH6 at the sites of DNA double-strand breaks.* Normal human IMR90 fibroblasts were treated with Cr(VI) and then simultaneously immunostained for MSH6 and 53BP1. 53BP1 is a protein involved in repair of DNA double-stranded breaks and serves as a marker of these lesions

Amino acid-Cr-DNA cross-links could be formed only with mononucleotides but not nucleosides, further supporting a critical role of the phosphate group in the formation of this biologically important class of DNA damage. Nucleotide-level mapping of binary Cr-DNA adducts and three Cr-DNA cross-links found a nearly uniform distribution along a long stretch of duplex DNA, and shielding of DNA phosphates with Mg^{2+} ions blocked Cr-DNA binding. However, all mutagenic events induced by Cr-DNA cross-links targeted G/C pairs with flanking purine bases (Zhitkovich et al. 2001; Quievryn et al. 2003), pointing to the potential presence of additional Cr-base interactions, possibly at N-7 position. Unlike ternary adducts, binary Cr(III) adducts are also formed with dG, indicating that binding at N-7 with possible microchelation at O-6 could be quite strong. An alternative cause for G/C pair specificity of Cr mutagenesis could be biological, such as properties of DNA polymerases and/or selectivity of DNA repair. These and structural factors impacting Cr adduct–induced mutagenesis have been discussed in detail elsewhere (Zhitkovich et al. 2001). At this time, the possibility of microchelate formation between phosphate group and N-7 of purine bases in mutagenic Cr-DNA adducts is neither proven nor disproven.

Indirect Mechanism of Cell Death and Chromosomal Breakage by Cr-DNA Adducts

The presence of ternary ascorbate- or amino acids-Cr-DNA cross-links but not binary Cr-DNA adducts strongly inhibits replication of plasmids in human cells (Zhitkovich et al. 2001; Quievryn et al. 2003). However, in vitro experiments with purified DNA polymerases found no significant replication-blocking activity of ternary Cr-DNA cross-links (O'Brien et al. 2003), indicating that genotoxicity of Cr-DNA damage was not a direct consequence of their direct polymerase-arresting activity. Genetic studies utilizing mutant mouse and human cells identified DNA mismatch repair (MMR) as a cause of cellular toxicity of ternary Cr-DNA adducts (Peterson-Roth et al. 2005). A normal function of MMR is to detect and correct DNA polymerase errors arising during replication. Consistent with plasmid replication experiments in cells, only ternary but not binary adducts were recognized by MMR proteins in binding experiments in vitro, and the loss of MMR components rescued replication blockage of cross-link-containing vectors in cells (Reynolds et al. 2009). "Mistaken identity" binding of ternary Cr-DNA cross-links by MMR proteins followed by aberrant processing causes highly toxic DNA double-stranded breaks, as determined by genetic approaches and confirmed by the presence of MMR proteins at the sites of DNA breakage (Fig. 4). The identification of MMR as a main cause of cell death and chromosomal breakage led to the formulation of the selection model of chromate carcinogenesis involving outgrowth of Cr(VI)-resistant cells with inactivated MMR (Peterson-Roth et al. 2005; Salnikow and Zhitkovich 2008). Cells lacking MMR rapidly accumulate mutations in cancer genes due to their inability to repair spontaneous replication errors. In agreement with this model, the majority of lung cancers among chromate workers are known to lack functional MMR.

Cross-References

▶ Chromium, Physical and Chemical Properties
▶ Hexavalent Chromium and Cancer
▶ Hexavalent Chromium and DNA, Biological Implications of Interaction
▶ Trivalent Chromium

References

Blankert SA, Coryell VH, Picard BT et al (2003) Characterization of nonmutagenic Cr(III)-DNA interactions. Chem Res Toxicol 16:847–854

Macfie A, Hagan E, Zhitkovich A (2010) Mechanism of DNA-protein cross-linking by chromium. Chem Res Toxicol 23:341–347

O'Brien TJ, Ceryak S, Patierno SR (2003) Complexities of chromium carcinogenesis: role of cellular response, repair and recovery mechanisms. Mutat Res 533:3–36

Peterson-Roth E, Reynolds M, Quievryn G et al (2005) Mismatch repair proteins are activators of toxic responses to chromium-DNA damage. Mol Cell Biol 25:3596–3607

Quievryn G, Peterson E, Messer J et al (2003) Genotoxicity and mutagenicity of chromium(VI)/ascorbate-generated DNA adducts in human and bacterial cells. Biochemistry 42:1062–1070

Reynolds M, Stoddard L, Bespalov I et al (2007) Ascorbate acts as a highly potent inducer of chromate mutagenesis and clastogenesis: linkage to DNA breaks in G2 phase by mismatch repair. Nucleic Acids Res 35:465–476

Reynolds MF, Peterson-Roth EC, Johnston T et al (2009) Rapid DNA double-strand breaks resulting from processing of Cr-DNA crosslinks by both MutS dimers. Cancer Res 69:1071–1079

Salnikow K, Zhitkovich A (2008) Genetic and epigenetic mechanisms in metal carcinogenesis and cocarcinogenesis: nickel, arsenic and chromium. Chem Res Toxicol 21:28–44

Zhitkovich A (2005) Importance of chromium-DNA adducts in mutagenicity and toxicity of chromium(VI). Chem Res Toxicol 18:3–11

Zhitkovich A, Song Y, Quievryn G et al (2001) Non-oxidative mechanisms are responsible for the induction of mutagenesis by reduction of Cr(VI) with cysteine: role of ternary DNA adducts in Cr(III)-dependent mutagenesis. Biochemistry 40:549–560

Chromium Carcinogenicity

▶ Chromium Toxicity, High-Valent Chromium

Chromium Chloride

▶ Chromium and Human Nutrition

Chromium Complexes

▶ Chromium and Nutritional Supplement

Chromium Dietary Supplement

▶ Chromium and Nutritional Supplement

Chromium Toxicity, High-Valent Chromium

Peter A. Lay and Aviva Levina
School of Chemistry, University of Sydney, Sydney, NSW, Australia

Synonyms

Chromium and cancer; Chromium and diabetes; Chromium carcinogenicity; Hexavalent chromium

Definition

Chromium(VI) (Cr(VI), hexavalent chromium) is an established human carcinogen and a major occupational and environmental hazard. Toxic properties of Cr(VI) are related to its ability to enter cells easily and to form reactive high-valent (Cr(VI), Cr(V), and Cr(IV)) intermediates on the way to stable Cr(III) products. In recent years, significant progress has been made in the understanding of chemical properties and biological roles of such intermediates, using physicochemical techniques such as electron paramagnetic resonance and X-ray absorption spectroscopies. The potential of biological oxidation of Cr(III) to Cr(VI) (via Cr(IV) and Cr(V)) intermediates has also been recognized as a likely link between the toxicity of Cr(VI) and antidiabetic activities of Cr(III), which highlights the potential danger of long-term use of Cr(III) nutritional supplements.

Chromium(VI) Toxicity

Hexavalent chromium (Cr(VI)) compounds, which are well-established (class I) human carcinogens, are among the most common occupational chemical hazards (for a review, see Levina et al. 2003). The most dangerous forms of Cr(VI) exposure are inhalation of insoluble chromate particles ($MCrO_4$, where M = Ca, Sr, Pb, or Zn) and inhalation of chromate fumes from stainless steel welding and electroplating baths, which pose a high risk of respiratory cancers (see section ▶ Hexavalent chromium and cancer (Costa) for details). The health hazards associated with ingestion of Cr(VI)-contaminated water have been extensively studied following the lawsuit against Pacific Gas and Electric, California, in 1993 (Levina et al. 2003). Although micromolar concentrations of Cr(VI) in drinking water are not highly carcinogenic, they are likely to sensitize the organism to other carcinogens, such as UV radiation, and recent studies suggest that oral ingestion of soluble chromates can lead to a range of cancers (section ▶ Hexavalent chromium and cancer (Costa)). A crucial recent finding is the ability of both water-insoluble ($PbCrO_4$) and water-soluble ($K_2Cr_2O_7$) chromates to cause malignant transformation of noncancerous human lung epithelial cells in culture (Xie et al. 2007; Costa et al. 2010). Significant progress has also been achieved in understanding of the roles of altered DNA repair mechanisms and disruptions of cell signaling in Cr(VI)-induced carcinogenicity (see sections ▶ Chromium binding to DNA (Zhitkovich) and ▶ Hexavalent Chromium and DNA, Biological Implications of Interaction (Wise)). This section concentrates on the links between chemical properties and toxicities of Cr(VI) and its metabolic products (Levina et al. 2003; Levina and Lay 2008).

Exposure routes and physical properties of Cr(VI) compounds are among the leading factors that determine their carcinogenic potential (Levina et al. 2003). For instance, inhaled particles of insoluble Cr(VI) salts adhere to the surface of bronchial epithelial cells, where they dissolve slowly, leading to prolonged exposure of cells to small doses of Cr(VI) (Xie et al. 2007). By contrast, ingested soluble forms of Cr(VI) are rapidly reduced to much less toxic Cr(III) in acidic media of the stomach (see section ▶ Chromium, Physical and Chemical Properties (Van Horn) for a review of chemical properties of various oxidation states of Cr), although this protective mechanism can be overwhelmed by exposure to large doses of Cr(VI) (Levina et al. 2003). Absorbed or in vivo-generated Cr(VI) (mainly in the form of $[CrO_4]^{2-}$) has a sufficient lifetime in the blood and other extracellular fluids at pH = 7.4 to be delivered to target cells (Levina et al. 2003). A framework for understanding of Cr(VI) interactions with cells has been provided by the uptake-reduction model (Fig. 1), first proposed by Wetterhahn and coworkers in the 1980s and updated by other researchers (for a review, see Levina et al. 2003).

The well-known ability of cultured mammalian cells to accumulate large amounts of Cr(III) when exposed to low concentrations of Cr(VI) is due to a combination of the following factors (Fig. 1) (Levina et al. 2003; Levina et al. 2007): (1) efficient uptake of $[CrO_4]^{2-}$ through anion channels (based on its structural similarity to essential anions, $[SO_4]^{2-}$ and $[HPO_4]^{2-}$); (2) intracellular reduction of Cr(VI) to Cr(III) (the likely reductants include glutathione, ascorbate, and NAD(P)H-dependent enzymes), which increases the chemical potential (Cr(VI) concentration gradient) for further Cr(VI) uptake; and (3) strong binding of the resultant Cr(III) species to biological macromolecules, which stores Cr(III) in inert forms that are difficult to remove by normal metal efflux mechanisms. Although recent data point to extracellular dissolution of $PbCrO_4$, or other weakly soluble chromates, as the predominant mechanism for intracellular uptake (Xie et al. 2007), their uptake by phagocytosis, followed by intracellular dissolution (assisted by acidic lysosomes that engulf such particles, Fig. 1) cannot be excluded (Levina et al. 2003). Formation of highly reactive Cr(V), Cr(IV), and organic radical intermediates during the cellular reduction of Cr(VI) (Fig. 1) is crucial for its genotoxic action, irrespective of the actual genotoxic agents, since chromate alone does not damage DNA under physiologically relevant conditions (Levina et al. 2003). In addition, biological reoxidation of Cr(III) to Cr(VI) through similar intermediates is also possible (see below). Details of

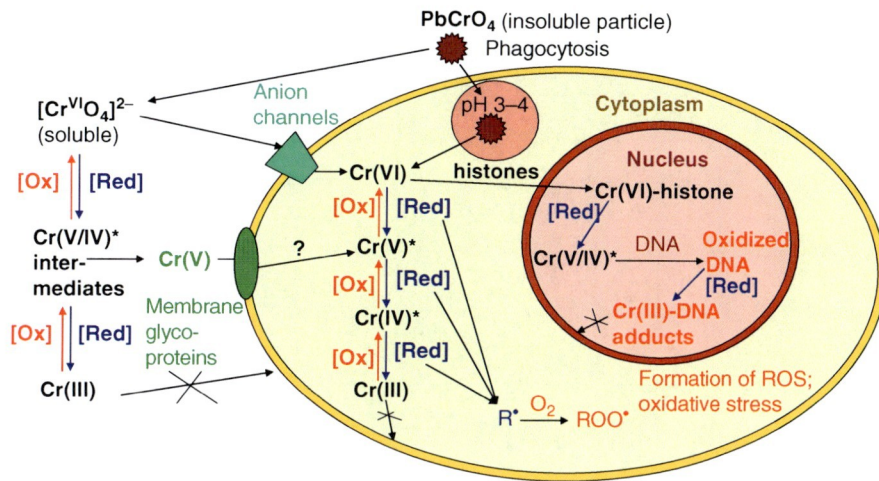

Chromium Toxicity, High-Valent Chromium, Fig. 1 Uptake-reduction model of Cr(VI) metabolism in mammalian cells (based on data reviewed in Levina et al. (2003), Levina and Lay (2008), Aitken et al. (2011))

[Red] = glutathione, ascorbate, catechols, tocopherols, flavonoid enzymes.
[Ox] = oxidase systems (e.g., xanthine + xanthine oxidase + O_2), H_2O_2, ClO^-.

* Reactive Cr(V/IV) intermediates are stabilized by intra- and extracellular ligands, including carbohydrates, glycoproteins, peptides, and 2-hydroxycarboxylates.

biochemical transformations of Cr(VI) within the cell (including the likely structures of reactive intermediates and products that are shown in Fig. 2) were deduced using a combination of electron paramagnetic resonance (EPR) spectroscopy (Levina et al. 2003; Levina et al. 2007), X-ray absorption spectroscopy (XAS) (Levina et al. 2007; Levina and Lay 2004; Aitken et al. 2011), electrospray mass spectrometry (Levina and Lay 2004), and electronic absorption spectroscopy (Levina et al. 2010; Zhitkovich et al. 2001), as well as X-ray fluorescence microscopy (XFM) (Aitken et al. 2011) and kinetic studies that incorporate these techniques, in both cultured cells and model cell-free systems.

Biological thiols, including glutathione, metallothioneins, and cysteine residues of protein tyrosine phosphatases (PTPs), are among the most likely intracellular reactive sites for Cr(VI) (Levina and Lay 2004; Levina et al. 2010). Chromate ($[CrO_4]^{2-}$, **1** in Fig. 2) rapidly reacts with thiols with the formation of five-coordinate Cr(VI) intermediates (**2** in Fig. 2), that have been characterized by XAS (Levina and Lay 2004). Structure **2** is similar to the proposed transition state of phosphate bound to the active sites of PTPs (**3** in Fig. 2, based on X-ray crystallographic studies of a V(V) analog) and is likely to be responsible for the inhibition of PTPs by Cr(VI) (Levina and Lay 2004; Aitken et al. 2011), which can affect the cellular signaling pathways in Cr(VI)-exposed cells. Involvement of thiols in cellular reduction of Cr(VI) is consistent with the EPR signals of thiol-containing Cr(V) species observed shortly after the exposure of cultured cells to Cr(VI) (Levina et al. 2007). Due to the high sensitivity and selectivity of EPR spectroscopy toward Cr(V) species, this method is widely used for the detection and characterization of these highly reactive and potentially DNA-damaging intermediates (Levina et al. 2003). The structures of Cr(V) species corresponding to the two EPR signals observed in Cr(VI)-treated cultured cells or live animals (g_{iso} = 1.986 and 1.979, **4** in Fig. 2; reviewed in (Levina et al. 2003)) were proposed on the basis of detailed EPR spectroscopic and various other kinetic studies of Cr(VI) reactions with model thiols in the presence of Cr(V)-stabilizing ligands (1,2-diols) (Levina et al. 2010).

Efficient stabilization of Cr(V) by ubiquitous biological 1,2-diols, such as carbohydrates (including sugar residues of ribonucleotides, **4a, b** in Fig. 2), sialoglycoproteins (**4c**), or ascorbate (**4d**) (Levina et al. 2003), means that significant amounts of these highly reactive species will persist in cells as long as the supply of Cr(VI) lasts or is regenerated by redox recycling of Cr(III) products (Levina et al. 2010).

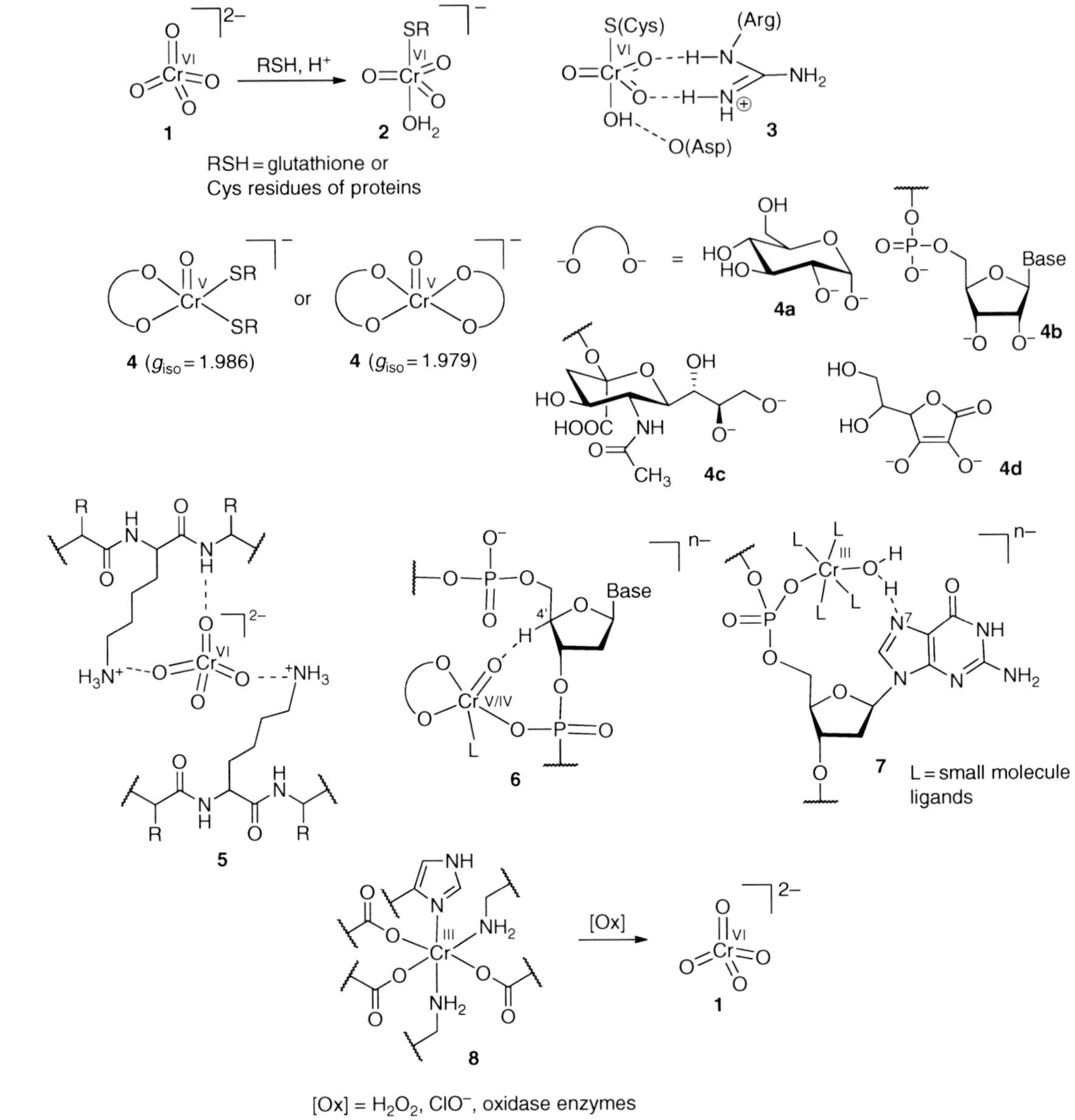

Chromium Toxicity, High-Valent Chromium, Fig. 2 Proposed structures of intermediates and products of cellular Cr(VI) metabolism (see text for details)

Ascorbate, along with glutathione, is among the most likely of the cellular reductants for Cr(VI) (Levina et al. 2003) (see also section ▶ Chromium binding to DNA (Zhitkovich)), so its role in the stabilization of Cr(V) intermediates is particularly notable. Extracellular reduction of Cr(VI) in the vicinity of the cell is likely to lead to the formation of Cr(V) species bound to sialoglycoproteins on the cell surface (Fig. 1) (Levina et al. 2003). Such binding may be particularly important given that alterations of the cell surface and surrounding areas (extracellular matrix) play a crucial role in malignant transformation of bronchial epithelial cells treated with Cr(VI) (Costa et al. 2010).

By contrast with the biospectroscopic EPR studies on Cr(V), there is currently no reliable technique for the detection of small amounts of Cr(IV) intermediates formed during the reduction of Cr(VI) in biological systems (several literature reports on the observation of such species are erroneous) (Levina et al. 2003), apart from multiple-linear regressions of Cr K-edge XAS from biological samples (Levina et al. 2007), which is not sensitive to small amounts of Cr(IV) in the presence of much higher concentrations of other Cr oxidation states. A likely reason for lesser biological significance of Cr(IV) compared with Cr(V) is its inability to form stable complexes with carbohydrate ligands, although stabilization by 2-hydroxycarboxylato and peptide ligands is possible (Levina et al. 2003). Model Cr(IV) 2-hydroxycarboxylato complexes cause oxidative DNA damage to a similar extent as Cr(V) complexes with the same ligands, and evidence was gained for Cr(IV) peptide complexes as being more reactive toward DNA than their Cr(V) analogs (Levina et al. 2003).

A significant proportion of Cr(VI) taken up by cells reaches the nucleus and binds to genomic DNA (Levina et al. 2003; Levina et al. 2007) (in the form of Cr(III), Fig. 1, see section ▶ Chromium binding to DNA (Zhitkovich) for details). By contrast with intact cells, isolated cell nuclei do not take up $[CrO_4]^{2-}$, so an active transport mechanism is likely to exist for the nuclear uptake of Cr in live cells (Levina et al. 2006). One possibility is the electrostatic binding of $[CrO_4]^{2-}$ to the positively charged residues of histones (**5** in Fig. 2, characterized by XAS) (Levina et al. 2006) that are synthesized in the cytoplasm and then transported across the nuclear membrane via the protein transport mechanism (Fig. 1), as is the case with phosphate and sulfate (Levina et al. 2006). Once in the nucleus, Cr(VI) is likely to dissociate from the histone protein due to thermodynamically more favorable histone binding to DNA polyanions during the formation of chromatin (Levina et al. 2006). Reduction of Cr(VI) in the vicinity of DNA can lead to the formation of Cr(V/IV) species bound to the phosphate backbone of DNA (**6** in Fig. 2) (Levina et al. 2003). Structure **6** was proposed based on kinetic and product studies of the reactions of isolated DNA with model Cr(V/IV) complexes (for a review, see Levina et al. 2003). Intramolecular electron transfer in **6**, involving the C4′ atom of the sugar ring (Fig. 2), will lead to the formation of stable DNA-bound Cr(III) species (due to the kinetic inertness of Cr(III)) (Fig. 1) (Levina et al. 2003). Such species can further interact with the N-donors of nucleic bases via linkage isomerization reactions or cross-linking, particularly with the N7 atom of guanine (**7** in Fig. 2), which would lead to distortions in DNA structure and to mutagenicity (Zhitkovich et al. 2001).

The mutagenic potential of the ternary DNA-Cr(III)-L$_n$ adducts, such as **7** (where L are small molecule ligands, such as ascorbate, glutathione, or cysteine), has been demonstrated experimentally (Zhitkovich et al. 2001), but the roles of more bulky DNA-Cr(III)-DNA and DNA-Cr(III)-protein cross-links remain unclear (section ▶ Chromium binding to DNA (Zhitkovich)). An alternative route to Cr(III)-DNA intermediates involves the formation of relatively labile low-molecular-weight Cr(III) species during the reduction of Cr(VI) in the nucleus, followed by their binding to DNA (Levina and Lay 2008; Zhitkovich et al. 2001), but the structures of the final products are likely to be similar. In addition to Cr(III)-DNA binding, oxidative DNA and protein damage by peroxo radicals (ROO•) formed as by-products of cellular Cr(VI) reduction (Fig. 1) is also possible, but it is more likely to be repaired by the existing cellular protection mechanisms (Levina et al. 2003).

Detailed XAS studies of Cr(III) products formed in cultured human lung epithelial cells treated with Cr(VI) (Levina et al. 2007) have shown that a major proportion of the Cr(III) is protein-bound, with a likely average coordination environment represented by the structure **8** (Fig. 2). The XAS results indicated that this environment was independent of whether the cells were ascorbate-deficient or ascorbate-saturated (with glutathione or ascorbate as the main Cr(VI) reductants, respectively, as shown by EPR spectroscopy) (Levina et al. 2007), which means that binding of the primary reductants to final Cr(III) products is insignificant. This work also highlighted the problem of aquation, hydrolysis, and redistribution of Cr(III) (as well as of other metal ions) during the cell lysis and subcellular fractionation, which greatly complicates the biospeciation studies of metal ions (Levina et al. 2007; Aitken et al. 2011). The use of XAS gives a unique opportunity to compare the metal coordination environments in intact cells and tissues with those in subcellular fractions, and thus detect if any such changes take place (Levina et al. 2007; Aitken et al. 2011). Given this information, Cr(III)-biomolecule adducts isolated from Cr(VI)-treated cells can be separated by gel electrophoresis and further analyzed by a combination of XAS and

XFM (Aitken et al. 2011); this work is currently in progress. Another highly promising research direction is the study of spatial distribution of Cr in organelles of single cultured Cr(VI)-treated cells, using XFM with submicron lateral resolution, and the determination of Cr coordination environments in single cells by micro-XAS (Aitken et al. 2011).

Biological oxidation of Cr(III) to Cr(VI) has long been considered highly unlikely, but recent studies (reviewed in Levina and Lay 2008) have shown that this process can occur in blood factions, particularly in the presence of the relatively high concentrations of H_2O_2 and ClO^- produced by macrophages at the sites of inflammation. These results may be crucial to explain the antidiabetic action of some Cr(III) complexes, such as a popular nutritional supplement, Cr(III) picolinate (see sections ▶ Chromium and Nutritional Supplement (Nair) and ▶ Chromium and human nutrition (Stoecker); for the discussion of biological roles of Cr(III)) (Levina and Lay 2008). Ligand-exchange reactions of this compound with blood serum proteins lead to the formation of Cr(III)-protein complexes similar to **8** (characterized by XAS) (Levina and Lay 2008; Aitken et al. 2011), which can be more readily oxidized to Cr(VI) by biologically relevant concentrations of H_2O_2 at pH = 7.4 (Fig. 2; Cr(III) picolinate itself is relatively unreactive toward H_2O_2) (Levina and Lay 2008). Such oxidation is likely to be facilitated in patients and animals with poorly controlled diabetes who suffer from chronic oxidative stress (Levina and Lay 2008; Aitken et al. 2011). The $[CrO_4]^{2-}$ oxidation product is efficiently taken up by cells, while cellular uptake of Cr(III) is several orders of magnitude slower (Fig. 1) (Levina et al. 2003). If the amounts of Cr(VI) product are small, Cr(VI) likely to be reduced by thiol-containing molecules (including PTPs) in the cell membrane and cytoplasm, which can lead to potentiation of the insulin signaling cascade (Levina and Lay 2008; Aitken et al. 2011). On the other hand, prolonged use of high doses of Cr(III)-containing supplements could possibly overwhelm the cellular protective mechanisms and lead to Cr(VI)-induced mutagenicity and/or other toxic effects (Levina and Lay 2008). Recent studies by a combination of XAS and XFM techniques (Aitken et al. 2011) revealed the formation of intracellular Cr(V)- and Cr(VI)-containing hotspots in cultured adipocytes (fat cells, crucial for insulin signaling) treated with another Cr(III) antidiabetic (trinuclear Cr(III) propionate), which means that reoxidation of Cr(III) can also occur within the cells (Fig. 1). In summary, toxicity of Cr(VI) and antidiabetic activity of Cr(III) are likely to arise from the same biochemical mechanisms involving the formation of reactive Cr(V) and Cr(IV) intermediates (Fig. 1), which enhances the current concern over the safety of Cr(III) nutritional supplements (Levina and Lay 2008).

Acknowledgments The authors are grateful for support from two Australian Research Council (ARC) grants to PAL for an ARC Professorial Fellowship (DP0984722) and for an ARC Senior Research Associate position for AL (DP1095310).

References

Aitken JB, Levina A, Lay PA (2011) Studies of the biotransformations and biodistributions of metal-containing drugs using X-ray absorption spectroscopy. Curr Top Med Chem 11:553–571

Costa AN, Moreno V, Prieto MJ, Urbano AM, Alpoim MC (2010) Induction of morphological changes in BEAS-2B human bronchial epithelial cells following chronic subcytotoxic and mildly cytotoxic hexavalent chromium exposures. Mol Carcinog 49:582–591

Levina A, Codd R, Dillon CT, Lay PA (2003) Chromium in biology: toxicology and nutritional aspects. Prog Inorg Chem 51:145–250

Levina A, Harris HH, Lay PA (2006) Binding of chromium(VI) to histones: implications for chromium(VI)-induced genotoxicity. J Biol Inorg Chem 11:225–234

Levina A, Harris HH, Lay PA (2007) X-ray absorption and EPR spectroscopic studies of the biotransformations of chromium (VI) in mammalian cells. Is chromodulin an artifact of isolation methods? J Am Chem Soc 129:1065–1075

Levina A, Lay PA (2004) Solution structures of chromium(VI) complexes with glutathione and model thiols. Inorg Chem 43:324–335

Levina A, Lay PA (2008) Chemical properties and toxicity of chromium(III) nutritional supplements. Chem Res Toxicol 21:563–571

Levina A, Zhang L, Lay PA (2010) Formation and reactivity of chromium(V) thiolato complexes: a model for the intracellular reactions of carcinogenic chromium(VI) with biological thiols. J Am Chem Soc 132:8720–8731

Xie H, Holmes AL, Wise SS, Huang S, Peng C, Wise JP (2007) Neoplastic transformation of human bronchial cells by lead chromate particles. Am J Respir Cell Mol Biol 37:544–552

Zhitkovich A, Song Y, Quievryn G, Voitkun V (2001) Non-oxidative mechanisms are responsible for the induction of mutagenesis by reduction of Cr(VI) with cysteine: role of ternary DNA adducts in Cr(III)-dependent mutagenesis. Biochemistry 40:549–560

Chromium(III) and Immune System

Jerry W. Spears and Shweta Trivedi
Department of Animal Science, North Carolina State University, Raleigh, NC, USA

Synonyms

Chromium; Cytokines; Immune responses; Stress; T & B lymphocytes

Definition

Chromium (III) is an essential trace mineral that potentiates the action of insulin. A number of different chemical forms of chromium have been utilized as dietary chromium supplements. Various studies have reported beneficial effects of supplemental chromium on the immune system function as well as exacerbation of disease pathogenesis. It is well documented that chromium exerts its effects on immune responses either by immunostimulatory or by immunosuppressive mechanisms. These mechanisms have been shown to be mediated by its effects on T and B lymphocytes, antigen-presenting cells like macrophages, or via cytokine production.

Introduction

The common valance states (III and VI) of chromium differ widely in their biological activity. Trivalent chromium is the form of chromium supplemented to humans and animals and is the form that occurs naturally in foods. Chromium (III) is considered an essential trace mineral for mammals and functions by potentiating the action of insulin. In chromium deficiency, glucose metabolism is impaired due to insulin-sensitive tissues becoming resistant to insulin. Chromium (III) is poorly absorbed from the diet, and is relatively nontoxic. In contrast, chromium (VI) is produced almost entirely by industrial processes and is highly toxic to mammalian cells.

Numerous studies in both animals and cultured immune cells have indicated that chromium can affect an array of immune responses. Many studies evaluating the effects of dietary chromium on immunity and disease resistance have involved animals exposed to various types of stress. Stress increases losses of chromium from the body and also impairs immune responses. Reports in the early to mid-1990s indicating that chromium supplementation could increase immunity and reduce incidence of respiratory disease in stressed calves generated considerable interest in the role of chromium in immunity and disease resistance (Spears 2000). The mechanism(s) whereby chromium alters immune responses is unclear. Chromium effects on immune responses may relate to increased insulin sensitivity by immune cells or the ability of chromium to minimize physiological responses to stress.

Immune responses to chromium supplementation have been variable with some studies showing no response to chromium supplementation. The control diet fed in some studies may have contained adequate bioavailable chromium prior to chromium supplementation. Supplemental chromium sources can differ in their absorption and postabsorptive utilization. A number of different chromium sources including chromium chloride, chromium picolinate, chromium nicotinate, chromium yeast, chromium propionate, chromium amino acid chelate, and chromium nanoparticles have been studied in regard to their ability to modulate immune processes. All of these chromium sources have enhanced immune response in one or more studies. In studies that have compared different chromium sources, organic chromium sources have generally been more effective than inorganic chromium chloride in enhancing immune response.

Overview of the Immune System

A brief overview of the immune system will be presented before discussing the effects of chromium on immunity. The immune system protects host via a two-tiered defense system that involves innate and acquired or adaptive immune responses. Innate immunity is the first line of defense against foreign pathogens and includes physical barriers such as skin, cellular mediators (complement, cytokines), and phagocytic cells (neutrophils, monocytes, and macrophages) that are capable of ingesting and killing pathogenic microorganisms by complement fixation or reactive oxygen and reactive nitrogen species. Neutrophils and monocytes function in blood, while macrophages function in tissues to engulf and subsequently destroy pathogens. The innate immune system is nonspecific for a particular pathogen and functions to prevent the entry of foreign pathogens into the body, and to attempt to rapidly eliminate organisms that do enter the body. This system does not confer long-term immunity against a given pathogen.

The acquired immune system involves T and B lymphocytes which exhibit a high degree of specificity against a particular antigen. An antigen is a foreign molecule/substance present on pathogens or other foreign invaders that causes activation and proliferation of lymphocytes, resulting in the production of cytokines or antibodies. T lymphocytes are involved in providing cell-mediated immunity, while B lymphocytes generate humoral immunity.

Cell-mediated immunity is conferred by T lymphocytes and is involved in destruction of intracellular pathogens and tumor cells, assisting with activating and directing other immune cells and antibody class switch. T cell receptors bind to fragments of antigen presented to them by macrophages or immune cells, and also to antigen fragments on the surface of infected body cells.

Humoral immunity results from the ability of B lymphocytes to proliferate and produce antibodies against a specific antigen. Antibodies produced by B lymphocytes neutralize pathogens by binding them and preventing their attachment to host cells. Antibodies are also involved in activating complement proteins that enhance the destruction of bacteria by phagocytic cells. Humoral immunity is an important defense mechanism against extracellular pathogens and their secreted toxins.

The innate immune response occurs rapidly, while the acquired immune response is slower to develop after activation by a foreign substance but is of sustained duration. The persistence or sustained duration of the acquired immune response results in immunological memory, resulting in an enhanced immune response on reexposure to the same pathogen or antigen.

A functional immune system is dependent on communication among cells within the acquired immune system and also between cells in the innate and acquired immune system. For example, macrophages process antigen and present it to effector cells in the cell-mediated immune system. B lymphocytes also require T helper cell help to start producing antibodies. A variety of immune cells also produce small soluble cell-signaling proteins called cytokines that promote inter- and intracellular communication among various immune cell types, resulting in immunomodulation. Interleukin 1 (IL-1), IL-6, and tumor necrosis factor α (TNF-α) are important cytokines produced by monocytes and macrophages. These cytokines stimulate T and B lymphocyte proliferation, activate phagocytic cells to kill microorganisms, and initiate production of a number of other cytokines. However, these cytokines are also responsible for many of the clinical signs of infectious disease including fever, anorexia, and inflammation as their effect is not only local but also systemic in nature.

Chromium and Immunity

The effect of chromium supplementation on adaptive immune responses has been explored utilizing various serologic- and immune-based assays. Several important studies are discussed here in an attempt to understand the role of chromium in potentiating humoral (B lymphocyte) and cell-mediated (T lymphocyte) adaptive immune responses against viral or bacterial pathogens.

Chromium supplementation to animal diets has increased humoral immune responses in a number of studies. Humoral immune response has been assessed in these studies by measuring antibodies produced following vaccination against viral and bacterial organisms or after injection of a foreign protein. In chicks, chromium has increased antibody responses following immunization against Newcastle virus and influenza virus. Chromium supplementation has increased antibody responses following vaccination against tetanus toxoid and infectious bovine rhinotracheitis virus in cattle. Chromium has also increased humoral immune response following injection of pigs with sheep red blood cells and cattle with ovalbumin or human red blood cells.

The ability of lymphocytes, isolated from blood or spleen of animals, to proliferate in vitro following mitogen stimulation has been used to evaluate the effect of dietary chromium on cell-mediated immune response. Mitogens stimulate DNA synthesis and induce lymphocyte transformation and cell division by mitosis. The plant-derived mitogens concanavalin A (ConA) and phytohemagglutinin (PHA) stimulate

T lymphocytes, while pokeweed mitogen stimulates T and B lymphocytes. Lipopolysaccharide from the cell wall of gram-negative bacteria is a mitogen that stimulates only B lymphocytes. Blood lymphocytes isolated from cattle supplemented with 0.4–0.5 mg chromium/kg of diet had greater proliferation responses to T cell mitogens than lymphocytes obtained from cattle not supplemented with chromium. Addition of blood serum from chromium-supplemented cattle to lymphocytes isolated from control cattle also increased ConA–induced lymphocyte proliferation. In pigs, the addition of 0.2 mg chromium/kg diet enhanced pokeweed mitogen–induced proliferation. Chromium supplementation, from chromium nanoparticles, at levels of 0.15–0.45 mg/kg diet increased responsiveness of T and B lymphocytes, isolated from blood and spleen, to mitogen stimulation in rats exposed to heat stress (Zha et al. 2009).

Adding chromium directly to lymphocyte cultures has also increased T lymphocyte proliferation from nonstimulated and ConA–stimulated cells. Chromium addition (from chromium chloride) at a concentration of 0.0045 µg/ml of culture medium increased proliferation of blood lymphocytes from cattle not receiving supplemental chromium. Adding a combination of chromium (0.005 µg/ml) and insulin (0.05 or 0.5 ng/ml) to culture medium also increased ConA–induced proliferation of blood lymphocytes from cattle. Fetal calf serum is added to medium used to grow lymphocytes in culture. However, when lymphocytes are grown in serum-free media, insulin addition stimulates mitogen-induced proliferation.

Glucose is the major energy source used by immune cells. Energy requirements of resting (nonactivated) immune cells are very low. Activation of immune cells enhances cell proliferation by a factor of several magnitudes, thus greatly increasing their glucose demand. Insulin receptors are found on the surface of resting and activated monocytes and B lymphocytes. T lymphocytes do not have insulin receptors in a resting state, but activation of T lymphocytes causes development of insulin receptors, especially in cells exposed to high glucose concentrations (Stentz and Kitabchi 2005). Glucose uptake by cells occurs via glucose transporters (GLUT). Glucose transporter 1 (GLUT 1) is not affected by insulin, while GLUT 4 and GLUT 3 are insulin responsive. Human monocytes and T and B lymphocytes express GLUT 1, GLUT 3, and GLUT 4, and expression of all three glucose transporters is increased when cells are activated (Maratou et al. 2007). Insulin addition to culture medium increases expression of GLUT 3 and GLUT 4 in monocytes and B lymphocytes, both in a resting and activated state. Insulin increases expression of GLUT 4 in activated T lymphocytes but not in resting T lymphocytes. Increased expression of GLUT 4 and GLUT 3 by insulin is consistent with a role for insulin in enhancing glucose uptake by activated immune cells when energy demands are high for cell proliferation. Chromium may increase lymphocyte proliferation and antibody production by increasing responsiveness of immune cells to insulin.

The effect of dietary chromium on cell-mediated immunity has also been assessed in vivo using intradermal injection of PHA (Kegley et al. 1996). Following administration of PHA, swelling response at the injection site is measured over 24–48 h as an indicator of the inflammatory response. Chromium supplementation has increased inflammatory response following intradermal administration of PHA in a number of animal species.

From the studies discussed here, we can conclude that chromium appears to assist the immune system by increasing antibody production against foreign agents and by driving the proliferation and cytokine production by T lymphocytes. Chromium may increase proliferation of T and B lymphocytes by enhancing their responsiveness to insulin.

Chromium and Cytokine Production

Chromium has been shown to alter cytokine production both in vivo and in vitro. Mononuclear cells isolated from cattle supplemented with 0.5 mg chromium/kg of diet had lower production of TNF-α, IL-2, and IFN-γ following stimulation with ConA than cells from nonchromium-supplemented animals (Burton et al. 1996).

Addition of 100 µM of chromium (from chromium chloride) to activated monocytes from a human cell line reduced TNF-α secretion. Exposure of monocyte cultures to high glucose concentrations results in increased secretion of proinflammatory cytokines, such as TNF-α, and increases oxidative stress. Oxidative stress is caused due to an imbalance in the production and removal of the reactive oxygen species which leads to protein, lipid, and DNA damage within the host. Activated monocytes, neutrophils, and macrophages produce high quantities of reactive oxygen species during the destruction of pathogens. Oxidative stress is usually assessed by measuring malondialdehyde, an end product of lipid peroxidation. Jain et al. (2007a) compared the effects of different chromium sources and levels on oxidative stress and secretion of IL-6 and IL-8 from monocytes exposed to high glucose concentration. Chromium chloride, chromium picolinate, and chromium niacinate were evaluated at chromium concentrations of 0.5, 1, and 10 μM. While all concentrations of chromium reduced IL-6 and IL-8 secretion, the 10 μM level was generally most effective. Chromium niacinate was also more effective than chromium chloride or picolinate in reducing IL-6 and IL-8 secretion in this study. Oxidative stress of monocytes exposed to high glucose was reduced by chromium chloride and chromium niacinate but not by chromium picolinate. Concentrations of chromium used in monocyte culture studies probably exceed physiological concentrations. Chromium is poorly absorbed and is usually found in blood and tissues at nM concentrations.

Monocytes and other immune cells can be exposed to elevated glucose concentrations in humans with diabetes. Diabetes is often associated with elevated circulating levels of proinflammatory cytokines and increased oxidative stress. Proinflammatory cytokines and oxidative stress are markers of vascular inflammation in diabetic patients. Diabetic rats have higher plasma levels of TNF-α, IL-6, and malondialdehyde than normal rats. Streptozotocin administration was used to induce diabetes. Supplementation of 400 μg of chromium/kg of body weight from either chromium niacinate or chromium picolinate for 7 weeks prevented the elevation of plasma TNF-α, IL-6, and malondialdehyde in diabetic rats (Jain et al. 2007b). The level of chromium supplemented in this study would be considered pharmacological relative to doses of chromium that would be supplemented to humans or animals.

Lipopolysaccharide (E. coli) injection has been used to study the effect of dietary chromium on release of proinflammatory cytokines in pigs. Lipopolysaccharide stimulates mononuclear phagocytes, causing a large elevation of TNF-α concentration within 2 h post–LPS dosing. The increase in TNF-α rapidly stimulates mononuclear cells to produce and release IL-1 and IL-6 into the blood. In pigs injected with a low dose of LPS (20 μg/kg body weight) intravenously, chromium picolinate supplementation (0.3 mg chromium/kg diet) reduced plasma TNF-α concentration at 1 and 2 h postinjection. Plasma IL-6 concentrations were not affected by chromium in this study. Injection of high doses of LPS can cause death due to septic shock which results from extremely high production of TNF-α and other proinflammatory cytokines. Supplementing 0.4 mg chromium/kg diet (from chromium propionate) reduced mortality in pigs injected intravenously with 200 μg LPS/kg body weight from 83% to 42%. This suggests a protective effect of chromium against septic shock.

Chromium supplementation at 0.4 or 0.8 mg/kg diet reduced liver malondialdehyde and plasma malondialdehyde, TNF-α, and IL-6 concentrations in Japanese quails housed at thermoneutral (22°C) or heat stress (34°C) environmental temperatures (Sahin et al. 2010). Responses to chromium were greater in heat-stressed quails. Reduced levels of proinflammatory cytokines in chromium-supplemented animals could indicate that chromium curtails the subsequent inflammatory signaling cascades that might occur. IFN-γ mRNA expression was enhanced at 1 and 3 days following vaccination against Newcastle disease virus in chicks supplemented with 0.5 mg chromium/kg diet. Supplementation with higher concentrations of chromium (1.0 or 1.5 mg/kg diet) downregulated IFN-γ mRNA compared to nonchromium-supplemented chicks. IFN-γ is the key cytokine that interferes with viral replication, and chromium may assist with initiating the antiviral cytokine responses following vaccinations.

Interaction Between Chromium, Stress, and Immunity

Humans and domestic animals are exposed to various stressors during their lifetime. Acute stress results in the release of adrenocorticotropin hormone (ACTH) from the pituitary gland which subsequently increases release of cortisol from the adrenal gland. Common stressors in humans include exercise, emotional stress, trauma, and surgical procedures. Stressors that animals may encounter include heat, cold, weaning from their dams, and shipping from one location to another. In humans, stress associated with physical trauma, acute exercise, and lactation increases urinary losses of chromium and may lead to chromium deficiency (Anderson 1994). A positive correlation exists between serum cortisol concentrations following exercise stress and urinary chromium excretion in humans.

Corticosteroid treatment in humans also increases urinary chromium excretion, and corticosteroid treatment can lead to impaired glucose tolerance due to insulin resistance. Chromium supplementation has been reported to improve glucose tolerance in humans with steroid-induced diabetes. This suggests that chromium deficiency is at least partially responsible for insulin resistance observed in steroid-induced diabetes.

It is well documented that cortisol and other glucocorticoids suppress a variety of immune responses. Depletion of chromium due to increased urinary losses during stress may further compromise immune responses. Studies with calves that have been stressed due to weaning from their dams, shipping to a new environment, and feed restriction during transportation indicate that chromium supplementation can increase immunity and health. The incidence of respiratory disease is frequently high in stressed calves. In calves stressed by weaning and shipping, chromium supplementation at 0.2–0.5 mg/kg diet has reduced morbidity in some studies (Spears 2000). Calves supplemented with 0.4 mg/kg diet (as chromium nicotinate or chromium chloride) tended to have lower body temperature at certain time points after intranasal inoculation with infectious bovine rhinotracheitis virus followed by intratracheal inoculation with *Pasteurella haemolytica* 5 days later (Kegley et al. 1996). In guinea pigs, chromium supplementation reduced mortality during pregnancy. Chromium addition to diets of laying hens has also reduced mortality rate.

Providing adequate dietary chromium can reduce blood cortisol concentrations during stress. In calves stressed via weaning and shipping, chromium supplementation has decreased serum cortisol in some studies. Following a respiratory disease challenge, calves supplemented with chromium had lower serum cortisol concentrations than control calves (Kegley et al. 1996). Chromium supplementation has also reduced circulating cortisol concentrations in heat-stressed rats and LPS-challenged pigs.

Cross-References

▶ Chromium and Human Nutrition
▶ Chromium and Insulin Signaling
▶ Chromium and Nutritional Supplement
▶ Chromium(III), Cytokines, and Hormones

References

Anderson RA (1994) Stress effects on chromium nutrition of humans and farm animals. In: Lyons TP, Jacques KA (eds) Biotechnology in the feed industry, Proceedings of the Alltech 10th symposium, University Press, Nottingham

Burton JL, Nonnecke BJ, Dubeski PL, Elsasser TH, Mallard BA (1996) Effects of supplemental chromium on production of cytokines by mitogen-stimulated bovine peripheral blood mononuclear cells. J Dairy Sci 79:2237–2246

Jain SK, Rains JL, Croad JL (2007a) High glucose and ketosis (acetoacetate) increases, and chromium niacinate decreases, IL-6, IL-8, and MCP-1 secretion and oxidative stress in U937 monocytes. Antioxid Redox Signal 9:1581–1590

Jain SK, Rains JL, Croad JL (2007b) Effect of chromium niacinate and chromium picolinate supplementation on lipid peroxidation, TNF-α, IL-6, CRP, glycated hemoglobin, triglycerides, and cholesterol levels in blood of streptozotocin-treated diabetic rats. Free Radic Biol Med 43:1124–1131

Kegley EB, Spears JW, Brown TT (1996) Immune response and disease resistance of calves fed chromium nicotinic acid complex or chromium chloride. J Dairy Sci 79:1278–1283

Maratou E, Dimitriadis G, Kollias A, Boutati E, Lambadiari MP, Raptis SA (2007) Glucose transporter expression on the plasma membrane of resting and activated white blood cells. Eur J Clin Invest 37:282–290

Sahin N, Akdemir F, Tuzcu M, Hayirli A, Smith MO, Sahin K (2010) Effects of supplemental chromium sources and levels on performance, lipid peroxidation and proinflammatory markers in heat-stressed quails. Anim Feed Sci Technol 159:143–149

Spears JW (2000) Micronutrients and immune function in cattle. Proc Nutr Soc 59:587–594

Stentz FB, Kitabchi AE (2005) Hyperglycemia-induced activation of human T-lymphocytes with de novo emergence of insulin receptors and generation of reactive oxygen species. Biochem Biophys Res Commun 335:491–495

Zha L, Zeng J, Sun S, Deng H, Luo H, Li W (2009) Chromium (III) nanoparticles affect hormone and immune responses in heat-stressed rats. Biol Trace Elem Res 129:157–169

Chromium(III) and Low Molecular Weight Peptides

John B. Vincent
Department of Chemistry, The University of Alabama, Tuscaloosa, AL, USA

Synonyms

Chromodulin; Low-molecular-weight chromium-binding substance (LMWCr)

Definition

Low-molecular-weight chromium-binding substance is a carboxylate-rich peptide isolated from the urine and tissues of higher animals that tightly binds Cr^{3+} ions. The peptide has been suggested to have a role in chromium detoxification and to potentially have a role in improving insulin sensitivity when animals are supplemented with high doses of chromium(III) compounds.

Chromium is absorbed by passive diffusion and then binds to the iron-transport protein in the blood. Transferrin delivers chromium to the tissues where it binds to a low-molecular-weight organic species that is cleared from the tissues. The chromium is ultimately eliminated from the body in the urine as a low-molecular-weight organic complex. This organic complex is called low-molecular-weight chromium-binding substance (LMWCr) and has also been coined chromodulin (Vincent and Stearns 2011).

Discovery of LMWCr

LMWCr was first reported by the toxicology group of Osamu Wada in 1981 (Yamamoto et al. 1981), who investigated the compound for about a decade (Vincent and Stearns 2011). Although other higher molecular weight species that bound chromium were present, a low-molecular-weight chromium compound(s) was identified by size exclusion chromatography of the cytosol of liver cells of male mice injected with a single dose of potassium dichromate (50 or 200 μg/kg body mass). In contrast to the higher molecular weight molecules that bound chromium, LMWCr could be detected for 7 days. A similar low-molecular-weight compound was found in the feces and urine and 2 h after injection in the plasma. The researchers suggested that a low-molecular-weight chromium-binding substance (LMWCr) was formed in the liver and participates in retention and excretion of chromium in the body. The material from the livers of rabbits treated similarly with dichromate (200 μg/kg body mass) was partially purified and found to apparently be an anionic organic-chromium complex containing amino acids with a molecular weight under 3,000 and a component that absorbed ultraviolet light at 260 nm (Yamamoto et al. 1981).

Subsequently, LMWCr was found to occur in urine normally, although the amounts were greatly increased after rats were injected with chromate (Wu and Wada 1981). Normal human and rat urine LMWCr was found not be saturated with chromium; addition of Cr^{3+} to urine would result in increased amounts of the chromium-containing species that eluted in a particular band upon size-exclusion chromatography. The LMWCr was believed to be similar to that of the liver and other organs of rabbits and dogs and to be involved in removing excess chromium from the body. Wada and coworkers (Yamamoto et al. 1984) followed the initial studies by examining the distribution of LMWCr. LMWCr was found in liver, kidney, spleen, intestine, testicle, brain, and blood plasma, with the greatest amount in liver followed by kidney. The organs were obtained from mice 2 h after injection with potassium dichromate. Supernatants of homogenates of the organs were found to possess more chromium bound to LMWCr when dichromate was added to the homogenate than when the mice were injected with dichromate. The time course of chromium binding to LMWCr after injection of dichromate was also examined. Chromium was found to be associated with liver and kidney LMWCr in just 2 min after injection and reached a maximum of 1–2 h after treatment. Repetitive treatments of mice with dichromate (150 μmol/kg daily for each of 4 days) had no effect on LMWCr levels in the liver; thus, no induction of formation of LMWCr was observed. LMWCr had greater affinity for chromium than transferrin or albumin. LMWCr was proposed to play a role in chromium detoxification.

In these studies, LMWCr was identified by its elution behavior in size exclusion chromatography and its Cr-binding ability. In other words, to identify LMWCr, chromium had to be added to the animal, tissue homogenate, or body fluid, and then the fraction of appropriate molecular from designated solution has to be separated by size exclusion chromatography. Thus, LMWCr refers to a low-molecular-weight organic species containing amino acids and able to bind chromium. These initial studies also leave questions as to whether LMWCr is a single species as no single well-characterized species was isolated, purified, and characterized, bringing back memories of the artifact first proposed as the biologically active form of chromium, glucose tolerance factor (Vincent and Stearns 2011).

Isolation and Characterization of LMWCr

Organic Composition

Efforts to isolate and characterize LMWCr continued. LMWCr has been isolated and purified from alligator liver, chicken liver, rabbit liver, bovine liver, porcine kidney, and porcine kidney powder and partially purified from dog and mouse liver. Inclusion of protease inhibitors in buffers during the isolation of bovine liver LMWCr does not affect the amount of oligopeptide isolated, suggesting it is not a proteolytic artifact generated during the isolation procedure. The materials from rabbit and dog liver were loaded with Cr by injection of the animal with chromate (or Cr(III) which provides lower yields). For the other tissues, chromate was added to the homogenized liver or kidney or suspended kidney powder. Cr(III) could also be added to the bovine liver homogenate to load LMWCr with chromium, but the loading was not as efficient as when chromate was utilized. A Cr-loading procedure is required so that the material can be followed by its chromium content during the isolation and purification procedures. The isolation procedures are similar involving an ethanol precipitation, anion exchange chromatography, and finally, size exclusion chromatography (Vincent and Stearns 2011).

The first purification was from rabbit liver by Wada and coworkers (Yamamoto et al. 1987). LMWCr was found to be an anionic organic chromium compound with an approximate molecular weight of 1,500 and to be composed of a peptide of glutamate (and/or glutamine), cysteine, glycine, and aspartate (and/or asparagine) and chromium (four Cr: amino terminus). Interestingly, the addition of purified LMWCr to isolated rat adipocytes in the presence of insulin (5 μU/mL) stimulated the conversion of glucose into carbon dioxide and the incorporation of hydrogen from glucose into lipids in a concentration-dependent manner. Increased incorporation of hydrogen into lipids also occurred in a concentration-dependent fashion in the absence of insulin, although to a lesser degree. Activation required LMWCr in Cr concentrations of 0.2–6 μM (Yamamoto et al. 1987). The biological activity is dependent on the chromium content. Removal of chromium decreases the production of glucose and incorporation of hydrogen from glucose into lipids, while most of the activity can be restored by addition of chromium (Yamamoto et al. 1989).

The amino acid composition of the LMWCr from mammalian sources is presented in Table 1. From the table, LMWCr appears to be an oligopeptide composed of glycine, cysteine, aspartate, and glutamate with the carboxylates comprising more than half of the total amino acid residues (Table 1). The amino acid composition data for the rabbit liver LMWCr (injected with dichromate) and bovine liver (dichromate added to homogenate) are extremely similar, indicating that the type of Cr-loading procedure utilized is probably not critical to the composition of the isolated material. Amino acid sequence data has only appeared most recently. Treatment of LMWCr from bovine liver, human urine, chicken liver, and alligator liver with trifluoroacetic acid in an attempt to remove the chromium in a more gentle fashion than used previously resulted in the production of a heptapeptide fragment of LMWCr with a mass/charge ratio of 802 as determined by matrix-assisted laser desorption ionization time-of-flight (MALDI-TOF) and electrospray ionization (ESI) mass spectrometry (MS) (Vincent and Stearns 2011). Postsource decay (POD) MALDI-TOF MS suggested the sequence was either pEEEEGDD (where pE is pyroglutamate) or pEEEGEDD. Use of collision-induced dissociation (CID) MS/MS or MS/MS/MS experiments allowed the sequence of the heptapeptide from the treatment of the bovine liver LMWCr to be confirmed as pEEEEGDD by comparison of spectra with those of synthetic peptides of the proposed sequences. The composition of the isolated heptapeptide was also confirmed by amino acid analysis. The treatment with trifluoroacetic acid in addition to removing the chromium cyclized the terminal glutamate residue while also dissociating some cysteine and glycine from the isolated LMWCr. The sequence of the heptapeptide explains why LMWCr could not be sequenced by Edman degradation as exposed N-terminal glutamates tend to cyclize under the Edman conditions; previous efforts at Edman sequencing had observed a low yield of glutamate at the N-terminus and essentially no yield at subsequent steps. Searches of sequence databases reveal a small number of possible sources for the heptapeptide; however, none have cysteine and glycine residues adjacent to the sequence. The ability of cysteine and glycine to be cleaved by the acid treatment and the lack of an appropriate sequence suggest these residues are

Chromium(III) and Low Molecular Weight Peptides, Table 1 Amino acid composition data for isolated LMWCr

Source	Glycine	Glutamic acid	Aspartic acid	Cysteine
Rabbit liver	3.22	3.91	1.98	1.75
Bovine liver	2.47	4.47	2.15	2.19
Bovine colostrum	1.98	5.0	4.12	0.93
Porcine kidney	1.45	4.05	2.31	0.622

associated with the heptapeptide through a nonstandard linkage (Vincent and Stearns 2011). Unfortunately to date, LMWCr has not proven to be particularly antigenic, preventing its presence to be detected using immunological techniques. Antibodies raised against the heptapeptide will recognize bovine liver LMWCr; however, the binding of the antibodies is not sufficient for use in techniques such as Western blotting (Vincent and Stearns 2011).

Three laboratories have reported not being able to isolate bovine liver LMWCr. One made the statement as a footnote in an article; one reported isolating instead a material they identified as containing Cr(IV) or Cr(V). Much of the work in this paper has been refuted, and the existence of such a Cr(IV) or Cr(V) species with the stability necessary to be isolated under these conditions is difficult to fathom. Another observed at least four species that bound chromium and isolated an approximately 15.6-kDa protein. As would only be expected by adding chromate to a liver homogenate, these researchers suggested that chromium binding was nonspecific. The other three Cr-containing species were not characterized (Vincent and Stearns 2011).

Metal-Binding Sites

Despite its small size (approximately 1,500 molecular weight; 1438 by MALDI-TOF mass spectrometry for bovine liver LMWCr), the molecule from rabbit and bovine liver tightly binds four equivalents of chromic ions. The binding is quite tight (K_a approximately 10^{21} M^{-4} for bovine liver LMWCr) and highly cooperative (Hill coefficient, $n = 3.47$); thus, the large value of the Hill coefficient indicates essentially only apoLMWCr and holoLMWCr (Cr_4-LMWCr) coexist in solution. The binding of chromium to synthetic heptapeptide pEEEEGDD is tight and cooperative, with four chromic ions binding with an essentially identical binding constant and Hill coefficient (Vincent and Stearns 2011).

Spectroscopic and magnetic studies suggest that the chromic ions comprise an anion-bridged multinuclear assembly supported by carboxylates from the oligopeptide. Electronic spectroscopic studies reveal that the Cr bound to LMWCr exists in the trivalent oxidation state; for the bovine liver material, 10Dq and the Racah parameter B were found to be 1.74×10^3 and 847 cm^{-1}, respectively, an indication of predominately oxygen-based coordination. In the ultraviolet region, the spectra of LMWCr possess a maximum or shoulder at approximately 260 nm; this feature may arise from a disulfide linkage. Paramagnetic ^{1}H NMR spectroscopy of the bovine liver reveals a downfield-shifted resonance at approximately +45 ppm, suggestive of the protons of a methylene carbon bound adjacent to a carboxylate bridging two chromic centers. The presence of a bridging ligand suggests the existence of a multinuclear assembly. Charge balance also suggests the existence of an anion-bridged multinuclear assembly. Four Cr^{3+} have a combined charge of +12, while holoLMWCr is anionic. The organic components cannot provide sufficient negative charge to compensate, indicating the need for additional anionic components. Also, the four chromium(III) centers require six coordination. How the organic composition could fill these 24 sites is difficult to envision (Davis and Vincent 1997b).

X-ray absorption spectroscopic studies on the bovine liver LMWCr have shown that the chromium atoms are surrounded by six oxygen atoms at an average distance of 1.98 Å and are consistent with a lack of sulfur-based ligands. A long Cr...Cr interaction at approximately 3.79 Å is present, and another such interaction may be present at 2.79 Å. This is also consistent with the presence of a multinuclear chromium assembly. As holoLMWCr can be prepared simply by addition of chromic ions to aqueous solutions of apoLMWCr, anionic bridges to the chromium assembly are probably hydroxide ions; the X-ray absorption studies failed to detect any short Cr-oxo interactions. The nature of the assembly has been narrowed to a few possibilities from electron paramagnetic resonance (EPR) spectroscopy to variable temperature magnetic susceptibility measurements. X-band EPR studies indicate that at least three chromic ions are coupled to give a species with an $S = 1/2$ ground state giving rise to a broad signal at $g \sim 2$; this signal appears to be broadened by interaction with another chromium species, giving rise to a complex EPR signal centered about g ~ 5. The g ~ 2 EPR signal sharpens as the temperature is raised from 5 to 30 K, suggesting that dipolar coupling exists between the two chromium species giving rise to the EPR signal. Finally, magnetic susceptibility studies are consistent with the presence of a mononuclear chromic center and an unsymmetric trinuclear chromic assembly (Jacquamet et al. 2003).

Put together, the spectroscopic and magnetic data on bovine LMWCr suggest the occurrence of a Cr_4 assembly in LMWCr. The chromium environment is mostly, if not exclusively, composed of O atoms, and the assembly is comprised of a single chromic ion and a trinuclear unit. Additionally, the sulfur atoms of the two cysteine residues of LMWCr appear to be involved in a disulfide linkage and not to be involved in binding chromium. Similarly, the N-terminal amine group can be derivatized, suggested it is not coordinated to chromium. Thus, oligopeptide-provided ligands appear to be limited to carboxylates from the side chains of the aspartate and glutamate residues and possibly the carboxy terminus.

Function

As described above, insulin dose–response studies using rat adipocytes have indicated a potential intrinsic biological function for LMWCr. Isolated rat adipocytes in the presence of LMWCr and insulin display an increased ability to metabolize glucose to produce carbon dioxide or total lipids; this increase occurs without a change in the insulin concentration required for half-maximal stimulation. This lack of change in half-maximal insulin concentration suggests a role for LMWCr inside the insulin-sensitive cells after insulin binds externally to the insulin receptor. The stimulation of glucose metabolism by LMWCr is proportional to the chromium content of the oligopeptide.

Because the primary events between insulin binding to its receptor and glucose transport are signal transduction events, a role for LMWCr in these events has been probed. LMWCr has been shown to activate the tyrosine kinase activity of insulin-activated insulin receptor and to activate a membrane phosphotyrosine phosphatase in adipocyte membranes. For example, the addition of bovine liver LMWCr to rat adipocytic membranes in the presence of 100-nM insulin results in a concentration dependent on up to eightfold stimulation of insulin-dependent protein tyrosine kinase activity, while no activation of kinase activity is observed in the absence of insulin. The dependence of the kinase activation on the concentration of LMWCr can be fit to a hyperbolic curve to give dissociation constants (K_m) of approximately 875 pM, indicating extremely tight binding. Blocking the insulin-binding site on the external β subunit with antibodies whose epitope lies in this region results in the loss of the ability to activate insulin receptor kinase activity. Examining the potential activation of isolated rat insulin receptor by bovine liver LMWCr in the presence of insulin indicates that LMWCr can amplify the isolated receptor protein tyrosine kinase activity by approximately sevenfold with an apparent dissociation constant of approximately 250 pM, suggesting that the receptor is the site of interaction with LMWCr (Davis and Vincent 1997a). The site of LMWCr binding on insulin receptor can be further refined. Studies with a catalytically active fragment (residues 941–1,343) of the β-subunit of human insulin receptor (which does not require insulin for kinase activity) reveal that LMWCr can stimulate kinase activity threefold with a dissociation constant. Thus, LMWCr apparently binds at or near the kinase active site.

As noted above, chromium plays a crucial role in the in vitro activation of insulin receptor kinase activity by LMWCr. ApoLMWCr is unable to activate insulin-dependent tyrosine kinase activity in the rat adipocyte membranes. Titration of apoLMWCr with Cr^{3+} results in the restoration of the enhancement of kinase activity; approximately four Cr^{3+} per oligopeptide are required for maximal activity, consistent with the number of chromium (four per oligopeptide) reported to be bound to holoLMWCr from liver sources. This reconstitution is specific to chromium. Transition metal ions other than chromium which are commonly associated with biological systems (V, Mn, Fe, Co, Ni, Cu, Zn, and Mo) are ineffective in potentiating the ability of apoLMWCr to activate kinase activity. In fact, all the ions except Cr^{3+} resulted in loss of activation potential relative to apoLMWCr (Davis and Vincent 1997a). Thus, the ability of LMWCr to potentiate the effects of insulin in stimulating the insulin-dependent protein tyrosine kinase activity of insulin receptor is specific to chromium and is directly dependent on the chromium content of LMWCr.

The activation of a membrane-associated phosphotyrosine protein phosphatase (PTP) by LMWCr has been little explored. Studies with isolated LAR and PTP1B have shown that chromodulin has no effect on these phosphatases (Vincent and Stearns 2011).

Based on these results, LMWCr has been proposed to function as part of a unique autoamplification system for insulin signaling (Fig. 1) and a new (and shorter) name, chromodulin, has been put forward (Vincent 2000). In this mechanism, apoLMWCr is stored in insulin-sensitive cells. In response to increases in blood insulin concentrations (as would result from increasing blood sugar concentrations after a meal), insulin binds to its receptor, bringing about a conformation change which results in the autophosphorylation of tyrosine residues on the internal side of the receptor. This transforms the receptor into an active tyrosine kinase and transmits the signal from insulin into the cell. In response to insulin, chromium is moved from the blood where it is maintained in the iron transport protein transferrin to insulin-sensitive cells. Here, the chromium flux results in the loading of apoLMWCr with chromium. The holoLMWCr then binds to the receptor, presumably assisting to maintain the receptor in its active conformation, amplifying its kinase activity. When the signaling is to be turned off, a drop in blood insulin levels facilitates relaxation of the conformation of the receptor, and the holoLMWCr is excreted from the cell into the blood. Ultimately, LMWCr is efficiently excreted in the urine. The basis of the alternative name chromodulin is the similarity of the proposed mechanism of action to that of the calcium-binding protein calmodulin. Both bind four equivalents of metal ions in response to a metal ion flux; however, the four calcium ions which bind to the larger protein calmodulin rest in mononuclear sites. Both holoproteins selectively bind to kinases and phosphatases, stimulating their activity (Vincent 2000).

Noting that all the studies on LMWCr and insulin receptor are in vitro studies is important. This proposed mechanism of action of chromium needs to be supported by in vivo studies. In other words, does LMWCr bind to insulin receptor in vivo and is insulin receptor kinase activity affected in vivo by interaction directly with LMWCr in vivo? As recent studies indicate that chromium is probably not an essential trace element, then LMWCr would not normally have a role in carbohydrate and lipid metabolism. However, studies using pharmacological doses of Cr^{3+} have generated beneficial effects on insulin sensitivity and blood lipid parameters in rodent models of diabetes (Vincent and Stearns 2011). This raises the question of whether pharmacological doses of chromium, orders of magnitude above the amount of chromium obtained in the diet, could result in a loading of LMWCr with chromium above normal levels, allowing for holoLMWCr to interact with insulin receptor to an appreciable effect and to increase insulin sensitivity.

The other suggested role for LMWCr is in detoxification of chromium. As noted above, chromium is absorbed by passive diffusion and maintained in the blood bound to transferrin. The movement of transferrin to tissues results in the transfer of chromium to LMWCr, and the holoLMWCr is eliminated in the urine. Thus, LMWCr appears to provide an efficient mechanism for elimination of chromium from the tissues and ultimately the body. ApoLMWCr is maintained in the soluble portion of tissue cells examined to date and in the nucleus of at least hepatocytes. ApoLMWCr levels are not increased in response to multiple intravenous administrations of chromium. However, the levels appear to not be diminished by administration of a chromium load; consequently, levels of apoLMWCr have been suggested to be under homeostatic control (Vincent and Stearns 2011). This is not typical for biomolecules whose primary role is detoxification. The efficiency of this clearance can be observed from the LD_{50} and mean tubular reabsorption rates of LMWCr. The mean tubular reabsorption rate for LMWCr of 23.5% in contrast to rates of 85.7% and 92.5% for chromate and chromium chloride, respectively; this is probably also responsible for the extremely high LD_{50} for LMWCr injected into mice of 135 mg/kg body mass (Yamamoto et al. 1984). Given a role for LMWCr under physiological circumstances does not appear to have a role in carbohydrate metabolism; a detoxification role would appear to be its natural function. Given that dietary intake of chromium is very low (approximately 30 μg/day), and lack of toxic from Cr^{3+} intakes several orders of magnitude higher, perhaps an inducible detoxification system for chromium, has not proven necessary.

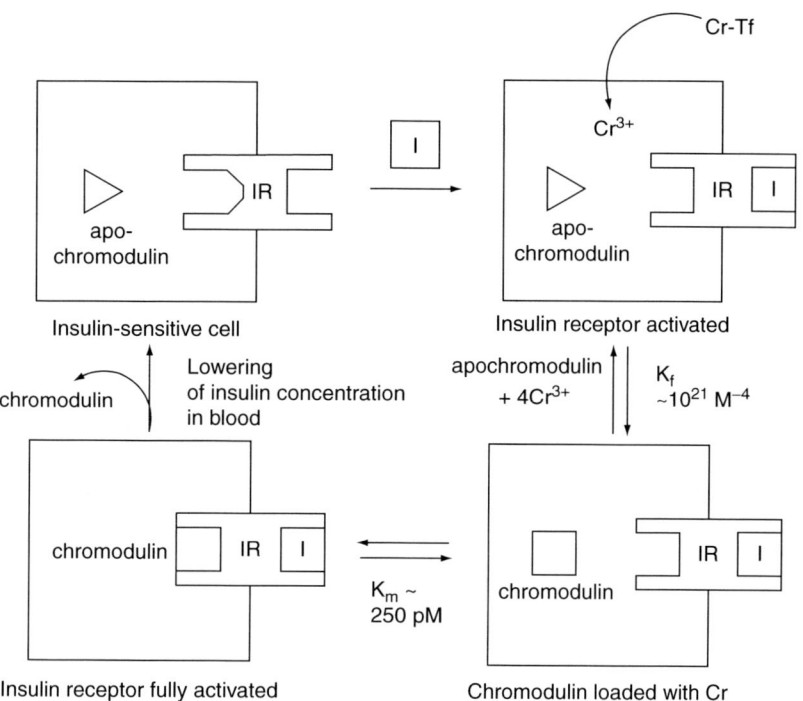

Chromium(III) and Low Molecular Weight Peptides, Fig. 1 Proposed mechanism for the activation of insulin receptor kinase activity by chromodulin in response to insulin (Vincent 2000). The inactive form of the insulin receptor (*IR*) is converted to the active form by binding insulin (*I*). This triggers a movement of chromium (presumably in the form of chromium transferrin, *Cr-Tf*) from the blood into insulin-dependent cells, which in turn results in a binding of chromium to apochromodulin (apoLMWCr) (*triangle*). Finally, the holochromodulin (*square*) binds to the insulin receptor, further activating the receptor kinase activity. Apochromodulin is unable to bind to the insulin receptor and activate kinase activity. When the insulin concentration drops, holochromodulin is released from the cell to relieve its effects (Reproduced with the permission of the copyright holder)

Bovine Colostrum LMWCr

A related chromium-containing oligopeptide from bovine colostrum (M-LMWCr) is comprised of the same amino acids but in distinctly different ratios and also stimulates insulin-dependent glucose metabolism in rat adipocytes. Whether the oligopeptide is present in other forms of milk is unknown. The significance of these differences between the liver and colostrum oligopeptides is essentially unexplored. The existence of multiple forms of LMWCr in cows raises concerns about identifying LMWCr in tissues or body fluids from only its apparent molecular weight and chromium-binding ability (Vincent and Stearns 2011). For example, are blood and urine LMWCr fractions comprised of one or more of these oligopeptides?

Conclusion

Overall, these studies show that chromium is transferred to a low-molecular-weight species in tissues and is subsequently lost in the urine as a low-molecular-weight species. These species are similar, if not identical, and can serve as a detoxification mechanism for chromium. An isolated species from liver and urine treated with chromium has been characterized and called LMWCr or chromodulin. While it is logical and likely that this species that tightly binds chromium is the same species that occurs naturally in tissues and urine, the chromium-loading process required to isolate the biomolecule does not allow this to be determined with certainty as it could represent an artifact generated during chromium loading. As chromium is probably not an essential element,

LMWCr (if the natural chromium-binding biomolecule) should not have a role in insulin signaling during normal conditions. The activity observed in test tube studies could reflect a mechanism for altering carbohydrate and lipid metabolism when excess chromium is present, as when rodents are treated with high doses of chromium(III) supplements. However, this will require in vivo evidence. Additional research on LMWCr and on the molecules in insulin signaling affected by chromium in vivo is needed to address these questions.

Cross-References

▶ Chromium and Glucose Tolerance Factor
▶ Chromium and Insulin Signaling
▶ Chromium(III) and Transferrin

References

Davis CM, Vincent JB (1997a) Chromium oligopeptide activates insulin receptor tyrosine kinase activity. Biochemistry 36:4382–4385

Davis CM, Vincent JB (1997b) Isolation and characterization of a biologically active form of chromium oligopeptide from bovine liver. Arch Biochem Biophys 339:335–343

Jacquamet L, Sun Y, Hatfield J et al (2003) Characterization of chromodulin by X-ray absorption and electron paramagnetic spectroscopies and magnetic susceptibility measurements. J Am Chem Soc 125:774–780

Vincent J (2000) The biochemistry of chromium. J Nutr 130:715–718

Vincent JB, Stearns DM (2011) The bioinorganic chemistry of chromium: essentiality, therapeutic agent, toxin, carcinogen? Wiley-Blackwell, Chichester

Wu GY, Wada O (1981) Studies on a specific chromium binding substance (a low-molecular-weight chromium binding substance) in urine. Jpn J Ind Health 23:505–512

Yamamoto A, Wada O, Ono T (1981) A low-molecular-weight, chromium-binding substance in mammals. Toxicol Appl Pharmacol 59:515–523

Yamamoto A, Wada O, Ono T (1984) Distribution and chromium-binding capacity of a low-molecular-weight, chromium-binding substance in mice. J Inorg Biochem 22:91–102

Yamamoto A, Wada O, Ono T (1987) Isolation of a biologically active low-molecular-mass chromium compound from rabbit liver. Eur J Biochem 165:627–631

Yamamoto A, Wada O, Manabe S (1989) Evidence that chromium is an essential factor for biological activity of low-molecular-weight Cr-binding substance. Biochem Biophys Res Commun 163:189–193

Chromium(III) and Transferrin

John B. Vincent
Department of Chemistry, The University of Alabama, Tuscaloosa, AL, USA

Synonyms

Siderophilin; Transferrin; β_1-metal-combining protein

Definition

Transferrins comprise a class of proteins with molecular weights of approximately 80,000 that reversibly bind two equivalents of metal ions. The proteins selectively bind ferric ions in a biological environment because the metal-binding sites are adapted to bind ions with large charge-to-size ratios. Transferrin is a blood serum protein, a β-globulin, although other forms of the protein are found in milk (lactoferrin) and avian egg white (conalbumin or ovotransferrin). Transferrin is the major iron transport protein in the bloodstream, while lactoferrin and conalbumin are believed to have antibacterial roles by depriving bacteria of iron. Because of its similarity in size and charge to ferric ion, Cr^{3+} is transported and stored in the bloodstream by transferrin.

Transferrins are a class of proteins of approximately 80 kDa (kilodaltons) that reversibly bind two equivalents of metal ions (Brock 1985). The protein exhibits amazing selectivity for Fe^{3+} in a biological environment because the metal sites are adapted to bind ions with large charge-to-size ratios. Transferrin is a blood serum protein, a β-globulin, although other forms of the protein are found in milk, lactoferrin, and avian egg white (conalbumin or ovotransferrin). Transferrin is the major iron transport protein in the bloodstream, while lactoferrin and conalbumin are believed to have antibacterial roles by depriving bacteria of iron. In humans, the protein is present at a concentration of approximately 3 mg/ml in serum and is normally about 30% saturated with iron (Brock 1985), allowing it to potentially bind and transport other metal ions. Consequently, transferrin has been proposed to serve as a chromium(III) transport agent, given the similar charge and relatively similar size of this ion to iron(III) (Vincent and Stearns 2011).

The transferrin molecule is composed of two lobes with approximately 40% sequence homology; the three-dimensional structures of the lobes are nearly superimposable (Baker 1994). Each lobe possesses an iron-binding site. Each Fe^{3+} binds concomitantly with a synergistic anion, usually bicarbonate or carbonate. The iron coordination is essentially identical in each site, being distorted octahedral and composed of two tyrosine residues, a histidine residue, an aspartate residue, and a chelating (bi)carbonate ion. The presence of the anion is essential for iron binding. The transferrin molecule undergoes a significant conformational change when binding and releasing iron. The apoprotein possesses a more open confirmation (Baker 1994). In the metal-loaded confirmation, transferrin binds to transferring receptor, a transmembrane protein of the cell membrane. Transferrin is brought into the cell by endocytosis. Acidification of resulting endosome releases the Fe^{3+}, and subsequent fusion of the endosome with the cell membrane releases and recycles the apotransferrin (Brock 1985).

Physical Characterization of Chromium Transferrins

In vitro studies have shown that Cr^{3+} readily binds to the two metal-binding sites of transferrin and concomitantly also binds two equivalents of bicarbonate, resulting in intense changes in the protein's ultraviolet spectrum. The changes in the ultraviolet spectrum suggest that each chromic ion binds to two tyrosine residues, suggesting that chromium binds specifically in the two iron-binding sites. The involvement of tyrosine ligands has been confirmed by Raman spectroscopy (Ainscough et al. 1980). Human Cr_2 transferrin has been described as pale blue in color with visible maxima at 440 and 635 nm (Aisen et al. 1969), while Cr_2 lactoferrin has been described as gray-green with maxima at 442 and 612 nm ($\varepsilon = 520$ and 280 M^{-1} cm^{-1}, respectively) (Ainscough et al. 1979). The visible spectra are typical for Cr(III) centers in a pseudooctahedral environment. The oxidation state of the bound chromium has been confirmed by variable temperature magnetic susceptibility studies, whose results are consistent with the presence of $S = 3/2$ centers, and by electron paramagnetic resonance (EPR) studies (Aisen et al. 1969). The two Cr-binding sites can readily be distinguished by EPR (frozen solutions at 77 K) (Aisen et al. 1969). At approximately pH 7.7, chromium binds to both sites on the protein. At pH 4.8–5.9, chromium only binds to one site. This tighter binding site possesses an EPR signal at $g = 5.43$. At near neutral pH, the Cr^{3+} in the tighter binding site that binds chromium at the lower pH cannot be displaced by Fe^{3+}, while Fe^{3+} readily displaces Cr^{3+} from the other site (Ainscough et al. 1980). The weaker binding site Cr^{3+} gives rise to EPR signals at $g = 5.62$, 5.15, and 2.42 (Ainscough et al. 1980).

The effective thermodynamic binding constants for chromium transferrin, actually using conalbumin, have been determined (Sun et al. 2000). The addition of chromic ions to apoconalbumin was monitored by following the enhancement of the intensity of the ultraviolet absorption bands at circa 240 and 290 nm. The value of $\Delta\varepsilon$ increased rapidly upon the initial additions of chromium but rapidly levels off after the addition of approximately 1.5 chromic ions, indicating the chromium was occupying both metal-binding sites. At a given pH and carbonate concentration, effective equilibrium constants can be written such that

$$K_1 = [CrT_f]/([Cr][T_f]) \quad (1)$$

and

$$K_2 = [Cr_2T_f]/([Cr][CrT_f]), \quad (2)$$

where [Cr] represents the concentration of all chromium not bound to transferrin. Based on these equations, the total concentration of Cr and of transferrin becomes

$$[Cr]_{total} = [Cr] + K_1[Cr][T_f] + 2K_1K_2[Cr]^2[T_f] \quad (3)$$

and

$$[T_f]_{total} = [T_f] + K_1[Cr][T_f] + K_1K_2[Cr]^2[T_f] \quad (4)$$

Using these equations, the $\Delta\varepsilon$ at any point of the titration, $\Delta\varepsilon_{calcd}$, can be calculated as

$$\Delta\varepsilon_{calcd} = \left(\Delta\varepsilon_{Cr}K_1[Cr] + 2\Delta\varepsilon_{Cr}K_1K_2[Cr]^2\right)/\left(1 + K_1[Cr] + K_1K_2[Cr]^2\right). \quad (5)$$

where $\Delta\varepsilon_{Cr}$ is the molar absorptivity per bound chromium at 245 nm. $\Delta\varepsilon_{Cr}$ can be determined by the slope of the linear portion of the curve in Fig. 2 at low Cr-to-transferrin ratios and is 7.94×10^3 M^{-1} cm^{-1}. Fitting the data to (5) gave the effective binding constants of $K_1 = 1.42 \times 10^{10}$ M^{-1} and $K_2 = 2.06 \times 10^5$ M^{-1}. This gives the overall effective binding constant $K = K_1 \times K_2$ to be 2.92×10^{15} M^{-2} (Sun et al. 2000). The difference in the binding constant for the two metal-binding sites is consistent with previous studies with chromium and transferrin. Given the binding constants for Cr^{3+}, the inability Fe^{3+} to displace Cr^{3+} from one of the binding sites, and that transferrin is maintained on average only 30% loaded with ferric ions, the protein appears to be primed to be able to transport chromium through the bloodstream.

In Vivo Studies

The first demonstration of the potential importance of transferrin in the transport of chromium resulted from in vivo administration of chromic ions to mammals; this results in the appearance of chromic ions in transferrin. $^{51}CrCl_3$ given by stomach tube to rats resulted in $\geq 99\%$ of the chromium in blood being associated with noncellular components (Hopkins and Schwarz 1964). Ninety percent of the Cr in blood serum was associated with the β-globulin fractions; 80% immunoprecipitated with transferrin (Hopkins and Schwarz 1964). In vivo and in vitro studies of the addition of chromium sources to blood or blood plasma also result in the loading of transferrin with Cr(III), although under these conditions, albumin and some degradation products also bind chromium; in fact, in vitro, more Cr may bind to albumin than transferrin. One must be careful to distinguish experimental design when examining chromium binding to serum proteins. When given orally, absorbed chromium appears in the blood essentially only as transferrin and a low-molecular-weight species; when given intravenously or added in vitro to blood or blood serum or plasma, nonphysiologically relevant binding of chromium to other species occurs. The results of the latter studies must be viewed with caution (Vincent and Stearns 2011).

Recent reports on the effects of insulin on iron transport support transferrin being the major physiological chromium transport agent. Plasma membrane recycling of transferrin receptors is sensitive to insulin as increases in insulin result in a stimulation of the movement of transferrin receptors from vesicles to the plasma membrane. The most detailed studies of Cr transferrin movement have been reported in the last few years using ^{51}Cr-labeled transferrin administered intravenously to rats. Injection of ^{51}Cr-labeled transferrin into the bloodstream resulted in a rapid and insulin-sensitive movement of chromium into the tissues as Cr transferrin; greater than 50% of the chromium is transported to the tissues within 30 min. Tissue levels of chromium were maximal 30 min after injection; decreases in tissue chromium with time were mirrored by increases in urine chromium. Thus, transferrin, in an insulin-dependent fashion, can transfer Cr to tissues from which Cr is excreted in the urine (Clodfelder and Vincent 2005).

Approximately 50% of the ^{51}Cr appeared in the urine within 360 min of injection of Cr transferrin into the tail vein of rats in the absence of added insulin; insulin treatment concurrent with injection of ^{51}Cr-labeled transferrin results in approximately 80% of the label appearing in the urine within 180 min. The removal of ^{51}Cr from the blood was faster than the appearance of ^{51}Cr in the urine; the lag in time indicates that the Cr transferrin in the blood and chromium in the urine are not in direct equilibrium and that intermediates in the transport of chromium must be involved. Separation of the urine components by G-25 size exclusion chromatography revealed that chromium occurred in the urine as apparently a single low-molecular-weight species, assumed to be low-molecular-weight chromium-binding substance (LMWCr, also called chromodulin). LMWCr when added to the urine comigrated with the urine chromium (Clodfelder and Vincent 2005).

When the species of chromium in the blood plasma as a function of time were examined by S-200 size exclusion chromatography, two primary features were observed (Fig. 1). The first was Cr transferrin, which disappeared quickly from the bloodstream. With time, a low-molecular-weight species, also proposed to be LMWCr, appeared. (Two species of intermediate molecular weight are also observable in Fig. 1, but they account for <10% of the applied chromium and probably were degradation products of transferrin) (Clodfelder and Vincent 2005).

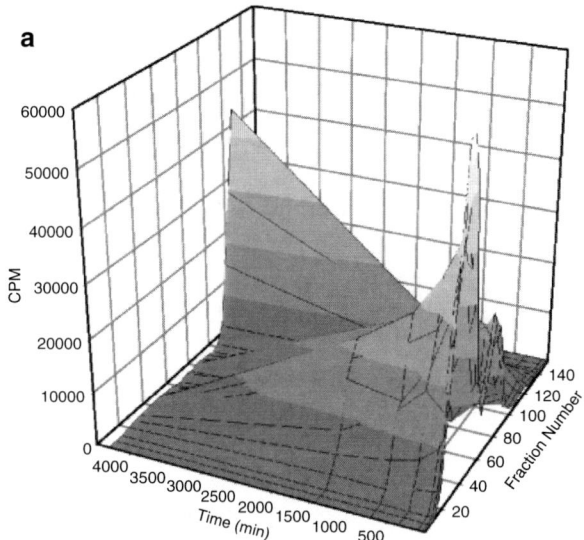

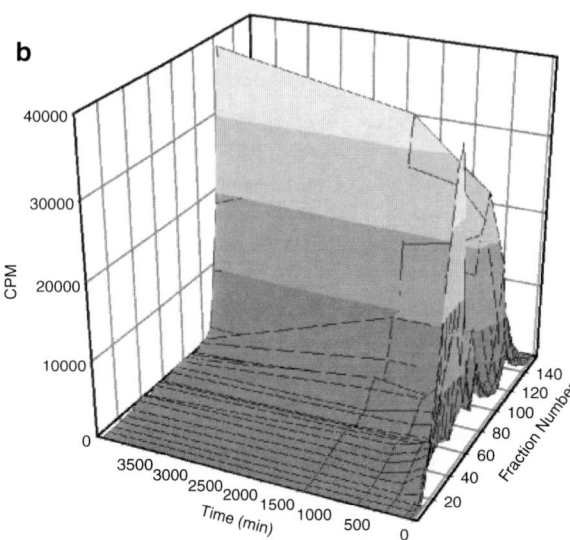

Chromium(III) and Transferrin, Fig. 1 Elution profiles of ^{51}Cr in blood plasma from adult rats from an S-200 column as a function of time. (**a**) Rats not receiving insulin. (**b**) Rats receiving insulin (Reproduced from Clodfelder and Vincent 2005 with the permission of the copyright holder)

This work established a clear pathway of transport of chromium starting from transport by transferrin from the bloodstream into the tissues, followed by release and processing in the tissues to form a low-molecular-weight chromium-binding species, excretion into the bloodstream, rapid clearance of the low-molecular-weight species or a similar species into the urine, and ultimately excretion as this species.

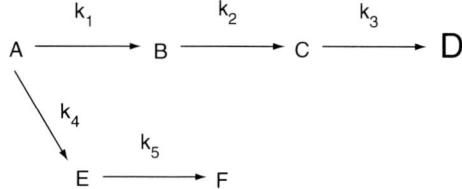

Chromium(III) and Transferrin, Fig. 2 Proposed mechanism of distribution of chromium from Cr transferrin (Reproduced from Clodfelder and Vincent 2005 with permission of the copyright holder)

Insulin stimulates the processing of chromium in the tissues. The rates of chromium movement were estimated, and on the basis of these results, a kinetic model for the movement of chromium from transferrin in the blood to LMWCr-like species in the urine was proposed (Fig. 2). The model assumes the presence of six major types of chromium: (a) Cr transferrin in the blood plasma, (b) chromium in the tissues, (c) LMWCr-like species in the plasma, (d) LMWCr-like species in the urine, (e) the larger unidentified species in the plasma, and (f) the smaller unidentified species in the plasma. Two pathways were required to fit the movement of Cr from transferrin to the urine and the Cr-binding species in the bloodstream. The first involves the transport of chromium by transferrin to the tissues (k_1), followed by the release of chromium and production and release of LMWCr-like species into the blood (k_2), and the movement of chromium as LMWCr-like species from the blood to the urine (k_3). The presence of LMWCr-like species in the tissues and in the urine necessitates the presence of LMWCr-like species in the blood plasma, although its presence appears to be masked by the smaller unidentified species. Hence, the LMWCr-like species is assumed to be kept at low steady-state levels in the blood, and k_3 is assumed to be much larger than k_1 and k_2. The second pathway includes the appearance of the larger unidentified species (k_4), which in turn is either metabolized to generate the smaller unidentified species or gives up its Cr to the smaller species (k_5). This kinetic model fit the experimental data well at the early time points but failed to simulate the gradual loss of chromium from the tissues to the urine. This failure arose because the model did not incorporate the processing and loss of chromium from the various tissues occurring at different rates. This manifested itself, for example, in a requirement for at least a biphasic function to fit to the appearance of LMWCr-like species in the urine as a function of time (Clodfelder and Vincent 2005).

Transferrin and LMWCr

While the studies of the movement of chromium from Cr transferrin to the tissues and ultimately the urine indicated that Cr chromium from transferrin was transferred to the tissues and then to a low-molecular-weight species for clearance in the urine, the relative ability of transferrin, LMWCr, and other biomolecules have been examined. Yamamoto and coworkers have previously investigated the ability of equal molar amounts of LMWCr, transferrin, and albumin to compete for chromic ions and found that LMWCr accepted more chromium than the two serum proteins (Yamamoto et al. 1984). These investigators also examined the ability of Cr_2 transferrin to donate chromium to apoLMWCr and for LMWCr to donate chromium to transferrin. These workers reported that approximately 10% of the chromic ions in Cr_2 transferrin were transferred to apoLMWCr and that 25% of the chromic ions in LMWCr were transferred to apotransferrin (Yamamoto et al. 1984). However, the data were collected after a single time interval (60 min) at an elevated temperature (37°C). LMWCr is highly susceptible to hydrolysis, particularly at elevated temperature. In light of this, the exchange of Cr^{3+} between transferrin and LMWCr has been reexamined. Within detection limits, LMWCr with its full complement of four chromium does not release its chromium to apotransferrin over greater than 25,000 min incubation (in pH 6.5 buffer at $\sim 4°C$) (Sun et al. 2000). However, over the same time period, chromium migration from Cr_2 transferrin is significant (approximately half the chromium), consistent with the larger chromium-binding constant of apoLMWCr versus halfapotransferrin. Although a direct transfer of chromium between transferrin and LMWCr is unlikely to be physiologically relevant (as metal-containing transferrin enters the cell through receptor-mediated endocytosis and releases metal during the lowering of pH that occurs with endosomal maturation), these results illustrate the ability of LMWCr to sequester chromic ions in the presence of other metal-binding species (Sun et al. 2000).

Significance of Chromium Transport by Transferrin

While chromium has generally been believed to be an essential trace element for approximately 50 years, recent research has not supported a role for chromium (Vincent and Stearns 2011). This has implications for the significance of chromium transport by transferrin. The ability of transferrin to bind chromium in vivo and transport chromium to tissues, particularly muscle, in an insulin-sensitive fashion is supportive of suggestions that chromium played a role in optimization of insulin signaling. Alternatively, transferrin could serve a role in detoxification of chromium. Transferrin, in such a role, could be viewed as scavenging absorbed dietary chromium from the bloodstream and delivering the metal to the tissues where it is ultimately bound to a low-molecular-weight peptide and rapidly cleared from the body via the urine.

Cross-References

▶ Chromium and Human Nutrition
▶ Chromium and Insulin Signaling
▶ Chromium(III) and Low Molecular Weight Peptides

References

Ainscough EW, Brodie AM, Plowman JE (1979) The chromium, manganese, cobalt and copper complexes of human lactoferrin. Inorg Chim Acta 33:149–153

Ainscough EW, Brodie AM, Plowman JE, Bloor SJ, Sanders Loehr J, Loehr TM (1980) Studies on human lactoferrin by electron paramagnetic resonance, fluorescence, and resonance Raman spectroscopy. Biochemistry 19:4072–4079

Aisen P, Aasa R, Redfield AG (1969) The chromium, manganese, and cobalt complexes of transferrin. J Biol Chem 244:4628–4633

Baker EN (1994) Structure and reactivity transferrins. Adv Inorg Chem 41:389–463

Brock JH (1985) Transferrins. In: Harrison PM (ed) Metalloproteins, Part 2. Verlag Chemie, Weinheim

Clodfelder BJ, Vincent JB (2005) The time-dependent transport of chromium in adult rats from the bloodstream to the urine. J Biol Inorg Chem 10:383–393

Hopkins LL Jr, Schwarz K (1964) Chromium(III) binding to serum proteins, specifically siderophilin. Biochim Biophys Acta 90:484–491

Sun Y, Ramirez J, Woski SA, Vincent JB (2000) The binding of chromium to low-molecular-weight chromium-binding substance (LMWCr) and the transport of chromium from transferring and chromium picolinate to LMWCr. J Biol Inorg Chem 5:129–136

Vincent JB, Stearns DM (2011) The bioinorganic chemistry of chromium: essentiality, therapeutic agent, toxin, carcinogen? Wiley-Blackwell, Chichester

Yamamoto A, Wada O, Ono T (1984) Distribution and chromium-binding capacity of a low-molecular-weight, chromium-binding substance in mice. J Inorg Biochem 22:91–102

Chromium(III), Cytokines, and Hormones

Sushil K. Jain
Department of Pediatrics, Louisiana State University Health Sciences Center, Shreveport, LA, USA

Synonyms

Inflammation; Micronutrients; Trace metals

Definition

Chromium (III) is a transition metal, and its trivalent state is the form most prevalent in organic complexes (Vincent 2004; Cefalu and Hu 2004). Chromium (III) supplements are widely consumed worldwide. Most chromium (III) in the diet is chromium (III), and any hexavalent chromium in food or water is reduced to chromium (III) in the acidic environment of the stomach.

Background

Various studies have reported lower levels of chromium (III) in the blood, lenses, and toenails of diabetic patients compared with those of the normal population (Rajpathak et al. 2004). Thus, subclinical chromium (III) deficiency may be a contributor to glucose intolerance, insulin resistance, and cardiovascular disease, particularly in aging populations or in populations that have increased chromium (III) requirements because of high sugar diets. Epidemiological data concerning chromium (III) intake and the risk of CVD are limited. Results from two case–control studies suggest an inverse association between chromium (III) levels in toenails and the risk of myocardial infarction in the general population. Similarly, a recent report within the Health Professionals Follow-up Study has found lower levels of toenail chromium (III) among men with diabetes and CVD compared with healthy control subjects (Rajpathak et al. 2004).

Chromium Content of Food

Foods with high chromium (III) concentrations include whole grain products, green beans, broccoli, and bran cereals. The chromium (III) content of meats, poultry, and fish varies widely since chromium (III) may be introduced during transport, processing, and fortification of foods. Not only are foods rich in refined sugars low in chromium (III), they actually promote chromium (III) loss. Based on the chromium (III) content of well-balanced diets, adequate intake values for chromium (III) in adults have been established at 35 µg/day in men and 25 µg/day in women. Although there are no national survey data available on chromium (III) intake, a study of self-selected diets of US adults indicates that the chromium (III) intake of a substantial proportion of subjects may be well below the adequate intake values; similar results have been shown in the United Kingdom, Finland, Canada, and New Zealand.

Cytokines and Chromium (III)

Cytokines interact with cells of the immune system to regulate the body's response to disease and infection. The cytokines locate target immune cells and interact with receptors on the target immune cells by binding to them. Overproduction of certain cytokines, such as interleukin-6 (IL-6) and tumor necrosis factor-alpha (TNF-α), is involved in inflammation and insulin resistance that can contribute to the development of CVD. The levels of pro-inflammatory cytokines are elevated in the blood of many subjects with CVD and diabetes (Jain et al. 2007a; Jain and Kannan 2001). An increase in circulating levels of TNF-α and IL-6 decreases insulin sensitivity. Elevated circulating levels of TNF-α and IL-6 can induce expression of adhesion molecules and thus monocyte-endothelial cell adhesion, now recognized as an early and rate-limiting step in vascular inflammation and the development of vascular disease.

Chromium (III) supplementation decreased secretion of IL-8, IL-6, and TNF-α in U937 monocyte cells exposed to high levels of glucose (Jain and Kannan 2001). Similarly, secretion of TNF-α and IL-6 was inhibited when isolated monocytes from human volunteers were used. These studies provide evidence that chromium (III) supplementation inhibits the increase in pro-inflammatory cytokine secretion and oxidative stress levels caused by exposure to high levels of glucose or ketosis (mimicking diabetes) in cultured monocytes (Jain and Kannan 2001; Jain et al. 2007b). Myers et al. (1997) determined the effect of growth hormone or chromium picolinate (CP) on swine metabolism and inflammatory cytokine production after endotoxin challenge and showed that IL-6 was not affected by CP or recombinant porcine somatotropin (PST) treatments. However, both CP and PST lowered the TNF-α response to lipopolysaccharide (LPS), similar to results observed in isolated monocytes exposed to high glucose levels in a cell culture model.

Studies using cobalt chromium alloy particles did not show any effect on TNF-α, IFN-α, IL-6, or IL-12 secretion or on mRNA expression of cytokines in a J774A.1 cell line exposed to chromium. In contrast, A549 human lung carcinoma cells pretreated with chromium showed inhibition of TNF-α stimulated expression of IL-8 and NFkB. Chromium (III) did not inhibit the TNF-α-stimulated IKB-alpha degradation or the translocation of the NFkB-binding protein to the nucleus. Both chromium chloride and CP showed a beneficial effect in reducing oxygen radical production and increasing glucose uptake and phagocytosis of *Escherichia coli* in the presence and absence of insulin using pulmonary alveolar macrophages.

The inhibitory effects of chromium (III) on pro-inflammatory cytokines were also observed in vivo in diabetic rats supplemented with chromium (III) compounds (Jain et al. 2007a, 2010). The levels of TNF-α, IL-6, IL-8 and MCP-1, and ICAM-1 were significantly lower in diabetic rats supplemented with chromium (III). It appears that the effects of chromium (III) on the secretion or expression of different cytokines may be dependent on the chromium (III) concentration and the specific cells. The effect of chromium (III) supplementation on lowering pro-inflammatory cytokines in the blood was associated with a beneficial effect in lowering elevated blood levels of glycemia, cholesterol, and triglycerides, as well as lowering oxidative stress in streptozotocin-treated diabetic rats. These animal studies suggest that chromium (III) supplementation can lower the level of vascular inflammation associated with diabetes.

The influence of chromium (III) on cytokines and inflammatory markers in cell culture and in vivo studies may be mediated by different mechanisms. One potential mechanism is the antioxidant effect of chromium (III). Several investigators have reported that chromium (III) supplementation lowers the blood levels of oxidative stress markers in an animal model as well as in diabetic patients (Jain et al. 2010). Chromium (III) activates glutathione reductase activity in a red blood cell model. Thus, it is possible that chromium (III) reduces oxidative stress by increasing the detoxification of oxygen radicals and the maintenance of cellular GSH. Whether chromium (III) supplementation has beneficial effects on oxidative stress, on pro-inflammatory cytokines, or on GSH in all of the cells crucial to atherogenesis, such as monocytes and endothelial cells, is not known. Chromium (III) may influence the immune system by modulating anti-inflammatory cytokines.

Hormones and Chromium (III)

Insulin is a major hormone that is critical to the metabolism and storage of carbohydrate, fat, and protein in the body. The role of chromium (III) in glucose metabolism has been known since 1955 when Metz identified chromium (III) containing glucose tolerance factor (GTF). GTF is a complex of chromium (III), glutathione, and nicotinate. Glutathione, a tripeptide of glutamate, glycine, and cysteine, is known to bind insulin tightly. The oligopeptide known as low-molecular-weight chromium (III)-binding substance is composed of glycine, cysteine, aspartate, and glutamate coordinated by chromic ions (Vincent 2004). The effects of chromium (III) on Zucker diabetic fatty (ZDF) rats, a genetic model of type 2 diabetes, was examined using 1,000 μg chromium (III)/kg body mass (Bennett et al. 2006). Treatment with chromium (III) caused significantly lower concentrations of blood insulin along with low levels of cholesterol and triglycerides. Similarly, a diet containing 5 mg chromium (III)/kg diet for 10 weeks caused a nearly 16% reduction in blood insulin concentrations in normal Wistar rats. It has been proposed that chromium (III) insulin complexes enhance the action of insulin on glucose metabolism and lower blood sugar levels. The crystallographic analyses of chromium (III)-Salen-soaked cubic insulin crystals revealed B21 glucose to be binding site for chromium (III) (Sreekanth et al. 2008). Bennett et al. (2006) also found a significant decrease in blood concentrations of leptin in ZDF rats supplemented for 4 weeks with chromium (III) (250–100 μg/kg BW).

The effects of a variety of dietary chromium (III) sources on adrenal steroid hormones have been investigated, but results have been inconclusive. Dietary chromium (III) decreased circulating adrenal steroids in stressed young calves, in epinephrine challenged lambs, in stressed dairy cows, and in postmenopausal women. Other studies have shown no effect of chromium (III) supplementation on adrenal hormones. Chromium (III) has a significant effect on insulin secretion from the islets of Langerhans and on catecholamine secretion from bovine medullary cells. Chromium (III) supplementation decreased the secretion of both cortisol and dehydroepiandrosterone in H259 R cells. The inhibition of steroid genesis by chromium (III) may explain the decreased blood cortisol concentration in chromium (III)-supplemented stressed calves and cows. However, other studies did not report any change in cortisol levels in chromium (III)-supplemented immune-stressed or thermally stressed pigs. The discrepancy of cortisol responses in chromium (III)-supplemented animals may be attributed to the dosage, source of chromium (III) or duration of dietary treatment, source and intensity of stress, and animal species. However, blood levels of cortisol were consistently lower when high-dose supplementation with chromium (III) (>0.4 mg/kg) was used in different studies.

Transthyretin (TTR) binds and transports thyroxin (T4) in human plasma. In vitro studies demonstrate that chromium (III) increased the T4-binding capacity of wild-type and amyloidogenic V30M-TTR. Chromium (III) and T4 cooperatively suppressed in vitro fibril formation due to the stabilization of wild-type TTR and V30M-TTR. Thus, transthyretin amyloid fibril formation, which is triggered by the dissociation of tetrameric TTR, is suppressed by chromium (III) in vitro (Sato et al. 2006). Other studies have also reported that the distribution of chromium (III) in the body is influenced by thyroid hormone activity but not by calcitonin or parathyroid hormone. This suggests that thyroid hormone controls cellular chromium (III) transport. Recent studies suggest that supplementation with chromium (III) nanoparticles significantly lowers serum concentrations of insulin, cortisol, insulin-like growth factor 1, and IgG after chromium (III) supplementation (150–450 μg/kg BW for 8 weeks in Sprague–Dawley rats.

The studies of Vincent et al. (2004) have shown that LMW chromium (III) is stored in the cytosol of insulin-sensitive cells in an apo (unbound form) that is activated by binding four chromium (III) ions. This activation is the result of a series of steps stimulated by insulin signaling. LMW chromium (III) potentiates the action of insulin once insulin has bound to its receptor. This insulin potentiating or autoamplification action stems from the ability of LMW chromium (III) to maintain stimulation of tyrosine kinase activity. Once insulin is bound to its receptor, LMW chromium (III) binds to the activated receptor on the inner side of the cell membrane and increases the insulin-activated protein kinase activity by eightfold. There is also evidence that the autoamplification effect of LMW chromium may be enhanced by the inhibition of phosphotyrosine phosphatase, which inactivates tyrosine kinase. LMW chromium (III) actually activates membrane-associated phosphotyrosine phosphatase in insulin-sensitive cells. As insulin levels drop and receptor activity diminishes, LMW chromium (III) is transported from the cell to the blood and excreted in the urine.

Trivalent chromium, the form found in foods and dietary supplements, is believed to be safe. The most popular complexed form is chromium picolinate, although chromium niacinate and chromium citrate are also used as nutritional supplements. Absorption of chromium (III) is low, as is common for other polyvalent minerals, ranging from less than 1% for chromium (III) chloride to above 1% for chromium (III)-niacinate to 1–3% for chromium (III) picolinate in rat studies. Human absorption from food is estimated at about 2–3% and 5–10% from brewer's yeast. In animal studies, the retention of chromium (III) from supplementation occurred chiefly in the kidneys, followed by liver and heart muscle. The order of retention was chromium (III) picolinate > chromium (III) niacinate > chromium (III) chloride, which varies from the order of absorption. No disturbances in the organ systems showing high concentrations of chromium (III) have yet been detected.

In conclusion, trivalent chromium can lower oxidative stress by improving glucose and lipid metabolism. It has been proposed that chromium (III) supplementation increases the amount of a chromium (III)-containing oligopeptide present in the insulin-sensitive cells that bind to the insulin receptor, markedly increasing the activity of the insulin-stimulated tyrosine kinase and phosphorylation of insulin receptor substrate-1 and glucose transporter (Jain et al. 2010). Oxidative stress associated with diabetes activates NFkB, which then activates the insulin resistance cascade. Hyperglycemia is known to increase reactive oxygen species generation and oxidative stress in diabetic rats and patients. Oxidative stress is a known activator of NFkB, which undergoes nuclear translocation and serine phosphorylation at residue 276 in its p65 subunit. It then associates with surrounding chromatin components and binds with DNA, which promotes the transcription of pro-inflammatory cytokines that mediate the insulin resistance cascade. Thus, the effect of chromium (III) on glucose and lipid metabolism is likely to be mediated by lowering levels of the pro-inflammatory cytokines TNF-α, IL-6, and MCP-1, which are known to cause insulin resistance.

Concluding Remarks

Chromium (III) supplementation in the form of commercially available chromium dinicotinate (CDN) or chromium picolinate (CP) is widely used by the public. Subclinical chromium (III) deficiency may contribute to insulin resistance and CVD, particularly in aging and diabetic populations. Although not in all studies, some studies of diabetic patients and diabetic animals have reported decreased blood glucose or decreased insulin requirements after chromium (III) supplementation (Cefalu and Hu 2004). A number of animal studies have demonstrated beneficial effects of chromium (III) supplementation on circulating levels of cytokines and hormones, which in turn may influence biological functions and diseases such as diabetes and cardiovascular disease (CVD). However, studies on the effect of chromium (III) supplementation on cytokines and hormones in humans have yet to be done.

Acknowledgments The author is supported by grants from the NIDDK, the Office of Dietary Supplements of the National Institutes of Health (RO1 DK072433), and the Malcolm Feist Endowed Chair in Diabetes. The authors thank Ms. Georgia Morgan for excellent editing of this manuscript.

Cross-References

▶ Chromium and Diabetes
▶ Chromium and Glucose Tolerance Factor
▶ Chromium and Insulin Signaling
▶ Chromium and Nutritional Supplement

References

Bennett R, Adams B, French A, Neggers Y, Vincent JB (2006) High dose chromium (III) supplementation has no effects on body mass and composition while altering plasma hormone and triglycerides concentrations. Biol Trace Elem Res 113:53–66

Cefalu WT, Hu FB (2004) Role of chromium in human health and in diabetes. Diabetes Care 27:2741–2751

Jain SK, Kannan K (2001) Chromium chloride inhibits oxidative stress and TNF-α secretion caused by exposure to high glucose in cultured monocytes. Biochem Biophys Res Commun 289:687–691

Jain SK, Rains J, Croad J (2007a) Effect of chromium niacinate and chromium picolinate supplementation on fasting blood glucose, HbA1, TNF-α, IL-6, CRP, cholesterol, triglycerides and lipid peroxidation levels in streptozotocin-treated diabetic rats. Free Radical Biol Med 43:1124–1131

Jain SK, Rains J, Jones K (2007b) High glucose and ketosis (acetoacetate) increases, and chromium niacinate inhibits IL-6, IL-8, and MCP-1 secretion in U937 monocytes. Antioxid Redox Signal 9:1581–1590

Jain SK, Croad JL, Velusamy T, Rains JL, Bull R (2010) Chromium dinicocysteinate supplementation can lower blood glucose, CRP, MCP-1, ICAM-1, creatinine, apparently mediated by elevated blood vitamin C and adiponectin and inhibition of NFkB, Akt, and Glut-2 in livers of zucker diabetic fatty rats. Mol Nutr Food Res 54:1371–1380

Myers MJ, Farrell DE, Evock-Clover CM, McDonald MW, Steele NC (1997) Effect of growth hormone or chromium picolinate on swine metabolism and inflammatory cytokine production after endotoxin challenge exposure. Am J Vet Res 58:594–600

Rajpathak S, Rimm EB, Li T, Morris JS, Stampfer MJ, Willet WC, Hu FB (2004) Lower toenail chromium in men with diabetes and cardiovascular disease compared with healthy men. Diabetes Care 2004(27):2211–2216

Sato T, Ando Y, Susuki S et al (2006) Chromium (III) ion and thyroxine cooperate to stabilize the transthyretin tetramer and suppress in vitro amyloid fibril formation. FEBS Lett 580:491–496

Sreekanth R, Pattabhi V, Rajan SS (2008) Molecular basis of chromium insulin interactions. Biochem Biophys Res Commun 369:725–729

Vincent JB (2004) Recent advances in the nutritional biochemistry of trivalent chromium. Proc Nutr Soc 63:41–47

Chromium(VI), Oxidative Cell Damage

Young-Ok Son, John Andrew Hitron and Xianglin Shi
Graduate Center for Toxicology, University of Kentucky, Lexington, KY, USA

Synonyms

Apoptosis, programmed cell death; Ascorbic acid, vitamin C; Aspirin, acetylsalicylic acid; Deferoxamine, desferrioxamine B, desferoxamine B, DFO-B, DFOA, DFB, desferal; Electron spin resonance (ESR), electron paramagnetic resonance (EPR); Proline oxidase, proline dehydrogenase; Sodium formate, formic acid

Definition

Hexavalent chromium (Cr(VI)) is a highly reactive metal capable of causing cellular oxidative damage through the generation of intracellular reactive oxygen species (ROS). ROS refers to a diverse group of reactive, short-lived, oxygen-containing species such as $O_2^{\bullet-}$, H_2O_2, and $^{\bullet}OH$. Overproduction or decreased removal of ROS leads to oxidative stress in tissues and cells. Cr(VI) can be reduced by various cellular reductants to its lower oxidation states, such as Cr(V) and Cr(VI). During the reduction process, molecular oxygen is reduced to $O_2^{\bullet-}$, which reduces to H_2O_2 upon dismutation. H_2O_2 reacts with Cr(V) or Cr(IV) to generate $^{\bullet}OH$ radicals. Thus, Cr(VI) is able to generate a spectrum of ROS. Biological systems are normally protected against oxidative injury caused by free radical reactions due to enzymatic and nonenzymatic antioxidants. When the balance between prooxidants and antioxidants shifts in favor of prooxidants, chromium-induced oxidative cell damage occurs.

Introduction

Cr(VI) compounds are known to cause serious toxic and carcinogenic effects. Intensive epidemiological studies of industrially exposed chromium workers have identified chrome plating, chrome pigment manufacturing, leather tanning, and stainless steel production as sources of potential exposure to this metal. Because of its wide industrial application, environmental contamination is considered to be an additional source of human exposure to Cr(VI). However, the biological mechanisms responsible for the initiation and progression of diseases resulting from exposure to Cr(VI) are not fully understood. A growing body of evidence reveals the correlation between Cr(VI)-induced generation of reactive oxygen species and cytotoxicity, or carcinogenesis.

Hexavalent chromium (Cr(VI)) and trivalent chromium (Cr(III)) are two stable chromium oxidation states found in nature. Cr(VI) is able to enter into cells through an anion transport system. If Cr(VI) enters cells, it is reduced by cellular reductants to its lower oxidation states, pentavalent chromium (Cr(V)) and tetravalent chromium Cr(IV). These intermediate states of chromium are reactive and capable of producing ROS. ROS are known to cause oxidative damage including DNA strand breaks, base modification, and lipid peroxidation (Ding and Shi 2002; Yao et al. 2008). Thus, it is generally believed that Cr(VI) can induce cell death by oxidative stress (Son et al. 2010).

Recent studies have suggested that the reduction of Cr(VI) to its lower oxidation states and related free radical reactions play an important role in Cr(VI)-induced carcinogenesis. Cr(VI) has been demonstrated to induce a variety of DNA lesions, such as single-strand breaks, alkali-labile sites, and DNA-protein cross-links. In contrast, most Cr(III) compounds are relatively nontoxic, noncarcinogenic, and nonmutagenic due to Cr(III)'s difficulty in entering the cells. Since Cr(VI) does not react with isolated DNA, the reduction of Cr(VI) by cellular reductants to lower oxidation states has been considered an important step in the mechanism of Cr(VI)-induced carcinogenesis (Kortenkamp et al. 1996).

Cr(VI)-Induced ROS Generation

In general, there are two pathways in the mechanism of Cr(VI)-mediated ROS generation. Cr(VI) is able to directly generate ROS during its reduction and subsequent reaction with cellular small molecules such as glutathione (GSH) and H_2O_2. A second, indirect pathway involves the stimulation of cells by Cr(VI). Cr(VI)-stimulated cells may increase the activity of NAD(P)H oxidase and generate ROS.

Direct Generation of ROS

Reaction of Cr(VI) with GSH generates glutathione-derived thiyl radical (GS$^\bullet$). An increase in GSH concentration enhances the GS$^\bullet$ generation. Reaction of Cr(VI) with cysteine or penicillamine also generates a corresponding thiyl radical (Shi et al. 1994a). The reaction equation is

$$Cr(VI) + GSH \rightarrow Cr(V) + GS^\bullet \quad (1)$$

The thiyl radicals generated by this reaction may cause direct cellular damage. These radicals may also react with other thiol molecules to generate a $O_2^{\bullet-}$ radical (Ding and Shi 2002; Yao et al. 2008) as

$$GS^\bullet + RSH \rightarrow ESSR^{\bullet-} + OH^- \quad (2)$$

$$ESSR^{\bullet-} + O_2 \rightarrow RSSR + O_2^{\bullet-} \quad (3)$$

The generation of $O_2^{\bullet-}$ radicals leads to the formation of H_2O_2 through a dismutation reaction. $O_2^{\bullet-}$ is able to cause additional oxygen radical generation by reducing Cr(VI) to Cr(V). Cr(V) can react with H_2O_2 to generate $^\bullet$OH radical through a Fenton-like reaction. Reaction of Cr(V) or Cr(IV) with H_2O_2 generates $^\bullet$OH radicals (Shi and Dalal 1992):

$$Cr(IV) + H_2O_2 \rightarrow Cr(V) + {}^\bullet OH \quad (4)$$

$$Cr(V) + H_2O_2 \rightarrow Cr(VI) + {}^\bullet OH \quad (5)$$

These reactions are similar to the Fenton reaction Fe(II) + H_2O_2 → Fe(III) + $^\bullet$OH; the reactions of Cr(IV) or Cr(V) with H_2O_2 are called Fenton-like reactions. Using xanthine and xanthine oxidase as a source of $O_2^{\bullet-}$ radicals, it has been demonstrated that Cr(VI) can be reduced to Cr(V) by $O_2^{\bullet-}$ (Shi and Dalal 1992):

$$Cr(VI) + O_2^{\bullet-} \rightarrow Cr(V) + O_2 \quad (6)$$

A combination of reaction Eqs. (5) and (6) leads to

$$O_2^{\bullet-} + H_2O_2 \xrightarrow{Cr(VI)/Cr(V)} {}^\bullet OH + O_2 + OH^- \quad (7)$$

Reaction Eq. (7) is similar to the Fe(II)/Fe(III) Haber-Weiss reaction, and thus reaction (7) can be called a Haber-Weiss-like reaction. This reaction could become particularly significant during phagocytosis, when macrophages and other cellular constituents generate large quantities of $O_2^{\bullet-}$ radicals in the so-called respiratory burst. It has been reported that a significant portion of oxygen consumed by phagocytes is first converted to $O_2^{\bullet-}$ radicals (Yao et al. 2008). However, further conversion of $O_2^{\bullet-}$ to $^\bullet$OH is too slow to be physiologically significant, unless a suitable metal ion is present as a Haber-Weiss catalyst. The finding that Cr(VI) can function as a Haber-Weiss catalyst may provide a basis for the known critical role of molecular oxygen in the genotoxic and carcinogenic reaction of Cr(VI)-containing particles. Using electron spin resonance (ESR) spin trapping, it has been demonstrated that both $^\bullet$OH and Cr(V) were generated in Cr(VI)-stimulated cells (Wang et al. 2000). The Cr(V) generated was identified as a Cr(V)-NADPH complex. Addition of NADPH enhanced the generation of these two kinds of reactive species. Addition of H_2O_2 enhanced the $^\bullet$OH generation, while catalase inhibited it, indicating that a Fenton-like reaction was involved.

Indirect Generation of ROS

Although the mechanisms of Cr(VI)-induced $^\bullet$OH generation may involve the direct interaction of Cr(VI) with cellular small molecules, such as GSH and glutathione reductase, an indirect mechanism may also occur. This indirect mechanism may involve activation of a certain signal transduction pathway to upregulate certain ROS-generating enzymes, such as NADPH oxidase. Treatment of cells with Cr(VI) increases oxygen consumption and increases the level of intracellular $O_2^{\bullet-}$. NADPH oxidase is one of the major enzymes that consumes most of the oxygen and converts it to $O_2^{\bullet-}$ during respiratory burst (Shi and Dalal 1990). It is the enzyme that catalyzes the transfer of one electron from NADPH to oxygen, leading to the formation of $O_2^{\bullet-}$.

Oxidative DNA Damage

Oxidative DNA lesions are one of the primary factors underlying Cr(VI)-induced apoptosis and carcinogenesis, the topics of the following two sections. Cr(VI) is able to cause DNA damage in both cell-free and cellular system. Studies utilizing the λ Hind III DNA digest have found that DNA damage was induced by free radical generation system consisting of a mixture of Cr(VI) and ascorbate, with and without H_2O_2 (Shi et al. 1994b). A significant amount of DNA strand breaks occur when DNA is incubated with Cr(VI) and ascorbate. The amount of DNA strand breaks is dependent upon the relative concentrations of Cr(VI) and ascorbate, which facilitates Cr(VI) reduction to Cr(V). Addition of H_2O_2 has been found to drastically enhance the DNA damage. Alternatively, addition of Mn(II) reduces DNA damage through the removal of Cr(IV) and inhibition of the Cr(IV)-mediated Fenton-like reaction.

The amount of DNA strand breaks present following Cr(VI) administration correlate with the amount of free radicals generated. ·OH radical can interact with guanine residues at several positions to generate a range of products, of which the most studied one is 8-hydroxy-deoxyguanosine (8-OHdG). The formation of this adduct is considered a marker to implicate ROS in the mechanism of toxicity and carcinogenicity of a variety of agents. Using single-cell gel electrophoresis, Cr(VI) has been shown to cause DNA damage in the human prostate cell line, LNCaP. Cr(VI)-induced ROS generation and, accordingly, DNA damage, is stronger in Ras protein–overexpressing LNCaP than in wild type (Liu et al. 2001). This indicates that Cr(VI)-generated ROS are capable of generating characteristic oxidative DNA lesions, including 8-OHdG.

Cr(VI)-Induced Apoptosis by Oxidative Stress

Apoptosis is a well-recognized form of cell death with some typical hallmarks, such as changes in nuclear morphology, chromatin condensation, and fragmentation of chromosomal DNA. During the last decade, there has been an overwhelming interest in apoptosis and the elucidation of mechanisms controlling this process. Apoptosis is an essential process required for development, morphogenesis, immune regulation, tissue remodeling, and some pathological reactions.

Under normal circumstances, the cell cycle proceeds without interruptions. However, when damage occurs particularly to DNA, most normal cells have the capacity to arrest proliferation in the G1/S or G2/M phase and then resume proliferation after the damage is repaired. The cell cycle controls the onset of DNA replication and mitosis in order to ensure the integrity of the genome. Lack of fidelity in DNA replication and maintenance can result in mutations, leading to cell death or, in multicellular organisms, cancer. Using flow cytometric analysis of DNA content, Cr(VI) is able to induce cell cycle arrest at the G2/M phase in human lung epithelial A549 cells; while at relatively low concentrations Cr(VI) causes cell cycle arrest, at relatively high concentrations Cr(VI) induces apoptosis, and ROS generated by Cr(VI)-stimulated cells are involved in Cr(VI)-induced cell cycle arrest and among the ROS H_2O_2 plays a key role (Zhang et al. 2001).

Since apoptosis is an important factor influencing the malignant transformation of cells, the regulation of cell apoptosis may be critical in metal-induced carcinogenesis. Cr(VI) itself is incapable of reacting with macromolecules such as DNA, RNA, proteins, and lipids (Shi et al. 2000). Instead, Cr(V) or Cr(III), intermediates of Cr(VI) reduction, can form covalent interactions with DNA and other macromolecules, a process that activates DNA-dependent protein kinases (DNA-PKs) and induces subsequent p53 activation and cell apoptosis (Singh et al. 1998). The DNA-damaging effect of Cr(VI) may also be through ROS generated during Cr(VI) reduction.

Activation of the p53 tumor suppressor protein is considered to be a major step in apoptosis induced by Cr(VI) (Son et al. 2010; Wang et al. 2000). p53 is considered an oxidative stress response transcription factor and can be activated in response to a variety of stimuli, such as UV, γ radiation, and nucleotide deprivation. Several mechanisms are involved in Cr(VI)-induced p53 activation. First, direct DNA damage by Cr(VI) or ROS generated during cellular Cr(VI) reduction activates upstream kinases, including DNA-PK, ATM, ATR, and others, for p53 phosphorylation and activation (Yao et al. 2008). Second, the p53 protein contains several redox-sensitive cysteines critical for the DNA binding activity of p53 (Meplan et al. 2000).

Evidence indicates that Cr(VI)-dependent p53 activation is ROS-mediated. Several recent studies have highlighted the importance of H_2O_2 and hydroxyl radical in Cr(VI)-induced p53 activation and cell death (Son et al. 2010; Wang et al. 2000; Ye et al. 1999;

Ye et al. 1995). SOD has been shown to increase p53 activity by enhancing the production of H_2O_2 from $O_2^{\bullet-}$. Alternatively, catalase, a H_2O_2 scavenger, inhibits p53 activation through elimination of $^{\bullet}OH$ radical generation. Sodium formate and aspirin, $^{\bullet}OH$ radical scavengers, also suppress p53 activation. Deferoxamine, a metal chelator, inhibits p53 activation by chelating Cr(V) to make it incapable of generating radicals from H_2O_2. NADPH, which accelerates the one-electron reduction of Cr(VI) to Cr(V) and increased $^{\bullet}OH$ radical generation, enhances p53 activation. Thus, $^{\bullet}OH$ radicals generated from Cr(VI) reduction are primarily responsible for Cr(VI)-induced p53 activation. The activation of p53 is at the protein level instead of the transcriptional level (Wang and Shi 2001).

As a transcription factor, p53 is able to upregulate the expression of genes involved in either ROS production or metabolism, including quinone oxidoreductase (Pig3), proline oxidase (Pig6) homologues, glutathione transferase (Pig12), and glutathione peroxidase (GPx) (Polyak et al. 1997). Moreover, p53 also activates the expression of several genes that directly control or regulate the process of apoptosis.

It has been shown that Cr(VI)-derived ROS initiate early apoptosis prior to activation of p53 protein (Ye et al. 1999). Alternate pathways have been implicated in activation of Cr(VI)-induced apoptosis, including MAPK-dependent signaling pathways and direct mitochondrial damage (Son et al. 2011). Although p53 is not necessarily required for initiation of Cr(VI)-induced apoptosis, it has been shown to significantly enhance late-stage apoptosis by transcriptional activation of p53-mediated apoptotic cascades. Therefore, Cr(VI) induces apoptosis through both p53-dependent and p53-independent pathways through a common ROS-mediated pathway.

Cr(VI)-Induced Carcinogenesis by Oxidative Stress

Cr(VI) is a known human carcinogen. The mechanisms underlying Cr(VI)-induced carcinogenesis are multifold. These include direct and indirect DNA damage, as discussed previously, as well as activation of oncogenic cell signaling proteins. These oncogenic proteins, or oncoproteins, typically promote cell survival, growth, and proliferation. Tumor suppressors, as their name suggests, repress carcinogenesis and tumorigenesis through causing cell senescence and/or apoptosis. These oncogenic pathways will be discussed in detail next.

NF-κB

NF-κB is considered to be a primary oxidative stress response transcription factor. NF-κB promotes cell survival by stimulating the transcription of a variety of cell survival genes. Cr(VI) stimulates NF-κB activation in vitro, an effect which is attenuated by the inhibition of IκB kinase (IKK), an upstream promoter of NF-κB phosphorylation. The reduction of Cr(VI) to lower oxidation states is required for Cr(VI)-induced NF-κB activation. The cotreatment of aspirin, an antioxidant, has been shown to attenuate NF-κB activation by Cr(VI). This indicates that hydroxyl radicals generated by Cr(V)- and Cr(IV)-mediated Fenton-like reactions likely play a prominent role in the mechanism of Cr(VI)-induced NF-κB activation. The inhibition of IKK, an upstream NF-κB activator, attenuates NF-κB activation by Cr(VI). This indicates that ROS acts as the ultimate upstream signal regulating NF-κB activation by Cr(VI), with IKK acting as a downstream terminal kinase.

NF-κB binding sites serve as an enhancer element in c-myc, a gene associated with the formation of Burkitt's lymphoma (Ji et al. 1994). Cr(VI) could induce expression of c-myc proto-oncogene via NF-κB activation. It is possible that NF-κB activation and a subsequent expression of NF-κB-regulated proto-oncogenes, such as c-myc, may play a role in the induction of neoplastic transformation by Cr(VI).

AP-1

Another important transcription factor whose activity is stimulated by Cr(VI) is AP-1. AP-1 is a multimeric protein consisting of Jun (c-Jun, JunB, and JunD) and Fos (c-Fos, FosB, Fra1, and Fra2) subunits. AP-1 binds to the TRE/AP-1 DNA response elements and regulates many kinds of early response gene expression (Munoz et al. 1996). Activation of AP-1 results in the overexpression of c-Jun and other proto-oncogenes. A number of mitogen-activated protein kinases (MAPK) members participate in the activation of AP-1 hierarchically through divergent kinase cascades. MAPKs, such as c-Jun-N-terminal kinase (JNK) and p38, are activated by a specific MAPK kinase (MAPKK) through phosphorylation of conserved threonine and tyrosine residues. In turn, a MAPKK is activated by a specific MAPKK kinase (MAPKKK) through phosphorylation of conserved threonine and/or serine residues.

Cr(VI) is capable of inducing AP-1 activation (Chen et al. 2000). The induction of AP-1 by Cr(VI) is associated with phosphorylation of the MAPKs p38 and JNK but not that of the extracellular-signal-regulated kinase (ERK). Cotreatment with aspirin has been shown to attenuate AP-1 activation by Cr(VI). Inhibition of p38 also decreases Cr(VI)-induced AP-1 activation. These results suggest that ROS serves as the ultimate upstream signal initiating activation of AP-1, with p38 acting as the downstream executive kinase.

HIF-1 and VEGF

Hypoxia-inducible factor 1(HIF-1) is a heterodimeric basic helix-loop-helix transcription factor, composed of HIF-1α and HIF-1β/ARNT subunits. HIF-1α is unique to HIF-1 and is induced exponentially in response to a decrease in cellular O_2 concentration. In contrast, HIF-1β is identical to the aryl hydrocarbon nuclear translocator (ARNT) that heterodimerizes with an aryl hydrocarbon receptor and is not regulated by cellular oxygen tension.

HIF-1 regulates the expression of many genes including vascular endothelial growth factor (VEGF), erythropoietin (EPO), heme oxygenase 1, aldolase, enolase, and lactate dehydrogenase A. High levels of HIF-1 activity in cells are correlated with tumorigenicity and angiogenesis in nude mice. HIF-1 is induced by the expression of oncogenes, such as v-Src and Ras, and is overexpressed in many human cancers. HIF-1 activates the expression of VEGF gene at the transcriptional level.

Vascular endothelial growth factor (VEGF) is an essential protein for tumor angiogenesis. VEGF plays a key role in tumor progression and angiogenesis. Inhibition of VEGF expression and function of its receptor dramatically decreases tumor growth, invasion, and metastasis in animal models. Tissue hypoxia is a major inducer for the expression of VEGF in tumors. Somatic mutations, such as oncogene Ras activation and tumor suppressor gene p53 inactivation, also increased VEGF expression.

Cr(VI) induces HIF-1 activity through the specific expression of HIF-1α, but not the HIF-1β subunit, and increases the level of VEGF expression in DU145 human prostate carcinoma cells (Gao et al. 2002). To dissect the signaling pathways involved in Cr(VI)-induced HIF-1 expression, p38 MAP kinase signaling was required for HIF-1α expression induced by Cr(VI). Neither PI3K nor ERK activity was required for Cr(VI)-induced HIF-1 expression. Cr(VI) induced expression of HIF-1 and VEGF through the production of ROS in DU145 cells. The major species of ROS for the induction of HIF-1 and VEGF expression is H_2O_2. These results suggest that the expression of HIF-1 and VEGF induced by Cr(VI) may be an important signaling pathway in Cr(VI)-induced carcinogenesis.

Tyrosine Phosphorylation

Tyrosine phosphorylation is an important step in the regulation of many key cellular functions. It is involved in control of cell proliferation, differentiation, cell-cycle regulation, cell signal transduction, metabolism, transcription, morphology, adhesion, ion channels, and cancer development. Cr(VI) increased tyrosine phosphorylation in human epithelial A549 cells in a time-dependent manner (Qian et al. 2001). N-acetyl-cysteine (NAC), a general antioxidant, inhibited Cr(VI)-induced tyrosine phosphorylation. Catalase (a scavenger of H_2O_2), sodium formate, and aspirin (scavengers of •OH radical) also inhibited the increased tyrosine phosphorylation induced by Cr(VI). H_2O_2 and •OH radicals generated by cellular reduction of Cr(VI) are responsible for the increased tyrosine phosphorylation induced by Cr(VI).

Summary

Cr(VI) is able to generate ROS in various biological systems. Reduction of Cr(VI) to its low oxidation states, such as Cr(V) and Cr(IV), is an important step. Fenton-like and Haber-Weiss-type reactions are common pathways for Cr(VI)-induced ROS generation. Cr(VI) may also be able to generate ROS through stimulation of the cells and upregulation of certain ROS generating proteins, such as NADPH oxidase. When the ROS present in the cellular system overpower the defense systems, oxidative stress will occur. The persistent oxidative stress caused by Cr(VI) exposure may play a key role in activation of transcription factors NF-κB, AP-1, p53, and HIF-1; regulation of cell cycle; and induction of apoptosis. All of these processes could be involved in the Cr(VI)-induced carcinogenic activation.

Cross-References

▶ Chromium, Physical and Chemical Properties
▶ Chromium Toxicity, High-Valent Chromium
▶ Hexavalent Chromium and Cancer
▶ Trivalent Chromium

References

Chen F, Ding M, Lu Y, Leonard SS, Vallyathan V, Castranova V, Shi X (2000) Participation of MAP kinase p38 and IkappaB kinase in chromium (VI)-induced NF-kappaB and AP-1 activation. J Environ Pathol Toxicol Oncol 19:231–238

Ding M, Shi X (2002) Molecular mechanisms of Cr(VI)-induced carcinogenesis. Mol Cell Biochem 234–235:293–300

Gao N, Jiang BH, Leonard SS, Corum L, Zhang Z, Roberts JR, Antonini J, Zheng JZ, Flynn DC, Castranova V et al (2002) p38 Signaling-mediated hypoxia-inducible factor 1alpha and vascular endothelial growth factor induction by Cr(VI) in DU145 human prostate carcinoma cells. J Biol Chem 277:45041–45048

Ji L, Arcinas M, Boxer LM (1994) NF-kappa B sites function as positive regulators of expression of the translocated c-myc allele in Burkitt's lymphoma. Mol Cell Biol 14:7967–7974

Kortenkamp A, Casadevall M, Da Cruz Fresco P (1996) The reductive conversion of the carcinogen chromium (VI) and its role in the formation of DNA lesions. Ann Clin Lab Sci 26:160–175

Liu K, Husler J, Ye J, Leonard SS, Cutler D, Chen F, Wang S, Zhang Z, Ding M, Wang L et al (2001) On the mechanism of Cr (VI)-induced carcinogenesis: dose dependence of uptake and cellular responses. Mol Cell Biochem 222:221–229

Meplan C, Richard MJ, Hainaut P (2000) Redox signalling and transition metals in the control of the p53 pathway. Biochem Pharmacol 59:25–33

Munoz C, Pascual-Salcedo D, Castellanos MC, Alfranca A, Aragones J, Vara A, Redondo JM, de Landazuri MO (1996) Pyrrolidine dithiocarbamate inhibits the production of interleukin-6, interleukin-8, and granulocyte-macrophage colony-stimulating factor by human endothelial cells in response to inflammatory mediators: modulation of NF-kappa B and AP-1 transcription factors activity. Blood 88:3482–3490

Polyak K, Xia Y, Zweier JL, Kinzler KW, Vogelstein B (1997) A model for p53-induced apoptosis. Nature 389:300–305

Qian Y, Jiang BH, Flynn DC, Leonard SS, Wang S, Zhang Z, Ye J, Chen F, Wang L, Shi X (2001) Cr (VI) increases tyrosine phosphorylation through reactive oxygen species-mediated reactions. Mol Cell Biochem 222:199–204

Shi XL, Dalal NS (1990) NADPH-dependent flavoenzymes catalyze one electron reduction of metal ions and molecular oxygen and generate hydroxyl radicals. FEBS Lett 276:189–191

Shi XL, Dalal NS (1992) The role of superoxide radical in chromium (VI)-generated hydroxyl radical: the Cr(VI) Haber-Weiss cycle. Arch Biochem Biophys 292:323–327

Shi X, Dong Z, Dalal NS, Gannett PM (1994a) Chromate-mediated free radical generation from cysteine, penicillamine, hydrogen peroxide, and lipid hydroperoxides. Biochim Biophys Acta 1226:65–72

Shi X, Mao Y, Knapton AD, Ding M, Rojanasakul Y, Gannett PM, Dalal N, Liu K (1994b) Reaction of Cr(VI) with ascorbate and hydrogen peroxide generates hydroxyl radicals and causes DNA damage: role of a Cr(IV)-mediated Fenton-like reaction. Carcinogenesis 15:2475–2478

Shi X, Leonard SS, Wang S, Ding M (2000) Antioxidant properties of pyrrolidine dithiocarbamate and its protection against Cr(VI)-induced DNA strand breakage. Ann Clin Lab Sci 30:209–216

Singh J, McLean JA, Pritchard DE, Montaser A, Patierno SR (1998) Sensitive quantitation of chromium-DNA adducts by inductively coupled plasma mass spectrometry with a direct injection high-efficiency nebulizer. Toxicol Sci 46:260–265

Son YO, Hitron JA, Wang X, Chang Q, Pan J, Zhang Z, Liu J, Wang S, Lee JC, Shi X (2010) Cr(VI) induces mitochondrial-mediated and caspase-dependent apoptosis through reactive oxygen species-mediated p53 activation in JB6 Cl41 cells. Toxicol Appl Pharmacol 245:226–235

Son YO, Hitron JA, Cheng S, Budhraja A, Zhang Z, Shi X (2011) The dual roles of c-Jun NH2-terminal kinase signaling in Cr(VI)-induced apoptosis in JB6 cells. Toxicol Sci 119:335–345

Wang S, Shi X (2001) Mechanisms of Cr(VI)-induced p53 activation: the role of phosphorylation, mdm2 and ERK. Carcinogenesis 22:757–762

Wang S, Leonard SS, Ye J, Ding M, Shi X (2000) The role of hydroxyl radical as a messenger in Cr(VI)-induced p53 activation. Am J Physiol Cell Physiol 279:C868–C875

Yao H, Guo L, Jiang BH, Luo J, Shi X (2008) Oxidative stress and chromium(VI) carcinogenesis. J Environ Pathol Toxicol Oncol 27:77–88

Ye J, Zhang X, Young HA, Mao Y, Shi X (1995) Chromium(VI)-induced nuclear factor-kappa B activation in intact cells via free radical reactions. Carcinogenesis 16:2401–2405

Ye J, Wang S, Leonard SS, Sun Y, Butterworth L, Antonini J, Ding M, Rojanasakul Y, Vallyathan V, Castranova V et al (1999) Role of reactive oxygen species and p53 in chromium (VI)-induced apoptosis. J Biol Chem 274:34974–34980

Zhang Z, Leonard SS, Wang S, Vallyathan V, Castranova V, Shi X (2001) Cr (VI) induces cell growth arrest through hydrogen peroxide-mediated reactions. Mol Cell Biochem 222:77–83

Chromium, Physical and Chemical Properties

J. David Van Horn
Department of Chemistry, University of Missouri-Kansas City, Kansas City, MO, USA

Synonyms

Cr, Element 24, [7440-47-3]

Definition

Chromium is a metallic element in the first transition series, Group IVB (6).

General Chromium Chemistry and Biochemistry

Chromium (Cr, atomic weight, 51.996; atomic number, 24) is an abundant metallic element usually found associated with iron-containing minerals and ores; chromite ($FeCr_2O_4$) is the most important source. The name "chromium" is derived from the Greek, χρώμα (color), referring to the intense coloration of many Cr compounds, such as $PbCrO_4$ (crocoite), a yellow pigment. Cr metal resists corrosion due to the formation of an oxide layer that passivates its surface (*cf.* aluminum). Cr metal is used in iron and stainless steel alloys for its hardening and anticorrosion properties; it is also used to electroplate surfaces.

Cr belongs to the first transition series, possesses an [Ar] $3d^5 4s^1$ electronic configuration, and can form compounds in every oxidation state from −2 to +6. The biologically important oxidation states include those from +2 to +6. The most prevalent oxidation states encountered are Cr(III) and Cr(VI); Cr(III) is the most thermodynamically stable state. While the highly charged Cr(VI) ([Ar] $3d^0$) metal center is unstable, it is kinetically stabilized by the presence of the oxo (O^{-2}) ligands. Thus, the chemistry of the higher oxidation states is dominated by the tetrahedral oxyanions of Cr(VI), namely, chromate (CrO_4^{-2}), hydrogen chromate ($HCrO_4^{-1}$), and dichromate ($Cr_2O_7^{-2}$). These three oxyanions are in equilibrium in a pH- and concentration-dependent manner. Therefore, oxyanions of Cr(IV) exist and are soluble over the entire range of pH. The structures and oxyanion chemistry of chromium trioxide (CrO_3), chromate, and dichromate are analogous to those of sulfur trioxide (SO_3), sulfate (SO_4^{-2}), and pyrosulfate ($S_2O_7^{-2}$), respectively. The Cr(VI) oxyanions are important as strong oxidants in analytical chemistry and in industry, and because of the health hazards they pose to biological organisms. The ability of chromium to pass through various biological compartments as a −2 oxyanion contributes to its hazardous properties. Cr(IV) and Cr(V) compounds readily undergo disproportionation, which may impede their characterization. Cr(III) and the higher oxidation states are hard Lewis acids and prefer oxygen, fluoride, or other hard donors; Cr(II) and lower states are softer acids and may accept nitrogen, carbon, and sulfur donors.

Chromium, Physical and Chemical Properties, Table 1 Comparison of M(III) hydroxide solubility products; a larger pK_{sp} indicates a more favorable reaction (Lide 2009)

M(III) hydroxide	K_{sp}	pK_{sp}
$Fe(OH)_3$	2.79×10^{-39}	38.5
$Ga(OH)_3$	7.28×10^{-36}	35.1
$Al(OH)_3$	3×10^{-34}	33.5
$Cr(OH)_3$	6.7×10^{-31}	*30.2*
$Nd(OH)_3$	1.1×10^{-26}	25.9
$Ce(OH)_3$[a]	2×10^{-20}	19.7

[a]Ce(III) hydrolysis is complicated by the extraordinarily stable Ce(IV) oxidation state

The extensive redox chemistry of chromium is reminiscent of that of manganese, but Cr does not play such an extensive and essential biological role as Mn. The standard reduction potentials of Cr(III) to Cr metal are −0.77 and −1.33 V in acidic and in basic conditions, respectively. The corresponding reduction potentials of Mn(II) to Mn metal are −1.18 and −1.56 V, respectively. The standard reduction potential for $Cr_2O_7^{-2}$ to Cr(III) (*aq*) in acidic solution is 1.38 V; the analogous three-electron-step reduction of permanganate (MnO_4^{-1}) to manganese dioxide (MnO_2) is 1.70 V. Redox reactions of the Cr(V) and Cr(IV) states are more typically observed in acidic conditions.

Cr(III) coordination chemistry is characterized by substitutional inertness and an essentially octahedral (O_h) geometric preference. In addition to extensive mononuclear examples, the types of Cr(III) compounds include numerous multinuclear clusters, such as $Cr_3O(O_2CCH_3)$ and $[Cr_8(OH)_8(O_2CCH_3)_{16}]$ (Eshel 2001). Many Cr(III) complexes are stable and soluble at neutral to high pH but are susceptible to hydrolysis and may precipitate as pure or mixed Cr(III) hydroxides. Cr(III) coordination and redox chemistry is sometimes compared to that of Fe(III). However, the biological redox chemistry of Fe is limited to the +2 and +3 states, and the comparable reduction reactions are quite different: The M(III) to M(II) potentials for Cr and Fe are −0.424 and 0.771 V, respectively. Chemical differences for Cr(III) may also be compared using the M(III) hydroxide solubility products (Lide 2009) of a range of +3 cations (Table 1). Thus, Cr(III) is less susceptible to hydrolysis than Fe and is more comparable to Al(III) or the

Chromium, Physical and Chemical Properties, Table 2 Comparison of Cr(III) to Fe(III), Mn(III), and cations with similar charge/ionic radius ratios. Charge/I.R. is proportional to cation surface charge (Shannon 1976)

Cation	Coordination number	Ionic radius (Å)	Charge/I.R. (Å^{-1})
Fe(III) or Mn(III) (high spin)	6	0.645	4.65
...			
Hf(IV)	8	0.83	4.82
Ga(III)	6	0.62	4.84
In(III)	4	0.62	4.84
Cr(III)	*6*	*0.615*	*4.89*
Co(III) (high spin)	6	0.61	4.92
Sn(IV)	8	0.81	4.94
...			
Fe(III) or Mn(III)	5	0.58	5.17

lanthanides, e.g., Nd(III). A comparison of ionic radii, and charge-to-radius ratios (Table 2) also indicates some metal ions that are more similar to Cr(III) than Fe(III) or Mn(III) (Shannon 1976).

While the [Ar] d^0 closed-shell configuration of Cr(VI) should not be colored due to lack of d electrons for d–d transitions, chromate and dichromate are colored yellow and orange, respectively, due to charge transfer between the metal center and oxo ligands. Most other Cr compounds are colored due to d–d orbital electronic transitions and thus available to spectroscopic analysis.

Chromates and chromium compounds are amenable to a wide variety of spectroscopic analysis and standard methods of metal analysis, including titrimetry, atomic absorbance, neutron activation, and X-ray fluorescence. Potassium dichromate ($K_2Cr_2O_7$) is a primary standard for redox titrimetry. Electron paramagnetic resonance is utilized in characterizing paramagnetic Cr(I, III, or V) compounds and in following the stages of redox reactions of biochemical interest from Cr(VI) to Cr(III); EPR analysis is typically done in the X band (~9.5 GHz) at liquid N_2 or He temperatures. In radiochemistry, Cr-51 is a gamma-emitting radioisotope with a half-life of 28 days; it is used as a radiotracer in analytical and biochemical experiments.

Cr(III) has not been identified in any biological redox system or enzyme, and a role for Cr(III) as a trace element in biological systems, especially in humans, has not been definitively described for healthy individuals. Some Cr(III) compounds have been evaluated as possible insulin mimetic agents or therapeutic agents for glucose metabolism or lipid maintenance in Type II diabetes and associated cardiovascular disease (Levina 2008). Cr(III) compounds generally do not pose an acute hazard due to the inability of Cr(III) salts, e.g., $CrCl_3$, to adequately pass through biological membranes. The oral bioavailability of Cr(III) as the chloride salt is less than 1%. Other compounds, notably, Cr(III) picolinate, have an oral bioavailability of about 10%. Chronic disease caused by Cr(III) has not been definitively determined, though the compounds that may access two-electron Cr(II)/Cr(IV) redox cycles may pose a potential hazard related to having Cr(III) present in the human body.

Cr(VI) is a significant anthropogenic environmental hazard in certain locations and in industrial settings and represents an acute or chronic health hazard. Cr(V) and Cr(IV) compounds participate in direct mechanisms of carcinogenesis. The most affected organs are the lungs and associated airways. Ingestion and absorption through the skin are the major routes of Cr(VI) contamination from contaminated water supplies or from contact with compounds or solutions in industrial settings. Other diseases caused by high-oxidation-state Cr compounds include gastroenteritis, liver damage, and acute renal failure.

Cross-References

▶ Chromium and Allergic Reponses
▶ Chromium and Diabetes
▶ Chromium and Glucose Tolerance Factor
▶ Chromium and Human Nutrition
▶ Chromium and Insulin Signaling

- Chromium and Leather
- Chromium and Membrane Cholesterol
- Chromium and Nutritional Supplement
- Chromium Binding to DNA
- Chromium: Chromium(VI)
- Chromium Toxicity, High-Valent Chromium
- Chromium(III) and Low Molecular Weight Peptides
- Chromium(III) and Immune System
- Chromium(III) and Transferrin
- Chromium(III), Cytokines, and Hormones
- Chromium(VI), Oxidative Cell Damage
- Chromium(III), Cytokines, and Hormones
- Diabetes
- Glucose Tolerance Factor
- Hexavalent Chromium and Cancer
- Insulin Signaling
- Intracellular Signaling
- Leather
- Transferrin
- Trivalent Chromium

References

Eshel M, Bino A (2001) Polynuclear chromium(III) carboxylates Part 2. Chromium(III) acetate – what's in it? Inorg Chim Acta 320:127–132

Levina A, Lay P (2008) Chemical properties and toxicity of chromium(III) nutritional supplements. Chem Res Toxicol 21:563–571

Lide D (ed) (2009) CRC handbook of chemistry and physics, 90th edn. Boca Raton, CRC Press

Shannon R (1976) Revised effective ionic radii and systematic studies of interatomic distances in halides and chalcogenides. Acta Crystallogr A32:751–767

Chromium: Chromate

- Hexavalent Chromium and DNA, Biological Implications of Interaction

Chromium: Chromium(VI)

- Hexavalent Chromium and DNA, Biological Implications of Interaction

Chromium: Cr(VI)

- Hexavalent Chromium and DNA, Biological Implications of Interaction

Chromium: Cr^{6+}

- Hexavalent Chromium and DNA, Biological Implications of Interaction

Chromobindins

- Annexins

Chromodulin

- Chromium(III) and Low Molecular Weight Peptides

Chronic Arsenic Intoxication Impairs Glucose Homeostasis

- Arsenic-Induced Diabetes Mellitus

Chronic Arsenicosis ("Arsenicosis" Needs to Be Differentiated from "Acute Arsenic Poisoning")

- Arsenicosis

Chronic Beryllium Disease (CBD)

- Beryllium as Antigen

Chronic Obstructive Pulmonary Disease

- Polonium and Cancer

Cisplatin

▶ Platinum Anticancer Drugs

Cisplatin Uptake and Cellular Pharmacology

▶ Platinum Interaction with Copper Proteins

Cluster Dynamics

▶ Iron-Sulfur Cluster Proteins, Ferredoxins

Clusters

▶ Gold Nanoparticles, Biosynthesis

Cobalamin

▶ Cobalt Proteins, Overview
▶ Cobalt-containing Enzymes

Cobalt Proteins, Overview

Masafumi Odaka[1] and Michihiko Kobayashi[2]
[1]Department of Biotechnology and Life Science, Graduate School of Technology, Tokyo University of Agriculture and Technology, Koganei, Tokyo, Japan
[2]Graduate School of Life and Environmental Sciences, Institute of Applied Biochemistry, The University of Tsukuba, Tsukuba, Ibaraki, Japan

Synonyms

Cobalamin; Methionine amino peptidase; Nitrile hydratase; Non-corrin cobalt; Radical rearrangement; Vitamin B_{12}

Definition

Cobalt is relatively rare among essential trace elements but plays important roles in various living organisms. For instance, cobalamin cofactor comprising a cobalt ion coordinated in a substituted corrin macrocycle is known as vitamin B_{12}, and its deficiency may cause severe damage to mammals. In this entry, the structure and function of vitamin B_{12}-dependent enzymes and non-corrin cobalt enzymes are reviewed.

Introduction

Cobalt is the 27th element in the periodic table and a transition metal with seven d electrons. It can take oxidation states from $-I$ to $+IV$ and preferably exists as Co^{2+} and Co^{3+}. Cobalt is relatively rare among essential trace elements, and its concentrations in seawater and the cytosol are 10^{-11} and 10^{-9} mol dm^{-3}, respectively. Geochemistry of trace metals suggests that the cobalt ion existed in ancient sea in relatively high concentrations, but it decreased after the generation of molecular oxygen through photosynthesis started. This view is consistent with the fact that most cobalt-containing enzymes are found in Archaea and Bacteria. Among the first-row transition metals, cobalt is less frequently used in a metalloenzyme than other elements such as iron, copper, zinc, and manganese. However, cobalamin cofactor, comprising a cobalt ion coordinated in a substituted corrin macrocycle, is essential for mammals, and a deficiency of it may cause severe damage to the brain and nerve system and lead to pernicious anemia (Matthews et al. 2008; Randaccio et al. 2007). Cobalamin is also known as vitamin B_{12}, and the biochemistry of the B_{12}-dependent enzymes has been extensively investigated. In addition, there are some enzymes as well as proteins that have been indicated to possess non-corrin cobalt cofactors (Kobayashi and Shimizu 1999). In particular, a mononuclear non-corrin cobalt enzyme, nitrile hydratase, has been biotechnologically very important because of its industrial use for acrylamide and nicotinamide production. This entry first overviews the biochemistry of and recent progress regarding B_{12}-dependent enzymes and then those for other cobalt-containing proteins.

Vitamin B$_{12}$-Dependent Enzymes

The biosynthesis of vitamin B$_{12}$ compounds is performed only by some microorganisms and algae (Escalante-Semerena and Warren 2008), and thus, mammals have to take them up via food (Matthews et al. 2008; Randaccio et al. 2007). Figure 1 shows the structure of adenosylcobalamin (AdoCbl). The cobalt is six-coordinated. Four ligands are provided by a corrin macrocycle, and the fifth ligand is the nitrogen atom of a 5,6-dimethylbenzimidazole nucleotide substituent covalently bonded to the corrin ring or a histidine from the enzyme. The sixth ligand is the C1 methylene group of 5′-deoxyadenosine which is a very unique example of a naturally occurred organic metal compound. The other B$_{12}$ cofactor is methyl cobalamin (MeCbl), in which the sixth ligand is substituted by a methyl group. Usually, B$_{12}$ compounds are taken as aquacobalamin, hydroxocobalamin, or cyanocobalamin through specific import and transport systems and then transformed into AdoCbl or MeCbl after binding to the catalytic site of the B$_{12}$-dependent enzymes.

There are two B$_{12}$-depndent enzymes in humans (Randaccio et al. 2007). One is methionine synthase (MS), which catalyzes the transfer of a methyl group from N-methyl-tetrahydrofolate (CH$_4$-H$_4$Folate) to homocysteine (Hcy), which produces methionine. Human MS is pathologically important because a dysfunction of it results in homocysteine accumulation in the blood, which causes homocystinuria. Although the structure of human MS has not been determined, the crystal structures of the N-terminal and C-terminal halves of methionine synthases (MetH) from *Thermotoga maritima* and *Escherichia coli* provide significant insights into the catalytic mechanism (Fig. 2). Human MS and bacterial MetHs share four modular structures. The N-terminus two modules are the Hcy-binding and CH$_4$-H$_4$Folate-binding domains (Hcy and Fol domains, respectively). The third module is composed of the Cap and Cob (B$_{12}$-binding) domains, and the C-terminus module comprises the AdoMet domain, which contains S-adenosyl-L-methionine (AdoMet). When MetH is activated, a strong nucleophile, cob(I)alamin, in the Cob domain extracts the methyl group from CH$_4$-H$_4$Folate to yield methyl-cob(III)alamin. Then, the Cob domain interacts with the Hcy-domain to

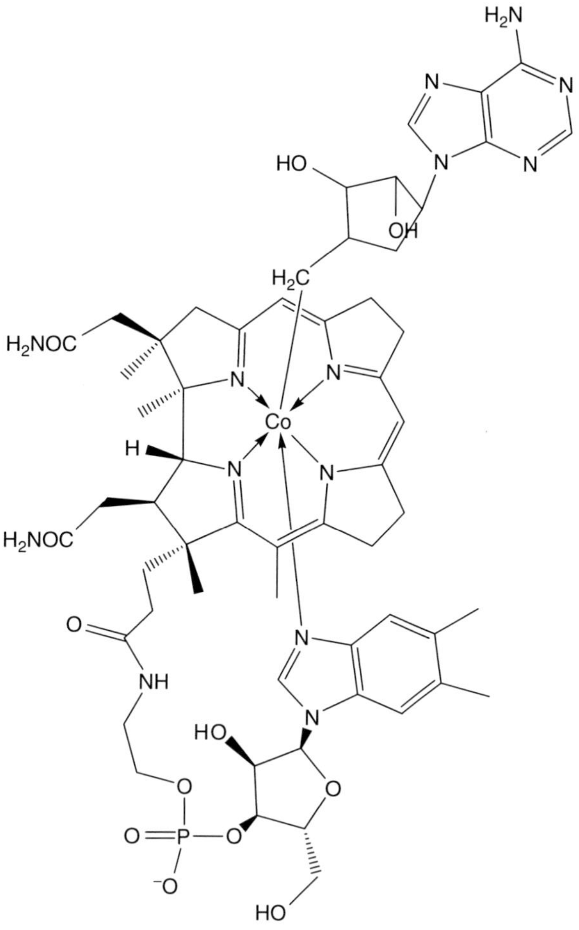

Cobalt Proteins, Overview, Fig. 1 Structure of adenosylcobalamin

methylate Hcy into methionine. It is known that both human MS and bacterial MetHs lose their catalytic activity in every 1,700–2,000 turnovers because the highly reactive cob(I)alamin is occasionally oxidized into an inert cob(II)alamin form under aerobic conditions. Therefore, organisms have intrinsic reactivation systems. The cob(II)alamin in MetH is reduced to cob(I)alamin by FAD-dependent ferredoxin-NADP$^+$ reductase (FNR) with flavodoxin, while that of human MS is reduced by methionine synthase reductase, which is the fusion protein of FNR and flavodoxin. The regenerated cob(I)alamin accepts the methyl group from an AdoMet molecule in the AdoMet domain to become methyl-cob(III)alamin again. Recent biochemical studies indicated that methionine synthase reductase not only reactivates

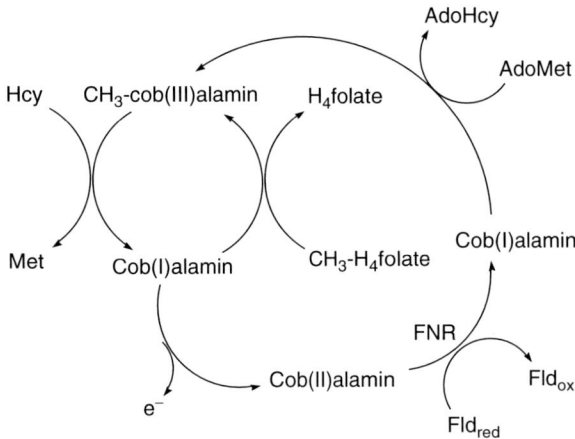

Cobalt Proteins, Overview, Fig. 2 Scheme of proposed reaction mechanism of methyltransferase. *Hcy* homocysteine, *AdoHcy* adenosylhomocysteine, *AdoMet* adenosylmethionine, *Fld* flavodoxin, *FNR* FAD-dependent ferredoxin-NADP$^+$ reductase

the cob(II)alamin in MS but also enhances the incorporation of cobalamin into the apo-form of MS through a specific protein-protein interaction (Wolthers and Scrutton 2009).

Some anaerobic organisms such as hydrogenogenic, acetogenic, and methanogenic bacteria and Archaea can convert CO and CO_2 into acetyl-CoA via the reductive acetyl-CoA pathway, the methyl group being transferred from CH_4-H_4Folate to the cob(I)alamin of the corrinoid iron-sulfur protein (CoFeSP) by its partner enzyme, methyltransferase (MeTr), and then, using the methyl group, acetyl-CoA is synthesized from CO and CoA by acetyl-CoA synthase (Matthews et al. 2008). Very recently, Goetzl et al. (2011) elucidated the crystal structures of CoFeSP and MeTr from *Carboxydothermus hydrogenoformans* and proposed a catalytic mechanism based on their conformational flexibility. Methanol:coenzyme M methyltransferase transfers a methyl group from methanol to CoM through a similar mechanism (Matthews et al. 2008). The enzyme is composed of three subunits, MtaA, MtaB, and MtaC. The cob(I)alamin in MtaC nucleophilically attacks the C-O bond of methanol activated by the zinc (II) ion of MtaB and cleaves it heterolytically to produce methyl-cob(III)alamin. Then the methyl group is transferred to CoM by MtaA. MtaC is structurally related to the cobalamin-binding domain of MetH.

Another human B_{12}-dependent enzyme is methyl malonyl-CoA mutase (MUT), which possesses AdoCbl at its catalytic site (Randaccio et al. 2007). MUT catalyzes isomerization of methylmalonyl-CoA into succinyl-CoA. The catalytic activity of MUT is explained as 1,2-radical rearrangement (Fig. 3). When a substrate binds to the enzyme, the Co(III)-C bond is cleaved homolytically to produce a Co(II) radical and an adenosyl radical. The adenosyl radical subtracts the hydrogen atom from the C1 site of the substrate to generate the substrate radical (reaction A). The product radical generated on the transfer of the X group from C2 to C1 subtracts the hydrogen atom from 5′-deoxyadenosine to yield the product and to regenerate the adenosyl radical (reaction B). The activities of the bacterial AdoCbl-dependent enzymes, 1,2-diol dehydratase, glycerol dehydratase, and ethanolamine ammonia lyase, can be explained similarly. 1,2-diol dehydratase and ethanolamine ammonia lyase produce 1,1-diol and 1-hydroxy-1-amine radicals as intermediates, respectively. Because of their chemical instabilities, the intermediates self-react immediately to produce water and ammonia in addition to the corresponding aldehydes.

A rare genetic disorder, isolated methylmalonic aciduria, mainly results from a deficiency of one of the three genes, *cbl*A, *cbl*B, and *mut*, which correspond to mitochondrial proteins MMNA, MMNB, and MUT. MMNB is an ATP: cob(I)alamin adenosyltransferase (ATR) that converts the inactive cob(II)alamin form into the active AdoCbl and delivers it to MUT. Banerjee and his colleagues (Padovani and Banerjee 2009) found that MeaB from *Methylobacterium extorquens* AM1, the ortholog of human MMNA, is a GTP-binding protein that functions as a chaperone editing the release of the inactive cob(II)alamin from MUT and gating the insertion of AdoCbl into the corresponding MUT. Since MUT is occasionally inactivated through loss of the deoxyadenosine moiety at the end of catalysis, MeaB is likely to rescue the removal of cob(II)alamin from the inactivated MUT. Recently, the crystal structures of a GTP-bound form of human MMNA and apo-, holo-, and substrate-bound forms of human MUT were determined (Froese et al. 2010). By combining crystal structural and kinetic studies, the authors demonstrated the guanine nucleotide-dependent interaction of human MMNA and human MUT and the gating function of human MMNA as to AdoCbl binding to human MUT.

Cobalt Proteins, Overview, Fig. 3 Scheme of proposed reaction mechanism of AdoCbl-dependent enzymes. Reaction A represents the hemolysis of adenosylcobalamin. Reaction B is the scheme of the radical rearrangement reaction of AdoCbl-dependent enzymes

Methionine Aminopeptidase

Methionine aminopeptidase (MetAP) is a ubiquitous enzyme that catalyzes the cleavage of the N-terminal methionine from newly translated polypeptide chains in eukaryotes as well as prokaryotes (Lowther and Matthews 2000). MetAP from *Escherichia coli* was reported as the first non-corrin cobalt protein whose crystal structure had been determined. Pseudo twofold-related N-terminal and C-terminal domains form the "pita bread" structure and contribute to hold a dinulcear cobalt center. Two cobalt ions are coordinated with His171 and Glu104 as monodentate ligands and Asp97, Asp108, and Glu235 as bidentate ones (residue numbers are based on MetAP from *E. coli*) (Fig. 4). A water ligand bridging two cobalt ions is considered as the nucleophile in the catalysis. These structures are conserved among all structurally known MetAPs.

Based on amino acid sequence similarities, MetAP is classified into two subfamilies. MetAP type 1 (MetAP1) is found in bacteria, while type 2 (MetAP2) is found in Archaea. MetAP2 has an inserted region that forms a helical surface subdomain. Eukaryotes possess both types of MetAPs. But each type has a specific N-terminal extension domain. Namely, eukaryotic MetAP1s have zinc-finger domains,

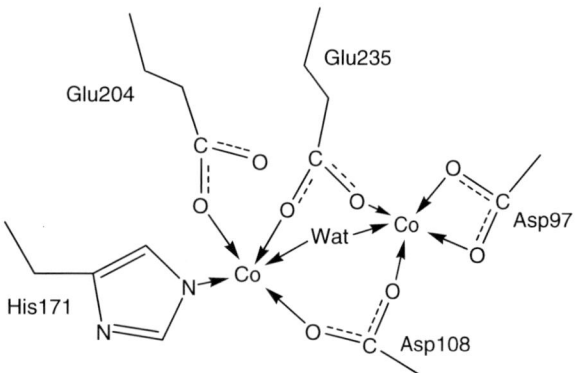

Cobalt Proteins, Overview, Fig. 4 Structure of the bimetallic catalytic center of MetAP from *Escherichia coli*

whereas eukaryotic MetAP2s have polybasic and polyacidic regions. These N-terminal extensions are not necessary for the catalysis because the truncated enzymes retain the catalytic activities. Interaction between the N-terminal regions and DNA and/or RNA has been suggested, but their biological functions remain unclear.

Deletion of both types of MetAPs in *Saccharomyces cerevisiae* is lethal. Since deletion of either of the two MetAPs results in slow growth, each enzyme can partly compensate for the biological function.

Also, MetAP is important as a target not only for antimicrobial agents but also for antitumor agents. Human MetAP2 was identified as the target of fumagillin family compounds that act as potent angiogenesis inhibitors. A synthetic analog of fumagillin, TNP-470, has entered clinical trials for a variety of cancers (Satchi-Fainaro et al. 2005). The crystal structure of hMetAP2 in complex with fumagillin has been determined. Fumagillin is covalently bound to the imidazole ring of His231 via C2, the carbon atom in the epoxide group in fumagillin, and occupies the substrate-binding pocket. The higher affinities of fumagillin compounds to MetAP2 rather than MetAP1 can be explained by the size of their substrate-binding pockets. TNP-470 induces the activation of p53 and p21CIP/WAF in endothelial cells to inhibit their proliferation in the G1 phase. Downregulation of MetAP2 by RNAi results in inhibition of the proliferation of germ cells in *Caenorhabditis elegans*. These findings suggest that MetAP2 plays an important role in the proliferation of specific cell lines. On the other hand, an inhibitor specific for MetAP1 causes a significant delay in the G2/M phase (Hu et al. 2006). Also, silencing of the human MetAP1 gene induces a delay in the G2/M phase in cell cycle progression (Hu et al. 2006). Although many of the details of the function of each type of MetAP remain unknown, it will serve as a promising target for drug designs.

MetAPs coordinate metals loosely because native MetAP1 from *Salmonella typhimurium* contains no metals when it is purified. *S. typhimurium* MetAP1 is activated by Co^{2+}, but not by Mg^{2+}, Mn^{2+}, or Zn^{2+}. But, generally, MetAP is sensitive to divalent cations such as Co^{2+}, Mn^{2+}, Zn^{2+}, and Fe^{2+}. Although MetAP1 from *E. coli* was determined to be a cobalt-containing enzyme, studies on the inhibitory effects of metal-selective MetAP inhibitors in vivo suggested that Fe^{2+} is the likely metal used by MetAP in *E. coli* and other bacterial cells (Chai et al. 2008). Likewise, hMetAP2 is suggested to function as a manganese enzyme in vivo. Thus, the physiologically relevant metals in MetAPs have not firmly established.

Nitrile Hydratase

Nitrile hydratase (NHase) catalyzes hydration of aromatic and small aliphatic nitriles into the corresponding amides (R-CN + H_2O → R-$CONH_2$). It is composed of two distinct subunits, α and β, and was originally identified as a nonheme iron enzyme. *Rhodococcus rhodochrous* J1 possesses two kinds of NHase genes, and both enzymes contain a non-corrin cobalt center in place of the nonheme iron one as the catalytic center. NHase is industrially very important because it has been used for kiloton scale production of acrylamide. Physiologically, NHase is considered to be involved in the aldoxime-nitrile pathway (Oinuma et al. 2003). Briefly, aldoxime compounds are converted into nitriles by aldoxime dehydratase. Then, the produced nitriles are hydrated into amides by NHase. Finally, the amides are hydrolyzed into the corresponding carboxylic acids and ammonia by amidase. The details of the characteristics of NHases are dealt with in a separate section.

Structural and functional studies have mostly concentrated on iron-type NHases. Nagashima et al. determined the crystal structure of *Rhodococcus* sp. N774 NHase at 1.7 Å resolution. The combination of high-resolution structure and mass spectrometric analyses revealed the unusual structure of the catalytic center at the α-subunit. Two main chain amide nitrogens and three cysteine sulfurs are coordinated to the iron, and two out of the three cysteine ligands are posttranslationally modified to cysteine sulfenic acid (Cys-SOH) and cysteine sulfinic acid (Cys-SO_2H), respectively. NHase is the first example of a metalloenzyme that has two cysteine ligands with different oxidation states. The crystal structure of the cobalt-type NHase from *Pseudonocardia thermophila* JCM 3095 is very similar to that of the iron-type one including two posttranslational oxidations of cysteine ligands. The sixth ligand site is occupied by a solvent water ligand. Some difference was observed in the β-subunit. In particular, the region of β111–β125 formed a helix and interacted with a helix in the α-subunit (α36–α49) in the cobalt-type NHase. The additional interaction is likely to contribute to the thermostability of the cobalt-type NHase.

Based on the crystal structure, three reaction mechanisms are proposed (Fig. 5). (a) A coordinated water molecule (or a hydroxide ion) provided nucleophilicity from the metal attacks the nitrile in the substrate-binding pocket. (b) A water molecule in the pocket is activated by the coordinated water molecule (or hydroxide ion) and makes a nucleophilic attack on the substrate. (c) A nitrile substrate activated through coordination to the metal is attacked by a water molecule in the pocket. Various studies

Cobalt Proteins, Overview, Fig. 5 Proposed catalytic mechanism of nitrile hydratase

involving model complexes mimicking the metallocenter of NHases, site-directed mutagenesis, specific inhibitors, and theoretical calculations have been conducted to understand the reaction mechanism of NHase. One of the authors has performed time-resolved X-ray crystallography of Fe-type NHase using a novel substrate, *tert*-butyl isonitrile (tBuNC). The nitrosylated inactive iron-type NHase from *Rhodococcus* sp. N771 soaked in tBuNC in the dark was activated by light illumination, and then its crystal structure was determined. After 120-min illumination, tBuNC coordinated to the sixth site of the iron center. Then, the shape of the electron density for the substrate changed, suggesting that a solvent water molecule activated by the oxygen atom of the cysteine sulfenic acid makes a nucleophilic attack on the nitrile carbon.

Thiocyanate Hydrolase

Thiocyanate hydrolase (SCNase) catalyzes the hydration and subsequent hydrolysis of thiocyanate (SCN$^-$) to produce carbonyl sulfide and ammonia (SCN$^-$ + $2H_2O \rightarrow COS + NH_3 + OH^-$). SCNase was identified in a sulfur-oxidizing bacterium, *Thiobacillus thioparus* THI115, isolated from active sludge for the treatment of factory wastewater for the gasification of coal. Based on amino acid sequence similarities, SCNase was found to belong to the same protein family as NHases. SCNase is composed of three subunits, α, β, and γ. The γ-subunit corresponds to the NHase α-subunit, while the α- and β-subunits correspond to the C-terminal and N-terminal halves of the NHase β-subunit. The crystal structure of SCNase shows that four αβγ-hetero-trimers comprise the dodecameric structure and that each γ-subunit possesses one non-corrin cobalt center. The structure of the αβγ-hetero-trimer of SCNase is very similar to that of NHase including two posttranslationally modified cysteine ligands, Cys-SOH, and Cys-SO_2H, respectively. The structural as well as biochemical characterization of SCNase will be discussed in a separate section.

Other Cobalt Proteins

Prolidase There are few proteases that cleave a peptide bond adjacent to a proline residue. This is because the cyclic structure of proline puts a conformational constraint in the polypeptide chain. Prolidase can hydrolyze dipeptides containing proline at the C-terminus, Xaa-Pro. Prolidase is widespread in nature from bacteria to mammalian tissues. Prolidase from *Pyrococcus furiosus* is homo-dimeric enzyme, and each subunit has a bimetallic catalytic center (Ghosh et al. 1998). The two sites exhibit distinct metal-binding affinities. The tightly binding site is occupied by a Co^{2+} ion. The second site binds Co^{2+} or Mn^{2+} (with a 25% decrease in activity) with a K_d of 0.24 mM, but does not bind Mg^{2+}, Fe^{2+}, Zn^{2+}, Cu^{2+}, and Ni^{2+}. The crystal structure of Pfprolidase has been determined. Unexpectedly, the enzyme in the crystal contains two Zn^{2+} ions, not Co^{2+} ones (Maher et al. 2004). This may be due to metal exchange during its crystallization. The fold of Pfprolidase is similar to those of two functionally related enzymes, aminopeptidase P and creatinase. Interestingly, Pfprolidase exhibited the highest activity when the metals were substituted by Fe^{2+}. Recently, human prolidase was shown to possess a Zn^{2+} ion and a Mn^{2+} one at the catalytic site (Besio et al. 2010). The difference in their metal composition may affect their substrate selectivities.

D-Xylose Isomerase D-Xylose isomerases catalyze conversion of D-xylose and D-glucose into D-xylulose and D-fructose, respectively, and have attracted considerable interest as to the production of high-fructose corn syrup. D-Xylose isomerase is dependent on a bivalent cation, Co^{2+}, Mg^{2+}, or Mn^{2+}. D-Xylose isomerases bind two metal ions per monomer with different binding affinities. The binding constant of the high-affinity site for Co^{2+} was estimated to $> 3.3 \times 10^6$ M^{-1} and that for the low-affinity site to be 4×10^4 M^{-1}. The crystal structure of D-xylose isomerase from *Streptomyces diastaticus* No. 7 strain M1033 showed that it had a homo-tetrameric structure. Two Co^{2+} ions exist in the catalytic site, revealing that *Sd* D-xylose isomerase prefers Co^{2+} ions to Mg^{2+} ones (Zhu et al. 2004).

Methylmalonyl-CoA Carboxytransferase Methylmalonyl-CoA carboxytransferase (MMCT) is a biotin-containing enzyme and also is known as transcarboxylase. MMCT from *Propionibacterium shermanii* is a 26S huge multi-subunit enzyme of 1.2 million Da, which is composed of three kinds of subunits, 1.3S, 5S, and 12S. 1.3S has twelve 12-kDa biotnylated linkers; 5S, six catalytic 116-kDa dimmers; and 12S, a catalytic 336-kDa hexameric core, respectively. The overall transcarboxylation reaction consists of two half reactions. In the first half reaction, 12S transfers CO_2^- from MMCoA to biotin to generate propionyl-CoA on 1.3S. 5S transfers the CO_2^- from the 1.3S biotin to pyruvate to produce oxaloacetate in the second half reaction. The 5S subunit is a dimer of $\beta_8\alpha_8$ barrel monomers (Hall et al. 2004). Each monomer contains one Co^{2+} ion at its catalytic site. Although the 5S subunit had been thought to contain one Co^{2+} and one Zn^{2+} per subunit, no Zn^{2+} ion was observed. The Co^{2+} is octahedrally coordinated by two imidazole groups of His215 and His217, one carboxyl group of Asp23, a solvent water molecule, and the CO_2^- from the carbamylated Lys184. The coordination is likely to be tight because the average ligand distance is 2.15 Å. The coordination sphere is unchanged in a complex with the substrate, pyruvate, but when the product, oxaloacetate, is co-crystallized, the carboxylate group of oxaloacetate is coordinated to Co^{2+} in place of the carbamylated Lys184.

Aldehyde Decarboxylase The final step of fatty acid synthesis is the decarboxylation of aldehyde catalyzed by a membrane protein, aldehyde decarboxylase (AD). The enzyme is attracting increasing attention because it will be applicable to biofuel production. However, information on its biochemical characteristics has been rather limited because of the difficulty in its solubilization and purification. AD purified from a green algae, *Botryococcus braunii*, exhibits a visible absorption spectrum characteristic of a porphyrin and contains 1.37 mole of cobalt per enzyme. This is the first example of a non-corrin cobalt-containing protein in a plant. Later, AD was purified from a higher plant, *Pisum sativum*. Interestingly, the higher plant AD contains a copper ion, probably with a porphyrin, despite that the enzyme reconstituted with cobalt showed about 70% of the activity of the wild type. Most recently, soluble aldehyde decarboxylase (cAD) was purified from cyanobacteria, *Prochlorococcus marinus* MIT9313 (Das et al. 2011). Surprisingly, cAD does not have a porphyrin-like cofactor, but has a nonheme di-iron center. AD may use a variety of prosthetic groups as the catalytic center.

Bromoperoxidase-esterase Haloperoxidase is classified into three groups: eukaryotic and bacterial heme-type, eukaryotic vanadium-containing-type, and bacterial nonmetal-type enzymes. Based on amino acid sequence homology, bromoperoxidase-esterase (BPO-EST) from *Pseudomonas putida* is grouped as a nonmetal-type enzyme, but its BPO activity was activated by ca. 300% in the presence of Co^{2+} (Itoh et al. 2001). The interaction between the enzyme and Co^{2+} is rather weak because Co^{2+} is easily eliminated on dialysis against the buffer. Interestingly, the EST activity was inhibited by Co^{2+}. Despite detailed kinetic analyses, the enzyme reaction mechanism and the role of Co^{2+} remain unclear because its crystal structure has not been determined.

Concluding Remarks

Cobalt ions are used much less frequently in living organisms than other transition metals. As overviewed here, however, most cobalt-containing enzymes catalyze very unique reactions. Therefore, they are very important from medical aspects as well as from industrial ones. Metal ions such as Mn^{2+}, Fe^{2+}, Ni^{2+}, Cu^{2+}, and Zn^{2+} involved in enzymes can often be substituted by Co^{2+} as in MetAP. In fact, cobalt substitution is one of the useful tools for studying the functions of other metal-containing proteins.

The concentrations of transition metals are strictly regulated in living organisms. The free cobalt ion is highly toxic because it may produce reactive oxygen species and because it may occupy the binding sites of proteins containing other metals. Therefore, the concentration of free cobalt ions is kept very low in cytosol. It is very important to understand the transport system controlling the uptake and transfer of cobalt. Future research will provide major clues as to the biological functions as well as to the industrial applications of cobalt proteins.

Due to the page limit, all of the related papers cannot be referenced. Please check them on PubMed using the corresponding keywords.

Cross-References

▶ Nitrile Hydratase and Related Enzyme

References

Besio R, Alleva S, Forlino A et al (2010) Identifying the structure of the active sites of human recombinant prolidase. Eur Biophys J 39:935–945

Chai SC, Wang WL, Ye QZ (2008) FE(II) is the native cofactor for *Escherichia coli* methionine amino peptidase. J Biol Chem 283:26879–26885

Das D, Eser BE, Han J et al (2011) Oxygen-independent decarbonylation of aldehydes by cyanobacterial aldehyde decarbonylase: a new reaction of diiron enzymes. Angew Chem Int Ed Engl 50:7148–7152

Escalante-Semerena JC, Warren MJ (2008) Biosynthesis and use of cobalamin (B12). In: Böck A, Curtiss R III, Kaper JB, Karp PD, Neidhardt FC, Nyström T, Slauch JM, Squires CL (eds) *EcoSal-Escherichia coli* and Salmonella: cellular and molecular biology. ASM Press, Washington, DC, Chapter 3.6.3.8

Froese DS, Kochan G, Muniz JRC et al (2010) Structures of the human GTPase MMAA and vitamin B_{12}-dependent methylmalonyl-CoA mutase and insight into their complex formation. J Biol Chem 285:38204–38213

Ghosh M, Grunden AM, Dunn DM et al (1998) Characterization of native and recombinant forms of an unusual cobalt-dependent proline dipeptidase (prolidase) from the hyperthermophilic archaeon *Pyrococcus furiosus*. J Bacteriol 180:4781–4789

Goetzl S, Jeoung J-H, Hennig SE et al (2011) Structural basis for electron and methyl-group transfer in a methyltransferase system operating in the reductive acetyl-CoA pathway. J Mol Biol 411:96–109

Hall PR, Zheng R, Antony L, Pusztai-Carey M, Carey PR, Yee VC (2004) Transcarboxylase 5S structures: assembly and catalytic mechanism of a multienzyme complex subunit. EMBO J 23:3621–3631

Hu X, Addlagatta A, Lu J et al (2006) Elucidation of the function of type 1 human methionine aminopeptidase during cell cycle progression. Proc Natl Acad Sci USA 103:18148–18153

Itoh N, Kawanami T, Liu JQ (2001) Cloning and biochemical characterization of Co^{2+}-activated bromoperoxidase-esterase (perhydrolase) from *Pseudomonas putida* IF-3 strain. Biochim Biophys Acta 1545:53–66

Kobayashi M, Shimizu S (1999) Cobalt proteins. Eur J Biochem 261:1–9

Lowther WT, Matthews BW (2000) Structure and function of the methionine aminopeptidases. Biochim Biophys Acta 1477:157–167

Maher MJ, Ghosh M, Grunden AM et al (2004) Structure of the prolidase from *Pyrococcus furiosus*. Biochemistry 43:2771–2783

Matthews RG, Koutmos M, Datta S (2008) Cobalamin- and cobamide-dependent methyltransferases. Curr Opin Struct Biol 18:658–666

Oinuma K, Hashimoto Y, Konishi K et al (2003) Novel aldoxime dehydratase involved in carbon-nitrogen triple bond synthesis of *Pseudomonas chlororaphis* B23. Sequencing, gene expression, purification, and characterization. J Biol Chem 278:29600–29608

Padovani D, Banerjee R (2009) A G-protein editor gates coenzyme B12 loading and is corrupted in methylmalonic aciduria. Proc Natl Acad Sci USA 106:21567–21572

Randaccio L, Geremia S, Wuerges J (2007) Crystallography of vitamin B12 proteins. J Organomet Chem 692:1198–1215

Satchi-Fainaro R, Mamluk R, Wang L et al (2005) Inhibition of vessel permeability by TNP-470 and its polymer conjugate, caplostatin. Cancer Cell 7:251–261

Wolthers KR, Scrutton NS (2009) Cobalamin uptake and reactivation occurs through specific protein interactions in the methionine synthase–methionine synthase reductase complex. FEBS J 276:1942–1951

Zhu XY, Teng MK, Niu LW et al (2004) Structure of xylose isomerase from *Streptomyces diastaticus* No. 7 strain M1033 at 1.85 Å resolution. Acta Crystallogr Sect D 56:129–136

Cobalt Schiff Base

▶ Zinc, Metallated DNA-Protein Crosslinks as Finger Conformation and Reactivity Probes

Cobalt Transporters

Thomas Eitinger
Institut für Biologie/Mikrobiologie, Humboldt-Universität zu Berlin, Berlin, Germany

Synonyms

Metal transporter; Transition metal uptake

Definition

Import systems that transport the transition metal ion across cell membranes in order to provide it for synthesis of coenzyme B_{12} and for incorporation into cobalt-containing enzymes.

Background

Cobalt is a trace nutrient for prokaryotes and utilized for biosynthesis of the cobalt-containing coenzyme B_{12} (see also ▶ Vitamin B_{12}), and for incorporation into noncorrin ▶ Co-containing enzymes. Animals must take up vitamin B_{12} with their diet or from intestinal prokaryotic producers. In higher plants, physiological roles for corrin and noncorrin cobalt enzymes have so far not been established. Nevertheless, beneficial effects of Co^{2+} for plant growth are known (Pilon-Smits et al. 2009). These effects may be due to stimulating growth of bacteria in the rhizosphere and of root-nodulating bacterial endosymbionts.

The focus of this short survey is on import systems in prokaryotes that transport Co^{2+} ions with high affinity and specificity under physiologically relevant conditions into the cells. Many of the transporters discussed here have closely related counterparts among the ▶ nickel transporters that function in the uptake of Ni^{2+} ions. From a bioenergetic point of view, cobalt importers can be classified into primary and secondary active transport systems (Eitinger et al. 2005; Rodionov et al. 2006) (Table 1). The latter include the long-known nickel/cobalt transporter (NiCoT) family and the weakly related HupE/UreJ systems. Primary active cobalt importers (CbiMNQO) belong to the more recently identified energy-coupling factor (ECF) transporter family of micronutrient importers and are driven by ATP hydrolysis. These systems are widespread in bacteria and archaea and represent a novel type of ABC (ATP-binding cassette) transporter (Rodionov et al. 2009; Eitinger et al. 2011).

CorA proteins form a huge group of Mg^{2+} channels in both prokaryotes and eukaryotes. It has long been known that CorA – in addition to Mg^{2+} – can transport Co^{2+} (and Ni^{2+}), however only at unphysiologically high concentrations of the transition metal ions in the micromolar range, underlining the primary role of CorA as a Mg^{2+} transporter. Very recently, however, CorA of the hyperthermophilic bacterium *Thermotoga maritima* was identified as a selective Co^{2+} transporter (Xia et al. 2011).

Another recent observation is energy-dependent transport of cobalt ions across the outer membrane of Gram-negative bacteria, and available data is included in the short overview.

Cobalt Transporters, Table 1 Classification and distribution of cobalt-uptake systems

System	Mechanistic type	Occurrence
CbiMNQO	Primary, ECF type	Prokaryotes
NiCoT	Secondary	Prokaryotes, fungi
HupE/UreJ	Secondary	Prokaryotes
CorA	Channel	Prokaryotes, eukaryotes
TBDT	TonB-dependent outer-membrane transporter	Gram-negative bacteria

NiCoT

Members of the NiCoT family are known since the early 1990s. They were originally identified as nickel transport systems in ▶ NiFe hydrogenase- or ▶ urease-producing bacteria. A Co^{2+} transporter within the NiCoT family was identified during sequence analysis of the gene cluster encoding a cobalt-containing ▶ nitrile hydratase in the actinobacterium *Rhodococcus rhodochrous* J1. This transporter, named NhlF, was later shown to transport both Ni^{2+} and Co^{2+} ions but to prefer the latter (reviewed in Eitinger et al. 2005). Similar properties, i.e., the capability to transport the two transition metal ions but to prefer one or the other, were reported for a couple of additional NiCoTs (reviewed in Eitinger et al. 2005).

Functional genomics proved a powerful method to predict the substrate preference of NiCoTs. The analyses uncovered that colocalization and/or coregulation of NiCoT genes with genes implicated in nickel or cobalt metabolism is a reliable indicator of a NiCoT's preference for Ni^{2+} or Co^{2+} (Rodionov et al. 2006).

Analyses of NiCoT sequences predict 2 four-helix segments formed by the N-terminal and C-terminal parts of the proteins (Fig. 1). The two halves are connected by a large and highly charged cytoplasmic loop. A couple of conserved sequence motifs mainly located within the TMDs have been found to be essential for transport activity. Of special importance is TMD II harboring the +HAXDADH (+, R or K; X, V, F, or L) motif which serves as the signature sequence for NiCoTs. This segment interacts spatially with a histidine- or asparagine-containing region in TMD I, together forming a central part of the selectivity filter that controls velocity and ion selectivity of the transport process (Degen and Eitinger 2002). Site-directed mutagenesis was applied to investigate the consequences of amino acid replacements introduced into a cobalt- and a nickel-preferring NiCoT. These experiments showed that an increase of transport velocity results in a decreased specificity. The data indicated that predictions of the preference of a NiCoT for either Co^{2+} or Ni^{2+} ion, solely based on the primary structure, are not possible (Degen and Eitinger 2002).

HupE/UreJ

As indicated by the designation HupE/UreJ, those proteins represent a family whose members – in many cases – are encoded within NiFe hydrogenase (*hup*, "hydrogen uptake") or urease (*ure*) gene clusters. Those proteins mediate the uptake of Ni^{2+} ions. HupE/UreJ are integral membrane proteins with six TMDs in the mature state. Many HupE/UreJ proteins are produced as precursors with a predicted N-terminal signal peptide which is cleaved off at a conserved site releasing an N-terminal histidine residue of the processed proteins. TMD I of mature HupE/UreJ proteins contain a HPXXGXDH motif which resembles the signature sequence of NiCoTs (Eitinger et al. 2005).

Another set of *hupE* genes, mainly found in cyanobacteria, is genomically unlinked to nickel metabolism. Bioinformatic analyses predicted a role of those HupE proteins as transporters for Co^{2+} ion in the metabolic context of biosynthesis of the cobalt-containing coenzyme B_{12} (Rodionov et al. 2006; Zhang et al. 2009).

CbiMNQO

CbiMNQO is the prototype of a novel type of micronutrient importers in prokaryotes, named energy-coupling factor (ECF) transporters (Rodionov et al. 2009; Eitinger et al. 2011). The first such system was mentioned upon sequence analysis of the coenzyme B_{12} biosynthesis gene cluster of *Salmonella enterica* serovar Typhimurium which contains a set of *cbiMNQO* genes. Since CbiO represents a typical nucleotide-binding protein of ATP-binding cassette (ABC)

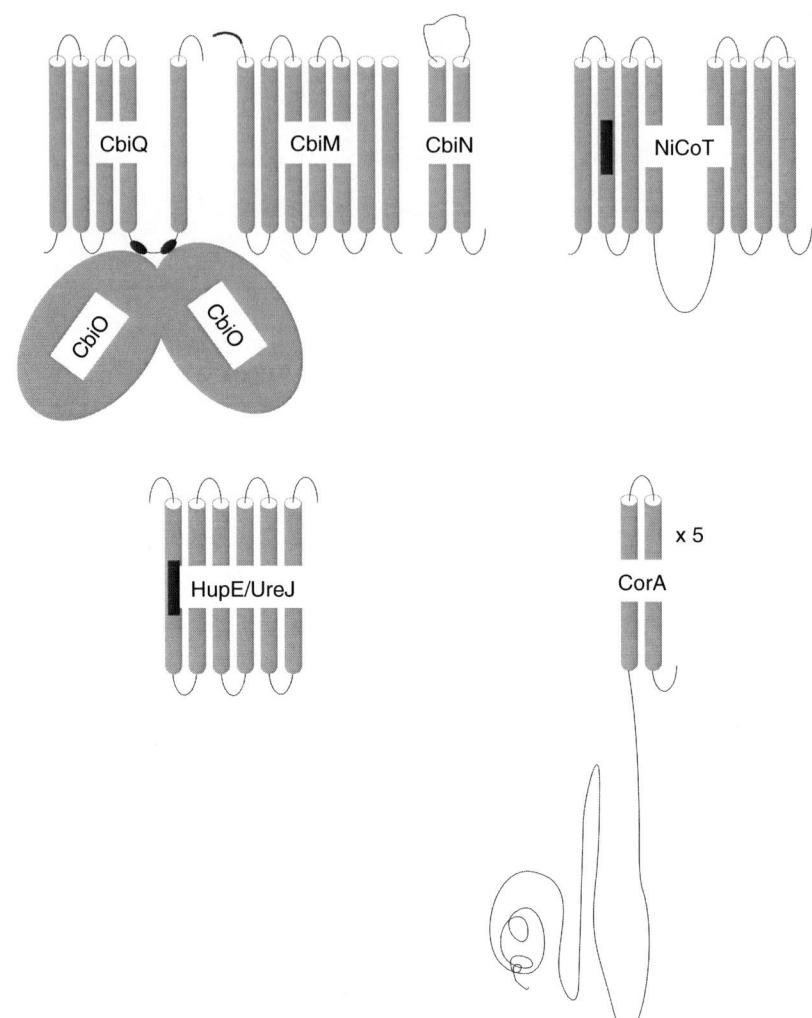

Cobalt Transporters, Fig. 1 Characteristics of microbial cobalt-uptake systems. See the text for details. Transmembrane helices are indicated as cylinders. The contact sites in the T unit CbiQ with the ABC ATPases of ECF-type Co^{2+} transporters are depicted. The strongly conserved N-terminus of CbiM proteins is highlighted. *Black bars* indicate the signature sequence in TMD II of NiCoTs and related sequences in TMD I of HupE/UreJ. The secondary structure of a CorA monomer is illustrated. CorA has a homopentameric quaternary structure. The N-terminal regions of the CorA pentamer form a funnel-like structure in the cytoplasm, and the C-termini contain a total of 10 transmembrane domains. Outer-membrane cobalt transporters of Gram-negative bacteria are not shown

transporters, it was proposed that CbiMNQO acts as an ABC transporter for Co^{2+} ions. The original conclusion neglected the fact that canonical ABC-type importers in prokaryotes strictly depend on extracytoplasmic solute-binding proteins, but no such protein was encoded in the *cbi* cluster (summarized in Rodionov et al. 2006).

Energy-coupling factors of ECF transporters consist of a conserved transmembrane protein (T component, CbiQ in the case of Co^{2+} transporters) and pairs of typical ABC ATPases (A components, 2x CbiO). Substrate specificity is conveyed through an S component which is made up of a single integral membrane protein (in the case of most ECF transporters) or multiple membrane proteins, e.g., CbiMN in the case of the Co^{2+} transporters. In contrast to canonical ABC importers, all ECF systems are devoid of extracytoplasmic soluble solute-binding proteins. The list of ECF transporters includes uptake systems for vitamins, for intermediates of salvage pathways and for Ni^{2+} or Co^{2+} ions (Eitinger et al. 2011).

Many components of ECF transporters in prokaryotic or even plant genome sequences are misannotated as cobalt transporters because the presence of *cbiQO*-like genes or of a solitary *cbiQ*-like gene (the latter occurs in plant genomes) is considered as an indicator. These approaches ignore the fact that the highly diverse S units are responsible for substrate specificity, and thus, the S units rather than homologous T and homologous A units should be used for assigning substrate specificities to ECF transporters (reviewed in Eitinger et al. 2011). Cobalt-specific ECF transporters can be identified by sequence alignments of the CbiM and CbiN proteins and by the fact that many *cbiMNQO* operons are colocalized and/or coregulated with genes for coenzyme B_{12} synthesis (Rodionov et al. 2006; Zhang et al. 2009; Eitinger et al. 2011).

Two bacterial ECF-type Co^{2+} transporters have been analyzed in some detail. Deletion analyses of the *cbiMNQO* operons of *S. Typhimurium* (Rodionov et al. 2006) and *Rhodobacter capsulatus* (Siche et al. 2010) revealed that in the absence of CbiQO, CbiMN has a basal Co^{2+} transport activity. The fact that the solitary CbiM and a tripartite CbiMQO system lacking CbiN are completely inactive underscores the essential role of the small transmembrane protein CbiN. Nevertheless, this component is only loosely bound to its partners and copurifies neither with CbiM nor with stable CbiMQO complexes. Construction of a Cbi (MN) fusion protein of the *R. capsulatus* system led to a functional S unit that interacted with CbiQO in vivo (Siche et al. 2010). The Cbi(MN) fusion was used to characterize the extremely conserved MHIMEGYLP N-terminal sequence of CbiM located on the extracytoplasmic side of the membrane (illustrated in Fig. 1). A total 16 Cbi(MN) variants was constructed in which individual N-terminal residues were replaced, deleted, or inserted. Among those, only the forms with the M_4A or M_4S replacement, which represent naturally occurring variations, were active. The position of the N-terminus of the peptide chain relative to the histidine residue at position 2 (in the natural situation) seems to be essential. Insertion of an additional N-terminal methionine residue or an alanine residue between M_1 and H_2 completely inactivated the core transporter.

As a distinctive feature between active and inactive variants, a significant percentage of active Cbi(MN) variants forms a recalcitrant structure that is not resolved in the presence of dodecyl sulfate. The molecular basis of this structure that obviously correlates with activity has not yet been unraveled (Siche et al. 2010).

CorA

CorA is the major transport system for Mg^{2+} uptake in prokaryotes, and functional homologs are known in eukaryotes (reviewed in Niegowski and Eshaghi 2007; Moomaw and Maguire 2008). As indicated by its name, CorA was originally identified during screens of cobalt-resistant clones of enterobacterial species. Enterobacterial CorA transports Mg^{2+} and Co^{2+} with affinities in the range of 20 μM, and Ni^{2+} with 10–20-fold lower affinity. Since micromolar concentrations of transition metal cations are toxic and considered to be unphysiological, the primary role of CorA is presumed to be the transport of Mg^{2+}.

Crystal-structure analyses of CorA of the hyperthermophilic bacterium *Thermotoga maritima* revealed a very unusual organization as a homopentamer. Each monomer of the channel consists of a cytosolic N-terminal domain and a C-terminal domain with two transmembrane helices. In the pentameric state, the latter form the pore in the membrane and the former a funnel-like structure in the cytoplasm.

The universal role of CorA as a Mg^{2+} transporter has been questioned recently when *T. maritima* CorA was shown to function as a Co^{2+} transporter that strongly prefers Co^{2+} over Mg^{2+} (Xia et al. 2011). This study suggests that variants of CorA proteins (e.g., in hyperthermophilic prokaryotes) may have different functions including the role as a selective transporter for Co^{2+} ions.

Other Co^{2+} Importers in Prokaryotes

NikABCDE and UreH represent canonical ABC-type and secondary transporters for Ni^{2+} ions, respectively. In a few cases, *nikABCDE* and *ureH* genes are under control of coenzyme B_{12}-dependent riboswitch regulatory elements or are colocalized with coenzyme B_{12} biosynthesis genes. Those NikABCDE and UreH systems may function in Co^{2+} uptake.

Based on bioinformatics, a couple of additional transmembrane proteins (CbtAB, CbtC, CbtD, CbtE, CbtF, CbtX) have been implicated in Co^{2+} uptake (summarized by Zhang et al. 2009). Like in the cases of hypothetical Co^{2+} transporters within the UreH and NikABCDE families, experimental evidence in support of this role has not been reported. Very recently, a canonical ABC-type cobalt transporter was characterized in the root-nodulating bacterium *Sinorhizobium meliloti* (Cheng et al. 2011).

TonB-Dependent Outer-Membrane Transporters (TBDT)

Until recently, active transport of solutes across the outer membrane of Gram-negative bacteria was only known for Fe^{3+} ions in complex with organic siderophores and for coenzyme B_{12}-related cobalamins. Recently, however, functional genomics and biochemical analyses identified a large number of putative outer-membrane transporters for alternate solutes (Schauer et al. 2008). Outer-membrane transporters consist of a C-terminal β-barrel domain and an N-terminal plug domain, and they bind their substrates with high affinity. They depend on ExbB-ExbD-TonB, a complex of proteins located in the cytoplasmic membrane. TonB contains an extensive periplasmic segment that interacts with the outer-membrane receptors.

Genes for outer-membrane transporters and *exbBD-tonB* clusters implicated in cobalt and/or nickel metabolism were identified by functional genomics in proteobacteria (Rodionov et al. 2006). Experimental analyses of methylotrophic bacteria correlated Co^{2+} deficiency and the resulting cobalamin deficiency with lowered expression of a TBDT gene located adjacent to a coenzyme B_{12} biosynthetic gene (Kiefer et al. 2009). These data corroborate the bioinformatic findings and support the hypothesis that Co^{2+} ions may cross the outer membrane of Gram-negative bacteria by a TonB-dependent active transport mechanism.

Cross-References

- ▶ Cobalt-containing Enzymes
- ▶ Nickel Transporters
- ▶ [NiFe]-Hydrogenases
- ▶ Nitrile Hydratase
- ▶ Urease
- ▶ Vitamin B_{12}

References

Cheng J, Poduska B, Morton RA, Finan TM (2011) An ABC-type cobalt transport system is essential for growth of *Sinorhizobium meliloti* at trace metal concentrations. J Bacteriol 193:4405–4416

Degen O, Eitinger T (2002) Substrate specificity of nickel/cobalt permeases: insights from mutants altered in transmembrane domains I and II. J Bacteriol 184:3569–3577

Eitinger T, Rodionov DA, Grote M, Schneider E (2011) Canonical and ECF-type ATP-binding cassette importers in prokaryotes: diversity in modular organization and cellular functions. FEMS Microbiol Rev 35:3–67

Eitinger T, Suhr J, Moore L, Smith JAC (2005) Secondary transporters for nickel and cobalt ions: theme and variations. Biometals 18:399–405

Kiefer P, Buchhaupt M, Christen P, Kaup B, Schrader J, Vorholt J (2009) Metabolite profiling uncovers plasmid-induced cobalt limitation under methylotrophic growth conditions. PLoS ONE 4:e7831

Moomaw AS, Maguire ME (2008) The unique nature of Mg^{2+} channels. Physiology 23:275–285

Niegowski D, Eshaghi S (2007) The CorA family: structure and function revisited. Cell Mol Life Sci 64:2564–2574

Pilon-Smits EAH, Quinn CF, Tapken W et al (2009) Physiological functions of beneficial elements. Curr Opin Plant Biol 12:267–274

Rodionov DA, Hebbeln P, Gelfand MS, Eitinger T (2006) Comparative and functional genomic analysis of prokaryotic nickel and cobalt uptake transporters: evidence for a novel group of ATP-binding cassette transporters. J Bacteriol 188:317–327

Rodionov DA, Hebbeln P, Eudes A et al (2009) A novel class of modular transporters for vitamins in prokaryotes. J Bacteriol 191:42–51

Schauer K, Rodionov DA, de Reuse H (2008) New substrates for TonB-dependent transport: do we only see the 'tip of the iceberg?'. Trends Biochem Sci 33:330–338

Siche S, Neubauer O, Hebbeln P, Eitinger T (2010) A bipartite S unit of an ECF-type cobalt transporter. Res Microbiol 161:824–829

Xia Y, Lundbäck A-K, Sahaf N, Nordlund G, Brzezinski P, Eshaghi S (2011) Co^{2+} selectivity of *Thermotoga maritima* CorA and its inability to regulate Mg^{2+} homeostasis present a new class of CorA proteins. J Biol Chem 286:16525–16532

Zhang Y, Rodionov DA, Gelfand MS, Gladyshev VN (2009) Comparative genomic analyses of nickel, cobalt and vitamin B_{12} utilization. BMC Genomics 10:78

Cobalt, Physical and Chemical Properties

Fathi Habashi
Department of Mining, Metallurgical, and Materials Engineering, Laval University, Quebec City, Canada

Physical Properties

Pure metallic cobalt has few applications, but its use as an alloying element and as a source of chemicals makes it a strategically important metal. End uses of cobalt-containing alloys include superalloys for aircraft engines, magnetic alloys for powerful permanent magnets, hard metal alloys for cutting-tool materials, cemented carbides, wear-resistant alloys, corrosion-resistant alloys, and electrodeposited alloys to provide wear and corrosion-resistant metal coatings. Cobalt chemicals, among their many applications, are used as pigments in the glass, ceramics, and paint industries; as catalysts in the petroleum industry; as paint driers; and as trace metal additives for agricultural and medical use. About 36% of the worldwide annual production of cobalt is converted to chemicals, whereas high temperature and magnetic alloys account for 41% and 14% of the consumption, respectively.

Atomic number	27
Atomic weight	58.93
Relative abundance in Earth's crust, %	2.3×10^{-3}
Density, $g \cdot cm^{-3}$	8.90
Liquid density at m.p., $g \cdot cm^{-3}$	7.75
Melting point, °C	1,495
Boiling point, °C	2,927
Heat of fusion, $kJ \cdot mol^{-1}$	16.06
Heat of vaporization, $kJ \cdot mol^{-1}$	377
Specific heat capacity at 25°C, $J \cdot mol^{-1} \cdot K^{-1}$	24.81
Crystal structure	Hexagonal
Magnetic ordering	Ferromagnetic
Electrical resistivity at 20°C, $n\Omega \cdot m$	62.4
Thermal conductivity, $W \cdot m^{-1} \cdot K^{-1}$	100
Thermal expansion at 25°C, $\mu m \cdot m^{-1} \cdot K^{-1}$	13.0
Young's modulus, GPa	209
Poisson ratio	0.31
Mohs hardness	5.0
Vickers hardness, MPa	1,043
Brinell hardness, MPa	700

The γ-rays emitted in the decay of ^{60}Co have energies of 1.17 and 1.33 MeV, and these taken with the 5.3 years half-life of the isotope, provide a widely used source of radioactivity for use in food sterilization, radiography, and radiotherapy as an external source. The isotope is also used in chemical and metallurgical analysis and in biological studies as a radioactive tracer.

Cobalt exists in two allotropic modifications, a close-packed hexagonal ε-form stable below ca. 400°C and a face-centered cubic α-form stable at high temperature. The transformation temperature is 421.5°C.

Chemical Properties

Cobalt is much less reactive than iron. It is stable to atmospheric oxygen unless heated. When heated, it is first oxidized to Co_3O_4 and then, above 900°C, to CoO. The metal does not combine directly with hydrogen or nitrogen, but it combines with carbon, phosphorus, and sulfur on heating. The reaction with sulfur is influenced by the formation of a low melting eutectic (877°C) between the metal and the Co_4S_3 phase; the reaction between cobalt and sulfur is rapid above this temperature. Below 877°C, a protective layer of sulfide scale is formed. In an atmosphere of hydrogen sulfide, cobalt also forms a scale of sulfide, but in air containing sulfur dioxide, a mixed oxide sulfide scale is formed.

The main oxidation states of cobalt are Co^{2+} and Co^{3+}. In acid solution and in the absence of complexing agents, Co^{2+} is the stable oxidation state. Cobalt is often removed from solution as its sulfide.

Cobalt is a transition metal having the electronic configuration 2, 8, 15, 2. It dissolves in ammoniacal solutions in presence of oxygen forming ammine complex:

$$Co + 1/2 O_2 + 2NH_3 + H_2O \rightarrow [Co(NH_3)_2]^{2+} + 2OH^-$$

Cobalt powder is precipitated from cobalt sulfate solution by hydrogen at high temperature and pressure:

$$Co^{2+} + H_2 \rightarrow Co + 2H^+$$

Cobalt sulfide undergoes oxidation in neutral medium at ambient conditions to $CoSO_4$:

$$CoS + 2O_{2(aq)} \rightarrow CoSO_4$$

In acidic medium, however, elemental sulfur forms:

$$CoS + 1/2 O_{2(aq)} + 2H^+ \rightarrow Co^{2+} + S + H_2O$$

Cobalt sulfide is oxidized at high temperature to CoO and sulfur dioxide:

$$CoS + 1 1/2 O_2 \rightarrow CoO + SO_2$$

Cobalt chemicals are used to correct cobalt deficiencies in soils and in animals. Soil treatments usually involve top dressings containing cobalt sulfates, whereas treatment of ruminant animals involves the use of either salt licks containing ca. 0.1% of cobalt as

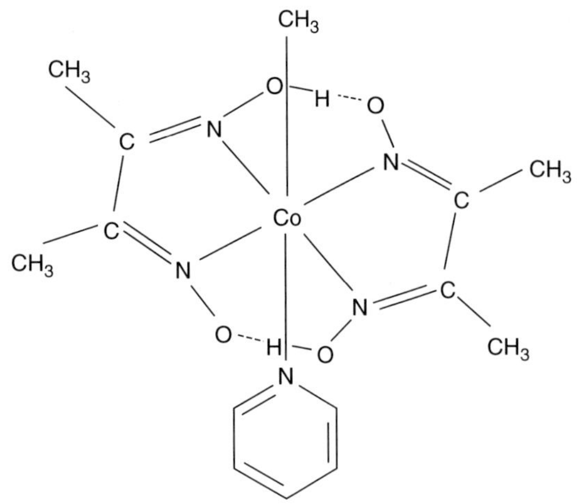

Cobalt, Physical and Chemical Properties, Fig. 1 Cobalt in vitamin B12

sulfate, or concentrated feeds, or pellets of cobalt oxide bound in an inert material such as china clay. The medicinal uses of cobalt are dominated by the use of vitamin B_{12} (Fig. 1).

References

Donaldson JD (1997) Chapter 18. In: Habashi F (ed) Handbook of extractive metallurgy. Wiley-VCH, Heidelberg

Habashi F (1999) Textbook of hydrometallurgy, 2nd edn. Métallurgie Extractive Québec, Québec, distributed by Laval University Bookstore

Habashi F (2002) Textbook of pyrometallurgy. Métallurgie Extractive Québec, Québec City, distributed by Laval University Bookstore

---

Cobalt-containing Enzymes

Todd C. Harrop[1] and Pradip K. Mascharak[2]
[1]Department of Chemistry, University of Georgia, Athens, GA, USA
[2]Department of Chemistry and Biochemistry, University of California, Santa Cruz, CA, USA

Synonyms

Cobalamin; Corrin; Hydrolases; Isomerases; Nitrile hydratase; Ribonucleotide reductase

Introduction

Among the metals of the first transition series, ▶ cobalt (Co) is next to iron (Fe) and exhibits three oxidation states (+1, +2, and +3) much like its congener iron (+2, +3, and +4). However, unlike iron, it is present in a relatively small number of enzymes and cofactors (Kobayashi and Shimizu 1999). In addition, it does not participate in oxygen activation, a process that iron frequently takes part in. In the +3 oxidation state, ▶ cobalt enjoys high crystal field stabilization energy (CFSE) in several coordination geometries. Significantly, high stability arises when Co^{3+} exists in spin-paired (low-spin) electronic configuration within an octahedral coordination environment of strong ligands. Such stability often makes the Co^{3+} center resistant to substitution processes such as exchange of a bound water molecule for the substrate. Such a center however can serve as a strong Lewis acid and promote activation of free water molecule(s) and subsequent nucleophilic attack of OH^- on substrate both present in the active site pocket (as in the enzyme nitrile hydratase). Although Co^{2+} is susceptible to oxidation, especially in aqueous environment, in many enzymes, Co^{2+} centers participate in catalytic processes without any tendency toward oxidation (as in various isomerases described below). Ligand fields arising from biological ligands around Co^{2+} centers in such proteins raise the oxidation potential and protect the metal centers from oxidation with dioxygen. Finally, cobalt-carbon bonds are readily formed and broken with concomitant formation of carbon-based radicals in certain biological pathways. This organometallic chemistry in an aqueous environment is quite unusual and is a unique aspect of vitamin B_{12} chemistry.

In biomolecules (proteins and cofactors), Co centers are bound to (a) amino acid side-chain residues such as carboxylato-O and imidazole-N, (b) carboxamido-N from the peptide frame, and/or (c) substituted corrin macrocycles. For example, in coenzyme B_{12}, an important cofactor, cobalt is bound to a corrin macrocycle and a carbon atom of a 5′-deoxyadenosyl moiety (or a CH_3 group), while in methionine aminopeptidase from *E. coli*, two Co^{2+} ions are bound to His-N, Glu-O, and Asp-O centers (Fig. 2). In general, Co^{2+} centers show preference for imidazole-N and carboxylato-O donors (methionine aminopeptidase is an example) while strongly σ-donating ligands like carboxamido-N and thiolato-S donors prefer Co^{3+} centers (as in nitrile hydratase, Fig. 1). In this chapter, the cobalt enzymes have been classified into two groups depending on whether cobalt is bonded to a corrin moiety (corrin-containing proteins) or not (non-corrin cobalt proteins). Since the corrin-containing enzymes include cobalt in the cobalamine cofactor, the non-corrin enzymes with cobalt bound directly to the peptide framework are discussed first.

Non-corrin Cobalt Enzymes

Nitrile hydratase (NHase). Among the non-corrin cobalt proteins, ▶ nitrile hydratase (NHase) has recently drawn special attention for its successful use in the industrial scale production of acrylamide, used in plastic and paper pulp industry (Kobayashi and Shimizu 1998). This soluble metalloenzyme is involved in the microbial degradation/assimilation of organic nitriles in nature. The heterotetrameric $(\alpha\beta)_2$ protein contains either a Fe^{3+} or a Co^{3+} ion at the active site depending on the organism and catalyzes the conversion of nitriles into amides (Fig. 1). Structural studies have revealed quite a few unusual features

$$R-C\equiv N \xrightarrow{NHase} R-\overset{O}{\underset{\|}{C}}-NH_2 \quad (1)$$

in the coordination sphere around the M^{3+} center in the functional αβ unit of both Fe- and Co-NHase. In Co-NHase from *Pseudonocardia thermophila* JCM 3095, the Co^{3+} center is ligated to four amino acid residues of the α subunit (αCys108, αCys111, αSer112, and αCys113) and strongly interacts with at least two arginine residues of the β unit (βArg52 and βArg157) (Miyanaga et al. 2001). The most unusual feature is the coordination of two carboxamido-N donors of the αCys111-αSer112-αCys113 portion of the peptide frame (Fig. 1). In addition to these two carboxamido-N donors, two S donors of αCys111 and αCys113 are also bound to the Co^{3+} center in the equatorial plane. Interestingly, these two S centers are posttranslationally oxygenated further to cysteine sulfinic (-SO_2H) and cysteine sulfenic (-SOH) acid, respectively (Fig. 1). The axial coordination is completed by the αCys108-S and a water molecule.

Cobalt-containing Enzymes, Fig. 1 *Top panel*: schematic structure of the cobalt-containing active site of Co-NHase from *Pseudonocardia thermophila* JCM 3095. *Bottom panel*: proposed mechanism of nitrile hydrolysis at the Co^{3+} site in Co-NHase

Although the α and β subunits of NHases do not show homology in amino acid sequence, in all known NHases, each subunit has highly homologous amino acid sequences. In particular, the Cys cluster sequence of the α subunit that binds the Fe^{3+} or Co^{3+} is highly conserved in addition to the two Arg residues of the β subunit. Strong interactions between the two subunits appear necessary for the function of the enzyme. The role of the unusual coordination of the carboxamido-N donors has been elucidated by modeling studies (Harrop and Mascharak 2004; Mascharak 2002). Coordination of the strongly σ-donating carboxamido-N centers stabilizes the +3 oxidation state of the Fe center in Fe-NHase, and as a consequence, the Fe^{3+} center exhibits no tendency to bind and/or activate dioxygen. Rather, it acts as a Lewis acid and promotes hydrolysis of nitriles into amide. In case of Co-NHase, the highly stabilized Co^{3+} center also behaves similarly. The posttranslational oxygenation of the equatorial Cys-S centers is essential for the enzymatic activity. Such modification most possibly alters the pK_a of the bound water at the active site.

The mechanism of nitrile hydrolysis has yet to be firmly established. Since low-spin Co^{3+} centers are substitutionally inert, direct binding of nitriles at the metal site and subsequent release of the amides following hydrolysis at a fast rate appear quite unlikely. It is therefore believed that a metal-bound hydroxide activates a free water molecule at the active site (via deprotonation), and nucleophilic attack by the resultant OH⁻ on the C atom of RCN (nested at the active site pocket) leads to the formation of amide (Fig. 1).

Thiocyanate hydrolase (THase). Closely related to Co-NHase is the protein thiocyanate hydrolase (THase) which catalyzes thiocyanate (SCN⁻) hydrolysis to carbonyl sulfide (COS), ammonia, and hydroxide ion (2). Interest in this enzyme stems from its role in

$$SCN^- + 2H_2O \rightarrow COS + NH_3 + OH^- \quad (2)$$

remediation of industrial wastewater. THase isolated and purified from *Thiobacillus thiocapsa*, an obligate chemolithotrophic eubacterium, has been structurally characterized (Arakawa et al. 2006, 2009). The heterododecameric structure $(\alpha\beta\gamma)_4$ is composed of four $\alpha\beta\gamma$ heterotrimers, each containing one low-spin Co^{3+} ion within the γ subunit. The close evolutionary relationship between THases and NHases is readily recognized by the conserved amino acid sequence and overall structure of the core domain of the $\alpha\beta\gamma$ trimer which resembles those of Co-NHase very closely. For example, the γCys128-γCys133 locus of the γ subunit of THase resembles that of the α subunit of NHase. In the equatorial plane of Co^{3+}, two carboxamido-N donors from the peptide backbone (of γSer132 and γCys133) and two sulfur atoms of γCys131 and γCys133 are coordinated. Also, the γCys131 and γCys133 are posttranslationally oxygenated to γCysSO₂H and γCysSOH, respectively; a modification once again is necessary for enzymatic activity. The only difference in the active site structure of these two proteins is the fact that in the THase active site no tightly bound water is present, i.e., the Co^{3+} is 5-coordinate. It is quite possible that the catalytic mechanism involves direct binding of SCN⁻ to the cobalt site.

Methionine aminopeptidase (MA). Methionine aminopeptidase (MA) is a ubiquitous enzyme that cleaves the N-terminal methionine from newly translated polypeptide chains in both prokaryotes and eukaryotes and plays important roles in both protein turnover and functional regulation. Although MAs examined so far contain variable amounts of metals and can be

Cobalt-containing Enzymes, Fig. 2 Schematic structure of the cobalt-containing active site of methionine aminopeptidase from *E. coli*

activated by metal ions such as Mn^{2+}, Co^{2+}, and Zn^{2+}, careful purification and structural characterization of MA from *E. coli* have been shown to be a true cobalt protein. *E. coli* MA is a monomeric protein (29 kDa, 263 residues) that binds two Co^{2+} ions at its active site. The dimeric site is situated between two double β-antiparallel sheets, and the two Co^{2+} ions are ligated to the highly conserved amino acid residues Asp97, Asp108, His171, Glu204, and Glu235. As shown in Fig. 2, the dicobalt site (Co–Co distance = 2.9 Å) is held by two bridging carboxylates (Asp 108 and Glu 235), and the metal centers have open site(s) for further binding. It is interesting to note that the active site structure of MA shares common features with other metalloproteases such as bovine leucine aminopeptidase with two zinc (Zn) ions. Although the catalytic mechanism of MA has not been firmly established, it is believed that activation of a Co^{2+}-bound water (either at one Co^{2+} site or in the bridge) and its subsequent attack on the substrate bound at the other Co^{2+} site leads to the peptidase action.

Prolidase. Prolidases are dipeptidases that selectively cleave pro-containing peptides and are widely distributed in nature (from bacteria to human). They are involved in degradation of intracellular proteins and recycling of proline. The enzyme from the hyperthermophilic archaebacterium *Pyrococcus furiosus* is a homodimer (39.4 kDa per subunit) that contains two Co^{2+} ions per subunit. One of the Co^{2+} ions is tightly bound, while the second Co^{2+} could be replaced by Mn^{2+} (resulting in a decrease in activity), but not by other divalent metal ions such as Ni^{2+}, Cu^{2+}, or Zn^{2+}. It is therefore evident that prolidases belong to the general class of metallohydrolases with two metal ions at the active site. Although the amino acid sequence of *P. furiosus* prolidase exhibits no significant homology to that of methionine aminopeptidases (MAs), all five cobalt-binding residues are conserved in the former enzyme (Asp209, Asp220, His280, Glu313, and Glu327). In addition, the two-domain polypeptide fold of *P. furiosus* prolidase is similar to that noted in MAs.

Corrin-Containing Cobalt Enzymes

B_{12}-Dependent enzymes. ▶ Vitamin B_{12}, first identified as the antipernicious anemia factor in 1925, is a cobalt-containing cofactor in biology. The cobalt is coordinated to a corrin ring and a 5,6-dimethylbenzimidazole ligand that is covalently linked to the corrin frame (Fig. 3). The sixth ligand in ▶ Vitamin B_{12} is cyanide (CN^-). In coenzyme B_{12}, the sixth ligand is 5′-deoxyadenosine (Fig. 3), while in alkyl cobalamine, a methyl group (CH_3) occupies this position. These two different forms of the B_{12} (cobalamine) cofactor, with cobalt bonded to a methyl group or to a 5′-deoxyadenosine (Fig. 3), are utilized by the various B_{12}-dependent enzymes.

Methyl Transferase (MT)

B_{12}-dependent methyltranferases catalyze transfer of a methyl group from different methyl donors (such as methyltetrahydrofolate) to acceptor molecules (Mathews 2001). For example, the enzyme methionine synthase transfers a methyl group from methyltetrahydrofolate to homocysteine to form methionine and tetrahydrofolate (3).

(3)

In such reactions, the Co(III)–CH_3 bond of methylcobalamine (MeCbl) is cleaved *heterolytically*, leaving both bonding electrons (i.e., a reductive elimination) on the cobalt (forming a Co(I)alamin), and formally, CH_3^+, a carbocation, is transferred to the

Cobalt-containing Enzymes, Fig. 3 Structure of coenzyme B_{12} (5′-deoxyadenosylcobalamine, dAdoCbl) an organic cofactor (axial ligation is shown as slanted to avoid cluttering). In methylcobalamin (MeCbl), the 5′-deoxyadenosine group in the axial position is replaced with a methyl (–CH_3) group

substrate. These enzymes play important roles in amino acid metabolism in many organisms including humans. The catalytic process involves three protein components, each of which is localized on different polypeptide domains. The first protein component, MT1, binds the methyl donor and transfers it to the B_{12}-containing protein, leading to the formation of the organometallic intermediate with the Co(III)–CH_3 moiety. The third component (MT2) catalyzes the transfer of CH_3^+ cation to the substrate Y^- to form CH_3–Y. The MT2 protein components in general are all zinc proteins that bind and activate the thiolate methyl acceptor (Y^-). Methionine synthase from *E. coli* is the best-characterized protein in this class.

During the catalytic cycle, B_{12} cycles between Co(III)–CH_3 and Co(I). In case of oxidative inactivation of the Co(I) center to Co(II), other ancillary reductases (methionine synthase reductase in human) prime the site back to Co(III)–CH_3 via CH_3 transfer from the methyl donor S-adenosyl methionine (AdoMet) which in turn binds at a different location on the peptide frame.

B_{12}-Dependent Isomerases

The B_{12}-dependent isomerases (mostly found in bacteria) employ 5′-deoxyadenosylcobalamin (Fig. 3) as the cofactor (Banerjee and Ragsdale 2003). The primary function of this unit is to act as a stable entity in the storage and generation of free radicals catalyzing rearrangement and isomerization reactions (e.g., glutamate mutase, methylmalonyl-CoA mutase, diol dehydratase, ethanolamine ammonia lyase), methyl group transfer (e.g., methionine synthase), and deoxyribonucleotide biosynthesis (class II ribonucleotide reductase). In these proteins, the first step is *homolytic cleavage* of the Co(III)–CH_2dAdo bond to form Cob(II)alamine and the 5′-deoxyadenosyl radical (dAdo-CH_2•). Specific amino acid residues around the cofactor at the B_{12}-binding domain of these enzymes initiate this *different mode* of cleavage of the cobalt-carbon bond (Ludwig and Mathews 1997). Although the cobalt-carbon bond in the Co(III)–CH_2dAdo unit is stable in water, it is inherently labile with a bond dissociation energy of ~30–35 kcal mol^{-1}. The B_{12}-dependent isomerases utilize the 5′-adenosyl radical (dAdo-CH_2•) to effect radical-based rearrangements (Fig. 4, Banerjee 2003). The high-energy dAdo-CH_2• radical abstracts an H atom from the substrate to generate a substrate-centered radical intermediate (step 2 in Fig. 4). Following the 1,2 rearrangement (step 3), this intermediate captures the H atom back from dAdo-CH_3 and affords the final rearranged product (step 4).

The corrin ring of 5′-deoxyadenosylcobalamin provides some support to the facile Co(III)-C bond homolysis (step 1, Fig. 4) in the enzyme-bound cofactor, a key feature of the B_{12}-dependent isomerases. This macrocycle is analogous to the porphyrin ring found in hemes with some notable exceptions. For example, the corrin ring is sufficiently reduced with predominantly sp^3-hybridized carbons along the outer ring that are all chiral in nature and carries a total 1 charge (versus 2- in porphyrin). These features along with the strained C–C link between rings A and D of the corrin result in a more distorted basal plane coordination unit. The structural distortion could further worsen upon binding of the substrates and could weaken the Co(III)–C bond promoting bond homolysis.

Methylmalonyl-CoA mutase. Methylmalonyl-CoA mutase (MMCM) is the only dAdoCbl-dependent enzyme present in both mammals and bacteria where it catalyzes the reversible rearrangement of (2R)-methylmalonyl-CoA (MMCoA) to succinyl-CoA (4) through hydrogen atom exchange with the methyl group for the carbonyl-CoA group on the adjacent carbon of MMCoA (Banerjee 2003). The most frequently studied enzyme was isolated from *P. shermanii* and consists of a heterodimer with subunits of 69.5 and 80.1 kDa. The catalyzed reaction is a crucially important step in the metabolism of odd-

$$\text{HOOC}-\underset{\underset{H}{|}}{\overset{\overset{H}{|}}{C}}-\underset{\underset{\text{CO-SCoA}}{|}}{\overset{\overset{H}{|}}{C}}-H \rightleftharpoons \text{HOOC}-\underset{\underset{H}{|}}{\overset{\overset{\text{CO-SCoA}}{|}}{C}}-CH_3$$

SuccinylCoA MMCoA

(4)

chain fatty acids, and the absence of a functional protein is the basis for the human metabolic disease methylmalonic acidemia. As is the case with many B_{12}-dependent enzymes, the mechanism of the rearrangement involves the conversion of substrate radical (S•) to a product radical (P•). Many polar residues such as arginine and tyrosine line the active site pocket to neutralize the carboxylic acid group of the MMCoA substrate. Additionally, the thioester-O of the MMCoA is within H-bonding distance of a crucial histidine-N that when mutated results in a 10^2–10^3-fold decrease in k_{cat}. It has been proposed that this residue promotes rearrangement chemistry by serving as a general base to facilitate proton transfer in the transition state of this enzyme (Thomä et al. 2000). The detailed mechanism of this reaction remains inconclusive, but two proposals involve the formation of a cyclopropyl oxy radical intermediate or

Cobalt-containing Enzymes, Fig. 4 General mechanism for 5′-deoxyadenosylcobalamine-dependent isomerases

acrylate intermediate via a fragmentation-recombination mechanism.

Ethanolamine ammonia-lyase. Another dAdoCbl-dependent bacterial enzyme that utilizes radical-initiated isomerization reaction is ethanolamine ammonia-lyase (EAL) which converts ethanolamine to acetaldehyde and ammonia (5). In EAL from *E. coli*, dAdoCbl binds at the interface of the α and β subunits of the $\alpha_6\beta_6$ enzyme (Shibata et al. 2010), and formation of the dAdo-CH$_2$• radical (and Cob(II) alamine) occurs upon substrate binding. Rotation of the ribosyl moiety around the glycosidic linkage brings the radical center close to the substrate C1, forming the substrate-based radical.

$$H_2NCH_2CH_2OH \xrightarrow{EAL} CH_3CHO + NH_3 \quad (5)$$

Following amino group migration via a cyclic transition state, the radical is transferred to C2 which then captures the hydrogen back from dAdoCH$_3$. Finally, the rearranged product 1-amino-1-ethanol undergoes elimination of ammonia to form the final product acetaldehyde.

Adenosylcobalamin in DNA biosynthesis. ▶ Ribonucleotide reductases (RNRs) are allosterically regulated enzymes responsible for the conversion of nucleotides to deoxyribonucleotides (6) in all organisms and represent the central players in DNA replication and repair by providing the four basic monomeric deoxynucleotide precursors (Nordlund and Reichard 2006). In all organisms, the RNR-catalyzed synthesis is achieved by the reduction of the corresponding ribonucleotide via radical-based chemistry mediated by a neighboring cysteine thiyl radical formed at a transition metal site (Stubbe and van der Donk 1998). RNR enzymes have been further classified based on the metal cofactor required for the radical initiation process, and these comprise the three distinct RNR classes I, II, and III.

$$(6)$$

The Co-dependent RNRs are found in bacteria, algae, and archaea and make up the class II RNRs, which require dAdoCbl for enzyme activity. The function of the Co center in class II RNR is slightly different than other B$_{12}$-dependent enzymes. In such RNR, the overall radical product responsible for catalysis is not dAdo-CH$_2$• but a cysteine thiyl radical originating from the peptide backbone; the enzyme in essence utilizes dAdo-CH$_2$• to generate the thiyl radical required for ribonucleotide reduction. The structure of a class II RNR from *L. leichmannii* reveals some similarities with other RNRs (Sintchak et al. 2002). In contrast to class I and III RNRs, the Co-dependent RNR is a monomeric protein (76 kDa) but contains a similar effector protein interface structure as class I RNRs. Despite its structural simplicity, class II RNRs contain a very similar site for catalysis, namely, the conserved 10-stranded $\alpha\beta$ barrel with a finger loop that houses the active Cys-S at the tip. The enzyme accelerates the homolytic Co–C bond cleavage rate

$\sim 10^{11}$-fold faster than the uncatalyzed reaction corresponding to a transition state stabilization of 15 kcal/mol. The radical character of the intermediate dAdo-CH$_2$• radical is then transferred to the active cysteine residue forming the thiyl radical which finally abstracts the H atom at the 3′ position of the ribose substrate moiety to form the deoxyribose product. Transfer of radical character from dAdo-CH$_2$• to cysteine is supported by epimerization of (5′R)-[5′-^{2}H] adenosylcobalamin observed when the critical cysteine was mutated to serine or alanine. The distance between the Co(II) center and the thiyl radical has been estimated to be in the 5–8 Å range.

Conclusion

Taken together, the role of Co centers in biology, while not quite as profound as Fe or copper (Cu), is multifaceted in nature. Cobalt either takes on a primary role in the catalytic process effecting direct chemical transformation or indirectly as a cofactor that initiates a reaction but remains a spectator after the reaction begins. In particular, Co(III), although generally thought of as kinetically inert, directly participates in the hydrolytic cleavage of relatively high-energy RC≡N bonds (where R = aliphatic or aromatic-C). Indeed, the high Lewis acidity of Co(III) provides an ideal platform to generate such a potent Co(III)–OH nucleophile analogous to Zn(II)–OH at the active sites of hydrolytic zinc enzymes. Since these centers are inert to oxygen, they are perfect choices to perform hydrolytic chemistry without the possibility of undesirable oxidation reactions that would occur at both Fe and Cu centers. As a cofactor, the Co center facilitates isomerization/methylation/redox processes through the formation of carbon-based free radicals to initiate chemical transformations. In this case, the overall facile nature of the Co(III)–C bond toward homolytic cleavage initiates and propagates the free radical chemistry performed at these unique active sites. Collectively, Co-containing proteins, though limited in number with respect to other first-row transition metals like Fe, partake in diverse roles in biology that impact many metabolic processes. ▶ The overview of Co-binding proteins supports this fact in a succinct way.

Cross-References

▶ Cobalt, Physical and Chemical Properties
▶ Cobalt Proteins, Overview
▶ Nitrile Hydratase
▶ Ribonucleotide Reductase

References

Arakawa T, Kawano Y et al (2006) Structure of thiocyanate hydrolase: a new nitrile hydratase family protein with a novel five-coordinate cobalt(III) center. J Mol Biol 366:1497–1509

Arakawa T, Kawano Y et al (2009) Structural basis for catalytic activation of thiocyanate hydrolase involving metal-ligated cysteine modification. J Am Chem Soc 131:14838–14843

Banerjee R (2003) Radical carbon skeleton rearrangements: catalysis by coenzyme B$_{12}$-dependent mutases. Chem Rev 103:2083–1094

Banerjee R, Ragsdale S (2003) The many faces of vitamin B$_{12}$ catalysis by cobalamin-dependent enzymes. Annu Rev Biochem 72:209–247

Harrop T, Mascharak P (2004) Fe(III) and Co(III) centers with carboxamido nitrogen and modified sulfur coordination: lessons learned from nitrile hydratase. Acc Chem Res 37:253–260

Kobayashi M, Shimizu S (1998) Metalloenzyme nitrile hydratase: structure, regulation, and application to biotechnology. Nat Biotechnol 16:733–736

Kobayashi M, Shimizu S (1999) Cobalt proteins. Eur J Biochem 261:1–9

Ludwig M, Mathews R (1997) Structure-based perspectives on B$_{12}$-dependent enzymes. Annu Rev Biochem 66:269–313

Mascharak P (2002) Structural and functional models of nitrile hydratase. Coord Chem Rev 225:201–214

Mathews R (2001) Cobalamin-dependent methyltransferases. Acc Chem Res 34:681–689

Miyanaga A, Fushinobu S et al (2001) Crystal structure of cobalt-containing nitrile hydratase. Biochem Biophys Res Commun 288:1169–1174

Nordlund P, Reichard P (2006) Ribonucleotide reductase. Ann Rev Biochem 75:681–706

Shibata N, Tamagaki H et al (2010) Crystal structure of ethanolamine ammonia-lyase complexes with coenzyme B$_{12}$ analogs and substrates. J Biol Chem 285:26484–26493

Sintchak M, Arjara G, Kellogg B, Stubbe J, Drennan C (2002) The crystal structure of class II ribonucleotide reductase reveals how an allosterically regulated monomer mimics a dimer. Nat Struct Biol 9:293–300

Stubbe J, van der Donk W (1998) Protein radicals in enzyme catalysis. Chem Rev 98:705–762

Thomä N, Evans P et al (2000) Protection of radical intermediates at the active site of adenosylcobalamin-dependent methylmalonyl-CoA mutase. Biochemistry 39:9213–9221

CO-Dehydrogenase/Acetyl-CoA Synthase

Stephen W. Ragsdale, Elizabeth Pierce and Gunes Bender
Department of Biological Chemistry, University of Michigan Medical School, Ann Arbor, MI, USA

Synonyms

Carbon monoxide dehydrogenase/acetyl-CoA synthase; CODH/ACS

Definition

Carbon monoxide dehydrogenase (CODH) is a Ni–Fe$_4$S$_4$-dependent enzyme that catalyzes the interconversion of CO and CO$_2$. Acetyl-CoA synthase (ACS) is a Ni–Ni–Fe$_4$S$_4$ dependent enzyme that synthesizes the central metabolite, acetyl-CoA, from CO, a methyl group from CFeSP, and CoA. CODH and ACS form a tightly bound bifunctional enzyme in *Moorella thermoacetica*.

Background

Carbon monoxide dehydrogenase (CODH) allows microbes to extract CO from the air at the low levels present in the environment and use it as a sole source of carbon and energy (Fig. 1). Besides being a toxic gas, CO is a remarkably potent source of reducing equivalents; thus, the CODHs can couple CO oxidation to the reduction of various cellular redox systems (ferredoxin, hydrogenase, pyruvate synthase, etc.). Two types of CODH have been described: one harbors a molybdopterin/copper active site and the other contains a nickel–iron–sulfur cluster catalytic core. Given the focus on nickel of this volume, we will focus on the Ni–CODH, which uses a bimetallic mechanism to catalyze the oxidation of CO to CO$_2$ (Fig. 1 and (1)) In some anaerobic microbes, the Ni–CODH couples to another Ni-enzyme called acetyl-CoA synthase (ACS). In this situation, they function as the major players in a microbial pathway of CO and CO$_2$ fixation called the Wood-Ljungdahl (or reductive acetyl-CoA) pathway (Ragsdale and Pierce 2008). When CODH and ACS are coupled, CODH catalyzes the reduction of CO$_2$ to CO, and CO is then channeled to ACS to catalyze the reaction of CO with two other substrates (CoA and a methyl group bound to a corrinoid iron-sulfur protein, CFeSP, CH$_3$–Co as shown in Fig. 1) to generate the central metabolite, acetyl-CoA (2). The Wood-Ljungdahl pathway is unusual in that it generates CO as a metabolic intermediate, channels it as a gaseous substrate and uses complex metal clusters and organometallic intermediates to fix CO and CO$_2$ into cellular carbon. This review is aimed at covering the contemporary literature related to the Ni-containing CODH and ACS and placing recent research results in perspective. The review is geared for advanced undergraduate and graduate students, researchers, instructors, and professors in the areas of biochemistry, cell biology, and genetics and for whom this may not be their primary area of expertise.

$$CO + H_2O \rightleftharpoons CO_2 + 2e^- + 2H^+ \quad \Delta E_o' = -540 \text{ mV} \quad (1)$$

$$CO + CH_3 - Co(III) - CFeSP + CoAS^- \rightarrow CH_3 - CO - SCoA + Co(I) - CFeSP \quad (2)$$

CODH

General Characteristics of Ni–CODH

Early studies on CODH (reviewed in Ragsdale and Pierce 2008) showed that Ni is necessary for CO oxidation by acetogenic bacteria and methanogenic archaea. CODHs catalyze reaction (1), which allows organisms to interconvert CO and CO$_2$ (Ragsdale and Pierce 2008). In the forward direction, this reaction allows microbes to grow on CO as the sole source of carbon and energy. The fate of the electrons determines the metabolic substrates utilized or end products produced; for example, CO oxidation can be coupled to the reduction of protons, metals, sulfate, etc. (Techtmann et al. 2009). The microbial coupling of CO oxidation to H$_2$ formation is linked to proton translocation. Similarly, purified CODH and hydrogenase have been adsorbed onto conducting graphite platelets to generate a device that can produce either

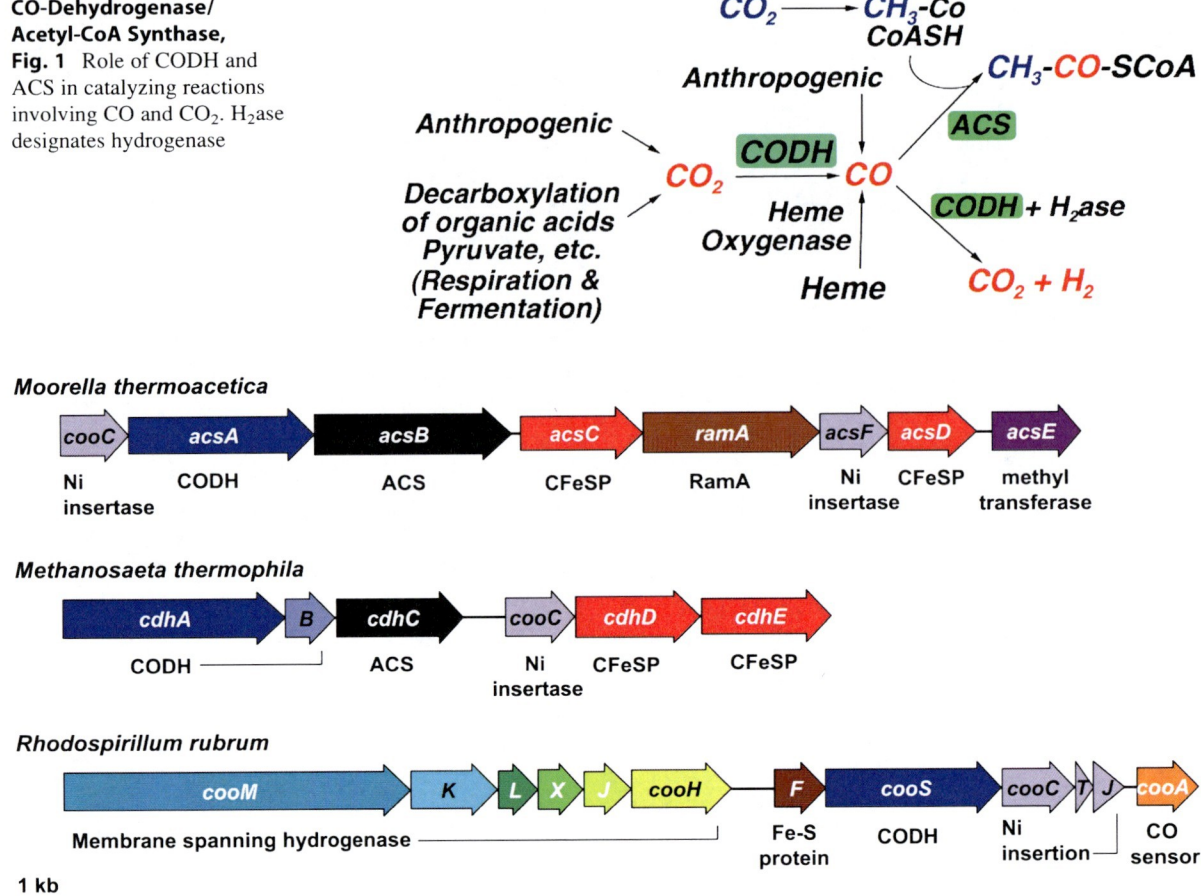

CO-Dehydrogenase/Acetyl-CoA Synthase, Fig. 1 Role of CODH and ACS in catalyzing reactions involving CO and CO_2. H_2ase designates hydrogenase

CO-Dehydrogenase/Acetyl-CoA Synthase, Fig. 2 Gene clusters encoding the Type III CODH/ACS from an acetogen, a methanogen the uses the Wood-Ljungdahl pathway in reverse, and a Type I CODH from *R. rubrum* that is coupled to a membrane bound hydrogenase

current or H_2 (Lazarus et al. 2009). Equation 1 also can be run in reverse, enabling organisms to grow on CO_2 as an electron acceptor coupled to the utilization of reducing equivalents produced by various fermentative reactions, e.g., H_2 oxidation and oxidative decarboxylation of pyruvate and aromatic acids. In the Wood-Ljungdahl pathway, CO_2 is reduced to CO, which reacts with CoA and the methylated CFeSP to generate acetyl-CoA, as shown in Fig. 1 (Ragsdale and Pierce 2008).

The genes encoding Ni-CODHs (*acsA*, *cdhA*, and *cooS*) are highly homologous and are found in different contexts, depending on the physiological function of that particular CODH (Fig. 2). For example, *acsA* encodes a CODH that forms a tight complex with ACS (*acsB*). AcsA and *acsB* are in the same cluster as the genes encoding the two subunits of the corrinoid iron-sulfur protein (CFeSP) and methyltransferase (MeTr), which are the component enzymes of the Wood-Ljungdahl pathway (Ragsdale and Pierce 2008). In methanogens, an analogous gene cluster (called *cdh* in *M. thermophila*) encodes the proteins involved in utilization of acetyl-CoA as an energy and carbon source. Indicating the function of the protein that it codes for, *cooS* is often found in a gene cluster that includes a membrane-bound hydrogenase.

There are several Ni–CODH subfamilies, with sequence identity as low as 30% between the most distantly related pairs of sequences across these subfamilies. In fact, *Carboxydothermus hydrogenoformans* encodes five CODHs (Techtmann et al. 2009), named CODH I–V. CODH I and CODH II are loosely associated with the inner side of the cytoplasmic membrane. The proposed physiological role of CODH I, encoded by a gene in a cluster like the one from *Rhodospirillum rubrum* in Fig. 2, is to couple CO oxidation to proton uptake by a membrane-associated hydrogenase, thus generating a transmembrane H^+ gradient for ATP synthesis. CODH II appears to be involved in coupling CO oxidation to reduction of cellular electron carriers.

CODH III shares ~50% sequence identity with CODH I and II; however, its sequence is most similar to that of the AcsA enzyme from *M. thermoacetica*, which catalyzes both CO_2 reduction to CO for acetyl-CoA synthesis and oxidation of CO to provide reducing equivalents for cellular redox carriers (Ragsdale and Pierce 2008). In *M. thermoacetica* the AcsA CODH is isolated in a complex with ACS. In some methanogens, a much larger protein complex called acetyl-CoA decarbonylase/synthase (ACDS) is assembled, which includes CODH, ACS, and both subunits of the CFeSP. This complex functions to catalyze the Wood-Ljungdahl pathway in reverse, with CODH oxidizing the CO that is generated during acetyl-CoA utilization (Ragsdale and Pierce 2008).

CODH IV appears to be involved in protecting against oxidative stress, while the sequence of CODH V differs most from the other CODH sequences and its gene neighborhood does not indicate a physiological role. Neither CODH IV nor CODH V have been purified.

The Involvement of a CO Sensor in Induction of CODH

A heme-containing transcriptional regulator called CooA, which is part of the *cooS* gene cluster, induces expression of CODH and the other proteins in the *coo* gene cluster (Techtmann et al. 2009). CooA is closely related to the cAMP receptor protein, containing a domain that binds ligand (heme instead of cAMP) and a DNA-binding domain that recognizes a promoter sequence similar to that targeted by CRP. Binding of CO to the heme elicits a conformational change that promotes association of CooA with DNA. Another CO-responsive transcriptional regulatory protein called RcoM has been identified that binds heme within a PAS domain (Techtmann et al. 2009).

Maturation of the Ni–CODH

While the various proteins involved in assembly of the active site of hydrogenase and urease are rather well defined, comparatively little is known about maturation of the Ni–CODH (Lindahl 2002). Many of the CODH gene clusters in bacteria contain two small genes, *cooC* and *acsF*, that appear to play a role in active site maturation. In *R. rubrum*, *cooC* is found in a *cooCTJ* gene cluster and deletion of portions of this gene cluster increases the concentration of Ni needed for CO-dependent growth of *R. rubrum*. Enzymatic experiments also indicate a role for CooC in the ATP-dependent insertion of Ni into CODH. It is not known if chaperones are involved in metal incorporation or maturation of the A-cluster of ACS. There is some evidence that the product of the *acsF* gene (Fig. 2) facilitates insertion of Ni into the A-cluster.

Metal-Centered Cofactors of the Ni–CODH

Crystal structures have been determined for the *R. rubrum* CO-induced type I CODH, CODH II from *C. hydrogenoformans*, the CODH/ACS complex from *M. thermoacetica* (Fig. 3) and the CODH component of the ACDS complex of *Methanosarcina barkeri*, as described in a recent structural paper (Kung and Drennan 2010). The bacterial CODH structures are very similar and overlay with root mean square deviations of <1.0 Å, while the methanogenic (*M. barkeri*) CODH contains an additional domain that harbors two additional Fe_4S_4 clusters and tightly associates with a small subunit whose function is unknown. All of the CODHs crystallize as a homodimer and each monomer contains a unique Ni–Fe_4S_4 cluster (the C-cluster) at its active site and another Fe_4S_4 cluster (the B-cluster). A third Fe_4S_4 cluster (D-cluster) bridges the two subunits by binding to cysteines from separate monomers in a manner similar to that in the Iron Protein, which transfers electrons to nitrogenase. The B- and D-clusters associate with the N-terminal helical domain and form a redox wire that transfers electrons to the C-cluster, which binds to two Rossmann domains.

The C-cluster is where CO oxidation is catalyzed (Fig. 3). It is conserved in all Ni CODHs and contains a Fe_3S_4 cluster that is ligated by three cysteine residues and is connected to a binuclear Ni–Fe site. The Fe of the Ni-Fe site, called ferrous component II (FC II), is ligated by a histidine and a fifth cysteine ligand that

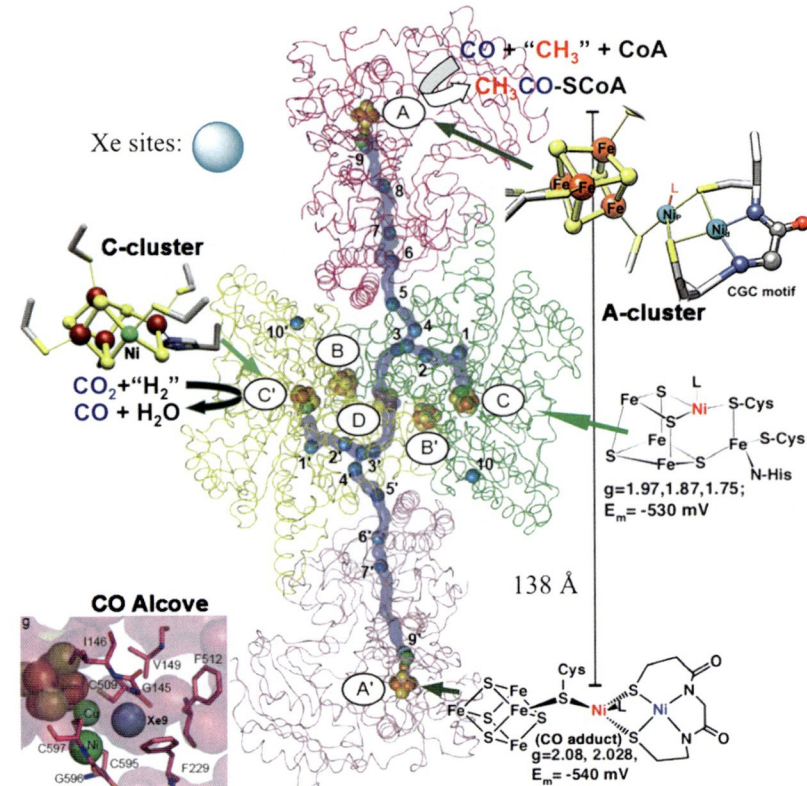

CO-Dehydrogenase/Acetyl-CoA Synthase, **Fig. 3** Structures of CODH/ACS and the metallocenter active sites, the C-cluster and A-cluster. The CO channel is shown in *blue*, with the *blue spheres* representing the Xe sites located by crystallographic studies. A site near the A-cluster, termed the CO alcove, is shown in the *lower left hand corner*. EPR g-values and midpoint redox potentials (E_m) of the C- and A-clusters are also indicated. The homodimeric CODH component of the complex is shown in *yellow* and *green* in the *middle*, while the two ACS subunits are located on opposite sides of the CODH dimer. The structures of the ACS-associated type III CODHs are very similar to those of the CooS-type I (or II) CODHs

forms a μ_3-S coordination at one corner of the cubane center, while the Ni, which is approximately planar, is bridged to two sulfurs of the Fe_3S_4 moiety and coordinated by a fourth cysteine. There has been some discussion about whether there is an additional bridge between Ni and FC II (Kung and Drennan 2010). In the first structure of the *C. hydrogenoformans* CODH II, an additional inorganic S was modeled at this bridging position, while in the *R. rubrum* CODH structure, the sulfur of a Cys residue appeared as the bridge. In the *M. thermoacetica* CODH/ACS structure, this cysteine could be modeled in different positions, one in which it coordinates Ni and another in which it coordinates FC II. In the most recent structures of CODH containing CO_2, water, and cyanide bound to the C-cluster, the bridging sulfur is absent and is replaced by one of the different ligands used.

Bimetallic Ni–CODH Mechanism

Determination of the enzymatic activity of CODH is straightforward and involves adding enzyme to a buffer solution containing CO and an electron acceptor. A low-potential one-electron acceptor, methyl viologen, is often used; however, the Ni–CODH is rather promiscuous, so most redox dyes can be employed. Methyl viologen is colorless in the oxidized state and is converted to a deep violet color with a high extinction coefficient in the one-electron reduced state. The natural electron acceptor is considered to be ferredoxin.

As described in Fig. 4, CODH uses a ping–pong mechanism (Ragsdale and Pierce 2008). In the first half-reaction, CO reduces CODH, thus forming CO_2 and, in the second half-reaction, an electron acceptor binds and reoxidizes CODH. Spectroscopic and crystallographic studies indicate that Ni and FC II are both involved in binding ligands during catalysis (Ragsdale and Pierce 2008; Lindahl 2002). The C-cluster can equilibrate among at least three different oxidation states: C_{ox}, which is EPR-silent, C_{red1} (with g values of 2.01, 1.81, and 1.65) and C_{red2} (with g values of 1.97, 1.87, and 1.75). One-electron reduction of C_{ox} generates C_{red1} (following a midpoint potential of -220 mV) and lowering the redox potential to around -530 mV leads to the formation of C_{red2}. As studied by rapid freeze-quench EPR experiments, when CODH is

CO-Dehydrogenase/Acetyl-CoA Synthase, Fig. 4 CODH mechanism. The individual steps 1–5 of the catalytic mechanism and the slow binding cyanide inhibition are described (Based on Kung and Drennan 2010)

treated with CO, C_{red1} is rapidly converted into C_{red2} at a diffusion-controlled rate. Next, the B-cluster undergoes reduction at a rate constant (60 s^{-1} at 5°C) that is similar to the k_{cat} for CO oxidation (47 s^{-1} at 5°C).

Step 1 of the CODH mechanism involves CO binding to the Ni site of the C-cluster, as suggested by X-ray crystallographic and spectroscopic studies of the complexes of CODH with CO and with CN, a competitive inhibitor with respect to CO (Ragsdale and Pierce 2008; Kung and Drennan 2010). Apparently there are two modes of CO and CN binding to the enzyme: one with an acute Ni-C-N/O angle and another that is linear. The existence of two Ni–CN conformations is consistent with kinetic studies of CN$^-$ inhibition, which demonstrate that CN is a slow binding inhibitor, indicating that CODH forms a complex with CN that is rapidly reversible followed by a separate complex that releases very slowly (Ragsdale and Pierce 2008). *Step* 2 involves the deprotonation of bound water, which apparently involves base catalysis by Lys or His residues near the C-cluster. In *Step* 3, OH$^-$ from the Fe-hydroxide attacks Ni–CO to form a carboxylate that bridges the Ni and Fe atoms. In *Step* 4, elimination of CO_2 is coupled to two-electron reduction of the C-cluster (Ni$^{"0"}$) and the binding of water. The two-electron reduction could generate a true Ni^0 state (Lindahl 2002) or, more likely, the electrons delocalize into the Fe and S components of the C-cluster. In *Step* 5, electrons are transferred from the reduced B- and D-clusters to an external redox mediator, e.g., ferredoxin. Each of the steps above can occur in reverse to catalyze the reduction of CO_2.

Electrochemical and Photochemical Studies of the Ni–CODH

Electrochemical methods have been used to help understand the redox chemistry associated with CO oxidation and CO_2 reduction. For example, spectroelectrochemical studies (monitoring changes in the spectra as a function of potential) of the interconversion between the different states of CODH defined the midpoint potentials for the C_{red1} and C_{red2} states, which are linked to catalysis (Ragsdale and Pierce 2008; Lindahl 2002). Several direct electrochemistry experiments have also been described in which cyclic voltammetry was performed with CODH directly attached to a pyrolytic graphite edge or glassy carbon working electrode (Lazarus et al. 2009). The direct reduction of CO_2 to CO in the presence of the mediator, methyl viologen, occurs at a rate that is half maximal at a redox potential near the midpoint potential for the CO_2/CO couple. The enzyme was active in both directions with the electrocatalytic reaction being dependent on the pH and CO/CO_2 concentrations. At low pH values, the rate of CO_2 reduction surprisingly exceeds that of CO oxidation. The high catalytic efficiency of CO_2 reduction at the thermodynamic potential for the CO_2/CO couple appears to be due to its ability to undergo rapid successive two-electron transfers coupled to proton transfer, unlike nonenzymatic catalysts that reduce CO_2 through a high-energy $\bullet^-CO_2$ anion radical. Electrochemical studies also revealed a one-electron oxidative inactivation of CODH at -50 mV and reductive reactivation at -250 mV.

Electrochemistry is also a valuable tool to drive the direct reduction of CO_2 and to produce electricity and H_2 from CO (a major component of syngas). For example, when CODH was adsorbed on graphite platelets with the *Escherichia coli* hydrogenase, H_2 was produced, thus, coupling CO oxidation to proton reduction to H_2 (essentially by (2)) (Lazarus et al. 2009). CODH could also be co-adsorbed with an inorganic ruthenium complex onto TiO_2 nanoparticles, allowing the photoreduction of CO_2 to CO using very mild reductants (Woolerton et al. 2010).

Energy Conservation Linked to CO Oxidation

Physiological data clearly show that microbial CO oxidation and CO_2 reduction are linked to energy conservation and ATP synthesis by formation of ion gradients. For example, acetogenic bacteria can grow autotrophically on H_2/CO_2 or CO using the Wood-Ljungdahl pathway (Ragsdale and Pierce 2008; Drake et al. 2008). Through this pathway, *M. thermoacetica* generates acetyl-CoA, which is converted to acetate and ATP through the actions of phosphotransacetylase and acetate kinase. However, because one ATP is required for the conversion of formate to formyltetrahydrofolate in the methyl branch of the Wood-Ljungdahl pathway, there is no net ATP synthesis by substrate-level phosphorylation. Therefore, for autotrophic growth by this pathway, ATP synthesis must occur through some type of chemiosmotic mechanism.

It has been proposed that CO oxidation in *Moorella thermoautotrophica* is coupled to reduction of membrane-bound *b*-type cytochromes in a process that is linked to the reduction of methylenetetrahydrofolate to methyltetrahydrofolate by methylenetetrahydrofolate reductase (Muller 2003). The following electron transport chain has been proposed: oxidation of CO coupled to reduction of cytochrome b_{559}, which then reduces methylenetetrahydrofolate or menaquinone and cytochrome b_{554}, which would finally reduce rubredoxin. A proton-pumping F_1F_0 ATP synthase from *M. thermoacetica* has been characterized and, when *M. thermoacetica* membrane vesicles are exposed to CO, they generate a proton motive force.

While formation of a proton gradient through a cytochrome-based pathway can be linked to CO oxidation, some acetogens like *Acetobacterium woodii* lack cytochromes. Such organisms have been suggested, by analogy to the corrinoid-containing, Na^+-pumping methyltetrahydromethanopterin: coenzyme M methyltransferases of methanogenic archaea, to use a sodium gradient involving a sodium-requiring methylenetetrahydrofolate reductase for energy conservation (Muller 2003). Furthermore, acetogenesis in *A. woodii* is coupled to Na^+ import and the *A. woodii* Na^+-dependent ATP synthase has been purified and characterized.

Mechanisms that are being considered for coupling CODH to generation of a transmembrane ion gradient include the Rnf-type NADH dehydrogenase complex, which has been recently shown to couple H_2 oxidation or caffeate reduction to ATP synthesis in *A. woodii* (Biegel et al. 2009). This complex translocates Na^+ while coupling the oxidation of ferredoxin to NAD^+ reduction. The NADH dehydrogenase genes are homologous to *E. coli* NADH dehydrogenase and to components of *R. rubrum* and *C. hydrogenoformans* hydrogenases that are coupled to CO oxidation.

ACS: Acetyl-CoA Synthase

Coupling CO_2 Reduction by CODH to Acetyl-CoA Formation by ACS

During growth of various substrates via the Wood-Ljungdahl pathway, CO is generated by CODH as a metabolic intermediate. For example, when acetogens grow on pyruvate, PFOR catalyzes the oxidative decarboxylation of pyruvate to acetyl-CoA and CO_2, CODH catalyzes the reduction of CO_2 to CO and ACS utilizes the CO (in its reaction with the methyl donor and CoA) to generate a second molecule of acetyl-CoA (Ragsdale and Pierce 2008). Given the very low midpoint potential for the CO_2/CO couple, the generation of CO is a significant investment by the organism. For example, there is over a 200 mV difference in redox potential between the NAD^+/NADH and CO_2/CO couples, which allows for the CO-coupled reduction of NAD^+ to yield about 40 kJ/mol of free energy, which is sufficient for synthesis of one mol of ATP.

Enzymatic and X-ray crystallographic studies revealed the existence of a gas channel connecting the active site of CODH, which produces CO, to the active site of ACS, which utilizes it (Ragsdale and Pierce 2008). When CODH/ACS crystals were subjected to high pressures of Xenon gas, Xe atoms were found at 19 discrete sites in the protein, showing the path of the CO tunnel between the active sites. Mutations of residues to block the movement of CO in the channel led to enzyme variants with much lower CODH and ACS activities. A similar gas channel was observed in the crystal structure of the acetyl-CoA decarbonylase/synthase complex of aceticlastic methanogens. The CO channel could confer several selective advantages. After the large energetic investment represented in CO synthesis, it would be important to sequester the produced CO, thus maintaining a relatively high local concentration and not having it diffuse away from the enzyme. It has been shown that CO_2 is a better donor of CO for acetyl-CoA synthesis than CO in solution, which is in relatively low concentration in the environment. In addition, CO is a potent inhibitor of various metalloenzymes like hydrogenase. Finally, the channel may help coordinate activity of the C- and A-clusters.

The Synthesis of Acetyl-CoA by ACS

The Structure of ACS and its Active Site A-Cluster

ACS catalyzes a remarkable reaction that is key to the Wood-Ljungdahl pathway (Ragsdale and Pierce 2008). The enzyme binds its three substrates, CO (generated by CODH and arriving from the channel just described), a methyl group (donated by the CFeSP) and CoA. Then, the enzyme connects together the methyl and CO groups to make an acetyl intermediate. Finally, it attaches the acetyl group to the thiol group of CoA to generate the high-energy and ubiquitous cellular feedstock, acetyl-CoA. This series of reactions involves the formation of methyl-Ni or Ni–CO and acetyl-Ni organometallic intermediates. ACS consists of three domains that exhibit large conformational changes to accommodate the three very differently sized substrates (a 15 Da methyl group on an 88 kDa CFeSP, a 28 Da CO molecule, and 770 Da Coenzyme A) and to coordinate the reaction steps. For example, to transfer the methyl group to the A-cluster, the C-terminal domain of ACS must open to bring the A-cluster and Co(III)-CH_3 moiety within bonding distance; however, it must coordinate this opening with closure of the CO channel to prevent the release of CO into solution. Thus, ACS must precisely accomplish a series of intriguing reactions to allow organisms to utilize the Wood-Ljungdahl pathway.

A-Cluster: Structure and Formation of the A-Cluster

As shown in Fig. 3 and reviewed recently (Ragsdale and Pierce 2008), the A-cluster consists of a Fe_4S_4 cluster that is bridged by a cysteinyl sulfur to a dinuclear Ni center. As in most Fe_4S_4 clusters, four Cys residues (Cys506, Cys509, Cys518, and Cys528, numbering based on the *M. thermoacetica* sequence) coordinate the iron sites in this component of the A-cluster. Cys 509 also bridges the Fe_4S_4 unit to the proximal Ni (Ni_p), nearest to the cluster. Two other Cys residues (Cys595 and Cys597) bridge Ni_p to the distal Ni (Ni_d), which is additionally coordinated by the backbone amide groups of Cys595 and Gly596, forming a square planar coordination environment for Ni_d.

One complicating feature of Ni_p is its lability and susceptibility to substitution by Cu and Zn (Ragsdale and Pierce 2008). In fact, Cu was present at high occupancy in the first crystal structure (with Ni in the distal site) and activity appeared to correlate with the Cu content, suggesting that the active form of ACS might contain a binuclear Cu–Ni center. However, various experiments, including activity measurements over a very wide range of metal concentrations, finally provided conclusive evidence that the Ni–Ni form of the A-cluster is active and the Cu–Ni form is not.

ACS exhibits conformational flexibility, as exhibited by large differences between the structures of the "open" Ni–Ni form of ACS versus the closed state of the Zn–Ni and Cu–Ni enzyme (Ragsdale and Pierce 2008). This plasticity may allow for alternative coordination and oxidation ($Ni^{1+, 2+, 3+}$) states during the catalytic cycle, e.g., adopting a closed conformation to react with CO as it exits from the channel and an open conformation to allow reaction with the much larger substrate, methylated CFeSP.

Assays and Mechanism of Acetyl-CoA Synthesis

The most convenient assay for ACS is an isotopic exchange reaction between the ^{14}C labeled carbonyl group of acetyl-CoA and unlabeled CO in solution (Ragsdale and Pierce 2008). This assay does not require any of the other Wood-Ljungdahl pathway proteins. In this assay, one incubates ACS or CODH/ACS with commercially available [1-^{14}C]-acetyl-CoA in a CO atmosphere and follows the decrease in radioactivity of the acetyl-CoA over time. Another relatively convenient assay is an isotopic exchange between radioactively labeled CoA or dephospho-CoA and acetyl-CoA. All activities have been shown to exhibit redox dependence and most of the data indicate an $n = 1$ (one-electron) process with an activation potential between -520 and -540 mV. One can also determine the activity of ACS by quantifying the relative (spins/mol enzyme) amount of an EPR signal that develops when ACS is treated with CO. The CO-treated reduced enzyme exhibits a characteristic EPR spectrum with g-values at 2.074 and 2.028. Studies of this EPR signal provided the first indication that ACS contains a heterometallic Ni–Fe–S cluster. When ^{13}CO was used and when the enzyme was labeled with ^{61}Ni and ^{57}Fe, hyperfine splittings were observed, indicating CO binds to a Ni-iron-sulfur cluster containing more than two iron atoms and that there is extensive delocalization of the unpaired electron spin among these components of the cluster; therefore, this signal was named the NiFeC signal.

It appears that the labile Ni_p in the A-cluster is the site to which substrates directly bind. Pulse-chase experiments indicate that CO and the methyl group bind randomly to ACS to form the acetyl-ACS intermediate that reacts with CoA to form acetyl-CoA. This is consistent with experiments indicating that methylated and carbonylated forms of ACS can both be intermediates during the catalytic cycle. Thus, as shown in Fig. 5, productive binary complexes can form between ACS and either CO or the methylated CFeSP to form viable ACS-CO or methyl-ACS intermediates in the overall synthesis of acetyl-CoA.

The active state of the A-cluster has been alternatively described as a Ni_p^0, a Ni_p^{1+}, or a Ni_p^{2+} species, as well as a spin-coupled center in which Ni_p^{1+} is coupled to a $[Fe_4S_4]^{1+}$ cluster. Based on the results of electrochemical, isotope chase, and transient kinetic experiments, the authors favor Ni_p^{1+} as the active starting state. Stopped flow IR and freeze quench EPR studies indicate that the only ACS-CO species that forms at catalytically relevant rates is the paramagnetic Ni_p^{1+}-CO species. Because either CO or the methyl group can bind to ACS in the first step in acetyl-CoA synthesis, the Ni_p^{1+}-CO intermediate must be a viable intermediate in the pathway. Transient kinetic studies following formation and decay of the EPR-active Ni^{1+}-CO species also have established its catalytic competence as an intermediate in acetyl-CoA synthesis. Furthermore, the midpoint potential for formation of the NiFeC species (-540 mV) is consistent with the values for reductive activation of ACS in the CO/acetyl-CoA and CoA/acetyl-CoA exchange reactions.

According to the "paramagnetic" mechanism (Fig. 5), *Step 1* of acetyl-CoA synthesis involves reductive activation of Ni_p^{2+} to the active Ni^{1+} state, which can bind CO ("a" branch in red) or the methyl group ("b" branch in blue). In *step 2a*, Ni_p^{1+} binds CO to form a paramagnetic Ni_p^{1+}-CO intermediate. This step can be reversed by photolysis of Ni_p^{1+}-CO, generating Ni_p^{1+}; however this is a highly labile species that must be maintained at very low temperatures, otherwise, it quickly recombines with CO. The CO used in this step is generated in situ by CODH and channeled to the active site of ACS. In *step 3a*, ACS binds the methyl group to form acetyl-Ni^{3+}, which is rapidly reduced (*step 4a*) by a proposed internal redox shuttle to form acetyl-Ni^{2+}. Finally, in *step 5a*, CoA reacts with acetyl-ACS to release acetyl-CoA in a reaction that leads to transfer of one electron to the

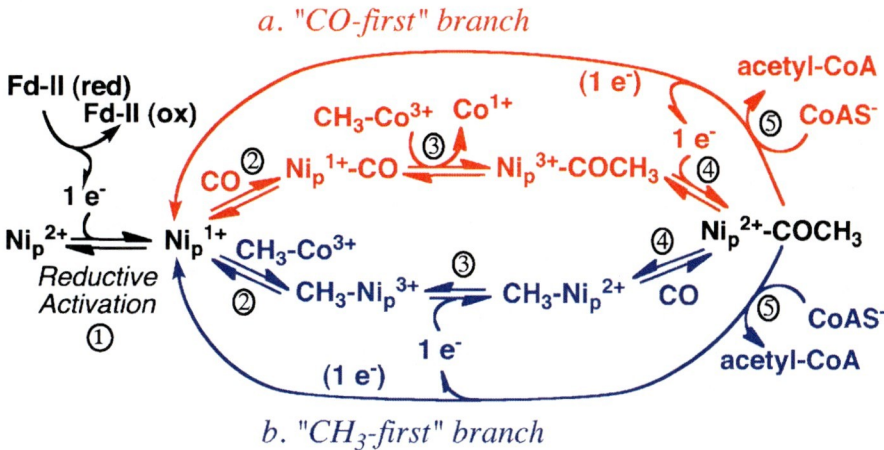

CO-Dehydrogenase/Acetyl-CoA Synthase, Fig. 5 Random paramagnetic mechanism. This mechanism indicates that ACS can first bind either the methyl group or CO (see the text for details). CH$_3$–Co represents the methylated CFeSP. When CoAS$^-$ reacts, two electrons are introduced, one of which is proposed to go to an internal electron shuttle to reduce Ni^{3+} to Ni^{2+} in the next catalytic cycle and the other is proposed to be retained at the Ni center (depicted by the 1 e$^-$ in parenthesis) to generate a transient unstable Ni(I) state

A-cluster to regenerate the Ni^{1+} starting species and the other to the electron shuttle, which can reduce acetyl-Ni^{3+} in the next catalytic cycle. In the lower "b" branch, reductive activation is coupled to methylation (*step 2b*). Transfer of the methyl group to ACS occurs by an S$_N$2 pathway and the methyl-ACS product appears to not be a species with an S = ½ EPR signal i.e., methyl-Ni(III). Thus, in *step 3b*, it is proposed that one-electron is transferred to methyl-Ni^{3+} to generate methyl-Ni^{2+}. Then, in *step 4b*, CO binds to form the acetyl-ACS intermediate, which reacts with CoA (5b), as in the "a" branch. Chemical modification studies implicate the involvement of arginine and tryptophan residues in binding CoA. Chemical modification and CoA protection experiments indicate that Trp418, which is in the middle domain of ACS, is involved in interacting with CoA.

The diamagnetic mechanism (Lindahl 2002) can be related to the lower path (b) except that it includes a two-electron reductive activation to generate Ni^0 or a spin-coupled species with Ni and the cluster in the 1+ states, methylation to generate methyl-Ni^{2+}, subsequent carbonylation to generate acetyl-Ni^{2+}, and nucleophilic attack by CoA to regenerate the active Ni$^{"0"}$ state. The diamagnetic mechanism considers Ni^{1+}-CO to be an off-pathway state and, due to the two-electron reduction, does not require an internal electron shuttle.

The methyl group of chiral isotopomers of CH$_3$–H$_4$folate is converted to acetyl-CoA with retention of configuration. Both the paramagnetic and diamagnetic mechanisms satisfy the stereochemical requirements because they involve two successive nucleophilic attacks, i.e., by Co(I) on methyl-H$_4$folate and by Ni(I) on the methylated CFeSP, which would lead to net retention of configuration.

Synthetic Analogs of the Anaerobic Ni–CODH and Ni–ACS Active Sites

Model complexes have been synthesized that mimic different aspects of the structure of CODH C-cluster and/or catalyze either CO oxidation or its reverse. Most of these are covered in a review by Evans (Evans 2005). Significant effort has been devoted to development of catalysts that can catalyze CO$_2$ reduction. The ultimate CO$_2$ reduction catalyst would be relatively inexpensive and have the ability to capture CO$_2$ from the atmosphere and reduce it to a useful fuel. Ideally, the catalyst would also mimic photosynthesis in coupling solar energy to convert a mild reductant to a powerful electron donor. Similarly, various models of the ACS active site have been synthesized. The ACS reaction is essentially the Monsanto or Reppe process.

The review by Evans also gives a succinct survey of model compounds related to the ACS active site.

Final Perspective

Over the past decade, many significant findings have been revealed related to the metabolism of CO_2 and CO by CODH and ACS. The structures of these enzymes have been determined and with these structures, surprising atomic level descriptions of novel metal-centered redox cofactors, including unusual nickel–iron–sulfur clusters, have been revealed. These structures provide an architectural framework for mechanistic studies, which are revealing novel ways that metals function in biology. Perhaps, further research on these processes will return important practical benefits, such as development of processes to more efficiently utilize CO_2 as a chemical feedstock and generate energy-rich compounds. While current nonbiological catalysts are relatively unselective and require significant overpotentials to drive CO_2 reduction, the enzymatic reactions are fast and operate at the thermodynamic redox equilibrium. These studies are expected to provide insights into the Wood-Ljungdahl pathway, which has been proposed to have been key to the emergence of life on earth. Studies of this pathway also have revealed novel mechanisms involving metallocenters that act as nucleophiles to form organometallic intermediates and catalyze C–C and C–S bond formation, the latter generating the high-energy thioester bond of acetyl-CoA.

Cross-References

- ▶ Nickel in Bacteria and Archaea
- ▶ Nickel Ions in Biological Systems
- ▶ Nickel, Physical and Chemical Properties
- ▶ Nickel-Binding Sites in Proteins
- ▶ [NiFe]-Hydrogenases

References

Biegel E, Schmidt S, Müller V (2009) Genetic, immunological and biochemical evidence for a Rnf complex in the acetogen *Acetobacterium woodii*. Environ Microbiol 11:1438–1443

Drake HL, Gossner AS, Daniel SL (2008) Old acetogens, new light. Ann N Y Acad Sci 1125:100–128

Evans DJ (2005) Chemistry relating to the nickel enzymes CODH and ACS. Coord Chem Rev 249:1582–1595

Kung Y, Drennan CL (2010) A role for nickel-iron cofactors in biological carbon monoxide and carbon dioxide utilization. Curr Opin Chem Biol 15:276–283

Lazarus O, Woolerton TW, Parkin A et al (2009) Water–gas shift reaction catalyzed by redox enzymes on conducting graphite platelets. J Am Chem Soc 131:14154–14155

Lindahl PA (2002) The Ni-containing carbon monoxide dehydrogenase family: light at the end of the tunnel? Biochemistry 41:2097–2105

Muller V (2003) Energy conservation in acetogenic bacteria. Appl Environ Microbiol 69:6345–6353

Ragsdale SW, Pierce E (2008) Acetogenesis and the Wood-Ljungdahl pathway of CO_2 fixation. Biochim Biophys Acta 1784:1873–1898

Techtmann SM, Colman AS, Robb FT (2009) 'That which does not kill us only makes us stronger': the role of carbon monoxide in thermophilic microbial consortia. Environ Microbiol 11:1027–1037

Woolerton TW, Sheard S, Reisner E et al (2010) Efficient and clean photoreduction of CO_2 to CO by enzyme-modified TiO_2 nanoparticles using visible light. J Am Chem Soc 132:2132–2133

CODH/ACS

▶ CO-Dehydrogenase/Acetyl-CoA Synthase

Coenzyme-B Sulfoethylthiotransferase

▶ Methyl Coenzyme M Reductase

Collagenase

▶ Zinc Matrix Metalloproteinases and TIMPs

Colloidal Arsenic

▶ Arsenic in Pathological Conditions

Colloidal Gold

▶ Gold Nanoparticles, Biosynthesis

Colloidal Silver

▶ Silver, Pharmacological and Toxicological Profile as Antimicrobial Agent in Medical Devices

Colloidal Silver Nanoparticles and Bovine Serum Albumin

Vinita Ernest, N. Chandrasekaran and
Amitava Mukherjee
Centre for Nanobiotechnology, VIT University,
Vellore, Tamil Nadu, India

Synonyms

Adsorption studies; BSA-coated AgNPs; Functionalization of AgNPs

Definition

Purposely modified silver nanoparticles (AgNPs) using bovine serum albumin (BSA) is a biofunctionalization technique to obtain biocompatibility in metals where the surface of the metal nanoparticle is altered for various biomedical applications. Among the serum albumins, the most abundant proteins in plasma and with a wide range of physiological functions, BSA has made it a model protein for biofunctionalization. BSA-coated AgNPs are being increasingly used in the recent past as sensors for environmental pollutants, targeted drug delivery, diagnostics, biosensors owing to exceptional biocompatibility, and optoelectronic properties. Besides functionalizing the nanoparticles for such uses, it highly stabilizes the nanoparticle, thereby preventing from aggregation and agglomeration, which is a major limitation for the application of nanoparticle in biological systems.

Principles and Role of BSA Interaction with AgNPs

Nanoparticles and proteins, in the formation of biocompatible defensive conjugates, have promising uses in targeted delivery (Di Marco et al. 2010). Serum albumins are the most abundant proteins in plasma (Carter 1994). BSA has exceptional property to bind reversibly a huge number of compounds. As the major soluble protein constituent of the circulatory system, albumin has a wide range of physiological functions involving binding, transport, and delivery of fatty acids, porphyrins, bilirubin, steroids, etc. (Olson and Christ 1996). The sulfhydryl groups present in BSA are scavengers of reactive oxygen and nitrogen species which plays an important role in oxidative stress (Valanciunaite et al. 2006 and Hansen 1981). BSA has been selected as a protein model due to its water-soluble nature which is important for interaction studies (Valanciunaite et al. 2006). It contains 582 amino acid residues with a molecular weight of 69,000 Da and two tryptophan moieties at positions 134 and 212 as well as tyrosine (Tyr) and phenylalanine (Phe) (Sklar et al. 1977 and Hansen 1981). BSA serves well as a bio-receptor since the molecule is known to the human antigen recognition system. The interaction of nanoparticles with the proteins present in plasma is of vital importance in biomedical applications of nanoparticles and also in the biosafety concern of nanomaterials. Moreover, biocompatibility of the nanoparticles is one of the prerequisites to biosensing applications. Over a decade, AgNPs have been recognized for its excellent optoelectronic properties (Mariam et al. 2011). As one of most popular surface enhancement Raman scattering (SERS) active substrates, AgNPs have been used to obtain millionfold enhancement in Raman scattering, which provides a highly sensitive tool for trace analysis and even for probing single molecules (Nie and Emory 1997). Coating of BSA or any other protein molecule reduces the toxicity levels of the nanoparticles to a larger extent.

Methods Adopted

UV–Vis Spectral Study

Depending on the shape, size, and size distribution of the nanoparticles, metal AgNPs are known to exhibit their unique surface plasmon resonance (SPR) upon excitation thorough light. UV–Vis spectra of AgNPs (50 µg/ml) and varying concentrations of BSA were recorded in the UV–Vis spectrophotometer (Shimadzu UV-1700, Japan). The recorded spectral range was between 200 and 600 nm.

Adsorption Isotherm Studies

For adsorption studies, concentrations of BSA ranging between 0.05% and 0.85% were interacted with AgNPs (50 μg/ml) for 4 h in a rotary shaker at 300 rpm. The interacted sample was centrifuged and the supernatant was carefully collected and the absorbance was recorded at 280 nm to measure unreacted BSA concentrations. Different graphs were plotted to find the mode of adsorption of BSA over the Ag nanoparticles. In our case, BSA and AgNPs were the adsorbate and adsorbent, respectively.

Linear form of Langmuir isotherm is:
$$C_e/q_e = C_e/q_m + 1/K_a q_m \quad (1)$$

Linear form of Freundlich isotherm is:
$$log\, q_e = 1/nf\, log\, C_e + log\, KF \quad (2)$$

where C_e (mg/l) is the amount of adsorbate in solution at equilibrium; q_e (mg/mg) is the amount of adsorbate adsorbed per gm of adsorbent; nF, KF, K_a, q_m are constants.

FTIR Studies

AgNP dispersion was centrifuged at 10,000 rpm, and the pellet was interacted with 0.85% BSA at pH 7, 10, and 12 to study the interaction of colloidal Ag nanoparticles with BSA. These measurements were carried out on a PerkinElmer Spectrum One instrument in diffuse reflectance mode at a resolution of 4 cm^{-1} in KBr pellets.

X-Ray Diffraction Studies

The colloidal Ag nanoparticles were lyophilized and subjected to interaction with BSA (0.85%) for 4 h and then subjected to XRD analysis.

Atomic Force Microscopic Analysis

Atomic force microscopy (Nanosurf Easy Scan 2, Nanosurf Inc; USA) was carried out to study the 3-D structures of nanoparticles in order to analyze the topography and size of engineered nanoparticles. The z dimension of individual particles was used for the histogram analysis and the average particle diameter calculation. Ag nanoparticles of 10 μg ml^{-1} were dispersed in Milli-Q water by sonicating for 15 min using a 750-W (20 kHz) ultrasonic processor (Sonics Corp., USA). A drop of nanoparticles dispersion was placed on to the cover slip and spread evenly in order to get a thin film and dried in hot air oven at 60°C for 30 min, and then the slides were subjected to microscopic analysis.

Results

Preliminary Characterization by the Atomic Force Microscopic Analysis

AFM measurements had been carried out for procured Ag nanoparticles to find out the surface topography and the 3-D structure of nanoparticles. Figure 1 below shows a characteristic AFM image of nano Ag. Nanoparticles were observed to be polydisperse condition; topography and shape of the particles were shown in the image with particles, ranging from 100 to 120 nm in diameter. The particles were nearly spherical to oval in shape.

UV–Vis Spectral Study

The λ_{max} value for colloidal Ag nanoparticles (50 μg/ml) was at 425 nm. A decrease in intensity at 425 nm was observed for lower concentrations of BSA, and after a certain concentration (i.e., 0.45%), the blue shift toward the lower wavelength is known to possess better stability. The colloidal AgNPs were stabilized by the negatively charged BSA at higher concentrations, which prevents aggregation. The lowest λ_{max} value was noted at 410 nm. Thus, the colloidal Ag nanoparticles have been capped by BSA as a function of protein concentration.

Surface Changes Recorded by FTIR over pH

The amide A band was obtained at 3,500 cm^{-1}, and amide I band obtained between 1,600 and 1,700 cm^{-1} represented carboxyl group. These peaks were found in all the three interacted mixtures studied at different pH (7, 10, and 12) which potentially proves the interaction of BSA on the AgNPs irrespective of pH. At pH 7, since the FTIR studies revealed only the bands corresponding to the protein (i.e., BSA), it is a proof that the AgNPs are completely coated by BSA (Ravindran et al. 2010). The similar bands

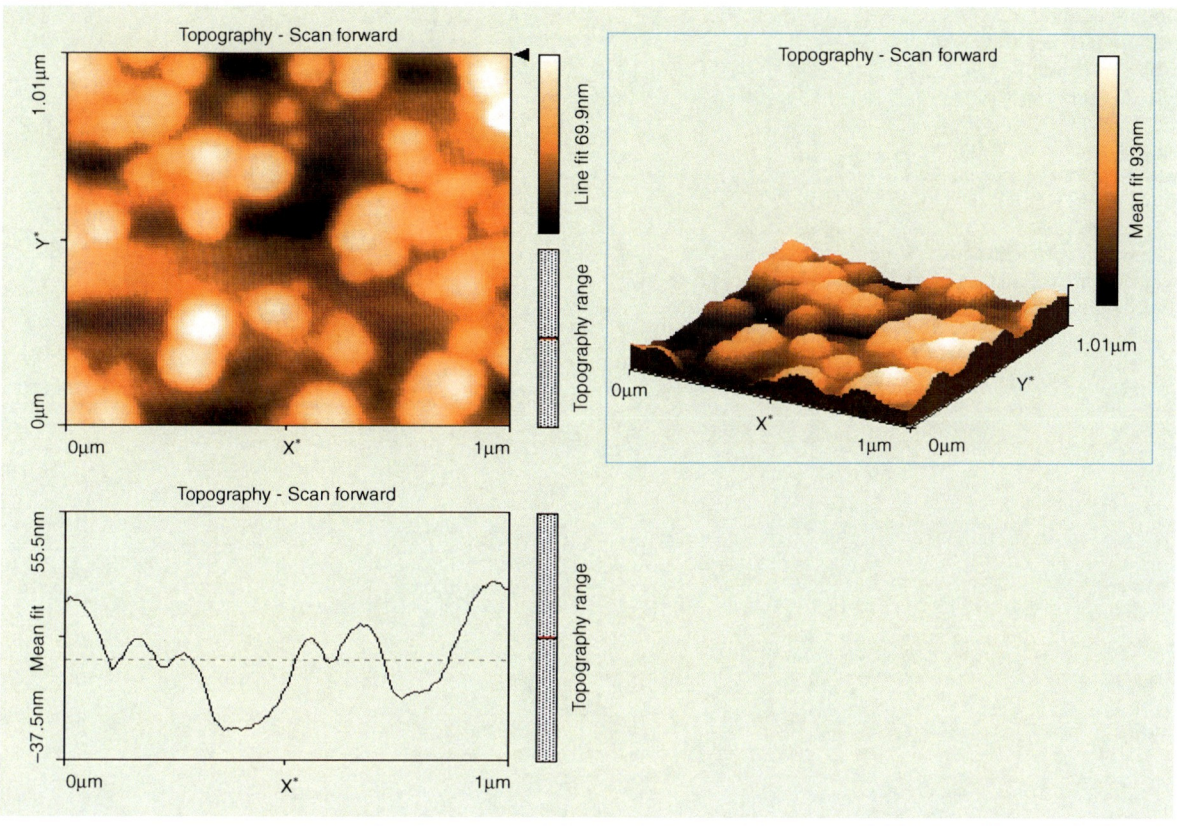

Colloidal Silver Nanoparticles and Bovine Serum Albumin, Fig. 1 Atomic force microscopic image of procured Ag nanoparticles

representing protein adsorbed were observed for elevated pH values (at pH 7 and 10) also as shown in Fig. 2a and b.

Adsorption Isotherm Studies

The adsorption data was fitted to the linearized forms of both Langmuir and Freudlisch isotherms to find the optimum isotherm relationship. The isotherms are plotted as a function of BSA/Ag and equilibrium concentration of BSA. Linear regression analysis was used to determine the best-fit isotherm, and the method of least squares was used in finding the parameters of the isotherm. From the R^2 values of both the isotherm equations, it is calculated that the Freudlisch isotherm was more appropriate for the interaction. Moreover, Freudlisch isotherm is based on multilayer adsorption with interaction between adsorbed molecules. The calculated regression values revealed that BSA formed multiple layers on the nanoparticle surface, thereby stabilizing the surface of the particles.

X-Ray Diffraction Studies

The X-ray diffraction image also proved the adsorption of BSA on the surface of Ag nanoparticles. The characteristic crystallite facets of AgNPs were not revealed. The absence of AgNP facets in the XRD spectra upon interaction with BSA confirmed the complete coverage of BSA on the surface of AgNPs (Fig. 3).

Concluding Remarks

In conclusion, the UV–Vis absorption properties of colloidal Ag nanoparticles coated with BSA were observed to behave different from the uncoated

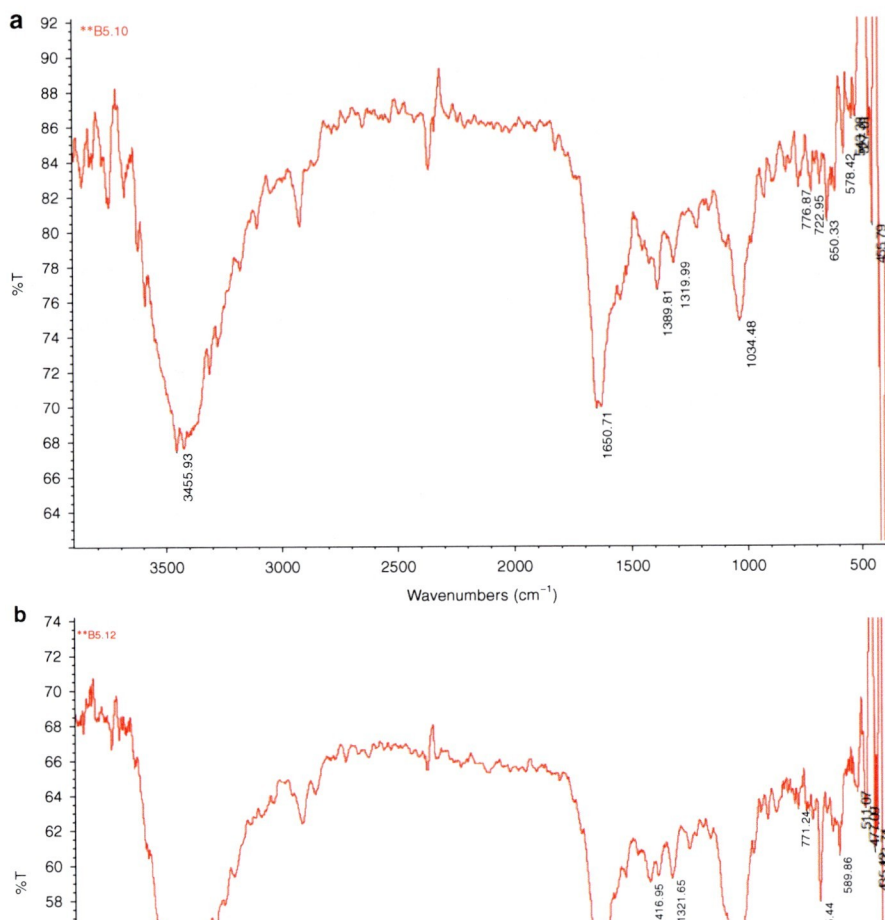
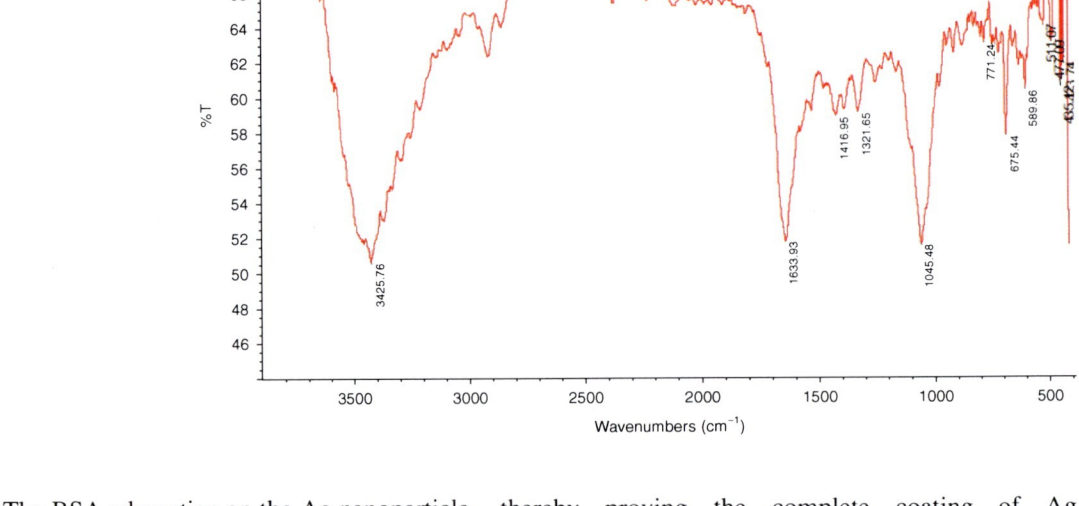

Colloidal Silver Nanoparticles and Bovine Serum Albumin, Fig. 2 (a) FT-IR spectra of BSA-coated silver nanoparticles (0.85% BSA with 50 ppm Ag nanoparticles) at pH 10 (b) FT-IR spectra of BSA-coated silver nanoparticles (0.85% BSA with 50 ppm Ag nanoparticles) at pH 12

particles. The BSA adsorption on the Ag nanoparticle surface prevented them from aggregating in solutions of pH > 5. With increasing concentrations of BSA, initially, a decrease in intensity of the absorption spectra was noticed without change in plasmon peak. The decrease in absorbance was corresponding to a stepped adsorption layer behavior. Beyond 0.45% of BSA concentration, a blue shift from 425 to 410 nm in the spectrum was noticed. The results of FT-IR spectroscopy also revealed the prominent peaks of proteins, thereby proving the complete coating of Ag nanoparticles by BSA. The equilibrium adsorption data was fitted better to the Freudlisch isotherm plot than the Langmuir curve. A strong hydrophobic interaction is between the colloidal AgNPs and BSA. Tailoring the concentration of BSA and pH of the medium, it is possible to reduce. The biomolecules conjugated with nanoparticles have received the most attention in the field of clinical diagnosis and drug delivery because of their high stability, good

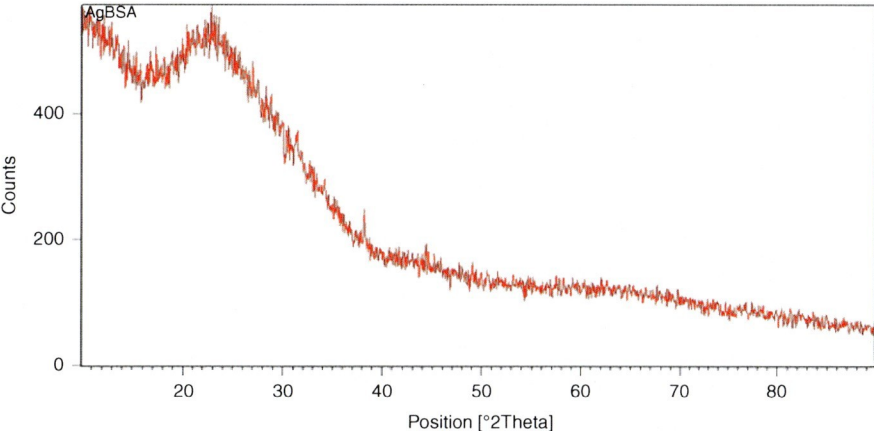

Colloidal Silver Nanoparticles and Bovine Serum Albumin, Fig. 3 XRD pattern of BSA (0.85%) coated over colloidal Ag nanoparticles where the characteristic facets for silver are not pronounced

biocompatibility, and high affinity for biomolecule. The colloidal Ag nanoparticles and BSA interact at the surface level and lead to the formation of protein "corona" around nanoparticles which highly characterize their biological identity as well as their potential usage for various applications. Lynch and Dawson suggested the importance of the "protein corona" as the vehicle and the biological identity of a nanoparticle for its transport through cell membranes (Di Marco et al. 2010). Hence, functionalizing the surface of the nanoparticles with biomolecules would pave a way for the use of these particles in various biomedical applications.

Advantages of Functionalization Approach

1. A major drawback of using the procured or synthesized nanomaterials due to their very responsive agglomeration and aggregation behavior has been overcome by functionalizing the surface with bioreceptive protein molecules or recognizable molecules such as antibodies, enzymes, starch, dextrin, polymers that are an added advantage for the usage of nanoparticles.
2. Toxicity levels of the BSA-coated Ag nanoparticles have also been considerably reduced with the proven seed germination and phytotoxicity tests.

Potential Applications

Biosensing is performed using BSA as a stabilizer on Ag nanoparticles for protein detection. Upon binding of proteins to the silver nanoparticles, the changes in both the intensity and the wavelength of the particle make an easy way for sensing the presence of a particular protein of interest from a heterogeneous sample. They are widely used in for imaging applications due to the easy recognition of the biomolecules such as BSA inside human system. They are also used as effective antimicrobial agents in the form of hydrogels and silver colloids which are safe on skin.

Research in the field of interaction of nanomaterials to biological conditions especially to human system is of a potential use in the development of nano-drugs and nano-sensors for the diagnostics and pharmacotherapy. Ag nanomaterials are widely synthesized with BSA as a bioreceptor for various biomedical applications such as piezoelectric sensors and antibody-mediated drug delivery systems.

Cross-References

▶ Gold and Nucleic Acids
▶ Gold Nanoparticles and Proteins, Interaction
▶ Silver in Protein Detection Methods in Proteomics Research

References

Carter DD, Ho JX (1994) Structure of serum albumin. Adv Protein Chem 45:153–203
Di Marco M, Shamsuddin S, Razak KA, Aziz AA, Devaux C, Borghi E, Levy L, Sadun C (2010) Overview of the main methods used to combine proteins with nanosystems:

absorption, bioconjugation, and encapsulation. Int J Nanomedicine 5:37–49

Hansen UK (1981) Molecular aspects of ligand binding to serum albumin. Pharmacol Rev 33:17–53

Mariam J, Dongre PM, Kothari DC (2011) Study of interaction of silver nanoparticles with bovine serum albumin using fluorescence spectroscopy. J Fluoresc 21:2193–2199

Nie S, Emory SR (1997) Probing single molecules and single nanoparticles by surface-enhanced Raman Scattering. Science 275:1102–1106

Olson RE, Christ DD (1996) Plasma protein binding of drugs. Annu Rep Med Chem 31:327–336

Ravindran A, Singh A, Raichur AM, Chandrasekaran N, Mukherjee A (2010) Studies on interaction of colloidal Ag nanoparticles with Bovine Serum Albumin (BSA). Colloids Surf B Biointerfaces 76:32–37

Sklar LA, Hudson BS, Simoni RD (1977) Conjugated polyene fatty acids as fluorescent probes: binding to bovine serum albumin. Biochemistry 16:5100–5108

Valanciunaite J, Bagdonas S, Streckyte G, Rotomskis R (2006) Spectroscopic study of TPPS4 nanostructures in the presence of bovine serum albumin. Photochem Photobiol Sci 5:381–388

Colloids

▶ Gold Nanoparticle Platform for Protein-Protein Interactions and Drug Discovery

Combined Toxicity of Arsenic and Alcohol Drinking

▶ Arsenic and Alcohol, Combined Toxicity

Complex – Coordination or Organometallic Compound

▶ Osmium Complexes with Azole Heterocycles as Potential Antitumor Drugs

Complex, Adduct

▶ NMR Structure Determination of Protein-Ligand Complexes using Lanthanides

Complexation of Actinides by Proteins

▶ Actinides, Interactions with Proteins

Conformation: Fold

▶ Monovalent Cations in Tryptophan Synthase Catalysis and Substrate Channeling Regulation

Connective Tissue

▶ Calcium and Extracellular Matrix

2nd Conserved Domain of Protein Kinase C

▶ C2 Domain Proteins

Contact Allergy to Beryllium

▶ Beryllium as Antigen

Coordinate

▶ Calcium-Binding Protein Site Types

Coordination Geometry: Bonding Geometry

▶ Monovalent Cations in Tryptophan Synthase Catalysis and Substrate Channeling Regulation

Coordination of Actinides to Proteins

▶ Actinides, Interactions with Proteins

Copper Amine Oxidase

Judith Klinman[1] and Albert Lang[2]
[1]Departments of Chemistry and Molecular and Cell Biology, California Institute for Quantitative Biosciences, University of California, Berkeley, Berkeley, CA, USA
[2]Department of Molecular and Cell Biology, California Institute for Quantitative Biosciences, University of California, Berkeley, Berkeley, CA, USA

Synonyms

Amine oxidase (copper-containing); Primary amine oxidase; Primary amine:oxygen oxidoreductase (deaminating)

Definition

Copper amine oxidase refers to a ubiquitous family of enzymes that catalyze the oxidative deamination of primary amines, concomitant with the two-electron reduction of molecular oxygen to hydrogen peroxide. To accomplish this reaction, these enzymes utilize the redox-active cofactor 2,4,5-trihydroxyphenylalanine quinone (TOPA quinone or TPQ).

Classification

Up until recently, copper amine oxidases (CAOs) were classified within the Enzyme Commission classification EC 1.4.3.6, distinct from the flavin-containing amine oxidases (monoamine oxidase, EC 1.4.3.4) (Boyce et al. 2009). This distinction was based on cofactor content rather than reaction catalyzed, however, leading to some confusion in the literature. For clarification, EC 1.4.3.6 has now been deleted and replaced with two distinct entries: EC 1.4.3.21 and EC 1.4.3.22 (Boyce et al. 2009). EC 1.4.3.21 (accepted name primary amine oxidase) refers to enzymes that oxidize primary monoamines but have little or no activity toward diamines, such as histamine, or toward secondary or tertiary amines (Boyce et al. 2009). EC 1.4.3.22 (accepted name diamine oxidase), on the other hand, refers to a group of enzymes that oxidize diamines such as histamine and also some primary monoamines but have little or no activity toward secondary and tertiary amines (Boyce et al. 2009). This entry will deal with the primary amine oxidases from EC 1.4.3.21.

Mechanism of TPQ Biogenesis

Historically, it had been thought that CAOs contained either a pyridoxyl phosphate cofactor or pyrroloquinoline quinone (PQQ), a bacterial vitamin (DuBois and Klinman 2005). It was later discovered that CAOs instead contained the novel quinocofactor TPQ, derived from a precursor tyrosine in the protein backbone (Janes et al. 1990). The discovery of TPQ led to the identification of a series of new quinone cofactors, including lysine tyrosylquinone (LTQ) in lysyl oxidase, cysteine tryptophylquinone (CTQ) in quinohemoprotein amine dehydrogenase, and tryptophan tryptophylquinone (TTQ) in bacterial methylamine dehydrogenase (Scheme 1) (reviewed in Davidson 2007).

The biogenesis of TPQ, formally a monooxygenation, hydroxylation, and two-electron oxidation of tyrosine, occurs without the aid of any auxiliary enzymes or external reducing equivalents (DuBois and Klinman 2005) (Scheme 2). This autocatalytic processing depends solely on the presence of molecular oxygen and a functional copper center (DuBois and Klinman 2005). Thus, CAOs are able to catalyze both cofactor formation and, subsequently, the oxidation of amines within a single active site. The active site Tyr that leads to TPQ (Y405 in the CAO from *H. polymorpha*, HPAO) is absolutely conserved among CAO family members and is contained within the consensus sequence T-X-X-N-Y-D/E (Guss et al. 2009). Other strictly conserved residues include three histidine residues (H456, H458, and H624 in HPAO) that form the three protein ligands of the copper ion and a second Tyr (Y305 in HPAO) that is located at the active site near the O-4 position of the TPQ cofactor (Fig. 1) (DuBois and Klinman 2005).

Prior spectroscopic and structural studies have led to a working mechanism for TPQ biogenesis, summarized in Scheme 2 (DuBois and Klinman 2005). When Cu(II) is added to apo-enzyme anaerobically, an absorbance at 380 nm forms immediately and then decays to

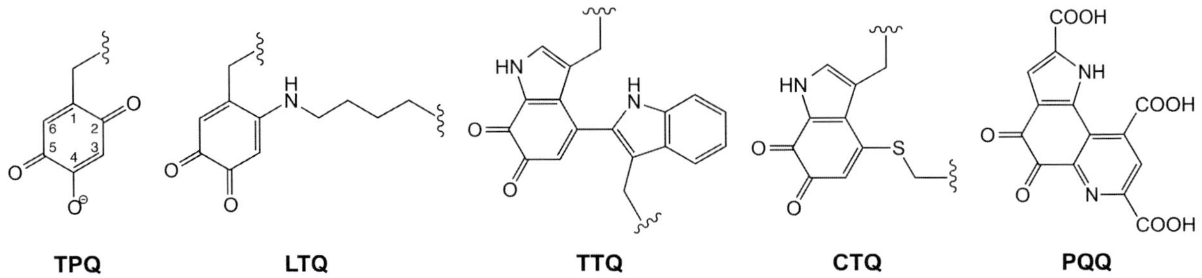

Copper Amine Oxidase, Scheme 1 Structures for protein-derived (TPQ, LTQ, TTQ, CTQ) and peptide-derived (PQQ) cofactors

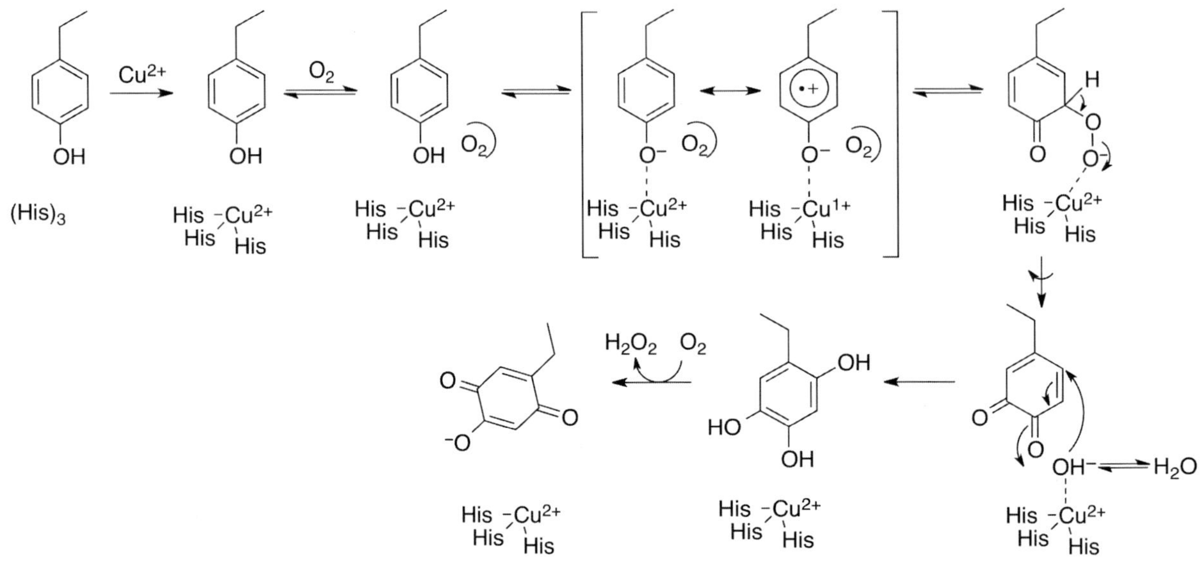

Copper Amine Oxidase, Scheme 2 Proposed mechanism for TPQ biogenesis (DuBois and Klinman 2005)

baseline. This absorbance has been attributed to a transient ligand-to-metal charge transfer between the Cu(II) and a residue en route to the active site. It is then proposed that binding of O_2 at the active site near the precursor tyrosine causes a conformational change that promotes the deprotonation of the tyrosine and ligation of the resulting tyrosinate to Cu(II) (DuBois and Klinman 2005). Next, dioxygen reacts with the "activated" precursor tyrosine in the form of a tyrosinate-Cu(II) ligand-to-metal charge transfer (LMCT) complex (bracketed species in Scheme 2) (DuBois and Klinman 2005). Due to the nature of the oxygen-copper bond in this complex, partial radical character is expected to be imparted onto the tyrosyl ring, thus circumventing the spin-forbidden nature of direct reaction between singlet tyrosine and triplet dioxygen. Experimental evidence for the LMCT complex comes from the observation of a spectral intermediate ($\lambda_{max} = 350$ nm) that forms rapidly upon aeration of HPAO prebound with Cu(II) and decays rapidly concomitantly with the appearance of TPQ (DuBois and Klinman 2005). The 350-nm intermediate represents the species that reacts with the first mole of oxygen, the rate-limiting step in Cu(II)-catalyzed TPQ biogenesis (DuBois and Klinman 2005). It should be noted that this 350-nm intermediate has only been seen in HPAO and not other well-studied CAOs from *E. coli* (ECAO) and *Arthrobacter globiformis* (AGAO). This is most likely due to the considerably slower rate of TPQ biogenesis in HPAO, which allows for the detection of the intermediate using a conventional UV-vis spectrophotometer. In the other two enzymes, stopped-flow techniques would likely be necessary to detect any intermediates.

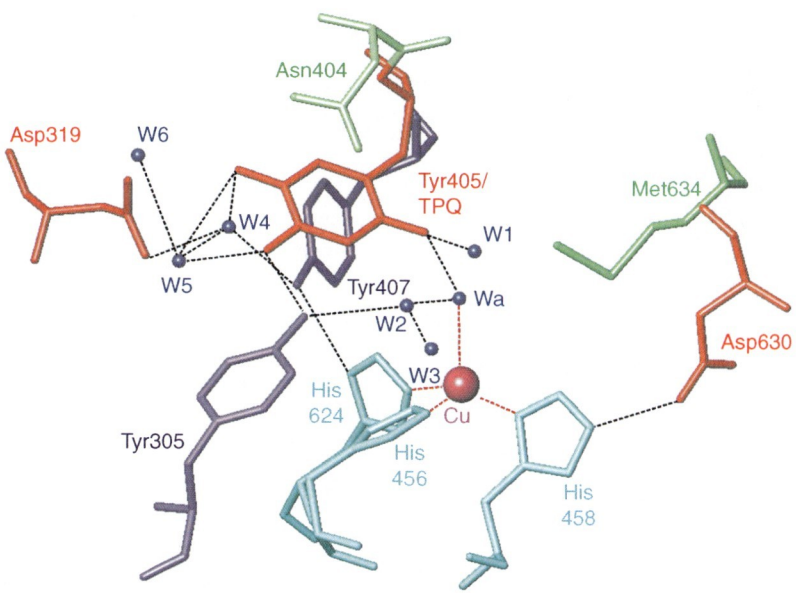

Copper Amine Oxidase, Fig. 1 Mature HPAO active site with TPQ in the reactive conformation. Acidic chains are in *red*, basic in *cyan*, aromatic in *purple*, neutral hydrophilic in *lime*, TPQ cofactor in *red*, and Cu(II) in *crimson*. A network of six active site water molecules (W1–W6) and one axial copper-coordinating water (Wa) that forms hydrogen bonds to TPQ and side chain atoms are shown in *blue*. Hydrogen and metal coordination bonds are shown as *black* and *red dashed lines*, respectively (Reprinted with permission from DuBois JL and Klinman JP 2006. Copyright 2006 American Chemical Society)

A second strictly conserved Tyr in the active site distinct from the TPQ precursor residue also plays an important role in TPQ biogenesis. This residue is Y305 in HPAO and hydrogen bonds to the O-4 of TPQ in the mature enzyme (Fig. 1). The role of Y305 in cofactor biogenesis has been studied by analyzing the properties of Y305A and Y305F mutant proteins (DuBois and Klinman 2006). It was found that the Y305F mutant gave rates for TPQ production that were close to that of WT (reduced ~3-fold), while the Y305A mutant was significantly more hindered, with a rate ~30-fold slower than WT (DuBois and Klinman 2006). In biogenesis, the fairly high rate for Y305F in relation to Y305A implicates a role for steric bulk and hydrophobicity at this position in steering the position of Y405 and its oxygenated intermediates that form en route to TPQ. It is also possible that this Tyr residue acts as a general acid to help facilitate breakdown of the aryl-peroxide intermediate (Scheme 2) or as a general base to abstract a proton from C-3 of the cofactor ring to drive breakdown of this intermediate (DuBois and Klinman 2006).

Furthermore, an additional product was detected that absorbed at ~400 nm in the case of Y305A and at ~420 nm for Y305F when Cu(II) was added to apo-enzyme (DuBois and Klinman 2006). The formation of this product, in relation to the amount of TPQ (λ_{max} ~480 nm), increased with pH (DuBois and Klinman 2006). The two peaks form independently without an isosbestic point, indicating that the alternate product is not a direct intermediate leading to TPQ. This has implicated two parallel pathways during cofactor biogenesis in the case of Y305 mutants, with one pathway leading to the expected TPQ product and the other producing a new, uncharacterized species. In the case of Y305F, the alternate species forms fourfold more quickly than TPQ at pH 7, facilitating its preferential crystallization and characterization by X-ray crystallography (Chen et al. 2010). In this crystal structure, a novel 2,3,4-trihydroperoxo, 5-hydroxo derivative was detected (Fig. 2) (Chen et al. 2010). The pathway for the formation of this new species is postulated to involve the partitioning of the normal, aryl-peroxy intermediate (E-I) between concomitant proton abstraction and O-O bond cleavage to generate dopaquinone and ultimately TPQ (top of Scheme 3a), and proton loss and ring aromatization to yield the first, off-path hydroperoxo-product (bottom of Scheme 3a). During the normal biogenesis pathway, E-I is expected to undergo a 180° rotation, placing the O-O bond

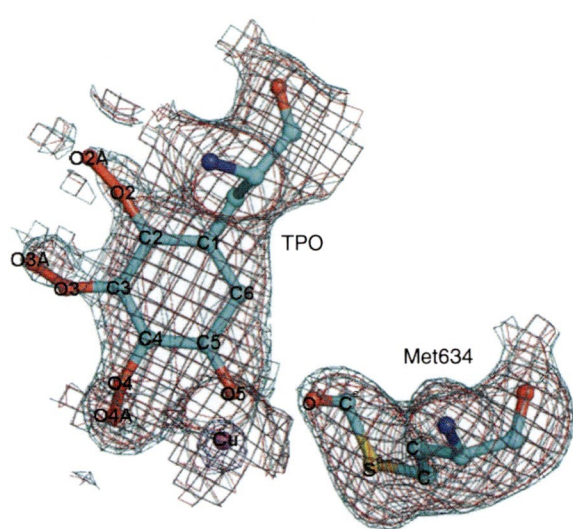

Copper Amine Oxidase, Fig. 2 Simulated annealing omit maps of the Y305F mutants of HPAO computed using coefficients $(F_o - F_c)$ where F_c were calculated from models in which the side chain of residue 405 was absent, and F_o were the structure factors observed from crystals of the *E. coli*–expressed mutant. The maps are contoured at σ levels of 2.5 (*cyan*), 3.0 (*crimson*), and 10.0 (*purple*). Atoms are colored *blue* for nitrogen, *red* for oxygen, and *magenta* for copper; for the residue 405 cofactor, the carbon atoms are in *cyan* (Reprinted with permission from Chen et al. 2010. Copyright 2010 American Chemical Society)

undergoing cleavage in close proximity to the hydroxyl group of Y305. Insertion of Phe in place of Tyr at position 305 creates, instead, a hydrophobic patch, precluding the normal 180° rotation and leading to the energetically more favorable deprotonation coupled to ring aromatization.

Proceeding further, it can be seen from the electron density map that the oxygen chemistry has not stopped at a single hydroperoxide insertion, indicating instead three such side chains (Fig. 2). A proposed pathway for the production of the 3,4-dihydroperoxo-product, designated TPO-3,4, is detailed in Scheme 3b. An analogous scheme can be written in which a 2,4-dihydroperoxo-product (TPO-2,4) is obtained (Scheme 3c). These results implicate a runaway oxidative process following the replacement of a single, absolutely conserved side chain.

Catalytic Mechanism

The reaction catalyzed by CAOs proceeds via a ping-pong mechanism involving a covalently bound redox cofactor, TPQ, and a copper ion, Cu(II) (Mure et al. 2002). The catalytic cycle consists of two half-reactions: (1) a reductive half reaction, in which the substrate is oxidized to an aldehyde product via a Schiff base intermediate and the cofactor is converted to a reduced, aminoquinol form; and (2) an oxidative half reaction in which dioxygen is reduced to hydrogen peroxide and the oxidized TPQ is regenerated following release of an ammonium ion (Mure et al. 2002) (Scheme 4).

Reductive Half Reaction

In the first half reaction, TPQ initially undergoes a two-electron reduction, serving as a "storage site" for reducing equivalents that are eventually transferred to O_2 to form hydrogen peroxide in the oxidative half reaction. The starting oxidized form of TPQ gives CAOs their characteristic absorbance around 480 nm and distinct pink color (Mure et al. 2002). TPQ exists in the oxidized state due to the acidity of the 4-hydroxyl group (pK_a = 4.1 in model compound) (Mure et al. 2002). Charge is delocalized between oxygens on C-2 and C-4, which decreases the electrophilicity at these sites and directs nucleophilic attack of amine substrates exclusively to the C-5 position. This nucleophilic attack is aided by an active site base that deprotonates the substrate (Asp319 in HPAO) (Mure et al. 2002). Addition of the substrate amine to TPQ leads to the formation of the first stable intermediate, the substrate Schiff base (Mure et al. 2002). This intermediate is proposed to form through a carbinolamine intermediate that has not been observed directly. The next step is the formation of the product Schiff base, proceeding through base-catalyzed proton abstraction at C-1 of the substrate (Mure et al. 2002). This involves the transfer of two electrons from the substrate amine to the TPQ cofactor and is driven by the gain of aromaticity in the cofactor (Mure et al. 2002). The final step in the reductive half reaction involves hydrolysis of the product Schiff base, releasing aldehyde product and generating the aminoquinol form of the cofactor (Mure et al. 2002).

Oxidative Half Reaction

In the oxidative half reaction, molecular oxygen binds to a hydrophobic pocket in the active site. Then, an electron is transferred from the aminoquinol form of TPQ to O_2, generating the semiquinone form of

Copper Amine Oxidase, Scheme 3 (a) Postulated mechanism for the branching of a biogenesis intermediate to TPQ and TPO. (b) Model to explain 3,4-dihydroperoxo-product. (c) Model to explain 2,4-dihydroperoxo-product

TPQ and superoxide, respectively (Mechanism II, Scheme 4) (Mure et al. 2002). Superoxide then moves onto the Cu(II), allowing for transfer of a second electron and two protons from the semiquinone, resulting in the formation of hydrogen peroxide and an iminoquinone form of the cofactor (Mure et al. 2002). Finally, hydrolysis of the iminoquinone results in the release of ammonium ion and regeneration of the oxidized form of TPQ, ready for another round of catalysis (Mure et al. 2002). Alternatively, release of ammonium from the iminoquinone can proceed by a transamination

Copper Amine Oxidase, Scheme 4 Proposed catalytic mechanism of CAOs (Mure et al. 2002)

reaction with substrate amine to form the substrate Schiff base (dashed arrow, Scheme 4) (Mure et al. 2002).

An alternative mechanism has also been proposed for the oxidative half reaction. The chief difference is that the first electron is transferred from the aminoquinol TPQ to Cu(II), generating a semiquinone radical and Cu(I) (Mechanism I, Scheme 4) (Mure et al. 2002). Molecular oxygen then combines with Cu(I) to form Cu(II) superoxide (Mure et al. 2002). The transfer of, a second electron and two protons yields hydrogen peroxide and iminoquinone. Finally, the iminoquinone is hydrolyzed, releasing ammonium and regenerating the oxidized form of TPQ. Evidence for this pathway comes from the accumulation of TPQ semiquinone/Cu(I) when the enzyme is reacted with amine substrate in the absence of O_2 (Mure et al. 2002). The level for accumulation ranges from 0 % to 40 %, depending on the source of the enzyme (Mure et al. 2002). In the case of HPAO, however, substitution of the active site cupric ion by a cobaltous ion yields an enzyme form with k_{cat} similar to the native enzyme. Since the redox cycling of Co(II) to Co(I) is prohibited by a very low redox potential, the metal substitution reaction provides strong support for Mechanism II (Mure et al. 2002).

Structural Properties of Copper Amine Oxidases

Structural studies have contributed greatly to the interpretation of the wealth of kinetic and spectroscopic data in the literature on CAOs. As previously mentioned, CAOs are ubiquitous and found in bacteria, plants, fungi, and animals. Structures of CAOs have been solved from a variety of sources, including *Escherichia coli* (ECAO), pea seedling (PSAO), *Hansenula polymorpha* (HPAO), bovine serum (BSAO), *Arthrobacter globiformis* (AGAO), *Pichia pastoris* (PPLO), and human (human vascular adhesion protein-1) (reviewed in Guss et al. 2009). The overall three-dimensional structure for all solved CAO structures is similar (HPAO shown in Fig. 3) (Brazeau et al. 2004). In all cases, the CAOs are homodimers of ~70 kDa subunits that contain a carboxy-terminal domain that folds into a β-sandwich with twisted β-sheets (Brazeau et al. 2004). The overall shape of the dimer resembles a mushroom cap. The active site is buried deeply within each subunit and contains a TPQ cofactor, which is not directly liganded to the Cu but rather has hydrogen bonding between O-2 and a water molecule and between O-4 and another

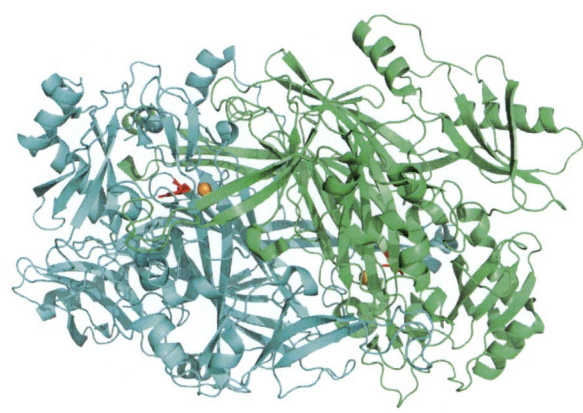

Copper Amine Oxidase, Fig. 3 Ribbon diagram of HPAO (PDB entry; 2OOV). One monomer is colored *cyan*, the other *green*. Copper ions are shown as *orange spheres*. TPQ is colored *red*

conserved Tyr in the active site (Y305 in HPAO) (Fig. 1). In addition, O-5 of the cofactor is oriented close to the catalytic base (Asp319 in HPAO) in the amine substrate binding pocket (Brazeau et al. 2004). The Cu(II) is liganded in a distorted square pyramidal geometry with three histidine sidechains (H456, H458, and H624 in HPAO) providing N ligands ~2.0 Å away, an axial water at a distance of ~2.4 Å, and sometimes a labile equatorial water ligand ~2.0 Å away (Brazeau et al. 2004).

Biogenesis Intermediates

Biogenesis of TPQ can be studied structurally by preparing enzyme in the absence of copper, as TPQ formation is a self-processing event. What has been particularly insightful is the identification of intermediates along the pathway by freeze-trapping techniques (Kim et al. 2002, reviewed in Brazeau et al. 2004). In the first of these structures, that of apo-enzyme, the His metal ligands along with the TPQ precursor Tyr residue are arranged in a tetrahedral geometry, poised to bind copper (Kim et al. 2002). When Cu is added anaerobically, another structure can be trapped that is virtually identical to that of apo-enzyme, with Cu now coordinated by the three His residues and TPQ precursor Tyr residue (Kim et al. 2002). After exposing these anaerobic crystals to oxygen for a short period (10 min), the next intermediate that is trapped is that of 3,4-dihydroxyphenylalanine (or the oxidized quinone form) (Kim et al. 2002). In this structure, an oxygen atom of either water or hydroxide is located in an equatorial position to Cu. When anaerobic crystals were instead exposed to oxygen for a longer period (100 min), a different, later-stage intermediate is observed, which now has three oxygens and can be modeled as either the reduced or oxidized form of TPQ (Kim et al. 2002). This structure also shows that the structure has rotated 180° about the Cβ-Cγ bond. Evidence that this structure is the reduced cofactor comes from single crystal microspectrophotometry, which shows no evidence of a 480 nm absorbance characteristic of oxidized TPQ (Kim et al. 2002). To visualize the final step of TPQ biogenesis, anaerobic crystals were aerobically soaked in copper solutions for a week, which resulted in pink crystals indicative of oxidized TPQ formation (Kim et al. 2002). In the final structure, the TPQ is clearly in an "off-copper" conformation, with the O-4 pointed away from the Cu.

Structural Data and Catalysis

Structural studies have also been insightful in studying intermediates of both the reductive and oxidative half reactions. The structure of a substrate Schiff base with the inhibitor 2-hydrazinopyridine (2-HP) has been solved (Brazeau et al. 2004). Hydrazines irreversibly inhibit CAO activity and can be used to derivatize the TPQ cofactor for spectral analysis (e.g., phenylhydrazine generates a strongly absorbing hydrazone). In this structure, 2-HP is covalently bound to the C-5 that is also the site of nucleophilic attack by natural amine substrates (Brazeau et al. 2004). This structure also shows N-2 of 2-HP hydrogen bonded to the carboxylate side chain of the catalytic base (Asp383 in ECAO). This position is analogous to the methylene carbon of a primary amine substrate where a proton is abstracted by the active site base in the conversion of substrate Schiff base to product Schiff base. To examine the aminoquinol form of the enzyme, crystals were prepared anaerobically and subsequently reduced with substrate (2-phenethylamine). In this structure, the aminoquinol form of the cofactor is ordered in the same position as the oxidized form of TPQ.

In order to study the oxidative half reactions, studies were conducted in which crystals were reduced with substrate anaerobically, reduced with substrate in the presence of nitric oxide (NO), or aerobically trapped. In the substrate reduced structure, previously mentioned with respect to the reductive half reaction, there is clearly the presence of product phenylacetaldehyde in the active site (Brazeau et al. 2004).

This is consistent with the ping-pong kinetics of the enzyme in which irreversible formation of product occurs prior to oxygen binding. In the structure with NO, NO takes the place of the axial water, and is situated between the copper and the O_2 of the aminoquinol (Brazeau et al. 2004). Thus, the dioxygen mimic NO appears capable of interaction with both the copper ion and the aminoquinol, consistent with a mechanism of electron transfer from either Cu(I) or aminoquinol to O_2. In the aerobically trapped structure, oxygen occupies the same site as NO did, with an electron density compatible with the peroxide product (Brazeau et al. 2004). The fact that these crystals lacked any visible absorbance is consistent with the presence of a hydrogen-bonded iminoquinone (Brazeau et al. 2004).

Cross-References

▶ Ascorbate Oxidase
▶ Biological Copper Transport
▶ Catechol Oxidase and Tyrosinase
▶ Copper, Biological Functions
▶ Copper, Mononuclear Monooxygenases
▶ Copper, Physical and Chemical Properties
▶ Copper-Binding Proteins
▶ Monocopper Blue Proteins
▶ Ribonucleotide Reductase

References

Boyce S, Tipton KF, O'Sullivan MI, Davey GP, Gildea MM, McDonald AG, Olivieri A, O'Sullivan J (2009) Nomenclature and potential functions of copper amine oxidases. In: Floris G, Mondovì B (eds) Copper amine oxidases: structures, catalytic mechanisms, and role in pathophysiology. CRC Press, Boca Raton

Brazeau BJ, Johnson BJ, Wilmot CM (2004) Copper-containing amine oxidases. Biogenesis and catalysis; a structural perspective. Arch Biochem Biophys 428:22–31

Chen ZW, Datta S, DuBois JL, Klinman JP, Mathews FS (2010) Mutation at a strictly conserved, active site tyrosine in the copper amine oxidase leads to uncontrolled oxygenase activity. Biochemistry 49:7393–7402

Davidson VL (2007) Protein-derived cofactors. Expanding the scope of post-translational modifications. Biochemistry 46:5283–5292

DuBois JL, Klinman JP (2005) Mechanism of post-translational quinone formation in copper amine oxidases and its relationship to the catalytic turnover. Arch Biochem Biophys 433:255–265

DuBois JL, Klinman JP (2006) Role of a strictly conserved active site tyrosine in cofactor genesis in the copper amine oxidase from *Hansenula polymorpha*. Biochemistry 45:3178–3188

Guss JM, Zanotti G, Salminen TA (2009) Copper amine oxidase crystal structures. In: Floris G, Mondovì B (eds) Copper amine oxidases: structures, catalytic mechanisms, and role in pathophysiology. CRC Press, Boca Raton

Janes SM, Mu D, Wemmer D, Smith AJ, Kaur S, Maltby D, Burlingame AL, Klinman JP (1990) A new redox cofactor in eukaryotic enzymes: 6-hydroxydopa at the active site of bovine serum amine oxidase. Science 25:981–987

Kim M, Okajima T, Kishishita S, Yoshimura M, Kawamori A, Tanizawa K, Yamaguchi H (2002) X-ray snapshots of quinone cofactor biogenesis in bacterial copper amine oxidase. Nat Struct Biol 9:591–596

Mure M, Mills SA, Klinman JP (2002) Catalytic mechanism of the topa quinone containing copper amine oxidases. Biochemistry 41:9269–9278

Copper and Prion Proteins

Glenn L. Millhauser
Department of Chemistry and Biochemistry, University of California, Santa Cruz, Santa Cruz, CA, USA

Synonyms

Creutzfeldt-Jakob disease; Mad cow disease; Prion; PrP^{Sc}; Transmissible spongiform encephalopathy

Definition

Prion diseases are infectious, neurological disorders that arise from accumulation of a misfolded form of the endogenous prion protein. These disorders affect a wide range of mammalian species and present an ongoing public health threat. The function of the normal prion protein is not well characterized, but recent research suggests that it is linked to the protein's ability to bind copper.

Prions, Copper Coordination, Function, and Disease

The prion protein, discovered by Stanley B. Prusiner and colleagues, is a ubiquitous component of tissues in

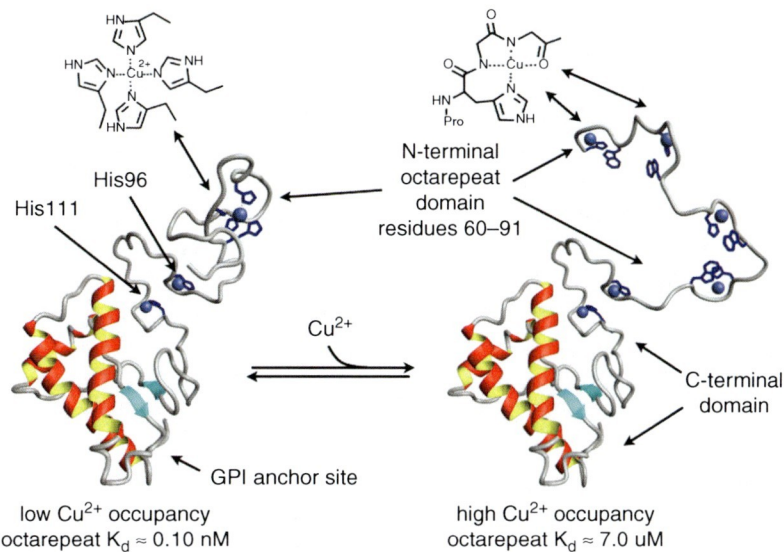

Copper and Prion Proteins, Fig. 1 Structural features of PrP^C showing copper coordination modes at different copper concentrations. The secondary structure of the folded C-terminal domain is primarily composed of three α-helices. Copper (Cu^{2+}) binds to two distinct regions. Coordination within the octarepeat (OR) domain depends on the ratio of copper to protein. At low copper levels, Cu^{2+} coordinates to imidazole side chains (*left*). At high levels, each OR segment takes up a single Cu^{2+} equivalent. Respective coordination features are shown at the *top* of the figure. In addition to OR binding, copper is also taken up at sites located to histidine 96 and histidine 111

both mammals and avian species that is found at high levels in the central nervous system (CNS). The normal form is referred to as cellular PrP and denoted PrP^C. PrP^C is a glycoprotein of approximately 210 amino acids, with an unstructured N-terminus and a folded C-terminus containing three α-helices (Fig. 1). The final amino acid links to a glycophosphatidylinositol (GPI) group that anchors the protein to extracellular membrane surfaces. The precise physiological function of the cellular prion protein remains unknown. It is thought to have a neuroprotective role, perhaps by suppressing apoptosis. PrP^C is also linked to transmembrane signaling.

The transmissible spongiform encephalopathies (TSEs), or "prion diseases," are infectious neurodegenerative disorders that include mad cow disease (bovine spongiform encephalopathy, BSE), chronic wasting disease (CWD) in deer and elk, scrapie in goats and sheep, and the four human disorders Creutzfeldt-Jakob disease (CJD), Gerstmann-Sträussler-Scheinker syndrome (GSS), fatal familial insomnia (FFI), and Kuru. TSEs are marked by spongiform pathology and progressive loss of neuronal function. Within the affected tissues, the prion protein is found in a β-sheet-rich misfolded form, referred to as a prion or PrP^{Sc} (scrapie form). Most cases of prion disease are sporadic, with no known cause, while approximately 10% are inherited and arise from gene insertions, deletions, or single point mutations. A very small percentage of prion diseases are caused by transmission of infectious PrP^{Sc}. These infectious routes are responsible for the 1980s outbreak of BSE in the United Kingdom and the current spread of CWD in the Rocky Mountain regions of North America. Approximately 200 individuals in the UK, who consumed infected meat, developed variant CJD (vCJD). Through the middle part of the twentieth century, cannibalistic practices of the Fore people of Papua New Guinea led to widespread transmission of Kuru. In Western Society, prions have been spread through infected cadaver-derived tissues and hormones, and improper surgical operating room procedures.

As demonstrated in 1997, PrP^C binds copper (Cu) in vivo (Brown et al. 1997). The connection to copper was initially suggested by mass spectrometry studies showing a high affinity for binding to the flexible PrP

N-terminus. Copper, one of the most abundant metal species of the CNS, is found both within cells and in the extracellular spaces. It is essential for numerous enzymatic processes including cellular respiration and neurotransmitter synthesis. Although copper has two oxidation states, Cu^+ and Cu^{2+}, most investigations into the interaction between copper and PrP examined Cu^{2+}, the dominant species in the extracellular space. There are two adjacent copper binding regions in PrP^C (Fig. 1) (Millhauser 2007). The first is in the N-terminal octarepeat (OR) domain, residues 60–91 in human PrP, comprised of tandem repeats of the fundamental sequence PHGGGWGQ (Pro-His-Gly-Gly-Gly-Trp-Gly-Gln). The precise way in which Cu^{2+} coordinates to this 32 amino segment depends on the ratio of copper to protein (Chattopadhyay et al. 2005). A single equivalent of Cu^{2+} coordinates to the imidazole side chain groups of the four OR histidine residues. The affinity is high, as reflected by a subnanomolar dissociation constant (K_d) of approximately 0.1 nM (Walter et al. 2006). This K_d is substantially lower than the estimated concentrations of extracellular copper, thus supporting an in vivo interaction between copper and PrP^C. The OR domain is also able to bind Zn^{2+} with a similar coordination mode, although with much lower affinity than that found for Cu^{2+} (Walter et al. 2007).

At higher Cu^{2+} concentrations, each HGGGW module within an individual repeat takes up a single copper equivalent. In this high occupancy state, a histidine imidazole, the backbone nitrogens from the two glycines immediately following the histidine and a backbone carbonyl oxygen, coordinate each Cu^{2+} (Burns et al. 2002). The K_d for this coordination mode is approximately 7.0 μM (Walter et al. 2006), comparable to the highest estimates of the Cu^{2+} levels in the synaptic space. Progressive copper uptake exhibits negative cooperativity, consistent with the increase in K_d from the low to the high occupancy state.

The second copper-binding region encompasses residues 94–111 (human sequence) with histidines at 96 and 111 (Fig. 1) (Burns et al. 2003). Although less studied than the OR domain, this segment is thought to take up two Cu^{2+} equivalents, each one localized to a histidine. Coordination at each site is from four nitrogens originating from the histidine imidazole and exocyclic nitrogen, and the backbone amide nitrogens from the two residues preceding each histidine. The dissociation constant is approximately 0.1 nM and does not vary significantly between the His96 and the His111 sites.

There are physiological connections between PrP^C and its ability to take up copper. In cell culture, Cu^{2+} levels in excess of 200 μM stimulate PrP internalization through endocytosis (Pauly and Harris 1998). Moreover, copper binding to the *PRNP* promoter region stimulates PrP expression, which then increases copper levels at extracellular membranes. Whole brain imaging finds that PrP expression correlates with copper levels in regions adjacent to the lateral ventricles (Pushie et al. 2011). The course of Wilson's disease, a genetic disorder associated with copper accumulation, is affected by PrP allele.

Many emerging functional studies consider the role of the copper binding domains. Early experiments suggested that PrP^C may function as a superoxide dismutase (SOD), converting superoxide to peroxide. This line of research has been controversial, and to date there is still no consensus. Another possibility is that copper complexation by PrP^C modulates copper's intrinsic redox activity thus protecting cellular components from radicals produced through Fenton-like reactions (Millhauser 2007). In this scenario, PrP^C functions as an antioxidant. Although this specific function remains unproven, it is noteworthy that the cerebral spinal fluid lacks many of the copper-binding proteins found in blood serum that serve a protective role outside of the CNS. Copper may also participate in PrP processing, releasing N-terminal fragments with downstream activities such as transmembrane signaling and caspase regulation.

Copper is also implicated in the development of prion disease. The histidine 96 and 111 sites are within a region thought to misfold in the formation of PrP^{Sc}. Kinetic studies demonstrate that synthetic prions form more slowly in the presence of high copper concentrations (Bocharova et al. 2005). A spectrum of rare genetic disorders also link copper binding to prion disease. Approximately 30 families and 110 individuals have been affected by prion diseases arising from expansions in the PrP OR domain, with modular inserts of one to nine PHGGGWGQ segments. Beyond a threshold of four to five inserts, prion disease switches from late to early onset, with some individuals diagnosed in their teen years. At the same insert number threshold, the PrP OR domain loses its ability to respond to increasing copper loads and does not readily transition to high occupancy binding mode where each repeat takes up a single Cu^{2+} equivalent.

Although the initial connection between copper and PrP was established nearly 15 years ago, the precise biological role of the copper sites is unknown, as is the function of PrP itself. The vibrant research activity driving toward a defined function will enhance the understanding of metal ion homeostasis in the CNS and may also clarify the molecular details that contribute to neurodegenerative disease.

Cross-References

▶ Copper Transport Proteins
▶ Copper-Zinc Superoxide Dismutase and Lou Gehrig's Disease
▶ Zinc in Alzheimer's and Parkinson's Diseases
▶ Zinc-Binding Sites in Proteins

References

Bocharova OV, Breydo L et al (2005) Copper(II) inhibits in vitro conversion of prion protein into amyloid fibrils. Biochemistry 44(18):6776–6787
Brown DR, Qin K et al (1997) The cellular prion protein binds copper in vivo. Nature 390:684–687
Burns CS, Aronoff-Spencer E et al (2002) Molecular features of the copper binding sites in the octarepeat domain of the prion protein. Biochemistry 41:3991–4001
Burns CS, Aronoff-Spencer E et al (2003) Copper coordination in the full-length, recombinant prion protein. Biochemistry 42(22):6794–6803
Chattopadhyay M, Walter ED et al (2005) The octarepeat domain of the prion protein binds Cu(II) with three distinct coordination modes at pH 7.4. J Am Chem Soc 127(36):12647–12656
Millhauser GL (2007) Copper and the prion protein: methods, structures, function, and disease. Annu Rev Phys Chem 58:299–320
Pauly PC, Harris DA (1998) Copper stimulates endocytosis of the prion protein. J Biol Chem 273:33107–33119
Pushie MJ, Pickering IJ et al (2011) Prion protein expression level alters regional copper, iron and zinc content in the mouse brain. Metallomics Integr Biomet Sci 3(2):206–214
Walter ED, Chattopadhyay M et al (2006) The affinity of copper binding to the prion protein octarepeat domain: evidence for negative cooperativity. Biochemistry 45(43):13083–13092
Walter ED, Stevens DJ et al (2007) The prion protein is a combined zinc and copper binding protein: Zn2+ alters the distribution of Cu2+ coordination modes. J Am Chem Soc 129(50):15440–15441

Copper Enzyme

▶ Nitrous Oxide Reductase

Copper Enzymes

▶ Copper, Biological Functions

Copper Metalloproteins

▶ Copper-Binding Proteins

Copper Nitrite Reductase

▶ Nitrite Reductase

Copper Trafficking in Eukaryotic Cells

▶ Biological Copper Transport

Copper Transport Proteins

▶ Platinum Complexes and Methionine Motif in Copper Transport Proteins, Interaction

Copper Transporting Proteins and Resistance of Platinum Anticancer Drugs

▶ Platinum Interaction with Copper Proteins

Copper, Biological Functions

Edward D. Harris
Department of Nutrition and Food Science, Texas A&M University, College Station, TX, USA

Synonyms

Copper enzymes; Micronutrient; Nutritionally essential metal; Trace metal

Definition

To understand the biological functions of copper is first to realize that the advent of copper (and iron) into the biosphere was timed with an enrichment of O_2 in the atmosphere. Consequently, many copper enzymes use O_2 as a substrate and in so doing have endowed living systems with the means to cope with a potentially toxic gas.

Background

In biological systems, copper is primarily a catalytic metal. The scarcity of copper in the system puts it in the category of a micronutrient, or more specifically a trace metal. Foremost in its actions is to allow the system to deal with iron. Other studies have linked copper in animals and humans with the synthesis of neurotransmitters, pituitary hormones, biopigments, and the establishment of a firm connective tissue network, functions that are manifested through copper-dependent enzymes. Copper in plants in confined mainly to the photosynthetic system of the plant. Through studies of copper deficiency, it is clear that ▶ Copper's physical and chemical properties allow the metal to serve as an enzyme cofactor, as an oxygen activator, and as an indispensible factor in ▶ iron homeostasis and transport in biological systems.

Historical Significance

By the mid-nineteenth century, copper was known only to be a constituent of blood, but of more importance to benefit people (mainly women) suffering from chlorosis, an anemic-like condition characterized by a grayish or yellow color of the skin, irregular menstruation, and breathlessness. A seminal discovery in 1928 showed that both copper and iron was required to fully restore blood hemoglobin in iron-deficient rats. An unexpected interdependence between the two metals was thus forecast. In time, dietary deficiencies in copper were shown to cause aneurysms and blood vessel ruptures in pigs and chicks, thus showing a need for copper for a sound cardiovasculature. Biochemically, copper was identified as cofactor for an enzyme that catalyzed the formation of cross-links in elastin and collagen in the blood vessels and in a connective tissue network in general. Further insights came with investigations of Wilson disease. First described in 1912 by an English physician, Wilson disease was characterized by copious amounts of copper accumulating in brain and liver. Sufferers had low levels of the blood copper protein ceruloplasmin and were unable to excrete copper through the bile. A second human disorder, Menkes disease, was more akin to a copper deficiency. Sufferers displayed arrested mental development, tortuous blood vessels, alterations in bone structure, and unusual hair texture referred to as "kinky." This X-linked disease, fatal to infants, was eventually traced to defective intestinal copper absorption. Both Wilson and Menkes diseases provided major insights into the pathologies that can occur if copper intake or metabolism is disrupted.

Copper Enzymes

In deciphering copper's necessity in biological systems, enzymes that require copper as a cofactor become a focal point. The catalytic function depends on copper; no other metal can substitute at the ▶ copper binding site. Many were discovered by identifying the biochemical factors underlying deficiency symptoms or genetic mutations. Some of the more notable ones are listed in Table 1. A brief description of their function and mechanism is listed below. A more complete understanding of their role in the system is described under "Functions of Copper in Organs and Tissues."

(a) ▶ Ascorbate Oxidase: Found in the cell walls of plants the enzyme catalyzes the oxidation of the

Copper, Biological Functions, Table 1 Biological functions of copper

Function	Copper factor
Ascorbate metabolism	Ascorbate oxidase
Antioxidant activity	Cu_2Zn_2 superoxide dismutase
Aerobic metabolism	Cytochrome c oxidase
Bone strength	Lysyl oxidase
Cardiovascular integrity	Lysyl oxidase
Iron metabolism	Ferroxidase, hephaestin
Lipid metabolism	Fatty acyl desaturase
Oxygen transport	Hemocyanin
Neurological functions	Dopamine β-monooxygenase
Neurohormone biosynthesis	Peptidyl-α-amidating monooxygenase
Pigmentation	Tyrosinase
Angiogenesis	Unknown

L-ascorbate to dehydroascorbate as shown in the reaction below. Although

$$2\text{L-ascorbate} + O_2 \rightarrow 2\text{ dehydroascorbate} + 2H_2O$$

the function of ascorbate oxidase has not been clarified, as an oxidoreductase, the enzyme is considered to be a component in the interaction between redox signaling and light-modulated control of ascorbic acid in apoplasts of plants.

(b) ▶ Copper-Zinc Superoxide dismutase: The copper/zinc form of the enzyme protects cells from oxidative damage by destroying superoxide anion, an aqueous phase free radical by-product of dioxygen. Otherwise known as SOD-1 or CuZnSOD, the enzyme is present in high concentrations in erythrocytes and the cytosol of most cells. The reaction catalyzed is shown below:

$$2O_2^- + 2H^+ \rightarrow H_2O_2 + O_2$$

Copper's role as an antioxidant is primarily through this enzyme.

(c) Cytochrome c oxidase: Arguably, the most important enzyme in aerobic metabolism, cytochrome c oxidase (CcO), is the terminal complex in the electron transport chain in the mitochondria. CcO catalyzes a four-electron transfer from cytochrome c to O_2. O_2 is tethered to the copper in the ▶ Cu-binding site in the protein. The reaction of CcO is shown below:

$$4 \text{ cytochrome } c(Fe^{2+}) + O_2 + 4H^+ \rightarrow$$
$$4 \text{ cytochrome } c(Fe^{3+}) + 2H_2O.$$

The transfer of electrons to O_2 occurs concomitantly with the establishment of a proton gradient whose energy is used to form ATP from ADP and inorganic phosphate.

(d) Ferroxidase: Ferroxidase is another name for *ceruloplasmin*, the major copper protein in plasma. The name recognizes the enzymes ability to oxidize ferrous iron to ferric, a reaction that occurs with the transfer of four electrons to dioxygen as shown below. Note the similarity between ferroxidase and the reaction catalyzed by cytochrome c oxidase. Ferroxidase, however, is believed to act on free ferrous ions, not a protein-bound form of the metal:

$$4Fe^{2+} + O_2 + 2H^+ \rightarrow 4Fe^{3+} + 2H_2O.$$

(e) Lysyl oxidase: The formation of the cross-links in collagen and elastin requires the oxidative deamination of select lysine residues in the soluble precursor proteins. Lysyl oxidase catalyzes the reaction shown below:

$$\text{Peptidyl-L-lysyl-peptide} + H_2O + O_2 \rightarrow$$
$$\text{Peptidyl-allysly-peptide} + NH_3 + H_2O_2.$$

The protein-bound aldehydes become a nucleus for forming cross-links at defined intervals along the chain. These cross-links support the architecture of blood vessels and other soft connective tissue. Since collagen also makes up most of the organic matter of bone, a lysyl oxidase failure can weaken bone structure and make bone prone to fracture.

(f) Peptidyl-α-amidating monooxygenase (PAM)
Many peptides synthesized by the pituitary gland have a glycine residue at the C-terminus. The biochemical rationale behind such structural similarity became clear with the discovery of PAM. PAM converts the C-terminal glycine of the peptide to an amide group which effectively activates the peptide. L-ascorbate provides the electrons for the reaction:

$$\text{Peptidylglycine} + \text{L-dihydroascorbate}$$
$$+ O_2 \rightarrow \text{peptidyl(2 hydroxyglycine)}$$
$$+ \text{L-dehydroascorbate} + H_2O$$
$$+ \text{desglycine peptide amide} + \text{glyoxylate}.$$

At least nine pituitary hormones undergo amidation as discussed under "Copper and Brain."

(g) Dopamine β-monooxygenase (DMO)

Dopamine is both a neurotransmitter and a precursor of norepinephrine. Dopamine β-monooxygenase (DMO) catalyzes the reaction of dopamine to norepinephrine. As with PAM, ascorbate supplies the electrons for the reaction:

$$3,4\text{-dihydroxyphenylethylamine (dopamine)} + \text{L-dihydroascorbate} + O_2 \rightarrow \text{norepinephrine} + H_2O + \text{dehydroascorbate}$$

(h) Tyrosinase

This enzyme present in melanocytes is responsible for the formation of melanin and other pigments (pheomelanin, eumelanin). The oxidation of tyrosine to L-dopa begins the cascade as shown below:

$$\text{Tyrosine} \rightarrow \text{L-dopa} \rightarrow \text{dopaquinone} \rightarrow \text{dopachrome} \rightarrow 5,6\text{-dihydroxyindole} + 5,6\text{-dihydroxyindole-2-}carboxylate \rightarrow\rightarrow \text{pheomelanin and eumelanin.}$$

Copper Deficiency

By far, the clearest insights into defining copper's role in biology have come through studies of copper deficiency. Feeding young pigs, chickens, and rat diets lacking copper produces pathological changes in animals. Correlating the symptoms with a specific biochemical factor has allowed workers to pinpoint a specific site that is affected. Such strategies lead to the discovery of biochemical factors that lose function when copper is missing or in slow supply. Other strategies have used animals with genetic mutations or young animals raised on forage grown in indigenous soil lacking copper. The symptoms that appear when an animal, human, or plant is deficient in copper or mismanages its metabolism are signs impinging directly on copper's role in the system. Some of the more important ones are shown in Table 2. Each disorder has a specific biochemical factor responsible for the symptom.

Copper, Biological Functions, Table 2 Symptoms of copper deficiency in animals and humans

1.	Stunted growth and arrested development
2.	Swayback (sheep and goats)
3.	Iron-resistant anemia
4.	Hypercholesterolemia
5.	Aortic rupture (falling disease in cattle)
6.	Loss of pigmentation
7.	Connective tissue weakness
8.	Inflammatory distress
9.	Neutropenia
10.	Neurological impairment

Functions of Copper in Organs and Tissues

Copper in Brain

Two key enzymes in brain require copper: dopamine β-monooxygenase (DMO) and peptidyl-α – amidating monooxygenase (PAM). Together, the two play a major role in neurotransmitter biosynthesis and pituitary hormone activation, respectively. DMO has both a soluble and membrane-bound form, the latter in the chromaffin granules of the adrenal medulla, which is also a rich source of the ascorbate required in the reaction. Like DMO, PAM also shows wide distribution appearing in the adrenal medulla, pituitary, pancreas, and the atria of the heart. The reaction catalyzed by PAM occurs in two steps, the first step requiring two copper atoms. The product of the first step is converted to a peptide amide by a lyase enzyme as shown in Fig. 1.

In the reaction the, COOH group and alpha carbon are removed as glyoxylate. This reaction is necessary to give functional status to the hormones shown in Fig. 2.

Whereas DBM and PAM catalyze distinctly different reactions, both enzymes appear to have structural similarity in their catalytic domain, which has led Southan and Kruse to suggest the two are homologous and may represent a family of copper type II ascorbate-dependent monooxygenases.

Copper in Connective Tissue

A lack of copper in the diet of new born chicks and pigs predisposes the animals to leg weakness, aneurysms, and death via aortic rupture. Rendering collagen and

Copper, Biological Functions

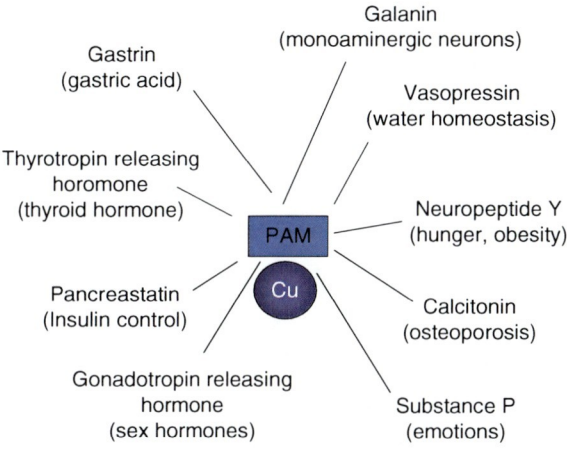

Copper, Biological Functions, Fig. 1 Reactions of peptidyl-alpha-amidating monooxygenase

Copper, Biological Functions, Fig. 2 Pituitary hormones that depend on PAM for activation

elastin capable of withstanding the rigors placed on connective tissues such as tendon, ligaments, and bone requires a posttranslational modification of the tropocollagen and tropoelastin molecules to a highly cross-linked network of proteins. For elastin, cross-linking provides the anchors that allow the resilience property typical of expanding and contracting blood vessels to manifest. As a prelude to their formation, lysyl oxidase, a monocopper oxidase, catalyzes the oxidation of peptidyl lysyl residues in collage and elastin forming peptidyl aldehydes which then condense through Schiff base and aldol condensation reactions to form cross-link (Fig. 3). If cross-links do not form, the precursor proteins are soluble and can easily be extracted by neutral salt solutions. Thus, high salt solubility of collagen is a biomarker of lysyl oxidase dysfunction. The multitude of connective tissue functions that rely on the enzyme are shown in Table 3. It is of some interest to note that in laying hens, an elastin-like protein coats the ovalbumin in the inner core of an egg prior to the application of the eggshell. A copper deficiency will weaken the structure, allowing the core to expand and give rise to oversized eggs. These in turn will have week shells and in some cases no shell at all.

Copper in Lung

Normal functioning of lung aveoli provides another example of the need for copper in connective tissue. The outer surface of the aveoli (air sacs) is supported by an extensive network of elastin protein. The role of the elastin is to expel air from the expanded aveoli when exhaling. In the absence of copper, lysyl oxidase in lung is compromised which results in the erosion of the elastin in aveoli and an emphysema-like condition.

Copper-Iron Interactions

The hematopoietic system's dependence on copper was one of the earliest signs of a biological need for copper. The efforts to uncover the iron-copper link, i.e., where the two metals merge functionally have led to the discovery of two copper-dependent oxidase enzymes, ceruloplasmin and hephaestin. These two copper enzymes are designed to oxidize the ferrous iron to its ferric oxidation state. The simple reaction is critical for a number of reasons. Ferric iron (Fe^{3+}) is the only form that will bind to the iron transport protein transferrin for transport to the tissues. A copper deficiency or genetic mutation affecting ceruloplasmin will result in an iron overload in the liver and other tissues. Hephaestin, a second mammalian copper oxidase/ferroxidase, is responsible for transferring iron from the intestine into the system. A defect in the hephaestin gene gives rise to a sex-linked (x-chromosome involved) anemia in mice. In yeast, Fet3p is the copper oxidase/ferroxidase essential for iron to engage a ferric ion transporter in the membrane of yeast cell. These observations have helped understand why copper is critical to the safe utilization of iron by higher animals as well as microorganisms.

Copper as an Antioxidant

The enzyme ▶ copper-zinc superoxide dismutase (SOD-1) has both a copper and zinc site in each of its two subunits. As noted above, the primary function of SOD-1 is to rid the system of superoxide anion that form in the cells interior. This two-step reaction occurs first with the transfer of the unpaired electron of O_2^- to a type II Cu at the active site of the enzyme. In so

Copper, Biological Functions, Fig. 3 Reaction of lysyl oxidase leading to cross-linking of collagen and elastin

Copper, Biological Functions, Table 3 Connective tissue functions dependent on lysyl oxidase

1.	Lung elasticity
2.	Ligamentum nuchae elasticity
3.	Bone and cartilage strength and stability
4.	Blood vessel integrity
5.	Eggshell shape and size

doing, the copper is reduced to Cu I. A second O_2^- then becomes a recipient of the electron whose passing gives rise to hydrogen peroxide and restores the copper to its Cu II form. The Cu_2Zn_2 enzyme is present in all tissues as a 32-kDa homodimer. Four histidine residues make up the type II copper binding site. A deficiency in copper will suppress the activity of the enzyme, but not the enzyme protein which stays at a near constant level.

Copper and Respiration

Respiration is the act of taking up O_2. As a cofactor for the respiratory enzyme cytochrome c oxidase, copper plays a central role in this event. The enzyme exists as a complex situated on the distal end of the electron transport chain in inner membrane of the mitochondria. The ultimate goal of the chain is to transfer electrons from reduced cofactors at the proximal end to O_2. In the process, protons are pumped from the interior of the mitochondria to the intermembrane space, and their return to the interior drives ATP formation. Three of the enzyme's subunits are synthesized by mitochondrial genes. These subunits hold both copper and iron in distinct coordination spheres. The ▸ copper A in cytochrome oxidase has one coordinate open to bind the O_2. Through this enzyme, copper is linked to the major energy-generating system in the cell.

Copper and Genetic Regulation

A number of studies have supported the understanding that copper has the capacity to regulate genetic expression. The genes targeted code for ▸ metallothioneins that bind and store copper, proteins that regulate its metabolism or respond to oxidants. The discovery of metal-responsive elements (MRE) in the DNA in the promoter region of genes coding for these proteins is testament to this understanding. First observed for *CUP1*, a gene that codes for a copper storage protein in yeast, ACE1 (activation of CUP1 expression), a 11-kDa cysteine-rich protein, was the first eukaryote copper-dependent transcription factor identified. ACE1 is able to engage the MRE only when Cu^+ (or Ag^+) is bound to the protein. The gene that codes for the antioxidant protein *SOD-1* likewise is regulated by copper through the ACE1 transcription factor. A mammalian counterpart for ACE1 has not been found.

Summary and Conclusions

The multifaceted roles of copper in biological systems center around a need to handle O_2 and its by-products. Copper metalloenzymes are designed to use O_2 as a substrate. Such enzymes vary from the simplest monooxygenases to the more complex multisubunit multicopper oxidases (▶ Copper-Binding Proteins). Nearly all copper-dependent systems exploit the redox properties of the metal. The cardiovascular system, respiration, soft and hard connective tissues, central and peripheral nervous system, and hair and integument all rely on copper for normal function or appearance and all can be altered metabolically by a deficiency. Many copper-dependent biochemical factors have come to light and through these factors one has gained an understanding of the molecular basis for copper's necessity. One must consider that the safe handling of iron and oxygen were early needs for copper in biology. A deficiency in copper has a serious impact on iron absorption, transport, and metabolism. As pointed out in this chapter, however, the role of copper in biology extends beyond a cofactor function for copper-dependent enzymes, although this is perhaps the most visible role for the metal. Traditionally, copper has been regarded as a passive element, i.e., available when needed. In the last 20 years, however, considerable interest has been directed at learning an active role for copper at the genetic level; functions that casts copper in the role of a regulator of genetic expression. Copper-dependent transcription factors have been identified in microorganisms such as yeast; however, the expected crossover to higher animals has not been fully realized. Thus, there is still much to be discovered with regard to the functions copper and to uncover all of the biological properties unique to this metal.

Cross-References

▶ Ascorbate Oxidase
▶ Cobalt Proteins, Overview
▶ Copper, Physical and Chemical Properties
▶ Copper-Zinc Superoxide Dismutase
▶ Iron Homeostasis in Health and Disease
▶ Metallothioneins and Mercury

References

Bousquet-Moore D, Mains RE, Eipper BA (2010) Peptidylglycine α-amidating monooxygenase and copper: a gene-nutrient interaction critical to nervous system function. J Neurosci Res 88:2535–2545

Easter RN, Qilin C, Lai B, Ritman EL, Caruso JA, Zhenyu Q (2010) Vascular metallomics: copper in the vasculature. Vasc Med 15:61–69

Fox PL (2003) The copper iron chronicles: the story of an intimate relationship. Biometals 16:9–40

Krupanidhi S, Sreekumar A, Sanjeevi CB (2008) Copper and biological health. Indian J Med Res 128:448–461

Southan C, Kruse LI (1989) Sequence similarity between dopamine beta hydroxylase and peptide alpha amidating enzyme: evidence for a conserved catalytic domain. FEBS Lett 255:116–120

Copper, Mononuclear Monooxygenases

Marius Réglier[1] and Catherine Belle[2]
[1]Faculté des Sciences et Techniques, ISM2/BioSciences UMR CNRS 7313, Aix-Marseille Université Campus Scientifique de Saint Jérôme, Marseille, France
[2]Département de Chimie Moléculaire, UMR-CNRS 5250, Université Joseph Fourier, ICMG FR-2607, Grenoble, France

Synonyms

Copper-containing hydroxylase; Copper-containing monooxygenase

Definition

Copper mononuclear monooxygenase is a unique class of metalloenzymes present in eukaryotes where they catalyze very important biosynthetic reactions in neurotransmitter and hormone pathways. Composed by dopamine β-monooxygenase (DβM), also named dopamine β-hydroxylase (DβH), tyramine β-monooxygenase (TβM), peptidylglycine α-hydroxylating monooxygenase (PHM), and monooxygenase X (MOX), this class of metalloenzymes is characterized by the presence of two independent T2 ▶ Cu-Binding Sites where only one copper atom is involved in the hydroxylation reaction.

General Background

Mononuclear copper monooxygenases catalyze the stereospecific oxygen atom insertion in organic substrates and are ascorbate (reducer) and dioxygen-dependents (Klinman 1996).

Dopamine β-monooxygenase (DβM, EC 1.14.17.1) is a mammalian enzyme involved in the catecholamine biosynthetic pathway where it transforms dopamine into norepinephrine (Scheme 1a) (Stewart and Klinman 1988). Tyramine β-monooxygenase (TβM), the insect homolog of DβM, transforms tyramine into octopamine in a similar way (Scheme 1a) (Hess et al. 2008).

Peptidylglycine α-hydroxylating monooxygenase (PHM) functions in vivo toward the biosynthesis of α-amidated peptide hormones in mammals and insects. PHM is a catalytic domain of peptidylglycine α-amidating monooxygenase (PAM, EC 1.14.17.3), a bifunctional enzyme involved in the posttranslational amidation of C-terminal glycine-extended peptides (Kulathila et al. 1999). This posttranslational modification is a key step in the activation of several neuropeptides and peptide hormones (substance P, oxytocin, neuropeptide Y, adrenomedullin...). PAM is encoded by a complex single-copy gene that is subject to tissue-specific and developmentally regulated alternative splicing. The isolated domains, separated either through endoproteolytic cleavage or through independent expression, retain their enzymatic activities. The NH$_2$-terminal domain of PAM, the PHM domain, catalyzes the stereospecific hydroxylation of the C-terminal glycine residue into hydroxyglycine, which in turn is transformed into amidopeptide and glyoxylic acid by the peptidyl-α-hydroxylglycine α-amidating lyase domain (PAL) of PAM (1b). Both enzymatic activities have broad substrate specificity since peptides with all 20 amino acid amides have been isolated.

Analysis of the sequence of the monooxygenase X (MOX) allows to group this mammalian enzyme in the mononuclear copper monooxygenase family. However, its catalytic activity has not yet been characterized and its function is unknown (Xin et al. 2004).

Structural and Functional Features

Dopamine β-Monooxygenase (DβM) and Peptidylglycine α-Hydroxylating Monooxygenase (PHM)

Despite a large difference in macroscopic features (DβM is a tetrameric glycoprotein of 75 kDa/monomer; PHM is active as a 35 kDa monomer), DβM and PHM exhibit strong similarities. First of all, (1) both enzymes catalyze the cleavage of dioxygen to form hydroxylated products and water, (2) both enzymes require two copper ions per subunit for full activity, and (3) both enzymes are believed to use ascorbate as the in vivo two-electron donor. In addition, DβM and PHM share a 28% sequence identity extending through a common catalytic domain of 270 residues, which includes the conserved copper ligands. Although there is no crystal structure yet, structural data exist for DβM. Extended X-ray absorption fine structure has been used to characterize the ligand environment of the two copper atoms in both oxidized and reduced forms. Without any evidence for backscattering between the two copper sites confirmed in the electron paramagnetic resonance (EPR) spectrum by the absence of any spin coupling, the distance between the two coppers has been estimated to exceed 4 Å. These findings have provided early evidence against a reactive binuclear center and, instead, have suggested separate functions for the two copper centers. The Cu$_A$ (Cu$_H$ in PHM), surrounded by three histidine residues and a water molecule, has been assigned as the electron transfer site where ascorbate binds and delivers two electrons, whereas Cu$_B$ (Cu$_M$ in PHM), liganded by two nitrogen atoms from histidine residues and a water molecule together with an elongated bond to the sulfur atom from a methionine residue, has been assigned as the substrate binding and hydroxylation site. The crystal structures of oxidized and reduced PHM forms with or without a bound peptide substrate N-α-acetyl-3,5-diiodo-L-tyrosylglycine (IYG) (Fig. 1) have confirmed the conclusions of many of the earlier spectroscopic data (Prigge et al. 1997, 1999). The important features determined from the crystal structures of the oxidized or reduced forms of the PHM catalytic core (PHMcc) are (1) a two-domain

Copper, Mononuclear Monooxygenases, Scheme 1 Reactions catalyzed by mononuclear copper monooxygenases. (**a**) Dopamine β-monooxygenase (DβM) and tyramine β-monooxygenase (TβM), (**b**) peptidylglycine α-hydroxylating monooxygenase (PHM)

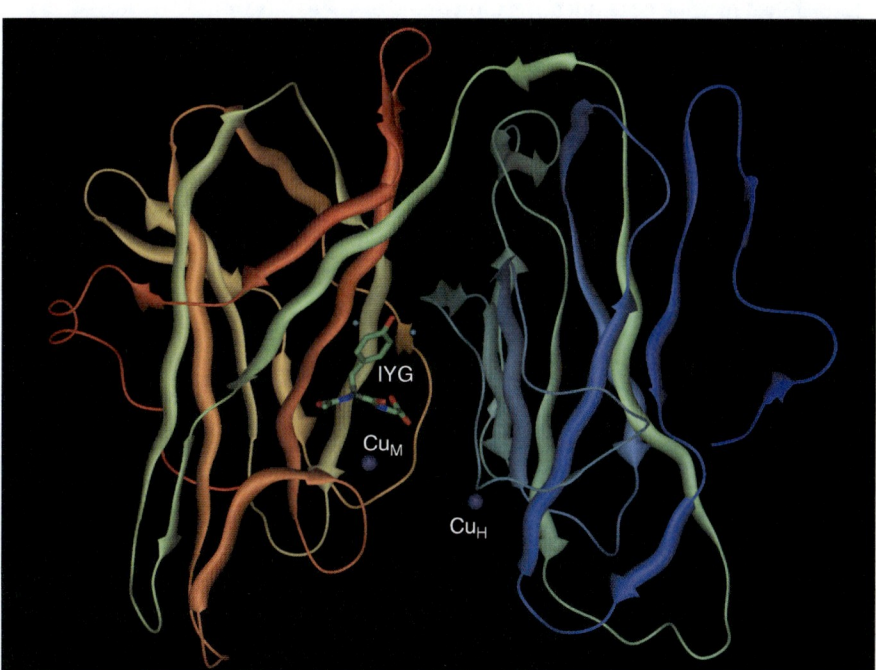

Copper, Mononuclear Monooxygenases, Fig. 1 Structure of the catalytic core of PHM (PHMcc) from Protein Data Bank (code 1OMP) showing the PHMcc fold with the binuclear active site and the bound substrate N-α-acetyl-3,5-diiodotyrosyl-glycine (IYG)

structure in which each domain binds a single copper atom, (2) a distance of 10.59 Å between the two copper sites, (3) the absence of closure in the copper binding domains in either enzyme form studied, and (4) the identification of a water-filled cavity that is located at the solvent interface and "links" the two copper binding domains. Several structures of oxidized and reduced forms of PHMcc with bound carbon monoxide, azide, and nitrite are now available on the Protein Data Bank and confirm a differential reactivity between the two copper sites (Chufán et al. 2010).

Comparison of the kinetic parameters for DβM and PHM with substrates of comparable reactivity indicates the same intrinsic H/D isotope effect for the

C–H activation step. Additionally, similar ^{18}O isotope effects decreasing with substrate deuteration observed for these two enzymes imply a chemical mechanism for substrate oxidation that is likely to be identical for both enzymes. Studies on kinetics indicate that both DβM (in the presence of the fumarate dianion activator) and PHM proceed in a preferred ordered mechanism with substrate binding to the enzyme before dioxygen interaction. Thus, all available data infer that DβM and PHM can be regarded as interchangeable with respect to mechanism and active site structure.

Regarding the pathway for the long-distance electron transfer between the copper sites and the nature of the various copper-oxygen species involved in dioxygen activation, an extensive and open debate has taken place in the recent literature. In early studies on DβM with either substrates or substrate analogs, it has been concluded that functionalization of the substrate involved hydrogen atom abstraction to yield a free radical intermediate. Identification of the oxygen species catalyzing hydrogen atom abstraction has proven to be far more elusive (Blain et al. 2002). The observation of pH-dependent isotope effects for DβM provided evidence for the involvement of a single proton in the chemical conversion process. This led to the hypothesis of a copper hydroperoxide (Cu(II)–OOH) as the reactive oxygen-centered intermediate. However, a detailed analysis of the effects of substrate structure and deuterium on ^{18}O isotope effects was found to be inconsistent with this hypothesis. Instead, a reductive cleavage of this intermediate to generate the copper-oxo (Cu(II)–O$^{\bullet}$ ↔ Cu(III)–O^{-}) as the species responsible for the hydrogen atom abstraction was suggested. In this proposal, the oxidation of both copper centers occurs before substrate activation and leads to the accumulation of a partially reduced form of dioxygen.

Recently, it was reported on two experimental probes of the activated oxygen species in DβM (Evans et al. 2003). First, the capacity of a substrate analog that cannot be functionalized (β,β-difluorophenethylamine) to induce reoxidation of the prereduced copper sites of DβM upon mixing with dioxygen under rapid freeze-quench conditions was examined. This experiment failed to give rise to an EPR-detectable copper species, in contrast to a substrate with an active C–H bond. This indicates either that the reoxidation of the enzyme-bound copper sites in the presence of dioxygen is tightly associated with the C–H activation or that a diamagnetic species Cu(II)–O$_2^{\bullet}$ has been formed. In the event of an active site that is open and fully solvent-accessible, as seen for the homologous peptidylglycine α-hydroxylating monooxygenase, the accumulation of a reduced and activated oxygen species in DβM prior to C–H cleavage would be expected to produce an uncoupling of dioxygen and substrate consumption by analogy to cytochrome P-450. For this reason, the degree to which dioxygen and substrate consumption are related in DβM was examined using both end point and initial rate experimental protocols. With substrates that differ by more than three orders of magnitude in rate, there was a strong correlation between dioxygen uptake and product formation. This has led to the conclusion that there is no accumulation of an activated form of dioxygen before C–H abstraction in the DβM and the formation a copper-superoxo species (Cu(II)–O$_2^{\bullet}$) was proposed to be responsible for the abstraction of the hydrogen atom of the dopamine substrate (Evans et al. 2003).

At the same time, a copper–dioxygen complex was trapped in PHM by freezing protein crystals that had been soaked in the presence of dioxygen with ascorbate and slow substrate N-α-acetyl-3,5-diiodo-tyrosyl-D-threonine (IYT). The X-ray crystal structure of this precatalytic complex, determined to 1.85 Å resolution (PDB code 1SDW), showed that dioxygen binds to one of the copper atoms in the enzyme with an end-on geometry (Fig. 2a, b). Given this structure, it is likely that dioxygen is directly involved in the electron transfer and hydrogen abstraction steps of the PHM reaction (Prigge et al. 2004). These findings suggested a mechanism in which a diamagnetic Cu(II)–O$_2^{\bullet}$ complex formed initially at very low levels, abstracts a hydrogen atom from the substrate to generate Cu(II)–OOH and a substrate-free radical as intermediates. Subsequent participation of the second copper site (Cu$_H$) per subunit completes the reaction cycle, generating hydroxylated products and water (Scheme 2).

Tyramine β-Monooxygenase (TβM)

Tyramine β-monooxygenase (TβM) catalyzes the synthesis of the neurotransmitter, octopamine, in insects (Scheme 1). TβM shares 39% identity and 55% similarity with the mammalian DβM. In addition, as indicated by kinetic isotope effect, the TBM oxidation seems to proceed in agreement with the mechanisms proposed for the mammalian enzymes

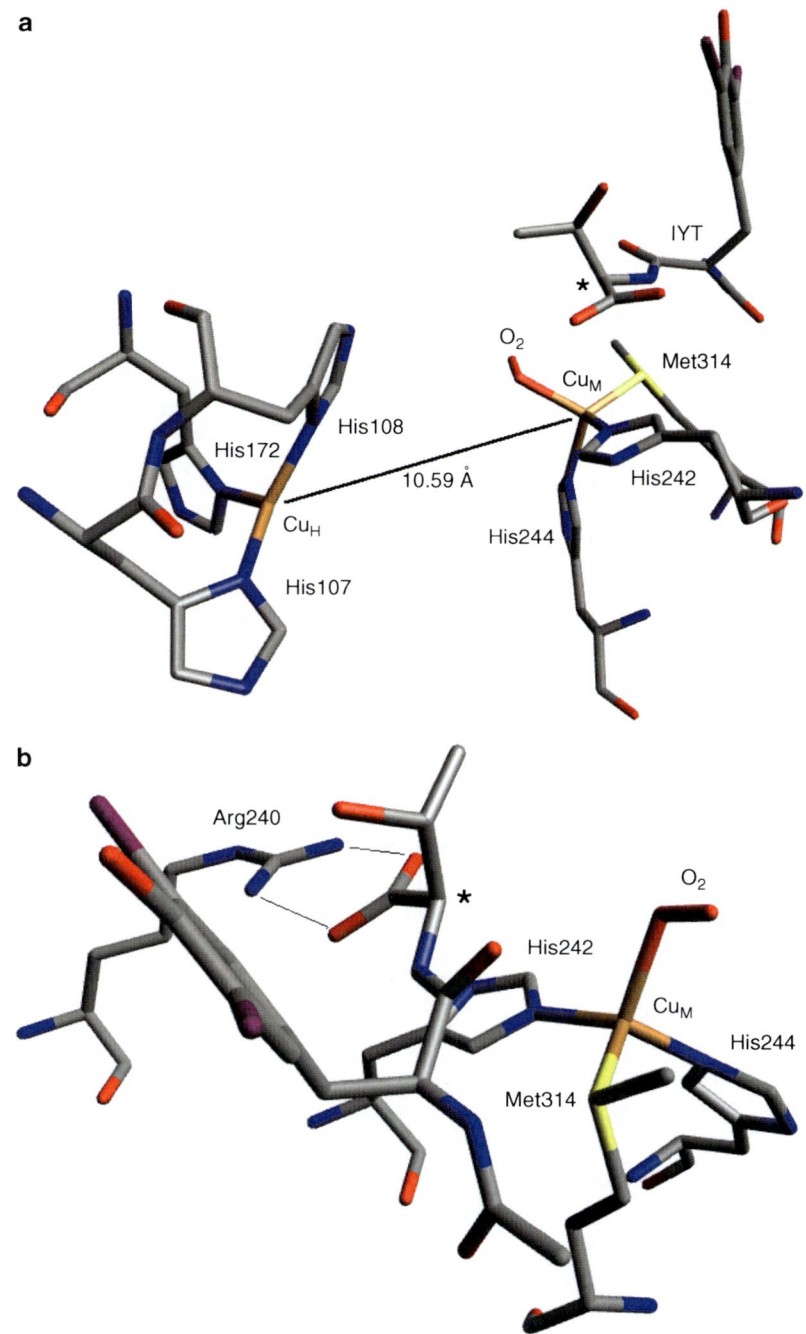

Copper, Mononuclear Monooxygenases, Fig. 2 (a) Structure of the active site of the PHMcc from Protein Data Bank (code 1SDW) showing the two Cu_H and Cu_M centers with the bound slow substrate N-α-acetyl-3,5-diiodotyrosyl-D-threonine (IYT) and end-on coordinated dioxygen, (b) focus on the Cu_M center showing the interaction of the carboxylate group of the substrate IYT with Arg240 and in *asterisk* the D-threonine Cα which is the subject of hydroxylation

(PHM and DβM). However, a distinctive feature of TβM is the very strong substrate inhibition that is dependent on the level of the cosubstrate, dioxygen, and ascorbate as well as substrate deuteration (Hess et al. 2008). This feature has led to a model in which the substrate tyramine can bind to either the Cu(I) or Cu(II) forms of TβM, with substrate inhibition increased at very high ascorbate levels. The rate of ascorbate reduction of the Cu(II) form of TβM is also reduced at high tyramine level, leading to propose the existence of a binding site for ascorbate to this class of enzymes and that it was never observed with DβM or PHM (Hess et al. 2010). These findings may be relevant to the control of octopamine production

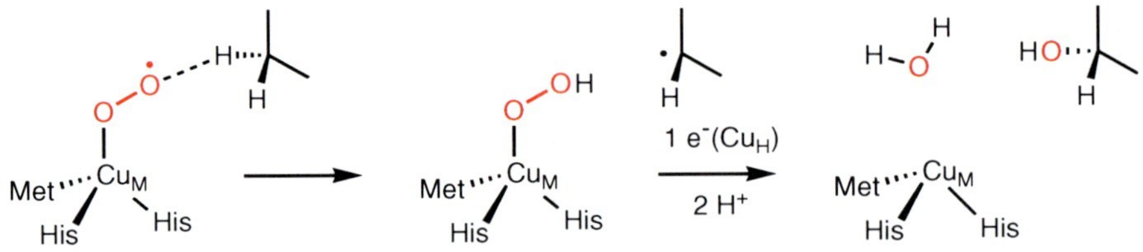

Copper, Mononuclear Monooxygenases, Scheme 2 Proposed mechanism for hydroxylation reaction catalyzed by mononuclear copper monooxygenases involving the formation of a copper-superoxo species Cu(II)–$O_2^{\cdot-}$

in insect cells. Recent studies have demonstrated that tyramine and octopamine have antagonistic effects and suggest that behavioral regulation may depend on the balance of these two hormones. Thus, TβM is exquisitely sensitive to small shifts in cellular conditions. Overall, the TβM kinetics data imply tighter regulation of neurotransmitter levels by the insect enzyme than in the mammalian homologue.

Physiological Roles

The monoamine and amidopeptides neurotransmitters are involved in the control of a wide variety of neuronal functions (psychomotor function, reward-driven learning, arousal, processing of sensory input, memory, appetite, emotional stability, sleep, mood, vomiting, sexual behavior, and secretion of anterior pituitary and other hormones). Since DβM and PHM play an important role in controlling the levels of the neurotransmitters/hormones, the literature on the physiological role of DβM and PHM is abundant.

Mammalian genomes encode only a small number of copper-containing enzymes. The many genes involved in coordinating copper uptake, distribution, storage and efflux make gene/nutrient interactions especially important for these copper-containing enzymes. Copper deficiency and copper excess both disrupt neural function. Using mice heterozygous for PAM, it was identified alterations in anxiety-like behavior, thermoregulation, and seizure sensitivity. Dietary copper supplementation reversed a subset of these deficits. Wild-type mice maintained on a marginally copper-deficient diet exhibited some of the same deficits observed in PAM[+/−] mice and displayed alterations in PAM metabolism. Altered copper homeostasis in PAM[+/−] mice suggested a role for PAM in the cell-type-specific regulation of copper metabolism. Physiological functions sensitive to genetic limitations of PAM that are reversed by supplemental copper and mimicked by copper deficiency may serve as indicators of marginal copper deficiency (Bousquet-Moore et al. 2010).

DβM deficiency is a very rare form of primary autonomic failure characterized by a complete absence of noradrenaline and adrenaline in plasma together with increased dopamine plasma levels. DβM deficiency is mainly characterized by cardiovascular disorders and severe orthostatic hypotension (Senard and Rouet 2006).

Cross-References

▶ Copper-Binding Proteins

References

Blain I, Slama P, Giorgi M, Tron T, Réglier M (2002) Copper-containing monooxygenases: enzymatic and biomimetic studies of the oxygen atom transfer catalysis. Rev Mol Biotechnol 90:95–112

Bousquet-Moore D, Prohaska JR, Nillni EA, Czyzyk T, Wetsel WC, Mains RE, Eipper BA (2010) Interactions of peptide amidation and copper: novel biomarkers and mechanisms of neural dysfunction. Neurobiol Dis 37:130–140

Chufán EE, Prigge ST, Siebert X, Eipper BA, Mains RE, Amzel LM (2010) Differential reactivity between two copper sites in peptidylglycine α-hydroxylating monooxygenase. J Am Chem Soc 132:15565–15572

Evans JP, Ahn K, Klinman JP (2003) Evidence that dioxygen and substrate activation are tightly coupled in dopamine β-monooxygenase: implications for the reactive oxygen species. J Biol Chem 278:49691–49698

Hess CR, McGuirl MM, Klinman JP (2008) Mechanism of the insect enzyme, tyramine β-monooxygenase, reveals differences from the mammalian enzyme, dopamine β-monooxygenase. J Biol Chem 283:3042–3049

Hess CR, Klinman JP, Blackburn NJ (2010) The copper centers of tyramine β-monooxygenase and its catalytic-site methionine variants: an X-ray absorption study. J Biol Inorg Chem 15:1195–1207

Klinman JP (1996) Mechanisms whereby mononuclear copper proteins functionalize organic substrates. Chem Rev 96:2541–2561

Kulathila R, Merkler KA, Merkler DJ (1999) Enzymatic formation of C-terminal amides. Nat Prod Rep 16:145–154

Prigge ST, Kolhekar AS, Eipper BA, Mains RE, Amzel LM (1997) Amidation of bioactive peptides: the structure of peptidylglycine hydroxylating monooxygenase. Science 278:1300–1305

Prigge ST, Kolhekar AS, Eipper BA, Mains RE, Amzel LM (1999) Substrate-mediated electron transfer in peptidylglycine α-hydroxylating monooxygenase. Nat Struct Biol 6:976–983

Prigge ST, Eipper BA, Mains RE, Amzel LM (2004) Dioxygen binds end-on to mononuclear copper in a precatalytic enzyme complex. Science 304:864–867

Senard JM, Rouet P (2006) Dopamine β-hydroxylase deficiency. Orphanet J Rare Dis 1:1–4

Stewart LC, Klinman JP (1988) Dopamine β-hydroxylase of adrenal chromaffin granules: structure and function. Annu Rev Biochem 57:551–592

Xin X, Mains RE, Eipper BA (2004) Monooxygenase X a member of the copper-dependent monooxygenase family localized to the endoplasmic reticulum. J Biol Chem 279:48159–48167

Copper, Physical and Chemical Properties

Fathi Habashi
Department of Mining, Metallurgical, and Materials Engineering, Laval University, Quebec City, Canada

Copper is an ancient metal has a characteristic red color, was used to make statues and coins, later mixed with tin to make bronze and with zinc to make brass. Bronze was cast in statues, bells, cannons, and other objects. Brass was an alloy that looked like gold. Copper occurs in nature in the metallic state in certain locations such as in Lake Superior region but mostly as sulfide and to a minor extent as oxide and silicate. Some of its minerals when pure have beautiful blue color and are used as gem stones.

Physical Properties

Atomic number	29
Atomic weight	63.55
Density, g/cm^3	8.89
	(continued)
Melting point, °C	1,083
Boiling point, °C	2,595
Heat of fusion, J/g	210
Heat of vaporization, J/g	4,810
Vapor pressure at m.p., Pa	0.073
Specific heat capacity, J g^{-1} K^{-1}	
At 20°C and 100 kPa	0.385
At 957°C and 100 kPa	0.494
Average specific heat, J g^{-1} K^{-1}	
0–300°C at 100 kPa	0.411
0–1,000°C at 100 kPa	0.437
Coefficient of linear thermal expansion, K^{-1}	
0–100°C	16.9×10^{-6}
0–400°C	17.9×10^{-6}
0–900°C	19.8×10^{-6}
Thermal conductivity at 20°C, W m^{-1} K^{-1}	394

The most important property of copper is its high electrical conductivity; among all metals, only silver is the better conductor. Both electrical conductivity and thermal conductivity are connected with the Wiedemann–Franz relation and show strong dependence on temperature.

Chemical Properties

Copper has the electronic configuration: 2, 8, 18, 1. It is considered a less-typical metal because when the outermost electron is lost, then an 18-electron shell will be exposed and not an inert gas structure. Copper in dry air at room temperature slowly develops a thin protective film of copper(I) oxide. On heating to a high temperature in the presence of oxygen, copper forms first copper(I) oxide and then copper(II) oxide, both of which cover the metal as a loose scale. In the atmosphere, the surface of copper oxidizes in the course of years to a mixture of green basic salts, which consists chiefly of the basic sulfate, with some basic carbonate. Such covering layers protect the metal.

Copper has a high affinity for free halogens, molten sulfur, or hydrogen sulfide. It is not attacked by nonoxidizing acids, such as dilute sulfuric, hydrochloric, phosphoric, or acetic, and other organic acids. Copper is not attacked by alkali-metal hydroxide solutions. Dissolution of copper is possible either by oxidation or by formation of complexed copper ions. Thus, copper is soluble in oxidizing acids, such as nitric acid, hot concentrated sulfuric acid, and chromic acid, or in nonoxidizing acids containing an oxidizing agent such as oxygen or hydrogen peroxide. The other method of dissolving copper is through formation of complex ions. The best reagents for this purpose are aqueous solutions of ammonia (Fig. 1) and ammonium salts or alkali-metal cyanides.

Copper dissolves in alkaline cyanide solution in presence of oxygen:

$$2Cu + 1/2 O_2 + 4CN^- + H_2O \rightarrow 2Cu(CN)_2^- + 2OH^-$$

Copper(I) ion precipitates from acid medium by HCN to form CuCN a white powder. Copper(II) ion precipitates from acid medium by H_2S to form CuS. Metallic copper is precipitated from $CuSO_4$ solution by hydrogen at high temperature and pressure:

$$Cu^{2+} + H_2 \rightarrow Cu + 2H^+$$

Copper sulfide undergoes oxidation in neutral medium at ambient conditions to $CuSO_4$:

$$CuS + 2O_{2(aq)} \rightarrow CuSO_4$$

In acidic medium, however, elemental sulfur forms

$$CuS + 1/2 O_{2(aq)} + 2H^+ \rightarrow Cu^{2+} + S + H_2O$$

Copper(I) and copper(II) can coexist in a single compound known as Chevreul salt: $Cu_2SO_3 \cdot CuSO_3 \cdot 2H_2O$.

Molten copper sulfide undergoes a conversion reaction when oxygen is blown through

$$Cu_2S + O_2 \rightarrow 2Cu + SO_2$$

This is the basis of copper metallurgy. The solubility of gases in molten copper follows Henry's law: the solubility is proportional to the partial pressure. Oxygen dissolves in molten copper as copper(I) oxide up to a concentration of 12.65% Cu_2O (corresponding to 1.4% O). Sulfur dioxide dissolves in molten copper and reacts:

$$6Cu + SO_2 \rightarrow Cu_2S + 2Cu_2O$$

Hydrogen is considerably soluble in liquid copper, and after solidification some remains dissolved in the solid metal, although copper does not form a hydride. The solubility follows Sievert's law, being proportional to the square root of the partial pressure because the H_2 molecules dissociate into H atoms on dissolution. Nitrogen, carbon monoxide, and carbon dioxide are practically insoluble in liquid or solid copper. Hydrocarbons generally do not react with copper. An exception is acetylene which reacts at room temperature to form the highly explosive copper acetylides Cu_2C_2 and CuC_2; therefore, acetylene gas cylinders must not be equipped with copper fittings.

Although copper is toxic in exceedingly low concentrations to certain lower life forms, notably

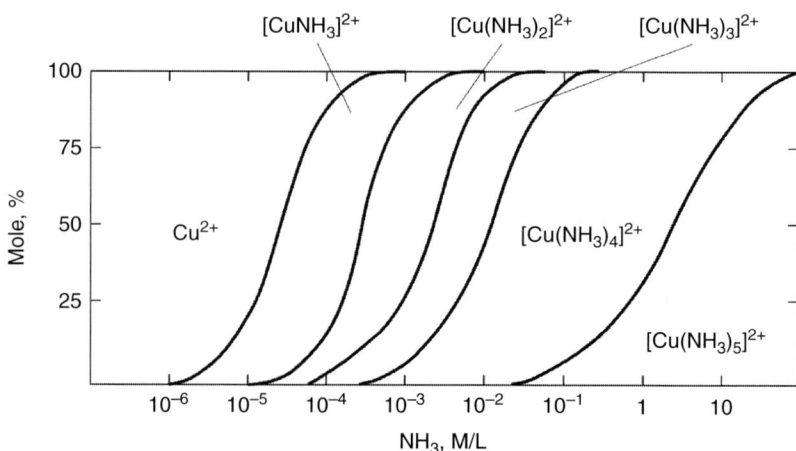

Copper, Physical and Chemical Properties, Fig. 1 Copper ammine complexes

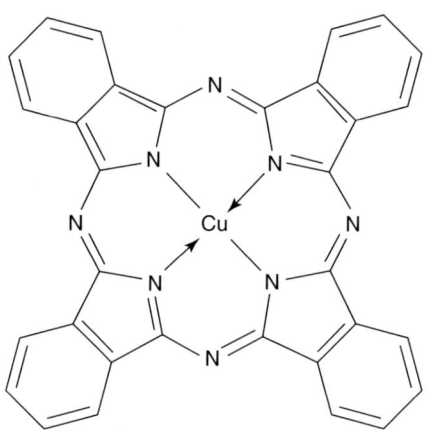

Copper, Physical and Chemical Properties, Fig. 2 Copper phthalocyanine

fungi and algae, it is a necessary constituent of higher plants and animals. It plays a necessary role as an oxidation catalyst and aids plants in photosynthesis and other oxidative processes. In higher animals, it is responsible for oxidative processes and is present in many proteins such as phenolase, hemocyanin, and galactose oxidase.

Copper Phthalocyanine

Copper phthalocyanine (Fig. 2) is analog of two natural porphyrins: chlorophyll and hemoglobin. It is a blue pigment.

References

Fabian H (1997) In: Habashi F (ed) Handbook of extractive metallurgy. Wiley-VCH, Heidelberg, Chapter 8

Habashi F (1999) Textbook of hydrometallurgy, 2nd edn. Métallurgie Extractive Québec, Laval University Bookstore, Québec City

Habashi F (2002) Textbook of pyrometallurgy. Métallurgie Extractive Québec, Laval University Bookstore, Québec City

Copper, Zinc Superoxide Dismutase

▶ Zinc in Superoxide Dismutase

Copper-Binding Proteins

Charles T. Dameron
Chemistry Department, Saint Francis University, Loretto, PA, USA

Synonyms

Copper metalloproteins; Cuproproteins

Definition

Copper-binding proteins specifically incorporate the metal into their structure for catalytic and structural purposes. Noncatalytic, structural sites are found in copper sensing proteins involved in the regulation of copper metabolism and in copper sequestering peptides and proteins involved in protection against copper intoxication. The sensing proteins regulate all aspects of copper metabolism, including the uptake, intracellular use, detoxification, and export of copper. Commonly, the regulation is affected at the transcriptional level, and the binding of copper by the protein is the switch that modulates the sensor protein's structure and function. The binding in these sites takes place through cysteinyl thiolates.

Noncatalytic Roles of Copper in Proteins

In noncatalytic proteins, copper serves a structural-regulatory role, a role mediated by its specific binding to sensory proteins where a change is elicited in the activity of the protein. Generally, copper's role in proteins is catalytic and is circumscribed by the metal's redox properties. In these copper enzymes, the metal repetitively cycles from a cupro (Cu^{+1}) to a cupric (Cu^{+2}) state as part of the catalytic cycle. It is, however, those same properties that dictate the need to control its availability in cells. Copper and other redox active metals can participate in the catalytic production of oxygen and other radical species capable of damaging biomolecules (Ercal et al. 2001). Copper bound to inappropriate binding sites in proteins has been postulated to play a role in many diseases through the production of radicals (Ercal et al. 2001).

When copper is specifically incorporated into enzyme and sensory protein structure, the redox capability of the ion is controlled.

Limiting the intracellular concentration is the principle function of copper sensing proteins as opposed to increasing the concentration. There are multiple ways by which copper can be brought into cells, both specific and nonspecific. Exportation and sequestration mechanisms are used to decrease the intracellular concentrations of the element. Exportation mechanisms, the most frequently used method in bacteria, utilize copper ATPases to translocate the ion out of the cell (Lu et al. 1999). Sequestration methods are varied and include methods in which the metal is imported into vacuoles, in some case as peptide complexes, or chelated by peptides and proteins. Regardless of which mechanism is utilized to relieve or prevent intoxication, the copper ion itself or a chelate of it becomes specifically bound to the sensor protein that controls the exportation or sequestration method. Acting at a DNA-binding level, the sensor proteins function as either negative repressors or positive-type transcription factors. For example, in the repressor case, copper binding lessens its affinity for DNA and, thereby, enables the transcription of the exporting ATPases when the protein-DNA complex dissociates (Cobine et al. 1999, 2002b). In the transcription factor case, copper binding increases the affinity of the protein for DNA and the active recruitment of the transcription complex (Dameron et al. 1993; Dobi et al. 1995).

Sensory Proteins Repressors

CopY, a Zn(II)-requiring but Cu(I)-regulated repressor from *Enterococcus hirae*, a gram-positive bacterium, is the archetype copper-regulated repressor (Strausak and Solioz 1997). CopY and the *Lactococcus lactis* homologue CopR are in the penicillinase repressor BlaI/MecI/CopY family of winged-helix repressors that bind DNA as dimers (Cantini et al. 2009; Portmann et al. 2006). Each of these members has a characteristic helix-turn-helix DNA-binding motif that is responsible for binding to DNA. Monomers in this family have a limited affinity for DNA (Gregory et al. 1997). CopY is the key regulatory component of the *E. hirae cop* operon and all related operons. The *E. hirae* operon contains CopY, a repressor; CopZ, a copper chaperone; CopA, a copper ATPase (import); and CopB, a copper ATPase (export) (Lu and Solioz 2001; Wunderli-Ye and Solioz 1999). The minimum requirement for a *cop* type operon is a CopY homologue and a CPX copper ATPase (Solioz and Vulpe 1996). Some *cop*-type operons do not contain CopZ and may also not have one of the ATPases. Analogous copper importing ATPases that use methionine sequences to bind copper are found in Ctr1 of ▶ *Saccharomyces cerevisiae* (Rubino et al. 2010) and hCtr1 of humans (Sharp 2003).

When in excess of the cell's needs, copper, as Cu(I), displaces the Zn(II) in CopY, leading to a decrease in the affinity for the promoter (Cobine et al. 1999, 2002a). CopY isolates with a Zn(II) to protein stoichiometry of 1:1 (Cobine et al. 2002b). The metal binding site at the C-terminus of the protein contains a single –CxCxxxxCxC-site. Zn(II) binding is essential for the formation of the DNA-binding active dimers (Cobine et al. 2002b). The dimerization takes place through the interaction of the two metal binding sites, but adjacent sequences containing regularly spaced hydrophobic residue repeats also contribute to the overall affinity (Pazehoski et al. 2008, 2011). Only the C-terminal 38 residues that contain the 10-residue metal-binding motif and the aliphatic repeats are needed for dimerization of CopY or other proteins to which it is fused (Pazehoski et al. 2011). The aliphatic repeats do not take the form of a leucine zipper. The formal stoichiometry is $Zn(II)_2(CopY)_2$. Similarly, when copper(I) displaces the Zn(II) from CopY, the final copper to protein stoichiometry is 2:1, but since the protein is still a dimer, the formal stoichiometry is $Cu(I)_4(CopY)_2$. The displaced Zn(II) does not copurify with the protein. In *E. hirae*, Cu(I)CopZ transfers Cu(I) to the Zn(II)CopY dimer (Cobine et al. 1999, 2002b). Cu(I)CopZ will also interact with and transfer Cu(I) to a homologous metal-binding domain of CopA (Banci et al. 2003). The displacement of one 4-coordinate Zn(II) by two 3-coordinate Cu(I) ions is postulated to promote the change in CopY's structure and, thereby, decrease its affinity for DNA. Derepression leads to the increased synthesis of the exporting copper ATPases and elevated export of copper (Lu et al. 1999, 2003; Solioz and Stoyanov 2003). As the intracellular copper concentration drops, newly synthesized Zn(II)CopY binds to the promoter and the synthesis of the ATPase is decreased and export slows. There is also evidence that high intracellular copper increases the turnover of Cop proteins through the increased synthesis of proteases (Solioz 2002).

Transcription Factors

The ▶ *Saccharomyces cerevisiae* transcription factor Ace1 is the archetype copper-regulated transcription factor. Copper(I) binding by Ace1, and close homologs such as Amt1, increases its affinity for the promoter. The distinct differences between CopY and Ace1 is that the copper binding to CopY causes it to lose affinity for DNA while copper binding to Ace1 causes it to have an increased affinity for DNA. In both situations, there is an increased synthesis of proteins involved in the copper detoxification process. Once bound to the DNA, the acidic N-terminal domain of Ace1 supports the recruitment of other transcription factors leading to an increase the synthesis of the yeast CUP1 (▶ Metallothioneins and Mercury, ScMT) which can ameliorate copper intoxication (Dameron et al. 1991; Winge et al. 1994). ▶ Metallothioneins (MT) are small cysteine-rich proteins that bind copper, and other transition metals, very tightly, making the metal less accessible (Coyle et al. 2002). Aside from their sequestration function, the MTs are proposed to be involved in a number of other metal homeostatic processes including serving as ▶ metallochaperones (Coyle et al. 2002). There are analogous proteins and pathways by which mammalian (human) MT production is increased, but the mechanism is more complex and not fully delineated (Samson and Gedamu 1998). The increased synthesis of the MT is only one of several ways that the yeast protects itself from copper intoxication. Export of copper is the primary protective mechanism. Ace1 and Amt1 are also involved in responses to oxidative stress, such as could be caused by adventitiously bound copper or organic oxidizers (Thorvaldsen et al. 1993; Zhou and Thiele 1993). The Ctr family of ATPases is transcriptionally regulated by an analogous protein called Mac1 (Xiao et al. 2004).

Copper-Binding Motifs

Cysteine-rich sequences are the hallmark of the metal-binding motifs in regulatory proteins such as CopY, Ace1, and Mac1, as well as the sequestration peptides and proteins such as the phytochelatins and ▶ metallothioneins. Ligation of the metals in these proteins is accomplished entirely through the cysteine residues. The cysteine content of the sequences within the metal-binding domains of these proteins is frequently between 20% and 30% and can be higher (Winge et al. 1994). The metals are bound in polythiolate clusters with a mixture of terminal and bridging cysteinyl thiolate ligands (Brown et al. 2002; Dameron et al. 1993; George et al. 1986; Zhang et al. 2008). X-ray absorption spectroscopy (XAS), crystallography, and NMR studies have been used to characterize these sites with a variety of metal ions in them. Zn(II), as expected, is bound in a 4-coordinate manner in sites with tetrahedral geometry (Cobine et al. 2002b). EXAFS indicates the average Zn-S distance in CopY to be 2.349 ± 0.005 Å, a typical value for complexes of this type. In biological systems, copper can be cuprous, Cu(I), a d^{10} electronic configuration, or cupric, Cu(II), a d^9 configuration. The proportion of Cu(I) versus Cu(II) is very dependent on ligands and anions present and solvent interactions (Cotton et al. 1999). In catalytic sites, the ligand structures are arranged to facilitate the redox changes between the Cu(I) and the Cu(II) which are typically part of the enzyme's catalytic cycle. In the sensor proteins copper is bound in the cuprous (Cu(I)) form, which has a preference for softer ligands like S. The distinction between Cu(I) and Cu(II) is significant because it profoundly affects the spectral properties; Cu(I) is diamagnetic so EPR spectroscopy cannot be applied to its complexes to determine the ligand structure around the metal. Also absent are the d-d transitions that lead to the blue color of Cu(II) complexes and are useful in characterizing the structure and electronic properties of the metal binding site. Cu(I)-thiolates do have weak-charge transfer bonds, and if the clusters are sufficiently shielded from solvent, they are weakly luminescent (Green et al. 1994). The principal tools for the study of these sites are XAS (XANES and EXAFS) and structural determinations by x-ray crystallography and NMR. Based on the analysis of a number of Cu(I)-thiolate centers by XAS and crystallographic studies, the Cu(I)s in these sensor proteins are bound in a 3-coordinate manner with trigonal planar geometry (Pickering et al. 1993). Detailed comparisons of the XAS of these proteins with control compounds containing mixed 2- and 3-coordination suggest that the site can have mixed coordination (Pickering et al. 1993). Cu(I)-thiolates can form 2-, 3-, and 4-coordinate complexes (Cotton et al. 1999). EXAFS analyses show that the metals ions in these clusters are close to van der Waals contact; the average Cu-Cu and Cu-S distances were 2.685 ± 0.0085 Å and 2.256 ± 0.0072 Å in CopY (Cobine et al. 2002a).

The clusters are very compact and thermodynamically stable, but isotope exchange studies show that the metals are kinetically labile; the metals can exchange rapidly. A number of computational studies have been applied to these clusters to investigate their properties (Ahte et al. 2009; Sivasankar et al. 2007).

Cross-References

▶ Metallochaperone
▶ Metallothioneins and Mercury
▶ Zinc Storage and Distribution in S. cerevisiae

References

Ahte P, Palumaa P, Tamm T (2009) Stability and conformation of polycopper-thiolate clusters studied by density functional approach. J Phys Chem A 113:9157–9164

Banci L, Bertini I, Ciofi-Baffoni S, Del Conte R, Gonnelli L (2003) Understanding copper trafficking in bacteria: interaction between the copper transport protein CopZ and the N-terminal domain of the copper ATPase CopA from *Bacillus subtilis*. Biochemistry 42:1939–1949

Brown KR, Keller GL, Pickering IJ, Harris HH, George GN, Winge DR (2002) Structures of the cuprous-thiolate clusters of the Mac1 and Ace1 transcriptional activators. Biochemistry 41:6469–6476

Cantini F, Banci L, Solioz M (2009) The copper-responsive repressor CopR of *Lactococcus lactis* is a 'winged helix' protein. Biochem J 417:493–499

Cobine P, Wickramasinghe WA, Harrison MD, Weber T, Solioz M, Dameron CT (1999) The *Enterococcus hirae* copper chaperone CopZ delivers copper(I) to the CopY repressor. FEBS Lett 445:27–30

Cobine PA, George GN, Jones CE, Wickramasinghe WA, Solioz M, Dameron CT (2002a) Copper transfer from the Cu(I) chaperone, CopZ, to the repressor Zn(II)CopY: metal coordination environments and protein interactions. Biochemistry 41:5822–5829

Cobine PA, Jones CE, Dameron CT (2002b) Role for zinc(II) in the copper(I) regulated protein CopY. J Inorg Biochem 88:192–196

Cotton FA, Wilkinson G, Murillo CA, Bochmann M (1999) Advanced Inorganic Chemistry. Wiley, New York

Coyle P, Philcox JC, Carey LC, Rofe AM (2002) Metallothionein: the multipurpose protein. Cell Mol Life Sci 59:627–647

Dameron CT, Winge DR, George GN, Sansone M, Hu S, Hamer D (1991) A copper-thiolate polynuclear cluster in the ACE1 transcription factor. Proc Natl Acad Sci USA 88:6127–6131

Dameron CT, George GN, Arnold P, Santhanagopalan V, Winge DR (1993) Distinct metal binding configurations in ACE1. Biochemistry 32:7294–7301

Dobi A, Dameron CT, Hu S, Hamer D, Winge DR (1995) Distinct regions of Cu(I)-ACE1 contact two spatially resolved DNA major groove sites. J Biol Chem 270:10171–10176

Ercal N, Gurer-Orhan H, Aykin-Burns N (2001) Toxic metals and oxidative stress part I: mechanisms involved in metal-induced oxidative damage. Curr Top Med Chem 1:529–539

George GN, Winge D, Stout CD, Cramer SP (1986) X-ray absorption studies of the copper-beta domain of rat liver metallothionein. J Inorg Biochem 27:213–220

Green AR, Presta A, Gasyna Z, Stillman MJ (1994) Luminescent probe of copper-thiolate cluster formation within mammalian metallothionein. Inorg Chem 33:4159–4168

Gregory P, Lewis R, Curnock S, Dyke K (1997) Studies of the repressor (Bla1) of beta-lactamase synthesis in *Staphylococcus aureus*. Mol Microbiol 24:1025–1037

Lu ZH, Solioz M (2001) Copper-induced proteolysis of the CopZ copper chaperone of *Enterococcus hirae*. J Biol Chem 276:47822–47827

Lu ZH, Cobine P, Dameron CT, Solioz M (1999) How cells handle copper: a view from microbes. J Trace Elem Exp Med 12:347–360

Lu ZH, Dameron CT, Solioz M (2003) The *Enterococcus hirae* paradigm of copper homeostasis: copper chaperone turnover, interactions, and transactions. Biometals 16:137–143

Pazehoski KO, Collins TC, Boyle RJ, Jensen-Seaman MI, Dameron CT (2008) Stalking metal-linked dimers. J Inorg Biochem 102:522–531

Pazehoski KO, Cobine PA, Winzor DJ, Dameron CT (2011) Evidence for involvement of the C-terminal domain in the dimerization of the CopY repressor protein from *Enterococcus hirae*. Biochem Biophys Res Commun 406:183–187

Pickering IJ, George GN, Dameron CT, Kurz B, Winge DR, Dance IG (1993) X-ray absorption spectroscopy of cuprous-thiolate clusters in proteins and model systems. J Am Chem Soc 115:9498–9505

Portmann R, Poulsen KR, Wimmer R, Solioz M (2006) CopY-like copper inducible repressors are putative 'winged helix' proteins. Biometals 19:61–70

Rubino JT, Riggs-Gelasco P, Franz KJ (2010) Methionine motifs of copper transport proteins provide general and flexible thioether-only binding sites for Cu(I) and Ag(I). J Biol Inorg Chem 15:1033–1049

Samson SL, Gedamu L (1998) Molecular analyses of metallothionein gene regulation. Prog Nucleic Acid Res Mol Biol 59:257–288

Sharp PA (2003) Ctr1 and its role in body copper homeostasis. Int J Biochem Cell Biol 35:288–291

Sivasankar C, Sadhukhan N, Bera JK, Samuelson AG (2007) Is copper(i) hard or soft? A density functional study of mixed ligand complexes. New J Chem 31:385–393

Solioz M (2002) Role of proteolysis in copper homoeostasis. Biochem Soc Trans 30:688–691

Solioz M, Stoyanov JV (2003) Copper homeostasis in *Enterococcus hirae*. FEMS Microbiol Rev 27:183–195

Solioz M, Vulpe C (1996) CPx-type ATPase – a class of P-type ATPases that pump heavy metals. Trends Biol Sci 21:237–241

Strausak D, Solioz M (1997) CopY is a copper-inducible repressor of the *Enterococcus hirae* copper ATPases. J Biol Chem 272:8932–8936

Thorvaldsen JL, Sewell AK, McCowen CL, Winge DR (1993) Regulation of metallothionein genes by the ACE1 and AMT1 transcription factors. J Biol Chem 268:12512–12518

Winge DR, Dameron CT, George GN (1994) The metallothionein structural motif in gene expression. Adv Inorg Biochem 10:1–48

Wunderli-Ye H, Solioz M (1999) Copper homeostasis in *Enterococcus hirae*. Adv Exp Med Biol 448:255–264

Xiao Z, Loughlin F, George GN, Howlett GJ, Wedd AG (2004) C-terminal domain of the membrane copper transporter Ctr1 from *Saccharomyces cerevisiae* binds four Cu(I) ions as a cuprous-thiolate polynuclear cluster: sub-femtomolar Cu(I) affinity of three proteins involved in copper trafficking. J Am Chem Soc 126:3081–3090

Zhang L, Pickering IJ, Winge DR, George GN (2008) X-ray absorption spectroscopy of cuprous-thiolate clusters in *Saccharomyces cerevisiae* metallothionein. Chem Biodivers 5:2042–2049

Zhou P, Thiele DJ (1993) Copper and gene regulation in yeast. Biofactors 4:105–115

Copper-Containing Hydroxylase

▶ Copper, Mononuclear Monooxygenases

Copper-Containing Monooxygenase

▶ Copper, Mononuclear Monooxygenases

Copper-Containing Nitrite Reductase

▶ Nitrite Reductase

Copper-Sulfur Biological Center

▶ Nitrous Oxide Reductase

Copper-Thioneins

▶ Metallothioneins and Copper

Copper-Zinc Superoxide Dismutase

▶ Copper-Zinc Superoxide Dismutase and Lou Gehrig's Disease

Copper-Zinc Superoxide Dismutase and Lou Gehrig's Disease

Herman Louis Lelie and Joan Selverstone Valentine
Department of Chemistry and Biochemistry,
University of California, Los Angeles, CA, USA

Synonyms

ALS; Amyotrophic lateral sclerosis; Copper-zinc superoxide dismutase; Lou Gehrig's disease; Motor neuron disease; SOD1. UniProtKB/Swiss-Prot P00441 (SODC_HUMAN)

Definition

Copper-zinc superoxide dismutase (SOD1) is an antioxidant metalloenzyme that is present throughout the body. It catalyzes the conversion of two molecules of superoxide into hydrogen peroxide and molecular oxygen in a reaction that involves alternate oxidation and reduction of the active site copper ion. Amyotrophic lateral sclerosis (ALS, Lou Gehrig's disease) is a fatal neurodegenerative disease characterized by progressive loss of motor neurons, leading inevitably to paralysis and death, usually within 3–5 years. Onset of disease is usually after age 50, and there is no cure. Most cases have no known genetic component, but in a small fraction of cases, ALS is caused by point mutations in SOD1. This linkage of SOD1 and ALS, first published in 1993, provided the first experimental route to understanding the cause of the disease.

Amyotrophic Lateral Sclerosis (Lou Gehrig's Disease)

Amyotrophic lateral sclerosis (ALS), commonly referred to as Lou Gehrig's disease because of the famous New York Yankee's baseball player whose career was ended by the disease, is the most common adult motor neuron disease. Originally described by the French neurobiologist and physician Jean-Martin Charcot in 1869, ALS is characterized by the progressive degeneration of spinal motor neurons leading to atrophy of voluntary muscles and spasticity, resulting in paralysis and relatively rapid death (Cleveland and Rothstein 2001). Some cognitive impairment is found in about 50% of cases, but ocular and sensory neurons are generally unaffected by disease. The average survival time is 3–5 years although much longer disease courses have been reported. ALS affects most patients in middle to late life with an incidence rate of 2 per 100,000 but a lifetime risk of 1 in 1,000.

Research efforts directed toward the understanding of ALS during most of the twentieth century involved epidemiological studies, which, although numerous, were overall inconclusive. The more recent breakthroughs in the genetics of ALS have driven the bulk of the research during the past two decades. However, despite tremendous efforts in research, there is still no cure for ALS, and treatments and clinical trials so far have only provided marginal beneficial effects.

The progressive uncovering of epidemiological and genetic factors has had the greatest impact in the area of ALS research. Some of the better known studies addressed the causes of an ALS epidemic on the island of Guam, where in the 1950s, it was discovered that the ALS frequency was 50 times higher than that of the rest of the world (Lelie et al. 2011). While a genetic factor was first hypothesized, it was later disproven when follow-up studies showed that the incidence rate rapidly decreased coinciding with changing environmental conditions including diet. Potential factors suspected to be involved include environmental toxins such as heavy metals and the neurotoxin beta-N-methylamino-alanine found in local food sources.

In 1991, geneticists identified a linkage to ALS in a gene somewhere on chromosome 21, and 2 years later, this gene was identified as the gene that encodes for SOD1 protein. The majority of ALS research since 1993 has focused on SOD1 in two ways: as a model of the disease in the form of genetically altered mice overexpressing mutant SOD1 and in efforts to determine how mutations in the SOD1 protein alter its properties and ultimately lead to ALS.

Copper-Zinc Superoxide Dismutase (SOD1)

SOD1 was first isolated from red blood cells in 1938 and, due to its characteristic copper binding property and ubiquitous presence in erythrocytes and liver tissue, it was initially termed haemocuprein, hepatocuprein, or erythrocuprein, depending upon its tissue source. In 1969, scientists J.M. McCord and I. Fridovich published their discovery that it catalyzed the disproportionation of superoxide, whereby two molecules of superoxide are converted into to hydrogen peroxide and dioxygen, and they named the enzyme superoxide dismutase. The substrate superoxide is classified as a reactive oxygen species (ROS), a reactive molecule derived from molecular oxygen that can lead to cellular oxidative damage either directly or indirectly. Superoxide is produced in small amounts as a byproduct of mitochondrial respiration and by some other reactions in vivo. Superoxide can exert its toxicity directly by inflicting damage to labile iron sulfur clusters of iron-containing proteins such as aconitase. However, its role as the starting point for the generation of other, more reactive ROS, such as hydroxyl radical or peroxynitrite, is probably more important (Valentine et al. 2005). The toxicity of superoxide is exploited by the immune system, which intentionally produces superoxide in macrophages using NADPH oxidase enzymes to destroy invading pathogens. Superoxide dismutase is included in a group of protective enzymes induced by various pro-oxidant stimuli and toxic compound such as 2,3,7,8-tetrachlorodibenzo-*p*-dioxin (TCDD), *para*-methoxyamphetamine (PMA), and ultraviolet radiation. Superoxide dismutase is regarded as a first line of defense against the damaging effects of reactive oxygen species and is present in every cell in the body.

Copper-zinc superoxide dismutase has been well characterized since its first description in 1938. SOD1 is an abundant enzyme, being present at concentrations up to the hundreds of micromolar in human cells, with a well-conserved sequence preserved across all eukaryotes. The human protein monomer is a 153-amino-acid polypeptide, and the protein is predominantly found in the cytosol, nucleus, peroxisomes, and mitochondria intermembrane space. To reach its fully mature homodimeric state, SOD1 undergoes several posttranslational steps besides the usual removal of its initiating methionine residue and N-terminal acetylation, SOD1 also acquires a copper and a zinc ion, forms an intrasubunit disulfide bond, and its two subunits dimerize.

SOD1 and ALS

The discovery of the link between SOD1 and ALS in 1993 was very exciting and has helped propel research on both the protein and the disease. In particular, it led to the first animal model of the disease in the form of transgenic mice and rats that overexpress mutant forms of SOD1 and exhibit ALS-like symptoms. Ten percent of all ALS cases occur by a genetic component, and approximately 20% of these familial cases (fALS) are linked to mutations in the SOD1 protein. Over 100 different mutations occurring along the length of the protein have been identified, mainly from single amino acid substitutions, but also including deletions, insertions, and C-terminal truncations. Interestingly, humans are not the only species to develop a myelopathy from SOD1, mutations in canine SOD1 have also been associated with ALS in dogs. The initial hypothesis for why mutations in SOD1 lead to disease was simply that a loss of dismutase activity resulted in toxicity due to excess ROS. However, soon after the genetic discovery, several studies demonstrated that the disease is not due to a loss-of-function but rather a gain-of-function (Cleveland and Rothstein 2001). First, the mutations are genetically dominant, meaning that the disease occurs even though at least 50% of the normal activity of SOD1 is present. In addition, most of the mutated proteins do not lose activity. Secondly, transgenic mice lacking all SOD1 do not develop the disease, while mouse models overexpressing mutant SOD1 fall ill with disease phenotypes even though they retain their own endogenous wild-type mouse SOD1. Thirdly, mouse models overexpressing both the dismutase-active and dismutase-inactive mutant proteins induce the disease phenotype with similar onsets and survival time.

The creation of these mouse models was invaluable not only because it established the gain-of-function feature of SOD1 mutations but also because it provided the first experimental platform to investigate the causes of ALS, particularly at early stages of the disease, to delineate the features of ALS pathology and to test possible treatments. The mouse models were used to investigate further the nature of the gain-of-function property of SOD1 as well as to elucidate the cellular pathologies that lead to motor neuron degeneration. Several hypotheses have been proposed for the gain-of-function property, including a pro-oxidant mechanism resulting from aberrant copper chemistry; however, other studies showed that a subset of mutants do not even bind copper, making this hypothesis less popular. Currently, the most widely accepted hypothesis centers on the misfolding, aberrant oligomerization, and aggregation of SOD1 mutants, bringing ALS into the larger family of neurodegenerative diseases that are due to aberrant protein aggregation (Valentine and Hart 2003). The central basis for this hypothesis comes from the appearance of large SOD1 inclusions in the ventral horn of both human fALS and transgenic mice spinal cords. To gain insight into how SOD1 aggregates, researchers have narrowed in on its folding, structure, and function, and how mutation affects these steps. This entry reviews the central aspects of SOD1 structure, function, and maturation as they relate to the disease and then highlights the pertinent pathological features learned from studies using animal models.

Structural Insights

SOD1 is an extremely stable dimeric protein that has a conserved eight-stranded Greek key beta-barrel structure, one zinc and one copper ion bound per subunit, and a single intrasubunit disulfide bond (Fig. 1). The zinc ion and disulfide bond are structurally important, and the copper ion is critical for enzymatic activity. The arrangement of the beta strands is shown in Fig. 2a where the beta-barrel structure, normally rolled up to form a closed barrel, is shown flattened out (Chattopadhyay and Valentine 2009). In the case of SOD1, there are eight antiparallel beta strands, which form a hydrophobic core with a flattened beta-barrel-like structure (Fig. 2b). Two main loop elements, the electrostatic and zinc loops (shown in teal and orange, respectively, in Fig. 1 and purple and light blue in Fig. 2), are important in metal binding and enzymatic activity. The zinc loop is held in place by the zinc ion, which is bound through one aspartyl side chain, Asp83, and three histidyl side chains, His80, His71, and His63 (shown in red in Fig. 1). His63 is a shared ligand and forms an imidazolate bridge to the copper ion. Thus, the copper ion remains in close proximity to the zinc but binds to SOD1 via three histidyl side chains located on beta

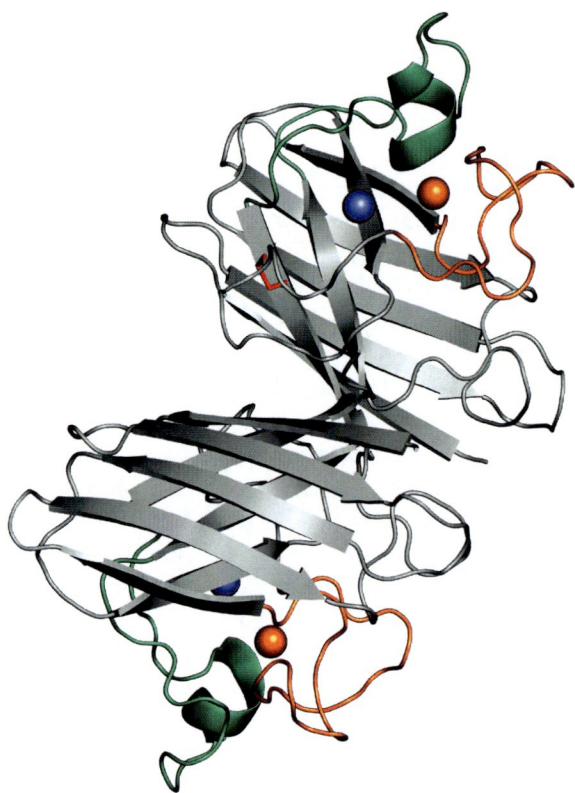

Copper-Zinc Superoxide Dismutase and Lou Gehrig's Disease, Fig. 1 Structure of dimeric metal bound SOD1. Copper is shown as a *blue sphere* and zinc as an *orange sphere*. The electrostatic loop is in *teal* and the zinc loop is in *orange*. The intrasubunit disulfide bond is shown as a *red connecting line*

strands 4 and 7, His46, His48, and His120 (shown in green in Fig. 1). The electrostatic loop serves two purposes: to provide a charged surface that guides the superoxide anion toward the catalytic copper center and to strengthen and stabilize the metal-binding region by creating a hydrogen bonding network between the metal-coordinating residues and the electrostatic loop residues Asp124, Gly141, and Arg143. The intrachain disulfide bond links residues Cys57 and Cys146, pinning the loop containing Cys 57 to beta-barrel strand 8 and stabilizing residues important for formation of the dimer interface. The same Arg143 that stabilizes the metal-binding site in the electrostatic loop also retains a hydrogen bond with Cys57, hence linking the important structural features of disulfide bonding, metal binding, and dimerization. Folding of the protein, binding of metal ions, formation of the disulfide bond, and dimerization are critical maturation events for the formation of the active enzyme (see below for more information). The structural interrelationships between these features lead to interesting features, many of which contribute to the extreme stability of the protein (Valentine et al. 2005). For example, the apoprotein is able to dimerize only if the disulfide bond is formed or if a metal is bound. Additionally, copper or zinc binding stabilizes the disulfide bond, making it less prone to reduction, which is vital in the reducing environment of the cytosol. Similarly, the disulfide bond contributes to enzymatic activity by strengthening the metal-binding region. The different structural features of SOD1 work in a synergistic manner to produce an extraordinarily stable protein.

Mature human SOD1 is among the most stable enzymes in the body. In its fully metallated, disulfide-oxidized, dimeric form, wild-type SOD1 (WT) has a melting temperature above 90°C and is resistant to 8 M urea, 1% sodium dodecyl sulfate, and proteolytic degradation. It retains enzymatic activity even after 1 h in 4% sodium dodecyl sulfate and is believed to have a very long half-life in vivo (Valentine et al. 2005). A central factor in driving this stability is the aforementioned posttranslational events, as is apparent by the 48°C difference in melting temperature between the disulfide-oxidized holo form and the apo (metal-free) disulfide-reduced form, which melts (unfolds) at 42°C (Shaw and Valentine 2007).

Compromising the intricate structural interrelationships of these posttranslational modifications leads to an overall less stable protein, providing a possible explanation for how mutations might lead to protein misfolding and consequent toxicity. Initially, many findings supported this hypothesis. For instance, many familial ALS mutant proteins, such as A4V, I113T, G93C, G37R, G41D, and G85R, have significantly reduced half-lives relative to the WT protein in cell culture systems. Additionally, many mutants, particularly those that have impaired metal binding, such as S134N, D125H, D124V, and H46R, have much lower melting temperatures, as measured by differential scanning calorimetry (DSC), because they are not properly stabilized by metal ions. Another class of mutants, termed "wild-type-like" (WTL) mutants, have similar metal-binding properties to wild-type SOD1 and are relatively stable when metallated but still exhibit reduced stability in their apo states relative to the wild-type protein. However, certain mutants do not follow these trends, for instance, D101N has

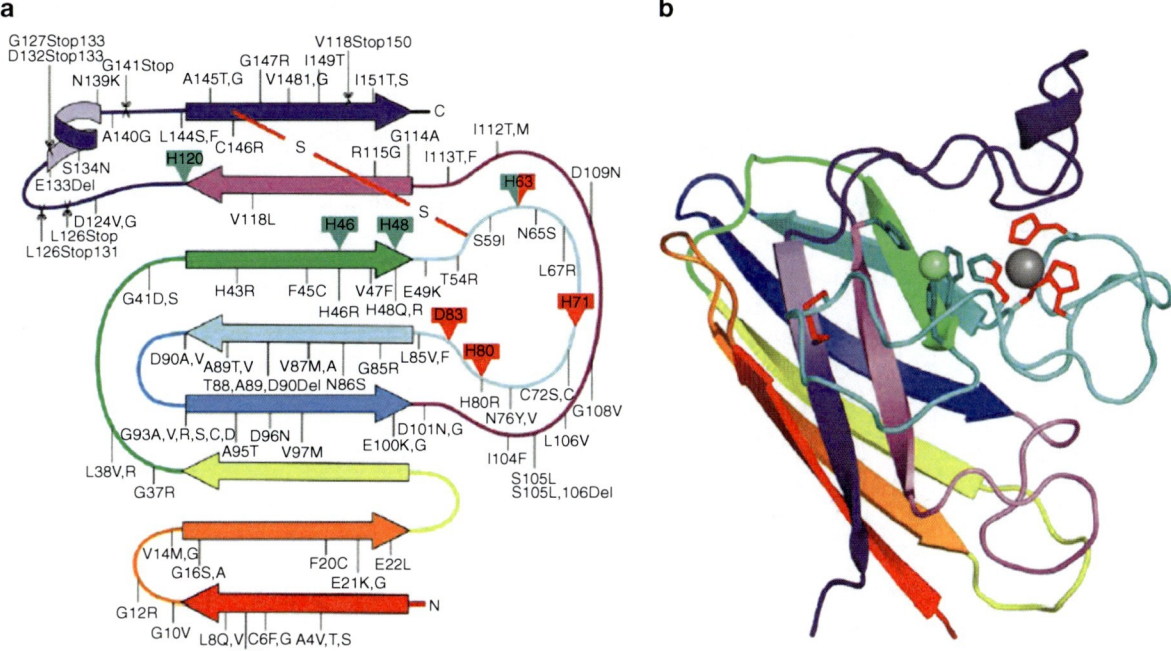

Copper-Zinc Superoxide Dismutase and Lou Gehrig's Disease, Fig. 2 Secondary structure representation and colorcoded structure of one subunit of SOD1. (**a**) The secondary structure diagram shows the beta barrel unrolled. The location of many of the ALS-causing SOD1 mutations is indicated, as are other important residues. The disulfide bond is shown as a red line. Zinc-binding residues are highlighted in red, and copperbinding residues are highlighted in green. (**b**) Backbone structure of one subunit of the dimeric SOD1 protein, color coded to match the flattened Fig. a. Copper and zinc are shown as green and grey spheres, respectively. The metal-binding ligands are fully drawn, and the disulfide bond is indicated in red

a melting temperature that is very similar to the wild type in the various states of metallation and disulfide status. Attempts to classify the mutants in terms of stability, activity, and metal-binding properties have helped define the effects of each mutation but have yet to elicit an obvious common misfolding mechanism. Taken together, it remains unclear if all mutations affect a common critical aspect of SOD1 structural stability that causes them to form aggregates (Valentine et al. 2005). More recent hypotheses for how mutations can lead to misfolded SOD1 include alterations in protein net charge, aberrant oxidative modifications, decreased metal-binding affinities, impaired folding and/or maturation (see below), or a combination of these four factors (Shaw and Valentine 2007).

Functional Insights

SOD1 functions as an important modulator of reactive oxygen species. SOD1 is expressed ubiquitously in human cells, and within these cells, it is found in the cytosol, nucleus, and mitochondrial intermembrane space, where it catalyzes the disproportionation of two superoxide anions into hydrogen peroxide and dioxygen.

SOD Reaction

(1) $O_2^- + Cu(II)ZnSOD \rightarrow O_2 + Cu(I)ZnSOD$
(2) $O_2^- + Cu(I)ZnSOD + 2H^+ \rightarrow H_2O_2 + Cu(II)ZnSOD$
(3) $Sum : 2O_2^- + 2H^+ \rightarrow H_2O_2 + O_2$

Copper is at the heart of the enzymatic ping-pong mechanism. In the two-step reaction, first, a superoxide anion reduces the cupric ion to form dioxygen (step (1)). Copper makes a 1.3 Å shift, breaking its bond with the bridging histidine, and the resulting cuprous ion is reorganized from a distorted square planar geometry, coordinated by four ligands, to a nearly trigonal planar three-coordinate configuration (Fig. 3). A second molecule of superoxide reoxidizes the cuprous ion to form hydrogen peroxide

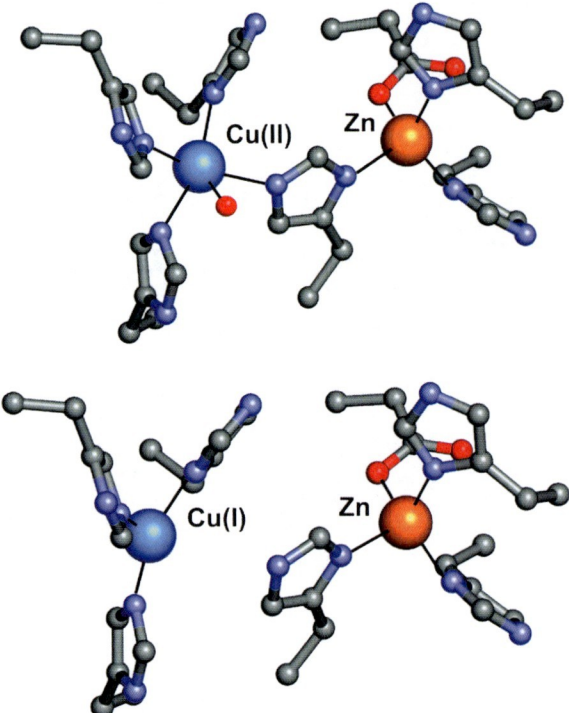

Copper-Zinc Superoxide Dismutase and Lou Gehrig's Disease, Fig. 3 Metal-binding-site geometries. The *top* shows copper in the oxidized state (Cu^{2+}) with an intact imidazolate bridge and a five coordinate square planar geometry to four histidyl side chains and a water. The *bottom* shows that the reduction of copper to Cu^{1+} breaks the bond to the bridging histidine, shifting the copper 1.3 Å *left* to form a nearly trigonal planar geometry

Using *Saccharomyces cerevisiae* (yeast) as a eukaryotic model system, SOD1 was knocked out to investigate the cellular effects of excess superoxide. The observed characteristic phenotypes for these SOD1Δ yeast include sensitivity to millimolar concentrations of zinc, diminished growth rate compared to wild-type yeast in nonfermentable carbon sources, sensitivity to 100% oxygen, and air-dependent amino acid auxotrophies for lysine and methionine. Interestingly, zinc sensitivity cannot be restored even with the introduction of MnSOD but is restored in the presence of an inactive mutant SOD1 (H46C), suggesting that yeast SOD1 might have an additional nondismutase function in zinc metabolism. Human SOD1 has also been suggested to play a role in zinc metabolism (Lelie et al. 2011).

The Maturation of SOD1

The steps along the maturation pathway for SOD1 have garnered increased interest with regard to understanding the mechanism of SOD1 misfolding and aggregation in ALS. Zinc acquisition by SOD1 is believed to be an early and important event that stabilizes the nascent polypeptide as it comes off the ribosome, but the detailed mechanism is unknown. Copper acquisition by SOD1 is better understood. It is mainly delivered to SOD1 specifically by the copper chaperone for SOD1 (CCS), which also serves to catalyze the formation of the intrasubunit disulfide bond.

The intramolecular disulfide bond in SOD1 serves an important role in stabilizing the quaternary structure by anchoring the loop from Glu49 to Asn53 thereby stabilizing the dimer interface. Interestingly, it has been reported that disulfide bonds in SOD1 mutants are more susceptible to reduction than those in WT, suggesting a possible mechanism whereby mutants are more prone to misfolding than the WT (Valentine et al. 2005).

It is unknown whether SOD1 dimerization occurs before or after the other posttranslational steps since it can occur in both the apo, disulfide-oxidized protein or in the partially metallated disulfide-reduced protein. Thus, there is a possibility that dimerization can occur at more than one point along the maturation pathway but it is probably dependant on the local concentration of SOD1. While the similarity in the overall thermodynamic stability of mutant and WT SOD1 does not reveal an obvious misfolding

(step (2)) and reestablishes the copper bond to the ε-nitrogen of His63. The reaction is highly efficient, occurring at near diffusion-controlled rates at physiological pH.

Despite the efficiency of SOD1 as a superoxide dismutase, other chemistries have also been associated with the protein. For example, hydrogen peroxide can also aberrantly react with SOD1, resulting in oxidation of the copper histidine ligands and leading to metal ion loss, loss of activity, and decreased protein stability. This mechanism may represent the basis of a possible inactivation and degradation pathway for SOD1 but may also provide an additional route by which the protein might misfold and aggregate (Valentine et al. 2005).

While the biochemical function of SOD1 in reducing intracellular superoxide concentrations has long been known, recent studies have helped shed new light on mechanisms of superoxide toxicity.

mechanism, the multiple steps along the way to mature SOD1 constitute a series of events that could, due to altered kinetics, lead to a greater sampling of off-pathway intermediates and ultimately a route to misfolding.

Misfolded SOD1

The misfolding of SOD1 and its consequent oligomerization and aggregation are now widely believed to underlie the toxic property of mutant SOD1 that causes ALS. Several key findings support this theory: Protein inclusions containing aggregated SOD1 have been identified in familial ALS patients and all mutant SOD1 transgenic mice. Also, formation and accumulation of aggregated protein occurs coincidently with the onset and progression of disease in fALS-model transgenic mice, and detergent-insoluble aggregates containing SOD1 are a unique property of mutant SOD1 but not of WT SOD1 when it is overexpressed in cell culture or in transgenic mice. Proteomic analysis on aggregates isolated from mice indicates that these aggregates contain some ubiquitin and neurofilaments but are dominantly composed of unmetallated full-length SOD1 (Lelie et al. 2011). However, animal studies suggest a role for WT SOD1 in accelerating the disease progression when it is coexpressed with an ALS-mutant SOD1 (Chattopadhyay and Valentine 2009). WT SOD1 may also play a role in sporadic ALS since spinal cord tissue samples from sALS patients consistently contain proteinaceous inclusions that have recently been shown to contain misfolded WT SOD1. Interestingly, immature WT SOD1 has been observed to aggregate and form amyloid fibrils under mild conditions in vitro. Together, these data highlight the relationship between SOD1, aggregation, and ALS.

While it is clear that mutant SOD1 can aggregate, it remains a mystery whether aggregation is truly the cause of toxicity and, if so, what form of aggregate (oligomer, amyloid, etc.) is toxic and how is toxicity exerted. In general, aggregates might be toxic in a number of ways: by interfering with normal cell function, depleting essential chaperones, overwhelming the proteasome, or inhibiting mitochondrial function (Cleveland and Rothstein 2001). A major portion of current research focuses on understanding the aggregation paradigm. This includes elucidating the folding, structural, and functional consequences of mutations in SOD1 that can possibly lead to misfolding in the cell. Extensive biophysical characterization of many SOD1 mutants has not revealed an obvious single converging pathway toward misfolding. Though the misfolding theory has the strongest support, it remains imperative to investigate all possible angles of disease including other pathological features present in tissues, which are not directly related to misfolded SOD1. In this light, the mutant SOD1 transgenic mice models have rendered tremendous insight into pathogenic mechanisms of ALS.

SOD1 and Sporadic ALS

Cases of familial ALS linked to SOD1 constitute only ~2% of all ALS cases, raising two questions: What is (are) the cause(s) in the other 98% of cases? What is the role of SOD1, if any, in these other cases? Recent evidence suggests that WT SOD1 might play a role in sporadic ALS cases. Misfolded forms of WT SOD1 are found in sporadic ALS tissue sections, but not controls, using antibodies that selectively recognize misfolded forms of SOD1. Furthermore, immature forms of WT SOD1 can readily fibrillate under physiological conditions in vitro suggesting that in vivo WT SOD1 could play a role in sporadic ALS. Additionally, oxidatively damaged WT SOD1 can misfold and acquire toxic properties similar to mutant SOD1 suggesting that oxidized SOD1 could be involved in sporadic ALS (Lelie et al. 2011). These findings provide hope that the studies on the SOD1 ALS transgenic mice may someday translate into therapeutics for all ALS patients.

ALS Pathology

The pathological clues gained from studies on ALS rodent models and actual human patient tissues have provided a greater understanding of neurodegenerative mechanisms involved and have helped pave the way for the creation of possible therapeutics (Bruijn et al. 2004). Mutations in SOD1 can trigger the series of deleterious events that lead to this midlife disease. Among the events that have been identified and characterized in one or another of the fALS-SOD1 mouse models are oxidative stress, apoptosis, excitotoxicity, neurofilament disorganization, mitochondrial dysfunction, proteasome inhibition, impaired axonal transport, reactive astrocytes, and protein aggregation (For a comprehensive review of these pathological features, please refer to Bruijn et al. 2004; Cleveland and Rothstein 2001; Turner and Talbot 2008). However, many questions remain regarding the significance of these events. For instance, it is unclear which event is the main upstream trigger that leads to the demise of motor neurons. Furthermore, the specific role of mutant SOD1 in each of these pathological abnormalities is unclear but has been attributed to its role as an antioxidant, its aggregation potential, and its partial mitochondrial localization. Therapeutics that target each of these pathological conditions have not provided the hoped-for benefit, confounding the issue of what the upstream event(s) may be. Additionally, the variability of disease manifestations in the different mutant mouse strains (expressing different mutant SOD1 proteins and/or at different levels) makes it difficult to ascertain the significance of each event (Turner and Talbot 2008). An updated theory suggests that ALS possesses a complex etiology reliant on a multifactorial convergence of pathways, all of which can be related to developing disease. The misfolding of SOD1 is regarded as the most pertinent of these factors due to its very early appearance in mice models, and understanding how mutations lead to aggregated protein remains a major priority for the SOD1-ALS field.

Direction of Research

Recent genetic studies have helped expand the ALS field to focus on two new ALS-associated proteins, TDP-43 and FUS. They both play roles in RNA metabolism, although this may be irrelevant to their roles in ALS. Interestingly, TDP-43 is involved in aggregation, and comparing its function and toxicity with that of ALS-SOD1 may help clear up the questions generated by the SOD1 studies regarding (an) underlying pathological mechanism(s).

The link between ALS and SOD1 has provided a tremendous impetus to research in both the disease and the protein. The focus on ALS has stimulated a number of studies leading to much better understanding of the biophysical properties of mutant SOD1. Conversely, the identification of mutations in SOD1 as a cause of fALS led to the first animal models for the disease and a great increase in our knowledge of that disease. While a cure for ALS remains stubbornly elusive, potential treatments are beginning to emerge largely due to the contributions of the ALS-SOD1 mouse studies. A central focus is determining the toxic form of mutant SOD1 and barring any major paradigm shifts, which includes understanding what drives mutant SOD1 to misfold and aggregate. By determining the underlying mechanism involved in SOD1-mediated ALS, researchers will hopefully be able to figure out what causes all forms of this devastating illness.

Cross-References

▶ Biological Copper Transport
▶ Copper-Binding Proteins
▶ Zinc in Superoxide Dismutase

References

Bruijn LI, Miller TM et al (2004) Unraveling the mechanisms involved in motor neuron degeneration in ALS. Annu Rev Neurosci 27(1):723–749

Chattopadhyay M, Valentine JS (2009) Aggregation of copper-zinc superoxide dismutase in familial and sporadic ALS. Antioxid Redox Signal 11(7):1603–1614

Cleveland DW, Rothstein JD (2001) From charcot to lou gehrig: deciphering selective motor neuron death in als. Nat Rev Neurosci 2(11):806–819

Lelie HL, Borchelt DR et al (2011) Metal toxicity and metallostasis in amyotrophic lateral sclerosis. In: Milardi D, Rizzarelli E, Milardi D, Rizzarelli E (eds) Neurodegeneration: metallostasis and proteostasis (Rsc drug discovery). Royal Society of Chemistry, Cambridge, IN PRESS

Shaw BF, Valentine JS (2007) How do ALS-associated mutations in superoxide dismutase 1 promote aggregation of the protein? Trends Biochem Sci 32(2):78–85

Turner BJ, Talbot K (2008) Transgenics, toxicity and therapeutics in rodent models of mutant SOD1-mediated familial ALS. Prog Neurobiol 85(1):94–134

Valentine JS, Hart PJ (2003) Misfolded CuZnSOD and amyotrophic lateral sclerosis. Proc Natl Acad Sci 100(7):3617–3622

Valentine JS, Doucette PA et al (2005) Copper-zinc superoxide dismutase and amyotrophic lateral sclerosis. Annu Rev Biochem 74(1):563–593

Corrin

▶ Cobalt-containing Enzymes

Cotransporters

▶ Sodium/Glucose Co-transporters, Structure and Function

Cowrins

▶ Zinc Metallocarboxypeptidases

CPG₂, Carboxypeptidase G₂ from *Pseudomonas Sp.* Strain RS-16

▶ Zinc Aminopeptidases, Aminopeptidase from Vibrio Proteolyticus (Aeromonas proteolytica) as Prototypical Enzyme

Cr, Element 24, [7440-47-3]

▶ Chromium, Physical and Chemical Properties

Crayfish Small-Molecule Proteinase

▶ Zinc-Astacins

CrCl₃

▶ Chromium and Human Nutrition

CrCl₃.6H₂O

▶ Chromium and Human Nutrition

Cr-DNA Adduction

▶ Chromium Binding to DNA

Creutzfeldt-Jakob Disease

▶ Copper and Prion Proteins

CSD, Cambridge Structural Database

▶ Zinc Aminopeptidases, Aminopeptidase from Vibrio Proteolyticus (Aeromonas proteolytica) as Prototypical Enzyme

CSQ

▶ Calsequestrin

Cu,Zn-SOD

▶ Zinc in Superoxide Dismutase

CuA Center

▶ Nitrous Oxide Reductase

Cu-Containing Respiratory Nitrite Reductase

▶ Nitrite Reductase

Cupredoxin

▶ Plastocyanin

Cuprein

▶ Zinc in Superoxide Dismutase

Cuproproteins

▶ Copper-Binding Proteins

CusCFBA Copper/Silver Efflux System

Megan M. McEvoy and Alayna M. George Thompson
Department of Chemistry and Biochemistry,
University of Arizona, Tucson, AZ, USA

Synonyms

Heavy metal efflux; Metal homeostasis; Metal-resistant bacteria

Definitions

Tripartite efflux complexes are protein export systems that span both membranes in Gram-negative bacteria and are involved in efflux of compounds from the cell. These systems are composed of three proteins: an inner membrane protein, an outer membrane protein, and a periplasmic adaptor that connects the two membrane proteins.

RND proteins are Resistance-nodulation-and cell division proteins. These are membrane proteins that expel compounds from bacterial cells using the protein gradient as an energy source. In Gram-negative tripartite efflux complexes, RND proteins are the inner membrane component.

MFPs are membrane fusion proteins; they function as the adaptors in the tripartite efflux complexes. They are also called periplasmic adaptor proteins.

Periplasm is the region between the inner and outer membranes in Gram-negative bacteria.

Introduction

Several transition metals are necessary as trace elements to carry out reactions inside of living cells. However, transition metals can be extremely reactive and cause cellular damage or even death when present in high concentrations. This toxicity allows the use of metals as biocides in clinical or industrial settings. In particular, copper and silver are commonly used to control microbial populations in hospital settings such as burn wards. Bacteria have developed elaborate protein systems that allow for the maintenance of homeostasis in response to copper or silver stress. This entry focuses on the CusCFBA system in *Escherichia coli*, which mediates Cu(I) homeostasis and Ag(I) resistance.

Copper ions are used in many cellular processes such as respiration and oxidative damage response. The ability of copper to cycle between the Cu(I) and Cu(II) oxidation states allows it to be used as an electron carrier and redox center inside of proteins. This reactivity, while useful for biology, can cause cellular damage when copper is present in excess of the required amounts (Mealman et al. 2012). Silver is found primarily as Ag(I) and is not biologically utilized. Thus when silver is present in cells it acts as a toxin by inappropriately binding to biomolecules and disrupting their function. Because of chemical similarities between copper and silver, the cellular machinery maintaining copper homeostasis is often used to confer silver resistance.

In *E. coli*, copper homeostasis is mediated by two complementary protein systems, the Cue and the Cus systems (Mealman et al. 2012). This entry focuses on the CusCFBA transport system, which is composed of CusCBA, a tripartite efflux system and CusF, a metallochaperone (Fig. 1). In addition to functioning in copper homeostasis, the Cus system is the main determinant of silver resistance for *E. coli* (Franke et al. 2003).

CusCBA contributes to silver and copper resistance by transporting excess metal out of the cell to the extracellular space. This tripartite system is an active transporter that uses the proton gradient across the inner membrane to provide an energy source to move substrates up their concentration gradient. The tripartite CusCBA system is similar in many ways to multidrug resistance systems, such as AcrAB-TolC. The main difference between metal and multidrug transporting systems is substrate specificity. CusCBA provides resistance to Cu(I) and Ag(I), while multidrug transporters export many different substrates. This difference makes comparing the two systems tricky. Although they assemble similarly and use proton gradients to actively transport substrates out of bacterial cells, the way in which these pumps handle their substrates is different due to the systems' specificity and chemically different nature of the substrates.

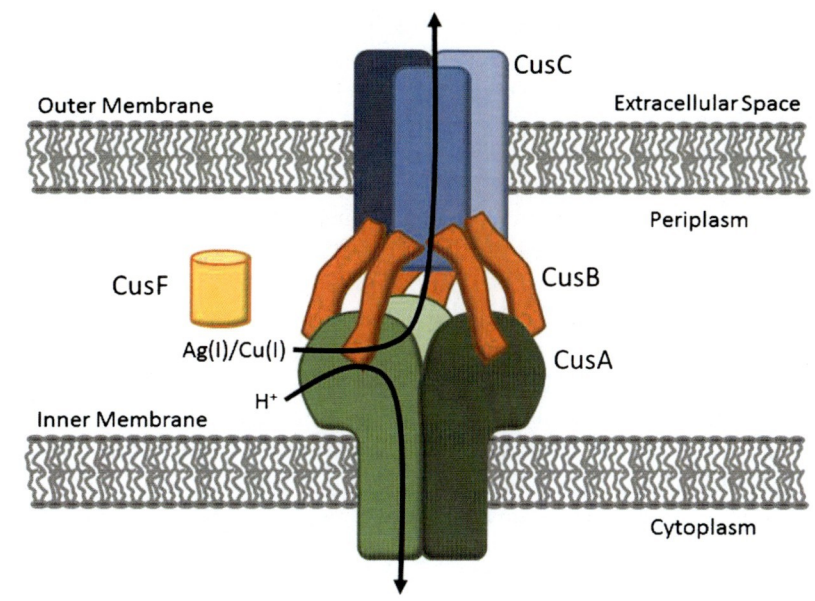

CusCFBA Copper/Silver Efflux System, Fig. 1 *The CusCFBA efflux system.* Trimeric CusA resides in the inner membrane and trimeric CusC resides in the outer membrane. Six CusB monomers interact with the inner and outer proteins. CusF is a soluble metal-binding protein that resides in the periplasm. Proton movement across the inner membrane is coupled with substrate export to the extracellular space

The CusCFBA system of *E. coli* has been studied genetically, biochemically, and structurally. Many studies have focused on the isolated proteins in this system, and those findings are reviewed below. More recent work has addressed how these proteins work together to export metal ions. What is known to date is summarized below, and some of the remaining outstanding questions are addressed.

CusA

CusA is the trimeric inner membrane transporter that drives efflux through the CusCBA complex. It is vital for Cu(I) and Ag(I) efflux, as deletion of the gene encoding CusA abolishes all efflux by the CusCBA complex (Franke et al. 2003).

CusA is a member of the RND family of transporters and is related to the multidrug transporters AcrB and MexB. RND proteins form homotrimers in the inner membrane and couple substrate export from the cell to the transport of protons into the cytoplasm. Experiments with AcrB show that RND proteins cycle between three major conformations to transport their substrate (Seeger et al. 2006; Murakami et al. 2006). RND proteins have a periplasmic substrate binding site and a channel that opens to the outer membrane protein, though how substrates move through RND proteins is not known.

In CusA, it was hypothesized that three conserved methionine residues, M573, M623, and M672, form the Cu(I)/Ag(I) binding site (Franke et al. 2003). A crystal structure of CusA was recently solved showing Cu(I) or Ag(I) bound to these methionines, confirming this is the periplasmic substrate binding site (Long et al. 2010).

At this point, there are uncertainties as to the path of metal ions through CusA. The periplasmic binding site has been confirmed, but how do Cu(I) and Ag(I) get to the periplasmic binding site? Does CusA acquire metal from the periplasm directly, or does it possibly interact with other metalloproteins? Recent work has also suggested that CusA can acquire metal from the cytoplasm as well as the periplasm (Long et al. 2012). Additional studies are needed to clarify these issues.

CusB

CusB is the membrane fusion protein (MFP) of the CusCBA complex. CusB is a soluble periplasmic protein that binds both CusA and CusC. As demonstrated in other systems, the membrane fusion protein stabilizes the formation of the tripartite complex (Nehme and Poole 2007). It is a necessary component of the CusCBA system, since deletion of *cusB* abolishes CusCBA efflux activity (Franke et al. 2003).

In addition to its role in stabilizing the CusCBA complex, CusB is also a metalloprotein (Bagai et al. 2007). This is unlike the multidrug efflux MFPs, which generally do not bind their substrates (Mealman et al. 2012). The fate of the metal bound to CusB is still a matter of investigation. It could be that the metal bound to CusB is transferred to CusA for efflux. Alternatively, metal binding to CusB could serve as a switch to activate the CusCBA complex for efflux (Kim et al. 2011). Both of these mechanisms are possible, and no current evidence completely rules out one or the other.

CusB binds Cu(I) and Ag(I) through three conserved methionine residues, M21, M36, and M38 (using the numbering of the mature protein sequence, Bagai et al. 2007). When the CusB crystal structure was solved, the N- and C-termini were missing, so the metal-binding site is not structurally characterized (Su et al. 2009).

From biochemical and structural characterizations, it has been shown that CusB is structurally flexible (Bagai et al. 2008; Long et al. 2012). In the crystal of CusB and a co-crystal with CusA, four separate conformations of CusB were crystallized (Su et al. 2009; 2011), further demonstrating the structural flexibility of this protein. The function of this flexibility is not clear. It could aid in assembling the CusCBA complex or the different conformations could allow CusB to aid in metal efflux, or serve as a regulator of CusCBA activity.

CusC

CusC is the outer membrane protein of the CusCBA complex. Three CusS monomers assemble to form a large aqueous channel that connects the periplasm with the extracellular space. CusC is not essential for CusCBA-mediated metal efflux; deletion of *cusC* decreases silver resistance but does not abolish it (Franke et al. 2003). It may be that one of the multidrug outer membrane proteins can substitute for CusC, but that protein has not been identified.

The crystal structure of CusC shows that it is a hollow channel with an inner diameter of $\sim$25 Å (Kulathila et al. 2011). No metal-binding sites have been identified in CusC, so potentially Cu(I) or Ag(I) are solvated by the extracellular medium once in the CusC channel. Because CusC may be partially substituted by an unidentified outer membrane protein, this lends support to the idea that CusC has no functional metal-binding sites.

CusF

CusF is a novel protein partner to the CusCBA complex. It is a soluble periplasmic protein that binds Cu(I) and Ag(I). Homologs of CusF are not seen in multidrug efflux systems, but are only found in other Cu(I)/Ag(I) tripartite efflux systems. Deletion of *cusF* decreases, but does not eliminate, copper or silver resistance, suggesting that CusF contributes to CusCBA efflux, but is not required (Franke et al. 2003).

The structure of CusF has been solved in both apo and metal-bound states (Mealman et al. 2012). CusF coordinates a metal ion in a two methionine, one histidine binding site with cation-pi interaction from a tryptophan (Xue et al. 2008; Loftin et al. 2007).

CusF is believed to be highly upregulated in response to copper or silver shock (Kershaw et al. 2005), so it is theorized to sequester metals and act as a molecular sponge. As a soluble periplasmic protein, it can bind excess toxic metal ions, thus protecting the cell from the damaging effects of free metal ions.

CusCBA Complex Formation

Since CusCBA forms a complex, the protein-protein interactions between components are crucial to understand function of the pump. CusA and CusC are trimeric membrane-bound proteins (Long et al. 2010; Kulathila et al. 2011). From a co-crystal of CusA and CusB, it is now known that two CusB monomers bind to each CusA monomer, such that the overall stoichiometry of C:B:A is 3:6:3 (Long et al. 2012).

From a CusBA crystal structure, it is shown that CusB interacts with the periplasmic domain of CusA. CusB also extends upward from the top of CusA to form a hexameric channel with an average internal diameter of 37 Å. This channel may form the beginning of the pathway out of the CusBA complex (Long et al. 2012).

To date, the interactions between CusB and CusC are unknown. However, the protein models can be docked onto each other using simulations. CusC has an external diameter smaller than the internal diameter of the CusB channel, and simulations show that CusC could fit inside the upper channel of CusB. It is not known if CusA and CusC interact directly, but simulations show that it may be possible for CusB to hold CusA and CusC close enough for them to come in contact (Long et al. 2012).

CusF-CusCBA Interactions

CusF is the soluble periplasmic protein in the Cus system. CusF has been shown to transiently interact with CusB and transfer metal ions to CusB (Bagai et al. 2008). Whether CusF interacts with or transfer metal to CusA is still under investigation. The interactions of CusF with the other Cus components are functionally important, and thus CusF has roles that are still being defined beyond that of a metal ion chelator.

CusCFBA Prevalence in Microorganisms

Cus protein homologs are found in bacteria besides *E. coli*. A BLAST-search found CusF and CusB homologs distributed throughout the Proteobacterial phylum (Kim et al. 2010). No CusF or CusB homologs were found in other phyla.

CusF homologs are found in a variety of genomic contexts. In the Enterobacteriales order, *cusF* was found with *CusCBA* efflux genes (Kim et al. 2010). However, in other systems, *cusF* was found as a fusion gene with *cusB*-like genes. Occasionally, *cusF*-like genes were found in close proximity to copper-responsive genes besides the *cusCBA* efflux genes. These genomic data suggest that *cusF* can function in Cu(I) and Ag(I) resistance through two pathways: one dependent on CusCBA, the other CusCBA independent. In CusCFBA in *E. coli*, though CusF has the ability to interact and transfer metal to CusB, it is not known how this contributes to metal resistance. The prevalence of CusF throughout bacteria necessitates its further study to understand its role in Cu(I) and Ag(I) resistance.

Conclusion

The CusCFBA system provides an important function in removal of excess Cu(I)/Ag(I) from cells to enable survival under conditions where these metal ions are in overabundance. From structural, genomic, and biochemical studies, much progress has been made to understand the working of this system. Future studies will reveal more fully the complex interplay between the components of the system and how they function dynamically to remove toxic metal ions from cells.

Cross-References

▶ Biological Copper Transport
▶ Copper, Biological Functions
▶ Magnesium, Physical and Chemical Properties

References

Bagai I, Liu W, Rensing C et al (2007) Substrate-linked conformational change in the periplasmic component of a Cu(I)/Ag(I) efflux system. J Biol Chem 282(49):35695–35702

Bagai I, Rensing C, Blackburn NJ et al (2008) Direct metal transfer between periplasmic proteins identifies a bacterial copper chaperone. Biochemistry 47(44):11408–11414

Franke S, Glass G, Rensing C et al (2003) Molecular analysis of the copper-transporting efflux system CusCFBA of *Escherichia coli*. J Bacteriol 185:3804–3812

Kershaw CJ, Brown NL, Constantinidou C et al (2005) The expression profile of *Escherichia coli* K-12 in response to minimal, optimal and excess copper concentrations. Microbiology 151:1187–1198

Kim EH, Rensing C, McEvoy MM (2010) Chaperone-mediated copper handing in the periplasm. Nat Prod Rep 27:711–719

Kim EH, Nies DH, McEvoy MM et al (2011) Switch or funnel: how RND-type transport systems control periplasmic metal homeostasis. J Bacteriol 193(10):2381–2387

Kulathila R, Kulathila R, Indic M et al (2011) Crystal structure of *Escherichia coli* CusC, the outer membrane component of a heavy metal efflux pump. PLoS One 6:1–7

Loftin IR, Franke S, Blackburn NJ et al (2007) Unusual Cu(I)/Ag(I) coordination of *Escherichia coli* CusF as revealed by atomic resolution crystallography and X-ray absorption spectroscopy. Protein Sci 16:2287–2293

Long F, Su CC, Zimmerman MT et al (2010) Crystal structures of the CusA efflux pump suggest methionine-mediated transport. Nature 467:484–488

Long F, Su CC, Lai HT et al (2012) Structure and mechanism of the tripartite CusCBA heavy-metal efflux complex. Philos Trans R Soc B 367:1047–1058

Mealman TD, Blackburn NJ, McEvoy MM (2012) Metal export by CusCFBA, the periplasmic Cu(I)/Ag(I) transport system of *E. coli*. Curr Top Membr (in press)

Murakami S, Nakashima R, Yamashita E et al (2006) Crystal structures of a multidrug transporter reveal a functionally rotating mechanism. Nature 443:173–179

Nehme D, Poole K (2007) Assembly of the MexAB-OprM multidrug pump of *Pseudomonas aeruginosa*: component interactions defined by the study of pump mutant suppressors. J Bacteriol 189:6118–6127

Seeger MA, Schniefner A, Eicher T et al (2006) Structural asymmetry of AcrB trimer suggests a peristaltic pump mechanism. Science 313:1295–1298

Su CC, Yang R, Long R et al (2009) Crystal structure of the membrane fusion protein CusB from *Escherichia coli*. J Mol Biol 393:342–355

Su CC, Long F, Zimmerman MT et al (2011) Crystal structure of the CusBA heavy-metal efflux complex of *Escherichia coli*. Nature 470:558–562

Xue Y, Davis AV, Balakrishnan G et al (2008) Cu(I) recognition via cation-p and methionine interactions in CusF. Nat Chem Biol 2:107–109

CuZ Center

▶ Nitrous Oxide Reductase

Cyclic Adenosine Monophosphate (cAMP)

▶ Calcium Sparklets and Waves

Cyclosporin a (CsA)

▶ Calcineurin

Cysteine Hydrolases

▶ Silicateins

Cysteine Protease Activity Inhibited by Terpyridine Platinum(II)

▶ Platinum(II), Terpyridine Complexes, Inhibition of Cysteine Proteases

Cysteine Sulfenic Acid

▶ Nitrile Hydratase and Related Enzyme

Cysteine Sulfinic Acid

▶ Nitrile Hydratase and Related Enzyme

Cyt19

▶ Arsenic Methyltransferases

Cytochrome C (Cyt C): Small Heme Protein

▶ Palladium Complex-induced Release of Cyt c from Biological Membrane

Cytochrome c Oxidase, CuA Center

Peter M. H. Kroneck[1] and Martha E. Sosa Torres[2]
[1]Department of Biology, University of Konstanz, Konstanz, Germany
[2]Facultad de Quimica, Universidad Nacional Autonoma de Mexico, Ciudad Universitaria, Coyoacan, Mexico DF, Mexico

Synonyms

Binuclear mixed-valent electron transfer copper center

Definitions

CuA is a binuclear, mixed-valent [Cu(1.5+)-Cu(1.5+)], spin $S = 1/2$ electron transfer center present in respiratory chains of numerous organisms, specifically in cytochrome *c* oxidase (CcO) which reduces O_2 to two molecules of H_2O and nitrous oxide (N_2O) reductase (N2OR) which reduces N_2O to N_2 and H_2O. The core structure of CuA consists of a Cu_2S_2 rhomb with unique spectroscopic features, specifically the seven-line electron paramagnetic resonance spectrum observed at low temperatures.

Structure and Function of CuA

Occurrence

CcO (ferrocytochrome c:oxygen oxidoreductase; cytochrome aa_3; complex IV; EC 1.9.3.1) is the key enzyme of cell respiration in all eukaryotes and many prokaryotes. In bacteria, CcO is located in the cell membrane, whereas in eukaryotic cells, the enzyme resides in the inner mitochondrial membrane (Wikström 2010; Yoshikawa et al. 2011). Mammalian and "classical" bacterial CcOs carry three Cu atoms: the binuclear mixed-valent CuA electron transfer center and CuB coupled to heme a_3 forming the dioxygen-reducing site. The enzyme catalyzes the terminal reaction in aerobic respiration, the reduction of dioxygen to water. This energy-conserving process is coupled to the generation of a proton gradient across the membrane (Babcock and Wikström 1992), with CcO acting as a redox-driven proton pump (1):

$$4Cyt-c-Fe^{2+} + 4H_c^+ + nH^+_i + O_2 \rightarrow \\ 4Cyt-c-Fe^{3+} + nH^+_o + 2H_2O \quad (1)$$

Hereby, H_i^+ and H_o^+ refer to protons being translocated from the inner (i) to the outer (o) side of the membrane. This process is also called protonpumping, usually with a stoichiometry of one proton per electron ($2H^+/O$). However, a lower stoichiometry, $1H^+/O$, has been reported for several microorganisms. H_c^+ refers to so-called chemical protons which are taken up from the inside for the formation of water.

By comparison, the copper enzyme ▶ N2O Reductase (N2OR), a head-to-tail homodimer (6Cu/monomer), is a soluble protein located in the periplasm of denitrifying bacteria. It is involved in the transformation of the greenhouse gas N_2O to water and dinitrogen (2) (Zumft and Kroneck 2007):

$$2H^+ + 2e^- + N_2O \rightarrow N_2 + H_2O \quad (2)$$

Recently, a nitric oxide reductase (NOR) in *Bacillus azotoformans* was identified as the third CuA-containing enzyme. In addition, artificial CuA sites have been engineered by loop-directed mutagenesis of ▶ Monocopper Blue Proteins (Savelieff and Lu 2010). These small proteins, such as ▶ Plastocyanin, azurin, or amicyanin, host the so-called blue type-1 Cu center. The similarity between the CuA-carrying enzymes N2OR and CcO becomes evident at the level of the amino acid sequence in a short region constituting the CuA-binding domain (Zumft and Kroneck 2007).

The cytochrome c oxidases belong to the superfamily of heme-copper oxidases, a group of structurally and functionally related enzymes. Within this superfamily, there exist two major branches: (1) cytochrome *bo* quinol oxidases with ubiquinol as primary electron donor and (2) cytochrome c cytochrome oxidases which use cytochrome c (Wikström 2010).

Overall Protein Architecture and Cupredoxin Fold

The "classical" CcO isolated from bacteria consists of three to four subunits depending on the organism, as illustrated for the enzyme isolated from the soil bacterium *Paracoccus denitrificans* (Fig. 1). Subunit I is located in the membrane; it constitutes the largest subunit with 12 transmembrane helices, and it contains the heme a_3-CuB dioxygen-reducing active site. Subunit II is attached to the periplasmic side of the membrane, anchored by two N-terminal helices. It has a large globular domain at the periplasmic side with a fold very similar to that of type-1 copper proteins. Subunit II harbors the mixed-valent CuA center which functions as the primary electron acceptor; the role of the other subunits remains unclear. In contrast, CcO isolated from bovine heart has 13 subunits, with subunits I and II carrying the canonical copper and heme iron redox sites (Iwata et al. 1995; Tsukihara et al. 1995). Note that the CuA center is absent in subunit II of the functionally related cytochrome *bo* quinol oxidases.

Comparisons of the overall folds and the copper binding sites have suggested a common evolutionary ancestry (Adman 1995). Notably, the key enzymes ▶ Nitrite (NO_2^-) Reductase, nitric oxide (NO) reductase, and nitrous oxide (N_2O) reductase of the denitrification pathway, which constitutes an important branch within the biogeochemical nitrogen cycle, share common structural elements with CcO (Castresana et al. 1994; Zumft and Kroneck 2007). This observation led to the hypothesis that aerobic respiration preceded photosynthesis during evolution (Wikström 2010). The cupredoxin family includes copper proteins carrying the mononuclear type-1 Cu as well as the binuclear CuA electron transfer centers. This can be clearly illustrated by circular dichroism (CD) spectroscopy of the CuA domain in CcO which

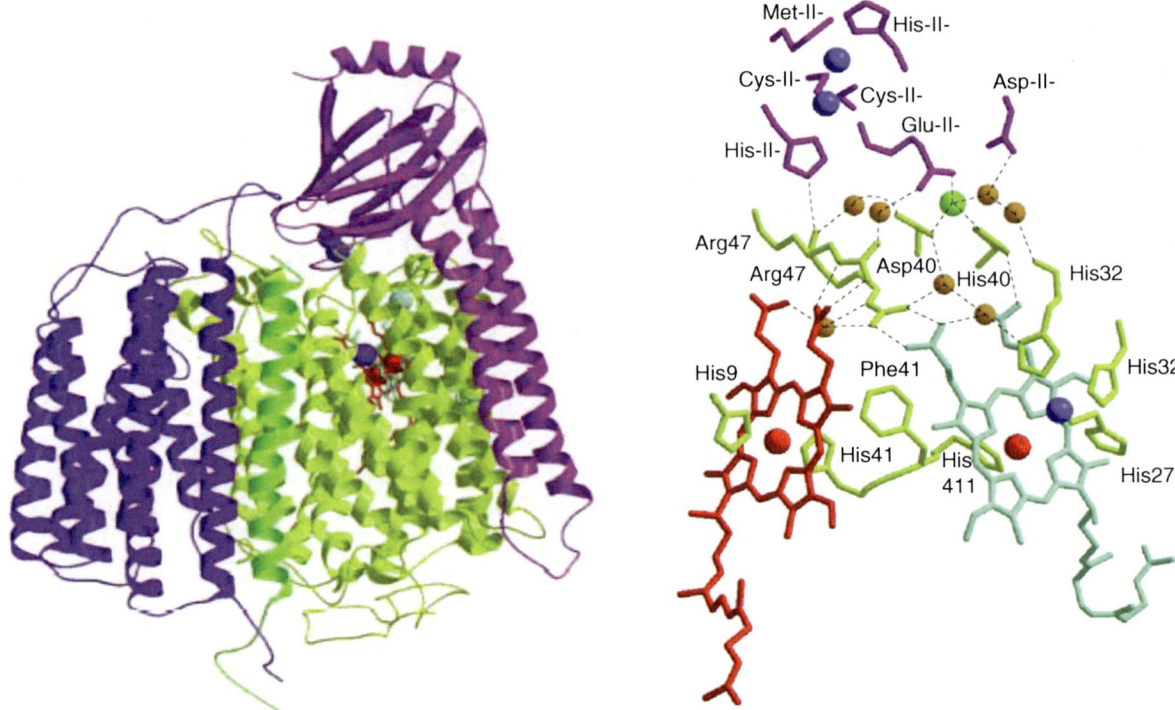

Cytochrome c Oxidase, CuA Center, Fig. 1 Structure of a "classical" cytochrome c oxidase. *Left*: Overall structure of CcO isolated from *Paracoccus denitrificans* showing the functionally essential subunits I (*magenta*) and II (*yellow*). *Right*: Redox metal centers (copper in *blue*, heme iron in *red*) (PDB code 1QLE) (Redrawn from Kannt and Michel 2001)

documents the presence of the so-called cupredoxin fold, a Greek key ß-barrel and common structural motif in small type-1 Cu proteins of plants and bacteria (Adman 1995). This structural relationship helped to engineer, through loop-directed mutagenesis, the purple CuA binuclear center into monocopper blue proteins (Fig. 2) (Saveleiff and Lu 2010). On the other hand, the CuA center of CcO from *Paracoccus denitrificans* could be transformed into a mononuclear type-1 Cu center by exchange of one bridging cysteine ligand (C216) into serine.

CuA Structure and Spectroscopy

CuA is a S_{Cys}-bridged electron transfer center, [Cu(1.5+)-Cu(1.5+)]. Interest in CuA is directly related to its unique spectroscopic properties. Numerous investigations by electron paramagnetic resonance (EPR), nuclear magnetic resonance (NMR), magnetic circular dichroism (MCD), resonance Raman, and X-ray absorption spectroscopy were key to establish a highly resolved picture of the electronic properties of the CuA site and the interaction of the metal atoms with neighboring amino acids (Zumft and Kroneck 2007). UV/visible, MCD, and EPR spectra of the binuclear CuA center differ significantly from spectra typical of the blue mononuclear type-1 copper site. The intense purple color of CuA is the result of mainly S(Cys) → Cu charge transfer bands at ~480 nm (~21,000 cm^{-1}), 530 nm (~19,000 cm^{-1}), and a class III mixed-valence charge transfer band in the near-IR region around 800 nm (~13,400 cm^{-1}) (Zumft and Kroneck 2007).

CuA is paramagnetic, with one unpaired electron resulting in a total spin S = 1/2. It comprises two Cu ions: one in the Cu(1+) $3d^{10}$, S = 0 configuration and one in the Cu(2+) $3d^9$, S = 1/2 configuration, respectively. However, the unpaired electron is fully delocalized over both Cu atoms within the $Cu_2(SCys)_2$ rhomb as documented by the typical seven-line electron paramagnetic resonance spectrum observed at low temperatures (Fig. 3). Consequently, a mixed-valent [Cu(1.5+)-Cu(1.5+)] S = 1/2 configuration has been assigned to CuA (Zumft and Kroneck 2007). The distance between the two copper atoms in CuA is in the range of 2.5 Å, similar to the Cu–Cu distance observed in metallic copper. As a consequence, the existence of

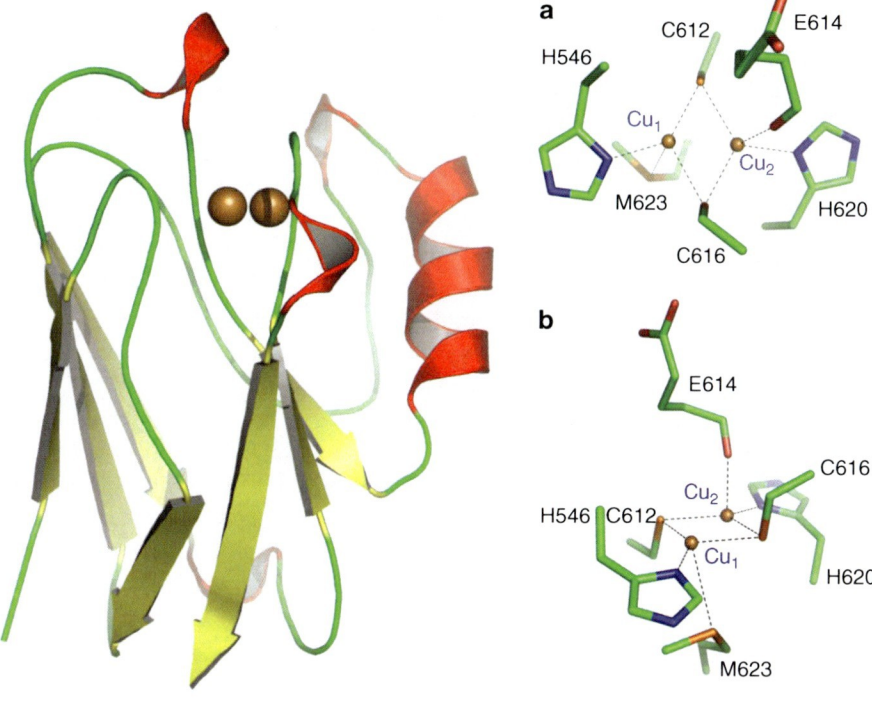

Cytochrome c Oxidase, CuA Center, Fig. 2 Structure of a CuA center engineered by loop-directed mutagenesis into a *blue* type-1 copper protein. *Left*: Ribbon representation of showing the cupredoxin fold. *Right*: Schematic drawing of the Cu_A site, view of the Cu_2S_2 rhomb (**a**), and side view of the Cu_A center (**b**). PDB 1CC3 (Redrawn from Zumft and Kroneck 2007)

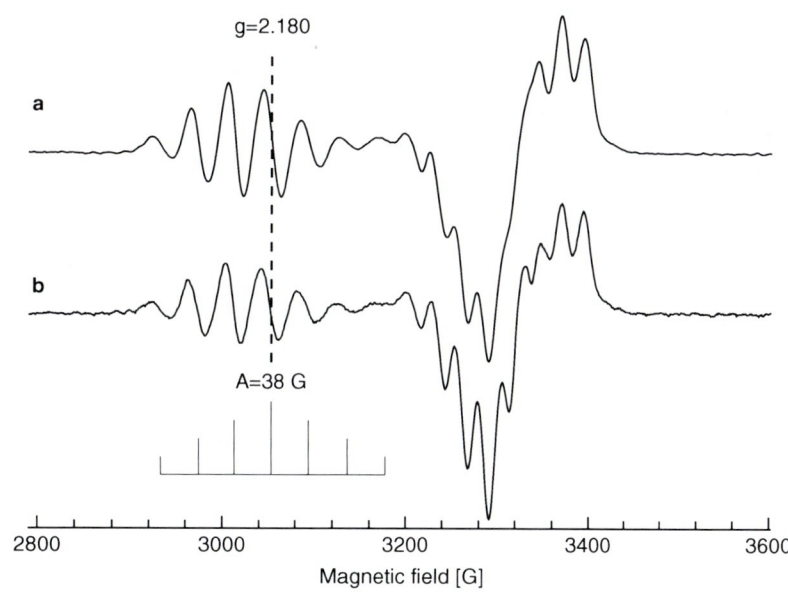

Cytochrome c Oxidase, CuA Center, Fig. 3 Electron paramagnetic resonance spectra (X band, second harmonic display) of the CuA site in nitrous oxide reductase from *Pseudomonas stutzeri* showing the characteristic seven-line hyperfine splitting in the $g_\parallel$ region at 2.18. (**a**) ^{65}Cu-enriched and (**b**) ^{65}Cu,^{15}N-histidine-enriched enzyme (Redrawn from Neese 1997)

a Cu–Cu metal bond has been proposed. Two cysteine sulfur atoms function as bridging ligands between the metal ions, and two histidines are terminal ligands of each of the two copper ions. The overall structure of the Cu_2S_2 rhomb is rather symmetrical, in agreement with the spectroscopic data, properties of variants obtained by site-directed mutagenesis, and theoretical calculations of the CuA binding site (Zumft and Kroneck 2007; Savelieff and Lu 2010). The purple [Cu(1.5+)-Cu(1.5+)] center will accept one electron upon reduction leading to the colorless reduced [Cu(1+)-Cu(1+)] state, with a virtually identical arrangement of the two coppers. The fully oxidized [Cu(2+)-Cu(2+)] state has not been detected so far.

By recombinant DNA technology, it was possible to express only the water-soluble fragment of CcO subunit II which contained the CuA center. As both hemes a and the bimetallic heme a_3-CuB catalytic site are located in the membrane part of CcO (Fig. 1), important spectroscopic information, without the spectral contributions of these metal sites, could be obtained to elucidate the electronic properties of CuA. Clearly, the unraveling of the binuclear and mixed-valent character of the CuA center in N2OR by spectroscopic techniques, most importantly by electron paramagnetic resonance spectroscopy (Fig. 3), was crucial for identifying the same structure in CcO (Beinert 1997). Later, high-resolution crystal structures of both bacterial and mammalian CcOs confirmed the binuclear nature of CuA (Iwata et al. 1995; Tsukihara et al. 1995).

Another distinct physical property of the mixed-valent CuA site is its unusually fast electron relaxation, which allowed extensive investigations of soluble CuA-containing protein fragments by ^{1}H-NMR spectroscopy and assignment of the individual ligands of the two Cu atoms.

Electron and Proton Transfer

The reduction of molecular oxygen catalyzed by CcO (Eq. 1) is highly exergonic, with a change in free energy $\Delta G = \approx -192$ kJ mol^{-1}. A major portion of this energy is conserved as an electrochemical proton gradient across the membrane (Wikström 2010). In "classical" CcO, CuA accepts electrons from cytochrome c which are then transferred via heme a to the coupled heme a_3-CuB dioxygen-reducing site (Fig. 1). As mentioned above, despite its binuclear composition, CuA will only transfer one electron at a time; it shuttles between the two biologically relevant redox states [Cu(1.5+)-Cu(1.5+)] (CuA$_{ox}$) and [Cu(1+)-Cu(1+)] (CuA$_{red}$). The water-soluble cytochrome c is located on the outer surface of the membrane; it docks to a specific site of CcO subunit II. Similar to the blue type-1 Cu ($\lambda_{max} \approx 600$ nm), the purple CuA ($\lambda_{max} \approx 530$ nm) is a very efficient electron transfer center. The reasons for utilizing a binuclear electron transfer center in CcO and N2OR are suggested to be its unidirectional electron transfer through the site or the lower energy of reorganization. In studies on intramolecular electron transfer in a purple azurin variant carrying an engineered CuA center, the rate constant of the intramolecular process, k_{ET}, was almost threefold faster than for the same process measured for the wild-type azurin carrying the mononuclear type-1 Cu (Zumft and Kroneck 2007).

The production of two molecules H_2O from one molecule O_2 requires the transfer of four electrons and four protons (so-called chemical protons) to the O_2 molecule which are coming from opposite sides of the membrane. This corresponds to the translocation of four electrical charge equivalents across the membrane per molecule of O_2 (Wikström 2010; Yoshikawa et al. 2011). Furthermore, in addition to the translocation of chemical protons, CcO translocates up to four protons across the membrane per O_2 reduced (Eq. 1). Thus, CcO is a redox-driven proton pump, with CuA acting as the primary electron acceptor. Uptake of protons into CcO, either to be used in $O_2 \rightarrow H_2O$ reduction or translocation across the membrane, occurs via defined pathways. Mutagenesis experiments in combination with X-ray crystallography helped to identify these pathways including their entrance and exit gates. In summary, CcO represents a remarkably efficient energy-transducing machine involved in the production of adenosine triphosphate (ATP) in aerobic life (Wikström 2010).

Cross-References

- Monocopper Blue Proteins
- Nitrite Reductase
- Nitrous Oxide Reductase
- Plastocyanin

References

Adman ET (1995) A taste of copper. Nat Struct Biol 2:929–931

Babcock GT, Wikström M (1992) Oxygen activation and the conservation of energy in cell respiration. Nature 356:301–309

Beinert H (1997) Copper A of cytochrome c oxidase, a novel, long-embattled, biological electron-transfer site. Eur J Biochem 245:521–532

Castresana J, Lübben M, Saraste M, Higgins DG (1994) Evolution of cytochrome oxidase, an enzyme older than atmospheric oxygen. EMBO J 13:2516–2525

Iwata S, Ostermeier C, Ludwig B, Michel H (1995) Structure at 2.8 Å resolution of cytochrome c oxidase from *Paracoccus denitrificans*. Nature 376:660–669

Kannt A, Michel H (2001) Bacterial cytochrome c oxidase. In: Messerschmidt A, Huber R, Poulos T, Wieghardt K (eds) Handbook of metalloproteins. Wiley, Chichester, pp 331–347.

Neese (1997) Electronic structure and spectroscopy of novel copper chromophores in biology. PhD thesis, Universitaet Konstanz

Savelieff MG, Lu Y (2010) CuA centers and their biosynthetic models in azurin. J Biol Inorg Chem 15:461–483

Tsukihara T, Aoyama H, Yamashita E, Tomizaki T, Yamaguchi H, Shinzawa-Itoh K, Nakashima R, Yaono R, Yoshikawa S (1995) Structures of metal sites of oxidized bovine heart cytochrome c oxidase at 2.8 Å resolution. Science 269:1069–1074

Wikström M (2010) Cytochrome c oxidase. In: Encyclopedia of life sciences. Wiley, Chicester, pp 1–10

Yoshikawa S, Muramoto K, Shinzawa-Itoh K (2011) Proton-pumping mechanism of cytochrome c oxidase. Ann Rev Biophys 40:205–223

Zumft WG, Kroneck PMH (2007) Respiratory transformation of nitrous oxide (N_2O) to dinitrogen by Bacteria and Archaea. Adv Microb Physiol 52:107–225

Cytocuprein

▶ Zinc in Superoxide Dismutase

Cytokines

▶ Chromium(III) and Immune System

Cytosolic Signaling

▶ Calcium in Biological Systems
▶ Calcium-Binding Proteins, Overview

Cytotoxic Potential of Gold(III) Complexes

▶ Gold(III) Complexes, Cytotoxic effects

Cytotoxicity

▶ Tin Complexes, Antitumor Activity

CzrA

▶ Zinc Sensors in Bacteria

Printed by Books on Demand, Germany